Science and Earth History—
The Evolution/Creation Controversy

Science and Earth History—
The Evolution/Creation Controversy

Arthur N. Strahler

Prometheus Books
Buffalo, N.Y.

Published 1987 by Prometheus Books
700 East Amherst Street, Buffalo, N.Y. 14215

91 90 89 88 87   4 3 2 1

Library of Congress Cataloging-in-Publication Data:

Strahler, Arthur Newell, 1918-
    Science and earth history: the evolution/creation controversy
  Arthur N. Strahler.
        p.   cm.
    Bibliography: p.
    Includes index.
    ISBN 0-87975-414-1:
    1. Creationism. 2. Evolution.
I. Title.
BS653.S77   1987   87-21169
231.7′65—dc19   CIP

Credits and permissions:

All quoted passages from copyrighted publications of the Institute for Creation Research,
Santee, California, are used by permission of Dr. Henry M. Morris, President of the
Institute for Creation Research.

Cover designed by Lois Darling. God is from the ceiling of the Sistine Chapel. Darwin is
from *The Science of Life* and *Bird* written and illustrated by Lois and Louis Darling.

Designer, illustrator, compositor: Arthur N. Strahler

"There is grandeur in this view of life, with its several powers, having been originally breathed into a few forms or into one; and that, whilst the planet has gone cycling on according to the fixed law of gravity, from so simple a beginning endless forms most beautiful and most wonderful have been, and are being, evolved."

From *The Origin of Species*, First Edition, 1859

H.M.S. *Beagle* leaving England, December 1831. The five year voyage around the world with Charles Darwin aboard was to become one of the most consequential of all time. Drawn and copyrighted by Lois Darling.

Preface

At first glance, readers may think that they have come upon a very strange book. I certainly would have said as much had I seen one like it ten years ago. There are, however, science educators and research scientists to whom this book will seem not only familiar but purposeful. These people are creationist watchers. I became an active creationist watcher only quite recently, but exposure to this absorbing occupation dates back to 1977, when I received from Professor Preston Cloud a copy of his paper titled "'Scientific Creationism'--A New Inquisition Brewing?" It had appeared in a popular journal called The Humanist early in that year. A distinguished biogeologist, Cloud had for many years been searching for fossils and other indications of the oldest life forms on earth. I was on his mailing list to receive copies of his journal articles, which arrived in welcome bunches at odd intervals. But an article on "creationism"--what's that all about? And what is a "humanist"? It seems that a year earlier Cloud had been in a public debate with two creation scientists, Henry M. Morris and Duane T. Gish; his Humanist paper was an outgrowth of that encounter. Although I read Cloud's excellent paper with interest, at that time it evoked no special enthusiasm to investigate further.

Then, in 1981, I began to take an interest in short news reports and letters appearing in such popular sources as Science News and Science telling of the activities of the creationists. In 1982 I attended a session of the Pacific Division of the American Association for the Advancement of Science, held on the campus of the University of California at Santa Barbara. The program included three sessions under the title "Evolutionists Confront Creationists." I listened with rising interest as representatives from both mainstream science and creation science presented their positions. It was my first exposure to real, live creation scientists (i.e., Duane T. Gish and Harold S. Slusher of the Institute for Creation Research) and to those mainstream scientists who considered creationism a threat sufficiently ominous to warrant carefully studied arguments supporting evolution through vast spans of geologic time. But why expend so much effort to stand up for monumental scientific achievements that had long been accepted without question? Surely, organic evolution (descent with modification) by natural selection needs no public defenders a century after Darwin, nor should anyone have to get up on the lecture platform to defend the radiometric age of four and one-half billion years for our planet and more than three and one-half billion years for its oldest rocks.

Perhaps it was the Arkansas bill requiring the teaching of creation science in the state's public schools on an equal basis with mainstream science, that really got to me. Then came the Louisiana bill of 1981, of which I obtained a copy from the New Orleans office of the American Civil Liberties Union. Clearly, a social epidemic of such documents was in progress.

Perhaps even more disturbing were the reports of textbook censorship in Texas and California. As a textbook author, I felt insult and revulsion at such degrading activities, even though my own texts were not directly involved. Perhaps the final prod came with the news that reputable textbook publishers (none were publishers of my books) had yielded to the demands of the censors by eliminating the subject of evolution from their biology texts.

So, here today is my very strange book, unlike any I have ever authored. For the creation watchers and anticreation activists, it contains much that is familiar, for it was preceded by several excellent books with like content and a journal (Creation/Evolution) devoted entirely to covering the same ground. Included in this group of authors and editors, from whose books I have drawn much substance, are Frank Awbrey and William M. Thwaites, Norman D. Newell, Dorothy Nelkin, Laurie R. Godfrey, Chris McGowan, and Philip Kitcher. Those of their works that are collections of papers include the contributions of other prominent anticreation activists, whose work is cited throughout this book.

The attempt by fundamentalist Christians to blend science and religion into a coherent view of the universe is what this book examines. As long as the blender is running at full speed, the mixture looks palatable enough, but it still has a strangely disturbing odor and taste. My thesis as a creation watcher is that when the blender is switched off, science and religion separate into two stable layers, neither of which will tolerate the presence of the other.' Creation scientists seek to emulsify two cognitive fields that not only resist union but crave separation.

Well, we do have a lively debate on our hands. I wrote a strange book to counter what to me looks like a strange, internally inconsistent view of the universe. There's nothing more fascinating than a good argument between groups deeply divided on an issue that both consider of vital importance to society.

Arguments rarely proceed for long with cold impersonality. Throughout this book I express some personal views and evaluations, although, for the most part, I let others speak for themselves in carrying on the argument. I have not always been kind to the mainstream scientists, particularly where they show unbecoming arrogance. On those few occasions when the creation scientists score a good point, I try to give them credit. I have tried to quote and paraphrase fairly and in accurate context, even when those around me are not doing the same. Arguments, especially those most heated, invite humor. If you find tongue-in-cheek here and there, please enjoy it.

On a more serious note, this book has a mission, which is primarily to explain what natural science is all about and how scientists go about doing it. The debate between creation scientists and mainstream scientists is an ideal vehicle for carrying out this mission. Mainstream science exposes its corporeal substance as a cognitive field most clearly when put on the spot to defend its specific findings, theories, and speculations. I hope those readers, if there be any, who follow the whole text through from cover to cover will say to me, "Thanks for letting me in on how the natural sciences really work."

Arthur N. Strahler

Acknowledgments

The person to whom I owe thanks for getting started on this book is philosopher Peter A. Angeles of the faculty of Santa Barbara City College. He had been most helpful in reviewing the manuscript of a short paper I wrote in 1981 on the evolution/creation conflict. During our discussions on creationism, secular humanism, and religion, Professor Angeles suggested that I go to work on a book embracing these subjects. Ever since, he has been a source of encouragement and guidance.

I am also indebted to a former graduate student of mine and a longtime geological colleague, Mark A. Melton, for his valuable criticism of my early drafts of chapters on the nature and philosophy of science.

During the early stages of this book, I developed an extensive three-way correspondence with two sympathetic individuals, each of whom has an expertise in important areas of the evolution/creation controversy--areas about which I know very little. John R. Armstrong is a geologist who also holds a Master of Divinity degree from Wycliffe College, Toronto School of Theology; besides having professional competence in stratigraphy and paleontology, he is well versed in the history of science, the church, and the Old Testament. Ronnie J. Hastings, with a Ph.D. degree in physics, teaches high-school science and math in Texas, where the textbook controversy has raged and creationist activity is ever close at hand. As vice-president of the Texas Council for Science Education, he is active in improving science textbooks in his state. John and Ronnie volunteered a critical reading of the entire manuscript, a feat carried out with distinction and with immeasurable benefit to the text. My sincere thanks to both of you.

I received invaluable assistance from over a dozen reviewers, each of whom read a particular chapter or group of chapters in the area of his professional competence. Some also generously gave me permission to use illustrations and to quote extensively from their books and journal articles. These reviewers are listed below in alphabetical order with the subject areas of their expertise:

Stephen G. Brush (Physics, History of science)
Bernard Campbell (Anthropology)
Edwin H. Colbert (Vertebrate paleontology)
Joel L. Cracraft (Evolutionary biology)
Thomas N. Cronin (Paleontology)
G. Brent Dalrymple (Radiometric dating)
William L. Donn (Meteorology, Geophysics)
Steven I. Dutch (Geology)
Stanley Freske (Physics, Astronomy)
Glen Kuban (Ichnology)
Chris McGowan (Vertebrate paleontology)
John W. Patterson (Thermodynamics)
David M. Raup (Paleontology)
Stanley A. Schumm (Fluvial geomorphology)
William M. Thwaites (Biology)
Stanley L. Weinberg (Biology)

The reviews offered by these specialists greatly improved the quality of the text, but it goes without saying that full responsibility for the content of this book rests with me alone.

A book that purports to present the case for creation science fully and fairly must quote extensively from the publications of the creation scientists. I thank Dr. Henry M. Morris, President of the Institute for Creation Research, for promptly and forthrightly granting me permission to quote from publications of the ICR.

A special note of thanks is reserved for my wife, Marge, who spent untold hours at the word processor, aided by her faithful and well-trained mouse, transcribing to floppy disks hundreds of typescript pages and thousands of index entries.

About the Author

Arthur N. Strahler holds the Ph.D. degree in geology from Columbia University. He was appointed to the Columbia University graduate faculty in 1941, serving as Professor of Geomorphology from 1958 to 1967 and as Chairman of the Department of Geology from 1958 to 1962. A Fellow of the Geological Society of America, his published research has dealt with processes and forms of fluvial erosion, morphometry and statistical analysis of landforms, and systems-theory applied to geomorphology and hydrology. He is the author or coauthor of textbooks on physical geology, the earth sciences, physical geography, and environmental science.

Abbreviated Contents

PART VIII The Rise of Man and Emergence
of the Human Mind

PART IX The Origin of Life on Earth--
Naturalistic or Creationistic?

Expanded Contents

PART I

Science and Pseudoscience

Introduction

A major debate takes place throughout the chapters of this book. In its broadest aspect the dispute is over the relative merits of two very different ways of viewing the universe and its contents. By "universe" I mean everything that can be observed and described by humans with reasonable assurance and general agreement that what is being observed exists as some recognizable form of matter or energy. The conflict we shall examine is not so much over the question "What's in the universe?" as it is a question of "How did the universe come about?"

Expanding and rewording this second question, it breaks up into such queries as "How did it all originate?" "What caused it?" and "When did it start?" Is it, then, a question of origins of things that we are debating? I think perhaps this is the crucial point of the debate. Science has one viewpoint on origins; various forms or brands of pseudoscience (false science) offer alternative views. Of these alternative views, we single out one for major confrontation: the claim of recent and sudden creation by a supernatural agency. From here on, I designate that particular view of the universe as creationism and describe it by the adjective creationistic.

As to science, its view of the universe can be described as naturalistic, using an adjective that has its historical roots far back in philosophy as explaining all phenomena by strictly natural categories--as opposed to explanations invoking supernatural forces. I could have just as easily used mechanistic as the adjective, but that is a harsh word, suggesting the actions of a machine and the work of an inventor. Another choice would have been materialistic, but for most persons that adjective carries a negative association in terms of values.

Taking the creationistic view first, it is simply that the universe was created from nothing--ex nihilo, that is --by a divine creator in ways and for reasons unknowable to humans except, perhaps, through revelation. The second, or naturalistic view, is that the particular universe we observe came into existence and has operated through all time and in all its parts without the impetus or guidance of any supernatural agency. The naturalistic view is espoused by science as its fundamental assumption. The creationistic view is espoused and interpreted in various ways, degrees, and levels. The version that concerns us here is a particular and rigid view based on fundamentalist Christianity: it can be called recent creation, for it specifically accepts the literal meaning of the words of the book of Genesis as providing the true six-day scenario for the origin of the universe. The creationistic view can also take other, broader forms of expression, and for some theistically oriented persons the creationistic and naturalistic views are dualistically worked into a complete explanation of the universe.

Ordinarily, as over the past two centuries or so, the creationistic view has cohabited with the naturalistic view on the understanding that the former is a religious belief, whereas the latter is not based on religious belief. In other words, they exist in separate realms of reality. Under this cohabitation there would not be even the slightest suggestion that a religion--Christianity or Buddhism, for example--is a form of pseudoscience. But what has happened in recent decades is something quite different. Creationists whose views of the universe are based in religion are now claiming that their view is a scientific one. Peaceful cohabitation has been abandoned and the new breed, calling themselves creation scientists, seeks to evict traditional science from its house and to move in as the new occupant and owner. The counterclaim of the traditional scientists is that the creationist view of the universe, now labeled creation science, is in essence pseudoscience. What was formerly religious doctrine has now become a secular ism, i.e., creationism.

That being the situation, broadly described, it is clear that our first objective must be to investigate both science and pseudoscience from various angles in order to find some reliable criteria whereby one can be distinguished from the other. This is a major task of the first part of my book.

There is much to be gained by such a comparative analysis beyond establishing a set of distinguishing criteria. Perhaps the most important gain is an in-depth understanding of the nature of science. Years ago, one of our university students, applying for admission to graduate school after a term of overseas service in the Peace Corps in a remote third-world country, was asked what he had learned from that experience. His reply: "I learned what the United States is really like." Recently, I read the same statement by another Peace Corps returnee. Perhaps it has been made innumerable times and is only a cliche. We learn most about ourselves by comparing our state with another state that is very different.

I have not written this book with the hope of achieving the conversion of creation scientists to the naturalistic view of the universe. That would be naive on my part. Isms hold their adherents in a firm grip of belief, well waterproofed against reason. True, every now and then one of them defects, but no mass defection is to be expected. Creation scientists in their writings make frequent references to conversions of the reverse kind experienced by their staff scientists, some of whom held the naturalistic viewpoint during their training and early period of scientific research, only to be born again to join the fundamentalist fold. So I suppose there is a flux of defectors in both directions, but that should surprise no one.

What I do hope to achieve is a stronger bargaining position of naturalistic science in the school boards, legislatures, and courts of our nation. Scientists can speak with a much louder voice than they have thus far in saying "No" to giving equal time to the creationistic view in science classes in our public schools. Scientists must come forward with reasoned answers to every distorted and unsupported claim from creationism and from any and all pseudoscientific isms that seek to undermine and weaken science. I can't agree that this task should be sidestepped on grounds that the issues are merely philosophical in content. There is a large bonus for scientists in this attention to details. Science will benefit greatly by critical reexamination of every law and theory that we scientists tacitly accept as "true" or "valid."

How should we react when we are confronted by the following creationist assertions? "The speed of light was radically different in the past than it is now." "The so-called constants of radioactive decay have not been constant through time." "The fossil record has no evolutionary forms transitional between the genus Homo and any other of the primates." We scientists, when we take notice at all, usually respond: "Rubbish!" But does this response enhance the image of science held by parents of school children, by budgetary committees of legislatures, and by voters on local and state referendum issues? Arrogance begets hostility. Consider for a moment that supremely arrogant pronouncement issuing from the mouths of distinguished biologists and geologists in recent months: "Evolution is a fact!" Judging from recent public opinion polls that sort of statement is making little headway in reversing the majority opinion on human origins (favoring recent creation) alleged to be held by young people of college age. Well, is evolution a fact, or isn't it? If you can persist through many chapters you may come across my own opinion, and you may not like it.

When I began to write, I envisioned my prospective group of readers as mostly nonscientists—outsiders beyond the walls of the science establishment. Then, as I progressed into the more abstruse problems of the philosophy of science—horrendous subjects such as "reductionism in biology"—it gradually dawned on me that very few research scientists and probably not very many college teachers of science have ever given much thought to the nature of science. I quickly became aware of how little I knew of the philosophy of science, despite some dabblings in that area early in my career. Mostly, it was cut-and-dried positivistic stuff with lots of lip service to testing, falsification, and verification. Operational rules for actually doing scientific research were relatively simple and, once the requirements of the science community were understood, we could get on with the really important job, which was to get some graduate students, adequate contract funding, and keep the output of journal articles coming.

Science philosopher John Ziman (1980, pp. 38-39) has expressed some sadness that most practicing scientists exhibit no fervor in philosophical analysis of what they are actually doing. He nevertheless shows tolerance for their dereliction and takes a pragmatic view of the situation, for he writes:

> One can be zealous for Science, and a splendidly successful research worker, without pretending to a clear and certain notion of what Science really is. In practice it does not seem to matter. Perhaps this is healthy. A deep interest in theology is not welcome in the average churchgoer, and the ordinary taxpayer should not really concern himself about the nature of sovereignty or the merits of bicameral legislatures. Even though Church and State depend, in the end, upon such abstract matters, we may reasonably leave them to the experts if all goes smoothly. The average scientist

will say that he knows from experience and common sense what he is doing, and so long as he is not striking too deeply into the foundations of knowledge he is content to leave the highly technical discussion of the nature of Science to those self-appointed authorities the Philosophers of Science. A rough and ready conventional wisdom will see him through.[1]

I was pleased to find that Professor Ziman had some more to add, taking a harder line with the "rough and ready" scientists:

> Yet in a way this neglect of—even scorn for—the Philosophy of Science by professional scientists is strange. They are, after all, engaged in a very difficult, rather abstract, highly intellectual activity and need all the guidance they can get from general theory. We may agree that the general principles may not in practice be very helpful, but we might have thought that at least they would be taught to young scientists in training, just as the medical students are taught Physiology and budding administrators were once encouraged to acquaint themselves with Plato's Republic. When the student graduates and goes into a laboratory, how will he know what to do to make scientific discoveries if he has not been taught the distinction between a scientific theory and a non-scientific one? Making all allowances for the initial prejudice of scientists against speculative philosophy, and for the outmoded assumption that certain general ideas would communicate themselves to the educated and cultured man without specific instruction, I find this an odd and significant phenomenon.[1]

Perhaps some of us who came on the science scene a halfcentury ago would have paid more attention to the philosophy of science if it had seemed to relate to the natural and historical sciences in which we chose to work. Instead, we were turned off by philosophers who dealt almost exclusively with formal science and pure physics. The worst part of it was that these masters of the abstruse did not even recognize biology and geology as being science in the first place. I well remember a futile exchange of letters with Ernest Nagel, a leading philosopher of science on the same campus, exploring some ideas on the nature of historical science. Upon rereading those letters, I see now that he had no idea what I was trying to say and I, in turn, found nothing of relevance or value in his response. Things have changed in recent years. Contemporary philosophers such as Philip Kitcher (1982) are talking about things in the range of my experience, which is in geology that includes the history of the planet and the evolution of its life forms. These are the topics we deal with most in the debate with the creationists over origins. Perhaps with the emergence of a philosophical analysis of historical science, more young professionals in the natural sciences will get interested in the nature of their science.

---

[1]From John M. Ziman, Public Knowledge, Cambridge University Press, New York, pp. 5-27. Copyright © 1968 by and reproduced by permission of The Cambridge University Press.

Chapter 1

Science—A Preview

Before I can engage in a debate with the creationists, I must try to explain what science is all about. There are positive features to be considered as well as misconceptions to be straightened out. Many of my academic colleagues in science and philosophy could do the job more authoritatively than I; many of them have already done so. But, after reading their presentations of the subject in both popular and academic styles, I'm not sure they have actually represented science to the average literate adult in a way that makes clear essential differences between science and the various other forms of knowledge on which it constantly impinges. What is more, many accounts of science as an activity seem to be telling us what scientists are supposed to be doing, rather than how they actually do things.

One difficulty is that the giants in science and philosophy are prone to restrict the scope of science rather severely, usually to physical science, and even as narrowly as to theoretical physics alone. Our debate, on the other hand, concerns a large area of natural science quite far removed from the ideal behavior of matter on a subatomic scale. Besides dealing with the origin and physical evolution of the universe--a field that does indeed rest largely on principles of pure physics--we must investigate the geological and biological evolution of our own planet Earth over time spans of billions of years. Here we will find extremely complex aggregations of matter that have long and involved histories of development. Scientists who investigate these historical areas of knowledge need to adopt specialized views of science. Be prepared for exposure to an unfamiliar set of notions of what science is and how it works.

Science as Information

Acquiring a full appreciation of science is much like climbing a high, rugged mountain peak--it can only be done in steps, sweated out one by one. The beginning may be the very worst part, as it is for all difficult undertakings. Where to start?

The mountain of science is one of many mountains consisting of knowledge. Having said that, we are immediately in direct confrontation with a concept so elusive that about all we can do at this point is to find an excuse to get out of it and on with something more tangible. Situations like this are often handled by making a circle of words--it's called a tautology--in which, as you move around the circle, each word means the same thing as the word before it and the word that follows it. Try this out on "knowledge." Knowledge of something is acquired by knowing; to know is to have cognition; cognition is a condition of awareness; awareness is the knowledge of something.

A major preoccupation of philosophers from classical times to the present is to break out of that neat little circle; they make a market in it under the name of epistemology. One group of scientists, the neurobiologists, approaches the question of knowledge through a study of the brain and the way in which it reacts with and processes sensory stimuli it receives. A few pages further on in this chapter we will firmly tackle the meaning of knowledge. For the moment you should simply say: "I know what it is to know, so tell me something I don't already know."

Science philosopher Mario Bunge of McGill University thinks of the various mountains of human knowledge as cognitive fields (1984, pp. 37-38). This substitution of words may not seem like a significant accomplishment, but it does lead to a first attempt to organize and classify the mountains of human knowledge. Here is how Professor Bunge gets started on that organization:

> We shall characterize a science, as well as a pseudoscience, as a cognitive field, genuine or fake. A cognitive field may be characterized as a sector of human activity aiming at gaining, diffusing, or utilizing knowledge of some kind, whether this knowledge be true or false. There are hundreds of cognitive fields in contemporary culture: logic and theology, mathematics and numerology, astronomy and astrology, chemistry and alchemy, psychology and parapsychology, social science and humanistic sociology, and so on.[1]

Professor Bunge has not told us what knowledge is, but he has told us something new and important, something of great substance. Notice the pairing of the cognitive fields he has listed. The first member of each pair belongs to science; the second to pseudoscience, or false science. Not that all my readers will agree with the professor on the things he has put in the pseudoscience group; for if you are a believer in astrology or parapsychology, you may feel outraged.

Nevertheless, a salient point has been made here. The mountains of knowledge can be separated into two mountain ranges, between which lies a great gulf. So there are two kinds of cognitive fields. One, Bunge says, consists of belief fields, in which the knowledge rests on belief--belief in something that cannot be observed to exist. He cites religions and political ideologies as examples. He also puts pseudoscience in with the belief fields. The other, Bunge says, consists of the research fields, in which knowledge rests solely on observation of the real world. He puts science--both basic and applied varieties--in this category along with the humanities. Bunge gives us one distinguishing feature that clearly separates the two fields: "Whereas a research field changes all the time as a result of research, a belief field changes, if at all, as a result of controversy, brute force, or revelation" (p. 38).

Already, we are well launched on our climb up the mountain of science, since we have been given a valuable piece of information about our mountain and others of the same chain: it is in continuous change, so we must select our route carefully. The old trail shown on the map may be washed out in places. Others who climb after us may not find our trail passable and will choose another one that looks better. Those who like things neat and easy may want to climb a mountain in the other range, where well-worn trails are permanent and secure, amply fitted with strong handholds and guide ropes.

Perhaps we have jumped ahead a bit too fast and far, but three purposes have been served. First, we have escaped, temporarily, from a tautology. Second, we have put all forms of knowledge into two distinct compartments, or boxes. Humans seem to feel satisfaction in classifying things into compartments with solid barriers, particularly when each item fits only one compartment. Third, we have sown the seeds of conflict, and conflict guarantees interest and purpose. Human adherents to the knowledge in one compartment hate the guts of those subscribing to the kind of knowledge in the other one--at least some of them do. Some, of course, divide their own brains into two compartments, one for each form of knowledge, and seem to get along fairly well. George Orwell's word for this strategy is "doublethink." In any case, there is more than enough hostility to support a major ideological war, and that war is in progress. If you plan to enlist on either side, you would do well to prepare yourself. The war is being fought with words and ideas, so fill your mind with as much information about science (and pseudoscience) as it will hold.

The Scope of Science

If science is a fund or body of collected knowledge, we need first to decide what scope that knowledge covers or deals with. Let us switch from "knowledge" (that which is known) to a synonym, "information"; it is a word that seems better to remove the knowledge that resides in the brain to an external location where it is more readily accessible. We can then observe that science consists of a body of information that can be stored as well as freely communicated by humans. Information entering one person's brain, analyzed and temporarily stored there, can be transferred to the brain of another human only by use of symbolic statements in oral, written, or pictorial forms. In that form, information can also be available in a general storage pool.

If science is a fund of information, what is the information and what is the information about? About what does science inform us? The question takes us into philosophy, because our reply is very sweeping: Science gathers, processes, classifies, analyzes, and stores information on anything and everything observable in the universe. We like to think that science deals only with that which is real in the sense that it is observable and identifiable as matter or energy. Physicists who study the structure of matter say that all matter has mass, and that matter is composed of atoms. Protons and neutrons that form the nucleus of the atom possess most of the atom's mass, whereas the electrons that surround the nucleus have very small mass. Electrons in high-speed motion possess a great deal of energy. Atoms in motion also possess energy. But there is also a major form of pure energy in electromagnetic radiation; it consists of fast-moving particles called photons, and these have no mass. There is also another package of almost pure energy, the neutrino, which like the photon moves at the speed of light. Thus, physicists have some very definite information about the nature of matter and energy within the universe. We should also include antimatter with matter. It has even been possible to demonstrate in the laboratory that matter can be annihilated and converted into pure energy. These ideas find wide application in postulating how the universe came into existence and reconstructing the events that took place in the earliest moments of that creative process. (See Chapter 14.)

The area of science I have just described, dealing with matter and energy in all their forms and functions, is usually called empirical science; it deals with the content of experience, i.e., with objects of nature established by actual observation. The word "empirical" simply means "relying on experience or observation alone" or "based on observation and experience" (Webster's Ninth New Collegiate Dictionary). Separate from empirical science is an area of knowledge often called formal science; it consists of logic and mathematics. Logic and, to a considerable degree, also mathematics are claimed by philosophy as its concerns. Both have minimal need for examining facts of nature based on observation. Instead, they construct hypothetical or theoretical situations and work from these to conclusions by logical deductions. As Wartofsky puts it, these are "linguistic systems of deductive inference" (1968, p. 99). Of course, empirical science must follow rules of sound logic and must make use of mathematics.

The scope of empirical science is incredibly broad, poking its nose into every nook and cranny of the universe and into almost every facet of our lives. Physical science includes physics and chemistry; it deals with the structure of matter and the nature of energy. Then there is natural science; it includes such well-known subjects as biology, geology, and astronomy. An interesting aspect of natural science is that it has to deal with the history of events happening over vast spans of time, such as the history of our solar system and the evolution of life on earth. The ways in which the human race organizes itself, behaves, and functions individually and in groups are also a legitimate area for scientific inquiry through the social sciences; they include economics, cultural anthropology, psychology, and sociology.

Are there any areas of human thought and action off limits to science? Empirical science cannot partake of the arts, ethics, or religion, for the products of these fields cannot be evaluated by empirical science. The arts create new, imaginative structures of ideas and sensory experiences that appeal to the emotions. Aesthetic judgments, involving as they do matters of taste, preference, and cultural conditioning, cannot be challenged by science. In ethics, moral values that are chosen as good or right are not subject to challenge or verification by science. Religious tenets, which usually involve belief in supernatural entities, are also beyond the limits of scientific appraisal, but science can examine religion as a form of human activity, i.e., as a real phenomenon.

Supernatural forces, if they exist, cannot be observed, measured, or recorded by the procedures of science--that's simply what the word "supernatural" means. There can be no limit to the kinds and shapes of supernatural forces and forms the human mind is capable of conjuring up from "nowhere." Scientists therefore have no alternative but to ignore claims of the existence of supernatural forces and causes. This exclusion is a basic position that must be stoutly adhered to by scientists or their entire system of evaluating and processing information will collapse. To put it another way, if science must include a supernatural realm, it will be forced into a game where there are no rules. Without rules, no scientific observation, explanation, or prediction can enjoy a high probability of being a correct picture of the real world. With no rules, anyone can play the game and it becomes chaos, generating a highly unreliable product.

Mind and Body

The pursuit of science requires highly complex and advanced forms of mental activity. We cannot hope to understand how scientists evolve their hypotheses and laws without analyzing human mental processes. Mental activity is a function of the human brain. The mind, as we use the word here, is the functional and subjective aspect of the living brain (Campbell, 1974, p. 332). Or as biologist Gunther Stent puts it, the mind is the "epiphenomenon" (1975, p. 1055) of the brain, i.e., the mind is a secondary phenomenon that accompanies and is controlled by the brain.

In accordance with that definition we must take a position on a longstanding philosophical argument, that of the mind/body relationship. At the risk of greatly oversimplifying the issues, consider two rather straight-forward alternatives: (a) Mind and body are one and the same basic form of reality; (b) Mind and body occupy two separate and distinct realms of reality. In philosophy the study of the nature of reality is known as ontology. The first alternative is said to represent a form of monism; it is a monistic ontology. Under the second alternative, a dualism exists; it represents a dualistic ontology. In the monistic view, "mind" is in the same physical realm with "body," so both are within the purview of science.

The monistic view held by science can also be labeled mechanistic materialism, since it assumes that the functioning of the brain can be explained by the same scientific principles used to explain the sensory activities of the eye and ear, or the contractions of a muscle, or any other organic function. The billions of interconnected nerve cells that make up the working brain carry out all mental activity, including thinking, imagining, re-membering, and emoting. "Mental activity," says anthropologist Bernard Campbell, "means, broadly, the neuronal mechanisms that operate between stimulus and response" (1974, p. 342). Thus, images produced within the brain are also physical phenomena and can be treated by scientists along with any other activities of the brain. This materialistic view of mind as physiologically one with body is today the universal view of scientists.

Perhaps the most widely celebrated version of mind/body dualism was that of René Descartes (circa 1640). His total package, Cartesian philosophy, regarded mind (mental states or events) as being in a nonphysical realm, a totally separate realm from that of the physical or material world. Beyond saying that "mind" is an unsolved mystery, Descartes did little to enlighten his listeners on the nature of mind. He considered mind so completely removed from the physical realm that its activation requires the intervention of a Deity, acting through a third agency, the "soul." This suggests that Descartes did not regard "mind" as supernatural in the same sense that God is supernatural.

The dualistic mind/body concept has recently resurfaced, this time at the suggestion of a scientist of high reputation: George Wald, a 1967 Nobel Laureate. As reported in Science News (Thomsen, 1983), Wald, in discussing his recent studies of the visual systems of frogs as compared with higher animal forms, suggests that there is something he calls "consciousness" existing "outside the parameters of space and time." As Wald describes it, consciousness pervades the universe as a kind of supernatural "force" capable of directing the development of the material universe and even of having guided the course of cosmology. To find a reputable scientist proposing a theory of supernatural force is disturbing to the community of scientists. If the realm of matter and energy with which scientists work is being influenced or guided by a supernatural force, science will be incapable of explaining the information it has collected; it will be unable to make predictions about what will happen in the future, and its explanations of what has happened in the past may be inadequate or incomplete.

Perception and Conceptual Thought

If we agree that the human brain is the control center of scientific thought and scientific activity, it is worthwhile to follow the pathways of information into and

out of the brain and the processing and integration of that information.

We must elect an arbitrary starting point; let it be the sensory input from environment to the brain--from the outside to the inside. Most systems descriptions begin with the input from external sources. This input information is sensory; in the simplest case it requires direct use of the human sensory mechanisms--visual, auditory, olfactory, or tactile. The sensing process itself needs no further elaboration here. The process is continuous and entirely unconscious, often with no conscious purpose. What happens to the sensory stimuli when they reach the brain is quite another matter, for here a selective process is at work, as well as an integrative process making use of stored, previously processed information. What we observe through sensory perception from the outside, according to Professor Marx Wartofsky, a specialist in the philosophy of science, "is largely a function of intent and context, and depends to a great extent on frame of mind, attention, and what we know to look for" (Wartofsky, 1968, p. 101). In other words, "observing" or "seeing" is a guided process.

Bernard Campbell analyzes the processing of sensory data as follows (1974, p. 332). The brain's observation, which is its own mental image of the external environment, can be called a percept; it is based on two kinds of information: (a) input from the senses, and (b) memory of previous experience. The two forms are unconsciously and continuously combined as long as the input of sensory information continues.

Perception, then, is the formulation of percepts; it is a very different thing from the raw sensory data upon which it is partially based. Perception is an activity common to many animal species, for it requires only the sensory apparatus and a certain amount of memory. Nevertheless, it is the basic mechanism for the development of a much more complex brain function found in humans, that of formulation of a concept, or conceptualization. A concept is "an abstraction from the particular to the class" (Campbell, 1974, p. 334).[2] The abstraction is not a conscious activity, but it forms the material used in conceptual thought, or thinking. Thinking uses imagination, the ability to conceive of acts, events, or artifacts that have not yet been realized. Is conceptual thought a unique capability of the human species? This is a question we examine in Chapter 51. Conceptual thought increased in power in humans along with the increase in size and surface of area of brain and was accompanied by the invention and growth of language (Campbell, 1974, p. 336).[2]

Getting back to perception of the environment, the perceived image, or percept, is completely private within the brain of the perceiver. It is a form of individual knowledge, but of no scientific value unless it can be communicated to other brains. This communication can only be made by use of language. Perception, then, is formulated in language: "To perceive something seems to come to saying to oneself or to someone else, 'This is a so-and-so, or such-and-such.' So intimately tied to the framework of language is our perception that our identification of things and properties of things in the language may in effect influence what we see and fail to see" (Wartofsky, 1968, p. 104).

A perception (a percept) transformed to words (language) can be called an observation. To distinguish the noun "observation" from the verb form, we should perhaps call it an observation statement. An observation statement is expressed as a claim that "some proposition, P, is true." The object of an observation statement is "what's out there in the environment." The observation

---

[2]From: Bernard Campbell, HUMAN EVOLUTION, Second Edition. Copyright © 1974 by Bernard Campbell. (Aldine Publishing Company, New York.) Adapted by permission.

statement is perhaps the most commonplace kind of statement made by humans. The observation statement is also the unit building brick of the structure that is science. To meet the special needs of science, those that are scientific statements must conform to a special standard of quality, both in the manner in which they are arrived at and in the language by which they are transmitted.

Science and Language

Scientific data and theory, consisting of observation statements and abstracted concepts, are communicated by language; they are also stored in the form of language. Language, according to Professor Campbell, exerted a strong positive force in the evolutionary development of conscious conceptualization in humans. He writes:

> But the conscious concept did not appear unaided. It seems probable that it was finally evoked by the use of symbols, gestures perhaps, but more often words, which make up language. The concept "bird" could be brought into full consciousness only by its identification with the word symbol "BIRD." The word symbol was the twin of the conscious concept, and it seems probable that they were born together and grew together. (1974, p. 336)[2]

You might say that language enables the "private" concepts that originate in one human mind to "go public," to become exposed to the light of close examination by anyone who chooses to do so. Transferral of private observation statements and percepts to the public domain requires statement in words and sentences or, in the case of mathematical language, by special symbols. Those special symbols are, however, definable in words. We may also wish to include drawings and other forms of graphic expression as ways to transmit and store information. A drawing or map made by a scientist using visual observation of a subject can be taken as a kind of scientific observation statement, but words will ultimately be required to express the content of that drawing or map in a useful or meaningful way.

An important operational rule is that each word in a scientific statement must carry exactly the same meaning to all scientists, at least to all who practice in a given field or area of science. This rule requires that all words be precisely defined. Scientists must be very fussy about definitions, even if that seems painful to others. For each branch of science, one can usually find a glossary that attempts to define all terms found in published works dealing with that branch. Most glossaries are published through a scientific society that is strongly representative of the particular branch of science. If a term is used in two or more meanings, the glossary will make a special point of explaining each.

But now we come to a real problem. A particular word must be defined by using a set of different words. A nice word like albedo probably means nothing to most persons, even though it is a fairly short word and looks simple. It might be a good calling name for a dog! The words we use to define "albedo" must be familiar to most persons with a reasonably good education and trained to use a dictionary. To define "albedo," a scientist needs to use these words: ratio, electromagnetic, radiation, reflection, incident. They too are words of science, but they are widely used in science and most scientists do not have to search for their meanings. Persons who are not scientists may need to take each term by itself and ferret out its meaning in everyday, commonsense language. Only then can a person who uses "albedo" be absolutely sure that he or she is communicating the following message to another person: Albedo is the ratio of the amount of electromagnetic radiation reflected by a body to the amount incident to it (American Meteorological Society,

Glossary of Meteorology, 1959, p. 21). The same definition can be written in mathematical symbols, but this also requires that each symbol be defined in words.

What we see here is a kind of hierarchy of language frameworks. At the top is the special scientific framework needed for a particular branch of science. Below it lies a much more general framework of language that is common ground for all science. At the very base is a framework of words and meanings common to most literate humans in the daily affairs and activities of living. The process of expressing one framework in terms of the more general one that lies below it is known to philosophers of science as linguistic reduction. Reduction proceeds from the special to the more general. The reduction of scientific language to common language (commonsense meanings) is what gives science its credibility. What we have here in practice is "a chain of so-called coordinating definitions or reduction sentences, which lead from the complex theoretical formulation to the basic predicates" (Wartofsky, 1968, p. 116).

Scientists set up their own theoretical framework of observational procedures and the language by which observations are communicated and stored within their own circle. They agree on the framework--indeed, they must agree on it--although they may disagree on the interpretations of the observational statements and concepts themselves. By linguistic reduction, the language of that small, elite circle can be brought to broader and simpler statements capable of being more widely understood.

Professor Wartofsky gives a rather useful analogy to explain what linguistic reduction means and how it works (1968, pp. 119-120). Consider our monetary system. The scientists' theoretical language system can be likened to a banknote of large denomination, printed on a piece of paper of very little intrinsic worth. What gives the banknote value is that the holder can take it to a bank and redeem it in "coin of the realm," accepted by everyone in payment of a debt. (The analogy has, unfortunately, lost most of its value, in the United States at least, as the government has substituted base metals of little market value in coins that once had a substantial content of silver or gold.) Wartofsky says: "This notion that theoretical terms in science, like atom or magnetic field may be reduced, or translated into basic coin leads us finally to the outcome of the argument for reduction to basic predicates." The value of the basic coin, in turn, rests in the strength of "the whole system that underwrites the exchange--the monetary system and the public agreement that upholds it" (p. 120). At this point, one can perhaps think of some appropriate words of advice for the scientific community: "Keep one foot on the ground at all times. If the public doesn't understand what you are saying, they will lose confidence in you." It does look as if science stays alive at the pleasure of the commonsense persons who make up the bulk of human society, and not the other way around. A measure of humility on the part of scientists seems called for, in any case.

What Are the Facts of Science?

Can science give us the truth? Try looking up the word "fact" in a dictionary. You will probably find that a "fact" is something that is "actual" (i.e., an "actuality"). Try cross-checking to see what "actual" means and you will learn that something that is actual is a "fact" (or is "factual"). This is a neat little circle. Webster's Ninth New Collegiate Dictionary (1985) manages to throw down a smokescreen with its final definition of "fact" as "a piece of information presented as having objective reality." One of the main concerns of philosophers over the centuries is to ponder the nature of "objective reality." "Objective" simply means that your mind is supposed to go outside itself and perceive some material thing or event as it "really is" (i.e., "that which is factual"). Again, we go into a circular maneuver ending in self-defeat, but the experience is chastening. Every first-year student of philosophy gets a good dose of the painful strivings of philosophers to get to the very heart of the problem of reality. To confuse the situation further, let me introduce another word with which we can play the circles game-- "truth." Truth is that which is a fact, which is an actuality, which is a truth, which is a fact . . . ad infinitum. When is an observation statement a fact? When is it true?

May I suggest that in our search for insight into the nature of science we set some strict limits to how we use the words "fact," "actuality," and "truth"? Let us vow never again to say "Scientists discover the truth." Observation statements that are the building blocks of science and of all knowledge within the research fields are designed only to minimize the probability of failing to make a true statement. Let us admit that the human mind or brain will never be privy to truth, but rather agree that the special kind of observation statement that is a scientific statement, despite being put forward as being true, actually contains a certain probability of being in error. This is a concept that needs to be developed in depth, for it is the very essence of empirical science and what has come to be called the scientific method.

Closely linked with what the previous paragraph says is another statement we should vow never again to say-- or even to think in the privacy of our minds: "Scientists believe that . . . (such-and-such is what happened)." Webster's Ninth New Collegiate Dictionary says "to believe" is "to have a firm conviction as to the reality or goodness of something" and also "to have a firm religious faith." "Belief" is defined as "conviction of the truth of some statement or reality of a fact, especially when well grounded." Shouldn't we admit "belief" to the circles game? Now there are four players: fact, actuality, truth, and belief. I propose instead that we leave belief in the realm of religion where it belongs, along with all other nonempirical concepts such as questions of ethics and morality, which may be at least partly religious in origin. A scientist is free, of course, to believe in God and to be a religious person, but forbidden to express belief in any of the scientific observation statements that make up the body of science. There always is the possibility, no matter how small it may be, that a scientific statement is false.

In everyday life scientists, like everyone else, treat many phenomena as if they were facts--the "facts of life." We would be foolish indeed to take a chance that at the precise moment we step off a forty-story building the phenomenon of gravity will be inactivated. Yet a scientist would also be foolish to assert flatly that the gravitational attraction between two masses separated by a certain distance is, always was, and forever will be exactly the same.

I can anticipate several of my readers coming up with a challenge to my flat assertion that there is no place in science for such concepts as a fact, an actuality, or a truth. You will put it to me that some events are facts, pure and simple. Consider an act of terrorism committed before the eyes of hundreds or thousands of persons and documented by motion pictures and television. Take the case of the 1981 attempt upon the life of Pope John Paul II in the crowded courtyard of the Vatican. How can I claim that there is some doubt that a particular individual fired bullets that struck the pope? I yield to the obvious. Hundreds of times a day, we observe and react to events that are factual beyond reasonable question. Descriptions of such events can be called statements of fact because we find no alternative statement acceptable for serious consideration.

I would like to limit the definition of a "scientific statement" to one that is subject to a finite probability, no matter how small, of being in error when it is asserted to be correct. To use the word "error" implies that there exists an alternative statement (or several alternatives)

that can be advanced for serious consideration. Take, for example, the assassination of John F. Kennedy in Dallas on November 22, 1963. That the president received a fatal wound is a statement of fact, but there is some shadow of doubt that Harvey Oswald's rifle was the one and only rifle to fire a bullet at the president. To state that Oswald was indeed the only person to fire a weapon can be classed as a scientific statement, because there is a certain degree of probability that the statement is false. There is also an alternative statement to complement it: namely, that a second assassin fired from a position on the grassy knoll. Each of these statements seems to have carried a substantial probability of being in error; for otherwise the lengthy deliberations of a national commission of inquiry would not have received widespread interest and support.

Taking a closer look at statements of fact, we can recognize that they take much the same form. Mostly, they consist of a noun or pronoun as the subject, a verb, and a predicate that is commonly an adjective or adverb. I shall use examples from elementary mineralogy and geology:

The streak of hematite is red.

The crystal form of aragonite is different from that of calcite.

A major earthquake struck Los Angeles on the morning of April 7, 1962.

A landslide dammed the Madison River, forming a lake.

For the most part, a statement of fact describes a tangible substance, a structure, or a unique event as perceived by the senses and, therefore, subject to being recorded independently by use of commonplace devices, such as the camera or microphone. The statement can be "verified" by a multiplicity of observers and by reference to permanent records, such as photographs, sound tapes, or video tapes. In each case, the suggestion that the statement may be in error is not spontaneously forth-coming (although an observer can be accused of lying). We would not think of countering the first statement by another reading: "The streak of hematite is blue." We would not think of negating the second statement thus: "The crystal form of aragonite is not different from that of calcite."

To summarize, a statement of fact is an observation statement accepted pragmatically and operationally as being true (a) because to substitute a predicate of opposite meaning would be absurd (would violate common sense) and (b) because the statement is subject to verification by many observers and by mechanical means of documentation, with no expressed alternative statements (i.e., no dissent). One feature of the statement of fact is that, by itself, it is not very interesting. Interest arises when the statement leads to anticipation of some con-sequence that is of interest. In the courtroom, statements of fact lead to the interesting possibility that the accused will be convicted and punished. In science, statements of fact are grist for the mill of conceptualization, where the scientist's interest is concentrated.

Statements of fact and all observations based on perception are arrived at by a mental process said to be inductive, i.e., the process of induction. The primary information is sensory, flowing from the object to the brain, where it is made into a percept. Perhaps this will be easy to remember because the first syllable of "induction" is "in"--that which first flows into the scientist's brain. Simple induction is a large part of our mental activity in daily life and we could not survive without it. Induction yields information we want or ask for, as well as a great deal we do not need and would prefer not to receive.

Scientific activity is the investigation of that which is to some degree unknown or uncertain. If this were not so, science would be highly uninteresting, to say the least. It is the element of something unknown or uncertain that attracts and intrigues the human mind, as we know from the lure and excitement we feel as a murder mystery begins to unfold. And so, of course, the investigation of a homicide may also be a bona fide scientific investigation --forensic science, you might call it.

A note of caution needs to be inserted here. What I have said in foregoing paragraphs is intended to apply only to empirical science. It does not apply in the formal sciences of logic and mathematics. Both of those areas of knowledge deal in absolutes, using such words as "prove or disprove," "correct or incorrect," "verify or falsify." Keep in mind that formal science starts with a premise-- from inside the brain--rather than from that which is perceived from the outside. The investigation typically begins with such statements as "Suppose that we take two things, A and B . . ." or "Assume that two variable quantities, x and y, are related in a way that . . ." The initial conditions are defined in absolute terms and manipulated in strictly regulated fashion. The end result is a decision that the initial proposition is correct or in error. Thus, formal science is rigorous. Unfortunately, the absolutes of rigorous formal science tend to spill over into the thinking and expression of the empirical sciences. Philosophers of science are especially prone to blur the distinction between formal science and empirical science. For example, Sir Karl Popper, whose admonitions about evaluating scientific statements we will refer to on later pages, talks about when to declare a scientific statement to be false (Popper, 1959). We will find it necessary to transform Popper's logic-language into the language of uncertainty.

Let us next turn to scientific statements and the way they are evaluated and organized during scientific investigations.

Scientific Observations

Scientific statements can include informational packages ranging from the extremely simple to the highly complex. Those statements based on observation are initially arrived at through the same general inductive process we have already assigned to formulation of statements of fact. The simplest class of scientific statements consists of singular observations, meaning that they are unique observations, not repeated. The singular observation may describe an attribute of some object or substance or an event with a fixed place in space and time. Observations commonly consist of single measurements of mass, length, time, or temperature. Examples:

Meteorite A-345 weighs 5.32 kilograms.

Elapsed time of fall of the boulder was 2.09 seconds.

Maximum recorded acceleration during the earthquake measured 0.92 g.

Crystallization was first observed at a temperature of 303 C.

Simple as they seem, measurements are subject to errors, both those committed by the observer and those induced by the equipment used and by variations in the environment. Consequently, statistical theory is applied to the distribution and magnitude of the errors. It is customary to add an estimate of the range of error that can be expected when a measurement is repeated many times. This "probable error" is a statement to the effect that, if we perform the measurement repeatedly (or many different observers each make one measurement), a

certain percentage of the observed values will, on the average, fall within a specified range. There is also a statement of the probability that a single observation will fall outside a stated range. That probability diminishes rapidly beyond the stated range but never falls to zero. This is why we must never refer to the single measurement as a "fact." Generally, we refer to collections of such observations as raw data. With proper precautions based on long experience, scientists may judge the data to be "good," "sound," or "reliable" and proceed with an investigation as if they were dealing with facts. You might put it this way: "There is safety in numbers." With large numbers of repeated measurements carried out by many observers, using many different sets of apparatus, the probability of error has been reduced to an acceptable level.

From what I have just stated, it looks as if scientists are professional gamblers, highly skilled in estimating the odds of winning or losing! Indeed, they must be familiar with the "laws of chance" and the mathematics that goes into those so-called "laws."

At the risk of seeming to harp endlessly on a favorite theme, I have more to say about the use of "fact" and "truth" in scientific writing. When we say "this is a fact" and "this is true" we are speaking in absolutes. One cannot say "this is an almost-fact" or "this is an almost-truth." On the other hand, as with any concept of the absolute, one can indicate an approach to it, thus: "This seems close to being the truth." "This is almost certain to be a fact."

If we glibly say that "science is made up of facts" and "science contains the truth," we have already trapped ourselves in an untenable position, for we have also said implicitly that scientific knowledge, once put in place, cannot be discarded or replaced. Lacking that capability for self-correction and internal improvement, scientific knowledge could only grow in bulk, like a brick wall to which we can add more bricks, but replace none of those already set in mortar. To reject one fact in favor of another and contradictory fact is a also a contradiction in logic since, if the first one were vulnerable to being found later to be a nonfact, it could not have been a fact in the first place.

This is why, I suggest, the words "fact" and "truth" belong to formal science--logic and mathematics--but not to empirical science, which starts with observation of nature and, therefore, does not have an unmodifiable premise. In this sense empirical science is open-ended at both ends. In examining the religious doctrine of recent creation by a Creator, we will observe that the creationist doctrine is not open-ended at the front, for it begins with an assumption asserted to be a fact, to be true, and never to be altered or abandoned.

Of course, scientists and philosophers of science will continue to talk freely of "facts" and their use in research. The more perceptive ones simply change the common meaning of "fact" to one that is not absolute. I discovered this dodge in the writing of a leading light in the philosophy of science, Thomas S. Kuhn, where he speaks of determining facts with "greater precision" (1962, p. 25). In connection with examples of measurements of such values as positions of stars and boiling points--of physical constants, that is--Kuhn refers to "attempts to increase the accuracy and scope with which these facts are known." Clearly, he implies that "fact" is not an absolute term and that a statement of fact should not be equated to a statement of truth.

The Scientific Method

We have been discussing methods used by scientists to obtain observational data. You may have heard it said that science is actually nothing more than a special method for obtaining information. Should we define science as a special kind of body of information, as suggested on earlier pages, or as the method itself? Perhaps science is both of these things. I propose definitions for each.

Scientific knowledge: the best picture of the real world that humans can devise, given the present state of our collective investigative capability. By "best" we mean (a) the fullest and most complete description of what we observe, (b) the most satisfactory explanation of what is observed in terms of interrelatedness to other phenomena and to basic or universal laws, and (c) description and explanation that carry the greatest probability of being a true picture of the real world. Scientific knowledge represents the harvest of human endeavor; it is an artifact and its makers are fallible. Therefore, scientific knowledge is imperfect and must be continually restudied, modified, and corrected; it will never achieve static perfection.

Scientific method: the method or system by which scientific knowledge is secured. It is designed to minimize the commission of observational errors and errors of interpretation. The method uses a complex system of checks and balances to offset many expressions of human weakness, including self-deception, narrowness of vision, defective logic, and selfish motivation.

To these terms, perhaps we should add a third. Scientific community: the collection of humans applying the scientific method. As we show in Chapter 8, the scientific community has a rather distinct set of characteristics as a society in itself. In other words, there is a sociology of science to be considered in developing a full understanding of how scientists work. Believe it or not, scientists are humans and behave much the same as any other collection of humans.

Laws of Science

Everyone has heard of "laws of science," which are passed off as eternal truths about the real world. In the role of unforgiving tyrants, the laws of science dictate exactly what will happen as energy is transformed and matter is moved about or changed in various ways. There is a law of gravity that tells how rapidly an object falling in a vacuum will accelerate its speed. Laws of motion tell us what happens when one object strikes another. A law of frictional resistance limits the top speed of an automobile.

The use of the word "law" for these scientific statements has been criticized, and perhaps rightly so, because we also use "law" to mean an order or directive set up by humans (or by God). It has been pointed out by geologist James H. Shea that the often-used metaphorical phrase "laws that govern or control nature" (1982, p. 458) is an anachronistic concept. Its usage probably derives from the early period of modern science in which even the most independent scientists believed that God created the universe and set up immutable laws of behavior of matter and energy. It seems more apt today to recognize self-imposed laws that govern or control the scientist who practices science.

The word "law" is firmly entrenched in science and we will make use of it. A law of science is a form of scientific statement with special attributes. First, it is a very general statement in the sense that it is applicable over a wide range of time and space; it is a statement that applies over and over, countless times, in countless situations. This is often referred to as a universal statement. Second, the statement of the law, once formulated, does not vary with repetition. Third and most important, the statement enjoys an extremely small probability of being in error because it has been tested in application countless times by innumerable investigators without once having failed its tests. Wherever and whenever the law has been applied, it has successfully predicted the outcome of the events it explains or the experiment set up to test it.

Yet another characteristic feature of laws of science is

their interrelatedness and interdependence, as explained by philosopher John Hospers:

> The laws of science are not viewed in independence of one another. Together they form a vast body or system of laws, with each law fitting into a system including many other laws, each mutually reinforcing the others. The laws that scientists are most loath to abandon are those that form such an integral part of a system of laws that the abandonment of the one law would require the abandonment or alteration of a large number of other laws in the system. Thus an observation that directly confirms one law indirectly confirms a group of laws, because of the interconnection of the laws in a system. . . . Whether or not something is called a law, then, depends to a large extent on how deeply embedded it is in a wider system of laws (1980, p. 110).

Although we will refer often to laws of science, it is always with the thought in mind that they are not absolute truths, perhaps because there is no absolute certainty that a law as now stated has always applied in the past and will apply through all future time, or that it applies in other realms of space than that in which it has been tested.

The concept of unavailability of absolute truth in empirical science is put this way by Professor Herbert Feigl, a specialist in the philosophy of science:

> The knowledge claimed in the natural and the social sciences is a matter of successive approximations and of increasing degrees of confirmation. Warranted assertibility or probability is all that we can conceivably secure in the sciences that deal with the facts of experience. It is empirical science, thus conceived as an unending quest (its truth-claims to be held only "until further notice"), which is under consideration here. Science in this sense differs only in degree from the knowledge accumulated throughout the ages by sound and common sense. (1953, p. 10)

Philosopher John Ziman observes: "Our experience both as individual scientists and historically, is that we only arrive at partial and incomplete truths; we never achieve the precision and finality that seem required by the definition" (1980, p. 38). At the great risk of giving the scientific creationists a beautiful quote in support of their contention that mainstream science is really not science, I add this line from Ziman's paragraph: "Thus, nothing we do in the laboratory or study is 'really' scientific, however honestly we may aspire to the ideal." The creationists may also want to quote this line by Ziman on the same page as the above: "Many philosophers have now sadly come to the conclusion that there is no ultimate procedure which will wring the last drops of uncertainty from what scientists call their knowledge."[3]

Credits

1. From Mario Bunge, What is Pseudoscience? Skeptical Inquirer, vol. 9, no. 1, pp. 36-46. Copyright © 1984 by the Committee for the Scientific Investigation of Claims of the Paranormal. Used by permission of the author and publisher.

2. From: Bernard Campbell, HUMAN EVOLUTION, Second Edition. Copyright © 1974 by Bernard Campbell. (Aldine Publishing Company, New York.) Adapted by permission.

3. From John M. Ziman, Public Knowledge, Cambridge University Press, New York, pp. 5-27. Copyright © and reproduced by permission of The Cambridge University Press.

Chapter 2

The Scientific Hypothesis: I.
Formulation and Prediction

Scientific statements that describe and explain specific but complex phenomena and can be considered in opposition to one or more alternative statements are known as hypotheses. A particular hypothesis in its simplest form arises in the human mind, using percepts based on sensory inputs, as well as previously collected statements of fact and/or scientific statements that possess only a small probability of being in error. In some manner that we do not fully understand, the mind spontaneously organizes the bits and pieces of information it has received into chains, networks, or other configurations that may involve a major cause and its effect or effects, or many interacting causes and their effects. In other cases the information may be organized into a static description of a physical structure of matter.

At the outset, let us clear up a question that is perhaps trivial, but that can cause minor irritation if not attended to. What is the distinction, if any, between a hypothesis and a theory? Based on answers I have read and heard, my own conclusion is that the two words may be used interchangeably without causing confusion or consternation. Some of my colleagues have decided that a theory is a major hypothesis, one of great scope and importance, that has successfully met all or most challenges put to it. In that view, a theory is a hypothesis that has achieved special distinction and has been given a fancy title, much as knighthood or an honorary doctorate degree is conferred on a person in recognition of meritorious achievement. I would prefer to use "hypothesis" as the more general term, irrespective of the rank of the statement. A lot depends on context. I use "theory" in many places where that agrees with general usage, as we find in the case of the Big Bang theory of the creation of the universe and the Darwinian theory of evolution by natural selection.

Another consideration is that the word "theory" has at least two meanings in common use. Besides referring to the hypothesis itself as a structured statement about a specific phenomenon, the word can mean the theoretical aspect of a widely repeated phenomenon. For example, the flow of heat through any substance can be expressed as a set of mathematical equations having the status of laws, which we refer to collectively as "the theory of heat flow." "Theoretical" is the appropriate adjective for that meaning.

Having disposed of something trivial, we turn to something of really serious concern. In Chapter 1 we noted that science aims to provide the best possible description and explanation of the phenomena it observes. Does it follow, then, that a scientific hypothesis must always include both a description and an explanation? The concept of "explanation" is complex and quite elusive when examined in depth, as we do in Chapter 4. For the moment, hang on to the commonsense view of a scientific explanation as an answer to the question "How does it happen?" Can a perfectly sound hypothesis describe a phenomenon without offering any explanation? One need only look back into the history of science to show that the answer is a clear "yes." Take, for example, the Copernican hypothesis, or theory that the planets revolve about the sun. Called the heliocentric theory, it revived an ancient view of followers of the Greek philosopher Pythagoras and was placed by the Polish astronomer Nicholas Copernicus in direct confrontation with the prevailing geocentric theory, which put the earth at the center. The heliocentric theory was greatly strengthened by Galileo's telescopic observations of the the phases of Venus and the revolution of the moons of Jupiter. Then Johannes Kepler worked out empirically--from direct astronomical observations provided by Tycho Brahe--three fundamental "laws" of planetary motion. These were actually in the nature of a more detailed description of what actually happens.

Yet, during all this time there was no real explanation of the heliocentric model in terms of fundamental or universal laws of the behavior of massive objects in motion. It remained for Sir Isaac Newton to furnish what we would regard as an explanation of the heliocentric model. In that case, a set of universal laws of gravitation and motion was brought forward as a form of explanation. The hypothesis of plate tectonics was initially a descriptive statement of the interaction of lithospheric plates. Even today, with many proposed and tentative explanations of those plate motions, a fully satisfactory understanding in terms of deep mantle motions and their driving forces has not been achieved.

Certainly, then, a scientific hypothesis can consist of description without explanation. In such cases, however, acceptance of that hypothesis as valid science is implicitly conditional upon the eventual achievement of an explanation. In later chapters we show how the fundamentalist creationists attack science in general, and organic evolution in particular, by asserting that a scientific hypothesis must be explanatory or it cannot be acceptable as a scientific statement. They then assert further that the Darwinian theory of evolution contains no explanation; therefore, that it must be rejected on grounds of not being within the fold of science. We can, of course, refute that assertion by showing that hypotheses of excellent quality explaining evolution are already on the books and strongly corroborated by hard evidence. However, the essential point is that the creationists' assertion as to the necessity of explanation is a fundamental misconception in the first place.

The Hypothesis as a Creation

The mental process of hypothesis-making has been described as one of inspiration, as distinct from pure logical reasoning from information. J. T. Davies observes that a hypothesis "comes from an intuitive leap of the imagination, from inspiration, from induction, or from a conjecture" (1973, p. 12). How remarkable it seems that the most crucial process in all of science is so shrouded in mystery! In this process, the mind is creating a model to fit as best it can the prototype, which is the external reality. The model is made complete by inserting imaginary pieces of information to fill gaps in the available supply of input information. Once conceived, the model is transported out of the mind by translation into an organized scientific statement--the hypothesis--where it becomes available to perception by other minds and where it can be stored more or less permanently.

Astronomers find it most difficult, if not impossible at this time, to observe directly the birth of a star. The event is always obscured by a dark veil of cosmic dust and gas that absorbs all light from the newborn star. Only when the veil is later dissipated can the star be seen. (The existence of a veiled star can, however, be detected by other forms of radiation that can pass through the veil.)

The origin of something, whether of a great scientific hypothesis or of a star, is thus the most elusive part of the scientific quest. Evolution that follows origin is usually open to observation and can often be understood in remarkable detail by the scientific method. I would urge you, however, not to succumb to the belief that creation of a hypothesis or the birth of a star is a supernatural act, incapable of being analyzed by science and therefore not a part of science itself.

Without doubt, those scientists who have achieved a rating of greatness have had extraordinary talent for constructing new hypotheses of tremendous depth and consequence. Their names adorn the walls of the world's science museums and universities--Copernicus, Newton, Darwin, Einstein, et al. Those who study the history and philosophy of science, in collaboration with psychologists, have made many attempts to analyze the creative process in science and to list criteria that set apart the most creative scientists. Despite all their efforts the nature of that creativity remains obscure.

The views of Sir Karl Popper are interesting in this connection. As one of the most influential contributors to the philosophy of science in the middle decades of our century, his ideas on the origin of scientific hypotheses, as well as the testing and corroboration of hypotheses, have attracted many followers. Popper was born in Vienna in 1902, educated in the University of Vienna, and taught philosophy at Canterbury University in New Zealand during the World War II period. He then transferred to the faculty of the London School of Economics and was knighted in 1964. Popper argued strongly against the inductive method of arriving at general hypotheses of universal importance (1959, pp. 27-32). He considered that approach illogical and offered the following alternative:

> However, my view of the matter, for what it is worth, is that there is no such thing as a logical method of having new ideas, or as logical reconstruction of this process. My view may be expressed by saying that every discovery contains "an irrational element," or "a creative intuition," in Bergson's sense. In a similar way Einstein speaks of "the search for those highly universal laws from which a picture of the world can be obtained by pure deduction. There is no logical path," he says, "leading to these . . . laws. They can only be reached by intuition, based upon something like an intellectual love ("Einfuhlung") of the objects of experience." (P. 32)

> (Note: Exact source of Einstein's statement is given in a footnote on p. 32. Popper states that the English translation of Einstein's German text gives "sympathetic understanding of experience" as an alternative meaning for "Einfuhlung.")

Because Popper so positively rejects the inductive process as capable of producing an important scientific hypothesis, he might seem to be arguing for creation of ideas ex nihilo (out of nothing) or in vacuo (in a vacuum). I feel sure that Popper would not allow such an interpretation. Surely, any idea or hypothesis has an experiential basis drawing on prior knowledge of

something. Note that Einstein refers to "the objects of experience." Consequently, I would not eliminate inductive input, even if the hypothesis itself does not follow logically from that information.

After reading Popper's statement, including that by Einstein, it would worry me that readers might infer that the hypothesis arrives by Divine revelation, put in the scientist's brain by the Creator for a purpose, and thus God guides human understanding of the cosmos. The fundamentalist creationists seem not to have hit on this interpretation, which would bring the great influence of both Popper and Einstein to their support. Could not the creationists say: "The theory of naturalistic evolution came to Darwin as a divine revelation and, that being the case, it is clear that evolution is religion rather than science." They could not stop there, however, but would need to go on to say "all science, revealed to humans by a Creator, is religion." And, of course, evolution along with all other science would be accepted as the truth, because God has revealed it and God's word is infallible. (Please consider this suggestion as a footnote for future reference in our consideration of creationism in Part 2.)

And so, I return to the inductive theory of the origin of a scientific hypothesis, which we can then modify to produce a more realistic explanation for the development of a scientific hypothesis in place of the choice of it being either induced from observation alone or by creative intuition alone.

Induction or Deduction?

Formulation of a hypothesis requires a supply of information obtained by the inductive process. The small fragments of information collected by the observer are organized and ordered into a single scientific statement that accounts for all of the information available at the moment. It is rare that a scientist formulates a hypothesis entirely from new information that no other observer has previously received. There are, however, occasional occurrences of a totally unpredicted exposure to new information. One example might be the scene that greeted scientists inside the deep-sea submersible vessel, Alvin, when they reached the floor of the Pacific Ocean in the Galapagos Islands to encounter a troughlike depression. From a chimneylike tube of mineral matter attached to the rocky floor of the depression there streamed upward a jet of black water, as if from a smokestack. It proved to be hot water rich in dissolved mineral matter. Surrounding the "smoker" was a community of strange living forms, including giant clams and enormous wormlike "things" attached to the bottom and waving to and fro in response to water motions. What a shock! The hypothesis the Alvin scientists immediately sought to formulate was that a community of unique life forms is in some way being sustained by nutrients emitted by the smoker, all in a cold environment with no light whatsoever.

Purely inductive processes of formulation of a hypothesis can also arise in space exploration of planets and telescopic observations of the universe. Happening upon something totally unexpected and wholly unfamiliar forces the inductive process to take place, but even so, the scientist cannot help but refer to knowledge already gained in different situations to guide the design of a new hypothesis. This form of guidance by analogy is commonplace.

Let us pause to deal with an example of the formulation of a major geologic hypothesis--a model of the behavior through time of the earth's outer rock layer, or lithosphere. By the late 1950s, there had accumulated a body of scientific statements relating to the possibility that the earth's crust is actively spreading apart along a great rift (a crack) that lies at the crest of the mid-oceanic ridge. The remarkable symmetry of ridge topography and of ages of sediments deposited on the basaltic crust, in addition to documentation of high heat flow and an outpouring of basaltic magma (molten rock) along the axial rift, strongly suggested a spreading apart of the crust.

The hypothesis we will examine took the form of a statement that, as seafloor spreading occurs and new oceanic crust is continually formed, the ocean basins widen, increasing in areal extent, while at the same time the earth as a whole is expanding in volume. Thus, the earth's total surface area is increasing at a rate sufficient to match the increase in area of new crust. It was obvious that while the scenario of crustal spreading and growth of new crust has only a rather modest probability of being wrong, the concept of an expanding earth is speculative and is initially lacking in support from scientific statements with a low probability of being wrong. To postulate that the earth is expanding ("balloon hypothesis") requires a great deal of imagination-- something that carries a high risk of being wrong. The hypothesis as a whole thus has a substantial probability of being in error, even though some of its parts, standing by themselves, can pass the test of low error probability.

Where high-quality statements are not immediately available in support of the hypothesis, you, as the investigator, are forced into an inventive role. You must engage in the process of deduction, a predictive process. The meaning of "deduction" is easier to grasp if we think of it as meaning anticipation, a reasoning process by which an investigator is led to anticipate the finding of a relationship or event not previously observed or identified. Actually, what is anticipated may have already been observed and placed on record, perhaps noted by someone else in a different investigation, but has not yet come to the attention of the inventor of the hypothesis.

Statement of the hypothesis is directed or aimed at providing an explanation and an interrelationship of a specific collection of statements of facts and simple scientific statements. Thus, a scientific deduction drawn from that hypothesis starts with the specific terms of the hypothesis--the "givens," you might say--then turns to a general statement or law, or a group of such laws, as providing the underpinnings for a rather specific conclusion (Nagel, 1961, p. 23). The deductive process results in a prediction of missing parts of the explanation contained in the hypothesis. Prediction of the specific is thus guided and supported by laws of science that have an extremely small probability of being in error.

Take this example. A geologist is trying to imagine (visualize) the changes that take place in a body of magma (molten rock) as it cools and solidifies several kilometers below the surface. Let us first bring forward the general or universal laws that will apply:

Law: A fluid exerts upon a body immersed in it a buoyant force equal to the weight of the fluid displaced by the body.

Law: An immersed body of solid matter of higher average density than the fluid will sink because the weight of the body (downward force of gravity) is greater than the buoyant (upward) force exerted by the fluid.

The initial conditions are now stated. These may take the form of scientific statements based on observation and experiment; they are inductively obtained:

From laboratory experiments with igneous rocks heated above the melting point at various pressures, it can be stated that the density of the magma in question lies in the range of 2.8 to 3.0 g/cc. Experiments also reveal that the mineral olivine, with a density ranging between 3.3 and 4.4, is the first common mineral to crystallize into

solid grains as the temperature of the melt falls during experimental cooling.

The deduction we can now make is that in the magma body, crystals of olivine will be formed in abundance at an early stage in the cooling and will sink through the magma until they reach the bottom of the magma chamber; therefore, we can expect to find a basal layer of igneous rock rich in olivine. The geologist now goes into the field to attempt to confirm this deduction. If the base of the igneous rock body can be found exposed or a drill core can be taken, the deduction can perhaps be confirmed.

In looking back over the steps in this example, we note that the strength or value of the final deduction depends upon the general laws that back it up. In practice, a scientist does not actually repeat the laws that apply to a deductive chain of reasoning. Those laws are learned early in the scientist's career, are accepted by other scientists, and remain unspoken unless a scientific statement violates one of them. It is assumed that the hypothesis first formulated by inductive processes also conforms to the laws of science. Always, those laws are in control of what the scientist says and does.

Getting back to the expanding-earth (balloon) hypothesis, if you wish to raise it to a level of high quality, you must begin a search for some consequences that are attached to it. In this case, you may reason that, if the mass of the entire earth has not changed, volume expansion to a larger sphere must be associated with a decrease in density of at least a substantial portion or zone of the earth. Perhaps your first thought is that if some of the earth's mass is displaced outward, farther away from the axis of rotation, the rate of rotation will correspondingly decrease. Now you have a lead to follow. What independent evidence has already been gathered that bears on the possibility of change in rotational velocity over the past 100 to 200 million years? You are forced to search for some structure or property of rock that carries within it the record of length of a day.

You come across a published study in which the lines of banding on a single coral structure are interpreted as records of growth. Each line represents a day, or one rotation of the earth. Duration of one year can also be read from a rhythmic change in the thickness of growth lines in a much longer cycle, representing the year. Thus, the number of days in the year is established, and the duration of the day becomes known. The age of the rock in which the fossil coral is embedded can be determined independently. The investigation shows that, indeed, length of day has been increasing rather steadily for more than 400 million years. This is the confirmation of your deduction! But wait, you learn that a much simpler answer is available to explain the evidence. Tidal friction that the moon exerts upon the rotating earth can be expected to slow earth rotation. Calculations can be made of the rate of slowing from this cause. The calculations seem to fit well with the evidence from the corals. There is no need to associate the lengthening day with slowing due to earth expansion. Your deduction, though confirmed, can be explained easily by a natural process that does not require an expanding earth. Your entire hypothesis remains weak because independent evidence is lacking for an expanding earth.

The emergence of a newer hypothesis of the lithosphere, based in part on the older one, was not long in coming. It retained the requirement of seafloor spreading and the formation of new oceanic crust as two platelike masses pulled apart along the axial rift. Meantime, evidence favoring seafloor spreading was getting even stronger.

In the middle 1960s, new statements of low-error probability could be made in support of a scenario of seafloor spreading continuously for as far back as 200 million years. This was evidence of numerous reversals of magnetic polarity recorded in stripelike zones parallel with the mid-oceanic ridge axis and arranged in mirror-image patterns with respect to the axis. Ages of the polarity reversals were determined by independent means and spreading rates were estimated. A short time later another statement of low error probability emerged from the observation that the first crustal motion of earthquakes generated along transverse (transform) faults offsetting the ridge axis is always in agreement with a spreading motion. The probability that this statement is in error had been reduced to a very low value by application of established principles of seismology.

The newer hypothesis accepted the high-quality evidence of seafloor spreading, but turned to a mechanism by means of which the oceanic crust could be disposed of as fast as it was formed, but without requiring the earth's volume and surface area to expand. The suggestion had been made many years earlier that a line of deep oceanic trenches close to the continental margins indicated that the crust was being deeply downbent. This configuration suggested that perhaps the oceanic crust could move sharply downward into the underlying soft mantle. For decades seismologists had been recording the presence of many earthquakes originating at great depth in a slanting zone extending from the trenches far under the continents. This finding led to the suggestion that perhaps the oceanic crust, down-tilted as a single slab, is being pushed down far beneath the continents and that internal friction is setting off earthquakes. The pattern of deep earthquakes was discovered as early as 1930 and continued to be reinforced in the ensuing decades, but the information remained largely unused until the mid-1960s, when it became a key element in the newer hypothesis.

The newer hypothesis proposes that the lithosphere, a thick, strong rock layer of which the oceanic crust is the uppermost layer, moves horizontally as a unit away from the mid-oceanic spreading axis. At the margin of an adjacent continent the lithosphere bends down sharply and descends into the soft, highly heated zone below, called the asthenosphere. The descent of the lithosphere is called subduction. The plunging slab becomes heated in contact with the surrounding hot rock and gradually softens or melts and is thus absorbed to become a part of the asthenosphere. In this manner, lithosphere is recycled into the earth at the same rate that it is formed at the spreading axis. (See explanation and figures in Chapter 20.)

At this point in formulating the newer hypothesis, some key deductions were possible and could be tested. Because the down-plunging slab must be comparatively cold and brittle, its descent should be accompanied by a great deal of minor fracturing (cracking and slipping of the rock), giving off many earthquakes. Existing plots of deep earthquakes were already available to confirm this deduction. Further analysis of earthquake records showed that the orientation of stresses that produce the earthquakes is just what would be expected under the kinds and directions of stresses that the slab would experience. Thus, statements of low error probability became available to support the mechanism of subduction (in itself a single hypothesis) and to support the entire hypothesis as well.

The total scheme became known as the hypothesis of plate tectonics. It has found broad-based support among geologists as many new deductions have subsequently been made and found to be in agreement with observations. Any scientific hypothesis carrying the strength that the total plate-tectonic model now enjoys is not likely to be discarded in total and replaced in the future by an entirely different model. It will more likely be continually revised and restructured to decrease the overall probability that it is incorrect. In this way, science gains ground in understanding and explaining the structure and

behavior of all parts of the universe. Despite revisions and restructuring, large areas of science enjoy a certain level of permanence because the odds are extremely small that a cataclysmic overturning of scientific statements will occur on a grand scale.

Quality of the Hypothesis

A pervasive principle of science, one that we as students exercised in our graduate seminars and debates, applies in the formulation of hypotheses and in setting up scientific explanations of any kind. Commonly called the principle (or law) of parsimony, it instructs us not to bring in unneeded explanations or causes, if those we already have on hand are adequate to do the job. The law has been around a long time, for it is dubbed "Occam's razor," after William of Occam (circa 1285-1349), a Franciscan and philosopher who taught the concept at Oxford for a number of years. In the language of his time, the admonition goes: "What can be done with fewer assumptions is done in vain with more." Most college students of my time memorized the equivalent in Latin: Entia non sunt multiplicanda praeter necessitatem. Now that Latin has gone out of style, we can try to remember the English translation: Entities must not be multiplied beyond necessity.

A good example of the principle of parsimony occurs in the development of our balloon hypothesis of seafloor spreading and an expanding earth. As first formulated, it contained two explanations; first, the production of new oceanic crust by upwelling of magma along a mid-oceanic spreading rift; second, an expanding earth to accommodate the increase in global surface area. The second one is in a sense contrived to explain the first (i.e., to make the first possible). If we could apply Occam's razor to get rid of the second explanation, the hypothesis would be more parsimonious and, in that way, would become much stronger. This action was taken when we showed that plate subduction can dispose of the oceanic crust as fast as it is being formed, thus eliminating the need for an expanding earth. A single, unified mechanism suffices, and the hypothesis is internally coherent and consistent.

To summarize the concept of quality of a scientific hypothesis and to give it graphic expression that may make it easier to grasp, I have drawn a rising line with a set of dots (Figure 2.1). Think of it as a "ladder of excellence" on which each "rung" denotes a tenfold increase in quality. We must take into account two complimentary statements of probability:

P_T: Probability that the hypothesis is true.
P_F: Probability that the hypothesis is false.

The sum of these two probabilities is unity; therefore, as P_T increases, P_F decreases. However, we must keep in mind that neither value can actually reach either unity or zero--those limits can only be approached, so our ladder has no upper or lower end. Gambling odds on the hypothesis actually being a true statement of nature are given as the ratio of P_T to P_F. I have suggested adjectives to describe the quality of the hypothesis, but these are quite subjective because, as with people in general, scientists differ among themselves as to the risks they are willing to take in a given situation. At some point high on the ladder, the hypothesis may take on the status of a law of science.

I gather from reading Karl Popper that there may be problems with the concept of a hypothesis being evaluated on the basis of the probability that it is false or true. One problem that seems genuinely serious was raised by Popper in the middle 1950s. It concerns the comparison of hypotheses that are of quite different levels of importance in terms of their power to relate diverse phenomena and

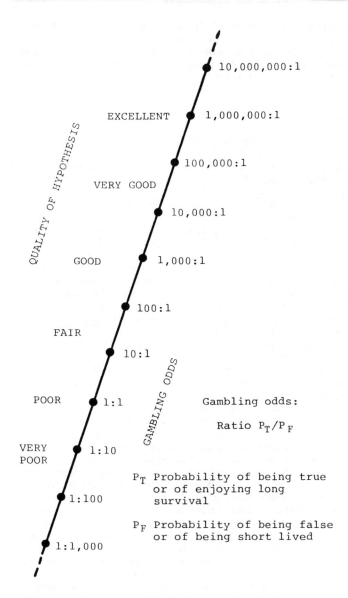

Figure 2.1 The ladder of quality of a scientific hypothesis. (A. N. Strahler.)

hypothesis on a different basis. We can think, instead, of a certain probability that the hypothesis will survive through time under continued testing and growth in complexity. We can think of some scale of value that expresses accurately the internal strength of the hypothesis. Consider that the ability of a hypothesis to survive will increase as does its ability to unify diverse phenomena under a single set of basic statements and its flexibility to adapt to the results of testing. Survival is then viewed in the context of actuarial statistics. What are the odds of the hypothesis enjoying strong support 20 years from now? or 50 years? or 100 years? This actuarial viewpoint takes into account not only the present strength of the hypothesis, but also its potential for internal change to meet stresses imposed by massive inputs of new information and the impacts of rival hypotheses. In any case, the concept remains one of probabilities, rather than one of absolute acceptance or rejection. We seem to have latched on to the evolutionary concept of survival of the fittest--the fittest hypothesis, that is.

Hypothesis and Prediction

From our example of a working hypothesis and its deduced consequences, leading to its partial rejection and the substitution of a new working hypothesis, it is evident that the scientific method is made complex by feedback mechanisms of both empirical observations and deductions.

In almost every field of science, explanatory hypotheses have evolved through numerous stages of deduction, testing, and revision. The scientist becomes familiar with accumulated hypotheses and empirical knowledge from student days--the content of professors' lectures, textbooks, journal articles and treatises, laboratory exercises, and finally, the master's essay and the doctoral dissertation. Research early in one's career is commonly an extension of research carried on for years by the student's supervising professor. It is almost impossible to say which comes first in scientific research: empirical data arrived at by induction from experiment and observation, or an imaginative and creative hypothesis. Some modern interpreters of philosophy of science follow Sir Karl Popper in placing emphasis on the latter phase; they characterize the scientific method as being a hypothetico-deductive process. Professor Francisco J. Ayala, a distinguished geneticist who has done extensive evolution research using the fruit fly (Drosophila), describes the hypothetico-deductive process of science as follows:

> Science is a complex enterprise that essentially consists of two interdependent episodes, one imaginative or creative, the other critical. To have an idea, advance a hypothesis, or suggest what might be true is a creative exercise. But scientific conjectures or hypotheses must also be subject to critical examination and empirical testing. Scientific thinking may be characterizeed as a process of invention or discovery followed by validation or confirmation. One process concerns the acquisition of knowledge, the other concerns the justification of knowledge. (1977d, p. 478)[1]

Ayala clearly gives the initial position to the creative act of formulating the hypothesis; validation through the deduction of consequences and their testing by empirical methods follows. But surely the hypothesis could not have been formulated in total absence of a rather large supply of knowledge, a supply that could only have flowed into the scientist's brain from outside sources, prior to the imaginative act.

Ayala pursues his point at greater length, but then, finally, in his last sentence he gives inductive activity its just credit:

to yield important deduced consequences. Popper's statement of the problem is as follows:

> I pointed out that the probability of a statement (or set of statements) is always greater the less the statement says: it is inverse to the content or the deductive power of the statement, and thus to its explanatory power. Accordingly every interesting and powerful statement must have a low probability; and vice versa: a statement with a high probability will be scientifically uninteresting, because it says little and has no explanatory power. (1980, pp. 31-32)

If my interpretation of Popper's statement is correct, it says that what I have described as a simple statement of fact would rate highest on the scale of quality; whereas a powerful hypothesis such as plate tectonics, with its enormous explanatory power and capability of producing testable deductions, would automatically get a much lower rating. Popper says that "we do not seek highly probable theories but explanations; that is to say powerful and improbable theories" (p. 32).

How can this difficulty be circumvented? I suggest that we can salvage the probability idea by evaluating the

Hypotheses and other imaginative conjectures are the initial stage of scientific inquiry. It is the imaginative conjecture of what might be true that provides the incentive to seek the truth and a clue as to where we might find it (Medawar, 1967). Hypotheses guide observation and experiment because they suggest what to observe. The empirical work of scientists is guided by hypotheses, whether explicitly formulated or simply in the form of vague conjectures or hunches about what the truth might be. But imaginative conjecture and empirical observation are mutually interdependent processes. Observations made to test a hypothesis are often the inspiring source of new conjectures or hypotheses. (P. 478)[1]

Professor Ayala has set down a list of four different activities that are involved in evaluating the quality of a scientific hypothesis (p. 479). First, the hypothesis must display internal consistency. It must be logically well formed and must not contain self-contradictory statements. Second, it must have explanatory value. The hypothesis must call upon causes or forces that are at work to produce what is observed to happen. Merely rephrasing the problem by use of synonyms (a tautology, that is) is not acceptable as an explanation. Third, the hypothesis must be evaluated to see if it really makes an advance in the state of knowledge as compared with existing hypotheses. For example, it must explain some observations that other hypotheses have not been able to explain, while at the same time being always consistent with basic laws of science that pertain. Fourth, the hypothesis must be vulnerable to being rejected through the use of empirical tests. The hypothesis must have the power to make "predictions" that can be put to such tests. Let us look into this fourth requirement in more detail, for its meaning must be understood.

Philosophers of science are prone to repeat that a hypothesis must carry the power "to make predictions." At first, I found it difficult to understand just what they meant by a "prediction." Most of us think of a prediction as a statement of some event that has yet to happen (i.e., a prognostication). Suppose that a scientist has devised this hypothesis: An earthquake is caused by a sudden slippage along a surface between two large masses of rock, releasing a large quantity of energy that has accumulated as a buildup of elastic strain, storing energy in increasing amounts over a long period of time. Let us suppose that testing of the hypothesis has produced scientific statements in agreement with the deduced consequences of the hypothesis, based on observations taken before and after several earthquakes. Using the hypothesis a scientist is then able to predict an earthquake in the future by pointing out those changes that can be observed as precursors to the event, for example, observed strain (bending) accumulating in rock on both sides of a fault.

Science philosophers have another meaning for the word "prediction"; it is the ability of a hypothesis to lead to deductions of scientific statements that were not anticipated when the hypothesis was formulated. This meaning of prediction we have already dealt with in some detail in our account of the deduction of consequences by means of which a hypothesis can be strengthened or weakened. Those deduced consequences do not refer to future events; they deal with phenomena that are presently observable or have already occurred. Although this is what the science philosophers mean by "prediction," the word can mislead the general public. Most persons don't associate "prediction" with a statement of what has already happened. The dictionary defines "predict" as "to declare in advance, to foretell." What the science philosophers really mean by "predict" is to foretell what the scientist will find if he or she goes out and looks for what is deduced as a consequence of the hypothesis. When the prediction is an inference of a past event or state of things that has yet to be discovered or verified, it is sometimes called a retrodiction (see, for example, Root-Bernstein, 1984, p. 11.)

The question is, must one or more deduced consequences always be associated with a scientific hypothesis? If the answer is "yes," we must say: A statement that cannot produce a deduced consequence is not a scientific hypothesis. Sir Karl Popper has made this pronouncement and it has been repeated by his students and followers for many years. In published arguments over whether organic evolution is a scientific hypothesis, we find the word "predictability" kicked around like a ball between opposing soccer teams. In another chapter, discussing organic evolution under attack by fundamentalist creationists, we will need to review these arguments and try to discern what the opposed groups are really trying to say. For now, it is certainly fair to conclude that a scientific statement sufficiently complex to be called a hypothesis is almost always capable of yielding a number of deduced consequences that should be investigated by empirical methods.

Credit

1. Francisco J. Ayala, Chapter 16, pp. 474-516 in T. Dobzhansky, F. J. Ayala, G. L. Stebbins, and J. W. Valentine, Evolution, W. H. Freeman and Company, San Francisco. Copyright © 1977 by W. H. Freeman and Company. Used by permission.

Chapter 3

The Scientific Hypothesis: II.
Falsification and Rejection

In reading various articles and published letters about the nature of science and how it differs from religion, I frequently came across the statement that a scientific statement or hypothesis must always be subject to the possibility of being falsified (whereas a statement of religious belief cannot be falsified). I immediately took the word "falsified" to mean that a scientist could put forward a scientific paper containing false statements--falsehoods, that is--in support of a hypothesis. The scientist, I assumed, could invent and record fictitious data from a laboratory experiment or a field study. I was puzzled because I felt sure that a religious leader or prophet could just as easily invent a false religious experience to gain support of followers and make new converts.

My puzzlement quickly came to an end when I realized that the verb "to falsify" can be misleading as used in the papers I was reading. Although Webster's Ninth New Collegiate Dictionary does define "falsify" as "to prove or declare false," it also gives alternative definitions: "to represent falsely (misrepresent)" and "to tell lies." It was the latter meaning that I was reading into the text.

Falsification and Corroboration

A typical statement on the requirement of falsifiability is this one given by Professor Ayala:

> The critical element that distinguishes the empirical sciences from other forms of knowledge is the requirement that scientific hypotheses be empirically falsifiable.

> If the results of an empirical test agree with the predictions derived from a hypothesis, the hypothesis is said to be provisionally corroborated, otherwise it is falsified.

> A hypothesis that is not subject, at least in principle, to the possibility of empirical falsification does not belong in the realm of science. (1977d, p. 479)[1]

Ayala is evidently a faithful disciple of Sir Karl Popper, the philosopher largely responsible for publicizing these concepts of prediction and falsifiability (Popper, 1959). Ayala goes on to state:

> The requirement that scientific hypotheses be falsifiable rather than simply verifiable may seem surprising at first. It might seem that the goal of science is to establish the "truth" of hypotheses rather than attempt to falsify them. (1977d, p. 479)[1]

My complaint with the Popper terminology arises from the observation that the words "verify" and "falsify" are absolutes. This quality of absolute certainty about a statement being correct or incorrect is exactly what I tried to throw out in my earlier discussion of the words "fact," "actuality," and "truth." Instead of such

absolutes, I proposed a probability concept--the odds of being in error when a statement is declared to be true or false. I stick to that concept here, for it would be most brash and nonobjective for a scientist to declare a scientific hypothesis to be false or true, no matter how decisive the outcome of the empirical tests might appear.

Popper's requirement of possibility for falsification of a scientific hypothesis raises questions much more serious than merely one of definitions. I argue that, just as a scientific statement cannot be proven to be true, neither can it be proven to be false. The most that can be done as science is to evaluate or estimate the probability that a statement is false or that it is true. The two values of probability are complementary. If the probability that the statement is false is valued at 1 percent, the probability that the same statement is true is 99 percent.

We have here two complementary values: in Popper's terminology, they are falsification and verification. In my probability terminology, they are, respectively, a large probability of being in error versus a small probability of being in error. My system is open-ended. Popper's is closed. (More about this distinction follows.) I have no doubt his terminology will prevail; it is already deeply embedded in the literature of philosophy of science.

Questions of terminology aside, there is an important point to be made about the alternatives with respect to a given hypothesis: shall it be accepted or rejected? These two alternatives are not symmetrical in power. If a deduced consequence should, on empirical examination, be identified as predicted, the hypothesis itself would not stand as proven correct. The reason is that some alternative hypothesis might also stand to be verified by the same test data. On the other hand, if the deduced consequence should be shown to differ from observation, the hypothesis from which it derives would automatically be shown also to be incorrect. (Note that I have used absolutes--"correct," "incorrect"--for the sake of the logic of the thing. We continue, using the language of logic.)

For example, Hypothesis A leads to the formulation of four deduced consequences, C_1, C_2, C_3, and C_4. The consequences are investigated and found to exist as deduced. That the consequences are all true does not prove that the hypothesis is true, because there may be a second but quite different hypothesis, Hypothesis B, for which the same deduced consequences apply. Thus, tests that are passed by Hypothesis A simply permit it to stay alive. On the other hand, if a fifth deduced consequence, C_5, would prove not to exist when tested empirically, Hypothesis A would be destroyed. We conclude that a hypothesis can be shown false (can be falsified), but can never be shown true.

Converting back from "absolutes" to "probables," we can say that a deduced consequence affirmatively shared by two or more conflicting hypotheses can do little or nothing to improve the quality of those hypotheses. If, however, a particular deduced consequence required of a favored hypothesis is incongruous to other rival hypotheses, that particular consequence is crucial; it sets up the test whose result can drastically increase or decrease the probability that the favored hypothesis is correct. Crucial tests that are successfully passed will tend to downgrade or to eliminate rival hypotheses for which a different test outcome is predicted. Thus, one hypothesis can improve in quality as compared with rival hypotheses.

Scientists are sometimes tempted to push a strong hypothesis beyond the limits of relative and probable validity to an absolute statement, such as "corroborated (or established) beyond reasonable doubt" or "shown to be a fact." See how Professor Ayala pushes toward this limit in evaluating the hypothesis of organic evolution:

> The larger the variety of severe tests withstood by a hypothesis, the greater its degree of corroboration. Hypotheses or theories may thus become established beyond reasonable doubt. The hypothesis of evolution, that new organisms come about by descent with modification from dissimilar ancestors, is an example of a hypothesis corroborated beyond reasonable doubt. This is what is claimed by biologists who state that evolution is a fact rather than a theory or hypothesis. In ordinary usage, the terms "hypothesis" and "theory" sometimes imply a lack of sufficient corroboration. The evolutionary origin of organisms is compatible with virtually all known facts of biology, and has passed a wide variety of severe tests. (1977d, p. 481)[1]

Ayala's comments on the status of the hypothesis of organic evolution through geologic time apply directly to the current debate over two opposing views on the origin of living forms on earth: scientific evolution vs. fundamentalist creationism. The creationists believe that the earth and all its life forms were created by God in a six-day period somewhere in the time range of 6,000 to 10,000 years ago. They seek to repudiate the biologists' entire reconstruction of an evolutionary sequence of life forms over a span perhaps as long as 3 to 3½ billion years. Evidence that evolution took place through geologic time rests largely on fossil remains of plants and animals in rock strata whose ages can be determined independently. The evolutionary "tree" establishes the order of appearance of each class of life forms as a "branch" of the tree. To date, all fossils that have been found can be represented by a single-trunked tree with numerous branches.

The creationists claim that the established sequence of evolutionary changes cannot be shown to be false, and therefore evolution is not a scientific hypothesis. The biologists point out, however, that the possibility always exists that fossils of a given life form will be discovered in the same strata with other forms that have heretofore always been limited to a different relative position and age. For example, it is conceivable that a scientist will some day discover human bones among dinosaur bones in such a relationship that it is judged highly likely that humans and dinosaurs lived at the same time. Such a finding would deal a crushing blow to the widely favored hypothesis of a unique evolutionary sequence. In Popper language, the hypothesis of evolution would be falsified. So far, no incongruous relationship of this type has been found, but there remains the remote possibility that such a find will be made and, if so, its impact would be to greatly increase the probability that the hypothesis of evolution is incorrect.

Geologists and biologists don't seem very worried that a downgrading of the evolutionary model will occur, and some have gone so far as to state bluntly to the public and press that "evolution is a fact." When arrogance like that begins to take over from skepticism among scientists, they are inviting trouble. As one scientist has remarked, "Science is based upon skepticism" (Root-Bernstein, 1981,

p. 1446). The best way to practice skepticism is to look at every scientific hypothesis as having, from the moment it is formulated, a built-in probability of being in error. With such an outlook, the scientist is much less likely to fall in love with his or her hypothesis and become blinded to whatever defects may appear in it from time to time.

I will discuss the controversy over evolution in some detail in Part 2 and ensuing parts. The controversy is much more complicated than I have sketched it here. The major scientific uncertainties that beset the evolutionists from within their own group concern the processes by which evolutionary changes came about and the tempo at which those changes proceeded. The evolutionary tree itself is by no means a perfect and whole structure. Here and there, sections of branches and trunk are missing but the whole tree is not likely to be toppled in decades to come. Perhaps a little pruning and grafting will make it an even better tree in terms of survival!

In recent years, the principle of falsification as the absolute criterion that every scientific hypothesis must satisfy has come under fire from philosophers of science. This topic is reviewed by science philosopher Philip Kitcher of the University of Vermont in his book _Abusing Science_ (1982). His discussion is prompted by the fundamentalist creationists' claim that the theory of evolution is incapable of being falsified, and thus cannot be accepted as empirical science. In Part 2, I deal with this argument in some detail. Here we need to reexamine Sir Karl Popper's rather firm and inflexible declaration that for a statement to be a scientific hypothesis it must always be falsifiable. When couched in strict language of formal science--logic, that is--Popper's principle may seem obviously sound. Kitcher points to pitfalls in application of the now-traditional falsifiability criterion in empirical science (1982, pp. 42-44). He says:

> The time has come to tell a dreadful secret. While the picture of scientific testing sketched above continues to be influential among scientists, it has been shown to be seriously incorrect. (To give my profession its due, historians and philosophers of science have been trying to let this particular cat out of the bag for at least thirty years. See, for example, Hempel 1941; Quine 1952.) Important work in the history of science has made it increasingly clear that no major scientific theory has ever exemplified the relation between theory and evidence that the traditional model presents. (P. 42)[2]

(Note: References for Hempel and Quine are given by Kitcher and are not repeated here.)

The subject can be extremely difficult to grasp, unless one is highly skilled in the language of logic, but Kitcher provides a clear example of a naive application of the concept of falsifiability that could have done irreparable damage to Newtonian laws of gravitation and motion. Newtonian celestial mechanics proved remarkably successful in predicting the orbits of the planets--successful, that is, until observations were improved in precision to make it obvious that the orbit of Uranus consistently failed to follow the calculated orbit. Uranus was at that time the outermost known planet of the solar system. Using the naive falsificationist criterion, it would have been necessary to declare the Newtonian laws false, this in spite of their excellent predictive performance for all the other known planets. A less drastic alternative was, however, available. Another as yet undiscovered planet could be postulated to exist in orbit beyond

---

[2]From Philip Kitcher, _Abusing Science: The Case Against Creationism_, The MIT Press, Cambridge, Mass. Copyright © 1982 by the Massachusetts Institute of Technology. Used by permission.

Uranus; its orbit could be calculated in such a way as to account for the seeming discrepancy in the observed orbit of Uranus. Here, then, was a prediction independent of the major hypothesis itself, but embodying the same laws and principles. Not long after, the missing planet was spotted by telescope. Named Neptune, its observed orbit agreed with the calculated orbit. The main hypothesis was saved.

Perhaps the game of chess can serve as a device to show how science can stave off the fate of falsification. Your opponent has put your king in check (meaning that your king is vulnerable to being taken off the board in opponent's next move). Opponent, seeing no way for you to get out of the situation, calls "Checkmate." Momentarily, you see no escape and seriously consider accepting defeat. Suddenly you see a way out: moving your knight between your king and opponent's queen can take you out of check. A few moves later, your opponent maneuvers you into another check; again you move a piece and save the situation. That move may result in loss of one of your pieces. This form of evasion can go on more or less indefinitely. In a similar manner, a complex scientific hypothesis involves many "chess pieces," each of which can be manipulated independently. Falsification can be avoided, perhaps indefinitely, by successive manipulations of only one piece at a time, or simply allowing one piece at a time to be taken off the board. As in the game of chess, science allows the hypothesis to be restructured, within certain rules, without limit. Thus naive falsification is circumvented and it may be futile to attempt to achieve it. But there is usually a final situation of checkmate in the game of chess. Is not an absolute and final falsification also possible during testing of a scientific hypothesis?

Kitcher asks us to consider the possibility that individual laws of physical science, treated apart from other laws, are neither testable nor falsifiable. He argues: "On their own, individual scientific laws, or the small groups of laws that are often identified as theories, do not have observational consequences" (1982, p. 44).[2] On the other hand, he does not explain to his readers just why this should be so. Take, for example, the essential assertion of the law of gravitation--that masses attract each other--and consider how you might wish to test it on the tentative assumption that it is a hypothesis. You can, of course, repeatedly raise a paperweight to shoulder height and release it, noting that it always drops to the floor. This act would not, however, qualify as a test, because the predictive statement "it will drop every time" is not an independent hypothesis; it is merely a special case of the statement of the hypothesis itself. Why is the experiment not an independent test (i.e., an observational consequence) of the hypothesis? Perhaps, in terms of logic, the answer is that the law itself was derived as a generalization of repeated observations that unsupported small objects always fall through the air toward the earth's center (more or less) unless other directed forces of greater intensity (such as strong updrafts of air) exceed the downward pull of the earth's mass upon the small object. Granting that, as Popper and Einstein seem to be telling us, Newton's brilliant generalization "all masses attract each other" may have come as sheer free-floating inspiration, it is more realistic to suppose that observation of nature drove Newton to his conclusion. In that case, to repeat the observations that initially forced the universal statement creates a circularity of reasoning. As long as one cannot get out of the tight little circle--observation, induction, deduction, and return to the same observation--it is perhaps impossible to falsify the universal law.

This may be a good point at which to attempt further distinctions between universal laws, on the one hand, and hypotheses, on the other. "Laws," it is sometimes said, "are discovered, not made" (Hospers, 1980, p. 106). The first point is, therefore, that a particular law "tolerates

no alternative," so to speak. Hypotheses are synthesized within the human mind. In contrast to a law, a given hypothesis is an artifact that can be matched off to an alternative hypothesis (or several such alternatives) describing the same set of observations.

Second and more important, perhaps, is that hypotheses most commonly are propositions designed to describe specific classes of events or states of matter. If relating to an event--a "happening"--the hypothesis tells us details of the specific succession of events that occurred. If relating to a specific state or structure of matter--the atomic structure of the diamond crystal, for example--the hypothesis describes a unique crystal lattice of carbon atoms. It may also tell us how the crystal lattice came into existence; that, too, was a event or happening. A third difference, yet to be presented, is that universal laws explain nothing, whereas hypotheses usually include explanation, and the explanation ultimately reaches down to the universal laws. Having thus further elucidated (or obfuscated?) the status of laws as distinct from hypotheses, we are ready to return to Kitcher's thesis that complex and powerful theories (hypotheses) are not easily, if ever, completely falsified. He says:

> This crucial point about theories was first understood by the great historian and philosopher of science Pierre Duhem. Duhem saw clearly that individual scientific claims do not, and cannot, confront the evidence one by one. Rather, in his picturesque phrase, "Hypotheses are tested in bundles." (1982, p. 44)[2]

> (Note: No reference to Duhem is cited. Pierre Duhem, 1861-1916, ranks as one of the founders of physical chemistry.)

Most major scientific hypotheses, Kitcher says, are "large bundles" of statements. Certainly this is true for organic evolution and plate tectonics. They include many smaller or lesser statements that are the prime targets for testing as independent hypotheses. If a lesser or secondary statement appears to be falsified by its disagreement with a test result, only that secondary statement need be eliminated or revised. We need not throw out the entire "bundle," for to do so would be to lose most of the knowledge it has gained for science. Successful testing of the larger portion of the bundle can preserve this knowledge.

Kitcher and those who share his views about falsifiability seem to be telling us to play down the Popper principle of falsification, which usually turns out in practice to be an unsupportable naive falsification, and to concentrate instead on something positive. Although logic can indicate that a hypothesis can never be proved true as a result of tests that turn out to be positive, surely the testing does more than just keep the hypothesis alive. Successful tests, involving as they do independent hypotheses, draw in new information. Added to the main hypothesis, the successes of test hypotheses increase the total scope and explanatory power of the main hypothesis. Popper grasped this idea in his "positive theory of corroboration" (1959, pp. 265-69). When a hypothesis stands up to numerous rigorous tests, it acquires a degree of "corroboration," which is a quality of strength independent of considerations of whether it is falsifiable. Popper ties in the degree of positive corroboration with the level of logical probability, for he says: "we can also say that an appraisal of corroboration takes into account the logical probability of the statement in question" (p. 269).

As the knowledge content of the entire hypothesis is increased, the hypothesis will tend to unify that knowledge under a single set of laws and principles. Thus

---

[2]See footnote 2.

both organic evolution and plate tectonics have great internal unity and are remarkably successful in explaining highly diverse phenomena by a single set of basic statements. Kitcher points to another positive feature of successful major hypotheses--their fecundity, shown in the capacity to reveal new areas of research suggested by the independent tests.

Kitcher, in reviewing the newer views of science philosophy, has emphasized three positive characteristics of "successful science":

> Independent testability is achieved when it is possible to test auxiliary hypotheses independently of the particular cases for which they are introduced. Unification is the result of applying a small family of problem-solving strategies to a broad class of cases. Fecundity grows out of incompleteness when a theory opens up new and profitable lines of investigation. Given these marks of successful science, it is easy to see how sciences can fall short, and how some doctrines can do so badly that they fail to count as science at all. A scientific theory begins to wither if some of its auxiliary assumptions can be saved from refutation only by rendering them untestable; or if its problem-solving strategies become a hodgepodge, a collection of unrelated methods, each designed for a separate recalcitrant case; or if the promise of the theory just fizzles, the few questions it raises leading only to dead ends.(1982, p. 48)[2]

The Method of Multiple Hypotheses

Pursuing a single hypothesis and standing up for it under attacks from fellow scientists can have some unwanted and unpleasant side effects. As the inventor of the hypothesis, you are not very receptive to alternative hypotheses; you may develop some blind spots, which is to say that you tend to lose objectivity in carrying on your investigation. This is something seen all too often in the past as scientists, pitted against each other in defense of a favorite hypothesis, debated publicly. Often one debater threw in a personal attack on the character of the opponent, just as candidates for political office are prone to do.

To curb the tendency to lose objectivity, a scientist will do well to formulate as many different, reasonable hypotheses as possible to apply to a given input of scientific data. Consequences are deduced for each hypothesis and tested accordingly. This approach has been referred to as the method of multiple hypotheses. It can be successful only if the scientist is genuinely devoted to problem solving, as compared with developing a public image as a warrior in defense of a cause. Scientists working as a team can practice this method by sharing and testing the hypotheses its members contribute.

The method of multiple working hypotheses was powerfully presented around 1900 by Thomas Crowder Chamberlin, a professor of geology at the University of Chicago. He is most widely known today for having collaborated with an astronomer, F.R. Moulton, to produce a rather unique hypothesis of the origin of the solar system. That hypothesis enjoyed a good deal of popularity, perhaps because it was the first strong contender to replace the then-ruling hypothesis formulated by the French astronomer, Laplace, over a century earlier.

In 1904, Chamberlin reviewed his ideas about the method of multiple hypotheses in an address given before the International Congress of Arts and Sciences in Saint Louis (Chamberlin, 1904). We can imagine him as a formidable figure, wearing cutaway coat and striped trousers as befitted a university professor in the

European tradition. His language was appropriate to the formalities and excrescences of the Victorian era he had outlived.

Chamberlin first denigrated what he called "the method of the ruling theory," in which a scientist uses an array of what are deemed to be "facts" to arrive by an inductive process at a theory to explain those facts. Quoting from his address: "as soon as a phenomenon is presented, a theory of elucidation is framed. Laudable enough in itself, the theory is liable to be framed before the phenomena are fully and accurately observed" (1904, p. 68). He went on to show how such a theory begins to dominate the thoughts and actions of its inventor, who tends to ignore new lines of evidence that should be used to test the theory:

> Soon also affection (for the theory) is awakened with its blinding influence. The authorship of an original explanation that seems successful easily begets fondness for one's intellectual child. This affection adds its alluring influence to the previous tendency toward an unconscious selection. The mind lingers with pleasure upon the facts that fall happily into the embrace of the theory, and feels a natural indifference toward those that assume a refractory or meaningless attitude. Instinctively, there is a special searching-out of phenomena that support the theory; unwittingly also there is a pressing of the theory to make it fit the facts and a pressing of the facts to make them fit the theory. When these biasing tenencies set in, the mind soon glides into the partiality of paternalism, and the theory rapidly rises to a position of control. Unless it happens to be the true one, all hope of the best results is gone. The defects of this method are obvious and grave. (P. 68)

Chamberlin mentioned no examples, but perhaps he had in mind a ruling theory prevalent in his time. Lord Kelvin (Baron William Thompson), a leading English physicist, strongly endorsed the theory put forward by astronomers and philosophers of the eighteenth century that the earth as a planet came into existence from a mass of high-temperature gas that condensed from a gaseous state into a molten state, then cooled to reach a solid state. Kelvin calculated the age of the earth on the basis of this ruling theory and concluded that an age of 20 to 40 million years was a reasonable figure. This pronouncement was a devastating blow to Charles Darwin, whose ruling theory of evolution required much more time than Kelvin's could allow. The story of how Kelvin's theory was demolished by the discovery in the early 1900s of the phenomenon of radioactivity is particularly fascinating because that discovery not only revolutionized all thinking about the age of the earth and the forces that power internal geologic processes, it also gave new life to Darwin's scenario of the evolution of species.

The method of the ruling theory has not gone out of style, despite Chamberlin's oration against it. We have a good example of it in the field of modern anthropology. An American school of cultural anthropology, led by Franz Boas, emerged during the 1920s and flourished for decades thereafter. Boas formulated a theory to the effect that each primitive group of humans develops its particular form of social behavior, including customs, mores, and religious practices, entirely independently of heredity; that culture is not under biological control of genes. Of those who sought evidence to support this "nurture theory," the person best known to Americans was Margaret Mead. Through her personal observations of a primitive society in Samoa, begun in the 1920s, she attempted to document Boas's theory of culture. You should know that the nurture theory had arisen to replace an extreme theory of genetic control over human behavior that flourished in the early 1900s. That biological theory

---

[2]See footnote 2.

turned sour when some of its proponents advocated the improvement of the human species by elimination of persons supposed to be genetically inferior.

The ruling nurture theory of Boas was strongly challenged in 1975 by Harvard biologist Edward O. Wilson, a specialist in insect societies. In a book titled _Sociobiology; The New Synthesis_ (Wilson, 1975) he revived the long dormant "nature theory" that biological forces acting through genetics control (to some extent at least) the culture patterns of any given primitive group of humans (Rensberger, 1983). Human behavior thus comes under biological guidance by an adaptive process of evolution forced by changes in culture. Adding fuel to the fire of controversy, field research methods used by Margaret Mead were strongly attacked in 1983 by anthropologist Derek Freeman and the debate was in full swing (Marshall, 1983). Will a new ruling theory displace the current one? Will "nature" replace "nurture" as the paradigm of modern cultural anthropology?

As Professor Chamberlin continued his address on methods in science, he turned to the method of the multiple working hypotheses. Here his eloquence surged to new extravagances of expression. We quote selectively from his text:

> The effort is to bring up into distinct view every rational explanation of the phenomenon in hand and to develop into working form every tenable hypothesis of its nature, cause or origin, and to give to each of these a due place in the inquiry. The investigator thus becomes the parent of a family of hypotheses; and by his paternal relations to all is morally forbidden to fasten his affections unduly upon any one. In the very nature of the case the chief danger that springs from affection is counteracted.

> The investigator thus at the outset puts himself in cordial sympathy and in the parental relations of adoption, if not of authorship, with every hypothesis that is at all applicable to the case under investigation. Having thus neutralized, so far as may be, the partialities of his emotional nature, he proceeds with a certain natural and enforced erectness of mental attitude to the inquiry, knowing well that some of the family of hypotheses must needs perish in the ordeal of crucial research, but with a reasonable expectation that more than one of them may survive. . . . Honors must often be divided between hypotheses. (1904, pp. 69-70)

Is it conceivable that if the culture school to which Margaret Mead belonged had used the method of multiple hypotheses, we would have seen the simultaneous testing by that school of two or more working hypotheses, each with its own set of deduced consequences to guide programs of field research? Must science continue to be, as in the past, a succession of title bouts in which one boxer after the other takes to the ring to dethrone a champion? Perhaps it is this very combatorial system that brings new blood into the study of science. If the champion of a ruling theory had, by chance, not become a scientist, would he or she have instead become a champion in the combat ring of politics or industry? Would the evenhanded application of the method of multiple hypotheses favor as its recruits those with less drive for success and less ability to achieve it?

Scientific Revolutions

We have seen that, as explained in Philip Kitcher's description of the fate of scientific theories, a widely held theory may begin to wither in the glare of new information that it lacks the power to explain. When this happens, it becomes vulnerable to dramatic collapse in the face of a new and more powerful theory. The king is dead; long live the king! The community of scientists then rallies around the new theory, giving it their almost undivided support and research effort.

Such is the scenario envisioned by science philosopher Thomas S. Kuhn, whose seminal work, _The Structure of Scientific Revolutions_, was published in 1962 and quickly caught the attention of the scientific community. Kuhn brought into widespread use a word that had little previous exposure among scientists and, for that matter, throughout the public in general, whereas today it is a buzzword freely tossed about in erudite discussions on almost any subject.

Kuhn chose _paradigm_ (rhymes with "pair-o'-dimes") to designate a scientific theory of great breadth and explanatory power within a particular area of science. A paradigm equates to what we have referred to earlier as a ruling theory. The word carries strong implications of "pattern" in the sense that it is the general plan to be followed in more specific applications. In _Webster's Third New International Dictionary_, the word is illustrated by anthropologist Margaret Mead's phrase "mistaken the paradigm for the theory." This suggests that paradigm and theory are on two levels, the former more general than the latter; yet this distinction disappears when Kuhn cites as examples of paradigms such great scientific theories of the past as geocentricism (earth at the center of the universe), replaced by heliocentricism (sun at center of planetary orbits). Other examples of paradigms given by Kuhn are the oxygen theory and the electromagnetic theory (1962, p. 150); these are highly specific theories. I suggest that you do not worry too much about fine distinctions between "paradigm" and such more easily understood terms as "dominant theory" and "ruling hypothesis."

As Kuhn sees the history of science, it is punctuated by crises that are resolved by _scientific revolutions_. The crisis is preceded by a period of decline in the then-current paradigm, as more and more new observations appear that are not accommodated by that paradigm. Conflicting observations of this type are described by Kuhn as _anomalies_. Despite these indications of approaching moribundity, the scientists remain loyal to their paradigm. In Kuhn's words:

> Though they may begin to lose faith and then to consider alternatives, they do not renounce the paradigm that has led them into crisis. They do not, that is, treat anomalies as counterinstances, though in the vocabulary of philosophy of science that is what they are.(P. 77)

Kuhn then goes on to say that "once it has achieved the status of a paradigm, a scientific theory is declared invalid only if an alternate candidate is available to take its place." Kuhn's use of the words "lose faith" and "renounce" lead to a new insight into his intended meaning of "paradigm"; it is a theory that has gradually come to involve belief (i.e., the theory carries the truth). Emotion has become involved in holding on to the dying paradigm. Ultimately, however, it is relinquished. Kuhn explains:

> The decision to reject one paradigm is always simultaneously the decision to accept another, and the judgment leading to that decision involves the comparison of both paradigms with nature and with each other. (P. 77)

This sounds like a tough and objective approach to the decision problem, the way we like to think scientists act. But elsewhere Kuhn seems to be emphasizing that belief and emotion are heavily involved in solving a crisis:

> The transfer of allegiance from paradigm to paradigm is a conversion experience that cannot be

forced. Lifelong resistance, particularly from those whose productive careers have committed them to an older tradition of normal science, is not a violation of scientific standards but an index to the nature of scientific research itself. The source of resistance is the assurance that the older paradigm will ultimately solve all its problems, that nature can be shoved into the box the paradigm provides. Inevitably, at times of revolution, that assurance seems stubborn and pigheaded as indeed it sometimes becomes. But it is also something more. That same assurance is what makes normal or puzzle-solving science possible. And it is only through normal science that the professional community of scientists succeeds, first, in exploiting the potential scope and precision of the older paradigm and, then, in isolating the difficulty through the study of which a new paradigm may emerge. (P. 150)

I see in the last sentence the idea that the new paradigm always arises from the corpus of the previous one; it could not arise independently from some outside field of knowledge. Previous knowledge is not simply thrown out as garbage. Instead, it is largely recycled and restructured.

What Kuhn says about crises and revolutions in science certainly applies to my own field, geology. Plate tectonics was the new paradigm, replacing one holding that crustal movements through geologic time have been predominantly vertical (up and down). We thought of mountains as moving more or less straight up along with upward movements of molten magma. Crustal blocks could also move down in fault basins. Erosion constantly tended to lower all exposed surfaces. The paradigm of the time had no name, but today we call it "vertical tectonics" to distinguish it from the "horizontal tectonics" of plate motions. True, there were some very nasty problems to fit into vertical tectonics. Crumpling of once-horizontal strata into tightly folded structures in the European Alps and elsewhere clearly required great horizontal motions.

The old paradigm endured for over a century, but it never worked very well. When Alfred Wegener proposed his theory of continental drift, the precursor of a revolution began to appear, but in it there was not really a new paradigm to contest the old one. As new information poured in about the crust of the ocean floors, the pressure for a new paradigm began to intensify beyond description. The information logjam was set free in remarkably short order by a coherent new paradigm, which we described on earlier pages. Rather than encountering slow and reluctant acceptance of new paradigms, plate tectonics swept the fields of geology and geophysics with mass conversions of its scientists. With incredible dispatch, these scientists fitted almost every category of secured knowledge of the old paradigm into the new one, and amazingly enough, nearly all of that knowledge was retained in its restructured setting.

Think about Darwin's theory of evolution through natural selection in the context of crises and revolutions. Surely the theory of evolution was a new paradigm, but did it displace an earlier one? Evolution was the naturalistic alternative to catastrophism, whether as a series of creations alternating with catastrophes, or a single creation followed by the catastrophic Flood of Noah. But the older view was a program of divine creation and extinction as revealed through the Scripture. The creation/catastrope view of life on earth was not science in the sense that we define it today (i.e., empirical science); instead, it was religious dogma, depending on faith in the supernatural.

We can conclude that in the knowledge field of the origin and development of diverse life on earth, Darwin's theory was the one and original paradigm of what Kuhn calls "normal science." For every field of what is now

normal science there was a first paradigm; before it was the "pre-paradigm" period, as Kuhn calls it (1962, p. 162). Kuhn says that whereas individuals practice science in the pre-paradigm era, "the results of their enterprise do not add up to science as we know it." Individuals were practicing good descriptive science in the field of biology long before Darwin's theory hit the world in 1859. Plants and animals, both living and fossil, had been described and classified in great detail in the previous hundred years. What was lacking, Kuhn suggests, was evidence of progress. By that, I presume he means absence of the coherent and universal explanation that the first paradigm was to provide. Lacking a paradigm--which is to say, lacking a powerful and fecund hypothesis or theory--the mechanism for deducing consequences and performing tests was lacking. Without such stimulus the science could not easily grow in new directions.

While Kuhn's view of crises, revolutions, and paradigms had enormous impact among science philosophers and upon the practitioners of science, there were some sticky side effects, arising from some of Kuhn's characterizations of the way the science community carries on its business. Kuhn's picture of the way the system works is not exactly flattering to scientists (p. 163). Working in isolation, directing efforts solely to an audience of colleagues, and selecting only problems that appear readily solvable are among the features cited that make science look more self-serving than perhaps it should be to maintain a pure and lofty public image.

What seems to have happened is that critics of science have used Kuhn's description of the scientific establishment itself (as distinct from the theory of crises, revolutions, and new paradigms) to attack the scientists in areas of morality and ethics, the implication being that scientists as a group are generally behaving in a socially undesirable manner. Philip Kitcher has sensed this reaction and seeks to correct its effects:

> Thomas Kuhn's book The Structure of Scientific Revolutions has probably been more widely read-- and more widely misinterpreted--than any other book in the recent philosophy of science. The broad circulation of its views has generated a popular caricature of Kuhn's position. According to the popular caricature, scientists working in a field belong to a club. All club members are required to agree on main points of doctrine. Indeed, the price of admission is several years of graduate education, during which the chief dogmas are inculcated. The views of outsiders are ignored. Now I want to emphasize that this is a hopeless caricature, both of the practice of scientists and of Kuhn's analysis of the practice. Nevertheless, the caricature has become commonly accepted as a faithful representation, thereby lending support to the Creationists' claims that their views are arrogantly disregarded. (1982, p. 168)[2]

We seem to have strayed a bit from the historical matter of paradigms and scientific revolutions into the sociology of science, a fascinating subject dealt with at some length in Chapter 8. In our next chapter we will be right back on course, investigating science.

Credits

1. From Francisco J. Ayala, Chapter 16, pp. 474-516 in T. Dobzhansky, F. J. Ayala, G .L. Stebbins, and J. W. Valentine, Evolution, W. H. Freeman and Company, San Francisco. Copyright © 1977 by W. H. Freeman and Company. Used by permission.

2. From Philip Kitcher, Abusing Science: The Case Against Creationism, The MIT Press, Cambridge, Mass. Copyright © 1982 by The Massachusetts Institute of Technology. Used by permission.

Chapter 4

Physical Science and Natural Science

Science, like an oriental carpet, has great strength because of the common threads that form the warp and woof. What we see, however, is a great variety in the patterns that result from the use of various kinds and lengths of fibers held in place by those common threads. Each branch or area of science has its special character, leading to different patterns of investigation and different modes of statement.

Physical science is usually set apart from natural science on grounds that the former seeks to understand the basic nature of matter and energy and their interactions; whereas the latter studies complex and unique systems of matter and energy that occupy specified positions in time and space. Physical science, consisting largely of physics and chemistry, is a search for general (universal) laws that can be expressed mathematically or by other symbols. Physical science is strongly quantitative, meaning that it collects numerical data arising from closely controlled experiments and expresses laws in the form of mathematical equations. Astronomy (including cosmology) is often said to be a physical science, but I am inclined to challenge that assignment.

Natural science includes the life sciences (biology), the earth sciences (geology, atmospheric science, ocean-ography), and, in my opinion, should also include astronomy. The physical structures with which the natural sciences deal are enormously complex aggregations of atoms and molecules. A single organism may contain millions or billions of molecules, while many kinds of organic molecules contain thousands of atoms. Even a single living cell is many orders of magnitude more complex than one of the atoms or molecules it contains. A chunk of rock contains millions or billions of atoms, often of many different elements, locked into geometrically complex systems of chains, layers, and lattices. The same statement about complexity applies to huge objects that make up the universe: stars, galaxies, gas clouds, and black holes.

The Role of History in Natural Science

Besides dealing with highly complex structures, the natural sciences investigate complex time sequences of events, something we do not find in basic physics and chemistry. Examples are the life cycle of an individual of a plant or animal species, the evolution of life on earth over hundreds of millions of years, or the birth and life history of a star or a galaxy. Thus the quality of history pervades the natural sciences.

History in natural science commonly takes the form of a sudden initiation of a system, followed by a much longer period of progressive change of the system (evolution). We are, of course, using the word "initiation" in a purely naturalistic sense, entirely devoid of any suggestion that a supernatural agent is involved. We can think of initiation as a relatively rapid reorganization of matter into a new system, derived from the substance and energy of an existing system but nevertheless uniquely

different. An example would be the formation of a star by the rapid gravitational collapse of a diffuse mass of dust and gas. There follows evolutionary change in the star over vastly longer spans of time than its initiation required. As used here, "evolution" carries no implication that the change is directed to some ultimate purpose or goal; the change simply occurs through the action of physical laws. A planet, such as our Earth, seems to have experienced a rapid initiation by gravitational collapse of dust and gas to form a solid sphere. This event is estimated to have taken about 200 million years, whereas the physical evolution of its internal structure has lasted over more than 4 billion years. An old star may end its long life by a cataclysmic event--an explosion--that initiates a new kind of structure, a neutron star, and a spreading cloud of gas and dust.

In biological evolution, an important hypothesis states that a new class of life--an order, family, or species--appears rather rapidly or even suddenly. This brief initiation phase is followed by a much longer period of smaller or minor changes, as organisms respond more or less gradually to environmental influences and other factors that control changes in gene composition. Among geneticists, this style of evolutionary tempo is called punctuated equilibrium. The term was introduced rather recently by paleontologists Niles Eldredge and Stephen Jay Gould, but expresses an idea that has been around since the 1940s (see Chapter 35). I like the wording used by Professor G. Ledyard Stebbins to describe this tempo. Referring to the evolution of the human brain, he says that increase in brain size took place in "quantum jumps followed by periods of relative stability" (Stebbins, 1982, p. 357).[1] He also uses the term "quantum bursts" (p. 358) for the events of rapid evolutionary change.

Although laws of physics and chemistry describe the kinds of activity associated with change in pure forms of matter and energy, the scientific statements that represent those laws are not uniquely positioned in time. Professor Walter Bucher, an American geologist, referred to this form of knowledge as being timeless (Bucher, 1941, p. 1). In contrast, knowledge of the origin and evolution of systems studied by natural science he described as timebound.

Looked at in a somewhat different light, Bucher's timeless knowledge consists of statements that describe events or relationships with an extremely large probability of being repeated. The radioactive disintegration of the nucleus of the uranium atom occurs somewhere in the universe in exactly the same way countless times each hour, minute, and second. Countless streams of photons are emitted in exactly the same way throughout the universe; they travel through space constantly, all with the same speed. When, however, we look at the evolution of life on earth, as displayed in the fossil record, we are viewing a long, complex chain of events that took place in a fixed span of time. As an entirety, the chain is unique in its structure; it is almost inconceivable that an identical chain of structures and events will ever occur on any planet in the universe, or has ever occurred. What we are really saying is that the probability of the particular evolutionary chain being repeated in exactly the same way is close to zero (approaches zero). I find satisfaction in this application of the theory of statistical probability to the distinction between physical science and natural science. It focuses attention on the uncertainty that attaches to every scientific statement, whether it be simple or complex, general or specific.

The Role of Explanation in Science

Most carefully composed definitions of science include its two basic aims: to describe and to explain. (A third aim often cited is to predict.) We have already delved into the nature of scientific description and how it makes use of special language and observational techniques. Now we need to devote some more attention to the process of explanation in science. What do we mean when we say that science explains the things it observes? Is explanation carried out in the same way for timeless knowledge as for timebound knowledge?

In Chapter 2 we made clear that science can consist of description without explanation, because a hypothesis can contain the description of something observed to exist or occur, but for which no explanation has yet been provided. Nevertheless, most descriptive statements evoke inquiring responses: Tell me more. Fill in the details. How did it come about? What forces caused it to happen? How is it controlled?

George Gaylord Simpson gives a rather straightforward and comprehensible analysis of scientific explanation that bears directly on the differences between physical and natural sciences (1963, pp. 33-35). He thinks of kinds of explanations in terms of kinds of questions we might want to ask. First is the question "How do things work?" It applies to timeless knowledge, which deals with how processes operate in nature. Specific questions might be phrased as follows: How do streams erode valleys? How do animals digest food? Processes of physics and chemistry are involved and the explanation consists of relating the appropriate process to the specific natural environment or setting in which it takes place. In explaining how a stream erodes a valley, we need to call on laws of mechanics of fluids, and these include reference to the laws of gravitation and motion. In explaining the digestion process, we need to set down the chemical composition of the nutrient substances as well as those of the reagents that act upon them in the alimentary canal of the animal. We then set up the chemical equations that show the reactions and products of those reactions. Scientists call this form of explanation "scientific reduction"; it is a topic we will examine later.

Simpson then turns to a second kind of explanation (p. 34), one that is appropriate to timebound knowledge: How did it come about? In short: How come? A geologist views a great mountain range with highly complex internal rock structure. How did this great mass we see here today acquire its size and shape, its internal composition and structure? The explanation is primarily historical and must be directed at reconstructing the chain of events that led to the final product. But the explanation must also deal with the natural processes that were involved in each step. For example, crustal compression occurred at one stage; rise of molten rock (magma) at another stage, and so forth. Thus the answer to "How come?" includes answers to repeated questioning as to "How do things work?"

Simpson refers to a third kind of explanation (p. 34). The appropriate question is "For what purpose?" or "What is the use of it?" The question has no meaning in physical science and in the purely physical areas of natural science. We would never think of asking "For what purpose does a volcano erupt?" We can, however, ask that kind of question about living organisms. For what purpose did warm-bloodedness develop in mammals?

Because "purpose" can easily suggest design with a purpose, which in turn can imply a designer with supernatural powers, we should rephrase this question to replace "purpose" with "function." Thus: What useful function does warm-bloodedness perform in mammals? The role of this kind of purpose in the evolution of life forms is something we will discuss on later pages.

One form of explanation in science, broadly viewed, consists of breaking down complex or unique phenomena into simpler and more readily repeated phenomena. This process comes to rock bottom when the explanation simply restates the most fundamental or universal laws of physics. How then can we explain those laws of physics? Philosophers of science have observed that such laws tell nothing about the cause of the relationships they describe. The law of gravitation asserts that an attractive force exists between two masses, but it leaves unanswered the question "What is the cause of the attractive force?" There are laws of magnetism, but what causes magnetic attraction and repulsion? Until recently, physicists had no explanation of the various forces that attract and repel bits of matter in a wide range of sizes. Today, however, physicists have a unified theory that can explain three of the four fundamental physical forces. Explanation of the fourth force, gravitation, is currently under development. (This subject is discussed in Chapter 15; see "grand unified theory.") These explanations require the existence of subatomic particles that are the carriers of the forces. Nevertheless, our conclusion here is that laws of physics need not offer an explanation for the relationships they describe.

At this point I would like to develop a philosophical point about what we really mean when we talk about laws of science. I propose for consideration the following description of a law of science, with special reference to physics: A law of science is a statement describing the relationship between or among variables that can be shown to co-relate always in exactly the same manner when a set of initial or boundary conditions is precisely established and defined. This definition may seem to lie in the area of formal science (logic and mathematics) and to be built on the absolutism of logic. On the other hand, such laws are based in empirical science and conform closely with what is reproducible in the laboratory or substantiated by repeated observations of natural phenomena outside the laboratory.

Nature can never follow precisely any formally stated law of science as defined here and, by the same token, we could conclude that no law can ever be proven true in the absolute sense by empirical science. Even in physics, which claims to discover laws fundamental to all science, a law is only a model of reality. No matter how precise the experimental equipment set up to "prove" or "validate" the model, and no matter how many times the experiment is replicated, the observational data can do no more than converge upon the terms of the law itself. In all of empirical science, whether it be physics or biology, we must cope with inherent inexactness in the form of the randomness that permeates the only real world we can observe.

Another class of statements, often called laws, consists of qualitative (verbal) assertions that some relationship or change involving matter and energy is always the same or always occurs in the same manner or in the same direction. An example might be as follows: "Any unsupported object at the earth's surface tends to fall along a line directed to the earth's center of gravity." Another example would be Newton's first law of motion: "Assuming the absence of any interaction with the rest of the universe, an object either remains at rest or moves in a straight line with constant velocity." We shall encounter this kind of statement later in Chapter 13, where we examine the so-called laws of thermodynamics.

The proposed definition allows us to include under laws of science classes of statements that are essentially definitions of terms, for example, Newton's second law of motion: "Force is the product of mass and acceleration." Perhaps, however, we should exclude simple qualitative statements of asserted fact, such as: "Light consists of photons."

Does Natural Science Recognize Laws?

Does natural science also have laws? For some scientists a "No" answer would be sufficient to banish natural science entirely from the fold of science. We must consider the possibility of there being laws for either or both of the two phases of natural science we have already identified: (1) timeless knowledge and (2) timebound knowledge. As to timeless knowledge, we are dealing with how things happen over and over in nature in almost the same way, but never in exactly the same way. In some cases a natural process can be approximated rather closely by laboratory experiments under controlled conditions. Both in biology and geology statements called "laws" have been made about the typical outcome of a specific process. That such statements should be considered laws comparable to those in physics is a debatable question. Ernst Mayr has evaluated the status of so-called laws in biology and, particularly, in evolution (1982, pp. 37-38). He refers to them as generalizations that describe regularities or trends but stops short of referring to them as "laws." The reason, he notes, is that these statements tell only what is most commonly the case or most likely to occur: "Generalizations in biology are almost invariably of probabilistic nature. As one wit has formulated it, there is only one universal law in biology: All biological laws have exceptions" (pp. 38-39). Similar statements would apply to generalizations made about repetitious processes and events in geology and other branches of inorganic earth science. The element of chance variation is always present in the real world of nature and usually it is a very important element. We will discuss this subject on later pages. What we need is a probabilistic model of nature that accommodates the range of chance variations we expect to find.

As to timebound knowledge, referring to complex sequences of events and changes that actually occurred in a specified segment of time, the possibility of formulating laws is rather more formidable than for timeless knowledge. If something happened only once in all of cosmic time can it be described by a law? Not likely, unless you intend to predict that there will be repetitions of the same thing in the future. George Gaylord Simpson, as a paleontologist, has one foot in biology and the other in geology; he takes a dim view of historical laws:

> The search for historical laws is, I maintain, mistaken in principle. Laws apply, in the dictionary definition "under the same conditions," or in my amendment "to the extent that factors affecting the relationship are explicit in the law," or in common parlance "other things being equal." But in history, which is a sequence of real, individual events, other things never are equal. Historical events, whether in the history of life, or recorded human history, are determined by the immanent characteristics of the universe acting on and within particular configurations, and never by either the immanent or the configurational alone.
>
> It is further true that historical events are unique, usually to a high degree, and hence cannot embody laws defined as recurrent, repeatable relationships. (1963, p. 29)

(Note: For Simpson, immanent is equivalent to "timeless"; configurational to "timebound" or "historical.")

I'm not altogether happy with Simpson's rejection of historical laws, even though I accept the logic of his position. In all of natural science, whether the content be timebound (historical) or timeless, the search must be made for generalizations that have a reasonably good probability of being repeatedly applicable. If we fail to identify such generalizations in the study of natural history--if our published accounts are no more than descriptions of the unique--we are vulnerable to the charge that we are not doing science at all. Even if that charge be withdrawn, we can be charged with doing a lower-grade or less-meaningful brand of science than science that produces laws or probabilistic generalizations.

Let me refer back to the geologic hypothesis used as an example in earlier pages. In modern geology, the basic mechanical activities of the earth's outer layer (the lithosphere) are encompassed by the general hypothesis of plate tectonics. Evidence is strong that the continents of North and South America, Eurasia, and Africa were at one time joined together in a single, great continent called Pangaea. About 200 million years ago, this single continent began to split apart along great fractures, known as "rifts." As the rift valleys widened, they became seaways, which expanded into deep ocean basins. New crust was continually being formed by rising molten rock (magma) along a central oceanic rift. Today a great Atlantic Ocean basin separates the American continents from those of Africa and Eurasia. Consider, now, that 200 million years is not a very long span of time in comparison with the age of an earth formed over 4 billion years ago. The opening of the Atlantic basin occupied only the last one-eighth fraction of the total earth history. Have continents split apart before in that earth history? Evidence is strong that it has happened several times before. There is also evidence that ocean basins close up and are eliminated by the underthrusting mechanism of subduction, which we used in our exemplary hypothesis. Subduction can cause an entire ocean subfloor to disappear into a deep, hot mantle region below the lithosphere. When this happens continents slam together like halves of a double sliding door--continental collision, it's called. Evidence is good that the closing of an ocean basin is a repeated event in geologic history; it alternates with the process of opening of a basin. A distinguished Canadian geologist, J. Tuzo Wilson, outlined a synthesis of alternate opening and closing of ocean basins; it is widely recognized today as the Wilson cycle.

In general terms, the sequence of events in one Wilson cycle has probably been very much like that in other cycles throughout at least three billion years. If only one cycle has been in operation at a given time on the globe, there has been time for perhaps six complete cycles. Does this give enough repetitions to let us assign to the idealized Wilson cycle the status of a scientific law? No, you say? Then are we permitted to call it a principle of geology? Perhaps so. In any case, we have enough geologic evidence to permit us to sketch a model cycle that embodies the most likely features of a natural cycle (i.e., the important features most instances share in common). The model is a hypothesis of a particular kind: it is a historical hypothesis that asserts what is likely to happen repeatedly in nature, including time yet to come. This historical hypothesis differs from a unique hypothesis that is a description and perhaps also an explanation of a single (singular) event.

Prediction--Logical or Temporal?

The possibility of a fundamental distinction between physical science and natural science, over and above the distinction I have already pointed out, has been put forward by philosophers and seized upon by the fundamentalist creationists as a weapon of attack on organic evolution through geologic time. Professor

Francisco Ayala deals with this issue, but without mentioning the creationists:

> Some philosophers of science have claimed that evolutionary biology is a historical science that does not need to satisfy the requirements of the hypothetico-deductive method. The evolution of organisms, it is argued, is a historical process that depends on unique and unpredictable events, and thus is not subject to to the formulation of testable hypotheses and theories.(1977d, p. 486)[2]

Ayala strongly disputes the claim that the hypothesis of evolution cannot be tested; he says that claim is based on "a monumental misunderstanding." He then points out that biological evolution has two scientific phases. One deals with causal questions; in other words, how biological processes or mechanisms work. This phase corresponds with the "timeless knowledge" of Bucher, and with my category of events or relationships with a very high probability of being repeated. The other phase of biological evolution is historical--the tree of life as it grew through geologic time with its various branches. This historical aspect corresponds with Bucher's "timebound knowledge," and with my category of evolutionary sequences having an infinitely small probability of being repeated elsewhere in space or time. It is with the historical phase that the dispute is concerned. Creationists claim that the history of biological evolution as it has been reconstructed by biologists is not a scientific hypothesis because it has no power of prediction.

Recall my account on earlier pages of how I was confused by the intent of the meaning of the word "prediction." It finally dawned on me that two meanings are in common use by scientists, but that the authors usually do not specify which one they intend. Confirmation and clarification came when I happened on the following statement by Professor Ernst Mayr of Harvard University, an eminent scientist in the field of evolutionary biology:

> The word prediction is being used in two entirely different senses. When the philosopher of science speaks of prediction, he means logical prediction, that is, conformance of individual observations with a theory or a scientific law. . . . Theories are tested by the predictions which they permit.
>
> Prediction, in daily usage, is an inference from the present to the future, it deals with a sequence of events, it is temporal prediction. (1982, p. 57)

George Gaylord Simpson uses the word "prediction" in the temporal sense (1963, p. 36). In discussing the physicists' narrowly circumscribed view of science, he notes that "some philosophers and logicians of science have concluded that scientific explanation and prediction are inseparable." He elaborates:

> Explanation (in this sense) is a correlation of past and present; prediction is a correlation of present and future. The tense does not matter, and it is maintained that the logical characteristics of the two are the same. They are merely two statements of the same relationship. This conclusion is probably valid as applied to scientific laws, strictly defined, in nonhistorical aspects of science. . . . But we have seen that there are other kinds of scientific explanations and that some of them are more directly pertinent to historical science. It cannot be assumed and indeed will be found untrue that parity of explanation and prediction is valid in historical science. (P. 36)

The temporal sense of prediction has been used more

recently by Ernst Mayr. During an interview with science writer Roger Lewin, Mayr observed that evolution, as a historical process, cannot predict future evolutionary changes. He is quoted as saying:

> If you had stood on the earth at the beginning of the Cretaceous (135 million years ago) and seen dinosaurs all over the place, you could not have predicted that the miserable little things* that came out only at night would eventually take over when Cretaceous came to an end. You can predict the next appearance of Halley's comet, but you can't predict changes in biological diversity. Such uncertainty is typical of evolution. (Lewin, 1982a, p. 719)

(*Shrewlike ancestors of the mammals.)

Creationists would latch onto a statement like that in an instant, for it supports a theme they have been trying to put across to the public for years. Richard K. Turner, an attorney for the creationists in their 1981 suit before the California Superior Court to force teaching of the biblical creation version in state schools, gave us a perfect example of this tactic. As reported by science writer William J. Broad (1981, p. 1332), Turner misinterprets Karl Popper's use of "prediction" (as deduction of consequences) to mean that a scientific hypothesis must have the ability to predict what will happen in the future. Switching from logical to temporal meaning of "prediction" allows the creationists to claim that because evolutionary theory cannot predict what specific evolutionary changes will occur in the future, the biological hypothesis of evolution is not scientific in character. From there, the creationists go on to argue that the hypothesis of evolution is actually a religious belief, and hence forms the basis of a religion.

Professor Ayala does not buy the creationists' version. Instead, he maintains, "even the study of evolutionary history is based on the formulation of empirically testable hypotheses" (1977d, p. 486). He gives an example from the evolutionary tree that includes hominids (humans) and the great apes (chimpanzee, gorilla, orangutan). The ways in which the branches of that tree can be drawn differ according to two current hypotheses. Using the logical sense of "prediction," he says: "A wealth of empirical predictions can be derived logically from these competing hypotheses" (p. 486). "These alternative predictions provide a critical empirical test of the hypotheses" (p. 487).

Philosophers of science have disagreed as to whether logical prediction (the deduction of some past event in support of a historical hypothesis) can be accepted in the absence of any means of making temporal predictions (what will happen in the future) from a given historical hypothesis (Hempel and Oppenheim, 1953). They have argued that both past and future must be predictable from the same hypothesis. This is a principle of logical symmetry that certainly must apply to laws of physical science that are in the timeless category.

But it has also been argued that, because of the extreme complexity of most historical hypotheses (timebound knowledge), prediction of what will happen in the future is neither required nor feasible (Scriven, 1959). That argument is based in formal logic, which I will not attempt to go into here. Scientific temporal prediction depends for its success on recurrence or repeatability, which is the nature of timeless knowledge (Simpson, 1963, p. 38). The kinds of events that characterize timeless knowledge can "move about freely in time," so to speak, from past through present to future. While it is true that a complex historical hypothesis contains both timeless and timebound classes of information or knowledge, the timeless constituents-- common events, that is--are not what make the hypothesis unique. It is the uniqueness of the succession of different kinds of events in a given history that counts--events that could not have been predicted if an observer had been there at the time, and can therefore not be predicted by today's observer looking into what is, to us, the future.

I have read statements to the effect that different branches of science use basically different means to formulate hypotheses and accumulate knowledge. I would like to think that such statements are generally unwarranted. Special ground rules for special groups is not something with which I can be comfortable. Excluding formal science (logic and mathematics), all of empirical science ("factual" science) must play by the same rules. From what I have reviewed of biologists' assessments of the scientific nature of the study of evolutionary history, I am satisfied that a single set of rules will work for all. Ernst Mayr has described the common aims of science in these words: "All sciences, in spite of manifold differences, have in common that they are devoted to the endeavor to understand the world. Science wants to explain, it wants to generalize, and it wants to determine the causation of things, events, and processes. To that extent, at least, there is a unity of science" (1982, p. 32).

The Uniqueness of Living Matter

In the interview with science writer Roger Lewin (1982a), referred to in an earlier paragraph, Professor Mayr discussed some ideas he has covered in his book, The Growth of Biological Thought (1982). Mayr has been disturbed by the misunderstandings that many physical scientists have about the field of evolutionary biology. He feels that physical scientists frequently do not appreciate important differences between the living world and the inanimate world. It is not that the same physical laws are not followed in both physics and biology; the basic laws are common to both fields. As Mayr puts it, "There isn't a process in a living organism that isn't completely consistent with any physical theory. Living organisms, however, differ from inanimate matter by the degree of complexity of their systems and by the possession of a genetic program" (p. 719). Of these two differences, the greatest level of uniqueness attaches to the second point --the possession of a genetic program.

One distinctive quality of the living organism, not found in the inorganic environment in which it lives, is the ability of the life form to achieve extraordinary molecular complexity, accompanied by a capacity to store energy in chemical forms. Living matter is organized into systems, which are usually rather clearly defined by a tangible boundary. The single cell of a plant or animal is perhaps the simplest example; another would be a single body organ, such as the kidney or liver. The total individual is also a clearly defined single system. We can go on to consider a community of plants or animals (or a mix of both) in their environment as a single system--an ecosystem. These organic systems import and export both energy and materials through their boundaries; they are open systems. Open systems are identifiable in all other areas of natural science, but these are purely inorganic systems (Strahler, 1980).

Consider a single plant cell as an open system of energy and matter. Nutrients enter through the cell wall as ions or molecules, while waste products of metabolism leave the cell through the same wall. Energy needed to power the biochemical reactions enters the cell as radiant light energy. The dominant process of chemical synthesis in the plant cell is photosynthesis, in which the light energy is used to combine carbon dioxide (a gas) with water molecules to form carbohydrate (compounds made up of carbon, hydrogen, and oxygen). In this process, solar energy is stored in the carbohydrate molecules as a form of chemical energy. Some of this, in turn, is used in the

synthesis of more complex organic molecules--proteins, for example. Again, energy goes into storage in those larger molecules. Among the most complex organic molecules synthesized in the cell are those of DNA and RNA; their complexity of structure and function are almost beyond comprehension. At the same time that complex molecules are being constructed, others are being broken down into simpler components that are released to the outside environment. The reverse process to photosynthesis is respiration; it releases heat energy to the environment.

Life Systems and Energy

Analysis of the organic cell as an open system was presented to scientists of North America by Ludwig von Bertalanffy, an Austrian biochemist, in a now-classic paper titled "The Theory of Open Systems in Physics and Biology" (von Bertalanffy, 1950). Upon first seeing his material in 1950, I was struck by how important his systems analysis could be for all areas of natural science, and I incorporated it into one of my papers on natural systems of water erosion, published a year later. The basic energy equation given by von Bertalanffy is a bit on the difficult side for most readers since it uses the differential calculus. At the risk of submerging many readers, I can state the equation in words as follows: The time-rate of change of energy concentration within a very small element of an open system is equal to the sum of the rate of change of production of energy (produced in biochemical reactions) and the net rate of outflow of energy from the element. By "element" is meant any very small unit volume in the cell; you can think of it as a tiny cube or sphere. The same statement could apply equally well to the entire cell as an open system. For a cell, three energy states are possible with this equation. First, the cell system can be experiencing a net gain in stored chemical energy. Second, the system can be experiencing a net loss in stored energy. Third, the quantity of stored energy within the system can be held constant with time; this represents a steady state of the open system. Biologists call the condition of steady state homeostasis.

An important point to be noted in connection with the theory of open systems and steady states in biology is that it seems--on the surface, at least--to violate one of the most sacred laws of physics: the second law of thermodynamics. This law includes the dictum that in isolated systems the tendency is for the matter inside the system always to go in the direction of greater disorder. A concept of ordered states and disordered states is involved here, something not previously mentioned in our description of the cell system. In thermodynamics, the complex organic molecules in the cell are considered to represent a high level of order (orderliness) of the atoms. In contrast, the collection of rather simple inorganic molecules and ions that the cell uses to construct the complex molecules is viewed as representing a low level of orderliness. The property of lack of order is disorder. Therefore, it can be said that within the growing cell, matter is being transferred from a state of disorder to a state of order, but this direction of change is forbidden by the second law. Creationists use this conclusion as the basis for claiming that the theory of organic evolution violates the second law; therefore, that evolution is false. Because the creationists think this argument is one of the strongest they have working for them, it must be thoroughly analyzed and repudiated, a task we undertake in Chapter 13.

For the moment, it is enough to be aware that the laws of thermodynamics forbid creation of new energy supplies from nothing (ex nihilo). Moreover, in biological processes, energy cannot be derived from matter by particle collisions. We must, instead, interpret the storage of chemical energy in cells by means of an energy cycle, or energy balance, in which all forms of energy entering and leaving the system are accounted for, much as you account for the flow of money in and out of your bank account.

Synthesis of large organic molecules in a living plant cell uses energy from outside the cell, transforming the sensible heat from solar sources into chemical energy that can be temporarily stored. During a stage of increase in stored energy, external energy is consumed by the cell, but a part of that energy is expended in the storage process itself. The important point is that for energy to be stored in a form capable of doing useful work, a great deal of energy must also be "burned" (i.e., lost as heat dissipated into the environment). A steam locomotive, in pulling a heavy train of cars up a long, steep grade, burns a great deal of coal or wood and dissipates a great deal of heat into the atmosphere; yet, at the top of the grade, the train has gained a large supply of stored energy (as potential energy). A cell, its energy input diminished, undergoes a degradation (chemical breakdown) of its complex, energy-storing molecules, releasing heat energy and resulting in a collection of raw materials of lower chemical energy.

The ratio of energy stored to energy consumed in open systems, such as a living cell, has been intensively studied. (See Strahler and Strahler, 1974, pp. 19-23, Appendix 2.) Charles A. Coulomb, the French scientist (1736-1806), was interested in evaluating the mechanical power capability of the human male. This was a subject of general interest at a time when men provided a great deal of the power of industry, using devices such as treadmills that would turn wheels to grind grain, lift water, or operate a crane. Coulomb observed that a porter who brought firewood up to his apartment, a vertical ascent of about 40 ft (12 m), had a maximum work capacity of about six wagonloads per day. He made 66 trips per day up the stairs, carrying an average load of 150 lbs (68 kg). Suppose that the porter carried only one or two sticks of wood per trip. He could have made many more trips per day, but since much of his work consisted of lifting his own body weight, the ratio of stored potential energy (stored work) to energy expended in moving his body (processing work) would have been quite small. On the other hand, if he attempted to lift a much heavier load, his progress up the stairs would have been painfully slow, and the total energy stored would be small by the end of the day. Somewhere in between was the optimum ratio of load to body weight, such that the maximum stored work was done in ratio to processing work. Evidently, the porter had learned from experience the optimum load of firewood that would get the largest total weight upstairs in one day.

Without going into an explanation, we can simply state here that, as a rule of thumb, the maximum rate at which energy can be placed in storage is achieved when 50 percent of the energy is stored and 50 percent is expended in processing work needed to accomplish the storage function. This 50/50 division of power into the two functions for optimum storage is known as the Darwin–Lotka law. The law is named for Sir Charles Darwin and A. J. Lotka; the latter in the early 1920s analyzed quantitatively the role of energy expenditure and storage in organic evolution (Lotka, 1922; Odum and Pinkerton, 1955). From an evolutionary viewpoint we simply comment that an organism capable of storing energy at the fastest possible rate would have advantages over other individuals of the same species with lesser abilities, since the food supply might be severely limited and it would be advantageous to be able to convert it into stored energy in the least possible time.

Genetic Programs and Feedback Systems

Professor Mayr's second point of uniqueness of living matter is the genetic program that directs the formation of each individual: "The genetic instructions packaged in an embryo direct the formation of an adult, whether it be a tree, a fish, or a human. The process is goal-directed, but from the instructions in the genetic program, not from the outside. Nothing like it exists in the inanimate world" (Lewin, 1982a, p. 719). But Mayr goes on to explain, the genetic package is much more than just a set of instructions; it is the product of a descent through evolution. (See also Mayr, 1982, pp. 55-56.) In this sense, the genetic program is history--millions upon millions of years of evolutionary history following a continuous chain of small steps. As I have emphasized earlier, the probability of such a chain being duplicated at any place in the universe, at any point in past or future time, is almost infinitely improbable. It is in this sense that the genetic program is unique.

Essentially the same point has been made by George Gaylord Simpson, who stresses that no two organisms are exactly alike, not even identical twins. He explains:

> Each is the product of a history both individual and racial, and each history is different from any other, both unique and inherently unrepeatable. These aspects of biology deal not with the immanent, the inherent and changeless characteristics of the universe, but with contingency, its states, fleeting and in ceaseless change, each derived from everything that went before and conditioning everything that will follow. (1969, p. 10)[3]

A unique feature of vertebrate animals, setting them apart as a class from other organisms and all inorganic natural systems and structures, is that they have built-in feedback systems, specifically developed to transmit information from sensory organs through neurons to a central clearinghouse--the brain--where a response is triggered and sent through neurons to the appropriate body organs for action (Simpson, 1963, p. 26).

The term "feedback" needs to be understood in each of two meanings intended in science. In purely mechanical systems in nature we can often recognize a self-adjusting or self-correcting mechanism that serves to keep the system on an even keel--a steady state, that is. An example would be an ocean beach of sand and pebbles, shaped by the action of breaking waves that generate the alternating uprush of a sheet of turbulent water and its downslope return by gravity. As many of you know from experience, the beach changes seasonally in form and position. In summer an embankment (a berm) is built seaward by relatively weak waves and swells; in winter the berm is eroded away by waves of high energy and the beach assumes a different profile. Geologists who study this cycle recognize that seasonal changes in beach form are responses to seasonal changes in the energy input; they refer to the self-adjustment process as "feedback," although it has nothing whatsoever to do with information. In the higher plants (the vascular plants) feedback can be recognized as carried out by physical and chemical processes, but not involving information as such.

The type of feedback we are talking about in animals is designated cybernetic feedback; it is found in biological, psychological, and social systems, as well as in a host of mechanical and electronic devices invented by humans (Strahler, 1980, pp. 25-26). Webster's Third New International Dictionary gives as one definition of feedback: "a return to the input of a part of the output of a machine, system, or process and leads to a self-correcting action." The word "cybernetic" refers to the control system itself. A familiar example in mechanical systems is the thermostatic control mechanism of a home heating unit. The neurological feedback system in vertebrate animals, evolved over hundreds of millions of years, is vastly more complex and efficient than any cybernetic feedback system invented by humans, electronic computers included. The human brain is also the control center for elaborate social feedback systems in which appropriate human responses are concepetualized to regulate society and keep it functioning on an even keel.

We can add a final unique quality or property of living organisms; they exhibit numerous behavioral properties and physical forms, functions of body, and organs that serve specific purposes in remarkably successful ways. Eyes are wonderfully adapted to the purpose of seeing, ears to hearing, hands to grasping. The process of organic evolution makes possible adaptations of form and function of the organism; these are genetically controlled features and they have undergone changes through time. Changes that increase the chances of the individual to survive and to reproduce tend to be preserved in the species. This is the natural selection process.

When we, as humans, observe these remarkable adaptations in our species and in other species, we think immediately of purpose being fulfilled. We think of a particular body organ as being designed for a useful purpose. Such thoughts are described by the term teleology, which simply means "to explain something in terms of fulfillment of purpose." A natural teleological explanation applies correctly only to features of living organisms. In the case of humans, it may be extended to human artifacts--a knife or fork, for example--which are deliberately designed to serve a purpose. This is artificial teleology. We are not particularly interested here in artifacts (including culture in general), but rather in attributes of organisms that are genetically determined.

Before Charles Darwin, natural teleology was, in the Hebrew/Christian tradition, closely tied to theistic religion. God created man and all other organisms exactly as they are today, along with their physical environments. All we humans needed to do was to sit back and admire the handiwork of the omniscient and omnipotent Creator. True, there was that bothersome organ, the appendix, that seemed to serve no good purpose and often dealt a lethal blow to its owner. Creationist fundamentalists still bask in the wonderment of divine creation. For them, nothing so marvelous as the human species could have reached such perfection through the blind and unthinking process of organic evolution. Such, at least, is one of their arguments in favor of divine creation.

The Question of Reductionism in Science

A philosophical issue currently undergoing vigorous debate concerns the relationship between natural sciences and physical sciences. The assertion is made that the underpinnings of natural sciences are those laws and relationships established in the physical sciences--physics and chemistry. Therefore, it is argued, each branch of natural science can be reduced to statements of physics and chemistry. Or, moving in the opposite direction, it is argued that all complex theories and laws of the natural sciences can be derived from simpler, more general theories and laws of physics and chemistry. The philosophical concept involved is called reductionism, defined as "a procedure or theory that reduces complex data or phenomena to simple terms" (Webster's Ninth New Collegiate Dictionary).

Offhand, reductionism sounds like a great idea. To make things simpler makes them more easily understood by more people. Natural sciences dealing with inorganic substances (in any case, with nonliving matter) lend themselves to being reduced to descriptions of physics and chemistry. The form of a mineral crystal reduces very nicely to an arrangement of atoms in a three-dimensional lattice. A flowing river easily becomes an exercise in the mechanics of flow of a viscous fluid. The Big Bang in cosmology is nothing but an intense flux of elementary particles, such as photons and neutrons.

When it comes to the study of live organisms, however, the process of reduction becomes a bit tacky. Is a living cell simply an aggregation of atoms and molecules carrying out processes of pure physics and chemistry? Is that all there is to life? Or is there some special force, quality, or substance in a living cell that lies apart from the domain of physics and chemistry? A "Yes" answer to the last question has been voiced for centuries, starting with Aristotle, by a class of philosophers called vitalists; their doctrine was vitalism. Vitalism was opposed by mechanism, which rejected the idea that there is "something more" in living cells than in nonliving matter. Vitalists postulated the existence of a mysterious, unique component of life described in such terms as "entelechy" (a perfecting principle), élan vital (vital force), or "radial energy." The vitalist/mechanist controversy raged from the time that Réné Descartes (about 1640) opted for the mech-anistic view, through to the early 1900s, when vitalism was put to rest. The reason for its demise is fairly obvious: vitalism invoked a supernatural concept, or at least a nonempirical concept, that cannot be tested scientifically. Vitalism therefore rested on faith alone.

The place of vitalism as an answer to the question "Is that all there is to life?" was taken by ontological reductionism. Recall that ontology is the branch of philosophy that asks "What is reality?" When a living cell is broken down into its material components, they prove to be nothing but atoms. Energy that the cell cycles through its metabolic processes proves to be no different from forms of energy known to physics. This is a purely naturalistic or mechanistic picture. As Professor Ayala says: "Ontological reductionism also implies that the laws of physics and chemistry fully apply to all biological processes at the level of atoms and molecules" (1977d, p. 488). He goes on to say:

> Ontological reductionism does not necessarily claim, however, that organisms are nothing but atoms and molecules. The idea that because something consists of "something else" it is nothing but this "something else" is an erroneous inference, called by philosophers the "nothing but" fallacy. Organisms consist exhaustively of atoms and molecules, but it does not follow that they are nothing but heaps of atoms and molecules.(Pp. 488-89)[2]

What Ayala seem to be getting at is that ontological reductionism does not concern itself with the question: How did the organism get to become the complex system it is? If we took all the individual atoms of which the cell is composed and dropped them into a beaker, what would we do next to persuade those parts to form a complete living cell? Relying on natural processes, aided or unaided by humans, the probability that all the components would spontaneously come together to form a cell is infinitely remote. Even if we assembled all the atoms into complete molecules, no cell would spontaneously form. Ayala sums it up: "Living processes are highly complex, highly special, and highly improbable patterns of physical and chemical processes" (p. 489). Emphasis should be given the word "improbable"; better to say "infinitely improbable by chance alone." We mean, of course, "improbable that it would happen now." The cell has a long and complex history of evolution from simpler states of matter.

If we turn to consider the properties of a living cell as compared with the properties of the component atoms, the two sets of properties are not alike. The cell has its own unique set of properties, referred to as emergent properties. Now the philosophical question takes on a new twist: Can the laws and theories that account for the behavior of the cell (a complex system) be logically derived solely from the laws and theories that govern the behavior of the atoms or the molecules as separate component parts? One particular brand of reductionism, called epistemological reductionism, considers the question as to whether "the laws and theories of biology can be shown to be special cases of the laws and theories of the physical sciences" (Ayala, 1977d, p. 491). Ayala gives a full explanation as follows:

> The connection among theories has sometimes been established by showing that the tenets of a theory or branch of science can be explained by the tenets of another theory or branch of science of greater generality. The less general theory (or branch of science), called the secondary theory, is then said to have been reduced to the more general or primary theory. Epistemological reduction of one branch of science to another takes place when the theories or experimental laws of a branch of science are shown to be special cases of the theories and laws formulated in some other branch of science. The integration of diverse scientific theories and laws into more comprehensive ones simplifies science and extends the explanatory power of scientific principles, and thus conforms to the goals of science. (P. 492)[2]

Several major successes have been achieved in epistemological reduction in the history of science. For example, much of what was the science of chemistry (especially about how compounds were formed from elements) prior to about 1900 yielded to a new and revitalized chemistry making use of knowledge the physicists furnished about the way in which electrons move in orbitals about the nucleus of the atom. More recently, a large part of what was the theory of genetics prior to the 1960s succumbed to explanation by organic chemistry--the structure and function of the DNA molecule. Dramatic as these reductions were, they have not been judged as completely successful, for there is always some unexplained content of the more complex system that fails to yield to the more general system.

Whether the processes of natural sciences will ever be fully explained by reduction remains a debatable question. Successes will doubtless continue in that phase of empirical science that concerns processes and mechanisms (timeless knowledge). I cannot see reduction having any success whatsoever in the historical aspects of natural science--historical organic evolution, evolution of the earth's crust, and stellar-galactic-cosmic evolution. These systems developed over a long, unique succession or chain of physical/chemical reorganizations. Once broken, such a chain could almost never be reproduced. Thus, while we can explain the general behavior of the complex natural system by basic laws of physics and chemistry, we cannot actually derive the more complex system from the more general one.

Biologists are particularly skeptical of the possibility that epistemological reduction can succeed in their area of science. Can the natural teleological phenomena seen in life forms be completely reduced to processes of physics and chemistry? Professor Simpson does not think so: "In physical sciences it is not legitimate, indeed it is downright silly, to ask what things are for or what good they are" (1969, pp. 10-11). Physics and chemistry do not ask what possible purpose is served by, say, oxygen combining with hydrogen to produce water. We would not think of saying that this chemical union occurs in order to produce a liquid compound that slakes the thirst of vertebrate animals. "But in biology it is not only legitimate but also necessary to ask and answer questions teleological in aspect, concerning the function or usefulness to living organisms of everything that exists and occurs in them." He goes on:

> The structures and processes of organisms are useful, they perform functions, and they would not exist or occur if that were not true. They can never be understood or explained by the most complete and exact specification of the chemical reactions involved. Those reactions themselves are meaningless except as they relate to the organisms and populations and ecosystems in which they occur. (P. 11)[3]

Professor Mayr, in his interview with science writer Lewin, had some cogent remarks to make about reductionism in science; he is skeptical of the dominant role assigned to it by physicists. He said to his interviewer: "This reductionism has led to what David Hull (no reference given) calls the arrogance of physicists" (Lewin, 1982a, p. 719). (See Mayr, 1982, p. 33.) "They say, yes, you biologists deal with complex things, but the ultimate explanation will be supplied by the level at which we study" (Lewin, 1982a, p. 719). Mayr doesn't think that an understanding of particle physics will provide an understanding of everything else in the real world. Instead, Mayr thinks "Complex systems have to be studied at high levels of complexity. New properties turn up in systems that could not have been predicted from the components, which means you have to study things hierarchically."

This chapter has stressed fundamental differences between natural science and physical science. As theoretical physics delves deeper and deeper into the nature of the fundamental forces, setting as its goal the achievement of a grand unified theory, physical science seems to drift farther and farther apart from the natural/historical sciences. Our next chapter seeks an area of common ground that serves to unify rather than to fragment science as a whole: it is the pervasive role of blind chance in the world of nature, encountered whether we are looking at the motions of individual atoms of a gas or at differences between individual organisms within an animal species.

Credits

1. From G. Ledyard Stebbins, Darwin to DNA, Molecules to Humanity, W. H. Freeman and Company, New York. Copyright © 1982 by W. H. Freeman and Company. Used by permission.

2. From Francisco J. Ayala, Chapter 16, pp. 474-516 in T. Dobzhansky, F. J. Ayala, G. L. Stebbins, and J. W. Valentine, Evolution, W. H. Freeman and Company, San Francisco. Copyright © 1977 by W. H. Freeman and Company. Used by permission.

3. From George Gaylord Simpson, 1982, Biology and Man, Harcourt, Brace & World, New York, pp. 10-11. Used by permission of Harcourt Brace Jovanovich, Inc.

Chapter 5

Determinism, Randomness, and the Stochastic View

One of the controversies in science most difficult to explain in simple language has smoldered for decades. It concerns the basic manner in which natural systems operate or, at least, how they are imagined to operate. Suppose that we had in our possession a book containing all physical laws needed to explain anything and everything in the universe. This volume would represent the ultimate achievement of epistemological reduction, discussed in Chapter 4. The question before us now is: How useful are these laws in describing and explaining systems of matter and energy that involve countless individual parts or particles, each of which is engaged in its own form of activity and in interactions with other parts or particles? The book of laws would be extremely useful in solving single-body problems, such as the trajectory of an intercontinental ballistic missile. Knowing with considerable accuracy the initial conditions at the instant of launch, we could use the laws to predict with a high level of accuracy the point of impact. But can the laws be applied to make an accurate description of all the trajectories of, say, 100 billion gas atoms in a closed container at a given instant, and a prediction of their trajectories at a future instant? As another example, is it possible, using the laws of physics, to predict in exact detail the complete branching system of air passageways in the lung of a particular adult human, given a description of the embryonic system of the same lung at a certain fetal stage?

The Deterministic Model

Let us approach the problem of the gas atoms in a container, using a similar but much simpler mechanical system for demonstration purposes. It is a billiard table (a carom billiards table with no pockets) and a set of billiard balls. Let us use six balls for this experiment. Place the balls in more or less random positions over the table. Let six players step forward, each holding a cue and taking up a position to strike a ball. Aiming in any convenient direction will be allowed. At a given signal, the six balls are struck in unison. The balls collide with each other and rebound from the cushioned edge of the table. Because of frictional resistance with the cloth and the air, the balls quickly slow down and finally come to rest. To cope with this difficulty, we must imagine that there is no friction in the system, so all the energy stays in the form of kinetic energy, with no energy loss to the outside. It is a closed energy system following the instant of impact.

The physicist now asks this question: Knowing exactly the starting position of each ball on the board and its initial velocity (speed and direction), can we predict the exact position of each ball after a lapse of exactly ten seconds? or one hundred seconds? A physicist who is a determinist will answer "Yes." Yes, because the laws of motion govern every move made by every ball; it is only a matter of plugging the initial positions and velocities into a set of mathematical equations and solving for the desired elapsed time. The system activity is precisely determined at all points in time by this unique set of

conditions and the laws of motion. The same answer could be given for a real billiard-table situation, provided that the exact values of frictional resistance and all other disturbing effects could be entered into the calculation. The system we have envisioned is a deterministic system; the way it is viewed as a scientific phenomenon is called determinism.

Determinism in empirical science was first expressed in the early 1800s by the French astronomer and mathematician Marquis de Laplace. This was about 130 years after Newton had formulated the laws of gravitation and motion. Other scientists had in the meantime developed these laws and applied them to many common mechanical phenomena. As translated from the French, Laplace wrote in 1812:

> Let us imagine an Intelligence who would know at a given instant of time all forces acting in nature and the position of all things of which the world consists; let us assume further that this Intelligence would be capable of subjecting all these data to mathematical analysis. Then it could derive a result which would embrace in one and the same formula the motions of the largest bodies in the universe and of the slightest atoms. Nothing would be uncertain for this Intelligence. The past and the future would be present to its eyes. (Wartofsky, 1968, p. 298)

On this cosmological scale, a deterministic solution would be far from practical. It works fairly well for orbiting objects in the solar system and allows us to predict eclipses and tides far in advance and with a remarkable degree of accuracy. It is not, however, practical in analyzing most ordinary natural systems we can observe quite closely on our planet. For example, in the billiard table case, the surface of the table has many tiny bumps and hollows we can scarcely see or feel, but these affect the motions of the balls. These disturbing effects are distributed more or less at random over the surface of the table. Given plenty of time and some fancy instrumental equipment, a scientist could measure all the irregularities and enter their effects into the equations. Even so, the corrective effort would be incomplete and some errors would remain to upset final conditions. The effect of the small initial errors tends to grow cumulatively as the experiment continues to run, so prediction would be subject to greater error as time passes. (See Smart, 1979, p. 652.)

Analyzing formally the nature of a strictly deterministic physical system, Professor Marx Wartofsky writes:

> Given the laws of the system, the values of the state variables at any time t determine uniquely the values of the state-variables at any other time t' (predictively or retrodictively, because time simply comes to nothing but a change in the values of the variables in accordance with the laws of the system). Thus, nothing comes into existence which is not already "contained" in the state description

of the system at any time, given the laws of the system. . . . A physical law would then be a universal statement giving the rule for generating the values of the state variables for any value of the time (or given our preceding account of time, generating the values of all of the state variables, given the change of values in any one of them). (1968, p. 298)

Uncertainty and Probability

Determinism encountered major difficulties in practice in many areas of physics. One was in the problem of predicting exactly where a particular electron orbiting the atomic nucleus will be located and how fast it will be traveling at any given instant. If it is agreed that an electron is a particle, that particle must, at a given instant of time, be located at a definite point in space and must have a certain velocity. Physicist Werner Heisenberg attempted to solve the problem of determining both the location and velocity of an atomic electron by experimental means. In 1927 he came up with a rather startling conclusion: it is simply not possible to measure precisely both the location and the velocity. (Strictly speaking, "momentum" should be substituted for "velocity.") In theory, either could be measured accurately, but not both. The solution to the problem is said to be inde- terminate (i.e., it cannot be determined). The principle involved here is called the Heisenberg uncertainty principle. At the heart of the principle lies the idea that the location of the electron in space at a given instant of time is subject to chance variation. There is a particular location that is its most probable position, and other locations that are less probable. A similar statement can be made about the velocity of the electron at a given instant: there is a most probable velocity, and numerous other velocities that are less probable. The electron, then, seems to follow mathematical laws of probability, rather than laws that lead always to a single exact solution. Let me try to clarify the concept of probability with a simple mechanical model.

Imagine that a rifle is securely locked into position on a test bench in a long straight tunnel. The rifle is fired repeatedly at a target 500 meters distant. The target is a sheet of graph paper--a grid of horizontal and vertical lines. We don't know where the bull's-eye should lie--that is something to be ascertained from the results and not in advance. Because the bullets make rather large holes in the target, it will soon be badly shredded in an area where the bullets tend to concentrate, so we use a fresh sheet of graph paper each time a bullet is to be fired. The exact position of the bullet hole is determined in terms of X and Y coordinates on the graph field and this information is fed into computer storage. On a computer screen the shots are displayed as sharp points as they accumulate. The points are at first scattered in a heterogeneous sort of pattern, but as the number of points increases, the points show clustering about a common center. The flight of each bullet is subject to aberrations from several sources. The bullets are not exactly uniform in mass, internal density distribution, and external form. Some, because of their shapes, will veer to the right or left, and up or down. The powder charge is not exactly the same in each shell, so the muzzle velocity of the bullet varies from one to the next. The Coriolis force that deflects to the right in the northern hemi- sphere increases in proportion to the speed of the bullet, so variations in speed will cause deflections to left or right. Air in the test tunnel is in turbulent motion, with innumerable eddies that form and dissolve. They act as if they were forces exerted on the bullet as it travels, pushing or pulling at right angles to its path, or forward and backward in its direction of flight.

For each bullet fired, the various aberrations lead to a single final effect, which determines the point at which the bullet impacts the target. With no such imperfections and disturbing forces whatsoever, all bullets would arrive at exactly the same point. With respect to that ideal point--the bull's-eye--we can describe the magnitude and direction of the aberrational effects. The direction of the effect is a radial line drawn from the bull's-eye (like the spoke of a wheel); the magnitude of the effect is the radial distance out from the ideal point. In theory, the most probable location for the individual bullet hole is exactly on the bull's-eye. The probability that the bullet hole will lie at a given distance from the bull's-eye diminishes with increasing distance from the bull's-eye. This probability distribution will begin to show on the computer screen as more and more shots are recorded. The clustering around a central region will become more smoothly graded and the cloud of points will show an outward thinning in density. There will, however, appear occasionally a rather isolated shot as far out as the screen permits.

Ultimately, the cloud of points in the central region becomes so dense that it appears solid. We must locate this central point of highest density, because it will be needed as the common center of a set of concentric circles we will draw to indicate levels of equal probability in deviations of the bullet holes. To help overcome this difficulty, the computer has been programmed to give us a three-dimensional picture of the points it has in storage. The two-dimensional graph serves as the horizontal base on which to erect perpendicular columns whose heights are proportional to the number of points falling within each square on the graph. We now have before us the image of a mound or hill, approximately circular at the base. The summit consists of a nest of four rectangular columns of nearly equal height.

Our next step is to use a grid with smaller squares and increase the number of shots fired in the experiment. The hill then becomes smoother in surface form and its summit point is more clearly defined. Finally, we go through the process of fitting an ideal smooth, continuous surface to the hill of tiny columns. This ideal envelope appears to have the shape of a bell--the Liberty Bell in Philadelphia, for example. Its side steepens upward to a middle region of maximum steepness, then lessens in steepness toward a summit that is broadly rounded. The base of the bell tapers outward to what seems to be zero thickness, but actually it extends outward to approach infinity in all directions.

What we have derived here is a particular kind of probability distribution. The central summit point of the bell is the average position, or statistical mean of the probability distribution; it is "the most likely position." Each bullet that was fired is called a statistical variate; the total number of bullets fired makes up the statistical sample. We could never fire enough shots to produce the perfect bell form, but it can be arrived at by mathe- matics. In theory, the shots we fired came from a statistical population of possible shots whose number approaches infinity. Another concept of mathematical statistics is most important in our experiment. Each shot must be thought of as having been drawn at random; meaning that from the box of shells containing all the bullets we fired, each shell has an equal probability of being drawn out of the box on each draw. For you smartalecks who are way ahead of me, let me add that the shells would all have to be identified by a number; each shell as it is drawn would have to be listed in serial order by that number, then returned to the box and mixed in well in preparation for the next draw. You could achieve the same randomness more easily by drawing the serial numbers at random, using a table of random numbers readily available to all scientists. The principle is that each variate drawn to become part of a sample of variates has exactly the same probability of being drawn as every other variate in the sample. Thus we have a random sample of variates.

The next thing we can ask our computer to do is to slice the bell surface with a vertical plane passing through the summit point, or mean. Let the computer display this surface as a simple curve. What we have here is a special kind of probability distribution known as the "Gaussian distribution," or the "normal curve of error." It is said that the curve was formulated by one Abraham de Moivre in 1733, based on probabilities encountered in games of chance. It later took the name of Karl Gauss, the mathematician, who derived the curve in an attempt to idealize the distribution of measurement errors. In science, when a controlled experiment involves numerous repetitions to arrive at a measurement of some dimensional property--such as mass, length, or time, or a product of those dimensions--the individual measurements (variates) tend to fall to the left or right of the mean value with a frequency in numbers proportional to the height of the Gaussian curve.

The great importance of the Gaussian distribution (along with certain other mathematical formulations of probability distributions) lies in its ability to describe the natural variation of things that are measured by scientists. For example, the body weight of an individual organism at a given age or stage in development can be sampled and fitted with a Gaussian curve. This gives us a general model of the relationship of individuals to the population average--extremes are rare, near-average values are common. Examples of the kinds of objects that are measured in science and treated in this manner are almost without limit. In geology, measurements are made of the dimensions of crystal grains in a rock or of sand grains in a beach or dune. Use of the Gaussian curve enables an investigator to make estimates of the reliability of scientific statements that serve as tentative conclusions based on measurements. Reliability is evaluated in terms of the percentage probability of being in error when a statement is asserted to be true.

The Null Hypothesis

To get a better idea of how the Gaussian model of distribution can be used in testing a scientific hypothesis, consider the following imaginary case. A physical anthropologist is studying two isolated groups of aboriginal humans separated from each other and from contacts with other populations by mountain barriers. The question under investigation is whether the two groups have important physical differences that would suggest that they migrated to their present sites from different source areas. The conservative scientific hypothesis reads thus: "There is no significant difference in the physical characteristics of the two groups." This is known as the null hypothesis. The anthropologist now takes measurements of various dimensions of individuals in each group. Suppose that one of these is the girth of the skull. The sample size is limited by the small total population of humans in each group. A Gaussian curve is fitted to each sample. When the two curves are drawn on the same graph, it is easy to see that although their two means differ by a small amount, the two curves overlap a great deal. This means that the range of variation within each group is large, larger in fact than the difference between the two means. A rigorous statistical test is performed and leads to the statement that, in adopting the null hypothesis, the chances of being in error in holding to it are very small--say, less than one percent. In view of this outcome the wise course of action is to accept the null hypothesis, at least for the time being.

Figure 5.1 may help to explain the use of the null hypothesis. It is the same "ladder of excellence" shown in Figure 2.1, but now the probability scale is reversed. Going up the ladder, the probability decreases by powers of ten; this is the probability that the scientific hypothesis is false when it is asserted as being true. Arrows point to a probability of 1 percent, which is a

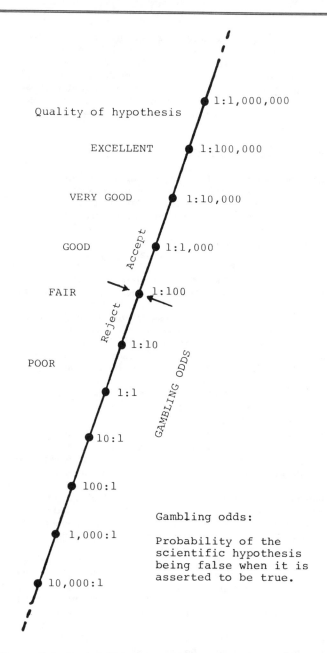

Figure 5.1 Probabilities derived from a statistical test serve as a basis for accepting or rejecting a scientific hypothesis. Compare with Figure 2.1. (A. N. Strahler.)

rather conservative number to use in deciding whether to accept or reject the scientific hypothesis. If, from the sample data, the test of significance of differences in the two means had yielded a probability smaller than 1 percent, we would have rejected the null statistical hypothesis and accepted the scientific hypothesis (i.e., the two populations of aborigines are actually different). As it happened, the probability was much larger than one percent and we were forced to adopt the opposite conclusion. The use of the null hypothesis in the test may seem an unnecessary complication and a waste of time, but it is an application of Occam's razor. It keeps saying to us: "Don't read into the observations a conclusion that complicates things in nature, unless there is a compelling reason to do so."

No scientific hypothesis can be either proven or disproven by mathematical statistics rigorously carried out on a sound mathematical basis. All that you will get out of it is a statement of odds of being in error when stating your hypothesis or proposition. Forget the politicians' favorite slogan: "You can prove anything with statistics."

In empirical science you can't prove anything with mathematical statistics.

Have we drifted far from the notion of deterministic versus nondeterministic systems? Not too far, I think. The hypothetical examples of bullet holes in a target and skull girths of aborigines deal with indeterminate solutions. As for the bullet holes, we can also adopt a rather strict deterministic approach. Each time a bullet is fired at the target, its course is fully determined by the forces that act upon it. Through every millimeter of its flight, the bullet moves in conformance with the laws of mechanics. In that sense, there is nothing at all "chancy" in the behavior of a particular bullet on a particular flight. All that we really need is the services of Laplace's "Intelligence" to work out the necessary equations to satisfy all the initial conditions and the transient conditions en route. It would be far too difficult for humans to actually do this job.

In practical terms, then, the deterministic approach simply has no scientific value because to subject it to testing is close to being impossible. This conclusion would probably be looked on favorably by physicist Percy W. Bridgman, a Harvard professor for many years and a Nobel laureate as well. Bridgman expressed a rather practical or hard-nosed view of what is useful or meaningful in scientific endeavor. His "operational method" (Bridgman, 1936, p. 10) in physics considers that a hypothesis has meaning only if it can be tested by accepted scientific procedure. Deterministic models of a complex system have essentially the status of untestable hypotheses--in the absence, that is, of Laplace's Intelligence.

The Stochastic System and Natural Laws

At this point we can no longer defer the introduction of a horrendous new term: stochastic. It is an adjective we can use to replace "indeterminate" as the alternative choice to "deterministic." You can think of "stochastic" as meaning "related to chance"; it is a system activity that involves chance variations in some parts of that system. A stochastic system has built into it a random process continually at work. Gambling machines illustrate the stochastic process very nicely. The roulette wheel, if it is not tilted or rigged in some way, delivers its winning or losing numbers in a random sequence. Drawing the numbers of military draftees or lottery tickets from a large goldfish bowl is a stochastic process, for it exploits randomness. (Don't forget to put the ticket back into the bowl!)

The stochastic nature of empirical science investigations lies in two basic sources of randomness in the sequence of numbers it delivers: (1) errors of observation (measurement), and (2) natural variations in the individual objects or properties that make up the variates of the statistical sample. Scientific investigation must deal in stochastic models or hypotheses because there is no practical alternative. This operational method in no way substitutes "laws of chance" for "laws of physics." Laws of physics can, by the deterministic process, explain fully why the roulette wheel stopped on a particular number on a particular turn of the wheel. No goddess of chance put her finger on the wheel to make it stop at a point other than where it would have stopped solely because of frictional loss of energy.

The reason why I place so much stress on this subject is there is prevalent a misconception to the effect that randomness explains physical phenomena; and, because that is the case (the argument goes), we have no further need of the basic laws of physical science. The misconception is the notion that when we have identified random variations in natural phenomena, we have therefore also explained what we observe. I will try to develop this concept in an area of natural science with which I am personally familiar: the science of rivers and streams of

water that flow in channelways over the land surface (Smart, 1979). If you have taken a plane flight over our arid western lands in clear weather you must have perceived the branching patterns of river systems, extending upstream to include all the smallest streams and even those dry channels that have flow only in wet weather. The impression is that of a leafless tree silhouetted against a bright sky. We can analyze the river system as if it were a tree, starting at the extreme ends of the smallest twigs and observing how each twig joins with another of approximately equal size to form a single larger twig or branch. Each branch in turn joins with another to form a still larger branch. A small twig with a free tip is sometimes joined to a much larger branch.

Geologists who study stream networks, as they are called, analyze the geometry of the network by dividing it up into individual segments, like so many matchsticks or toothpicks laid out upon a table in such a way as to touch ends. These segments are called "links." A single link is a segment of channel that extends from one stream junction to the next junction downstream (or from the end tip to the first junction). Given a box of 100 matches or toothpicks, you can lay out a variety of link networks to resemble a real stream network. Of course the resemblance is highly superficial. This simple task preoccupies at least a few geologists who are perhaps too lethargic to get outdoors and look at the real thing. One of these "linkers" decided to program a computer to generate networks of links in a random manner. When compared with actual stream networks having the same number of links, the randomly generated network showed no essential difference from the natural network so far as the pure geometry of the link layout was concerned. The investigator's conclusion: natural systems of streams are governed by a random process. The mistaken implication is that randomness acts as a natural cause, or, shall we say, randomness is creative. In this respect, randomness is viewed as part of a system of stochastic laws capable of replacing mechanical laws of physics in the explanatory function of science (i.e., the stochastic model can replace the deterministic model). If so, all we need to know is how to program our computers to produce stochastic models. So, who needs to take courses in physics or chemistry?

What, if anything, is wrong with the above reasoning? In focusing their attention upon the random nature of the geometric variations in arrangements of stream links, the linkers overlooked the strongly deterministic character of the system as it actually occurs in nature. There are constraints on the way a stream network must operate in nature. First, the water must flow downhill throughout the entire system in response to the law of gravitation. Two stream segments (links) must always converge to meet at their downstream ends; a single link cannot contribute to two links at its lower end. This convergence is a physical requirement to increase the efficiency of the stream as it moves to lower levels. Also, real streams are not just straight lines on paper; they are water masses in motion, occupying troughlike channels adjusted in width, depth, and form of bed to carry that water in the most efficient manner at the same time that they carry along rock particles, some of which have to be dragged along the bed. Potential energy is converted to heat energy through friction. None of these essential forms and functions of real streams are of any concern whatsoever to the linkers. They have explained by a random model a superficial geometric variation that comes close to being of no importance or value.

Let's set up the stochastic link game in an honest way. We need a box of matches. Let the match heads show which end of the link is the downstream end of the link. Put the matches in a bowl, shake them up well, then fling them into the air and let them fall as they will on the surface of a table. Now we have a true random system, free of constraints. I can almost guarantee that no matter how many times you repeat this toss, you will never get anything that resembles a stream network. You need to generate a complete network of matchsticks in which two sticks meet with their heads touching and fall so that they also touch the upper (unheaded) end of a third stick, and so forth, to completely incorporate all the matches in a converging network. There is a remote probability that a single toss will achieve this result-- maybe it will happen once in each 10^{36} tosses, on the average. We can't rule out that possibility, but who could stay alive long enough to complete that number of tosses?

Very similar to the above example is the case of the "random-walkers." These are members of the scientific community who can reproduce certain patterns of nature by generating a random walk. The procedure is to move out from an arbitrary starting point on a flat field on the basis of numbers drawn from a table of random numbers. One set of numbers gives the direction of each move; a second prescribes the distance of travel to find the next point. Following a random walk gets you nowhere in the long run, but it can tire you out just the same. I have used the random walk as a device to locate points on the ground at which to take random samples (samples of soil, for instance). It assures objectivity by overcoming your natural reluctance to take a sample in the heart of a prickly-pear cactus plant. Suppose, however, that you are plotting the movements of a predatory organism as it searches for and seizes its victims. You are impressed by the organism's track as you have plotted it on paper. It looks like a random walk. Indeed, a statistical test shows that the pattern is not significantly different from a random walk. That's the answer! The predators pere- grinations are governed by pure chance. Forget about determinism and deterministic models! But wait, did you stop to think that the organisms that are its prey might also be randomly located on the field of travel? Was the predator being deterministically guided by its senses in a way that would lead it to the most conveniently located free meal?

The Stochastic Model and Prediction

It looks as if a purely stochastic system always has a stochastic beginning, in which initial quantities, such as energy, momentum, speed, acceleration, and direction, arise at random from a population that is described by a probabilistic model. Now, take a second look back at the deterministic model and ask this question: does it have a stochastic requirement somewhere in the system operation?

In a strict deterministic analysis, there is no actual physical explanation of the specific states of the system at specific times. The only place where explanation comes into the picture is in the general laws. Is it not obvious that the conditions we specify at a given instant of time arise by chance in the first instance, if we could trace the state of the system back to an absolute zero point in time? In the case of the billiard balls, the zero point in time was the instant that all six balls were struck by the six different players. The balls had been placed in a random pattern to start with and it can be assumed that the directions in which the players aimed were also randomly distributed over a 360-degree range.

Let us now apply the stochastic model to the more general case of atoms of a gas in a closed container, perfectly insulated from its surroundings. The atoms of the gas are assumed to be composed of only one element (nitrogen or helium, for example). The individual atom is visualized as a perfect elastic sphere; all the spheres are exactly alike (as in our example of the billiard balls). These elastic spheres are continually in motion, flying through the vacuum in straight-line paths at high speeds, perhaps on the order of 5 kilometers per second (about 10,000 miles per hour). A particular atom will strike either another atom or the wall of the container and will

change course or rebound, as do the billiard balls. The paths of free flight are, on the average, of great length compared with the diameter of the individual atom. Weak forces of attraction between the atoms can be disregarded. Constant bombardment of the walls of the vessel by atoms accounts for the pressure the gas exerts on those walls.

The atoms of a gas cannot actually be observed as flying objects; they are much too small and move much too fast. To apply a deterministic model to the entire container of gas is ruled out as a practical procedure because of the impossibility of actually measuring the initial and sequential positions and velocities of all (or any) of the atoms in the container. This impracticality in no way detracts from the validity of the mechanical laws that govern the activities of each and every atom. The only physical variables that can actually be measured in this experiment apply to the entire population of atoms treated as a uniform substance. These variables are the temperature of the gas and the pressure that the gas exerts on the container walls. Temperature can be read from a thermometer that penetrates the wall of the container; pressure can be read from a gauge that is connected with the gas chamber by means of a small tube. Limited to these observations, we turn to a <u>stochastic model</u> that will relate what the gas atoms are doing collectively to the observed temperature and pressure of the gas.

The stochastic model views the gas atoms as individual variates in a statistical population. The speeds of the atoms at any given instant can range from as low as zero to very high values. Thus there is a statistical population of speeds. This being the case, there must be an average speed that characterizes the entire population of speeds. We can also envision a probability distribution of speeds; it will be a peaked curve, perhaps something like the Gaussian curve. Physicists think of it as being sharply peaked, so that the speed of most of the atoms is not far from the average speed. Physicists use a relatively simple mathematical equation to demonstrate to their satisfaction that average speed of the population is related to the temperature of the gas. The equation requires that the mass of the individual atom be known; it has, in fact, been established independently. Without going into further details, it can be said that it is a relatively easy task to establish the statistical mean speed of the gas atoms. The actual speeds of individual atoms can be estimated only in terms of probability. If individual atom speeds could be sampled at random, the probable speed of an atom would be indicated by the probability distribution. There is no way that an actual speed in kilometers per second can be assigned to a particular atom drawn at random.

Under the stochastic model of behavior of gas atoms, the exact positions in space of individual atoms at a given instant cannot be stated, but the distribution is assumed to be described by a particular probability distribution. The average distance separating one atom from its neighbors can be calculated from the number of atoms in the container and the volume of that container. Here, again, an ideal probability curve could be assumed to predict the likelihood of any two atoms being separated by a certain distance. As for the travel direction in space of a single atom at a given instant, that can be assumed to follow a random model--all directions are equally probable when individual travel paths are randomly sampled from the population.

In the stochastic model of behavior of gases there is provision for events of extremely low probability to occur. The stochastic model we have outlined describes the most probable state of the system in terms of energy. But because the probability curves on which it is based have "tails" extending toward infinity in either direction (or approaching zero), extreme values are possible. There is always a possibility, no matter how remote, that some very strange arrangement of the gas atoms will occur for at least an instant of time. For example, there is a remote possibility that we might observe an instant in which all the atoms spontaneously became crowded together in one corner of the container (see Chapter 13). Just don't wait around for this remarkable event to occur. The subject of extremely improbable events provided for by the stochastic process will come up again. The fundamentalist creationists have used the idea in arguments to support their contention that life could not have come into existence on earth on the basis of chance alone.

It may distress some of my readers that scientists must resort to stochastic models to describe natural systems. The rules of stochastic procedures say that no statement can be proved either correct or incorrect; therefore that absolute truth is unobtainable. The same rules present a scientist with a scale of betting odds. The scientist must select in advance the particular odds that will lead to tentative rejection or retention of the hypothesis being tested. Can you think of a better way to investigate complex natural phenomena? The world is waiting for your improved method.

For an appropriate comment with which to conclude this chapter, I defer to Alan Lightman, a contributing editor to Science 83, published by the American Association for the Advancement of Science. In a brief essay on probability in science, "Weighing the Odds," he wrote:

> Most people, I suspect, have a deepseated reluctance to welcome probabilities into their private lives. At least since the Greeks, mankind has harbored a passion for knowing some things with certainty. Probabilities, by definition, shimmer in a mist of uncertainty. Einstein contributed little to science in the last three decades of his life, in large part because he could never accept the probabilistic nature of the emerging quantum physics. "God does not play dice," he insisted. (1983, p. 22)

Chapter 6

Science and Religion—Do They Mix?

Judging from the current vitriolic debates between scientists and fundamentalist creationists, science and religion don't mix. It might seem to be a case of "either/or," but the creationist position is only one expression of religion out of many alternatives. Alternatives exist not only within the group of Western monotheistic religions--Judaism, Christianity, and Islam-- but also with respect to many other religions of the world. Scientists also come from the ranks of believers of the Hindu, Buddhist, Confucian, Shinto, and Taoist religions. We need to examine the relationship of science and religion on much broader terms than appear in the "creation/evolution" controversy.

Ontological Models Relating Science to Religion

The philosophical field of ontology, which examines the nature of reality, provides for a number of different models that can be helpful here.[1] Four such models are pictured in Figure 6.1. Each model must concern itself with the origin of the universe (cosmology) and, more specifically, with the event of initiation of life on earth (biopoesis), and with the diversification of that life into numerous forms, including the human species (referred to here as "Man," the English translation of the genus Homo). Each model must deal with the nature of reality and the role of time. The role of purpose or design may also be included, but only where required by the model.

The ontological models are presented as schematic diagrams in which certain definitions and rules are imposed for the sake of clarity and uniformity. All models are based on two possible realms within which reality is presumed to reside. In each diagram a horizontal line of dots separates a natural realm from a supernatural realm. A given hypothesis or model may require both realms or only one. Where both realms are postulated to exist, their spatial realationship is not specified; they may occupy the same space.

The natural realm includes all physical reality that is amenable to analysis by empirical science. You will immediately identify this realm with the cognitive field we recognized in Chapter 1 that comprised the research fields of knowledge. We identified the supernatural realm in Chapter 1 as being one of the belief fields; it includes religion, which is the main concern here.

First, consider the classification of the models as monistic or dualistic. Ontological monism asserts that there is only one kind of reality. Here, ontological monism is taken to mean acceptance of either the supernatural realm or the natural realm, but not both. Ontological dualism asserts the reality of both realms. Usually, ontological dualism postulates some interaction or influence between realms, and usually the direction of flow of influence is from the supernatural to the natural. The case of a totally separate and independent existence of the two realms can simply be dismissed as uninteresting.

Model A, representing theistic-teleological dualism, recognizes the existence of God in the supernatural realm. In the philosophy of religion, the concept of God as a continuously active agent influencing the natural realm is known as theism. In the natural realm we show a succession of physical constructs produced through cosmic and geologic time. They follow the current, or conventional, scientific opinion or consensus, which in later chapters I like to call mainstream science, to distinguish it from "creation" science. First is the origin of the present universe, some 10 to 20 billion years ago (-10 to -20 b.y.), possibly by the Big Bang mechanism. The existence of God and/or universal matter prior to the cosmic explosion is not specified here. The Christian Church seems to hold to the belief that whereas God is eternal and has always existed, the universe was created by God out of nothing (ex nihilo). On the other hand, the model allows us to suppose that matter and energy have always existed, i.e., they are eternal. This alternative follows historically the Greek atomists, who postulated that the universe consists of a finite number of atoms that are permanent and eternal, without a beginning or end (Angeles, 1980, p. 32).

The mainstream science view is that about -3.6 to -3.7 b.y. at the earliest, biopoesis occurred on our planet. Biopoesis is the formation of living matter from nonliving matter, a topic we cover in Chapters 53 and 54. An important feature of the theistic-teleological model is the belief that God caused biopoesis, i.e., God created life from nonlife. In this model, divine creation provided the starting point of organic evolution, which then proceeded under pervasive divine guidance toward a higher goal. For some persons, this goal is believed to have been the rise of the human genus, or Man. (Other versions of divine intervention include numerous events of creation of life forms, but we leave such possibilities for later consideration.)

Notice that in the theistic-teleological model, the flow of influence across the boundary of the two realms is in one direction only. The same is true for models B and C. If we subscribe to the principle that the sum total of energy and matter in the universe cannot be increased or decreased (law of conservation of matter and energy), the control exerted by God on the natural realm can carry no flux of either energy or matter. This conclusion is consistent with our characterization of the supernatural realm as that which is not amenable to analysis by the scientific method. Just how God controls the natural realm is a mystery that rests totally in faith or religious belief.

While the theistic-teleological model is probably satisfactory for large numbers of educated and devoutly religious persons of the Judeo-Christian tradition (excepting the fundamentalists), it grates rather harshly on the sensibilities of the mechanistic (nontheistic) scientists. Under the theistic model, science cannot explain all of nature, because divine intervention is unfathomable. In ways unknown and unknowable the stochastic process in nature is being stacked to achieve an unknown purpose.

Model B, deistic-mechanistic dualism, follows the same program of physical constructs in the natural realm as in Model A. The essential difference is that the deistic model shows God's creative work to have ceased with biopoesis so, from that point to the present, organic evolution has proceeded on its own in a completely mechanistic manner.

The cessation of divine causation and guidance following creation is referred to in philosophy of religion as deism (Angeles, 1980, p. 52). The concept is seen in Cartesian philosophy and may have been adopted by none other than Charles Darwin, who wrote:

> There is grandeur in this view of life, . . . having been originally breathed by the creator into a few forms or into one; and that, whilst this planet has gone cycling on according to the fixed law of gravity, from so simple a beginning endless forms most beautiful and most wonderful have been, and are being evolved. (The Origin of Species, 1859)

One has to be cautious here in reading Darwin's inner mind. In an statement by Ashley Montagu including the above quote (1984, p. 14), I came across a footnote explaining that the words "by the Creator" were not in Darwin's first edition, but were inserted in later editions "as a concession, probably, to his wife's religious views."

Once life was started by God, biopoesis never again occurred, either by divine creation or by natural (mechanistic) processes. Thus a totally mechanistic system has prevailed over approximately the past 3.6 b.y. This deistic model obviously has some major advantages for the mechanistic scientists, since it keeps God "out of their hair" while they are investigating evolution and geologic history; they are also free from being "bugged" by teleological suggestions. Any evolutionary changes that might look like the long-range strivings of organisms toward a distant goal must be accounted for by physical circumstances alone. What's more, attributing biopoesis to God gives the geologist and biologist an aura of respectability with the religious community; any charge of atheism can be protested. Perhaps for some scientists, this belief assuages guilt for having excluded God from performing any useful role in evolution. At the same time the supreme mystery of God's breath of life can be enjoyed and even savored as a numinous experience.

An important variation of the deistic-mechanistic model limits divine activity to the single act of creation of the universe. Biopoesis then becomes just one of many purely mechanistic developments taking place billions of years after the universe was formed. This variation of the model places scientific astronomy on a purely mechanistic basis, free of any supernatural influences, while at the same time allowing astronomers to express a belief in God and His/Her great act of special creation. Thus the astronomers and cosmologists too escape being branded as atheists.

Model C, fundamentalist creationism, is another dualistic model. It represents (more or less) the doctrine of recent creation held by Christian fundmentalists. We shall give full attention to this model in Part 2. It is unique in dividing the content of knowledge into two divisions, one succeeding the other in time. The first body of knowledge, dealing with recent creation and the Flood of Noah--the question of origins, that is--falls into the belief field as religion; the second, into the research field as conventional (mainstream) science. For a believing scientist, this epistemological dichotomy or dualism removes the stress of having to cope with possible divine intervention in the present world of nature. A pure temporal deism may not accurately represent the working beliefs of the creationists. As we shall explain later, their tenets seem to allow room for miracles performed by God in post-Flood time.

Model D, mechanistic monism or naturalistic materialism, eliminates the supernatural realm entirely. This is pure ontological monism, for only the natural realm exists. In this sense, it appears atheistic, or at least nontheistic. It does, however, incorporate God into the natural realm under a mechanistic concept: God is anthropogenic; God is created by Man.

With this brief description of four ontological models as a framework for viewing Western religion in reference to science, we turn to some further considerations and implications they evoke.

Theistic Science--A Contradiction in Terms?

Turning now to the group of scientists professing to be theists, we raise the question: Is there any inner conflict in holding this dualistic position? We ask them: Does such belief inhibit your scientific activity in any way? I feel concerned that some level of conflict within the individual may exist, but not necessarily for those in all branches of pure science, and perhaps not at all for those in the applied or normative sciences.

For most theistic scientists there seems to be no problem with God's role in initiating the Big Bang, postulated by many cosmologists to have started the universe. Science has no means of probing farther back in time than that instant. There seems to have been no problem in gaining acceptance by the religious establishment (fundamentalist creationists excluded) of the scientific estimates of the age of the universe and its components, such as galaxies and stars. There has been general acceptance of the geochemists' determinations of the age of the earth as a planet, of rocks that form the earth's crust, and of fossils contained in those rocks. A large number of well-educated persons within the Catholic, Protestant, and Jewish communities, including their clergy and leaders, give broad support to the scientists' time scale and to the descent with modification that is shown in the fossil record.

As to the position of the Christian theologists (creationists excepted), I quote from a paper written by a scientist, Rev. James W. Skehan, S.J.: "the creation stories in Genesis 1-2 are theological reflections on the origin of the Universe and on humanity as the apex of creation" (1983, pp. 307-308). Further on, Skehan writes: "Among the biblical theologicians there is wide agreement that the story of the creation of the earth and of mankind in the first chapters of Genesis, is presented to recount the beginnings of the religious history of the people of Israel, and is not a scientific analysis to establish either the age or mode of origin of the earth."

Although this theological view of the significance of the Genesis account may seem to enable the scientist to practice without inhibition, there actually remains a large body of religious belief that attributes an active and directive role to God. Members of the Christian faith believe that God is a living force, for they pray to God for strength and for the enactment of beneficial events, such as an end to poverty, cure of a disease, or for peace on earth. They believe in miracles--events that cannot be explained by science and must therefore be acts of God. Moreover, God has always had a purpose or plan, directing all changes and events toward a goal. Study of such divine direction and its revelation in a design or pattern is teleological, but much more far-ranging in time than the natural teleology we have already referred to in biology.

Thus, in the historical sciences, the theistic scientist is faced with the problem of deciding which events or portions of the record represent God's work and which remain as naturalistic phenomena. To conduct research as if God had played no part is a rejection of one's own religious belief; surely this would be traumatic. On the other hand, if one proceeds on the assumption that God's hand was at all times into everything, historical science becomes nothing more than the documentation of God's work. Since God also created all laws of science, reductionist explanations invoking underlying laws are also attempts to understand God's plan. Indeed, documenting God's glorious and perfect work was held by

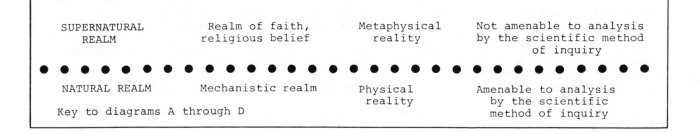

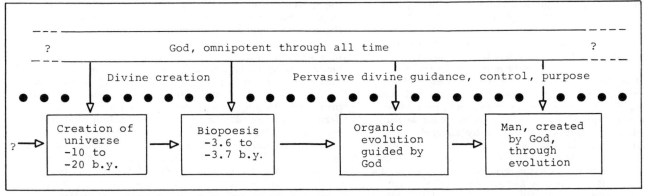

A. Theistic-Teleological Dualism

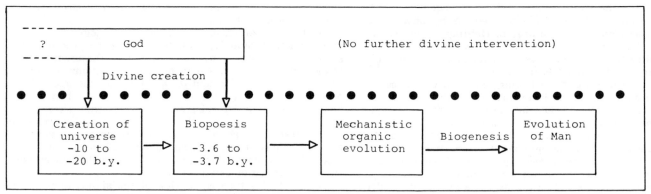

B. Deistic-Mechanistic Dualism

Figure 6.1 Schematic diagrams of four ontological models extant in philosophies and religions of Western cultures. (Adapted from A. N. Strahler, 1983, Jour. of Geological Education, vol. 31, p. 88, Figure 1. Used by permission.)

the early naturalists to be the purpose and justification of natural science. That same view is expressed in one form or another by many modern theistic scientists; perhaps for them it successfully bridges the two realms of reality.

Under theism the initial formation of life on earth from nonliving matter was a creative act of God. Many biologists and geologists who are not theists can live with that assumption. It was, after all, a singular event. But what about the evolution of life through at least three billion years since life started? Scientists must investigate evolution as a naturalistic phenomenon. Can divine intervention be ruled out if God has the capability for forcing change?

The theistic scientist relies on the body of statements we have called laws of science, judged by long experience to be in such low probability of being in error that they must be respected as forming the foundation of science.

First, there is the fundamental principle or law that states that the sum total of energy and matter in the universe is constant. If God set things up this way, cannot God also subvert the law? How can we be sure that at times God has not added a substantial quantity of matter or energy to the universe? There is a law of gravitation and there are laws of motion, which scientists dare not ignore. These and many other scientific statements are accepted as working principles by scientists and woven into their hypotheses of evolutionary change. Hypotheses that attempt to explain how new species of plants and animals come into existence must, according to the Darwinian theory, be wholly naturalistic (mechanistic). Yet, while formulating such hypotheses and attempting to strengthen them by testing of deductions, does there not constantly linger in the minds of the theistic scientists the thought that perhaps the formation of a new species was forced by a supernatural act in fulfillment of God's unknown and unknowable design? My own reaction would be to feel a sense of futility in continuing my research under such circumstances. The probability of my scientific hypothesis being false is perhaps being increased to a very high level by supernatural forces that cannot be evaluated.

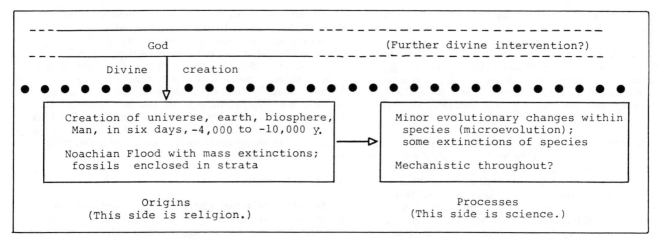

C. Fundamentalist Creationism (dualistic)

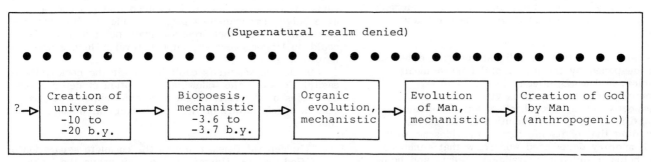

D. Mechanistic Monism (Naturalistic materialism)

In this context, consider the remarkable linkage of religion and science in the practice of seismology (earthquake phenomena) by members of the Roman Catholic order the Society of Jesus, or the Jesuits, as they are commonly called. In their universities and far-flung observatories, the Jesuits have contributed greatly to the development of the science of seismology and for decades they were the major gatherers of seismological records on a global basis. As avowed theists, do Jesuits feel insecure or uneasy about possible divine intervention in seismic phenomena? Would it worry them that God might consider it desirable to set off an earthquake long before enough mechanical stress had accumulated to cause the event under natural circumstances? Could God choose to alter the velocity of seismic waves in a particular event? One would need to have faith that God is not capricious, that God's laws are strictly observed by God on a voluntary basis to prevent chaos.

The question of miracles in possible conflict with the practice of strictly materialistic science has intrigued me for years, and this may explain the length to which I have gone to discuss it. To carry the matter further, I decided to write to a distinguished Jesuit scientist for an opinion. Accordingly, I sent some of the relevant pages of my manuscript to Rev. James W. Skehan, whom I quoted earlier. Reverend Skehan holds the Ph.D. in geology from Harvard University and is Professor of Geology and Director of the Weston Observatory, Department of Geology and Geophysics, Boston College. He is also a theologian with degrees in sacred theology and has made a special study of the origin and significance of the earliest religious ideas expressed in the book of Genesis. Much to my gratification, Reverend Skehan replied with a letter in which he set down his position as a scientist and as a theologian dealing with the question of possible divine intervention into the naturalistic realm with which scientists are concerned. His analysis includes the question of divine response to intercessory prayer. With

Reverend Skehan's kind permission, I quote several paragraphs from his letter, dated July 15, 1986:

I undertake my scientific research with the confident assumption that the earth follows the laws of nature which God has established at creation. Moreover, I should note in passing that my geological research is carried out using the same measuring instruments that other geologists in my field of structural geology use in their field and laboratory analysis. On the basis of such observations, I develop models or working hypotheses to explain how such structures have been produced by various kinds of earth movements. Moreover, my work is performed with the idea of testing the relatively new theory of plate tectonics and developing new models. In this respect the framework for my scientific studies is the same as that of other scientists worldwide.

At the same time, however, my religious belief, supported by theological analysis, is that miracles can occur and, indeed, have taken place from time to time in a religious context, and that they can be recognized as such. Additionally, I believe that God can and does answer intercessory prayer. There is strong biblical to contemporary support for miracles and for the efficacy of prayer in the Old and New Testaments and up to the present. The Apostles, noting that Jesus often went up to a mountain to pray, asked Him to teach them to pray. And in response He taught them "The Lord's Prayer."

I accept the possibility and the reality of miracles because God, as the "Architect" of the Universe who "taught" creation the laws by which it must operate, can intervene for an important religious purpose. However, such intervention can be recognized as such. In no way does my acceptance of the reality of miracles interfere with my ability to carry out scientific investigations because my studies are performed with the confidence that God will not capriciously confound scientific results by "slipping in" a miracle! Such an anomaly is a non-problem as far as I am concerned; nor have I ever known a Jesuit scientist who was bothered by such a possible problem.

Nor is intercessory prayer and God's response to it a problem to my equanimity as a practicing scientist (except on such occasions as I may not have kept up my practice of prayer or meditation!). Theologically this is a non-problem because God lives in the timelessness of eternity, as I and others who believe in an afterlife expect to do after death, and from the beginning He could foresee and provide for each individual prayer and His response. Far from generating confusion or tensions, I find that these two spheres of my life, the scientific and religious, are mutually complementary and provide reinforcing perspectives from which to view the earth and universe.

Basically I find that my religious beliefs, enhanced and deepened by theological and philosophical studies, add a magnificent dimension to my appreciation of the earth as viewed from the perspective of my branch of geology. Conversely my scientific study of the evolution of the earth, including that of its life forms, provides me with a contemporary data-based worldview that enhances and helps to enliven my religious beliefs and helps situate my theology in a contemporary setting. I feel that I am immensely privileged to have been provided the opportunity to view creation through two different telescopes whose overlapping images yield a synthetic view of the world and universe of unparalled beauty and grandeur.

This great richness and beauty, derived from my dual perspective as a scientist and as a Jesuit-Catholic-Christian with Judaic religious roots, grows out of my theological, philosophical and scientific studies filtered through my personal experience. These have led me to the conviction that the supernatural realm, though intangible, is indeed a magnificent and substantial reality. Thus, I believe that God, the Spirit of Wisdom, is pervasively present throughout the universe and to each person. There is a large segment of the world's population which shares this basic belief though it may be formulated in various terms. I believe personally that there is interaction between God, the Spirit of Wisdom, and individual humans that is at once real, non-manipulative, and which can have a beneficial effect in one's life and activities, including the quality of one's scientific research. One form of this interaction is in prayer, either verbal or non-verbal (meditative prayer). It is my impression that the level of this interaction is according to one's openness to this interaction. Thus I believe that prayer is necessary to nourish my religious life. Furthermore, not only does prayer not interfere with my ability or desire to carry out scientific research but I find that it supports and provides motivation and thus adds a level of personal fulfillment to my teaching, research and publication activities. I think this outlook is compatible with that saying of the founder of the Jesuit Order, Ignatius of Loyola: "Trust in God as if everything depends on Him, but work as if everything depends on you."

We will return to the subject of theistic science in Chapter 12, where the discussion is narrowed to theistic evolution and its negative reception by the fundamentalist creationists. What seems to have emerged thus far in our examination of theistic science is the great diversity of relationships between God and Nature conceived within the general model of theistic-teleological dualism. Accommodations are made, tailored to suit each individual's concept of God and how that God works. Nothing I have encountered thus far leads me to think that the integrity of science is being compromised by such individualized religious philosophies.

Deism and Science

Scientists of today who favor the deistic-mechanistic model can recognize that life was initiated by divine forces; they can set the approximate date by reference to scientific determinations of the ages of rocks in which the first recognizable fossils are found. This would be about 3.6 to 3.7 billion years ago, but the dates are subject to revision. Once life began, it evolved under a completely materialistic (mechanistic) program. One can broaden this statement to read that from a certain point in time, God could no longer exert any supernatural influence upon the natural realm.

But deistic scientists can't be let off the hook quite so easily! Consider the following description of deism given by Professor Peter A. Angeles, a philosopher who has made a special study of the God concept in Western theology:

> In general Deism holds to the complete separation of God and the Universe. God transcends the Universe and is different in His Nature from the Universe. He has set up the Laws governing it according to His Idea. Occasionally He suspends the Laws of Nature and produces a beneficent event. This is called a miracle.(1980, p. 52)[2]

Those last two sentences certainly take the serenity out of deism so far as the research scientist is concerned. The possibility of intervention by God may not be as serious a problem for the deistic scientist as for the theistic scientist, but it is still there. All of you have watched in a leaky faucet the very slow growth of a water drop, enclosed in its surface-tension "skin," gradually forming a drooping sac and suddenly detaching to fall free as a sphere and land with a sharp "splat" or "plink." The phenomenon of an occasional miracle leaking through the otherwise impervious roof of deism is something I have elsewhere referred to as leaky deism (Strahler, 1983, p. 91).

For deistic scientists it will be of crucial importance to recognize a miracle when it occurs, so that they can say: "Chalk that one off to God--and thank God, I don't have to try to explain it!" The problem is that recognizing miracles leaves the scientist with only the routine events of nature to handle. The really interesting and challenging problems of science, such as the origin of life on earth and the origin of Homo sapiens as a remarkable and unique species, are turned over to God and are off limits to study.

I would ask these deistic scientists to set down their criteria for identifying an event as a miracle. To simply say that the event is a miracle because no explanation is immediately available in terms of basic laws of science is totally unacceptable. The entire history of science shows that phenomena that seemed at the time to have no explanation in scientific laws proved later to be fully explicable by such laws. Perhaps we could compromise by allowing God one miracle per each one-billion-year period of geologic time. This would accommodate the major events of cosmology and earth history. Would God agree to this compromise?

Creation Ex Nihilo?

Whether one's view of Western theology in the Hebrew/Christian tradition is theistic or deistic, there is agreement in the belief that God created the universe. When this happened is a subject of major disagreement. On the one hand, the fundamentalist creationists of Protestant denominations put the moment of creation at some point around 6000 years before present, essentially according to the Genesis version. On the other hand, there are those Protestant and Jewish adherents who are willing to let the astronomers set whatever date seems to be indicated by the latest available information (i.e., -10 to -20 b.y.) The second position is, of course, essentially identical with that taken by scientists who are atheists, so it is here that theism, deism, and atheism converge to common ground. From an operational standpoint it matters little whether God created the universe and then immediately withdrew into seclusion, became impotent, or died, or whether God never existed in the first place. What is disturbing here is that some rather distinguished scientists, among them astronomers, cosmic physicists, and geologists, have felt the urge to bring God into events leading up to the moment of universal creation, even though their description of the Big Bang itself is strictly mechanistic. Is this being done to try to remove the onus that conservative religious persons place on a science that claims no need for God? Is it done to invite personal approval from the religious community or perhaps to sell a scientist's popular writings on cosmology?

Astrophysicist Robert Jastrow has delved into this sensitive area, as we read in his popular article, "Have Scientists Found God?" (1980). He stresses that scientists cannot answer questions about reality at or prior to the cosmic singularity. He reasons that "the shock of that instant must have destroyed every particle of evidence that could have yielded a clue to the cause of the great explosion." To pursue scientific inquiry farther back in time we must (Jastrow explains) place ourselves in the company of the theologians, who have been pondering such questions for centuries.

Preston Cloud, a highly respected and much-honored geologist who has searched for the earliest fossil life forms, lends an aura of respectability to the belief in divine creation of the universe. He writes:

> Given a ball of neutrons at the beginning, scientists can think of naturalistic explanations of varying degrees of probability and testability for all subsequent events. There is, however, no scientific explanation for where the primordial ball of neutrons might have come from. In fact, there is no certainty that there was only one and not several balls of neutrons, or even that the universe didn't emerge from one or several black holes, or from a deity. And, of course, we have no idea what such a deity may have been like, or from where it (or She or He) may have come. That is the problem of first causes. Science has no answers to the problem of first causes, although it can place limits on what kinds of answers are permissible. Science does not contradict the idea of a divine origin for the embryonic universe, during which it acquired those characteristics we designate as natural laws, whose unfolding underlies all later events. It does, however, have something to say about the permissible time framework and the composition of primordial materials. (1977, p. 10)

Christian theologians, starting with Saint Anselm (1033-1109), have gone to great lengths to rationalize the existence of God, using the methods we find in formal science, which is to say, the use of logic that argues from the assumed base. These arguments deal with the necessity of God, God's role as creator of the universe, and God's purpose in that creation. Scientists who are interested in publishing their personal views on the possibility that God created the universe should, I suggest, review the arguments from theology and the criticism of those arguments under examination by modern philosophy. Christianity holds that God created the universe out of nothing (ex nihilo). Is this belief compatible with the first law of thermodynamics? Can you, as a scientist, feel comfortable with this belief? Or would you be inclined to favor an updated version of the view held by the Greek atomists, that the universe consists of a finite number of kinds of "atoms" that are eternal? How does the idea of a universe entirely predetermined by an omniscient and omnipotent God fit into your model of a changing lithosphere and an evolving biosphere?

If you are interested in these fundamental questions, I strongly recommend Professor Peter Angeles's book, titled The Problem of God (1980). The author slices through the anatomy of the theological argument with a relentless microtome, revealing layer upon layer of the inner structure of that argument. Under high magnification and strong illumination, each proposition in the theological argument is scrutinized with an objectivity that should appeal strongly to the sensibilities of any scientist.

Is Mechanistic Monism the Answer?

Finally, we return to those scientists who espouse naturalistic materialism as their grand hypothesis of the nature of reality and the workings of everything in the universe. Unlike simple atheism, which summarily denies the existence of God and all other supernatural phenomena, naturalistic materialism has a place for the God-phenomenon within the natural realm. God is incorporated into the natural realm under a purely mechanistic concept: God is anthropogenic, meaning that God is created by humans. "So Man created God in his own image, in the image of Man created he God" (Gen. 1:27, rephrased). God exists in ideas, and ideas are physical realities produced in the brain by neuro-physiological processes capable of being investigated and explained by scientific study.

Naturalistic materialism recognizes that Homo sapiens-- the human species--has become differentiated into many cultures, evolved over thousands of years. God takes many forms and displays a wide variation in function. For some cultures, God is multiple--a collection of deities, as in Hinduism. For other religions God is one entity, but the God of each religion differs in specifications from that of other religions.

The anthropogenic hypothesis of God (including with God any other entities that are assigned to the super-natural realm) has the strong logical position of accounting for all forms of God and accepting them all as reality. It is important to make clear that physical reality attaches to the ideas and images of God and that they exist in the natural realm (i.e., the model-image is real). In contrast, the prototype is nonreal (nonexistent). Mechanistic naturalism is not atheistic; it does not deny the existence of God. Quite to the contrary, it has a rational position for all God-models within the total mechanistic system.

Mechanistic naturalism maximizes the opportunity for humans to encompass under a single general hypothesis or theory all physical and biological forms and processes of the cosmos, because it has a rational accommodation for all phenomena capable of identification and description. Under mechanistic naturalism such diverse disciplines as psychology, sociology, ethics, religious philosophy, and political science can be covered by the same general systems theory that applies in the natural and physical sciences. This is surely the ultimate in reductionism. In this context, religion becomes amenable to analysis by science and the supposed conflict between two particular religions or between a given religion and a given naturalistic philosophy (such as secular humanism) is restructured into a case of two diverse phenomena capable of coexistence, each in its own habitat--a human mind.

Credits

1. Portions of this section of text are taken from the author's earlier work: Arthur N. Strahler, 1983, Toward a broader perspective in the evolutionism-creationism debate, Journal of Geological Education, vol. 31, pp. 87-94. Used by permission.

2. From Peter A. Angeles, The Problem of God: A Short Introduction, Prometheus Books, Buffalo, N.Y. Copyright © 1980 by Peter A. Angeles. Used by permission of author and publisher.

Chapter 7

Pseudoscience: I. Three Scenarios

In pursuing the main theme of this book, which is the conflict of naturalistic and creationistic views of the universe, one of the specific issues will be the question: Is creation science bona fide science or is it pseudoscience? If the former, creation science belongs with the research fields of knowledge; if the latter, it belongs with the belief fields (using the classification of cognitive fields explained in Chapter 1). Accordingly, we devote two chapters to pseudoscience. In this chapter, three examples are used to help us discern the criteria for recognition of pseudoscience. The second chapter summarizes the criteria of distinction and considers the historical and sociological factors in the phenomenon of pseudoscience.

Pseudoscience--A Checklist

Pseudoscience is information that is promulgated as science, but that fails to meet the qualifications for admission to empirical science--it is false science and, as as such, is not science at all. Don't confuse pseudoscience with science fiction; they can sound almost alike in places. Science fiction is a legitimate branch of literature, dealing in fantasy in the established tradition of such famous writers as Edgar Allan Poe and Jules Verne. A fairly reliable criterion is this: science fiction makes many people happy; pseudoscience makes many people angry--angry, that is, with those who ridicule them. Science fiction has its devotees, among them many excellent scientists; pseudoscience has its cults and the followers in those cults can get pretty nasty. In showing this kind of behavior, the pseudoscience cultists act much like the fundamentalists of the New Right, who find the scientific theory of evolution abhorrent and seek to have it expunged from the minds of school children and replaced with their own product, which they call "creation science."

Listed below are some topics that may or may not qualify as being partly or wholly in the category of pseudoscience. How many of them are familiar to you? Place a "P" in front of those you recognize as pseudoscience, an "S" in front of those that belong in science. I have enough sense not to give you my own ratings--I just couldn't cope with the calumny that would be aimed at my head! For the answers, see Martin Gardner's excellent books: Fads and Fallacies in the Name of Science (1957) and Science: Good, Bad, and Bogus (1981).

Bermuda Triangle

Big Foot (Sasquatch); Abominable Snowman (Yeti); Loch Ness Monster

Flat Earthers; Koreshanity; Hollow Earthers; Atlanteans (Lost Atlantis); Lemurians

Shroud of Turin

Jupiter Effect

Pyramid Power; Pyramidology

Astrology

Lysenkoism

UFOs, Flying Saucers, ETI

Worlds in Collision (Velikovsky)

Chariots of the Gods (von Däniken)

Exobiology

Dianetics; Biorhythms

Homeopathy; Naturopathy; Osteopathy; Chiropractic

Acupuncture

Laetrile, Krebiozen

Here's another one to add to the list: A new and rather startling hypothesis has been making its way into geology and biology at this time. An asteroid impacted the earth with great violence some 65 million years ago. The effects of this explosive impact were lethal to many forms of life and caused mass extinctions of entire species, families, and even some entire orders of animals. On land, the dinosaurs were completely wiped out; in the shallow seas,the ammonites all but disappeared. This scenario is now being studied and debated in reputable science journals and books by some of the best paleontologists and biochemists we have in our universities. The science community regards the asteroid hypothesis as good science, even though it is on the frontier of science. The scenario is as weird as almost any you would find in science-fiction magazines and movies. On what grounds is it science and not pseudoscience? I hope to give you the criteria by which you can reach your own conclusions.

My review of pseudoscience is developed along the following lines. I have selected three major scenarios put forward just after World War II. All three are reconstructions of supposed events; their content falls in the category of timebound knowledge. All three consist of assertions or allegations that particular events occurred at particular times; all three introduce agents responsible for those events; in all three the agents are asserted to be extraterrestrial, but do not require the intervention of the God of contemporary theistic religions. All are based, in one way or another, on testimony given and recorded in a language by humans following sensory experience. In some cases the testimony is in the form of symbols on artifacts and in the nature of the artifacts themselves. The events are of a singular nature, not capable of being duplicated on demand for direct observation by either their authors or independent investigators. Additional common characteristics of the three scenarios relate to their social impact. Each quickly caught the imagination of millions of persons because of rapid and widespread dissemination through the popular media. Large followings of believers were generated from among persons outside the scientific community, and cults have arisen in support of each scenario. In each case the scientific community reacted with skepticism and strongly expressed its disapproval. Within each cult, believers have attempted to draw support from the body of empirical science in an effort to claim that the phenomena described are explicable in a naturalistic manner and do not require supernatural intervention.

Three Scenarios to Consider

The three scenarios are (1) Immanuel Velikovsky's reconstruction of world history of the first and second millennia B.C., (2) Erich von Däniken's claims of visitations by extraterrestrial beings from the time of earliest recorded human history to modern times, (3) alleged visitations by beings in spaceships from outer space in the period following World War II. The last item is closely tied to but not to be equated with the total content of cases of Unidentified Flying Objects (UFOs).

The word "scenario" is used with the implication that the succession of singular events contained in each may be in some degree fictional; indeed, the entire scenario may be an artifact. The scientific community views all three scenarios as being outside the realm of empirical science (i.e., they belong in pseudoscience). No one of the scenarios poses a serious threat to the security of the human race, because the first two deal with ancient history and the third is viewed by many of its believers as a benign phenomenon. In this light, the intensity of conflict between scientists and cultists seems all out of proportion to its importance in terms of content.

We cannot escape from the realization that the highly emotional exchanges between scientists and cultists in public debates are symptoms of a deeply rooted social malaise. The conflict between science and the forces of antiscience or cultism is one in which the stakes are much greater than for mere refutation or confirmation of the particular scenario. The malaise is not new but seems to have taken a new form adapted to an age of space technology.

My suggestion for your consideration is that in earlier decades of this century throughout the culture of western Europe and North America, highly dogmatic Christian beliefs were being abandoned by increasing numbers of persons. Some sectors of Protestantism, and to some extent Catholicism as well, became sufficiently liberalized to accept the phenomenon of organic evolution occurring through an enormously long period of geologic time; this trend was perhaps accelerated by the impact of the Scopes trial. Freethought was on the rise and secular humanists were advertising the merits of a purely rational approach to human problems, using science as a guide to a new ethics. Perhaps the population of agnostics, atheists, and those simply having little interest in religion was steadily growing larger. At the same time, science in the area of nuclear physics had shown its enormous power to create destruction through the atomic bomb while developing its theoretical concepts far beyond any possibility of being understood by the vast majority of citizens.

Perhaps for vast numbers of persons the need was growing for something both new and irrational to believe in--something that, like a religion, required no formal education and did not tax the intellectual faculties, but at the same time carried an air of mystery and romance. The arrival on the scene shortly after World War II of Velikovsky's scenario and a rash of reports of flying saucers almost instantly filled that growing vacuum and, at the same time, polarized the science and antiscience factions as never before in the century. Not only was the

vacuum filled, but there was provided a public spectacle in which the ivory tower of science was to be struck repeatedly with roughhewn two-by-fours, wielded by folk heroes. However one may wish to explain why these scenarios of pseudoscience took off like huge rockets, the evidence is clear that they did.

A somewhat similar explanation of the recent resurgence of pseudoscience is given in John T. Omohundro's article, "Von Däniken's Chariots: a Primer in the Art of Cooked Science." (1981, pp. 307-309)

Some surveys and census statistics presented in a recent paper by William S. Bainbridge and Rodney Stark (1981) are supportive of this scenario. These authors zero in quite closely on the major group that has been most ready to accept pseudoscience. It appears from the results of questionnaires in 1963 and 1973 that the religious liberals and the irreligious persons are more likely to accept the "new superstitions" (p. 50), such as occultism and pseudoscience, than are fundamentalists. The latter group, professing in the questionnaire to be born-again Christians, dislike far-eastern cults and occult literature and take a dim view of UFOs being spaceships from other worlds. They are, however, strongly opposed as a group to the theory of evolution (p. 48). Evolution is most strongly supported by those with no religious preference and those admitting to be agnostics and atheists.

According to 1970 U.S. Census Bureau figures quoted by Bainbridge and Stark (p. 54), there were about 115 million church members in a total population of 205 million, the average rate being 560 church members per thousand. This leaves 90 million persons unaccounted for. Some are children or incompetent adults who would not participate in a response to pseudoscience, and some may be excluded as holding other faiths that disfavor pseudoscience, but there must still remain an enormous audience available for participation in the cultist response to pseudoscience (exclusive of creation science). Bainbridge and Stark make a significant comment on our possible misconception of the character of this susceptible group:

> It would be an equally great mistake to conclude that religious liberals and the irreligious possess superior minds of great rationality, to see them as modern personalities who have no need of the supernatural or any propensity to believe un-scientific superstitions. On the contrary, they are much more likely to accept the new superstitions. It is the fundamentalists who appear most virtuous according to scientific standards when we examine the cults and pseudosciences proliferating in our society today. (P. 50)

The three scenarios I have selected are presented, argued, and discussed in thousands of pages of published books and magazine articles. There is no way this mountain of information (and misinformation) can be adequately digested and covered here. Keeping in mind that we are interested in gaining insights into the workings of science, we must settle for a very brief statement of the contents of each scenario--just enough to provide some targets and examples as we try to identify the essential characteristics of pseudoscience and its modus operandi. The crucial questions we will examine lie mostly in the areas of philosophy, psychology, and sociology. To find answers to these questions, we must try to stand clear of the arguments themselves. In any case, neither side will concede anything to the other, regardless of the merits of the arguments.

Velikovsky's Worlds in Collision

Immanuel Velikovsky's first major work, Worlds in Collision published in 1950, is essentially a recon-struction of world history spanning a period that began when human myths and legends were first being recorded in writing. His references to events earlier than the beginning of the second millennium B.C. (2000 B.C.) are vague; they include a novalike explosion that caused the Flood of Noah. Let us begin this brief review with a catastrophic event occurring about 1450 B.C., when Venus made its first major encounter with Earth. Velikovsky states that this encounter was recorded by many ancient cultures, but is particularly noted for causing the plagues associated with the Exodus of the Hebrews from Egypt (Ransom, 1976, p. 21). At that time Venus was a comet; it had been born of material ejected from the great outer planet Jupiter. The comet's trajectory was erratic for a period of 800 years, during which time it terrorized Earth by near-collisions. In Velikovsky's words, when a comet makes contact with a planet, "the planet slips from its axis, runs in disorder off its orbit, wanders rather erratically, and in the end is freed from the embrace of the comet" (1950, p. 156). The Jews during Exodus experienced unusual phenomena including a fall of red dust, which "turned rivers to blood"; falls of either hail or meteorites (or perhaps both); a fall of petroleum droplets from the sky; fall of vermin from the sky; a period of several days of darkness; and great earthquakes.

Current astronomical knowledge suggests that the dense core of a typical comet consists of a mixture of icy particles that may include substantial amounts of such compounds as water, ammonia, and methane, and of particles of solid mineral matter similar in composition to materials found in meteorites that have reached the earth's surface. In popular parlance, the typical comet is a "dirty snowball." The diversity of a comet's composition can perhaps be held to account for the diversity of substances alleged to have been transferred from a cometary Venus to Earth.

Other phenomena associated with this first encounter with Venus included the parting of the Red Sea, gigantic tides, and an enormous electrical discharge between Venus and Earth (Ransom, 1976, p. 30). Great geologic events occurred; they included growth of innumerable great volcanoes, rise of mountain ranges, and large-scale crustal rifting. Later, food of some sort also fell from the sky. For Hebrews in the desert, the food was manna; in other parts of the world it was described variously as ambrosia, heavenly bread, and honey (pp. 31-32). Following the encounter there was a period of several years of dense cloud cover and gloom (p. 34).

Meantime, comet Venus continued on a highly elliptical orbit that eventually brought it back to a second encounter with Earth. Ransom describes that encounter in these words:

> The second near encounter was not as close or as destructive as the first. The Earth was not engulfed in the extended atmosphere of Venus although numerous meteorites hit the Earth. Earthquakes were also common. (P. 35)

At this time, the Israelites had crossed over the Jordan River, entered the Promised Land, and encountered the city of Jericho. The collapse of that city may have been accomplished by one of the earthquakes referred to above. During the battle of Gibeon, the sun "stood still" (held its apparent position in the sky) for a whole day, enabling Joshua to complete the slaughter of the enemy at Beth-horon, begun by a rain of huge hailstones sent down by Yahweh (Josh. 10: 1-15). This astronomical event is interpreted by Velikovsky as a mechanical coupling effect with Venus, causing earth rotation to cease, but then to start up again to resume its previous rotational rate.

In what we might want to label as "Act 2" of the Velikovsky scenario, Mars walks on stage as the principal troublemaker. Venus, much greater in mass than Mars,

severely disrupted the orbit of Mars, causing that planet on two occasions to pass close to Earth. The Mars encounters with Earth were not as violent as were the Venus encounters. The first came in 747 B.C., about the time of the founding of Rome. It produced violent earthquakes and altered the earth's orbit, increasing its path length. Instead of 360 days in each year, the number increased to the present 365 days (approximate). Earth's axis was tilted by an additional ten degrees. The second encounter occurred in 687 B.C. It produced a great "thunderbolt" or "blast from heaven" that Veilovsky claims obliterated the Assyrian army led by Sennacherib as it lay camped near Jerusalem (Pénsee Editors, 1966-76, p. xii). The angle of tilt of the earth's axis was restored to its value prior to 747 B.C.

In the closing scene of the Velikovsky account, Mars engaged Venus in a final dramatic event, in which Mars was thrown outward (with respect to the sun) into an orbit beyond that of Earth, while Venus became a planet and assumed a nearly circular orbit between Earth and Mercury. In that location, it became a bright planet seen only as an evening star and morning star. The end of the "battle of the gods" ushered in a new age of serenity for planet Earth.

Immanuel Velikovsky was born in Russia in 1895, obtained the M.D. degree from Moscow University in 1921, and later migrated to Palestine, where he became a general medical practitioner. After study in Vienna, he became a psychoanalyst. In 1939 he and his family migrated to the United States, settling in New York City. For nearly a decade thereafter he spent much of his time in the library of Columbia University, searching out ancient writings of both western and eastern cultures. Worlds in Collision was published in 1950 by the Macmillan Company. It immediately drew heavy fire from the scientific community, and in particular from astronomers. One of the latter was Harlow Shapley of Harvard University, held in great distinction for his pioneering work on galaxies. Summaries of Velikovsky's scenario flourished in popular magazines, including Harper's, Collier's, and Reader's Digest. Under a second publisher, Doubleday, Worlds in Collision was reprinted numerous times and was followed by four more Velikovsky books.

The ranks of Velikovsky supporters grew rapidly and his scenario quickly became the basis of a loosely organized, but highly vocal cult. In using the word "cult," I have taken my cue from Martin Gardner, who refers to "the rise of the Velikovsky cult" (1981, p. 382). Devotees of the cult have attempted to demonstrate the scientific validity of events in the Velikovsky scenario. See, for example, a volume of such articles titled Velikovsky Reconsidered (Pensée Editors, 1976). It contains reprints of papers appearing between 1966 and 1976 in Pensée, the magazine of the Student Academic Freedom Forum, published in Portland, Oregon. Conspicuous among the contributors to this collection of articles is Lynn E. Rose, a professor of philosophy at the State University of New York, Buffalo. (See also Ransom, 1976, The Age of Velikovsky.)

In the early 1970s, scientists concerned about the popular persistence of Velikovsky's claims began to organize a symposium within the American Association for the Advancement of Science. It was held in 1974 and included a paper delivered by Velikovsky himself. Among the scientists participating, perhaps the best known to the public was astronomer Carl Sagan, who had been highly vocal in opposing Velikovsky's versions of planetary behavior. The principal papers of that symposium (sans Velikovsky's contribution) were published in a 1977 volume entitled Scientists Confront Velikovsky (Goldsmith, 1977). Velikovsky died in 1979; he was eighty-four years old.

Although few persons on either side of the debate are aware of it, Velikovsky's scenario was preceded by at least three theories of comet-caused catastrophes (see

Gardner, 1957, p. 41). One was proposed by William Whiston, a British clergyman and mathematician, in a book titled New Theory of the Earth, published in 1696. Whiston had the solar system formed from the tail of a comet. Visited somewhat later by a second comet, the earth experienced a great flood (Flood of Noah) caused by precipitation coming from water vapor in the comet's tail.

Of considerably greater importance relative to the Velikovsky scenario is a massive tome published in 1913 by Hans Hörbiger, a Viennese mining engineer, under the title of Glazial-Kosmogonie (Glacial Cosmogony). An English account of the major points of Hörbiger's scenario was written by a British follower, Hans Bellamy, and published in the early 1930s. The main point of the story is that a small stray planet was captured by Earth, to circle Earth at close range. As this moon orbited Earth, it drew closer and closer, exerting increasing gravitational pull on Earth and raising an enormous equatorial bulge of the oceans. Ultimately, earth's attraction overcame the cohesion of the moon and it began to fall apart. Great chunks of ice broken from the moon's surface melted and the water was drawn to the earth as rain and hail. There followed a rain of rocks, as the moon disintegrated entirely. Relieved of the moon's attraction, the equatorial tidal bulge subsided and the ocean spread over the higher latitudes to produce the Flood of Noah. A second capture gave Earth its present moon in a catastrophic event that resulted in the Ice Age, with the growth of great continental ice sheets and the foundering of the continent of Atlantis. The latter capture is said to have occurred about 13,500 years ago. "Racial memories" of the event are "buried in our subconscious." The present moon is also spiraling closer to Earth, and a new doomsday scenario can be anticipated.

What is most interesting about the Hörbiger scenario is that it became increasingly popular among anti-intellectual persons in the German Nazi movement, perhaps in part because conventional German astronomers spoke out against the scenario with such vigor. According to Martin Gardner (1957, p. 37), the fantastic reconstruction of world history soon acquired millions of followers and became a cult known as WEL (initials of Welt-Eis-Lehre, translated as Cosmic-Ice-Theory); the term was coined by Hörbiger. Gardner cites Willy Ley, the well-known German rocket scientist, as saying the WEL "functioned almost like a political party. It issued leaflets, posters, and publicity handouts. Dozens of popular books were printed describing its views, and the cult maintained a monthly magazine. . . . Disciples often attended scientific meetings . . . to interrupt the speaker with shouts of Out with astronomical orthodoxy! Give us Hörbiger!" (p. 37). Hörbiger even wrote to Ley "either you believe in me and learn, or you must be treated as an enemy." I dare say few Americans have even heard mention of Hörbiger and his WEL. I, for one, was totally unaware of them. Charles Fair states that "Hörbiger's disciples tried to ally the WEL movement with the Nazis, as their official scientific arm, but Hitler would have none of it!" (1974, p. 185)

Von Däniken's Chariots of the Gods?

Erich von Däniken's major works are Chariots of the Gods? (1969), Gods from Outer Space (1971), and The Gold of the Gods (1973). The main thesis he presents in Chariots is that during the period of early human civilizations, the earth was visited repeatedly by highly intelligent, technologically advanced individuals who resembled humans and were transported in spaceships from some unknown planet in outer space. In Von Däniken's text, these astronauts are called "gods," which is the way they were regarded by the earthlings who felt their impacts. The date of the earliest visitation is not made clear, but visits were taking place at the time of the earliest Egyptian pyramid builders, and possibly as far back as 45,000 years before the present.

Von Däniken postulates that at some prehistoric time the visiting gods found a population of apes or of hominids that predated Homo sapiens. They are not identified in the conventional anthropological sequence, but let us suppose they were members either of H. erectus, or of Neanderthal man (now considered a subspecies of H. sapiens). The gods artificially inseminated females of this indigenous species and were successful in breeding H. sapiens. The gods were evidently experts in hybridization and may have understood the genetic code. Later they destroyed the remaining population of parent hominids.

The scientific and technical capabilities of the gods were particularly noteworthy in areas of navigation, metallurgy, and construction of stone buildings and monuments; they demonstrated this capability in many lands of the earth. Von Däniken hopes that modern humans will, as soon as feasible, make spaceships that can transport earthlings to some distant planet in another solar system, where they will find a primitive culture, for which they in turn will serve as "gods" and return the favor, so to speak.

Von Däniken's evidence is in the form of a collection of diverse artifacts. They include: an ancient world map (attributed to Piri Re'is) interpreted by von Däniken as an aerial view of the earth from a vantage point in space about over Cairo; lines on the Nazca plain of Peru, interpreted as landing strips made by pre-Inca inhabitants to facilitate the landing of spaceships; huge mountainside drawings that served as signals to spaceships; an ancient astronomical calendar at Tiahuanaco in Peru, and assorted other astronomical data (Sumerian), for which the knowledge was supplied from extraterrestrial sources; a large assortment of buildings, idols, and monuments consisting of hewn stones far too large to have been carved and transported by earthlings--localities include Tiahuanaco, Sacsahuaman, Baalbeck, Egypt (pyramids), Easter Islands (stone idols), and Guatemala and Yucatan (pyramids); cuneiform texts and tablets from Ur telling of "gods who rode in the heavens in ships" and "gods who came from the stars, possessed terrible weapons, and returned to the stars"; cave drawings from many localities depicting beings from outer space dressed in space suits and wearing goggles; miscellaneous smaller objects from Lebanon, Egypt, Iraq, Peru, China, India, and elsewhere that required manufacturing technologies far beyond the capabilities of the earthlings (made of platinum, aluminum, and nonrusting iron); biblical references interpreted as describing gods from outer space (giants, angels, Ezekiel's spaceship). The total package of so-called evidence can be interpreted as making the following argument: Because so many of the artifacts and technologies of ancient civilizations are so very far beyond the capabilities of earthlings of that period to make or perform, we can only conclude that extra-terrestrial beings must have come to earth to supply the necessary advanced knowledge and technology.

Von Däniken seems to look with approval on the interpretation of certain sightings of unidentified flying objects (UFOs) in the present century as continuing visitations of gods from outer space (1969, p. 138), at least for surveillance of what is going on here on our planet.

Early in his first book, von Däniken gives a brief review of modern astronomy that shows his acceptance of the great age of the cosmos and the earth, and the appearance of Man about one million years ago. In this respect, he differs from Velikovsky, who confined his scenario to events within the time frame of literal interpretations of the book of Genesis. The result is that von Däniken immediately came under fire of the funda-mentalist creationists; whereas that group seems to have paid little serious attention to Velikovsky.

In von Däniken's third work, The Gold of the Gods (1973), we encounter a scenario quite unlike that of the first two books. Now the space visitors are seeking refuge after suffering defeat in a cosmic war. They build great tunnel systems for safety and set up a decoy on a fifth planet then orbiting in our solar system between Mars and Jupiter. The enemy space beings annihilate that planet, which is fragmented into the asteroids, and then leave the scene for good. Meantime, back on planet Earth, the newcomer "gods" reel from the effects of the planetary explosion, which has caused a great flood (Flood of Noah?). Recovering, they turn on earth "monkeys," representing the hominoids of that time, and manipulate their genes through controlled mutations to create intelligent humans. But then the gods become dissatisfied with their product and begin to punish and kill the humans, who seek shelter in underground hideouts.

Does this flip-flop mean that the von Däniken cultists must drop the Chariots version and substitute the Gold version? Will there appear a third scenario to replace the second? Is von Däniken getting in a little Velikovsky planetary catastrophism to increase book sales? I'm really confused! One nice feature of this switch is that we need no longer debate the merits of the Chariots evidence. If we wait a little longer, we may not need to debate the Gold evidence.

Erich von Däniken was born in Switzerland in 1935. His early education in a Catholic school was not followed by advanced schooling in either astronomy or archaeology, subjects that caught his interest and that he attempted to link to religion (Story, 1976, pp. 1-2). He became an avid amateur archaeologist and traveled to many sites of ancient civilizations. While employed on the staff of a Swiss hotel he wrote his first book, published by Econ-Verlag in 1968. It was serialized in a Swiss newspaper and attained great popularity that quickly spread to Germany, where within a year it was the number-one-selling book. The English translation, titled Chariots of the Gods? was published a year later in England, followed in 1969 by publication in the United States by G. P. Putnam's Sons. A German-made film version of the book was widely viewed and led to production of an American TV show based on the book, airing in 1973. As a result, U.S. sales of the book skyrocketed. An interesting sidelight is that in 1966, the year in which von Däniken was writing his book, two prominent astronomers--I. S. Shklovski of the USSR and Carl Sagan of Cornell University--published a book titled Intelligent Life in the Universe. According to Ronald Story, "This book contained many ideas that were later expressed (although some in a distorted form) in Chariots; it may well have given von Däniken the brainstorm to provide the world with a new set of gods to worship, to replace the traditional deity, who was being murdered by the poison pens of contemporary theologians" (1976, p. 5).[1] Reference in the last phrase of the quotation is to the God-is-dead movement among theologians at that time. Story also cites earlier publications by other authors that contained much the same archaeological "evidence" used by von Däniken. (1976, p. 5)

Von Däniken's chariots-of-the-gods scenario seems to have had a milder impact than Velikovsky's world's-in-collision on both the general public and the scientific community. If a cult of believers has formed, it has not been very vocal in the English-speaking world, at least. Reaction has been strongly negative from the creationists, who regard it as an affront to the literal interpretation of the Old Testament. The idea of numerous "gods" from

outer space replacing the functions of the one and only true and living God of all Creation is anathema. A denunciation from the creationists' side was not long in making its appearance. In 1972, Crash Go the Chariots appeared as a short book written by Clifford Wilson, Ph.D., "archeologist, authority in psycho-linguistics, education professor, and Bible scholar" (so the Foreword to his book reads). Goaded by von Däniken's put-down of the archaelogists as a group, Wilson scathingly attacks the gods and their chariots. He puts things back into their traditional Judeo-Christian framework with one Supreme God at the controls. Although originally published by Lancer Books, Crash was adopted by the Institute for Creation Research and reprinted by them as one of its first-line educational offerings to the fundamentalist sector.

I recommend the detailed critical analysis of von Däniken's scenario and interpretation of archaelogical materials in Ronald Story's The Space Gods Revealed (1976). It seems completely secular in tone and will appeal to scientists and humanists who might be turned off by the emotive style and ad hominem arguments that characterize Wilson's Crash. Story's book contains a Foreword written by Carl Sagan, offering some interesting speculations on the theological connotations of von Däniken's scenario. Sagan sees in popular acceptance of the shoddy scholarship of von Däniken's books a profound yearning of vast numbers of people to embrace extra-terrestrial beings for their superior support and guidance, a commodity no longer available through traditional religious beliefs. As we will see, this same yearning may have found satisfaction in belief in contemporary visitations by benign extraterrestrial beings arriving in flying saucers.

UFOs and UFOlogy

Of the three scenarios, that of unidentified flying objects (UFOs) is by far the most difficult to analyze--it has a literature of incredible proportions throughout which confusion, contradiction, and calumny have proliferated unchecked. For a definition of UFO, I quote from Allan Hendry's tightly organized UFO Handbook:

> UFO (Unidentified flying object): Any anomalous aerial phenomenon whose appearance and/or behavior cannot be ascribed to conventional objects or effects by the original witness(es) as well as by technical analysts who possess qualifications that the original observer(s) may lack. (1979, p. 4)

Is there any American capable of reading newspapers or comprehending a television or radio program who has not been exposed to stories of UFO "sightings" in the past few years? Most stories tell of one or more individuals seeing strange objects in the form of luminous bodies--commonly cigarshaped, spherical, or disklike--and showing a wide variety of behaviors, including rapid travel, hovering, abrupt change in direction, change in shape and size, and sudden disappearance. The highly luminous forms, emitting light from an internal source and displaying various colors, are usually of indistinct outline or outer boundary; those that seem to reflect light brilliantly are often described as solid objects of metallic luster with sharp outlines. Various kinds of sounds can accompany the antics of the object. A few of the objects, when close to the ground, exert physical effects on things in the vicinity; for example, they may cause automobile engines to misfunction, radios to emit static, or TV pictures to break up. In a very few cases, distur-bances of the ground surface, including burning or charring of vegetation, are reported. Finally, in a very few reports some sort of living creatures are said to have emerged from the object--in this case a "flying saucer" or "space vehicle" of some kind--and even to have spoken to

---

[1]From Ronald Story, The Space Gods Revealed: A Close Look at the Theories of Erik von Däniken. Harper & Row, Publishers, Inc., New York. Copyright © 1976 by Ronald Story. Used by permission of the publisher.

the observer(s). In the most bizarre reports, the observers are kidnapped and later released.

According to Hendry, a 1978 Gallup poll stated that 9 percent of adult Americans "believe they have seen a UFO" (1979, p. 1). This extrapolates to about 13 million sightings. Upon careful scrutiny by persons trained in UFO evaluation, a large proportion of the reported incidents--as high as 90 percent--can be resolved as explained phenomena and relabeled as IFOs (Identified Flying Objects). For example, the object has turned out to be an aircraft or an astronomical object. Some reports are hoaxes. Also under consideration is the possibility that what is reported as an object is a kind of hallucination, originating within the mind. Of those events remaining in the UFO category, only a few are so strongly supported in terms of the quality of the witnesses and their allegations that organizations engaged in serious study of the phenomena accept them as real physical phenomena worthy of careful further analysis.

The enormity and intensity of public interest in UFOs, together with a sustained interest over some forty years, results in a great diversity in groups of interested persons and a vast literature. I have read a statement that several hundred books have been published on the subject. Lists of organizations devoted to UFOs suggest that they are international in distribution and number at least in the dozens. Two major reasons for the magnitude of the social phenomenon are apparent. First, there is (or has been) concern that the UFOs include surveillance tools of the Soviet Union and are thus a threat to national security, both because of the information they can gather and the manifest existence of advanced Soviet technologies as yet unknown to the Western democracies. Second is the concern that some of the UFOs are space vehicles bearing life forms from other solar systems and visiting our planet for reasons unknown.

Because of the first type of concern, UFOs attracted the attention of the U.S. military establishment, the Central Intelligence Agency (CIA), and numbers of politicians. Because of the second type of concern, UFOs gave rise to livid stories in the news media and attracted an enormous following of "believers" who form a cult of quasi-religious nature, organized into numerous societies for the culture of "UFOria." This movement is now worldwide. Charles Fair, in his 1974 book titled The New Nonsense, divides the cultists into two groups (1974, p. 137). The angel school views the visitors from outer space as benign and desirous of heading off the great nuclear holocaust or Armageddon. The visitors' superior intelligence may include the ability to save earthlings from this fate. The devil school sees the space visitors as evil and threatening, with the power to wreak a catastrophe on our planet.

A small handful of qualified scientists has taken a strong interest in the UFO problem and considers it worthy of serious scientific study. Fair states that these individuals can be divided into two groups (pp. 140-41). The skeptics are hardnosed individuals who tend to work from a strong bias against any unconventional proposal unsupported by physical evidence. An example is Edward U. Condon, a distinguished physicist who, as we shall see, headed up a committee to investigate UFOs. Another, but quite different example is Carl Sagan, noted astronomer, who has evaluated the possibility of contacting extraterrestrial intelligence (ETI). A third example, also an astronomer, is David H. Menzel of Harvard University; he has offered an alternative physical explanation for one class of UFOs. I should also include Philip J. Klass, a graduate electrical engineer who has proposed his own scientific hypothesis, based on principles of plasma physics, to explain a large proportion of UFOs (1968).

There are a very few qualfied scientists who are either believers in the possibility that some UFOs are manifestations of ETI, or are neutral on that issue to the extent of being agnostic. One of the most puzzling of these (my reaction) is James E. McDonald (deceased) of the Institute of Atmospheric Physics at the University of Arizona. His statements can be interpreted as indicating a certain measure of belief that ETI has reached our planet through UFOs and thus created a situation of grave scientific concern. A second person whose activities and statements suggest the possibility of belief is Dr. J. Allen Hynek (deceased), a former Chairman of the Department of Astronomy and Director of the Dearborn Observatory of Northwestern University. He became Director of the Center for UFO Studies and for some years was a scientific consultant to the Air Force in their program of documentation of UFO sightings. For many years a skeptic, Hynek became less sure of that position. He stated that from his long experience, most reported UFO cases were misrepresentations but a "relatively small residue of UFO reports . . . were so well attested and so compellingly strange that the chances were overwhelmingly great that they could not be ascribed to collective misidentification, hoax, or hallucination" (Hendry, 1979, pp. ix-x). He goes no further than to leave open the possibility that UFO cases may be manifestations of ETI.

The beginning of the UFO era is usually placed in June 1947, with the first publicized sighting of an aerial object and its being dubbed by a newspaper reporter as a "flying saucer." Equally strange luminous objects, called "foo fighters," had been the subject of numerous accounts by military aircraft pilots in World War II, but those incidents did not precipitate a public reaction. Reports of strange aerial objects go back to the age of early hot-air balloons and to dirigibles of World War-I-vintage; some of these were described in lurid terms in the newspapers. Evidently, the immediate post-WW-II period had generated a national psychology unusually favorable for a mass response in the United States to the UFO phenomenon. Tensions with the USSR were increasing with the start of a nuclear-weapons race and there were many rumors of Soviet spying. Perhaps because several UFO sightings occurred near the Pentagon in Washington, D.C., and many other sightings were described by trained aircraft pilots, including personnel of the U.S. Air Force, the military establishment began to take the phenomenon seriously as a possible security threat. This concern led to the setting up in 1947 of Project Blue Book, a registration and documentation of UFO sightings maintained by the Air Force in Washington and continued until 1969. The material was classified and closed to public access, perhaps contributing to the early charges from the public sector that the Air Force was withholding secret evidence that could prove that Soviet spying on the nation or extraterrestrial visitations were actually taking place.

In 1966, the Congress stepped into the investigatory picture. The House Armed Services Committee directed the Pentagon to set up a civilian inquiry into the UFO phenomenon to be based in part on a review and restudy of the accumulated data of Project Blue Book. The result was funding of a national committee to study UFOs. The committee was set up to do its work at the University of Colorado and requested to submit its report to the National Science Foundation, which, in turn, was to release it to the public. The committee was headed by Dr. Edward U. Condon, at one time a professor at Princeton University, a past president of the American Association for the Advancement of Science, and widely known for his research in nuclear physics and his role in developing the atomic bomb. Later he served as Director of the National Bureau of Standards and in advisory positions related to atomic-energy development and nuclear-weapons testing. With a staff of scientists that included both physical scientists and social scientists, Condon's committee set to work to reexamine and evaluate UFO cases in the Blue Book file.

The Condon Committee released its report in 1969 (see

Hall, 1969; Condon, 1969). The report concluded that no independent physical evidence had been found to substantiate the allegations made in the case studies they had examined; that further scientific study of the matter was not justified because no threat could be implied. A possible exception to the first part of that statement lies in the interpretation of three small, metallic fragments alleged to have been collected from a shallow-water site off the coast of Ubatuba, Brazil, where a UFO was alleged to have fallen, exploded, and burned. Analysis done under the committee's direction showed the metal to be magnesium. Neutron activation analysis of the metal showed an elemental composition that has been interpreted by some to mean that the material is not of earthly origin. (It could be of meteoritic origin.) Details of the actual collection and handling of the material are confused, and its direct connection with the alleged UFO incident is much in doubt (Fair, 1974, pp. 149-50). In December 1969, a symposium sponsored by the American Association for the Advancement of Science aired the diverse opinions of a number of scientists on the significance of UFOs (Sagan and Page, 1972). One of the opinions expressed was that the Condon Committee report should not be accepted as final, and that further scientific study of UFOs was both desirable and important.

In considering the status of UFOlogy in terms of whether it is science or pseudoscience it is essential to be quite precise as to exactly what we are talking about. Hendry reminds us: "We only get to study reports of UFOs--not the UFOs themselves" (1979, p. 6-7). The reports consist only of allegations. Hendry puts it this way: "We would be dishonest with ourselves if we considered the reports as anything but allegations; this is unfortunate, but necessary, if we are to treat UFOlogy as a science" (p. 7).

It is generally agreed by both skeptics and believers that to date no physical object identifiable as a "flying saucer" or "spaceship" has been retrieved or captured for direct physical and chemical examination. Empirical science is reluctant to admit allegations as evidence in support of a working hypothesis that demands interpretation beyond the sensory perception itself.

A particular sighting of a UFO, even if corroborated by several witnesses, is a singular timebound event and cannot be repeated on demand. What can be done as acceptable science is to classify and catalog the percepts on record and to rate their reliability according to some reasonable uniform standards, and from those data reconstruct a general description of a phenomenon (or several different phenomena) about which some pertinent questions can be asked. It would be foolish to rule a priori that everything claimed to be perceived by humans is nonreal, for that would be to deny the existence of all commonsense knowledge essential to the conduct of human activities. What we can work with is a body of persistent, repetitious allegations that various objects were observed to have certain properties (such as shape, size, color, or texture), that they changed those properties in certain ways, engaged in certain types of motions, and that they emitted, besides visual light, some other form of electro-magnetic radiation or some kind of sound. In making the tentative assumption that these perceptions arise from real phenomena, it is quite permissible to formulate one or several hypotheses that could explain the typical or common perceptions through application of accepted laws and principles of empirical science. In such a program we would be simply carrying on a form of descriptive science typical of natural sciences in their formative periods.

If it be agreed by scientists that some common ground of observational reality exists within many UFO reports, hypotheses offered in explanation can be subjected to testing through the derivation of deduced consequences, which in turn may lead to the discovery of previously unnoticed relationships. For example, a certain class of statements might prove to be significantly associated with a particular time of day or night, or with a particular environmental setting. If an investigation of UFOs is to be taken seriously by scientists, the hypothesis will conform as closely as possible with the principle of parsimony. This means that where possible, the explanation makes use of natural phenomena already accepted through extensive prior observation and experimentation.

For example, a hypothesis requiring that the UFO is of extraterrestrial origin will be given relatively low status if another hypothesis can explain the same phenomenon in terms of well-known terrestrial phenomena. A case in point would be the plasma hypothesis advanced by Philip J. Klass (1968), who suggests that many of the perceived qualities and behavior patterns of UFOs are what might be expected of masses of strongly ionized air containing highly excited free electrons capable of emitting various colors of light and other forms of electromagnetic radiation. A commonplace example of this effect is the fluorescent light from gas-filled glass tubes through which an electrical current is passed. Related phenomena in nature include the corona (St. Elmo's fire), seen to emanate from solid objects on the ground or from ships at sea, and ball lightning, both of which are occasionally observed when intense fields of atmospheric electricity are present. Visible plasma effects are also known to occur on or near high-tension power lines. What Klass has done is to attempt to explain at least a large proportion of luminous UFOs as plasma phenomena. Numerous deduced consequences of this hypothesis are capable of being tested, and Klass has endeavored to do just that.

One might even wish to entertain as a working hypothesis from the field of psychology that most UFOs are hallucinations arising within the human brain. There might be difficulties in reconciling the allegations that the same hallucination occurred simultaneously in the brains of two or more observers, but the magnitude of those difficulties would be small compared with a hypothesis of UFOs as spaceships directed by ETI or occupied by extraterrestrial creatures. The ETI hypothesis has the fatal weakness of merely citing as "supporting evidence" the assertions of the hypothesis itself. The ETI hypothesis is thus not now capable of being dealt with by the methods of empirical science because no evidence independent of the hypothesis itself can be brought to bear on it. But, at the same time, its mere statement can be labeled as pseudoscience only when persons who advance it use in their argument unacceptable claims for supporting evidence. For example, the allegation that small creatures with large heads and spindly bodies emerged from the UFO cannot be accepted as evidence of ETI, because those little guys (or gals) are simply an extension of the hypothesis itself! The argument that "a particular UFO could not possibly be an artifact of earthlings because it represents a technology far beyond the capability of earthlings" fails because that assertion is also an elaboration of the hypothesis itself. It also fails on grounds of logic because to claim that another hypothesis cannot explain the alleged phenomenon in no way adds value to the ETI hypothesis. Persons who use such specious arguments are practicing pseudoscience; they are abusing a hypothesis that might at some future time become amenable to treatment by acceptable methods of scientific investigations, deriving support from naturalistic lines of evidence not presently available.

With the three chosen scenarios in mind, we must now attempt to decide whether they deal in science, pseudoscience, or some of both. To reach such a decision we need to probe into the general characteristics of pseudoscience and to set down some criteria of recognition to guide us.

Chapter 8

Pseudoscience: II. The Phenomenon Examined

To prepare ourselves for passing judgment on whether Velikovsky, von Däniken, and the priests of the ETI cult have dealt in pseudoscience, we will find it helpful to learn more about the nature of the scientific community. Do Velikovsky and von Däniken share the same set of characteristics common to the practitioners of mainstream science? Do they practice in the same manner? Do they share the same set of norms?

The Scientific Community--A Social System

Among the collected papers emerging from the A.A.A.S. Velikovsky symposium of 1974 (Goldsmith, 1977) is one by sociologist Norman W. Storer, a professor in Baruch College of the City University of New York. His paper is titled "The Sociological Context of the Velikovsky Controversy" (Storer, 1977). So far in this account of the workings of science I have not presented the view from sociology, but it is pertinent at this point, since there is a sociology of science as well as a sociology of pseudo-science. Storer has made a specialty of the sociological study of science and scientists. To give us that perspective, he begins with "a sketch of science as a community, or as a coherent social system" (p. 29). He finds a clear distinction between this community and other sectors of society; its principal product is "organized, certified empirical knowledge." As with the members of any social group, "the distinctive relationships found among a set of people occupying certain social positions are due to their sharing a set of norms--standards of proper behavior--that tell them how to behave with respect to each other" (p. 30). Storer draws upon the work of Robert K. Merton (1973) to identify four norms "that are central to the ethos of science."

First, the value attached to a scientific statement must in no way be connected with the personal characteristics of the scientist who makes that statement. Merton's term for this principle is <u>universalism</u>, but it does not seem to me to give any clue to the norm intended, namely, that the strength or weakness of a hypothesis proposed by a scientist must be considered strictly on its scientific content and supporting evidence. It should make no difference whatsoever that the scientist is of a certain race, religion, sex, age, political affiliation, and so forth. Perhaps we should call this norm "depersonalism" or "apersonalism." If the principle were extended to include disregard of professional qualifications, we would need to give unbiased and serious attention to a scientific statement made by anyone, scientist or not, and including Velikovsky and von Däniken. I doubt that Merton intended such inclusion.

Second, findings made by one scientist must be shared freely and openly with the entire scientific community. Publication of such findings is thus a moral obligation. This is the principle of <u>communality</u>. It doesn't help us to distinguish scientists from the practicioners of pseudo-science; the latter publish compulsively and could not be restrained from doing so--verbal diarrhea is their chronic disease.

Third, scientists must practice <u>organized</u> <u>skepticism</u>. Each scientist must scrutinize the publications of others in the same area of specialization and express his or her criticism in print, in journal articles, reviews, and letters, as well as orally from the floor of a meeting room or a seat on the debating stage. This activity is a form of mutual policing needed to sustain a high quality of published scientific information. Perhaps the most important part of the policing action occurs through peer reviews of articles submitted to scientific journals. Reviewers must take their job seriously; they must search closely for errors in observations and weaknesses in arguments. They receive no monetary reward for this service, which draws time from their own research programs, but it is to the mutual benefit of all.

The scrutiny of one's work by colleagues is a feature wholly lacking in the publication of pseudoscience literature. Velikovsky, von Däniken, and their publishers' editors never sought critical reviews from scientists familiar with those areas of astronomy, geology, and archaeology that form the skeletal structure of their scenarios. If those authors had submitted their manu-scripts to scientific journals, rejection notices would have been swift in coming. It looks, then, as if authors of pseudoscientific material shy away from the scientific community. Instead, they seek support in the nonscience community, and particularly from those persons having little higher education in any field of knowledge. <u>Prima facie</u> evidence of this audience selection lies in the fact that pseudoscience is published by those same publishing houses (or divisions within a publishing house) that handle fiction, science fiction, and the more sensational forms of biography and autobiography. You would not find a <u>Worlds</u> <u>in</u> <u>Collision</u> or a <u>Chariots</u> <u>of</u> <u>the</u> <u>Gods?</u> on a publisher's list of scientific textbooks and monographs.

A fourth norm recognized by Merton is what he calls <u>disinterestedness</u>, meaning that a scientist's research should not be guided by desire for personal rewards. He refers to such rewards as private economic gain, glory in the eyes of the nonscientific public, and even the honors and medals awarded by scientists to each other. We must be careful here to emphasize that such personal rewards are essentially excrescences or trappings that do not always accurately measure the quality of the scientific work of the individual. We of university experience know that nearly every senior professor has a following of former students who conspire to get the "old prof" a medal or prize. Award committees rely mostly on the number of nominating letters received in support of a candidate for the honor.

Of the four norms presented here, this last one is least likely to be observed within the scientific com-munity, and is so often flagrantly violated that it is perhaps little more than a sham. I can assure my readers unfamiliar with the academic profession that nearly every scientist seeks to maximize private economic gain in one way or another, and many try to get public exposure through the news media. Many (with thinly veiled understandings of reciprocity) encourage colleagues to come through with an honor.

As for the pseudoscientists, the norm of disinterestedness is simply not there, and no shame is to be incurred from violating such a norm. Emphasis is on rolling up the royalty earnings and fees from book sales, TV/motion picture adaptations, and lectures to lay audiences, on public exposure through media interviews, and on receiving expressions of adulation from fan clubs within the cult. Pseudoscience is big business and very little else!

Storer turns next to examine the driving force behind scientific research and publication (1977, p. 31). He follows Merton in identifying it as "the scientist's interest in acquiring professional recognition." Professional recognition has a meaning here quite distinct from the personal rewards listed above. Recognition is judged primarily in terms of acceptance of one's scientific reports for publication in journals operated by peers in one's own field of specialization. Peer reviews serve to let pass only the highest-quality products, while the excess in number of submitted manuscripts over the number capable of being accommodated makes competition severe. Journals that can be the most choosey confer the highest value upon the papers they publish. Thus, faculty committees who must evaluate a colleague for tenure appointment tend to place higher value on the candidate's articles that have appeared in the more prestigious journals. Another source of professional recognition comes from the citation of a scientist's published works in the texts and bibliographies of other scientists' works. A high frequency of citation is equated to high value of the product. The common practice of citing papers written by one's friends, students, or those of shared opinions, even when such citations are not essential to the content of one's paper, pays off in the favor being returned by getting more citations of one's own work in the works of other authors.

Professional recognition also comes from the general excellence of the chair that a professor occupies. The prestige of the university attaches to the chair and its occupant, as does the prestige that the particular science department enjoys, both nationally and internationally. The same effect operates in terms of one's position on the staff of a private or public research institution--for example, appointment to the Institute for Advanced Studies at Princeton, New Jersey.

Pseudoscientists as Exoheretics

As to the producers of pseudoscience, professional recognition within the science community is nonexistent. They are excluded in a very firm manner. Exclusion is then seized upon by the pseudoscientist and attached cult as an opportunity to indulge in paranoia, a subject we will come to shortly.

Writing in a Foreword to the A.A.A.S. Velikovsky symposium volume, Isaac Asimov has identified and described two kinds of scientific heretics (Goldsmith, 1977, pp. 8-15), distinguished in terms of their relationship to the scientific community we have just examined. One is the endoheretic, the other, the exoheretic. Endoheretics arise within the scientific community of which they are a part; exoheretics arise from outside the scientific community. As examples of endoheretics, Asimov cites Galileo and Charles Darwin. Both were well qualified as scientists according to the prevailing standards; both were opposed by orthodox members of the science community who attempted to discipline them and bring them into line with the orthodox views. (In Galileo's time, Asimov points out, both scientific and religious orthodoxies were one and the same, so that pressure to recant came from the Inquisition.) In the long run their hypotheses won out. A striking example of an endoheretic of modern times was Alfred Wegener, a German geophysicist who proposed the drifting apart of the continents to the accompaniment of a widening of the intervening oceans. Geologists of the time persecuted Wegener mercilessly, and it was not until the 1960s--long after his death--that his hypothesis (greatly revised) prevailed.

Asimov suggests that there may have been at least fifty endoheretics whose hypotheses failed to every one whose hypothesis was eventually accepted. The losers simply dropped out of sight and the public knew little or nothing of their attempts. The successful endoheretics won out because the self-correcting methods of the scientific community, directed by norms described in earlier paragraphs, continually sort the wheat from the chaff, retaining that which has value; but the process may be painfully slow--painful, particularly, to the endoheretic.

The exoheretic is an outsider to science and is not schooled in practice of the norms of the scientific community, or in the methodology and language of science. The manner in which the exoheretic attacks some area of orthodox science strikes those relatively few scientists who take note as being strange, unintelligible, or otherwise deviant, often to the point of causing amusement or derision. Knowing that this will happen, the exoheretic directs expression to the general public. Those persons educated primarily in the arts and humanities will receive the strange ideas as worthy expressions of the human imagination and will respond to the romantic overtones. The exoheretic finds public response deeply satisfying as well as lucrative. The public in turn will turn against the scientists and goad them into denunciations, if possible. Goading was not needed for Harlow Shapley, the Harvard astronomer; he attacked Velikovsky savagely even before Worlds was published, basing his information on prepublication reviews. Velikovsky cultists have made a big thing of these early attacks, which were successful in forcing the original publisher to turn the book over to another house. Seen in retrospect by members of the science community today, the concerted attack on Velikovsky was a tactical blunder, for it quickly brought the public to Velikovsky's side, making him a martyr, while doing nothing to cause his readers to reject the bizarre astronomical scenario.

You might want to question the classification of Velikovsky as an exoheretic, since he had obtained a medical degree and practiced psychoanalysis. That area of knowledge, however, is normative science and quite unrelated to the planetary science on which Worlds depends. Velikovsky's competence as a biblical scholar seems not to have been questioned. As to von Däniken, there seems to be little contest over his candidacy for exoheretical status. He admits to being an amateur. When it comes to UFOlogy and the question of ETI, both exo- and endo- types have gotten into the act; they must be sorted out (as we have done) according to their levels of skepticism or belief, as well as their professional qualifications and the quality of their writing.

Martin Gardner has set down what he regards as characteristics that most pseudoscientists share in common (1957, pp. 8-15). "First, and most important is that cranks work in almost total isolation of their colleagues" (p. 8). This certainly describes both Velikovsky and von Däniken. They not only worked in isolation from the scientific community, but from other persons of any description. It is said that Velikovsky "opened and closed the doors of the Columbia University Library" for several years, probing into the literature of the Fertile Crescent in biblical times. Von Däniken wrote in the solitude of his quarters in a Swiss hotel. Some successful scientists have worked for years as loners in the sense of having secured no support from colleagues, but that is another matter;

they had full access to the scientific community and its body of empirical knowledge.

Second, the self-inflicted isolation of the pseudo-scientist goes hand-in-hand with paranoia, which manifests itself in a sense of personal greatness, enabling him or her to stand firm in defiance of the recognized scientists. Gardner lists "five ways in which the sincere pseudo-scientist's paranoid tendencies are exhibited."

> (1) He considers himself a genius. (2) He regards his colleagues, without exception, as ignorant blockheads. (3) He believes himself unjustly persecuted and discriminated against. (4) He has strong compulsions to focus his attacks on the greatest scientists and the best-established theories. (5) He often has a tendency to write in a complex jargon, in many cases making use of terms and phrases he himself has coined. (Pp. 12-14)

I'm not so sure all five of these criteria can be detected in Velikovsky and von Däniken--that is something best left to the psychiatrists. The second, third, and fourth criteria are clearly visible in the writings of the more outspoken supporters of both men, and especially virulent in the cultist authors who are believers in UFOs with ETI connections.

Science and Pseudoscience as Cognitive Fields

In recent years philosophers of science have become increasingly aware that a single criterion for distinguishing between science and pseudoscience cannot be fully effective. This problem has been treated by Professor Mario Bunge of McGill University, whose analysis of cognitive fields we introduced early in Chapter 1. Bunge has made an in-depth study of epistemology, presented in two recent works: Exploring the World (1983a) and Understanding the World (1983b). In a paper titled "What is Pseudoscience?" (1984, pp. 37-38), Bunge singles out the problem of comparing the two cognitive fields mentioned in Chapter 1: (a) belief fields that include religions, political ideologies, and pseudoscience; (b) research fields that include humanities, mathematics, and the pure and applied sciences. His opening paragraph reads as follows:

> Most philosophers have attempted to characterize science, and correspondingly pseudoscience, by a single feature. Some have chosen consensus as the mark of science, others empirical content, or success, or refutability, or the use of the scientific method, or what have you. Every one of these simplistic attempts has failed. Science is far too complex an object to be characterizable by a single trait--and the same holds for pseudoscience. Just as we must check a number of properties in addition to color and brilliance in order to make sure that a chunk of metal is not fake gold, so we must examine a number of features of a field of knowledge to ascertain whether it is scientific. (P. 36)[1]

Bunge begins his analysis by listing ten essential characteristics, or components, of a cognitive field; they are all nouns--"things" that every cognitive field possesses. I list them in the order and with the names used by Bunge, but with some minor rephrasing of his explanatory phrases:

C Cognitive community. The purveyors of the knowledge category (e.g., the scientists or the pseudoscientists themselves).

S Society hosting the cognitive community. The society of which the cognitive community is a part.

G General outlook. The world-view, or philosophy of the cognitive community.

D Domain. The universe of discourse of the cognitive field; the objects of study or inquiry of the particular field (i.e., the subject matter itself).

F Formal background. The logical and mathematical tools employable by the cognitive field.

B Specific background. The set of presuppositions about the domain (D) borrowed from other fields of knowledge.

P Problematics. The set of problems the cognitive field may handle or treat.

K Specific fund of knowledge. The knowledge accumulated by the cognitive field.

A Aims. The goals of the cognitive community in cultivating its cognitive field.

M Methodics. The collection of methods utilizable in the cognitive field.

Figure 8.1 is my sketch of the ten characteristics or components, attempting to show their relationships to one another. The cognitive community (people) has a particular domain of reality in which it operates and within which lie its problems, its accumulated knowledge, and its aims. Drawn into the domain from outside it are essential inputs in the form of methods or information. This diagram may be going beyond or astray from what Bunge had in mind, but it may help in getting some sort of coherent mental picture of the composition and structure of a cognitive field.

Bunge's next step is to describe and define the concept of science in terms of twelve conditions, all of which must be satisfied. If any cognitive field fails to satisfy all twelve conditions, it will be judged as nonscientific. Examples of nonscientific fields named by Bunge are theology and literary criticism. Nothing in such a designation implies the nonscientific field to be of lesser intellectual rank or value than science. Now comes the crucial point, which is: "any cognitive field that, though nonscientific, is advertised as scientific will be said to be pseudoscientific" (p. 39). In other words, a claim put forward by its adherents that a belief field is science is a fraudulent claim; it is a misrepresentation of the real nature of that belief field. Thus, our thesis in future chapters will be that when creationism, which qualifies as a nonscientific field (i.e., a belief field), is alleged to be science--"creation science"--a pseudoscience has been established.

So let us turn to the twelve conditions set down by Bunge (pp.38-39). In places I have edited or paraphrased parts of his statements, but mostly they are verbatim.

1. Every one of the ten components of the cognitive field changes, however slowly, as a result of inquiry in the same field as well as in related fields, particularly those supplying the formal background (F) and the specified background (B).

2. The research community (C) of the cognitive field is a system composed of persons who have received a specialized training, hold strong information links among themselves, and initiate or continue a tradition of inquiry.

3. The society (S) that hosts the cognitive community encourages or at least tolerates the activities of the ten components (A through S).

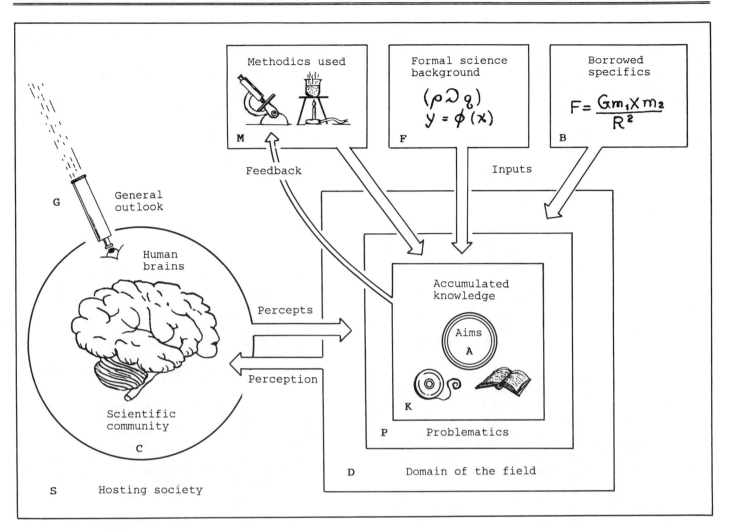

Figure 8.1 A fanciful presentation of the components of a cognitive field--with apologies to philosopher Mario Bunge. (A. N. Strahler.)

4. The domain (D) is composed exclusively of (certified and putatively) real entities (rather than, say, freely floating ideas) past, present, or future. In terms of the ontological models described in Chapter 6 (Figure 6.1), the domain is entirely in the mechanistic realm.

5. The general outlook or philosophical background consists of (a) an ontology according to which the real world is composed of lawfully changing concrete things (rather than, say, of unchanging, or lawless, or ghostly things); (b) a realistic theory of knowledge (rather than, say, an idealistic or a conventionalist one); (c) a value system enshrining clarity, exactness, depth, consistency, and truth; (d) the ethos of the free search for truth (rather than, say, that of the bound quest for utility or for consensus or for conformity with dogma).

6. The formal background (F) is a collection of up-to-date logical or mathematical theories (rather than being empty or formed by obsolete formal theories).

7. The specific background (B) is a collection of up-to-date and reasonably well confirmed (yet not incorrigible) data, hypotheses, and theories obtained in other fields of inquiry relevant to the cognitive field.

8. The problematics (P) consists exclusively of cognitive problems concerning the nature (in particular the laws) of the members of the domain (D), as well as problems concerning other components of the cognitive field.

9. The fund of knowledge (K) is a collection of up-to-date and testable (though not final) theories, hypotheses, and data compatible with those of the borrowed specifics (B) and obtained in the cognitive field at previous times.

10. The aims (A) include discovering or using the laws of the domain of the field, systematizing (into theories) hypotheses about the domain, and refining methods in the methodics used (M).

11. The methodics (M) contains exclusively scrutable (checkable, analyzable, criticizable) and justifiable (explainable) procedures.

12. The cognitive field is a component of a wider cognitive field, i.e., there is at least one other (contiguous) research field such that (a) the general outlooks, formal backgrounds, specific backgrounds, funds of knowledge, aims, and methodics of the two fields have nonempty overlaps, and (b) either the domain of one field is included in that of the other, or each member of the domain of one of them is a component of a system belonging to the other domain.

Professor Bunge turns next to the set of conditions that will identify a cognitive field as pseudoscience (pp. 39-40). Point by point, it corresponds with the list we have just presented. In reviewing this list, I try to be as brief as possible in quoting or paraphrasing Bunge, but adding comments in parentheses relating to the two historical scenarios we have reviewed, namely, Velikovsky's Worlds and von Däniken's Chariots.

1. Little change occurs in the components in the course of time; when it happens, it is forced by controversy and outside pressure rather than by the results of research. (Velikovsky held to a single scenario; von Däniken switched to a radically different one after the first was widely received; no changes were forced by research findings.)

2. There is no research community as such. Instead, the cognitive community consists of believers, who, although calling themselves scientists, conduct no scientific research. (Velikovsky's library study and von Däniken's travels to archeological sites may have in some sense been research into history, but neither conducted research on the astronomical or space-travel phenomena at the heart of the historical scenario.)

3. The host society (S) supports the cognitive community for practical reasons (because it is good business) or tolerates it while relegating it beyond the border of its official culture. (Most certainly, Worlds and Chariots were good business for publishers, and press, and cultist writers, while at the same time providing great entertainment and amusement for a large sector of the public.)

4. The domain (D) teems with unreal or at least not certifiably real entities, such as astral influences, disembodied thoughts, superegos, and the like. (Worlds and Chariots certainly rest on astral influences; the "Space gods" were superegos. None of the actors, whether aberrant comets, planets gyrating out of control, or space gods are certifiably real in the roles claimed for them.)

5. The general outlook (G) includes either (a) an ontology countenancing immaterial (nonmaterial) entities or processes, such as disembodied spirits, or (b) an epistemology making room for arguments from authority or for paranormal modes of cognition accessible only to the initiates or to those trained to interpret certain canonical texts, or (c) a value system that does not enshrine clarity, exactness, depth, consistency, or truth, or (d) an ethos that, far from facilitating the free search for truth, recommends the staunch sense of dogma, including deception if need be. (This is a long list, but the points are all antithetical to those listed for science. Perhaps the final point applies to Worlds and Chariots, both of which are delivered by their authors as dogma to be staunchly defended against the inquiries of science.)

6. The formal background (F) is usually modest. Logic is not always respected, and mathematical modeling is the exception rather than the rule. (This condition scarcely applies to the historical, or timebound, scenarios of Worlds and Chariots, but is important in other areas dealing with timeless knowledge, such as parapsychology.)

7. The specific background (B) is small or nil; a pseudoscience learns little or nothing from other cognitive fields. Likewise it contributes little or nothing to the development of other cognitive fields. (Surely this condition applies to all three scenarios we have reviewed in Chapter 7.)

8. The problematics (P) includes many more practical problems concerning human life (in particular how to feel better and influence other people) than cognitive problems. (This condition does not apply to Worlds or Chariots, but certainly applies to parapsychology.)

9. The fund of knowledge (K) is practically stagnant and contains numerous untestable or even false hypotheses in conflict with well-confirmed scientific hypotheses. And it contains no universal and well-confirmed hypotheses. (This condition certainly applies to all three scenarios we reviewed in Chapter 7.)

10. The aims (A) of the members of the cognitive community are often practical rather than cognitive, in consonance with its problematics. They do not include the typical goals of scientific research, namely, the finding of laws or their use to understand and predict facts. (This condition does not apply to the historical scenarios, but is certainly descriptive of extrasensory perception research, which has yielded no laws and nothing in the way of a testable explanation.)

11. The methodics (M) contains procedures that are neither checkable by alternative (in particular scientific) procedures nor justifiable by well-confirmed theories. In particular, criticism is not welcomed by pseudoscientists. (This condition would relate to extrasensory perception research, which has followed only a single mode of investigation for which no explanatory theory exists.)

12. No other field of knowledge, except possibly another pseudoscience, overlaps with the stated cognitive field. This means that the field is isolated and free from control of other cognitive fields. (The historical scenarios are, of course, isolated as unique conceptions.)

Finally, Professor Bunge presents us with a table comparing the attitudes and activities of scientists and pseudoscientists (p. 41). If you have found the foregoing analysis of diagnostic conditions a bit too heavy, perhaps this checklist will be rewarding in its brevity and directness (Table 8.1).

In appraising Bunge's analysis, keep in mind that it fits best into physical science. The historical sciences, such as geology, paleontology, and evolutionary biology, have some special qualities that must be taken into consideration, as explained in Chapter 4. For example, formal science (mathematics and logic) may not be directly used in unraveling the historical sequence of events through geologic time. Appropriate adjustments can, however, be made to accommodate the historical (timebound) aspects of science.

Protoscience

Always at the outer limits of the expanding frontier of scientific knowledge there lies a nebulous zone in which tentative or suggestive scientific statements are made, but for which supporting evidence is lacking. This is the zone of protoscience, also described as emerging science (Bunge, 1984, p. 44).

Perhaps a good example is the recent suggestion of certain astronomers that there may exist a companion star to our own sun. Now dubbed "Nemesis" by some (and just "George" by others), the companion is a very small dead

Table 8.1   Attitudes and Activities of Scientists and Pseudoscientists

Typical Attitudes and Activities	Scientist	Pseudo-scientist
Admits own ignorance, hence need for more research	Yes	No
Finds own field difficult and full of holes	Yes	No
Advances by posing and solving new problems	Yes	No
Welcomes new hypotheses and methods	Yes	No
Proposes and tries out new hypotheses	Yes	Optional
Attempts to find or apply new laws	Yes	No
Cherishes the unity of science	Yes	No
Relies on logic	Yes	Optional
Uses mathematics	Yes	Optional
Gathers or uses data, particularly quantitative ones	Yes	Optional
Looks for counterexamples	Yes	No
Invents or applies objective checking procedures	Yes	Optional
Settles disputes by experimentation or computation	Yes	No
Falls back consistently on authority	No	Yes
Suppresses or distorts unfavorable data	No	Yes
Updates own information	Yes	No
Seeks critical comments from others	Yes	No
Writes papers that can be understood by anyone	No	Yes
Is likely to achieve instant celebrity	No	Yes

From Mario Bunge, The Skeptical Inquirer, vol. 9, no. 1, p. 4, Table 1. Copyright © 1984 by the Committee for the Scientific Investigation of Claims of the Paranormal. Used by permission.

(or almost dead) dwarf star revealing nothing of itself through electromagnetic radiation or other measurable effects. Nemesis orbits far out beyond the solar system, but in an eccentric orbit that occasionally brings it into the region where comets or cometary materials reside. This region in itself is a hypothetical zone lying beyond the most distant of our known planets (Pluto). No one would have any reason to conjure up Nemesis were it not a possible mechanism for having a large number of cometary masses thrown out of their home region and sent hurtling toward the sun, on which course they can collide

with our Earth and produce devastating impacts. Such impacts in the past might have been the cause of mass extinctions, such as that which killed off the dinosaurs and many other animal groups at the close of the Cretaceous Period, some 65 million years ago. The possibility that Nemesis exists and has played its disruptive game is pure conjecture, but one that can hardly be classed as pseudoscience. In this case, not even the phenomenon itself has been observed but, as conceived, it agrees with our knowledge of the classes of stars (binary stars, dwarf stars) to which it is assigned; its orbit is postulated in accordance with laws of motion and gravitation. The speculation is conducted within the cognitive community of scientists and is discussed in their journals and debated just as is any scientific hypothesis. The entire idea may eventually be shelved or dropped by those who proposed it. Perhaps the time will come when a companion star of this kind is actually located and its orbit computed. A telescopic search for Nemesis by scientists of the Lawrence Berkeley Laboratory was begun in 1984, so we are not dealing here with pseudoscience.

Pseudoscience often consists of speculations that might seem at first glance to be allowable, but on further consideration must be assigned to pseudoscience. Bunge looks to parapsychology for examples. He would relegate clairvoyance, precognition, and psychokinesis to pseudoscience on grounds that they all conflict with physical laws. On the other hand, he thinks that the investigation of mental telepathy, which is thought transmission directly from one brain to another, may perhaps be protoscience. He argues that "if the thought transmission does exist, then it must be a physical process. So, if it were discovered, it would not confirm parapsychology, but would become a subject of ordinary scientific research. . . . Such discovery would be the coup de grace of parapsychology, just as the chemistry of Boyle finished off alchemy and Newtonian astronomy killed astrology" (pp. 44-45).[1]

Protoscience may also be distinguished from pseudoscience by using Asimov's criteria for distinguishing between endoheretics (within the scientific community) and exoheretics (outside the scientific community). In this context, protoscience can be said to be endoheresy (Bunge, 1984, p. 45). Endoheresy is not only tolerated by scientists but welcomed when it comes from a scientist highly respected by colleagues.

So that you may be better armed to take part in controversies involving science vs. pseudoscience, I call attention to another term, fringe science, that appears in writings on pseudoscience (Dutch, 1982, pp. 6-13). It seems that J. S. Trefil (1978), in a popular article in Saturday Review titled "A Consumer's Guide to Pseudoscience," classified scientific ideas into three categories: center, frontier, and fringe. The first two are genuine science, the third is often pseudoscience, but not necessarily so. From the definition that fringe science lies in "a region where ideas are highly speculative or weakly confirmed" (Dutch, p. 6). Dutch would place the Nemesis speculation in fringe science, but we have already argued for its correct place in science. Dutch would also place the Velikovsky and von Däniken speculations, along with creation science, psychic phenomena, and the Loch Ness monster, in fringe science "because they are supported by little data, appear to be untestable or are in conflict with better and more conventional interpretations."

I sense confusion through ambiguity in the term "fringe science," since it seems to encompass both science and pseudoscience in a common zone of overlap. No such overlap needs to be recognized if we apply the list of distinguishing criteria laid down by Bunge. To place the scientific speculation of a Nemesis in the same category as recent creation, Genesis style, leaves the science classroom door wide open to creation science, virtually guaranteeing the creationists the use of at least some fraction of the space and teaching time in that room.

How Pseudoscience Arguments Are Flawed

Flaws in arguments offered by pseudoscience are exposed in an article by Steven I. Dutch of the University of Wisconsin--Green Bay (1982a). Flaws include logical fallacies, often presented in subtle ways that can escape the attention of unwary readers.

We can start with perhaps the oldest logical fallacy known to orators--the straw-man argument. (In case some of these dummies were female, "straw person" is to be preferred, but I defer to tradition here.) The strategy is to generate false assumptions or postulates, then show that they are indeed false. In other words, the pseudo-scientist puts into the mouths of the scientists statements that no scientist has ever said or would have said. Actually, the pseudoscientists often pick up outmoded and discarded hypotheses that science has already repu-diated on the basis of newer evidence. In such cases a repudiated hypothesis can be quoted from obsolete scientific articles, making it seem all the more respectable as genuine science. The false assumption is swallowed by the listener as a genuine current product of science, and is easily and quickly demolished by the orator, using evidence from science itself. We shall encounter many such straw-man ploys in later chapters on creation science.

Another basic ploy, called by Dutch the residue fallacy (1982a, p. 30), is for the pseudoscience advocate to point out that there are many things in nature that science cannot explain, and therefore that all explanations offered by scientists are inherently suspect. It is very easy to glean from the scientific literature statements to the effect that an observed phenomenon is not as yet explained or is poorly understood. Selective quotation is a favorite tool of the pseudoscientist and we shall encounter it again and again in examining creation science. Frequently, the pseudoscientist deliberately ignores the existence of satisfactory, well-supported explanations readily available in the published scientific literature. After offering the selected statement, the false claim is then made that no explanations whatsoever are forthcoming from science. That assertion having been accepted by the gullible members of the public or the cult, there swiftly follows a fallacy of logic, the non sequitur, to the effect that because science has no explanation, the asserted scenario or hypothesis of pseudoscience must be valid. This is the well-known fallacy of "proof from ignorance."

Von Däniken repeatedly makes assertions that science has no explanation for this or that occurrence when, in fact, there do exist published articles containing carefully considered and reasonable explanations. Phrased as rhetorical questions--often many on a single page--his repeated calls for explanations are followed by exhor-tations to science to devote its full attention to the matter. In this way he tries to make fools of scientists and, of course, his audience loves it!

Dutch refers to a common practice of courtroom lawyers, which he calls the "stacked argument" (1982a, p. 11), a question so worded that you are given only two possible answers, either of which is incriminating. Example: Have you stopped distributing counterfeit bills? Answer "yes" or "no"! Dutch selects from von Däniken a question about the Great Pyramid; it is in the context of his general argument that humans are not physically capable of transporting and lifting great pieces of stone: "Who is so ingenuous as to believe the pyramid was nothing but the tomb of a king?" (von Däniken, 1969, p. 11). You have no choice in answering but to choose either (a) "Yes, I'm a naive scientist, like all the rest," or (b) "Not I! Gods from outer space must have built the pyramid." The kind of stacked question used by von Däniken is argument ad hominem, which attacks the person of the opponent rather than the opponent's argument. Scientists have not always been above that sort

of argument, but their bad acting does not make it acceptable.

Dutch describes another tactic of the pseudoscientists, which is to take aim at indirect scientific evidence and degrade its value (pp. 11-12). Geophysics, which studies the structure and composition of the earth's interior by analysis of seismic waves (earthquake waves), has presented strong evidence to support the hypothesis of a dense, probably metallic core, surrounded by a thick, rocky shell (the mantle). No human has ever seen this core, nor have samples of its material been extracted for laboratory examination. The geophysical evidence is nevertheless considered extremely good because it is based on laws of physics, has been repeated innumerable times, and is consistent with other categories of independent evidence (such as measurement of the earth's average density).

Perhaps you have read of the various "hollow earth" theories of pseudoscience, which assert that the earth's interior is hollow or has deep cavities and that the hollow space is inhabited by colonies of living creatures. In a modern version of one of these theories, its author will point out that scientific statements about what lies deep beneath the earth's surface are based only on con-jectures, guesses, and suppositions, and not on any hard evidence. In making this assessment, the author has downgraded to mere speculation indirect evidence from highly sensitive and reliable scientific instruments interpreted in strict conformity to laws of physics. The practice is, of course, a form of attempted deceit, but it is effective in convincing an unwary and unsophisticated audience. What the pseudoscience author is really saying is: "My theory of a hollow earth is just as good as your theory of a dense earth without cavities, because no human has ever been down there to look at the core and get samples of it."

This brings us to a general logical fallacy prevalent in pseudoscience. Dutch calls it relativism; he describes it as follows:

> The general thrust of all of the fallacies above is to create a fog in which all theories appear to be on the same level, none more probable than any other, and the consensus among scientists is made to appear much weaker than is actually the case. Amid the seeming confusion, the fringe theorist argues that his theory is just as likely to be valid as any other, and that experts don't agree, so the non-scientist is free to choose whichever alternative looks best. (P. 12)

The most devastating argument for segregating pseudoscience from science has not, I think, been brought forward strongly enough in the numerous writings of skeptics. In virtually every scenario of pseudoscience of which I am aware, the principle of parsimony is grossly violated. The pseudoscientists have postulated elaborate and complex events that are not necessary to explain the phenomena observed (or asserted to have been observed). In each case, a simpler explanation, reasonably well supported by the present state of empirical science, is available. For the Velikovsky case, the events described in ancient texts relating to the Flood of Noah and the exodus and the captivity in Babylon do not require a comet of unique dimensions and behavior to impact Earth and set off Mars into an orgy of destruction. All that is needed is an appreciation of the naturalistic manner in which humans, possessed of remarkable imaginative ability but of limited knowledge of the world of nature, were prone to construct myths and legends, some of which may have had a basis in real events that were not fully understood. Earthquakes, floods, tidal waves (tsunami), and volcanic explosions are examples of such events. As to the Velikovsky astronomy, the principle of parsimony suggests that, in the absence of any acceptable physical evidence to the contrary, we assume that the planets have maintained stable orbits for a vastly longer period of time than the few thousands of years Velikovsky asks for. Why was he compelled to read into vaguely worded, incomplete documents from ancient civilizations a sequence of catastrophic planetary phenomena? Perhaps he felt the same need to create myths that was felt by our ancestors in the dawn of civilization. Why should his mind work differently from theirs?

In addition to the principle of parsimony, but as a consequence of its abuse, the pseudoscientists share a common failing--they rely heavily on hearsay evidence, which is a body of allegations in oral or written language, passed from one person to another. As pointed out in the evaluation of UFOlogy, all that we have is a set of allegations, within which the distillate worthy of provisional admission as evidence is capable of a number of alternative physical explanations. Von Däniken, for example, bases his case on artifacts or what he claims to be artifacts, in addition to the texts and symbols of uncertain meanings in ancient languages. In this respect his scenario may demand more serious attention from science than those of Velikovsky and the UFOlogists. But, here again, those artifacts for which von Däniken asserts a requirement of space gods can be given reasonable alternative explanations commensurate with what engineering science can produce and, in fact, can reproduce. The quarrying, carving, transporting, and erecting of huge monoliths are operations capable of explanation in terms of human and animal power and the use of tools and machines available to humans who lived in those times.

Why Are Scientists Concerned about Pseudoscience?

If these and other forms of deceit and fallacious argument are so widely used and so easily exposed, why has not the scientific community quickly put down each pseudoscientific theory? Besides the obvious answer--that the believers within the cults of followers are totally oblivious to reasonable debate--there is another and more serious answer. If scientists were to dissect every unsupported theory proposed in popular books and magazines, their entire working capacity would be filled and they would have no time for their research. To undertake a complete refutation of the Velikovsky scenario, for example, would require that a complete textbook be written, covering all the pertinent areas of astronomy, physics, chemistry, archaeology, and engineering. The scientific community has no obligation to perform this function; it is the obligation of the pseudoscience author to document the scenario with high-quality evidence and present the whole package in scientific language with full referencing and in a form acceptable for publication as in a scientific journal or monograph series--acceptable, that is, after searching review by scientists familiar with the areas of science involved in the scenario. Unless material of such quality is prepared by the pseudoscientist, it deserves only summary dismissal. That the A.A.A.S. should have sponsored symposia to focus attention upon the evidence for Velikovsky's assertions or for cultists' assertions of the existence of visitors from outer space is a sign of concern that the unchecked growth of cultism and its unwarranted attack upon science may undermine the strength of the nation.

Professor Bunge addresses our question in somewhat different language and perspective well worth our attention:

Scientists and philosophers tend to treat superstition, pseudoscience, and even antiscience as harmless rubbish, or even as proper for mass consumption; they are far too busy with their own research to bother about such nonsense. This attitude is most unfortunate for the following reasons. First, superstition, pseudoscience, and antiscience are not rubbish that can be recycled into something useful; they are intellectual viruses that can attack anybody, layman or scientist, to the point of sickening an entire culture and turning it against scientific research. Second, the emergence and diffusion of superstition, pseudoscience, and antiscience are important psychosocial phenomena worth being investigated scientifically and perhaps even used as indicators of the state of health of a culture. Third, pseudoscience and antiscience are good test cases for any philosophy of science. Indeed, the worth of such philosophy can be gauged by its sensitivity to the differences between science and nonscience, high-grade and low-grade science, and living and dead science. (1984, p. 46)[1]

The Emotional Appeal of Pseudoscience

Why are the gods and their chariots needed? The answer may lie in psychology and sociology. Perhaps we need to pay more attention to how the mind of the author responded to the cultural environment and forces of history and less attention to questions of physical science. Ronald Story may have given us a clue to what is going on in his comments on von Däniken's "new mythology:"

Man's inability to rise to a high moral plane, even in times of dire need, is all too evident in the present world situation with its threat of atomic war, depletion of natural resources, and widespread hunger. Just as in the days of the contactees and the "Space Brothers," we are facing bad times and fears of the future--fears of an ever-advancing technology that seems to be running out of control.

What could be more appealing than the modern Space Brothers--the ancient astronauts of von Däniken, who are godlike in their technical knowledge (which is so threatening but means so much to us) and in their wisdom (so we assume), and who could direct us in the use of advanced knowledge for the ultimate good of mankind? Since the gods may be our salvation, we want to believe in them, whether we realize it or not. By identifying with these gods, we both comfort our fears and flatter our egos. (1976, pp. 18-19)[2]

(Note: The "Space Brothers" referred to are modern visitors from space in UFOlogy.)

Now, at the close of this section on pseudoscience, you may be wondering why I have not referred specifically to creation science--that body of claims in the name of science by which the fundamentalist creationists seek to support a hypothesis of recent cosmic creation, a hypothesis that happens to fit a religious belief in Divine Creation as described in the book of Genesis. I felt that inclusion of their arguments would be premature in view of what comes next. In Part 2 I will treat as a major topic the conflict between empirical science and creation science. I hope that with this examination of the nature and workings of science the stage will have been set and the tools of analysis explained--ready to be applied point-by-point to the arguments put forward by the creation scientists.

An Unfinished World View

My attempt to describe empirical science and how it works has required that we observe science from a variety of vantage points and perspectives. I hope I have given some insights that are new to many who are not scientists, and to remind some scientists of some principles they may have neglected because of lifelong confinement in the tiny niches of their specialties. As we turn to the conflict between fundamentalist creationists and evolutionary biologists, it may be well to let Professor Herbert Feigl state the guiding spirit of empirical science:

> Instead of presenting a finished account of the world, the genuine scientist keeps his unifying hypotheses open to revision and is always ready to modify or abandon them if evidence should render them doubtful. This self-corrective aspect of science has rightly been stressed as its most important characteristic and must always be kept in mind when we refer to the comprehensiveness or the unification achieved by the scientific account of the universe. It is a sign of one's maturity to be able to live with an unfinished world view. (1953, p. 13)

Credits

1. From Mario Bunge, What is Pseudoscience. Skeptical Inquirer, vol. 9, no. 1, pp. 36-46. Copyright © 1984 by the Committee for the Scientific Investigation of Claims of the Paranormal. Used by permission of the author and publisher.

2. From Ronald Story, The Space Gods Revealed, Harper & Row, Publishers, Inc., New York. Copyright © 1976 by Ronald Story. Used by permission of the publisher.

PART II

Creationism—Its Roots and Tenets

Introduction

The science community of the English-speaking world is under attack by a relatively small group of Christian fundamentalists purporting to speak for the large body of concerned members of their fundamentalist churches and sects. These attackers are trained in science and can be called creation scientists. Their alternative to the prevailing content of natural science, as it is taught in nearly all accredited colleges and universities of the Western world, they call creation science. For brevity, we use creationists to designate creation scientists, creationism for creation science.

Creationists focus their attack on the naturalistic theory of organic evolution through geologic time as originally proposed by Wallace and Darwin in the 1860s and subsequently expanded with many modifications and revisions into the modern general theory of evolution, often called "the new synthesis." The creationists' attack on science is by no means limited to organic evolution. Because the creationist view is one of recent special creation of the universe and everything in it, they are required to include as targets the prevailing scientific theories and findings of geology, astronomy, and cosmology.

From this point on I will use the terms mainstream science and mainstream scientists to mean the product and membership of the international community of scientists that adopts naturalistic organic evolution through geologic time as a major unifying hypothesis or theory. Mainstream science describes nearly all of the scientific literature that will be found in university and museum libraries the world over, the science curricula of those universities, and the subject matter of pure-science research in universities as well as private and public research organizations. "Mainstream" is a good word for this purpose, because the middle region of any symmetrical stream channel is where the water flows fastest and deepest--symbolizing the dynamics of scientific activity. In this region turbulence is intense, with numerous eddies of all sizes continually forming, dissolving, and reforming-- symbolizing the continual formulation and reshaping of scientific hypotheses, while scientific knowledge as a whole moves forward. I would avoid using such adjectives as "traditional" or "conventional," which imply dogmatism and stasis; those are adjectives that fit better the funda- mentalist creationists' Bible-based view of a stagnant postcreation world undergoing continual decay, increasing entropy, and increasing disorder.

Creation science takes its tenets from the Book of Genesis, literally interpreted by the creationists to mean that all of the universe was created out of nothing (ex nihilo) by an omnipotent and omniscient Creator. Supernatural creation occurred within a period of six days of mean solar time during one year approximately 6000 tropical years before present. The Creator carried out his Creation with a divine purpose. Each kind of plant and animal (what biologists would call a species or a genus) was specially created in complete form as we know it today. Not one of the original kinds has given rise to another kind since the Creation, but some kinds have since become extinct. Minor changes and variations within kinds have taken place following the principles of modern genetics, as we see in the case of guided development of varieties of food and ornamental plants and of breeds of domesticated animals.

We can call this total picture of creation of the universe recent creation, to distinguish it from other versions of divine creation that accept a time scale going back as far as 3 billion years (b.y.) for the creation of earliest life on earth, about $4\frac{1}{2}$ b.y. for creation of the planet Earth, and 10 to 20 b.y. for creation of the universe.

According to the creation scientists, the universe operates today within fixed natural laws. These laws were created by the Creator at the time of his Creation and are maintained constantly by the Creator. There is the possibility, always present, that the Creator may intervene to change or suspend his laws to provide a miraculous event.

As a consequence of the belief in fixed natural laws, creation scientists accept much of physical science-- physics and chemistry--as presently developed by the international community of scientists and taught in accredited colleges and universities the world over. This acceptance applies to physics and biology that underlie present-day practical or applied science in fields such as engineering and medicine, for example. Mainstream genetics and molecular biology are accepted, as are most basic fields in physical geology--mineralogy, seismology, and hydrology, for example.

Although the tenets of creation science obviously rest entirely on a religious base, according to the statements of the creationists themselves, it is their claim that creation science is not religion, but instead is science. Then, in an even more remarkable juxtaposition of the common meanings of words, they claim that organic evolution is not science, but instead is religion. On this basis, and invoking the principle of separation of church from state, they would like to see naturalistic evolution and most of historical geology excluded from the school science curriculum (because it is religion), while its place should be taken by creation science (which is not religion, but is science). Failing to achieve this one-on-one replacement, they express willingness to settle for the teaching of both views on an equal-time basis in the same natural science course.

It appears, then, that we must examine the claims of the creationists by using two avenues of exploration. First is from the position and methodology of sociology and psychology, i.e., creationism as a behavioral phenomenon. Second is from the position and methodology of analytical philosophy, including the problems of ontology, epistemology, and logic. Perhaps the first of these avenues of exploration will lead us to understand why there are creation scientists, why they are so strongly motivated, and why they are supported within the larger community of religious fundamentalists. Perhaps the second avenue will lead us to decide whether creation science is acceptable as a form of empirical science or whether it should be regarded as pseudoscience. If

scientific issues of substance remain after our behavioral and philosophical investigations are complete, we can turn to them on an individual-case basis.

Our program is to make first a historical review of the origins of scientific thought about evolution in relation to religion, leading to the rise of fundamentalist creationism in our century. Following this we will apply the philosophy of science to the tenets of creation science. Finally, we investigate each major science topic of the creationists' dispute with mainstream cosmology, astronomy, geology, and biology.

Chapter 9

The Roots of Creation and Evolution

The roots of controversy over the origin and development of life forms on earth go well back in church history. A brief review can give some useful perspectives, provided that we focus upon two questions of natural science: (a) What is the significance of fossils, those ancient plant and animal remains or impressions preserved by burial in sedimentary rock strata; (b) How old are fossils and the rocks in which they occur? The arena in which these questions were first examined critically was limited to Europe, and specifically to Western and Southern Europe and the British Isles, under domination of the Christian church, both Catholic and Anglican.

Fossils and the Flood of Noah

We need go back only about a century before the time of Bishop James Ussher (1581-1656) and publication in the 1650s of his chronological studies leading to the establishment of the Genesis creation as occurring in the year 4004 B.C. Throughout the Middle Ages and Renaissance fossils had no place in the scenario, and the idea that they might have once been living plants or animals was heresy. A challenge was to arise about A.D. 1500 in Italy, when canals were being dug into marine rocks of Cenozoic age (covering roughly the time span of 10 to 60 million years ago). Marine fossils, strikingly resembling living forms seen along the Mediterranean shores today, were uncovered in large numbers in the canal excavations. The fossils attracted the attention of Leonardo da Vinci, who had a strong interest in geology. He argued that the fossils were the remains of animals living in those rocks at the time of burial, when the rock material was soft sand and mud. His pronouncements set off a great fossil controversy between the liberal Renaissance thinkers and the Christian clergy. Supporters of the church claimed that the fossils were not of organic nature, but that they were the product of mystical forces, or even implantations by the devil to delude honest Christians. In the ensuing two centuries, evidence for the organic nature of fossils grew increasingly strong, and even the most orthodox churchmen could no longer escape the naturalistic interpretation of burial of life forms.

One important figure in promoting the organic theory of fossils was Robert Hooke (1635-1703), a member of the Royal Society, which at the same time included distinguished scientists Robert Boyle and Isaac Newton (see Faul and Faul, 1983, pp. 40-44). Hooke wrote this highly perceptive comment on the fossils he had examined: "There have been many other Species of Creatures in former Ages, of which we can find none at present; and that 'tis not unlikely also but that there may be divers kinds now, which have not been from the beginning" (1665, p. 291). Hooke encountered opposition both from scientists of the Royal Society who were conservative church members and from Anglican theologians.

In ensuing decades of the 1600s the proposal was made to explain fossils as the remains of animals and plants killed and buried during the Flood of Noah. Despite problems of scriptural interpretation, the Christian church eventually made the switch to the Flood-annihilation hypothesis. The name diluvialist school is usually given to believers in that hypothesis.

Fossils and Strata

What we need to look at next is the relationship of fossils to the rock layers, or strata, in which they are found enclosed. Strata are typical of sedimentary rock, a major rock class consisting of sediment composed of particles of mineral or organic matter. Those loose particles are transported by currents of water or air from some source region and deposited in layers on the surfaces of the continents or on the floors of rivers, lakes, or the oceans. Another class of sedimentary rock consists of mineral matter precipitated from chemical solution. Familiar kinds of sedimentary rock are sandstone, shale, and limestone. When sediment particles settle out of suspension in water or air, they form more or less horizontal layers that are successively older from the top down or, conversely, younger in age from bottom to top. This relationship is elementary from our sandbox experience in the kindergarten.

The age-layering principle is called the principle of superposition. It is attributed to a Danish physician, Nicolaus Steno, who worked out the concept while serving as physician to the duke of Florence. Steno studied sedimentary strata exposed along the walls of the Arno Valley, not far southeast of Florence. His findings were published in 1669; it was the first fruitful effort to work out the geologic history of a region by interpretation of strata. Although the principle of superposition is reliable, based as it is on the action of gravity at the earth's surface, strata deposited in the order of age sometimes get turned upside down in a later episode of mountain upheaval. Geologists have independent criteria for determining "which way is up" in a sequence of strata, regardless of how they are found today.

Using the principle of superposition, the relative ages of assemblages of fossils can be established within a stack of strata exposed to view by natural erosion, or in an excavation, or extracted from a drill hole. About the year 1800, an English civil engineer and geologist, William Smith, collected fossils exposed in canal excavations. He found fossils of like species to be present in all parts of a single group or rock layers (called a formation) that could be proved to be one and the same layer by tracing it horizontally in continuous exposure. Fossil species in formations above or below were found to be distinctively different, but always to occur in the same order in widely separated localities. The conclusion is inescapable through simple logic: Fossils are of the same kinds within a single formation, but of different kinds in formations of different ages. Notice that I cannot infer from that statement alone that the fossils in an upper formation are descended from (or evolved from) those in a lower formation. That inference can, however, be stated as a hypothesis, but there is an alternative hypothesis: The fossil species in

the lower formation were made extinct--wiped out, that is
--by a major catastrophe, and at a later time some
entirely new species were created to occupy the
environment of the upper formation. This extinction could
have alternated with creation to explain what is observed.

Catastrophism and Uniformitarianism

If the catastrophe were a worldwide flood of great
proportions, then there could have been several or many
floods, each following an interval in which an unique
assemblage of life forms enjoyed a period of survival over
many generations. This scenario is called catastrophism,
and it was developed in detail by a French naturalist,
Baron George Cuvier (1769-1832). A zoologist of
distinction, Cuvier became an authority on fossils, a
paleontologist, that is, and he is often called the "father
of paleontology." In the northern part of France, within a
region known as the Paris Basin, the land surface is
underlaid by sedimentary strata that have been slightly
warped into the form of a stack of nested saucers.
Erosion has exposed the edges of the saucers to form
"cuestas," and it is in those exposures that fossils can be
collected. By the early 1800s, the sequence of strata and
fossils they contained had been worked out in considerable detail. It is noteworthy that fossils in the
lowermost formation are very much unlike modern species,
and that upward into younger formations fossils show
increasing degrees of similarity to modern forms.
However, these changes occur in abrupt jumps from one
formation to the next rather than continuously within each
formation. Cuvier, who was devoutly religious, never
doubted that all fossils were of divine creation. It was
Cuvier who proposed a series of catastrophic extinctions
by sudden floodings of the region by ocean water.
Following each extinction, a new set of animals was
created. The final extinction was correlated with the Flood
of Noah, but this connection was established by an
Englishman who translated Cuvier's works. Catastrophism,
Cuvier-style, features heavily in the modern creationists'
arguments, and we will examine those arguments on later
pages.

But Cuvier did not have the French stage all to
himself. An associate of his at the Museum of Natural
History was Jean Baptiste Pierre Antoine de Monet de
Lamarck (1744-1829), a botanist who was also a
freethinker and thus not compelled by conscience to
subscribe to the creationist doctrine of the church.
Lamarck also studied the fossils and strata of the Paris
Basin, but he came to a different interpretation from
Cuvier's. Lamarck emphasized the general similarities in
life forms as they reappeared in higher, younger strata.
He concluded that members of a particular taxonomic
group--a particular order or family, that is--persisted
through time, undergoing change in response to
environmental pressures. What seemed to be an abrupt
jump from one set of fossils to a different set could be
explained if a time gap exists in the record--a time period
in which the Paris Basin region was uplifted and exposed
to erosion, causing removal of some strata. Life would
continue to thrive and to change in form in shallow ocean
water beyond the limits of the upraised area. Then the
land would sink and the ocean water invade and submerge
the region, bringing with it the changed marine fauna,
which would become preserved as fossils in the newer
strata. What we see here is a unified concept: continuous
evolution of life forms coupled with a continuous, rather
uniform process of erosion and deposition shifting its
activity in geographical location through time. No
catastrophe is required in this scenario; no need for mass
extinctions in cataclysms, and no need for divine creation
following each cataclysm. The question of origin of the
first life forms in this sequence does, however, remain
open, and there persists the option of divine creation
versus a naturalistic beginning.

While Lamarck was doing his thing in France, a major
revolution in geologic thought had already made
considerable progress in Scotland, but with emphasis on
how rocks are formed and changed, rather than on the
significance of fossils in strata. To understand this
revolution we need to go back a couple of decades and
look at a rather remarkable exchange of sparks between
Scots and Germans, going right over the heads of French
scientists of the museum in Paris. The revolution was
spearheaded by two geologists associated with the
University of Edinburgh, at that time enjoying a period of
vigorous scholarly activity and providing an ideal climate
for new ideas in various branches of science. The older
of the two geologists was James Hutton (1726-1797); his
younger friend and associate was James Hall (1761-1832).
Together they studied rocks exposed on the rugged coast
of their native land. What they found spurred them into
undertaking a two-man war against a totalitarian German
system of geology headed by Professor Abraham Gottleib
Werner of the Freiburg Academy. This powerful figure of
European geology preached a theory that all rocks,
whatever their compositions and properties, came into
existence quite suddenly by precipitation from the waters
of a single great world ocean. Werner's schedule of events
fitted the doctrine of the Christian church regarding
creation and was widely accepted. Werner and his
followers were called neptunists.

The neptunist theory was brought back to Edinburgh
University by Robert Jameson (1774-1854), who had
espoused neptunism in Europe. Between 1800 and 1820
Jameson lectured to Edinburgh students, and it was here
that neptunism came in direct person-to-person conflict
with the ideas of Hutton and Hall, developed over some
three decades and first set down in abstract form by
Hutton in 1785. The theme of this new view of geology
was that geologic processes such as those observed in
action today have been in operation for vast spans of
time, with no beginning point in time observable and no
end in sight. Those processes include stream floods,
moving glaciers, winds, breaking surf, ocean tides, and
tidal currents, all of which are processes that transport
sediment and lead to formation of new strata. Also
included are volcanic eruptions, earth movements
accompanying great earthquakes, and great landslips in
alpine mountains. This last group of events falls in the
category of catastrophic happenings in terms of threat to
human life. Today we would include among the catastrophic events the impacts of asteroids or comets from
outer space. Nevertheless, all such events are of natural
occurrence, and that is an essential point of the principle
of uniformitarianism, as Hutton's thesis came to be known.
It stood in contrast to the catastrophism of Cuvier and
the creation scene of the neptunists. Lamarck's view of
biological evolution complemented Hutton's view.

Uniformitarianism should be viewed as antithetic to any
and all theories invoking supernatural forces for the
origins of things geologic or biologic or of any other
natural-science phenomena. Uniformitarianism includes the
assumption that laws of physics and chemistry as
presently understood have applied to all past time and will
apply to all future time; they are assumed valid unless
compelling independent evidence should be found to
suggest that they may have differed in quality or
quantity at other points in time or space. In using that
assumption, all science operates today under the
uniformitarian principle. In cosmology, the study of the
structure and evolution of the universe, it is assumed
that the laws of physics are similiar throughout the entire
universe. Another principle implicit in uniformitarianism is
that all interpretations of the universe and its parts
should be as simple as possible, following the principle of
parsimony, or Occam's Razor, which we explained in
Chapter 2. Catastrophism, invoking special causes and
processes of supernatural origin, is viewed as violating
the principle of parsimony.

Modern creationists reject the universal principle contained in uniformitarianism. As did their predecessors in Hutton's time, they require supernatural creation of the universe and its present laws, followed by a brief but enormously cataclysmic event--the Flood of Noah. Only since that one-year catastrophe have God's laws been in full control of the universe. We have really not gotten very far in resolving this basic conflict that arose as far back as the late Renaissance.

Between Hutton and Lamarck, the trend of science had been launched in the direction of a concept of organic evolution through a vast span of time. Werner's neptunism persisted in Europe through the first three decades of the 1800s, but was soon to be phased out as another Scottish geologist, Sir Charles Lyell, came on the scene to champion Hutton's uniformitarianist picture of the world and, through his widely used textbook of geology, to make that picture the generally accepted standard of earth science. Principles of Geology, as Lyell's textbook was titled, appeared in 1830, was widely accepted, and went through many revisions. Captain Fitzroy of H.M.S. Beagle had obtained a copy for the ship's library, and his young staff scientist, Charles Darwin, set about to read it on the transatlantic leg of the five-year discovery voyage (1831-1836).

Fossils and Geologic Time

The 1830s saw great advances in determining the relative ages and fossil content of strata of the British Isles. Two field geologists, Adam Sedgwick and Roderick Murchison, are particularly worthy of mention in this connection. Within each formation they found a distinctive collection of fossils--a fossil fauna, that is. After having established a sequence of such faunas in a succession of formations at one locality, it was then possible to correlate the strata of other localities, using the fossil faunas themselves as the identifying features. In some instances, a single fossil proves to be uniquely limited to a narrow zone of strata. Called an index fossil, its presence in a formation on any continent establishes the position of the formation in the standard geologic time-scale of relative age.

Groups of formations sharing the same fossil faunas and index fossils were assigned to a system, belonging to a time unit known as a geologic period. For example, the name Silurian System was applied by Murchison to a group of formations he studied in Wales, in a region occupied in Roman times by a native tribe of that name. Strata of that system belong to the Silurian Period of geologic time. Once the sequence of systems was established in England, it could be applied to strata far removed in other continents. By means of their fossils, formations of Silurian age were identified in other parts of Europe and in North America. This activity is called stratigraphic correlation. Gradually the entire table of geologic time with its periods was constructed and filled in. From the earliest period in which invertebrate marine animals of complex forms were abundant--the Cambrian Period--to the present, ten geologic periods came to be recognized.

By comparing the fossil content of rocks of each period with that of the next younger period, a particular class of plant or animal could be traced through time from the period in which it first appeared to the time that it disappeared (became extinct) or to its persistence to the present. It could be seen that several classes of animals-- among them the corals, sponges, worms, and crustaceans --were present from the start of the Cambrian Period and have continued to the present day. However, in each case the representatives of those classes had undergone a change from one period to the next. Some classes of animals and plants enter the time scale "as if from nowhere." For example, the crinoids, starfish, and echinoids, all classes of marine invertebrates, are not found in the Cambrian Period, but appear at the outset of the next period, which is the Ordovician Period. Here we would have a choice of two explanations: (a) These new life forms were independently created at the time they appeared; (b) They arose by rapid change within another, already existing class of animals, which is to say that a new form evolved from an old one. The latter explanation represents organic evolution.

At this point in our inquiry, we are forced to look at a biological principle: All life derives from other, preexisting life; it is the principle of biogenesis. Of the two alternatives offered to explain the appearance in the geologic record of a new life form, the second, or evolutionary, explanation conforms with the principle of biogenesis. That principle is universally accepted by biologists today, but such was not always the case. Cuvier's catastrophism required that new life forms be created from nonliving matter following each extinction. The entire theory of organic evolution depends upon acceptance of the naturalistic principle of biogenesis, but when traced far enough back into geologic time, it leads us to a point where nonliving matter must have given rise to living matter. That event is called biopoesis, a rather strange term deriving from the Greek word poiesis (Latin, poesis), meaning "creation." (Note the same root for "poesy," a poem, which is a literary creation.) This is not the place to get into arguments over biopoesis, whether it was an act of God or occurred spontaneously in a naturalistic manner. We will get into that question in due time. The point here is that the concept of evolution depends on there being only one originating point in time and place for life on earth. In that case, evolution can be visualized as a tree, originating from one seed and growing as a single trunk to which many branches are attached (Figure 9.1).

The branching tree of life forms took shape through the middle years of the 1800s. In practical form, it is presented as a chart in which time-lines run horizontally across the page, while the lines of continuity of life forms run vertically upward. An interesting pictorial diagram of this type can be found in a geology text by David Page, published in 1860, where it is called an "oryctological chart" (Figure 9.2). Lamarck's ideas on organic evolution, which were quite consistent with the new uniformitarianism, included the suggestion that species of plants and animals might undergo change in response to environmental influences, and that changes might be passed on genetically. But Lamarck did not use the word "evolution," and he seems not to have had any well-developed hypothesis of the dynamics of an evolutionary process (Faul and Faul, 1983, p. 139). Much later, the neo-Lamarckian doctrine of inheritance of acquired traits was to be brought forward by anti-Darwinists, but that is a different story.

The idea of an evolutionary pattern generally from lower, simpler life forms to higher, more complex forms seems to have been gaining strength, since it soon became obvious that following marine invertebrate dominance in the Cambrian Period, the vertebrates appeared as fishes and sharks in the ensuing Ordovician, Silurian, and Devonian periods, followed in later periods by reptiles and birds and, much later, the mammals. Thus the essence of the evolutionary process was understood and embodied in the words descent with modification.

Darwin and Natural Selection

Among the naturalists of the early 1800s--one of them being Charles Darwin as a young student at Cambridge-- the Lamarckian evolution concept went under the name of transformism or, more commonly, transmutation. It was considered a highly revolutionary idea and was, of course, in direct opposition to the Cuvier scenario of repeated divine creation. Young Darwin held strongly to the prevailing religious view that each species had been separately created; he held it even after his return to

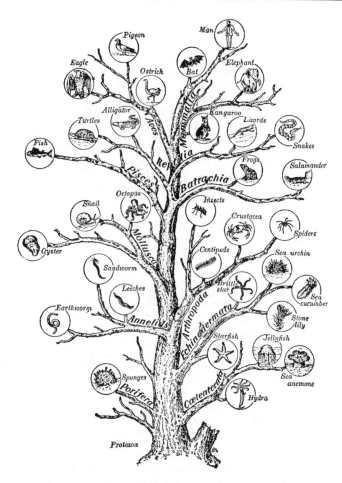

Figure 9.1 The tree of life in pictorial rendition. If a comparable tree of plant life were to be added, it would branch off near the base (shown here as a stump). (From Gruenberg, Elementary Biology, Ginn and Company.)

England from the Beagle voyage, this despite the great amount of information he had collected. Professor A. Lee McAlester of Yale University describes Darwin's change of viewpoint:

> The turning point came in 1837, a few months after his return to England. Reflecting on the experiences of the voyage, it occurred to Darwin that certain puzzling facts about South American fossil mammals and the distribution of living species on isolated islands could best be explained by the heretical idea of transmutation. The more he considered the idea, the more probable it seemed, yet there was one key still missing--an adequate mechanism to explain just how one species changes into another. (1968, p. 25)

After another two years of thought and speculation a new explanation for transmutation took root in Darwin's mind. Taking account of the random physical variation that is present within a population of individuals of a species, he realized that an individual that happened by chance to possess a set of characteristics giving it a superior opportunity to survive would also be likely to reproduce more offspring. It is important to realize that the characteristics referred to are genetically controlled from within the individual and are not those acquired by the individual in response to the environment during the lifetime of that individual. In competition with other individuals of the same species, with other species, and with environmental stresses, the better-endowed

individuals would, on the average, tend to survive longer and produce more offspring. The less favorably endowed individuals would tend, as a group, to survive for a shorter average lifespan and to reproduce fewer offspring. Thus, the average composition of the species in terms of its physical characteristics would undergo gradual change toward a better adaptation to the total environment. This change could lead to the emergence of a new species. Darwin described this evolutionary process as natural selection. The outcome of natural selection soon came to be known as the survival of the fittest. Darwin recognized that the process of inheritable change in a species can be induced artificially by breeders of plants and animals through deliberate selection of individuals to be retained; hence his use of the adjective "natural."

Darwin was occupied for twenty years with putting together the evidence needed to make a strong scientific case for his hypothesis of natural selection. In the meantime, another British naturalist, Alfred Russel Wallace, had independently devised the same explanation and made his discovery known to Darwin. They agreed that each should make public a brief statement of his own version simultaneously, which they did in 1858, but it seems to have attracted little notice at the time. Darwin went on to complete his major work, The Origin of Species by Means of Natural Selection; it was published a year later (Darwin, 1859).

The Impact of Darwinism

Darwin's work immediately attracted widespread attention and quickly gained numerous adherents from within the scientific community. It was widely accepted by the public as well. The work had surprisingly few detractors, considering that the hypothesis was totally naturalistic and demanded an enormous block of geologic time in which to operate.

Nevertheless, some scientists of the day strongly attacked Darwin's hypothesis. One was Louis Agassiz, the renowned Swiss zoologist who is perhaps best remembered today for his hypothesis of continental glaciation during the last Ice Age. Agassiz was devoutly religious and a strong supporter of Cuvier's scenario of repeated extinction and divine creation. Agassiz saw as the purpose of science to provide proof of the existence of God as a Creator giving purpose to observed progressive changes in life forms. For Agassiz, each species, as created, remained immutable. This is, of course, a tenet of the modern fundamentalist creationists. Criticism of Darwin's presentation came from several geologists and paleontologists, among them Adam Sedgwick, referred to in earlier paragraphs. They were displeased with what they considered the lack of evidence for Darwin's important points. Furthermore, they pointed out that transitional fossil forms are lacking between species asserted to have been in evolutionary sequence. This attention to "gaps" in the fossil record is today strongly pursued by the fundamentalist creationists, and we will need to give it due consideration.

As to the reaction from the supporters of the church, this paragraph by Dorothy Nelkin describes it nicely:

> Beyond its impact on traditional science, Darwinism was devastating to conventional theology. Religious traditionalists accused Darwin of "limiting God's glory in creation," of "attempting to dethrone God," of "implying that Christians for nearly 2,000 years have been duped by a monstrous lie."* Evolution theory violated traditional theological assumptions, and, above all, the assumed distinction between man and the animal world. (1982, p. 28)[1]

(*Quoted phrases are from Andrew Dickson White, 1896, A History of the Warfare of Science and Theology in Christendom.)

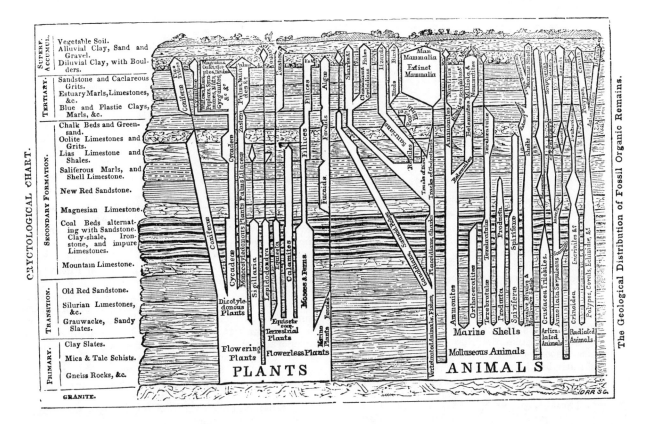

Figure 9.2 The original legend for this plate from an American geology textbook of Civil War vintage reads as follows: "The Oryctological Chart--This is a representation of the leading facts of Oryctology, exhibiting all the important organic remains of the different formations. The greater or less space occupied by the various tribes of animals and plants on the chart shows their comparative abundance or paucity. The branches designate the species. When one entirely disappears during some formation, but afterwards reappears, a line is drawn where it is wanting. The whole brings under a glance of the eye the rise, developement, ramification, and extirpation of the different tribes." (From David Page, 1869, Elements of Geology, A. S. Barnes and Burr, New York.)

Darwin himself seems to have retained as much of his religious faith as his hypothesis would permit, espousing the deistic view that God had created life on earth in its most primitive form but had thereafter withdrawn to allow the mechanistic course of evolution to proceed untended. For a person of his time, trained in theology and seriously considering it as a profession, he had traveled a long way from home territory, evidently with extreme reluctance.

According to Nelkin, acceptance of Darwinism by both scientists and theologians had increased markedly by the end of the century (1982, pp. 28-29). Attempts were made to reconcile the requirements of the process of natural selection over a long time span with the Genesis story. These alternatives will be a subject for discussion when we consider the various forms of creationism subscribed to today. Another phenomenon of the post-Darwin period was the rise of Social Darwinism, a vehicle by which human social and political behavior could be justified as a "natural" phenomenon through analogy with the concept of survival of the fittest.

In the United States, the period from about 1895 to 1920 saw a remarkably high level of acceptance of the Darwinian hypothesis of evolution, and it began to appear in high school and college textbooks, presented with great assurance and sometimes with the arrogance often associated with a ruling hypothesis having no current rival (Nelkin, 1982, pp. 29-30). But the American scene was soon to change with the arrival of a new wave of religious fundamentalism.

Credit

1. From Dorothy Nelkin, 1982, The Creation Controversy; Science or Scripture in the Schools, W. W. Norton & Company, New York. Coppyright © 1982 by Dorothy Nelkin. Used by permission of author and publisher.

Chapter 10

A Resurgence of Fundamentalism

Longstanding emphasis on literal interpretation of the Bible by groups within Protestantism--for example, in the Methodist revivalism of the early 1800s (Nelkin, 1982, p. 30)--was submerged only temporarily in the immediate post-Darwin period; it showed a sharp resurgence shortly after the turn of the century. According to Nelkin, fundamentalism began to direct its attack on Darwinism through "a series of pamphlets entitled Fundamentals."

> These pamphlets attacked "modernism" and, in particular, evolution theory; the idea that evolution involved discrete accidental changes determined by the circumstances of the moment shattered fundamentalist faith in planned and purposeful change. They attacked the theory with zeal and enterprise, and worried about its implications for Christian behavior. (Nelkin, p. 30)[1].

By the 1920s division was strong within the Protestant denominations over the issue of evolution. The fundamentalists denied the validity of the theory, while the modernists felt at ease with a reconciliation between all science, including even evolution, and the teachings of the Bible. The battle was actually waged over the issue of controlling education. The line was drawn between the predominantly fundamentalist inhabitants of the rural South and liberal intellectuals of the industrialized North. Thus, in the southern states, fundamentalism pretty much had its way in producing restrictive legislation relating to teaching in its schools. According to Nelkin, between 1921 and 1929 antievolution bills were introduced into thirty-seven state legislatures (1982, p. 31); those that passed were in southern states. In March of 1925, Tennessee passed its bill prohibiting teaching in the public schools of theories contrary to the accepted interpretation of the biblical account of human creation, and soon history was to be made in Dayton, Tennessee, scene of the Scopes trial of July 1925.

The Scopes Trial

In what has become known as "The Monkey Trial," John T. Scopes, a biology teacher, was put on trial for teaching Darwinism in the Dayton public school, in violation of the state law passed earlier in the same year. Because two nationally prominent figures came to Dayton to participate, the trial drew nationwide attention. Agnostic trial lawyer Clarence Darrow, fresh from success in the 1924 murder trial of Leopold and Loeb, came to serve as one of Scopes's attorneys. Coming to the aid of the prosecution was William Jennings Bryan, fundamentalist Presbyterian and in early life an unsuccessful Democratic candidate for the U.S. presidency. An orator with great popular appeal, he had previously addressed state legislatures to urge measures against the teaching of evolution.

The trial quickly turned into a contest between fundamentalism and Darwinism, with attention focused on the lively encounters between Darrow and Bryan. Bryan presented essentially the same argument we hear from the creation science advocates today: "What right have the evolutionists--a relatively small percentage of the population--to teach at public expense a so-called scientific interpretation of the Bible?" Under severe questioning by Darrow as to the literal interpretation of the Bible, Bryan fared badly, undergoing severe ridicule; he died in his sleep a few days later. Scopes was convicted and fined, but the decision was later overturned on a technicality by the state supreme court.

According to Nelkin the Scopes trial brought a temporary end to the fundamentalists' efforts to secure further legislation banning the teaching of evolution (1982, p. 32). Fundamentalists then turned much of their attention to support of the prohibition amendment. Nevertheless, fundamentalist sects continued to issue tracts denouncing evolution for its role in bringing moral decline to the nation.

Nelkin attributes the relative quiet on the fundamentalist side following the Scopes trial to the voluntary reduction or elimination of evolutionary biology from textbooks (1982, p. 33), a response by many publishers to unfavorable publicity. Some revised editions deleted the word evolution and the name Darwin, and some even added religious material. A nationwide survey of secondary school teachers in 1942 "indicated that fewer than 50 percent of high school biology teachers were teaching anything about organic evolution in their science courses" (p. 33). State statutes in the South continued to bear laws against teaching of evolution, and these persisted into the 1960s.

The Rise of Creation Science

Professor Ronald L. Numbers of the University of Wisconsin, Madison, has documented the beginnings of creation science in the 1920s. Numbers quotes a statement by a Baptist minister of the time, William Bell Riley, giving his reason for opposition to the theory of evolution: "The first and most important reason for its elimination is in the unquestioned fact that evolution is not science; it is hypothesis only, a speculation" (Numbers, 1982, p. 539). This view was shared by William Jennings Bryan, who was actively crusading against evolution in the post-World War I period. Numbers explains that their expressed view on evolution as nonscientific was in the context of an archaic and outmoded model of science, associated with Sir Francis Bacon, "that emphasized the factual, nontheoretical nature of science. By identifying with the Baconian tradition, creationists could label evolution as false science, could claim equality with scientific authorities in comprehending facts, and could deny the charge of being antiscientific" (p. 539). Creation scientists of today do not use this archaic concept of science in their argument; instead, they have latched onto Karl Popper's newer view of what science should be.

One self-styled antievolution scientist of the time was a Presbyterian minister, Harry Rimmer (1890-1952). By laboratory experimentation, he attempted to show that biological science and the Bible were not contradictory in their statements (Numbers, 1982, p. 539). A second antievolution scientist, destined to become by far the more influential of the two, was George McCready Price (1870-1963). In 1923 Price published his major work titled The New Geology. A fundamentalist of the Seventh Day Adventist sect, Price had held a post as professor of geology in a small Adventist college in Nebraska, followed by service on the faculty of Walla Walla College in Washington. Martin Gardner describes the 726-page college textbook in these words:

> It is a classic of pseudo-science. So carefully reasoned are Price's speculations, so bolstered with impressive geological erudition, that thousands of Protestant fundamentalists today accept his work as the final word on the subject. Even the sceptical reader will find Price difficult to answer without considerable background in geology. (Gardner, 1957, pp. 127-28)

Price's religious views are essentially the same as those of the present-day religious fundamentalists, and his geological arguments for recent creation and the Flood of Noah are repeated today in much the same form by the creation scientists. Strangely, Price seems to have considered his creationist view of the earth and life as a new theory--new, that is, with respect to the conventional geology and evolutionary theory of the late 1800s and early 1900s. He expressed the opinion that his version of geology would replace the outmoded view. He even went so far as to declare that organic evolution is dead (Gardner, 1957, p. 132). As a college student majoring in geology in the late 1930s, I knew of Price's work and discussed with others some of his lines of argument. But we did not take it seriously for, having recently been liberated from fundamentalism ourselves, we held such antiscientific views in contempt. Our textbook of historical geology was a standard work of the time: the third edition of Textbook of Geology, Part 2, by Yale professors Schuchert and Dunbar (1933). It has a full chapter on evolution and makes no concessions to religious views. In referring to Darwin's theory of evolution, those authors wrote:

> The theory met with bitter opposition, the echoes of which have not yet altogether died out, but its truth is now accepted as a working principle by the biologists of the world, who hold that "evolution is a fact."

> It is important to make a distinction between the general doctrine of evolution and the various theories propounded to explain how the modifications have been brought about. The former is now universally accepted by enlightened people, but the latter are still the subject of much controversy.

> The evolution concept is without doubt the grandest generalization of the nineteenth century, since it has not only transformed the method of study in Biology, Geology, and the social sciences, but has given a new point of view to all science and art, and even to progressive religion. (Schuchert and Dunbar, 1933, p. 29)[2]

I do not recall much about our biology text, but the course content was largely anatomy and physiology, with some taxonomy. The biology department chairman was active in genetics research, and there was no suggestion that any problem existed as to what should be taught.

I gather that in that time college science curricula received little or no attention from fundamentalists, at least in the North. The same situation seems to persist today. In Arkansas and Louisiana, where recent bills have been enacted, efforts of the creationists are directed primarily at precollege schooling. I am not aware that similar efforts have been directed at the science programs of Louisiana State University; at least nothing to that effect is implied in the wording of the recent Louisiana law. I suspect that the fundamentalists have thought better of locking horns with college and university science faculties. High school teachers of biology are in a much weaker position because they work in relative isolation from colleagues but in direct contact with parents of the students in their classes, only the local community being involved in what goes on.

An important trend of the late 1930s and early 1940s was the defection of some younger university-trained scientists from the fundamentalist ranks (Numbers, 1982, p. 541). The impact was traumatic for three fundamentalist organizations then active in fighting the theory of evolution. One was the Religion and Science Association, formed in 1935 under the leadership of a Wheaton (Illinois) College professor. The second was the Deluge Geology Society, formed in 1938 by George Price and colleagues. The third was the American Scientific Affiliation (ASA), formed in 1941.

The ASA was formed to investigate problems bearing on the relation between Christian faith and science (Nelkin, 1982, p. 77). The position of the ASA on the teaching of evolution and creationism is a moderate one, judged in comparison with the other organizations listed above. According to Nelkin, ASA members regard the theory of evolution as misleading and having "serious moral and social as well as theological implications" (p. 78). Nelkin states further:

> Because of diverse opinions among its membership, the organization has avoided taking a position that advocates teaching creation theory in public schools. ASA does, however, criticize the evolutionary emphasis in textbooks, arguing that the theory is extended beyond what is scientifically appropriate, and that it unnecessarily excludes consideration of alternative theories. (P. 78)[1]

One of the early members of the ASA was J. Laurence Kulp, a geochemist on the faculty of Columbia University. A graduate of fundamentalist-oriented Wheaton College, Kulp had obtained the doctor's degree from Princeton University. His pioneering research at Columbia included the development of improved radiometric methods of determining geologic age of rocks and fossils. As a departmental colleague of his, I was closely acquainted with Kulp and knew of his fundamentalist background and active interests in religious education. Despite his strong early belief in the literal interpretation of Genesis and its recent-creation scenario, Kulp felt persuaded to adopt the new geochronometry and the general scheme of geologic evolution through an enormously long span of time. He proceeded to advise his fundamentalist associates to turn away from the teachings of George Price and his New Geology and to adopt current views of geology based on established physical and chemical laws (Kulp, 1950). Evidently Kulp's effective influence helped to hasten the loss of some of the more extreme fundamentalist ASA members.

Among those fundamentalists who refused to heed Kulp's admonition to "get with it" was Henry M. Morris, a member of the Southern Baptist denomination and a strong believer in recent creation. He went along with Price's interpretation of geology and the Flood of Noah. As a teacher of civil engineering at Rice Institute he began to apply his knowledge of mathematics and physics to issues of creation versus evolution (Numbers, 1982, p. 541). He became particularly interested in the numerical probability that complex organisms could develop by pure chance, something that seemed to be a requirement of evolutionary biology. His first book on creation science appeared in 1946 under the title That You Might Believe (Morris, 1946; 1978). Late in the 1940s he joined the ASA, but attracted little notice until after he joined forces with fundamentalist theologian John C. Whitcomb, Jr., to publish in 1961 a book titled The Genesis Flood (Whitcomb and Morris, 1961). As we shall see next, the timing of this creation science book was such that it reinforced a new wave of fundamentalist activity directed against intervention of organized scientists in the school science curriculum.

Scientists Formulate a New Science Curriculum

The launching in 1957 by the Soviet Union of Sputnik, the first human-made orbiting earth satellite, can be taken as a turning point in the American attitude toward science education (Nelkin, 1982, p. 39). Realization that the nation had fallen far behind in science came sharply to those of us lucky enough to spot that faint point of light moving across the October sky. We needed to catch up in space technology, and that need could only be met with a revised and beefed-up program of science education in the public schools.

In the same year, a group of physicists from Cambridge, Massachusetts (where Harvard University and the Massachusetts Institute of Technology are located), organized the Physical Science Study Committee (PSSC) with the purpose of developing teaching films for use in physics courses in the high schools. The project was funded by the National Science Foundation (NSF) and was to be the first of several major science-education projects extending into mathematics, chemistry, biology, and the social sciences. By 1975, fifty-three projects had been funded at a cost of over 100 million dollars (Nelkin, 1982, p. 41).

For the first time, scientists in groups participated in shaping the science curricula, including determination of goals and methodology and the actual writing of text materials and teaching aids. The student was to be cast in the active role of the investigator, with emphasis on innovation and discovery, but with de-emphasis on memorizing facts and solving numerical problems. The teaching materials were to be complete to the extent that the learning could be carried out fully without requiring or even allowing interference from teachers and the local school board and community. This policy was soon to bring sharp negative reaction from parents and the general public who viewed it as an arrogant display by scientists and an attempt to dictate a science curriculum at a national level, but at the same time through federal funding at the expense of the taxpayer.

In physics and chemistry the new curricula were not likely to clash with fundamentalist religion, but for the biology program--the Biological Sciences Curriculum Study (BSCS)--it was quite a different matter, because of the need to include the topic of organic evolution. Another difference was that, whereas physics and chemistry were elective high school courses with limited participation, the basic biology course was nearly everywhere a required science. The American Institute of Biological Sciences, which initiated the new curriculum study, found public-school teaching of biology sadly deficient in modern areas of biology such as genetics and evolution and lacking in emphasis on problem solving. New textbooks prepared by the BSCS appeared in 1963 and immediately were met with hostility in southern states and their local communities.

Stiff opposition to the BSCS in Texas reflected the favorite theme of the Protestant fundamentalists, which is that the decline in religious belief manifested in the teaching of evolution has led to widespread immorality and criminality among the youth of the nation. Nelkin refers to an almost incredible tie-in of teaching of evolution with

the assassination of President Kennedy in 1963 in Dallas (1982, p. 47), namely, that the godless individual who perpetrated the murder fired his bullets from the state textbook repository--which, of course, contained books teaching evolution. Are we to infer that if the repository had contained no such atheist-inspired textbooks Kennedy might be alive today? In Texas the BSCS texts were approved, but "only after several changes in the books that softened their evolutionary emphasis" (Nelkin, 1982, p. 47)[1]

Although the BSCS textbooks and supporting teaching aids were eventually used by about half of the American high schools, "the major problem facing BSCS was less a matter of social protest than the inertia of high school teachers, who often failed to understand the materials and the methods of science sufficiently to convey the character and use of evolution theory in biology" (Nelkin, 1982, p. 47).

Curriculum reform in the social sciences also ran into difficulty with the fundamentalists. A study program titled "Man: A Course of Study" (MACOS) attempted to focus attention on fundamental questions of human behavior, using an open-ended approach in which students were urged to explore values freely (Nelkin, 1982, p. 49). Topics for discussion included religion, reproduction, aggression, and murder--obviously an open invitation to attack by religious fundamentalists.

Another Resurgence of Fundamentalism

Throughout the period of growth of the curriculum projects sponsored by the NSF, all public school textbooks were under close scrutiny of various individuals and groups in many states; these "textbook watchers" were particularly critical of all science texts. Of all the textbook watchers perhaps the most widely renowned are Mel and Norma Gabler of Longview, Texas. The Gablers for years scrutinized all new textbooks offered for use in their state, noting inclusion not only of evolution, but also of any material that might seem to have a negative effect on the morals of young people. They formed a nonprofit supporting organization to fight in various ways the battle for the minds of millions of students. I find it interesting that in 1973 the Gablers attempted to introduce the teaching of both evolution and recent creation in the Texas schools, with both to be elective subjects. I presume it was their plan that evolution would at the same time be removed from required science courses. This may be one of the first appearances of the "let's-make-a-deal" strategy of the creationists, which reads: "If you won't throw evolution out, at least let us come in on an equal-time basis." The Gabler proposal was rejected, but they were able to have introduced by legal amendment the requirement that textbooks "identify evolution as only one of several possible explanations of origins, and to clarify that this treatment is theoretical rather than factually verifiable" (Nelkin, 1982, p. 65).

On a national scale, efforts to reduce federal influence on public school curricula, to eliminate teaching of evolution or give creationism equal time, and in general to restore the teaching of fundamentalist morality instead of secular humanism were promoted throughout the 1960s by conservative organizations such as the Heritage Foundation and the Council for Basic Education. More recently, the emergence of the New Right, the Moral Majority, the Christian Broadcasting Network, and the Eagle Forum, using the mails and electronic evangelism, has set off an outright ideological war against the "Godless minority." According to Nelkin:

> The war focuses on diverse issues--among them are reinstating prayer in the schools, blocking the Equal Rights Amendment, prohibiting abortion,

banning books, influencing elections, and removing the civil rights of homosexuals. The educational system, as a source of values, is a critical target. (P. 67)[1]

The New Right's major target is secular humanism, which strongly endorses science and the scientific method, and along with it the theory of organic evolution through geologic time.

With this general review of the ups and downs of religious fundamentalism over the past four centuries we need to turn to a relatively new vehicle of fundamentalism that specializes in attempting to apply the scientific method to support their doctrine of recent and sudden creation of the universe. If biblical creation can be made respectable as science, it can bypass the exclusionary church/state principle and gain legitimate entrance to the public schools.

The Modern Creation Scientists

Creation scientists are those persons of Christian or Jewish faith who are unreservedly committed to belief in recent, sudden creation of the universe as described in Genesis and to the literal meaning of everything in the Old Testament, but who are also scientists in one field of specialization or another, and who seek from empirical science information that either (a) supports the biblical scenario, or (b) tends to weaken or destroy the naturalistic theory of organic evolution through geologic time. Let us assume that the adjective "scientific" describes the professional status of the individuals themselves, and in no way suggests that recent creation is a scientific hypothesis. If the latter were intended, the title of "scientific creationist" would be a contradiction in terms.

One of the first organizations of scientific creationists to appear in the United States has already been mentioned. It is the American Scientific Affiliation (ASA) (Nelkin, 1982, pp. 77-78). Formed in 1941, it seems to have confined its attention to pointing out supposed inadequacies and weaknesses in the theory of evolution and in protesting the manner in which that subject was being taught, i.e., in a dogmatic or authoritarian way. The ASA advocated the teaching of alternative views of creation, but not necessarily as part of the biology course. This policy came to be regarded by some of its members as overly moderate for even suggesting that evolution should continue to be taught.

As a result of this internal difference in opinion, some members of the ASA defected in 1963, forming with others the Creation Research Society (CRS) of Ann Arbor, Michigan. The objective of the CRS, as stated in their articles of incorporation, is "to publish research evidence supporting the thesis that the material universe, including plants, animals, and man are the result of direct creative acts by a personal God" (Nelkin, 1982, p. 78). Voting members of the CRS must have a postgraduate degree in science and must believe in the literal truth of the Bible. Upon application they sign a statement asserting that the Bible is the Word of God, its content historically and scientifically true, and that all life forms on earth were directly created by God, according to Genesis, and that the Flood of Noah was an actual event. It is essential for anyone studying the phenomenon of creation science to take note that this statement of belief, and others like it, constitutes the basic "scientific hypothesis" of creation science; but, of course, it can have no such status because (a) it cannot be modified or altered in the light of evidence, (b) it cannot be tested, and (c) it cannot be falsified by any evidence obtained by empirical science. Furthermore, because it invokes supernatural forces as the causative agent, it offers no explanation amenable to examination in the context of laws of empirical science.

An important product of CRS activity, so far as the school curriculum is concerned, was the publication in 1970 of a biology textbook titled Biology--A Search for Order in Complexity (Moore and Slusher, 1970). It was "prepared by the Textbook Committee of the Creation Research Society" (from title page) and its two editors were John N. Moore and Harold S. Slusher. Moore holds degrees of master of science (M.S.) and doctor of education (Ed.D.) and at the time was a professor in the Department of Natural Science of Michigan State University, East Lansing. Slusher, with the M.S. degree, was Professor of Geophysics and Astronomy and Director of Kidd Memorial Seismic Observatory, University of Texas at El Paso. Although Slusher's title indicates a specialization in seismology, he is not listed in the 1984 Directory of Members of the American Geophysical Union, the leading scientific organization that encompasses all aspects of earth physics. In addition to the above editors is a list of seventeen persons , who presumably are members of the textbook committee (p. xvi). The list includes several biologists holding the Ph.D. degree.

The CRS textbook was revised in 1974 and went through its eighth printing in 1981. Although the book outwardly resembles a typical high school biology text, the words "Creator" and "God" appear in a number of places in presenting the recent-creation view. This is a crucial point because the most recent legal strategy of the creationists is to avoid all mention of religion or religious entities in presenting recent-creation in the schools. Nevertheless, they cannot escape such published records as this of their religious doctrine and the religious basis of their view of science. If they meant it in the 1970s, they cannot protest that in the 1980s they do not mean it. They cannot simply say, "We take it all back; we didn't really mean it."

Creation Science Goes Academic

The CRS was, in turn, faced with internal dissension and schism that followed when, in 1970, several of its members formed the Creation Science Research Center (CSRC) of San Diego, California. Their stated purpose was "to reach the 63 million children in the United States with the scientific teaching of Biblical creationism" (CSRC in Nelkin, 1982, p. 79). The research program of the CSRC is intended "to clarify problems in the field of geophysics, oceanography and structural geology as well as Biblical and geological chronology." The organization strongly opposes the use of federal funds in public school teaching and in production of evolution-biased textbooks. Schism again occurred, this time in 1972, and some members of the CSRC formed the Institute for Creation Research (ICR), also located in San Diego.

The ICR began operation as the research arm of Christian Heritage College, founded in 1970 under the auspices of an independent Baptist church. The college has both undergraduate and graduate programs. Its science faculty is dedicated to opposing materialistic organic evolution. According to the 1974 catalog of the college, as quoted in Nelkin, each faculty member, both at the time of appointment and annually thereafter, must swear to agreement with the following statement:

> We believe in the absolute integrity of Holy Scripture and its plenary verbal inspiration by the Holy Spirit as originally written by men prepared for God for this purpose. The scriptures, both Old and New Testament, are inerrant in relation to any subject with which they deal, and are to be accepted in their natural and intended sense . . . all things in the universe were created and made by God in the six days of special creation described in Genesis. The creationist account is accepted as

factual, historical and perspicuous and is thus fundamental in the understanding of every fact and phenomenon in the created universe. (Nelkin, 1982, p. 81)[1]

Heading the ICR's academic program are Henry M. Morris, Ph.D., president and director, and Duane T. Gish, Ph.D., vice president and associate director. The ICR's first graduate school catalog appeared in 1981, offering the M.S. degree in science and science education. The list of faculty included nine individuals besides Morris and Gish, all holding doctorate degrees. One is a legal expert, Wendell R. Bird, with the J.D. degree from Yale University Law School. Scientific specialties represented include biology, geology, physics, astrophysics, biochemistry, hydrology, and hydraulics. A series of technical monographs, authored by the faculty and other ICR scientists, represents a serious attempt to support the recency of special creation and to vitiate conventional hypotheses of evolution through the multibillion-year schedule of geologic time.

For general consumption the ICR publishes a series of numbered pamphlets titled Impact: Vital Articles on Science/Creation. Through January of 1987, 163 Impact units had been published. They contain, collectively, all of the esssential arguments and evidence relating to creation science topics with which the ICR has involved itself, including its views on education and secular humanism. Promotional activities of the Institute for Creation Research, including descriptions of debates in which its members have participated, are covered in another set of pamphlets, titled Acts and Facts, also regularly issued. Numerous book titles and teaching aids, directed at all levels of learning and interest, are distributed through its marketing division, Master Books (P.O. Box 1606, El Cajon, California, 92022). The ICR products for popular consumption are attractive, highly professional in design, and moderately priced.

A new ICR headquarters building, located in Santee (near San Diego), was opened in 1986. It houses the Museum of Creation and Earth History and the ICR Research Library, as well as science laboratories and an education center. An interesting field project of the ICR is a continuing search for remains of Noah's Ark high on the slopes of Mount Ararat. A number of expeditions have been mounted by ICR under the direction of John D.Morris, Ph.D., the institute's director of field research. The Flood of Noah figures strongly in the creation science scenario because of the sedimentary deposits the flood left and the incorporation of animal and plant remains in those strata. Another project of special interest to the ICR is the investigation of alleged "mantracks" next to dinosaur tracks in strata exposed in the bed of the Paluxy River in Texas.

A number of other antievolution organizations and educational programs are in operation today, and these are reviewed by Nelkin (1982, pp. 83-84). The Genesis School of Graduate Studies in Gainesville, Florida, offers a Ph.D. in science-creation research. A few creationist courses are taught in isolation in the secular colleges and universities. The Bible Science Association of Caldwell, Idaho, has an active program of public education by mail and the electronic media. Other organizations of more limited scope within individual states support the work of the Bible Science Association.

The creation science movement has also appeared in England, where the Creation Science Movement (formerly, the Evolution Protest Movement) and the Biblical Creation Society are currently active. A third, the Newton Scientific Organization, has branches in Australia, New Zealand, and South Africa. In 1978, a coalition calling itself Creation Science Australia (CSA) began publication of Ex Nihilo, a quarterly journal for popular presentations of creation science. In 1985 the journal title was changed

to Creation Ex Nihilo. The member group behind CSA is the Creation Science Association, organized by Carl Weiland. An important feature of Creation Ex Nihilo is its series of technical articles, written in scientific journal style by certified scientists. In 1984, there first appeared the Ex Nihilo Technical Journal, devoted almost entirely to articles at the professional science level and including on occasion counterarticles by mainstream scientists, along with rebuttals by the creation scientists. This journal is clearly a strong bid for leadership in creation science on an international level.

Academic Affiliations of Creation Scientists

Some interesting observations on the qualifications and viewpoints of the principal leaders and participating scientists of the ICR and other creation science groups have been offered by Nelkin (1982, pp. 84-87). "Many of the creationists are from the applied sciences and engineering" (p. 86). For example, Henry Morris has served as a professor of hydraulic engineering and as chairman of a civil engineering department (Virginia Polytechnic Institute). Duane Gish, with a Ph.D. in biochemistry, was formerly employed as a member of the research staff of Upjohn and Company. Rocket scientist Wernher von Braun expressed belief in the guidance of a Creator and approved of teaching of alternate theories of the origin of the universe and life forms (Nelkin, 1982, p. 86). Three astronauts (Irwin, Borman, Mitchell) have supported the creationist scenario.

Much, if not all, of applied science, including engineering and technology, can be practiced with a narrow framework of specialization that does not come into relevant contacts of any kind with religious beliefs or, for that matter, with any intellectual activities. A person who designs dams, or computers, or aircraft is free to pick and choose from what the arts and humanities have to offer and to place those objects in a different mental compartment from that of the daily job. This dualism allows the applied scientist to carry on two independent thought programs, one that adheres to laws, principles, and constants derived from empirical science, and another that deals in a body of supernatural or transnatural concepts. The two realms are completely isolated and well insulated from one another, allowing each to be cultivated with dedication and fervor and in total unawareness of any logical conflicts between the two types of systems.

In contrast, the evolutionary biologist and paleontologist (often one and the same individual) focus on timebound knowledge with its inquiry into the origin of living matter, the origin of species, and questions of natural teleology. Scientific hypotheses about such questions are, from their very inception, in conflict with theistic religion of the Hebrew/Christian tradition as interpreted literally from its scriptures. For all but a very small handful of such scientists holding two flatly contradictory views in equal epistemological status, knowing that one resides in the naturalistic realm and the other in the transnatural (supernatural) realm is simply a logical impossibility.

Whether my interpretation is reasonable or not, so far as I can ascertain from my perusal of a highly diverse literature on creation science by many authors, most staff members or advisory board members of the creation science organizations have not achieved international status in the scientific community in basic research in such fields as biology, geology, paleontology, or geochemistry, which are the fields most closely tied in with naturalistic evolution. The creationists' lot has been to attack evolutionary theory from positions outside the specialistic community that is their target. Thus the attack upon evolutionary theory comes not from scientific peers, but from those unfamiliar through advanced research with the subject being attacked. The result is that scholarly arguments attempting to overthrow naturalistic evolution in favor of recent creation do not (or rarely) appear in the journals of the scientific community. Instead, the attack of the creationists upon evolution is routed through popular publications, such as independent magazines of general reader interest, electronic media presentations by both the news organizations and the religious fundamentalists themselves, and through privately financed books and tracts. The creation scientists, unable to find acceptability for their output within the science community, take their argument instead to those in the poorest position to evaluate it--the general public, and especially its young people. This is, of course, the situation we find in all pseudoscience, and I have dealt with it in Part 1 in my analysis of such pseudoscience cults as the Velikovsky and von Däniken scenarios and UFOs guided by extraterrestrial intelligence.

Credits

1. From Dorothy Nelkin, The Creation Controversy; Science or Scripture in the Schools, W. W. Norton & Company, New York. Copyright © 1982 by Dorothy Nelkin. Used by permission of author and publisher.

2. From Charles Schuchert and Carl O. Dunbar, A Textbook of Geology, Part II--Historical Geology, Third Edition, John Wiley & Sons, Inc., New York. Copyright, 1933, by Charles Schuchert and Carl O. Dunbar. Used by permission of the publisher.

Chapter 11

The Tenets of Creation Science

There seems little question that the fundamentalist creationists' attack upon naturalistic evolution through geologic time is under the firm command of the Institute for Creation Research (ICR): they account for most of the currently published materials and oral debates in the United States. Thus, at the outset of my inquiry into the philosophical and scientific issues in the debate, it seems appropriate to quote exactly the ICR text describing the tenets of scientific creationism.

Tenets of the Institute for Creation Research

The educational philosophy of the ICR is clearly presented in the <u>Graduate</u> <u>School</u> <u>Catalog</u> of that institute (Institute for Creation Research, 1981-82, pp. 12-15; see also Morris, 1984b, pp. 361-65). In quoting verbatim, we avoid the risk of misstating the ICR's position. We also have specific targets on which to focus a point-by-point analysis. With the scientific tenets we include a list of ICR tenets of biblical creation. This section of the ICR catalog begins with the following statements (starting with the second sentence of the text):

> The Institute for Creation Research bases its educational philosophy on the foundational truth of a personal Creator-God and His authoritative and unique revelation of truth in the Bible, both Old and New Testaments.

> This perspective differs from the evolutionary humanistic philosophy which has dominated most educational institutions for the past century, providing the more satisfying and meaningful structure of a consistently creationist and Biblical framework, and placing the real facts of science and history in the best context for effective future research and application.

More explicitly, the administration and faculty of ICR are committed to the tenets of both scientific creationism and Biblical creationism as formulated below. A clear distinction is drawn betwen scientific creationism and Biblical creationism but it is the position of the Institute that the two are compatible and that all genuine facts of science support the Bible. ICR maintains that only scientific creationism should be taught in tax-supported institutions, but that both should be taught in Christian schools.

Tenets of Scientific Creationism

1. The physical universe of space, time, matter and energy has not always existed, but was super-naturally created by a transcendent personal Creator who alone has existed from eternity.

2. The phenomenon of biological life did not develop by natural processes from inanimate systems but was specially and supernaturally created by the Creator.

3. Each of the major kinds of plants and animals was created functionally complete from the beginning and did not evolve from some other kind of organism. Changes in basic kinds since their first creation are limited to "horizontal" changes (variations) within the kinds, or "downward" changes (e.g., harmful mutations, extinctions).

4. The first human beings did not evolve from an animal ancestry, but were specially created in fully human form from the start. Furthermore, the "spiritual" nature of man (self-image, moral consciousness, abstract reasoning, language, will, religious nature, etc.) is itself a supernaturally created entity distinct from mere biological life.

5. Earth pre-history, as preserved especially in the crustal rocks and fossil deposits, is primarily a record of catastrophic intensities of natural processes, operating largely within uniform natural laws, rather than one of uniformitarian process rates. There is therefore adequate reason for investigating the many scientific evidences for a relatively recent creation of the earth and the universe, in addition to the strong scientific evidences that most of the earth's fossiliferous sediments were formed in an even more recent global hydraulic cataclysm.

6. Processes today operate primarily within fixed natural laws and relatively uniform process rates but, since these were themselves originally created and are daily maintained by their Creator, there is always the possibility of miraculous intervention in these laws or processes by their Creator. Evidences for such intervention should be scrutinized critically, however, because there must be clear and adequate reason for any such action on the part of the Creator.

7. The universe and life have somehow been impaired since the completion of creation, so that imperfections in structure, disease, aging, extinctions and other such phenomena are the result of "negative" changes in properties and processes occurring in an originally perfect created order.

8. Since the universe and its primary components were created perfect for their purposes in the beginning by a competent and volitional Creator, and since the Creator does remain active in this now-decaying creation, there do exist ultimate

purposes and meanings in the universe. Teleological considerations, therefore, are appropriate in scientific studies whenever they are consistent with the actual data of observation, and it is reasonable to assume that the creation presently awaits the consummation of the Creator's purpose.

9. Although people are finite and scientific data concerning origins are always circumstantial and incomplete, the human mind (if open to the possibility of creation) is able to explore the manifestations of that Creator rationally and scientifically, and to reach an intelligent decision regarding one's place in the Creator's plan.

Tenets of Biblical Creationism

1. The Creator of the universe is a triune God-- Father, Son and Holy Spirit. There is only one eternal and transcendent God, the source of all being and meaning, and He exists in three Persons, each of whom participated in the work of creation.

2. The Bible, consisting of the thirty-nine canonical books of the Old Testament and the twenty-seven canonical books of the New Testament, is the divinely-inspired revelation of the Creator to man. Its unique, plenary, verbal inspiration guarantees that these writings, as originally and miraculously given, are infallible and completely authoritative on all matters with which they deal, free from error of any sort, scientific and historical as well as moral and theological.

3. All things in the universe were created and made by God in the six literal days of the creation week described in Genesis 1:1-2:3, and confirmed in Exodus 20:8-11. The creation record is factual, historical and perspicuous; thus all theories of origins or development which involve evolution in any form are false. All things which now exist are sustained and ordered by God's providential care. However, a part of the spiritual creation, Satan and his angels, rebelled against God after the creation and are attempting to thwart His divine purposes in creation.

4. The first human beings, Adam and Eve, were specially created by God, and all other men and women are their descendants. In Adam, mankind was instructed to exercise "dominion" over all other created organisms, and over the earth itself (an implicit commission for true science, technology, commerce, fine art and education) but the temptation by Satan and the entrance of sin brought God's curse on that dominion and on mankind, culminating in death and separation from God as the natural and proper consequence.

5. The Biblical record of primeval earth history in Genesis 1-11 is fully historical and perspicuous, including the creation and fall of man, the curse on the creation and its subjection to the bondage of decay, the promised Redeemer, the worldwide cataclysmic deluge in the days of Noah, the post-diluvian renewal of man's commission to subdue the earth (now augmented by the institution of human government) and the origin of nations and languages at the tower of Babel.

(Note: Items 6 and 7 have been deleted because they deal with theological subjects such as sin, redemption, resurrection, forgiveness of sins, and eternal life. These have no bearing on the creation scenario.)

Comparison with the Views of Mainstream Science

It may be helpful next to abstract from each of the nine "tenets of scientific creationism" the essence of the statement and compare it with the largely prevailing equivalent hypothetical statement of mainstream science. My intention here is to apply principles of philosophy of science, covered in Part 1, to the creationists' statements, leaving the science content itself for later discussion.

1. Supernatural creation of the universe ex nihilo was done by a Creator. Modern scientific cosmology, to the extent that it is completely materialistic or mechanistic, must simply ignore the question of existence of a supernatural Creator, but it is obliged to express an opinion about the coming into existence of something (energy, matter, space, time) from absolutely nothing. That suggestion could be put forward as a purely materialistic hypothesis in itself, but could serve no useful purpose when we realize that evidence in science requires that something exist to be observed. If, then, we postulate that prior to some initial point in time nothing existed, it follows that neither could evidence of nothing exist, so we cannot prove the existence of nothing. The scientific absurdity of the ex nihilo hypothesis is obvious through logic.

Philosopher Peter A. Angeles has thoroughly discussed the problem of creation ex nihilo (1980, pp. 45-71). He notes that theologians have argued that the Creator's logic may be different from ours but, upon examination, this ploy proves only to be a definitional self-contra-diction. Angeles shows that because creation ex nihilo is a contradiction in terms, it "would be a logical impossibility even in God's Logic or any reality. Self-contradictions have no existence in reality. Self-contradictions can only appear in the realm of language or definitions which is the only possible place for them since they are meanings that exclude any possibility of their being the case in external reality" (p. 64).[1]

An alternative suggestion that at least allows a possibility of rational speculation about reality is that the total content of the universe, whether it be in the form of pure energy, or of some combination of matter and energy, has remained more or less constant far back into time, with time approaching negative infinity. This assumption allows an evolving (changing) universe to be projected back to a point in time when the universal package of matter and energy existed in an extremely dense state and occupied an extremely small volume.

Under what is currently the favored cosmic scenario of events from an arbitrary initial time-zero--the Big Bang hypothesis--the initial package underwent explosive expansion to the accompaniment of a succession of changes in the forms assumed by energy and matter, ultimately leading to the formation of elements in a dispersed state and their subsequent aggregation into denser states that we see today in the form of clouds of gases and dust, galaxies, and stars. Scientific speculation can even be carried back into time before the singularity that marked the start of the present universe. This speculation is to the effect that the dense state of the singularity was itself the end point in the condensation of a previous universe from its earlier state of greatest dispersion. Thus the evolutionary process can be imagined as cyclic, with endlessly repeated cycles. The oscillation between diffuse and dense states of matter has analogs in observable phenomena, such as star formation by condensation ending in explosive destruction to produce supernovae and a new body of gas and dust. This imaginary view of a universe without beginning or end can be formulated so as to conform with existing laws of conservation and states of energy and matter.

The point to be made is simply this: The statement of divine creation ex nihilo is merely an assertion of a singular event and does not qualify as a scientific hypothesis because it neither describes how that creation was achieved nor explains how creation took place in terms of the application of laws and principles of empirical science. Thus tenet number 1 has no scientific content or merit whatsoever.

2. The initiation of life (correctly termed biopoesis) is a supernatural act of the Creator. Empirical science has formulated hypotheses, general and speculative as they may be, for the transition in a naturalistic manner from inorganic molecules to organic molecules of increasing complexity, and for the organization of those complex molecules into living systems capable of reproduction. Unquestionably, these hypotheses, which deal with historical events that cannot be replicated in the laboratory and cannot be observed today through natural replication, enjoy poor status in terms of quality, but they invoke only forms of matter and energy observed to exist now, and they postulate only physical and chemical processes observed in action today in conformance with laws accepted as universal. In this respect, the scientific speculations about biopoesis at least have scientific content, whereas the supernatural agent and acts proposed by scientific creationism have no scientific status whatsoever.

3. Each "kind" of organism was separately created from nothing by the Creator in complete form as it exists today. Denial of evolution of one kind from another is a logical corollary of that statement. According to creation science, changes occurring in any kind since its creation are genetic variations of minor consequence, not leading to formation of new kinds. The term "kind" is of biblical origin (Gen. 1:21,25) and has no place in scientific biology; the term is ambiguously used by creationists and its intended meaning is not clear (see Chapter 37). It will be quite unnecessary to repeat here that the assertion of creation of kinds from nothing by a Creator has no merit as a scientific hypothesis.

Empirical science in the form of the hypothesis of organic evolution traces the development of all life-forms continuously (but not uniformly in rate) from the initial life-form of the event of biopoesis to the present time. The thread, or chain, of life is postulated to have remained unbroken from biopoesis to the present time for forms now living, but to have ended in many of its branches as extinctions. Many consequences can be deduced from that hypothesis, and all are capable of being tested and of possibly being falsified. Supernatural separate creation--or any creation--by a Creator is untreatable by the scientific method.

4. Supernatural creation by the Creator of the human species was a special event. This is merely a special case of the assertion of separate creation of every species. The hypothesis of organic evolution provides for the evolution of primates, hominoids, and hominids in exactly the same manner as for all other species. Comments are the same as for tenets 2 and 3.

5. (a) Natural catastrophism was the dominant geologic activity in producing the earth's rocks and fossils, and included a great flood of more recent date. (b) Creation of the earth and universe took place at a recent date. Points (a) and (b) require individual and separate consideration, since neither requires the other. As to point (a), the substitution of a naturalistic catastrophism for uniformitarianism can be entertained as a scientific hypothesis. The creationists' concept of both catastrophism and uniformitarianism is sadly archaic and does not correspond with modern geologic thought. This topic will be developed in depth on later pages. As to point (b), the meaning of the word "recent" must be sought for in other creationist writing. It is a relative time concept and is not acceptable in scientific writing unless a specific point in time is identified with respect to which some event can be described as being more (or less) recent. When spelled with a capital "R," Recent time has a generally accepted definition in geology as synonymous

with Holocene time, the latest of the Cenozoic epochs. The Holocene Epoch follows the Pleistocene Epoch, which ended approximately 10,000 years ago with the rapid disappearance of the great continental ice sheets.

The ICR elsewhere expresses its philosophical position with respect to a question we examined in depth in Part 1: Does historical (timebound) information fall within the realm of empirical science? Recall that the argument on the negative side is to the effect that events of cosmic and geologic time cannot be repeated for scientific examination, cannot produce predictions, and are therefore untestable and hence not falsifiable. I think we showed in Part 1 that logical prediction is possible for such past events as geologic history and organic evolution, and that the criteria of empirical science are fulfilled.

In the ICR's textbook, Scientific Creationism, Public School Edition (Morris, 1974a) we read: "It should be remembered, however, that real history is available for only the past few thousand years" (p. 131). The text continues: "To keep this problem in its proper perspective, one should remember that no one can possibly know what happened before there were people to observe and record what happened. Science means 'knowledge' and the essence of the scientific method is experimental observation" (p. 131). Clearly, this is a philosophical position with which the modern natural sciences--astronomy, geology, and biology--take issue; they simply could not function on the basis that no knowledge is possible of what occurred before there were humans to observe and record what actually happened.

To show the logical impossibility of the creationists' view of science, let us apply their provision to the Genesis creation scenario. The first human, Adam, appeared late in the sixth day of creation, after all else had been created. Assuming that Adam was endowed by the Creator with the ability to record what he, Adam, could observe, the instant of Adam's creation must be the earliest time for which any scientific knowledge could be obtainable. It follows that all earlier events of creation are incapable of being known to humans and are not within the domain of science. Well, then, does not the creationists' own view of science effectively dispose of nearly all of creation science? What is left as creation science, except human history? Answer: "Nothing."

In all fairness, and playing by the standards of the scientific community, we are bound to forgive the creationists for their logical self-immolation and proceed on the basis that empirical science can explore pre-human history, propose historical hypotheses of merit, make logical predictions from those hypotheses, and test them accordingly. Having given back to the creationists the dueling foil that so unceremoniously flew out of hand, we can resume the argument.

In the ICR textbook we find this statement:

> As a matter of fact, the creation model does not, in its basic form, require a short time scale. It merely assumes a period of special creation sometime in the past, without necessarily stating when that was. On the other hand, the evolution model does require a long time scale. The creation model is thus free to consider the evidence on its own merits, whereas the evolution model is forced to reject all evidence that favors a short time scale.
>
> Although the creation model is not necessarily linked to a short time scale, as the evolution model is to a long scale, it is true that it does fit more naturally in a short chronology. Assuming the Creator had a purpose in His creation, and that purpose centered primarily in man, it does seem more appropriate that He would not waste aeons of time in essentially meaningless caretaking of an incomplete stage or stages of His intended creative work. (P. 136)

Disregarding the fallacious logic expressed in the first paragraph, these statements cast doubt that recency of creation is really a tenet of scientific creationism. First the author offers to abandon the recency requirement, then reads the Creator's mind to come up with the necessity of retaining the requirement. I had never before realized that the Creator was so rational and pragmatic!

The ICR's textbook states: "According to the evolution model, man has been on earth for at least a million years, whereas the creation model postulates probably only a few thousand years, corresponding to the approximately 4000-5000 years of recorded history" (p. 167). Because the creation of the human species occurred almost simultaneously with all other species and the entire universe (six days of mean solar time), we can perhaps safely assume that the creationists' hypothesis intended in point (b) of tenet number 5 should read: creation of the universe and earth occurred somewhere in the time span of 4,000 to 10,000 years before present. To make this statement acceptable as a scientific hypothesis, we would need to eliminate the words "creation of" (previously asserted to be the work of a Creator) and substitute the words "initial event in formation of." In that form, the hypothesis can be tested in competition with the prevailing hypothesis of an age of 10 to 20 billion years for the present universe and 4.5 to 4.7 billion years for the earth as a solid planet. This is a topic requiring full discussion at a later point.

6. The Creator can intervene at any time to work a miracle, in which laws of nature are suspended or modified by the Creator to accomplish the miracle. This tenet is religion, pure and simple; it has no place in a discussion of science.

7. The Creator made everything in the universe perfect, but since then it has been deteriorating in quality. The "imperfections" cited here are given a negative value. The creationists evidently view aging of an individual and affliction with a disease as bad or evil in the moral sense. It scarcely needs pointing out that such value judgments are outside the bounds of science. This creationist viewpoint is clearly religious, not scientific. In pure science, aging is viewed as a natural process, neither good nor evil, when studied as a biological or psychological phenomenon. This tenet should be eliminated or transferred to the list of religious tenets.

8. The Creator designed the universe and everything in it with an ultimate purpose or destiny in view. This, too, is a religious concept and has no scientific meaning. As we found in Part 1, natural teleology is recognized in evolution to the extent that the changes within organisms provide improved means of coping with the environment, but in no sense does a genetic change come about in response to conscious formulation based on ability of the individual of a species to perceive a goal. This creationist tenet, too, should be dropped from our further consideration.

9. If we use the brains given to us by the Creator we will adopt the doctrine of recent supernatural creation. No comment is necessary.

After eliminating from the stated tenets of scientific creation those assertions that are outside the scope of empirical science because they invoke a supernatural agent, we are left with very little to examine. We most certainly will not entertain as a scientific hypothesis the flat claim that someone else's hypothesis is erroneous. We need entertain only a positive suggestion that something exists or occurs in a certain form or manner vis-à-vis an alternative positive suggestion. Falsification must be demonstrated as the product of testing a hypothesis, not merely asserted without evidence. On that basis, scientific creationism offers precious little for science to debate.

I suppose that we could simply drop all mention of the role of a Creator or any other supernatural agent and merely assert that suddenly, not too long ago, everything in the universe came to exist, whereas prior to that instant in time nothing existed. In that case the assertion would fail as a scientific hypothesis simply because it does not describe what happened, how it happened, or what caused it all; nor could such a bare assertion lead to logical predictions. After all, we have general agreement that the function of science is to describe, to explain, and to produce logical predictions. If an assertion performs none of these functions, it is simply not a scientific hypothesis.

Why not stop here? Is there any point in going on? Unfortunately, the advocates of creation science, as with purveyors of all brands of pseudoscience, devote their major efforts to negative forms of activity. Promoters of pseudoscience declare that rival hypotheses arising from within the scientific community, where they are scrutinized by peer scientists and tested in every conceivable way, are in error. Pseudoscience promoters then claim that the scientists cannot prove the validity of their hypotheses, and therefore those hypotheses are false. The purveyors of pseudoscience are well aware of the method of multiple hypotheses and the continual debating that goes on within science to find and expose weaknesses in rival hypotheses. But, instead of recognizing this adversary process as one that strengthens the body of science, they use it only as an indication of the inadequacy and weakness of a scientific hypothesis they despise. Quoting out of context, they try to show that a particular scientist "does not believe in" evolution--or the radiometric method of dating rocks, or whatever the hypothesis is about. They interpret the meanings of words to suit themselves, and they refer to hypotheses long ago discarded as if these were current fare in order to place their own neat scenario upon a contrasting background of uncertainty and confusion.

It is because some creationists use these debating tactics and direct their expressions to school teachers and to school children and their parents that the creationists' arguments must be investigated and exposed, and that the scientists' own point of view be also directed to the same audience at an appropriate level. Of course, as you know, I am not writing for that level of audience, but perhaps some of you can, in turn, deal with problems at that popular level in the schools and in law-making bodies.

Are Origins Off Limits?

In the area of science philosophy, the creationists make other assertions that deserve closest scrutiny. One of these relates to what creationists call "origins." They seem to be referring to the initial appearance of a unique structure of matter such as the universe, a star, a planet, a species. Of course, the creationists' doctrine is that all such initial appearances are the work of the Creator. They state flatly that scientific proof of such "origins" is impossible. This topic is presented early in the content of the ICR textbook, Scientific Creationism (Morris, 1974a, pp. 4-5). Both naturalistic evolution and supernatural creation are stated to deal with "origins," and it is considered of vital importance to study the subject of "origins" (p. 4). "At the same time," they state, "it must be emphasized that it is impossible to prove scientifically any particular concept of origins to be true" (p. 4). The text continues as follows:

> This is obvious from the fact that the essence of the scientific method is experimental observation and repeatability. A scientific investigator, be he ever so resourceful and brilliant, can neither observe nor repeat origins!

> This means that, though it is important to have a philosophy of origins, it can only be achieved by faith, not by sight. That is no argument against it, however. Every step we take in life is a step of faith. Even the pragmatist who insists he will only

believe what he can see, believes that his pragmatism is the best philosophy, though he can't prove it! He also believes in invisible atoms and in such abstractions as the future. As a matter of observation, belief in something is necessary for true mental health. A philosophy of life is a philosophy, not a scientific experiment. A life based on the whim of the moment, with no rationale, is "a tale told by an idiot, full of sound and fury, signifying nothing."

Thus, one must believe, at least with respect to ultimate origins. However, for optimally beneficial application of that belief, his faith should be a reasoned faith, not a credulous faith or a prescribed faith. (Pp. 4-5)

Let us start with the assertion that it is impossible to prove scientifically the truth of any concept of "origins." In discussing the nature of science in Part 1, I placed strong emphasis on a fundamental concept in empirical science: no proposition can be proved to be true or false. The most that can be done is to reduce to a small value (or increase to a large value) the probability that the proposition is false when declared true. As I explained earlier, the words "proof" and "true" apply only in the area of formal science, namely, logic and mathematics. Thus, for the creationists to say that no hypothesis of "origins" can be proven to be true simply does not apply to the natural and physical sciences. When applied to recent creation by a Creator, which is a religious statement invoking the supernatural, science and the scientific method are simply not relevant.

Next, notice the use of the words "believe" and "faith." Early in Part 1, I suggested that the word "belief" and the concept behind it (faith without evidence) have no place in any form of science, whether formal or empirical. "Belief" is appropriate only in matters and concepts that are in either a supernatural realm (in reference to religious concepts) or in a transnatural realm that involves moral or aesthetic values. The quoted ICR text is clearly an invitation to believe in a supernatural creation and purpose, not an appeal to reason. ("Reasoned faith" is another contradiction in terms.)

Science does concern itself with causes (i.e., with explanations) and that concern is legitimate. Historical geology and biology search for causes (explanations) of everything and anything material that can be examined today, especially explanations of the record of the past preserved as fossils and of the enclosing rock. That the study of events in past geologic time can be carried out in strict observance of the requirements and criteria of the scientific method is agreed on by most geologists and biologists. Even Sir Karl Popper, who in his earlier writings seemed to say that evolution was not a scientific hypothesis, later reversed his position to accept the prevailing modern view. (See below.) Historical geology and biology must consider the origin of everything observable. Every geologist who examines a particular rock asks: "What is the origin of this rock?" And so do the scientific creationists. When they look at the same rocks they search for the origins of those rocks--they postulate that the rocks originated in a great biblical Flood. They even attempt to prove that postulate by interpreting in their own way the features of the rocks and the fossils they contain. Creationists cannot have it both ways. They cannot first assert that origins cannot be proven and then go right out into the field in search of geologic evidence by which to prove the asserted origin. Such self-contradiction is an essential part of the modus operandi of creation science. Moreover, assertion simultaneously of contradictory statements can be a form of self-deception.

The ICR text continues with more details on creation and evolution as the two contrasting concepts of origins (Morris, 1974a, p. 5). It asserts that neither can be proved. Granting that special creation cannot be proved, or even rationally examined by science, we need to take a look at reasons offered by the creationists for the assertion that evolution cannot be proved.

"If evolution is taking place today, it operates too slowly to be measurable and therefore is outside the realm of empirical science" (p. 5). As already pointed out above, this argument is in line with the creationists' specious claim that nothing that happened before there were humans to observe the happening can be investigated by science, or even become a part of human knowledge. The argument about slowness of evolutionary change will be examined in our later discussion of scientific topics.

Evolution Viewed as Dogma

"Evolution is a dogma incapable of refutation" (Morris, 1974a, p. 6). The ICR textbook inserts two quotations, one from Paul Ehrlich and L. C. Birch (1967, p. 352), the other from Peter Medawar (1967, p. xi). Both statements reiterate an earlier position, held by Sir Karl Popper, that organic evolution cannot be tested by making predictions that have the possibility of bringing falsification upon the theory. In terms of temporal prediction, this may be true but, in terms of logical prediction, deductions are readily available with which to test hypotheses concerned with the mode and tempo of evolution. The same can be said for the general hypothesis of organic evolution through geologic time. We discussed this question in Part 1. For example, the entire evolution hypothesis could be effectively falsified, or at least thrown into serious disarray, by the finding of human fossils in company of, say, dinosaur fossils, in such a way that the contemporaneity of the two life forms is beyond reasonable doubt. This finding has not occurred to date, but it is logically possible.

Notice that the creationists, in each of the two 1967 citations, have selected statements expressing a philosophical position that has since been reversed by the philosopher responsible for it originally--Popper himself-- and that would not be shared by the majority of evolution scientists today. Passing off obsolete and discarded views and opinions as current fare is standard operating procedure of the creation scientists, as with purveyors of pseudoscience in general. Popper's change of position needs to be documented here, so that it cannot be said I offer it as mere hearsay. John R. Cole, writing on the subject of the use of misquotations by creationists, refers to this incident in detail (1981, pp. 43-44). (See also Kofahl, 1981, p. 873; Zeisel, 1981, p. 873.) He repeats an oft-quoted statement by Popper: "Darwinism is not a testable scientific theory, but a metaphysical research programme--a possible framework for testable scientific theories" (Popper, 1976, p. 168). In evaluating this statement, you need to know that the word metaphysical has two meanings in philosophy, and both are in common use. On our side of the Atlantic, metaphysics is usually associated with supernatural phenomena and beliefs; in Europe (where Popper did his lecturing and writing), it is rigorously used by philosophers to refer to basic philosophical questions "such as ontology, the body-mind problem, the problem of origins . . . the problem of natural versus supernatural," and "other questions of ultimate significance" (Ferm, 1936, p. 92). I think it would be a serious mistake to assume that Popper intended to mean that Darwin's theory lies in the supernatural realm. Creationists would like to use that meaning to support their claim that "evolution is religion."

Popper's change of position was given in published form: "I have changed my mind about the testability and logical status of the theory of natural selection; and I am glad to have an opportunity to make a recantation" (Popper, 1978, p. 344). (Be sure you understand the distinction between logical prediction and temporal prediction, as explained in Part 1. Even the best scientists and science-news reporters have confused and abused the meanings, as have the creationists. Historical hypotheses, such as organic evolution, are capable of producing logical predictions, whereas temporal predictions are, as a general rule, impossible. It is in the sense of logical prediction that Popper accepts the testability of Darwin's hypothesis.)

Creationists actually do look for evidences of human beings in the same place and time as dinosaurs. In the rocky bed of the Paluxy River in Texas, they find what they interpret as human footprints on the same rock surface that also bears numerous tracks of dinosaurs. Creationists make a big thing of the Paluxy tracks. In so doing, they attest to the falsifiability of the hypothesis of naturalistic evolution. Because geologists set the age of dinosaurs as limited to the Mesozoic Era (245 to 66 million years before present) and the first appearance of hominids at about 5 to 17 million years before present (at most), the Paluxy tracks, if genuinely human and if unquestionably contemporary with dinosaurs, would effectively serve to falsify the evolutionary tree of life in relation to relative ages of strata and determinations of absolute ages of those strata.

Again, I say, creationists cannot have it both ways. They cannot assert that naturalistic evolution is incapable of falsification through testing based on logical prediction, while at the same time they put forward what they consider to be empirical evidence that the hypothesis of evolution is proven false by those same criteria. To assert both positions simultaneously is self-contradictory; it may mislead students, their parents, and legislators unskilled in use of logic to detect self-contradiction.

"Evolution is an authoritarian system to be believed" (Morris, 1974a, p. 7). This assertion is followed by quotations from three authors who are or were evolution scientists. (Full references to the three quotations are contained in footnotes in the ICR textbook. I have not checked them out and I have no idea of the contexts in which they were originally made.) The first, by G. A. Kerkut (1960), includes this sentence: "It is premature, not to say arrogant, on our part if we make any dogmatic assertion as to the mode of evolution of the major branches of the animal kingdom." This word of caution, intended for Kerkut's scientific colleagues, seems reasonable enough. No scientific hypothesis should be put forward as a dogmatic assertion. On the other hand, supernatural creation by a Creator can only be asserted in a dogmatic manner. The second quotation is by D. Dwight Davis (1949): "But the facts of paleontology conform equally well with other interpretations . . . e.g., divine creation, etc., and paleontology by itself can neither prove nor refute such ideas." A good statement, it might be broadened to read that anything and everything that we can observe is fully accounted for by the doctrine of divine creation and that no proof of divine creation is possible by either philosophy or science. The third quotation is from Thomas Huxley (1903); it reads: "creation in the ordinary sense of the word, is perfectly conceivable. I find no difficulty in conceiving that, at some former period, this universe was not in existence; and that it made its appearance in six days . . . in consequence of the volition of some pre-existing Being." My comment would be: Neither do I find difficulty with conceiving of that scenario. Every adult human is possessed with adequate imagination to handle the story of six-day creation. Why should we be less able to imagine that creation story than those who imagined it originally?

Do Both Models Enjoy Equal Status as Science?

The ICR textbook continues by developing the idea of two models of origins enjoying equal status: "Neither evolution nor creation can be either confirmed or falsified scientifically" (Morris, 1974a, p. 9). "Furthermore, it is clear that neither evolution nor creation is, in the proper sense, either a scientific theory or a scientific hypothesis. . . . This is because neither can be tested" (p. 9). I have already gone over the science-philosophers' argument that naturalistic evolution through geologic time has all the necessary attributes of a scientific hypothesis, and that it is capable of being tested. I have also given the reasons for dismissing divine creation as a scientific hypothesis. What, then, is the debating strategy being pursued by the ICR in denouncing both models in the same terms? Apparently, their reasoning runs that, since it must be conceded that recent creation by a Creator has no chance whatever of being passed off as a scientific hypothesis, every attempt must be made to demote the evolution hypothesis to that same level of impotence. Perhaps they reason that, if neither is scientific hypothesis and neither can be tested or proved or falsified, then both deserve equal time and treatment in the classroom. My own conclusion from reasoning by the same token is that neither model would deserve any time in the classroom. The "what's-good-for-the-goose-is-good-for-the-gander" strategy is a mainstay of the creationists' attempts to secure equal time in the schools, and they seem quite willing to substitute "bad" for "good" in that saying, if only it will bring recognition of equality.

Having given up the possibility of gaining acceptance of both creation and evolution as scientific hypotheses, the ICR textbook shifts to another tack, namely, that even though the two scenarios are not in themselves scientific, they can be treated in a scientific manner. In this way, they can both be insinuated into the science classroom:

> All of these strictures do not mean, however, that we cannot discuss this question scientifically and objectively. Indeed, it is extremely important that we do so, if we are really to understand this vital question of origins and to arrive at a satisfactory basis for the faith we must ultimately exercise in one or the other. (Morris, 1974a, p. 9)

Apparently, the authors had forgotten their earlier assertion that the question of origins is beyond the limits of understanding (see pp. 4-5). Understanding as a basis of faith seems to be another one of those contradictions in terms. Is not "faith" simply "belief without understanding"? To understand is to know how something works. With understanding in hand, there is no need for faith.

If creation and evolution are not scientific hypotheses, what can they be called in order to justify their treatment by the scientific method? The answer: both are models:

> A more proper approach is to think in terms of two scientific models, the evolution model and the creation model. A "model" is a conceptual framework, an orderly system of thought, within which one tries to correlate observable data, and even to predict data. When alternative models exist, they can be compared as to their respective capacities for correlating such data. When, as in this case, neither can be proved, the decision between the two cannot be solely objective. Normally, in such a case, the model which correlates the greater number of data, with the smallest number of unresolved contradictory data, would be accepted as the more probably correct model. (Morris, 1974a, p. 9)

I am pleased to find the word "model" used as I have defined it in Part 1, that is, as a structured concept within the mind, in distinction from an external reality that may, or may not, exist in nature. I have proposed that all such conceptual models are real, whether there is an external prototype or not. The creation model will correlate perfectly with anything and everything in the external naturalistic realm. In that sense, neither model is capable of being proved. The real difference, not mentioned in the ICR text, is that, whereas testing of the evolution model has the potentiality of leading to its falsification, the creation model cannot even be tested and, that being the case, is not falsifiable. (I use the absolute term, falsifiable, in the logical sense.) The last sentence of the quotation is patent absurdity when we realize that the creation model explains all "data" and thus creation could not include any "contradictory data." This problem is, however, recognized in a later paragraph: "There is no observational fact imaginable which cannot, in one way or another, be made to fit the creation model" (p. 10). For the evolution model, the same is also held to be true, because "When particular facts show up which may seem to contradict the predictions of the (evolution) model, it may still be possible to assimilate the data by slight modification of the original model" (p. 10). Thus (they quote Ehrlich and Birch, 1967): "Every conceivable observation can be fitted into it." So the ICR text does, in fact, recognize the basic difference in the two models. The evolution model, as a scientific hypothesis, is open to modification--it is open-ended. The strength of science lies in its flexibility and in its ability to introduce self-correction, and thus to improve the probability that a scientific hypothesis is not in error. The creation model, derived from divine fiat, encompasses everything while it explains nothing, and therefore requires no modification. It requires no modification because it is perfect from its inception. That perfection demands no proof: it demands only belief.

I hope I have been able to put across a central thesis, that naturalistic evolution and supernatural creation are not fundamentally alike but are, instead, fundamentally very different--different, that is, in the philosophical sense of both ontology (nature of reality) and epistemology (nature of the knowledge) within each. No contortions and distortions of logic can make the two models alike in kind and substance. To bring both into the same classroom under the same teacher for equal treatment could only result in total confusion.

Credit

Chapter 12

Conservative Religious Alternatives
to Recent Creation

Before we get into questions of biology and geology, it will be useful to consider that there are theological alternatives to the strict line of modern fundamentalist creationism. I have outlined basic religious alternatives briefly in Part 1, in discussing the common dualistic ontological models that include both a natural realm and a supernatural realm; they are forms of theism and deism.

In this review the alternatives to recent creation are arranged in order from most conservative to most liberal. We begin with attempts to reinterpret the words of Gen. 1: 1-2, to give some accommodation to the diluvialists' needs for more time. In describing these alternatives, I have relied rather heavily on the ICR's textbook, Scientific Creationism, General Edition (Morris, 1974a). This edition contains a chapter on "Creation According to Scripture" (pp. 203-255), not included in the Public School Edition. The general edition has (as of this writing) gone into its twelfth printing, dated 1985. Thus it can, with some confidence, be taken as the official commentary of the ICR on matters of biblical interpretation.

The Gap Theory

The gap theory of biblical creation separates the first and second verses of Genesis 1 by a large time gap. Here are the two verses, according to the King James Version:

> Verse 1. In the beginning God created the heaven and the earth.
> Verse 2. And the earth was without form, and void; and darkness was upon the face of the deep. And the Spirit of God moved on the face of the waters.

The creation account is contained in the remaining verses of Chapter 1. Now it seems that a great deal of interest attaches to the possibility that the Hebrew word for the word "was" in verse 2 should be translated as "became" to make the verse read "And the earth became without form." According to Bernard Ramm, this wording change was suggested as early as 1791 (1954, p. 135); it made possible the insertion of a great time gap between an initial creation and a later creation. Time was what the diluvialists of England and France needed most, and this was where they found it. Cuvier, Buckland, Sedgwick, and others of that diluvialist group could fit into the time gap their entire program of successive catastrophic extinctions, each ending a long period of deposition of fossiliferous strata.

Those who subscribe to the gap theory assume that the creation of the universe and earth, described in verse 1, was complete and perfect. In contrast, the earth described in verse 2 is in terrible condition. Biblical scholars can interpret the Hebrew words for "without form, and void" as also connoting "ruined and empty." This interpretation leads to the supposition that the Creator's perfect world was subjected to a gigantic cataclysm, leaving a shattered, uninhabited earth. As a result, it was necessary for God to restore the earth by means of a six-day "re-creation." For this reason, the gap theory is also known as the creation-ruination-re-creation theory, or restitution theory (Ramm, 1954, p. 135).

For those of you who want to get more deeply involved in this kind of free-wheeling reconstruction of the Scriptures, look at what happened more than a century later. In 1909, American biblical scholar Reverend C.I. Scofield, D.D., published the first edition of his Reference Bible, incorporating the gap theory and giving it widespread credence. By that time Darwin's theory of evolution had come to dominate the scientific world and the gap theory had proved a godsend through its granting of adequate time for evolution to have taken place. In Scofield's 1917 edition, page 3, we find wedged in between verses 1 and 2 the notation "Earth made waste and empty by judgment" and a reference to Jer. 4: 23-26. In a footnote, Scofield says that the Jeremiah reference, as well as two in Isaiah (24:1 and 45:18) "clearly indicate that the earth had undergone a cataclysmic change as the result of a divine judgment. The face of the earth bears everywhere the marks of such a catastrophe."

Scofield's influence seems to have been enormous. Ramm writes that the gap theory "accumulated to itself all the veneration and publicity of that edition of the Bible" (1957, p. 135). Ramm tells us that the gap theory was espoused by a leading spokesman of American Protestant hyperorthodoxy, Harry Rimmer. As a result, says Ramm:

> The gap theory became the standard interpretation throughout hyper-orthodoxy, appearing in an endless stream of books, booklets, Bible studies, and periodical articles. In fact, it has become so sacrosanct with some that to question it is equivalent to tampering with Sacred Scripture or to manifesting modernistic leanings. (P. 135)

But "hyperorthodoxy" is a relative term, translating to "modernism" in the minds of another group. The gap theory proved most unpalatable to the Christian fundamentalists of the South, whose resurgent anti-evolution movement started around 1900 and moved into high gear during the 1920s (Nelkin, 1982, pp. 30-33). So today, the fundamentalist creationists will have none of the gap theory. There is no such gap, they say. The perfect world created in 4004 B.C. endured through a millenium and a half of Edenic bliss, secured under a vapor canopy, until God released the great Deluge that exterminated life and left its one-year record of all the world's fossiliferous strata.

There are some variations of interpretation of the life forms that belong in the gap and their relationship to life created in the re-creation. Henry Morris explains these options:

> Many holding this theory, though not all, have found it convenient to place the fossils of dinosaurs and ape-men and other extinct forms of life in this great gap, hoping thereby to avoid having to

explain them in the context of God's present creation. Others have tended to postulate only a partial pre-Adamic cataclysm, allowing plant seeds from the pre-world to survive and even certain pre-Adamite hominids to survive in order to provide a wife for Cain (Genesis 4:17) and mothers for the "giants" (Genesis 6:4). For the most part, however, expositors advocating the gap theory believe the cataclysm to have devastated the whole world, leaving it completely waste and empty. (1974a, p. 232)

Today, the gap theory pleases no large group, least of all the creationists. For them, the gap has been trumped up merely to reconcile Bible and evolutionary geology. They are puzzled as to why God would go through the entire process of guided evolution, resulting in plants and animals closely resembling living species, only to destroy that life entirely and "restock" the biosphere with a closely similar fauna and flora. The modern fundamentalists regard evolution through extended geologic time as an evil and cruel process causing much suffering and death before culminating in emergence of humans. God would not want to engage in such activity. In contrast, God's six-day creation was humane and free of evil. Fundamentalists just do not want any "geologic ages" in the scenario, whether they conflict with the Bible or not.

The Day-Age Theory

The day-age theory, which has a number of variations, is based on the expansion of each "day" of the six creation days into a much longer time interval, making possible the evolutionary sequence and the great accumulations of strata that go with it. Authority for stretching the mean solar day as required is often found in 2 Peter 3:8, which reads: "But, beloved, be not ignorant of this one thing, that one day is with the Lord as a thousand years, and a thousand years as one day." The basis of this verse may lie in Psalm 90:4: "For a thousand years in thy sight are but as yesterday when it is past, and as a watch in the night." It is obvious to anyone reading the epistle that the verse is taken out of a very different context. Henry Morris makes a good point of this and takes the opportunity to preach a mini-sermon against quoting out of context (1974a, pp. 226-227).

The ratio of one day per 1,000 years yields only 6,000 years for Creation, not even a drop in the bucket for the Darwinists. But a little manipulation is possible with the same ratio. If all time since creation is taken as 7,000 tropical years, which contain 2,555,000 mean solar days, and we multiply by 1,000, we get 2.555 billion years (Morris, 1974a, p. 227). Not bad at all for a planet now regarded as about 4.6 b.y. old, but not good enough for a universe 10 to 20 b.y. old. This maneuver still has creation of all living forms compressed into the first 6,000 years, leaving more than 2.5 b.y. of recorded human history to follow to the present. It just won't work. So we have on our hands another theory that pleases no one, least of all the uncompromising creationists.

One variant of the day-age theory separates each of the six days of creation by a large time gap sufficient for big blocks of geologic time and for organic evolution. This can be called the isolated-day theory. The fundamentalists nevertheless insist on creation being accomplished entirely within each of the separated six days of mean solar time, so the intervening time spans are unnecessary and simply represent stagnation. A second variant, the revelatory-day theory, states that the six days of creation were actually days in which God revealed creation to some recipient (unnamed). Both variants are rejected by the creationists and neither one is of any scientific interest.

Progressive Creation

Progressive creation is an acceptance of the scheme of organic evolution through geologic time, but with the provision that God intervened at intervals to create new lines of life forms. A strong supporter of this theory was Bernard Ramm (1954, p. 76, 191). Essentially it is a form of deism. Examples given in the ICR textbook are God's intervention to create the first of the line of horses, Eohippus, which then evolved through the Cenozoic Era, and in like manner the early hominoid (ancestral ape) that then evolved into the hominids and modern Man (Morris, 1974a, p. 220). The idea seems to be that evolution would not generate new lines such as new orders or classes, so that to keep the evolutionary tree growing, God needed to step in and graft on (as it were) a new branch from time to time. Creationists reject progressive creation both because it does not agree with the literal meaning of the six-day creation and it limits God's intervention to a sporadic activity. Evolutionists will resent the idea that their theory cannot account for new branches on the tree of life. Also, the dualistic mechanism runs contrary to the basic principle of parsimony.

Progressive creation seems to be a relative newcomer as compared with the gap and day-age theories. It was promoted in the 1950s as one of the alternatives considered by the American Scientific Affiliation. A more recent version of progressive creation has been put forward by Davis A. Young, a professor of geology at Calvin College in Grand Rapids, Michigan (Young, 1977, 1982). A stout fundamentalist so far as belief in the Bible is concerned, Young is also convinced of the great length of geologic time required by geologic evidence. He also accepts organic evolution but with the belief that God can and does perform miracles, one of which is the creation of Man. He accepts the naturalistic evolution of the hominoids and earlier hominids and the similarity of Homo sapiens to those older species. It does not seem to bother Young that modern Man may seem to have evolved from more primitive species of the genus Homo, as the mainstream evolutionists theorize. He reasons that God would simply have created Man in such a way as to have suggested this evolutionary connection. The creationists of ICR have denounced Young (Morris, 1978), and he probably has few supporters from the side of the more liberal theistic scientists.

Gosse's Omphalos

Before turning to the serious business of theistic evolution, we should take time out for something in a lighter vein. It is from the pen of one Philip Henry Gosse (1810-1888), an experimental zoologist who was a contemporary of Charles Darwin, with whom he was personally acquainted and exchanged correspondence during the 1850s (Morowitz, 1982, p. 20). As a lay preacher for a Protestant sect with extreme fundamentalist views (The Plymouth Brethren), Gosse was obsessed with the need to reinforce the biblical doctrine of special creation against the rising mountain of geologic evidence favoring transmutation and a continuous evolutionary tree of life. In a book bearing the title Omphalos: An Attempt to Untie the Geological Knot (1857), Gosse successfully defended the Creator's entire Creation from all possible attack, rendering it completely invulnerable to any and every objection that might be brought against it. And how did he do this?

Omphalos describes a stone of religious significance, shaped in the form of a navel; it was used in cultist rites in religious practices of the Greeks and Romans. The Greek word means "navel" or "umbilicus." The omphalos residing in the Temple of Delphi was the most famous of all: it represented the center of the earth. But it was another navel--Adam's navel in particular--that raised embarrassing questions for the fundamentalists from the first time it was seen--seen, that is, in the eyes of religious artists. As all of you know, Michelangelo's Creation-of-Adam scene on the Sistine Chapel ceiling shows a half-reclining Adam, in the buff and displaying a magnificent umbilicus, no doubt about it. If Adam was created from scratch, out of nothing, how come he had a navel? The longstanding argument of theologians was that Adam had a navel because God wanted him to look as if he had developed in utero, like all other humans (except Eve) to follow. Eve is also shown with a navel, as for example in the expulsion scene as rendered by Jacopo della Quercia (ca. 1425) and Masaccio (ca. 1427). In the navel-art competition, consider also Jan van Eyck's rendition of both Adam and Eve in the Ghent Altarpiece (ca. 1430). It was a good decade for navels! Martin Gardner develops this idea further:

> This is not as ridiculous as it may seem at first. Consider, for example, the difficulties which face any believer in a six-day creation. Although it is possible to imagine Adam with a navel, it is difficult to imagine him without bones, hair, teeth, and fingernails. Yet all these features bear in them the evidence of past accretions of growth. In fact there is not an organ or tissue of the body which does not presuppose a previous growth history. (1957, p. 125)

The number of examples is as great as the objects in nature that reveal growth or change through time. A stately tree in the Garden of Eden must have shown numerous annual growth rings appropriate to its species and the prevailing climate, all emplaced in the trunk at the instant of its creation. We can suppose (as Gosse did) that Adam had partially digested food and its residue in his alimentary canal, food that he had never eaten. With impeccable logic, Gosse carried his principle to its logical conclusion: The Creator created everything in the universe to look as if it had an antecedent existence that flowed smoothly into the time stream of the new universe. Fossils, for example, were created in place in rock strata, designed to look as if they were once living forms, which they never were. The Creator thus built into the universe a complete line of things that never were.

Although Gosse deserves full credit for developing the Omphalos scenario to its fullest limits of absurdity, the same idea was voiced in 1802 by the distinguished French writer, Francois René de Chateaubriand (1768-1848), in a work titled Genius of Christianity. The topic is mentioned by Andrew White in his discussion of attempts at compromise between orthodox church views of creation and the evolutionary paleontology of Lamarck and burgeoning uniformitarian geology of Hutton and Lyell. The theologians' compromise was to attribute fossils to the great Deluge of Noah--they were destroyed and buried in the flood. This doctrine is firmly held by today's creationists and remains for them a problem of awesome proportions. Andrew White quotes from Chateaubriand in the following paragraph:

> By large but vague concessions Cuvier kept the theologians satisfied, while he undermined their strongest fortress. The danger was instinctively felt by some of the champions of the Church, and typical among these was Chateaubriand, who in his best-known work, once so great, now so little--the Genius of Christianity--grappled with the questions of creation by insisting upon a sort of general deception "in the beginning," under which everything was created by a sudden fiat, but with appearances of pre-existence. His words are as follows: "It was part of the perfection and harmony of the nature which was displayed before men's eyes that the deserted nests of last year's birds should be seen on the trees, and that the seashore should

> be covered with shells which had been the abode of fish, and yet the world was quite new, and nests and shells had never been inhabited." (White, 1896, p. 231)

All findings of science that relate to geologic age of the earth and the universe are completely explained by such "false records of a non-existent past" (Morowitz, 1982, p. 20). Take, for example, the ratio of uranium-238 to lead-206, the latter an isotope produced by the spontaneous decay of the former at a constant rate. When the geochemist today analyzes the ratio of these parent and daughter isotopes in a particular rock, the age of the rock is revealed, and it may extend back 3 or 4 billion years in the past. For Gosse the answer would have been obvious: the Creator determined the isotopic ratio when He created the rock and did so with the intent to lead geologists into assigning a spurious age to the rock, but one fully consistent with all other indications of its age (also placed there by the Creator). And what of the red-shifts in the spectra of galaxies, a phenomenon that is interpreted to assign ages of several billion years to a single galaxy and an enormous radial velocity in an expanding universe? The Creator took care of that, too, with an appropriate Doppler effect. One of the nicest things about the Omphalos scenario is that scientists can search for centuries to turn up new kinds of fossils, minerals, and rocks containing evidences of their ages, and in all cases the results will be perfectly consistent with the table of geologic time as it is so far constructed--or will it? Suppose that, somewhere, the Creator has arranged for dinosaur bones to be intermixed with human bones--what then? Knowing that the Creator is always consistent and humane, we do not need to worry about that ever happening.

And, of course, there is no reason why the event of Creation should be assigned to the year 4004 B.C., give or take a couple of thousand years. Why not in the year 1 A.D., or in the year 1900, precisely at Universal time 0001 hours? Gosse, himself, pointed out this logical extension of his Omphalos view of Creation (Gardner, 1957, p. 126); he suggested that Creation may have occurred only a few minutes ago and we would be unaware of it! We realize that the Creator would have also provided all humans with a complete memory of all events of life prior to Creation, and there would exist newspaper files complete through that date, along with libraries and all other necessary artifacts. Because the instant between nothing and something was just that--an instant--there would be no way it could have been detected or recorded. Don't forget, the Creator would have also set up the complete line of ancient documents and inscriptions--Dead Sea Scrolls, included--to place the false creation date at 4004 B.C., but how could one detect that? The entire Bible would have been created by God, complete with English and other language translations; family bibles would have been created complete with handwritten insertions of records of family births and deaths. Thus it would be impossible for humans to know when Creation occurred, and it could have occurred at any time within the past 5000 to 6000 years, as the fundamentalists require. The total scheme of Divine Creation based on the Omphalos principle might be called indeterminate creation, setting it apart from theories we have already reviewed, such as progressive creation and the day-age and gap theories.

Did contemporaries of Gosse welcome God's perfect compromise as the ultimate expression of design? Gardner says that it was not accepted in Gosse's time (1957, p. 126). Using quotations from a biography written by Gosse's son, Edmund, Gardner writes:

> Nevertheless, Omphalos was not well received. "Never was a book cast upon the waters with greater anticipation of success than was this

curious, this obstinate, this fanatical volume,"
writes the younger Gosse in his book Father and
Son." . . . He offered it, with a glowing gesture,
to atheists and Christians alike. . . . But, alas!
atheists and Christians alike looked at it and
laughed, and threw it away . . . even Charles
Kingsley, from whom my father had expected the
most instant appreciation, wrote that he could not
. . 'believe that God has written on the rocks one
enormous and superfluous lie'. . . . a gloom, cold
and dismal, descended upon our morning tea cups."
(Pp. 126-127)

Charles Kingsley, referred to above, was a minister
and friend of Gosse who had been asked to review
Omphalos. According to Garrett Hardin, Kingsley wrote to
Gosse accusing the latter of having given a great boost to
the theory of evolution (1984, pp. 164-65). For the text
of that letter, see Hardin (pp. 164-65); it contains the
phrase quoted in the above paragraph from Edmund's
biography.

Finally, this summation by Harold Morowitz is given in
a short piece titled "Navels of Eden":

> The case of Philip Gosse is a most poignant example
> of an individual who was devoted to science yet
> brought to his subject an inflexible prior commit-
> ment to an absolute truth. This left him in the
> position of holding contradictory views on a number
> of subjects. Since logicians have long known that
> two contradictory premises can lead to any
> conclusion whatsoever, it is not unexpected that
> Philip Gosse produced a very enigmatic book. The
> reason for wiping the dust off this seldom read
> volume is to remind present-day fundamentalist
> scientists of the difficulties of assuming that
> reading an English translation of a very old Hebrew
> book takes precedence over reading much older
> rocks if we wish to establish the most accurate
> possible history of our planet and the life thereon.
> In a free society one can, of course, accept
> contradictions and conclude anything. That
> procedure has never led to particularly good
> science--or good theology either, for that matter.
> Our sympathy for the pathos of Philip Gosse, the
> man, cannot extend to a sympathy for a scientist's
> gross intellectual errors about belly buttons in
> Paradise. (Morowitz, 1982, p. 22)

Gosse's Omphalos theory is not identified specifically in
the ICR's basic textbook in which an entire chapter is
devoted to "creation according to the scriptures" (Morris,
1974a, pp. 203-60). There is, however, a brief section on
"appearance of age," in which Morris states that the
whole universe had the appearance of age from the very
start (pp. 209-10). He clarifies the point in these words:

> Note that this concept does not in any way suggest
> that fossils were created in the rocks, nor were any
> other evidences of death or decay so created. This
> would be the creation, not of an appearance of age,
> but of an appearance of evil, and would be contrary
> to God's nature. (P. 210)

Morris is perhaps making reference to the great fossil
controversy of the Renaissance period, when fossils were
explained by some persons as having been created in the
nonliving state by God (or by the devil) within the rocks
in which they are now found (see Chapter 9). This view
was gradually superseded by the diluvial theory in which
living organisms buried in the Flood of Noah became
fossilized. Perhaps Morris's reason for steering clear of
any specific reference to the Omphalos theory is because
Gosse had recognized the perfect conciliation between
Darwinism and creationism.

For naturalistic scientists, the Omphalos theory can
simply be ignored (or accepted) as religion. As for
scientists holding fundamentalist religious views, it allows
them to continue research into the geologic past, secure
in the belief that they are seeing God's work as He
intended it to be. If there be seeming deception in God's
creative work it is not malicious in any sense, but a
means of promoting harmony and peace within all
humanity. The creation of a relict past universe as it
could have been, but never was, presents us with a
marvelous continuity of past with present, ameliorating
the psychologically disturbing trauma of sudden
catastrophic creation.

Theistic Evolution

A program of theistic evolution has been sketched in
general terms in Chapter 6 as a dualistic ontological model
(Model A in Figure 6.1). It is given the title "theistic-
teleological dualism." The diagram shows the geologic
timetable of universal creation, biopoesis, and a
naturalistic evolution leading to Man. Divine guiding
influence is pervasive and continuous. In a short
quotation from a paper by Rev. James Skehan, I
presented the theistic evolutionists' view that the biblical
account of creation is not a scientific statement, but
rather a story that recounts "the beginnings of religious
history of the people of Israel." This interpretation of
the Genesis story accepts it as a statement of great
spiritual value, but not of scientific value. Skehan's
letter, which I quoted on the question of acceptance of
miracles by a theistic scientist, also describes very well
the general view of theistic evolution. The theme is one of
compatibility and complementarity of two realms as giving
the fullest possible meaning to humanity.

The fundamentalist creationists' outspoken denunciation
of mechanistic evolution scientists and a return of
hostility by the latter to the former have prompted a
number of protesting views from the liberal scientific
group. Some of these are worth quoting to develop
further the position of theistic evolution. W. H.
Hildemann, a biologist in the Department of Microbiology
and Immunology of the University of California, Los
Angeles, in a letter to Science (a journal of the American
Association for the Advancement of Science) explains:

> Many evolutionary biologists appear to be
> responding in an uncompromisingly hostile manner
> as if no compromise were conceivable in teaching
> about the origins of life. Overlooked is the fact that
> many of us teaching life sciences in the universities
> and high schools are both Christians and
> evolutionists. The view has long been held among
> many, if not most, educated Christians that
> evolution is God's awesome method for achieving the
> creative process--in other words, adaptive diversity
> of species. One need only look at the relatively
> rapid appearance of new variants of animals and
> plants or pesticide-resistant and antibiotic-resistant
> strains of organisms to realize that this process
> continues unabated. The sadness of the rigid
> reasoning of spokesmen for the Institute for
> Creation Research is in considering creation and
> evolution as irreconcilable. Many biologists who also
> believe in a supreme being governing an orderly
> universe of marvelous design deplore the efforts of
> these "creationists" to force their literal religious
> views into the curriculum. One may also object to
> the attitude of intellectual arrogance among certain
> evolutionists who push their view that the original
> forms of life appeared entirely by accident or that
> matter itself sprang from nothing. The evidence of
> evolution does not and cannot reveal the source of
> the basic chemical elements or the primal source of
> life. (1982, p. 1182)

Consider the viewpoint of a distinguished English scientist who has recently been elected Archbishop of York. The Right Reverend John Habgood, Ph.D., served on the faculty of the medical school in Cambridge, then chose to become an Anglican priest and head of a theological college. In an article titled "Evolution and the Doctrine of Creation" he expresses his strong support of biological evolution as a leading theory of science, then turns to the reconciliation of evolutionary theory with Christian theology (Habgood, 1982). He asks: "How can this evolving world as disclosed by science, bear a theological interpretation?" He goes on to suggest three ways in which it can:

> The first is very general. The world that we explore turns out to be an intelligible world and the more intelligible we find it, the more successful our science, the more I believe it makes sense to see it as ultimately related to a God of order, a God whose mind is not totally foreign to our mind.

> The model that emerges as we make sense of it, is a dynamic model, a model where we see a universe in constant process of change and development in ways which are not predetermined and yet which are discovered to be intelligible once they have happened. As I see it, such a model of a universe in process of change is very much more compatible with the notion of the Living God than that of a universe created whole, perfect and entire at one moment in time.

> Secondly, what we find as we look at the universe, is the emergence of meaning. We see a hierarchy of forms developing which give rise to the possibility of meaning. I don't believe that the interpretation of the universe which sees humanity as in some sense the crown of creation is too hopelessly anthropocentric. But as you look at the process of evolution and ask why things have developed in the way that they have, then the trends towards the greater complexity of living forms, the advantage given to creatures which are more and more adaptable, more and more exploratory, these begin to make some sort of sense in terms of the whole process. (Pp. 4-5)[1]

This second point is suggestive of the strong teleological philosophy of Teilhard de Chardin, whose views I describe below. In developing this point of increasing complexity and intelligence, Habgood seems to get into sociobiology, a subject I treat in Part 8. He sums up the second point, that evolution is directed toward a purpose in these words:

> In a nutshell, I would say that the whole process through which some kind of intelligent responsiveness to God develops is inherent in the way that the world is, and therefore it is not totally chance or an accident that we have come to be what we are. (P. 5)

Habgood's third point refers to a major argument that the scientific creationists make against evolution, namely, that mutation and natural selection operate by blind chance alone, and that such an unguided process could not possibly lead to development of more highly intelligent animals, culminating in Man. Habgood explains that the random element in the genetics of evolution should not be construed to mean that the total process of evolution by natural selection is a random process. He says:

> If we were to say that the process of evolution is random, full stop, then I think one would have to go on to say it is ultimately meaningless. But it is

not random, full stop. The engine which drives evolution is the successful selection from a whole enormous series of chance variations, of those elements in life which give an advantage to their possessor. What I think the element of chance does in evolution is to throw about all the possibilities which life might pursue. Those possibilities which are selected, which go on to develop, are then dependent upon the properties of matter, the kind of environment in which the selection has taken place and also the behaviour of the living things in so far as they have reached the stage of development in which talk about behaviour makes sense. I believe that this element of chance at the heart of things, far from being a counter-argument to belief in a loving God, is in some way a symbol of the sheer gratuitousness and fecundity of God's creation. What we see as we look around nature is an enormous overflow of energy--Creation as it were pouring out of this marvellous phenomenon of life which explores every possible means of being alive. We don't have to prove that on the basis of evolutionary theory. We have only to look around the world through our ordinary unscientific eyes and see the enormous wastage in nature, when it takes millions of acorns to produce a single oak, and billions and billions of human beings with only one Christ. (P. 5)

Another illustration of a highly thoughtful and analytical approach to theistic evolution is seen in the philosophy of Pierre Teilhard de Chardin (1881-1955), a French paleontologist and Jesuit. Teilhard seems to have concentrated upon the purposefulness of evolution throughout all the cosmos, culminating in Man as the supreme goal. Based upon reading one of Teilhard's posthumously published works, Man's Place in Nature (1966), I judge that he accepted a largely mechanistic origin of life but considered that transnatural teleological forces became progressively more important through organic evolution and finally dominated in the rise of Man. Thus, concerning evolution he states: "we must assume the existence and influence, underlying it, of some powerful dynamism" (p. 32).

Teilhard regards the phenomenon of Man and civilization as too complex to be accommodated in deterministic natural selection (p. 72). The phenomenal growth of human brain power, which he calls "hominisation," is considered unique: "Hominisation, a mutation that, in its development, differs from all the others" (heading on p. 72). A direct reference to a divine driving force seems limited to his final sentence:

> And it is at this point, if I am not mistaken, in the science of evolution . . . that the problem of God comes in--the Prime Mover, Gatherer and Consolidator, ahead of us, of evolution. (P. 121)

And, of course, theistic evolution is thoroughly unacceptable and repugnant to the creation scientists, as to most right-wing Protestant fundamentalists. They argue that theistic evolution is a theological compromise with which to placate theists while allowing the false doctrine of naturalistic evolution to be upheld. Randy L. Wysong, a creation scientist, refers to theistic evolution as a "hybrid," a "baptized evolution" (1976, p. 63). Whatever role in evolution a more liberal theologian attempts to ascribe to the Creator, it is declared patently false if it departs in the slightest from the literal interpretation of Genesis.

Just how many, or what proportion, of the members of the science community hold to some version of the general theme of theistic evolution is very difficult to determine or even to estimate. The position itself is a satisfactory working philosophy to those who embrace it and, as such, it is not likely to produce a strong evangelistic movement. Nontheistic (atheistic, agnostic) scientists seem to feel little urge to argue with the theistic scientists, perhaps feeling that the mixture of theism and naturalism it requires could be something of a contradiction, but of no consequence if it remains a personal matter. I, myself, wonder about the ability of a scientist to "serve two masters" while carrying out research in the historical areas of biology, geology, and astronomy; but as long as no mention of miracles appears in published research in the scientific journals, I see nothing to worry about. The important point is that the theistic scientist wants no intrusion of religion into the science classroom and this view unites the theistic scientist with the nontheistic (materialistic) scientist in opposing the fundamentalist creationists' efforts to get their anti-evolutionary doctrine into the classroom. Humanists, even though some of them subscribe to a religious form of humanism, are highly vocal in opposing teaching creation science in the schools, but it must be remembered that secular humanism is considered by fundamentalists to be an evil consequence of Darwinism. Perhaps what science needs is a highly vocal united front with responsibilities shared by both humanists and theistic scientists.

Credit

1. John Habgood, 1982, Evolution and the Doctrine of Creation, Insight, no. 13, pp, 1-9. (Published by Wycliffe College, Toronto, Canada.) Used by permission of the author and publisher.

PART III

Two Views of Cosmology and Astronomy

Introduction

In their book <u>Scientific Creationism</u>, prepared for public school use, creation scientists of the Institute for Creation Research (ICR) take a forthright position on universal laws of the physical universe:

> Creationists obviously would predict that the basic laws, as well as the fundamental nature of matter and energy, would not now be changing at all. They were all completely created--<u>finished</u> in the past, and are being <u>conserved</u> in the present. (Morris, 1974a, p. 18)

The text then goes on to suggest that mainstream science is prepared to consider the alternate possibility that the so-called "universal laws" could, if the universe as a whole changes, also change in an unpredictable manner. This view is expressed in a paragraph quoted from Professor W. H. McCrea (1968, p. 1297), who at the time was research professor of theoretical astronomy at the University of Sussex, England. McCrea's remark was made in the context of his questioning the naive idea that physical laws existed in complete form at, or prior to, the time of the origin of the universe. He remarked: "Actually it seems more natural to suppose that the physical universe and its laws are inter-dependent." The point has a logical base, namely, that what we call laws are merely descriptions of what we think best describes that which happens in nature; if things have happened in different ways as time has passed, the descriptive laws must necessarily be adapted to fit what we accept as reality.

So far as I am aware, however, mainstream science has always constructed its hypotheses on the first assumption that universal laws are truly universal in time and space. Postulates to the contrary, when required, are part of the hypothesis itself and are to be regarded with considerable skepticism unless testable, for, if untestable, such postulates are transempirical.

Thus, when the creation scientists claim that the "evolution model" predicts that matter, energy, and laws have evolved (changed) and are still evolving (Morris, 1974a, p. 18), they have created a fictional image of mainstream empirical science for the sole purpose of obtaining an image that contrasts with theirs. Actually, the principle of postcreation invariancy in the nature of matter and energy and the laws that describe their behavior fits closely the views accepted by mainstream science. Professor McCrea, quoted by the ICR as possibly questioning the invariance of physical laws, says elsewhere in his paper: "Various recent investigations are interpreted as showing that the constants of physics are the same throughout the observable universe. This supports the hypothesis that the universe obeys the same laws throughout" (1968, p. 1297). The ICR text goes on to list and discuss ten concepts that the creationists accept, just as they are universally accepted in all of science. In subscribing to all of the accepted laws and constants of mainstream science, the creation scientists would seem to have agreed to a set of ground rules for debate. For example, they should stand committed to the constancy of the speed of light from the instant of creation of the universe, and they should refrain from postulating that the speed of light diminishes as light travels through space. Their allegiance to the principle of invariance of laws should not permit them to postulate that a decay constant of radioactivity was different in the past from what it is now.

Actually, an important line of attack used in recent years by the creation scientists against a universe several billion years in age is to attempt to infer or demonstrate that such "constants" as the speed of light and the decay constant have not been the same in the past as today. Thus, promoting their model of recent creation, they seem to have contradicted and abandoned the very position stated by Morris in 1974.

Chapter 13

Entropy and Universal Decay

Turning now to the arguments by which creation scientists attempt to invalidate the structure of mainstream science and, at the same time, to validate their own doctrine of recent creation, it might be best to start with the most fundamental or universal questions and work down through more specific or limited questions. So we start with the creationists' view of the universe as a thermodynamic system. As luck would have it, we have picked out one of the most abstruse topics in classical physics.

Thermodynamics, the study of the relationships between heat and work, had its beginnings during the Industrial Revolution. It was a period in which the coal-fired steam-engine powered textile factories and mine pumps. The first practical steam engine had appeared in 1689, and those that followed it made use of the principle that a chamber or cylinder filled with steam could be cooled suddenly by a water jet to produce a vacuum, which then forced a piston to move under atmospheric pressure. The Newcomen engine of 1705 was one of these and it was capable of lifting water from great depths in mines. Most persons know of James Watt's steam engines of the late 1700s; they represented major improvements in efficiency. Scientific study of such machines was undertaken not only to improve their efficiency, but also to prove (if possible) that perpetual-motion machines are an impossibility. Thermodynamics did that job effectively but seems never to have succeeded in completely quashing public interest in such miraculous machines. The honor of being the founder of thermodynamics has been given to Sadi Carnot, a French physicist whose 1824 treatise on heat contained the description of an ideal heat engine that makes use of expanding gas to do work. The Carnot cycle, carried out by this imaginary engine, leads to the conclusion that no heat engine can convert into mechanical work all of the heat energy supplied to it.

In its classical early phase, thermodynamics took no account of the atomic nature of the gases, liquids, and solids with which the subject dealt. Bringing into consideration the atomic (kinetic) theory of gases was the accomplishment of Ludwig Boltzmann (1844-1906), an Austrian physicist. A younger contemporary of Charles Darwin, whose theory of evolution he warmly supported, Boltzmann produced a statistical analysis of the behavior of gas atoms, putting thermodynamics into a new framework, much as we find it today. Presentations of thermodynamics and its laws to students of physics invariably draw on the atomic theory of gases, which is what we shall do here.

Kinds of Systems

Classical thermodynamics is usually presented in terms of the isolated system, a kind of system quite different from the open system described in Chapter 4 in our discussion of the cell as a biological system. In outline form, we can classify all systems as follows:

I. Isolated systems framed in formal science (logic and mathematics); neither energy nor matter can be transported across the system boundary.

II. Open and closed systems framed in empirical science and identifiable in nature. They are of two kinds:

A. Open systems in which matter or energy or both can be transported across the system boundaries. (Examples: living cell, thunderstorm, river system.)

B. Closed systems in which matter cannot be transported across system boundaries, but the matter inside can be driven by external energy sources. (Examples: refrigerator, air conditioner.)

The isolated system typically presented in classical thermodynamics is unreal in every aspect. It is typically pictured as a cylinder whose walls provide perfect insulation (impossible), containing a piston that moves in frictionless contact with the cylinder walls (impossible), and in which the motion of the piston and the changes in pressure or temperature of the enclosed gas take place at a rate approaching zero (impossible). To make this impossible machine "do something," heat is somehow injected into the cylinder or removed from it, or a weight has to be inserted on top of the piston (or removed from that position); yet these operations are forbidden by the very premise of total isolation. No wonder the average person finds it almost impossible to understand the laws of thermodynamics based on such a strange machine!

Class II systems deal in reality. They can be observed in nature and often replicated and controlled in model form in the laboratory. Their inputs and outputs of energy and matter can be easily imagined in terms of sensory input. In short, these systems fall within the range of human experience or analogs of human experience. Their operation can be understood without using the formal terminology of the laws of thermo-dynamics, even though they conform with those laws. Nevertheless, we need to review those laws and know something of the kinetic theory of gases to understand their meaning.

The Laws of Thermodynamics

The first law of thermodynamics can be stated in various ways. In classical terms, it is stated as follows: The quantity of heat added to an isolated system is always equal to the increase of energy within the system. If it bothers you that you are not allowed to add heat to an isolated system in the first place, try this statement: The total quantity of energy within an isolated system remains constant. The important message is that energy can be neither created nor destroyed; thus the first law is a way of stating the principle of conservation of energy. Now, the only possible truly isolated system of which we can conceive is the universe itself, provided that the universe is finite in extent. This leads to another way to phrase the first law: The energy inventory of the universe is constant.

To take into account that matter can be changed into energy and vice versa in such phenomena as radioactive decay and matter/antimatter interactions, the statement of the first law needs to be broadened, thus: The total quantity of energy and matter existing in the universe is constant. So stated, it is commonly known as the law of conservation of energy and matter. Everyone, including the creationists and the mainstream scientists can live with this definition, and it is universally accepted. Unfortunately, the first law is really of no help in understanding the second law.

To understand the second law, we need to review briefly a few principles of the behavior of gas atoms in a closed container. (These were stated in Chapter 5.) Gas atoms (and gas molecules) are constantly moving at high speed. Atoms are visualized as perfect elastic spheres, the distance separating the atoms being very great in comparison with the atom diameter, but frequent collisions occur. Atoms following a free path must impact the wall of the container, from which they will rebound at the same angle as that at which they strike the wall. The atoms thus obey the Newtonian laws of motion. Impacts of atoms on the container wall provide the outward pressure of the gas. It is assumed that the gas consists entirely of single, pointlike atoms of the same element (i.e., the gas is monatomic and homogeneous), and that weak interactive forces such as gravitation that may also be in action are small enough to be neglected. Also assumed is that the gas behaves according to the well-known gas laws, the names of which are familiar to students of physics (Boyle's law, Charles' law); these relate temperature, pressure, and volume of the gas. The heat present in the gas is in the form of the kinetic energy (energy of motion) of the gas atoms. With this brief introduction, we are ready to tackle the statistical aspect of the atom motions.

As explained in Chapter 5, within the gas the distribution of the atoms in space is perfectly random in the statistical sense. The travel paths of the atoms are uniformly distributed throughout all possible directions. However, the travel speeds of the atoms are not all the same. The range of speeds is from zero to very high values. Since it would be impossible to measure individual speeds of large numbers of atoms, an average, or mean, value of speed is used to represent the gas as a whole. A constant average speed is associated with a constant gas temperature. Temperature is always referred to the absolute scale, degrees Kelvin. At zero K, called "absolute zero," the atoms have no motion whatsoever and therefore the matter has no kinetic energy. The average kinetic energy of the gas, dependent on the average speed, is always proportional to the absolute temperature. The total population of speed values is described by a probability distribution that is strongly peaked, the summit of the peak being the mean value. Values fall off rapidly in both directions. Having this brief recitation of the behavior of gas atoms in a container, we are ready to launch into the concept that lies at the heart of the second law of thermodynamics.

Drawing upon a device used by physics professor Kenneth R. Atkins in his popular college textbooks, imagine that we could somehow arrange the atoms in a container into a single line running down the length of the center line of the container (which may be thought of as having flat, parallel sidewalls), as in Figure 13.1A. Let alternate atoms start moving in opposite directions-- half of them heading toward one wall, the other half toward the opposite wall. Each atom, when it reaches the

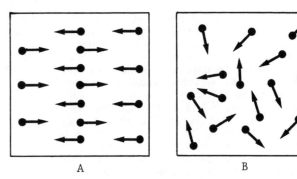

A   B

Figure 13.1 Order and disorder of gas atoms. (A) Imagined state of perfect order. (B) Random motions in disorder. (From Kenneth R. Atkins, <u>Physics--Once Over --Lightly</u>, p. 98, Figure 5-11. Copyright © 1972 by John Wiley & Sons, Inc. Reprinted by permission of John Wiley & Sons, Inc.)

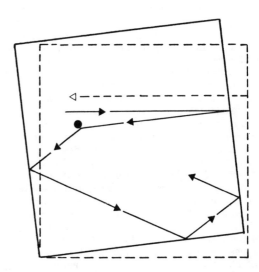

Figure 13.2 Tilting the container shown in Figure 13.1 so as to induce disorder. (From Kenneth R. Atkins, <u>Physics --Once Over--Lightly</u>, p. 99, Figure 5-12. Copyright © 1972 by John Wiley & Sons, Inc. Reprinted by permission of John Wiley & Sons, Inc.)

wall, rebounds so as to retrace its path exactly. All the atoms are shuttling back and forth at high speed (the same speed for all the atoms). Left undisturbed, they can continue to do so indefinitely. We have now created a total situation of perfect orderliness, described in thermodynamics simply as one of <u>order</u>. Next, we generate a very trivial disturbing factor into the situation. Professor Atkins suggests that we jolt the container ever so slightly, tilting it a bit off its original stance, as shown in Figure 13.2. Now the atoms, which maintain their paths in space, will strike the walls at a slight angle, throwing them into paths that soon lead to collisions with each other. The collisions quickly bring about the randomness of motions normally present in the gas (Figure 13.1B). Randomness, in the view of thermodynamics represents a state of <u>disorder</u>; it is the most probable state of the system. The second law requires that spontaneous changes in an isolated system always go in the direction from order to disorder. Can this statement be challenged, or is it always true? (When using formal science, based on logic, a conclusion can be declared true or false.)

For those of you interested in the probability aspect of order and disorder, I offer the following example from the writings of physicist Philipp Frank (1946, pp. 24-26). Let us imagine a container of monatomic gas subdivided into a large number of small space- compartments, or cells, all of the same volume. (The imagined cell walls have no effect on motions of the atoms.) Writes Frank, if we could trace and record the path of a single atom over a time of years or centuries, we could establish the specific fraction of the total time that the atom spends in one particular cell--that would be the "dwelling time," expressed as a ratio. The dwelling time, p, represents the probability that the atom will be found in the particular cell. For a system of 100,000 cells, p = 1/100,000. A second gas atom also has the same probability, p, of being in that same cell, i.e., 1/100,000. Now let us assume that the total number of atoms in the container, N, is one million. We can calculate the time that we can expect to find all the atoms in the same cell; it is equal to p raised to the Nth power times the total observation time, T:

$$T (p^{1,000,000}) = T (1/100,000)^{1,000,000}$$

$$= T (1/10^{5,000,000})$$

If T is, say, a billion years, the average dwelling time during which all atoms are in one cell "will be much less than the billionth of the billionth part of a second" (1946, p. 25). He adds: "Moreover, if we are lucky observers of such an event, we can bet that it will soon disappear and not reappear for a billion generations to come."

Perhaps easier to grasp is a similar example based on a deck of playing cards. We start with a new deck in which there is perfect order--two through ace in each of the four suits. Repeated shuffling quickly destroys that initial order and replaces it with a random distribution that represents disorder. The random distributions we continue to obtain are rarely exactly the same, but the level of disorder represented by the randomness is overwhelmingly the more probable. The probability of recovering the initial ordered distribution with any one deal is about 10^{-68} (Patterson, 1984, p. 141).

Now comes the most difficult concept in thermodynamics. The term entropy is a mathematical concept related to laws of probability (Atkins, 1972, p. 101), but we can best handle it here by relating it to order and disorder. The dictionary definition of the word typically reads: "A measure of the unavailable energy in an isolated system." We can simply say that the greater the disorder in the system, the greater is the entropy. This relationship allows us to state the second law of thermodynamics as given by Professor Atkins: "When a system containing a large number of particles is left to itself, it assumes a state with maximum entropy, that is, it becomes as disordered as possible" (p. 100).

The state referred to is the equilibrium state, which is the final stable condition. Entropy of the system increases as the absolute temperature of the system increases. This means that the system is more disordered at higher temperatures. It also means that as the temperature is lowered, the entropy decreases, while at the same time the system becomes more ordered. "When the system is perfectly ordered, it is not possible to go any further and the absolute zero of temperature has been reached" (Atkins, 1984, p. 101). This statement leads us to the third law of thermodynamics, stated by Atkins as follows: "A system in equilibrium at the absolute zero of temperature is in a state of perfect order and has zero entropy" (p. 101).

Another way of looking at entropy is through heat flow. We are asked to think of an imaginary isolated system in the configuration of a dumbbell, as shown in Figure 13.3. The contents consists of two bodies, one hot and one cold. Initially, they may be separated by a thermal barrier, but this is now removed and heat flows from the hot body to the cold body. Eventually, the equilibrium isothermal state is reached, and when this has

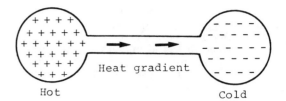

Figure 13.3 An imagined isolated system in which heat flows from a hot body into a cold body, with resulting increase in total system entropy. (After Kenneth R. Atkins, Physics--Once Over--Lightly, p. 102, Figure 5-13. Copyright © 1972 by John Wiley & Sons, Inc. Reprinted by permission of John Wiley & Sons, Inc.)

happened it will be found that the total entropy of the system has increased; the change in entropy is described as positive. In reaching the equilibrium state, "downhill" flow of heat has occurred (hot to cold), but total entropy and disorder have increased.

In the state we first assigned to the system shown in Figure 13.3, it had the capability of undergoing an increase in entropy, i.e., it possessed the ability to gain entropy. In the words of physicist Stanley Freske, it initially had an entropy deficiency (1981, p. 8). We define entropy deficiency as the difference between the system's entropy capacity and the amount of entropy it is actually holding. (Capacity means the maximum amount possible.) Freske explains further:

> This deficiency may also be referred to as negentropy (short for negative entropy)--a concept which, had it been generally adopted, might have been less confusing than entropy. Negentropy, then, has been defined as a measure of order, information, lack of equilibrium, and the availability of energy for doing work. But most fundamentally, negentropy--or entropy deficiency--is a measure of the improbability of a system being in a given state. (P. 8)[1]

Entropy deficiency (negentropy) has great importance in open systems, whether they are biological or physical, because such systems have mechanisms for creating entropy deficiences and maintaining them. The concept of storage of chemical energy in biological open systems, referred to in Chapter 4, is identical with the concept of developing an entropy deficiency. Keeping this concept in mind will prove most useful in analyzing the creationists' arguments on the subject of open systems.

The only possibility we have of a truly isolated system is the entire universe (if it is finite in extent). When it originated, we suppose, the degree of order was high and the entropy low, but the system's entropy capacity was enormous. In the billions of years that have elapsed since the Big Bang with which it started, the total entropy has been increasing and so has the disorder. But the system is still changing and equilibrium has by no means been reached. It is this very state of change--a state of flux-- that allows local and temporary reversals of the entropy change from positive to negative to occur, accompanied by corresponding local changes in the direction toward increasing order. Let us turn next to consider that possibility.

A schematic view of the universe is shown in Figure 13.4. Within the limits of the universe are many open subsystems within which entropy change can be negative to the accompaniment of the construction of highly ordered states of matter. The largest of the open subsystems are galaxy clusters and galaxies; within them are smaller subsystems--luminous gas clouds and stars. Many stars have planetary systems, and within each solar system the individual planets and other orbiting objects

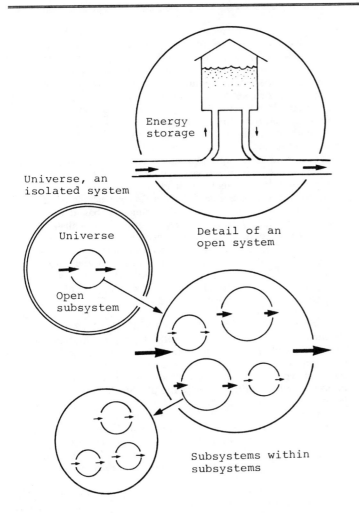

Figure 13.4 The universe visualized as an isolated system within which are nested sets of open subsystems. (A. N. Strahler.)

form another set of subsystems. Each planet, in turn, includes numerous subsystems. Associated with a typical energy subsystem is a subsystem of matter that may be either closed or open. Subsystems of all types and scales continue to be formed. After the growth stage of decreasing entropy is completed, the system may enter into a steady state, and this may be followed by a decay state in which system entropy increases along with increasing disorder, until the system itself goes out of existence. Meantime, the universe as a whole proceeds on its general course toward increasing entropy and increasing disorder. With this general picture in mind, we can turn to the creationist's view of thermodynamics, entropy, and the order/disorder phenomenon.

Thermodynamics and the Scriptures

Fundamentalist creationists consider that the laws of thermodynamics originate through God, the Creator of the universe, and that this fact is clear from the Scriptures (Morris, 1974a, p. 211-13). They point out that, as originally created, the universe was perfect and that perfection applied to the earth and all living things on earth, including Man. However, God also endowed Man with a certain amount of freedom of choice, including freedom to love God or not, to obey God or not, and to take responsibility for such choices. When Adam and Eve in the Garden of Eden, using this prerogative, chose (in eating of the fruit of the Tree of Knowledge) to doubt and reject the Word of God, a sin was committed, and with that sin came God's curse of death to the world. In

cursing Adam, God said "cursed is the ground for thy sake" and "In the sweat of thy face shalt thou eat bread, till thou return unto the ground; for out of it wast thou taken: for dust thou <u>art</u>, and unto dust shalt thou return" (Gen. 3:17-19). Morris explains that the dust of the ground refers to the basic physical elements (1974a, p. 212). Thus we can infer that God cursed all matter of the universe.

But if the laws of thermodynamics are not clear from Genesis, they emerge clearly in the New Testament. In Rom. 8:20-22 we read:

> For the creature was made subject to vanity, not willingly, but by reason of him who hath subjected <u>the same</u> in hope.
>
> Because the creature itself also shall be delivered from the bondage of corruption into the glorious liberty of the children of God.
>
> For we know that the whole creation groaneth and travaileth in pain together until now.

Creationists see in "bondage of corruption" ("corruption" is more literally translated as "decay") the statement of the second law. Ever since God cursed Adam and the ground beneath him, the universe came into a regimen of continual decay, meaning that the total supply of entropy was on the increase.

As to the first law, the revelation is clear from Gen. 2:2: "And on the seventh day God ended his work which he had made; and he rested on the seventh day from all his work which he had made." Because God was through making things, he made nothing more, and the total universal supply of matter and energy was complete. If this Genesis passage seems to require stretching meanings, revelation is perfectly clear in Eccles. 3:14: "I know that, whatsoever God doeth, it shall be forever; nothing can be put to it, nor anything taken from it."

Henry Morris makes a significant statement to the effect that empirical science has demonstrated the universality of the laws of thermo- dynamics, but thus far no scientific explanation has come forth as to why they work (1974a, p. 212). Morris shows from several scriptural passages that the origin of those laws in the Creator's hands is implicit in the Bible and has been available to humans through revelation for thousands of years.

The creationists now apply the second law in an argument designed to prove that God created the universe. Quoting from an undated issue of <u>Impact</u> (1973?, no. 3) Henry Morris writes:

> Thus, the Second Law proves, <u>as certainly as science can prove anything whatever</u>, that the universe had a beginning. Similarly, the First Law shows that the universe could not have begun itself. The total quantity of energy in the universe is a constant, but the quantity of <u>available</u> energy is decreasing. Therefore, as we go <u>backward</u> in time, the available energy would have been progressively greater until, finally, we would reach the beginning point, where available energy equalled total energy. Time could go back no further than this. At this point both energy and time must have come into existence. Since energy could not create itself, the most scientific and logical conclusion to which we could possibly come is that: "In the beginning, God created the heaven and the earth."

Granting the logic of projecting the state of the universe backward in time toward a point in time when the available free energy of the system was at a maximum (or, at least, greater than at any time since then), the remainder of the argument does not follow. There is no reason to infer that time did not exist beyond that point. Moreover, it can be postulated that in the model of an alternately expanding and contracting universe, the initial state of maximum order and minimum entropy could have marked the end of a prior contracting phase in which the total universal matter and energy had become concentrated into an extremely small volume. Nothing in the initial postulates of Morris's argument leads to the conclusion that "At this point both energy and time must have come into existence." His final sentence then is a total _non sequitur_ and becomes nothing more or less than a statement of religious belief, which has no scientific status whatever.

Cosmic Creation as a Unique Event

In discussing science and religion in Chapter 6, I noted that the instant of creation or initiation of the present universe is also a point of convergence of ontological systems. Under theism, whether fundamentalist creationism or a more liberal theism that includes evolution through geologic time, the role of initiation of the universe can be ascribed to a Creator and viewed as a supernatural event. The same applies for a deistic view that terminates the Creator's role at the instant of creation. For the mechanistic-materialistic (nontheistic) scientist there is no way that events preceding the instant of creation--the singularity--can be examined by empirical science. Speculations can be made, based on the assumptions of physical laws that are accepted as applying to the present universe from its inception, but as speculations with no possibility of being tested, they have no scientific substance. Thus, the question of the causation of the present universe stands entirely independent from questions of origin and possible causation of the contents of the universe, including the cosmic structures of outer space, the solar system and its orbiting objects, the planet Earth in particular, and life on that earth. The question of reality of organic evolution through extended geologic time by natural selection, toward which the major thrust and energy of the creationists' negative argument is directed, can be examined independently of considerations of origin of the universe.

But consider, then, that there is also another and later point of ontological convergence: the question of origin of living matter on earth. For both theists and deists who subscribe to the geological time scale in which the initial appearance of life goes back to something on the order of 3.5 billion years, or perhaps more, and who also consider organic evolution by natural selection to be a strong working hypothesis, it is permissible to hold the belief that God created the initial living substance. For the mechanistic materialists, it is permissible to postulate a naturalistic, spontaneous mode of biopoesis (origin of living matter) based on what is known today about chemical and physical processes and materials. In other words, this is one of the singular points in universal history when both dualistic and monistic ontologies come up with exactly the same result, which is the unique "seed" or "seedling" of the tree of life. Whichever origin is preferred, one being a religious belief and the other a naturalistic hypothesis of science, the question of continuous evolution of life thereafter can be examined in full independence of the question of its origin.

Fundamentalist creationists thus have two independent target hypotheses with respect to life on earth: (a) the naturalistic origin of living matter; (b) continuous organic evolution of life forms from the most primitive forms known to the most advanced. The argument from the laws of thermodynamics is used by creation scientists against both targets. Their general thesis is that both naturalistic biopoesis and continuous organic evolution run counter to the second law because they require a decrease in entropy and a flow from a disordered state to an ordered state. Because the second law is inviolate, they say, life cannot arise from nonlife spontaneously, nor can living forms evolve from simpler to more complex forms.

In Chapter 4 we especially emphasized the ability of living organisms as open systems to place energy temporarily in storage in complex molecules and to release energy as required by breakdown of those molecules. We explained that this localized decrease in entropy in no way violates the laws of thermodynamics, because the conversion and storage processes require the expenditure of even more energy as waste heat than the quantity of energy that is chemically stored.

In reading the arguments of mainstream scientists against the creation scientists' arguments based on thermodynamics, I find one persistent theme (already developed fully in Chapter 4): creationists ignore the existence of open energy systems that can locally and temporarily reverse the universal trend toward increasing entropy and increasing disorder. As a typical case of the exercise of this theme, I cite John W. Patterson's chapter, "Thermodynamics and Evolution," in a volume of essays edited by Laurie R. Godfrey under the book title _Scientists Confront Creationism_ (Godfrey, 1983; Patterson, 1983). It is an excellent chapter in terms of presenting clearly the meaning of the second law and in showing that, although "water runs downhill" is a general rule, there are specific situations in which water can actually "run uphill." The principle of the hydraulic ram is explained, showing how the kinetic energy of downhill flow of water can be used to lift a small quantity of water to a storage tank many feet above the base level. Other illustrations follow, and the argument is strongly reinforced. Nevertheless, Patterson's concluding remarks turn to the debating theme referred to above:

Because the second law of thermodynamics is nonintuitive and because few people have studied it in depth, it is ideally suited to the apologists' favorite techniques of obscurantism. Moreover, the second law does provide a criterion for determining if certain processes are impossible in nature. Hence, by misinterpreting the second law, whether by ignorance or deliberate deception or both, the creationists are able to convince unwitting audiences that evolution is impossible.

In reality, however, the "uphill" processes associated with life not only are compatible with entropy and the second law, but actually depend on them for the energy fluxes off of which they feed. Numerous other kinds of backward processes in simpler, nonliving systems also proceed in this way and do so in complete accord with the second law. The creationists' second law arguments can only be taken as evidence of their willingness to bear false witness against science itself. It is a sad testimonial to the community of professors, engineers, and scientists that so many have ignored their professional responsibilities in failing to expose the creationist thermodynamics apologetic. (Pp. 114-15)

(Note: See Patterson's more comprehensive treatment of the entire subject in a paper titled "Thermodynamics and Probability," 1984, pp. 132-52.)

Creation scientists are by no means unaware of open systems and their characteristics, since the subject has been aired numerous times in debates with mainstream scientists. Henry Morris responded in a 1976 issue of _Impact_:

By far the majority of evolutionists, however, attempt to deal with this Second Law argument by retreating to the "open system" refuge. They maintain that, since the Second Law applies only to isolated systems (from which external sources of information and order are excluded), the argument is irrelevant. The earth and its biosphere are open systems, with an ample supply of energy coming in from the sun to do the work of building up the complexity of these systems. Furthermore, they cite specific examples of systems in which the order increases--such as the growth of a crystal out of solution, the growth of a seed or embryo into an adult plant or animal, or the growth of a small Stone Age population into a large complex technological culture--as proof that the Second Law does not inhibit the growth of more highly-ordered systems. (1976b, p. ii)

Morris considers such arguments as specious. He claims that creationists have emphasized for years that the second law is applicable only to open systems and it can be proven that no truly isolated system exists. He then goes on to develop a restatement of the second law:

Creationists have long acknowledged--in fact emphasized--that order can and does increase in certain special types of open systems, but this is no proof that order increases in every open system! The statement that "the earth is an open system" is a vacuous statement containing no specific information, since all systems are open systems.

The Second Law of Thermodynamics could well be stated as follows: "In any ordered system, open or closed, there exists a tendency for that system to decay to a state of disorder, which tendency can only be suspended or reversed by an external source of ordering energy directed by an informational program and transformed through an ingestion-storage-converter mechanism into the specific work required to build up the complex structure of that system."

If either the information program or the converter mechanism is not available to that "open" system, it will not increase in order, no matter how much external energy surrounds it. The system will proceed to decay in accordance with the Second Law of Thermodynamics. (P. ii)

A clear distinction needs to be made between the statement of the second law as formal science and its application in empirical science. The assumption of an isolated system, while none may exist (other than the entire universe as a system), is perfectly permissible as a premise of an argument in the exercise of logic. It is a presumption that a perfect insulator exists with which to enclose the system. With or without that presumption, the second law is always valid and can be broadly stated as follows: The total system entropy cannot decrease in any process.

Stanley Freske (1981) has replied to Morris's attempt to rewrite the second law. Freske directs his fire at the three-condition second law as proposed by the creationists for open systems. He refers to it as "The Creative Trinity," a title given earlier by a creationist, R. G. Elmendorf (reference cited by Freske). Consider point by point the three terms of the revised law. Point 1 is that the tendency toward increase in entropy (and increase in disorder) can only be reversed if energy is supplied from outside the system. This statement is not necessarily true, although importation of energy typically occurs. Take, for example, a large storage tank holding water and equipped with a drainpipe exiting from the bottom of the tank. Let the drainpipe feed through a small turbine pump capable of lifting a small part of the outflow to a small storage tank several feet above the main tank. (A

hydraulic ram-jet can also perform this function.) We now open the drain and let the pump do its work. By the time all the water has drained from the tank, a small fraction of it will have entered the upper tank. Thus, mechanical work against the pull of gravity will have stored energy in a subsystem of the open system. No free energy was supplied from outside the system. As the system was "running downhill" it was able to perform a lesser amount of work "running uphill." Note, also, that at the conclusion of the work run, the system will continue to have an entropy deficiency, albeit a much smaller one than that it had at the outset.

Point 3 (taken out of order) is that the system must contain an energy conversion mechanism. Typically, an open energy subsystem does function by the conversion of energy from one form to another, including placing some of the energy in temporary storage. For example, in a thunderstorm, latent energy held in water vapor is drawn into the updraft, released as sensible heat, and later dissipated as longwave radiation. In a plant, photosynthesis is the mechanism for converting the energy of sunlight into chemical energy stored as carbohydrate. Granted that such energy conversion mechanisms exist, they may explain how entropy can decrease (or hold at steady state) within a subsystem, but the existence of an energy conversion mechanism is not required in the broad statement of the the second law, which is simply that the total system entropy cannot decrease in any process.

Point 2 is that the external energy source must be directed by an "informational program." Freske notes that creationists also refer to the "informational program" as "intelligence," or a "control system" (p. 10). Perhaps the creationists also would include cybernetic feedback, mentioned in Chapter 4. Morris gives as a specific example of a "directing program" (1976, p. ii) the case of a growing plant as an open system: it is the genetic code that directs the course of the growth. In this case the creationists follow strictly the principles of modern genetics and molecular biology. When it comes to the origin of the first living molecule (as an open system), the inference seems to be that the directing program is one of divine guidance by the Creator. Although not directly stated as such, the implication arises from the creationists' claim that there is no naturalistic directing program in the event of biopoesis, hence a supernatural direction is required.

That an informational control system is required of all open systems in order that they can reverse the trend to increase in entropy is certainly not the case in natural physical (inorganic) open systems in nature and cannot, therefore, apply to all open systems as part of a universal statement of a law. Freske concludes: "Creationists are not showing that evolution contradicts the second law of thermodynamics; instead, they are saying that the second law, as accepted by conventional science, is incorrect and insufficient to explain natural phenomena. They insist that something else of their own making must be added--namely, a divinely created directing program or a distinction between different kinds of entropy" (1981, p. 10).[1]

Order and Disorder--Creationist Style

Creation scientists, when discussing the second law of thermodynamics use the words "order" and "disorder" with great frequency, and in so doing tend to generate extreme confusion in the minds of listeners unfamiliar with the language of classical thermodynamics. They use these words in reference to organic evolution and the origin of living matter in ways that are hard for most people to associate with familiar, observable phenomena. Obviously, use of the word "disorder" ties the running-downhill process in the universe together with the biblical doctrine of decay of the universe willed by God following its creation and the expulsion of Adam and Eve. Stanley Freske cautions us in these words:

> A final warning: the word order in popular usage is highly ambiguous and should be scrupulously avoided in explanations of entropy for the benefit of anyone not already familiar with scientific jargon, lest it cause a great deal of confusion. (1981, p. 9)[1]

Let's look at a physics experiment and see how the words "order" and "disorder" are used. We follow the thermodynamic stages in transition from ice to liquid water to water vapor (Table 13.1). The unit used for entropy change doesn't need to be defined for our purposes. Notice that while the ice is melting and when it is passing from the liquid state to the vapor state, the temperature remains constant while the inputs of heat of fusion and heat of vaporization are in progress. We all realize that to melt ice or snow, heat needs to be applied. True to our earlier statement, system entropy increases with temperature. (The heat necessary for this process must be imported, which would be impossible in a truly isolated system, but we supply the heat anyway.) Ice represents the state with most order, gas the state of greatest disorder. If the ice had initially had a temperature of zero K, it would have started with perfect order and zero entropy.

Ice consists of water molecules neatly packed into a uniform geometrical arrangement called a crystal lattice. In ordinary pond ice or refrigerator ice we don't see evidence of that high level of structural order, but it is obvious in snowflakes seen under a low-power microscope. Snowflakes form by sublimation, a process in which gaseous molecules of water vapor become attached to a "seed" nucleus, which might be a smaller ice particle or a mineral particle. As the snowflake grows, it develops a six-sided symmetry (hexagonal symmetry) and can become extremely ornate. Order in terms of symmetry, perfection, and ornate design is extremely high.

Does the growth of the snowflake actually represent a reversal of the thermodynamic trend? If so, the growth of the crystal structure must involve the giving up of internal energy; it is the heat of fusion in reverse, and because that is indeed the case, the crystal itself, apart from its surrounding medium, can be considered a subsystem that has achieved decreased entropy while achieving increased order.

Crystal growth is an important topic in creationist writing because it is a purely inorganic process that seems on first thought to mimic organic growth to the extent that a highly ordered structure is created from molecules previously in random disorder. Creationists have erroneously attempted to liken the formation of ice crystals to organic growth by the observation that both involve increasing order and a negative change in entropy. But, as explained by Stanley Freske, there is an important difference between the growing crystal and a living system:

> In the crystal, the entropy is always at the maximum. In other words, while it is true that the entropy decreases as the liquid changes into a solid, this happens because the entropy capacity of the system decreases. The living system, on the other hand, contains an entropy deficiency, and this deficiency increases as the system grows or evolves. It should now be obvious that a debater who tries to draw too close a parallel between crystals and living systems will be in trouble. (1981, p. 11)[1]

I can discern another fundamental difference between crystals and living cells. Both show a high level of order but are very different in level of complexity. The ice

Table 13.1   Increase in Entropy with Changes in State of Water

Change	Entropy increase
Ice to liquid water at 273 K	+5.27
Liquid water from 273 K to 373 K	+5.66
Liquid water to vapor at 373 K	+26.00
Overall, from ice to vapor	+36.93

Data source: John B. Fenn, 1982, Engines, Energy, and Entropy; A Thermodynamics Primer, W. H. Freeman and Company, p. 223.

crystal includes only two kinds of atoms formed into a very simple molecule; the unit cell of its crystal lattice is really very simple in structure and never varies. Molecules of DNA and RNA, in contrast, are orders of magnitude greater in complexity than water molecules. The double helix of DNA is a long chain of molecular units, the nucleotides, each of which is much more complex than a water molecule. The four kinds of nucleotides are arranged in various sequences to provide the genetic code. To use the ice crystal as an analog of the DNA molecule in terms of complexity is highly misleading.

Creationists who engage in debating on questions of thermodynamics do not always have their principles straight, as Freske points out in reporting the following incident involving the text of a creationist author, R. G. Elmendorf, in his 1978 book titled How to Scientifically Trap, Test, and Falsify Evolution:

> Nevertheless, creationists have expended a great deal of effort attempting to explain the entropy decrease inherent in crystal growth. Elmendorf claims that there is no decrease in entropy, because liquids are more orderly than crystals (1978).* When I pointed out to him in an exchange of letters that gases turn into liquids by a similar removal of heat, he decided that gases are the most orderly of all. I might have asked him why we observe changes of state in nature which proceed in the opposite direction by means of the simple addition of heat, such as snowflakes melting, however, I did not pursue the matter any further. (Freske, 1981, p. 11)[1]

(*Reference given by Freske, p. 16.)

Perhaps this has been a difficult chapter for many of you. If you have grasped the essential points you are reasonably well equipped to refute some of the creationists' favorite arguments against evolution and in support of special creation--arguments they have used with some success in public debates, because the subject of classical thermodynamics is unfamiliar to the audience even when it contains persons who have had college courses in physics, chemistry, and biology. We move next to confront the creationists on their arguments in favor of an extremely young universe.

Credit

1. From Stanley Freske, Creationist Misunderstanding, Misrepresentation, and Misuse of the Second Law of Thermodynamics, Creation/Evolution, vol. 2, no. 2, pp. 8-16. Copyright © 1981 by Stanley Freske. Used by permission of the author.

Chapter 14

How Mainstream Science Views the Universe

Creation scientists of the Institute for Creation Research (ICR) are well equipped to understand and discuss the technical and mathematical aspects of the fields of science that they, as individuals, elect as areas of specialization. Their arguments often seem highly erudite and make reference to numerous published journal articles of mainstream science. For you to follow their arguments about cosmology and astronomy, a basic knowledge of the contents of the universe--galaxies, cosmic clouds, stars, and planetary systems--is essential. While some of you have a fairly good knowledge of modern astronomy, the likelihood is that many of you do not. It's of little help to suggest that you take time out to read a recent astronomy textbook (or textbooks of biology and geology).

As in sections of previous chapters, I resort to a sort of first-aid treatment, which is to provide a descriptive summary of the current topics and terms in each science area, hoping it will enable you to follow the debate through. In this chapter I offer a brief overview of prevailing astronomical and cosmological concepts and hypotheses. Recapitulating the history of science, it begins with objects closest to our planet and therefore most familiar, working outward to distant objects recorded only as faint light images on photographic plates or as fluctuating signals transmitted through infrared and radio waves or as streams of X rays.[1]

Celestial Objects of Antiquity

Astronomers of Mediterranean lands in the several centuries before and after the birth of Christ could identify with the unaided eye celestial objects of several distinct classes. First, of course, was the Sun, unique because of its great luminosity. Obviously, the Sun was some sort of ball of fire. Then there was the Moon, unique because of its appearance as a solid circular disk or sphere with a fixed set of surface markings and a pale luminosity. Astronomers observed that the Moon's outline changed through a regular monthly cycle as its position changed in the sky with respect to the Sun. These changes we know as the Moon's phases. Another class of celestial objects were the stars--pinpoints of light fixed on a black background that turned as a unit through a daily cycle. As in a modern planetarium, the stars appear to lie on the inner surface of a hollow sphere--the celestial sphere. Then there were several bright objects that moved slowly against the background of fixed stars. Like the stars, these planets appeared as points of light. There were visible only five planets--Mercury, Venus, Mars, Jupiter, and Saturn. They moved through the constellations of stars along about the same track as the Sun and Moon, but each planet had its own period of motion. (The Greek root of "planet" means "wanderer.") Venus was observed to be extremely brilliant at times and to shine brightly as an evening star or morning star

during twilight, when no other celestial objects except the Moon could be seen. Jupiter was also a brilliant planet; Mercury was seen only close to the Sun and moved back and forth in a rather strange way (hence our adjective "mercurial"). We might also mention that these ancient astronomers observed the coming and going of comets on occasion and were familiar with the faint light trails of meteors (falling stars) and the rare arrival of a fireball that sometimes brought strange stones crashing into the Earth's surface. Of considerable interest to modern astronomers are the ancient records of isolated stars that flared up with astonishing brilliance, only to dim slowly and finally disappear from sight after only a few months. These supernovae we now know to be exploding stars; they have played a major role in evolution of the universe.

The ancient astronomers held that the Moon and planets, along with the Sun, moved in more or less circular orbits around the Earth--a geocentric model of the universe (Figure 14.1). We now accept the heliocentric model of a solar system in which Earth, itself a planet, and the other planets move in orbits about the Sun. We now know that the Sun is a star, consisting of a great mass of highly heated incandescent gas, whereas the members of the solar system--planets, moons (satellites), comets, and other kinds of objects--are solid bodies, cold and very small in comparison with a typical star. Perhaps the most significant finding of modern astronomy was that the stars (other than the Sun) lie at enormous distances from the solar system and from each other.

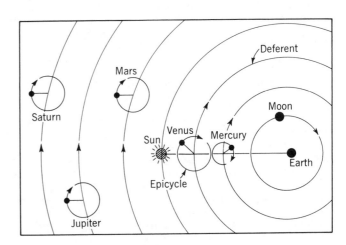

Figure 14.1 The Ptolemaic system required epicycles superimposed on deferents to explain the retrograde motions of the planets. Notice that the epicycles of Mercury and Venus remain fixed on a line connecting Earth and Sun. (A. N. Strahler.)

Stars

The closeness of our Sun enables astronomers and physicists to learn a great deal about a typical star. Like most stars, it is a spherical body of gas bound together under the enormous inward pressure of gravitational attraction. The Sun's diameter is 140,000 km, or more than 100 times Earth's diameter. The mass, or quantity of matter, contained in the Sun is more than 330,000 times that of Earth; the Sun's volume, 1,300,000 times that of Earth. Like other stars of its class, the Sun has an internal energy source that gives it enormously high temperatures, both in its interior and on its surface. The internal temperature is estimated to be 13 to 18 million Kelvin degrees (kelvins, K), a figure far beyond the imagination of humans, but one that can be reproduced by humans during the instant of detonation of a hydrogen bomb. The Sun's surface temperature is from 4,000 to 6,000 K, causing it to radiate an enormous quantity of energy in all directions into outer space.

The Sun is composed largely of two elements--hydrogen and helium. Hydrogen makes up about 90 percent and helium about 8 percent of the total mass. The heat-producing process that takes place in the Sun's interior is called nuclear fusion; it consists of the conversion of hydrogen into helium, a process involving the conversion of part of the mass of each hydrogen atom into energy. Over long periods of time the quantity of a star's mass lost by the entire nuclear conversion process is an extremely small part of the original hydrogen mass. It can be calculated that at the present rate of energy production, our Sun will diminish by only one-millionth part of its mass in 15 million years. Thus the Sun will continue to burn its nuclear fuel for a life span of several billion years.

Looking over the entire field of stars that can be studied by means of telescopes, we find a great range in mass, diameter, interior density, surface temperature, and luminosity.

Taking the mass of the Sun as unity (one solar mass), we find that some stars have a mass as small as 1/10 the Sun's mass, whereas the largest have masses as great as 20 to 30 times that of the Sun. Stars also show a great range in diameter. For example, a small companion star to Sirius has a diameter only 1/30 that of the Sun, whereas the diameter of Antares is almost 500 times greater than that of the Sun.

Density of a star refers to the degree of concentration of its mass within a given volume of space. Taking liquid water at the Earth's surface as the standard of density with a value of one gram per cubic centimeter ($1 \text{g}/\text{cm}^3$), the average density of the Sun is about 1.4 g/cm^3 or only slightly greater than the density of water. Stars show a truly enormous range in density, from less than a millionth that of the Sun to more than 100 million times as great. The tiny companion star to Sirius, referred to above, has a mass almost equal to that of the Sun, and consequently its density is 35,000 times that of water at the Earth's surface!

Surface temperature of stars ranges from below 3,500 K to more than 80,000 K. A star's color is closely related to its surface temperature. The hottest stars are blue; those only a little cooler are white. At progressively lower temperatures star color ranges from yellow through orange to red. Our Sun, for example, is described as yellow in color; Antares, mentioned above, is a large red star with surface temperature only in the range of 2,000 to 3,000 K. Rigel, with surface temperature in the high range of 11,000 to 25,000 K, is a blue-white star.

A basic principle of physics is that the rate at which a surface radiates energy increases greatly as surface temperature goes up. More precisely, the total quantity of radiation energy for a unit area of surface (one square centimeter) in a unit of time (one second) varies as the fourth power of the temperature in kelvins. (This law,

called the Stefan-Boltzmann law, assumes that the emitting surface is a perfect radiator, called a blackbody.) Star luminosity is a measure of the total radiant energy output of a star as if it were being measured at the star itself. From the fourth-power law, we realize that luminosity depends in part on the star's surface temperature. However, luminosity will also depend on the star's total surface area, which increases with its diameter. This means that great luminosity may in one case be largely the result of extremely high temperature, whereas in another case it can result from a combination of low temperature with extremely great diameter. Luminosity is usually scaled in reference to the Sun's luminosity taken as unity (Sun's luminosity = 1.0). The observed range in luminosity among stars is from as low as one-millionth that of the Sun to as high as 10,000 times that of the Sun. Actually, luminosities less than one ten-thousandth (0.0001) are rare. Figure 14.2 is a graph showing the relationship between a star's mass and its luminosity. Except for the white dwarfs, the plotted stars fall close to a smooth curve in which luminosity rises with increase in mass.

Before going further with an investigation of the origin and evolution of stars, we need to get a handle on the dimensions of space in which those stars are located. Determining distances to stars and other objects distant in space is a difficult subject to explain, and we shall not give attention to measurement techniques used by astronomers. The main point will be to try to get across a feeling for the vastness of space and to learn the unit of length used to state the distance from Earth to a given star or between two stars.

Our Sun lies at an average distance of about 150 million (150×10^6) km from Earth. Solar electromagnetic radiation, which includes X rays, light rays, infrared rays, and radio waves, travels through space at a uniform speed of about 300,000 km/sec (186,000 miles per second). Thus it takes a particle of radiation energy (a photon, that is) about 8 1/3 seconds to travel from the Sun to the Earth. The average distance between Earth and Sun, precisely given as 149.6×10^6 km, is taken as a distance unit called the astronomical unit (a.u.). Pluto, the most distant planet of our solar system, lies at a distance of just over 39 a.u. from the Sun.

Consider next the distance to the nearest star beyond our Sun: it is Alpha Centauri and its distance from the Sun is about 300 a.u. To gain some comprehension of this distance, consider that our Voyager space vehicles required about four years to travel from Earth to Saturn, a minimum direct distance of 8.55 a.u. Let us round off that figure to 8 a.u., for an average speed of 2 a.u. per

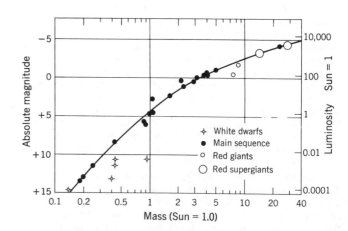

Figure 14.2 Mass-luminosity diagram. (From T. G. Mehlin, _Astronomy_, p. 45, Figure 2-3. Copyright © 1959 by John Wiley & Sons, Inc. Reprinted by permission of John Wiley & Sons, Inc.)

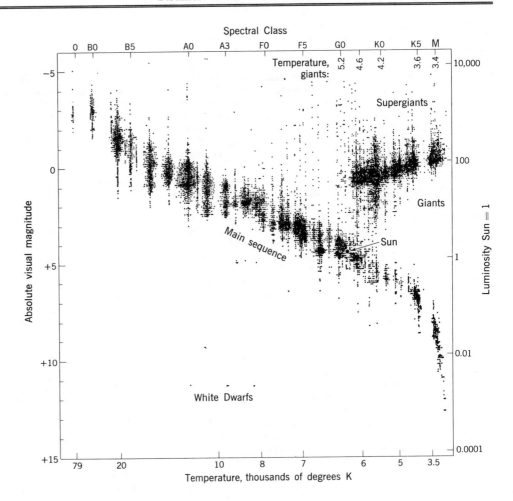

Figure 14.3 The Hertzsprung-Russell spectrum-luminosity diagram. Each dot represents a star. Altogether, a sample of 6700 stars is recorded on the diagram. (The Yerkes Observatory, University of Chicago.)

year. At that speed, a space vehicle would require 150 years to travel to the nearest star. It does not seem very likely that astronauts aboard that vehicle would live to reach Alpha Centauri, unless some strange biological effect caused normal human aging to slow down dramatically!

To state distances to stars more conveniently, astronomers quickly adopted a much larger unit, the light-year: it is the distance traveled by light in one year's time. Multiplying the speed of light (300,000 km/sec) by the number of seconds in a year we obtain a figure of approximately nine million million kilometers (9 x 10^{12} km). (A more nearly exact value of the light-year is 8.898 x 10^{12} km.) Of the fifteen brightest stars, all lie no more than 400 light-years away, but this is an insignificantly small distance compared with that to most objects we shall be investigating throughout the universe. Do you think you can cope with a distance of one billion light-years? or five billion?

Classes of Stars

With such enormous ranges in star diameters, masses, densities, temperatures, and luminosities, how can we make out some sort of meaningful system of star classes or types? Actually, not all possible combinations of size, luminosity, and temperature are represented. A pattern emerges when two properties, temperature and luminosity, are plotted as points on a simple diagram, or graph (Figure 14.3). Star temperature is plotted on a logarithmic horizontal scale with values declining from left to right. Star luminosity is plotted on the vertical axis, increasing upward on a logarithmic scale. Each dot on the graph represents one star, and several thousand stars were plotted to make the pattern show. Most of the stars

cluster in two bands; one runs diagonally downward across the graph from left to right; the other is a short horizontal band in the upper right-hand part of the graph. The first of these bands contains stars of the main sequence, in which we find our Sun. At the upper left end of the main sequence are the blue giants, which are both large and extremely hot. At the lower right end of the main sequence are the red dwarfs, which are both small and cool. The second band consists of very large stars that range in temperature from moderate to cool. There are the supergiants and giants, with colors ranging from yellow to red. In the lower part of the graph, near the center from left to right, are a few dots forming an isolated group. These are white dwarfs: stars that are relatively hot but of extremely small size. They are also stars of extremely great density, an example being Sirius's tiny companion star, referred to earlier.

The graph reproduced in Figure 14.3 is called the Hertzsprung-Russell diagram, or simply H-R diagram. It is named for two astronomers, working independently about 1910, who first plotted star luminosity against temperature. It has been used as the base not only for classifying stars into types, but also for formulating an evolutionary sequence of star development.

Formation and Evolution of Stars

You will notice that we are approaching the problem of origin and evolution of the universe in a series of steps that leads us from things closer at hand to those farther distant and from simpler objects to more complex objects. The approach path we are following is also that followed by modern astronomers of the twentieth century, based on the increasing capability of their telescopes to explore more distant and complex objects of the universe. One of the problems that is best approached early in a study of the universe is the creation and evolution of stars as individual objects.

Besides the stars themselves, astronomers observe in the space between and surrounding stars--in interstellar space, that is--some enormous glowing clouds of dust and gas, called nebulae. The word nebula is from Latin and Greek roots that mean "cloud" or "mist." As seen in the first telescopes, a nebula appeared as a faintly luminous film or mist that dimmed and diffused the images of bright stars lying beyond the cloud. As a matter of explanation, we now assume that the bright nebulae get their luminosity from clusters of young stars embedded in the clouds. Gas molecules in a cloud absorb radiation from those stars and reradiate that energy as light. The young stars within a bright nebula are thought to have formed from the substance of the cloud itself, but these new stars cannot be seen by optical telescopes and the earliest stages of stellar evolution could not, until recently, be directly observed. The presence of the hidden stars can now be detected by their infrared radiation. Later, after they are formed, radiation from the hidden stars blows away the surrounding cloud, or the stars may travel out of the cloud. Thus the clusters of young visible stars complete their formative process in comparative secrecy.

Besides the luminous nebulae astronomers observe larger interstellar clouds of dust and cold gas (mostly hydrogen) that form black patches or blobs among the stars. These clouds appear black because the dust particles block the passage of light from stars behind them, and the clouds themselves emit no light.

The interstellar clouds of dust and gas, both luminous and dark, are interpreted as the remains of numerous massive stars that have exploded, dispersing the pulverized and gaseous remains of the former stars far into surrounding space. Called a supernova, this type of star explosion has been seen on the average of about once per century. Throughout centuries following an explosion, an expanding bright nebula surrounds the point in the sky where the supernova occurred. For example, the remains of a supernova seen in A.D.1054 as a brilliant star is today the Crab nebula in the constellation of Taurus.

According to a prevailing hypothesis, star formation depends on the gravitational attraction of every gas molecule and dust particle in a nebular cloud for every other molecule and dust particle. The mutual attraction causes contraction of the entire cloud. But this can begin to happen only when outside forces have caused the cloud to be compressed to a density of about 1,000 hydrogen atoms in each cubic centimeter of space. There is good reason to suspect that the original outside force that compresses the cloud comes from a nearby supernova. The expanding remains of the supernova, consisting of minute grains of mineral matter, move out into space with a sharply defined front. Upon reaching a diffuse cloud of interstellar gas and dust, the moving front of grains impacts the cloud as a shock wave, then completely surrounds the cloud and compresses it into a smaller volume within which cloud density is sufficient to initiate gravitational collapse. Once it begins, gravitational collapse is rapid and the cloud experiences a rise in its internal temperature. It has now become a stellar nebula shaped like a thick lens and rotating about an axis perpendicular to the broad dimension of the lens (Figure 14.4). In the center of the nebula is a dense, ball-shaped mass of highly heated gas that we can call the protostar. For a star of the mass of our Sun, its protostar would initially have had a diameter of at least 10,000 times the Sun's diameter. Up to this time its internal temperature is entirely the result of gravitational compression of gas into a shrinking volume. There comes a point in the compression-heating process that nuclear fusion reactions begin near the center of the protostar. Nuclear fusion

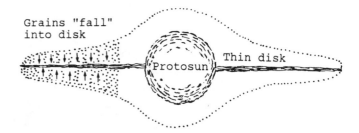

Figure 14.4 Schematic diagram of a stellar nebula with a protostar at its center. (A. N. Strahler.)

greatly increases its internal temperature and causes the new star to become highly luminous.

Fusion begins when temperatures within a contracting star exceed 4 million K. In detail, the fusion process consists of several forms of atomic reactions involving hydrogen with the elements lithium, beryllium, and boron, and ending with the production of helium. This set of reactions is found in stars the size of the Sun. For much larger stars, with internal temperatures over 20 million K, another series of complex nuclear reactions occurs in which hydrogen is involved with carbon and nitrogen to yield helium.

Hypothetical stages in the evolution of a star can be related to the Hertzsprung-Russell diagram in a schematic way, as shown in Figure 14.5. The diagonal shaded band shows the position of the main sequence stars. The chain of arrows entering from the right represents the evolutionary time-path of a single star about the same size as our Sun. The path moves horizontally across the graph toward the line of the main sequence, indicating a rise of temperature. This horizontal portion of the line is traversed rapidly and represents the stage of gravitational contraction of the gas and dust cloud as a protostar. When the protostar begins to consume its hydrogen by fusion, further contraction under gravity is

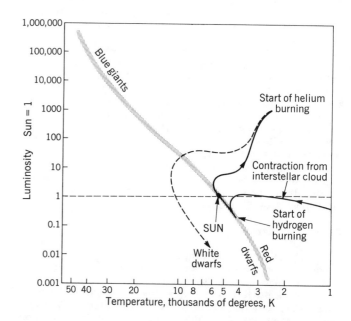

Figure 14.5 Simplified Hertzsprung-Russell diagram showing the inferred evolution of an average star. (From Otto Struve, The Universe, p. 55, Figure 36. The MIT Press, Cambridge, Mass. Copyright © 1962 by The Massachusetts Institute of Technology. Used by permission.)

halted. Now the mass is located on the line of the main sequence and its life as a star has begun. It may then brighten somewhat as its temperature increases and may move upward along the main sequence track, stabilizing in one position for several billion years.

The next stage in the life history of a star of about the Sun's mass is entered when its hydrogen supply is seriously depleted. Nuclear fusion ceases first in the central region of the star, which then contracts. Nuclear activity continues in a surrounding zone that gradually moves outward from the central region toward the surface. As this happens, the star may expand greatly. Although the luminosity increases greatly, the surface temperature falls, and the star becomes more reddish in color. As the track on the graph shows, the star has now arrived at the region of the red giant stars, becoming one of them for a time. At this point the star's helium is being burned in a fusion process that converts helium to carbon. The demise of the star, which is a hypothetical event, is suggested by a steep downward path leading to the area of the white dwarfs. This change occurs because the helium fuel is exhausted. The star contracts to an extremely small object of enormously great density and its luminosity correspondingly declines to a very low level.

Depending on their masses, stars of the main sequence have greatly differing life spans while maintaining their stable configurations during hydrogen burning. Stars of small mass--those near the lower right end of the sequence--are stable at comparatively low temperatures and pressures and therefore produce nuclear energy at a relatively slow rate. They are stars of faint luminosity and are red dwarfs. Such small stars will have extremely long lives, because the burning of the hydrogen supply takes place so very slowly. Their life spans will be vastly longer than that of our Sun, and they may be expected to endure for as long as thousands of billions of years. On the other hand, those stars at the high-temperature and large-mass end of the main sequence--blue giants, that is--are converting their hydrogen supplies into energy at an extremely fast rate. Their life expectancies will be short. For example, a star of mass 10 times that of the Sun will radiate energy about 10,000 times more rapidly than the Sun. The life of such a large star must therefore be on the order of 1 percent of the life of our Sun, or as short as 100 million years. As a blue giant nears the end of its hydrogen-burning period, its position on the diagram is shifted directly toward the right, bringing it into the group of reddish giants.

Red giants whose mass is about that of our Sun reach a stage of nuclear exhaustion in which the central region of the star consists mostly of carbon and oxygen, and this is surrounded by an outer envelope of hydrogen and helium. Temperatures in this outer region are not high enough to sustain fusion of hydrogen and helium. Now a series of explosions takes place within the star, blowing off the hydrogen and helium envelope, but this leaves most of the mass of the star intact, and it contracts into a white dwarf. In contrast, red giants having a mass on the order of five times that of the Sun reach a stage in the burning of their nuclear fuel at which they are ripe for a violent explosion, producing the supernova we mentioned earlier. When the bomblike detonation occurs, most of the star's mass is propelled outward into surrounding space in the form of gas and dust. Now the remaining mass, which is about the same as the Sun's mass, evolves into a white dwarf.

Neutron Stars and Pulsars

Although the final explosion of some stars produces a supernova and its remains become a white dwarf, there is another possible outcome for the final explosion. The remaining mass may be too large and too dense to become a white dwarf. Instead, it shrinks into a ball of matter so dense that it consists entirely of neutrons in close contact. Called a neutron star, it may be no larger than 10 km in diameter, but its density may be on the order of one million billion (10^{15}) times greater than water at the Earth's surface. To gain some idea of what such an enormous density means, think of a cube of matter the size of a sugar cube but weighing 100 million tons under the Earth's surface gravity!

A neutron star is thought to have an enormously strong magnetic field--perhaps a thousand billion times stronger than the Sun's magnetic field. Thus it is capable of emitting a powerful beam of radiation in the form of rays of light or radio waves. Because the neutron star rotates on an axis, the beam of light or radio waves sweeps through space much like the light beam from a lighthouse or rotating beacon light.

Actually, neutron stars had been predicted since the early 1930s, but for over three decades none had been observed. Then in 1967 astronomers at the Lick Observatory discovered a star that flashed "on" and "off" with surprising regularity; they called it a pulsar. The light pulse was accompanied by a radio beam following the same rhythm. Discovery of other pulsars followed rapidly. Their light pulses range in frequency from one pulse per four seconds in the slowest rhythm to as fast as thirty pulses per second. Although varied explanations were brought forward for a pulsar's behavior, evidence soon centered upon the hypothesis of a rotating star emitting a beam of radiation. A measurable slowing of pulse rate in these stars suggests that their energy output is steadily declining. Then, in 1969, the central star of the Crab nebula (the remains of a supernova) was shown to be a pulsar, fitting nicely into the hypothesis that the supernova explosion of A.D. 1054 had left behind sufficient mass to contract into a neutron star.

Galaxies

Galaxies are huge aggregations of stars, bunched comparatively close together and separated from one another by large volumes of seemingly empty space. Our solar system lies within a galaxy known as the Milky Way galaxy. Nearly all the stars and bright nebulae we can observe with the unaided eye and with small telescopes are members of our own galaxy.

The Milky Way galaxy has the form of a great disk or wheel, with a marked central thickening at the hub (Figure 14.6). If it could be seen from an outside vantage point, our galaxy would probably be quite similar to the Whirlpool nebula (galaxy M51), and to the Great Spiral galaxy (M31) located in the constellation of Andromeda.

Our Sun occupies a position more than halfway out from the center toward the rim of the galaxy (Figure 14.6). As we look out into the plane of the disk, we see the stars of the galaxy massed in a great band, the Milky Way, that completely encircles the celestial sphere. Astronomers estimate that the Milky Way galaxy contains about 100 billion stars. The diameter of the galaxy is about 100,000 light-years; its thickness ranges from 5,000 to 15,000 light-years. The galaxy has a system of spiral arms, comparable to those seen in the Andromeda and Whirlpool spirals. The arms consist of individual aggregations of stars, known as star clouds, each having dimensions of 5,000 to 20,000 light-years.

The Milky Way galaxy also contains clouds of gas and dust, which we have already described as nebulae. Concentrations of these clouds are particularly heavy in the plane of the galactic disk. Surrounding the disk is a vast halo of stars widely scattered above and below the disk, and among them are globular star clusters. Globular star clusters tend to be more or less spherical and are free of clouds of gas and dust (except perhaps in the very center of the cluster). A typical cluster may have a diameter of 30 light-years and may contain 100,000 stars.

A

Perseus arm

Local arm

Sagittarius-
Carina arm

Cygnus arm

(Unnamed arm)

10,000 parsecs

B

Figure 14.6 Schematic diagrams of the Milky Way galaxy:
(A) As viewed from a point in the plane of the spiral.
Large spots represent stars clusters; small dots represent
stars. (B) As viewed from a point above the center of the
disk, showing four major spiral arms. The black dot
represents our Sun. Open circles represent older, fainter
stars, surrounded by giant molecular-cloud complexes.
(A. From Otto Struve, The Universe, p. 76, Figure 44.
The MIT Press, Cambridge, Mass. Copyright © 1962 by
The Massachusetts Institute of Technology. Used by
permission.) (B. Modified from Leo Blitz, Scientific
American, vol. 246, no. 4, p. 94. Copyright © 1982 by
Scientific American, Inc. Used by permission.)

Star clusters also are present within the galactic disk,
but these are smaller and more open than the globular
star clusters. The process of star formation, already
reviewed, seems to produce new stars in clusters, and
when the clouds of dust and gas are blown away, the new
stars appear as open clusters within the disk.

Spiral galaxies such as ours and the Andromeda galaxy
rotate slowly about the central axis or hub. The central
region rotates more rapidly than the more distant outer
regions, and this accounts for the development of curved
spiral arms. At the position occupied by our Sun, a full
cycle of rotation would require about 200 million years.
The velocity of our solar system in the circular path due
to rotation is about 800,000 km/hr. Besides this general

rotational motion, individual stars have motions in various
directions with respect to the plane of the disk and to
one another.

The Andromeda galaxy, a very near neighbor as
galaxies go, can be seen with the naked eye, although
only the dense central part is visible and it appears as
just another star. It lies at a distance of about 20 million
light-years from us and is one member of a small cluster
of galaxies known as the Local Group. The group consists
of about sixteen galaxies, but most are small ones, known
as dwarf galaxies. Several of these are at distances of
only 200,000 to 800,000 light-years. These small galaxies
lack a distinct spiral structure. Individual stars and dust
clouds can be recognized within the nearer galaxies.

Most galaxies of the universe lie at enormous distances
from us and each other, for example, 300 million or 3
billion light-years. To the most distant limits of telescopic
penetration our universe contains galaxies, of which an
estimated 10 billion can now be observed. The total extent
of possible telescope observation of light from distant
galaxies is estimated to be 10 billion light-years. Within
this theoretical maximum radius of observation there may
be as many as 100 billion galaxies. Although galaxies are
found in all regions of the visible universe, they tend to
form groups called galaxy clusters. A typical cluster has
from 5 to 20 members, but a few very rich clusters each
contain more than 100 major galaxies.

When the positions of galaxy clusters are plotted on a
complete sky map, their distribution appears more or less
uniform over the entire map area, excepting that portion
of the map on which the Milky Way lies. (Density of stars
and gas clouds in our galaxy prevents observation of
distant space that lies beyond it.)

Galaxies fall into several classes, according to their
forms. One major type consists of the spiral galaxies,
exemplified by the Milky Way, Whirlpool, and Andromeda
galaxies. Another important class of galaxies are the
barred spirals, in which two arms uncoil from a central
bar. Equally important as a group are the elliptical
galaxies, which are ellipsoidal or spherical objects having
a high degree of symmetry. This form suggests that the
elliptical galaxies, like the spirals, are rotating. Clouds of
dust and gas that are typical of the spiral galaxies seem
to be absent from the elliptical types. In addition, there
are galaxies of highly irregular shapes, but large galaxies
of this kind are relatively few.

Edwin P. Hubble (1889-1953), an American astronomer
who did much of the pioneering work in galactic
investigation, suggested an arrangement of galaxy types
that could represent an evolutionary sequence through
time. It begins with almost spherical elliptical galaxies,
which then progress to more flattened elliptical types;
these in turn evolve into either spirals or barred spirals
(Figure 14.7).

When more became known about the galaxies and the
ages of stars in them, another American astronomer,
Harlow Shapley (1885-1972), suggested that the
evolutionary sequence might well begin with the irregular
galaxies, developing in turn into spiral systems in which
the nucleus moves tightly into the arms with increasing
age. Then, as the stars age and the interstellar clouds of
gas and dust were eliminated, the spirals might evolve
into elliptical galaxies. This hypothesis does not,
however, explain why some spirals take the normal form
and some become barred.

In speculating upon how an individual galaxy is
formed, we can take our cue from the hypothesis we have
already presented for the formation of a single star from
a cloud of gas and dust. This time, however, it is
necessary to postulate that we are dealing initially with a
gas cloud containing enough matter to produce 100 billion
stars! For reasons that will become clear later, the cloud
consisted only of gas, and most of that gas was hydrogen
and perhaps a small proportion of helium. (According to
this hypothesis, when the first galaxies were formed,

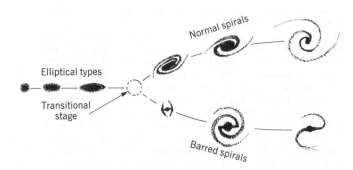

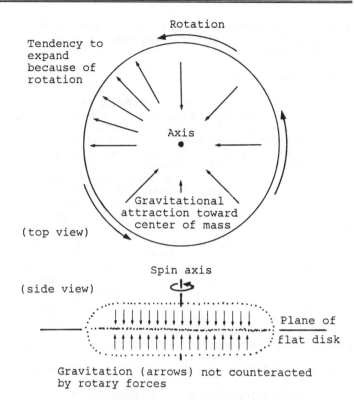

Figure 14.7 Diagram of a sequence of galaxy types as arranged by Edwin P. Hubble. No galaxy has been recognized in the transitional stage, which is hypothetical. (From E. Hubble, The Realm of the Nebulae, p. 45, Yale University Press, New Haven, Conn. Copyright 1936 by the Yale University Press. Used by permission of the publisher.)

Figure 14.8 Schematic diagrams of gravitational force in relation to rotary forces, tending to cause expansion in a rotating gas cloud. (Based on data of Fred Hoyle, 1975, Astronomy and Cosmology, W. H. Freeman & Company, p. 273, Figures 6.8 and 6.9.)

elements heavier than helium had not yet been formed.)

Our hypothesis requires that the huge gas cloud began to contract under the mutual gravitational attraction of all its gas atoms. (A possible cause for such contraction to occur is discussed later.) As contraction occurred, the entire cloud began to spin about an axis of rotation. Because the average radius of the cloud was decreasing, the rate of spin was steadily increased. This familiar principle of physics is sometimes called spin-up. But the same rotation was also bringing into play a set of forces acting radially outward at right angles to the central axis of rotation (Figure 14.8a). This outward rotary force increases as spin-rate increases, so that a time must be reached when the opposed sets of forces are in balance, and contraction toward the spin axis ceases. At the same time, however, the atoms of gas are attracted by gravitation to the central plane of the disk, but their response to this attraction is not met by an opposing force due to the rotation (Figure 14.8b), Thus the gas cloud could develop into the form of a flattened disk, such as that of a spiral galaxy.

In addition to purely mechanical forces, the spinning galactic gas cloud was strongly influenced by magnetic fields in which lines of magnetic force must have radiated outward from the central core to the periphery of the disk. The force lines were deformed into outward-spiraling bands. The effect of this magnetic field was to transfer kinetic energy to the outer part of the spinning disk, while at the same time reducing the rate of spin of the central core. The same principle is used in explaining the early history of our solar system, which is a rotating system of extremely small dimensions in comparison with a galaxy.

Contraction of the galactic gas cloud would result in a rise in the internal temperature of the gas, which would tend to expand the gas and thus resist the contraction. Here again, the result would be a cessation of further contraction were it not for another process that comes into play. As the gas temperature increases, the intensity with which energy is radiated outward into space from the cloud must also increase. This energy loss would tend to cool the cloud and allow contraction to continue. (This is quite a different result from that for individual stars forming as protostars within a cloud of gas and dust, for such a cloud would be opaque to radiation and would hold the heat energy of compression within the enclosing cloud.)

How did the first stars form within a contracting galactic cloud? Earlier, our hypothesis attempted to explain the birth of a star, or group of stars, through the shock effect of the front of a dust cloud produced by

a supernova, but here we cannot think in terms of a previously existing star. We are limited to deriving stars from primeval hydrogen gas alone. The suggestion has been made that, early in the history of the universe, energy traveled through the dispersed gas of space in the form of pressure waves. The passage of a pressure wave through the gas would locally increase the density of the gas in the wave "crest" and decrease the density in the "trough" to the rear of the wave. Gravitational forces would now draw the denser gas closer together and the final contraction process would set in. The contracting mass would fragment into many separate contracting centers, each becoming a critical point, hydrogen fusion would start, and a star would come into existence. Based on reasons of physics, that we cannot go into here, it is considered likely that the clouds that formed the first stars must have had masses thousands to hundreds of thousands of times greater than the Sun. Such stars would have been formed in great clusters, each one having thousands to hundreds of thousands of stars. The globular star clusters seen today could thus represent the first class of stars to come into existence.

We can imagine numerous star clusters to be forming in a spinning glactic cloud prior to the time it collapsed into a flattened disk. Stars lying far above or below the equatorial plane of the galaxy would tend to remain near the location where they formed. Later, when the bulk of the gas had been drawn into the thin disk, these early stars would become the globular star clusters that are seen today surrounding tightly wound spiral galaxies. Under this scenario, globular star clusters consist of very old stars; they must also be of comparatively small mass in order to have had very long lives. Of course, many very large stars were also formed in the early period, but they would have had short lives and would have soon exploded into supernovae. The dust and gas of these explosions would have moved into the thin disk. To explain the scattered halo stars in this outer region, we would need to speculate that they bacame separated from their parent clusters and drifted far away to assume isolated positions.

Origin of the Elements

Throughout the development of chemistry as a quantitative science in the latter half of the nineteenth century it was assumed that all matter consisted of atoms that were indestructible. When and how the atoms had originally been created were beyond the limits of useful speculation, and it could be assumed that atoms had always existed in the varieties in which they are found today. This principle made chemistry a precise and orderly science, and it certainly ruled out the possibility that latter-day alchemists could ever transmute lead into gold.

The discovery of radioactivity by Henri Becquerel in 1896 and subsequent research by Marie and Pierre Curie led to the first suspicion that it was possible for atoms of one element to change spontaneously into atoms of another element. This was what appeared to be happening to uranium--it was somehow changing into radium. Further research led to the conclusion that a long series of such spontaneous changes (radioactivity) end up in the production of lead, which is a stable form of matter. The transformation of one element into another was observed to be accompanied by the release of some rather strange form or forms of energy.

Within only a few years after the discovery of radioactivity, increased knowledge of the process led to startling new concepts in both astronomy and geology. As we explain on later pages, chemists used the principles of radioactivity to show that some rocks of the Earth's crust are at least 2 billion years old. This conclusion conflicted sharply with Lord Kelvin's previous calculation that the Sun could not be more than 20 million years old. In 1909, John Joly, a chemist, brought forward the hypothesis that heat generated by radioactivity within the Earth's crust was the energy source of volcanic activity and of various upheavals and deformations of crustal rocks. Astronomers were quick to suggest that radioactivity might be the source of energy in stars, which could thus be vastly older than had previously been thought. Unfortunately, this suggestion encountered serious objections.

In 1920, Sir Arthur S. Eddington (1882-1944), a leading British astronomer and physicist, proposed an entirely different process by which one kind of atom might be changed into another kind of atom. He visualized that within the Sun and other stars the nuclei of hydrogen atoms were being fused together to form helium atoms, and he was able to show that this nuclear fusion process would yield energy. Eddington's proposal must surely mark the point of liberation of modern stellar astronomy from its straitjacket of dependence on mechanical laws of physics that characterized the Victorian era of science. Astrophysicists have a great deal of confidence in the hypothesis that elements are presently being created within our own galaxy and in other galaxies, and that this process has been going on for billions of years.

Nucleosynthesis is the scientific term for the creation of elements by forming new groupings of protons and neutrons in the nucleus of the atom. The list of elements so far identified by chemists numbers 103, but of these 11 have only been produced by nucleosynthesis in the laboratory and are not known to exist in nature. That leaves 92 elements to account for by natural processes acting somewhere within the universe. (We assume under the Big Bang theory that no elements, as such, existed at the instant the cosmic explosion was set off.)

Based on information obtained from the radiation spectra of stars, galaxies, and bright nebulae, estimates have been made of the relative cosmic abundance of the naturally occurring elements. The elements are numbered in integer sequence from 1 through 92. This number is the atomic number and it is equal to the number of protons in the nucleus of the atom. It happens that elements with atomic numbers 1 and 2 (hydrogen and helium, respectively) are first and second in order of cosmic abundance. Hydrogen is the lightest element, with one proton as the nucleus and one electron orbiting the nucleus. Helium has four times the mass of hydrogen, because its nucleus consists of two protons and two neutrons. The ratio of cosmic abundance of these two elements is about on the order of $14\frac{1}{2}$ parts hydrogen to 1 part helium (Table 14.1). Hydrogen alone makes up close to 93.4 percent of all cosmic matter; helium, 6.5 percent; the two combined, 99.9 percent.

The third most abundant element, oxygen, and the fourth, carbon, are roughly equal in abundance but only on the order of 1/100 that of helium and 1/1000 that of hydrogen. The next two elements, nitrogen and neon, rank fifth and sixth respectively and are each on the order of 1/10,000 the abundance of hydrogen. There follow in order magnesium, silicon, aluminum, iron, sulfur, argon, sodium, calcium, and nickel, to complete the fifteen most abundant elements. Below this point, abundances fall off rapidly. Table 14.1 carries the list through twenty-six elements.

In our earlier explanation of the nuclear activity in stars, we noted that the dominant process in stable stars of the main sequence is the conversion of hydrogen into helium by fusion. It appears, then, that the matter of the early universe consisted entirely, or almost entirely, of hydrogen. According to one version of the Big Bang theory, the formation of hydrogen atoms occurred rather suddenly, beginning when the universe was about 300,000 years old. By then the temperature had dropped to about 4,000 K, enabling one electron to combine with one proton in a stable relationship. As the universe expanded rapidly, the hydrogen gas cooled rapidly and its temperature fell steadily as a result of the expansion.

Table 14.1 Cosmic Abundances of the Elements

Rank	Name	Chemical symbol	Atomic number	Abundance*
1	Hydrogen	H	1	31,800
2	Helium	He	2	2,210
3	Oxygen	O	8	21.4
4	Carbon	C	6	11.8
5	Nitrogen	N	7	3.6
6	Neon	Ne	10	3.44
7	Magnesium	Mg	12	1.06
8	Silicon	Si	14	1.00
9	Aluminum	Al	13	0.85
10	Iron	Fe	26	0.83
11	Sulfur	S	16	0.50
12	Argon	A	18	0.12
13	Calcium	Ca	20	0.07
14	Sodium	Na	11	0.06
15	Nickel	Ni	28	0.05
16	Chromium	Cr	24	0.013
17	Phosphorus	Po	15	0.0096
18	Manganese	Mn	25	0.0093
19	Chlorine	Cl	17	0.0057
20	Potassium	K	19	0.0042
21	Titanium	Ti	22	0.0028
22	Fluorine	F	9	0.0025
23	Cobalt	Co	27	0.0022
25	Zinc	Zn	30	0.0012
26	Vanadium	V	23	0.0003

*Abundance is given relative to number of silicon atoms expressed as unity.

Data of A. G. W. Cameron, 1970, Space Science Reviews, vol. 15, pp. 121-46.

In discussing the origin of galaxies, we noted that the first stars must have been derived from clouds of gas that consisted almost entirely of hydrogen. When nuclear fusion began in these first stars, helium was produced. In large stars, those of masses some twenty times the mass of the Sun, the fusion of helium was followed by the burning of helium to yield carbon, and of carbon to yield oxygen, of oxygen to yield magnesium, and of magnesium to yield silicon. In the largest stars, fusion would have led to the formation of iron in the core of the star. By then the star was unstable and ready to explode as a supernova. Thus in its final stage, the star would have consisted of concentric layers of the elements we have listed, with silicon or iron forming the inner core. Explosion of the star as a supernova would have released the elements synthesized by fusion as part of the resulting cloud of gas and dust. In this way the original hydrogen clouds of the early universe would have become enriched by heavier elements.

Synthesis of the heavier elements--those with atomic numbers greater than that of iron--requires enormously high temperatures and pressures for nucleosynthesis. Astronomers visualize the sudden collapse of the core of the star to form a dense neutron mass as the source of a shock wave that moves rapidly out through the surrounding shells of silicon, magnesium, oxygen, carbon, and helium, creating intensely high temperature as it travels. At the same time the collapsing core radiates neutrons that bombard the nuclei of the elements of the outer layers. The combination of high temperature and neutron bombardment causes new and larger atomic nuclei to be formed by fusion. In this way all of the remaining elements, for a total of ninety-two varieties, together with their isotopes, came into existence and were dispersed into interstellar clouds of gas and dust, only to be re-formed into new stars. In the atmosphere of our Sun, sixty-six elements have been identified by spectroscopic analysis, and doubtless most of the others are present but in quantities too small to detect. The richness of element composition of the Sun shows that it is a second-generation star formed from the debris of earlier supernovae explosions.

It is supposed that nucleosynthesis of the heavier elements was very rapid during the time of the formation of the earliest stars of the galaxies. Stars of very large mass were then probably much more numerous than now. With short lives, these massive stars lived for only a few millions of years before exploding. The frequency of supernovae was then perhaps 100 times greater than in our galaxy today, where we have been observing supernovae about once per century. Thus nucleosynthesis continues on a reduced scale, and the elements it produces will be recycled through succeeding generations of stars.

How Far Away Are Galaxies?

Major advances in science are often won only through enormous inputs of hard, tedious work, extending over many years. An outstanding example of this principle was the measurement of distances to stars and galaxies. Distance measurement proceeded by a series of steps, from the closest stars to the more distant ones. Each step proved more difficult than the previous one. Regrettably, this great scientific campaign can't be traced here in detail, for it involves some highly complex concepts and would take several pages to explain.

The campaign took a great step forward with the establishment of a foolproof method of measuring distances to stars of a certain class: they are variable stars, whose luminosity rises and falls in a regular cycle with a period of from 1 to 45 days. For this class of variable stars astronomers use the name classical Cepheids. It is established that the luminosity of any particular member of the classical Cepheids depends on the length of its period--the longer the period, the greater the luminosity. This relationship in itself allowed astronomers to measure the distance to groups of classical Cepheids, not only in our Milky Way galaxy, but also in galaxies as far distant as several million light-years. This was possible because, in the bright phase of their cycles, these stars are extremely brilliant. Measurement of distances to other galaxies was a major breakthrough in astronomy. It was followed by a chain of somewhat similar measurement procedures in which the luminosity of bright nebulae was used and also the luminosity of giant elliptical galaxies and clusters of those galaxies. Although the chain of calculations is subject to a certain percentage of error, and the error can be assumed to increase with distance, we now have a reasonable scale of distances extending out to very distant galaxies.

The Redshift

To understand the astronomers' next giant step into the distant universe we must draw on a principle of physics that can be explained in rather simple terms using your ears as test instruments. Suppose you are waiting in your auto at a railroad grade-crossing while a fast-moving diesel locomotive bears down upon you, sounding a steady note on its powerful horn. The instant the locomotive passes you, now moving rapidly away, the pitch of the horn drops sharply to a lower note. This is the Doppler effect. It results from the fact that during the period of approach of the sound source, the frequency of the sound waves is increased in proportion to the speed of the source relative to the observer whereas during the period of recession, the frequency of the sound is decreased in a similar proportion.

A very simple analogy may help to illustrate the Doppler principle. Imagine yourself to be standing beside a long, horizontal conveyor belt and that you place a pebble on the belt at uniform intervals of time--say, once every second. If the belt speed is constant, the pebbles will be uniformly spaced. Next, as you place a pebble on the belt each second, walk slowly in the direction in which the belt is moving; the pebbles will now be spaced closer together. If you turn about and walk in a direction opposite to the belt motion, continuing the same activity at one-second intervals, the pebbles will be spaced farther apart on the belt than when you are standing still.

To carry the Doppler effect into outer space requires an understanding of basic principles of electromagnetic radiation and its spectrum and the principle of the spectroscope. Take the case of a star emitting a given light spectrum. If the star is moving earthward it appears to us that the frequencies of vibration constituting the light rays are all increased slightly, resulting in a slight change in the star's dominant color. This change occurs because the color we observe is determined by the frequencies of light, and these have been increased by the star's motion in the direction that the rays are traveling. In this case a reddish star would appear slightly less red and more yellow. If the star is moving away from the Earth, a reverse effect occurs: the frequencies of the vibrations are all slightly reduced. Thus a yellow star would appear slightly more reddish.

In applying the Doppler effect, astronomers measure the displacement in wavelength of a given absorption line in the spectrum of the star. (Wavelength and frequency are inversely related; as frequency increases, wavelength diminishes.) For example, the spectrum of star 61 Cygni shows that the H line of calcium is displaced toward the blue end of the spectrum in an amount equal to 0.86 Angstrom unit. By use of a simple formula, the approach velocity of the star along the line of sight (the radial velocity, that is) can be calculated and proves to be 65 km/sec. Of course, the star's true motion is most likely to be in a path at some angle to the line of sight. The

Doppler method only yields the radial velocity, whether toward us or away from us.

For distant galaxies, the spectrum of the entire galaxy is photographed as if it were a single point source of light. After spectra of many galaxies were photographed and displacements of the spectral lines measured, a remarkable conclusion emerged: the shift of the spectral lines was always toward the red end and, moreover, the amount of that shift was always in direct proportion to the distance of the galaxy. The phenomenon is called simply the redshift. Once the relationship of amount of redshift to distance was established, the redshift itself could be used to estimate the approximate distance to any galaxy within the range of telescopic observation. Figure 14.9 shows the spectra of five galaxies ranging in distance from 78 million light-years to nearly 4 billion light-years. Arrows show the amount and direction of shift of the H and K lines of calcium.

An Expanding Universe

If we assume that the physical principle of the redshift as a real Doppler effect holds valid to the greatest distance that it is observed in galaxies, an inescapable general conclusion follows. Nearly all distant galaxies are moving radially outward relative to our position of observation, and their speed of recession increases proportionately with their distance.

If all galaxies seem to be receding from us at speeds proportional to their distances, must we conclude that we are located at the center of the universe where all matter was originally located? The answer does not need to be "yes." If the universe is expanding equally in all directions, any vantage point of observation we might choose to occupy within the universe would produce the same recessional phenomenon we observe. The concept of a universe expanding uniformly in all directions impacted cosmology with tremendous force and quickly led to the formulation of the Big Bang theory.

Figure 14.10 shows the radial velocities of Figure 14.9 plotted against their distances. True proportionality would be represented by a straight line originating at the zero point on both scales and slanting upward to the right. The points representing the five galaxies lie close to such a line. The relationship of radial velocity to distance has been called Hubble's law; it can be expressed by the following simple equation:

$$V = HD, \text{ where}$$

V is radial velocity in km/sec,

D is distance in millions of light-years (m.ly), and

H is a numerical constant, or constant of proportionality, in units of kilometers per second per million light-years (km/sec/m.ly).

The redshift of galaxies was discovered by V. M. Slipher, an astronomer who did his work at the Lowell Observatory in Flagstaff, Arizona. Between the years 1912 and 1917 he measured the redshifts of more than twenty galaxies. This work was carried on by Edwin M. Hubble and a colleague, Milton Humason. Their accumulated data revealed the consistency of the equation we stated above, and it has since become a cornerstone of cosmology. The constant of proportionality, H (the Hubble constant), has been since revised toward lower values. Hubble used a constant of 160 km/sec/m.ly, as shown in Figure 14.10, and this led him to the conclusion that the age of the universe would be 2 billion years (2 b.y.). This figure turned out to be much less than the calculated ages of ancient rocks in many parts of the continents. The geological calculations had been based on rates of

Distance in millions of light years:	Red-shifts: (km/sec)

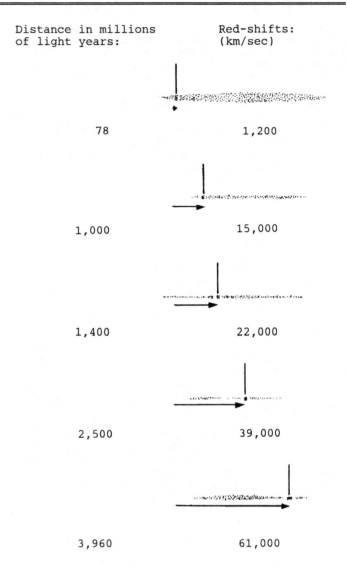

78	1,200
1,000	15,000
1,400	22,000
2,500	39,000
3,960	61,000

Figure 14.9 The spectra of five galaxies, showing the relationship of redshift to distance. An arrow on each spectrum shows the shift of calcium lines H and K. (Data from a Palomar Observatory Photograph. Used by permission.)

radioactive decay of minerals and were then, as now, highly regarded as reasonably accurate ages. Even to astronomers of that time, the age of 2 b.y. seemed much too low an estimate to accommodate the evolution of galaxies.

In 1952, data obtained from the 200-inch (500-cm) reflector telescope at the Mount Palomar Observatory required that Hubble's constant be revised to 65 km/sec/m.ly, and this revision placed the age of the universe at nearly 5 b.y. It was not long, however, before the constant was revised to 32 km/sec/m.ly for an age of 10 b.y., and by the late 1970s, the age was raised even higher, to about 17 b.y., or even as great as 20 b.y. The age is still under debate, however, and various values are given by different authors. Independent evidence of the age of our galaxy, based on calculations of the production of radioactive elements in our galaxy, have led to the conclusion that it has an age at least as great as 8.5 b.y. and probably close to 12 b.y. Thus an age for the universe itself ranging from 10 to 20 b.y. is not unreasonable.

An interesting thought is that the radial velocity of an extremely distant galaxy might approach and even exceed the speed of light (300,000 km/sec). The corresponding distance in light-years (indicated on Figure 14.11) would

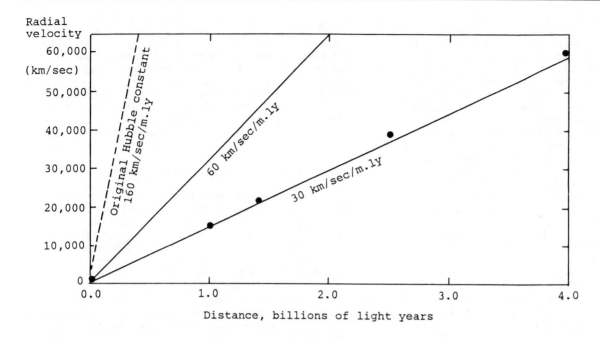

Figure 14.10 Plot of radial velocity versus distance for the five galaxies shown in Figure 14.9. (A. N. Strahler.)

constitute the limit of the universe, beyond which we could receive no light or radio waves from the emitting sources. (The special theory of relativity requires that no object in motion have a speed greater than that of light, for as that limiting speed is approached, the mass of the object increases to infinitely large values.)

Galaxies are known with measured redshifts so great as to indicate a radial velocity one-third of the speed of light (or even more). From this observed redshift and an assumed Hubble constant, we infer that they were 5 billion light-years distant at the time the light now entering our telescopes left the galaxies. In such a case, a galaxy can be said to have minimum age of 5 b.y. However, that galaxy may have formed much earlier in the history of the universe and without some other evidence, independent of the redshift, we can't even guess just how much longer the galaxy had been in existence before it sent out the signals we are now receiving. If our selected Hubble constant leads to an age of the universe of 20

b.y., the signals from the galaxy were sent out 15 b.y. after the universe came into existence. That leaves a lot of blank time to fill in.

By the same token that we observe a particular galaxy as it was 5 billion years ago, we can have no knowledge of what has happened to it since. For that matter, it may no longer exist and, if so, its fate will never be known to humans.

Because the speed of light is constant (or if we choose to assume that it is) and because interstellar and intergalactic distances are so vast, each star and galaxy presents itself to us in reference to a different point in past time. We may see one star as it was 10 million years ago, and next to it another star as it looked only 100 years ago. A telescopic examination of this same part of the sky might reveal a galaxy located between those two stars, and its photographic image shows it as it existed, say, 3 billion years ago. This concept is hard to grasp as a reality, because in the familiar surroundings of our planetary environment everything our eyes survey is a true image of the present instant of time.

The Big Bang Theory

The discovery of Hubble's law quickly led to a new theory of origin of the universe. It was proposed by Abbé George Lemaitre, a Belgian, who referred to it as a "fireworks theory." Uniform expansion of the universe carried backward in time must lead us to a zero point in time--perhaps 10 to 20 billion years ago--when all matter and energy was concentrated into a very small volume of space. This dense cosmic nucleus had exploded as if it were a powerful bomb and its fragments moved outward in all directions. Although we have earlier referred to this theory as the Big Bang, a more conservative title is evolutionary theory. Lemaitre's view of the expanding universe was that it is finite but unbounded. Such a model universe is described as an open universe, in contrast to a model that has an outer boundary and is a closed universe. The evolutionary theory was based on Albert Einstein's general theory of relativity and perhaps for that reason rapidly gained strong support.

The Big Bang theory and others that have been seriously considered by modern cosmologists adhere to a general hypothesis known as the cosmological principle.

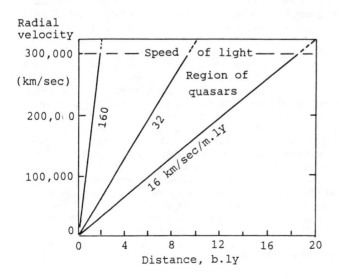

Figure 14.11 Radial velocity, distance, and the speed of light for three values of the Hubble constant. (A. N. Strahler.)

The principle rests on the assumption that the laws of physics as determined from observations made by humans on Earth apply to the entire universe throughout its entire history. Without such an assumption all sorts of wild speculations could be made about the behavior of the universe in other parts and at other times. Under the cosmological principle the distribution of galaxies is considered to be uniform in all directions through space, so that an imaginary observer occupying a position in any galaxy would have the same general picture of the universe. The uniformity referred to here is a general or average condition, for we know from observation that galaxies are clustered into small clusters and great clusters. An absolutely uniform distribution of galaxies is an idealized concept, but is taken to be the basis of any theory of origin and evolution of the universe. Operating under the cosmological principle there is room for other theories of cosmology quite different from the Big Bang theory, and we now turn to one that has been its strongest rival in past years.

In 1948 three astronomers, Hermann Bondi, Thomas Gold, and Fred Hoyle came forward with the steady-state theory of the universe. To the cosmological principle as stated above, they added another provision: the universe is changeless in both space and time. This did not imply that there was no physical activity or change going on within stars and galaxies, but only that the general composition and structure of the universe was sustained in a certain unchanging condition. They recognized and accepted the redshift of distant galaxies as evidence of an expanding universe, but reasoned that such expansion would be accompanied by a steady decrease in average density of matter in the universe. To offset the density increase, they postulated that matter in the form of hydrogen atoms was being continually created in intergalactic space. They held that the new atoms were being created from nothing. From this new matter new galaxies were continually being formed. Galaxies pass through an aging cycle, and as they become old they separate from one another and are replaced in space by young galaxies.

Perhaps your immediate reaction is that the steady-state hypothesis violates the most strongly held postulate of mainstream science--matter cannot be created from nothing (ex nihilo). All well and good if the new matter is derived from preexisting energy or from matter in another form, but not creation from nothing! The authors of the new theory pointed out that the rate of creation of hydrogen is at the rate of one atom per billion years in a volume of space of about one liter. This rate is much too slow to be measured by any known means, but it would be fast enough to produce new galaxies as needed to keep constant the average density of matter in space. The steady-state theory implies a universe with no beginning in time and no ending. Its age could be infinite and so could its duration into future time. Thus it conveniently avoids the problem of how the universe began, what created it, and what preceded it.

Cosmologists have thought of several ways in which the steady-state model might be tested. One point to consider is that under this theory a cluster of galaxies would have the same proportion of new and old galaxies, no matter where in the universe that group might be situated or at what age it is observed. The argument might be illustrated by considering the ages of individuals in an ecologically stable group of animals, such as a herd of elephants. In the herd we should expect to find very young and very old individuals present, because new individuals are being born yearly, all the individuals are aging, and each year as many die as are born. With the population in a steady state in terms of numbers and ages of individuals, an elephant herd observed in the year 1700 would appear about the same as one observed in 1800, or one observed in 1900. Observations to date fail to show a wide range of ages of individual galaxies from young to old within a single cluster. Instead, not only are all galaxies within a single cluster of roughly the same age, but all contain old stars. This evidence seems to favor a scenario in which all galaxies and galaxy clusters formed at about the same general stage in time, which under the Big Bang theory was perhaps within the first 5 billion years following the Big Bang itself. Several other arguments seem unfavorable to the steady-state theory, but we shall not go into them here. Later, we will introduce some evidence that strongly favors the Big Bang theory to the exclusion of the steady-state theory.

Radio Galaxies, Quasars, and Black Holes

Part of the electromagnetic radiation spectrum--that in the extreme longwave region of the spectrum--consists of radio waves. In the range of wavelengths between about 1 cm and 20 m, radio waves from outer space can pass freely through our atmosphere and can be received by radio telescopes. These instruments use a huge bowl-shaped antenna of parabolic cross-section that can be aimed at a distant emanating source.

Thousands of radio-emitting sources have been discovered and their positions plotted, but only a few can be identified with stellar objects that actually appear on photographs. Some sources of radio emission lie within our Milky Way galaxy; others are in distant galaxies, called radio galaxies. Radio galaxies are extremely powerful sources of radio emission.

In the early 1960s a discovery of major importance shook the field of astronomy. It was the finding of some extremely small sources of intensely powerful radio emission not associated with any surrounding galaxy. These objects were first named quasistellar radio sources but this term was quickly reduced to quasars. By zeroing in on the exact point of radio emission, the optical image of the quasar could be identified as a faint pinpoint of light. The spectra of many of these starlike objects showed an extreme redshift--much greater than observed in the most distant galaxies. This observation was interpreted to mean that quasars are extremely distant objects. The light from the most distant quasars began to emanate from them as long ago as 15 b.y., which can have been only a few billion years after the Big Bang had initiated the universe.

Getting back to the radio galaxies for the moment, the quantity of energy they emit as radio waves may exceed the total light energy emitted by all the stars in the same galaxy. For radio waves to be produced in such great intensity, it is necessary that very strong magnetic fields exist within the galaxy. Free electrons are propelled by the magnetic fields at speeds approaching that of light and move in tight spiral loops within which the radio waves are generated. Physicists refer to this form of radio emission as synchrotron radiation.

It is found that the source of radio emission, which can be plotted in detail on a sky map of the galaxy, often does not coincide with the visible portion of the galaxy. Instead, it may come from a pair of radio lobes, one on either side of the visible galaxy (Figure 14.12). It is thought that the radio lobes are the result of internal explosions within the galaxy that send out in opposite directions two clouds of highly charged particles that are the emitters of the radio waves.

Astronomers now speculate that the source of energy within the core of the galaxy may be a black hole--a concentration of mass so dense and large that its gravitational force pulls the surrounding matter and all radiation into itself. You might say that a black hole can "swallow its own light" and, in so doing, make itself invisible. The streams of charged particles that form the radio lobes are not part of the black hole but are surrounding gas clouds energized by the enormous gravitational force of the black hole. Most radio galaxies have large redshifts and are among the most distant of

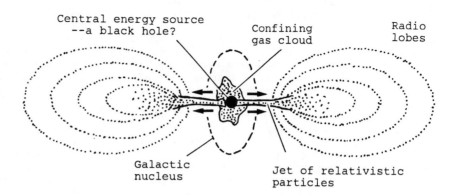

Figure 14.12 Schematic diagram of the twin-exhaust model of a radio galaxy. (From The Big Bang: The Creation and Evolution of the Universe, by Joseph Silk, p. 208, Figure 12.5, W. H. Freeman and Company, New York. Copyright © 1980. Used by permission.)

the galaxies. Thus, they may be tied in with the even more distant quasars in terms of their sources of energy.

The redshift observed in many quasars is so great that some astronomers have expressed doubt it has the same significance as for galaxies--namely, that the shift results directly from the radial velocity of the quasar. However, most astronomers seem to be willing to accept quasar redshift as a result of recessional velocity, and this leads them to regard quasars as the most distant class of objects of our universe. (We discuss this redshift question in a later section.)

Although the first quasars to be discovered showed strong radio emissions, many others have been found more recently that are weak radio sources. Many of these emit ultraviolet radiation, an activity that enables them to be detected by the use of photographic film highly sensitive to ultraviolet rays. Many quasars are also emitters of intense X rays. The luminosity of quasars has been observed to vary greatly in time periods of a few years. In their brightest phases, some quasars exceed the energy output of the brightest known galaxies. Some quasars are comparatively close to our galaxy, but their spectral characteristics are much the same as the distant ones.

While the origin and inner workings of a quasar remain extremely puzzling to astronomers, speculations tend to favor two basic conclusions. First, because of their great distance, quasars may represent one of the most primitive forms of galaxies. Second, quasars may be the massive, dense cores or nuclei of such galaxies, cores that emit intense synchrotron radiation. Perhaps frequent collisions of dense stars in a contracting galactic nucleus furnish their great energy output. Even more attractive is the speculation that a black hole lies at the dense nucleus of the aging galaxy. As in the black-hole model for a radio galaxy, the powerful inward suction of matter and radiation into the black hole would produce intense radiation from hot gas in the surrounding region.

Events of the Big Bang

Equipped with an adequate amount of descriptive basic information on the contents and structure of the universe, we are prepared to follow the simplified script of a hypothetical Big Bang creation scenario. The events described here are based on a detailed account presented by Professor Joseph Silk (1980) of the University of California in a book written to explain cosmology to persons outside the professional field of astronomy. In this brief review of Professor Silk's account, we can only touch on some of the evidence on which it is based, because to understand that evidence you would need a professional knowledge of particle physics, including the advanced mathematics that goes with it.

The absolute zero-point in time in the schedule of creation of the universe is known as the singularity. How long ago that instant in time is to be placed depends upon what Hubble constant is adopted. We will use a date of

-20 b.y. for that instant, knowing that others may prefer a shorter time span, such as -15 b.y. or -10 b.y. The concept of the singularity is impossible to define or describe. Instead, cosmologists are prepared to describe the possible state of the universe at an instant known as Planck time, when the universe was 10^{-43} seconds old. For all practical purposes we can take the instant of Planck time as marking the birth of the universe. What ensued constitutes the Big Bang. At the instant of Planck time all matter of today's universe was contained in a sphere having a radius of only 1/1000 cm. We are talking now about a mass equivalent to about 10^{20} solar masses, so its density must have approached 10^{90} kg/cm^3. Temperatures within this minute speck of a universe were enormously high--the figure of 10^{31} K has been mentioned.

The nature of the universe at the Planck instant was quite unlike that of matter we are familiar with. It may have consisted largely of neutrinos, along with comparatively small numbers of electrons, protons, and neutrons. A neutrino is a particle that moves with the speed of light but has no mass and lacks an electrical charge; thus, it does not interact readily with atoms and is extremely difficult to detect.

Besides the particles mentioned above, the universe of Planck time was filled with radiation. Radiation was in the form of photons, massless particles of energy that travel at the speed of light. From this description, you might think that photons are about the same as neutrinos. One important difference is that photons interact readily with atoms, so they are easily detected. We are familiar with photons through the phenomenon of visible light, which is easily reflected by a solid surface that intercepts light rays.

The energy carried by a photon can be vastly greater than the energy of an ordinary photon of the visible light spectrum. The radiation we are speaking about as being present in the early universe consisted of high-energy photons called gamma rays; they have much higher frequencies in the electromagnetic energy spectrum than light rays.

Photons that have the highest level of energy are produced by a phenomenon of particle physics known as the annihilation of matter. Various kinds of particles of matter have counterparts known as antiparticles. For example, the antiparticle of an electron is a positron, which is positively charged (an electron has a negative charge). The antiparticle of a proton (positively charged) is an antiproton, having a negative charge. A particle and its corresponding antiparticle can come together in such a way that both are annihilated and become energy. For example, annihilation of a proton and an antiproton can yield gamma radiation. Physicists also argue that gamma radiation can produce particles in pairs of matter and antimatter, such as protons and antiprotons, or mesons and antimesons. Thus there may have been a brief period in the early universe when annihilation of matter was producing gamma rays, while at the same time gamma rays were being transformed into matter-antimatter particle pairs. But because the temperature was rapidly dropping, there came a point that the annihilation of matter went on almost to completion and there remained mostly radiation.

It seems, however, that there were originally a few more particles than antiparticles, so that when annihilation was complete, some particles remained, and they were destined to constitute all of the matter we observe in the universe. The universe was now one second old and the remaining matter consisted largely of protons and neutrons, which are the heavy particles that were to become the nuclei of atoms. The temperature of the universe had now dropped below 10 billion K. As expansion of the universe continued and the temperature continued to drop, a new critical point was approached. At the age of about one minute, when the temperature had dropped to one billion K, nuclear fusion began and the nucleosynthetic era had arrived. Neutrons and protons were brought together in a series of steps to create helium, which has a nucleus consisting of two protons and two neutrons; its mass number is four.

The abundance of protons was six times greater than neutrons, and because of the relatively small proportion of neutrons available for helium fusion, the abundance of helium atoms in the universe turned out to be one-tenth as great as that of hydrogen atoms (produced at a much later time). That ratio persists throughout the universe so far as we can observe. As already noted, nucleo-synthesis of elements heavier than helium occurred much later in supernova explosions of large stars. Nucleosynthesis at this time could also have produced deuterium, which is one of the variant forms of hydrogen --it is an isotope of hydrogen. The nucleus of deuterium contains one proton and one neutron, whereas the ordinary hydrogen nucleus consists of a single proton. Both deuterium and ordinary hydrogen have only a single electron and are thus the same kind of atom, but their atomic masses differ; for hydrogen the mass number is one, for deuterium it is two. Deuterium is extremely rare in nature. Throughout the universe there is only about one deuterium atom for every 30,000 ordinary hydrogen atoms.

Physicists suppose that the nucleosynthesis of helium occurs in stages by addition of single neutrons or protons to the atomic nucleus. The first stage would have been the capture of a proton by a neutron to form the nucleus of a deuterium atom. The next stage would have been the addition of another neutron, forming the nucleus of tritium, another isotope of hydrogen. Finally the addition of another proton would have completed the synthesis of the helium nucleus. Thus it seems possible that the deuterium in our universe had its origin, along with helium, during the early nucleosynthetic era of the Big Bang. One reason that deuterium is interpreted as having been formed in this early stage of the Big Bang instead of later is that deuterium does not survive from the helium fusion process that goes on constantly in the interior of stars.

At the age of only a few minutes nucleosynthesis had largely ceased in the expanding universe; it now entered a stage known as the radiation era. The universe as it emerged in this era has been called the "primeval fireball."

Throughout the radiation era, the universe was expanding in volume and its temperature was falling. This would require that the quality of radiation was undergoing a corresponding change. Recall that the first radiation was in the form of high-energy photons of gamma radiation. The effect of expansion was to reduce the average energy of the photons to lower and lower levels. The radiation spectrum of a radiating object or substance is determined by its temperature. For a given temperature the spectrum has a particular peak--the higher the

temperature, the higher the frequency and the lower the wavelength of the energy peak in that spectrum. Thus in its earliest stage, the radiation spectrum of the universe peaked in the extreme high-frequency end of the spectrum where the radiation takes the form of gamma radiation. Here the wavelength ranges from 10^{-10} to 10^{-8} cm. As temperature falls, the blackbody radiation spectrum undergoes two forms of change: (1) the spectrum is stretched out over a wider band of frequencies; (2) the peak of maximum energy intensity is reduced and is shifted toward lower frequencies (Figure 14.13). Actually this phenomenon is a redshift occurring with the passage of time. The radiation peak thus passes from the gamma-ray band into the X-ray band, then through the ultraviolet band to the visible-light band. Next it moves into the infrared band and from there into the band of radio waves.

A good question now arises. At the present time, some 20 b.y. after the Big Bang, what is the radiation spectrum of the universe like? Physicists had predicted that the radiation of the early universe still persists, but it must have very low frequency and a very small level of energy. They estimated that its spectrum would be essentially identical with the blackbody radiation spectrum for a temperature of only a few K degrees--perhaps 10 K or less. This would mean a spectrum of radio waves peaking in the band of microwaves in the wavelength range from 0.01 to 10 cm. They could also predict that this cosmic background radiation would be reaching us from all directions in space with equal intensity. You might imagine this radiation to resemble a low-pitched sound reverberating from the confining walls of a great auditorium. You cannot trace the sound to any one source area, and it seems to be equally intense no matter in what direction you turn your head.

The story of the discovery of the cosmic background radiation is one of the most exciting in all physical science, because it strongly corroborated the Big Bang theory as an adequate explanation of the origin of the universe. In 1965 two radio astronomers--Arno Penzias and Robert Wilson at the Bell Laboratories in New Jersey --began constructing a strange-looking trap for cosmic microwaves. The device looked like a huge cornucopia--an expanding horn of square cross-section. It could be pointed over a wide range of directions into outer space to trap radio signals coming in from a very narrow travel path, and it excluded radio waves from other sources. What they got was "noise"! Amplified as sound through a radio loudspeaker, it is continual hiss; but the noise was the same from all points in space. Analyzed into the component frequencies, the noise proved to fit the radiation spectrum of a blackbody with a temperature of 2.9 K. This is very close to the 3 K curve shown in Figure 14.13. Princeton physicist Robert Dicke and his colleagues were quick to interpret the radio signals collected by Penzias and Wilson as cosmic radiation left over from the Big Bang, and this interpretation soon gained wide acceptance among astronomers. Subsequent observations by other scientists added details to confirm the first observations. For their pioneering research, Penzias and Wilson received the Nobel Prize, and cosmology has never been quite the same since. An alternative theory, the steady-state model we mentioned in our introduction, was particularly hard hit by the discovery of cosmic background radiation.

We now continue with the narrative of events early in the history of the universe. Throughout the radiation era the universe continued to expand, so that the density of protons and electrons decreased. This meant that photons collided less frequently with electrons. There came a point at which the photons, traveling always at the speed of light, became more or less independent of the presence of electrons. Physicists say that radiation had now become independent of matter: i.e., radiation was "decoupled"

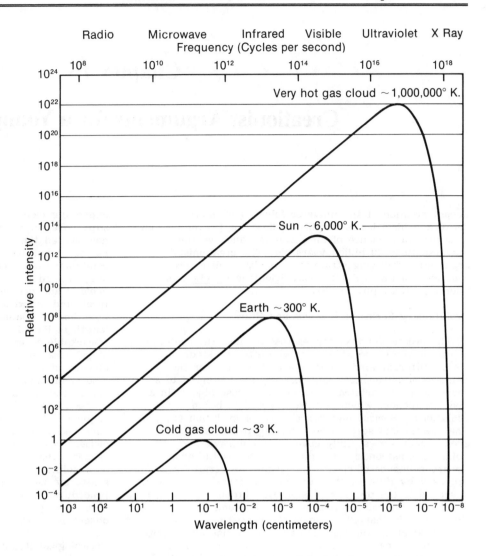

Figure 14.13 Curves of black-body radiation at four different temperatures. (From <u>Astronomy</u> <u>and</u> <u>Cosmology</u>: <u>A</u> <u>Modern</u> <u>Course</u>, by Fred Hoyle, p. 670, Figure 16.3, W. H. Freeman and Company, New York. Copyright © 1975. Used by permission.)

from matter. Thus the <u>decoupling</u> <u>era</u> began. The temperature had now dropped to about 4,000 K, permitting electrons to combine with protons on a one-to-one basis to form hydrogen atoms. This process is thought to have started when the age of the universe was about 300,000 years. Photons now had less energy and were unable to break apart the hydrogen atoms upon impact. The universe was now transparent to radiation, which could move freely throughout the whole universe.

As time passed, the wavelength of radiation increased and its frequency decreased, following the schedule we have already outlined. Most of the matter of the universe was now in the form of stable atoms of hydrogen and helium in the ratio of about 10 hydrogen atoms to 1 helium atom. We can think of the universe as consisting of cold gas rather uniformly distributed through space. We have now arrived at a point when the dispersed atoms were being acted upon by pressure waves that began to cause density fluctuations leading to gravitational instability, and resulting in the collapse of clouds of matter. From this point on, our scenario follows what we have already said about the formation of star clusters to build the first galaxies, and the evolution of those stars to cause galaxies to evolve in terms of their structure and contents.

Although the Big Bang theory of cosmology dominates the scientific scene today, it has a number of variations in the interpretation of important matters. Thus there are several current models of the Big Bang theory, and these

are the subject of a great deal of scientific debate. Keep in mind that the Big Bang scenario does not include a description or explanation of the instant of singularity, nor can it deal with anything that might have existed before that instant. Whether the expansion of the universe will continue forever or be halted and reversed to cause a collapse of the universe are questions still open to debate. The Big Bang theory is thus incomplete-- it does not account for the very beginning; neither does it have a satisfactory closing scene.

If the principle of the redshift, on which explanation of the universe now rests, should be demolished, the Big Bang theory would fall with it. If gravitation can be shown to have been a highly variable force through cosmic time, the Big Bang theory may be hard hit and require major restructuring. In cosmology particularly, highly respected scientists have come up with some radically new and different proposals as to the nature and behavior of matter and energy on cosmic scales of time and distance. We cannot dismiss lightly the possibility that a new and coherent theory of cosmology will appear as a successful challenger to the Big Bang theory.

Credit

1. Parts of this chapter are extracted or adapted from the author's textbooks of geology and the earth sciences published by Harper & Row, Inc., New York. Copyright © 1971, 1972, 1976, and 1981 by Arthur N. Strahler.

Chapter 15

Creationist Arguments for a Young Cosmos

Recent creation of the universe followed by increasing decay and disorder is weighed against the prevailing view of mainstream science of an expanding and evolving universe 10 to 20 billion years old. The creationists' arguments, following strategies already familiar, are contained in the ICR textbook, Scientific Creationism (Morris, 1974a, pp. 24-29).

Evolutionary Series of Galaxies and Stars

The creationist text argues that because the contents of the universe are widely varied--different kinds of stars, different kinds of galaxies--it is easy for the mainstream astronomers and cosmologists (referred to as "evolutionists") arbitrarily to arrange these objects in a model sequence, which is then asserted to be an evolutionary series, and hence proof of evolution (p. 24). The creationists seem to be referring to two kinds of evolutionary series, fully covered in our review of mainstream astronomy (Chapter 14). One is the evolutionary sequence of classes of galaxies--a series first proposed by Hubble, then reversed in order by Shapley. A second is the program of stellar evolution depending on location in the main sequence. The creationists' position in this matter is that all varieties of celestial objects were created at about the same instant with a distinct purpose for each and have not changed in the slightest since then. This position is completely nonscientific in outlook, since it does not seek organizational relationships nor does it look for explanations in terms of universal physical laws.

In comment, it should be pointed out that arrangements of observed entities, such as star types, into series according to a continuous range of some physical property, such as temperature, mass, density, or luminosity (which need not be assigned a time sequence of change to be ordered), is a perfectly legitimate form of scientific investigation. Two such varying physical properties can be plotted against each other to reveal natural order in the universe. Assignment of a particular time sequence to such data involves the formulation of scientific hypotheses, followed by testing and further observation in search of corroboration. When an evolutionary time series is postulated, it must be based on physical laws relating to processes and, in that sense, is not arbitrary. The creationists, in accusing the mainstream scientists of arbitrary ordering of observations without a physical basis, are, in effect, impugning the methods of all empirical science.

The ICR textbook goes on to say that even if a particular model--such as one for galactic or stellar evolution--seems reasonable, it cannot be tested by observation because no changes can actually be observed in a single star or galaxy (Morris, 1974a, p. 24). Specifically, it states that as long as humans have been observing stars, those stars have remained the same. Such a statement is utter nonsense. Humans have observed novae and supernovae over the centuries. Astronomers observe and record the tangential motions of expanding clouds of gas emanating from the site of a supernova for which records exist. Astronomers have documented the diminishing frequency of the energy pulses of pulsars, showing them to be slowing in rotational rate. The creationists' argument is a general one, applied to all historical natural science including geology and organic evolution. But we have seen that this timebound knowledge can be the subject of scientific hypotheses capable of being tested, as discussed at some length in Part 1. In astronomy, however, there is a remarkable access to the time dimension in the evolution of galaxies and stars. The varying distances of these objects allow them collectively to be observed over a large time span, and thus the observations, while singular in age as individuals, are ordered according to age, independently of any arbitrary assumptions. The age sequence of the observations can thus be correlated with observed physical differences in the object itself, such as element composition, mass, density, or luminosity. In this respect, that ordering of stellar objects by age--relative or absolute--is analogous with the ordering of sedimentary strata by use of the principle of superposition. Both procedures depend on physical laws for their validity; those laws preclude arbitrary age ordering by human observers in purely subjective ways.

Cosmological Hypotheses and Thermodynamics

The ICR textbook goes on to consider the laws of thermodynamics, which creationists accept as proven laws (Morris, 1974a, p. 24). Strangely enough, the text states that the second law is actually transforming the universe from one of order to one of disorder (p. 25). This change or transformation is inevitable, they say. And yet, on the same page they have made the statement that the creation model is perfect and nothing in it can change. This contradiction is glaring and defies logic. I have already noted that the idea of a changeless universe stands vitiated by observation, so that leaves us to consider the second law as a guiding principle--one that is accepted by both creation scientists and mainstream scientists. We have discussed this topic in considerable detail in Chapter 13 and it does not need to be repeated here.

The ICR text continues with a largely religious argument supporting creation of the universe by "a Cause transcendent to itself" (p. 26). This material has been discussed in our earlier section on entropy and does not bear restatement here. The text next restates the two cosmological hypotheses most seriously considered a decade ago, when the book was written: steady state and Big Bang. About these the text states:

It is obvious by definition that neither the big-bang theory nor the steady-state theory has any observational basis. In fact, they contradict both Laws of Thermodynamics. Therefore, they are philosophical speculations, not science, secondary assumptions to avoid the contradictions implicit in the evolution model.

The creation model, on the other hand, in effect predicts the two Laws of Thermodynamics, as noted before. A special creation of space, matter and time, by an omnipresent, omnipotent, eternal Creator is the only logical conclusion to be drawn from the two most certain and universal laws in science. (P. 26)

These paragraphs repeat the familiar creationist theme; that mainstream science (evolution) is not scientific, whereas the creation model is logical, predictive, and therefore good science.

The ICR text next raises what is considered a very important unanswered problem concerned with the origin of galaxies and stars (p. 27). A statement by James Clerk Maxwell, the Scottish physicist, written in 1873 is quoted in a paragraph from W. H. McCrea's paper, referred to in our Introduction to Part 3, (McCrea, 1968, p. 1298). Maxwell was evidently much impressed by observational evidence, such as was then available, that all of the stars, including those most distant, are composed of the same molecules found on earth, and he found no theory of evolution to account for that similarity of matter. McCrea's point was that modern observations continue to tell us that "all electrons are everywhere the same, all protons are the same, and so on." McCrea suggests that a sophisticated theory of cosmology should explain why this is so. (We now have such theories.) The creationists use such statements to their advantage with the familiar anti-science argument: If scientists cannot explain something essential to their hypothesis, even though there is general agreement that it is something observed as reality, their entire hypothesis is false. Of course, the creationist text follows immediately with the claim that the creation model does explain the observation--the Creator made it that way! This ploy is what Steven Dutch calls "explanation by default" (1982a, p. 11).

The final argument in the ICR text is that the Big Bang hypothesis does not account for the initial quantity of matter and energy present at the instant the cosmic explosion began (Morris, 1974a, p. 28). (This topic was discussed in Chapter 6.) Recall that scientists Robert Jastrow and Preston Cloud admitted that what may have been in existence prior to the singularity can only be an object of speculation and cannot be handled by empirical science. Theistic scientists may wish to suppose that a Creator was responsible for the initial conditions, and many of them hold that belief. That precreation reality has long seemed inaccessible to science in no way supports a religious belief that seeks to provide an answer in a supernatural realm.

The New Inflationary "Nothing" Universe

My heading is the title of a short article in Science News written by Dietrick E. Thomsen (1983) to update readers on a new development in cosmology known as the new inflationary theory. It concerns what may have happened very shortly after Planck time, which we mentioned in Chapter 14 as fixed at 10^{-43} second following the singularity, or zero-time. Not only is this new theory a veritable bombshell dropped in the midst of a cosmology already replete with incomprehensible events--the Standard Big Bang theory--but it has also proven a windfall for the scientific creationists.

No doubt many of you saw in the news media in 1983 the photo of a very youngish man pointing to a blackboard loaded with mysterious mathematical statements. That was Alan H. Guth at the time (1982) he presented his new theory to the Eleventh Texas Symposium on Relativistic Astrophysics. Today Guth is a tenured professor on the faculty of the Massachusets Institute of Technology--a meteoric academic rise for one who was a postgraduate scholar in 1979, when he put together his theory. Other specialists in cosmology and particle

physics have lent a hand in debugging Guth's original inflationary theory, and it is today considered a "satisfying" explanation of what may have happened in an inflationary era lasting only about 10^{-32} second--about one trillion-trillionth of a second.

By way of introduction, we should note that the new theory arose from a major development in theoretical physics occurring in the 1970s--the attempt to arrive at a grand unified theory (GUT) that can describe in a single mathematical framework three of the four fundamental physical forces. These are (a) the weak force (enabling the neutron to decay into a proton), (b) the strong force (between protons and neutrons), and (c) the electromagnetic force (between charged particles). Gravitation, the fourth force, is not yet included in GUT but continues to be described under the theory of general relativity. (Efforts to include gravitation in GUT are in progress.) Actually there are several or many versions of grand unified theory; they are collectively designated "GUTs." Success in establishing GUTs not only laid the foundation for the new cosmological advance but supplied the mathematically capable physicists to work it out.

The new inflationary model does not simply supplant the entire Standard Big Bang (SBB) theory that we described in Chapter 14; it adds a specific scenario to fill out details of the inflationary era and the time between that era and Planck time, i.e., between 10^{-43} and about 10^{-30} seconds. Changes that occured in this time interval relate to temperature, density, and size of the universe. It begins with a universe having a radius of only 10^{-53} cm, rather than one of 10^{-3} cm called for in the SBB. This is smaller by a factor of about 10^{50}. Temperature--some 10^{30} K at Planck time--follows the same downward path in both scenarios until the inflationary era begins. Whereas in the SBB temperature falls steadily, it does a sudden dip followed by sudden rise within the inflationary era. I suggest you think of this temperature fluctuation as the kind of ride you would experience in a jet aircraft beginning its descent for a landing and encountering first a sudden downdraft, then a sudden updraft. You might drop like a stone through several thousand feet, level off for a few seconds, then shoot up to an altitude even higher than before, after which the correct approach path would be resumed. If your seatbelt was unfastened, you might have been anatomically and emotionally changed by this terrifying experience. Similarly, our universe experienced fundamental changes that were to determine its behavior in all time to come.

What happened was that the sudden temperature drop followed by sudden rise released an enormous quantity of energy of a sort called "vacuum energy" that caused the sudden expansion to a universe about the size of a softball (about 10 cm). This is expansion by a factor of some ten trillion-trillion times (X 10^{50}). During expansion the original vacuum energy was, in the words of Science writer M. Mitchell Waldrop, "converted into a hot, dense firestorm of electrons, photons, quarks, and neutrinos. . . . The violently expansive vacuum energy had spent its fury, and the universe was crammed with a hot, uniform plasma of newborn particles. The universe lay poised for . . . the destruction of its antimatter, the creation of the elements, and the formation of stars, galaxies, planets, and us" (1984, p. 51).

Crucial in the creationists' interpretation of the new inflationary theory is the question of what the universe was like before the inflationary era began. Under the SBB, the universe at Planck time consisted of matter with a mass equivalent of about 10^{20} solar masses. As we stated in Chapter 14, the nature of that matter is subject only to speculation; it may have consisted largely of neutrinos, along with comparatively small numbers of electrons, protons, and neutrons. As to the new theory, the nature of the universe prior to the inflationary era seems not to be specified. Waldrop, apparently quoting Guth in interview, has Guth saying: "The universe's initial state is arbitrary, since the predictions of inflation are independent of any assumptions you make once the inflation begins" (Waldrop, 1984, p. 50). Waldrop goes on to state:

> All that was needed to begin inflation was for one region of the chaos, the forerunner of what we now call "the universe," to have held at least 20 pounds of matter and for that region to have been exceptionally hot and expanding.

"Twenty pounds"--about 10 kg mass--is virtually infinitesimal in compared with 10^{20} solar masses, mentioned as the mass of the universe at Planck time under the SBB. Keep this point in mind as we turn to the creationists' interpretation of the new inflationary theory.

In a 1984 update of the creationist position on the origin of the universe, Henry M. Morris uses an entire article in the ICR Impact Series (No. 135) to attack mainstream cosmologists through their own statements. Several quotations picked up from articles published in 1982, 1983, and 1984, are used by Morris to suggest that many modern cosmologists are close to being convinced that the universe may have been created from nothing (ex nihilo), or from almost nothing.

Morris begins by selecting a statement by Alan Guth from a 1984 Scientific American article titled "The Inflationary Universe" and coauthored with Paul J. Steinhardt. The quotation from these authors refers to their suggestion that during the inflationary era the universe, which had been of a size much smaller than a single proton, almost instantaneously underwent a dramatic expansion to become an object about the size of a softball. Morris states that concepts such as these are strictly mathematical and impossible even to visualize, let alone test. Morris wants to convince us that Guth's inflationary scenario is absurd or ridiculous; he says:

> Now 10^{-35} second is one hundred millionth of a billionth of a billionth of a billionth of a second, whatever that can possibly mean. These inflationary cosmogonists are telling us that, at the beginning, the entire universe (of space, time and matter) was concentrated as an infinitesimal particle, with all force systems (gravity, electro-magnetic, nuclear and weak forces) unified as a single type of force. This "universe" somehow went through an inconceivably rapid inflationary stage, reaching grapefruit size in 10^{-35} second, by which time the four forces had become separate forces, the heterogeneities had been generated which would eventually become expressed in the heterogeneous nature of the expanded universe, and the universe was ready to enter the "normal" phase of its big bang. (Morris, 1984a, p. ii)

Morris wants to know where the initial "point universe" came from. He asks:

> This amazing infinitesimal particle which contained the entire universe and, in principle, all its future galaxies, planets and people--how do we account for it? Now, if one thinks that the scenario up to this point has been enchantingly preposterous, he will surely think the rest of it is simply a creationist plot to make evolutionists look ridiculous. Readers should certainly check this out for themselves! (P. iii)

Morris follows with a quotation from a 1984 article in New Scientist by Edward P. Tryon, one of the active modern cosmologists speculating on the earliest events of the Big Bang: "So I conjectured that our Universe had its physical origin as a quantum fluctuation of some pre-existing true vacuum, or state of nothingness." Morris has now nearly arrived at his goal, which is to pin down the alleged belief of the cosmologists that the universe was created from nothing, for he concludes: "So our vast, complex cosmos began as a point of something or other which evolved as a fluctuation from a state of nothingness!" (1984, p. iv) To cinch the matter, Morris quotes again from Guth and Steinhardt:

> From a historical point of view probably the most revolutionary aspect of the inflationary model is the notion that all the matter and energy in the observable universe may have emerged from almost nothing. . . . The inflationary model of the universe provides a possible mechanism by which the observed universe could have evolved from an infinitesimal region. It is then tempting to go one step further and speculate that the entire universe evolved from literally nothing. (Guth and Steinhardt, 1984, p. 128)

Mission accomplished! Morris offers his final thought:

> Regardless of the sophisticated mathematical apparatus leading the inflationary-universe cosmogonists to their remarkable statement of faith in the omnipotence of nothingness, there will continue to be a few realists who prefer the creationist alternative: "In the beginning God created the heaven and the earth." (Morris, 1984a, p. iv)

Morris has successfully made the classic switch: Mainstream science is based on faith; creationism, in contrast, is based on reality. Science has become supernaturalism; the supernatural has become the empirical.

How can one comment on Morris's argument, based as it is on what seem to be outright confessions of cosmologists that the universe may have materialized ex nihilo? Perhaps the following paragraph from Guth and Steinhardt contains a much better answer than any I could possibly think of:

> Recently there has been some serious speculation that the actual creation of the universe is describable by physical laws. In this view the universe would originate as a quantum fluctuation, starting from absolutely nothing. The idea was first proposed by Edward P. Tryon of Hunter College of the City of New York in 1973, and it was put forward again in the context of the inflationary model by Alexander Vilenkin of Tufts University in 1982. In this context "nothing" might refer to empty space, but Vilenkin uses it to describe a state devoid of space, time and matter. Quantum fluctuations of the structure of space-time can be discussed only in the context of quantum gravity, and so these ideas must be considered highly speculative until a working theory of quantum gravity is formulated. Nevertheless, it is fascinating to contemplate that physical laws may determine not only the evolution of a given state of the universe but also the initial conditions of the observable universe. (1984, p. 128)

Whatever qualifications the physicist might attach to the meaning of "nothing," I think we must accept the above statements as an almost unequivocal acceptance of scientific speculation on the initiation of the universe ex nihilo, at least as the average intelligent layperson would interpret those statements. Does such a speculation thereby place the entire Big Bang cosmology in jeopardy? Not in the least! Not any more so than the speculation (as belief) of a theistic scientist that God performed the initial creative act at time zero.

So what is all the fuss about? Perhaps the creationists are unhappy that mainstream science can postulate its own version of creation ex nihilo in competition with their own six-day model. Mainstream science is increasingly invading the creationists' turf, much to their confusion and discomfiture. We will see this invasion occurring in the adoption by mainstream science of catastrophism on a colossal scale in the latest scientific hypotheses of devastating impacts on our planet by space objects, such as asteroids or comets.

The Dark-Sky Paradox

Although the ICR textbook, Scientific Creationism (Morris, 1974a) uses quasi-religious arguments in favor of recent creation and against the mainstream hypotheses of cosmology and can scarcely be described as rigorous in the context of cosmology of the 1980s, the creation scientists do have substantive arguments to offer. Physicist Harold S. Slusher, a faculty member of the University of Texas at El Paso and former staff scientist of the ICR, has undertaken to challenge the mainstream cosmologists on certain technical matters and to attempt to show that the universe is much younger than 10 to 20 billion years. Let us next consider these arguments.

Slusher asks us to consider a question posed to all students of astronomy: Why is the sky dark at night? (1978, pp. 16-18; 1980, pp. 21-23) The question refers to the blackness of the celestial sphere between planets, stars, and luminous gas clouds. Most of us would give a rather straightforward reason, such as that in the black areas (a) there are no light sources, or (b) stars and galaxies in those areas are obscured by clouds of cold dust and gas. The question might never have arisen except for the musings of a German astronomer, Heinrich Olbers (1758-1840), who undertook a chain of reasoning based on Sir Isaac Newton's hypothesis that the universe is infinite in extent. In 1826 Olbers presented his conclusion: in an infinite universe filled with stars the entire sky would be extremely bright--as bright as the sun itself. Olbers's argument is clearly described by astronomer Michael Zeilik of the University of New Mexico:

> Olbers tackled a seemingly simple question: what is the total amount of starlight striking the earth? He assumed that (1) the universe was infinite; (2) it was uniformly populated by stars similar to the sun; (3) no interstellar material blocks out any of the stellar light; (4) the light diminishes as the inverse square of the distance; and (5) the universe (and the stars in it) does not evolve. Because this static universe is filled with stars of the same luminosity, an observer sees differences in the brightnesses of stars only if they are at different distances from him or her. Recall that the brightness depends only on the inverse square of the distance. But as you look at fainter stars, you see deeper into space, and the number of stars increases directly as the square of the distance. The diminishing intensity (inverse-square law) is just balanced by the increase in number of light sources (direct-square law). Net result: The total light striking the earth is equivalent to that from one average star placed so close that it completely fills the sky.

> Another way to see it: Every line of sight must end at a star if space is infinite and the stars are uniformly scattered. All fused together, these individual images blend into a uniformly bright surface. (Zeilik, 1982, p. 529)

In this way Olbers presented us with a paradox, now known as Olbers's paradox. Slusher develops Olbers's assumptions mathematically in terms of successively larger concentric shells of stars and by integration arrives at a total amount of light (1980, pp. 21-22); it approaches infinity. Slusher notes that if the self-shadowing effect, in which nearer stars obstruct the path of light from more distant stars, is taken into account, the brightness of the sky should be only that of a typical star. Slusher then lists some attempts that have been made to resolve Olbers's paradox. It does no good, he says, to introduce curved space instead of the conventional Euclidean space. He then considers the argument that interstellar dust may absorb the starlight; he rejects this effect because in a very old universe the dust would come into radiative equilibrium with the radiation sources and would emit as much light as was absorbed. Finally, Slusher says we may be led to one of three conclusions: (a) at great distances either the luminosity of the stars diminishes or their distribution in space is less dense; (b) physical constants vary with time (presumably the speed of light might be one of these); and (c) distant galaxies receding at velocities approaching the speed of light would have their spectra so strongly redshifted that the photons would have near-zero energy (p. 22).

In reference to argument (a) Slusher states that it would hold if the universe were very young, so that stars would have been radiating for only a short period of time (p. 22). This statement seems to have no meaningful relationship to argument (a). Slusher rejects argument (b), as would mainstream cosmologists, on the grounds that there is no evidence in its support. Slusher does not find fault with the principle of argument (c), but questions the hypothesis that the universe is expanding, as the Big Bang scenario requires; this is to imply that the redshift has some cause other than radial velocity.

Although Slusher does not say it in so many words, the impression is given that Olbers's paradox cannot be resolved if the universe is very old (10 to 20 b.y.) but may possibly be resolved by a very young universe. His final conclusions from consideration of Olbers's paradox are stated as follows:

> The universe having a dark night sky seems in the consideration of other arguments persuasive evidence of a young universe. The presence of very bright objects seems indicative of a low entropy state and a short time that the universe has been running down since its beginning. These bright objects (and there is an infinite number of them, practically speaking) have not been in existence long enough to give a bright night sky. (Slusher, 1980, p. 23)

This conclusion does not seem to have any relevance to the paradox and its assumptions. Instead, I read in his conclusion an argument about the second law of thermodynamics based on the presence of highly luminous celestial objects: they could only be luminous if they are very young; if very old, they should no longer be luminous. Mainstream cosmology calls for star formation more or less continuously through the entire history of the universe following the condensation of hydrogen and helium into galactic clouds. Bright stars of large mass burn intensely and have short lives, but this does not mean the universe is no older than those bright stars. Slusher seems to have fallen into a non sequitur: existence of young stars requires that the entire universe be young. (The existence of infants in the human population requires that the human race can have existed no longer than two years.) That spurious relationship follows logically upon the creationists' assertion that the increase of entropy can never be reversed. In Part 1 we discussed this claim at length, showing that entropy can, and is, decreased locally within open systems. Star formation is one of those open systems in which gravitational energy is stored in a decreasing volume of space to produce a high-temperature object capable of internal nuclear fusion.

And now, back to Olbers's paradox. Is there a way out? After all, the night sky is indeed black. Steven I. Dutch, in examining Olbers's paradox, has made calculations of sky brightness based on the assumption of a finite universe (1982b, p. 29). He adopts a model universe 2×10^{23} km in radius. Using a value of starlight based on typical nearby stars, he calculates that total starlight would be equal to that from 5 billion first-magnitude stars. This amounts to "only about four percent of the brightness of the Sun, or roughly the equivalent of a first-quarter moon for each square degree of sky. Very bright, to be sure, but nowhere near infinite." He follows that analysis with one greatly refined on the basis of a nonuniform distribution of stars and comes up with a much lower light value. He states: "There is no need for exotic explanations to resolve Olbers' paradox, and no conflict with an ancient universe. In fact, Olbers' paradox is no longer a paradox and its relevance to the age of the universe is tenuous at best."

Those readers wishing to delve more deeply into the demolition of Olbers's paradox will enjoy reading an article by E. R. Harrison, published in Mercury magazine (Harrison, 1980).

The Big Bang theory also avoids the paradox by the alternative assumption that the universe is not infinite. It also rejects the assumption of a static universe. Outward radial motion of the galaxies at speeds that increase with distance provides for reduction in energy of photons in proportion to the distance. This principle is stated by Slusher, but he is unwilling to accept redshift as a function of radial velocity. Clearly, we need to move the battlefield from Olbers's assumed infinite, static universe replete with identical stars to one of a finite expanding universe. Can the redshift be explained by an effect other than recessional velocity? Is there independent evidence of great age of the universe?

The Cosmological Redshift--Is It Reliable?

In Chapter 14 we noted that the Big Bang theory of cosmology is almost totally dependent on the assumed significance of the redshift--the shift of the light spectrum of a distant celestial object (star or galaxy) is a measure of its radial velocity. This mechanism of redshift is usually termed the cosmological redshift. Is it possible that the observed shift of the spectrum has a cause other than the optical Doppler effect? Could one or more other causes of spectral shift also act in the universe?

One might suppose that the scientific creationists would make a major bone of contention of the integrity of the cosmological redshift. They might wish to exploit two possibilities: (a) the speed of light may not be a constant value throughout all parts of the universe throughout all time; (b) with the speed of light assumed constant, the energy of the photons in a beam of light might in some manner be depleted as the light passes through space.

The redshift question is addressed by Harold S. Slusher in an ICR monograph published in 1978 under the title "The Origin of the Universe: An Examination of the Big-Bang and Steady-State Cosmogenies." But in his monograph "Age of the Cosmos," published in 1980, the redshift question does not appear. I will review the arguments in the former work (Slusher, 1978, pp. 13-16). Slusher states that many interpretations of the redshift other than the classical Doppler effect have been proposed. He describes four interpretations.

The first hypothesis has elsewhere been dubbed the "tired light" effect. Assuming that the galaxies are stationary with respect to one another (a nonexpanding universe), as light from the galaxies travels through space, the photon beam loses energy. Its speed remains constant, but energy reduction increases the wavelengths throughout the entire spectrum, which is shifted toward the red end in an amount proportional to the distance of travel. The redshift would then correctly measure distance to the galaxies if they were static light sources. Slusher states that among mechanisms proposed for light becoming tired, one invokes photon-neutrino interactions, another requires photon-photon interactions, but he admits that neither has been observed in the laboratory. Other related effects that might be deduced from either mechanism have not been observed. Thus tired light may be shelved as an unlikely hypothesis.

The second hypothesis is seriously considered by mainstream astronomers; it is the gravitational redshift (Slusher, 1979, p. 14). A lucid description of the phenomenon is given by Michael Zeilik in his college textbook of astronomy (1982, p. 456). We have already seen that quasars are noteworthy for the large redshifts many of them display. Some astronomers have questioned the great distances these large redshifts require by use of the cosmological redshift. Perhaps, instead, the quasars are mostly rather close in, so that some other phenomenon is responsible for the redshifts. Under Einstein's theory of relativity, a photon has an equivalent mass and is subject to gravitational attraction by another mass. Thus gravitational attraction can bend the path of light--a phenomenon already observed to occur. Imagine photons streaming from a small quasar of large mass. The photons will be pulled upon by a gravitational force opposed to their direction of travel, their energy depleted, and the spectrum will be redshifted. The principle of gravitational redshift seems well accepted by mainstream scientists, but to find convincing evidence that it occurs is quite another matter. Slusher rejects the hypothesis because it is based on the theory of general relativity (1978, p. 14), which is not accepted by the creation scientists.

An entirely different explanation for the very large redshifts of many quasars is based on simple Newtonian mechanics. Suppose that a quasar has been shot out of a galaxy by a powerful explosion--moving like a rifle bullet --and is headed on a course that includes motion away from an observer's location. Its radial velocity relative to the galaxy will be added to the radial velocity of the galaxy itself, and the result will be an abnormally large redshift for the actual distance involved. While some evidence has been collected to show that this effect may occur, other evidence fails to support the hypothesis. One rather strong argument against this cause of redshift in quasars is that at least a few quasars would be moving toward us after the explosion that set them in motion, but no quasars with blue shifts of the spectrum have thus far been found (Zeilik, 1982, p. 475).

The third hypothesis covered by Slusher (1978, p. 14) was proposed in 1972 in a paper published in Nature (Pecker, Roberts, and Vigier, 1972). Its authors make use of a provision of the relativity theory, that photons have a small mass and are thus capable of speeds less than the speed of light. Mutual interactions among photons can thus occur. Collisions between photons could reduce their speeds. In a dense cloud of colliding photons a redshift would be generated, and this would increase with increased temperature of the photon cloud. Slusher says that one consequence of the hypothesis is that, if corroborated, "it would explode modern cosmology by explaining the red shifts of the most distant objects, the alleged evidence for the expanding universe and the rest of modern cosmology, as the result of scattering of their light by the universal background radiation, the so-called three-degree blackbody radiation" (1978, p. 15). Slusher reports that the authors of the hypothesis proposed an experimental test. In Slusher's 1980 monograph no mention appears of this hypothesis, and one might suppose that it has simply dropped by the wayside, as is so often the case in mainstream science.

A fourth item, not actually an alternative hypothesis, is a reference by Slusher to an observation by astronomer W. G. Tifft in 1973, in which an apparent anomaly in the redshifted spectra of galaxies in the Coma cluster seemed to the author unexplained by the cosmological redshift (Slusher, 1978, p. 16). The reference given by Slusher is to a brief report in Science News (1976, vol. 110, p. 6). The report states that Tifft "does not deny that the effect may be due to the sparsity of data." Slusher offers no comment of his own on Tifft's observations.

The cosmological redshift seems to be surviving the various attacks directed at it from within mainstream science and continues to be considered a very strong working hypothesis--the "hypothesis of choice." So far, the creation scientists of the ICR have only called attention to alternative hypotheses arising from within the community of mainstream scientists. From this source the alternatives usually deserve serious consideration, and it has been given to them by members of that community. The self-policing activity of mainstream science is always in full swing and needs no prodding from the outside. Nevertheless, the arguments aired by the creation scientists can serve a valuable purpose if they force science teachers and their students to study carefully the evidence for and against all hypotheses that seriously attempt to describe the nature of the real world.

Galactic Clusters and the Hidden Mass

Harold Slusher interprets observations of galaxies and clusters of galaxies as evidence that they appear to have originated rather recently (1980, pp. 7-16). He notes that members of pairs of galaxies or of clusters are sometimes joined by bridges of luminous matter, and that the galaxies themselves are moving at high radial velocity. This he interprets to mean that the moving objects are not gravitationally bound and that this is made possible by the inadequate quantities of mass they contain. This situation leads Slusher to the conclusion that the galaxies and clusters originated recently, since otherwise after many billions of years they would be widely dispersed.

The crux of the problem is the concept of hidden mass (Silk, 1980, pp. 188-91). The mass of a galaxy can be estimated if its rotational velocity can be determined, and this can be done by observing the shift of the light spectrum--toward the red region on one side of the center of the galaxy, toward the blue region on the opposite side. Calculations based on such observations lead to the estimate of the mass within the galaxy, and this proves to be much larger than the total measured mass of all luminous objects in the galaxy. The observations on rotational velocity seem to require a mass several times greater than what is observed.

This seeming discrepancy has been a major puzzle for astronomers; it is just the kind of situation capitalized on by the creation scientists, whose argument is that if something should be found and has not been found, it does not exist in the first place, and therefore a hypothesis that requires its presence is untenable. Creation scientists refer to the hidden mass required to be present as the "missing mass," implying that it is actually nonexistent. The history of science shows a long succession of cases in which the presence of something not as yet observed was a requirement of a general hypothesis or theory. An example from astronomy was the prediction of a planet orbiting the Sun beyond Uranus, which was then the most distant planet known. (We referred to this example in Chapter 2 in discussing the problem of naive falsification.) That, too, was a case of "hidden mass" versus "missing mass." If hidden, it might be found; if missing, the theory of Newtonian celestial mechanics would need to be discarded.

Astronomers have hypothesized that the hidden mass in galaxies may be in the form of (a) stars of such small mass that they emit almost no radiation and cannot be detected, or (b) black holes that can emit no radiation. So far there is no good evidence that either form of matter is present in sufficient quantity to account for the hidden mass. A third possibility remains open: there exist large clouds of extremely cold gas, too cold to emit measurable radiation. The gas, largely hydrogen, may be in either the atomic form or the molecular form (or in both forms). Recently, clouds of molecular hydrogen have been identified in our galaxy (Blitz, 1982, p. 84; Blitz et al, 1983). These molecular clouds have been found to be much colder and much denser than clouds of atomic hydrogen. Their temperature is about 10K. Although the hydrogen molecules cannot be observed because they do not emit radiation, other molecules contained within the cloud do radiate electromagnetic energy and can be observed. Most important in this function are molecules of carbon monoxide (CO), present in very small proportions in the cloud. About 4,000 such molecular hydrogen clouds have so far been identified in the galaxy. Each cloud has a mass 100,000 to 200,000 times greater than the Sun; the clouds are therefore the most massive objects in the galaxy (Blitz, 1982, p. 87). The clouds are thought to have formed from supernova explosions and are comparatively young--perhaps no more than 100 million years (Blitz, 1982, p. 90).

Thus, despite the creation scientists' denial of the existence of hidden mass in the galaxies, progress is being made in finding it. Also being observed today are intergalactic hydrogen clouds. As reported in Science News in 1983, one such cloud observed by its radio emissions has a mass one billion times greater than our Sun, and its length dimension is three times the dimension of our galaxy (Chen, 1983, p. 148). Discoveries such as these are important in weighing the intriguing cosmological question of whether our universe in its entirety contains enough mass to gradually reduce its rate of expansion, stop expanding, and then begin to contract (closure of the universe), or whether it must continue to expand forever.

And now, to get back to Slusher's interpretation of luminous bridges between galaxies and high-speed radial motions of galaxies, these observations also agree with an interpretation of such youthful features as meaning that rapid changes continue to occur even in an extremely old universe. Despite the creationists' claims that the second law of thermodynamics forbids postcreation changes, we observe many changes in celestial objects such as stars and luminous gas clouds. We have already shown that the very young age of one member of a population of individuals does not require the conclusion that the entire population is also very young. It's that same non sequitur being handed out again by the creation scientists!

Slusher refers to the spiral galaxies, those with curved radial arms of luminous stars:

> A galaxy is an assemblage of stars that does not rotate as a rigid body; the inner parts revolve more rapidly than the outer. An enormous difficulty which all hypotheses face that propose a long age for the universe is that any spiral arm structure will be almost completely wound up into a circle in one to a few (at most) revolutions of the galaxy-- 200 million to 1,000 million years--because of the Keplerian motion with which the parts of these galaxies rotate. However, there is still a huge number of spiral-arm galaxies in the universe. If they had a common origin in time, how could they still be around for supposed billions of years?

> The magnetic field which runs through the gases in a spiral arm is not strong enough to give the arm appreciable coherence against the dragging effect of the rapid rotation of the nucleus. Further, the stars in the arms are not coupled to this magnetic field. In other words, the galaxy will wrap itself up in a short time, relatively speaking. This analysis does not, of course, determine the age of the universe; but certainly does establish limits which are far below evolutionists' estimates of the age of

the universe, since there are very many spiral galaxies with the arms in well-defined existence. (1980, pp. 15-16)

Slusher's argument is answered by Steven Dutch (1982, pp. 27-28). Dutch introduces comparison photos from a Scientific American article by Stephen and Karen Strom (1979, p. 77). The spiral structure is seen in photographs taken in blue light, which is emitted by young blue stars that are very hot. Seen in near-infrared light, the same galaxy appears as a smooth disk in which the older, redder stars are rather uniformly spread, and with only a faint spiral structure. The rather obvious conclusion is that the galaxy as a whole is very old, whereas the spiral structure is confined to a phenomenon associated with the formation of new stars. Dutch refers to a hypothesis of spiral galaxy structure outlined in 1973 by astronomer Frank H. Shu in which the spiral arms are explained as the effect of spiral density waves. These are the same waves that cause compression of clouds of gas and dust to initiate star formation (Shu, 1973, pp. 525-27). Dutch concludes:

> In summary, there is every reason to believe star formation is still going on and that the formation of new stars is largely responsible for maintaining the spiral form of galaxies. In galaxies where star formation is not occurring, spiral structure has been largely obliterated. Spiral galaxies and galactic clusters are entirely consistent with the accepted evolutionary view of the Universe. (P. 28)[1]

Formation of Interstellar Grains

Modern hypotheses of star formation require that atoms and molecules of interstellar gas come together to form grains of solid matter with diameters on the order of about one hundred-thousandth centimeter (10^{-5} cm). This is about the diameter of smoke and haze particles suspended in our atmosphere. Such grains of interstellar matter must then clot rapidly together into larger units to form the body of the star. Professor Martin Harwit of the Astronomy Department of Cornell University, in his 1973 volume titled Astrophysical Concepts, discussed the formation of interstellar grains (Harwit, 1973, pp. 394-97). His introductory sentences read: "One of the puzzles about interstellar grains is their origin. The density in interstellar space is so low that grain formation appears impossible there." Harwit's calculations show that the time required to form a grain would be on the order of 3 b.y. Harold Slusher has seized upon that statement and the calculations that Harwit uses to back it up to claim that star formation as envisioned by mainstream astronomers could not happen within the maximum time available (some 20 b.y.) since the Big Bang (Slusher, 1980, pp. 17-19). To achieve this result Slusher has resorted to decreasing to one-tenth the value of one of the constants in the equation (the sticking coefficient). Slusher considers the smaller value to be "more realistic." This reduced value yields 30 b.y. as the time required for grain formation. Slusher (following Harwit) then discusses possible destructive effects (collisions and vaporization) that would tend to destroy grains, making the probability of their formation even more remote. Slusher summarizes as follows:

> From a consideration of the various effects involved, it would seem that grain formation is well nigh impossible. If it is granted that they would form, we are faced with the inconsistency for the evolutionist that the time of formation of the grains is greater than the alleged age of the universe. If it takes as long as indicated to form such a simple body as an interstellar grain as the calculations indicate under the most optimistic conditions (which

actually do not seem to exist), how can the huge ages for the stars and galaxies postulated by the evolutionists have any credibility and thus be taken seriously? (P. 19)

What Slusher does not tell us is that Harwit states that there are regions in space (the Orion nebular region, for example) where the density of atoms is greater and grains could perhaps form in about 3 million years (Harwit, p. 395). Whereas Harwit and Slusher used a density of one atom per 1,000 cm^3, recent estimates of the density of cold molecular hydrogen clouds (to which I referred in earlier paragraphs) run as great as 100 to one million atoms per cm^3 (Dutch, 1982b, p. 28, citing Gammon, 1978). This would bring the time of grain formation down to a few million years at most. Nothing that Slusher has written in his chapter on formation of interstellar grains relates to creation of the universe in less than 10,000 years. Dutch appraises Slusher's argument in the following words:

> The irony in this line of argument is that Slusher is using a presumed long time for grain formation to discredit the general view of the age of the Universe. He argues that the grains would take longer to form than the accepted age of the Universe, whereas one might just as easily argue that his model of grain formation shows that the Universe is older than generally thought. On the other hand, proof that grains form in a short time could be used by creationists to prove that the Universe could be young, as Morris (1977) does with a long list of short-term "age indicators." If the grains are old, creationists argue that they could not have formed naturally, and if they are young, creationists might well argue that the universe as a whole is young. (1982b, pp. 28-29)[1]

The Travel of Light in Space--In Circles?

Over the decades of the 1900s the astronomers' working information on the form and structure of the universe expanded in proportion to estimates of the distances of celestial objects from our point of observation. Distances to those objects are enormously greater than the creationists want them to be, if the speed of light is to be accepted as a constant value throughout all past time and space. To deny the distance values accepted by virtually the entire community of mainstream astronomers puts the creationists in an unenviable position, because that denial must also negate a vast body of scientific information along with the laws and principles that underlie it.

Although not all creation scientists have expressed acceptance of the constancy of the speed of light through past time, Harold S. Slusher does accept this assumption in his chapter titled "Travelling of Light in Space" (1980, p. 25). He understands that if an object can be assigned a distance of one billion light years, light from that object that we record today on a photographic plate must have begun its journey one billion years ago. Thus the universe must be at least one billion years old. How can that age be reduced to 10,000 years or less while retaining as valid both the speed of light and the measured distance to the object? Creation scientists of the ICR think they can accomplish this reduction.

Slusher's chapter gives a detailed review of the methods of measuring distances to stars and galaxies. It begins with stellar parallax and moves out in steps that include the Cepheid variables and apparent brightnesses within galactic clusters. Slusher then draws attention to the point that in all these methods of distance measurement, light is assumed to travel in straight lines, and the solutions use ordinary Euclidean geometry (p. 32). He is now prepared to bring forward an alternative description of the geometry that applies to the trajectories of light.

To prepare the background for an alternative hypothesis, we need to review a principle of Albert Einstein's special theory of relativity, formulated in 1905. The theory requires that for two objects in motion in space at uniform speeds (nonaccelerating objects), one of which emits light and the other observes the light, the numerical value of the speed of that light is always exactly the same, even when the two objects are moving relative to each other. This principle had been suggested years earlier by experiments intended to reveal the existence of "ether" in space, through its possible effects in altering the speed of light. The experiments had revealed no measurable difference in the speed of light, and thus cast doubt on the existence of ether.

Although Einstein's special theory of relativity gained almost universal acceptance among physicists in the years that followed, including the invariant speed of light and the relationship between mass and energy ($E = mc^2$), attempts to disprove the theory continued to be made for many years. Evidently one such attempt was that made in 1953 by Parry Moon of MIT and Domina E. Spencer of the University of Connecticut (Moon and Spencer, 1953). To understand what Moon and Spencer were trying to show, we need to look back some years to find the frame of reference of their explanation. It seems that about the same time Einstein was formulating his special theory of relativity, a Swiss physicist, Walther Ritz, was formulating another theory of relativity that postulated that the speed of light varied according to the speed of a light-emitting object, and also according to the speed of the receiving object. This was, of course, in direct contradiction to Einstein's postulate. A Dutch astronomer and mathematician, Willem de Sitter (1872-1934), took a keen interest in this conflict of views about the speed of light, because de Sitter had proposed his own theory of the universe combining certain of Einstein's ideas with his own. In an attempt to devise a test of Ritz's theory, de Sitter set up the terms of an astronomical experiment, which involved measurement of light coming from a gravitationally linked pair of stars (a binary star) rotating about a common center of gravity. I will not attempt to describe the experiment, but a full account is given by Slusher (1980, pp. 33-35). If successful, the experiment would result in recording multiple images of the two stars. Results of this and similar experiments did not produce evidence that the speed of light varies as Ritz's theory required.

Moon and Spencer in 1953 came into this picture with a suggestion that would make possible the correctness of Ritz's theory, despite the negative experimental results. They proposed that instead of straight-line travel of light in a Euclidean geometrical frame, a curved-space geometry should be used. One such system of curved-space geometry is that proposed by a German mathematician, Georg Riemann, and is known as <u>Riemannian</u> <u>space</u>. (Einstein used Riemannian space in his 1915 general theory of relativity in which light rays are bent into curved patterns by strong gravitational fields in space.) When an object moves in a curved path--an automobile rounding a curve, for example--the "sharpness" of the curve is measured by the length of the radius of a circle fitted to the curve. In Einstein's general theory of relativity, the radius of curvature is on the order of billions of light years. Moon and Spencer selected a radius of curvature of only five light years in their revision of Ritz's theory, for in so doing, they could explain why the de Sitter astronomical experiment could not produce multiple images of the double stars.

In 1980, Slusher revived Moon and Spencer's device to reach an astounding conclusion: the light from even the most distant celestial object in the universe cannot require more than 15.71 light years to reach the earth! Slusher gives full details of the mathematical calculations leading to this conclusion (1980, p. 36). When the "Euclidean distance" of an object goes to infinity, the "Riemannian distance" levels off at a limit of 15.71 years. Perhaps a simple physical model will help to get across the idea expressed here: If your automobile is driven with the steering wheel securely lashed in position for a tight turn, you will turn endlessly in a circle of constant radius. Light rays in the Riemannian universe also are restricted to a circular path, but it can be followed on the inside of an imagined hollow sphere, with Earth at the center. Slusher gives conversion values from "Euclidean distance" to "Riemannian distance" (1980, p. 36, Table 1). At a Euclidean distance of only 10,000 light years the Riemannian distance at 15.7 light years is almost at the limiting distance. What Slusher is saying is that no fault is to be found with the conventional methods of arriving at distances to stars and galaxies, but the assumed straight travel paths are unreal and must be converted to curved paths to give true distances in light years. Slight adjustments downward in the assumed radius could easily bring down the limiting value to, say, 6,000 light years. Thus light that began its journey at the instant of creation, no matter where the point of origin, took no more than a few thousand years to reach the earth. This conclusion does not prove the age of the universe as being only a few thousand years, but it permits that hypothesis to be held. It also permits the mainstream astronomers to continue to talk about Euclidean distances of billions of light years, and corresponding ages of billions of years for distant galaxies. Everyone gets what everyone wants.

How does Slusher's argument strike you? Is it unflawed? Steven Dutch has this to say:

> There are several fallacies in this approach. In the first place, the cosmological usefulness of Riemannian geometry derives from relativity, which in turn hinges on the notion that the speed of light is independent of the observer. Abandon that notion, and there are no longer any grounds for postulating a Riemannian space at all. In the second place, there are a great many data indicating that relativity theory is correct and that Moon and Spencer are simply wrong; data that have nothing to do with evolution.
>
> The central fallacy of this argument is the idea that the universe can be both Euclidean and Riemannian at the same time. The only way we can know where the stars are is by observing the light from them. If starlight travelled Riemannian paths no longer than 15.7 light years, it would diminish in brightness according to $1/r^2$ in Riemannian space. No star would appear to be more than 15.7 light years away. All stars of a given type would be about equally bright and for the most part, of naked-eye brightness. With such a small radius of curvature, we should be able to see images of the same star in many different directions. Granted, the universe may be Riemannian with a radius of curvature of billions of light years, but assuming a radius of only five light years rapidly leads to absurdities. (Dutch, 1982b, p. 29)[1]

Stanley Freske has also commented on Slusher's use of Moon and Spencer's material (1980, p. 37). Freske considers it ironic that Einstein's general theory of relativity uses Riemannian curved space with a radius of curvature of billions of light years--this in enormous contrast to the very small value arbitrarily assumed by Moon and Spencer. Freske continues:

> Perhaps the most sidesplitting assumption in the theory of Moon and Spencer is the size of their radius: 5 light-years. Why 5 light-years? Because this is large enough so that the curvature can not

be detected by any experiments performed in the solar system, yet small enough to take care of all the binary stars studied! What we have here is nothing more than a mathematical trick specifically designed to make things appear just the way Moon and Spencer wanted them.

One has to entertain the possibility that the article was written and published as a joke. But creationists certainly don't take it as a joke; to them it must seem like a godsend. With light travelling in a Riemannian space having a radius of only 5 light-years, the time it would take to reach us from any source no matter how distant would never exceed 16 years! Again we see the Gosse Hypothesis in all its glory. The creator decreed that light, and only light, should travel in a Riemannian space with a 5-light-year radius, again for the purpose of making the universe appear to be much older than the actual 10,000 years. As an added bonus, it made us poor fools accept the preposterous notion of relativity! (Freske, 1980, p. 37)[2]

Speedy Light--How Fast Was It?

If light running in a tight little circle in Riemannian space proves a fatal maelstrom for creation scientists, the reasonable course of action is to devise an alternative hypothesis, one in which light travels in Euclidean straight lines but has not always traveled at the same speed in the past as now. This possibility we mentioned earlier in the chapter, noting that ICR's Harold Slusher preferred not to adopt it. Not so, a brave Australian creation scientist, Barry Setterfield, at one time a lecturer with the Astronomical Association of South Australia in Adelaide. He threw down the Aussies' challenge glove to the American group in a "technical" paper appearing in Ex Nihilo in 1981.

As for all creation scientists, Setterfield finds unacceptable the resort to the Omphalos-type cop-out, namely, that the Creator made both the star and its light beams at the same instant, thus making all stars immediately visible to observers. While he admits to finding the mathematics of Riemannian space fascinating, he agrees that confirmation of that idea is lacking (Setterfield, 1981, p. 39). This is his new hypothesis:

There is a third alternative which to date has not been explored, but which I believe solves not only many of the observational problems of astronomy and Genesis creation, but has wide ranging implications for the whole of the physical sciences.

The basic postulate of this article is that light has slowed down exponentially since the time of creation. This thought is radical and at first looks outside of confirmation. However, there are at least 40 observations of the speed of light since 1675 which support this suggestion. (P. 39)

Setterfield lists these observations in chronological sequence noting the estimated, or standard, error of observation associated with each. Table 15.1 gives some samples, selected for historical interest and to show a general trend, which is generally toward lower values. They are grouped by method of instrumentation. Speed is given in kilometers per second.

Since about 1950 the values seem to hold nearly constant, although a downtrend of about 0.3 km/sec/yr has been noted in that period (Setterfield, 1981, p. 39). Notice also that the standard error has fallen greatly during the entire 300-year period, as would be expected of any such series of determinations in experimental physics begun so long ago. Setterfield places great

importance on the observation that "the observed change in c (speed of light) over 3 centuries is far greater than the margins of error in the observations." He goes on: "The above results compel me to formulate a new theory of light."

Actually, Setterfield has no new theory to propose in the sense of process or explanation, but he has fitted various kinds of mathematical curves (regression equations) to the forty determinations in his table in an attempt to find one that is the best fit. (Does this mean the best fit to the data or to the year of creation?) He reports that power curves, polynomials, logarithmic and hyperbolic functions were all tried but were not successful (p. 40). He did, however, coax the computer to come up with a log-sine curve that seemed to fit very well. The general form of this equation is $\log c = A + B (\log \sin T)$, T being time in years. The point is that this plotted curve steepens as it is is plotted back in time until, at 4040 B.C. ± 20 yr, it is essentially vertical (asymptotic) on the graph. (Could you have possibly guessed that would happen?) Moreover, close to the ultimate point, the speed of light is enormous: about 1.5 x 10^{17} km/sec or roughly 5 X 10^{11} times faster than today--that's 500 billion times faster than now! Setterfield adds:

I will assume that this value held from the time of creation until the time of the fall, as in my opinion the Creator would not have permitted it to decay during His initial work. At the same time, I propose that this initial high value of c would have produced the appearance of great age to the Universe in that one week (to those who look with eyes and minds fixed on the current value of c).

Integration over the curve shows that our initial problem of light travelling millions of light years in only 6000 years, is solved and that one major problem associated with the quasars is overcome also. The total distance travelled by light since creation would be about 12 X 10^9 light years. (P. 44)

That last figure comes to 12 billion light years, which is in good conformity with the range assigned by the mainstream cosmologists to the age of the universe.

By 1983, in response to questions and criticisms directed toward his hypothesis, Setterfield reworked his data and undertook a second session of computer analysis. This resulted in a simplified equation in which the logarithms of the variables were removed and the cosecant-squared was substituted for the sine: $C = A \csc^2 (kT)$, where A is the minimum value for the speed of light (last figure on the above list). This refinement moved the date of origin from its original estimate of 4040 B.C. to 4082 B.C. (-6058 yr). For that year's curve, the correlation coefficient, r^2, was 0.998, the highest of a range of years for which the solution was run.

At this point we should look at the mathematical phenomenon involved in Setterfield's analysis, i.e., the extent to which extrapolation has been carried. He has points only for the final 300-year period of more than 6000 years of plotted curve. He has extrapolated back over twenty times the time span of his data. As we shall find in connection with the creationists' handling of data on the intensity of the earth's magnetic field, extrapolation in time carried to outrageous lengths is one of their means of arriving at a very young age of the universe. Creationists argue that this extreme extrapolation is not nearly as bad as that used by mainstream physics in arriving at the working conclusion that the radioactive decay constants have been invariant over several billions years.

In mainstream physics, the general decrease in experimentally derived values of the speed of light, as

Table 15.1   Selected Observations of the Speed of Light

Optical Methods: (Km/sec)

Roemer	1675	301,300 ± 200
Bradley	1728	301,000
Cornu-Helmert	1874	299,990
Michelson	1879	299,910 ± 50
Newcomb	1882	299,860 ± 60
Perrotin	1906	299,880
Mittelstaedt	1928	299,778 ± 10
Huttel	1940	299,768 ± 10

Cavity resonators:

Essen	1947	299,797 ± 3
Hansen	1951	299,789 ± 1

Geodimeter:

Bergstrand	1949	299,796 ± 2
Edge	1956	299,792.4 ± 0.11

Radio interferometer:

Froome	1958	299,792.5 ± 0.1

Tellurometer:

Wadley	1957	299,792.6 ± 1.2

Radio:

I.T.T. Staff	1970	299,793
C.C.D.M.	1975	299,792.458 ± 0.004

Data source: B. Setterfield, 1982, _Ex Nihilo_, vol. 1, no. 1 (Int.), p. 56, Table 3.

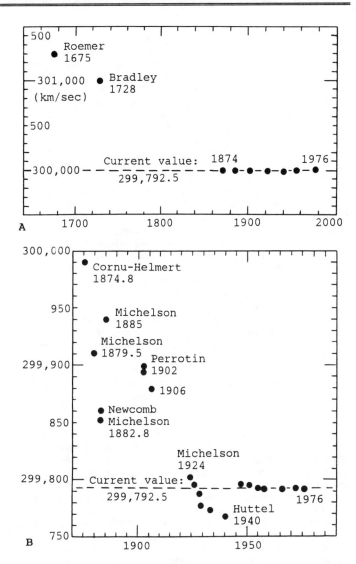

Figure 15.1 Selected observed values of the speed of light plotted on different vertical and horizontal scales. A. Three centuries of observations. B. Observations of the modern era of physics. (Based on data of Barry Setterfield, 1981, _Ex Nihilo_, vol. 4, no. 1, Tables 1 and 2.)

shown in Table 15.1 and in the accompanying graph (Figure 15.1), would not be immediately assigned to a time-decrease in the actual value of _c_. Instead, explanations relating to the methods used would be first examined. The earliest calculations, based on telescopic observations within the solar system, would be scrutinized for possible sources of error in the ability of observers to measure the starting and stopping times of the phenomena observed. Second, physicists would want to look into physical causes that would lead to erroneously larger speeds of light where the rays pass through various media of transmission. As every student of elementary physics knows, light rays passing through a prism are bent, or refracted, so as to split up into a rainbow spectrum. Astronomical observations would be affected by refraction caused by the earth's atmosphere, causing errors in the reading of ray angles.

Physicist Ronnie J. Hastings has studied Setterfield's published reports and calls attention to a major weakness in the curve-fitting procedure. In a manuscript yet to be published, Hastings writes:

> His entire premise rests mainly upon two very early optical measurements made by Roemer in 1675 and Bradley in 1728. The former was made by observing the eclipsing of Jupiter's Galilean moons behind the planet, and the latter by observing stellar aberration, both methods subject to atmospheric interference and observational error every modern observer wants to avoid, though Setterfield wants us to believe in the precision of Roemer's instruments. (Hastings, 1987)[3]

Further details of the problem with the Roemer value are given by Edward D. Fackerell, an associate professor

in the Department of Applied Mathematics in the University of Sydney, Australia (Fackerell, 1984, pp. 90-91). It seems that Setterfield has used a _c_ value of 301,300 km/sec taken from a popular article in _Sky and Telescope_ (June, 1973), in turn referring to an article in the _Astronomical Journal_ (Goldstein, Trasco, and Ogburn, 1973) in which observations made by Roemer (and Picard) at the Paris Observatory from 1668 to 1678 were restudied by the three authors named above. Pushing past many details, available in Fackerell's text we come to the significant conclusion: Roemer's data yield a value of _c_ that does not differ by 0.5 percent from the current value of about 299,270. That is a statistical statement of probable error, i.e., "plus-or-minus 0.5 percent." Fackerell states that Setterfield, without any justification, changed the statement of the three authors to read that the speed of light in 1675 was 0.5 percent higher than the present value (Fackerell, 1984, p. 91). This adds the full 0.5 percent (1,500 km/sec) to the present value to give Setterfield's value of 301,300. Inserting today's value into the 1675 position makes a dramatic reduction in the correlation coefficient; this step, writes Fackerell (p. 91), has the ability to invalidate Setterfield's curve. If this, the earliest value of _c_, is not significantly different from the present value, there is good reason to reject the entire hypothesis that the speed of light was substantially greater in the past.

We can turn now to examine the broader consequences of Setterfield's hypothesis. A change in the value of speed of light must, by reason of the interrelatedness of many laws of physics, also require corresponding changes with time of several other constants, among them the constants of mass, gravitation (G), and the half-lives of radioisotopes. Hastings points out that these related forced changes must lead us to conclude that the universe of several thousand years ago must have behaved very differently from how it does today (1987).

Consider, for example, the effect of an enormously high value of c upon the rate of production of heat by spontaneous radioactive decay in minerals of the earth's crust. Radioactivity is explained in Chapter 17, so we must be limited here to the predicted effect itself. In short, the energy released by radioactive decay in, say, the year 3,940 B.C. would have been 1,200 times greater than today. Hastings offers calculations to show that under those conditions it would take only about seven years for the world ocean to experience a temperature rise of one Celsius degree. A temperature rise of 100 C would thus require only seven centuries, assuming no radiational heat loss. But when the required reduction in the mass of the oceans is also taken into account, greatly reducing the specific heat of seawater, the time for a one-degree temperature rise is reduced to only about $2\frac{1}{2}$ minutes. Says Hastings: "At this rate it would take just over a day for enough heat to be generated to raise all the oceans 100 degrees Celsius and vaporize them into steam!" Hastings goes on to point out that such high rates of radioactive decay would mean ionizing radiation in deadly intensities in the earth's surface environment. Created life would have been destroyed by this intense radiation unless, for some mysterious reason, cells of plants and animals of that time were immune to the somatic damage that cells experience today under similar radiation dosages. The point is, of course, obvious. In devising an outrageous hypothesis to fit exactly the biblical scenario of recent creation, the creation scientists have painted themselves into a corner--perhaps several corners at one time--from which they cannot extricate themselves without postulating a universe preposterously different from that observed throughout recorded human history. Creationists are thus required to repudiate one of their stated tenets: that God initially created the laws of the universe as forever after valid and unchanging. Of course, there are creationists who adhere to that tenet and correspondingly reject Setterfield's tampering with the speed of light. Has Setterfield announced a bold new paradigm of creation science to be set up in place of the prevailing paradigm of rule of the universe under God's unchanging laws and their immutable constants? Is a great schism making its appearance to rend creationism into two warring doctrinal factions?

Setterfield continues to promote his hypothesis of the former enormous value of the speed of light. His later version in the 1984 issue of Ex Nihilo Technical Journal is accompanied by two papers criticizing it from the position of mainstream physics, Professor Fackerell's being one of them. Rebuttals by Setterfield are also included. Altogether, about seventy large-format pages in dense type are given to the debate in this single issue. The Australian creationists are to be commended for their inclusion of critical articles from the other side of the debate. No such evidences of free debate have as yet appeared in publications of the Institute for Creation Research.

Much of this chapter deals with what Ronnie Hastings calls "creation physics" (1987). It seems fitting, then, to close this chapter with the following lines from the final paragraph of his manuscript:

> "Creation physics" bears the mark of Christian fundamentalism as do all other facets of creationism. . . . Such faith is the substance of personal and even social inspiration, not science. Such faith is apparently inspirational enough to Setterfield to be very selective of his data so that it yields a "creation" date corresponding to fundmentalist cosmology, to be oblivious of gross inconsistencies with modern physics, and to cling to the Bohr model of the atom in his applications without reference to its shortcomings and without attempting to explain why physicists the world over consider the Bohr model to be better replaced with a quantum mechanical interpretation. It is true that science can be inspired by almost any human emotion or insight, including religious fervor, but the resulting insights into the universe retain neither the content of nor the justification from that emotion or insight. Only nature itself justifies scientific statements. Considering the methods and implications of "creation physics" such as Setterfield's, "creation physics" needs far more than nature to extricate itself from its problems. (Hastings, 1978)[3]

Credits

1. From Steven I. Dutch, 1982, A Critique of Creationist Cosmology, Journal of Geological Education, vol. 30, pp. 27-33. Used by permission.

2. From Stanley Freske, Evidence Supporting a Great Age of the Universe, Creation/Evolution, vol. 1, no. 2, pp. 34-39. Copyright © 1980 by Stanley Freske. Used by permission of the author.

3. From Ronnie J. Hastings, 1987, "Creation Physics" and the Speed of Light. (Manuscript to be published.) Used by permission of the author.

Chapter 16

How Mainstream Science Views the Solar System

Creation scientists of the Institute for Creation Research (ICR) have presented a number of arguments favoring recent creation of the solar system. These relate to such topics as Newtonian mechanics of orbiting bodies, cosmic dust, comets, the Moon's orbit, and geology of the Moon. Perhaps most important are the creation scientists' claims that the methods of determining the age of Earth rocks, Moon rocks, meteorites, and the solar system itself are worthless. To follow the pros and cons of the various arguments, you need to have a general background in solar system astronomy and the various hypotheses of solar system origin, including that most widely favored today by mainstream astronomers, geochemists, geophysicists, and geologists.

As we did for astronomy and cosmology, we now offer a brief review of the solar system. If you are already familiar with this material, go immediately to Chapters 17, 18, and 19 dealing with the creationists' arguments.[1]

Our Solar System

We need first to get a good perspective on the relative size of our Sun in comparison with a single planet, such as the Earth. (We capitalize the names of Earth, Sun, and Moon to conform with capitalized names of the other planets and their satellites.) Our Sun, a huge sphere of incandescent gas, has a diameter of more than 1000 times that of Earth, a volume 1,300,000 times that of Earth, and a mass more than 330,000 times that of Earth (Figure 16.1).

Actually, the Sun contains about 99.84 percent of the total mass of the entire solar system. Most of the remaining 0.16 percent of mass is in two giant planets, Jupiter and Saturn. So planet Earth is merely a grain of sand--so to speak--compared with the enormous cauldron of hot, seething gases that make up the Sun.

We need to make one more point of comparison between planets and Sun before turning our minds to their origin. The Sun, as a star, has an internal energy source-- nuclear fusion--that gives it an enormously high temperature, both in its interior and on its surface. As in other similar stars, the Sun's internal temperature is estimated to be from 13 to 18 million Kelvin degrees (K), equivalent to 22 to 32 million Fahrenheit degrees (F), and this is a level of heat far beyond the imagination of any human being. The Sun's surface temperature is from 4,000 to 6,000 K (8,000 to 11,000 F), and this can be directly measured by scientists. The planets, in contrast, have no internal mechanism of generating energy to produce temperatures of millions of degrees. The Earth is thought to have a temperature of about 3,000 K (4,500 F) in its inner core--scarcely more than one-thousandth of the Sun's interior. Surfaces of all the planets are cold compared with the Sun's surface. Whereas the Sun radiates internal energy intensely from its hot surface, the planets depend upon receiving that solar radiation for most of their surface heat. Consequently those planets nearest the Sun have the warmest surfaces (Mercury,

Venus, Earth, Mars) while those farthest distant from the Sun are coldest (Jupiter, Saturn, Uranus, Neptune, Pluto). On Mercury, for example, temperature on a surface exposed to the Sun reaches nearly 420 C (800 F), high enough to melt tin or lead. On the surface of Neptune--the most distant large planet--temperature is around -200 C (-300 F).

What Science Must Explain about the Solar System

If you wish to begin speculating on the origin of the planets, keep in mind that they all move around the Sun in more or less circular (elliptical) orbits and approx- imately in one flat plane, and that all move in the same direction (Figures 16.2 and 16.3). Your hypothesis must explain these observations as well as many others. For example, it must explain why each of the inner four planets (Mercury, Venus, Earth, Mars) is composed largely of solid rock surrounding a large central core of iron or an iron-sulfur compound. It will be necessary to explain why the four great outer planets (Jupiter, Saturn, Neptune, Uranus) are so much larger than the inner group, and why they consist largely or in large part of hydrogen and helium and/or solid compounds of carbon, hydrogen, nitrogen, and oxygen. Besides accounting for satellites of the planets--moons, that is-- your hypothesis should also explain minor objects that orbit the Sun--the asteroids (minor planets), meteoroids, and comets. Many asteroids and meteoroids are composed of materials closely resembling the inner planets in composition. Comets contain compounds of carbon, hydrogen, nitrogen, and oxygen similar to those found in the atmospheres of the outer planets.

Taking this information into account, your speculations might include the following three general possibilities for the origin of the planets:

(1) Each planet was formed in a distant and different place, far from the Sun in some other part of our galaxy, and somehow each planet was brought to a position near the Sun and launched into an orbit around it.

(2) The material out of which the planets were formed was taken out of the Sun itself, formed into planets, and these were launched into orbits.

(3) The planets and Sun (and other objects such as moons, comets, and asteroids) were all formed at about the same time from a single, great parent- mass of diffuse matter in space.

The first possibility may be the poorest, because it requires many special events and forces. With space so enormous and distances so vast, it is very unlikely that a planet formed many thousands of millions of miles away from the Sun could be guided toward the Sun and arrive with precisely the angle and speed to be captured into

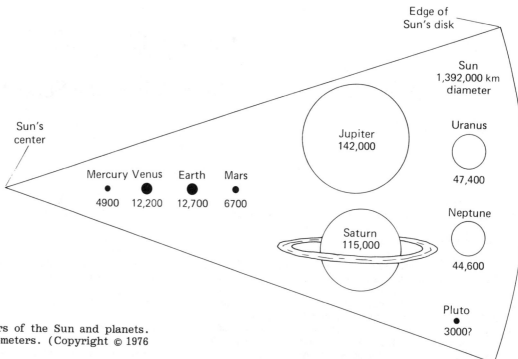

Figure 16.1 Relative diameters of the Sun and planets. Figures give diameters in kilometers. (Copyright © 1976 by Arthur N. Strahler.)

the Sun's gravitational field and remain in orbit. The second and third possibilities would be expected to receive favorable attention because they are quite simple in general plan. Each of the hypotheses has a certain basic unity about it.

Early Hypotheses of Catastrophic Origin

Not surprisingly the second possibility was taken up seriously by respected scholars as far back as the middle of the eighteenth century. The French philosopher and naturalist, George Louis Leclerc, Comte de Buffon, proposed in 1749 that a comet collided with the Sun, tearing loose the solar substance from which the planets were formed. Not a bad idea for his time! Collision of a comet with the Sun was long suspected because every now and then a comet that was observed to approach close to the Sun failed to reappear on the other side. In 1981 such an event was actually recorded in data sent back from a space vehicle. But the idea of planets being formed in this way is not given much likelihood, because comets lack the great mass needed for the kind of impact Count Buffon had in mind.

The idea of planets derived by violent action from the Sun was revived in 1905 by two highly respected scientists: T. C. Chamberlin, a geologist, and F. R. Moulton, an astronomer. Both were professors at the University of Chicago. They set up the hypothesis that a rapidly moving star once passed close to our Sun. The mutual gravitational attraction of these enormous bodies for each other raised great tidal bulges on the Sun's surface. One bulge faced the moving star, the other lay on the opposite diameter of the Sun. When the attractive force became too great, large amounts of gaseous matter were ejected from the Sun, forming two jets that reached far into space. These jets took the form of curved arms. They were similar to, but much larger than, bursts of gaseous matter that are frequently observed to be thrown upward from the Sun's surface and known as solar prominences. The total amount of matter thus ejected was many times greater than the planets, but most of it fell back into the Sun. That which did not fall back into the Sun traveled around the Sun in orbit. Gradually the gas cooled and condensed into small solid particles, and these

in turn joined together as they collided in space. The larger masses, called planetesimals, continued to grow by attracting the remaining smaller particles, and thus the major planets were formed. Gradually the space occupied by the planetary orbits was swept clear of most of the planetesimals.

Well, what, if anything, is wrong with the Chamberlin-Moulton hypothesis? First, it calls upon a rare special event that is most unlikely to begin with, considering the vast dimensions of space within our galaxy. Furthermore, later calculations by physicists showed that for the gravitational attraction between the two stars to have been sufficient to cause the matter to be drawn out from the Sun, the two stars would have had to come within a separating distance not greater than the diameters of the two stars themselves--a most unlikely event. Their closing speed would have to be about 5,000 km (3,000 mi) per second, much faster than a star would be able to attain within the galaxy relative to other stars.

The idea of planets forming as the result of a passing star was taken up again in 1917 by two English astronomers, Sir James Jeans and Sir Harold Jeffreys. They suggested that the passing star caused a single, long filament of gases to be drawn from the Sun. While the outer part of the filament escaped into space, the inner part fell back into the Sun. The middle part condensed into solid particles, and these came together to form the planets. This hypothesis was beset with most of the same difficulties as the Chamberlin-Moulton hypothesis. Scientists think now that any gaseous matter that could be drawn from the Sun by a passing star would quickly escape into space, and none would remain in orbit to form planets.

Laplace's Nebular Hypothesis

And so we come to the third possibility, that the Sun together with its entire solar system was formed from a single, great parent-body of matter. The idea is just about as old as Count Buffon's, for it was put forward in 1755 by the distinguished German philosopher, Immanuel Kant. He proposed that the solar system originated as a cloud of gas and dust, with a high concentration of matter in a central region that later became the Sun.

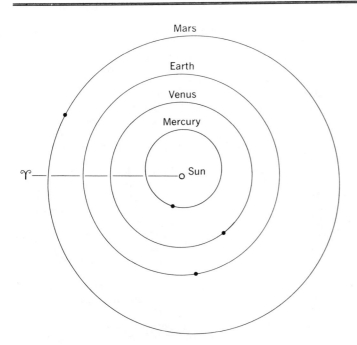

Figure 16.2 Orbits of the four inner planets. A black dot represents the point at which the planet is closest to the Sun. The orbits are drawn as circles and do not show the true elliptical form of each orbit. (Copyright © 1976 by Arthur N. Strahler.)

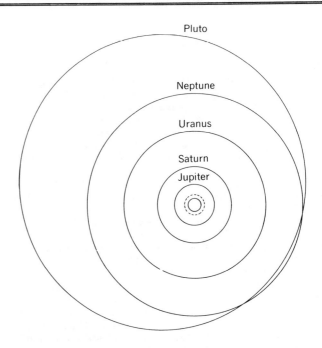

Figure 16.3 Orbits of the outer planets. The innermost circle represents Mars's orbit; the dotted circle represents the zone of asteroid orbits. Pluto's orbit is inclined more than seventeen degrees with respect to Neptune's orbit. (Copyright © 1976 by Arthur N. Strahler.)

Kant's idea was taken up a few years later by the French astronomer and mathematician Pierre Simon, Marquis de Laplace, who published his version in 1796 under the name of the nebular hypothesis. Laplace's reconstruction of the origin of the solar system begins with a hot, rotating mass of gas and dust--a nebula-- isolated in space. The nebula had a diameter larger than that of the largest planetary orbit. Because the finest solid particles and even the molecules of gas are attracted to each other by gravitation, the nebula began to contract into a smaller volume and took the form of a thin, flat disk rotating on a central axis. As contraction occurred, the speed of rotation (spinning) of the nebula increased. Spinning set up outward radial forces, which tended to act against the gravitational force that tended to cause contraction. At a certain critical point, these outward forces of rotation exactly balanced the gravitational force within an outer ring of the nebula. The outer ring condensed into solid particles of increasing size, and these fused together, eventually forming a planet. More rings were formed in succession in the same manner, and each became a planet. Laplace even had an explanation for the moons that now orbit the planets. He supposed that as each planet formed, it shrank in diameter and left concentric rings of gas and dust that became its moons. The great central mass of gas and dust contracted to become the Sun.

Laplace's nebular hypothesis enjoyed great popularity for more than a century, but as physics developed increasing levels of sophistication, it became all too apparent that if the nebular hypothesis were correct, the Sun should now rotate about 200 times faster than it does. Also, physics could furnish no physical reason for intermittent contraction of the nebular disk, rather than continuous contraction. Finally, it was concluded that the molecules of highly heated gas within the rings would have easily escaped from the gravitational hold of the central solar mass of the nebula. And so, by the year 1900, the hypothesis had been placed on the shelf of history to become merely a curiosity.

Like many outmoded theories of science, the nebular

hypothesis retained the seeds of a newer and better hypothesis that could explain new observations of astronomy. These seeds were lying dormant through the late 1800s and early 1900s, while astronomers were thinking in terms of ways that matter could be drawn out of the Sun to make planets. But astronomers were also hard at work studying the stars and with the aid of better telescopes were learning more and more about the various kinds of stars--blue giants and supergiants, red dwarfs and white dwarfs, and just ordinary medium-sized stars like the Sun. This new knowledge allowed astronomers to develop a general theory of the origin of stars and their typical life histories.

In our earlier section on astronomy and the origin of the universe (Chapter 14) we reviewed modern ideas on star formation, in which a cloud of interstellar gas and dust began gravitational collapse when impacted by the shock front from the expanding remains of a supernova. Rapid collapse produced a rotating stellar nebula in the form of a thick lens with a dense ball-shaped protostar at the center (see Figure 14.4).

The Condensation Hypothesis

Modern explanations of the origin of the planets favor the condensation hypothesis, in which the planets grew within the outer part of the solar nebula by condensation of gases into liquid and solid particles and by the clotting together of those particles. By numerous collisions the particles accumulated into larger objects, and these in turn joined together to form the planets and other objects of the solar system. Alternative titles for this general explanation are accumulation hypothesis and accretion hypothesis. There are many variations within the general hypothesis, and the details are strongly debated as to their merits and weaknesses.

Just prior to the time of the formation of the solid particles that were later to become the planets, our solar nebula was extremely hot--several thousand Kelvin degrees--but had not contracted enough to generate intense internal heat by the process of radioactivity. In other words, the Sun as a star had not yet formed. The dense inner part of the nebula can be called the protosun. During contraction, as we have already stated, the entire solar nebula began to spin, rotating upon a central axis (Figure 14.4). As the rate of spinning increased, the nebula took on a more flattened shape, shrinking along the dimension of the spin axis, but expanding in dimension of the circular rim, or "equator." As the rim of the nebula moved out farther from the protosun, temperatures began to fall within the disk because of an increase in distance of the gas from the central source of radiant heat. Ultimately, temperatures in the outermost zone of the nebula fell to only a few tens of Kelvin degrees, whereas in the inner part of the nebula, close to the protosun, temperatures remained high--over 1,000 K. Thus a complete gradation of temperature came to exist from a cold outer zone to a hot inner zone.

The basic concepts of the modern condensation hypothesis relate to what happened chemically and physically as temperatures decreased in various zones of the nebular disk. Wherever the nebular temperatures remained above 1,600 K, all nebular matter was in the gaseous state, but where the temperature fell below that level and continued to fall, condensation of matter from the gaseous state to the liquid or solid state followed a chemical sequence in which different substances were produced in succession from the nebula. The chemical sequence based on nebular temperature is shown in Table 16.1. The chemical changes and reactions shown here can actually be demonstrated by laboratory experimentation. (The critical temperature at which a particular change in state or chemical reaction occurs depends partly upon the gaseous pressure within the nebula and would have become lower outward through the nebula where gas density was progressively less.)

To understand what was going on within the nebular disk at a particular time and place, we need only ascertain the nebular temperature and read from the table the activity appropriate to that temperature. Because the nebular temperature had previously fallen to the specified value, those activities applying to higher temperatures had already occurred in that region, whereas the activities specified for lower temperatures would not yet have occurred.

Let us pick a point in time when the inner zone of the nebular disk, closest to the protosun, had temperatures in the range of 1,200 to 1,300 K. At the same time, temperatures in the middle zone of the disk were in the range of 400 to 500 K, and in the outer zone, from 100 to 200 K. We can now read from the table what was happening in each zone of the nebular disk, and we can also see what chemical changes had previously occurred in that same zone.

In the inner zone where temperature was about 1,600 K, oxygen was combining with calcium, aluminum, and titanium to form solid, dustlike grains of calcium oxide, aluminum oxide, and calcium titanium oxide. These oxides are described as refractory substances. Somethat farther out, in the zone at 1,300 K, the elements iron and nickel were able to condense into minute droplets of a nickel-iron alloy. A bit farther out, where temperature was around 1,200 K, a compound of silicon, oxygen, and magnesium was being formed. On Earth today we find this mineral in crustal rocks; it is called enstatite and belongs to a group of silicate minerals known as the pyroxene group.

Throughout the entire temperature zone from 1,200 to 490 K, metallic iron was uniting with oxygen to form iron oxide, and this in turn would have been reacting with enstatite grains already formed to yield olivine, a silicate mineral with magnesium and iron. (Much of the Earth's mantle is thought to consist of a mineral of the same general composition as olivine.)

In the nebular zone at 1,000 K, we read that condensation was allowing the formation of a group of silicate minerals called the feldspars. They consist of silicon oxide in combination with aluminum, sodium, and potassium. Feldspars are common in the Earth's crust. For example, a large proportion of ordinary granite consists of feldspars. At 680 K, gaseous hydrogen sulfide was combining with metallic iron to form an iron-sulfur mineral known as troilite. It has been suggested that the Earth's outer core consists of troilite, which is somewhat less dense than nickel-iron. Earth has a small inner core in the solid state, and this is thought to be nickel-iron.

Moving out into a somewhat cooler zone, we come to a temperature of 550 K, at which molecules of water in the gaseous state--water vapor, that is--were uniting chemically with calcium and magnesium to form the mineral tremolite. At 425 K, water unites with olivine to form serpentine. This chemical activity involving water is extremely important because it means that at a temperature far above its boiling point, water in large quantities could have become locked into abundant minerals of the inner planets.

We move out next to the vast region of the cold outer zone of the solar nebula, thinking now in terms of temperatures far below zero on both the Celsius and Fahrenheit scales. At 175 K (-100 C), under the very low pressures that would have prevailed in this part of the nebula, gas molecules of water-vapor were condensing into water-ice. At 150 K, gaseous ammonia was uniting with water-ice to form an ammonia-ice compound.

Finally, in the temperature zone of 120 K, methane gas was uniting with water-ice to form a methane-ice compound. Methane has been detected at the outer surfaces of the atmospheres of Uranus and Neptune, and it is possible that a considerable quantity of methane is present deeper in their atmospheres.

The remaining forms of gas condensation at still lower temperatures listed in the table are beyond the region of planetary formation. However, they point to an important principle: argon, neon, hydrogen, and helium would never have condensed from gaseous to liquid forms in the solar nebula. How, then, does it happen that the outer planets consist largely of hydrogen and helium? We will turn to this problem in a later paragraph.

Planetesimals and Protoplanets

Our next step is to set up a working hypothesis for the actual physical formation of the two groups of planets--inner and outer. Condensation resulted in minute liquid droplets or solid grains of microscopic dimensions. The solid grains now began to move under gravitational attraction to occupy a flat plane at the equator of the rotating nebula (Figure 14.4). Particles above that plane moved down to arrive at the plane; particles below the plane moved up to reach it. The particles thus began to form a very thin disk, which we can visualize as quite like the disk formed by the rings of Saturn.

Rapidly the solid grains of the disk began to clot together to form larger particles, a process called accretion. These particles quickly grew to the size of pebbles, then began to merge by collisions into larger masses. In a stage in which the larger masses had reached diameters of one to several kilometers, they became planetesimals (the same term invented earlier by Chamberlin and Moulton). Now the planetesimals within the flat disk moved in swarms in orbits that intersected to permit numerous collisions. Some of the more violent collisions shattered the planetesimals into small fragments. (This is one possible origin of meteorites, those fragments of iron or silicates, or mixtures of the two, that still move in planetary orbits and occasionally impact a planet.) Other collisions were relatively gentle, and the mutual attraction of the colliding particles caused them to join into a single mass. Numerous collisions of this type resulted in the rapid growth of much larger bodies, which became the ancestral planets, or protoplanets. Quickly the

Table 16.1   Stages of Condensation from the Solar Nebula

Nebular temperature, K		Chemical activity	Planetary environment	
Above 1600		All nebular matter in gaseous state.		
1600		Condensation of oxides of calcium and aluminum (refractory substances).	---	
1300		Condensation of nickel-iron alloy.	Mercury	
1200	Rock-forming silicate minerals	Condensation of enstatite, a silicate mineral consisting of silicon-oxide and magnesium.		Inner planets
1200 to 490		Metallic iron unites with oxygen to form iron-oxide, which in turn reacts with enstatite to form olivine, a silicate of iron and magnesium.		
1000		Feldspars, a silicate mineral class, form by combination of silicon-oxide with aluminum, sodium, and calcium.	Venus	
680	Water-rich silicate	Hydrogen-sulfide gas comines with metallic iron to form troilite, an iron-sulfur compound.	Earth	
550		Molecules of gaseous water (water vapor) unite with silicate minerals of calcium and magnesium to form a water-rich mineral, tremolite.	Mars	
425		Water molecules in vapor state unite with olivine to form a water-rich silicate mineral, serpentine.	--- (Asteroids) ---	
175		Molecules of water vapor condense into water-ice.	Jupiter	Outer planets
150	Ices	Ammonia gas unites with water-ice to form an ammonia-ice compound.	Saturn	
120		Methane gas unites with water-ice to form a methane-ice compound.	Uranus Neptune	
65	Gases	Argon gas and remaining methane gas condense to solid argon and solid methane.	---	
20		Condensation of gaseous neon and hydrogen.		
1		At close to absolute zero, helium gas condenses into liquid helium.		

Data sources: John L. Lewis, 1973, Technology Review, vol. 76, no. 1, pp. 21-35; 1974, Scientific American, vol. 230, no. 3, pp. 51-65.

protoplanets swept up from their orbital zones most of the remaining smaller fragments, and the growth of the primitive planets was largely completed.

We noted earlier that one reason the Laplace nebular hypothesis was discarded was that the Sun would be expected to now rotate much faster than it does. Another way to say this is that the angular momentum within the solar system is not correctly distributed between Sun and planets under Laplace's version. Angular momentum can be thought of as describing "quantity of motion in a curved path." (For an object moving in a circle about a center of gravity, angular momentum is the product of the quantity of mass, the velocity in the path, and the radius of the circle.) Now, the point is that our rotating Sun, with nearly 99.9 percent of the total solar system mass, possesses only about 2.75 percent of the total angular momentum, whereas Jupiter and Saturn together, with just over 0.1 percent of the solar system mass, possess over 85 percent of the total angular momentum. In some manner, by some mechanism not explained by simple laws of motion, an enormous amount of momentum was transferred from the Sun to the two giant planets.

Creation scientists of the ICR have used this problem as an opportunity to attack mainstream science. Duane

Gish of the ICR has called attention to the distribution of angular momentum as we have described it and flatly states that no plausible mechanism for this distribution has yet been proposed (Gish, 1974, p. i).

Contrary to Gish's statement, the answer to the problem of momentum distribution is now known with a high degree of certainty. The outer region of the solar nebula remained under the influence of the protosun through lines of magnetic force generated within the protosun and extending out through the entire rotating nebula. The lines of force coupled the outer nebula with the central protosun and had the physical effect of slowing the rotation of the protosun, while accelerating the rate of rotation of the outer rim of the nebula and its disk of solid particles. This outward transfer of energy increased the angular momentum of the great outer planets at the expense of the angular momentum of the protosun.

Internal Structure of the Planets

Let us now combine the principle of a chemical sequence of condensation based on temperature with the physical process of planetary accretion in an attempt to reconstruct the formation of the actual planets in terms of

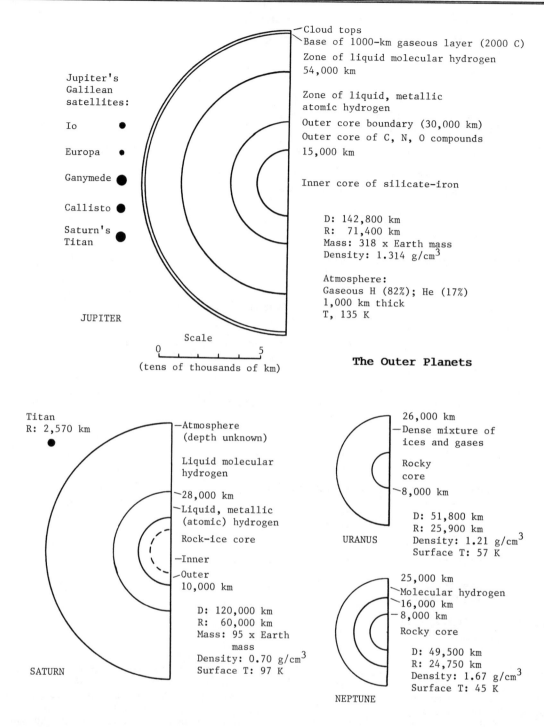

Figure 16.4 Structure and composition of the outer and inner planets compared. (A. N. Strahler.)

their chemical composition and internal structure. Reversing the usual order, we will start with the outermost planets and work inward.

In the region of Uranus and Neptune, as nebular temperatures fell, condensation of grains of the refractory oxides would have begun first, followed by condensation of nickel-iron. (See names of planets in Table 16.1 in column labeled "Planetary environment.") According to one modern version of planetary formation, called hetero-geneous accretion, these substances would have quickly developed into protoplanets, formed largely of iron, and possibly or probably in a molten state. Thus the dense innermost cores of Uranus and Neptune came into existence (Figure 16.4). With falling temperatures, the silicate minerals condensed, and silicate grains reaching

the disk formed a new generation of planetesimals that were swept up by the protoplanets to be added to each as a silicate layer enclosing the iron core. There followed condensation of the volatiles: water-ice, and the ammonia-ice and methane-ice compounds. These were added to the existing protoplanets as thick mantles of ice. Thus Uranus and Neptune emerged as fully developed planets that have undergone few changes since. However, a large number of planetesimals of the volatile substances escaped collisions with the two great planets. These were thrown into orbits beyond that of the planet Pluto and became the source of the comets. The remaining gases of the nebula--argon, neon, hydrogen, and helium--could not condense under the prevailing temperature conditions. Moreover, Uranus and Neptune were not large enough to capture hydrogen and helium in their gravitational fields.

To reconstruct the evolution of Jupiter and Saturn we have only to repeat what has been said for Uranus and

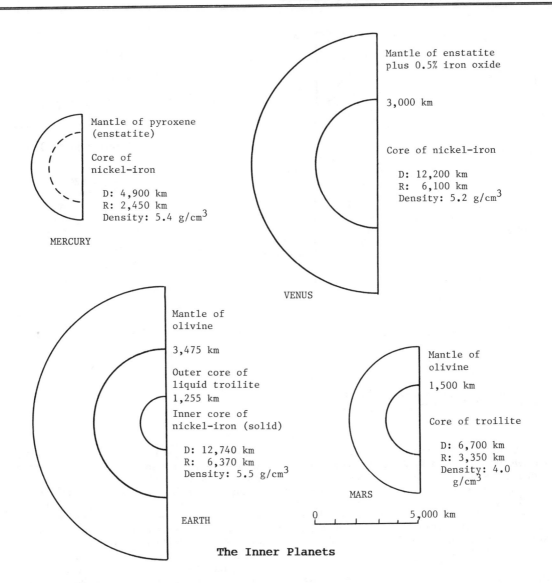

Mantle of pyroxene
(enstatite)

Core of
nickel-iron

D: 4,900 km
R: 2,450 km
Density: 5.4 g/cm³

MERCURY

Mantle of enstatite
plus 0.5% iron oxide

3,000 km

Core of nickel-iron

D: 12,200 km
R: 6,100 km
Density: 5.2 g/cm³

VENUS

Mantle of
olivine

3,475 km

Outer core of
liquid troilite
1,255 km
Inner core of
nickel-iron (solid)

D: 12,740 km
R: 6,370 km
Density: 5.5 g/cm³

EARTH

Mantle of
olivine

1,500 km

Core of troilite

D: 6,700 km
R: 3,350 km
Density: 4.0
   g/cm³

MARS

0   5,000 km

The Inner Planets

Neptune. This scenario would lead to two protoplanets very similar to the latter pair. However, we must assume that these newer protoplanets were much larger than Neptune and Pluto, because the nebula in the zone of Jupiter and Saturn was thicker and denser than in the outermost zone and could thus supply much greater masses of volatiles. Much more massive protoplanets were produced and these, in turn, could attract and hold hydrogen and helium in their gravitational fields. Thus the two greatest planets quickly grew into giants, each several times the mass of either Neptune or Uranus.

And now we come to the principal satellites (moons) of Jupiter and Saturn (Figure 16.5). Jupiter's four large moons--Io, Europa, Ganymede, and Callisto--are called the Galilean satellites, because they were discovered in 1610 by Galileo, using the first astronomical telescope. Titan is the single large moon of Saturn. (Figure 16.5 also includes Earth's Moon and planet Mercury for comparison.) Io and Europa are thought to be composed largely of silicate rock, but of the two, Io is of greater average density and may have a small core of troilite. Ganymede, Callisto, and Titan are of much lower average density and are thought to have massive mantles of ice surrounding large cores of silicate rock. The ice mantle is most probably water-ice in Ganymede and Callisto, but ammonia-ice and methane-ice compounds may be important forms of ice in the mantle of Titan.

The origin of the five large satellites of Jupiter and Saturn fits rather nicely into the same series of stages we outlined for the great planets themselves. Silicate cores were formed first, then the ice mantles were added. These large moons can be considered as small protoplanets that escaped destruction by collision. It is thought that they and some of the smaller moons formed within the cloud of dust and gas that surrounded each of the growing giant planets.

An interesting sidelight to the formation of planets in the cold outer region of the nebula is the possibility that comets originated here. Comets have a composition dominated by volatile substances--compounds of hydrogen, carbon, oxygen, and nitrogen--for example, carbon monoxide, carbon dioxide, ammonia, and methane. Some silicate mineral matter is also present as solid grains. It has been suggested by astronomer Fred Hoyle that comets originated as planetesimals by condensation and accretion in the region of Uranus and Neptune. Because of their wide-ranging orbits, the comets escaped being incorporated into the outer planets.

Between the orbits of Jupiter and Mars is a space gap occupied by large numbers of small, solid objects called asteroids. Their origin is discussed in a later paragraph, but we can note here that their compositions (which vary over a wide chemical range) are compatible with this temperature zone, 400 to 550 K.

The four inner planets, while forming a group characterized by dense, iron-rich cores and thick mantles of silicate rock, are by no means alike. Figure 16.4 shows cross sections of these planets to the same scale, and we immediately note their size differences. The two larger planets, Earth and Venus, are very closely matched in

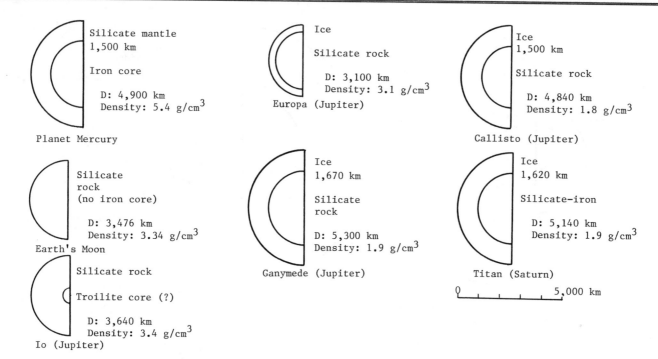

Figure 16.5 The large satellites compared with planet Mercury. All are drawn to the same scale. (A. N. Strahler.)

diameter. Mars is of intermediate size and Mercury quite small by comparison with the others. Actually, their sizes are not of concern to us in a discussion of origin, whereas their average densities will tell us a great deal.

Starting with Mars, we see that its average density of 4.0 g/cm³ is substantially lower than for the other three (5.2 to 5.5 g/cm³). This must mean there is an important chemical difference between Mars and the others. The Martian core is judged to have little or no nickel-iron alloy, but to be composed largely of the iron-sulfur compound troilite, which condenses at 680 K. The mantle is interpreted as being composed of olivine, a silicate mineral of iron and magnesium. Table 16.1 shows olivine to be formed by chemical reaction at temperatures below 1,200 K and extending down to 490 K. Thus the inferred composition of Mars seems compatible with an environmental temperature range between 600 and 700 K. In general, for Earth, Venus, and Mercury, the proportionate size of the metallic nickel-iron core increases in the same direction as the inferred increase in nebular temperature. Earth's olivine mantle gives way to enstatite mantles in Venus and Mercury, a change that is also in the direction of increasing nebular temperature shown in Table 16.1. Thus in a general way, the inferred compositions of the four planets fit into the condensation sequence based on nebular temperature.

The Sun Goes on Nuclear Power

You might reason that if temperatures in the nebula continued to fall, volatiles would ultimately condense in the region of the inner planets, and they too should have finally accumulated icy mantles. Our hypothesis should provide for one of two possibilities: One, the cooling trend was reversed before volatiles could begin to condense. Two, ice mantles were formed on the inner planets, but a temperature reversal set in and the ice mantles evaporated. Both of these possibilities are provided for in an event of great importance that occurred after the primitive planets were formed. The protosun continued to contract and, despite the steady cooling of the surrounding nebula, interior pressure and temperatures continued to rise. At a certain critical point, nuclear fusion began, rapidly raising internal temperatures to several million Kelvin degrees.

Fusion consists of the conversion of hydrogen into helium, a process involving the conversion of part of the mass of each hydrogen atom into energy. The process begins to occur when temperatures within a contracting star exceed 4 million K. Actually, the fusion process consists in detail of several forms of atomic reactions involving the elements lithium, beryllium, and boron, and ending with the production of helium. When internal temperatures exceed 15 million K, yet another series of complex nuclear reactions begin to occur in the process of transformation of hydrogen to helium. Over long periods of time the mass lost by the total conversion process is an extremely small part of the original hydrogen mass. With fusion in progress, the Sun's surface became highly luminous.

The Sun's surface now began to put forth a great stream of charged atomic particles; they were protons and electrons resulting from the ionization of hydrogen atoms. The particles flowed outward from the Sun as a <u>solar wind</u>. A solar wind exists today; it can be seen as the pearly <u>corona</u> visible during a total solar eclipse.

At the time the primitive planets had just been formed, the solar wind was probably much more powerful than now and was capable of forcing free gas molecules and microscopic dust particles to move with it into outer space. Thus the solar wind literally swept the entire planetary region free of the remaining gas and dust of the original nebula. Solar radiation was no longer blocked by dust suspended within a nebular mass and could reach each planet in full force, heating the planetary surface to a temperature proportional to its closeness to the Sun.

If mantles of ammonia-ice and methane-ice compounds had accumulated on the inner planets prior to this clearing out of the nebular dust, those ices would have been evaporated by surface heating and the gaseous matter would have been swept out by the solar wind. In the region of the outer planets, solar radiation falling on the planetary surfaces is much weaker, and the surfaces of those planets remain below the melting points of the ices that comprise their atmospheres and surface layers.

Actually, we see this vaporization process acting upon comets that occasionally pass close to the Sun. Solar radiation vaporizes the icy volatiles to produce gaseous tails that are swept up in solar wind.

It is easy to understand why Mercury, a small planet close to the Sun, is totally devoid of volatile substances. Because of its small size, it has weak gravity, enabling gas molecules to escape quite readily. It is also strongly heated by the Sun, with surface temperatures reaching 400 C where the solar rays are most direct in angle.

On the other hand, both Venus and Earth have dense atmospheres of gas molecules that, because of the strong planetary gravity, cannot easily escape into space. Moreover, Earth is unique in having an abundance of surface water in liquid and ice form, which it has retained since early in its history. Is Earth's surface water an inheritance from a layer of volatile substances condensed from the solar nebula? Conventional wisdom holds that the Earth's surface volatiles (water, nitrogen, oxygen, carbon dioxide, and many others) were originally locked in the mineral compounds of rocks of the mantle layer and were later released by volcanic action. We may also want to entertain the more exciting hypothesis that Earth's supply of volatiles was brought to it from far out in the solar system, as a large comet by sheer accident impacted Earth.

Asteroids and Meteoroids

In the early 1800s, astronomers discovered in the space gap between Mars and Jupiter a few small planetary objects, which appeared through their telescopes as points of light moving among the stars. Known as asteroids, the four largest of these later proved to be masses ranging in diameter from 450 to 1,000 km (300 to 600 miles), but most appear only as dots on photographic film and are only a few kilometers in diameter. More than 2,000 have been named and their orbits determined. The asteroids probably number in the tens of thousands, but their total combined mass is estimated at no more than 1/1000 that of Earth. Most asteroids have more or less circular orbits between Mars and Jupiter, but a few have rather odd, elongated elliptical orbits that repeatedly bring them close to Earth. For example, an asteroid called Icarus passed within $6\frac{1}{2}$ million km of Earth in 1968. This was close enough for astronomers to observe that it has an irregular shape, perhaps something like a badly formed Idaho potato, is no more than 2 km in width, and is probably composed of iron.

A new wrinkle in the scientific study of asteroids came in the late 1970s, when Viking spacecraft began to orbit the planet Mars and took excellent pictures of Mars's two moons, Phobos and Deimos. Phobos, about 25 km in diameter, is irregular in outline and has a heavily cratered surface. Deimos is also small and irregular in shape. It has occurred to astronomers that these strange moons may be asteroids that were captured by the gravitational field of Mars.

And now, a few observations about meteoroids, those fragments of solid matter that enter the Earth's atmosphere from outer space. Most are particles no larger than dust grains; they simply create a thin trail of light --a meteor--as they vaporize in the upper atmosphere. Swarms of these meteors that visit us on a regular yearly schedule are thought to represent the solid remains of comets that long ago lost their volatile substances. But, every now and then, a large meteoroid, even though partially vaporized by its passage through the atmosphere, reaches the Earth's surface to become an exotic sort of rock called a meteorite. Some of these objects arrive intact as one body; others explode before reaching the ground and are found as widely scattered fragments.

Of the meteorites collected and analyzed, about 94 percent are classed as "stones"; they are composed largely of silicate minerals, mostly olivine and pyroxene, but they contain a little nickel-iron as well. About $4\frac{1}{2}$ percent are composed almost entirely of nickel-iron alloy; they are called "irons." The remaining $1\frac{1}{2}$ percent are "stony-irons"--mixtures of the silicate minerals and nickel-iron. Right away, you are reminded of the materials scientists think comprise the bulk of the inner planets--the nickel-iron in planetary cores and their silicate mantles. Do the meteorites give samples of the interior of our own planet? We should add that stony meteorites of one particular class known as chondrites have an internal composition and structure unlike anything found in earthly or lunar rocks. They contain small rounded bodies of silicate mineral composition. In some cases the rounded bodies, called chondrules, have a large amount of the element carbon, which is almost unknown in Earth rocks composed of silicate minerals that have solidified from a molten state. These same carbonaceous chondrules are rich in water, bound up in the silicate mineral serpentine. So it seems that we have here a sample of the volatile substances that we have supposed were present in the solar nebula and condensed as ices on the outermost planets. The carbonaceous chondrites also contain iron-sulfide as troilite, which we have postulated to be an important part of certain planetary cores.

The message we read from the meteorites takes on special significance when we look at their radiometric age. Without exception and regardless of their class and composition, they all yield an age close to 4.5 billion years (-4.5 b.y.)! This age fits well with the age of the Earth as calculated from radioisotopes contained in ancient crustal rock. The most satisfactory interpretation of the age of meteorites is that they consist of solid matter condensed rapidly from the solar nebula in a time span of only a few million years. The carbonaceous chondrules are thought to be original spherical bodies, directly condensed from the nebula, and later incorporated into larger masses of silicate minerals without being remelted or otherwise changed.

Meteorites that have been collected and studied seem obviously to be fragments of much larger masses. For one thing, the crystalline grain structure of the silicate minerals requires slow cooling deep within a much larger mass. In some cases the nickel-iron meteorites show a crystalline structure that also required slow cooling in a large mass. What were these larger parent masses, and how large were they? Perhaps they were as large as the largest of the asteroids. Reflected light from asteroids has been analyzed to reveal much about their chemical composition. Some are of iron composition, others are of silicate composition. Some dark-colored asteroids show a composition almost exactly matching that of the carbonaceous chondrites.

And so the weight of all evidence so far collected points to a conclusion consistent with our general hypothesis of origin of the major planets. Planetesimals, many of them ranging in diameter from 100 to 1,000 km, formed rapidly by accretion of solid particles that were condensing out of the solar nebula. Within the larger planetesimals, melting took place followed by cooling and crystallization of the silicates and nickel-iron. The iron melt was concentrated in the core of the planetesimal while the silicates formed a surrounding mantle. Although most of the planetesimals eventually became incorporated into the nine major planets, some managed to escape that fate and were preserved as the asteroids in the space gap between Mars and Jupiter. There were, however, many severe collisions between large asteroids, fragmenting them into odd-shaped masses of all sizes, from small asteroids down to the smallest meteoroids. This fragmentation probably continued through much of the 4.5 billion years that followed.

To this scenario we can add the possibility, stated

earlier, that the comets are fragments of planetesimals formed largely of volatiles in the outermost region of the solar nebula, and that these bodies (or their fragments) have survived today because they were thrown into wide-ranging elliptical orbits that pass far out beyond the most distant planet, Pluto, and reach far inward to close proximity to the Sun. Traveling at high speeds in and out of the solar system, numerous comets have thus far escaped collision with the Sun or a major planet.

Origin of the Moon

Earth's Moon seems to be the most difficult to explain of all the planetary satellites. First, it is exceptionally large in mass in relation to its planet, with a mass about 1/81 that of Earth. The ratio of diameters is only about one-to-four. Astronomers have expressed the situation by saying that the Earth-Moon pair can be thought of as a binary planet (a double planet). How could so large an object as the Moon have come into existence as a satellite so close to a planet?

Scientists had hoped that information gained on the Apollo missions to the Moon's surface would help solve the basic problem of the Moon's origin. Rock samples all proved to consist of silicates of magnesium and iron, with some aluminum-silicates (feldspars). The composition of these rocks is generally similar to a class of rocks of the Earth's crust known as basalts, which make up the crust of the ocean floors and occur as one kind of volcanic lava on continents as well. However, there proved to be some subtle chemical differences between the minor ingredients of Earth basalts and those of lunar basalts. This observation suggested to some geologists that the Moon did not form in the same physical environment as Earth.

Other direct evidence relating to the Moon's origin came from determinations of the ages of lunar rock samples. Practically all samples revealed an age of mineral crystallization between -4.2 b.y. and -3.1 b.y. (Until recently, the oldest date on metamorphic and igneous Earth rocks was not older than -3.8 b.y. In 1983, zircon crystals in ancient sedimentary rocks in western Australia were dated as -4.1 to -4.2 b.y.) An important lunar find was a rock specimen containing several small fragments of olivine (magnesium-iron silicate) that proved to have an age of -4.6 b.y. This oldest material matches the meteorites in age and seems to assure that the substance of which the Moon is now composed was formed by condensation and accretion from the solar nebula to create planetesimals that grew into protoplanets.

But there remains an observation difficult to explain. The Moon's density of 3.34 g/cm^3 rules out the possibility that it has an iron or troilite core. If the Moon grew independently as a small planet in the same general region as the inner planets, it should have at least a small iron core.

Three hypotheses of origin of the Moon have had some degree of popularity at one time or another. One is that the Moon and Earth grew as a binary planet by accretion from the solar nebula and thus never existed as two independent planetary objects. The advantage of this hypothesis is that it avoids the difficulties of explaining how so large an object as the Moon could become trapped into orbit by the Earth. The Moon's lack of an iron core is the greatest single obstacle to the binary accretion hypothesis. Minor but distinct differences in the mineral chemistry of the two bodies have also been cited as evidence against their having formed simultaneously in close proximity.

The capture hypothesis supposes that the Moon was formed by accretion in a different part of the nebula, presumably more distant from the Sun than the region of the four inner planets. Accretion began with silicate

minerals, in some manner escaping the earlier stage of condensation and accretion of iron or iron sulfide. (The same can be said of the formation of Europa, Ganymede, Callisto, and Titan, whose densities do not allow for iron cores.) Even if the chemical problems connected with distant accretion can be explained, there remains the mechanical problem of capture of Moon by Earth during a close encounter. Some supporters of the capture hypothesis place the date of this event at about -3.7 to -3.6 b.y., based on the age of great outwellings of lava upon the Moon's surface. However, it can be argued that an encounter at such close range would have disrupted the Moon completely.

A third working hypothesis is actually one of the earliest. Called the fission hypothesis, it was first put forward in 1897 by Sir George Darwin, son of Charles Darwin and an authority on tides and tidal forces. He calculated that the Earth was formerly spinning on its axis much more rapidly than today. Outward forces of rotation produced a great bulge in the Earth's equatorial region. As these forces increased, the bulge broke away from the Earth and moved out into orbit to become the Moon. Darwin's hypothesis was rather popular in his time and there were those who guessed that the Pacific Ocean basin represents the cavity from which the lunar material was largely derived.

Revived in the early 1960s, the fission hypothesis again became popular because it could explain why the Moon lacks an iron core. Fission occurred after the Earth's core had been formed, so that rock removed from the Earth came from the silicate mantle. More recently, a new version of the fission hypothesis has called upon a tremendous impact by a planetary object with approximately the present mass of the Moon. Impact tore away a large chunk of the Earth's silicate mantle and both the impacting object and the mantle material were pulverized into dust. The orbiting dust cloud underwent accretion and formed the Moon. By introducing a foreign body into the scenario, minor differences in chemistry between lunar rocks and Earth rocks can perhaps be explained.

This impact hypothesis has gained greatly in strength within the past few years. Since it was proposed in 1975 by William Hartman and Donald Davis of the Planetary Science Institute in Tucson, the hypothesis has garnered some strong supporters (Kerr, 1984). Calculations by A. G. W. Cameron of the Harvard-Smithsonian Center for Astrophysics have shown the feasibility of such an impact for propelling an adequate mass into orbit (Taylor, 1985, p. 17).

Despite these signs of the emergence of a dominant hypothesis capable of achieving a consensus, or paradigm, of lunar origin, the various alternatives continue to be revived with new twists. Creation scientists have taken delight in what seems to have been little progress despite direct observations of the lunar surface and intense study of samples of its rocks. Frustration of the mainstream scientists is seen in their own facetious, off-the-record comments that the Moon is an "impossible object" and that perhaps it is an "imaginary object." In a more serious vein, they will insist that the Moon was formed by processes that adhere strictly to laws of physics and chemistry. They will continue to search for new evidence bearing on this subject, but the ending of the Apollo program may have postponed for decades a major scientific breakthrough.

Credit

1. Parts of this chapter are extracted or adapted from the author's textbooks of geology and the earth sciences published by Harper & Row, Inc., New York. Copyright © 1971, 1972, 1976, and 1981 by Arthur N. Strahler.

Chapter 17

Age of the Solar System and Earth

In our earlier discussion of the origin of the universe we talked about a universe 10 to 20 billion years old. That cosmic concept of age derives from two assumptions. First is the constancy of the speed of light through all time since the Big Bang. Second is the reliability of the Doppler shift of the light spectrum, or redshift, for all astronomical distances used in conjunction with an assumed Hubble constant. Estimates of the value of that constant have changed over the decades since Hubble in 1927 used a value that led to an age of 2 b.y., much less than the age of crustal rocks previously measured by radioactivity.

Before embarking on a point-by-point consideration of creationists' arguments favoring recent creation of the solar system, we should review the entire question of determining ages of the materials and structures that make up the earth, solar system, our galaxy, and distant galaxies. The question of duration of geologic time is vital to the fundamentalist creationists, since their religious philosophy requires creation of the universe in a six-day period not farther back in time than about 10,000 years. The general plan of the creation scientists of the ICR is twofold in concept. First, scientific arguments must be put forth to show that methods used by physicists and geologists to establish ages of rocks as great as millions or billions of years are in error and must be discarded. Second, scientific arguments must be put forward to show that all kinds of geologic features, ranging from rocks and minerals to surface features of the land (landforms), show evidence of very recent age.

Creation scientists make use of the history of geology to show that older methods of estimating the age of the earth and its rocks were poorly based on scientific evidence and yielded results that differed greatly from one method to another. This evaluation of earlier studies reiterates many of the negative arguments made by mainstream geologists for decades. We will start with a review of some of the early attempts to determine the absolute age or actual age (age in years before present) of rocks and the earth itself. The term actual age is preferred by many geologists, who point out that the word "absolute" suggests perfection in age determination. All age determinations involve errors, which can arise from several sources. We will follow this review with an explanation of modern methods of dating based on principles of radioactivity.

Early Estimates of the Earth's Age

The search for a means by which to establish the absolute age of an event in the earth's past history was for decades frustrating and, in retrospect, highly misleading. Most geological estimates of the ages of crustal rocks were based on two lines of calculation: (a) the salinity (degree of saltiness) of ocean water and total salt content of the oceans, and (b) the total accumulated thickness of sedimentary rock strata of all geologic time. It was assumed that to arrive at the age of the oceans (and hence an early time in planetary history), the estimated total weight of salt presently held in the oceans could be divided by the annual increment of salt, most of which is brought into the oceans from the lands by streams and rivers. Chemical analysis of the salt content of rivers could form the basis for estimating the annual contribution. Sodium and chlorine as dissolved common salt, the mineral halite (NaCl), together make up about 85 percent of the salt content of seawater. Toward the end of the 1800s the calculation was made that the total weight of sodium in the oceans is about 1.6×10^{16} tons, and that the annual increment is about 1.6×10^8 tons. Division yields a figure of roughly 100 million years (10^8 m.y.).

You can probably suggest at least one good reason to doubt the value of such a calculation. One is the possibility that not all sodium that has entered the ocean since it was formed remains there to the present day. Many of you know about the existence of great layers of rock salt deep beneath the continental surfaces. In the Permian Basin under Kansas and Oklahoma a few of the many layers of salt reach thicknesses of 90 to 120 meters. Good evidence is present to show that this salt was evaporated from seawater in shallow estuaries. The evidence thus indicates that a large, but unknown, quantity of sodium chloride was deposited in salt beds through geologic time, so that an estimate of the present quantity of sodium in the ocean is of little value in age calculation.

A second, and probably greater, source of error is to use the estimate of the present rate of salt entering the ocean as valid through millions of years of the past. Because of continual motions of lithospheric plates, the ancient continental shields have traveled widely over the globe throughout Precambrian and younger geologic time. Assuming that the general locations of belts of dry and wet climates have always been governed by an atmospheric circulation dominated by the earth's rotation on its axis, individual continents were at times located in the equatorial zone of high rainfall, at other times in one of the two belts of great tropical deserts, and even at times in a polar location under a great ice sheet. This history makes any extrapolation of present supply rates of sodium back over hundreds of millions of years far too risky to be attempted.

The second approach was to take the thickest known accumulation of sedimentary strata of each geologic period and total them for all periods into one imagined grand pile of strata--it was estimated as being 150 km thick! In one calculation, using a value of one meter per 650 years as an average rate of accumulation, an age of about 100 m.y. was obtained for the total elapsed time since sediment deposition began. Allowing for numerous, great time gaps almost certainly occurring in the record and requiring introduction of a correction factor of 15 times, the total age of the strata would come to 1.5 b.y. Between uncertainties as to rates of deposition and lengths of periods of nondeposition, the results were scarcely better than blind guesses. More than thirty estimates made between 1860 and 1917 ranged from roughly 3 m.y. to 15 b.y.

Late in the nineteenth century, the distinguished

British physicist, William Thomson (later to be known as Lord Kelvin), calculated the age of the earth using the premise that the earth cooled from a molten state and that the cooling rate followed simple laws of radiative and conductive heat loss. On this basis, he concluded that the earth could not be older than 100 m.y. and that an age of 20 to 40 m.y. was a reasonable figure. This figure was disappointingly short to the geologists, and also to Charles Darwin and his followers. The stratigraphic column seemed to require much longer spans of time. Lord Kelvin had moved his estimate far in the "wrong" direction, but it was difficult to find any flaw in his application of what were believed to be correct laws of physics.

Radioactivity

Dramatically, the dilemma over the age of the earth and the duration of periods of geologic time was resolved with the discovery of radioactivity. Using these principles, the first crude age determinations of rocks were made in 1905 and 1907; the oldest rocks analyzed proved to have ages of 2.2 to 2.4 b.y.

Radioactivity is the spontaneous breakdown of certain elements, leading to permanent changes in the atoms involved. Radioactivity is accompanied by the emission of energetic atomic particles and the production of heat. Another way of describing radioactivity is that it is a natural process of conversion of matter into energy. Let us review some of the principles of nuclear physics involved in radioactivity.

The nucleus, or dense core, of an atom consists of two types of particles, neutrons and protons. For a given element the number of neutrons varies, whereas the number of protons is fixed. Take, for example, an important radioactive form of the element uranium: In the nucleus of this form of uranium there are 146 neutrons and 92 protons. The total of neutrons and protons is therefore 238. This quantity is known as the mass number and is designated by a superscript before the symbol for uranium, thus: ^{238}U. This form of the element is also written as uranium-238, or simply U-238. The atomic number of an element is the number of protons contained in the nucleus. In the case of uranium the atomic number is 92. Although the atomic number is fixed for each named element, the number of neutrons in the nucleus is subject to some variation.

For uranium-238 the number of neutrons is 146, but there also exists another form of uranium with 143 neutrons. The latter form thus has a mass number of 235, and it is designated as uranium-235. These differing varieties of a single element are referred to as isotopes.

A key to the understanding of radioactivity is that certain isotopes are unstable. This instability can result in the flying off of a small part of the nucleus, reducing the mass number or the atomic number or both. The original element is thus transformed into a different element, having a different name. In this spontaneous breakdown, mass is converted into energy, released into the surrounding matter. The term radioactive decay covers the entire process.

Let us look further into the nature of the spontaneous radioactive process. Breakdown of the atomic nucleus may result in the emission of an alpha particle, consisting of two neutrons and two protons (same as the nucleus of a helium atom). As a result, in this case there is a decrease of four in the mass number and two in the atomic number. As the alpha particles travel outward through the atoms of the surrounding substance, their energy is transformed into the increased activity of the electrons of those atoms, imparting heat to the surrounding substance.

Another emission in the radioactive breakdown is the gamma ray, an energy form that acts similarly to electromagnetic radiation (sun's radiation). In gamma emission, the atomic number remains the same. Gamma rays are absorbed by the atoms of the surrounding substance. A third form of emanation is the beta particle, an electron traveling at high velocity. One form of beta emission (shown in Figure 17.1) raises the atomic number by one but does not change the mass number. (One of the neutrons is changed into a proton.) Beta particles also react with surrounding atoms and increase the quantity of heat.

The radioactive disintegration of one parent isotope may lead to the production of another unstable isotope, known as a daughter product. This product, in turn, may produce yet another unstable isotope, and so forth, until ultimately a stable isotope results and no further radioactivity occurs.

Take as an example the system of uranium-238 and its daughter products, leading finally to the stable lead isotope lead-206. Figure 17.1 shows steps in this decay series. Arrows show the direction of successive changes. Note that each step in the direction of the arrow to the left (alpha emission) marks a decrease in the mass number. Uranium-238 decays to produce thorium-234. This is followed by a succession of isotopes of different elements, listed along the bottom of the graph. The disintegration process achieves a steady rate, or equilibrium, with time.

Of prime importance in the dating of geologic events is a physical law that governs the rate of decay of a parent radioactive isotope (or radioisotopes). The law applies unless the supply of atoms is being replenished, because it is itself the daughter of another isotope. The law states that the ratio of decrease in the number of atoms of the parent isotope with each unit of time is a constant. In other words, the process can function as a perfect clock. This is a statistical law based on large numbers of atoms of the isotope. Each radioactive nucleus decays randomly on an individual basis.

Let us take the above example, U-238 and its final stable product Pb-206, and plot its decay against time, as shown in Figure 17.2. We can start at any point in time with a small quantity of pure U-238. Let the number of atoms of U-238 at time-zero be designated as unity, or 100 percent, as shown in the upper-left corner of the graph. After 4½ b.y. have elapsed, the number of atoms of U-238 will have been reduced to half the initial number, or 50 percent. The time span of 4½ b.y. is known as the half-life. In the first half-life the number of atoms of Pb-206, which was zero at time-zero, has risen to 50 percent, equal to the number of remaining atoms of U-238. After two half-lives have elapsed, U-238 will have diminished to 25 percent, while Pb-208 will have risen to 75 percent. This schedule of decrease with time is called exponential decay.

Different radioisotopes have different half-lives. For example, another radioisotope of uranium, U-235, has a half-life of about 0.7 b.y. For other radioisotopes the half-life may be very much longer; for example, the half-life of rubidium-84 is 48.8 b.y. In contrast, many radioisotopes have short half-lives; that of carbon-14 is about 5,730 years; for tritium, a radioisotope of hydrogen, the half-life is 12½ years. Both carbon-14 and tritium are useful in dating events of relatively recent occurrence (see Chapter 19).

An important assumption in using radioactivity to determine the age of rocks is that the half-life of each kind of radioisotope has been constant over at least the past five billion years. So far as can be determined by nuclear physics, and as predicted from theory, the spontaneous decay of the radioisotope is completely unaffected by the external environment of pressure and temperature or the chemical processes in which the radioisotope participates at or near the earth's surface. This immunity to the ranges in external conditions of our planet and other solar system satellites can be understood in a general sense through the model of the atom as a

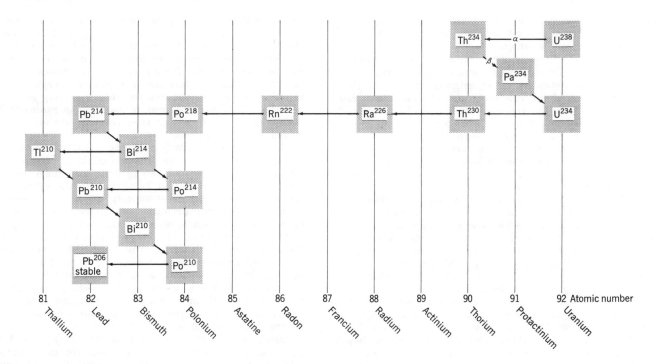

Figure 17.1 Radioactive decay series of U-238 to Pb-206. (Based on data of P. M. Hurley, 1959, <u>How Old Is the Earth?</u> Doubleday & Company, p. 62, Figure 9.)

central massive nucleus surrounded by a cloud of electrons that form an effective shield against all outside forces. There is, however, a small measure of uncertainty in the values measured for the half-lives. For example, the half-life of U-238 is considered accurate to within one percent of the value; it is presently established as 4.498 b.y.

Radiometric Age Determination

Basic principles of spontaneous decay of radioisotopes provide us with a method of determining the age of an igneous rock, a procedure of science known as geochronometry. Ages thus determined are referred to as radiometric ages.

At the time of crystallization of an igneous rock from its hot liquid state, minute amounts of minerals containing radioisotopes are trapped within the crystal lattices of the common rock-forming minerals, in some cases forming distinctive radioactive minerals. Assume for the moment that at this initial point in time there are present none of the stable daughter products that constitute the end of the decay series. (Initial daughter isotopes are usually present, but these can be evaluated.) As time passes, the new, stable end member of each series is produced at a constant rate and accumulates in place.

Knowing the half-life of the decay system, we can estimate closely the time elapsed since mineral crystallization occurred. An accurate chemical determination of the ratio between the radioisotope and the stable daughter product must be made. A fairly simple mathematical equation is used to derive the age in years of the mineral under analysis.

Take, for example, the uranium-lead series, uranium-238 to lead-206, with a half-life of $4\frac{1}{2}$ b.y. (Figure 17.2). Quantities of both uranium-238 and lead-206 are measured from a sample of uranium-bearing minerals or from a common mineral (usually zircon) enclosing the radio-isotopes. The ratio of lead to uranium is entered into the equation and easily solved for an age in years.

Similar age determinations can be made using the series uranium-235 to lead-207. Because both series of uranium-

lead isotopes are normally present in the same mineral sample, age analysis of one series can serve as a cross-check upon the other. It is possible to determine the absolute age of a sample of uranium-bearing mineral to within about 2 percent of the true value, and in some cases to within 1 percent. But this level of accuracy also assumes that none of the components in the decay series has been lost from the sample. Use of the uranium-lead systems for age determination can be applied to the oldest rocks known, as well as to meteorites. Age of meteorites is close to 4.5 b.y., about 0.7 b.y. older than the oldest rocks of the earth's crust that have thus far been dated. Isotopes of lead can be used to date the time of formation of the earth as a planet at -4.5 b.y. (billion years before present), a figure in good agreement with the age determined for meteorites and one that has been accepted by mainstream science for more than thirty years. Most lunar rocks fall in the age range of -4.2 to -3.8 b.y., but none has been dated as younger than -3.0 b.y.

Of great importance in age determination is the potassium-argon series, potassium-40 to argon-40, with a

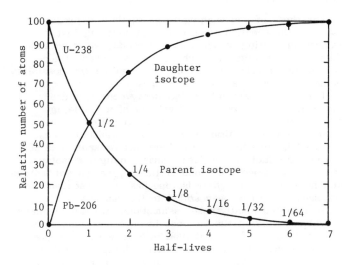

Figure 17.2 Relative decrease in parent isotope and increase in daughter isotope through seven half-lives. (A. N. Strahler.)

half-life of 1.25 b.y. It is particularly adaptable to use with the mica group (specifically, muscovite and biotite) and hornblende, all of which are widely present in igneous rocks. The potassium-argon series gives reliable minimum ages for fine-grained volcanic rocks (lavas), which cannot be dated by other methods. Moreover, the method can be used for relatively young rocks (less than 1 m.y.), as well as for the most ancient rocks. It has been a highly important tool for the geologist, particularly in deciding which of two rock groups is the older.

The rubidium-strontium decay series, rubidium-87 to strontium-87, has an extremely long half-life: 48.8 b.y. This series is of great value in dating both individual minerals and whole-rock samples.

Ideally, the above three dating systems--uranium to lead, potassium to argon, and rubidium to strontium-- should serve as cross-checks upon one another when all three are applied to mineral samples from the same rock body. In some instances the ages check out as closely similar, but there are instances in which moderate discrepancies are evident. Despite existing uncertainties, the radiometric ages given for various events in the timetable of the earth's history are now accepted by mainstream geologists as valid within small percentages of error. The success of the radiometric age determinations of rocks stands as a striking scientific achievement based upon the application of principles of nuclear physics to geology.

Errors in dating with uranium-lead and rubidium-strontium decay series, which have very long half-lives, decrease to small values only with great age of the sample, and their use is limited largely to the older crustal rocks, those of Precambrian age. As we noted above, the potassium-argon series, with a half-life of only a little over 1 b.y., is useful in rocks as young as 1 m.y. For rocks younger than 1 m.y., other radioisotopes must be used. One of these is thorium-230, which decays with a half-life of 75,000 years (y.). It has proved particularly useful in dating deep-sea sediments that have accumulated as far back as several hundred thousand years.

For even younger events--those of Holocene time (present to -10,000 y.) and very late Pleistocene time-- the radioisotope of choice is carbon-14, with a half-life of 5,730 y. Its use is limited to about -40,000 y. Thus the date of Divine Creation of the universe as interpreted from Genesis 1 is well within the scope of carbon-14 dating, and we need to know how the method works (see Chapter 19).

Methods of Checking the Integrity of Radiometric Ages

For those of you wishing to go further in preparing to evaluate the creation scientists' arguments against radiometric dating of minerals and rocks, we offer some details of the ways in which geochemists check their results for internal consistency and against the possibility of having arrived at erroneous ages.

Besides the assumption that decay constants are invariant through past time, validity of a single age determination depends on the assumption that at time-zero, which is the time at which the radioisotope was enclosed in its parent rock, there was present none of the end product of the decay series (daughter isotope). It is also assumed that none of the original parent isotope and none of the daughter isotope has since escaped from the sample (or has been added to the sample).

In thinking about these assumptions, we must visualize a system involving the isotopes in question. This system can be thought of as contained in a very small fragment of solid, crystalline mineral matter holding the isotopes in question. The system could, of course, be a much larger body of rock. In our earlier discussions of thermodynamics, we referred to open systems and closed systems, making the point that no true closed system of both matter and energy exists in nature, unless perhaps the universe as a whole is to be designated the one truly closed system. In that context, a closed system imports or exports neither matter nor energy. There are, however, natural systems that for all practical purposes are closed in terms of matter, but not in terms of energy. In that sense, the radioisotope system we are talking about can be closed only with respect to matter. As radioactive decay takes place, matter is converted into energy that leaves the system.

At this point you will probably want to stop me and recall that alpha-particle emission consists of the flying off of two neutrons and two protons into the surrounding region. Because matter is being lost to the system it cannot be described as absolutely closed. So it seems that we must revise our definition of "closed" in this context to mean that no whole atoms of the parent or daughter isotopes are imported or exported through the system boundary. In other words, the sum of the parent atoms and daughter atoms must remain constant. That is the assumption we make, but does it hold valid in nature?

In experimental physical science, which is what we are here dealing with in drawing and analyzing samples, the safety-in-numbers principle is put into practice. In determining the age of a particular body of granite, for example, several or many samples are analyzed and the ages compared. Small departures from the mean value relative to the quantity of the mean itself signal reliability and a certain satisfactory level of confidence. Another approach is to use two different methods on the same sample, methods that are independent in the sense that the one method does not influence the outcome of the other one. Let us turn now to methods of cross-checking the reliability of radiometric ages. In describing two methods of cross-checking radiometric ages, I have relied rather heavily on lucid descriptions by Dr. G. Brent Dalrymple of the U.S. Geological Survey (1984) and Professor Don L. Eicher (1976). The latter gives excellent coverage of the entire subject of geologic age determination in his book Geologic Time.

The first procedure to consider is the isochron method. A preliminary step in understanding this method is to learn what is meant by an isochron, beyond the rather obvious meaning as a line of "equal-time" (from the Greek isos, equal; chronos, time). The isochron is used in various ways in science, but here it is a straight, slanting line on a graph showing the relative proportions of parent isotope and daughter isotope of a decay series at a given point in time.

In its simplest form, the isochron can be derived from the two exponential curves shown in Figure 17.2. Figure 17.3 shows the percentages of parent and daughter isotopes plotted simultaneously for various half-lives; they lie on a diagonal line extending from the lower-right corner to the upper-left corner. Suppose, now, we say that instead of percentages, the Y and X scales can give the actual number of parent atoms in the sample at time-zero. For a sample half as large as that represented by 100 percent, we would plot a line of points starting at the 50 percent mark on the X-axis and ending at the 50 percent mark on the Y-axis. Ratios of parent to daughter remain the same for each half-life value. Next, we pass straight lines through our two sets of points and these all terminate at the origin (X = 0; Y = 0). The new set of lines are isochrons; they are labeled in terms of time elapsed from time-zero, using the artificial assumption that the half-life of the radioisotope is exactly 1.0 b.y. For a perfect closed decay system, the isochron begins at time-zero as a horizontal line (abscissa) and increases in slope as time passes, approaching the vertical (ordinate) as time approaches infinity.

Applying the isochron principle to the rubidium-strontium (Rb-Sr) method, let us imagine that three samples have been taken from the same mass of rock that, from its uniform physical and chemical characteristics,

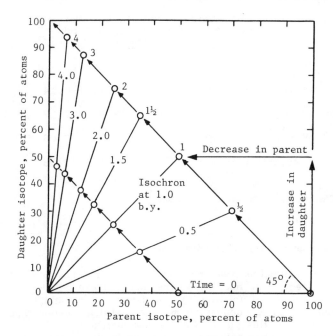

Figure 17.3 Arithmetic-linear plot of quantity of parent isotope against quantity of daughter isotope, defining isochrons. (A. N. Strahler.)

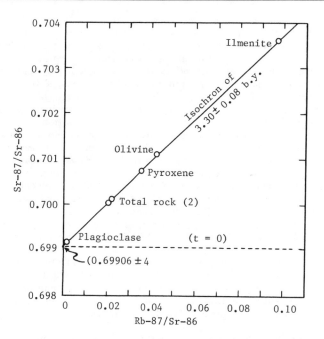

Figure 17.5 Isochron for rubidium-strontium ratios for mineral and rock samples from a lunar rock collected by astronauts of the Apollo-15 mission. (Based on data of Rama Murthy et al., Science, vol. 175, Figure 1, p. 420.)

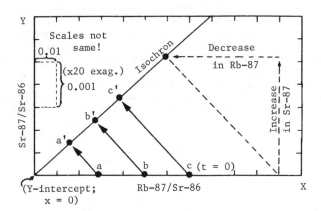

Figure 17.4 Isochron for ratios of rubidium and strontium isotopes. (A. N. Strahler.)

gives every indication of having had one and the same history of origin. Assume in this case that the rock is igneous--solidified from a high-temperature melt, or magma. The parent radioisotope is rubidium-87 (Rb-87); the daughter is strontium-87 (Sr-87); the half-life of Rb-87 is 48.8 billion years (b.y.).

Essential in the Rb-Sr isochron method is recognition of the presence in most rocks of initial strontium, Sr-86, that is not radiogenic. Because the quantity of Sr-86 remains constant, it provides a stable reference base and can be used to derive two ratios: Sr-87/Sr-86 and Rb-87/Sr-86. The two ratios can be plotted on a simple (arithmetic-linear) graph, as in Figure 17.4. At time-zero (t = 0) the ratio of the two forms of strontium (plotted on the ordinate, or Y-axis) is zero for all three samples, a, b, and c, that therefore fall on a straight horizontal line. As time passes, the Rb-87/Sr-86 ratios will decline (move left), while at the same time the Sr-87/Sr-86 ratios will increase (move up). The points a, b, and c thus move diagonally up and left with time, as shown by three parallel arrows. At a second point in time the three points will be represented by a', b', c', also on a straight line, the isochron, meeting the time-zero line at a common point

on the ordinate. That common point on the Y-axis establishes the initial Sr-87/Sr-86 isotopic ratio. Through time, the slope of the line has steepened, pivoting on the common point on the Y-axis. The steeper the slope angle, the older the rock sample. (Note that on this graph the horizontal and vertical scales are not the same; the vertical scale is stretched out by a factor of 20. This exaggeration is required by the greatly disparate sizes of the two isotopic ratios, given in Figure 17.5.)

The points a', b', and c' correspond with ratios actually measured in rock samples. Figure 17.5 is an example of the isochron method applied to a specimen of basalt rock collected by astronauts in Hadley's Rille on the moon. The plotted points come from different kinds of mineral matter within the rock fragment, which explains their spread along the line. One of the points represents the whole-rock analysis. The points lie very close to a straight line and define the appropriate isochron. The slope of the isochron gives the information needed to derive the age, which is 3.30 ± 0.08 b.y. In some cases, the validity of a Rb-Sr age can be corroborated by plotting the isochron for samarium/neodymium (Sm-147/ Nd-143) ratios. This cross-check has been effectively used in the age determination of meteorites (see Dalrymple, 1984, p. 74).

For the uranium-lead (U-Pb) method, a somewhat more complex method of cross-checking is used. It compares the ratios of two different decay systems of uranium and lead isotopes: U-238 to Pb-206 (half-life 4.47 X 10^9 yr) and U-235 to Pb-207 (half-life 7.04 X 10^8). Ratios used here are those of daughter-to-parent, i.e., Pb-206/U-238 and Pb-207/U-235. At time-zero these ratios are also zero, since there is no daughter product at time-zero. Figure 17.6 shows one ratio plotted against the other on a graph with arithmetic scales. Because the half-lives are quite different, the time-plot is curved convexly upward. Points on the curve show elapsed time in billions of years. This curve is known as the concordia. If ratios determined from samples actually fall on the concordia, their ages are said to be concordant and would mean that none of the daughter lead has been lost from the system.

What is commonly found, however, is that the observed ratios do not fall on the concordia, but lie instead below that curve and form a more or less straight line sloping down to the left. An example is shown in Figure 17.6. This relationship is described as discordant and a straight line fitted to the points is known as the discordia; it forms a chord that intercepts the concordia in two places. The discordia is interpreted to mean that some of the daughter lead was lost from the system in a single geologic episode occurring earlier in the decay history. It can be shown that the upper point of interception corresponds with the true or original radiometric age of the samples. The lower point of interception is commonly interpreted as marking the date of the episode of lead loss. In Figure 17.6 nine samples of a metamorphic rock (gneiss) from a locality in Minnesota define the discordia. The rock is just over 3.5 b.y. old, a near-record age for rocks in North America; the event at -1.85 b.y. was probably one of heating, causing lead loss.

Dalrymple evaluates the method in these words:

> The U-Pb concordia-discordia method is one of the most powerful and reliable dating methods available. It is especially resistant to heating and metamorphic events and thus is extremely useful in rocks with complex histories. (1984, pp. 75-76)[1]

The concordia-discordia method is often used in conjunction with the potassium-argon and rubidium-strontium isochron methods to give independent corroboration of the age of the rock. The example in Table 17.1 illustrates the excellent agreement among five methods, three by isochron and two by discordia. They show the radiometric ages of what is considered one of the oldest rocks on earth, the Amitsoq gneiss of western Greenland (Dalrymple, 1984, p. 97).

With multiple cross-checking giving such good agreement, it is small wonder that mainstream geologists and geochemists place a considerable degree of confidence in their age determinations. Dalrymple points out that "many tens of thousands of radiometric age measurements are documented in the scientific literature" (1984, p. 95). These have been made in laboratories numbering between 50 and 100 and distributed worldwide. While this richness of information has built an enormously strong level of confidence, it has also provided creation scientists with a vast literature from which to extract bits and pieces of information they can turn against the radiometric method.

Table 17.1   Radiometric Ages of the Amitsoq Gneisses, Western Greenland

Method	Age (b.y.)
Rb-Sr isochron	3.70 ± 0.14
Lu-Hf isochron	3.55 ± 0.22
Pb-Pb isochron	3.80 ± 0.12
U-Pb discordia	3.65 ± 0.05
Th-Pb discordia	3.65 ± 0.08
Weighted mean age	3.67 ± 0.06

Data of G. Brent Dalrymple, 1984, p. 97, Table 6, from several sources.

Radiometric Age of the Universe

Although the creationist publications that have come to my attention do not include the topics of age determination of the universe by radioisotope decay and the distribution of nuclides, it may be useful to include that evidence at this point. Nuclide is a term used by physicists when referring to a given kind of isotope in terms of the composition of its nucleus. All nuclides with the same atomic number answer to the same element name, but, as isotopes of that element, they differ from one another because of differences in mass number.

Can the radiometric methods used to date the oldest rocks and the earth itself also provide an age for the universe? If so, we would have an independent source of information with which to corroborate a cosmic age derived from the cosmological redshift and Hubble constant, i.e., the Hubble age.

The general subject of determining the age of the universe from the nuclei of the heavy elements has been given a fantastic name: nucleocosmochronology (Schramm, 1974). It is the companion science to geochronology. Astrophysicists consider it likely that all the elements heavier than bismuth originated in supernova explosions. The process is thought to occur in a region just outside the dense core that remains as a neutron star or a black hole. There the neutron flux would be extremely intense and the nuclei of the heavy elements could grow at a rapid rate. Some elements produced in this process are unstable and immediately begin radioactive decay. Two decay sequences have proved useful indicators of the age of formation of heavy elements and both have a large relative abundance: thorium-232/uranium-238 and uranium-235/uranium-238. Abundance ratios of the two members of each pair can be determined from meteorites and lunar rock samples. Half-lives are calculated theoretically. More recently, the pair rhenium-187/ osmium-187 with a half-life of 40 b.y. has been used. Depending upon what assumption is made as to the rate of production of supernovas during the early history of the universe, these observed abundance ratios yield an age of the universe of roughly 10 to 15 b.y. (Schramm, 1974, p. 77). A more recent appraisal of currently available data places the start of galactic star formation at about -11 b.y. and the age of the universe at -12 b.y.(van den Bergh, 1981, p. 825), the latter with an uncertainty of being as great as -15 b.y. or as small as -10 b.y.

Besides the Hubble age and nucleocosmochronological (Wow!) age, a third independent estimate of age of the universe is available and is in general agreement with the first two. Recall from our review of astronomy in Chapter 14 that the globular star clusters within the halo of the Milky Way galaxy are spherical in form and largely free of clouds of gas and dust. All of the stars of a given cluster are of about the same age and of similar chemical composition (Schram, 1974, pp. 70-71). The globular clusters are among the oldest celestial objects in our

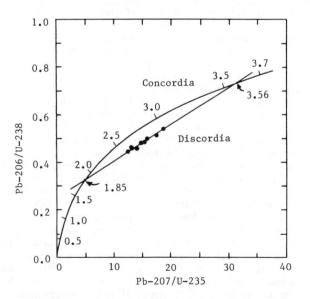

Figure 17.6 Uranium-lead concordia-discordia diagram for nine samples of the Morton Gneiss, Minnesota. (From G. Brent Dalrymple, 1984, p. 76, based on data of Goldich et al., 1970, Bulletin, Geological Society of America, vol. 81, pp. 3671-96.)

universe (van den Bergh, 1981, p. 825). Calculations made in 1980 yielded for clusters of metal-poor stars (which were made early in the history of the universe) an age in the range of 12 to 19½ b.y. These stars are thought to have been formed as early as within 1 b.y. of the Big Bang. Another similar but independent estimate gave an age range of -14 to -16 b.y. for the oldest metal-poor clusters.

So far as I have been able to determine from the published literature of the ICR scientists, they have not put forth objections to the second and third methods of age determination reviewed here. Although each method requires its own assumptions and has a substantial measure of uncertainty, the age brackets that each method supplies for the age of the universe show a large overlap. All three share a common age span between 10 and 15 b.y. (Schramm, 1974, p. 77). The extent of corroboration is thus very strong, and not likely to be repudiated or greatly altered by future observations.

Another line of argument for great age of the earth and universe has been offered by Stanley Freske, who makes use of available data on the half-lives of all known nuclides (1980). Of the radioactive nuclides occurring naturally, Freske tells us: "There are 47 nuclides with half-lives between 1,000 years and 50 million years." He continues:

> If the earth were only 10,000 years old, then there should be detectable amounts of all 47 in nature because 10,000 years is not enough time for them to decay totally. However, only 7 of these are actually found, and that is only because they are continually being generated: 4 of them are members of natural decay series; C-14 is generated by cosmic rays acting on nitrogen nuclei; Np-237 is produced by cosmic rays on the moon; and the seventh, U-236, is generated by slow neutron capture in uranium ore where neutrons are available. Creationists have to explain why the other 40 missing. What makes this significant is that all 17 nuclides with half-lives longer than 50 million years are found in nature. Simple calculations show that this division between nuclides which are absent and those that are present is exactly what would be expected if all the nuclides were generated (probably in some star) about 4.5 billion years ago. The longest-lived one among the 40 absentees is Sm-146 with a half-life of 50 million years. If it had existed for 4.5 billion years, only 8×10^{-28} of the original amount would remain today, which would explain why it has not been detected. The same would, of course, be true of those with even shorter half-lives. (Freske, 1980, pp. 37-38)[2]

Using the creationists' assumption that the earth is only 10,000 years old, Freske has calculated the probability of the 40 short-lived nuclides being absent and the 17 long-lived ones being present (as opposed to a random distribution between absence and presence) to be 7×10^{-15}. Thus the observed presence and absence of isotopes in terms of their half-lives strongly corroborate a very old earth, in agreement with a radiometric age of 4.5 b.y. One can anticipate that the creation scientists' arguments against Freske's calculation would center on the question of constancy of the half-lives of the radio-isotopes.

Creationists Challenge the Validity of Radiometric Ages

For answers to the creationists' arguments that radiometric ages are grossly in error, I have relied heavily on two highly qualified mainstream scientists. Both have gone into great detail to answer each of the creationists' specific points. One of these specialists is

Stephen G. Brush, holder of a Ph.D. degree in physics from Oxford University and presently a professor of history of science at the University of Maryland, College Park. He has worked as a physicist at the Lawrence Radiation Laboratory and has taught at Harvard University. One of his excellent articles is readily accessible to teachers and students in the Journal of Geological Education (Brush, 1982; see also Brush, 1983). The second scientist, G. Brent Dalrymple, mentioned on earlier pages, has answered the creationists' arguments in a paper prepared for the A.A.A.S. 1982 symposium on the subject "Evolutionists Confront Creationists" (Awbrey and Thwaites, 1984; Dalrymple, 1984). I limit my review to those aspects of their arguments that can be understood in general terms by persons unfamiliar with nuclear physics and its mathematics. As of the date this is written, creation scientists of the ICR have not replied to these published arguments supporting the integrity of radiometric ages.

First, take note of one of the creationists' commonly used strategies, that of selective citation and out-of-context quotation of bits of scientific information found in mainstream journal articles and treatises. In this case, the argument consists of assembling a list of "anomalous" radiometric ages that are pointed to as deviating far from "expected" values. Dalrymple describes a list of such asserted anomalous ages prepared by J. Woodmorappe and published in 1979 in the Creation Research Society Quarterly (Dalrymple, 1984, pp. 76-82). Woodmorappe's list consists of more than three hundred supposedly "anomalous" radiometric ages; his claim is that published radiometric ages are highly selected to agree with what is expected, ignoring or excluding ages that do not agree. It turns out that in each case discussed in detail by Dalrymple, a satisfactory explanation is available. Generally, the description of the incident offered by Woodmorappe turns out to be a misrepresentation or misstatement of what the original text actually stated. For example, in the case of a wide range in radiometric ages of "Hawaiian" basalts, it turns out that the deviating ages were from basalt samples on other islands of the Hawaiian chain. The lengthy chain was formed over a long time span as the Pacific plate moved over a mantle hot spot. The radiometric ages actually increase progressively northwestward from island to island in the chain, as would be predicted for a lithospheric plate moving at uniform speed.

In another case of citation of assertedly "anomalous" radiometric ages, creationist Ralph D. Matthews refers to "selective data publication" as the reason for general agreement among scientists on radiometric ages found in the conventional table of geologic time (Matthews, 1982, p. 43). Self-incriminating quotations by radiochronologists are offered to show that their house is in serious disarray.

The culling of scientific journals for discrepancies reminds one of being offered a bushel basket of rotten apples culled from among many thousands of sound apples that passed through a packing house in a given day. We are being asked to believe that (a) all apples are rotten and (b) apples should be avoided as food for humans. Most persons would simply suggest that we avoid eating the rotten apples and that they were eliminated for good reason. Dalrymple has this to say about the creationist's conclusion, drawn from supposed "anomalous" ages, to the effect that we should therefore abandon the radiometric methods:

> This argument is specious and akin to concluding that all wrist-watches do not work because you happen to find one that does not keep accurate time. In fact, the number of "wrong" ages amounts to only a few percent of the total, and nearly all of these are due to unrecognized geologic factors, to unintentional misapplication of the techniques, or to

technical difficulties. Like any complex procedure, radiometric dating does not work all the time under all circumstances. Each technique works only under a particular set of geologic conditions and occasionally a method is inadvertently misapplied. In addition, scientists are continually learning, and some of the "errors" are not errors at all but simply results obtained in the continuing effort to explore and improve the methods and their application. There are, to be sure, inconsistencies, errors, and results that are poorly understood, but these are very few in comparison with the vast body of consistent and sensible results that clearly indicate that the methods do work and that the results, properly applied and carefully evaluated, can be trusted. (1984, p. 76-77)[1]

We turn next to creationists' arguments on theory and methodology designed to show that radiometric age determinations cannot be accepted. Basic arguments of this group are presented in the ICR textbook, Scientific Creationism, edited by Henry M. Morris (1974a, pp. 131-49). The claim is made that rocks are not dated radiometrically by mainstream geologists (p. 133). Instead, says Morris, geologists established the ages of rocks long before radioactivity was known, and they did it by interpreting the positions of fossils in strata. As I tried to make clear in Chapter 9, geologists of the 1800s established the relative ages of strata by applying the principle of superposition, itself based on the physical principle of downsinking of particles through a fluid medium under the attraction of gravity. During all that time of field study, the absolute (actual) age in years of any fossil or rock layer was beyond scientific determination; it was only the subject of speculations to which we have already referred. Morris continues in stride to assert that geologic ages of rocks are determined by the index fossils they contain (1974a, pp. 134-37). As I explained above, the sequence of strata had first been determined from direct field observation, giving relative age by the principle of superposition. Afterward, the fossil flora and fauna were studied and certain assemblages were consistently found in formations of the same relative age. This observation allowed a "tree of life" to be constructed; but nothing in this inductive process implied a dependence on any theory of evolution or descent with modification. (In fact, relative age had been reasoned a generation or two before Darwin's theory by investigators who never suggested evolution.) The observations could equally apply to a theory of repeated special creation, such as that held by Cuvier. Nothing in this chain of reasoning has anything to do with assigning actual ages in years to any rocks. Although Morris quotes several paragraphs from writings of geologists in support of his assertion, none of them makes a clear reference to actual age of rocks or fossils.

We next transfer attention to the publications of Harold S. Slusher of the ICR, who has undertaken to present the creationists' arguments against a great age of the earth and universe. He holds the M.S. degree in physics and astronomy from the University of Oklahoma. Slusher in 1981 was Assistant Professor on the faculty of the Department of Physics at the University of Texas at El Paso, and also served as Research Associate in Geoscience and Astronomy with the ICR and as Adjunct Chairman of the Department of Physical Sciences at Christian Heritage College. He has authored several technical monographs of the ICR on various aspects of cosmology and astronomy. His arguments against a great age of the earth are contained in his ICR technical monograph titled "Critique of Radiometric Dating," first published in 1973 and updated by revision in 1981.

We can start with the question of constancy of the process of radioactive decay, since this is crucial to the validity of radiometric ages. Slusher claims that the assumed constancy of decay rates has been challenged by several forms of evidence, including studies of carbon-14 dating (which we discuss in Chapter 19) and of the phenomenon of "polonium halos" in minerals (discussed later in this chapter) (1981, pp. 24-26). Elsewhere, Slusher has claimed that there is excellent laboratory evidence that external influences can cause change in the decay rates" (1976, p. 283).

The principle of constancy of decay rates is defended by Dalrymple, who asserts: "A great many experiments have been done in attempts to change radioactive decay rates, but these experiments have invariably failed to produce any significant changes" (1984, p. 88). He notes that decay constants have been found to be uniform under temperatures ranging as high as 2,000 C and as low as -186 C, and through pressures as low as a virtual vacuum to as high as several thousand atmospheres. Tests have also been conducted with differing gravitational and magnetic fields. Except for slight observed deviations in three radioisotopes (C-14, Co-60, Ce-137), the decay constants for the three principal forms of decay used in age determination adhere strictly to the ideal random distribution predicted by theory.

Dalrymple goes on to describe a fourth type of decay that is known to be very slightly affected by physical and chemical conditions. It is electron capture (e.c.), in which an orbital electron is captured by the nucleus of the atom and a proton is converted into a neutron. The only important case of e.c. in radiometric dating is the decay of potassium-40 (K-40), the parent isotope in the K-Ar method.

Specific causes for variable or uncertain decay rates are put forward by both Morris and Slusher. Let us look at three of their suggestions.

Morris suggests that a supernova explosion might have released a burst of cosmic rays that would have greatly increased the decay rates, or that a flux of neutrinos or of neutrons would have had a seriously disturbing effect (1974a, pp. 142-43). Such possibilities are mere speculations about special events and processes not known to have occurred. That cosmic rays do not influence the rate of radioactive decay has been tested by observing the decay of polonium in a deep mine shielded from cosmic rays; no significant change in rate was observed (Brush, 1983, p. 66).

Morris has suggested that reversals of the earth's magnetic field may have been accompanied by changes in the decay rates (1974a, p. 142). The principle here is that at times of low or zero magnetic field strength, the shielding force-lines of the earth's external magnetic field (the magnetosphere) would have disappeared, allowing high-energy particles (cosmic rays) to bombard the earth's surface. This same reasoning has been used by paleontologists to allow formulation of a hypothesis of events of rapid evolutionary change through brief episodes of greatly increased rates of mutation. Actually, the primary shield to cosmic rays is the earth's atmosphere. If the magnetic field were to collapse completely, the cosmic-ray influx would increase about 14 percent at the equator and none at all at the poles (Dalrymple, personal communication, 1986). In the absence of any evidence that cosmic rays can influence decay rates, Morris's suggestion cannot be pursued further, but it is interesting to note that another ICR creation scientist, Thomas Barnes, maintains that no reversals of the earth's magnetic field have occurred (see Chapter 19).

As to Morris's claim that free neutrons might change decay rates, Dalrymple comments that "his arguments show that he does not understand either neutron reactions or radioactive decay" (1984, pp. 88-89). Dalrymple continues:

Neutron reactions do not change decay rates but, instead, transmute one nuclide into another. The result of the reaction depends on the properties of the target isotopes and on the energy of the penetrating neutron. There are no neutron reactions that produce the same result as either beta or alpha decay. (Pp. 88-89)[1]

Dalrymple points out that there aren't enough free neutrons in nature to affect any of the isotopes used in radioactive dating.

With regard to Morris's suggested neutrino flux, the idea is taken from one of the regular columns contributed to the journal Industrial Research by Frederick Jueneman, director of research for the Innovative Concepts Association. The author titled his column "Scientific Speculation." His suggestion was that the neutrino flux from a supernova explosion might have had "the peculiar effect of resetting all our atomic clocks" (Jueneman, 1972). "This," he says, "would knock our Carbon-14, Potassium-Argon, and Uranium-Lead dating measurements into a cocked hat! The age of prehistoric artifacts, the age of the earth, and that of the universe would be thrown into doubt." Dalrymple points out that Jueneman does not propose that the neutrino flux would change decay rates, nor does he offer any explanation as to how the neutrino flux could "reset" atomic clocks. Neutrinos have no charge and little or no mass; they do not interact with matter and are close to impossible to detect for that reason. Jeuneman has provided a sweet morsel for the creationists, but it has no nutritional value.

The neutrino-flux idea is taken up by Slusher (1981, pp. 20-23). He goes to considerable length to make use of a most unusual general theory of physical science put forward by H. C. Dudley (1976, pp. 52-55). Brush discusses this theory in some detail; it rejects Einstein's theory of relativity and attempts to revive the long-abandoned theory of the existence of "ether" in space (1983, pp. 66-68). Dudley's theory requires rejection of quantum mechanics, a drastic and unwarranted move in the face of massive corroborative evidence for quantum mechanics. Brush emphasizes the severe consequences to all physical science of throwing away much of the foundation of modern physics merely to cast doubt on radiometric dating. It turns out that Dudley himself does not endorse the conclusions Slusher drew from the Dudley hypothesis. Dudley would limit the effects of his hypothesis to making changes in the age of the earth by only a few percent, and this would still put the earth's age at more than 4 b.y. Brush quotes these words from a personal letter to him from Dudley: "The figure of 4.5 billion years for the earth's age seems to be a good ball park figure" (Brush, 1982, p. 51). Brush also comments:

> To attack the theory of radioactive decay by abandoning quantum mechanics seems almost suicidal; one can only suppose that the creationists know nothing about modern atomic physics (in spite of their "qualifications") or that they hope no one will notice how absurd their position is. (1982, p. 51)

This is not the first instance we have observed in which the creation scientists have painted themselves into a corner, so to speak. Ill-advised hypotheses all too often carry the penalty of a long string of absurd consequences.

Another recent attempt by creationists to discredit the principle of constancy of half-lives has been made by Theodore W. Rybka, who cites two cases in which measurements show that the half-life is increasing with time (1982, p. ii). If such a trend is significant, the decrease backward in time can be extrapolated using any one of several mathematical models. If an exponential variation with time is assumed, as Rybka suggests be done, the half-lives of the radiometric nuclides are reduced by orders of magnitude. For U-238 the half-life

is reduced by a factor of 10^5. Therefore, states Rybka, the whole subject of radiometric dating should be rethought (p. iii).

Rybka's conclusion is challenged by Dalrymple, who points out that the measurements alleged to show increase with time during the historical period of measurements are carefully selected from the total list of measurements from earliest to most recent (1984, pp. 90-91). While pairs of measurements used by Rybka show such an increase, the full list, which includes much more recent and presumably more accurate determinations, reverses the supposed earlier trend. Taken altogether, however, the data show no significant change to either higher or lower values.

Dalrymple's closing paragraph on the subject of constancy of decay constants reads:

> In summary, the attempts by creation "scientists" to attack the reliability of radiometric dating by invoking changes in decay rates are meritless. There have been no changes observed in the decay constants of those isotopes used for dating, and the changes induced in the decay rates of other radioactive isotopes are negligible. These observations are consistent with theory, which predicts that such changes should be very small. Any inaccuracies in radiometric dating due to changes in decay rates can amount to, at most, a few percent. (1984, p. 91)[1]

On a related but somethat different tack, the creationists have challenged the accuracy of decay constants, arguing that rates are uncertain by a large factor (Morris, 1974a, pp. 142-49). Slusher asserts that the decay constant of the K-40/Ar-40 system is adjustable and subject to manipulation rather than being an accurate half-life (1981, p. 40). Dalrymple answers by pointing out that the cases cited by Morris and Slusher date back to the 1940s and 1950s, when accurate determinations had not been made (1984, p. 91). Today, however, "all constants for the isotopes used in radiometric dating are known to better than one percent. Morris and Slusher have selected obsolete information out of old literature and tried to represent it as the current state of knowledge" (p. 91).

Yet another line of attack used by creationists to undermine radiometric age determinations centers on the third assumption, that at the outset (time-zero) none of the final decay product (daughter isotope) was present and that none of that product was lost from the system. Slusher points out that not all lead isotopes in rocks are formed by radioactive decay (1981, p. 27). There may be present a supply of lead isotopes of nonradiogenic origin, and there is no way to distinguish them chemically from those produced in the decay system. Here, the creationists claim, is a possible source of major error in uranium/lead ages. A particular fly in the ointment seems to be the idea that nonradiogenic lead isotopes can be produced by "neutron reactions." Slusher states: "Neutron reaction corrections in the U-Th-Pb series reduce 'ages' of billions of years to a few thousand years because most of the Pb can be attributed to neutron reactions rather than to radioactive decay" (1982, p. 54).

Dalrymple analyzes this argument, showing where the idea came from and showing that it has no substance (1984, pp. 92-94). The discussion is highly technical, with mathematical equations awesome to the average person; I attempt only to state the gist of the matter. It seems that creationist Melvin A. Cook published several articles criticizing conventional radiometric dates (references cited by Dalrymple). The thrust of his argument appears to be that lead-208 (Pb-208) got into particular ore masses through neutron reactions upon lead-207 (Pb-207), and that the latter came into existence through neutron reactions with lead-206 (Pb-206). This neutron activity was supposed to have taken place in bodies of uranium ore at various localities on several continents. It would, therefore, have seriously altered results of the usual uranium-lead age determinations on samples of those ores. Cook's suggestion is that, with a large amount of lead of nonradioactive origin being mistaken for the daughter product of the decay series, a great age was arrived at, whereas the actual amount of true radiogenic lead is very small, and if correctly evaluated would yield a very young age.

Dalrymple finds several serious fundamental flaws in Cook's argument and calculations. In a specific instance, Cook seems to have misread a dash in a data table to mean "zero," when, in fact, it meant "not measured." Even if the data were not flawed, there remains the problem that there were far too few free neutrons available at the ore body sites of the supposed neutron activity to have been responsible for the alleged magnitude of the effect. This problem of free neutron deficiency was acknowledged by Cook himself (1966, p. 54). Dalrymple's conclusion is: "Cook's proposition and calculations, enthusiastically endorsed by Morris and Slusher, are based on data that do not exist and are, in addition, fatally flawed by demonstrably false assumptions" (1984, p. 94).

Halos Produced by Radioactivity

Creation scientists offer as evidence of past extreme variations in decay rates a phenomenon well known in mainstream science. When a thin-section (very thin slice) of biotite mica is observed at high magnification under a microscope, using transmitted light, there may be seen circular dark-brown patches, all of the same diameter. Surrounding the darkened disk can be seen several concentric rings. Known as <u>pleochroic halos</u>, these circular objects have at the very center a tiny mineral inclusion interpreted as consisting of the daughter products of radioactive uranium or thorium, originally enclosed in the mineral and its associated igneous rock at the time of formation. The halo has been formed by innumerable alpha particles, sent outward in all directions from the radioative particle and causing dislocations of the atoms of the mineral crystal lattice.

Each of the halo rings is interpreted as associated with alpha-particle emission of a particular isotope in the decay series. Looking back at Figure 17.1, you find that alpha particles are given off by eleven members of the series. Of the eleven, rings can be identified for eight: U-238, U-234, Th-230, Ra-226, Rn-222, Po-218, Po-214, and Po-210. Figure 17.7, upper-right quadrant, shows the rings for these eight members of the U-238/Pb-206 series. The radius of a particular ring agrees very well with the calculated path length of the alpha particle of the isotope for the particular medium through which it has traveled. The radius is directly related to the energy of the alpha particle. Energies in million electron-volts (MeV) are listed beside each isotope; that with the greatest energy is polonium-214, which produces the outermost ring.

Pleochroic halos were recognized as early as 1907 by John Joly and were quickly interpreted as records of radioactive decay through time (York, 1979, p. 617). Scientists who studied the rings hoped that they might provide a time-clock for determining the age of the host mineral, but this aim was never achieved because of technical difficulties. One of the scientists who studied pleochroic halos during the 1930s was G. H. Henderson of Dalhousie University in Nova Scotia, Canada. He classified the halos into several types and associated each with a radioactive decay series and its members.

Henderson was able to identify halo types produced by each of the three polonium isotopes or combinations of the three. The remarkable thing about this discovery is that polonium isotopes are extremely short-lived. For example Po-214 has a half-life of less than 200 microseconds! Polonium-218 has a half-life of about two minutes, and in

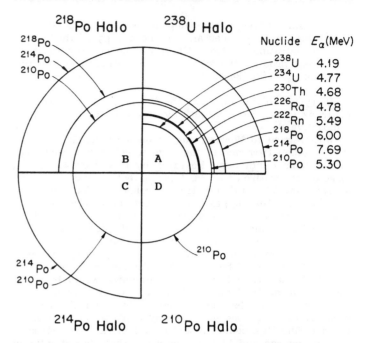

Figure 17.7 Pleochroic halos. (A) Complete U-238 halo. (B, C, D) Polonium halos. (From Robert V. Gentry, p. 48, Figure 7, in F. Awbrey and W. M. Thwaites, eds., 1984, <u>Evolutionists</u> <u>Confront</u> <u>Creationists</u>, Proceedings of the 63rd Annual Meeting of the Pacific Division, American Association for the Advancement of Science, vol. 1, Part 3. Used by permission of the A. A. A. S. and Robert V. Gentry.)

only one hour after time-zero the number of atoms initially present will have been become reduced to one-millionth of the original number. These polonium rings were found in complete isolation from rings of other members of the uranium decay series. The polonium itself must have been produced in a uranium decay series, as shown in Figure 17.1. How, then, did the polonium isotopes manage to become isolated as specks of matter in the crystal lattice of biotite mica in sufficient quantity to produce the halos? The difficulty in finding an answer to this question is elaborated upon by Derek York of the Department of Physics of the University of Toronto:

> If the Po isotopes are isolated from their parents, they will all have decayed away within a few hours. The situation Henderson had to face then was the following: Po halos must have been produced from a local concentration of Po in a mica. Since there are no uranium rings associated with Po halos, then the Po must have been separated from its supporting ancestors when it was being concentrated into the tiny volume which was to become the halo center. Since, as we have just seen, Po isotopes decay almost at the same moment as they are removed from their longer-lived ancestors, only a very limited number of separation and concentration processes may be envisaged. (York, 1979, p. 618)

Henderson and his colleague F. W. Sparks observed that in many slides of Po halos in mica, the halo centers seemed to be "strung together along obvious channels or microconduits in the cleavage planes" (Henderson and Sparks, 1939, as reported in York, 1979, p. 617). Cleavage is the natural parting found in certain minerals: in this case, perfect parting on a single parallel set of planes. Henderson suggested that as uranium-bearing hydrothermal (hot-water) solutions moved slowly through the rock mass, their movement was localized in the microchannels within the mica crystals. Po isotopes that would be present in the solution might, for specific reasons uncertain, have precipitated from the solution and built up at certain points along the flow path. Alpha emission would have taken place from those points. Accumulation would continue over a long period of time so that despite the extremely short half-lives of the Po isotopes, sufficient point-radiation would occur to produce the halos. Henderson's hypothesis was carefully conceived to conform with physical laws and existing knowledge of radioactivity and mineral structure. It would not be possible to reproduce the natural conditions in the laboratory; hence, the hypothesis would (like all hypotheses that ask "What happened in the distant past?") be difficult to test and corroborate.

At this point you will, if you are an experienced creationist watcher, see a perfect opportunity presenting itself to that group to leap in and seize the chance to promote recent creation instead of an obviously speculative scientific hypothesis. Polonium halos would have presented no difficulty to the Creator of the universe. No matter how short the half-lives, God would have emplanted specks of radioactive polonium isotopes in adequate quantities wherever halos are found today. The initial polonium was not formed in a decay series--it was simply created from scratch. Science, say the creationists, has no good answer to polonium halos, and that ignorance is an expected consequence of a false theory of a great age of the universe. It would be the creationists' argument from ignorance, now so familiar to my readers. Did the creationists get into the polonium act? Indeed they did!

Creationist Robert V. Gentry, a physicist on the faculty of Columbia Union College, Takoma Park, Maryland, has substantial qualifications in the field of radioactivity (Brush, 1983, p. 69). He has made an intensive study of polonium halos and has published his results in Science (journal of the A.A.A.S.) and in the British journal Nature. Other papers of his have appeared in EOS (journal of the American Geophysical Union) and Physics Today. Gentry was invited to participate in the 1982 A.A.A.S. symposium "Evolutionists Confront Creationists." His findings and arguments are contained in the proceedings volume of that title (Gentry, 1984), published under the editorship of Frank Awbrey and William Thwaites (1984).

Gentry has proposed that the polonium halos in so-called "Precambrian" granites were made almost instantaneously during the creation of the universe nearly 6,000 years ago (1984, pp. 38, 50-51). He refers to this creation event as a "singularity." It was followed by a brief period in which "uniformitarian" principles prevailed, and radioactivity followed the decay schedules now assigned to them by mainstream science. A second singularity was the Fall of man, ending the first uniformitarian period. A third singularity was the year of the Flood of Noah. During these brief second and third singularities the physical laws were suspended by the Creator in favor of special conditions in the form of enormously accelerated radioactive decay rates. The greatly intensified radioactive decay during the Flood produced an enormous quantity of sensible heat, which can explain the catastrophic volcanic and tectonic processes of that event. Between the second and third singularities, and following the third, were periods of uniformitarian rule, continuing to the present. Gentry also suggests a possible fourth singularity coinciding with the rifting apart of the continents. This singularity is to be used if plate tectonics is accepted as a working hypothesis.

It may be best to let Gentry tell us how uranium-lead ratios are to be interpreted in terms of the singularities:

> In this scenario U/Pb ratios are presently utilized as indicators of elapsed time since the last singularity. U-238/Pb-206 ratios are not used as time measures prior to this last singularity because

of conflicting evidence of very high Pb and He retention in natural zircons subjected to a prolonged high temperature environment in deep granite. Those results, discussed below, are consistent with a very young age of the earth, and suggest that the radioactive decay rate may have been enhanced (indeed, had to be if this creation model is correct) during any one of the three singularities. (Gentry, 1984, p. 51)[3]

The essential feature of Gentry's hypothesis thus seems to be that most of the daughter lead now found associated with uranium was generated in one or two or more very brief episodes of enormously high rates of radioactive decay. And what does this scenario for U/Pb ratios have to do with polonium halos? There is no direct connection, so far as I can see, because the primordial polonium was introduced into the primordial rocks (Precambrian granites) by the Creator in the first singularity (Gentry, 1984, p. 50). The decay rates of the three isotopes of polonium are so rapid in any case that the primordial polonium would have virtually disappeared early in the 6,000-year period available since creation.

Gentry cannot offer any substantiating evidence for his creation hypothesis of primordial polonium and granite. In this matter he seems to go along with the creationists' tenet that no event occurring before humans were on earth to observe it can ever be proven to have occurred. Instead, he has repeatedly (he says) thrown a challenge to the mainstream scientists to produce by laboratory synthesis a "hand-sized" specimen of granite or biotite in which the subsequent formation of polonium halos would be observed. If that attempt were successful, he says, his own hypothesis would be falsified. Gentry claims that his challenge has been ignored. He presses his point as follows:

> It is inescapable that these experiments should be successful if the uniformitarian principle is true. Thus, with so much at stake for evolution, I suspect the reason why my evolutionary colleagues have failed to achieve success is because the Precambrian granites never formed by the uniformitarian principle to begin with; hence, to attempt to utilize it now to produce a synthesized piece of granite is just a futile effort. The end result is that the uniformitarian principle is essentially falsified because of its failure to live up to its own predictions. But since the pieces of the evolutionary puzzle are glued together by this principle, we must now come to the same conclusion about evolution itself. (Gentry, 1984, p. 50)[3]

Perhaps nowhere else in the creationist literature has the fallacious argument from ignorance been so eloquently worded! The real reason why no geochemist has presented Gentry with a hand specimen of granite containing polonium halos is twofold. First, the difficulties of simulating the deep high-pressure, high-temperature environment of formation of granite from a magma are enormous. Even if the chemical constituents of that magma, including the original fugitive volatiles, were assembled, they could not be handled in a quantity capable of producing a chunk of granite the size of a baseball, or even a pea. To produce the biotite crystals would require extremely slow cooling, which in nature probably takes hundreds of thousands of years or more. Second, how does the laboratory scientist introduce the polonium (Po-218) into the crystallizing magma and have it become concentrated in isolated specks in the biotite mica? Gentry has presented mainstream scientists with a totally unrealistic proposal, then faulted them for their silence. "See," he as much as says, "you can't do it, therefore science is impotent and in error, therefore recent creation is true."

Derek York has called Gentry to task for dismissal of Henderson's naturalistic hypothesis of origin of polonium halos (York, 1979, p. 617). York observes that "Gentry's own observations and measurements of halos" (and other observations) "add support to Henderson's suggestion that types A, B, and C really are Po halos." York continues:

> Why, then, Gentry takes the position he does, totally ignoring the enormous mass of self-consistent geochronological data and theory, is difficult to understand. He presents no alternative theory and does not even consider why geochronology as presently practiced is so successful. (P. 617)

Gentry refers to Henderson's 1939 hypothesis of a secondary source of polonium in the movement of uranium-bearing hydrothermal solutions. Gentry faults that hypothesis on grounds that "only a negligible supply of Po daughter atoms would be available for capture at any one time" (1984, p. 47). He also cites as an objection the low diffusion rates of the those isotopes in minerals. Yet, surprisingly enough, Gentry states in the same paragraph that he himself used the migration/capture mechanism to explain Po-210 halos in coalified wood during the Flood singularity. He limits the process to the longest-lived of the three polonium isotopes. His argument is that the absence of Po-214 and P0-218 rings in the coal halos is explained by their extremely short half-lives.

Gentry has much more to say on pleochroic halos and other subjects related to radioactive decay products (retention of daughter lead and helium in deep granite cores) (1984, pp. 52-53), but we must bring this topic to a close. Gentry's scientific expertise and experience in radioactivity force mainstream scientists to examine his work carefully and evaluate his arguments on a point-by-point basis.

Gentry freely admits the scriptural basis of his hypothesis of a very young, instantaneously created universe (1984, p. 38). His own research, he says, leads to a scenario in accord with Ps. 33:6-9: "By the word of the Lord were the heavens made; and all the host of them by the breath of his mouth. . . . For he spake, and it was done; he commanded, and it stood fast." For those who ponder the workings of a creation scientist's mind and the seeming difficulties of simultaneously harboring in that mind two such different and mutually exclusive views

of the universe, there may be a clue in Gentry's last paragraph:

> In closing I wish to express my gratitude to those of my evolutionary colleagues who on so many occasions have assisted me, and on other occasions have collaborated with me in my research. Of one thing I am certain: Only in America could my research over the past two decades have been accomplished. I close by expressing gratitude to my Creator for allowing me the privilege of being an American. I submit this article to the scientific community not as an antagonist who purports to have the last word on the subject, but as a colleague who, in the spirit of free scientific inquiry, genuinely seeks a vigorous, critical response to the evidence presented herein. Perhaps a future "Evolutionists Confront Creationists" AAAS symposium would be the ideal forum for this exchange to occur. (1984, p. 63)[3]

Credits

1. From G. Brent Dalrymple, How Old is the Earth? A Reply to "Scientific" Creationism, pp. 66-131 in Frank Awbrey and William M. Thwaites, eds., 1984, Evolutionists Confront Creationists, Proceedings of the 63rd Annual Meeting of the Pacific Division, American Assoc. for the Advancement of Science, vol. 1, Part 3, San Francisco, 213 pp. Used by permission of G. Brent Dalrymple.

2. From Stanley Freske, Evidence Supporting a Great Age for the Universe, Creation/Evolution, vol. 1, no. 2, pp. 34-39. Copyright © 1980 by Stanley Freske. Used by permission of the author.

3. From Robert V. Gentry, Radioactive Halos in a Radiological and Cosmological Perspective, pp. 38-65 in Frank Awbrey and William M. Thwaites, eds., Evolutionists Confront Creationists, Proceedings of the 63rd Annual Meeting of the Pacific Division, American Assoc. for the Advancement of Science, vol. 1, Part 3, San Francisco, 213 pp. Copyright © 1984 by the Pacific Division of the American Association for the Advancement of Science. Used by Permission of the author and publisher.

Chapter 18

Creationist Arguments for a Young Age of the Solar System

Creationists look to the sun, planets, planetary satellites, and other orbiting objects of our solar system for indications of simultaneous creation only a few thousand years ago. No members of the solar system are too large or too small to be studied for evidences of recency of origin. Observations by astronomers over the past three or four centuries can, they say, be extrapolated backward in time to arrive at the date of creation. Even the earth's moon has participated in revealing the date of creation by slowing the period of the earth's rotation. The short lives of one group of comets are said to point to recent creation. Even the tiny specks of interplanetary dust are called on to bear witness to that event.

This is a motley collection of astronomical topics to be found assembled in a single chapter. They share one characteristic in common, however, in that they do not involve radiometric age determination. Instead, resolution of the arguments they involve rests on astronomical observations and physical principles that can be discussed independently of the validity of radiochronometric considerations.

A Shrinking Sun?

Writing in the ICR's Impact Series, creation scientist Russell Akridge titles his paper: "The Sun Is Shrinking" (1980). Akridge holds the Ph.D. in physics from Georgia Tech and is on the physics faculty of Oral Roberts University. He is referring to an abstract of a paper by John A. Eddy and Aram A. Boornazian, astrophysicist and mathematician, respectively, announcing that their examination of a century's record of solar observations (1863-1953) showed that size of the sun's disk has undergone a decrease amounting to 2.25 seconds of arc per century (1979). This would be a decrease of the horizontal dimension of the sun by about five feet per hour. The vertical dimension showed a decrease about one-third as great in the same period.

Akridge uses this rate, which is equivalent to 0.1 percent per century, to extrapolate far back into time. At the time of divine universal creation, 6000 years ago, the sun would have been only about 6 percent larger than now, not enough greater to constitute unfavorable environmental conditions on the newborn earth. However, when "evolutionists" project that rate to approximately 20 million years ago, the sun would have been so huge that the sun's surface would have touched the earth's surface. Going back still further, through much of geologic time the earth would have had to be located within the sun! Akridge notes that this extrapolation is conservative, because the rate of shrinkage would have been greater when its radius was greater. Furthermore, Akridge says, the rapid rate of shrinkage must be attributed to gravitational collapse; thus gravitational energy must account for part or all of the heating and radiation of the sun.

A 1979 report in Science News (vol. 115, June 30,

1979, p. 420) reviews the announcement by Eddy and Boornazian and makes this comment:

> The shrinkage does not apply to the entire solar mass, but rather to the sun's outer layers. Since the rate of shrinkage is so fast, the two researchers believe it is only a temporary contraction phase. "It's unrealistic to assume this will continue," Eddy told Science News. "It does seem to imply that the sun is oscillating in some way. However, going farther back into time to find an expansion will be difficult since the records get dimmer and dimmer."

In the same year, scientists of the NASA Goddard Space Flight Center and Louisiana State University reported the results of a similar kind of study of solar data (Sofia et al., 1979). They found that astronomical data collected between 1850 and 1937 show a decrease in the sun's diameter, but it amounted to only 0.25 seconds of arc, which was about the same as the value of the standard error of the observations. Thus their change rate, which is on the order of only one-tenth that found by Eddy and Boornazian, is of questionable significance as an actual change.

Another research group, using data based on direct ground observations of the limits of the path of totality of eclipses in 1715, 1976, and 1979, found that a diameter decrease of 0.34 seconds of arc was indicated, with a standard error of 0.2 seconds. Between 1976 and 1979, no significant difference was found. (Dunham et al., 1980). An apparent decrease at the rate of 0.3 seconds per century was independently determined by measurements made during twenty-three transits of the planet Mercury between 1736 and 1973 (Shapiro, 1980). The decrease was not considered significant. In 1981 a report on photoelectric solar radius measurements made at Mount Wilson Observatory between 1975 and 1981 showed a highly regular annual variation but no observable trend (Labonte and Howard, 1981).

Scientists of the University College London and Royal Greenwich Observatory calculated from transits of Mercury observations over the past three centuries a decrease in solar diameter of only 0.008 percent, with a standard error of 0.007 percent. This amount, they say, is negligible, but the data did reveal evidence that the diameter oscillates with a period of 80 years and with an amplitude of about 0.025 percent (Stephenson, 1982, p. 172).

In 1983, during the total eclipse of June 11 in which the path of totality swept across the island of Java, ground observations were again made of the width of the zone of totality. Preliminary results were reported in Science 84 (Gribbin and Sattaur, 1984, p. 56). The leader of the 1983 Java expedition, John Parkinson of University College, London, sees in the observations made by his expedition no support for the claim by Eddy and Boornazian that the sun has been shrinking rapidly and steadily since 1715. Parkinson's data, combined with that of other recent studies, are consistent with the hypothesis that solar changes are a pulsation with a rhythm of 76 years. Pulsation would not imply any net long-term gain or loss in the solar diameter.

The ICR scientist, Dr. Akridge, has followed a pattern of development of his thesis almost identical with that used by Thomas G. Barnes in the argument from the recent decline in the earth's magnetic dipole field (see Chapter 19). A rate of change derived from observations of an extremely short recent time span is extrapolated without good reason or evidence at the same rate over a vastly longer earlier period. Published articles of mainstream scientists indicate that the extrapolation carried out by Akridge is unwarranted and totally incongruent with their consensus derived by independent investigations using different methods. Statements available prior to the 1980 publication of Akridge's Impact article are on record expressing the opinion of Eddy and Boornazian that the change they observed is temporary and is appropriate to an oscillating system. Those who read only Akridge's account of the case are unaware of this alternative interpretation backed by other sets of data. For a scientist well qualified to understand the subject to omit alternative statements and data in such a presentation cannot be excused as a simple oversight. No mainstream scientific journal would accept Akridge's paper for publication without full citations of the pertinent literature and a discussion of differences in data and their interpretation.

The Lifetimes of Comets

In his ICR Technical Monograph titled Age of the Cosmos, Harold S. Slusher devotes a chapter to comets and their significance (1980, pp. 43-54). His argument for a young solar system, conforming with the requirements of recent creation, centers on the short-period comets. These comets have comparatively short lives--a few thousand years or so--according to the prevailing opinion of mainstream astronomers. Slusher's argument is that if these comets came into existence at the same time as the solar system, their short lives place an upper limit on the age of the solar system. This is a familiar argument, and it stems from the creationist doctrine that Divine Creation was complete in every respect and detail. This religious postulate forbids the formation of comets since the creation, for that would be a violation of the creationists' erroneous interpretation of the second law of thermo-dynamics. According to creationist religious dogma, decay and death are permitted by that law, but not the reforming of systems or objects toward higher states of internal order. This doctrine leads them to the conclusion that the universe can be no older than the youngest individual objects within it.

A brief review of the nature and composition of comets and their orbits will be useful in following the creation-ists' argument. (For background reading, see Comets, John C. Brandt, 1981.) The typical comet, as seen by the unaided eye or with a small telescope, consists of a brightly luminous head, called the coma, from which a luminous tail streams off in a direction pointing away from the sun. Both coma and tail consist of gas and fine dust driven off a small dense nucleus, which may appear in the telescope as a bright, starlike point embedded in the coma. During close approach to the sun, solid matter of the nucleus is vaporized and ionized under the intense solar radiation. Under the pressure of the solar wind this diffuse material is pushed away from the sun to form the tail. The nucleus consists of roughly equal parts of dust (mineral matter) and ices (volatiles). The ices include frozen methane, ammonia, carbon dioxide, and water. As we noted earlier, the comets are thus quite similar in general composition to the outermost planets and their satellites.

As a result of close passage to the sun, a comet may lose part of its mass or even be completely disrupted; it may simply disappear. Evidently, the debris of a disrupted comet continues in orbit as a meteoroid swarm. Several cases are on record in which the orbit of a meteoroid swarm is identical with the orbit of a previously known comet.

Orbits have been accurately determined for more than 600 comets. The orbits are highly elongated ellipses--very much more flattened than planetary orbits--and they show a wide range of inclinations with the plane of the earth's orbit (the ecliptic plane) (Figure 18.1). Most of the observed comets (about 500) have orbital periods longer than 200 years; these are classed as long-period comets. The 100 or so remaining comets have periods shorter than 200 years; many of them with periods of 3 to 8 years. Some 40 of these short-period comets have their point of greatest distance near Jupiter and are known as the Jupiter family, or Jovian comets. This relationship has led to a hypothesis that this group was captured by the strong gravitational field of that great planet. Halley's comet, with a period of 76 years, travels far out beyond the orbit of Neptune (Figure 18.1)

A considerable degree of support exists among mainstream astronomers for the hypothesis that the long-period comets come from a "storage" region, so to speak,

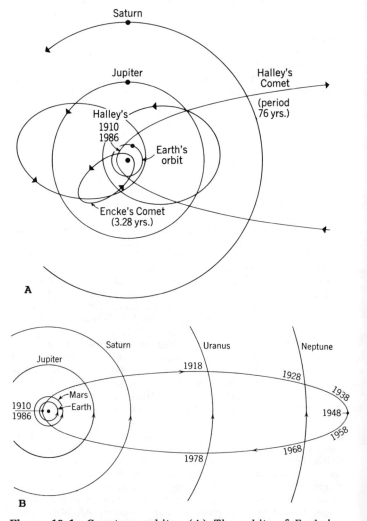

Figure 18.1 Cometary orbits. (A) The orbits of Encke's and two other Jovian comets. (B) The orbit of Halley's comet. (A. Copyright © 1976 by Arthur N. Strahler. B. From T. G. Mehlin, Astronomy, p. 336, Figure 13-2. Copyright © 1959 by John Wiley & Sons, Inc. Reprinted by permission of John Wiley & Sons, Inc.)

in the form of a spherical cloud around the sun. The cloud lies far beyond the orbit of Pluto, the outermost planet, and ranges from 10,000 to 100,000 astronomical units (AU) in radial distance from the sun. (The astronomical unit is the average distance between sun and earth.) The radius of the orbit of Pluto is a mere 40 AU, so the cometary cloud is extremely far out in comparison with the planetary system (Figure 18.2). The hypothesis of a distant cloud of comets was put forward in 1950 by the Dutch astronomer, Jan H. Öort, to explain the rate at which long-period comets appear. Comets within the cloud orbit the sun, as do the planets, but every now and then one gets thrown into an orbit bringing it close to the sun. A few of these later become trapped into orbits around both the sun and Jupiter or another outer planet.

Slusher discusses Öort's cometary-cloud hypothesis in great detail, together with objections advanced by English astronomer R. A. Lyttleton (Slusher, 1980, pp. 42-53). In evaluating this part of Slusher's chapter, Steven I. Dutch points out that Slusher has been rather selective in his choice of publications dealing with the cometary-cloud hypothesis (Dutch, 1982, p. 30). For example, Slusher does not refer to more recent papers by LePoole and Katgert (1968), Marsden and Sekanina (1973), and Marsden (1974). (References are given by Dutch.)

Slusher also reviews four other hypotheses of origin of comets: capture of short-period comets; comets derived from disrupted asteroids; comets volcanically erupted from Jupiter; accretion of planets. As Slusher shows, none of these modes of origin stands up effectively to close scrutiny. So the chapter ends on a negative note:

The failure to find a mechanism to resupply comets or to form new comets would seem to lead to the conclusion that the age of the comets and hence the solar system is quite young, on the order of just

several thousand years at most. Occam's razor should be followed in this matter. The obvious is that the solar system has been operating on a short time scale since its creation. (Slusher, 1980, pp. 53-54)

Need I repeat, as before, that Slusher's conclusion does not follow logically from his premise? Support for a positive empirical assertion--"the solar system is very young"--cannot be derived from false information or non-information. Slusher is saying all hypotheses thus far proposed are false, therefore recent creation is true. A second unacceptable argument is that the age of a group or collection of individuals can be no greater than the age of the youngest individual of the group. That reasoning is valid for an assumption that the entire group was created in the same instant and that no mechanism exists for individuals to be created since that instant. Under the cometary-cloud hypothesis one might be able to argue in this same creationist fashion as follows: All comets were created about -4.5 b.y. during the condensation process in the cold outer reaches of the condensing nebular disk; from time to time since then individual comets have been put into elongate orbits penetrating the region of the inner planets and sun; the term "short-period" applied to these objects does not signify their short life, for all comets are of the same age.

Science has not failed because one of its hypotheses has not yet been corroborated to the point of near-certainty. The cometary-cloud hypothesis remains viable and provides an explanation for the continual resupply of comets to replace those destroyed by close passage to the sun. Sooner or later new evidence for or against the distant cometary cloud will be forthcoming, perhaps sent back to us from one of our space vehicles passing through that zone.

Interplanetary Dust on the Earth and Moon

Slusher derives from the fall of cosmic dust upon the surfaces of the earth and moon an argument for a very young age of those bodies (1980, pp. 39-42). "Cosmic dust" has since been renamed interplanetary dust (IPD) to limit the meaning to particles within the inner solar system (Bradley et al., 1984, p. 1432). This term avoids confusion with dust clouds observed in galaxies as interstellar matter. IPD consists of extremely fine solid particles thought to enter the earth's upper atmosphere from outer space; they are considered to be composed of the same materials as meteorites and are therefore sometimes called micrometeorites.

Since 1981 NASA has been collecting IPD particles from the stratosphere at altitudes of about 18 km (29 mi). (See EOS, vol. 63, no. 11, March 16, 1982.) Called Brownlee particles, they are trapped on plates coated with silicon oil. The assumption is that the particles are extra-terrestrial and could not be derived from the earth's ground surface by rising air currents. The particles range in diameter from less than 0.001 cm (10^{-3} cm) to as large as 0.05 cm. Some of the particles are spherical, others angular. Their chemical composition includes iron, magnesium, potassium, calcium, aluminum, manganese, sulfur, chromium, nickel, silicon, and oxygen. Most of the particles are classified as mineral silicates, such as those found in stony meteorites. The iron and nickel could be of similar origin to that in the iron meteorites.

Slusher suggests that the infall of IPD can be used as sort of clock to judge the age of the earth (1980, p. 39). First he takes estimates of the quantity of IPD reaching the earth's surface in one year and extrapolates back over a five-billion-year period at the same rate. According to Slusher, the total thickness of dust that would now lie on the continental surfaces (if none had been removed) would be as small as 15 to 30 m (50 to 100 ft), or perhaps as great as 300 m (1,000 ft). He then

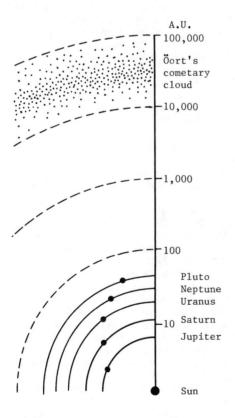

A.U.

100,000

Öort's cometary cloud

10,000

1,000

100

Pluto
Neptune
Uranus

10 Saturn

Jupiter

Sun

Figure 18.2 A schematic rendering of Öort's cometary cloud, thought to lie between 10,000 and 100,000 astronomical units from the sun. The radial scale is logarithmic. (A. N. Strahler.)

points out that in the normal process of land erosion (denudation) some of the nickel in the dust layer would have been carried into the oceans by streams, adding to nickel that arrived directly at the sea surface by fallout from the atmosphere above.

The process described seems reasonable enough, although it is quite unnecessary to calculate the thickness of an imaginary dust layer accumulated through five billion years. Based on reasonable calculations of the average rate of lowering of the continental land surface by erosion processes, that surface would have been reduced at a rate hundreds to thousands of times faster than IPD could accumulate, keeping the continental surfaces swept clean of dust.

We must not move on without pointing out that Slusher's figures for annual accumulation of IPD, based as they are on a 1960 estimate, are sadly outdated and badly in need of correction (see Awbrey, 1983, pp. 22-24; Dalrymple, 1984, pp. 108-11). Orbiting space vehicles have been used as platforms from which to sample IPD on a global basis, resulting in a great accumulation of particle density data. Results published in 1972 by J. S. Dohnanyi (and available to Slusher years before his technical monograph was published) put the annual influx of IPD at about 22,000 metric tons. At that rate a layer only about 8 cm thick would be expected to accumulate in 4.55 b.y., assuming none to be removed. More recent estimates place the annual input at 11,000 to 18,000 tons (Dalrymple, 1984, p. 109). This lower rate is confirmed by studies of IPD in deep-sea cores. Thus the best figures based on recent observations leave Slusher no basis for his argument. One wonders why Slusher chose to ignore the newer data, which are obviously of a much higher order of reliability than those gathered in 1960 from air samples collected at a single point--the summit of the Hawaiian volcano Mauna Loa.

Slusher's next step is to calculate the residence time of nickel in the oceans (pp. 40-41). Following the accepted methods of chemical oceanography (Goldberg, 1961, pp. 585-86), Slusher arrives at a residence time for nickel as 8,900 y. He then states that the nickel presently in the oceans could have accumulated in 8,900 y. from continental runoff alone. He does not state flatly that his calculation limits the age of the oceans to 8,900 y., but that happens to be quite close to the going age of the universe by Divine Creation (i.e., 6,000 to 10,000 y.).

In Chapter 19 we will go into the subject of "residence times" of elements in seawater. Elements that enter the world ocean as dissolved solids reside there until they are precipitated as mineral matter in ocean-floor sediments. Residence times range widely in duration, depending on which element is selected. If one were to select iron, titanium, or aluminum, with residence times between 100 and 200 years, it could be argued that the divine creation occurred during the nineteenth century. Residence times relate to chemical systems that are open systems in steady state. These systems simply process the elements they receive. Some elements leave quickly; others reside for long spans of time. Steven I. Dutch comments on Slusher's argument: "The fallacy lies in the fact that the oceans are demonstrably not closed systems and no oceanographer assumes they are. . . . The short residence time of nickel is in no sense a measure of the age of the Earth" (1982, p. 30).

Interplanetary dust on the moon requires a different sort of argument since there is no liquid ocean on the moon. Slusher calls attention to the discovery by Apollo astronauts of only a very thin dust layer in the landing sites (1980, pp. 41-42). This, he says, was a great surprise in view of predictions that cosmic dust accumulating on the moon's surface for 4.5 b.y. should have been found to be very deep and a source of peril, because the landing vehicle could have stuck in the layer of dust and been unable to leave the moon. Slusher, in asking what could have happened to that great layer of

dust that should have been there, implies that it was never there because of recent creation of the moon.

Slusher goes on to quote from a 1950 book written by astronomer R. A. Lyttleton of Cambridge University. Lyttleton speculated that the impact of unimpeded solar ultraviolet and X-ray radiation would have reduced exposed lunar rock to dust at the rate of a few ten-thousandths of an inch per year. Slusher is then able to calculate that in 4.5 b.y. a dust layer 45 km (28 mi) in depth would have formed--but of course the astronauts found no such layer.

Wherever the Apollo astronauts were able to observe cross-sectional exposures of lunar material below the general surface (as in crater rims and rilles), it proved to be a mixture of fine and coarse fragments formed by the churning action of repeated cratering. Even the largest of the rocklike boulders proved to be formed of rock fragments or clusters of fragments of older rock. In his recent volume on planetary geology, Professsor Billy P. Glass of the University of Delaware reviews our knowledge of the lunar regolith, based on the Apollo missions (1982, pp. 207-9). It is generally agreed that the layer of regolith, including the finest particles, was produced by meteorite impacts that pulverized the bedrock. Cratering was accompanied by the throwing out of fine particles that fell widely in layers surrounding the craters. For that reason thin layers of fine particles are found with large rock fragments embedded in them. A coring tube about 2.4 m (7 ft) long was used to take probes of the regolith. Average thickness of the regolith on the highlands has been estimated as about 10 m (30 ft). It has also been estimated that the thickness of the regolith increased by about 1.5 to 3 meters per one billion years at the Apollo site, but accumulation may have been ten times more rapid during an early period of intense cratering. The surfaces of regolith fragments show clear indications of erosion by micrometeoroids as well as damage from cosmic-ray and solar-flare particles, but this damage is on an atomic scale. The melting and vapor-ization of mineral matter accompanying the production of microscopic craters suggest that such impacts served to bind particles of regolith together. Thus, there has been an ongoing process of cementation of regolith, as well as one of fragmentation by impacts of large objects.

Slusher's review of pre-Apollo speculations about an enormous layer of lunar dust being a requirement of a 4.5 b.y. age of the solar system is irrelevant to what was actually found by the astronauts. Thick dust that "should have been there" is absent not because the moon is only a few thousand years old; it is not there because lunar processes of cratering have a comparatively shallow limiting depth, reworking the same layer over and over again. As Dutch aptly points out, if a farmer plows the same field to a depth of 20 cm each spring, even in a lifetime or several lifetimes the layer disturbed by plowing will be only 20 cm deep (unless accelerated erosion is in action) (1982, p. 30). On the moon, in the absence of an atmosphere and free water, cratering has simply moved the contents of a shallow surface layer from one point to another in a random manner over the past 3 b.y. or so.

The moon shows no evidence of deep-seated tectonic activity that might have produced deep basins to receive large amounts of sediment, even if processes of transportation had been available to move particles from highlands to such basins. (Broad, deep basins, called maria, were produced by enormous impacts and subsequently filled with basaltic lavas, but these are not basins of sediment accumulation.) Nothing actually observed on the moon by the Apollo astronauts and by laboratory analysis of materials brought back to earth suggests that the moon is only a few thousand years old. Instead, the complex layering and intermixing of the regolith and its enclosed rock masses suggest a very long geologic history. Radiometric ages on rock samples have yielded no age less than -3.1 b.y. for rock crystallization from magma. The greatest radiometric age found, -4.5 b.y., was of some minute mineral particles enclosed in a fragment of igneous rock. That age agrees with independent radiometric age determinations from meteorites found on earth--the age generally accepted for the origin of the planets.

Sweeping of Interplanetary Dust from the Solar System

Closely related to the subject of interplanetary dust attracted to the surfaces of the earth and moon is a question of the behavior of those dust particles in interplanetary space. These are the same kinds of micrometeorites that arrive in the earth's upper atmosphere, but in the interplanetary space, far from the gravitational fields of the planets, they orbit the sun. Each particle is a miniature planet in orbit. We can assume that there is a complete gradation of particle diameters upward into sizes like sand grains, pebbles, and boulders, and then into the class of full-sized asteroids. The important point is that particles smaller than about 0.01 cm are influenced by forces other than that of the sun's gravitational attraction. One such force we have mentioned is the pressure of the solar wind-- charged particles that continually stream from the solar corona. Another effect is that of sunlight striking the particles and influencing their orbits.

Named the Poynting-Robertson effect, this influence of solar radiation is the subject of a chapter by Harold S. Slusher (1980, pp. 55-64) in which he uses the effect to argue that the solar system cannot be more than a few thousand years old. Martin Harwit gives the basic equations for the Poynting-Robertson drag on a grain of dust orbiting the sun in interplanetary space (1973, pp. 176-77). The situation can be visualized in two ways. First, looking along the path of solar photons, we see that the grain absorbs that energy but also emits photons in all possible directions into surrounding space. It can be shown that the grain must lose orbital angular momentum and, as a result, will draw closer to the sun, following an inspiraling orbit, and eventually will fall into the sun. As viewed from the moving grain, solar radiation is attacking at a slightly advanced angle. An analogous situation is that of an automobile moving through vertically falling rain. Because of the auto's motion raindrops will strike vertical surfaces on the front of the auto. This can be understood as a slowing effect on the forward motion. The effect on a particle is the same, whichever analysis is used.

Based on the Poynting-Robertson effect alone, particles 0.001 cm in diameter located at a distance equal to that of the earth's distance from the sun (one AU) would spiral into the sun in about 19,000 years; particles 0.000,1 cm diameter would require less than 2,000 years. Slusher estimates that the typical lifetime for interplanetary particles is of the order of 10,000 years (1980, p. 60). But because small interplanetary particles are known to be present, the age of the solar system cannot be more than 10,000 years. For a very old solar system, all particles should have already been swept clean.

Slusher goes on to discuss the meteor swarms that probably have resulted from the disintegration of comets (pp. 60-62). He supposes that the Poynting-Robertson effect would by now have separated the particles of the swarm according to their sizes, but such separation has not been observed. This situation also favors a very young solar system.

Although Slusher's argument may sound reasonable it is seriously flawed by his omission of other effects that influence very small orbiting interstellar particles. Steven I. Dutch calls attention to this omission:

> Slusher's discussion is incomplete in several important respects. The Poynting-Robertson Effect requires that sunlight be absorbed and reradiated

in all directions. Reflected light from the particle will result in a radiation-pressure effect (Alfven and Arrhenius, 1976, p. 80) which will tend to push a particle outward. The net effect, since radiation intensity follows an inverse-square law, is to lessen the effective gravitational pull on the particle. Gravitational attraction will depend on the mass of the particle (cube of its radius) whereas radiation pressure will depend on its cross-sectional area which is proportional to the square of its radius. For very small particles (less than 6×10^{-4} cm in diameter), radiation pressure will actually tend to drive particles away from the Sun. (1982b, p. 31)[1]

(Note: The relationship between force of gravitational attraction, F_g, and radiation pressure force, F_p, can be derived as follows. F_g is directly proportional to particle mass, m, which in turn is directly proportional to the cube of particle radius, r^3:

$$F_g \propto m \propto r^3 \quad (1)$$

F_p is equal to the product of pressure, P, and area; A varies as r^2:

$$F_p = PA \propto Pr^2 \quad (2)$$

Combining (1) and (2):

$$\frac{F_p}{F_g} \propto \frac{Pr^2}{r^3} \propto \frac{1}{r} \quad (3)$$

In words, (3) reads: "The ratio of pressure force to gravitational force increases as particle radius diminishes.")

Dutch (p. 31) calls attention to another major point overlooked by Slusher. Particles spiraling in toward the sun must pass through the orbital zone of each planet in succession. Close encounter with a planet could result in the particle being propelled into an elongate orbit (such as that of a comet) carrying it farther out from the sun and delaying its eventual end in the sun. Another effect overlooked by Slusher is trapping of particles by gravitational resonances with the larger planets (Alfven and Arrhenius, 1976, p. 81). So trapped, particles could remain in stable orbits indefinitely. Dutch calls attention to the reasonable possibility that small particles are being continuously supplied to our region of the solar system from the region of the outer planets and the cometary cloud (if such a cloud exists), or by shedding of particles from asteroids. (1982b, p. 31)

In summary, Slusher's argument is defective, first, because it does not take into account known effects counter to the Poynting-Robertson effect; and, second, it assumes that no particles are furnished to the inner solar system from outer regions. In this respect, Slusher's argument follows the same pattern as that which he advances for the short-period comets: No external sources are available for replenishment of either the comets or the interplanetary dust particles; but the comets and particles are still here; therefore, the solar system is very young. Slusher's basic assumption is that of recent divine creation of the universe, which allows for no subsequent change other than entropy increase. That basic assumption is, of course, outside the scope of empirical science.

Changes in Saturn's Rings

Harold S. Slusher finds in Saturn's rings an argument for a young age of the ring system and the planet itself (1980, pp. 65-71). Saturn's rings consist of billions upon billions of solid particles ranging in size from as small as 1 mm in radius to as large as 2 to 4 m (Kerr, 1982, pp.

143-44). The particles lie in what seems to be an extremely thin disk--so thin that it is invisible from earth when viewed edge-on. Estimates of the thickness of the disk range from 10 to 20 m to no more than 200 m. A thickness of 30 to 50 m is commonly suggested (Kerr, p. 143). The disk of particles is formed into several rings, separated by narrow gaps. Prior to the Voyager 1 and 2 encounters with Saturn in 1980 and 1981, three rings were recognized; they were designated A, B, and C, in order from outside to inside. Voyager observations added a host of details to this basic picture, including two additional narrow rings, bandings within rings, spokelike lighter zones, tiny moons embedded within the rings, and various irregularities of the rings. The cause of the major gaps, particularly the Cassini gap located between rings A and B, has long been the subject of speculation. The presence and maintenance of the gaps seems linked to the presence of moons.

Three important hypotheses of origin of Saturn's rings and the particles that compose them are available (Pollock and Cuzzi, 1981, pp. 127-29). First, and oldest, is that the particles represent what remains of a single satellite body that formerly orbited the planet. That object was disrupted by internal stresses induced by tidal forces. This hypothesis is now considered unlikely because of the insufficiency of tidal stressing to disrupt such an object. A second hypothesis that conforms with the accretion hypothesis of origin of the solar system is that the ring particles are "leftovers" of the accretion process in which gaseous matter condensed on nuclei; many particles failed to grow any larger, although moonlets were formed by accretion from some of the matter. A third hypothesis, suggested by geologist Eugene Shoemaker, is that a single large moon orbiting Saturn collided quite accidentally with a giant meteoroid or an asteroid. The impact produced billions of particles of a wide range of sizes, and these eventually settled into a thin, flat disk.

Whatever hypothesis one chooses to favor, the consensus of mainstream scientists is that the rings date back to a time early in the origin of the solar system, i.e., about -4.5 b.y. Slusher, of course, takes exception to this age and attempts to present evidence that the rings and the planet itself are of very young age (Slusher, 1980, pp. 71-72). He argues that astronomer Otto Struve in 1852 noted that observations of Saturn's rings over the period from 1657 to 1851 show an increase in the widths of the rings and in the width of the gap between the planet and the inner edge of the B ring. The changes are interpreted to mean that the ring system is rapidly evolving and has not yet reached an equilibrium. If so, he argues, the ring system could not have been in existence for more than 100,000 to 500,000 years at the very most. Steven I. Dutch has evaluated Slusher's arguments and questions the observations interpreted as changes in the ring widths and distance from Saturn (1982b, pp. 31-32). Drawings by Huygens in 1659 and Cassini in 1676, according to Dutch, show the proportions of the rings essentially as they are known today. Considering the poor quality of the early telescopes and the crudity of the drawings, no significant change can be inferred with confidence. Dutch summarizes with the remark that "the present creationist position is based on faulty data and erroneous reasoning, and is simply irrelevant to the age of Saturn" (p.32).[1]

The Moon's Age and the Slowing of Earth Rotation

Creation scientists are very pleased with mainstream scientists' inability to get together in strong support of a single hypothesis of origin of the moon. To the confusing and unsatisfactory conclusions that I summarized on this subject in Chapter 16 on the origin of the solar system, the creationists say "Amen!". Creationist Donald B. DeYoung wrote:

The Lord knew that men would attempt to account for the moon by natural evolutionary mechanisms. Thus there are forty lunar references in Scripture, many of which declare the moon's supernatural origin and beneficial design. . . . The moon serves as a constant reminder of God's faithfulness and of the creation event. (1979, p. i)

DeYoung then asserts that the moon was created instantly out of nothing as a fully functioning satellite (p. ii). He cites six examples of unique features of the moon that show evidence of purposeful design. (1) The moon is the earth's only satellite and is exceptionally large in proportion to the earth. (2) The moon, because of its large size and nearness to earth, provides adequate night illumination. (3) The moon throughout history has provided man with an accurate time record. (4) The moon's orbit is stable, with no danger of the moon falling into the earth and burning up. (5) No hint of life has been found on the moon, because none was created there. (6) Several lines of evidence point to recent lunar creation on the scale of thousands of years, rather than billions of years.

On the last point, DeYoung gets into topics we have already covered. One of these is the lunar dust layer and regolith, and the radiometric ages of these materials.

In a 1982 article, ICR creation scientist Thomas G. Barnes refers to the slow recession of the moon from earth because of tidal friction. He introduces his paper with the following paragraphs:

It takes but one proof of a young age for the moon or the earth to completely refute the doctrine of evolution. Based upon reasonable postulates, great scope of observational data, and fundamental laws of physics there is proof that the moon and the earth are too young for the presumed evolution to have taken place.

There is an easily understood physical proof that the moon is too young for the presumed evolution- ary age. From the laws of physics one can show that the moon should be receding from the earth. From the same laws one can show that the moon would have never survived a nearness to the earth of less than 11,500 miles. That distance is known as the Roche limit. The tidal forces of the earth on a satellite of the moon's dimensions would break up the satellite into something like the rings of Saturn. Hence the receding moon was never that close to the earth. The present speed of recession of the moon is known. If one multiplies this recession speed by the presumed evolutionary age, the moon would be much farther away from the earth than it is, even if it had started from the earth. It could not have been receding for anything like the age demanded by the doctrine of evolution. There is as yet no tenable alternative explanation that will yield an evolutionary age of 4 billion years or more for the moon. Here is as simple a proof as science can provide that the moon is not as old as claimed. (Barnes, 1982, p. i)

Barnes's statements invite a skeptical response because they lack sufficient information and supporting evidence to stand by themselves. We must look to other creationist publications to find details of an argument in favor of a young moon based on its present rate of recession from the earth.

One creation scientist who has put forth this argument is Walter T. Brown, Ph.D., Director of the Midwest Center of the Institute for Creation Research at Naperville, Illinois. His argument is presented in a publication titled Evidence that Implies a Young Earth and Solar System. (See Thwaites and Awbrey, 1982, p. 22, for this undated citation.) Brown wrote:

> Atomic clocks, which have for the last twenty-two years measured the earth's spin rate to the nearest billionth second, have consistently found that the earth is slowing down at the rate of almost one second a year. If the earth were billions of years old, its initial spin rate would have been fantastically rapid--so rapid that major distortions in the shape of the earth would have occurred.

This statement has been thoroughly analyzed and refuted by San Diego University professors William M. Thwaites and Frank T. Awbrey, who find it contains a serious error that destroys the validity of the conclusion (1982, p. 19). Checking Brown's calculations and using the slowing rate he gives, Thwaites and Awbrey project back into time the conditions that would have prevailed 4.6 b.y. ago. At that time the earth would have been spinning 143 times faster than it does today and the day would have been only 10 minutes long. There would have been about 53,500 days per year. With this rate of spin the planet would have been severely deformed if not totally disrupted. Brown's conclusion is that the earth must be very much less than 4.6 b.y. old. Even if his conclusion seems valid, could it be derived from an erroneous rate of slowing? Let us look into the background of this argument to try to understand what it is all about.

Gravitational attraction exerted by the moon on the earth tends to deform the earth into a prolate ellipsoid (shaped like an American football) with its long axis along the line connecting moon and earth. As a result, there are two diametrically opposite centers upon which tide-producing forces converge. If the earth were covered by a single open ocean of uniform depth, the ocean water would rise to a summit, or bulge, at each of the tidal centers and would subside in a gentle trough girdling the globe in a plane passing through the polar zones.

If the earth did not spin on its polar axis, the tidal bulges would be fixed in position relative to the earth's surface. But, of course, the earth does spin on its axis, and the two tidal-force bulges thus travel endlessly around the earth's equatorial region. The ocean water tends to respond by rising and falling in unison with the passing of the tide-producing force centers, and thus the system of ocean tides is energized and its period or rhythm is controlled. The waters of the various ocean basins tend to oscillate in periods determined by their dimensions; the actual global pattern of tidal rise and fall is extremely complex. The main point here is that the back and forth tidal movements of ocean water encounter frictional resistance against the ocean bed, an effect that is particularly great in shallow waters. What we have here is an energy-dissipation system. The continuous work done by tides on the whole earth is estimated to be on the order of 3×10^{12} watts (two billion horsepower). This tidal friction tends to slow the earth's rotation, and it also has the effect of changing the angular momentum of the moon in its orbit. The reason for this effect is that the total angular momentum of the earth-moon system must remain constant, therefore, the slowing of earth rotation must be accompanied by an increase in angular momentum of the moon. This increase takes the form of an increase in the moon's distance, along with a decrease in the moon's linear velocity in its orbit. It is calculated that each year the distance to the moon increases by 3 to 4 cm. The distance increase has now been detected by laser measurements (Wahr, 1985, p. 44).

As to the slowing of the earth's rotation because of the effects of tidal friction, two methods have been used to estimate the rate (Goldreich, 1972, p. 47). One is based on observations of the longitudes of the sun, Mercury,

and moon with reference to the stars. The records extend over the past 250 years and are considered quite reliable. The second method uses historical records of ancient solar eclipses, but the reliability of such records has been seriously questioned (Newton, 1969).

The first astronomical method has yielded a rate of increase of length of the earth day of approximately 20 microseconds per year (0.000,02 seconds per year). It might seem wholly unwarranted for a scientist to extrapolate such a figure back in time as far as one billion or two billion years. After all, we have strongly criticized the creation scientists for that kind of extrapolation--witness the case of their claimed decrease in the speed of light (Chapter 15) and the claimed exponential decay of the earth's magnetic field, which we discuss in Chapter 19. There is, however, some completely independent evidence from paleontology suggesting that the estimated rate of slowing has been roughly as given above.

During the 1960s, Professor John W. Wells of Cornell University observed that the structure of fossil corals consisted of yearly growth bands, and that those bands consisted of daily growth ridges (Goldreich, 1972; Runcorn, 1966). The number of days per year could be estimated by counting the number of daily ridges per annual band. For corals in the Devonian period at about -380 m.y., the day length was judged to be 22 hours and the distance from earth to moon about 370,000 km. Determinations of the same kind were made for algal deposits (stromatolites) of the Upper Cambrian (-510 m.y.) (Pannella et al., 1968). Plots of the collected data for the entire time span from Recent back through the Paleozoic Era showed a nonuniform increase in days per month going back in time, and from this it is inferred that tidal friction has not been uniform in that period. There is some indication that the slowing was halted during the Mesozoic period when the great opening up of the Atlantic and Indian ocean basins was taking place.

Confusion was introduced and compounded in 1978 when two researchers, Peter G. H. Kahn and Stephen M. Pompea, paleontologist and astronomer, respectively, published in Nature an article proposing the chambered nautilus as a time clock (Kahn and Pompea, 1978). Studying the modern nautilus, they found an average of 30 growth lines on the outside of the shell between every two septa (partitions) that form the chambers inside the shell. They concluded that the nautilus produces one growth line per day for an average of about 30 lines for each synodic month of 29.53 days. (The synodic month is the period to which the phases of the moon and the tidal rhythm are adjusted.) They then counted growth lines for fossil ancestral nautilus shells (nautiloids) through geologic time ranging back to -420 m.y. One specimen from South Dakota, 69.5 m.y. in age (Late Cretaceous), had 22 lines per chamber. If this number is accepted as the number of days in the synodic month, the rate of slowing of earth rotation would be 17 times greater than the amount calculated from astronomical observations of the past 3,000 years. For a 326-m.y. nautiloid they counted 15 lines per chamber. The smallest count was 8 to 9 lines per chamber for nautiloids from the Upper Ordovician, age about -420 m.y.

Keep in mind that, as the rate of earth rotation increased backward in time, the moon's period of revolution around the earth--the synodic month--would also have decreased. Nine growth lines per septum thus works out to a day length of about 21 hours. Extrapolated back at that rate to -4.0 b.y., the day length is a little over 10 hours, whereas rates based on astronomical calculations of tidal friction (Runcorn, 1966) give a day length of about 11½ hours. Kahn and Pompea come up with a moon distance in Upper Ordovician time as about 0.4 that of the present distance. If we extrapolate that rate back to zero distance, the moon would have touched the earth about 750 m.y. ago. Even the more conservative data extrapolate to zero distance at -1.5 to -2 b.y. (Runcorn, 1966).

Within the scientific community, reaction to Khan and Pompea's nautilus data soon materialized (West, 1978). The coral geochronometry proposed earlier by Wells was already under strong criticism from other scientists and the same critics were quick to attack the nautilus geochronometry. They pointed out that there is no evidence that the nautilus produces one growth line per day or that it produces one septum per month. Actual observations on the living nautilus do not confirm the figures. Details of the argument are far too numerous and technical to present here. The point is that mainstream science is up to its old custom of greeting all new ideas with skepticism and subjecting those ideas to every possible form of intense scrutiny. As usual, the creationists see this as a weakness in science and its hypotheses. "Survival of the fittest"--in this case the fittest hypothesis--is a concept abhorrent to creationists. Their own theory of recent creation by a Creator is above and beyond any contest for survival conducted by humans. Yet, strangely enough, the creationists seek empirical support for recent creation. Does not this imply that they question the inerrancy of the Scriptures?

Creation scientist Donald B. DeYoung, a physics professor at Grace College in Indiana, used the nautilus data in his argument for a young moon and earth (DeYoung, 1979). He states, quite correctly, that the nautilus data show that the moon was 60 percent closer to the earth at about -400 m.y. than it is today. He does not interpret the figures explicitly, but as we have seen, they extrapolate backward to a close encounter of moon with earth in less than one billion years. That extrapolation has, however, been challenged in a more recent scientific study.

In 1982 a report published by Kirk Hansen of the University of Chicago described a revised model for calculating the position of the moon as far back as -4.0 b.y. (Kerr, 1983, p. 1166). The principle used is that, as earth rotation increased in rate back into time, the period of the tidal cycle decreased. This decrease would place the tidal period out of the range of the natural period of tidal oscillation of the ocean basins, reducing the coupling effect and the rate of energy dissipation and slowing the rate of change in rotation period. Extrapolating backward, Hansen's model gives a distance of 290,000 km at -4.0 b.y. and a minimum of 225,000 km at any earlier time. Hansen's argument provides a reasonable solution to the dilemma of hypotheses of the moon's origin that require that event to have occurred early in the history of the solar system.

So now let us get back to the question: How could Walter Brown have come up with a rate of earth slowing (one second per year) that leads to a -4.6 b.y. day-number over 60 times larger and a day-length 1/60 as great as the astronomical data produce?

Thwaites and Awbrey suggest that Brown made use of an entirely different kind of measurement of the "slowing" of earth rotation and "lengthening of day" (1982, pp. 20-21). I came across a note in Science News (vol. 100, p. 408, December 18, 1971) that seems to pinpoint the cause of confusion on the part of Brown. The news note described the coming need to add about one additional or "leap" second to atomic time (AT) because the second of universal time (UT) has been getting longer since 1900. The insertion of the leap second will help close the gap between AT and UT. The point is that atomic time is considered a base of absolute constancy, since atomic clocks are not known to vary appreciably. On the other hand, universal time is based on the earth's rotation, which is influenced by a variety of fluctuating causes. For example, there is an annual rhythm of meteorological origin, related to seasonal shifts in prevailing winds. There are irregular changes in speed of rotation for which no cause is known. There are probably also longer period changes caused by changes in the magnetic coupling between the earth's mantle and iron core. Records going back to 1820 show that there was a period of increase in rotation rate between 1860 and 1880 (Strahler, 1971, p. 39, Fig. 3.4). A strong decline since 1900 is obvious, but it has not been a steady decline. The accumulated loss in UT with respect to the time based on astronomical observations other than earth rotation has been nearly 50 seconds between 1900 and 1970. (Ephemeris time, ET, was used before the atomic clocks were installed.)

What Brown seems to have done is to confuse the need for occasionally adding a leap second with the long-term rate of slowing of earth rotation based solely on tidal friction. The two quantities are completely unrelated physically. (For details, see Thwaites and Awbrey, 1982, pp. 20-21). The subject of time measurement and time systems is extremely complex and can confuse even readers with good science training. The easiest mistake for a reader to make is to confuse constant rate of change (equivalent dimensionally to velocity) with acceleration. Use of the words "time per unit of time" puts both variables in the same dimension. Acceleration reads "seconds per year per year" and is a different quantity from "seconds per year." Fortunately, Awbrey and Thwaites make the distinction very clear and give examples for calculating cumulated deficits from acceleration figures. (For background reading on the subject of earth rotation, time, moon, and tides, see Strahler, 1971, Chapters 2, 3, 8, and 9; Wahr, 1985.)

Credit

1. From Steven I. Dutch, 1982, A Critique of Creationist Cosmology, Jour. of Geological Education, vol. 30, pp. 27-33. Used by permission.

Chapter 19

Creationist Arguments for a Young Earth

Creation scientists have used what they call uniformitarian principles of geology to examine the age of the earth. They do not, however, subscribe to uniformitarianism, which they consider a false doctrine held by mainstream geologists and evolutionists. By "uniformitarianism" they mean an assumption that geologic and biologic processes seen in action today have been of more or less uniform intensity and rate throughout all of geologic time. As we stated in Chapter 9, the extreme form of uniformitarianism was promoted by Sir Charles Lyell through his popular textbook Principles of Geology, first published in 1830. Taken out of its early historical context, that definition of uniformitarianism has no useful meaning today, because no mainstream geologist would want to assume, let alone declare, that the rates at which geologic processes operate today were the same throughout all past geologic time. Let us see how creation scientists carry the corpse of a discredited uniformitarianism before them to bear the brunt of their attack.

Rates of Influx of Solids into the Oceans

Creationists derive the age of the world ocean by dividing the quantity of a given element present in seawater today by the estimated annual rate at which that element is brought into the ocean by river flow from the lands. Mainstream geologists would, of course, consider the use of present rates for extrapolation far back into geologic time unreliable in the extreme and therefore inadmissible as a scientific method. The record of marine sediment deposition revealed in strata of past eras clearly shows that the influx of dissolved solids and rock particles into the oceans has varied greatly through geologic time. Despite such evidence, creation scientists of the Institute for Creation Research (ICR) insist on using the premise of uniformity to compile figures giving the age of the oceans varying from a few thousand years to over 200 million years (Morris, 1974b, pp. 153-56). In an early issue of ICR Impact Series (No. 17) titled "The Young Earth" (it bears no date of publication), Henry Morris lists seventy-six processes for calculating the age of the earth, many of which are based on the influx of a single element to the oceans by way of rivers. These estimates vary from as low as 200 y. (for aluminum) to as high as 260 m.y. (for sodium).

As pointed out earlier, the world ocean is an open system, so far as input and output of matter are concerned. Elements that enter the ocean also leave the ocean to become part of the accumulated sediment on the ocean floor. Each element has it own residence time. Generally speaking, the longer the residence time, the greater will be the quantity of the element present in storage, dissolved in seawater. Assuming a uniform rate of entry, elements with short residence times exit quickly and their relative quantities in seawater are proportionately small. The same principle would apply to the individual inhabitants of a boarding house, or a city. Morris, in his table of estimates, has failed to adjust his age calculations for residence times. Refer to Table 19.1

to see how this is evident. It looks very much as if Morris has simply listed the residence time as his "indicated age of the earth."

Quite apart from the possible misuse of residence time figures, and the inadvisability of extrapolating annual rates of the present century to all of geologic time, it is interesting to note that, of the 76 "indicated-age-of-earth" values in Morris's Table, 25 are in excess of 1 m.y. while only about 20 are 10,000 y. or less. Morris states that this great variation--from as low as 100 y. to as great as 500 m.y.--is the most obvious characteristic of the list of ages. All but perhaps one figure must be in error, since the earth can have only one age. Quite rightly, he attributes the variability to errors in the assumptions used in making the calculations. But then, Morris advances an argument in favor of the low age estimates: they are likely to be more accurate than the high values. One reason he gives for this conclusion is that the dangers of extrapolation are much less for a short period of time (a few thousand years) than for a long period of time (hundreds of millions of years).

Morris concludes that the weight of the scientific evidence is on the side of a young earth; that there was not enough time for organic evolution by natural selection; and that the data indicate the origin of everything by special creation. My own suggestion would be that for ages based on elements in seawater, much greater earth ages seem likely for two reasons. First, substantial amounts of the element have left the system and are now in long-term storage in crustal rocks. Second, today's rates of influx of the elements by rivers draining the land may be well above the average of the past 3 to 4 b.y. Either consideration, if taken into account, would increase the age estimate of the oceans, moving the estimates in the direction of radiometric ages.

A perusal of the list of seventy-six "processes" and the earth ages they are assigned shows that a wide

Table 19.1 Residence Times of Selected Elements in Seawater

Element	*Residence time, m.y.	*Concentration in seawater, percent	**Morris: age of earth, m.y.
Sodium	260	31	260
Magnesium	45	3.7	45
Potassium	11	1.1	11
Silicon	10,000	0.003	8,000

*Data of E.D. Goldberg, 1961, in Mary Sears, Ed., Oceanography, Washington, D.C., Amer. Assoc. for the Advancement of Science, p. 586.

**Henry M. Morris, 1974, ICR Impact Series, no. 17, Table 1.

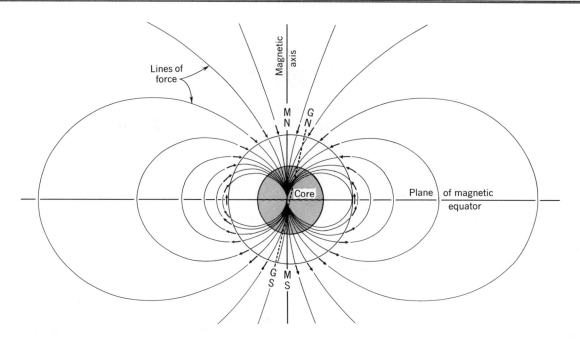

Figure 19.1 Lines of force in the earth's magnetic field shown in cross section passing through the magnetic axis. Letter M̲ designates magnetic; G̲, geographic. Arrows at the surface of the earth show the orientation of a dip needle. (Copyright © 1971 by Arthur N. Strahler.)

variety of physical processes is involved, and these include astronomical, galactic, geophysical, and geologic processes. Each different process requires separate evaluation as an independent indicator of age. To assemble them all in one table and from the total collection draw a conclusion as to which alternative is favored--small age or great age--is simply an exercise in futility that can lead only to confusion. Throwing an audience into extreme confusion is a debating tactic used by purveyors of pseudoscience. On later pages we shall, however, take a close look at certain of the processes listed in the table. For example, the alleged decay of the earth's magnetic field (Item 27) as indicating an age of 10,000 y. is well worth examining critically. Another is the accumulation of sediment, brought to the continental shores by rivers and spread by currents in layers over the ocean floors (Item 73), calculated by Morris to give an earth age as low as 20,000 y.

Earth Magnetism

Creation scientists of the ICR place considerable emphasis on a method they consider reliable in yielding a very young age for the earth; it is based on a set of scientific observations purporting to show that the earth's magnetic field has grown steadily weaker over the past 150 years. To understand the argument it is necessary that you know something about earth magnetism.

In its most simple aspect, the earth's magnetic field resembles that of a bar magnet located at the earth's center (Figure 19.1). The axis of the imaginary bar magnet is situated approximately coincident with the earth's geographic axis. At the points where the projected line of the magnetic axis, or geomagnetic axis, emerges from the earth's surface are the magnetic poles. Note that the earth's present magnetic axis forms an angle of about 20° with respect to the geographic axis. As a result, the magnetic poles today do not coincide exactly with the geographic poles. (Averaged over long periods of time the two poles would be coincident.)

Figure 19.1 shows lines of force of the earth's magnetic field in relation to the earth's core. The force lines pass through a common point close to the earth's

center. The magnetic axis is oriented vertically in this diagram. There exists a magnetic equator, lying in a plane at right angles to the geomagnetic axis and encircling the earth's surface approximately in the region of the geographic equator. Visualized in three dimensions, the lines of force of the earth's magnetic field form a succession of doughnutlike rings, suggested in Figure 19.1. The small arrows show the attitude that would be assumed by a small compass needle, free to orient itself parallel with the force lines close to the earth's surface. Force lines extend out into space surrounding the earth to distances as great as 150,000 km. This entire external field of magnetic effect is called the magnetosphere.

Earth magnetism, or geomagnetism, has been an object of scientific measurement and study since the early 1800s. Measurements of the strength of the magnetic field at the earth's surface were begun as early as 1835 by Carl Friedrich Gauss (1777-1855), the noted German mathematician, physicist, and astronomer. Since 1880 the measurements have been repeated about every 10 to 15 years (McDonald and Gunst, 1967). What the measurements showed was a persistent decrease in strength of the main magnetic field. In terms of percentage, the decline from 1835 to 1980 has been about 7 percent (Merrill and McElhinny, 1983, p. 45 and Figure 2.10a).

Dr. Thomas G. Barnes, who serves as professor of physics on the graduate staff of the ICR Graduate School, and who was formerly a professor of physics at the University of Texas in El Paso, in 1973 published a remarkable hypothesis of geomagnetism that led him to conclude that the observed decline in field strength referred to above can be projected back into the past to reveal a limit to the earth's age. That limit, he concluded, is much less than 20,000 y. and is probably less than 10,000 y. (Barnes, 1973, pp. 25, 38; 1981, p. iv).

The key to Barnes's limiting age lies in the method by which less than two centuries of observation is projected (extrapolated) backward many tens of thousands of years. Barnes postulates that the waning of field strength has been an exponential decay (similar in mathematical principle to the decay of radioisotopes). He made use of the two end values of the measurement series--1835 and 1965--leading to the conclusion that the half-life of the decay program is 1400 y. It is easy enough to reconstruct the decay curve backward into time as far as desired. Barnes made calculations as far back a 1 b.y. In Figure 19.2 I have constructed the Barnes curve back to the year 10,000 B.C., or -12,000 y. (See Barnes, 1973, Table

2, p. 37, for data shown as points on the curve.) At -12,000 y., the field strength is about 400 times greater than today. At that value, the enormous strength of the field would be associated with extreme heat within the earth's core, and the electric currents that produce the magnetism would have been extremely intense. Barnes figures that, at -20,000 y., the electric currents in the core would have been 50,000 times greater than they are today, and the heating of the core could have been 250 million times greater than at present. The core, he concludes, could not have held together at such a temperature; hence, a reasonable limiting earth age is -10,000 y. or less.

Barnes's thesis may seem, on first reading, to be logical and direct. Fortunately, there are mainstream scientists, well versed in principles of geophysics, who find a great deal to critize in Barnes's method and conclusions. One such critic is Dr. G. Brent Dalrymple, holding the Ph.D. in geology from the University of California at Berkeley. Since 1963 he has served as research geologist with the U.S. Geological Survey in its branches of theoretical geophysics, regional geophysics, and isotope geology. Dalrymple has undertaken to show "that Barnes' hypothesis is scientifically untenable and should not be taken seriously. . . . In the process of supporting his arguments and defending his conclusions, (however,) Barnes ignores or attempts to refute much of what is known about the magnetic field." (1983, p. 125)[1]

The Dipole Field

To understand what Dalrymple finds to be the trouble, we must begin by asking: What property of the earth's magnetic field is actually represented by the values of "magnetic strength" recorded between 1835 and 1965? The correct technical term for the quantity measured is the dipole moment; it describes a property of the earth's dipole field. A simple steel bar magnet has two poles--one that seeks the north magnetic pole, the other that seeks the south magnetic pole, as we find in the simple magnetic compass. The bar magnet (or a compass needle) is a dipole--it has two poles. You can think of the earth's dipole field as its main or principal magnetic field. It is derived by mathematical calculation to be a perfect model with perfect radial symmetry around the earth's magnetic axis. This model will have "magnetic meridians" analogous to the meridians of longitude that run north-south between the north and south geographic poles. There will be a "magnetic equator," lying in a plane at right angles to the magnetic axis. On this ideal magnetic globe, a compass needle is always aligned perfectly with a magnetic meridian and points exactly toward the magnetic pole.

The actual situation is quite different from the ideal model given by the dipole field. Carrying around your magnetic compass and recording the true geographic direction in which it points at many widely separated global locations, you would find many seemingly irregular variations in the directions the needle assumes. This means that the magnetic field as measured at the earth's surface has numerous irregularities, and you might think of these as taking the form of centers that have more than the average attraction for the compass needle, while other centers have less than the average attraction. When the values described by the ideal dipole configuration are subtracted from values actually measured over the earth's surface, there remains a residual magnetic field, or nondipole field, consisting of the centers of greater or lesser magnetic field strength. The entire nondipole field is slowly shifting westward around the globe, the rate of motion being such that the pattern would make a complete circuit of the globe every 2000 years. The features of the nondipole field are also continually changing in form. The point of all this description of two "magnetic fields" is this: The total strength of the earth's magnetic field measured at its surface is the sum of the dipole field and

the nondipole field. As Dalrymple points out: "It is important to understand that the dipole field is not a real field at all. Rather it is an idealized mathematical model that best fits the real field" (1983, p. 126).

Next, consider this possibility: If the dipole field is steadily becoming weaker, perhaps the nondipole field is at the same time becoming stronger, and if so, is it not possible that the total field strength is remaining constant? Barnes has created the impression in published presentations of his theory that the observed decrease in the strength of the dipole field represents a decrease in the strength of the earth's entire magnetic field. Dalrymple comments: "This is a serious error, because the dipole field as measured at the earth's surface is only one idealized component of the field and cannot be equated with either the total field strength or the total field energy" (1983, p. 128). Dalrymple cites the results of one study of the magnetic field data as showing that the decrease in the dipole field strength is being approx-imately balanced by an increase in the nondipole field strength (p. 129). There is, however, evidence of a very small annual decrease in the total field strength; it amounts to about 0.01 percent. Actually, this observed decrease may be of little significance, since most of the energy of the earth's total magnetic field lies deep in the core, where it cannot be measured.

Although it seems clear enough that Barnes's calculations of the age of the earth are based on an unwarranted assumption that the earth's total magnetic field is becoming weaker, it should also be pointed out that the assumption of an exponential rate of that supposed decrease is also entirely unwarranted. Looking at the uppermost graph in Figure 19.2, showing how the actual measurements of dipole strength have declined through the years, there is nothing to suggest that the points lie on a curve that steepens toward the left (as it would in an exponential curve, and as shown in the middle and lower diagrams, which are hypothetical). A straight line could just as easily be inferred as the best representation of the trend of the points. Actually, a straight-line fit is a somewhat better fit, as judged mathematically, but that is of little consequence. The point is that carrying back far into time either an exponential decay or a straight-line change is completely unwarranted by the observations themselves.

Using the slanting straight line in the uppermost graph, I figure that the dipole field strength would have been increased by a factor of only about 8 at -10,000 y., surely not enough to place much physical stress on the core. An equally reasonable hypothesis is that the dipole field strength, and perhaps also the total field strength, has actually fluctuated, perhaps cyclically, for tens of thousands, or perhaps hundreds of millions of years in the past. Dalrymple brings to our attention evidence already accumulated to support alternating increasing and decreasing dipole field strength. Analysis of the carbon-14 content in lava flows and in archaeological materials, such as bricks and pottery, shows the dipole strength to have been less than at present in a time period between about -4000 and -5000 y. Barnes flatly rejects such measurements as having any validity.

In August 1983, Barnes published a rebuttal to Dalrymple's criticism of the exponential decay theory. Barnes (1983) explains that in the language of electrical or communications engineering, the dipole field is the "signal," whereas the nondipole field is the "noise" that tends to obscure the signal. He points out that at times a severe magnetic storm can create so much "noise" (static) that long-distance radio communication is totally disrupted (which is indeed the case at times). Barnes insists that the magnetic age of the earth is indicated by the dipole field, which is the true "signal," and the nondipole field cannot carry any such information about the earth's age. It strikes me that Barnes has resorted to an incorrect analogy. The noise background created by a magnetic

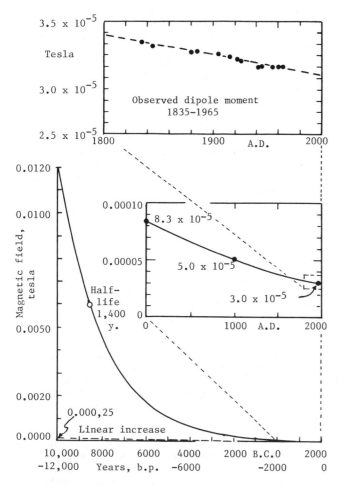

Figure 19.2 Observed dipole moment extrapolated back to 10,000 B.C. (-12,000 y.). (Based on data of Thomas G. Barnes, 1973, <u>Origin and Destiny of the Earth's Magnetic Field</u>, Technical Monograph no. 4, p. 37, Table 2. Institute for Creation Research, Creation-Life Publishers, San Diego.)

storm represents an influx of energy from a solar flare impacting the magnetosphere. It drowns out the comparatively weak radio signal powered by an electrical input at a radio transmitting station. The two energy sources are entirely dissimilar and independent. In the case of the earth's magnetic field, both the dipole and nondipole fields share the same kind and source of energy, which is simply divided up between them artificially by mathematical calculation. The "signal/noise" analogy is simply specious and irrelevant; it can scarcely be taken seriously except to recognize it as a desperate attempt to confuse unwary persons in the audience hearing the debate.

Cause of the Earth's Magnetic Field

In mainstream scientific study of geomagnetism, the prevailing explanation of the source of earth magnetism is known as the <u>dynamo theory</u>. This theory postulates that the liquid iron of the earth's core is in slow rotary motion with respect to the solid mantle that surrounds it. It can be shown by application of laws of physics that such fluid motion is capable of causing the core to act as a great dynamo, generating electrical currents. These currents at the same time set up a magnetic field. A single, symmetrical current system can thus explain the magnetic field as essentially resembling a simple bar magnet. Dalrymple states: "The reason the dynamo theory has gained near-universal acceptance is that it is the only

proposed explanation that can explain the observed features of the magnetic field" (1983, p. 130). Barnes rejects the dynamo theory and sets up an alternative theory of his own.

Note first, however, there is no disagreement about certain basic inferences concerning the earth's core. Studies of earthquake waves show that the core has a radius of 3470 km, and that a sharp boundary separates it from the enclosing <u>mantle</u>, composed of silicate rock. Based on calculations of the earth's average density and measurements of the compressibility of materials at high pressure, it is usually assumed that the core is composed largely of iron and some lighter element such as sulfur or oxygen. That the outer region of the core is in a liquid state is shown by the behavior of earthquake waves--the S-waves do not pass through the core. An inner region of the core may be in the solid state. Confining pressure in the liquid core is enormous (2000 to 3000 kilobars), and temperatures, which are between 2700 and 2800°K, are above the melting point of iron at the prevailing pressure. Obviously, conditions of pressure, temperature, and physical composition of the core cannot be directly observed or sampled, and this lack of specific information makes extremely difficult the exact solution of a dynamo theory. Barnes uses this lack of direct observational information as reason to categorically reject the dynamo theory. Referring to a work published in 1933 Barnes finds support for this conclusion Dalrymple discusses this reference in detail, showing that nothing in the statements of that earlier work preclude the operation of a dynamo in the core (1983, p. 130).

Dynamo theory provides for the polarity reversal of the magnetic field. Dalrymple states that "polarity reversals are a permissible, perhaps even an expectable, property of self-exciting dynamos" (1983, p. 131)[1] . He adds that "periodic reversals are probably an intrinsic feature of the large-scale dynamo process." Polarity reversal has been documented for the dipole field of the sun, which is a great ball of rotating dense gases. Between 1953 and 1958 a complete reversal of polarity was observed, including a period of a few years in which no field strength was detectable (Dalrymple, 1983, p. 131).

In that connection, it is interesting to note that sunspots, which are associated with intense magnetic fields, continued to operate even when the dipole field had almost disappeared. This observation suggests that there is no reason to suspect that a decay of the solar dipolar field is associated with a decay of the total magnetic field strength.

Despite repeated assertions by Barnes that there can be no energy source for a dynamo effect in the earth's core, several reasonable sources can be suggested. According to Dalrymple, these include cooling in the core, radioactivity in the core, precession of the earth, seismic energy, and gravitational energy (1983, p. 131). Energy presently stored in the core can be calculated as fully capable of having kept the dynamo in action over the past 3 b.y.

Barnes's theory is that the earth was supplied initially (during recent creation) with a supply of electrical energy that, through electric currents in the core, have produced the earth's dipole field. Since then the store of energy has been on an exponential depletion schedule. As the currents grow weaker, the magnetic field becomes weaker. Barnes attributes his theory to conclusions reached in 1883 by Sir Horace Lamb to the effect that the earth's magnetic field could be due to an original event (creation) and has decayed ever since (1973, p. viii). However, Dalrymple, after reviewing Lamb's papers cited by Barnes, finds in them no theory of the earth's magnetic field (1983, p. 130). Not only does Lamb not mention the earth's magnetic field, but he makes no mention of an "original event" or "creation." Barnes's theory of decay from an initial store of energy is in full accord with the creationists' religious philosophy: Since creation the earth as a system has been decaying--getting physically and morally worse, that is--in accordance with the second law of thermodynamics.

Rock Magnetism and Polarity Reversals

To follow further the exchange of views between Dr. Barnes and Dr. Dalrymple, you need to have a general knowledge of the ways in which the earth's magnetic field has become recorded in crustal rocks and sediments, allowing a history of geomagnetism to be extended far into the past. Perhaps this information will bear on whether Barnes's theory of exponential decay can be sustained, or whether the magnetic field has fluctuated repeatedly. One of the most abundant kinds of igneous rock is basalt, a dark, dense rock resulting from the outpouring of molten rock (magma), usually from long, narrow fissures (cracks) leading up from pockets of magma below. Basalt is the common rock of the Hawaiian Islands and excellent examples of the outpouring and solidification of basalt can be seen on the island of Hawaii, where the active volcanoes Kilauea and Mauna Loa are located. Basaltic lavas contain minor amounts of oxides of iron and titanium. Magnetite, the mineral of which "lodestone" is a naturally magnetic variety, is an example.

At the high temperatures in a basalt magma (around 1100 C), these iron and titanium minerals have no natural magnetism. However, as cooling sets in, each crystallized mineral passes a critical temperature, known as the Curie point (between 600 and 400 C), below which the mineral is magnetized by lines of force of the earth's field. The magnetism acquired by minerals on cooling resembles the permanent magnetic condition of the alnico magnet. In this way a permanent record of the earth's magnetic field is locked into the solidified lava.

In the study of rock magnetism, a sample of rock-- usually a small drilled core--is removed from the surrounding bedrock. Orientation of the specimen is carefully documented in terms of geographic north and horizontality. The specimen is then placed in a sensitive instrument, the magnetometer, that measures the direction and intensity of the permanent magnetism within the rock. Using a number of samples obtained from a single lava flow, the magnetic values are compared for consistency and averaged, yielding the compass direction and inclination of the magnetism. (Inclination refers to the down-pointing of a magnetic compass mounted on a horizontal axis so that it can "dip" in either direction.) Effects of lightning and the earth's present field are removed, and the magnetism is tested for stability in various ways. The term paleomagnetism is used for such locked-in magnetism dating far back into the geologic past. Paleomagnetism can be compared with present conditions and with the magnetic field at other locations and in different times in the geologic past.

As early as 1906, Bernard Brunhes, a French physicist, observed that the magnetic polarity of some samples of lavas is exactly the reverse of present conditions. He concluded that the earth's magnetic polarity must have been in a reversed condition at the time the lava solidified. You might wish to argue that the rock magnetism itself may have undergone a change in polarity; that is, self-reversal has occurred. However, that possibility has been excluded by careful experiments, and in recent years there has been general agreement among mainstream geoscientists that the rock magnetism is permanent and is a reliable indicator of the former states of the earth's magnetic field.

In addition to the magnetic data of the lava specimen, we need a radiometric determination of the age of the rock, giving the date of solidification of the magma. Extensive determinations of both magnetic polarity and rock age have revealed that there have been at least eighteen reversals of the earth's magnetic field in the last 5 m.y.

During this phenomenon of magnetic polarity reversal, reversal starts by a weakening (decay) of the geomagnetic field of force while the polarity and direction of the force lines remain the same. After the strength of the magnetic field has fallen to a low value, it exhibits an erratic directional behavior, reverses, and begins to build back to normal, but with the polarity in the opposite direction, i.e., reversed polarity. The polarity reversal can require some 500 to 1000 years to complete. The time intervals between polarity reversals seem to be irregular in length; the sequence does not follow a cyclic or rhythmic pattern.

Figure 19.3 shows a timetable of magnetic polarity changes. Polarity such as that existing today is referred to as a normal epoch; opposite polarity is called a reversed epoch. Each epoch is named for an individual or a locality. For example, the pioneer work of Bernard Brunhes is recognized in the present normal epoch, which began about -700,000 y. (-0.7 m.y.). An epoch of reversal named for the Japanese scientist Motonori Matuyama extends to -2.5 m.y.; it includes four shorter periods of normal polarity classified as magnetic events. A still older normal epoch, named in honor of the mathematician Karl Gauss (1777-1855), carries back the paleomagnetic record to about -3.5 m.y. and contains two reversed events. The oldest of the reversed epochs shown in the figure is named after Sir William Gilbert, who in 1600 pioneered in the study of terrestrial magnetism.

A large number of polarity reversals have been found throughout the geologic record, going back to nearly 150 m.y., which is through the entire Cretaceous Period and even beyond. As yet, the older reversals in this sequence cannot be dated with great accuracy, but the general pattern in time is known. Recall from Chapter 2 that the modern theory of plate tectonics had its beginnings in attempts to explain the presence of a great mid-oceanic ridge with its central or axial rift valley. In many places,

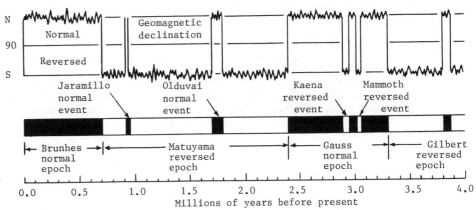

Figure 19.3 Magnetic-reversal polarity scale for the past four million years. (Based on data of E. A. Mankinen and G. B. Dalrymple, 1979, Journal of Geophysical Research, vol. 84, pp. 615-26.)

basaltic lavas emerge from the floor of the rift valley, and locally these accumulate to form islands, of which Iceland is the prime example. Seafloor spreading along the axial rift emerged as a concept well before the theory of plate tectonics was formulated as a complete system. If the axial rift valley is a line of upwelling of basaltic lavas, and if crustal spreading is a continuing process, the lava flows that have poured out in the vicinity of the axial rift will be slowly moved away from the rift. Lavas of a given geologic age will thus become split into two narrow stripes, one on each side of the rift. As time passes, these stripes will increase in distance of separation, as shown in Figure 19.4. The lavas can be identified and classified in terms of the epochs of normal and reversed magnetic field; these epochs will be represented by symmetrical striped patterns on either side of the rift.

Confirmation of the symmetrical magnetic stripes was gained in the course of oceanographic surveys made during the mid-1960s. We cannot take oriented core samples of lavas from the ocean floors. However, it is possible to operate a sensitive magnetometer during a ship's traverse of the mid-oceanic ridge. When this is done, it is found that there are minute variations in the value of magnetic field strength. These departures from a constant normal value are referred to as <u>magnetic anomalies</u>.

When several parallel lines of magnetometer surveys have been run across the mid-oceanic ridge, the magnetic anomalies can be resolved into a striped pattern. From a study of the anomaly pattern it is possible to identify the normal and reversed epochs, as we have done in Figure 19.4.

It has also been found possible to identify the normal and reversed magnetic epochs in core samples of soft sediment obtained from the ocean floor by piston-coring devices. Here, the epochs are encountered in sequence from top to bottom within the core. In this way numerous older epochs were discovered, extending back through the Cretaceous Period.

Finding the magnetic stripes on the ocean floor proved to be the key to the revolution in geology. There followed rapidly a series of magnetic surveys along many sections of the mid-oceanic ridge, all revealing similar striped patterns in mirror image. Magnetic evidence not only corroborated the phenomenon of seafloor spreading, but allowed the rates and total distances to be estimated as well.

Thomas Barnes simply denies that evidence exists for magnetic polarity reversals (Barnes, 1973, pp. 27, 49). Dalrymple points out that Barnes cites statements from papers or books published before a large body of corroborative evidence for reversals had been obtained (1983, p. 128). For example, one of Barnes's references is to a 1963 book by J. A. Jacobs, a leader in geomagnetic research, in which Jacobs expresses some reservations about the reliability of certain interpretations leading to establishing that a reversal has occurred (Jacobs, 1963). In ensuing years a great deal of positive evidence was collected and published. When in 1975 Jacobs published a second book, he took the new evidence into account and wrote as follows: "The evidence seems compelling that reversals of the Earth's field are the cause of the reversals of magnetization (in rocks), and this provides a further constraint on any theory of the origin of the field" (1975, p. 140). The "further constraint" would, of course, apply to Barnes's decay theory that permits of no reversals. (Jacobs' more recent book, published in 1984, bears the title <u>Reversals of the Earth's Magnetic Field</u>.)

I noticed in Barnes's ICR Impact Series article (1981) that his reference to J.A. Jacobs' 1963 book, titled <u>The Earth's Core and Magnetism</u>, published by Macmillan, omits the year of publication. Again, we have an example of a creationist practice that would not be permissible in the community of mainstream scientists--to cite statements

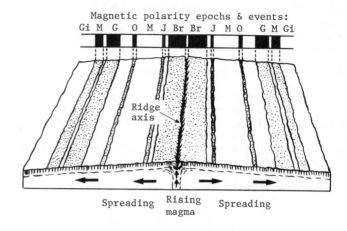

Figure 19.4 Schematic diagram of development of symmetrical pattern of magnetic polarity belts in ocean-floor basalts during seafloor spreading. (Copyright © 1971 by Arthur N. Strahler.)

made at a time when pertinent evidence had not been discovered, ignoring the current "state of the art" obvious in more recently published research. Creation scientists of the ICR, I feel sure, are well acquainted with all scientific publications relating to their specialties. All graduate university students pursuing the M.A. or Ph.D. degree are expected to search the published scientific literature and be cognizant of the very latest announcements of research findings and their bearing on current hypotheses. To omit such recent publications from consideration, whether through lack of diligence or for a purpose, would bring a sharp rebuke from the supervising professor and possible termination of the degree candidacy.

I recommend that you read Dalrymple's full discussion of the evidence for paleomagnetic polarity and its reversals (Dalrymple, 1983, pp. 126-28). It shows that three completely different geologic processes record the same sequence of polarity reversals with the same ages. Evidence from rocks of different types and from diverse global locations is in remarkably close agreement on the reversal schedule. Measurements have been made on thousands of rock specimens. Methods have been perfected to eliminate spurious data possibly caused by external factors, such as exposure to lightning or to chemical alteration. Particularly impressive is corroboration of the magnetic polarity observed in igneous rocks by an entirely different method of preservation of magnetic history. Sediments accumulated on the seafloor at great depths, where they remain undisturbed, contain magnetized mineral particles that have oriented themselves according to the magnetic force lines prevailing at the time the particles settled into place. This process takes place in a cold environment, totally unlike that of cooling of a molten magma, yet the program of polarity reversals it yields substantiates that obtained from lavas in other locations.

A Creationist's Rebuttal

Barnes, in his 1983 rebuttal to Dalrymple's arguments, continues to attack the dynamo theory on the grounds that as yet an exact model of the dynamo mechanism is not available. The general thrust of the argument is a familiar one: Because science has not yet solved all its problems, those of its hypotheses as yet uncorroborated by empirical evidence must be false. This is clearly an antiscience position, and it is one expressed repeatedly by the purveyors of pseudoscience. Actually, Barnes can cite no evidence in observation to support his theory of an original supply of magnetic field energy (1983, p. iii), but this does not deter him from asserting that the source of energy is clearly identified. Identified it is, as energy furnished by the Creator a few thousand year ago.

Barnes says the "evolutionists" have nothing but faith to support the dynamo hypothesis (1983, p. ii). Here another familiar creationist theme creeps in: Science depends on faith, therefore science is not actually science, but instead is religion.

Barnes, nearing the conclusion of his rebuttal, claims the dynamo theory "has no substantive theoretical basis" (p. iv). I am a little reluctant to apply the adjective "absurd" to that assertion, but it seems inescapable here. The dynamo theory makes use of a great body of the most fundamental and universal principles of physical science by which mainstream science has been guided and constrained for decades, and even for centuries. The substantive theoretical basis for the dynamo theory is massive and formidable, making use as it does of everything that is experimentally corroborated about fluxes of heat, electricity, and magnetism and the behavior of fluids. In his dismissal of the dynamo theory Barnes is dismissing almost the entire substance of modern science. This cavalier dismissal of a great accumulation of human knowledge about the nature of the real world seems to come rather easily to the funda- mentalist creationists, and the reason is quite obvious: Recent creationism really requires no science whatsoever, because it relies entirely upon the work of a Creator. With no need to understand the origin of the universe and everything in it, and no recognition of a cosmic history longer than 10,000 years, science is irrelevant and immaterial to a creationist. Mainstream science does not need to be debated and its hypotheses are patently false when they contravene the literal meaning of even one word in the book of Genesis.

We must now return to the carbon-14 method of radiometric dating, mentioned briefly in Chapter 17, because there is a connection between that radioisotope of carbon, produced high in the atmosphere, and the strength of the earth's magnetic field--the external field of force lines that arch out into space. We are not yet done with reversals of magnetic polarity, or with the creationists' claim that the earth cannot be more than 10,000 years old.

Radiocarbon Dating

Prior to 1950, archaeologists, anthropologists, and geologists could make only educated guesses as to the ages of plant and animal remains, artifacts, and ice-laid sediment deposits of the late Ice Age and the Holocene Epoch that has followed. Other radiometric methods we have described cannot be applied to such young organic compounds and sedimentary materials. About 1950, there came a great scientific breakthrough when a new radiometric method was developed by Willard F. Libby of the Institute for Nuclear Studies of the University of Chicago. His work won him a Nobel prize.

Libby's method made use of a radioisotope of carbon, carbon-14, that originates in the earth's upper atmosphere. At levels above 16 km, atoms of ordinary nitrogen (nitrogen-14) are subject to bombardment by neutrons created by highly energetic cosmic particles (cosmic rays) penetrating the atmosphere from outer space. Upon being struck, an atom of nitrogen-14 absorbs the impacting neutron and emits a proton. The nitrogen atom is thus transformed into carbon-14, which quickly combines with oxygen to form carbon dioxide (CO_2). Carbon-14 is radioactive and decays back to nitrogen-14 by emitting beta particles. The half-life of carbon-14 is 5730 ± 40 years.

The rate of production of carbon-14 in the upper atmosphere is first assumed to be constant. If so, atmospheric carbon dioxide that is taken up by plants and animals will contain a fixed proportion of carbon-14

relative to the total amount of ordinary carbon (carbon-12). From an initial point in time marked by the death of the organism, the proportion of carbon-14 in the organic structure declines steadily, following the exponential curve of decline. By making precision measurements of the extremely small amounts of carbon-14 in a sample of organic matter, the age in years of that matter can be estimated to within a fairly small percentage of error. There are about one trillion (10^{12}) atoms of C-12 to one of C-14. The very short half-life of carbon-14 makes it an excellent tool for age determinations in the last few tens of thousands of years. On the other hand, the uncertainty of measurement increases at such a rate that the present limit of usefulness is about -40,000 y.

Libby began by making age determinations of such materials as charcoal, shells, wood, and peat derived from archaeological sites and glacial deposits. Materials whose age was documented from other historical records served as a check upon the accuracy of the method. By 1952, Libby's laboratory had made age determinations of a large number of carefully selected samples. Other laboratories were soon set up, and the radiocarbon method became established as one of the most important research tools in geological and archaeological research.

A good example of the use of the radiocarbon method was to determine the ages of some tree trunks felled by the rapid advance of the ice sheet at a locality in Wisconsin. A forest had grown up in a brief mild period in late glacial time, but the advancing ice broke off the tree trunks and incorporated them into dense red clay. Radiocarbon analysis of the logs gave an age close to -12,000 y. This was a record of the last readvance of glacial ice before it finally disappeared from the region.

Dendrochronology and Carbon-14 Ages

As the years passed, discrepancies began to appear in the radiocarbon dates when they were compared with dates arrived at by other means. An alternate method of dating makes use of tree rings exposed in a sample cut at right angles to the tree trunk. Each growth ring represents one calendar year, and the age of a tree is obtained by simply counting the rings. A given tree trunk shows distinctive sequences of wide and narrow rings that are controlled by variations in climate from year to year. Trees whose life spans overlapped in time can be correlated by matching the distinctive ring sequences. In this way, logs and timbers used in ancient dwellings can be dated accurately. The method of precise tree-ring dating is called dendrochronology, developed by an American astronomer, A. E. Douglass, and extended by a number of collaborators and successors. Another approach has been to count the growth rings of long-lived species of trees and compare the ring counts with carbon-14 dates of the same wood. One tree in particular, the bristlecone pine (Pinus aristata), has an extremely long life span, some specimens living to an age of just over 5000 years.

The trunks of dead bristlecone pines could also be used for tree-ring analysis, and this made possible the extension of the total chronology to about 9000 years. (For highly readable reviews of this subject, see Ralph and Michael, 1974; Hitch, 1982.)

By the early 1970s, the carbon-14 ages had been obtained for the same tree rings used to establish the dendrochronology to -7300 y. Assuming that the tree-ring count gave an absolute age, it was clear that the C-14 age was subject to an error that first decreased with time to zero, then increased to a small maximum, again decreased to zero, and then again increased to reach a large error. Figure 19.5 is a graph on which the age difference is plotted against time as curve A. (Data of Ralph and Michael, 1974, Figure 2.) The tree-ring age (dendrochronological age) is subtracted from the C-14 age. Between 1700 and 1500 A.D. (-300 to -500 y.) the

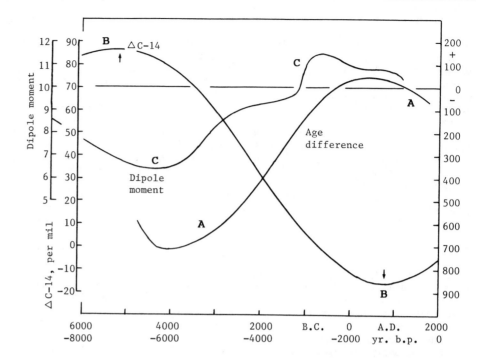

Figure 19.5 Relationship of C-14/ dendrochron age differences to relative atmospheric C-14 content and strength of the earth's dipole field. See text for data sources. (A. N. Strahler.)

C-14 ages run 50 to 100 years too low, but with considerable irregularity in trend. By about 1300 A.D. (-700 y.) the two sets of ages agree. Then the C-14 ages gradually give ages too large by some 200 to 400 years. This trend then reverses and by about 400 B.C. (-2400 y.) the two ages are again in agreement. From this point on, the C-14 ages begin to develop an error in which they are too small. The error increases steadily to about -6000 y., when the discrepancy amounts to about 700 years. The trend then reverses, but here the information runs out. The smooth curve on the graph has been fitted to the observed data, but the deviations from the individual observations are small throughout much of the time-span, so that the reliability of the cycle shown is high. The C-14 ages are calculated from the observed ratio of C-14 to ordinary carbon, C-12. Therefore a plot of the C-14/C-12 ratio against time should give a similar but inverse curve, showing the same cycle. The inverse curve is shown by curve <u>B</u> in Figure 19.5 (data of Suess, 1982). It extends farther back in time because it uses samples from rings in trunks of dead trees. The curve shown is a sine wave fitted to the individual observations, which fluctuate with a short-period cycle that may be an effect of the sunspot cycle. The long period shown by the sine curve is on the order of 6,000 years from maximum to minimum (12,000 y. per full cycle).

Effect of Variations in the Magnetic Field

What might be the cause of the 12,000-year cycle of variation in ratio of carbon-14 to carbon-12? Variation in the dipole field strength of the earth's magnetic field seems like a good candidate for the cause of the cycle. We can reason that when the dipole field increases, the magnetosphere is strengthened and is more effective in shielding the upper atmosphere from cosmic rays, and, under those conditions, C-14 is produced at a lower rate. When the dipole field weakens, increased bombardment by cosmic rays leads to a higher rate of production of C-14. Figure 19.5, curve <u>C</u>, shows the dipole field strength calculated from measurements of magnetism of lava flows and of artifacts such as pottery and bricks, whose age can be determined. The curve is roughly fitted to mean values determined about every 500 to 1,000 years (data of McElhinny and Senanyke, 1982, in Merrill and McElhinny, 1983, p. 105.). The curve is roughly 180 degrees out of phase with the C-14 curve.

The strength of the cause/effect relationship we have recognized lies in the nature of the two processes that are being correlated. The fluctuation of the earth's magnetic field is controlled from deep within the earth by fluid motions in the core, and these cannot be influenced by the cosmic ray bombardment from outer space. Neither can the amount of C-14 in the atmosphere be seen as a control mechanism over the magnetic field strength. Furthermore, at this time at least, one cannot postulate that the two processes are in turn the effects of a third independent physical cause that directly controls them both.

Creationists' Arguments against Radiocarbon Dating

As we have seen, the scientific creationists reject all evidence of a cyclic variation in the earth's magnetic field strength. No reason is given for that rejection. Neither do the creationists offer any evidence favoring an exponential decay of the magnetic field; they simply assert that such is the case. They have no scientific working hypothesis of an initial source of magnetism that went into effect about 10,000 years ago, more or less; the hypothesis they offer is a religious concept.

On the subject of carbon-14 as an indicator of the earth's early age, the creation scientists have a well-prepared set of arguments to make. The subject is given seven pages in their public-school textbook, <u>Scientific Creationism</u> (Morris, 1974, pp. 161-67). For answers to the creationists' arguments, see a paper by Christopher Gregory Weber (1982), published in <u>Creation/Evolution</u>. The ICR text states that the radiocarbon dating method makes use of some doubtful assumptions, some of which are serious enough to place doubt on ages greater than about 2000 to 3000 years. Four points are discussed:

1. Creation scientists point out that living mollusks have been found whose shells show C-14 ages as great as 2300 years. Reference here is to a paper by two mainstream scientific researchers who published their results in the journal, <u>Science</u>, under the title of "Radiocarbon Dating; Fictitious Results with Mollusk Shells." (Keith and Anderson, 1963). This result is absurd on the face of it and indicates to the creationists that many living systems are not in equilibrium for C-14 exchange. The authors of the paper present evidence that the mussels they sampled obtained much of the carbon used in their shells from limestone (calcium carbonate) in

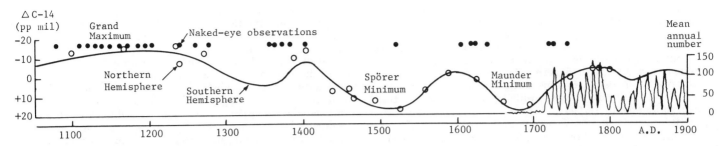

Figure 19.6 Relationship between sunspot frequency and deviation of carbon-14 from its normal value. The heavy line and open circles show deviation of C-14. Black dots show occurrences of sunspots recorded from naked-eye observation. Fine-line peaked graph at right is modern record of sunspot frequency, scaled in number of sunspots per year. (From J. A. Eddy, <u>Science</u>, vol. 192, p. 1195, Figure 5a. Copyright © 1976 by the American Association for the Advancement of Science. Used by permission.)

their habitat. The C-14 content of the limestone carbon is very low because of the great age of the rock. The authors intended to point out a source of major error in the dating method that could be avoided by having made sure that the carbon used by an invertebrate animal is of recent atmospheric origin. In this respect, plants that photosynthesize carbohydrates are reliable sources of ages because they withdraw carbon directly from the atmosphere as gaseous carbon dioxide.

2. Creationists say that the decay rate of carbon-14 may not have been constant in the past. This is the same argument presented by the creationists against the other forms of radiometric age determination, such as the uranium-lead decay series. The creationist text refers to an abstract of a paper presented to the American Chemical Society in 1971, in which the speaker (J. L. Anderson) described experiments allegedly showing that C-14 decay rates could have varied in the past to a degree that would render most radiocarbon ages invalid (Morris, 1974, p. 162). It is difficult, if not impossible, to evaluate such a claim, where no details of the experiments are made available, and no mechanism is suggested for varying decay rates. To date, the mainstream scientists who have been engaged in active research on the radiocarbon method have found no reason to question the basic theory of nuclear physics in which the decay constant of all natural radioisotopes is held to be independent of pressure, temperature, or the chemical state of the isotope. That radiocarbon ages agree so closely with tree-ring counts over at least 8000 years, when the observed magnetic effect upon the production rate of C-14 is taken into account, suggests that the decay constant itself can be assumed to be reliable.

3. Creationists claim that the amount of natural carbon present in the atmosphere as carbon dioxide may have varied in the past, and thus the ratio of C-14 to C-12, which is the basis on which the age of a sample is determined, will be affected by the total concentration of CO_2 in the atmosphere. Let us consider this effect.

Atmospheric C-14 comes into existence almost entirely from impacts on nitrogen atoms by neutrons produced by galactic cosmic rays (Lingenfelter, 1963, pp. 36-37). The production rate of C-14 can remain invariant only if both the cosmic ray flux and density of nitrogen molecules in the upper atmosphere remain constant. The independence of atmospheric molecular nitrogen concentration from CO_2 concentration seems a reasonable assumption.

The concentration of atmospheric CO_2 is known to have increased by about 13 percent by volume in the past 120 years, and this is attributed to the steadily increased combustion of fossil fuels. The contribution of C-14 from such combustion would be so small (because of the great age of the fossil fuels) as to be neglected as a source of C-14. Therefore, we would expect the ratio of C-14 to ordinary carbon to have diminished in the past 120 years from the increase in CO_2, the production of C-14 being

assumed constant. The diluting effect of addition of C-14 free CO_2 from fossil fuel burning is known as the <u>Suess effect</u>.

A group of scientists of the Lamont-Doherty Geological Observatory of Columbia University, headed by geochemist Wallace S. Broecker, reported in 1979: "The change in the atmospheric C-14/ordinary carbon ratio caused by this dilution has been recorded by tree rings. The decrease between 1850 and 1950 was about 2.4 percent" (Broecker et al., 1979, pp. 412-13). They attribute the smallness of the percentage (about one-fifth as great as expected) to mixing of the fossil-fuel carbon with a much larger carbon reservoir of the biosphere and the ocean. After 1950, large increases in atmospheric C-14 were recorded as a result of nuclear weapons testing, ending further possibility of detecting the effect on C-14 of increasing CO_2 from burning of fossil fuels.

Fluctuations in C-14 ratios observed for the past 500 years seem to fit quite closely with major sunspot deficiencies, such as the Maunder minimum, 1645-1715. The C-14 ratio increased substantially during that interval in which sunspots virtually disappeared. During that same Maunder period auroras were also extremely rare. An increase of the C-14 ratio coincided with the Maunder minimum and with an earlier minimum period (1400-1600). These effects are shown in Figure 19.6. Lack of solar activity during these extreme solar minimum periods is inversely related to the intensity of cosmic ray activity in the upper atmosphere. The same effect is observed on a smaller scale in connection with the 11-year sunspot cycle that has prevailed with considerable regularity since about 1720, following the Maunder minimum. Stuiver and Quay explain: "During intervals of low sunspot activity the magnetic shielding properties of the solar wind are such that a larger galactic cosmic-ray flux arrives in the upper atmosphere, whereas cosmic-ray fluxes are lower during periods when sunspot numbers are higher" (1980, p. 11).

Evidence of past history of C-14 concentration in the atmosphere is now available through the past 22,000 years, using ages of lake sediments in which organic carbon compounds are preserved. Reporting before a 1976 conference on past climates, Professor Minze Stuiver of the University of Washington found that magnetic ages of the lake sediments remained within 500 years of the radiocarbon ages throughout the entire period. He reported that the concentration of C-14 in the atmosphere during that long interval did not vary by more than 10 percent (Stuiver, 1976, p. 835).

Thus, the available evidence is sufficient to validate the radiocarbon method of age determination within an error of about 10 percent for twice as long a period as the creation scenario calls for. No evidence has been put forward by the scientific creationists to invalidate any of the radiometric methods or the chronology of magnetic polarity reversals.

4. Creationists claim that mainstream scientists are

making an unwarranted assumption that the ratio of C-14 to C-12 is in a steady state, in which the worldwide production of C-14 is balanced by worldwide decay of C-14 (Morris, 1974, p. 164). They point out that a period of about 30,000 years is required for steady state to be attained from an imagined zero point in time in which cosmic radiation began to impact the atmosphere. This is much too long a time to agree with Divine Creation of the universe at about -10,000 y. The Bible requires that galactic cosmic radiation was zero until the instant of creation, so it follows that the rate of production of C-14 must presently be much larger than the rate of decay. After all, 10,000 years is less than two half-lives of C-14, so no other conclusion could be tenable. The creation scientists try to prove a young age of the earth and universe by taking the presently observed rates of C-14 production and decay and extrapolating them back in time when the quantity of C-14 was zero. Creation scientist Melvin Cook (1968) has carried out this calculation and reaches the conclusion that time-zero was -10,000 y., which is the age of the atmosphere and probably for the earth itself (Morris, 1974a, p. 165). Another creation scientist, Robert L. Whitelaw (1968) uses an even larger ratio of C-14 to C-12, yielding a zero value at -5000 y. (Morris, 1974a, p. 166).

One authority from mainstream science, cited by Morris in support of the greater production than decay of C-14, is Richard E. Lingenfelter (Morris, 1975, p. 164). Lingenfelter states:

> On comparing the calculated value of the carbon 14 production rate, averaged over the last ten solar cycles, of 2.50 ± 0.50 C^{14} atoms per square centimeter per second with the most recent estimates of the decay rate of 1.8 ± 0.2 . . . and 1.9 ± 0.2 . . ., there is strong indication, despite the large errors, that the present natural production rate exceeds the natural decay rate by as much as 25 per cent. (1963, p. 51)

(Note: Citations of sources given by Lingenfelter have been deleted from the above paragraph).

What Morris does not tell the reader is that Lingenfelter continues his discussion by attributing the discrepancy between production and decay rates to possible variations in the earth's magnetic field, a subject we have explored in earlier pages. Lingenfelter's paper was written in 1963, before the cycles of C-14 variation we described had been fully documented. The point is that fluctuations in the rate of C-14 production mean that at times the production rate will exceed the decay rate, while at other times the decay rate will be the larger. As curves A and C in Figure 19.5 show, before about -3000 y., C-14 was decaying faster than it was being

formed, with the result that the C-14 ages are too young. The creation scientists ignore this newer information, which is readily available to the public in published journals; it was available in 1974 when the creationist textbook was published. They have made their calculations from the unsupportable assumption that the C-14 production rate has always been greater than the decay rate by the percentage observed today. That is an untenable assumption in the light of evidence to the contrary. Creationists have used the same form of unsupported extrapolation that we found in creation scientist Barnes's calculation of a limiting age of the earth by assumed constant decay rate of the earth's magnetic field.

It is interesting to see that the scientific creationists accept the authenticity of radiocarbon ages going back as far as about -3000 y. (Morris, 1974a, p. 167). In this period, they say, the historical methods of establishing ages are in general agreement with both the mainstream scientists' "equilibrium" model and their own "non-equilibrium" model. The point is that they are looking for confirmation of the ages assigned to biblical and other historical records of Jewish history of the first millennium B.C.

The creationists are interested in fitting radiocarbon age determinations into their cataclysmic model of events following creation (Morris, 1974a, p. 163). The "cataclysm" is, of course, the Flood of Noah. Prior to the Flood, the proportion of land surface to water surface was very great, so that "tremendous amounts" of vegetation were produced, as shown by the world's coal deposits. Of course, plants that produced that coal would have contained very little C-14, because C-14 production had scarcely begun; by now the coal will show no measurable amounts of C-14 (which is indeed the case for coals of Mesozoic and Paleozoic ages). Following the Flood, the land surfaces were small in extent (because of high water level and presence of ice sheets) and were denuded of vegetation. That being the case, the atmospheric C-12 content would have been small, with the result that radiocarbon ages of materials produced in that period would appear younger than true ages. This is the creationists' way of accounting for the observed discrepancy between radiocarbon ages and bristlecone ages between -6000 and -4000 y. Here, again, we see the creationists' acceptance of mainstream science information, when it fits their picture.

Credit

1. From G. Brent Dalrymple, 1983, Can the Earth Be Dated from Decay of Its Magnetic Field? Jour. of Geological Education, vol. 31, pp. 124-33. Used by permission of the author and publisher.

PART IV

Two Views of Geology and Crustal History

Introduction

The full scenario of fundamentalist creationism comes in two great acts. Act 1, which we have now reviewed in depth, is the six-day creation, which was perfect in every detail, and after which God rested. Act 2 is the Noachian Deluge, or Flood of Noah. This, too, was God's act, but it was punitive and destructive. The Flood was devoid of creation in the sense of the formation of wholly new forms of matter and energy, but, as a cataclysmic event, matter was rearranged, redistributed, and restructured on an enormous scale--or at least, such is the creationists' claim. The new geologic arrangements and structures are thus derivative, or secondary features, with respect to creation. These Flood-produced forms are, of course, limited entirely to the single planet Earth in this ultramyopic and wholly anthropocentric view of the cosmos.

The great significance of the Flood to natural science lies in the vast changes it accomplished. Creationists consider that all rocks that contain fossils are products of the Flood. Primitive life forms (algae) are recognized in rocks of Precambrian age, and these very ancient rocks are therefore recognized by the creationists as products of the Flood. For example, a technical article appearing in 1984 in the creationist publication Ex Nihilo recognizes as Flood products a great mass of severely deformed and altered (metamorphosed) rocks of the Mount Isa ore body in Queensland. One of the rock units, originally a shale formation, contains abundant remains of microorganisms, interpreted as blue-green algae (Snelling, 1984). Mainstream geology places the age of these rocks at more than one and one-half billion years. Thus all geological materials and structures that are observed in the Mount Isa locality must, according to the creationists, have been formed in about one year, that particular year being 2350 B.C. (-4,350 years before present, approximately).

If some gnarled Precambrian rocks in a remote spot in Australia carry little meaning to you, consider a scene closer to home--the Grand Canyon of the Colorado River. Standing on the South Rim you stare down into a mile-deep chasm in which there lie exposed some 900 m (3,000 ft.) of horizontal strata of Paleozoic age--all younger than 570 m.y., according to mainstream geologists. Fossils are found in strata of Cambrian age near the base of the pile and in various formations right up to the limestone caprock. Your creationist companion, should there be one at your side, will tell you that the whole sedimentary accumulation occurred in one year. As if that were not enough to stagger the imagination, a look farther to the right (east) into the depths of the canyon reveals a great wedge of tilted strata, and these contain some fossils of

very primitive types, so we must add about another 600 m (2,000 ft) of strata to the Flood accumulation. But wait, there's more to it than that! All geologists, creationists among them, will agree that the older strata of the wedge (the Algonkian rocks) were broken by major faults and tilted up at an angle after they were deposited, forming blocklike mountains. Then the block mountains were beveled by erosion and the region reduced to a low plain, which later sank below sea level to receive the younger marine Paleozoic layers. All of this crustal breakage (diastrophism) and continental erosion (denudation) has to fit into the one year available for the job. To make things more bizarre, take a journey due north into Utah. You will climb step by step over the exposed edges of younger strata that make up the walls of Zion Canyon and, finally, the Pink Cliffs of Bryce Canyon. This adds at least another 2,500 m (8,000 ft) of strata. Our total pile of strata is now up to 4,000 m, or 4 km (13,000 ft).

But how about those enormous canyons--Grand Canyon, Zion Canyon, and many others? If they were not started by river action before the waters of the Flood receded, they must have been carved in no more than 4,350 years. Photographs and detailed drawings of the walls of Grand Canyon and its inner gorge were made over a century ago (Dutton, 1882). In that time scarcely any change can be discerned in the details of the cliffs, pinnacles, and clefts, although here and there a small rock mass has fallen away. The same gigantic rapids that Major Powell and his party endured in their wooden hulls in 1869 are here today, scarcely changed. At this pace, does it make sense that the mile-deep gorge was carved in forty centuries? Your creationist companion urges that you believe it, indeed, that you must believe it.

Clearly, Flood geology, as we shall refer to the creationist view of geology, is up against a problem of awesome proportions: How can all of those things have happened in so short a time, affecting vast areas of the continents of the globe?

Creation scientists are well versed in the literature of geology. To follow their arguments and the counter-arguments of mainstream geologists, you need a good grounding in modern geology, including plate tectonics--the new paradigm unifying geology and earth history under a single grand hypothesis of the dynamics of the earth's outermost rigid shell, or lithosphere. If you have not had a basic geology course within the past decade or two, there is much that is new to be learned. To this end we begin Part 4 with a brief overview of modern geology from the standpoint of mainstream science.

Chapter 20

How Mainstream Geology Views Earth
Structure and Dynamics

Like the other three inner planets of our solar system--Mercury, Venus, and Mars--planet Earth is a spherical ball of rock consisting largely of compounds of the elements oxygen, silicon, magnesium, and iron.

The average radius of the earth sphere is about 6400 km (more exactly, 6370 km). Nearly all of this radius is taken up with two zones or regions: a core and a mantle (Figure 20.1). The spherical inner core, with a radius of 3470 km, is thought to consist of metallic iron mixed with a small proportion of nickel, and perhaps with a very small proportion of sulfur. The outer two-thirds of the core is in a liquid state, intensely hot and under enormous confining pressure. There is evidence that the inner portion of the core is solid. Surrounding the core is the mantle, an enormous rock shell nearly 2900 km thick. Thus, the mantle occupies nearly half the earth's radius. The dense mantle rock consists mostly of the elements silicon, oxygen, iron, and magnesium in a solid state, although it has a very high temperature and is under great confining pressure. [1]

Extremely thin in comparison with the mantle is the outermost earth shell, the crust, ranging in thickness from about 5 to 70 km. If spread evenly over the entire earth, its average thickness would be about 17 km. Such a thin layer cannot be shown to true scale on our cutaway drawing of the earth, Figure 20.1. Rock of the crust differs from that of the mantle in containing substantial proportions of the lighter metals: aluminum, sodium, calcium, and potassium, along with abundant silicon and oxygen. For this reason, crustal rock is less dense than mantle rock, just as a light metal such as aluminum is less dense than a heavy metal such as iron. Mantle rock is, in turn, less dense than the iron-nickel mixture of the core. Thus, our first generalization about the planetary structure is that it consists of shells that decrease in density outward from the center.

The three earth shells we have described--core, mantle, and crust--are different in chemical composition, the primary basis on which each is defined. Superimposed on the outer shells (the outer mantle and crust) are layers defined according to the physical state of the rock --whether the rock is hard and brittle or soft and plastic. Depending largely on temperature, the same rock can be in either of these conditions, just as a bar of cast iron is strong, hard, and brittle when cold, but becomes soft and plastic when heated to a high temperature.

Lithosphere and Asthenosphere

The earth has an outer layer of hard, brittle rock known as lithosphere. This layer includes the entire crust and a portion of the upper mantle. Thickness of the lithosphere ranges from under 50 to over 125 km, with a rough average of perhaps 75 km, as shown in Figure 20.2. Below the lithosphere lies the asthenosphere, a soft layer in the upper mantle. The word is derived from the Greek root asthenes, meaning "weak." The asthenosphere is in a soft condition because its temperature is high-- about 1400 C--and close to its melting point. The rock

behaves much like an ingot of white-hot iron that will hold its shape when resting on a flat surface, but is easily formed into bars or sheets when squeezed between rollers.

Temperature increases steadily inward from the earth's surface, so that the change from lithosphere to asthenosphere is gradual, rather than abrupt. We know from the behavior of earthquake waves as they travel through the outer mantle that the asthenosphere extends to a depth of about 300 km, below which the strength of the mantle rock begins to increase. The weakest portion of the asthenosphere lies at a depth of roughly 200 km.

The important concept you should derive from this information is that the rigid, brittle lithosphere forms a hard shell capable of moving bodily over the soft, plastic asthenosphere. This motion is exceedingly slow and is distributed through a thickness of many tens of kilometers. A simple model of the motion is a deck of playing cards resting on a tabletop, its top ten cards glued together to form a solid block, representing the lithosphere (Figure 20.3). A horizontal force against the edge of the deck will move the upper block horizontally, while the motion becomes a slipping between the free cards beneath it. The card at the bottom of the deck will

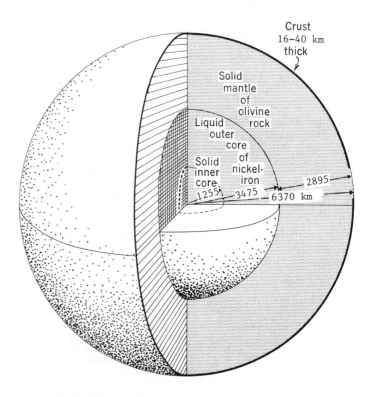

Figure 20.1 The earth's core, mantle, and crust. The crust is much too thin to be shown to true scale. (Copyright © 1975 by Arthur N. Strahler.)

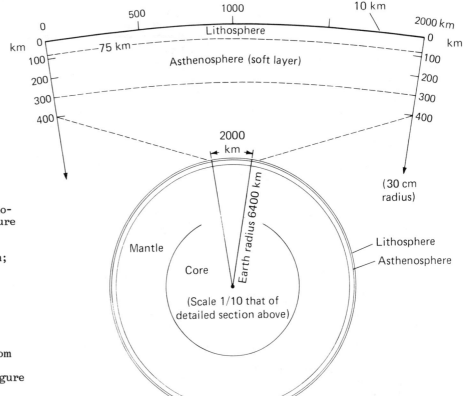

Figure 20.2 The lithosphere and astheno- sphere drawn to true scale. The curvature of the upper diagram fits a circle 30 cm in radius. The black line at the top is scaled to represent a thickness of 10 km; it will contain about 98 percent of the earth's surface features, including the ocean floors and high mountains and plateaus. The complete circle below is drawn on a scale one-tenth as great as the upper diagram. Seen in true scale, the lithosphere is a very thin shell compared with the mantle and core. (From A. N. and A. H. Strahler, Elements of Physical Geography, 3d ed., p. 215, Figure 12.3. Copyright © 1984 by John Wiley & Sons, Inc. Reprinted by permission of John Wiley & Sons, Inc.)

remain fixed to the table. Slippage on a great number of very thin parallel layers is described as underlineshearing. Layers involved in shearing are no thicker than atoms or molecules, so that we could not actually see one layer gliding over another. The entire asthenosphere seems to move by the kind of flow we observe in a tacky liquid, like thick syrup. But the behavior of earthquake waves shows that the asthenosphere is not a true liquid. It is difficult for us to understand how rock behaves under such conditions as exist in the asthenosphere, where temperatures are very high and the rock is under enormous confining pressure from the weight of the overlying rock.

Lithospheric Plates

If the earth consisted of a perfectly spherical lithosphere with no flaws or fractures, it is conceivable that the entire lithosphere could move as a whole with respect to the rest of the earth (middle and lower mantle and core). If that kind of motion actually occurred, it would carry with it as a unit all of the earth's continents and ocean basins. We can imagine a situation in which western Europe might be moving toward the earth's north pole, so that there would come a time when the north pole would be situated in Ireland. At the same time, New Zealand would be nearing the south pole, while the city of Cape Town, South Africa, would be approaching the equator.

It is more realistic to suppose that the lithosphere has tended to break into large sections, each on the order of size of a whole continent or a whole ocean basin. There is strong evidence to support the statement that the lithosphere is quite thin in some areas--notably beneath the deep ocean floors--and quite thick in other areas-- notably under the continents. Thin lithosphere might be expected to fracture quite easily, while thick lithosphere would tend to hold together much better.

In reality, breakup of the lithosphere has formed a number of lithospheric underlineplates, each of which has some freedom to move independently of the plates around it. Like great slabs of floating ice on the polar sea, lithospheric plates can be seen to be pulling apart in some places and colliding in others. In the case of floating ice, two plates pulling apart leave a widening gap of exposed water. This gap can be filled with new ice as the top layer of the exposed water freezes. When two ice plates collide, they often come together in crushing impacts that raise great welts, called "pressure ridges." These welts remind us of mountain chains found along the margins of continents. When collision occurs, one ice plate might be expected to be forced down beneath an adjacent plate. The down-diving plate would melt and finally merge with the surrounding water.

Because ice is less dense than water, an ice plate floats easily and resists being pushed down below the water surface. Thus, it would not be wise to carry the floating ice-plate model any further in search of a true

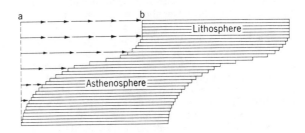

Figure 20.3 The shearing motion of soft rock in the asthenosphere resembles the slip of cards in a deck. We imagine the cards in the upper part of the deck to be glued together so as to move as a solid plate, representing the lithosphere. (Copyright © 1976 by Arthur N. Strahler.)

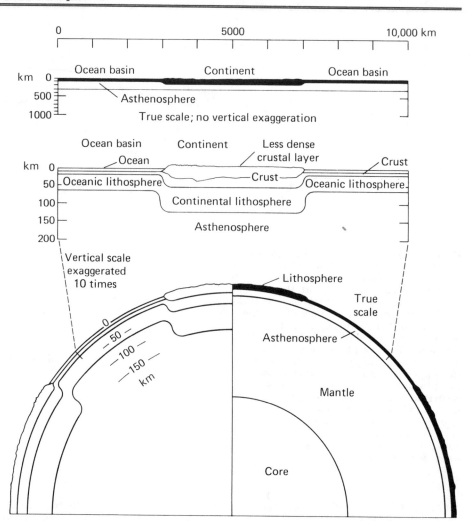

Figure 20.4 Continental lithosphere is thicker than oceanic lithosphere. The crust of the continents is also thicker than the crust of the oceans. For purposes of illustration, the continental lithosphere is shown as being twice as thick as the oceanic lithosphere. (Copyright © 1981 by Arthur N. Strahler.)

model of lithospheric plates. Unlike ice, a given kind of rock in the solid state is denser than the same rock in the molten condition, when the two are compared at the same pressure.

Continental Lithosphere and Oceanic Lithosphere

We know that continents rise high in elevation above sea level--if they did not, they would not be continents. We know that the floors of the ocean basins lie at an average depth of some 4 km below sea level. In this case, our zero reference is the surface of the ocean, which tends to form a nearly perfect sphere. Figure 20.4 shows the continental surfaces rising above sea level and the ocean basin floors below that level. Some good scientific evidence has now established that the lithosphere under the continents is thicker than the lithosphere under the ocean basins. In Figure 20.4 we show continents to be underlain by a thick layer of continental lithosphere and the ocean basins by a thin layer of oceanic lithosphere.

Besides being thicker, continental lithosphere is also different in chemical composition from oceanic lithosphere. Continental lithosphere includes a special upper crustal layer of rock having lower-than-average density (Figure 20.4). Continental lithosphere is, therefore, more buoyant than oceanic lithosphere, and this property contributes to the higher upper surface of a continent. To explain this principle in simple terms, consider what happens if we take two wooden blocks of exactly the same size and shape, one a block of oak (a dense wood), the other of balsa (a wood of very low density). Placed in water, the balsa block floats much higher than the oak block. Our schematic diagram of continental lithosphere (Figure 20.4)

includes an upper crustal layer of lower rock density. Notice that this crustal layer is missing from oceanic lithosphere. After we have completed a review of igneous rocks, it will be a simple matter to assign rock names to the layers that make up the lithosphere.

Plate Tectonics

Our next step is to visualize the lithosphere broken into lithospheric plates that move with respect to one another. Figure 20.5 shows some of the major features of plate interactions. In Diagram B we can see two plates, X and Y, both made up of oceanic lithosphere, pulling apart along a common plate boundary. This activity tends to create a gaping crack in the crust, but molten rock from the mantle below rises continually to fill the crack. The rising molten rock, called magma, solidifies in the crack and is added to the two edges of the spreading lithospheric plates. In this way, new solid lithosphere is continually formed. At the distant boundary of oceanic plate Y, the oceanic lithosphere is shown to be pushing against a thick mass of continental lithosphere, plate Z. Because of its greater crustal buoyancy, the continental plate remains in place, while the thinner, denser oceanic plate bends down and plunges into the asthenosphere. The process of downsinking of one plate beneath the edge of another is called subduction.

The leading edge of the descending plate is cooler than the surrounding asthenosphere--cool enough that this descending slab of brittle rock is denser than its surrounding asthenosphere. Consequently, once subduction has begun, the slab can be said to "sink under its own weight." Gradually, however, the slab is

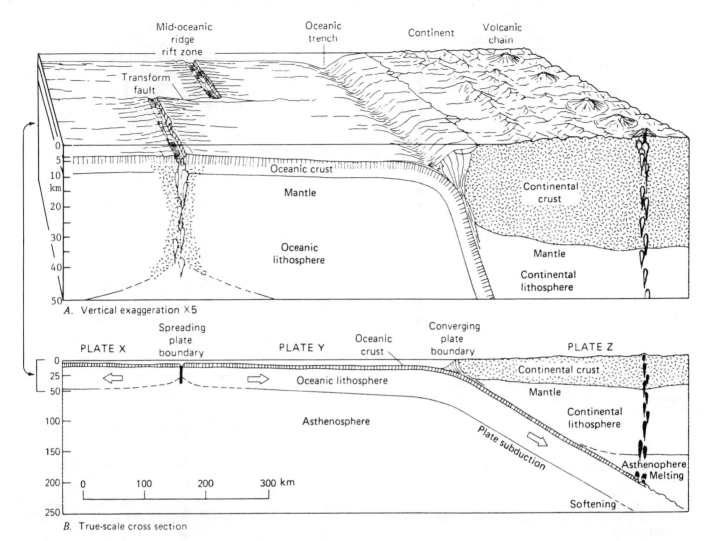

A. Vertical exaggeration ×5

B. True-scale cross section

Figure 20.5 Schematic cross sections showing some of the important elements of plate tectonics. Diagram A is greatly exaggerated in vertical scale so as to emphasize surface and crustal features. Only the uppermost 30 km is shown. Diagram B is drawn to true scale and shows conditions to a depth of 250 km. Here, the actual relationships between lithospheric plates can be examined, but surface features can scarcely be shown. (Copyright © 1981 by Arthur N. Strahler.)

heated by the surrounding asthenosphere and thus it eventually softens. The underportion, which is mantle rock in composition, simply reverts to asthenosphere as it softens. The thin upper crust, formed of less dense mineral matter, may actually melt and become magma, which tends to rise because it is less dense than the surrounding material. Figure 20.5 <u>A</u> shows some magma pockets formed from the upper edge of the slab. They are pictured as rising like hot-air balloons through the overlying continental lithosphere. Reaching the earth's surface, quantities of this magma build volcanoes, which tend to form a chain parallel with the deep <u>oceanic trench</u> that marks the line of descent of the oceanic plate.

Viewed as a unit, plate <u>Y</u> (Figure 20.5<u>B</u>), appears as a single lithospheric plate simultaneously undergoing <u>accretion</u> (growth by addition) and <u>consumption</u> (loss by softening and melting), so that the plate might conceivably maintain its size without necessarily expanding or diminishing. Actually, our model can also call for a plate of oceanic lithosphere either to grow or to diminish, and we also have models that allow for the creation of new plates of oceanic lithosphere where none previously existed. In this respect, the theory is quite flexible.

The general theory (or hypothesis) of lithospheric plates, their relative motions, and their boundary interactions is called <u>plate</u> <u>tectonics</u>. <u>Tectonics</u> is a noun meaning "the study of tectonic activity." <u>Tectonic activity</u>, in turn, refers to all forms of breaking and bending of rock of the lithosphere.

We have yet to consider a third type of lithospheric plate boundary. Two lithospheric plates may be in contact along a common boundary on which one plate merely slides past the other with no motion that would cause the plates to separate or to converge (Figure 20.6). The plane along which motion occurs is a nearly vertical fracture or shear zone, extending down through the entire lithosphere; it is

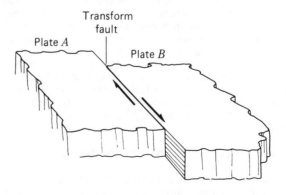

Figure 20.6 A transform fault involves the horizontal motion of two lithospheric plates, one sliding past the other. (Copyright © 1981 by Arthur N. Strahler.)

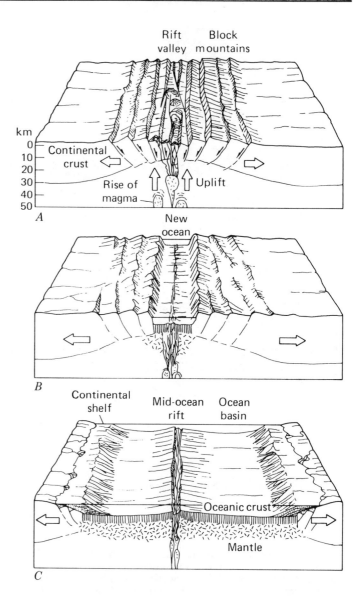

Figure 20.7 A schematic diagram of a single rectangular lithospheric plate with four boundaries. (Copyright © 1981 by Arthur N. Strahler.)

Figure 20.8 Schematic block diagrams showing stages in continental rupture and the opening up of a new ocean basin. The vertical scale is greatly exaggerated to emphasize surface features. (A) The crust is uplifted and stretched apart, causing it to break into blocks that become tilted on faults. (B) A narrow ocean is formed, floored by new oceanic crust. (C) The ocean basin widens, while the stable continental margins subside and receive sediments from the continents. (Copyright © 1981 by Arthur N. Strahler.)

called a <u>transform</u> <u>fault</u>. The word <u>fault</u> is used by geologists for any plane of fracture (a crack) along which there is motion of the rock mass on one side with respect to that on the other.

Thus, in summary, there are three major kinds of active plate boundaries:

<u>Spreading</u> <u>boundaries</u>. New lithosphere is being formed by accretion.

<u>Converging</u> <u>boundaries</u>. Subduction is in progress; lithosphere is being consumed.

<u>Transform</u> <u>boundaries</u>. Plates are gliding past one another.

Let us put these three boundaries into a pattern to include an entire lithospheric plate. As shown in Figure 20.7, we have visualized a moving rectangular plate like a window, set in the middle of a surrounding stationary plate. The moving plate is bounded by transform faults on two parallel sides. Spreading and converging boundaries form the other two parallel sides. Some familiar mechanical devices come to mind in visualizing this model. One is the sunroof top of an automobile; it has a window that opens by sliding backward along parallel side tracks to disappear under the fixed roof. Another is the old-fashioned rolltop desk. There are various different arrangements of plate boundaries, which can be curved as well as straight, while individual plates can pivot as they move. Thus, there are many geometric variations to consider.

Continental Rupture and New Ocean Basins

Plate tectonics provides for a most remarkable geologic event--a <u>continental rupture</u>, the splitting apart of a plate of continental lithosphere. When this occurs, an entire continent is split into two parts that begin to separate as shown in Figure 20.8. At first, the crust is

both lifted and stretched apart. Then a long narrow valley, called a <u>rift</u> <u>valley</u>, appears (Block <u>A</u>). The widening crack in its center is continually filled with magma rising from the mantle below. The magma solidifies to form new crust in the floor of the rift valley. Crustal blocks slip down along a succession of steep faults, creating a mountainous landscape. As separation continues, a narrow ocean appears; down its center runs a spreading plate boundary (Block <u>B</u>). Plate accretion then takes place to produce new oceanic crust and lithosphere. The Red Sea is an example of a narrow ocean formed by continental rupture. Its straight, steep coasts are features we would expect after much deformation. The widening of the ocean basin can continue until a large ocean has formed and the continents are widely separated (Block <u>C</u>).

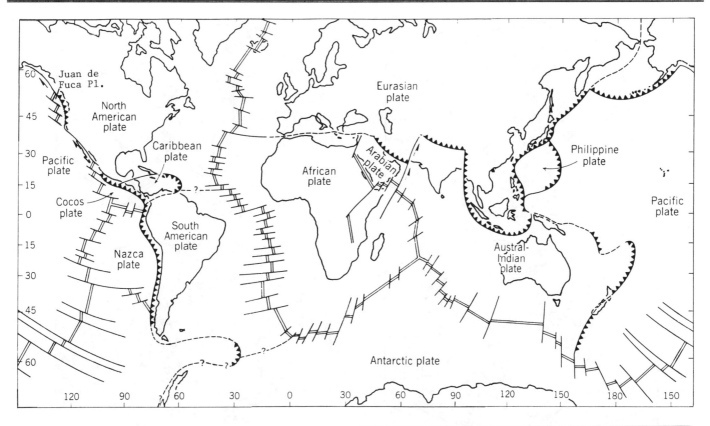

Figure 20.9 Generalized and simplified world map of lithospheric plates. (Copyright © 1981 by Arthur N. Strahler.)

Plate boundaries			
══════	Spreading	▬▬▬▬	Transform fault
▲▲▲▲▲▲	Subduction	‒ ‒ ‒ ‒	Uncertain or inactive

The Global System of Lithospheric Plates

The global system of lithospheric plates consists of twelve major plates, shown in Figure 20.9. Of the twelve, six are plates of enormous size; the remaining six range from intermediate to comparatively small in size. Geologists have identified and named several even smaller plates within the twelve major plates. Plate boundaries are shown by standard symbols, explained in the key accompanying the map. Keep in mind that the Mercator grid distorts the areas of the plates, making them appear greatly expanded in high latitudes.

The great Pacific plate occupies much of the Pacific Ocean basin and consists almost entirely of oceanic lithosphere. Its relative motion is northwesterly, so that it has a subduction boundary along most of the western and northern edge. The eastern and southern edge is mostly a spreading boundary. A sliver of continental lithosphere is included and makes up the coastal portion of California and all of Baja California. The California portion of the boundary is an active transform fault (the San Andreas Fault).

The American plate includes most of the continental lithosphere of North and South America as well as the entire oceanic lithosphere lying west of the mid-oceanic ridge that divides the Atlantic Ocean basin down the middle. For the most part, the western edge of the American plate is a subduction boundary; the eastern edge is a spreading boundary. The Eurasian plate is mostly continental lithosphere, but is fringed on the west and north by a belt of oceanic lithosphere.

The African plate can be visualized as having a central core of continental lithosphere nearly surrounded by oceanic lithosphere. The Austral-Indian plate takes the form of an elongate rectangle. It is mostly oceanic

lithosphere but contains two cores of continental lithosphere--Australia and peninsular India. The Antarctic plate has an elliptical shape and is almost completely enclosed by a spreading plate boundary. The continent of Antarctica forms a central core of continental lithosphere completely surrounded by oceanic lithosphere.

Of the remaining six plates, the Nazca and Cocos plates of the eastern Pacific are rather simple fragments of oceanic lithosphere bounded by the Pacific mid-oceanic spreading boundary on the west and by a subduction boundary on the east. The Philippine plate is noteworthy as having subduction boundaries on both east and west edges. The Arabian plate resembles the "sunroof" model shown in Figure 20.7; it has two transform-fault boundaries and its relative motion is northeasterly. The Caribbean plate also has important transform-fault boundaries. The tiny Juan de Fuca plate is steadily diminishing in size and will eventually disappear by subduction beneath the American plate.

Continental Drift--The Breakup of Pangaea

Although modern plate tectonics became scientifically acceptable within only the past two decades, the concept of the breakup of an early supercontinent into fragments that drifted apart to make the modern continents is many decades old. Almost as soon as good navigational charts showed the continental outlines, educated persons became intrigued with the close correspondence in outline between the eastern coastline of South America and the western coastline of Africa. In 1666, a French moralist, Francois Placet, interpreted the matching coastlines as proof that the two continents became separated during the Noachian Flood. In 1858, Antonio Snider-Pelligrini produced a map

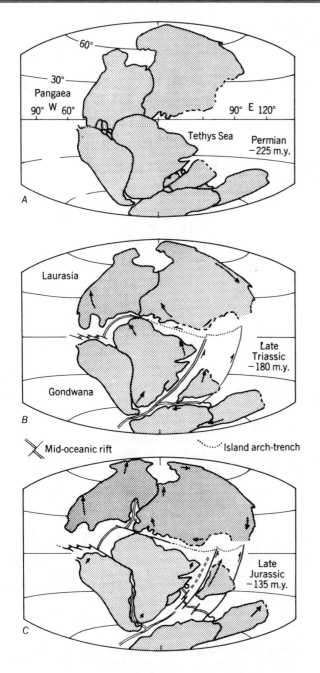

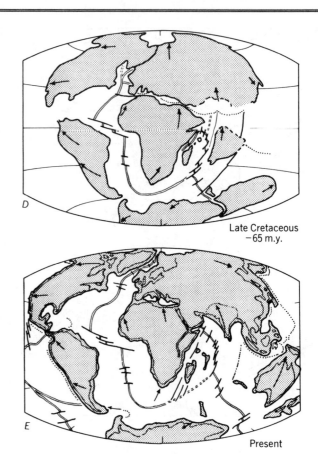

Figure 20.10 The breakup of Pangaea is shown in five stages. Inferred motion of lithospheric plates is indicated by arrows. (Redrawn and simplified from maps by R. S. Dietz and J. C. Holden, <u>Journal of Geophysical Research</u>, vol. 75, pp.4943-51, Figures 2 through 6. Copyrighted © 1970 by The American Geophysical Union. Used by permission.)

to show the American continents nested closely against Africa and Europe. He went beyond the purely geometric fitting to suggest that the reconstructed single continent explains the close similarity of fossil plant types in coal-bearing rocks in both Europe and North America.

In the early twentieth century, two Americans, Frank B. Taylor and Howard B. Baker, published articles presenting evidence for the hypothesis that the New World and Old World continents had drifted apart. Nevertheless, credit for a full-scale hypothesis of breakup of a single supercontinent and the drifting apart of individual continents belongs to a German scientist, Alfred Wegener, a meteorologist and geophysicist who became interested in various lines of geologic evidence that the continents had once been united. He first presented his ideas in 1912, and his major work on the subject appeared in 1922. A storm of controversy followed, and many American geologists denounced the hypothesis.

Wegener had reconstructed a supercontinent named <u>Pangaea</u>, which existed intact about 250 million years ago in the Permian Period. Figure 20.10 is a modern version of Pangaea and shows successive stages of its breakup. Wegener visualized the Americas as fitted closely against

Africa and Europe, while the continents of Antarctica and Australia, together with the subcontinents of peninsular India and Madagascar, were grouped closely around the southern tip of Africa. Starting about 180 million years ago, continental rifting began as Australia, Antarctica, and India pulled away from the rest of Pangaea. As these fragments rifted apart, separating from Africa and from each other, there began the opening up of the ancestral Indian Ocean basin. As the American continents drew away from Eurasia and Africa, the Atlantic Ocean basin widened. Thus enormous expanses of younger oceanic lithosphere came into existence.

Igneous Rocks and the Earth's Crust

It's not wise to go further with details of the theory of plate tectonics without introducing chemical differences in rocks, and learning the properties of the major kinds of rock that are recognized in terms of those chemical differences. Rocks are composed of minerals, which in turn are inorganic compounds. We need information about minerals and rocks in order to understand how the crust differs from the mantle, and how crust of the oceanic

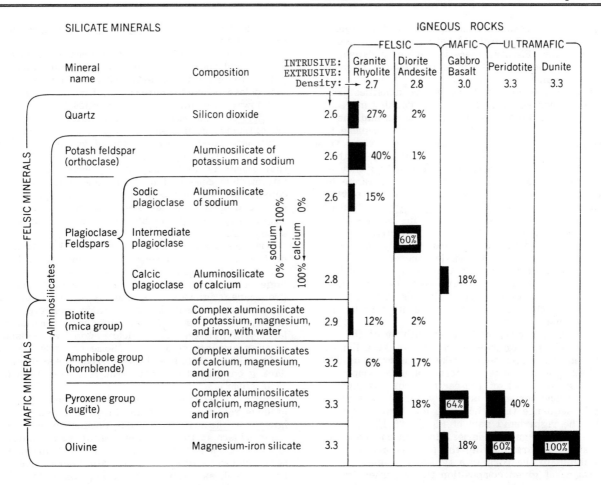

Figure 20.11 The principal igneous rocks and their major constituent silicate minerals. (Copyright © 1976 by Arthur N. Strahler.)

lithosphere differs from crust of the continental lithosphere. With this knowledge, you will be able to understand how lithospheric plates interact along converging boundaries, and what happens when lithospheric plates collide.

The <u>igneous rocks</u> are formed from magma that rises from source regions in the upper mantle or in the basal part of the thick continental crust. Igneous rocks fall into three major groups in terms of chemical composition, which is reflected in the kinds and proportions of mineral varieties they contain. Let us start, however, by going one step further back to examine a list of the most abundant elements in the earth's crust.

Table 20.1 lists the eight most abundant chemical elements in the earth's crust. The order of listing is according to percentage by weight. Several points are of interest in these figures. Notice, first, that the eight elements constitute between 98 percent and 99 percent of the crust by weight and that almost half of this weight is oxygen. Notice that silicon is in second place with about 28 percent, or roughly half the value of oxygen. Aluminum and iron occupy intermediate positions, while the last four elements--calcium, sodium, potassium, and magnesium--are each in the range of 2 to 4 percent.

The vast bulk of all igneous rock consists of mineral compounds containing the elements silicon and oxygen. Collectively these minerals are known as <u>silicates</u>. In a silicate mineral both silicon and oxygen are combined with one or more of the metallic elements listed in Table 20.1.

We can gain a good appreciation of the nature of igneous rocks as a class by noting the proportions of only seven silicate minerals, or mineral groups. These are shown in Figure 20.11. The mineral list begins with quartz, containing only silicon and oxygen. The next five

Table 20.1 The Most Abundant Elements in the Earth's Crust

Element	Percentage by weight
Oxygen	46.6
Silicon	27.7
Aluminum	8.1
Iron	5.0
Calcium	3.6
Sodium	2.8
Potassium	2.6
Magnesium	2.1
Total	98.5

silicate compounds all contain aluminum and can be designated aluminosilicates.

The first aluminosilicate is potash feldspar. Potassium is the dominant metallic element in potash feldspar. The mineral name for a common kind of potash feldspar is orthoclase.

Next come the plagioclase feldspars. They span a continuous range from the sodic plagioclase end of the series, with sodium making up most of the metallic element, to calcic plagioclase at the other end, with calcium making up most of the metallic element. Intermediate plagioclase contains about equal proportions of sodium and calcium. Pure quartz and the common varieties of feldspars are light in color.

Biotite is the dark-colored representative of the mica group of silicate minerals. Biotite is a complex aluminosilicate of potassium, magnesium, and iron, and with some water. Continuing down the list, we come to two more mineral groups. In each group are several closely related minerals, each with its own name and distinctive chemical composition. The amphibole group is represented by the mineral hornblende, the pyroxene group by augite. Both of these groups are complex aluminosilicates of calcium, magnesium, and iron. Minerals of both groups are usually very dark in color. Finally, there is olivine, a dense, greenish mineral that is a silicate of magnesium and iron, but without aluminum.

Now refer back to Figure 20.11, where densities are listed opposite each mineral. Notice that density increases from the top to the bottom of the list. Quartz and the feldspars range from 2.6 to 2.8 g/cm^3; they comprise the felsic group. "Felsic" is a coined work; it is derived from "fel" in "feldspar" and "si" in "silica." Not only do the felsic minerals have comparatively lower density; they are typically light in color. The remaining minerals range in density from 2.9 for biotite to 3.3 for augite and olivine. These denser minerals comprise the mafic group. "Mafic" is a word coined from the syllable "ma" in "magnesium" and "fic," a contraction of "ferric" (an adjective describing iron). The presence of iron largely accounts for the greater densities of the mafic minerals.

Looking ahead, we can reason that an igneous rock composed of felsic minerals will have lower density than an igneous rock composed of mafic minerals. This difference must influence the rearrangement of rocks into layers in response to gravity.

Igneous rocks are composed almost entirely of the seven silicate minerals or mineral groups we have listed. It is estimated that these seven make up as much as 99 percent of all igneous rocks. In terms of bulk composition, most igneous rock of the earth's crust belongs to the granite-gabbro series. Customarily these rocks are presented in sequence from the felsic end toward the mafic end. The right-hand part of Figure 20.11 lists rocks of the granite-gabbro series. In each column under the rock name are bars showing the proportions of minerals making up the rock in a typical example.

For a magma of given composition the solidified igneous rock will take one of two forms, depending on whether the magma solidifies completely surrounded by older rock or whether it solidifies in contact with air or water after emerging from a vent or conduit. In the first case, the rock is described as intrusive; in the second case, as extrusive. The intrusive igneous rocks solidify into rock bodies called plutons; they come in a wide range of sizes and shapes. Where the pluton is large, the magma has undergone very slow cooling and crystallization, with the result that the mineral crystals are comparatively large. The rock is then described as being "coarse-textured," and it will be possible to identify individual mineral crystals with the unaided eye or with a simple hand lens. The extrusive igneous rocks emerge as lava, the fluid magma itself, that rapidly solidifies and typically forms a tongue or sheet of rock called a lava flow. But extrusive igneous rock can also be blown out of surface vents under pressure of gases contained in the magma, and, in that case, takes the form of solid particles of various diameters (dust, ash, cinders, bombs) collectively called tephra. In lava flows and tephra rapid cooling has produced extremely small mineral crystals--too small to be distinguished without the aid of a microscope--or the magma may have solidified as a glass, lacking in crystalline structure. In Figure 20.11 two names appear for rock of a given mineral composition, one for the intrusive variety (plutons), another for the extrusive variety (lavas).

Granite is dominated in composition by the feldspars and quartz. Potash feldspar of the orthoclase variety is the most important mineral, while sodic plagioclase may be present in moderate amounts or absent. Quartz, which accounts for perhaps a quarter of the rock, reaches its most abundant proportions in granite. Biotite and hornblende are commonly present.

Granite is a light-colored igneous rock and is grayish to pinkish, depending upon the variety of potash feldspar present. Its density, about 2.7 g/cm^3, is comparatively low among the intrusive igneous rocks. The extrusive

equivalent of granite is rhyolite, a light gray to pink form of lava.

The granite-gabbro series progresses through transitional rocks not named here. During this transition potash feldspar and quartz decrease in proportion while plagioclase feldspar increases and moves from the sodic end toward the intermediate varieties.

Diorite is the most important plutonic rock on our list. Its extrusive equivalent, andesite, occurs very widely in lavas associated with one class of volcanoes. Its density is 2.8 g/cm^3. Looking at the bars in Figure 20.11, we see that diorite is dominated by plagioclase feldspar of intermediate composition, while quartz is a very minor constituent. At this point in the granite-gabbro series, pyroxene of the augite variety makes its appearance. Amphibole, largely hornblende, is also important, and some biotite is present.

Gabbro is an important plutonic rock, but not a great deal of it is seen exposed on the continents. Its extrusive equivalent is basalt. We find that basalt makes up a few huge areas of lava flows on the continents and is the predominant igneous rock underlying the floors of the ocean basins. Basalt is also a major rock type at the surface of the moon. Gabbro and basalt are composed largely of pyroxene and calcic plagioclase feldspar with varying amounts of olivine. (Some types lack olivine.) Gabbro and basalt are dark-colored rocks--dark gray, dark green, to almost black--of relatively high density (3.0 g/cm^3).

We can now apply the adjectives "felsic" and "mafic" to igneous rocks as well as to silicate minerals. Felsic igneous rocks are those rocks dominantly composed of felsic minerals; they include granite and diorite. Mafic igneous rocks are those rocks composed dominantly of mafic minerals. However, the mafic rocks include some types, such as gabbro and basalt, containing substantial amounts of calcic plagioclase feldspars. As shown in Figure 20.11, the felsic rocks have densities of less than 3.0 g/cm^3. The mafic rocks have densities of 3.0 or higher.

Continuing the igneous rock series depicted in Figure 20.11, we arrive at peridotite, a rock composed almost entirely of olivine and pyroxene. On the continents peridotite is seen only in relatively small bodies, but rock of peridotite composition probably makes up most of the upper mantle. Peridotite is a dark-colored rock of high density, 3.3 g/cm^3, and belongs to a group designated as ultramafic igneous rocks.

Finally we include a variety of igneous rock consisting almost entirely of the mineral olivine. This ultramafic rock is dunite. Like peridotite, it has a density of 3.3 g/cm^3. Although dunite is a rare rock at the earth's surface, it may be present in the mantle.

Continental Crust and Oceanic Crust

Our knowledge of the granite-gabbro series of igneous rocks can now be put to use to describe the composition of the earth's crust. Figure 20.12 shows details of the crust and its composition. The oceanic crust, an average of about 6 km thick, is composed of mafic igneous rock of the composition of basalt and gabbro. The upper part is basalt that solidified from lava reaching the seafloor in the spreading plate boundary between two oceanic lithospheric plates. The same magma solidifying in the lower part of the crust, below the basalt, is thought to consist of gabbro, the intrusive equivalent of basalt.

The continental crust averages some 30 to 35 km in thickness generally, but extends down to depths of 65 km or more beneath the high mountain belts. An upper felsic rock layer can be distinguished from a lower basal, mafic layer, but no sharp boundary separates these layers. The upper felsic layer, often described by the term granitic rock, has a chemical composition about that of granite with a density of about 2.7 g/cm^3 near the surface. Much

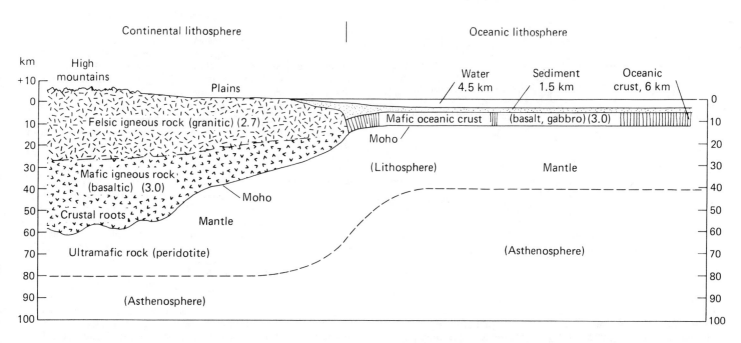

Figure 20.12 Schematic cross section of the crust beneath the continents and ocean basins. The vertical scale of the drawing is greatly exaggerated. (Copyright © 1981 by Arthur N. Strahler.)

of it is intrusive rock of granitic composition or metamorphic rock derived from granite. (Sedimentary rock that covers the granitic rock in many places makes up only a very small percentage of the crust.) Although little is known with certainty about the composition of the lower, mafic layer of the continental crust, it is usually considered to have a chemical composition similar to basalt and gabbro. The mafic layer should not, however, be visualized as a continuation of the oceanic crust, nor should it be thought to have had a similar origin.

Notice that the base of the crust is shown by a solid line under both oceanic and continental crust. This line represents a surface of physical discontinuity; it is labeled here as the moho. This strange word is simply the first two syllables of the name of a seismologist, Andrija Mohorovičić, who in 1909 discovered that earthquake waves passing through this surface underwent an abrupt change in speed. Beneath the moho lies the mantle, composed of ultramafic rock, essentially like peridotite. The moho thus marks a rather abrupt change in density from 3.0 g/cm^3 in the crust to 3.3 in the upper mantle.

In Figure 20.12, a dashed line shows approximately where the lithosphere has its base, and below which the asthenosphere sets in. As noted earlier, the lithosphere is, on the average, substantially thicker under the continents than under the oceans. Thus a lithospheric plate composed of oceanic crust at the top can be expected to be thinner than a lithospheric plate bearing continental crust. (Keep in mind that a single plate may bear both continental and oceanic crust, both moving as one unit.)

The Principle of Isostasy

We have emphasized that the oceanic crust, which is comparatively dense and thin, stands low relative to the continental crust, which is less dense and much thicker than oceanic crust. Apparently a condition of equilibrium prevails, and we alluded to this earlier in likening lithospheric plates--in a superficial way--to floating plates of sea ice. Equilibrium is possible because the soft, plastic asthenosphere has the capability of yielding by slow flowage and can move from place to place to establish

and maintain equilibrium. In this role the asthenosphere behaves like a true fluid. The continental lithosphere floats high because of the lower density of its thick crust. This is the same principle that applies to the buoyancy of all floating objects--a ship's hull, for example.

We are also familiar with the fact that a floating object comes to rest with a certain proportion of its bulk submerged and the remaining portion above the water line. The Archimedes principle requires that a floating object displace a volume of liquid having the same mass as the entire floating object. If an object has a density half that of the liquid, it will float at rest with half its volume below the water line and half above.

When applied to the lithosphere, the principle of a mass floating at rest in a supporting liquid is referred to as the principle of isostasy. This word comes from the Greek words isos, equal, and stasis, standing still. Isostasy requires that a particular column of the lithosphere must float in an equilibrium state on the underlying asthenosphere. Over most areas of the earth, including the ancient continental shields and the young ocean floors, the equilibrium required by isostasy exists or is closely approached. In other words, isostatic equilibrium prevails.

A physical model for isostasy was put forward about 1850 by Sir George Airy, Astronomer Royal of England. His intent was to explain what seemed at the time a very strange phenomenon encountered in geodetic surveys on the Indo-Gangetic plain. It will have to suffice here to note that the great Himalayan range standing nearby did not seem to be exerting the gravitational attraction it should be exerting in consideration of its great mass. Airy had concluded that a deep crustal root of less-than-average density must be situated under the mountain range.

Airy's hypothesis is nicely demonstrated by a simple physical model, shown in Figure 20.13. Suppose that we take several blocks, or prisms, of a metal such as copper. Although all prisms have the same dimensions of cross section, they are cut to varying lengths. Because copper is less dense than mercury, the prisms will float in a dish of that liquid metal. If all blocks are floated side by side in the same orientation, the longest block will float with the greatest amount rising above the level of the mercury surface, and the shortest block has its upper surface lowest. With all blocks now floating at rest, it is obvious that the block rising highest also extends to greatest depth.

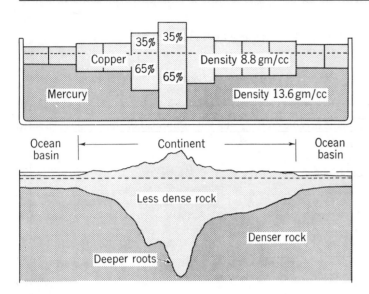

Figure 20.13 Prisms of copper floating in a dish of mercury can be used to illustrate the Airy hypothesis of mountain roots in conformity with the principle of isostasy. (Copyright © 1971 by Arthur N. Strahler.)

Airy supposed that the rock of which high mountains are composed extends far down into the earth to form roots composed of less dense rock. This crustal rock protrudes downward into a location normally occupied by denser mantle rock. Under a low-lying plains region, the root of less dense rock will be shallow. The thin oceanic crust consists of mafic (basaltic) rock, resting on ultramafic rock of the mantle. Because it is both a thin layer and of higher density than the upper felsic zone of the continental crust, its upper surface will take a position lower than the surface of the continental crust. This explains why the ocean floors are the lowest parts of the earth's surface.

Suppose, now, that we cut off a piece of the top of the longest prism in our model. The prism will rise and come to rest in a new position with the top surface lower than before and the bottom at a shallower depth. The ratio of volumes above and below the mercury surface level will remain the same. If, on the other hand, we should add a small prism of copper to the longest block it would sink and come to a new position of rest. The top of the longer prism would now stand higher than before, but the bottom would sink deeper.

The Rock Transformation Cycle

We have chosen to begin the study of rocks with igneous rock. This is about as good a starting point as any, because magma originates deep in the mantle and probably well beneath the continental crust; but is nevertheless an arbitrary starting point, because rock of the crust has been involved throughout geologic time in a great process of recycling, that we can call the rock transformation cycle. Figure 20.14 illustrates the cycle as a flow system of matter powered by two energy sources.

Essentially, there are two distinctly different environments of rock formation: (1) a deep environment, characterized by high temperature and great confining pressure, and (2) a surface environment of low pressure and low temperature. The deep environment has as its energy source the continuous production of heat by radioactive decay of a few kinds of naturally occurring radioisotopes, principally isotopes of uranium, thorium, and potassium. The surface environment obtains its energy from solar radiation, and, being in contact with the atmosphere, has direct accesss to water, oxygen, and

carbon dioxide used in chemical reactions affecting minerals.

Igneous rocks are formed of magma originating in the deep environment through melting or partial melting of mantle rock, or of rock that has been carried down into the mantle by plate subduction. Magma rises through the intermediate region and may solidify at comparatively shallow depths as intrusive igneous rock or may emerge upon the surface as extrusive igneous rock.

Sedimentary rocks originate through the disintegration, decomposition, or dissolution of existing rock of any one of the three rock classes. Living organisms can also produce organic and mineral compounds that become sediment. These processes operate in the surface environment or within close range of water and chemical reactants from the surface. The product is sediment that, broadly speaking, consists of solid particles of a great range of sizes and may include ions in solution. Sediment is eventually deposited in layers; it may undergo burial under newer overlying layers; it may experience hardening (lithification) to become sedimentary rock.

The metamorphic rocks are produced in the deep environment where, in addition to the presence of high temperature and high confining pressure, the rock mass may have been subjected to kneading in a plastic state-- the kind of action a mass of bread dough is subjected to before being formed into a loaf. Any one of the three classes of rock can be altered into a metamorphic rock. Soft sediment saturated with water may enter directly into this process when it is being dragged down into the mantle by a descending plate. Metamorphism may begin with the driving off of surplus water and continue with the formation of new internal structures, such as the development of thin parallel plates or scales. New minerals usually appear, being chemically reconstituted from mineral components already present.

Both intrusive igneous rocks, as plutons, and metamorphic rocks are found today exposed at the surface. This means that they have been uncovered as a result of removal of a thick layer of overlying rock. In the case of some large plutons the intrusion may have originally been covered by a rock layer probably at least one or two kilometers thick. Some kinds of metamorphic rocks may have been formed at depths as great as 10 to 40 km, requiring removal of an overlying rock layer of that thickness in order that they can lie exposed today at the surface.

Plate Tectonics and Igneous Activity

The principal forms of igneous activity are shown in an idealized cross section in Figure 20.15. As before, drawings of the same thing are shown on two scales. The upper diagram, A, has a vertical exaggeration of scale by a factor of five that allows details of the crust to be shown; the lower diagram is drawn to natural scale-- vertical and horizontal scales are the same--and this allows the full thickness of the lithospheric plates to be shown.

First, let us examine the igneous activity along the spreading plate boundary. It is concentrated in the rift that marks the center line of the mid-oceanic ridge. Basaltic magma that rises beneath spreading boundaries of oceanic lithospheric plates is now thought to originate far below in the mantle by the process of partial melting, meaning that certain of the mineral components melt, whereas others remain crystalline. We have described the composition of the upper mantle as that of peridotite-- mostly olivine and pyroxene. Included in this peridotite are minor amounts of calcic plagioclase feldspar that, together with pyroxene (and sometimes olivine), make up basalt. Partial melting of the peridotite would free the necessary pyroxene and calcic plagioclase feldspar; it would move slowly upward, accumulating in bodies of basalt/gabbro magma beneath the spreading centers. It is

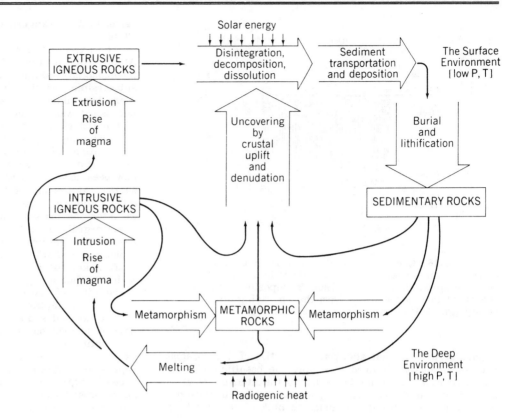

Figure 20.14 This schematic diagram of the rock transformation cycle shows how the three rock classes are related to the surface environment of low pressure and low temperature and the deep environment of high pressure and high temperature. (Copyright © 1971 by Arthur N. Strahler.)

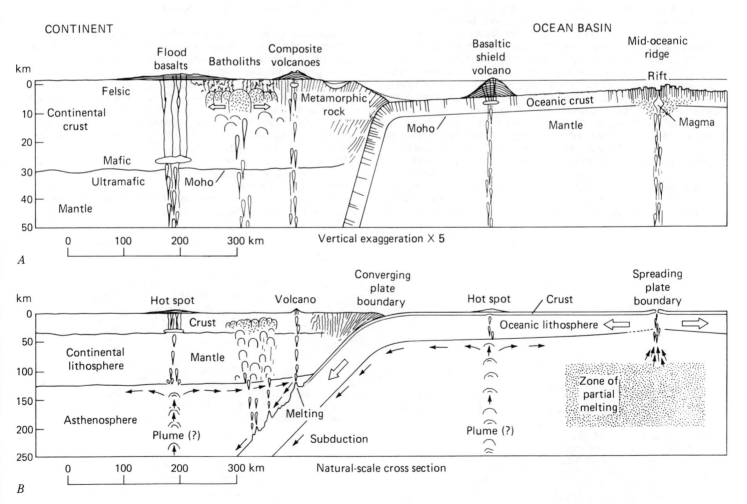

Figure 20.15 Schematic cross sections showing the relationship to plate tectonics of various forms of igneous activity. (A) Cross section with fivefold vertical exaggeration to show crustal and surface features of the uppermost 50 km. (B) Natural-scale cross section to show lithospheric plates and asthenosphere in the upper 250 km. (Copyright © 1981 by Arthur N. Strahler.)

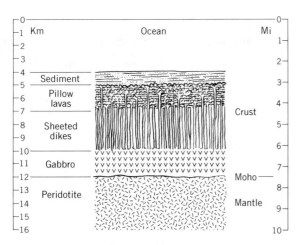

Figure 20.16 Composition and structure of oceanic crust, formed at an accreting plate boundary in a zone of seafloor spreading. The sediment is deposited later, after the crust has moved away from the spreading zone. (Copyright © 1977 by Arthur N. Strahler.)

important to note that enormous quantities of mafic magma are continuously required to supply the basaltic and gabbroic rock that must be added to the lithosphere at spreading boundaries. The hypothesis of partial melting meets this requirement, because the bulk of the mantle is so enormous.

At a few places along spreading plate boundaries of a mid-oceanic ridge, volcanic islands have been formed. Here, basalt lavas have accumulated faster than they could be moved away from the spreading boundary by plate motion. A particularly important example is Iceland, which lies on the spreading boundary between the North American plate and the Eurasian plate (see Figure 20.9). Iceland has several active volcanoes. Here, geologists can examine a surface exposure of the spreading axis and examine the rocks in detail. The spreading axis is marked by a great troughlike rift running through the center of Iceland. Several points of current and recent volcanic activity occur along this rift zone. In deeply eroded canyons, carved by glaciers on older parts of Iceland, the geologist can examine the structure of the spreading zone. Particularly striking is the presence of innumerable vertical basalt dikes, stacked side by side like a deck of

cards in an arrangement called sheeted dikes (Figure 20.16).

The axial rift of the mid-oceanic ridge of the North Atlantic Ocean (the Mid-Atlantic Ridge) has been studied by geologists taking observations from the deep-diving submersible vessel, Alvin. Freshly formed basalt observed there has taken on strange bulbous and tubelike shapes because it was extruded from narrow cracks in older lava of the rift floor. This form of flow is called pillow lava.

Combining the seafloor observations with the observations of sheeted dikes on Iceland, we can reconstruct an idealized section of the rock structure in new oceanic crust formed in spreading zones (Figure 20.16). At the top of the figure are pillow lavas, fed by dikes, above a zone of sheeted dikes. Still farther down we infer the presence of a layer of gabbro, a plutonic (coarse-grained) mafic rock of the same composition as basalt. The gabbro would represent magma bodies that cooled very slowly as they moved away from the spreading axis. Below the gabbro we would expect to find the moho, with a rather abrupt transition into ultramafic rock (peridotite) of the mantle. This sequence is confirmed by study of earthquake waves.

The layered sequence of mafic and ultramafic rocks we have described has been identified in rock exposures on the continental crust where it is called an ophiolite suite. Geologists now realize that these exotic occurrences represent fragments of oceanic crust that have somehow become caught in continental collisions and lifted in great slices or slabs along faults, eventually becoming part of the continental crust.

Basaltic Volcanoes and Hot Spots

Another class of volcanic islands rising from the oceanic lithosphere forms volcanic island chains or isolated groups of volcanoes that are within the oceanic lithospheric plates, rather than on plate boundaries. The Hawaiian Islands are an outstanding example. Because basaltic lava has low viscosity (i.e., flows easily) and is not heavily charged with gases under pressure, the lava flows accumulate as a broad, domelike structure with gentle side slopes; it is called a shield volcano (Figure 20.17). The lava wells up through long, narrow fissures that radiate from a central depression. The great bulk of the volcano lies concealed below sea level, for the ocean floor is 4 km or more deep.

Volcanic chains and groups of this class are thought to have been formed above a magma source called a hot spot.

Figure 20.17 (Above) Block diagram of a large basaltic shield volcano such as those making up the Hawaiian Islands. The vertical scale of the diagram is greatly exaggerated. (Below) A cross section drawn to true scale, showing the relationship between the volcano and the underlying crust and mantle. The crust has sagged under the load of the volcano. (Copyright © 1981 by Arthur N. Strahler.)

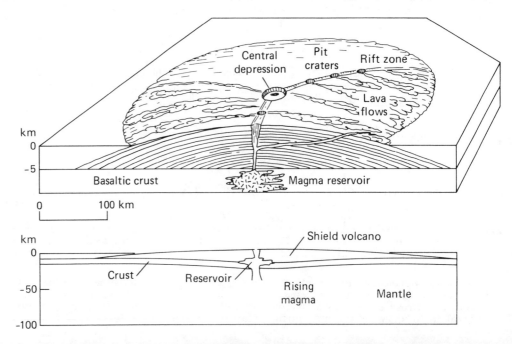

Some geologists think the hot spot is located over a slowly rising column of heated mantle rock known as a <u>mantle plume</u> (Figure 20.15). Whereas the mantle plume and its associated hot spot remain approximately fixed in position relative to the asthenosphere for long periods of time, the overlying oceanic lithospheric plate is drifting past it in continual motion. As a pulse of volcanic activity occurs above the hot spot, a volcanic island, or island group, comes into being but is then carried away with the moving plate. A short time later, a new pulse of volcanic activity occurs above the hot spot, forming a new volcanic island. In time, a chain of islands results.

Also related to hot spots is a form of isolated volcanic activity occurring within continental lithospheric plates, far from presently active boundaries. From a hot spot beneath the continental lithospheric plate, enormous volumes of basaltic magma may rise to the surface, emerging through fissures and pouring out upon the landscape as thick basalt flows (Figure 20.18). Called <u>flood</u> <u>basalts</u>, these outpourings continue until a total thickness of some thousands of meters of basalt layers have accumulated. Often cited as an example of a great expanse of flood basalts is the Columbia Plateau of Oregon, Washington, and Idaho, where basalts covering about 130,000 sq km have a total thickness ranging from 600 to 1200 m, and a volume of at least 250,000 cu km.

Continental Volcanic Arcs

Volcanic eruptions of felsic lavas (rhyolite, andesite) are largely concentrated directly above active subduction zones. Felsic lavas are highly viscous and tend to be explosive. The lava congeals close to the vent and is interlayered with tephra (volcanic ash). The product is a <u>stratovolcano</u>, which is steepsided and often beautifully shaped into a cone that steepens toward the summit, where the crater is located (Figure 20.19).

<u>Volcanic</u> <u>arcs</u> of great andesitic cones, many of them active today, occupy positions above subduction zones bounding the Pacific plate, forming a great "ring of fire" that borders the Pacific Ocean. Note in Figure 20.9 that the Pacific, Nazca, and Cocos plates consist almost entirely of oceanic lithosphere. In North and South America, oceanic lithosphere of the Pacific, Nazca, and Cocos plates is being subducted beneath continental lithosphere of the American plate. <u>Mountain</u> <u>arcs</u> of andesitic stratovolcanoes lie adjacent to these subduction zones in the Cascade Range, Central America, and the Andes Range. (The word "andesite" is derived from its abundance as a rock type in the Andes Range.) Other arcs of andesitic volcanoes follow the western boundary of the Pacific plate, making several <u>island</u> <u>arcs</u> extending south from the Japanese Islands through the Philippines, and into the southwestern Pacific. Another similar volcanic arc forms the Indonesian crescent, including the islands of Java and Sumatra. Here, the Austral-Indian plate is being subducted beneath the margin of the Eurasian plate. Rhyolite lavas are also present in volcanic eruptions of the "ring of fire," but in smaller volumes than the andesite lavas.

Magmas of the mountain arcs arise from a position above descending slabs of lithospheric plates, where subduction of oceanic lithosphere is in progress (Figure 20.15). It is thought that the upper surface of a descending plate becomes highly heated in contact with hot rock of the asthenosphere. The descending layer of basaltic oceanic crust, forming a thin cap of the plate, is rich in water because of its prior contact with seawater. It melts readily, releasing magma of basaltic composition. The depth at which this process begins is estimated to be about 120 km, but melting then extends to depths of 200 km or more on the upper surface of the plate.

A difficult question is how this process can result in extrusion of enormous volumes of andesite and rhyolite at the continental surface. One hypothesis states that the

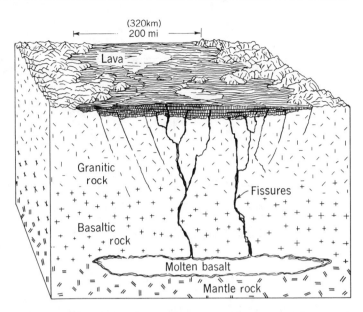

Figure 20.18 This schematic block diagram suggests the inferred relationship between flood basalts and a reservoir of basalt magma located near the base of the continental crust or in the uppermost mantle. (Copyright © 1971 by Arthur N. Strahler.)

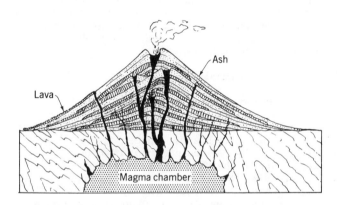

Figure 20.19 This idealized cross section of a stratovolcano shows feeders rising from a magma reservoir beneath. (Copyright © 1972 by Arthur N. Strahler.)

rising basaltic magma undergoes <u>magmatic</u> <u>differentiation</u>, a process of separation of mineral components of a magma. The more felsic mineral ingredients of the melt are less dense than the more mafic ingredients and rise more rapidly. In this way, magmas of felsic and intermediate composition become separated from mafic magmas. The mafic magmas solidify near the base of the continental crust whereas the felsic magmas solidify as plutons in the upper layer of the crust or are extruded to form volcanoes. The hypothesis of differentiation has major weaknesses and does not seem adequate to explain the enormous volume of felsic and intermediate igneous rock in the continental crust.

A hypothesis that appears more successful is based on a recycling principle. It says that the felsic and intermediate magmas come largely from the melting of preexisting rock of the continental crust. The role of the rising basaltic magma from the descending plate is that of a carrier of the heat necessary to melt the crustal rock. At the continental margin, where subduction has been in progress for many tens of millions of years, crustal rock consists largely of deformed, water-rich sediments that have been plastered against the continental plate above the descending plate (see Figure 20.26). Sedimentary rock as a whole has the necessary chemical components to

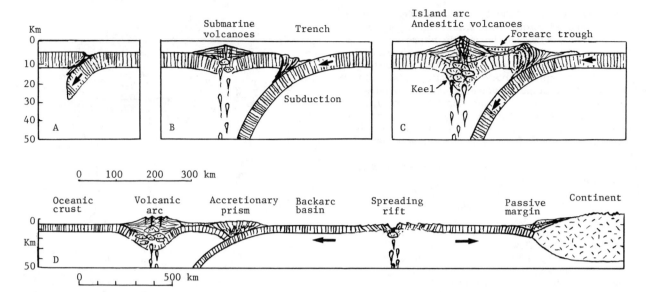

Figure 20.20 (A, B, C) Stages in the development of a volcanic island arc. (D) Relationship of an island arc to a passive continental margin. (A. N. Strahler.)

produce felsic magma. For the same reason, metamorphic rock formed from sedimentary rocks would, by melting, also yield felsic magmas. Finally the continental crust contains large masses of felsic, plutonic igneous rocks produced in earlier tectonic events. These plutons are available for remelting.

The Building of Volcanic Island Arcs

Volcanic island arcs, situated far out from the continental margins, are typical of the western Pacific Ocean basin. A good example is the Mariana arc lying far east of the Philippine Islands. Another is the Lesser Antilles arc of the Caribbean region; it contains the volcanic string of the Windward Islands. These arcs originated within the oceanic crust and lithosphere. The process is postulated to begin with a new break in the oceanic lithosphere, as shown in Diagram A of Figure 20.20. The broken plate edge sags down into the asthenosphere, creating a new subduction plate boundary. Once the downplunging plate has reached well into the asthenosphere, magmas begin to rise beneath the stationary plate some 150 to 200 km from the plate boundary. Some of the magma--that of andesitic composition--rises through the oceanic crust and begins to build submarine volcanoes. As these accumulate, the crust sags beneath the load, but eventually volcanoes reach the surface as islands (Diagram C). Rising magma of mafic composition forms plutons in the lower part of the crust, which is thickened, forming a keel beneath the volcanic arc. The entire structure is maintained in isostatic equilibrium.

Between the volcanic island arc and the continental margin is a backarc basin, floored by oceanic crust (Diagram D). Volcanic arcs play an important role in

continent building, for, as we shall see, they are later driven toward the continental margins, with which they collide, and are added to the continental crust.

Batholiths

The largest of the plutons is the batholith, consisting of granite or a felsic igneous rock of a related variety (e.g., granodiorite). A single batholith may be exposed over a surface area of several thousand square kilometers. Batholiths are shown in Figure 20.15 as forming in the upper part of the continental crust.

Because batholiths are formed at considerable depth within the crust, they can be seen only after erosion over millions of years has removed the overlying crustal rocks, exposing the mountain roots. As a batholith is uncovered, there appear first smaller bodies of the plutonic rock (Figure 20.21). Remnants of the country rock (older rock into which the igneous rock was intruded) will be found extending down into the batholith as roof pendants, but with continued erosion of the land surface a vast expanse of uninterrupted plutonic rock will appear.

Close examination of rock exposures of a batholith will often reveal the presence of irregularly shaped inclusions of country rock, which may be of almost any variety (Figure 20.21). These inclusions suggest that the entry of magma to form a batholith is in part, at least, a mechanical process, in which masses of the brittle country rock are wrenched loose and become incorporated into the magma. Batholiths are thick enough to prevent the bottom being exposed to observation.

One of the major problems in determining the origin of batholiths is to explain what happened to the country rock that was displaced by the invading magma--assuming that the magma was produced from other sources deeper than the country rock. If, instead, we envision the magma as produced directly by melting of the older country rock, the problem of finding room for the batholith largely disappears. Geologists have an alternative explanation for large plutons that relates to

Figure 20.21 A granite batholith exposed by erosional removal of the overlying country rock. Portions of the rock that projected down into the batholith remain as roof pendants, along with inclusions of the country rock surrounded by granite. (Copyright © 1981 by Arthur N. Strahler.)

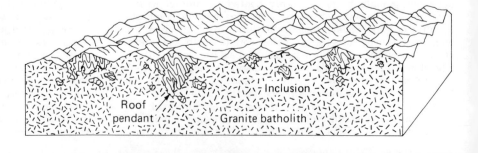

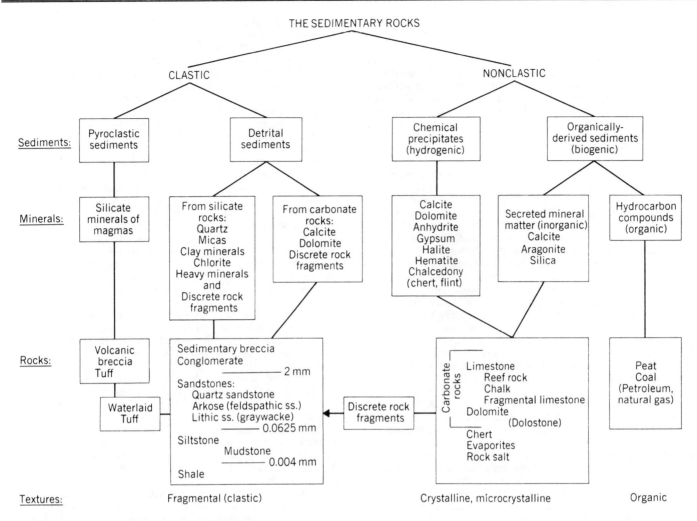

Figure 20.22 Composition and classification of the sedimentary rocks. (Copyright © 1976 by Arthur N. Strahler.)

plate tectonics. The crust above a descending lithospheric plate is stretched horizontally and tends to become thinner. Rising magma takes advantage of this mass motion, pushing the older country rock aside, rather than upward. Under this hypothesis, there is no limit to the volume of new plutonic rock that can be accommodated in the crust.

In this brief summary of the relationships between igneous activity and plate tectonics, we have not described all types of magma intrusions or all regions of volcanic activity. What we have covered is, however, enough to give the most important relationships on a global scale and to reinforce the bare outlines of plate tectonics already sketched.

The Sedimentary Rocks

Sedimentary rocks have formed in a wide range of surface environments ranging from continental locations to the deep ocean floors. These environments are largely present today and can be studied in order to document the processes that form the older strata. This is one of those areas of geology in which the classic uniformitarian principle of Hutton and Lyell works quite well: The present is the key to much of the past. Perhaps not all kinds of environments of sediment deposition that prevailed in the past are seen fully active today, but if not present on a grand scale, they may perhaps be seen locally in action on a small scale.

Sedimentary rocks are usually recognizable through the presence of distinct layers resulting from changes in particle size and composition during the period of deposition. These layers are termed strata, or simply beds. The planes of separation between layers are stratification planes, or bedding planes. The rock is described as being stratified, or bedded. In their original condition major bedding planes of strata deposited in quiet water are nearly horizontal, but they may have become steeply tilted or otherwise distorted by subsequent tectonic activity.

Sediment Types

Geologists classify sedimentary rocks in terms of the possible origins of the sediment constituting the rock (Figure 20.22). The first order of classification is into clastic and nonclastic divisions. Clastic sediment consists of particles individually broken away from a parent rock source. The clastic rocks are in turn subdivided into two groups. One rock is made up of pyroclastic sediments. This material was called tephra in earlier paragraphs. A second group consists of rocks made up of detrital sediments, mineral fragments derived by the disintegration of preexisting rocks of any classification.

The nonclastic division also includes two basic subdivisions, chemical precipitates and organically derived sediments. Chemical precipitates are inorganic compounds representing solid mineral matter precipitated from a water solution in which that matter has been transported. The organically derived sediment consists of both the remains of plants or animals and mineral matter produced by the activities of plants and animals. This includes, for example, the shell matter secreted by animals, which is a true mineral and constitutes an inorganic sediment.

On the other hand, accumulating plant remains, consisting of hydrocarbon compounds, form a truly organic sediment. We shall need to be careful to distinguish between organically derived mineral matter and organic sediment (hydrocarbon compounds). The most important solid hydrocarbon compounds occurring within sedimentary rocks are peat and coal. Petroleum, while not strictly classed as a mineral, is a closely related form of hydrocarbon deposit usually found in sedimentary rocks.

The Detrital Sediments

The most abundant particles of detrital sedimentary rocks consist of quartz, rock fragments, feldspar, and clay minerals. Fragments of unaltered fine-grained parent rocks can easily be identified in coarse sandstones by microscopic examination. Such fragments are typically second in abundance to quartz grains and are the chief component in the coarser grades of detritus. Mica and other minerals generally make up less than 3 percent of coarse-grained detrital rock. Clay minerals may be abundant in the finer-grained detrital sediments.

The clay minerals, as a group, are of enormous importance in sedimentary geology. Basically, clay minerals are derived from the silicate minerals of igneous rock under the attack of weak acid solutions present in rainwater and soil water. The chemical alteration that ensues consists in part of the addition of water to the lattice structure of the silicate mineral. This chemical process is not generally reversible in the surface environment. It produces a class of minerals characterized by being soft and plastic when moistened.

The clay minerals take the form of microscopic flakes or scales, so small that they remain in suspension indefinitely in water. In this respect they are classed as colloids. The clay particles have important electrochemical properties, since they bear surface electrical charges (negative) and can hold not only water molecules, but also ions (cations) of such elements as hydrogen, aluminum, calcium, sodium, potassium, and magnesium. Colloidal clay particles can become clotted together in comparatively large aggregations, a process known as flocculation. It occurs when fresh water of rivers, carrying clay minerals in suspension, enters the ocean and mixes with salt water. The aggregates (flocs) can then settle to the bottom to become layers of clay sediment.

Detrital Sedimentary Rocks

The common detrital sedimentary rocks are conglomerate, sandstone, siltstone, claystone, and shale. The distinctions among them are made in terms of the diameter of the particles comprising the rock. Table 20.2 gives the grades and their diameter limits. Gravel is a term applied to the three coarsest grades: boulders, cobbles, and pebbles.

A sediment accumulation that has become hardened into rock is described as lithified. The noun lithification refers to the hardening process. Usually, cementation by mineral matter is responsible for lithification of the coarser grades. Compaction and expulsion of water is the main cause of lithification of clays.

Table 20.2   Grade Sizes of Sediment Particles

Grade name		Diameter (mm)
Boulders		over 256
Cobbles	Gravel	64-256
Pebbles		2-64
Sand		0.06-2
Silt		0.004-0.06
Clay		Under 0.004

Conglomerate consists of pebbles, cobbles, or boulders, usually quite well rounded in shape, embedded in a fine-grained matrix of sand or silt. Rounding of the conglomerate pebbles is a result of abrasion (wearing action) during transportation in stream beds or along beaches. Commonly, then, conglomerates represent lithified stream gravel bars and gravel beaches.

Sandstone is composed of grains in the range from 2 mm to 0.06 mm (1/16 mm). Perhaps the most abundant and familiar form is quartz sandstone, in which quartz is the predominant constituent. Quartz sandstones contain minor amounts of other detrital minerals such as small flakes of muscovite mica and grains of feldspar.

The quartz sandstones commonly represent lithified sediment deposits of the shallow oceans bordering a continent or of shallow inland seas. The quartz grains have survived a long distance of travel. Finer particles have been sorted out and removed during the transportation process. Certain quartz sandstones were formed from large deposits of dune sands in ancient deserts on the continents. Also, some quartz sandstones are formed largely of recycled grains derived from preexisting sandstones.

Lithification of quartz sands requires cementation by mineral deposition in the interstices (open spaces) between grains. This cementation is accomplished by slowly moving ground water importing the cementing matter as ions in solution. The cementing mineral may be silica (silicon dioxide). In this case the sandstone is an extremely hard rock with great resistance to weathering and erosion. If the cementing material consists of calcium carbonate, a less durable rock results.

The compaction and cementation of layers of silt gives a compact, fine-grained rock known as siltstone. It has the feel of very fine sandpaper and is closely related to fine-grained sandstone, with which there is a complete intergradation.

A mixture of silt and clay with water is termed a mud, and the sedimentary rock derived from such a mixture is a mudstone. The compaction and consolidation of clay layers forms claystone.

Many sedimentary rocks of mud and clay composition are laminated in such a way that they break up easily into small flakes and plates. A rock that breaks apart in this way is described as fissile, and is generally called a shale. Shale is fissile because clay particles lie in parallel orientation with the bedding to form natural surfaces of parting.

Shales of mud and clay composition make up the largest proportion of all sedimentary rocks. They can be subdivided by color. The red shales owe their color to finely disseminated oxide of iron. Red shales are associated with red siltstones and red sandstones in enormously thick accumulations. Collectively known as red beds, these strata are interpreted as having been deposited in an environment of abundantly available oxygen. Favorable conditions are found on river floodplains and deltas of arid climates.

The gray and black shales, also found in great thicknesses, are interpreted as deposited in a marine environment in which oxygen is deficient. The dark color is due to disseminated carbon compounds of organic nature.

Referring back to the pyroclastic sediments, layers of fine volcanic ash and dust become lithified into tuff, a fine-grained sedimentary rock. The mineral composition of tuff is usually the same as for rhyolite and andesite, which are the lavas associated with explosive volcanoes.

Carbonates and Evaporites

Perhaps the most important minerals of the nonclastic sediment class are the carbonates. These are compounds of the calcium ion or magnesium ion, or both, with the carbonate ion. Calcium carbonate ($CaCO_3$) is the composition of one of the most abundant and widespread of minerals, calcite. Another important carbonate mineral is dolomite; it contains both magnesium and calcium.

Evaporites form a major class of chemical precipitates. There are highly soluble salts deposited from saltwater bodies when evaporation is sustained under an arid climate. We are all familiar with halite, or rock salt (NaCl), one of the commonest of the evaporites. Two sulfate compounds of calcium, anhydrite and gypsum, are also among the common evaporites.

Another common mineral important in nonclastic sedimentary rocks is hematite, an oxide of iron. Also important are forms of silica lacking obvious crystalline structure. These occur abundantly as nodules and layers in sedimentary rocks and are referred to as chert or flint. Most nonclastic sedimentary rocks are either carbonate rocks or evaporites (Figure 20.22). Chert is a third important class.

Of the carbonate rocks, the most important is limestone, a sedimentary rock in which calcite is the predominant mineral. Because either clay minerals or silica (as quartz grains or chert) may be present in consider-able proportions, limestones show a wide variation in chemical and physical properties. Limestones range in color from white through gray to black, in texture from obviously granular to very dense.

The most abundant limestones are of marine origin. Some of these are formed by inorganic precipitation; others are by-products of organic activity. The marine limestones show well-developed bedding and may contain abundant fossils. Dark color may be due to finely divided carbon. Many limestones have abundant nodules and inclusions of chert and are described as cherty limestones. An interesting variety of limestone is chalk, a soft, pure-white rock of low density. It is composed of the hard parts of microscopic organisms.

Important accumulations of limestone consist of the densely compacted skeletons of corals and the secretions of associated algae--they are seen forming today as coral reefs along the coasts of warm oceans. Rocks formed of these deposits are referred to as reef limestones. These limestones are in part fragmental, since the action of waves breaks up the coral formations into small fragments that accumulate among the coral masses or in nearby locations. Limestones composed of broken carbonate particles are recognized as fragmental limestone.

Dolomite is a rock composed largely of the mineral of the same name. (It is also called "dolostone.") Dolomite rock poses a problem of origin, since the mineral is not excreted by organisms as shell material. Direct precipitation from solution in sea water is not considered adequate to explain the great thicknesses of dolomite rock that are found in the geologic record. The most widely held explanation of the formation of dolomite rock is that it has resulted from the alteration of limestone by the substitution of magnesium ions of seawater for part of the calcium ions.

Environments of Sediment Deposition

Environments of sediment deposition are usually divided according to whether they are terrestrial (continental) or marine. Many terrestrial environments are obvious in the landscape. River floodplains and deltaic plains receive annual increments of gravel, sand, silt, and clay. Intermontane desert basins receive fluvial sediment from streams that originate in adjacent mountain ranges and terminate on dry lake beds where the sediment accumulates, together with salts left from evaporation of the stream water. Sand, bounced or dragged along the bare ground surface by wind, forms and reshapes thick accumulations of dune sands, both in interior desert areas and close to marine and lake shorelines. Continental ice sheets transport sediment and deposit it as layers of bouldery clay and as sand and gravel deposits of

meltwater streams at the margins. Dust carried high in the atmosphere in frontal storms settles in distant locations to build layers of fine silt over uplands.

Strata of the marine environments of deposition differ enormously from one another in such categories as depth below sea level, rate of accumulation, total thickness, and sediment composition. Two major subdivisions are (a) deep-ocean environments, and (b) continental margin environments.

The Deep-Ocean Environment

In the deep-ocean environment, far from the nearest large land area, the seafloor receives a rain of sediment particles from the ocean above. Some of it is the hard parts (tests) of microscopic organisms that live as plankton in upper water layers. Accumulating very slowly, these materials form deep-sea ooze, which may be either calcareous or siliceous in composition. Mineral sediment that rains down consists of dust particles of terrigenous sources (coming originally from continental surfaces); it includes brown and red clays and volcanic ash.

As new basaltic crust is produced at a spreading plate boundary, the rain of sediments begins to accumulate on the new rock surface, forming a deep-sea sediment layer that may consist of bedded limestones and bedded chert, as well as brown and red clays. In general, the sediment layer becomes thicker at greater distance from the active plate boundary. Over the oldest oceanic crust, that of Jurassic age, the sediment layer reaches a thickness of 300 to 500 m.

Passive Continental Margins

Along continental margins--such as those that border large stretches of the Atlantic and Indian oceans--we today have great sediment accumulations. These zones are referred to as passive continental margins because they do not coincide with contacts between moving lithospheric plates. This means that they have not experienced tectonic activity of any severity for tens of millions of years, or even longer. Typical of such margins is a great, tapering wedge of shallow-water marine sediments, thickening seaward to the outer limits of the continental shelf (Figure 20.23). The technical term for this wedge is miogeocline. (A large sediment wedge in which the strata are inclined in one direction only is called a geocline.) Sediments that accumulate here are clays, lime-muds, sands, and some pebble layers. The shallow marginal continental seas are rich in marine life, and the deposits of past geologic periods contain rich faunas of marine invertebrate animals. Reefs and platforms of coral have also accumulated in this environment in the geologic past.

Notice in Figure 20.23 that the continental shelf forms the floor of the shallow marginal sea in which the strata of the miogeocline are being deposited. The shelf terminates on the seaward side in a steeply descending slope, the continental slope. Whereas the edge of the shelf has a water depth on the order of 200 m, the foot of the continental slope is at a depth of 1500 to 3000 m, which is the deep-ocean, or abyssal environment. The floor of this deep zone is the continental rise; it slopes gently away from the continent and may show a depth of 5000 m where it finally merges into a flat abyssal plain. Beneath the continental rise is another great sediment wedge called a eugeocline. Notice that it tapers to a thin edge outward, away from the continent, which is just the opposite taper from that of the miogeocline. The eugeocline is made up of sediment that has been swept off the continental shelf and has made its way down the continental slope. As masses of mud on the brink of the shelf come loose and begin to slide, the mud becomes mixed with seawater to form a tonguelike turbidity current that flows rapidly down the slope. Upon reaching

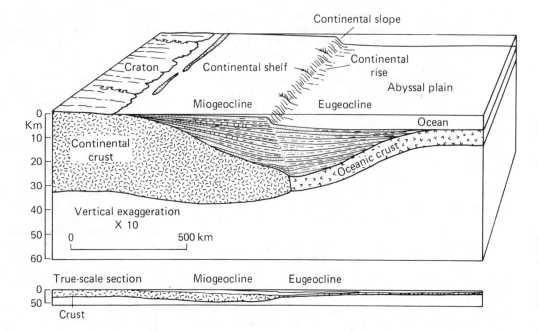

Figure 20.23 Schematic block diagram of geoclines on a stable continental margin. (Copyright © 1981 by Arthur N. Strahler.)

the rise, the current spreads into a broad sheet and flows for long distances before the sediment settles out.

The deposit of a single turbidity-current flow is represented by a distinctive unit layer of sediment called a turbidite. Within this turbidite is an arrangement of layers changing in texture from coarse at the bottom to fine at the top. Typically, the basal zone of the turbidite also shows a distinctive particle size arrangement known as graded bedding, characterized by a continuous upward gradation of sizes from coarse sand or gravel to fine sand. The arrangement is illustrated in Figure 20.24. You can simulate graded bedding by placing a clean mixture of fine pebbles, coarse and fine sand, and coarse silt in a tall glass cylinder filled with water. Upend the cylinder and shake it vigorously, then allow it to stand upright. The rain of particles reaching the base of the container will be assorted in the same manner as in graded bedding. Above the basal graded zone, the turbidite consists of sand in parallel and rippled laminations. Close to the top is fine sandy to silty clay. The topmost layer is made up of clay-sized particles.

Turbidites that make up the eugeocline have in some places accumulated to a thickness of several kilometers. Their presence was not even suspected a few decades ago, but today they are revealed by geophysical methods, in which seismic waves are used as a probe.

Notice in Figure 20.23 that the miogeocline rests on continental crust, whereas the eugeocline rests on oceanic crust. The entire mass of sediments has gradually subsided along with the crust beneath it. The addition of sediment load always tends to produce crustal subsidence, which is an expression of the principle of isostasy. At the same time, removal of rock from the adjacent continent (process of denudation) removes load from the continental crust, which will tend to rise. Figure 20.25 is a schematic diagram of the opposite effects of sediment deposition and denudation. Rising and sinking of the lithosphere requires a compensating flowage of plastic rock in the astheno-sphere. At first thought, this principle might seem to explain geoclines, because once sediment accumulation sets in along a shallow continental shelf, the trough would deepen steadily as more sediments arrive, and the water depth would remain shallow throughout the entire formation of the geocline. But on second thought, this conclusion appears false. Sedimentary layers have lower density than the deeper crustal rocks they displace to a lower level. Even a dense sedimentary rock, such as limestone, is not as dense as mafic rock of the lower crust, so the amount of sinking by isostatic response will

not be as great as the thickness of the sedimentary layer that accumulates. It can be shown that a downsinking of only 5000 m will follow the accumulation of 6000 m of compacted sediment--a ratio of five to six.

Now, if the lithosphere were affected only by the isostatic force of buoyancy, the shallow sea present at the onset of sediment accumulation would soon be filled and deposition would then change from marine to continental sediments. The surface of deposition would continue to rise in elevation above sea level, or the shelf would simply be built farther out into the ocean basin. Clearly, an independent mechanism of sinking must be sought to explain a geocline.

A plausible explanation in tune with plate tectonics lies in plate cooling accompanying the widening of the ocean basin. At the time of rifting, the continental plate would be relatively warm because of the rise of mantle rock. As the passive continental plate margins retreated from the spreading axis, the plate would have steadily cooled, increasing its density, and therefore causing slow crustal sinking. Added to the sinking required by isostasy as the sediment wedges thickened, the total effect would have been to maintain a shallow continental shelf for nearly 200 million years.

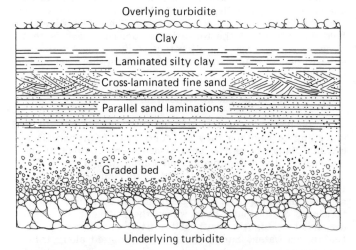

Figure 20.24 A schematic cross section through a single turbidite showing the typical structure. Above the basal layer of graded bedding are laminated sand and silt layers, topped by a clay layer. (Copyright © 1981 by Arthur N. Strahler.)

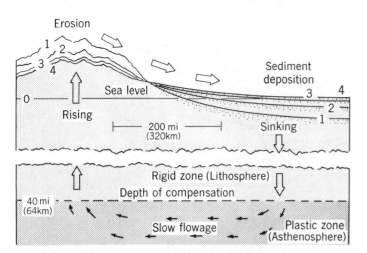

Figure 20.25 Isostatic compensation takes place within the crust when rock is either removed by denudation or added by sediment deposition. (Copyright © 1971 by Arthur N. Strahler.)

Platform Strata

At various times in the geologic past, shallow seas have widely flooded low-lying continental surfaces that subsided below sea level. Here, richly fossiliferous strata have covered areas of subcontinental extent. These are sometimes called platform strata, or simply platforms. They cover much of the interior region of the United States. Strata of Paleozoic and younger age in the Colorado Plateau are a fine example of a sedimentary platform. These strata were described in our earlier section introducing the concepts of Flood geology. Some of the Grand Canyon formations were essentially the marine sediments of ancient continental shelves that were exceptionally wide. Other formations in the sequence are clearly of terrestrial environments, such as delta plains, desert plains, and dune regions. We shall refer to these later.

Platform strata also are found in southern England and northern France, where strata ranging in age from Paleozoic through Mesozoic and Cenozoic ages were first studied and given names by such pioneers as Smith, Murchison, and Sedgwick. The early geologists and paleontologists recognized the alternations of drowning (transgression) and uncovering (regression) of the area by shallow seas, and it was here that Cuvier's multiple catastrophes and creations were envisioned.

Active Continental Margins

Active continental margins are largely those at which subduction is in progress. A deep oceanic trench is characteristic of a subduction boundary. It is there because the downward movement of the oceanic plate is creating a crustal depression that can be only partially filled with sediment from terrestrial sources. Figure 20.26 is an idealized cross section of a subduction zone, showing the surface features as well as structures at depth.

A high-standing volcanic arc can supply large quantities of detrital sediment to the coastline, from which it is swept across a narrow continental shelf by currents. At the outer edge of the shelf, turbidity currents are formed and carry the sediment in tonguelike surges down canyons in the steep inner slope of the trench to reach the trench floor. The movement and accumulation of terrestrial sediments are pictured in Figure 20.26, which also shows pelagic sediment of the ocean floor being carried toward the trench axis on the moving plate.

Arriving at the trench floor, terrestrial and pelagic sediments are subjected to intensely disruptive tectonic activity. The fixed edge of the continental plate acts as a gigantic scraper for sediments dragged against it by the downgoing plate. The activity is often described as offscraping, and is pictured as a process somewhat like the action of the blade of a snowplow. Although some of the snow at the bottom of the layer passes beneath the blade, most is pushed along in front of the blade, where it piles up in a succession of crumpled wedges. Every now and then, the blade may strike a bump in the pavement beneath, breaking loose fragments of the pavement, and these become mixed in with the snow. The pavement in this model might be represented in reality by the basaltic oceanic crust with "bumps" made by submarine volcanoes.

Figure 20.27 is a series of schematic cross sections showing the growth of a succession of thrust slices of crumpled trench sediments to form a rising pile. The process has been called understuffing. These imbricated thrusts are formed at a low angle, but are pushed upward by newer wedges and steepen in angle. The imbricate structure is revealed in detail by seismic reflection profiling. The wedges are pushed upward; they are also being buried by newer sediment layers. The newer sediments in turn become folded between the upper edges of the wedges. In this way, numerous small catchment basins are formed and filled with sediment on the inner trench slope.

The result of the understuffing process is an accretionary prism of deformed sediments. (As an alternate term, subduction complex may be used.) The density of the material in the prism is quite low and it tends to rise to form a tectonic crest. The crest then tends to form a barrier to terrestrial sediment arriving from the coastline, trapping part of it in a shallow depression that we may call a forearc trough (or forearc basin). Here, sediment layers accumulate, and the base of the deposit subsides because the combined load of the accretionary prism and the forearc trough sediments causes the moving plate to sink somewhat lower. At the same time, the accretionary prism grows in width, extending farther and farther out upon the moving plate. The process of growth of the accretionary prism and forearc trough can continue more or less interruptedly for tens of millions of years. In some cases, it continues as long as 100 million years, or about the duration of a major geologic period such as the Cretaceous Period.

To complete this survey of sedimentary deposits of subduction plate boundaries, we need to take a look at an East Indian volcanic arc that includes the islands of Sumatra and Java and a number of other islands to the east that form a curving chain between the Java Sea and the Indian Ocean. Cross sections are shown in Figure 20.28. The features of the trench are familiar-- accretionary prism, tectonic crest (tectonic arc), and forearc trough. On the opposite (Asiatic) side of the volcanic arc is another trough, the backarc trough. It is a zone of accumulation of sediment swept down the northerly slopes of the volcanic arc. In Java and Sumatra, the backarc trough now has its surface above sea level, but farther to the east it is a seaway with water depths exceeding 3 km. The great sediment troughs associated with subduction zones have been discovered and deeply probed only in the past two decades, using seismic reflection methods to reveal the bedding planes in the deposits.

The Metamorphic Rocks

Our description of the metamorphic rocks will be brief and will be largely limited to some generalizations about how this major class of rocks is related to plate tectonics. We consider only rock metamorphism related to intense tectonic activity in which pressures are applied to rock in the zone where two lithospheric plates are coming

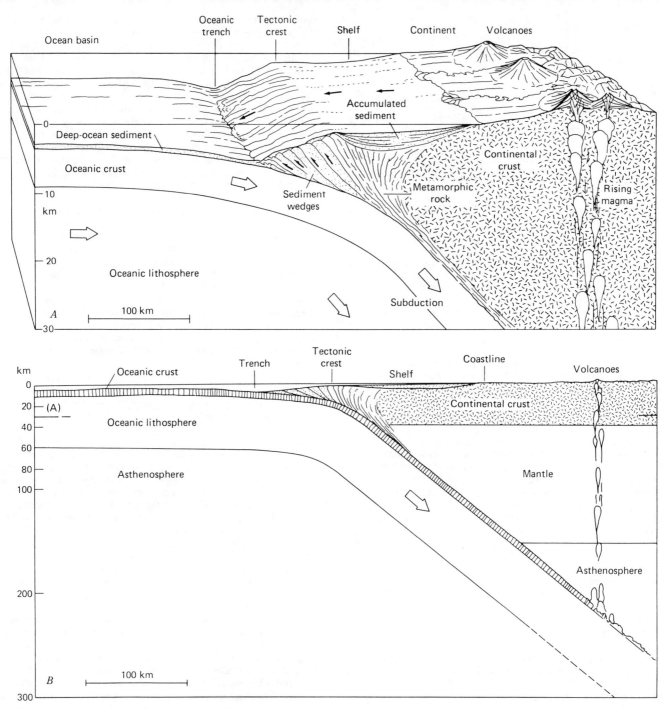

Figure 20.26 Some typical features of an active subduction zone. (A) Great vertical exaggeration is used to show surface and crustal details. Sediments scraped off the moving plate form tilted wedges that accumulate in a rising tectonic prism. Between the tectonic crest and the mainland is a shallow trough in which sediment brought from the land is accumulating. Metamorphic rock is forming above the descending plate. Magma rising from the upper layer of the descending plate reaches the surface to build a chain of volcanoes. (B) True-scale cross section shows the entire thickness of the lithospheric plates. (Copyright © 1981 by Arthur N. Strahler.)

together. One of these has already been described--the zone of plate subduction. The other is the zone of plate collisions, in which subduction ultimately brings masses of continental lithosphere into an impact relationship.

The kinds of metamorphism associated with tectonic activity on a large scale may affect rocks of both igneous and sedimentary origins, but it is on the latter type that the results are most dramatic, because the bedding of sedimentary strata is visibly crumpled and sheared on a large scale.

Metamorphism associated with compressional tectonic

activity is expressed in two ways. First, original minerals recrystallize and new minerals are formed. Second, a new set of structures is imposed on the rock and may replace or obliterate original bedding structures.

Two common kinds of metamorphic rock illustrate these changes. One is slate, familiar through its use as a roofing shingle and as flooring slabs for patios and walks. Slate is a very fine-grained rock that splits readily into smooth-surfaced sheets along natural parting surfaces, called cleavage surfaces. Slate is largely derived from fine-grained, clay-rich marine clastic sedimentary rocks

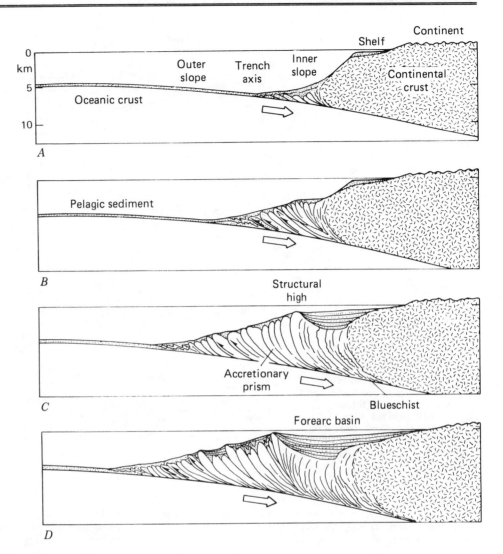

Figure 20.27 Stages in the construction of an accretion prism by offscraping and understuffing. (Based on data of W. R. Dickinson and D. R. Seely, <u>Bull.</u>, <u>Amer.</u> <u>Association</u> <u>of</u> <u>Petroleum</u> <u>Geologists</u>, vol. 63, no. 1., pp. 2-31. Copyright © 1981 by Arthur N. Strahler.)

(shale, claystone). The cleavage of slate is a new structure imposed by metamorphism and usually cuts across the original bedding.

Another class of metamorphic rock is <u>schist</u>, which is characterized by a structure called <u>foliation</u>--a crude layering along which the rock easily separates into irregular slabs. Foliation results from the parallel alignment of platy minerals such as mica. The reflecting surfaces of these minerals give a characteristic glistening sheen to the foliation surfaces. Schists have undergone a high degree of metamorphism, and their origin is not always clear. Most schists are interpreted as altered clastic sedimentary strata rich in aluminosilicate minerals. It is commonly inferred that slates represent an intermediate grade of metamorphism between shale and schist. This sequence is actually demonstrated in some localities by tracing the changes continuously from shale, through slate, to schist. Limestone and dolomite are metamorphosed into marble, which is typically a light-colored granular rock exhibiting a sugary texture on a freshly broken surface.

Metamorphism during Subduction

Development of the accretionary prism gives an opportunity for the formation of metamorphic rocks (see Figure 20.26). Deep in the subduction zone (deeper than 20 km), pressures are very high (greater than 6 kilobars). Here the environment is favorable for the formation of <u>blueschist</u> from turbidites. Blueschist is a metamorphic rock that requires high pressure combined with relatively low temperature. Temperatures close to the descending plate are comparatively low (200 C to 400 C) at this depth because the plate itself is cool in comparison to the surrounding mantle. At still greater depth, with increased pressure, conditions may be favorable for the formation of <u>eclogite</u>, a metamorphic rock of approximately the same chemical composition as basalt and gabbro.

The metamorphic rock is added to the continental crust and thus the continental margin is gradually extended. Denudation of the continental margin is accompanied by a rise of the continental crust, so that eventually these deep-seated metamorphic rocks may appear at the surface.

The intense tectonic activity that goes on within a growing subduction complex affects a wide variety of sediment types as well as oceanic crust. These diverse ingredients become mixed together in a remarkable kind of rock called <u>mélange</u>. As you know, the French word "mélange" refers to an incongruous mixture. Here and there, among the deformed rocks of ancient mountain ranges, occur exposures of rock that would best be described as a very coarse breccia. <u>Breccia</u> is a rock containing numerous angular blocks that are enclosed in a matrix of finer-grained detrital materials. What is remarkable about breccia is not only the great size of some of the blocks, but also the diversity of their composition and origin. Some of the blocks appear to have been torn off the downmoving plate and consist of basaltic oceanic crust; these are <u>exotic</u> <u>blocks</u>. Other blocks are fragments of brittle layers that were formed within the deformed trench sediments; they are native blocks. Enclosing the blocks is the matrix, consisting of intensely deformed sediments--a complex mixture of turbidites and pelagic sediment that was soft at the time of deformation.

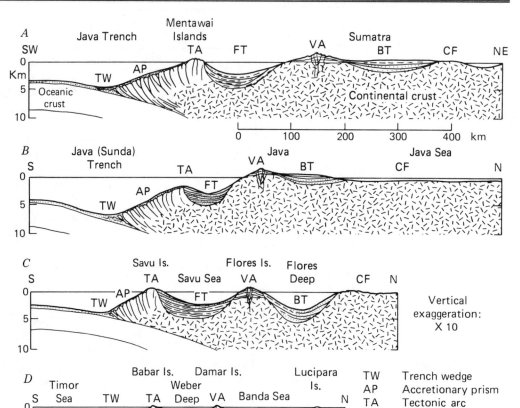

Figure 20.28 Cross sections of the Sumatra-Java volcanic arc. (A) Across central Sumatra. (B) Across central Java. (C) Flores and the Savu Sea. (D) Timor Sea to Banda Sea. (Copyright © 1981 by Arthur N. Strahler.)

TW	Trench wedge
AP	Accretionary prism
TA	Tectonic arc
FT	Forearc trough
VA	Volcanic arc
BT	Backarc trough
CF	Continental foreland

If the mélange is carried down to greater depths in the subduction zone, it may be subjected to high pressure, transforming the mélange into schist.

Metamorphism during Continental Collision

Metamorphism occurs on a vast scale during a continental collision, illustrated in Figure 20.29 by three schematic cross sections. In Diagram A the two types of continental margins are familiar: passive margin with geoclines on the left; active margin with subduction complex on the right. The ocean basin is steadily closing and the two marginal complexes are converging toward an inevitable collision. As the two continental masses converge, oceanic crust is fractured by diagonal slices, called imbricate thrusts. The surfaces of slippage are overthrust faults. The thrust slices begin to ride up on one another, in what is sometimes described as "telescoping." Sediments of both margins are caught in the squeeze and are intensely folded, while at the same time the overthrust faults penetrate the deformed sediments. As slices become more and more tightly squeezed, they are forced upward. The upper part of each thrust sheet, under the force of gravity, bends over into a horizontal position to form a nappe. One nappe is piled upon the other. As you recall from an earlier section, the oceanic crust/mantle sequence is an ophiolite suite. Metamorphosed slices of ophiolite are carried up with the nappes. Where later exposed, they mark the line of a continental suture, which is a permanent union of two masses of continental lithosphere. The entire mass of deformed rock is called an orogen.

Notice in the diagrams that as the pile of nappes rises, sediment eroded from the mountainous mass is carried to lower marginal zones by streams. In early stages, it may accumulate in shallow ocean troughs paralleling the mountain belt. Later, these basins are filled and the

sediment accumulates as a terrestrial deposit of coarse sand and gravel. This type of sediment accumulation is called a foreland trough and can be added to our collection of environments of sediment deposition.

The most recent major continental collision occurred in Cenozoic time along a line extending from the Alps in Europe, through the eastern Mediterranean Sea and Iran, and finally through the Himalayas of south Asia. The African and Indian plates collided with the Eurasian plate. Earlier, a seaway (the Tethys Sea) had separated these continental plates, but it had been gradually closing since the breakup of Pangaea began. These developments can be traced on the series of maps shown in Figure 20.10.

Continental collisions have produced orogens throughout all geologic time as far back as the middle Precambrian. Orogens remaining from these collisions are found in the ancient rocks of all the continents. Not all collisions were between two large bodies of continental lithosphere. Frequently, the collision was between an island arc and the main continent. In other cases, rifting away of small subcontinental fragments of a large continent was followed by a closing of the intervening ocean basin. Numerous smaller crustal masses thus came to be welded onto the main bodies of continental lithosphere. Each of these smaller masses is distinctive in the kind of rocks it contains, and these are often referred to as terranes. Some terranes consist of metamorphic rock, others of igneous rock.

In contrast, oceanic crust in place on the ocean floor is nowhere on the globe older than about 200 m.y. (Jurassic age and younger). Most of the continental crust ranges in age from 1.0 to 3.5 b.y. The reason for age difference is, of course, that oceanic lithosphere is not only continually forming, but also continually disappearing into the asthenosphere by subduction. The continental crust, in contrast, is thick and buoyant, so that no matter how severely it is treated by tectonic processes of

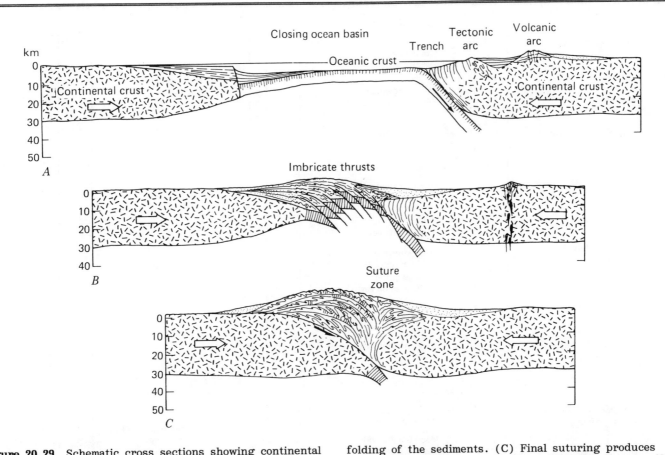

Figure 20.29 Schematic cross sections showing continental collision and the formation of nappes. (A) An ocean basin is closing, bringing two continental margins closer together. On one side (left) is a passive continental margin with geoclines; on the other side (right) is an active subduction boundary. (B) A series of imbricate thrusts breaks through the oceanic crust and causes folding of the sediments. (C) Final suturing produces nappes with ophiolites and metamorphosed sediments. Note that the continental crust on the left has been forced beneath the edge of the opposing continental crust along a surface of underthrusting. A foreland trough lies next to the suture zone. (Copyright © 1981 by Arthur N. Strahler.)

plate rifting the fragments "remain afloat." This is not to say that there is no process by which continental crust can be destroyed. Fragments of its mass can experience deep burial after being torn off by a descending plate during subduction. But the chances are that the less dense felsic rock that gets forcibly dragged down into the asthenosphere will later form granitic magma that will rise to reenter the continental crust as intrusive rock.

Early Earth History

Early earth history, which began about -4.6 b.y. with rapid accretion of the planet and ended about -3.8 b.y. (the age of the oldest known rocks), has been called Protoarchean time. Events of that 800-million-year time block invite speculation by igneous-rock specialists and geochemists who study the oldest rocks available in the continental crust and the composition of mafic and ultramafic magmas that issue from spreading plate boundaries and hot spots. These scientists also have the benefit of what is known of the moon's history and the chemistry of samples of lunar rock, all of which are older than the oldest earth rock. In the late 1980s there still is no consensus on Protoarchean history. Perhaps we should call each speculative statement "a tentative view of the early earth," as distinct from a full scientific hypothesis. At least five different views have been expressed by competent specialists who meet occasionally in conferences (see Geotimes, 1981, vol. 26, no. 6).

Scientific creationists are delighted with the diversity of views on early earth history; they are pleased that no rocks have yet been found older than -3.8 b.y. The miracle of recent Creation provides for this period of

mystery; no questions need be asked. This stultifying acceptance of origins as beyond the limits of science runs counter to the philosophy of mainstream science.

Referring back to the section on the origin of the solar system, recall that a hypothesis currently in good standing is that of heterogeneous accretion of the planets. In the case of the inner, earthlike planets, this calls for accretion of the iron core, followed by accretion of the silicate mantle. Only a few years earlier, the most popular hypothesis was quite different; it called for accretion of a more-or-less homogeneous planet, in which the iron was mixed in with the silicate rock. It was supposed that the planet underwent a period of mineral segregation, in which the iron settled to the core region, leaving the silicate rock to form a mantle. That history would have involved a great deal of internal melting and overturning.

Under heterogeneous accretion, with core and mantle essentially in place at the outset, there would be general stability of the earth's deep interior, but the surface zone would have been in the grip of "catastrophism" on a vast scale. (A good place to use the creationists' favorite word!) Using the history of the moon as a guide, it seems likely that the earth was impacted by numerous solid objects--meteorites--that churned up the outer 2 to 3 km of the earth's solid surface. A few of the impacting objects may have been asteroids as large as 1,000 km in diameter. Fragmented rock would have been thrown about and thoroughly mixed, while locally the heat of impact may have produced substantial lakes of lava. The moon seems to have experienced a period of major bombardment between -4.0 and -3.8 b.y., producing huge basins and causing the melting of enormous volumes of mafic lunar rock. Magma so produced filled the basins and made flat

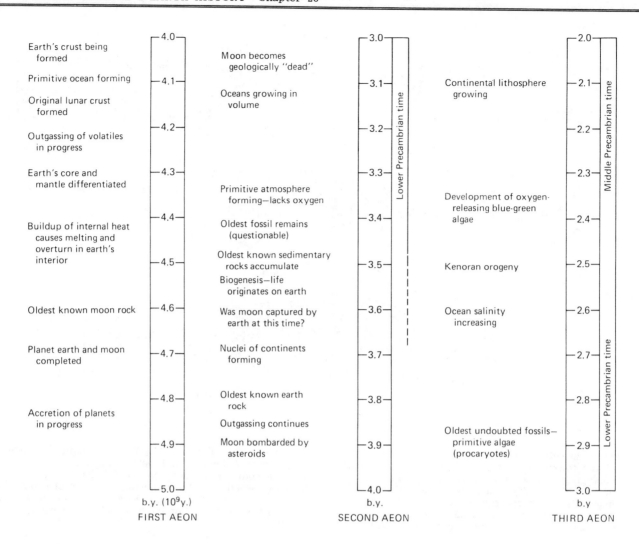

Figure 20.30   Five aeons of geologic time.   (Copyright © 1981 by Arthur N. Strahler.)

floors in the lunar maria. Speculation goes that if a similar sequence of events occurred on earth, the earliest crustal rocks to survive following cessation of the great bombardment would have been felsic magmas that rose to the surface to solidify as patches of continental crust. These high-standing crustal masses would have produced sediment and the first sedimentary rocks accumulating in shallow basins.

In the continental interiors, we find today patches of Archean rock that are mostly in the age range 2.3 to 2.7 b.y., but some of it as old as 3.3 to 3.5 b.y. Whether this older rock was produced by plate tectonics in subduction zones and by collisions, or by a primeval process of separation of magmas referred to above, is a subject of debate (see Alfred Kroner, 1985). These ancient continental nuclei have some unique chemical features that require explanation. They contain large masses of a plutonic rock called anorthosite, which is largely calcic plagioclase feldspar. Anorthosite intrusions do not seem to have been produced in later geologic time by igneous processes related to plate tectonics. The suggestion has been made that crystals of this variety of plagioclase would have tended to rise through bodies of partly crystalline magma.

It is not possible at this time to agree on a particular date at which lithospheric plates began to separate, move over an asthenosphere, and disappear by subduction. Plate tectonics began its operation early in earth history, but how early? One view is that it was in operation at the time of the oldest known rock; another is that it started much later--perhaps no earlier than -2.7 b.y. The argument revolves about the interpretation of what seem to be ophiolite suites in rocks as old as -3.6 b.y. in Canada and South Africa (see Richard Kerr, 1986). If plate collisions were occurring as early as -3.6 b.y., that would still leave about one billion years for formation of crustal rock by processes other than plate spreading and plate subduction.

When plate tectonics did start, it probably proceeded at a much faster pace than today, because the rate of production of radiogenic heat must have been much greater then than it is today. Production of continental crust by accretion and understuffing in the subduction zone, and by rising felsic magmas in volcanic arcs, would perhaps have produced continental crust at a much more rapid rate than today.

The formation of ancient continental crust has produced cratons, also often called shields, that underlie large areas of the continents. The production of such crust is often referred to as cratonization, which is simply the process of growth of continents. In a general sense cratonization by plate tectonics is an expression of uniformitarianism as Hutton and Lyell intended it to mean: an ongoing process, open to inspection by science today, that has been in operation throughout much of earth history. The intensity or rate of action of plate tectonics may have undergone episodes of sharp increase, with slower rates prevailing in long, intervening periods. In the qualitative sense only, uniformitarianism can be applied.

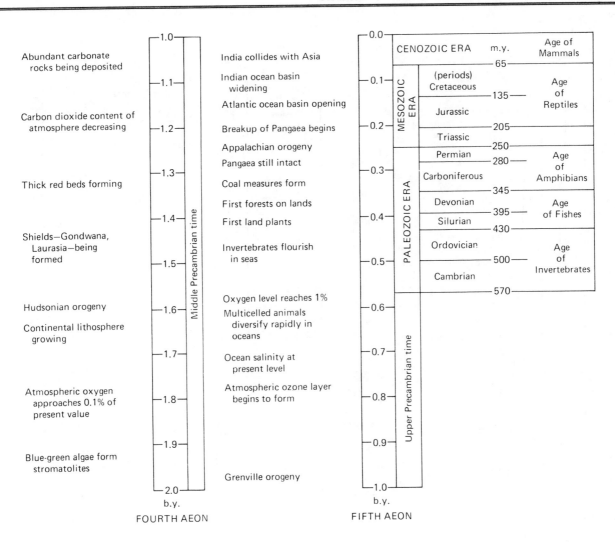

Abundant carbonate rocks being deposited

Carbon dioxide content of atmosphere decreasing

Thick red beds forming

Shields—Gondwana, Laurasia—being formed

Hudsonian orogeny

Continental lithosphere growing

Atmospheric oxygen approaches 0.1% of present value

Blue-green algae form stromatolites

Middle Precambrian time

India collides with Asia

Indian ocean basin widening

Atlantic ocean basin opening

Breakup of Pangaea begins

Appalachian orogeny

Pangaea still intact

Coal measures form

First forests on lands

First land plants

Invertebrates flourish in seas

Oxygen level reaches 1%

Multicelled animals diversify rapidly in oceans

Ocean salinity at present level

Atmospheric ozone layer begins to form

Grenville orogeny

FOURTH AEON

CENOZOIC ERA	m.y.	Age of Mammals
MESOZOIC ERA (periods) Cretaceous	65	Age of Reptiles
Jurassic	135	
Triassic	205	
PALEOZOIC ERA Permian	250	
Carboniferous	280	Age of Amphibians
Devonian	345	Age of Fishes
Silurian	395	
Ordovician	430	Age of Invertebrates
Cambrian	500	
	570	

Upper Precambrian time

FIFTH AEON

Creation scientists are reluctant to accept plate tectonics; they write very little on the subject. Their first negative comments made use of scientific papers published in the 1970s by the few mainstream scientists who held out against plate tectonics. Now even these dissenters have largely disappeared. As evidence accumulates on a massive scale of the workings of plate tectonics and the rocks it produces, creation scientists must fall back on their contention that radiometric ages of rocks are false, therefore all geologic history based on great spans of time is false, and that only a six-day Creation and a one-year Flood give a true history of our planet.

The Table of Geologic Time

The vastness of geologic time, which we can say starts with the accretion of the planets, is almost beyond human comprehension. This same vastness of time is one of the most fascinating aspects of geology, setting both geology and astronomy apart from other physical and natural sciences.

This is a good place to extend our introduction to early earth history into a sweeping view of all of geologic time. Figure 20.30 is a table of geologic time devised in such a way as to emphasize the enormousness of the five-billion-year span of our planet's history. For convenience, we begin five billion years ago (-5 b.y.) and recognize five <u>aeons</u> of time, each one enduring a billion years.

The formation of the planets of the solar system was in progress as the First Aeon began. The earth's growth to form a solid spherical body was complete about -4.7 b.y. There is no record of this event in earth rocks, but a moon rock called the Genesis Rock has been found to contain mineral particles of this age. Throughout the First Aeon, the earth's history is completely unknown, and it is not until the Second Aeon that the oldest earth rock (so far discovered) appears with an age of 3.8 b.y. It is speculated that the origin of living matter took place in shallow ocean water in the time span of -3.7 to -3.5 b.y. What may be the remains of colonies of bacterial forms (blue-green algae) have been discovered in rocks 3.5 b.y. old in Western Australia. The moon became geologically "dead" (inactive in terms of crustal movement or volcanic activity) at the end of the aeon.

The Third Aeon (-3.0 to -2.0 b.y.) saw the development of primitive one-celled organisms--forms of algae--and these are preserved in rocks of that aeon. The nuclei (oldest central parts) of the continents were growing in size. The ocean waters, which had been accumulating from water vapor emitted by the solid earth, were nearly at their present volume. The Fourth Aeon saw a continuation of much the same activity as in the previous aeon, with a continued growth of the continents and the continental lithosphere that underlies them.

Lithospheric plates were continually in motion in the Second, Third, and Fourth aeons, but we know very little about them. The metamorphic rocks that formed from repeated continental collisions are found in great

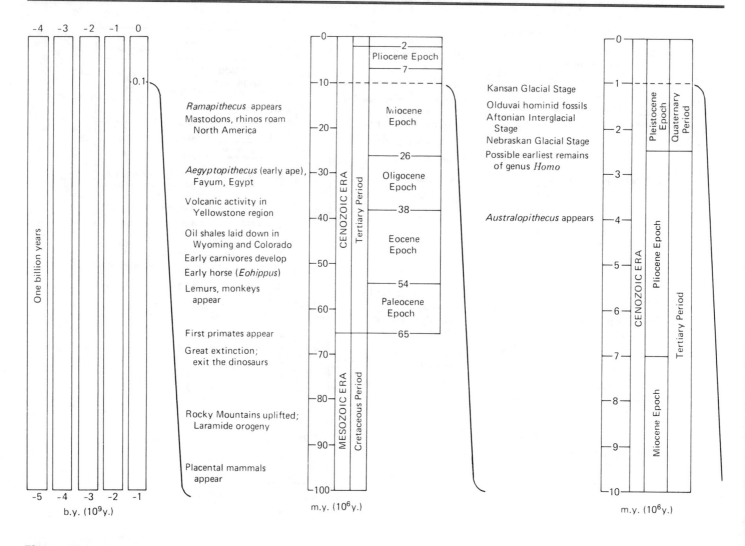

Figure 20.30 (continued) Five aeons of geologic time. (Copyright © 1981 by Arthur N. Strahler)

abundance in the continental interiors today.

The Fifth Aeon, covering the last billion years, witnessed the very rapid evolution of many complex life forms; much of that organic change was concentrated into a short period around -0.5 b.y. The second half of the aeon saw most of the evolution of advanced life forms as we know them today.

Figure 20.30 shows the Fifth Aeon subdivided progressively into smaller time units, each being the youngest one-tenth part of the previous one. Human evolution takes place largely within the final ten million years; human civilizations do not appear until the final 5000-year time block.

Credit

1. Text and illustrations in this chapter are largely extracted and adapted from the author's textbooks of geology and the earth sciences published by Harper & Row, Inc., New York. Copyright © 1971, 1977, 1981 by Arthur N. Strahler.

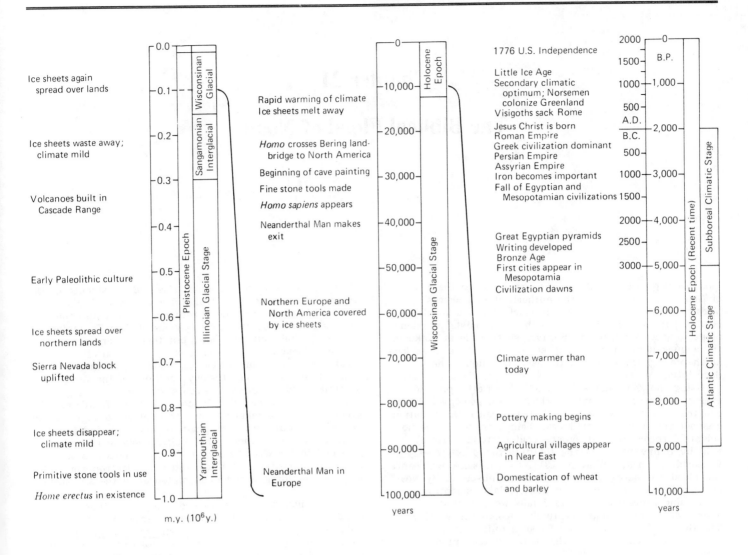

Ice sheets again
spread over lands

Ice sheets waste away;
climate mild

Volcanoes built in
Cascade Range

Early Paleolithic culture

Ice sheets spread over
northern lands

Sierra Nevada block
uplifted

Ice sheets disappear;
climate mild

Primitive stone tools in use

Home erectus in existence

m.y. (10⁶y.)

Wisconsinan Glacial
Sangamonian Interglacial
Pleistocene Epoch
Illinoian Glacial Stage
Yarmouthian Interglacial

Rapid warming of climate
Ice sheets melt away

Homo crosses Bering land-
bridge to North America

Beginning of cave painting

Fine stone tools made

Homo sapiens appears

Neanderthal Man makes
exit

Northern Europe and
North America covered
by ice sheets

Neanderthal Man in
Europe

years

Holocene Epoch

Wisconsinan Glacial Stage

1776 U.S. Independence

Little Ice Age
Secondary climatic
optimum; Norsemen
colonize Greenland
Visigoths sack Rome
Jesus Christ is born
Roman Empire
Greek civilization dominant
Persian Empire
Assyrian Empire
Iron becomes important
Fall of Egyptian and
Mesopotamian civilizations

Great Egyptian pyramids
Writing developed
Bronze Age
First cities appear in
Mesopotamia
Civilization dawns

Climate warmer than
today

Pottery making begins

Agricultural villages appear
in Near East

Domestication of wheat
and barley

B.P.
A.D.
B.C.

years

Holocene Epoch (Recent time)

Subboreal Climatic Stage

Atlantic Climatic Stage

Chapter 21

The Biblical Flood of Noah

Flood geology must conform in every detail to the words of Gen. 6-8 according to the authorized King James Version of 1611. The account is brief, but the words set certain parameters within which the creationists' version of geology is constrained. Because so much is at stake in the argument at hand, I propose that we take a close look at chapters 6 through 8 in Genesis, containing the description of the Flood.

Using the King James Version, I have attempted to extract the essential data on which Flood geology must be constructed. Note at the outset that because the Creation was not observed by any human being, there can be no scientific record of that event. That position is firmly held by Henry Morris of the Institute for Creation Research (Morris, 1974a, p. 131). By the same argument, the Flood was actually experienced and observed by Noah and his family; it is an event admissible to scientific study since, presumably, the details were related to Noah's descendants and eventually recorded in writing. The Gen. 6-8 account of the Flood is quite detailed and specific, as compared with the Gen. 1-2 account of creation.

The Genesis Account of the Flood

The sources of the Flood waters are two: (1) "fountains of the great deep" (Gen. 7:11; 8:2), and (2) "windows of heaven," from which came "rain from heaven" (7:11, 8:2). Both sources became activated on the same day (7:11). The rain continued for 40 days (7:12). Rising water level lifted the ark until it was free-floating (7:17; 7:18). All "high hills" were covered (7:19). Rise of the waters continued to a height of an additional "fifteen cubits" and "the mountains were covered" (7:20). The flood crest remained 150 days (7:4; 8:3), after which the "waters were abated" (8:3). The ark was grounded on Mt. Ararat on the seventeenth day of the seventh month (8:3); on the first day of the tenth month, mountain tops were seen (8:5). One year and 10 days after the rain started "was the earth dried." (Subtract data in 8:14 from 7:11). There is a little confusion here, because in 7:13 it is stated that the "face of the ground was dry" 10 months and 15 days from the start. We can simply settle for one full year from onset of rain to complete drainage.

In studying the Genesis account of the Flood, I came to notice a salient point: Nothing in the account refers to any geologic effects of the aqueous inundation. The only form of destruction mentioned is death by drowning of all life forms (Gen. 7:21-23). No mention is made of currents capable of scouring the solid bed beneath the waters. In 8:1 we read that God made a wind to pass over the earth, but this occurred when all the land surface was under water, and it would be foolhardy to speculate as to the intensity and duration of the wind. The entire description gives the impression of a quiet rise and subsidence of the water. (We discuss this "tranquil Flood" hypothesis later.) The water may have returned in part into the earth's interior by the same systems of deep fountains, reversing their outward flow, but that is mere speculation. Conspicuously lacking is any mention of diastrophism (crustal movement, including folding and faulting) or volcanism. Clearly, nothing in the Genesis account could be taken as even suggesting a geological catastrophe or cataclysm. It looks as if the fundamentalist creationists have taken it upon themselves to augment the Sacred Scriptures with imaginative processes and events of their own concoction. The typical geologic history revealed in the Grand Canyon, in which episodes of sediment deposition are interrupted by diastrophism and mountain building is not even hinted at in the statements in Gen. 7-8.

Let us now present the Flood story according to the creation scientists of the ICR. In Chapter 5 of their basic volume for public school use, Scientific Creationism, we find a full exposition of the subject (Morris, 1974a, pp. 91-130). For the moment, let us postpone discussion of the subject of catastrophism versus uniformitarianism, a subject that takes much of the first part of the chapter. We go straight to the description of the Flood:

> Visualize, then, a great hydraulic cataclysm bursting upon the present world, with currents of waters pouring perpetually from the skies and erupting continuously from the earth's crust, all over the world, for weeks on end, until the entire globe was submerged, accompanied by outpourings of magma from the mantle, gigantic earth movements, landslides, tsunamis, and explosions. The uniformitarian will of course question how such a cataclysm could be caused, and this will be considered shortly, but for the moment simply take it as a model and visualize the expected results if it should happen today.

> Sooner or later all land animals would perish. Many, but not all, marine animals would perish. Human beings would swim, run, climb, and attempt to escape the floods but, unless a few managed to ride out the cataclysm in unusually strong watertight sea-going vessels, they would eventually all drown or otherwise perish.

> Soils would soon erode away and trees and plants be uprooted and carried down toward the sea in great mats on flooding streams. Eventually the hills and mountains themselves would disintegrate and flow downstream in great landslides and turbidity currents. Slabs of rock would crack and bounce and

gradually be rounded into boulders and gravel and sand. Vast seas of mud and rock would flow downriver, trapping many animals and rafting great masses of plants with them.

On the ocean bottom, upwelling sediments and subterranean waters and magmas would entomb hordes of invertebrates. The waters would undergo rapid changes in heat and salinity, great slurries would form, and immense amounts of chemicals would be dissolved and dispersed throughout the seaways.

Eventually, the land sediments and waters would commingle with those in the ocean. Finally the sediments would settle out as the waters slowed down, dissolved chemicals would precipitate out at times and places where the salinity and temperature permitted, and great beds of sediment, soon to be cemented into rock, would be formed all over the world. (Pp. 117-18)

Some further geologic and hydrologic details are available on later pages (Morris, 1974a, pp. 124-25). These are introduced by a rhetorical question as to what would cause a global flood with accompanying igneous and tectonic activity.

What Caused the Flood?

Creationists place much stress on the significance of fossil content of strata over the entire earth, indicating a worldwide warm climate with no distinct climatological zones such as exist today (Morris, 1974a, p. 123). If you wonder what they are leading up to with such suggestions, you have a big surprise in store! They are looking for an enormous extraterrestrial water source to be let loose in the Deluge. Why the universal warm climate? Let them tell you:

The most likely explanation is that something outside the earth's surface so controlled the incoming solar energy as to maintain a global greenhouse-type environment. There are three components of the atmosphere which, in lesser measure, have this function today; namely, ozone, carbon dioxide and water vapor.

If one or more of these were a much more abundant constituent of the atmosphere prior to the cataclysm, there would indeed have been a universal "greenhouse effect." The most important is water vapor. If there were, in the beginning, a vast thermal blanket of water vapor somewhere above the troposphere, then not only would the climate be affected, but there also would be an adequate source to explain the atmospheric waters necessary for the Flood.

However, the postulated cataclysm also involves tectonic and magmatic upheavals, as well as tremendous hydraulic and sedimentary disturbances, on the bottom of the ocean. Thus a secondary source of water is postulated as existing in vast subterranean heated and pressurized reservoirs, perhaps in the primeval crust or perhaps in the earth's mantle itself, a situation similar to that existing at present but greater in quantity. The explosive release of those waters, accompanied by magmas and followed by earth movements, provides another cause of the cataclysm.

The primeval creation of those two vast bodies of water, one above the troposphere and the other deep in the earth's crust, would thus serve the dual purpose of providing a perfect environment for

terrestrial life and also for transmitting the energy for the universal cataclysm which later would destroy that life.

On the surface of the primeval world, it is postulated, there was probably an intricate network of narrow seas and waterways whose precise locations need yet to be determined. Though the uniform climate would inhibit air mass movements, as well as storms and heavy rains, a daily cycle of local evaporation and condensation would maintain an equable humidity everywhere. The favorable climate, aided by the highly effective radiation filter provided by the vapor canopy, would favor abundant plant and animal life, longevity of animal life, and growth of large-sized animal organisms. The trigger to unleash the stored waters and initiate the cataclysm might have been any one of a number of things. The simplest explanation would be to assume that the pressurized waters below the crust suddenly erupted at a point of weakness. Collapse at one point would cause a chain reaction leading to similar eruptions at many other points around the world.

The turbulence in the atmosphere which would result, together with immense amounts of dust blown skyward, would then initiate the condensation and precipitation of the vapor canopy. (Pp. 124-25)

What the Diluvialists Added

Where did the creationists get all the geologic and hydrologic ideas in this scenario? Certainly not from the account in Gen. 7-8. Did these latter-day creationists concoct this material or did they dig into the past to resurrect fantastic ideas expressed by the early Diluvialists, hardpressed at the time to reconcile the growing naturalistic view of fossils with the Scriptures? Turning to Henry and Carol Faul's history of geology, an answer was quickly forthcoming (Faul and Faul, 1983). The "diluvial doctrine" seems to have arisen in the Christian church in the 1500s. The Fauls attribute it to a Frenchman, Bernard Pallisy, about 1580, but in any case, "The diluvialist concept grew from these ancient roots through the Renaissance, flourished in the seventeenth and eighteenth centuries, and produced a tremendous literature" (p. 26). I struck pay dirt some pages later with the Fauls' description of a book written by a Cambridge divine and royal chaplain, Thomas Burnet (1636-1715), under the title of The Sacred Theory of the Earth (pp. 48-51). In this volume, published in English in 1680-1689, Burnet presented a novel interpretation of the Creation and Flood scenarios of Genesis. In the Fauls' words, this was Burnet's version of the Deluge:

Burnet argued that the paradisical world was perfectly smooth and spherical. It was composed of differentiated parts: a core of heavy particles surrounded by shells of liquid and air. The liquid differentiated into fat and oily liquids above and water below. Dust from the primeval atmosphere settled into the oily shell and eventually it became firm and habitable land. The Deluge resulted when this solid shell collapsed into the waters below and they gushed out to flood the Earth, leaving it in ruins. It is written in a rich, flowing style and still makes good reading, but many of the ideas it presents with such smooth assurance were debatable at the time and produced a strong response from theologians and scientists alike. It was not coincidence that the Theory was reprinted often and remained a best seller for decades. (P. 49)

Nothing in Burnet's geological hypothesis is suggested

by the Genesis account of the Flood. It is pure, original fabrication and it most certainly describes a tremendous global catastrophe. The notion of collapse of a thin, brittle crust to release great volumes of pressurized water seems almost identical in the Burnet and ICR versions, separated in time by three full centuries.

Burnet seems also to have been responsible for the modern creationists' idea of a moist, mild global climate during the "golden age" following the Creation. Burnet "concluded that before the Flood there was over the whole earth perpetual spring, disturbed by no rain more severe than the falling of dew" (White, 1896, p. 219). How like the climatic description in the paragraphs quoted above from the ICR textbook: "Though the uniform climate would inhibit air mass movements, as well as storms and heavy rains, a daily cycle of local evaporation and condensation would maintain an equable humidity everywhere" (Morris, 1974a, p. 125). In reviewing Burnet's account of the "golden age" following Creation, White makes a remarkable statement: "he (Burnet) now mixed the account of the Garden of Eden with heathen legends of the golden age" (p. 219). Does this mean that the modern creationists of the ICR are accepting a mixture of God's word and "heathen legends"? If so, the level of their desperation has sunken tragically low.

Modified Versions of the Flood of Noah

As with the Creation scenario, theologians and theistic scientists have attempted to modify the Flood scenario to more reasonable proportions in terms of stringent requirements--in the case of the Creation, by allowing greater time for it to occur and, for the Flood, to require less water depth and a more limited extent. One popular modification, called the "local flood theory" is discussed by the ICR (Morris, 1974a, pp. 250-54). A recent example, cited by Morris, is Bernard Ramm's version, published in 1954.

It is a rather simple matter for any geologist or hydrologist to adapt the words of Genesis to a description of a major river flood inundating a broad floodplain of extremely low gradient. In Noah's time, human concepts of the extent of the "world" must have been severely restricted--a flat world, in any case, and one of very small radius of direct knowledge. A flood crest of 8 to 10 m might easily top the natural levees bordering a meandering river channel, ponding large backwater areas between the river and distant terraces or bluffs. Full drainage of the ponded water might require weeks to months. To naturalize the biblical event does, of course, require that one abandon a strictly literal interpretation of the scripture. This liberalization is totally unacceptable to the creationists because, as the ICR text explains, if the magnitude and intensity of the Flood as stated in Genesis is strictly upheld, then the case for evolution collapses (Morris, 1974a, p. 251). Their text states:

> It is not easy in the academic world to maintain a so-called "flood theory of geology." There are, no doubt, certain geological problems in such a position, but a far more real problem is the "flood" of scholarly wrath and ridicule that descends upon those who hold it--that is no theory! The Genesis Flood is the real crux of the conflict between the evolutionist and creationist cosmologies, and evolutionists invariably concentrate their strongest attacks at this point. By the same token, this is where Christians should also marshall their strongest and most vigorous campaign. Unfortunately, their strategy until recent years has almost completely ignored it.

> If the system of flood geology can be established on a sound scientific basis, and be effectively promoted and publicized, then the entire evolutionary

cosmology, at least in its present neo-Darwinian form, will collapse. This in turn, would mean that every anti-Christian system and movement (communism, racism, humanism, libertinism, behaviorism, and all the rest) would be deprived of their pseudo-intellectual foundation.

> These are the stakes involved and it is no wonder that evolutionists have so opposed the historical fact of the global cataclysm known as the Genesis Flood. (P. 252)

The second modified version of the Flood is the "tranquil Flood theory," which holds that the rise and fall of the waters were placid events, lacking in forces capable of doing important geologic work, such as erosion, sediment transport, and sediment deposition. The flooding would not have been accompanied by volcanic or tectonic activity. The ICR text specifically notes the promotion of the tranquil flood theory by Dr. J. Laurence Kulp, a geochemist and geochronologist, formerly on the faculty of Columbia University (Morris, 1974a, p. 254; Kulp, 1950). In an earlier paragraph I remarked that the Genesis story of the Flood is quite devoid of mention of destructive forces accompanying the Flood. Interpretations to the contrary come from outside sources. We should not lose sight of the fact that God was fully capable of carrying out a worldwide deep inundation with no other destruction than the drowning of living forms. To suggest otherwise is an unwarranted presumption, infringing upon the divine prerogative and casting doubt on God's methods and purposes. If the words of Gen. 7-8 are taken literally, they most certainly describe a tranquil flooding from start to finish. For the ICR text to assert flatly that such a flooding is an absurd idea from the standpoint of hydrology and geophysics is clearly out of order (p. 255).

Creationists' Views on Catastrophism and Uniformitarianism

We should go no further in analyzing the creationists' Flood geology until we have exposed, once and for all, their distorted and inaccurate views on the modern mainstream version of two relics of past controversy in the history of geology as a science: catastrophism and uniformitarianism. Creation scientists endlessly parade the desiccated corpses of these maimed veterans of a past war to the accompaniment of ritual chants that, when closely examined, are meaningless in the late twentieth century. Let us put these two corpses into massive sarcophagi, encased in reinforced concrete, and let them lie in peace.

The context of the two contrasting views of geologic history has been covered in Chapter 9, in our section on the fossil controversy and its gradual resolution through a succession of rather heated debates. Catastrophism had a fairly simple and direct introduction through the linkage of fossils with a single global cataclysm, as witness the fantastic scenario put forward by Thomas Burnet in the 1680s. Less dramatic versions simply had the living organisms drown in the rising Flood of Noah and receive a simple, dignified burial in Noachian sediments. The one-step view of catastrophism, which is held as doctrine today by the creationists, became greatly complicated in the first two decades of the 1800s when Baron Cuvier proposed a succession of sudden catastrophes, each followed by creation of new life. As we noted earlier, Cuvier's multiple extinction/creation scenario caused a theological problem, and it was not really resolved by Robert Jameson's gratuitous suggestion that only the final extinction of the series was the Flood of Noah. Modern creationists do not like Jameson's suggestion either, because God had finished his work on the sixth day of Creation and chose to do no more. Besides, to accept Cuvier's plan they would have to disrupt the warm, humid softness of the golden era with a succession of catastrophes and creations.

To make things even more confused, Abraham Werner's view of geology was then in vogue in Germany, England, and the United States, and it could be construed as a kind of catastrophism. Faul and Faul note that Jameson, in translating Cuvier's work into English, "added a lot of notes that try to make Cuvier's statements appear in Wernerian light" (1983, p. 142). Will the real catastrophism please stand up! My suggestion is that we erect a barrier in time--say 1868--beyond which the use of "catastrophism" may not pass. I pick 1868 because it was the year in which Louis Agassiz, the last holdout from Cuvier's time, made his final statement against Darwinism and in support of Cuvier (A Journey in Brazil, coauthored by his American wife, Elizabeth Cabot Cary).

And what should we recommend for uniformitarianism? Should it, too, be declared dead? More importantly, why was it born and what made it viable? James Hutton, who gave it birth, and Charles Lyell, who raised it as an adopted child, had a single purpose--to counter a contrived catastrophism that required supernaturalism and to replace it with a naturalistic view of the world of nature. Unfortunately, but of necessity, Hutton's zeal led him to stress the continuity of process acting over spans of time that seem infinitely long to a human being. The Flood of Noah lasted only one year. A million years is to the mind, by comparison, infinity. Hutton reasoned that if the alternations of sea with land (transgressions and regressions of shallow seas) had occurred repeatedly and continuously, we should look at processes in action today for the clues to the nature of processes that acted in the distant past. But read Hutton's words for yourself:

> Hence we are led to conclude, that the greater part of our land, if not the whole, had been produced by operations natural to this globe.

> By thus proceeding upon investigated principles, we are led to conclude, that, if this part of the earth which we now inhabit had been produced, in the course of time, from the materials of a former earth, we should, in the examination of our land, find data from which to reason with regard to the nature of that world . . . .

> If we could measure the progress of the present land, towards its dissolution by attrition, and its submersion in the ocean, we might discover the actual duration of a former earth; . . . But, as there is not in human observation proper means for measuring the waste of land upon our globe, it is hence inferred, that we cannot estimate the duration of what we see at present, nor calculate the period at which it had begun; so that, with respect to human observation, this world has neither a beginning nor an end. (Abstract of a Dissertation, 1785, as given in Faul and Faul, 1983, p. 108)

Hutton's writing style did not lend itself well to popular reading, but it certainly raised the hackles of the religious fundamentalists of the time. Uniformitarianism underwent an explosive expansion, so far as public recognition was concerned, under Charles Lyell, whose first edition of Principles of Geology appeared in 1830. Lyell was a competent and widely traveled geologist in his own right; he was also an enticing writer and given more to subtle persuasion than vindictive argument. Somehow he got across a message that the uniformitarian view of geology could be held independently of one's traditional religious views. In placing geology on a naturalistic basis, Lyell seems also to have left a legacy that has plagued geology as a science even into modern times. This legacy can be summed up in now-hackneyed statements: "The present is the key to the past; geological processes have always acted through the past in the same manner and at the same rate as they do today." The impression of a gentleness of nature is coupled with an image of extreme slowness of result.

But, of course, no geologist of Hutton's and Lyell's time could shut out of mind the consideration of natural events of great suddenness and violence. The great Lisbon earthquake of 1755 was then fresh in the minds of many scientists, for it was a true human catastrophe. Thousands perished in buildings that collapsed to rubble. It was also a very real flood disaster, for the withdrawal of seawater from the harbor had enticed a large number of townspeople to examine the strange disclosure and many of them drowned as the ensuing seismic sea wave arrived in great force. Altogether, some 60,000 persons perished in a city of 235,000. Volcanoes were active from time to time in the Mediterranean region, and every geologist was familiar with Pliny's account of the great eruption of Vesuvius in A.D. 79, which entombed many humans under a fall of volcanic ash in Pompeii, while destroying Herculaneum in a volcanic mudflow. Lyell had visited Mount Etna, an active volcano on the island of Sicily. He concluded that the tremendous pile of volcanic rock had accumulated since the end of the Tertiary Period--a million or two million years in the making to be sure, but finite in age and comparatively young as crustal rocks go.

Certainly geologists of Lyell's time were also familiar with rapid trenching by great river floods, violent wave action that cut back the shore rapidly in times of storm, and all sorts of devastating avalanches and landslides in mountainous terrain such as the Alps. These are all processes seen in action today at one place or another around the world.

Lyell made clear his recognition of events of unusually large magnitude in terms of energy expended and geologic materials shifted about in short time spans. Today, we describe such events as episodic, meaning that they are brief episodes relative to the time that intervenes and implying that their occurrence is irregular in time. In 1982 the episodic view of geologic events was brought to the attention of geologists by Robert H. Dott, a leading figure in stratigraphy (the study of of strata). As reported in Geotimes (vol. 27, no. 11, pp. 16-17), Dott addressed his colleagues on the modern meaning of uniformitarianism. Is the record of the strata mostly continuous or is it episodic? He said that he had chosen the term "episodic" over "catastrophic" and that the latter should be purged from our vocabulary because it feeds the "neo-catastrophist-creationist cause." Dott referred to the principle of episodic events as episodicity. He remarked: "I hope I have convinced you that the sedimentary record is largely a record of episodic events rather than being uniformly continuous. My message is that episodicity is the rule, not the exception." More recently, the subject of episodicity in Lyellian uniformitarianism has been analyzed by geologist Donald H. Zenger of Pomona College. The Lyell quotations given below are from his paper (Zenger, 1986).

Of episodic great floods, Lyell wrote the following passage describing the ability of running water to transport large rock fragments in large quantities. He is describing a summer storm in Scotland:

> The floods extended over a space of about five thousand square miles . . . . All the rivers within that space were flooded, and the destruction of roads, lands, buildings, and crops along the courses of streams was very great. (Lyell, 1830, p. 174)

Lyell seems to have given a very clear picture of the episodic nature of both volcanic and tectonic processes--descriptions that we would accept today. On volcanoes, he wrote:

Again, if one of these igneous formations is examined in detail, we find it to be the product of many successive ejections or outpourings of volcanic matter. As we enlarge, therefore, our knowledge of the ancient rocks formed by subterranean heat, we find ourselves compelled to regard them as the aggregate effects of unnumerable eruptions, each of which may have been comparable in violence to those now experienced in volcanic regions. (Lyell, 1872, pp. 114-15)

Interestingly enough, Lyell recognized and accepted the conclusions of a fellow geologist (George Scrope) to the effect that extinct volcanoes of the Auvergne district of France had erupted in an episodic fashion, burying freshwater sediments with lava flows. In the company of Roderick Murchison, Lyell examined these alternating volcanic rocks and lake sediments. Their conclusion was that this complex history of erosion and deposition, punctuated by repeated volcanic events, falls within the Tertiary (Faul and Faul, 1983, p. 131). This is certainly the modern view of the episodic activity of great volcanoes--Mount St. Helens, Lassen, and Shasta of the Cascades, for example.

On the subject of tectonic activity leading to the rise of great mountain blocks by faulting, Lyell's description of the process is essentially that which we use today, namely, an episodic succession of fault slippages, each generating an earthquake, that add up over many centuries or millenia to a rise of thousands of feet a single fault block--the Sierra Nevada, for example.

Zenger has examined Lyell's writings to reveal what his position really was on the tempo of events in physical geology (1986). What we most commonly read about Lyell's uniformitarianism is his emphasis on the constancy and slowness of the action of agents of erosion, transportation, and deposition. This view of geologic processes and the course of organic evolution is now often referred to as <u>gradualism</u>. We find this view expressed in the second sentence of the following statement by Lyell appearing early in his first edition of <u>Principles of Geology</u>:

The form of a coast, the configuration of the interior of a country, the existence and extent of lakes, valleys, and mountains, can often be traced to the former prevalence of earthquakes and volcanoes. . . . On the other hand, many distinguishing features of the surface may often be ascribed to the operation at a remote area of slow and tranquil causes. (Lyell, 1830, p. 2, in Zenger, 1986, p. 11)

The idea of slowness and tranquility of geologic process (gradualism) is the part of Lyell's picture that seems to have been lifted from his writings to be transplanted widely into elementary textbooks of geology in the present century. Zenger rightly urges us to give Lyell just due for his recognition of the episodic nature of processes in all areas of physical geology, including sedimentology, volcanism, and tectonism (1986, pp. 11-12).

Zenger's main thrust, however, seems to be that we recognize what Lyell was attempting to do in the context of the uniformitarianist/catastrophist debate of his time. The violence of great storms and floods, avalanches, earthquakes, and volcanic explosions, impressively catastrophic as they may seem to humans witnessing them, is insignificant when compared with the immensity and intensity of the biblical cataclysm of the Flood, or of each of the successive cataclysms that were postulated in the multiple-catastrophe program of Cuvier and his followers. Writes Zenger: "Lyell described numerous episodic processes, recognizing them as such, but these would not have been in the least 'catastrophic' or 'castaclysmic' to a

catastrophist!" (1986, p. 11). That's the whole point--the scale of the Diluvialists' catastrophe is several or many orders of magnitude greater in intensity than any episodes of rapid change seen in action today. Lyell made the point clearly in the following passage, in which he contrasts a succession of tectonic episodes with a single genuine diluvialist catastrophe:

We know that one earthquake may raise the coast of Chile for a hundred miles to the average height of about five feet. A repetition of two thousand shocks of equal violence might produce a mountain chain one hundred miles long and ten thousand feet high. Now, should only one of these convulsions happen in a century, it would be consistent with the order of events experienced by the Chileans from the earliest times; but if the whole of them were to occur in the next hundred years, the entire district must be depopulated, scarcely any animals or plants would survive, and the surface would be one confused heap of ruin and desolation. (Lyell, 1830, pp. 79-80)

In this example Lyell assigns a century to the duration of a diluvialist catastrophe. Modern creationists give the great catastrophe of the Deluge only one year! Lyell's figures describe a diluvialist catastrophe 2,000 times more intense than his 200,000-year uniformitarian period required to get the same job accomplished. The creationists' one-year event increases the intensity to a factor of 200,000 over the uniformitarian schedule of one shock per century. The mountain mass would have to rise 10,000 feet in no more than one year, and that would mean more than five earthquakes of Richter magnitude 8-plus per day. Aftershocks included, the shaking might be almost continuous. And think of the enormous tsunamis sweeping back and forth over the world ocean that whole time. Poor Noah! Frivolity aside, this little exercise puts some perspective on Lyellian uniformitarianism compared with supernatural catastrophism. Never, says Lyell, would the action of internal forces of volcanism and tectonism "lay a whole continent in ruins" (1872, pp. 130-31).

Why, then, all the current fuss over uniformitarianism? My recommendation is the same as for catastrophism: let them both lie dead and buried in the context and time in which they arose. Let us designate these terms as relics of the past and salvage only the useful ideas about uniformitarianism persisting into the modern era of geology. This course of action seems now to be the consensus among mainstream geologists, but the fundamentalist creationists insist upon disinterring the corpses and parading them around to the accompaniment of the same meaningless chant. Let me go further with an appraisal of the current situation.

The term "uniformitarianism" remains in use, and it is futile for me (or Professor Dott) to try to lay it to rest. As long as the creationists insist on bringing a discredited view of uniformitarianism into the public school classroom, the word will be bandied about. The first thing we must do is to switch metaphors. From corpses we turn to "straw persons" or "straw dummies." What we are going to find is a classic example of the straw-person strategy: set up a fallacious image of the opponent's position and attack that image. It is the oldest and most successful method of pseudoscience, as we observed in Part 1 in our analysis of the anti-science movement. It is a successful strategy because it makes use of scientific statements beyond the abilities of the audience to verify or even understand. It is particularly effective when statements of mainstream scientists are quoted out of context. The straw person contrived by the creationists is simply an outmoded and discredited version of the meaning of uniformitarianism. The built-in advantage of reviving the outmoded model is that it has already been rejected by mainstream scientists, and their own words from scientific journals and textbooks can be cited as valid reasons against it.

The first signs of misrepresentation appear in the creationists' characterization of uniformitarianism as a doctrine essential to mainstream geology. For example, ICR creation scientist Henry Morris refers to mainstream historical geology as a "religion of evolutionary uniformitarianism" (1968, p. 12). The assertion that uniformitarianism is a religious concept carries with it the implication that it is therefore not scientific. This is a familiar diversionary theme. As we have explained, the writings of the early uniformitarian geologists--Hutton, Lyell, and others--clearly intended it as a naturalistic hypothesis designed to replace the orthodox religious diluvialist version of stratigraphy and paleontology.

The scientific creationists profit greatly from statements made by mainstream geologists. First, there are geology textbook authors (myself included) who, in the past, perpetuated the Huttonian concept that uniformitarianism requires geologic processes to have operated throughout all past time in the same manner and at the same rate as today. Most often, we have used Sir Archibald Geikie's motto, coined in 1882: "The present is the key to the past." Although such statements have appeared and may accurately report what Hutton had in mind, they are usually presented in direct contrast to the catastrophist version, particularly that of Cuvier. It is easy enough to pick up only the Huttonian line and call it to fault. Authors are now much more cautious in their statements. In my 1981 college textbook, <u>Physical Geology</u>, I present an interpretation of uniformitarianism that is much more closely in accord with the present consensus. After reviewing the Hutton-Lyell version in the context of its development, I continue with an updated version, as follows. (Note that I use "principle of uniformity" for the older term.)

Looked upon today in the context of modern science, the principle of uniformity means simply that all phenomena of geology must be explained through the laws of science we accept as valid. The fundamental laws of physics, chemistry, and biology must have applied uniformly from early Precambrian time to the present. No explanation of an event that occurred in the Cambrian Period or the Triassic Period can be considered acceptable if it violates a scientific law we apply today. Supernatural forces are ruled out of the physical explanations of deposition and deformation of strata.

The principle of uniformity does not, however, require that the intensity of each geologic process shall have been uniform throughout all time. To the contrary, we accept as a valid conclusion that deformation of strata and intrusion of plutons may have been much more intense at one point in time than another. . . . There are some good reasons to suppose that the overall total intensity of igneous and tectonic activity has been declining since early Precambrian time. The total release of heat energy from radioactive elements within crustal rocks has almost certainly been steadily decreasing. Nevertheless, we believe that the physical laws of radioactivity have not changed since earliest geologic time, for to think otherwise would violate the principle of uniformity. (Strahler, 1981, p. 152)

A second source of what we might refer to as "self-incriminating" statements by mainstream geologists is in their journal articles, particularly in the field of paleontology and evolutionary biology, where a heated controversy has grown up over the tempo of evolution. Henry Morris found good pickings from articles by paleontologist Stephen Jay Gould. Morris's article is designed to show that mainstream geologists are coming

over to the side of catastrophism while at the same time abandoning uniformitarianism (1976a, pp. i-ii). The misdirection in Morris's presentation is the suggestion that because "catastrophism" is being bandied about by mainstream geologists, they are becoming persuaded of the rightness of the biblical doctrine of the Creation and Flood geology. Actually, these geologists are talking about intense episodic events, meaning natural events and processes of catastrophic proportions in terms of the impact on life forms. Most striking of these hypotheses is one currently enjoying a great deal of attention--that an impact by an asteroid or a large comet caused extinctions on a worldwide scale at the close of the Cretaceous Period. Hypotheses of this kind should under no circumstances be characterized as expressions of a religion-based doctrine of catastrophism. In reviving the archaic term and giving it an entirely new meaning, these geologists and the science news reporters who follow them unthinkingly play into the hands of the creationists.

Having offered that commentary, I give you a statement of Gould's that Morris has included in his 1976 article:

Charles Lyell was a lawyer by profession, and his book is one of the most brilliant briefs ever published by an advocate. . . . Lyell relied upon true bits of cunning to establish his uniformitarian views as the only true geology. First, he set up a straw man to demolish. . . . In fact, the catastrophists were much more empirically minded than Lyell. The geologic record does seem to require catastrophes: rocks are fractured and contorted; whole faunas are wiped out. To circumvent this literal appearance, Lyell imposed his imagination upon the evidence. The geologic record, he argued, is extremely imperfect and we must interpolate into it what we can reasonably infer but cannot see. The catastrophists were the hard-nosed empiricists of their day, not the blinded theological apologists. (Gould, 1975, pp. 16-17)

Morris hastens to remind his readers that Gould is not a creationist or a biblical catastrophist, and he inserts another quotation in which Gould clears himself of any involvement with supernaturalism. Nevertheless, a distinguished evolutionary scientist has given the creationists a free lift. That Gould is referring to the catastrophism of Cuvier is of no matter; "catastrophists" were empirical scientists, and that is what Morris wants the readers to believe.

Indications now are that the extreme episodicity of catastrophism and the Lyellian expression of uniformitarianism (including both gradualism and moderate episodicity) are becoming united in a single view of the world of nature. At a conference on the dynamics of extinction, held in 1983, one scientist was quoted as saying, "It is a great philosophical breakthrough for geologists to accept catastrophe as a normal part of Earth history." Another scientist said, "We have to accept asteroid impacts as part of the uniformitarian process" (Lewin, 1983, p. 935). As long as we must keep the two words, why not fuse them into one? For example, we could take our choice of "uniformicatastrophism" or "cataformitarianism." The merger would put a crimp in the creation scientists' style and require some revision of their ICR textbook, <u>Scientific Creationism</u>. For example, on page 236 we read: "The whole dogma of modern geology has been built upon the dogma of uniformitarianism, not catastrophism. . . . One cannot have his cake and eat it! The geological strata can be explained in terms either of global catastrophism or of uniformitarianism, but not both together" (Morris, 1974a). To Dr. Morris we can now say: "Not any more; not since we got uniformicatastrophism!" A few lines later on the same page by Morris we read: "It should be strongly emphasized that orthodox geology has no place for worldwide cataclysms." Well, Dr. Morris, we do now! We will have more to say about the new view of natural history in looking at recent trends in thought about mode and tempo in organic evolution.

Under the updated statement of a useful principle of uniformitarianism it boils down essentially to affirmation of the validity of universal scientific laws through time and space, coupled with a rejection of supernatural causes. For the creationists, the first part of this principle is held as valid for, as the ICR textbook points out: "Creationists believe in general uniformitarianism as an evidence of the Creator's providential maintenance of the laws He created in the beginning" (Morris, 1974a, p. 91). The text goes on to say that "The creation model is fundamentally catastrophic because it says that present laws and processes are <u>not</u> sufficient to explain the phenomena found in the present world. It centers its explanation of past history around both a period of special <u>constructive</u> <u>processes</u> and a period of special <u>destructive</u> <u>processes</u>, both of which operated in ways or at rates which are not commensurate with present processes" (p. 92). Clearly, the periods referred to by Morris are the Creation and the Deluge, respectively. Post-Deluge processes are assumed to follow the updated principle of uniformitarianism, and on this point both sides seem to be in agreement. This means that a field party consisting of both creationists and mainstream geologists can harmoniously investigate the volcanic ash deposits of Pompeii, the moraine of the receding Columbia glacier in Alaska, and the recent Madison landslide in Montana. If some remains of humans, animals, or plants or some artifacts happened to be found enclosed in deposits formed during these events, they are not to be treated as fossils formed during the Flood. This gives us all a grace period of about 4000 years of past geologic history about which we all think alike. Radiocarbon dates, adjusted by tree-ring dating, are acceptable to both parties in this grace period. How sad to think that a radiocarbon date of -7000 years from a tree trunk buried in the moraine of an alpine glacier would divide our little group of geologists into two camps, no longer on speaking terms--separated now by a bottomless chasm of theological dogma.

Professor James H. Shea of the University of Wisconsin at Parkside has done a thorough analysis of the prevalent misconceptions as to the current meaning and working status of uniformitarianism. Quoted here is the major part of the abstract of his paper:

> Advocates of creationism claim that uniformitarianism is a substantive theory, that it includes concepts of identity of ancient and modern causes, constancy of rates, gradualism, a very old Earth, and various specific assertions about Earth processes. None of these ideas is, in fact, a part of modern uniformitarianism, but each of them is quite common in the geological literature, thus providing a basis for creationist attack on the validity of historical geology. Modern uniformitarianism has been shown to be no more and no less than the philosophical rule of simplicity. Contrary to creationist allegations, modern uniformitarianism makes no assertions about nature, but instead, tells scientists to choose the simplest hypothesis that both fits the observations and leads to greatest simplicity in overall theory. What the creationists attack, therefore, is not uniformitarianism as it is used by contemporary geologists, nor uniformitarianism as it has been clearly explained in several careful analyses published since 1965, but a false 19th-century uniformitarianism that has been abandoned. (Shea, 1983, p. 105)

By way of a summary, what can we say is the modern view of uniformitarianism? What is included in its definition and what is excluded? First, let's eliminate things that don't belong.

What Shea calls "substantive theory" does not belong in uniformitarianism. By "substantive theory" is meant the findings of science as to the nature of the real world. Uniformitarianism should be restricted to statements of the methods adopted in pursuit of natural science. One such guiding principle we have already stated is the assumption of past and present constancy of physical laws throughout the universe. Another is the importance of using the simplest possible explanation consistent with observation. Under the latter point, explanations of past events in the light of what we observe of present-day processes is logically the method of choice, where possible. In making this choice we serve the principle of simplicity. Perhaps this is the full extent of uniformitarianism. If so, it nevertheless constitutes a strong foundation for the conduct of natural science.

What are some of the substantive ideas that, while not being included in the statement of uniformitarianism, are nevertheless its products? One is the consensus about the great age of the earth and universe. Another is the admission to consideration of a wide range in rates at which phenomena operate. Hypotheses in this area include gradualism (very slow change), periodicity (regular rhythms of change), episodicity (irregularly timed, brief events), and even stasis (no change with time). Episodicity can span an enormous range in the intensity of an episode, from very minor events in deposition of sediment to great asteroid impacts. But even as evidence is being claimed for past occurrences of such world-catastrophic happenings, the very slow but steady motions of the great lithospheric plates cast a vote for gradualism on a colossal scale. Combinations of rates and modes of change are also admitted for consideration. For example, gradualism may alternate with episodicity. We see this combination in a popular current theory of the mode and tempo of evolution called "punctuated equilibrium." All such hypotheses concern observed phenomena or the empirical evidence of their having occurred.

Another category of substantive hypotheses that may be considered products of uniformitarianism are those that invoke explanations for which there are no known equivalents in recorded history to serve as examples. Illustrations are numerous, and without them planetary history would be devoid of much of its substance. Often cited is the evidence of enormous floods of glacial meltwater released by the sudden breakup of huge ice dams enclosing marginal glacial lakes. Global climatic conditions of the present do not allow the existence of large continental ice sheets that have spread far from their centers of accumulation. Thus the conditions for great meltwater floods do not presently exist. But the requisite physical processes can be observed today on a small scale along many marginal areas of glacial ice; hence, there is a tie established between the present and the past. Another example is the inferred existence in the past of vast shallow inland seaways in the continental interiors. In our present situation of high-standing continents with high freeboards, these seaways are not to be expected. Evidence that they did exist comes from the marine strata preserved as platforms in continental interiors. In yet another example, we postulate with much good evidence that for a time in the past there existed a single world continent--Pangaea--but today there are only several widely separated continents. In short, the present holds many keys to the past, but not all the keys for everything that needs to be explained. Our world does not contain a contemporary example of every geological and biological entity that ever existed.

The subject of uniformitarianism is complex; ideas expressed about it are often obscure and difficult to grasp. Divergent interpretations on substantive matters are strongly debated within the community of scientists, leaving the false impression that the meaning of uniformitarianism as a methodology is under fire. In ensuing chapters we will continue to hammer away at the basic concepts of modern uniformitarianism, exposing whenever necessary the creationists' attempts to confuse, confound, and misuse the issues that surround those concepts.

The Water Canopy Theory

We now return to some technical aspects of the postulated Flood of Noah, as described in Genesis and modified by extraneous postulates and assumptions from other sources. Creation scientists are not satisfied to accept the work of God in the Holy Scriptures as to the progress and duration of the Flood of Noah. They feel compelled to fit the simple and straightforward account with a mechanistic explanation based on laws of physics and chemistry as we know them today. God was in no way obliged to follow such laws in beginning and ending the Flood. Just as in the Creation, God's causation of the Flood could be carried out in ways of which humans have no knowledge, and no way ever to find out. Nevertheless, because the creationists want the Genesis account to be believed by all and to enjoy the status of natural science, thus making it a fitting subject for science classes in the public schools, they must supply a full and credible naturalistic explanation in conformity with laws of physics. This they have attempted in the water canopy theory.

The water canopy theory is fully described by Randy L. Wysong, a creation scientist whose major work, The Creation-Evolution Controversy, published in 1976, is intended as a textbook. It develops the scientific and technical aspects of creationism in greater detail and depth than the ICR textbook (Morris, 1974a). The canopy theory dates back to earlier creationist works, among them The Genesis Flood by Whitcomb and Morris (1961), a book I referred to in reviewing the history of the recent rise in fundamentalism. In 1984, ICR staff scientist and faculty member, Dr. Larry Vardiman reviewed the canopy theory (or "model") that has been significantly expanded with mathematical modeling by J. C. Dillow (1981; 1983). Vardiman holds the Ph.D. in atmospheric science from Colorado State University and is also chairman of the physical science department of Christian Heritage College.

The canopy theory or model requires that there existed in the golden age following Creation, but preceding the Flood, an envelope of water completely surrounding the earth in a zone suspended between the earth's surface and the ozone layer. According to diagrams in Wysong the water canopy lies about midway between the earth's surface and the ozone layer (1976, p. 389). The diagrams lack vertical scales and no altitudes are given in the text, but we can infer altitudes by reference to the diagram of post-Flood conditions. The ozone layer today lies largely in the altitude range of 20-35 km (12-21 miles), but it varies with both latitude and season (Strahler, 1971, pp. 190-91). It lies in the stratosphere, which is the layer above the troposphere. Wysong's diagram shows (if a linear scale is to be assumed) that the water canopy is in the upper troposphere, which is where the jet streams are found today and where upper-air waves and disturbances are most strongly developed. Because such a location seems unlikely, we shall assume that the water canopy in pre-Flood times was located about where the ozone layer is found today, and the ozone layer itself was pushed to a higher level. (It is regrettable that neither Wysong's or Vardiman's accounts meet the rigorous requirements of scientific presentation; their vagueness makes them difficult to evaluate in specific terms.)

Lest the reader be misled, the "canopy" consisted not of liquid water but of water vapor, which is water in the gaseous state. All precipitation as rain or snow occurs through the condensation of atmospheric water vapor and is invariably associated with cooling of the air through which the water vapor is diffused. Various mechanisms exist for the requisite cooling to occur. Under normal conditions, cooling leads first to formation of cloud particles, and these can grow rapidly into water droplets or snow crystals. The vapor canopy (as Vardiman correctly terms it) is postulated to have contained the equivalent of a water layer about 12 m (40 ft) in depth over the entire earth. It is described as "resting on top of the current atmosphere" (Vardiman, 1984, p. i). The Flood hypothesis calls for a rapid condensation of the vapor in the canopy over a 40-day period, causing it to fall to earth as the Flood, or universal Deluge.

The vapor canopy serves two purposes in the Flood scenario. First, it was a major source of floodwater, the other source being subterranean water emerging from springs. Second, the canopy caused the uniform warm, moist climate that is postulated to have prevailed over the entire globe in pre-Flood times. The latter condition is attributed to the well-known greenhouse effect on planets with atmospheres. Incoming solar radiation, reradiated as longwave radiation, is absorbed by the atmosphere and causes the air to be warmed. Generally speaking, the greater the density and thickness of the atmosphere, the stronger the effect and the higher the surface temperature of the planet. Venus, with a dense atmosphere made up mostly of carbon dioxide, has an atmospheric pressure at its surface about 100 times greater than the Earth's surface pressure; Venus's surface temperature holds at about 480 C (900 F) because of the very strong greenhouse effect.

The vapor canopy in pre-Flood times not only maintained a uniformly warm and moist climate over the globe, but it also caused sea-level barometric pressure to be about double what it is today. Creation scientists suggest that this greater pressure made it easier for animals to breathe. Thus longevity was greatly increased, along with general health benefits. Interestingly enough, the greater air density made it possible for the gliding winged reptiles, like the pterodactyls, to glide long distances despite the general calm that prevailed. The flying reptile, pteranodon, was able to launch itself from the ground in still air by flapping its wings, a feat that would have been impossible under today's lesser air pressure and density (Vardiman, 1984, p. iii). It was indeed a good time to be alive--oh, to have lived 969 years, as did Methuselah!

Creationists have made the vapor canopy theory a kingbolt of the Flood scenario. But does it "hold water" in terms of what is known of atmospheric science? I am concerned about the stability of such a system. It seems to me that the canopy could not exist for two reasons. First, water vapor, if not isolated or enclosed by an impervious barrier, quickly diffuses into surrounding regions of lower vapor density. Everyone who uses a vaporizer (humidifier) in the home or who boils a pot of water on the stove in a closed room knows this. Water vapor of a dense "canopy"--should such a canopy have momentarily come into existence for reasons unspecified-- would quickly diffuse downward and upward, until it attained a uniform distribution over the globe such that the proportion of water molecules would be in a uniform ratio to the other molecules of the atmosphere (largely nitrogen and oxygen). In other words, the water vapor would respond to the laws of behavior of gaseous molecules. The density of the gaseous mixture would quickly reach a distribution like that found today-- densest at the surface and diminishing upward exponentially to the upper limit of the atmosphere.

A second reason why the supposed "canopy" would not survive has to do with dynamics and the distribution of incoming solar energy. Two factors are involved. The energy input of the atmospheric system is the sun's rays, largely in the shortwave spectrum that includes visible light. The earth is a sphere turning under the sun's rays. Disregarding for the moment the tilt of the earth's axis with respect to the plane of the ecliptic, maximum input of energy per unit area is at the earth's equator; the input dimishes with increasing latitude and falls to zero at the poles, where solar rays parallel the surface. But the earth is also an emitter of energy, mostly in the longwave (infrared) portion of the spectrum, and must return to space exactly as much energy as it absorbs from the sun's rays. While this budgetary equation is correct for the globe as a whole, imbalances exist according to latitude: an energy surplus in low latitudes, a deficit at high latitudes. The imbalance requires energy transport poleward across the parallels of latitude and it occurs by flow of the atmosphere, i.e., a global circulation system. There is no way the atmosphere in pre-Flood times could be kept static and preserve a uniformly dense canopy over the entire earth, providing, of course, that the earth was spherical and was exposed to the parallel rays from the sun. It would not have mattered if the earth rotated more or less rapidly than today. On Venus, with a dense atmosphere and a rotation period of once each 243 days, there is a vigorous global circulation system with wind speeds up to 360 km (220 mi) per hour and turbulent convection cells also visible.

Earth rotation is also a major factor in determining the atmospheric circulation system. Rotation is responsible for the Coriolis effect, in which the gaseous flow is deflected to either the right or left of the path of motion, depending on which hemisphere is involved. The combination of energy imbalance and rotation results in a great overturning motion in low latitudes; it is called the Hadley cell. Heated air tends to rise in the equatorial belt, to travel poleward to a higher latitude, and to descend over the two subtropical belts where we find deserts. Hadley cell circulation would have destroyed a vapor canopy, if for any mysterious reason it had been momentarily formed. In higher latitudes, great upper-air waves form and re-form repeatedly, mixing warm air from low latitudes with cold air of polar sources. This mixing (known as advection) would also have acted to destroy a vapor canopy in high-latitude zones. The vapor canopy theory is thus absurd in the light of fundamental principles of science. Genesis says nothing of such a canopy. Robert J. Schadewald pinpoints the source of the idea:

> To account for the Deluge waters, some creationists propose that before the Flood the earth was surrounded by a canopy of water vapor. This basic idea was proposed in 1874 by Isaac Newton Vail, a Quaker Bible-scientist. Vail suggested that the planets evolve through a ringed stage (like Saturn) to a canopied stage (like Jupiter) to a final earthlike condition. Modern creationists throw out Vail's evolution and keep his canopy. (1983, p. 288)

Vail's theory appears to be completely secular, perhaps merely a bizarre concoction of his own imagination. Are there earlier sources of the canopy idea within the doctrines of the Christian Church?

Looking back in history to the pre-Diluvialist tenets of the Church, we find that the Ptolemaic system of the universe was long the accepted doctrine. The Greco-Egyptian astronomer Claudius Ptolemy (second century A.D.) envisioned a universe of crystal spheres. The spherical earth at the center was surrounded by concentric spheres of what were then known as the "elements"--water, air, fire, and ether, in that order-- and beyond them the seven astronomical zones, or "heavens." The "vapor canopy" of the creationists does not fit into this system, unless for some reason it substitutes for the fire sphere, and that certainly doesn't make sense.

At this point I thought it best to recheck the Scriptures on the subject of the Creation. In the King James version of Genesis 1, verses 6 and 7 read: as

6. And God said, Let there be a firmament in the midst of the waters, and let it divide the waters from the waters.

7. And God made the firmament, and divided the waters which were under the firmament from the waters which were above the firmament: and it was so.

Because the firmament clearly lies between the upper and lower waters, I sought a definition of "firmament." Webster's Third New International Dictionary tells us that the word comes from the Latin firmamentum, vault of the sky, a translation of the Greek word stereoma for a solid body, or a foundation. The first definition reads: "the vault or arch of the sky." Consulting next the Jerusalem Bible (English), I found a modern translation of the ancient Hebrew writings that was extremely enlightening. Here, verses 6 and 7 read as follows:

6. God said: "Let there be a vault in the waters to divide the waters in two." And so it was.

7. God made the vault and it divided the waters above the vault from the waters under the vault.

(Note: Footnote b reads as follows: "For the ancient Semites, the 'arch,' or 'vault' of the sky was a solid dome holding the upper waters in check.")

We realize, of course, that these ancient Semites conceived of their universe as a flat land surface, perhaps of circular outline at its limits, beneath the solid arch or vault that bore the celestial objects. The Scriptures, which must be taken literally under the creationists' beliefs, leave no question that the firmament was a solid barrier. A solid barrier makes sense in terms of the "windows of heaven" referred to in Gen. 7:11 and 8:2, as opening to release Flood waters from above. The idea of windows being installed in a tenuous vapor layer is nonsensical.

Figure 21.1 is a reproduction of an artist's conception of the world as described in Genesis. It shows a solid firmament, installed like a false ceiling beneath a roof, with waters filling the attic and sluices for releasing the waters. Note the "fountains of the deep" for releasing the "rivers of the nether world." There is even an omphalos, labeled as "navel of the earth." What a delightful legend this drawing portrays!

Next, I consulted the Scofield Reference Bible (1909, 1917) edited by C. I. Scofield with the aid of a group of consultants that appears to be strongly fundamentalist, since it included a faculty member from the Moody Bible Institute. Here, I struck pay dirt. The Genesis verses are identical with those of the King James Version, quoted above, but there is a notation attached to the word "firmament" in verse 6 reading: "Lit. expanse (i.e. of waters beneath, of vapour above)." This looks like the place where the Scofield committee adopted Vail's vapor canopy, substituting it for the upper water layer and at the same time dissolving the solidity of the firmament.

If I have traced the corruption of the Scriptures correctly, it exposes a conflict between the stated beliefs of the creationists and their practice. In the ICR textbook Scientific Creationism we read that the Gen. 1:1-2:3 Creation scenario must have been written directly by God Himself, since the Creation could not have been observed by any human being; it is God's personal narrative (Morris, 1974a, p. 206). The text comments as follows on the Creation story:

> It would be well not to try to explain away its historicity by calling it merely a literary device of some kind. Rather, man should bow before its Author in believing obedience, acknowledging that

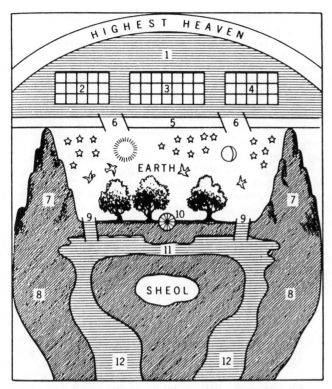

Figure 21.1 The world as described in the Old Testament; an artist's conception. (1) Waters above the firmament. (2) Storehouses of snows. (3) Storehouses for hail. (4) Chambers of winds. (5) The firmament. (6) Sluices. (7) Pillars of the sky. (8) Pillars of the earth. (9) Fountains of the deep. (10) Navel of the earth. (11) Waters under the earth. (12) Rivers of the nether world. (From The Interpreter's Dictionary of the Bible, vol. 1, pp. 702-09, Abingdon Press. Copyright © 1962 by the Abingdon Press. Used by permission.)

He has clearly spoken, in words that are easy to be understood, concerning those things which man could never discover for himself. (P. 206)

Need more be said? The creationists, in devising their own fictional story of a vapor canopy, have flagrantly put aside the clear word of God. The book of Genesis neither describes nor implies the existence of a vapor canopy, nor can that imagined vapor canopy evoke the slightest vestige of support from mainstream natural science.

The creation scientists have, in their canopy model, used a water-equivalent depth of 12 m (Vardiman, 1984, p. i). Just how much more water may have come from the subterranean sources is not mentioned, but a uniform addition of 12 m to existing sea level would not be much of a flood--certainly not if it were to land the Ark on Mount Ararat. Two professors of earth science at St. Cloud State University, Minnesota, Leonard Soroka and Charles L. Nelson, have made extensive calculations of the total quantity of water required to raise present sea level to the height of Mount Everest, so as to satisfy the Genesis provision of covering all mountaintops (Soroka and Nelson, 1983). The total requirement comes to 4.4 billion cubic kilometers. They conclude that to accommodate this volume as water vapor in the atmosphere is impossible. They give four reasons for that conclusion:

First, atmospheric pressure would be about 840 times higher than it is now. Second, the atmosphere would be 99.9 percent water vapor and it would be impossible for humans and other animals to breathe such an atmosphere. Third, such a mixture of gases could not even exist as gases at temperatures that humans could tolerate. Fourth, neither could the

water be accommodated as cloud droplets since there would be insufficient nitrogen and oxygen (less than .1 percent) to support them and since as clouds they would have prevented nearly all sunlight from reaching the surface. In short, such an atmosphere would not have allowed terrestrial life as we know it to exist on the surface of the earth. (P. 135)

Even more damaging to the theory of an atmospheric water source of such magnitude is the thermodynamic effect of condensation of the water vapor. Soroka and Nelson show that the rise in temperature caused by liberation of the latent heat of vaporization during a 40-day period would raise the atmospheric temperatures over the entire earth to over 3,500 C (6,400 F). Before this heat could have been dissipated by radiation into space, it would have set the ocean to boiling and would have cremated the Ark (p. 135). The same authors also evaluate the subterranean water source through springs, assuming that the subterranean source must have supplied most of the 4.4 billion cubic kilometers. If that much water were to be exuded from the crust of the continents and oceans, the porosity of the crust must be at least 50 percent (half solid mineral matter and half open voids). Actual porosity of crustal rock is much less than 1 percent because rock under the pressure that exists at depths of several kilometers is ductile in behavior and capable of closing any open pores that might exist.

Spring water that issues from the earth is commonly referred to as ground water; it originates as precipitation, percolates downward, where it may become heated, and returns to the surface. This recycling process also applies to the hot springs (hydrothermal springs) recently found to be common along the mid-oceanic spreading axis or rift of the deep ocean floor. The hot water is recycled seawater (brine) that made its way downward along faults and came in close contact with rising magma beneath the spreading rift. To suppose that a large volume of water could be supplied from greater depths than the crust is completely unreasonable in consideration of the physical state of the mantle material that lies below the crust. Information from earthquake waves shows that a soft or weak layer in the mantle consists of silicate rock close to its melting point, and perhaps partially melted in some places. These conditions preclude the existence of reservoirs of free water. In any case, if highly heated water did burst forth from such an environment into the ocean and atmosphere, the heat it would bring with it would raise the temperature of the atmosphere to a level in which the Ark could not survive (Soroka and Nelson, 1983, p. 136).

Soroka and Nelson also consider the possibility that the water for the Flood of Noah might have been obtained from a comet that impacted the earth (1983, pp. 136-37). A comet of water ice, sufficiently large to supply the necessary water, would on impact release so great an amount of heat as to raise the atmospheric temperature to over 6,800 C, insuring destruction of the Ark. The same authors also consider a comet-impact model in which small comets falling into the ocean would generate great standing waves (seiches) that would temporarily flood the continents to the requisite heights. This mechanism would not, however, cause a continuous inundation for 40 days, as the Genesis account requires. The authors accompany their text with sets of calculations for each of the models they consider.

Noah's Ark

A topic distinct in itself is the capability of Noah's Ark to carry out the functions required of it by the Genesis story. It is difficult to treat this subject in a coldly materialistic way, since what we are dealing with is a treasured story about animals caught up in a great adventure. To attempt to justify the Genesis story in terms of its logistics puts the fundamentalist creationists in the role of despoilers of a cherished childhood tale, but that is the path they have chosen. Those who refute the adequacy of the Ark can fall easily into the role of heapers of ridicule since, upon close examination, the Ark is indeed full of holes.

It seems hardly profitable to review this topic-- sometimes referred to scathingly as "Arkeology." It includes the fascinating story of reports of remains of the Ark on Mount Ararat and of the many expeditions mounted to find those remains. Creationists have been active in that search, and have written a great deal about it. One account is by Tim F. LaHaye and John D. Morris under the title The Ark on Ararat (1976). John Morris, a son of Henry Morris, holds the Ph.D. degree and is a professor of geological engineering at the University of Oklahoma; he also serves as Assistant Dean of the ICR Graduate School. As an active organizer and participant in recent expeditions to Mount Ararat, John Morris gives frequent progress reports in the ICR Impact Series (Nos. 22, 47, 116, 125). (For a brief review of the ICR expeditions, see Edwords, 1983, p. 35.)

Suppose that the Ark should be found on Mount Ararat, either intact or with a large portion of it remaining. What then? Would creation scientists submit to having it dated by the radiocarbon method? Would evolution have to be scuttled? A much more important question would remain, according to Robert A. Moore:

Suppose you picked up the newspaper tomorrow morning and were startled to see headlines announcing the discovery of a large ship high on the snowy slopes of Mt. Ararat in eastern Turkey. As you hurriedly scanned the article, you learned that a team from the Institute for Creation Research had unearthed the vessel and their measurements and studies had determined that it perfectly matched the description of Noah's Ark given in the book of Genesis. Would this be proof at last--the "smoking gun" as it were--that the earliest chapters of the Bible were true and the story they told of a six-day creation and a universal flood was a sober, scientific account?

Perhaps surprisingly, the answer is no. Even this sensational find is not enough to validate a literal reading of Genesis. Our continuing skepticism is in the tradition of philosopher David Hume, who wrote that "the knavery and folly of men are such common phenomena that I should rather believe the most extraordinary events to arise from their concurrence than admit of so signal a violation of the laws of nature." As we shall see, the story of the great flood and the voyage of the ark, as expounded by modern creationists, contains so many incredible "violations of the laws of nature" that it cannot possibly be accepted by any thinking person. Despite ingenious efforts to lend a degree of plausibility to the tale, nothing can be salvaged without the direct and constant intervention of the deity. (1983, p. 1)[1]

Moore's lengthy essay is titled: "The Impossible Voyage of Noah's Ark." A partial list of the headings and subheads will give you some idea of the richness of the subject: building the Ark, needs of animals, problems for the builders, accommodating all those animals, genetic problems, taxonomic problems, leaving some things behind, plants and seeds, animal migration, parasites and diseases, surviving the Flood, those that died, survival of the Ark, fate of the cargo, caring for the cargo, feeding the animals, special dietary needs, storage of food and water, sanitation and water disposal, ventilation,

landing on Ararat, release of the animals, survival and redistribution. An enticing list, indeed!

For a rebuttal (written earlier, hence, a "prebuttal") you might try "An Engineer Looks at Noah's Ark," Chapter 31 in LaHaye and Morris (1976). A much briefer account than Moore's, it attempts to make the Ark seem capable and its problems tractable.

When one is done with a reading of Moore's careful consideration of every conceivable technical and scientific aspect of the Ark and its mission, the creationists' literal adherence to the entire biblical scenario appears to have failed utterly to retain credibility--so, at least, it seemed to me. The problem of the Ark clearly goes much deeper than an exercise in technology. I can do no better than to let Moore present a modern philosophical perspective of the Noachian Deluge. Moore (a footnote on the opening page of his essay tells us) is a "a writer on religious subjects, has testified on church-state issues and is an experienced mountain climber (with no intention of joining an Ark expedition)." Here are some of Moore's concluding statements:

When one reads the story of the great flood in the book of Genesis, one is struck by the matter-of-fact style of the narrative. While it definitely has the larger-than-life flavor typical of legends, the reader would not suspect that he or she is dealing with the bizarre impossibilities we have detailed above. After all, the ancient Hebrews lived on a small, disc-shaped world with a dome overhead and waters above and below. There were only a few hundred known animals, and subjects such as ecology, genetics, and stratigraphy were not even imagined. The deluge was a mighty act of God, to be sure, but nothing that the ancient Hebrews would have found too extraordinary.

When, however, this same story is brought into the twentieth century and insisted upon as a literal account of historical events, a considerable change is observed. No longer a simple folk tale, it has become a surrealistic saga of fantastic improb-abilities. Events which seem relatively straight-forward at first glance--building a boat, gathering animals, releasing them afterwards--become a caricature of real life. The animals themselves are so unlike any others that they may as well have come from another planet; genetic Frankensteins with completely unnatural social, reproductive, and dietary behavior, they survived incredible hazards yet remained amazingly hardy and fecund.

How can we account for this transformation? Put simply, the tale of the ark grows taller in inverse proportion to the advance of science. Two centuries ago, when biology and geology were in their infancy, the theory of a worldwide flood as a major event in the earth's physical history seemed perfectly plausible and, in fact, was advocated by various scientists. But as geology progressed and as evolution gradually achieved a position of fundamental importance, the concepts of biblical literalists were shown to be untenable and were falsified. At the same time, the disciplines of biblical criticism, comparative religion, and archaeology uncovered the true origins of these stories and myths and showed that they were a natural part of the religious development of the Near East. (Pp. 37-39)[1]

A Question of Responsibility

In mainstream science, a scientist who puts forward a novel hypothesis is required to present it in full and in detail, explaining each feature or step precisely in accepted scientific language and with full documentation. The scientist must tell how the investigation was conducted, what methods were used, and exactly what was observed. Flood geology follows none of these procedures and offers none of the required information by which the hypothesis can be understood and evaluated. If creation scientists expect to have their product taught in the public schools on an equal footing with mainstream science, it is their responsibility to fulfill these requirements of science. In failing to do so, they fail in their responsibility to the students, to the students' parents, to the local communities, and to the nation as a whole. Mainstream science has no obligation whatsoever to attempt to refute Flood geology--a hypothesis vaguely and confusingly worded, lacking in completeness of statement, and nearly devoid of evidence.

Unfortunately, the going strategy of pseudoscience is to put forth vague and unsupported assertions and to defy science to disprove them. If science exercises its right to ignore the assertions, the purveyor of pseudoscience declares "See! Science has no answer for our theory, which is a true theory because the scientists cannot find fault with it." Then they add: "Because our theory is true, what mainstream science claims must be false." And then they can go the final step: "Evolution science, being false, must not be taught in the schools." Knowing that many voters and legislators are to be found among the hooting and stomping supporters of pseudoscience, the scientific community is moved by a sense of public duty to take upon itself the responsibility of answering. Perhaps some of the legislators will listen, and one can usually (maybe) count on the sanity and commonsense of the judiciary. So, back to the shop and on with the job!

Credit

1. From Robert A. Moore, The Impossible Voyage of Noah's Ark, Creation/Evolution, vol. 4, no. 1, pp. 1-43. Copyright © 1983 by Robert A. Moore. Used by permission of author and publisher.

Chapter 22

The Creationist Version of Flood Geology
Vis-à-Vis Plate Tectonics

Flood geology is divided into two consecutive time sequences. First is the single year of the Flood itself; second is the post-Flood period of some 4,350 years. Nearly all crustal rock features were generated in the Flood year, which was a time of enormously accelerated volcanic and tectonic activity, quite aptly described as cataclysmic, or catastrophic. The post-Flood period was, in contrast, one that even the creationists would recognize as dominated by uniformitarianism in the Huttonian/Lyellian sense that "the present is the key to the past." Although volcanism and tectonic activity persisted, these activities were rapidly diminishing in intensity and frequency, whereas the major role in shaping the landscapes of the continents was being carried out by glaciers, streams, waves, and wind-- agents whose action can be observed today. No really distinct time boundary in the geological sense separates the Flood year from the post-Flood period, despite the exactness of the termination of the Flood itself as marked by the grounding of the Ark and the liberation of its passengers.

In this chapter and the following three we concentrate on geologic processes and events of the year of the Flood. First, we take up the cataclysmic events involving wholesale dislocation of the crust, igneous intrusion, and volcanic extrusion. We follow with an account of the several kinds of deposits, largely sedimentary, that are attributed by creation scientists to the Flood year. As we proceed, we will state the creationists' arguments favoring extremely rapid occurrence of these varied phenomena, answering each directly with the findings of mainstream geology that would appear to demand vastly longer spans of time than a single Flood year.

Flood Chronology

Creation scientists of the Institute for Creation Research have attempted to correlate the Flood deposits and related events with the standard geologic timetable of mainstream science (Morris, 1974a, p. 129). Their correlation is shown in Table 23.1. The Flood itself encompasses all time since the beginning of the Proterozoic Eon, -2.5 b.y. The outstanding feature of this description of Flood stratigraphy is its vagueness, a characteristic of all Flood geology, emphasized by Philip Kitcher (1982, p. 129).

Creationists must tentatively acknowledge the established sequence of strata, their relative ages and names, and their fossil content, as derived by geologists working in all parts of the world over a total span of nearly two centuries under the principles of superposition and stratigraphic correlation. Using the same scientific method, the creationists would be obliged to start a new ordering of strata from scratch, based on a set of logical deductions (predictions) from a carefully stated general hypothesis. They would be free to rearrange the strata and their fossil content according to the hydrodynamic needs of the Flood hypothesis. Why have they not done so? They recognize that it would be a task of monumental

proportions, as Henry Morris states in the following paragraph:

> A great deal of research needs yet to be done, of course, to work out the details of this proposed revised geologic column. It should be remembered that the work of thousands of geologists for 150 years has all been described and classified in terms of the standard evolutionary column, so that the work of re-classifying this mass of material represents a monumental task which cannot be done overnight by a relatively small number of creationist geologists. (Morris, 1974a, p. 129)

It is the creationists' hope that in time this effort will succeed:

> In fact, teachers could render a real service to science and to their students and to mankind in general by encouraging significant numbers of their gifted students to prepare for careers in the earth sciences with this very goal in mind. We predict that viewing earth structure and history in terms of the more realistic model of creationism and catastrophism will eventually result in a much sounder understanding and utilization of earth's resources. (Morris, 1974a, p. 130)

For the present, then, their problem is to reconcile the accumulated data of mainstream geology with the limiting parameters of a one-year Flood.

Cataclysmic Events of Flood Geology

In Chapter 21 we reviewed the entire Flood scenario, quoting from the ICR textbook (Morris, 1974a, pp. 117-18). We saw that scarcely any of the happenings it relates can be identified with the Genesis account. Nearly all of the cataclysmic stuff is fabricated or borrowed from earlier diluvialists who fabricated it as far back as the original Burnet model. Thus we are being asked to focus attention on a fabricated secular scenario.

Where shall we begin? We have already dealt at length with the supposed "currents of water pouring perpetually from the skies and erupting continuously from the earth's crust all over the world" (Morris, 1974a, p. 117). Meteorology and geophysics show that part to be untenable. There could have been no vapor canopy and no vast reservoirs of free water in the deep crust or upper mantle, unless laws of physics did not prevail in the third millennium B.C. How about the "outpourings of magma from the mantle, gigantic earth movements, landslides, tsunamis, and explosions"? (P. 117)

Outpouring of basaltic magma can be observed somewhere on the earth in any given year. With the continual spreading apart of lithospheric plates along much of the 60,000-km mid-oceanic ridge, the rise of magma must also be more or less continuous, with basaltic magmas emerging at one point or another. Recent observations of the floors of the active rifts show fresh pillow lavas in the Mid-Atlantic rift floor and ropy lava surfaces on the floor of the Galapagos rift. Heat flow is at a high rate near the rift lavas, indicating magma at depth. The process is almost "uniformitarian" in the classical sense. It has nothing to do with a biblical Flood in any case.

"Gigantic earth movements" must mean tectonic activity, such as block faulting and overthrusting, that is a required ingredient of Flood geology. Creationists seem to accept the existence of faulting and folding as real phenomena affecting crustal rocks, although the Genesis account is devoid of mention of such activities. We will need to consider whether or not all documented deformation of Proterozoic and earlier rocks could be accomplished in a single year.

A landslide occurs somewhere on the earth at least once in every year, if one's definition of the phenomenon is not restricted to a special mass limit. We can assume that large numbers of landslides have occurred each year since the Flood, as well as during the year of the Flood. Tsunamis are generated by submarine earthquakes, and we can count on a tsunami to occur somewhere on the globe as often as once a year, if it does not have to be a really big one. As to the "explosions," we are not given the faintest idea what the creationists have in mind; nor is mainstream science about to put words into their mouths. Did God have a supply of fissionable isotopes? Genesis says nothing of "explosions."

The ICR version of the Flood describes severe, rapid soil erosion, which would not be unreasonable during protracted periods of torrential rainfall (Morris, 1974a, pp. 117-18). The problem here is that there is no description in Genesis of the pre-Flood landscape. Creationists do not supply any details. Under the vapor-canopy theory, the lower atmosphere was virtually stagnant and the implication is that atmospheric lift necessary to produce condensation was not operative; hence, there was no precipitation other than perhaps light drizzle. We have no way of knowing whether stream channel systems existed. We cannot guess the relief characteristics of the land surface, nor the extent to which chemical and mechanical weathering had produced a regolith of comparatively soft mineral matter over the Archeozoic bedrock. Fully lithified, hard, dense rock-- such as granite, gneiss, or quartzite and other kinds of igneous and metamorphic rocks that make up the Archean rocks as we know them--could withstand forty days and nights of torrential rainfall with scarcely a measurable quantity of erosional removal. Perhaps locally, chutelike channels might have formed under cavitation. Even on the assumption that a thick (100-meter) layer of decayed rock (saprolite) was available for removal and conversion into sediment, it would be woefully inadequate to supply the quantity needed to form all existing Proterozoic and younger sedimentary and metasedimentary rocks.

The ICR text refers to "upwelling sediments" on the ocean bottom (p. 117). No such process is known to science, unless the term might refer to the mineral matter suspended in the vents of hot water in rifts at the spreading plate boundaries. Actually, the "smokers" observed in the Galapagos rift serve to sustain invertebrate life, not "entomb" it.

The ICR text, after reciting the secular and largely imaginary description of the Flood cataclysm (see full quotation given earlier), adds this significant paragraph:

> The above of course is only the barest outline of the great variety of phenomena that would accompany such a cataclysm. The very complexity of the model makes it extremely versatile in its ability to explain a wide diversity of data (although, admittedly, this makes it difficult to test). (Morris, 1974a, p. 118)

Yes, indeed, a cock-and-bull story can be devised to explain just about anything. Why should those who are asked to listen to such a story be the ones required to test its validity?

Diastrophism and Volcanism during the Flood

I have pointed out on earlier pages that diastrophism and volcanism have been brought into the creationists' Flood scenario, whereas nothing of the sort is mentioned or implied in Genesis. Adopting the imagined Flood model of Burnet, the creationists have thrown caution to the winds and trumped up an incredible set of crustal displacements and volcanic action, all fitted into the single year available for the Flood. In this connection, ideas presented by Randy Wysong are of interest. Wysong holds the degree of doctor of veterinary medicine, so his views on geology are probably borrowed, but his statements are nevertheless straightforward. Wysong says that if, in addition to the floodwater supplied by rain, there had been such geological activity as "seismic fissures, volcanism, earthquakes, tidal waves, etc.," the universal nature of the Flood would be accounted for (Wysong, 1976, p. 391). Ocean basins present before the Flood would lift upward, which would cause ocean water to spill over the continents. At the same time, the continents would be dropping down to lower levels. This equalization, or leveling process, would allow the pre-Flood ocean water to cover the entire globe to a more or less uniform depth. Wysong also postulates a great deal of volcanic action during the Flood and this, he says, explains the great basaltic lava plateaus of the Deccan region of India and the Columbia Plateau of the northwestern United States. He adds that the volcanic ash spewed into the upper atmosphere from erupting volcanoes could have provided nuclei needed for intense condensation of water vapor.

Wysong has hit upon an interesting geologic device for causing a great flooding and worldwide inundation without needing very much, if any, water from either the atmosphere or subterranean springs. I have attempted to show the geometry involved in this explanation, using a highly schematic diagram in which the curvature of the spherical earth is flattened out (Figure 22.1). As the Flood begins (A), the floors of the ocean basins rise, while the continents subside. This tectonic activity takes place on vertical faults (normal faults) at the contacts of ocean basins and continents. When fault motion has equalized the surfaces of ocean basins and continents, the ocean water uniformly covers the solid earth surface (B). Meantime, a volcano--Mount Ararat--has been constructed at a continental location. It is a submarine volcano during formation. Late in the year of the Flood the tectonic process reverses: continents are up-faulted; ocean basins are down-faulted. The water pours back into the basins, the continents emerge again to stand high (C). The Ark has been grounded high on the slopes of Mount Ararat. Actually, Mount Ararat may have dated from pre-Flood times, along with other volcanic peaks and mountain ranges of the Creation, so volcanic construction of summits is not essential to the theory.

If the tectonic model suggested by Wysong seems to solve the problem of where the water came from, it runs into another problem from which there is no escape and must lead to abandonment of the whole scheme. What happens to the active faults at depth? Downward movement of the continental blocks must displace rock material in the mantle zone beneath. It is reasonable to suppose that the hot plastic mantle material of the asthenosphere would move laterally to a position beneath the rising ocean floors. This slow compensatory flowage of the mantle is permitted by physical theory--but not on the time scale of one year! The great viscosity of the mantle rock is such that its flowage motions must be incredibly slow by standards of human history. At any

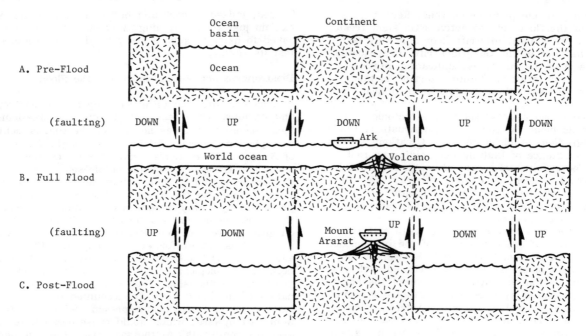

Figure 22.1 Schematic diagram of a tectonic model for the Flood of Noah. It requires no additional water sources. (A. N. Strahler.)

given point in the flow system, the speed of a mineral particle would be not more than a few centimeters per year. Moreover, as the difference in level of the individual crustal blocks diminished, the hydrostatic pressure differential in the mantle would have declined, decreasing the rate of flow.

By far the weakest element in Wysong's hypothesis of sinking continents and rising ocean basins is that no natural forces are available to cause such gross displacements. The crust is presently close to being in equilibrium--it is in isostatic equilibrium. The continents stand high because they are underlain by thick crust of relatively low density--about that of granite. The ocean basins lie low because they are underlain by thin crust of relatively high density--about that of basalt. The principle is that of flotation of a ship's hull in water. Neither type of crust is disposed to exchange its level with the other. Even with continual lithospheric plate motions in progress, including plate separation and subduction, the state of equilibrium is well preserved by continual compensation through slow vertical motions.

In this connection, it is worth noting that Whitcomb and Morris in their comprehensive volume The Genesis Flood take great liberties with geophysics in postulating wholesale crustal redistributions in total disregard of what is known of crustal and lithospheric structure and isostatic adjustment (1961, pp. 121-22). In attempting to find a place to put the floodwaters, they write:

Whatever the source of the Deluge rain, the mass of waters which descended to the earth could hardly have been elevated back into the heavens, because it is not there now. This can only mean that much of the waters of our present oceans entered the oceans at the time of the Flood. This in turn implies that the proportion of land area to water area was larger before the Flood, perhaps very much larger, than at present. Much of the present sea-bottom was once dry land. Very likely, in order to accommodate the great mass of waters and permit the land to appear again, great tectonic movements and isostatic adjustments would have to take place, forming the deep ocean basins and troughs and elevating the continents. (Whitcomb and Morris, 1961, pp. 121-22)

As in Wysong's hypothesis, and for the same reasons, it would be physically impossible for large parts of the continental crust to be suddenly and permanently depressed into deep ocean basins and hold that position against the tendency for the crustal mass to rise and reestablish isostatic equilibrium. In any case, no such changes could be accomplished by natural causes in the time of one year, or several tens of thousands of years.

To introduce tectonic activity (diastrophism), a subject not mentioned or inferred in the Genesis account, has led not to a solution but to impassable obstacles. The notion of volcanism and diastrophism as essential large-scale activities of the year of the Flood may be, as Andrew White suggests, part of the "heathen legends of the golden age" (1896, p. 219) that Thomas Burnet may have mixed in with the Genesis account in developing his monstrous scenario of a catastrophic Deluge. Legends, whatever their origin, have no place in science, and there is no justification for promoting such pseudoscience, however wellmeaning the intent may be. To depart from the actual words of Genesis has been folly and has led only to delusion.

According to the ICR text, aftereffects of the cataclysmic geologic activities of the Flood continued for hundreds of years and even persist to some degree at the present time (Morris, 1974a, pp. 125-30). The phenomena are under the aegis of "residual catastrophism" and include mountain making, volcanism, glaciation, and pluviation. These activities, the ICR text explains, have been decreasing asymptotically since the closing stages of the Flood, when they were in a phase of maximum intensity (p. 128). The suggestion is implicit in that schedule that the second law of thermodynamics is now fully in charge of the earth, but that the Flood represented a sudden and temporary reversal of the increase in entropy. Although not specifically stated in their text, we might wish to infer that the Creator reversed the universal trend by making a massive input of free energy to carry out the work of the Flood. While that work was destructive to life on earth, it also created a stratigraphic system carrying a high degree of order. Another hypothesis might be that the energy required for the Flood had been stored during the Creation. In any case, the uplift and construction of high mountains represent work done against the pull of gravity and can be seen as a case of reversal of increase in entropy.

Creationists' Arguments against Plate Tectonics

Creationists are, for good reason, unable to accept the modern theory of plate tectonics, which is highly uniformitarian (their version) in that plate motions are extremely slow but unceasing. Tectonic processes inferred from relationships seen today at the spreading, converging, and transform plate boundaries are envisioned by mainstream geologists as having been taking place in more or less the same way back through time for at least two and perhaps three billion years.

Creationists cannot even accept such contemporary plate activities as seafloor spreading, subduction, motion on transform fault boundaries, and continental rifting, even though seismic activity--which all humans can feel as earthquakes--reveals to seismologists the existence of such motions and their relative directions. Note that the religious doctrine of a biblical Flood does not prevent creationists from accepting the theory of ice-sheet growth and wastage overwhelming entire continents. The great scientist and creationist Louis Agassiz fought a hard battle for acceptance of his glacial theory, so latter-day creationists must respect his theory and weave it into Flood and post-Flood geology. Not so plate tectonics, which had no creationist sponsor. Instead, plate tectonics is to be viewed as a spinoff of Darwinism; it embodies the corrupt concept of evolution--slow evolution of continents by materialistic physical processes unguided by ulterior design but responding, instead, to senseless random variations in a mysterious system of convection currents deep down in the asthenosphere. For the creationists, it is imperative to stop those moving plates dead in their tracks.

Continental drift, Wegener style, posed no threat to the gurus of creation science prior to about 1960. George McReady Price could ignore Alfred Wegener's crazy notion because mainstream science had no use for it either (1923). That situation held for Whitcomb and Morris as they put together The Genesis Flood; they could with impunity write these words:

> In general, there are currently two main hypotheses of mountain-building. One depends on thermal contraction of the crust, the other on subcrustal convection currents. Another, the theory of continental drift, is at present running a poor third. None of them is based on present measurable processes, but solely on hypothetical speculations which may or may not be meaningful. Proponents of the two leading hypotheses have each advanced arguments showing the inadequacies of the other. (1961, p. 140)

Halcyon days for the creationists! No ruling hypothesis to contend with and no lack of scientists publicly beating each others' brains out and falling exhausted in mutual defeat. But that situation was soon to end. No longer could Whitcomb and Morris claim "that the mountain-making processes with all their associated phenomena--the faults, folds, rifts, thrusts, etc.--have been geologically active in recent times. But they are not active now, at least not measurably so!" (P. 142)

By 1967, Columbia University seismologist Lynn R. Sykes had published the evidence of first motion in earthquakes centered on mid-oceanic transform faults-- evidence that demonstrated beyond question the pulling apart of the crust along the Mid-Atlantic Ridge. By 1974, the publication date of the ICR textbook Scientific Creationism, the new paradigm of plate tectonics had become securely emplaced in mainstream geology and geophysics. The creationists' reaction to the new paradigm was negative, of course, but they brushed over the subject lightly:

> Until about 1960, the old idea of continental drift had been rejected and even ridiculed by practically all geologists, who were convinced they had worked out a complete explanation of earth history and the stratified rocks in terms of stable and permanent continents. Currently, however, the pendulum has swung and most geologists have become committed to the concepts of plate tectonics, sea-floor spreading and continental drift. All the older explanations, which they once dogmatically accepted as certainties, have now been discarded in favor of drift-centered concepts. There still remains a minority of outstanding earth scientists (Jeffries, Meyerhoff, the Russian geophysicists, et al) who oppose the idea of continental drift as geophysically impossible, and there are some signs that indicate the pendulum may be starting to swing back again. (Morris, 1974a, p. 128)

Morris may be confusing the older Wegener version of continental drift with the modern physical theory of lithospheric plate motion. Jeffries, the British geophysicist, found Wegener's drift mechanism to be "geophysically impossible" and so it remains today. If the Russian geophysicists upheld Jeffries calculations, that is no surprise, because so did the American geophysicists. The difficulty was overcome, as we explained in Chapter 20, by introducing an entirely different hypothesis, that allowed motion of a rigid lithosphere over a weak asthenosphere. Morris does not mention this important distinction between the two models.

As to mention of the name of Meyerhoff, that is a reference to a series of five papers authored by petroleum geologist Arthur A. Meyerhoff and published in 1970, 1971, and 1972 in the prestigious Journal of Geology. Coauthors in this series included Curt Teichert, Howard A. Meyerhoff, and R. S. Briggs, Jr., all of whom are well-qualified scientists. Much of the content of these papers is directed to what are considered inadequacies of paleomagnetic determinations to explain patterns of sediment deposition seen in continental platform strata. In the final paper of the series a novel hypothesis, called the "fracture-contraction model," is proposed as an alternative to plate tectonics. Also appearing in the same journal in those years were two other papers posing objections to plate tectonics. One, by M. L. Keith, proposed "ocean-floor convergence" as the alternative mechanism to sea-floor spreading (1971). The other, by Paul S. Wesson of Cambridge University, objected to plate tectonics on grounds that (1) continents have not moved with respect to each other; (2) convection is not active throughout the whole mantle; (3) if convection were active it could not account for drift of continents; (4) pole positions derived from paleomagnetism are subject to an unknown error and are too inexact for use in constructing plate motions (1972). Wesson's paper was later cited by creation scientists, as we shall see.

In the years 1970-1972 plate tectonics was barely emerging from its infancy. Xavier Le Pichon's seminal paper of 1968, the first to exhibit world maps of the lithospheric plates, their boundaries, and relative motions, was still fresh on the scene. The small flurry of papers raising objections to the theory is just what would be expected of the scientific community. Data were being gathered at an enormously rapid rate from various disciplines such as seismology, paleomagnetism, radiometric dating, submarine topography, sediment-core analysis, heat-flow analysis, and geochemistry. Confusing and conflicting sets of data were being studied, some of them from the early cruises of the deep-sea drilling vessel Glomar Challenger. Altogether, it was a case of indigestion from eating too much too fast. But we must not lose sight of the rapid strengthening that plate tectonics was undergoing at the same time. Basic points of the theory were being heavily reinforced and corroborated in this period.

As creation scientists began to look further into plate tectonics, they could no longer shrug it off as a bad dream. They began to probe for points of seeming weakness that might be exploited to their advantage. One result of this new look was a 1976 ICR Impact Series article by creationist Stuart E. Nevins, bearing the title "Continental drift, plate tectonics, and the Bible." He focused his attention on the following points: Numerous versions of a reconstruction of Pangaea have appeared, only one of which could possibly be correct, and most showing gaps, overlaps, and omissions. Magnetic anomaly "stripes" should be symmetrical, but are not. Dating of the magnetic anomalies is of questionable accuracy. Subduction is not consonant with the discovery of flat-lying trench sediments; earthquake first-motion studies show the opposite relationship to that expected of a down-plunging plate. Explanations offered for causes and driving forces for plate motions range from doubtful to impossible. Neverthless, concluded Nevins, there is validity to the concept of platelike slabs being deformed at their edges as a result of continental separation, seafloor rifting, and underthrusting in trenches. This deformation, he says, is best interpreted as a rapid process, not occurring today, and attributable to a catastrophic mechanism associated with the Flood of Noah. Separation of the continents may well have occurred during the Flood and, indeed, this would help to explain passages in Genesis to the effect that following creation the waters were "gathered together unto one place" (1:19) and that later "was the earth divided" (10:25). Some scholars suppose the scattering at Babel (11:8) to have included continental separation (Nevins, 1976a, p.iii).

I gather from Nevins's paper that some accommodation is now to be made with mainstream science on the question of plate tectonics. Evidence of plate motions is now much too strong to be flatly negated, but it can be fitted into the Flood catastrophe. In that position, of course, it faces the same horrendous difficulties as does most of Flood geology, namely, how to have it all accomplished in one year. Right away, you realize that the great basaltic/gabbroic expanses of oceanic crust forming the present floors of the Atlantic, Pacific, and Indian oceans must have been formed by magma extrusion at a fantastic rate --minimally, horizontal crustal growth at the rate of 3,000 km per year, which for a 300-day period requires accretion at the average rate of 10 km per day.

Let us now turn to more recent creationist writings to find out how plate tectonics is to be handled. The Australians are now making their bid for the prize trophy in creation science, led by Andrew Snelling, Ph.D., a geologist well equipped to analyze the professional literature on plate tectonics. Writing in Ex Nihilo, Snelling gives the latest position on these matters in answer to the question: "Have the continents really moved apart?" (1984a) For the most part, his points are almost identical with those given by Nevins in 1976. Snelling enlarges on a question of the reliability of potassium /argon dating of ocean floor basalts, that has come up with the conclusion that there is no evidence of increasing basalt age with increasing distance from the spreading axis. Snelling makes no mention of accommodating continental separation into the Flood year as a catastrophic event, as Nevins proposed. Instead, Snelling would recommend exterminating the beast: "The absence of sufficient mechanism(s) for plate motion, the uncertainty regarding palaeomagnetism (sic) and the existence of sea-floor spreading, and the doubts about subduction render the whole idea of continental drift and the theory of plate tectonics highly speculative and questionable" (p. 16). It looks as if the creation scientists may have a doctrinal schism on their hands. Should they crucify plate tectonics or adopt it? Dr. Morris: Please step in and straighten this matter out! Would it help to have an international committee on creationist doctrine--some kind of body that could set up

our targets in clear and unambiguous forms that minimize waste effort expended in beating dead horses or parading corpses?

We turn now to the points raised by Nevins and Snelling. The objections they raise are specific and of empirical nature, making possible their evaluation in the light of observational data.

Objections to Seafloor Spreading

Magnetic anomalies, translated into a symmetrical, mirror-image set of stripes of alternating magnetic polarity, were recognized in the mid-1960s and quickly interpreted as evidence of seafloor spreading. Nevins and Snelling fault this interpretation on three counts. First, they say, anomaly patterns may be caused, not by polarity reversals, but by other causes. Second, the anomaly pattern is normally asymmetrical, rather than symmetrical. For these points Nevins cites cites Meyerhoff and Meyerhoff (1972b, pp. 337-59). Referring to that source, Nevins says: "It has been argued that the linear patterns can be caused by several complex interacting factors (differences in magnetic susceptibility, magnetic reversals, oriented tectonic stresses)" (1976a, p. ii).

To get a better basis for understanding the nature of Nevins's complaint, we need to know exactly what is meant by "magnetic anomaly." As mentioned in Chapter 19, the variable quantity actually measured in shipboard magnetometer traverses is the strength of the magnetic field. The magnetometer is towed from the ship along a series of traverses at right angles to the spreading plate boundary. Figure 22.2 shows a typical magnetometer profile; its fluctuations represent departures from the mean value of total magnetic intensity measured in the direction of the earth's magnetic field. Below the observed profile is a theoretical profile constructed from the established scale of polarity reversals derived from basalt core samples in a terrestrial location, such as on Iceland, for example. If you run your eye back and forth from one profile to the other, you will be able to match them up, based on the grouping of the distinctive "wiggles." When several parallel profiles are laid out on an oceanographic chart, the anomaly pattern can be made two-dimensional, as shown in Figure 22.3.

The point is that the polarity epochs and events must be interpreted from a pattern of field-strength anomalies, and this requires a "fix" on the polarity scale based on another set of observations. Actual age of the ocean-floor basalt beneath a particular anomaly zone, or "stripe," can be determined (a) from the age of the bottom-most sediment layer resting on the basalt and/or (b) the radiometric age of the basalt itself. Many cores penetrating the sediment layer and entering the basalt beneath were obtained by deep-sea drilling from the Glomar Challenger. The sediment cores themselves gave a record of changes in magnetic inclination corresponding to changes in polarity, while fossil content of the sediment as well as radiometric age determinations allowed those polarity reversals to be identified.

These procedures brought together all the information needed to assign geologic ages to the anomaly stripes. In this way the reality of the anomaly patterns as reflecting polarity reversals was independently corroborated. If you care to examine a geologic map of the Pacific Ocean (Heezen and Fornari), you will find that deep-sea cores show consistently that the age of the oldest sediment layer and of the basalt below it agrees closely with the age assigned to the anomaly recorded above that point. References cited by Snelling to cast doubt on the meaning of the magnetic anomaly patterns date back into the mid-1960s, before the deep-sea drilling program provided the corroboration referred to above (Snelling, 1984a, p. 15).

As to the lack of symmetry of some anomaly patterns, the claim is a valid one, but the phenomenon can not only be explained, but it is also to be expected. Rate of

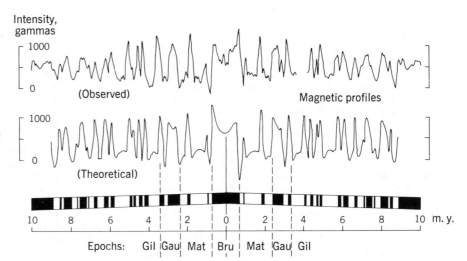

Figure 22.2 Observed profile of magnetic intensity (above) along a traverse of the mid-oceanic ridge at about latitude 60° S. in the South Pacific. Theoretical magnetic profile and time-scale of magnetic polarity are shown below. (From W. C. Pitman III and J. R. Heirtzler, Science, vol. 154, p. 1166, Figure 3. Copyright 1966 by the American Association for the Advancement of Science. Used by permission.)

motion of a plate is measured with respect to the rift axis of rising basaltic magma that is deeply rooted in the underlying mantle. Asymmetrical spreading has been documented (Rona and Gray, 1980, pp. 490-91). As Figure 22.4 shows, it is with respect to the axial reference line that one plate can move away faster than the opposing plate moves away. The faster-moving plate generates a wider set of stripes than the slower-moving plate. One might imagine a case in which the motion of one plate virtually ceases for a period of time. In that case, one or more anomaly bands might be missing entirely.

Under the most popular hypothesis of forces that cause motion of oceanic lithospheric plates, a major force urging a plate to move away from the spreading axis is induced by the lifting of the plate near the axis, giving it an outward gradient and leading to motion to a lower level under gravitational force. This force is often called ridge-push force. There also operates another

gravitational force; it acts on the down-plunging slab of the same lithospheric plate and is called slab-pull force. Figure 22.5 is a schematic diagram showing certain forces involved in motion of oceanic lithospheric plates. The moving plate exerts a mantle-drag force on the plastic asthenosphere beneath it.

There are various possibilities for explaining asymmetric spreading rates in terms of these forces. Greater distance from spreading axis to subducting boundary could increase the total mantle-drag force,

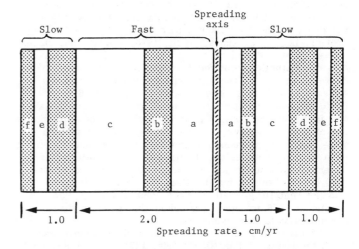

Figure 22.4 Asymmetrical plate spreading occurring during the production of anomaly stripes a, b, and c. (A. N. Strahler. Based on data of W. H. Menard, 1984, Geology, vol. 12, p. 178, Figure 3.)

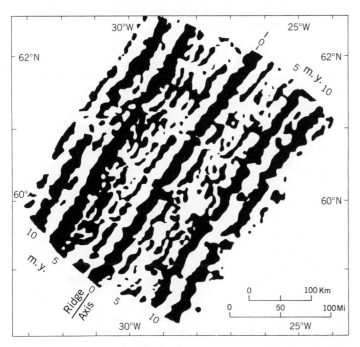

Figure 22.3 Magnetic anomaly pattern for Reykjanes Ridge, located on the Mid-Atlantic Ridge southwest of Iceland. Figures give approximate rock ages in millions of years. (From J. R. Heirtzler, X. Le Pichon, and J. G. Baron, Deep-Sea Research, vol. 13, p. 427. Copyright 1966 by Pergamon Press, Ltd., Used by permission.)

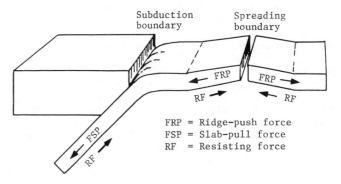

Figure 22.5 Major forces acting upon an oceanic lithospheric plate. (A. N. Strahler.)

reducing the spreading rate as compared with the opposing plate. There is, then, no validity in citing asymmetric spreading as an objection to plate tectonics. The phenomenon is "normal" in all respects; it represents expected behavior of opposite plates, not unexpected behavior. The creationists have made an unwarrranted assumption to begin with, namely, that asymmetrical spreading is an unexpected phenomenon. The specious nature of such an argument (setting up a straw person) is easily exposed but it can, nevertheless, be an effective debating ploy, particularly when the audience knows little of the scientific background information.

In their general argument against the validity of dates assigned to magnetic anomaly stripes, the creationists cite Paul S. Wesson, the Cambridge geophysicist mentioned above, as one who opposed plate tectonics (Wesson, 1972). This is what Wesson wrote:

> Direct dating of the sea floor (Fisher 1969) is contradictory: K/Ar ages, properly interpreted, show no evidence of increasing age with distance from the ridge system, the data from seamounts is equally uninformative. (P. 191)

First, we should eliminate data from seamounts, which are mostly volcanoes built later upon the basaltic oceanic crust. They can form from mantle hot spots far distant from spreading axes and their ages would not necessarily have any genetic relationship to the magnetic anomalies in the crust on which they stand.

In the year 1969, in which Fisher's article appeared in New Scientist, the Glomar Challenger had only begun to carry out deep drilling capable of reaching the basaltic ocean crust. Consistent results began to appear on Leg 9 of the drilling program when, in January 1970, several holes were drilled in a west-to-east traverse leading to the crest of the East Pacific Rise (see Deep Sea Drilling Project: Leg 9, Geotimes, April, 1970, pp. 11-13). Basalt was encountered and recovered in 8 of 9 drilling sites. Age determinations showed a remarkably consistent decrease in age as the axial rift zone was approached. Ages ranged from 21 m.y. (Early Miocene) to 6 m.y. (Late Miocene) not far from the ridge crest. The basalt was identified as being in the form of igneous sills, hence not of the pillow-lava form that elsewhere has yielded some confusing potassium/argon dates. The essential point here is that the creationists have drawn a statement from observations made prior to the great series of deep-sea drill cores obtained by the Deep Sea Driling Program (DSDP) that functioned between 1969 and 1981. For Nevins in 1976 to have overlooked this great body of consistent data is inexcusable.

The creationist authors refer to serious discrepancies in K/Ar dating of basalts (Morris, 1974a, pp. 146-47; Nevins, 1976a, p. ii; Slusher, 1981, p. 39; Snelling, 1984a, p. 15). These relate to ages of submarine pillow lavas a short distance east of the island of Hawaii. Although these lavas are obviously of recent age (200 yr), they gave K/Ar ages ranging to values as high as 22 m.y. Playing into the creationists' hands, two scientists of the Hawaiian Institute of Geophysics, who made the age determinations, urged caution in applying K/Ar dates from deep-ocean basalts on seafloor spreading ridges (Noble and Naughton, 1968, p. 265). This quotation was picked up by Morris who could not resist adding: "The comment about sea-floor spreading is most interesting, in view of the fact that the modern concept of continental drift, especially its very slow rate, is based mainly on similar potassium-argon dates in basalts at the bottom of the Atlantic" (1974a, p. 147). The scientists offered as an explanation of the spurious ages the suggestion that excess argon present in the fluid magma could have been incorporated into fluid and gaseous inclusions in the basalt at the time the pillows came in contact with the cold seawater (Funkhouser and Naughton, 1968, p. 4606).

Morris moved in quickly to press his advantage:

> The creationist does not question the fact that the anomalously high ages of the lava rocks noted above may well be due to incorporation of excess argon at the time of formation. Again, however, he points out that if this is known to have happened so frequently in rocks of known age, it probably also happened frequently in rocks of unknown age. Since there is no way to distinguish Argon 40 as formed by unknown processes in primeval times and now dispersed around the world, from radiogenic Argon 40, it seems clear that potassium-argon ages are meaningless in so far as true ages are concerned. (1974a, p. 147)

G. Brent Dalrymple considers Morris's argument to be a red herring (1984, pp. 80-81). Dalrymple, a specialist in radiometric age determination who has published his own research results on ages of historic basalt flows on Hawaii, gives us this explanation of the two 1968 pillow-basalt studies referred to above:

> Two studies independently discovered that the glassy margins of submarine pillow basalts . . . trap Ar-40 dissolved in the melt before it can escape. This effect is most serious in the rims of the pillows and increases in severity with water depth. The excess Ar-40 content approaches zero toward the pillow interiors, which cool more slowly and allow the Ar-40 to escape. . . . The purpose of these two studies was to determine, in a controlled experiment with samples of known age, the suitability of submarine pillow basalts for dating, because it was suspected that such samples might be unreliable. Such studies are not unusual because each different type of mineral and rock has to be tested carefully before it can be used for any radiometric dating technique. In the case of the submarine pillow basalts, the results clearly indicated that these rocks are unsuitable for dating, and so they are not generally used for this purpose except in special circumstances and unless there is some independent way of verifying the results. (Dalrymple, 1984, pp. 890-81)[1]

We have already explained that independent determinations of age of the oceanic basalts are possible from the sediments that lie directly on the basalt crust. Rubidium/strontium ages are available from the sediments, along with distinctive fossils whose geologic age has been established independently in rock sequences now exposed on the continents. Deeper drilling into the basalt is now able to reach the zone of sheeted dikes, which can be confidently dated by the K/Ar method because they are of plutonic (rather than extrusive) origin, intruded into the deeper zone without experiencing direct contact with cold seawater.

Snelling has one more point against the use of magnetic polarity stripes as evidence of seafloor spreading (1984a, p. 15). He refers to a 1979 Science article by J. M. Hall and P. T. Robinson pointing out that deep-sea drilling in the North Atlantic basin has given a confused picture of magnetic polarity conditions in the comparatively shallow holes drilled into the basalts near the ridge crests of spreading axes. Snelling quotes these authors as saying: "it is clear that the simple model of uniformly magnetized crustal blocks of alternating polarity does not represent reality." The authors say further that drilling revealed that downward in the holes "a variation of magnetization intensity occurs on several scales from centimeters to tens of meters, and there are no consistent trends with depth." They say further: "there is poor agreement between the sense of the effective magnetization in the drilled holes and the associated linear anomalies." Snelling could have reported an even more disturbing find by Guy M. Smith, which is that "both normal and reversely magnetized rocks have been found in the same vertical section of crust by the Deep Sea Drilling Project . . . as well as many samples with directions that differ significantly from the expected geocentric axial dipole value" (1985, p. 162; see Smith for references cited). Offhand, these findings might seem to knock the entire magnetic-anomaly idea into a cocked hat. But not all is lost!

Smith notes that the extreme hydrothermal alteration of the shallow basalts near a spreading axis can strongly influence the magnetic properties of those rocks (Smith, 1985, p. 162). It is also known that normal faulting of basaltic crust accompanying spreading has caused numerous dislocations of the original basalt layering. A slumping motion of such blocks could easily cause the juxtaposition or overlap of two zones of opposite polarity.

The general conclusion Smith and other scientists have reached is that the rock magnetism that is responsible for the anomaly patterns is rooted much more deeply in the oceanic crust. Confirmation of that inference came in 1984 from data of DSDP, Leg 83, which yielded up a core penetrating 300 m of the sheeted dike complex in oceanic crust just south of the Costa Rica Rift (Smith, 1985, p. 162.) Smith concluded that the transition zone above the sheeted dike zone "lacks sufficient magnetization to make a significant contribution to marine magnetic anomalies. The sheeted-dike complex, on the other hand, is sufficiently magnetized to be a part of the magnetic source. This high magnetization is at least 300 m thick and may extend to much greater depths" (p. 165). No more need be said to refute fully Snelling's attempt to completely undermine the basis of the magnetic-anomaly model. In all fairness, we should keep in mind that the new drilling data were not available to Snelling for his 1984 report.

Seafloor spreading is corroborated by other lines of evidence than those we have discussed above. From that hypothesis certain deductions can be made to serve as tests. If magma is rising from the mantle beneath the spreading axis, we should expect to observe fresh extrusive rock on the floors of the rifts. The presence of pillow lavas in the more slowly spreading rifts of the Atlantic basin and ropy lava surfaces of sheet flows in the faster spreading Pacific rifts have been repeatedly observed, sampled, and photographed from submersibles, such as the Woods Hole research vessel <u>Alvin</u>. Along with the pillow lavas, gaping fissures (open vertical fractures) have been observed paralleling the rift axis, an expected consequence of the pulling apart of the new lavas in the rift.

If magma is steadily rising from deep in the mantle, we should expect high rates of heat flow through the crust near the rift, but a diminishing rate of heat flow at increasing distances away from the rift. Scientists of the Lamont-Doherty Geological Observatory of Columbia University developed a model of the ideal distribution of heat in the oceanic lithosphere on the two sides of a spreading rift (Langseth, 1977, p. 42). The plate becomes cooler as it moves away from the rift, as the heat contours in Figure 22.6A show. Cooler lithosphere is also denser, so that under the principle of isostasy the plate should gradually subside to greater depth. A predicted curve of ocean depth versus distance is shown in Figure 22.6B; it is strikingly matched by points showing the observed depth. From the heat model the rate of heat flow can be predicted for increasing distance, as shown in Figure 22.6C. Here the observed values are much less than predicted in a 1000-km belt adjacent to the spreading axis, but match the prediction for greater distances. Creationists might want to zero in on this discrepancy as a defect of the hypothesis, but would perhaps want to back off when the scientists' explanation is given. As we shall explain in Chapter 25, circulation of seawater as

ground water through the young basalt near the rift is intense, removing much of the heat by outflow from submarine springs and resulting in abnormally low values of heat flow through the upper surface of the basalt (see Figure 25.6).

An entirely different line of evidence of seafloor spreading is one we mentioned briefly in earlier paragraphs. The mid-oceanic ridge throughout most of its length shows lateral offsets, in which leglike sections of rift are connected by active transform faults. This relationship is shown in Figure 22.7. Motion on the faults must be horizontal, as illustrated in Figure 20.6. The transverse faults were first interpreted from ridgelike topographic features connecting the ends of the offset axial rift. Seismic maps showed that shallow earthquakes of moderate to low intensity were generated on these transverse segments as well as in the rift axis itself. This observation led to the inference that the transverse ridges represent active faults.

While the evidence of the earthquake epicenters might

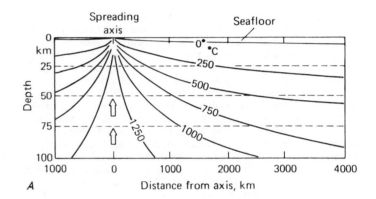

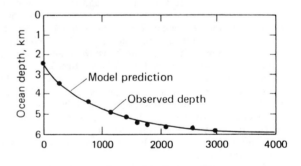

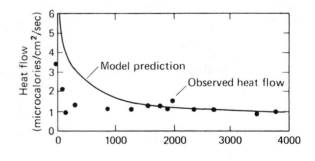

Figure 22.6 Temperature, depth, and heat flow in the vicinity of a spreading plate boundary. (A) Ideal distribution of temperatures in the lithosphere, as computed by model. (B) Observed and predicted depths of the ocean floor. (C) Observed and predicted rates of heat flow through the ocean floor. (Redrawn from M. G. Langseth, Lamont-Doherty Geological Observatory of Columbia University, <u>Yearbook</u> <u>1977</u>, vol. 4, p. 42.)

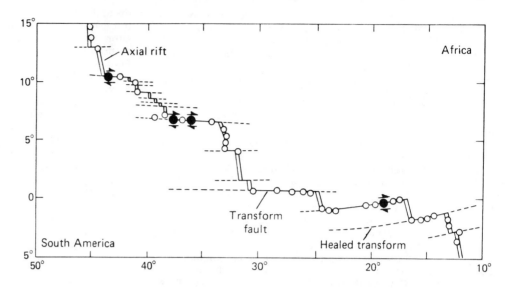

Figure 22.7 Sketch map of the Atlantic Ocean basin between Africa and South America, showing the locations of earthquake epicenters (open circles) with respect to the axial rift of the Mid-Atlantic Ridge and its offsets on transform faults. The solid circles with arrows are epicenters for which direction of first motion was determined, establishing that they lie on transform faults. (Simplified and stylized from L. R. Sykes, 1967, _Journal of Geophysical Research_, vol. 72, p. 2137, Figure 4. Copyrighted by The American Geophysical Union.)

seem strong enough proof that the transverse segments are active faults, more precise evidence was needed in the form of the relative direction of the fault movement. Confirmation came through the work of Professor Lynn R. Sykes, then a seismologist on the research staff of the Lamont-Doherty Geological Observatory of Columbia University. Examination of a seismogram will reveal the direction of the first motion; it is the initial impulse that generates a compressional wave (P-wave). Sykes was able to report in 1967 that the first motion along transverse faults between the offset ridge ends is just what would be expected of a transform fault lying between spreading plates. Arrows in Figure 22.7 show these directions. Such transform faults had been predicted by Canadian geophysicist J. Tuzo Wilson only a short time previously, but now his hypothesis was confirmed beyond any reasonable doubt. It was at this point that crustal spreading won full acceptance of nearly all members of the mainstream science community. Incidentally, analysis of first motion of earthquakes along the axial rift showed them to be normal faults, such as we would expect of gravity movements on crustal blocks breaking off on the two sides of the widening rift and settling down into the bottom of the rift valley. To the best of my knowledge, creationists have not tackled the first-motion evidence of active crustal spreading. Should they decide to make a try for it, they had better have a very carefully prepared case, because the seismologists will prove to be very tough customers.

There is more evidence to be presented of spreading motion at the mid-oceanic rift, carrying the newly formed oceanic lithosphere to points hundreds and even thousands of kilometers distant. If this hypothesis is correct, it should be revealed in the accumulation of fine sediment, called pelagic sediment, that continually rains down through the overlying ocean. First, the sedimentary layer should thicken with increasing distance from the

spreading axis, as shown in Figure 22.8. Thickening has been fully demonstrated by the deep-sea cores gained from the DSDP. In the oldest areas of oceanic crust, which are found in the western Pacific basin, sediment cores reach a thickness of more than 500 m.

But there is more than just thickness to consider. One common kind of pelagic sediment is deep-sea ooze, consisting of the tests (hard parts) of tiny planktonic organisms. One kind of ooze is calcareous in composition, meaning that the tests are composed of mineral calcium carbonate. Several kinds of one-celled animals produce calcareous tests, among them the foraminfera. As these tests sink, they arrive at a crucial depth, called the compensation depth, at which the calcium carbonate begins to dissolve. In a zone roughly between 3 and 4 km down, all calcareous tests are thus destroyed. Consequently, calcareous ooze will be limited to lesser depths, which are found on the flanks of the spreading rift axis, as shown in Figure 22.9. Now, as spreading continues and the water depth increases through plate subsidence, the initial layer of calcareous ooze will come into a zone where only tests of silica composition, along with some very fine mineral clay particles, will reach the ocean floor. Here a siliceous clay layer will accumulate and become thicker with greater distance from the spreading rift, as Figure 22.8 shows. This predicted arrangement is actually found in the deep-ocean cores, provided one looks at cores from abyssal plains well isolated from seamounts and land masses.

Finally, we look at a remarkable feature of the Pacific Ocean basin: chains of volcanic islands in which the radiometric age of the islands increases progressively along the chain. Figure 22.10 is a schematic diagram showing the prevailing hypothesis of island-chain formation. A submarine volcano forms over a hot spot in the mantle. Rising magma penetrates the crust and builds the volcano. Plate motion carries the volcano away from

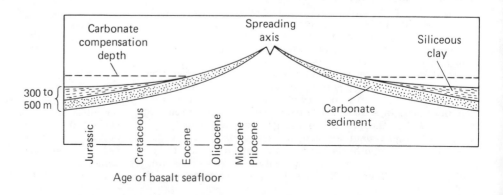

Figure 22.8 A schematic cross section showing the thickening of pelagic seafloor sediment away from a spreading plate boundary. The vertical scale (thickness) of the sediment layers is enormously exaggerated. (Copyright c 1981 by Arthur N. Strahler.)

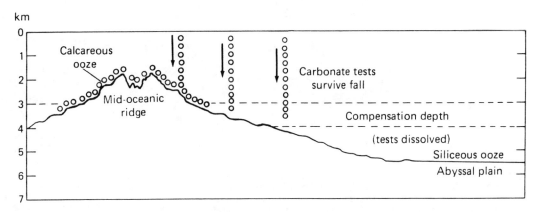

Figure 22.9 Schematic diagram showing that carbonate tests begin to dissolve at the compensation depth and will fail to reach the ocean floor below about 4 km. (Copyright © 1981 by Arthur N. Strahler.)

the hot spot, causing the volcano to become extinct. Wave action may then bevel off the top of the volcano. Crustal subsidence carries the truncated volcano below the surface, where it is now a seamount. Renewed rise of magma from the hot spot then produces another volcano, which follows the same history of extinction, truncation, and submergence. Repeated several or many times, this scenario generates a long seamount chain.

The classic example of such an island/seamount chain is the Hawaiian-Emperor chain, shown in Figure 22.11. It begins with active volcanoes on the island of Hawaii, easternmost of the Hawaiian Islands. Maui, the next island to the west, has extinct volcanoes, only partially destroyed by erosion. Volanoes on islands farther west are long extinct and even more deeply eroded. A chain of seamounts continues far westward, rising steeply from a broad submarine ridge. At the end of this straight chain lies Midway, some 2400 km from Hawaii. Ages of Hawaiian basalts have been determined, using (with caution) the K-Ar method. The ages consistently increase from Hawaii (recent to -0.5 m.y.) to Kauai (-4.5 to -5.6 m.y.). The tiny island of Necker, located much farther west, has a rock age of -11 m.y. Using this evidence, it is concluded that the Pacific plate has moved west-northwest at a speed of about 12 cm per year. At a point west-northwest

of Midway, another chain of seamounts begins--the Emperor Seamounts. They run in a direction just west of north for about 2400 km. Ages of the lavas of the Emperor Seamounts start with about -40 m.y. at the southern end and reach -75 m.y. at the northern end, where the chain meets the Aleutian Trench. The interpretation seems clear enough; the Pacific plate first moved almost due north, then abruptly shifted its direction to more west-northwest. Two other seamount /island chains, both in the South Pacific, show the same west-northwest direction of Pacific plate motion as does the Hawaiian chain. We might anticipate that the creationists will attack this interpretation on grounds that the K/Ar dates are unreliable, a matter we have already dealt with at some length. Careful sampling to avoid glassy outer zones of pillow lavas was doubtless well up front in the minds of the geochemists who determined the ages given above.

All-in-all, the evidence of plate motions away from the axial rift of the mid-ocean ridge system is massive in total and extremely strong because of the independence of the several tests which the hypothesis has successfully

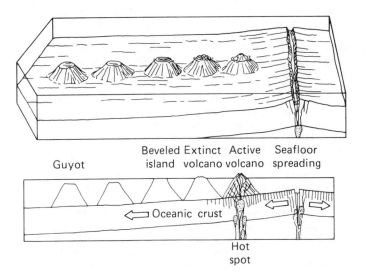

Figure 22.10 Schematic diagram of a hot spot producing a chain of volcanic islands as the oceanic plate moves away from a spreading zone. Beveled extinct volcanoes are submerged to become guyots. Ocean depth and height of volcanoes are enormously exaggerated. (Copyright © 1981 by Arthur N. Strahler.)

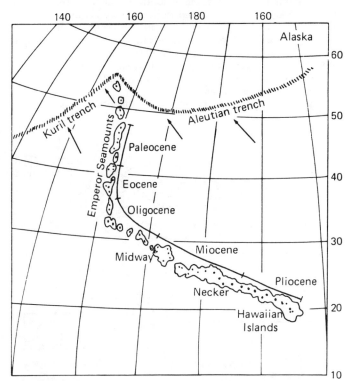

Figure 22.11 Sketch map of the North Pacific Ocean, showing the Hawaiian-Emperor seamount chains. Dots are summits; enclosing line marks the base of the volcanic piles. Geologic ages are given for volcanic rock of the seamounts. Present motion of the Pacific plate is shown by arrows. (Copyright © 1981 by Arthur N. Strahler.)

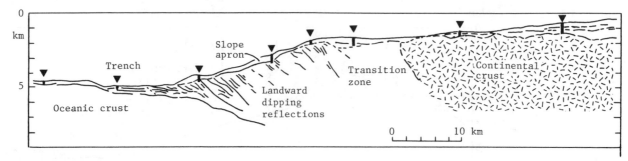

Figure 22.12 A composite cross section along the line of deep-sea drill cores (triangles), obtained on Leg 66 of the DSDP off the coast of southern Mexico. (From J. Casey Moore et al., 1982, Bull., Geological Society of America, vol. 93, p. 849, Figure 2. Used by permission.)

passed. We need to turn next to the creationists' objections to present and past lithospheric plate subduction along active continental margins and volcanic island arcs.

Objections to Plate Subduction

Creationists' objections to plate subduction at converging boundaries are twofold (Nevins, 1976, p. ii; Snelling, 1984, p. 16). First, they say, if subduction occurs as postulated, the soft seafloor sediments deposited in the trench should show deformation, compression, and thrust faulting, whereas observations in the Peru-Chile Trench and the Aleutian Trench show the trench sediment to be undeformed and flatlying. For this statement they cite two papers of 1970 and 1972 vintage appearing in the Bulletin of the Geological Society of America (Scholl, 1970; Van Huene, 1972). These sources are impeccable, of course, but the full story shows a far different picture.

The presence of undisturbed, flat-lying soft sediments in the trench floor in various localities is well known and fully documented. These sediments are typically turbidites, the deposits of turbidity currents racing down the inner trench slope and spreading out over the trench floor. The presence of severely deformed soft sediments immediately adjacent to the trench floor on the continentward side is also well known and fully documented in numerous cases.

The layering of the flat-lying sediments is clearly

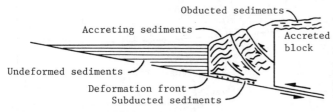

A. TRENCH

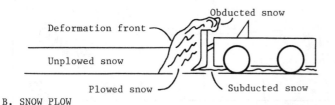

B. SNOW PLOW

Figure 22.13 Schematic diagram of undeformed and deformed sediments of a trench and subduction zone. The process is analogous to that of a snow plow. (From J. Helwig and G. A. Hall, 1974, Geology, vol. 2, p. 311, Figure 3. Reproduced by permission.)

revealed by seismic reflection profiling. Figure 22.12 is a drawing made from seismic reflection data. The locality is the Middle America Trench off the coast of Mexico, not far from Acapulco (Moore et al., 1982). The triangles show locations of drill holes on Leg 66 of the DSDP. Notice that reflections under the deepest part of the trench show the sediment layers to be essentially horizontal, whereas reflectors on the landward rising slope are steeply inclined, indicating thrust faults of an accretionary prism (see Figure 20.26).

Figure 22.13 is a cartoonlike drawing showing the general interpretation of commonly observed relationships in trenches (Helwig and Hall, 1974, p. 311). The undeformed strata are accumulating on the descending slope of the plate as it moves toward the continent. At a certain point, called the deformation front, the blocking effect of the continental plate manifests itself like the blade of a snowplow, pictured in the lower diagram. Here the soft sediments are severely deformed. What could be simpler as an explanation of the presence of both undisturbed and highly deformed sediments in the same subduction boundary zone?

What the creationists also neglect to tell us is that deep trenches have been discovered that contain no appreciable amount of soft sediment of any kind. They are known as barren trenches; a type example is the Marianas trench in the western Pacific basin. Here, the subducting plate is steeply downbent and carries down with it the thin sediment layer lying on the seafloor, as well as any sediment that is moved down from the adjacent volcanic island arc.

The kinds of simple relationships we have described here are widespread and typical, although a few different and puzzling arrangements of rocks have been found in subduction zones. The creationists' argument fails because it rests on a carefully selected statement that gives only a small part of the observed relationships. In view of the abundance of published detailed descriptions of trench sediments and their tectonics available to the creationist authors, their selectiveness cannot be excused through ignorance.

The creationists' second argument is that first-motion data for earthquakes centered within down-plunging plates commonly indicate tensional stress (extension, or stretching) within the plate, whereas data indicating compressional stresses are rare. Nevins cites as the source of this observation a 1973 paper by W. F. Tanner. To understand what is implied by the objection, we need to review the relationship between earthquakes and subducting plates.

Long before plate tectonics was put together, seismologists were accumulating data on earthquakes emanating from a zone adjacent to the great circum-Pacific trenches. Japanese seismologists in particular were giving intensive study to the seismic activity of their Japan Trench and its contiguous Kuril, Shikoku, and Izu trenches. Large numbers of relatively shallow-focus quakes, including some devastatingly great ones, are generated in a zone immediately adjacent to the trench on the landward side, as shown in Figure 22.14. Intermediate-focus quakes (around 200 km depth) are

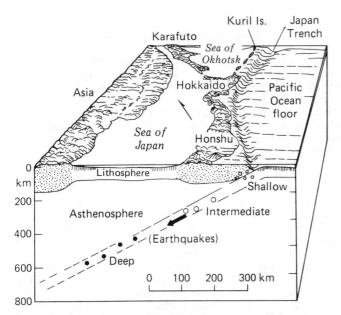

Figure 22.14 Block diagram of the Japan-Kuril arc showing how earthquake foci are distributed in the crust and mantle beneath. (Copyright © 1981 by Arthur N. Strahler.)

generated under the island belt, while those of deep focus (400 to 600 km depth) come from a zone some 400 to 500 km to the west, beneath the Sea of Japan and close to the Asian mainland. This slanting zone of earthquakes is also found on the mainland side of all the circum-Pacific trenches. Next to the Peru-Chile Trench, the deep-focus quakes occur directly beneath the east side of the Andes range. The slanting belt of earthquake foci was recognized as early as 1930 by a Japanese seisnmologist, K. Wadati, and shortly thereafter by an American seismologist, Hugo Benioff. In scientific publications of western seismologists, the feature became known as the Benioff zone, but it is now proper to refer to it as the Wadati-Benioff zone.

The existence of the Wadati-Benioff zone posed some very difficult problems. If, as seismologists were reasonably certain, most earthquakes are generated by stick-slip movements on fractures in brittle rock, the occurrence of deep earthquakes with foci far down in the hot, plastic mantle seemed particularly puzzling. In that zone, unequal stresses would be expected to be relieved by plastic flowage.

When the concept of a descending lithospheric plate was finally introduced in the 1960s, the problem seemed largely solved. The dense, brittle slab descends so

rapidly that its interior remains relatively cool, preserving the property of brittleness. Cracking in that cooler inner zone could take place even when the slab reached a depth as great as 600 to 700 km. I will skip over some important seismic information explaining how the internal physical properties of the slab can be evaluated, but the evidence is firm for a cool, brittle plate at great depth.

Now we come to the creationists' objections based on the stress directions within the plate. Nevins and Snelling are both in gross error when they imply that earthquake stresses in the Wadati-Benioff zone are "rarely compressional." Data collected by Japanese seismologists Honda and Ichikawa showed conclusively that deep foci are associated with compressive stress acting in the plane of the plate and in a direction directly down the slope of the plate (see Uyeda, 1978, pp. 139-40). This information agrees with the assumption that as the plate descends, its leading edge enters a zone of increasing strength of surrounding mantle, i.e., approaching the lower limit of the asthenosphere. The force of resistance to penetration is passed back up the rigid slab as a compressional stress.

Later, however, the same kind of seismic analysis revealed that in many subduction zones stresses causing intermediate-focus earthquakes near the top of the slab are of the tensional type. Apparently, the upper part of the slab is under tensional stress at the same time that the deep portion is under compressional stress. Why is the slab being "stretched" in its upper part? The answer offered by seismologists is that the slab is sinking under the force of gravity. As it moves into the low-strength zone of the asthenosphere, the plate tends to sink faster, causing a stretching force in the plate in this upper region. Figure 22.15 shows schematically the relation of stresses to depth as a plate descends into the asthenosphere. (For details of the entire subject of earthquakes and subduction, see Strahler, 1981, pp. 287-91).

In each of the two creationists' papers cited, their objection to subduction based on stresses requires only four or five lines of type. Their objection is groundless. It never should have been offered in the face of a voluminous scientific literature containing massive data that refute the objection. Yet this is their version of a scientific debate. It amounts to little more than making a charge that science is wrong about plate tectonics, but with no real evidence put forward to support that charge. The many lines I have written in only the briefest of explanations of the questions raised are not, however, an exercise in total futility. Consider the explanation as a learning experience--not about creationist debating tactics, but rather about natural science itself and about the scientific method.

The creationists conclude their papers with a parting shot alleging that no satisfactory system of driving forces

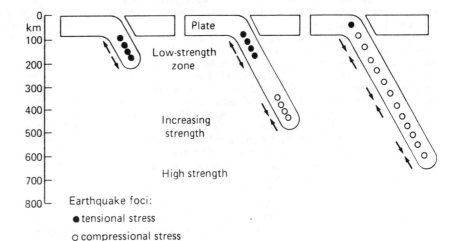

Figure 22.15 Schematic diagram of a lithospheric plate plunging into the asthenosphere. Locations of tensional and compressional stresses are shown by closed and open circles, respectively. (Based on data of B. Isacks and P. Molnar, 1969, _Nature_, vol. 223, p. 1121.)

can be found for lithospheric plate motions. Nevins (1976, p. iii) cites Wesson (1972, p. 187) stating that large-scale convection systems in the mantle are impossible. Nevin goes on to say that none of the plate-moving forces postulated so far are capable of overcoming the viscous drag at the base of the plate.

Many versions of a large-scale convection system in the mantle have been proposed. Over the years, the suggestion that currents in the mantle drive the plates through boundary drag on the undersides of the plates has been discarded. That the supposed convection would be incapable of causing plate motion seems to have been demonstrated to the satisfaction of geophysicists (Uyeda, 1978, p. 190).

One hypothesis attracting considerable favorable response today is that the plates themselves, moving under ridge-push and slab-pull forces, induce upper-mantle convection through the exertion of drag force by the plate on the mantle beneath it. This turns the older idea completely upside down. According to Seiya Uyeda, who has made a special study of the problem, it is the slab-pull force that is the most important driving force, whereas mantle drag has little influence on the velocity of the plate motion (1978, p. 197). In 1983 Wayne D. Pennington of the University of Texas made the interesting suggestion that at a depth of about 40 km, the subducting slab undergoes a change to a rock type of higher density (a phase change, it is called), increasing the downward body force and the tensional stress within the plate.

The problem of driving forces in plate tectonics is by no means solved, but reasonable hypotheses are available for examination. Science rarely provides an immediate and fully acceptable explanation of every important new phenomenon it observes. Earlier hypotheses of driving forces have failed and died; now they are exhumed by the creationists in order to follow a fallacious debating strategy, commonly described as beating a dead horse. How familiar that sounds!

Are Plate Motions Observed to Be Taking Place?

As a closing note in our discussion of plate tectonics, we make a prediction: On-going relative motions between lithospheric plates will soon be routinely measured with considerable precision. As early as 1979, when NASA set up its Crustal Dynamics Project, it was recognized that the precision of geodetic surveying using orbiting satellites was soon to reach a level at which lithospheric plate motions could actually be measured. The technique used is called Very Long Baseline Interferometry (VLBI) and involves analysis of radio signals emitted by quasars and other celestial objects (Geotimes, 1981, vol. 26, no. 12, p. 17). It was soon realized that, in combination with laser ranging techniques, crustal movements as small as one centimeter per year can be measured between widely separated stations. The slowest-moving lithospheric plates move at about that speed, while the fastest move at rates around 10 cm per year.

Observatories participating in the Crustal Dynamics Project are located in such diverse places as Australia, North America, South America, and Europe (see EOS, 1982, vol. 63, no. 19, p. 502). By 1985 it could be safely announced that the earth's continents are actually observed to be moving relative to each other at rates of a few centimeters a year. Researchers at NASA announced that the motions they recorded support the plate-tectonics theory. Scientist David E. Smith, head of NASA's geodynamics branch, was quoted as saying that the fastest-moving continent NASA has observed is Australia, which seems to be moving toward North America at a rate of 7 cm/yr (Earth Science, 1985, vol. 38, no. 1, p. 8). Note that a great-circle line between Australia and California crosses a subduction boundary, the New Hebrides Trench, so the observed closing of distance

between the two continents is to be expected. On the other hand, NASA reported that Australia seems to be moving away from South America at a rate of 2 to 3 cm/yr. A connecting line drawn from southeastern Australia to Nazca in Peru, where a geodetic station is located, crosses not only the Kermadec-Tonga Trench, but also the East Pacific Rise, a spreading plate boundary between the Pacific and Nazca plates, and the Peru-Chile Trench. While both lengthening through spreading and shortening through subduction are in progress on this line, the former rate seems to exceed the sum of the latter by a small amount.

VLBI data published in 1986 showed that measured rates of plate motions are remarkably close to rates previously estimated from paleomagnetic data (Carter and Robinson, 1986, p. 54). These data include closing rates across the subduction boundary between the Pacific plate and the Eurasian and North American plates and opening rates across the Mid-Atlantic spreading boundary between the North American and Eurasian plates. For the November, 1986, issue of The Journal of Geological Education, editor James H. Shea saw fit to select as his "Food for Thought" quote the following statement: "The most pronounced conclusion which can be drawn from the analysis of laser measurements taken on LAGEOS from 1979 through 1982 is that movement between the continental plates was measured and and is therefore confirmed. The motions we are measuring, albeit preliminary, agree overall in magnitude and direction with those found in the geological record . . . which reflect average plate movements over millions of years" (Christodoulidis et al., 1985, p. 9261).

While it may be a bit preliminary to press the point, it looks as if Drs. Whitcomb and Morris may have to eat their words (quoted on earlier pages) on tectonic processes: "But they are not active now, at least not measurably so!" (Whitcomb and Morris, 1961, p.142).

Igneous Intrusion--How Rapid?

The "doctrine of recency and rapidity," as we can refer to the creationists' basic tenet of Flood geology, applies to just about any and every kind of mineral and rock and to every configuration and structure of rock. Whether produced in the six-day Creation or the one-year Flood, production was accomplished with extreme rapidity and extreme recency, as compared with the schedules of mainstream geology. Among the earth materials referred to in the ICR textbook as produced recently and rapidly are igneous intrusive bodies (plutons), metamorphic rocks, sedimentary rocks, ore deposits, and accumulations of coal and petroleum (Morris, 1974a, pp. 101-11). We will take these categories in succession through the remainder of this chapter and in the following three chapters.

Let us begin with the igneous rocks. Already, we have covered in some detail the production of crustal igneous rock on an enormous scale by long-continued upwelling of molten mantle rock along the spreading plate boundaries located in mid-oceanic positions. According to the theory of plate tectonics, almost the entire oceanic crust, underlying about 59 percent of the earth's surface area to an average depth of about 7.5 km consists of mafic igneous rock--basalt in the form of lavas and sheeted dikes above, and gabbro as a coarse-textured plutonic layer below. All but the uppermost layer of lava is intrusive igneous rock.

Multiplying the areal extent of oceanic crust by its average thickness gives a total volume of just over two billion cubic kilometers of mafic igneous rock, the oldest of which is of Jurassic age. Thus, all of the present oceanic crust is 200 m.y. or younger in age. Using the figure of 2 billion km^3 of oceanic crust and a time of 200 m.y., we can derive an average production rate of 1 km^3 of igneous rock per 10 years, or 0.1 km^3 per y. Even if the rate were faster by a factor of 10 (1 km³/y.), it would still be an extremely modest rate.

For a Flood scenario in which the period of igneous intrusion and extrusion on spreading boundaries was, say, 100 days, the igneous rock would have had to be emplaced at the rate of 20 million km³ per day! That entire mass would have had to congeal by the final weeks of the Flood. The enormous quantity of heat delivered so rapidly to the ocean and atmosphere would have surely vaporized the ocean and killed all passengers on the Ark by steam cookery. Marine animals that had been left to survive the Flood on their own would have perished entirely.

Do the creationists have an escape from this terrible dilemma arising from their version of Flood geology? One way out would be to declare the formation of nearly all oceanic crust to be a product of the Creation. It might be argued that the igneous oceanic crust contains no fossils, hence is primeval. If God could have created the entire universe in six days, He would have had not the slightest difficulty in creating the oceanic crust within the same schedule, along with the nonfossiliferous Archean continental rocks and the plutons incorporated in them. There are fossils in the sediment layers overlying the oceanic crust, but these layers can be assigned to the Flood and post-Flood period. As we shall see in later pages, deep-sea sedimentation of itself presents enormous difficulties to the creationists, but at least the marine animals and the Ark passengers would not need to have perished by boiling and steaming, because sedimentation can be thought of as taking place in a thermal regime not much different from that existing today. The advantages of a tranquil Flood are obvious. Leave the catastrophic global volcanism and tectonism to the Creation, where it belongs!

We turn now to the continental crust and its plutons of igneous rock that are found in contact with older rocks of various kinds in such a relationship that the igneous rock has either (a) occupied widening fractures, filling the gaping space as it became available, or (b) removed the surrounding country rock bodily while simultaneously occupying the vacated space.

In the first case, it may not be clear whether the widening of the fracture occurs through tectonic forces or by hydraulic pressure of the magma. The igneous body takes the form of a slab that may be very thin in relation to its dimension along the widened fracture; it is called a dike. Dikes cut across various structures previously existing in the country rock.

In the second case, the nature of the process itself is not always clear, but the evidence of replacement of country rock by igneous rock is often clear beyond question. An example is the Sierra Nevada batholith of California, an enormous mass of coarse-grained granitic igneous rock. It is widely exposed at the surface and its contact with the older country rock can be examined in detail (see Figure 20.21). Portions of the "roof" of the batholith remain, and these project down into the body of the batholith. There are also inclusions of country rock within the igneous rock. It looks as if the process of intrusion of large plutons is, at least in part, a physical one in which pieces of the country rock were actually broken loose and incorporated into the magma. The country rock was in some manner "digested" by the magma. Perhaps it simply melted and mixed with the magma.

Igneous dikes of basalt might conceivably be produced rapidly, close to the surface, congealing in passageways (fissures) for lava outpourings (fissure flows). Igneous rock remaining in the fissures could perhaps solidify rapidly enough to fit into the Flood geology. But when we turn to consider the time factor in the emplacement and cooling of a batholith, the matter is very different. Batholiths come in enormous sizes. One gigantic example, the Idaho batholith, is exposed over an area of 40,000

square kilometers. The granite (or related type of igneous rock) that forms a batholith extends downwards at least a few kilometers and the base cannot be established or viewed directly. Laboratory studies show that samples of such igneous rock melt at temperatures of 700 to 1000 C. Basaltic lava emerging from Hawaiian volcanoes has a temperature of 1200 C. Within the asthenosphere, where partial melting probably supplies the original magmas or parent magmas, the temperature is probably in the range of 1000 to 1500 C.

How long must it take for an enormous body of magma, say, on the order of 10,000 km³, to rise from a source 100 km beneath the surface, make its way through overlying crustal rock, come to an equilibrium position, and cool sufficiently by heat conduction to the surface to completely crystallize into solid rock? The thickness of rock overlying the batholith is thought to have been as little as 1 to 3 km in some cases, but much greater in others.

Detailed field studies of great batholiths show that they are composed of smaller units, suggesting that magma rose in individual masses on the order of 50 to 100 km across (Bayly, 1968, p. 116). Magma formed at a depth of 20 to 30 km is less dense than the surrounding rock and tends to rise, as does a hot-air balloon through the lower atmosphere. A model of intermediate scale, working on this principle, is the salt diapir, a mass of rock salt that makes its way slowly upward through thousands of meters of strata in a stalklike column. Upon reaching the surface, these diapirs become salt domes, familiar from their occurrence on the Gulf Coast, where they are associated with accumulations of petroleum. A geophysicist, Hans Ramberg, carried out laboratory experiments simulating the rise of magma bodies (Ramberg, 1963). Using elaborate calculations based on known properties of the magma (density, vicosity), he came to the conclusion that the magma body of a small batholith, about 10 km across, would require 150,000 y. to rise to a near-surface location, after starting at a depth of 30 km. To complete a large batholith, consisting of tens to hundreds of such units, would obviously require a much longer time. Therefore we are dealing with a total time in the millions of years for a great batholith. Even the Gulf Coast salt diapirs have had comparable times of formation, or even longer, since the salt beds from which they arise are of Jurassic age and began their ascent about -120 m.y. According to a recent detailed study the maximum rate of rise was on the order of 400 to 530 m/m.y. (Jackson and Seni, 1983).

As to cooling rate of a batholith, principles of thermodynamics have been applied to the dissipation of heat from a magma body under a 5-km rock cover. Temperature near the contact of magma with the country rock will remain within 20 percent of its maximum temperature for at least one million years (Bayly, 1968, p. 113). The creation scientists may point to a few cases in which the contact of a batholith with the country rock shows that rapid chilling of the igneous rock occurred in a narrow zone. Don't be misled if they imply that the entire batholith experienced rapid chilling within the time framework of the Flood scenario. Solidification times of magma bodies have been calculated by Frank Spera of Princeton University. For a pluton with a radius of 10 km, the solidification time is about one million years. The presence of a small amount of water in magma greatly speeds its rate of solidification. For a pluton with radius of 5 km and a water content of 4 percent by weight, the solidification time is much less--only about 50,000 years (Spera, 1980, pp. 299, 301).

The creationists would have these igneous processes compressed into one year or less. The ICR text simply states in one sentence that "giant batholiths . . . must have formed quickly once the material emerged from the mantle" (Morris, 1974a, p. 101). How quick is "quickly"? Let the creation scientists show their calculations to demonstrate the clear possibility of a one-year history of all plutons. It is not the responsibility of mainstream scientists to do that job.

Metamorphic Rocks and Crustal Shortening

The ICR textbook has only a few lines on metamorphic rocks. Referring to how sedimentary rocks are converted to metamorphic rocks (i.e., to metasediments), they note that the process is poorly understood and is not taking place today (Morris, 1974, pp. 101-102). They refer to the process of granitization, in which metamorphic rocks are thought to be recrystallized into what appears to be granite, but without undergoing remelting. Thus, they recognize the existence of metasediments but can give no evidence that those of Proterozoic age or younger were all formed in a one-year Flood.

If we were to make a geologic map of North America in which platform strata--post-Cambrian in age--are imagined to be removed, and leaving only Precambrian to show, we would see areas with rock of Archean age (pre-Flood) over parts of Canada. There are also large areas in the age range -1.3 to -2.3 b.y. that would be Proterozoic in age. Creationists might want these early and middle Proterozoic rocks assigned to the Creation, rather than to the Flood, because they might, with some justification, argue that spherical and chainlike objects of microscopic dimensions, considered by many geologists to be fossil forms of algae, are not genuine organic forms. However, there are major belts on both the eastern and western sides of North America with metamorphic rocks of late to upper-middle Proterozoic age; they go by the name of the Grenville series. These most certainly contain fossils. The Grenville rocks are in large part Precambrian strata deformed in the Grenville orogeny that took place about -1 b.y. Corresponding late Proterozoic rocks are found in the Northern Rockies, where they go by the name of the Belt series, and they can be seen exposed in great cliffs in the national parks of the United States and Canada. Here, reefs of calcareous algae are exposed, and there are some tubelike structures that might possibly have originated as burrows. No fossil remains of multicelled animals (metazoans) have been identified with certainty in rocks older than -700 m.y., which is a time transitional into the Paleozoic era. However, the calcareous algae are enough to require their inclusion into Flood geology. (For those of you knowledgeable in historical geology, it is obvious that I am excluding from the Precambrian the Ediacaran fauna, sometimes assigned to an Eocambrian position, -600 to -700 m.y.; those organisms are clearly metazoans.)

Besides metasediments of late Proterozoic age that must be fitted into Flood geology, there are major belts of younger metasediments, produced in the early part of the Paleozoic Era. These are found in the Appalachian region in a band from Georgia to Newfoundland in which we recognize severely deformed rocks such as schist, slate, marble, and quartzite, as well as gneiss. The strata involved in this metamorphism are of the Cambrian and Ordovician periods, in which highly diverse marine life flourished. Even younger metamorphic belts were produced in the Appalachian orogeny in Carboniferous time.

Consider the enormous tectonic activity that must have

been associated with the metamorphic process. Strata were crumpled, sliced up, and thrust in one rock sheet over another. Figure 40.1 shows some restored recumbent folds, called nappes. Figure 40.6 shows a cross section of the Canadian Rockies in which fault slices, called overthrust faults, are numerous. A single great overthrust fault lies at the base of the slices. The point is that these forms of structures require crustal compression. Two reference points, originally separated horizontally by distances of several tens of kilometers, have been brought together in one place.

Modern theory of plate tectonics interprets these compressional tectonic structures as the result of collisions between island arcs and large continental lithospheric plates. Crustal shortening occurs when the island arc collides with the continental plate at a subduction boundary. Subduction is established by firm evidence as an ongoing process. The speed of plate motions can also be calculated from paleomagnetic data. The important point to be made is that tectonic crustal shortening cannot proceed at a rate more rapid than the rate at which a subducting plate moves under an adjacent plate. Relative motions of plates in such boundaries are as slow as 1 to 2 cm/y. to as rapid as 6 to 8 cm/y.

Suppose that the kinds of nappe and overthrust complexes that are observed in the Appalachians and Rockies are "unscrambled," meaning that we unfold the nappes and pull the thrust sheets back to their starting positions. We may well come up with a total shortening as great as 100 km, bringing us back to a time when the strata lay undisturbed on a passive continental margin. Let us take 10 cm/y. as the limiting rate of crustal shortening. For shortening in the amount of 100 km, one million years would be required. Clearly, this discrepancy --six orders of magnitude greater than the one-year Flood--puts the Flood geology entirely beyond the range of credibility. But to this time span we must add at least as much more time in those cases where a batholith has intruded the metamorphic rock. You might argue that the use of plate motions involves extrapolation far back into geologic time. Such is not really the case. The magnetic anomalies, from which spreading rates can be closely estimated in combination with radiometric ages, go back to Cretaceous and Jurassic time, which will encompass the time in which metamorphic rocks were produced by continental collisions along the western margins of North America and in a great collision belt extending from the European Alps to the Himalayas. The creation scientist will need to produce a mountain of contradictory evidence to convince anyone that metamorphic rocks can be formed in one year, or even in 1,000 years. (Metamorphism continuously going on in the accretionary prisms of active subduction zones is an entirely different process, unrelated to collision tectonics.)

Credit

1. From G. Brent Dalrymple, How Old Is the Earth? A Reply to "Scientific" Creationism, pp. 66-131 in Frank Awbrey and William M. Thwaites, Eds., 1984, Evolutionists Confront Creationists, Proceedings of the 63rd Annual Meeting of the Pacific Division, American Assoc. for the Advancement of Science, vol. 1, Part 3, San Francisco, 213 pp. Used by permission of G. Brent Dalrymple.

Chapter 23

Rapid Formation of Sedimentary Rocks during the Flood

The Institute for Creation Research textbook emphasizes two features of Flood stratigraphy: (a) it was rapid; (b) it was continuous (Morris, 1974a, pp. 94-117). They claim that various lines of evidence point to the rapidity of accumulation of sediments. They find no evidence of time-breaks, or time lapses in the record. These are specific claims that can be evaluated in terms of mainstream geology, because both creation scientists and mainstream scientists are looking at the same record. Field parties made up of representatives of both groups can go to the same exposure of strata and discuss their divergent views on the spot--as indeed they have done on the rocky bed of the Paluxy River in Texas, where creationists assert that human footprints are contemporaneous with dinosaur footprints.

Table 23.1 compares the creationists' timetable of Flood phases with the standard system of mainstream geology. More than one billion years of stratigraphic history under the latter chronology must be compressed into the single year of the Flood and a 4000-year post-Flood period. The creationists' claim that various kinds of sedimentary rocks contain evidence of a very rapid rate of formation includes references to sandstones, shales, conglomerates, limestones, dolomites, cherts, and evaporites.

Cementation and Lithification of Sediments

The creationists' case is made in terms of uniformitarianism and the rate at which sand and gravel can become cemented into their respective rock types-- sandstone and conglomerate. They argue that the kinds of deposits we see in process of formation today in stream channels and beaches could become cemented only under very unusual circumstances. Mainstream geology has ample provision for cementation through the action of ground water following burial of the sand or gravel under younger layers of the same material or under other kinds of sediments. Slowly circulating ground water contains carbonate matter in solution as calcium bicarbonate. Calcium ions are abundantly available through the chemical decomposition of igneous rocks and from previously formed carbonate rock (limestone). Calcium readily precipitates as calcium carbonate, a common cementing material. In some cases the circulating ground water carries and deposits silica (silicon dioxide), also abundant in igneous source rocks, and this, too, can form a cement for sandstone and conglomerate. Thus cementing substances and processes are abundantly available in natural environments, but the action is far too slow to fit into a one-year Flood.

The ICR text points out that concrete, a mixture of sand, water, and portland cement, sets hard in a few hours, as everyone knows (Morris, 1974a, p. 102). The implication is that a similar activity occurred during the Flood. Note, however, that portland cement is produced by firing of limestone and siliceous rock at high temperature. When the cement is exposed to water, it rapidly combines with water to achieve a chemically stable state. If some natural form of portland cement (such as a variety of volcanic ash called "pozzolan") were produced on a vast scale by volcanic eruptions during the Flood, it would react immediately upon immersion in floodwaters and would be so quickly dispersed that it would not remain capable of setting in a mixture with sand and gravel. Calling attention to industrial portland cement is no substitute for proposing a detailed and workable process by which sand and gravel could be cemented during some part of a one-year Flood.

The creationist text does not mention lime (calcium oxide), which is produced by "burning" (heating) of limestone. Upon being mixed with water, it forms hydrated lime--Ca(OH)$_2$. Lime mixed with sand is familiar as mortar used to plaster walls. Over long periods of time, the hydrated lime absorbs carbon dioxide from the air and is altered to calcium carbonate. Perhaps the creationists would want to postulate the occurrence of lime during the Flood as an explanation for carbonate cement in sandstones. (No charge for this suggestion.)

Sand Sheets of the Continental Interiors

The uniformitarian argument is applied by creationists to the observation that some sandstone formations occur over enormous continental areas. The formation the creationists pick as an example is the St. Peter Sandstone of Ordovician age (Morris, 1974a, p. 103). It is found distributed over an enormous area of the upper and middle Mississippi Valley (but not over the entire United States, as the creationists' text implies). It is a remarkably pure sandstone, consisting of quartz grains that are beautifully rounded. Upon close examination, the grain surfaces are seen to be frosted by microscopic fracture marks, typical of quartz grains found in active sand dunes. The frosting is the result of impacts between grains rebounding during high-speed travel under the impulse of wind. Geologists interpret the St. Peter Formation as a coastal deposit formed from coastal sand dunes reshaped by wave action. Creationists claim that such a widespread, uniform sheet of sand is unlike anything being formed today, and that it seems to be explainable only as the deposit of a continentwide inundation. Another widespread formation cited by the creationist text is the Shinarump Conglomerate of the Colorado Plateau, which, they say, extended over an area of more than 300,000 square kilometers (pp. 103-4). They ask, how could such a vast layer of gravel have formed, except by a continentwide flooding?

In both of these cases, the corpse of an outmoded and rejected concept of "uniformitarianism" is set up and flayed. If such great sheets of sand or gravel cannot be found in process of formation today, as uniformitarianism requires be the case, they must be the result of catastrophism--which, of course, is the Flood. Mainstream geologists do not claim that the world of today contains examples of recent formation of everything that has ever happened in all of geologic time. Clearly recognizable from the study of strata are periods of the past when large proportions of the continents had been reduced to low-lying plains, rising not much above sea level. The strata themselves tell a story of the gradual submergence of such surfaces (often called peneplains, a contraction of the two words "penultimate" and "plain"). As submergence occurred, the marine shoreline and its offshore depth zones migrated inland over distances of thousands of kilometers. Sediments of the beach and offshore zones were continually deposited, but the zone of deposition migrated into the continent. There is also a record of the reverse process in which the continents slowly emerged and the zones of marine sediment deposition migrated off the continental interior toward the margin. In either case, sheets of particular kinds of sediment would be formed, but the age of the sheet would not be the same throughout its extent. A simple analogy comes to mind. Consider how a roof is covered with shingles. The first row is laid at the eaves, followed by rows extending higher and higher until the roof is completely covered. The shingle cover is younger at the roof peak than at the eaves, is it not? Nobody would be so foolish as to explain a shingle roof as something that suddenly fell out of the sky in a single piece. Yet, that is just what the creationists are trying to say: the sheet of sand or gravel simply "fell" out of the floodwaters to blanket the ocean floor at a single instant.

The Shinarump Conglomerate is interpreted by geologists as a gravel-bar deposit left by streams that shifted from side to side over a sloping plain in an arid environment. Enclosed in the conglomerate are logs of trees, now silicified. Gravel deposits of this kind can be seen forming today in piedmont locations throughout the arid southwestern United States. Over a long period of time an extensive, continuous gravel sheet is formed, but the construction requires time and does not occur in a single flood.

The creationist text makes a similar argument about the widespread distribution of certain shale formations, and the argument can be answered by much the same interpretation as applies to the sandstone and conglomerate formations (Morris, 1974a, p. 103). The shale was deposited in a particular depth zone with reference to the land margins, and this zone shifted with time.

In this connction, the creationist text makes a point of the meaning of a typical succession of strata, starting at the base with conglomerate and sandstone and overlain by a shale formation. This, they say, is what is to be expected if a watery mixture of sediment of many size grades is allowed to stand. The coarse particles settle out first, followed by those of finer grade, and finally by clay particles that comprise the shale. The principle is entirely valid and applies in the case of turbidites, which were described in Chapter 20 (see Figure 20.24). On the other hand, the same succession from coarse to fine can be interpreted as the successive deposits of a deepening ocean, as a continent subsides and the shoreline transgresses the continent at a given location. The time sequence will consist of shoreline deposits of coarse sediments, followed in time by the clays and muds of a zone of greater depth that replaced the shoreline.

The depositional sequences worked out by mainstream geologists require no worldwide flood and no catastrophic deep flooding of all continents simultaneously. Under the general theory of plate tectonics, gradual inundations of low-lying continents can be explained by changes in rates of crustal spreading along the plate boundaries. Changes in spreading rate affect the average depth of the ocean basin and its ability to accommodate a given volume of water. When basin floors are deeper, the "freeboard" of the continents is greater; when basins are shallow, the continents can experience partial inundation. In comparatively recent (late Cenozoic) time the continents have tended to be emergent, so that shallow inland seas are largely lacking. During the Cretaceous Period, as an example, continents were widely submerged. Perhaps at that time plate spreading was rapid. Rapid spreading means more rapid production of new oceanic lithosphere that, being hotter than old lithosphere, is more buoyant and has a shallower seafloor depth. This relationship must be considered as hypothesis, but it is remarkably consistent with the overall workings of plate tectonics.

Table 23.1 A Creationist Flood Chronology

Age	Standard system	Corresponding stage of the Flood*
	Recent (Holocene)	Period of post-Flood development of modern world.
-10,000 y.	---------------	
	Pleistocene	Post-Flood effects of glaciation and pluviation, along with lessening volcanism and tectonism.
-1.6 m.y.	---------------	
	Tertiary	Final phases of the Flood, along with initial phases of the post-Flood readjustments.
-66 m.y.	---------------	
	Mesozoic	Intermediate phases of the Flood, with mixtures of continental and marine deposits. Post-Flood possibly in some cases.
-286 m.y.	---------------	
	Paleozoic	Deep-sea and shelf deposits of the early phases of the Flood, mostly in the ocean.
-570 m.y.	---------------	
	Proterozoic	Initial sedimentary deposits of the early phases of the Flood.
-2.5 b.y.	---------------	
	Archaeozoic	Origin of crust dating from the Creation Period, though disturbed and metamorphosed by the thermal and tectonic changes during the cataclysm.

*From Henry M. Morris, Ed., Scientific Creationism, Creation-Life Publishers, San Diego, Calif., p. 129. Copyright © 1974. Used by permission.

Eolian Sediments

While we are on the subject of marine transgressions and the regressions from the continental margins and interiors, it is appropriate to bring up the subject of formations within the geologic column that clearly formed directly beneath the atmosphere, and not under a continuous cover of water, be that water oceans or inland lakes of naturalistic origin or a divinely created worldwide Flood. Gravels deposited by shifting streams on sloping desert plains (including features called <u>alluvial</u> <u>fans</u>) are one such variety of <u>subaerial</u> <u>deposit</u>. Of greatest interest, however, are the great layers of hardened dune sand that are clearly recognizable in the record of strata exposed for all to examine in the walls of Grand Canyon and Zion Canyon.

Desert dunes occur in extensive areas today in the Sahara Desert and in areas of lesser extent in the United States--between Calexico and Yuma (Holtville region), California, and the Alamosa Valley of Colorado, for example. These desert dunes have distinguishing physical properties that set them apart from all other known forms of well-graded sand deposits. The dune surfaces are devoid of plant cover and are formed into great wavelike ridges with sharp crests and steep lee slopes. The sand, usually almost entirely of quartz composition, is extremely well graded in terms of size. The grains are spherical to a degree of perfection not found in water-transported sands. The grain surfaces are frosted by the force of intergrain impacts in free air, not subject to the cushioning effect that is found in water. Under prevailing strong winds, with dry conditions, the sand is carried up the windward dune slopes by low leaps and rebounds. Upon reaching the dune crest, the grains are projected into the air to fall in the comparative calm of the protected lee slope, where they build up the sand slope to a steeper angle of inclination. This slope is the slip face. At intervals, an unstable surface layer under gravity slides down the slip face until stability is resumed. This process, repeated innumerable times, gives the dune an internal structure consisting of long, steep sand laminae. The structure is called dune bedding or planar lamination (Figure 23.1). In a region in which sand is accumulating, such structures are buried, and if geologic conditions are generally favorable, a formation of dune sand is buried and later cemented into sandstone.

Exposed in the walls of Grand Canyon is the Coconino Formation of Permian age. It is about 90 m thick and qualifies in all respects as a dune formation. In the walls of Zion Canyon the Navajo Formation of Jurassic age, over 500 m thick, consists of cross-laminated dune sand. The highway approaching Zion from the east provides spectacular close views of the cross-lamination etched into relief by rock weathering. Restorations of the positions of continents in the Permian and Jurassic periods show this region to have been in the subtropical latitudes that contain persistent large deserts. Fossils are either totally lacking in these dune formations, or extremely rare. In the Coconino Formation quadruped tracks have been found on the lamination surfaces. The evidence is overwhelming in support of major periods of time in which this region was in an inland location and exposed to the atmosphere under dry conditions. This is an environment totally incongruous with the Flood scenario. According to Flood geology, while land surfaces were above the rising flood, they were subjected to almost continual heavy rain, which would prevent wind action and dune formation. When covered under the floodwaters, no such dunes could have formed.

The evidence of subaerial origin of the dune-sand formations is undisputed as to its significance by mainstream geology; in itself it is sufficiently weighty to totally discredit the biblical story of the Flood of Noah as a naturalistic phenomenon occurring in one year. In both examples, the dune formations are underlain and overlain

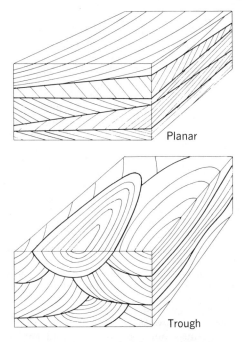

Figure 23.1 Two varieties of cross-laminations. The heavy line at the base of each set of laminations represents a surface of erosion, truncating older sets of laminations beneath. Upper block shows planar lamination, seen in dune deposits. Lower block shows trough variety, typical of streambed deposits. (Based on data of E. D. McKee and G. W. Weir, 1953, <u>Bull.</u>, <u>Geological</u> <u>Soc.</u> <u>of</u> <u>America</u>, vol. 64, p. 387, Figure 2.)

by thick sequences of marine strata, rich in marine fossils. There is no evidence whatsoever in this entire sequence of formations having been bodily overturned or seriously disturbed by any tectonic activity, other than minor faulting and gentle warping.

The Permian dune formation (Coconino Sandstone) is within the Paleozoic Era that, according to the creationists' chronology (Table 23.1), was an early Flood phase during which deep-sea and shelf deposits were formed. The Jurassic dune formation (Navajo Sandstone) is within the Mesozoic Era. According to the table, the Mesozoic strata are mixtures of continental and marine deposits, formed in intermediate phases of the Flood. But also included is the waiver that they may be post-Flood deposits in some cases. The last suggestion is absurd, because overlying Cretaceous marine strata in the same region of Utah, highly fossiliferous, were also Flood deposits.

I have found no mention of this evidence in the major creationist works, including Whitcomb and Morris (1961), Morris (1974a,), and Wysong (1976). That the subject is not known to creation scientists is most unlikely, since it is found in many widely used textbooks of physical and historical geology. Weber, writing about the dune formations and their significance and discussing specifically the tracks of quadrupeds on the dune laminations, says: "ICR geologist Dr. Steve Austin has taught the theory that amphibians resting between underwater dunes made the tracks. His theory is very interesting, but rather implausible since the Flood must have been violently dumping several meters' worth of sediment per day" (1980a, p. 26). Apparently, the creation scientists' argument, when expressed, is to the effect that the laminated sandstone formations referred to are not ancient terrestrial dunes.

Besides dune formations, the stratigraphic column contains many formations that seem beyond doubt to have been formed under conditions of persistent atmospheric exposure between surges of stream water across floodplains and deltaic plains, and tidal inundations of the latter. One such formation is the Moenkopi Formation of Triassic age, exposed in low cliffs north and east of Grand Canyon (Strahler, 1971, pp. 508-9). The Moenkopi Formation contains a variety of interesting features. Siltstones in thin slabs are in many places beautifully ripplemarked. This kind of ripple mark is made by water currents and suggests the action of shallow tidal waters. The same slab has shrinkage cracks, suggesting exposure to the air, perhaps at low tide. Other slabs show rain pits, the fossilized impressions of rain drops. There are also slabs with casts of salt crystals, suggesting evaporation of concentrated brine in shallow tidal lagoons. It is evident that at least some parts of the Moenkopi Formation were deposited in an intertidal zone (littoral environment) on mudflats alternately covered by shallow water and exposed to a dry atmosphere. Skulls and other bones of amphibians are found in these rocks, as well as reptile tracks. These fossil evidences also support the interpretation of partial exposure of mudflats at low water. Perhaps some of the deposits are actually those of broad river floodplains. All of the red mud and sand is considered to have been brought to the coastal belt of deposition from highland sources to the east, in what is now Colorado and New Mexico. Red beds, as continental sediment sequences such as the Moenkopi Formation are referred to, are widespread and numerous in the geologic record, especially so in the Mesozoic Era. Their presence is entirely unpredicted by and incongruous with the Flood geology program.

Christopher Gregory Weber sums up the whole idea in a nutshell: "You don't need a Ph.D. in geology to know that desert dunes and other desert deposits do not form under roaring flood waters. These require not only time, but also dry land. The Flood of Noah supplies neither" (1980a, p. 25).

Coal as a Flood Deposit

Coal layers, found in strata of most of the periods of geologic time since the first forests flourished in Middle Devonian time, are a subject of great interest to the scientific creationists, who see them as products of the Flood, formed rapidly and by catastrophic processes. That coal has been derived from plant remains accumulated in a water-saturated environment is a conclusion shared by creationists and mainstream scientists. Beyond this point, the explanations are widely divergent.

The picture of coal formation derived by mainstream geology begins in a swamp or bog environment where water saturation persists. Here plant remains accumulate faster than they can be destroyed by bacterial activity. Only partial decomposition occurs because oxygen is deficient in the stagnant water, and the organic acids released by the decay process inhibit further bacterial activity. The product of this environment is peat, a soft fibrous material ranging in color from brown to black. Although peat has formed in North America and Europe in bogs occupying former glacial lakes, that environment is not considered likely for the formation of coal. A second contemporary environment of peat formation is in salt marshes in the tidal zone. Because of the slowly rising sea level in Holocene time following disappearance of the ice sheets, tidal peat has accumulated in successive layers and in places reaches a thickness of 10 m or more. This particular scenario does not, however, prove useful in explaining the thickness and extent of peat layers capable of yielding coal seams of the geologic record.

In the geologic past, peat layers have become compacted first into lignite or "brown coal," a low-grade fuel intermediate between peat and coal. Lignite has a woody texture; its origin from plant matter is obvious from the organic structures it contains. Upon further compaction under the load of accumulating sediments lignite has been transformed into bituminous coal, or "soft

coal," that makes up the vast bulk of coals of the geologic record. In some areas, however, tectonic activity associated with continental collisions folded the strata in which bituminous coal was present, subjecting it to intense pressure and altering it to anthracite, a metamorphic product containing about 96 percent carbon. Bituminous coal, in comparison, has a substantial content of volatile hydrocarbon compounds.

Coal occurs in layers, known as seams, interbedded with sedimentary strata, which are usually thinlybedded shales, sandstones, and limestones. Collectively, such accumulations are known as coal measures. Individual coal seams range in thickness from a few millimeters to as much as 10 to 15 m in extreme cases. It is estimated that the production of a 1-meter coal seam required a 30-meter layer of peat, and that the accumulation of 1 meter of peat may have required 1000 years. On this basis, the great Mammoth coal seam of Pennsylvania, 15 m thick, would have required a peat layer about 450 m thick, in turn requiring a period of plant accumulation lasting over 450,000 years. Such figures are staggering to the imagination because of the necessity that the environment of accumulation of the peat must have been extraordinarily stable. This aspect of the problem is a juicy morsel for the creationists! How much simpler to let it be accomplished in hours or days in the great Flood. They must search for evidence of an extremely rapid rate of formation of both the coal seam and of the strata above and below it.

Prior to about 1960, which is about the time Whitcomb and Morris' major treatise, The Genesis Flood, was first published (1961), the great coal seams were viewed as originating in accumulations of forest vegetation in enormous freshwater swamps existing for extremely long spans of time on low-lying plains. The kind of environment that comes to mind in this connection is the swampy floodplain expanse of a great equatorial river, such as the Amazon or Congo, located in a region of perpetually warm climate and abundant rainfall well distributed throughout the year. Between the long periods of peat production, there occurred invasions of shallow seawater, in which marine shales and limestones were deposited, followed by withdrawal of the sea, exposure of the surface to erosion by streams--which scoured the marine strata--and deposition of a layer of continental (nonmarine) sediments. Of these, a sand layer was first deposited, followed by mud layers, and then another peat layer. This cycle consisted of withdrawal of the sea (regression) followed by flooding of the sea (transgression). The sediment sequence of a full cycle was given the name cyclothem. In coal measures of the Upper Carboniferous Period in North America, a typical cyclothem is about 30 m thick. This older version of depositional sequence in a single cyclothem is represented in Column A of Figure 23.2.

In the above cyclothem sequence, the nonmarine sequence begins with a basal sandstone having typical trough lamination of water currents such as occur in streams, including distributary streams on deltas (see Figure 23.1). Typically, the surface on which the sands are deposited is one of erosion, with channels carved into the marine shales of the underlying sequence. A scoured channel of this type is shown in Figure 23.2. The nonmarine sequence grades upward into silty shale, above which the coal seam occurs. These strata contain plant and animal remains that show they were deposited on land (after the manner of the Moenkopi deltaic red beds described earlier, but in a different climatic environment). Nonmarine strata would have been brought from inland locations by streams draining the land. This is a familiar activity seen today in many locations around the world. Transgression of marine waters would have caused the switch to marine sediments. The Flood hypothesis, on the other hand, explains these systematic alternations by alternating currents requiring the utmost in uniformity of

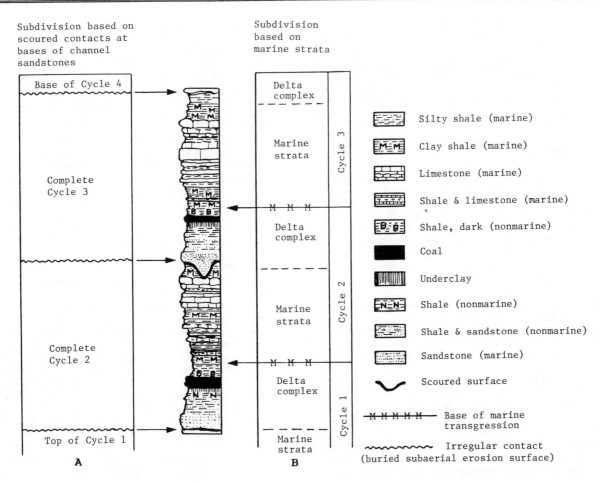

Subdivision based on
scoured contacts at
bases of channel
sandstones

Subdivision
based on
marine strata

Base of Cycle 4

Complete
Cycle 3

Complete
Cycle 2

Top of Cycle 1

A

Delta
complex

Marine
strata

Delta
complex

Marine
strata

Delta
complex

Marine
strata

Cycle 3

Cycle 2

Cycle 1

B

Silty shale (marine)

Clay shale (marine)

Limestone (marine)

Shale & limestone (marine)

Shale, dark (nonmarine)

Coal

Underclay

Shale (nonmarine)

Shale & sandstone (nonmarine)

Sandstone (marine)

Scoured surface

Base of marine
transgression

Irregular contact
(buried subaerial erosion surface)

Figure 23.2 Strata of a typical Pennsylvanian cyclothem interpreted in two different ways. (A) The older interpretation, based on the theory of inland floodplain deposition of peat. (B) The newer interpretation, based on alternating deltaic and marine environments. (From G.

M. Friedman and J. E. Sanders, <u>Principles</u> <u>of</u> <u>Sedimentology</u>, pp. 302-3, Figure 10.28. Copyright © 1978 by John Wiley & Sons, Inc. Reprinted by permission of John Wiley & Sons, Inc.)

action--a physical system entirely antithetical to a chaotic scenario of enormous water turbulence containing great eddies and continually disturbed by violent volcanic eruptions and tectonic movements on the ocean floors.

In Flood geology, the plant material from which coal was formed did not come from plants growing in place, but was transported long distances by Flood currents. It originated as forest growth on the pre-Flood land surfaces, from which it was stripped by torrential water flows and rafted in great floating mats far out into open ocean water, where it sank to the bottom. To avoid any possibility of inadequately or incorrectly presenting the creationists' hypothesis, let us read their own words:

> The physical evidence plainly and emphatically demonstrates the fact that the coal seams are water-laid deposits, in which great agglomerations of plants were rafted down on the surface of the Deluge rivers, then conveyed back and forth on the shifting currents until finally brought to rest in some basin of deposition, to be followed by a reacting current from another direction bearing nonorganic materials perhaps, then another current with a load of plant debris, and so on. (Whitcomb and Morris, 1961, p. 277)

We might wish to call this mode of origin the "alternating current hypothesis." Despite the enormous turmoil and turbidity of the floodwaters (as described by the creationists), currents coming from one direction alternated in just the right way with currents coming the other way to construct great numbers of cyclothems, each

remarkably similar to the next. According to the ICR textbook, the currents carried sand, silt, or lime mud (Morris, 1974a, p. 108). Obviously, an extremely complex flow system operating with great precision and order is required here, but the creationists provide no details whatsoever. Clearly, an issue is developed over whether the coal was formed from vegetation growing in place or was imported as plant material produced elsewhere.

True to their practice, the dispute over the origin of coal is introduced by the creation scientists by the ritualistic parading of a straw-person asserted to represent the prevailing view of uniformitarianism. How familiar is the plan! Whitcomb and Morris attack the peat-bog hypothesis as follows:

> This theory, which is purportedly uniformitarian in essence, is actually anything but that, as there is no modern parallel for any of its major features. The peat-bog theory constitutes a very weak attempt to identify a modern parallel, but it will hardly suffice. (1961, p. 163)

The authors then quote a paragraph from an authoritative work on coal to the effect that no known modern bog or marsh could supply sufficient peat to make a large coal seam (E. S. Moore, 1940, p. 146). Whitcomb and Morris continue:

> The Dismal Swamp of Virginia, perhaps the most frequently cited case of a potential coal bed, has formed only an average of seven feet of peat, hardly enough to make a single respectable seam of

coal. Furthermore, there is no actual evidence that peat is now being transformed into coal anywhere in the world. No locality is known where the peat bed, in its lower reaches, grades into a typical coal bed. All known coal beds, therefore, seem to have been formed in the past and are not continuing to be formed in the present, as the principle of uniformity could reasonably be expected to imply. (Pp. 163-64)

One page later, they say:

Regardless of the exact manner in which coal was formed, it is quite certain that there is nothing corresponding to it taking place in the world today. This is one of the most important of all types of geologic formations and one on which much of our supposed geologic history has been based. Nevertheless, the fundamental axiom of uniformity, that the present is the key to the past, completely fails to account for the phenomena. (P. 165)

There is no need to repeat the details of a full reply to such statements. The uniformitarianism they refer to has long since been rejected by mainstream geology, replaced by a common-sense actualism that is the basis of all natural science. The present does not hold all keys to the past.

The statement to the effect that no locality is known where a modern peat bed grades downward into coal is fully in conformity with prevailing views of coal formation based on physics and chemistry. Compaction of peat is essential to increase its density, drive out water and volatiles, and increase the relative amount of carbon. This process requires a substantial loading beneath accumulating overlying strata and/or application of heat. It would be remarkable indeed if a peat layer graded downward into lignite, then bituminous coal, and finally into anthracite, at a depth of only a few meters.

The creationist version of the origin of coal was reiterated in 1976 by Stuart E. Nevins, Professor of Geology and Archaeology at Christian Heritage College. In Chapter 22 we discussed his critique of plate tectonics. In an issue of ICR Impact Series (No. 41) titled "The Origin of Coal," Nevins enlarges on the topics covered by Whitcomb and Morris in 1961 and Morris in 1974, adding details of the fossil content of coal seams and related strata.

Nevins calls attention to the presence of marine fossils "such as fish, molluscs, and brachiopods in coal" (1976b, p. i). He also mentions the finding of small marine tubeworms (Spirorbis) attached to plants in Carboniferous coals. In an earlier paragraph, Nevins has also pointed out that fossil plants found in Carboniferous coal seams are interpreted as freshwater plants, and that some of them may have preferred well-drained ground rather than swampy conditions. Putting these two indications together --freshwater (terrestrial) plants in association with marine animals--Nevins finds a contradiction that flaws the prevailing early theory of coal formation, which we have reviewed. At the same time, the mixture of terrestrial and marine features is, according to Nevins, exactly what would be expected of the Flood scenario, which calls for rapid mixing of rafts of terrestrial vegetation with marine sediments.

In passing, we note that Nevins makes use of two ponderous and mystifying geological terms to describe the opposed versions of the origin of coal. The mainstream model of in-place growth of the plants that were to become peat and then coal is described as the autochthonous theory; the creationist model of rapid transport from the growth area to the distant deposit area is the allochthonous theory. While these adjectives have been applied to theories of coal formation, their principal use in geology has been to describe great masses of crustal

strata with respect to their place of origin: autochthonous strata remain in the same place of deposition; allochthonous strata have been imported from an adjacent region by the tectonic mechanism of overthrusting. Nothing is to be gained by incorporating Nevins's use of these terms in the context of coal formation or any other topic of Flood geology in other than the original tectonic meaning. In the event of having these terms thrown at you in a public debate, it may be wise to have been forewarned.

What we need to do next is to present a modern view of the origin of major extensive coal measures. This newer mainstream view of the origin of coal is very different from the old tropical-coal-swamp version with alternating marine regression and transgression that would keep the terrestrial and marine deposits in completely separate compartments, sequential in time. Let us take a second look at the cyclothem and an alternative interpretation of the sedimentary sequence it shows.

Since the 1960s, great strides have been made in probing the world's great deltas to determine how they are constructed and their past history. At the same time, newer, more detailed studies have been made of coal seams of various ages and locations. There has emerged a hypothesis of peat accumulation that is a major step in relating delta structure and composition to ancient coal measures. The deltaic origin of cyclothems and their coals has been generally adopted as a working hypothesis, replacing the previously prevailing reconstruction of vast floodplain swamps extending far into the ancient continental interiors. Good reviews of the newer version are given by Friedman and Sanders (1978, pp. 301-3) and Landis and Averitt (1978, pp. 165-66). Details are given by Fassett and Hinds (1971). Emphasis today is upon the former construction of an offshore barrier of marine sand, behind which extensive coastal marshes were sequestered, and in which marsh vegetation produced thick peat. Large deltas normally undergo continual subsidence, both because of compaction of sediment and the crustal-loading effect that requires isostatic compensation. Continued upbuilding of the sand barrier would have allowed peat accumulation to continue for long periods. Migration of the sand barrier toward the continent would have not only buried the peat layer, but also allowed its development to extend farther landward, resulting in an aerially vast coal seam in some cases. Column B of Figure 23.2 shows the cyclothem divided according to the deltaic/marine hypothesis.

Thus, delta-forming processes observed today can be used as a key to processes active in past geologic periods. A modern statement of uniformitarianism does indeed provide for that form of extrapolation. We must assume that any physical process observed to happen today might also have happened in the past. It does not, however, allow us to limit interpretations of past events, in terms of both intensity and magnitude, to events observed today. Observations of the present are permissive of interpretation, but not restrictive. Events that are indicated by the rock record as having occurred in the distant past do not, therefore, necessarily have to be part of contemporary process. The older hypothesis of coal formation required a process not seen in action today--an acceptable feature under modern uniformitarianism, but one that increases its vulnerability to attack.

It is interesting to make estimates of the rate of deposition of sediment, including coal seams, under the Flood scenario. Take, for example, the Pennsylvanian (Upper Carboniferous) system, in which coal measures were deposited in the Middle West and Appalachians. Using the mainstream geologists' time scale for proportional time allocation only, the Pennsylvanian system (35 m.y. duration) represents about 1.4 percent of the 2.5 b.y. that includes all time since the start of the Proterozoic. Assuming 300 Flood days of major depositional activity, the Pennsylvanian coal measures get about 4 days, or 96 hours, as their fair time share. For southern West Virginia, with 90 cyclothems in a total sediment thickness of about 1520 m (5000 ft), the average cyclothem thickness is 17 m. During the 96 hours available for total deposition, the average rate would have been about 16 m (50 ft) per hour. Roughly, this is one cyclothem per hour. Under mainstream geology a cyclothem in this area should have required about 400,000 years, for a depositional rate of one meter per 23,500 years. Kay and Colbert give an estimate of about 1 m.y. per cyclothem (1965, p. 248). This schedule seems ample to allow peat accumulation from plants growing in place over a sufficient time to produce a coal seam 1 to 10 meters thick.

Under the Flood schedule, the transporting currents would be required to alternate direction about every half hour. This would seem to require rather strict traffic control. There would be no provision for long-settling time of clays, but it can be supposed that they were clotted into larger particles by flocculation. Difficulties arise when we try to visualize individual thin sheets of turbid water (turbidity flows), each one carrying its respective load of sand, clay, or lime mud. (Graded bedding is absent from the cyclothems.) Assuming a depositional area 100 km across and 10 individual layers to be deposited in each direction, each layer by one flow, a single flow would have to streak across the entire area no less rapidly than 200 km/hr, assuming that each flow moving in one direction could start within seconds of the departure time of the previous flow. You can think of it as a case where 10 rolls of carpet are to be unrolled, one over the other, in rapid succession. One roll can closely follow the other by a short distance. Then 10 other rolls are unrolled in a sequential group in the opposite direction. We have made no time allowance for a great raft of matted plant matter to fall into place between strata of the two directions. The required current speed is excessive in view of estimates of turbidity current speeds of not over 30 km/hr (Shepard, 1963, p. 340).

Creation scientists make a strong point of the significance of horizontal, rootlike structures that are sometimes found beneath a coal seam within a dense plastic clay layer (fire clay) that is commonly present directly under the coal. Called <u>stigmaria</u>, they have been interpreted as root structures or rhizomes of trees that grew in the swamp forest, but this is questioned by sources quoted in Whitcomb and Morris (1961, p. 164). Elsewhere, in Carboniferous strata of Nova Scotia, upright tree trunks are exposed enclosed in sandstone (not coal) (Kay and Colbert, 1965, p. 259). At the base of the trunk, stigmaria are found attached. In those cases, the implication of rapid burial of live trees is strong, as the creationists point out (Morris, 1974a, pp. 107-8). It should suffice to note that evidence clearly pointing to rapid deposition of a single layer a few centimeters to several meters thick of a particular kind of sediment, such as sand, is quite irrelevant to the question of the time required for the accumulation of sequences of strata made up of many layers of various rock types. Finding a tree trunk enclosed in sandstone in no way supports the hypothesis that several thousand meters of coarse sediments must have accumulated in a few weeks or months.

Fossil Forests and Flood Tsunamis

Geologic phenomena that produce repetitive sedimentary sequences include a remarkable formation of "fossil forests" of the Yellowstone National Park. The sequence was discovered and first described by William Henry Holmes over a century ago (Holmes, 1878, p. 48). Exposed in the steep canyon walls of the Lamar River on the northern slope of Amethyst Mountain, he found "a multitude of the bleached trunks of the ancient forest." Here, "rows of upright trunks stand out on the ledges

like the columns of a ruined temple." On the lower slopes "petrified trunks fairly cover the surface." Holmes, who was a landscape artist as well as a geologist, made a detailed sketch of a similar exposure at Specimen Ridge; it gives the strong impression that mature forests were buried in succession under volcanic materials, i.e., tephra that rained down from nearby volcanoes or poured as mudflows down the volcano flanks (Figure 23.3). The exposed section is about 600 m thick and includes some 15 to 18 separate "forest" layers.

J. Laurence Kulp, a geochemist to whom we referred earlier as a theistic scientist of the American Scientific Affiliation, and who urged acceptance of the radiometric time scale, cited the Yellowstone buried forests as a repeating stratigraphic sequence that must have required much more time to complete than the single year of the Flood (Kulp, 1950, p. 10). He wrote:

> In Yellowstone Park there is a stratigraphic section of 2000 feet exposed which shows 18 successive petrified forests. Each forest grew to maturity before it was wiped out with a lava flow. The lava had to be weathered into soil before the next forest could even start. Further this is only a small section of stratigraphic column in this area. It would be most difficult for flood geology to account for these facts.

Whitcomb and Morris admit that this occurrence is spectacular, and perhaps difficult to accommodate in Flood

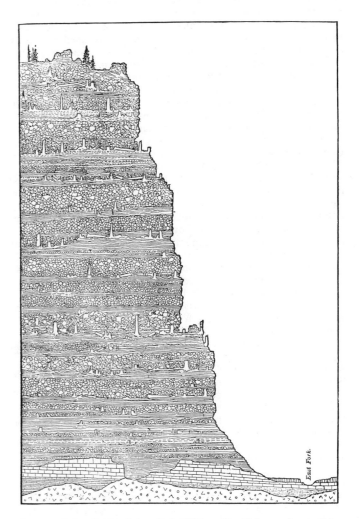

Figure 23.3 Geologist-artist William H. Holmes drew this composite sketch of the arrangement of fossil tree trunks interbedded with tephra in Specimen Ridge, Yellowstone National Park, Wyoming. (W. H. Holmes, 1878, p. 48.)

geology, but at the same time they see in the volcanic sequence a record of catrophic events such as would be expected in the Flood (1961, p. 419). They make their case by proposing an alternative scenario to one of forests being destroyed in situ. The fossil trees, they say, were transported by great water surges of the Flood from some distant area where they had grown. This is how Whitcomb and Morris describe those catastrophic events:

> The stumps give every appearance of having been in some manner sheared off by some overwhelming force (possibly tsunami-driven debris), then uprooted and transported and sorted out from other materials and then suddenly buried beneath a volcanic shower. Then came another wave of sediment and stumps (several layers of sediment, however, appear to be without any stumps), possibly resulting from the tsunami generated by the preceding eruption, then another volcanic shower, and so on. The whole formation, as does the volcanic terrain all over the Yellowstone region and the Pacific northwest, literally proclaims catastrophic deposition! (P. 421)

The creationists' picture of great surges of floodwater sweeping up over the higher land surfaces, perhaps in both rising and falling stages of the Flood, is a familiar one--as we saw in the case of the formation of coal measures by rafting of great masses of forest debris. In the case of the buried forests, the surges would be in late stages of the Flood. Whitcomb and Morris point out that many of the tree trunks lie prostrate and many lack long root systems (p. 419). Those that are erect probably assumed this position as they came to rest because of the low center of gravity of the tree base and its root mass.

Recent studies of the "fossil forests" show that the environment of deposition was indeed a complex one. William J. Fritz examined details of the formation in which the trees are incorporated. It represents a complex alluvial system in which contributions of layers were made both as mudflows and as beds of gravel and boulders (conglomerate) of braided streams (Fritz, 1980a, pp. 312-13). There are also layers of tuff (volcanic ash) of sand and silt grades, but much of this material was transported in water. A large proportion of the fossil trees are in the form of horizontal trunks. In some layers, all trunks are horizontal. This suggests accumulation in a logjam, either in a stream or watery mudflow. Some trunks are in a diagonal position with respect to the vertical. The upright (vertical) stumps tend to be short and to show evidence of trunk breakage before burial. Generally, the bark has not been preserved, although bark is found intact on some vertical trunks. Added to this physical evidence of transport and violent treatment is the observation that the species of trees represent a mixture of those of cold-temperate climate and those of tropical climate. This mixture could be explained by the intermingling of trunks swept down from high-altitude locations on the volcano summits with trunks of trees growing on the lower slopes. Some trees remained in place, becoming surrounded and buried in the moving sediment. These recent studies would seem to reinforce the creation scientists' claims for transport under catastrophic conditions, but otherwise the two scenarios are as unlike as day and night.

I noted that Fritz's paper, referred to above, was accepted for publication by the Geological Society of America on April 25, 1980. On May 18, 1980, Mount St. Helens produced its single, great eruptive event of catastrophic proportions. First, a large section of the upper volcano was blown out by the explosion of gases, generating a hot cloud of gas and ash that devastated a large area of forest. A conspicuous part of the landscape devastation that followed was accomplished by stream-floods, heavily charged with sediment and by mudflows.

Both snowmelt and heavy precipitation supplied the water, while the great quantities of loose ash and coarse debris of avalanches provided sediment (Meier et al., 1981, p. 625). Logs were swept into the flows and carried along to lower levels. These flows continued long after the eruption, as new breakouts of mud occurred and moved down the gradient of former stream valleys.

And who should step forward to observe the transport of logs and stumps at Mount St. Helens but William Fritz! During several visits to the area in the summer of 1980, Fritz "observed numerous stumps that have been deposited upright after being transported many kilometers by streams and mudflows resulting from the May 18, 1980 eruption" (1980b, p. 586). He went on to write: "These observations support the conclusion that mudflows and streams of the same variety transported and deposited stumps that are preserved in a vertical position in the Eocene Lamar River Formation."

What a remarkable illustration we have here of the principle of uniformitarianism as it is viewed today by mainstream science! Violent, destructive events that are surely catastrophic in human terms were observed by humans to happen in the present, producing stratigraphic deposits matching those of Eocene strata 40 to 50 million years old. Catastrophism of naturalistic causes joins with the uniformity and continuity of commonplace landscape-forming processes in a realistic view of the world of nature.

From what has been reviewed here, it looks as if Kulp's picture of forests, their trees rooted in thick, old soils, dying in place as a benign rain of volcanic ash buries them, as would a heavy snowfall, is too placid a scenario both for naturalistic science and Flood geology. After all, Kulp favored a tranquil Flood (if indeed there was a worldwide Flood), and this seems to come out in his reconstruction of the burial of forests in situ. But it turns out that the creation scientists, in refuting Kulp's interpretation, are beating a dead horse. Whitcomb and Morris have had ample time to revise their major work, The Genesis Flood, since the Mount St. Helens eruption. Instead, they chose to reprint the original book for a twenty-sixth time in August of 1982. What better proof would one need that the account of recent creation and the Flood, once formulated, has been frozen in time? What better proof would one need that creation science is not natural science, but is, in fact, pseudoscience?

And now, back to the original bone of contention. Could the buried forests of the Yellowstone region have been deposited in less than one year--the year of the Flood? Rapid deposition of one layer of debris produced by a single eruption can obviously be a rapid process, as witness Mount St. Helens and many other eruptions of stratovolcanoes in historical time. But, as we have said before, rapidity of an event does not mean recency of that event. An accumulation of many rapid events can in no way be construed to mean that the total collection of events was accomplished in a brief period. It seems reasonable that each depositional event in the construction of the sequence that makes up the Amethyst Mountain and Specimen Ridge exposures represents a major eruption of a single stratovolcano.

From what is known of the history of Cascade volcanoes in Holocene time, events such as hot or cold mudflows have highly irregular recurrence intervals for a given locality. For example, for Mount Shasta, a published table of events contains observations such as these: "Three or more mudflows . . . within the last 210 years." "A pyroclastic flow, a hot mudflow, and three cold mudflows . . . during the last 200 years." "Seven or more mudflows . . . during about the last 1000 years" (D. C. Miller, 1980, Plate 1). Cold mudflows could occur during periods of volcanic quiescence. A table of the eruptive sequence at Mount St. Helens during the past 1,500 years shows a repose interval of 123 years following the last reported eruption before 1980. This was preceded by four volcanic events between -123 and -180 y., and before that, a repose interval of about 170 years. (Hoblitt et al., 1980, p. 556). It would be futile to attempt to find a regular periodicity of events of a magnitude capable of producing single stratigraphic units such as those in the Yellowstone sequences. We can, however, be fairly certain that a sequence of 15 to 20 events would require a time span of, say, at least the same number of years as the number of events. This is reasonable because cold mudflows and streamfloods are likely to be annual events at the minimum periodicity, i.e., produced by melting snow and seasonal rains.

Other scientists who have studied the Yellowstone sequence have reported finding stumps well rooted in paleosols (ancient soils). Greg Retallack, in commenting on Fritz's 1980 articles, writes that "there are at least some cases of pertified tree stumps unquestionably in place" and "I examined these stumps carefully . . . and found that these stumps are firmly attached to extensive root systems in paleosols that, compared to previous accounts, are surprisingly well differentiated" (1981, p. 52). He goes on to describe A, B, and C horizons of the paleosol. But he cautions that laboratory data will be needed to accurately identify the soil. For our purpose it would be enough to know that the formation of a soil with differentiated horizons would require, at the very least, some tens of years. Then consider that if the fossil tree were occupying its original position in that soil, it would have required at least 100 years to grow to the size it had attained when terminated. Consider further that if there were only two or three such successions of soil formation and tree growth, the total time span is measured in centuries. Not by the greatest stretch of natural rates of growth and geologic change could that time span be fitted into one year--the Flood year.

Also lending support to the tree-in-place interpretation of the Yellowstone fossil forests is a more recent field study by Richard F. Yuretich, who reexamined the section at Specimen Ridge (1984). Although he agrees with Fritz that fluvial and mudflow deposits have in places buried the vertical trunks, he finds no evidence that the trunks were transported. Instead, he finds that, typically, the stumps are rooted in a fine-grained sandstone, but he did not recognize distinct soil horizons. He suggests that burial in place could have come about in one of three ways: by mudflows (producing conglom- erate), by lake sediments in dammed sections of valleys, and by ash-fall. If one accepts the burial-in-place hypothesis at eight stratigraphic levels, as reported by Yuretich, the case against the Flood hypothesis appears greatly strengthened. Perhaps judgment on this conclusion should be suspended pending further research. In any case, science deals with tentative judgments, not in final judgments. Perhaps we are permitted to say that as things are now understood, the case of a one-year completion of the Yellowstone stratigraphic sequence is extremely weak and fails to attract even the scantiest measure of corroboration. The level of unreasonableness required by the Flood version is simply too great to stand against existing rates of natural processes as they are constrained to occur by biological, chemical, and physical laws.

The eruption of Mount St. Helens and the sweeping of vast quantities of debris into Spirit Lake set the creation scientists off on a new tack: the formation of coal during the Flood occurred by the action of volcanic heat on those rafts of debris swept down from the water-deluged lands. Australian creation scientist John Mackay quickly picked up this opportunity for discussion in a 1983 Ex Nihilo article titled "Mt. St. Helens: Key to Rapid Coal Formation?" Intended for the "layman," his article is nevertheless on the same level of presentation as those in the ICR Impact Series.

Mackay injects something new in the line of theories of coal formation (1983, p. 6). He describes coal measures of

Swansea Heads, Australia, where the lower coal seam has vertical tree trunks jutting from the coal and protruding into a volcanic ash layer overlying the coal. The same trunks also penetrate upward into a second coal seam overlying the ash bed. Mackay contends that the information from tree stumps and sediments in Spirit Lake now provides confirming evidence of the same volcanic/ catastrophic origin for the Swansea coal seams: the vegetation swept into the standing water was converted almost instantaneously into coal by the heat of the ash flow that buried the vegetation layer. Mackay notes that the Spirit Lake wood fragments, leaf debris, and bark that sank to the bottom of the lake showed charring (p. 7). He makes no claim that true coal has been produced by the Mount St. Helens eruption but suggests a connection, for he writes:

> Laboratory experiments indicate that rapidly applied intense heat or vibrating pressure are mechanisms whereby woody or vegetable material can be converted to coal in a short time. This makes areas like Mt. St. Helens especially relevant to coal formation. They could supply both heat and shock. (P. 7)

My own conclusion would be that the situation at Spirit Lake, which shows no evidence of coal formation (albeit of charcoal produced by the blast of hot ash), is irrelevant to the question of formation of great coal seams of the geologic record. Perhaps aware that his case is practically without value, Mackay continues:

> The size of any deposits which may be produced in Spirit Lake will be insignificant compared to known coal beds and "buried forests." Many Australian coal beds occupy thousands of square kilometres. It should be obvious that any catastrophe which could produce these beds, would have been on a much greater scale than the impressive but relatively minor eruption which was Mt. St. Helens. But what large volcanic and watery catastrophe has the world experienced? (P. 8)

The answer to that last question is, of course, the Flood. That immense catastrophe would, Mackay says, have devastated the forests of the entire world, and there would have been ample volcanic activity to produce the coal. What we actually have in Mackay's argument is a failed attempt to produce coal by contemporary unifor- mitarian volcanic processes, leaving only the resort to a catastrophism that need not rely on scientific evidence for its embrace. And so we can say "Rubbish!" to Mackay's closing line: "All of which makes the flood of Noah a primary contender when it comes to the origin of many coal deposits."

In 1984, in the lead-off paper of the first issue of Ex Nihilo Technical Journal, Andrew Snelling joined with John Mackay in repeating the latter's arguments favoring Flood coal and beefing up the arguments on laboratory production of coallike substances by heating samples of lignin and cellulose in the presence of clays (Snelling and Mackay, 1984, pp. 22-25). They also flagellate the corpse of the discredited inland peat-swamp hypothesis. Here, they introduce an irrelevant observation, namely, that modern cold-climate peats lack the basal clay layer that is typically present beneath geologically ancient coal seams (p. 26). They see in this observation the conclusive indication that subarctic (tundra) environments are not suitable for formation of coal seams. We have already pointed out that this form of cold-climate bog peat has never been seriously considered as a form of parent matter for bituminous coals. An amusing suggestion is offered under the heading of "Industrial Applications": "exploration coal geologists should be examining areas of ancient explosive volcanism and tectonism for hitherto undiscovered coal measures" (p. 28).

As a postscript to all this ash-and-coal fantasy, I offer as an antidote the contents of a scientific abstract of a paper presented before the 1985 annual meeting of the American Association of Petroleum Geologists (see Association Round Table, p. 246). Its title: "Sixty-five Volcanic Events Recorded in a Single Coal Bed." Its author: Carol Waite Connor of the U.S. Geological Survey. The locality is the Big Dirty coal bed exposed in the Bull Mountain coalfield, northwest of Billings, Montana. This thick coal seam contains as many as 65 ash layers. Individual layers average 1.5 cm thick and are separated by an average 7.6 cm of coal, tuffaceous coal, or carbonaceous tuff. No wonder it's called the Big Dirty coal bed! The author cites the coal bed as an example of relatively rapid normal sedimentation that covers each ash fall before another fall occurs. Between ash falls swamp vegetation was able to produce a peat layer that later became coal.

The 1980 ash plume from Mount St. Helens left a layer of fine ash over a large area. It fell like snow and could be swept up as if it were a snowfall. There was no heat problem for those on whom that ash rained, and it charred nothing. Tuffaceous sediments in the Big Dirty coal bed were doubtless initially the fallout deposits of ash plumes, but the particles of tuff that arrived at the coal swamp were carried there by stream floods. Keep these considerations in mind if and when you read a creationists' claim that the coal of the Big Dirty coal bed was formed by intense heating of plant debris by catastrophic waves of intensely hot volcanic ash and tuff.

Coral Reefs--How Recent?

Whitcomb and Morris claim that coral reefs, both modern reefs and those enclosed in strata of early Paleozoic and younger age, could have been formed in short periods compatible with the Flood chronology (1961, pp. 408-9). They cite a 1950 work by a marine geologist (Ph. H. Kuenen) giving estimates of growth rate of modern coral-algal reefs as being in the range of 0.1 to 5 cm/yr. Data of Chave, Smith, and Roy (1972), quoted by F. P. Shepard yield as a reasonable growth rate 1 cm/yr, based on net calcium-carbonate production rate on modern reefs "where a well-covered coral surface exists with adequate food supply" (1973, pp. 347-48).

On the north coast of Papua, New Guinea a coral reef 8 m thick was dated by carbon-14 and thorium/uranium ratios at various sample points from base to top. The rate of upward growth of the reef averaged about 0.5 cm/yr; the most rapid rate was 0.8 cm/yr (Chappell and Polach, 1976, pp. 235, 239). These rates are in good agreement with that based on the modern rate of calcium-carbonate accumulation. We will use 1 cm/yr in our calculations.

Coral atolls of the Pacific Ocean are more or less circular reefs surrounding a central lagoon. Core borings on several atolls, including Funafuti, Eniwetok, Bikini, and Midway, have revealed thicknesses of 200 to 1,400 meters of reef rock resting on a platform of volcanic rock (Ladd and Gross, 1967, p. 1092). These observations have made virtually certain the origin of atolls through slow subsidence of volcanic islands. A volcanic island, following cessation of its constructional activity, was reduced by erosion and planed off by wave action to have a flattened summit, serving as a platform for reef growth. Continued slow subsidence allowed the reef to build upward and maintain itself. On Eniwetok, the carbonate rock has a thickness of 1,200 to 1,400 m (Figure 23.4). The lowermost 500 m is dated as Upper Eocene, with an age of about -24 m.y. (Saller, 1984, p. 219).

Using a rate of 1 cm/yr as the rate of upward growth, assumed continuous and constant, 1,300 m of reef rock would require 130,000 years to accumulate. This does not allow for compaction at depth by pressure solution and

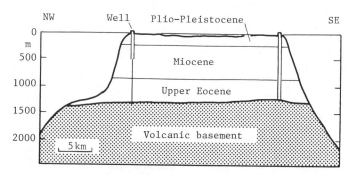

Figure 23.4 A cross section of Eniwetok Atoll, based on borings and seismic refraction soundings. (From Arthur H. Saller. Geology, vol. 12, p. 219, Figure 4. Copyright © 1984 by the Geological Society of America. Used by permission.)

fracturing or for periods of nondeposition, either of which would lengthen the total time. If, as the creation scientists assert, modern reefs were produced in the post-Flood time of 4,300 years, the available time is far too short to account for the Eniwetok carbonate deposit. Lesser accumulations, such as 200 m, would exceed the available time by a factor of five.

When it comes to great limestone reef structures of the older record, the available time of the Flood is less than one year. Consider, for example, the problem of fitting into the one-year Flood scenario the great mass of Permian reef limestone exposed in the Guadalupe Mountains of western Texas. Called the Capitan Limestone, the total thickness of reef rock is more than 500 m. To postulate that this enormous barrier grew to full height in only a few days or weeks through the growth of reef-building organisms is patent absurdity. How could hundreds of generations of coral colonies reproduce and grow massive skeletal structures in only days or weeks? Keep in mind that geologic events of the Flood are natural phenomena following the principles of physics, chemistry, and biology; they cannot be construed as miraculous works of God.

Whitcomb and Morris state: "Particularly during the Flood, the extensive reefs formed in the warm waters of the antediluvial seas would have been eroded and redeposited, often giving the appearance now of an ancient reef of great extent" (1961, p. 409). Even this time extension is of no avail, however, because the elapsed time between the Creation and the Flood is only about 1,650 years--considerably less than the post-Flood period of 4,350 years. Erosion and redeposition is ruled out in the case of the Capitan reef, for its original structure is still in place, along with other limestone strata into which it grades in what was formerly the deeper water in front of it and the shallow lagoon behind it. (See Kay and Colbert, 1965, Figure 14.7, p. 297, for restored cross sections.)

Notice that Whitcomb and Morris, in basing their argument on current rates of reef growth, have used the uniformitarian principle as the basis of their statement that the estimated current rate would account for most coral reefs of the world during the post-Flood period. Ordinarily, they would be scoffing at uniformitarianism and putting catastrophism in its place.

The discussion of coral reefs is followed by a section on deep-sea sediments (Whitcomb and Morris, 1961, pp. 409-12). It consists largely of a list of seafloor sediments and morphological features that seemed puzzling. The references cited are dated in the late 1950s, when seafloor exploration was still limited and many discoveries were yet to be made. For example, the finding of ripple marks on the abyssal floor was a surprise. Bottom currents now explain such ripples. Turbidity currents were then somewhat of a novelty, and their origin and limits were not yet fully understood. Further discussion of the creationists' limited set of observations would not be profitable in such an outdated context.

Chert Beds and the Flood

Chert, a dense nodular form of silica (SiO_2), is another sedimentary material that is singled out by the creationists for attention in their textbook <u>Scientific Creationism</u> (Morris, 1974a, p. 105). Bedded chert, which is a common marine sedimentary rock, seems not to be forming today, their text says. A widely used standard textbook on sedimentary rocks is the source of a quotation that "the origin of the bedded cherts is a very controversial subject" (Pettijohn, 1957, p. 442). There follows the creationists' conclusion that because bedded cherts are not forming today, their presence in the geologic record requires a catastrophic origin, namely, volcanic action during the Flood.

Bedded cherts formed on the deep ocean floor are now attributed to both organic and inorganic processes (Thurston, 1978, p. 122). Silica can be precipitated and reach the ocean floor as tests of radiolarians. Following accumulation, the silica is subject to diagenesis, i.e., to being dissolved and reprecipitated as chert. This process seems to be very widespread. The inorganic process, which is direct precipitation of SiO_2 in solution from seawater, requires a great increase in silica concentration over the normal value in seawater--from about 3 parts per million (ppm) to over 120 ppm. The creationist textbook is on the right track in suggesting the role of volcanism in this inorganic concentration and precipitation process.

What the creationist text does not say is that requisite conditions are present in the axial rift zones of oceanic spreading plate boundaries. Silica is liberated from mafic silicate rock acted upon by hot, circulating seawater and is brought up in the solutions that emerge in seafloor vents, including the smokers (Edmond and Von Damm, 1983, p. 84). The concentration of silica in this emerging water can be exceptionally great because of the high water temperature (350 C). Silica concentration reaches nearly 1,300 ppm in the emerging hot water and readily precipitates in the surrounding area as the water temperature falls. Layers of massive chert have been produced in this manner (Snyder, 1978, p. 743). Chert is thus a normal product of the hydrothermal system.

The hydrothermal system of the rift zone is volcanic in the sense that the heat is furnished by rising magma. The process is ongoing and probably has been more or less continuous through much of geologic time, albeit at varying rates. This uniformitarian aspect completely escapes the grasp of the creationists, but the hydro-thermal process has now been amply documented in many places on the rift systems. Thus, bedded cherts of both origins--organic and inorganic--seem securely established by mainstream science in terms of conditions of origin. Their existence does nothing to suggest their formation during the year of the Flood.

On the other hand, the radiolarian cherts require that radiolaria live as plankton in surface waters. During the one-year Flood the rate at which radiolaria normally produce silica would have had to be accelerated beyond all reason to produce all the radiolarian cherts of the geologic record in one year's time. As in the case of the reef corals, growth limits set by organism function (metabolic rates), population density, and nutrient supply constrain the organic production of silica (or calcium carbonate) to a rate that simply rules out Flood geology as a possibility. Thus nature's own organisms, by their very existence, deal a fatal blow to Flood geology, and they testify (so to speak) without any help from radiometric indicators of absolute age. Is net rate of organic production to prove the Achilles' heel of creation science?

Chapter 24

Evaporites and Rhythmites of the Flood

Creation scientists devote considerable attention to the subject of evaporites and rhythmites (banded sediments) in the sedimentary record. Evaporites belong to the general class of sediments that are of hydrogenic origin, meaning that they are directly precipitated from a water solution. Although in inland desert basins today evaporites are being formed in shallow lakes (playas) and in some deeper lakes occupying downfaulted basins (Dead Sea, for example), the great sequences of evaporites throughout the geologic record are explained as the result of sustained evaporation of seawater in various forms of restricted water bodies, such as large coastal lagoons and narrow or small deep ocean basins.

Evaporite Deposits--The Mainstream Consensus

Many varieties of evaporite minerals are produced from seawater; the great bulk of the deposits consists of the sulfates of calcium as gypsum ($CaSO_4.2H_2O$) and anhydrite ($CaSO_4$) and of sodium chloride as halite (NaCl). These compounds are commonly referred to as salts.

The evaporites occur in association with one another in sedimentary strata, usually with marine sandstones and shales and in some instances with chemically precipitated limestone and dolomite. Although details differ, most mainstream hypotheses of origin that try to explain the thick sequences of evaporites in the geologic record require a special set of environmental conditions. First, we assume a warm arid climate in which evaporation on the average exceeds precipitation. Such climates are widespread in tropical latitudes today and can be presumed to have been present in the past. Second, an evaporating basin is required--perhaps a large, shallow bay or lagoon cut off from the open sea by a barrier bar (Figure 24.1). A narrow inlet, through which ocean water can enter to replace water lost by evaporation, is necessary to account for great thicknesses of evaporite beds. A slow subsidence of the area of deposition also accommodates the accumulating beds.

When seawater is evaporated under laboratory conditions, there is a definite order of precipitation of salts, in reverse order of solubility. First, calcium carbonate is precipitated as the mineral calcite. It is followed in order by calcium sulfate as gypsum or anhydrite, sodium chloride as halite, complex salts of magnesium, chlorine, and sulfate, and finally by the compounds of potassium. Calcium carbonate formed in this way can be included among the evaporites. Calcium sulfate is precipitated as gypsum at temperatures below 35 C and as anhydrite at higher temperatures. Halite follows gypsum and anhydrite and is usually the last member of the series to be deposited in any large quantity. The more soluble salts are found in the geologic record very rarely, but they are common in playas.

Enormous accumulations of halite have formed at several points in geologic time. A good example is the accumulation of halite beds and associated layers of red shales and sandstones, gypsum, and anhydrite of the Permian basin in Kansas and Oklahoma. In the Permian Period, dozens of halite beds were deposited here, many of great thickness. A few individual salt beds 90 to 120 m in thickness are known. Thick anhydrite beds also occur in the series--one bed is over 400 m thick. Important salt beds of older geologic ages occur beneath Michigan, Ohio, and western New York.

A new concept of the origin of thick salt beds has been supplied by the theory of plate tectonics. Recall that in the early stages of continental rifting, a narrow ocean appears. It seems likely that this new ocean would, for long periods of time, be connected with the open ocean by narrow straits. These connecting straits would restrict interchange of water with the open ocean and might at times be closed, and thus isolate the narrow ocean. If the narrow ocean happened to lie in a dry tropical zone with a desert climate, the trapped ocean water would evaporate rapidly, leading to the precipitation of thick salt beds.

The Red Sea basin seems to fit rather well into this hypothetical description. Figure 24.2 shows steps in the development of the basin and the accumulation of evaporites (see Degens and Ross, 1976, p. 186). In Diagram A, the continental crust is being stretched and thinned. The upper part of the crust is broken by faulting, but semiplastic yielding in a lower (ductile) zone allows the crust to thin without major breaks occurring. In Diagram B the thinned zone has subsided, forming a narrow but shallow seaway, in which evaporites were deposited. About 3,500 m of halite were formed at this time. There followed a stage in which the seaway was open to adjacent oceans (Mediterranean and Indian oceans) allowing normal salinities to prevail. During this time pelagic marine sediments were deposited over the evaporites (Diagram C). Then, rather recently (about -2 to -3 m.y.), a deep rift appeared and an axis of seafloor spreading was initiated. Basalts emerged from the axial rift and the first oceanic crust appeared (Diagram D). It is in this deep rift that the hot brines are pooled today.

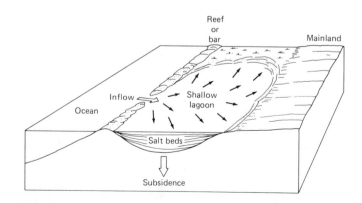

Figure 24.1 A schematic diagram of an evaporating basin in which evaporites are accumulating. (Copyright © 1981 by Arthur N. Strahler.)

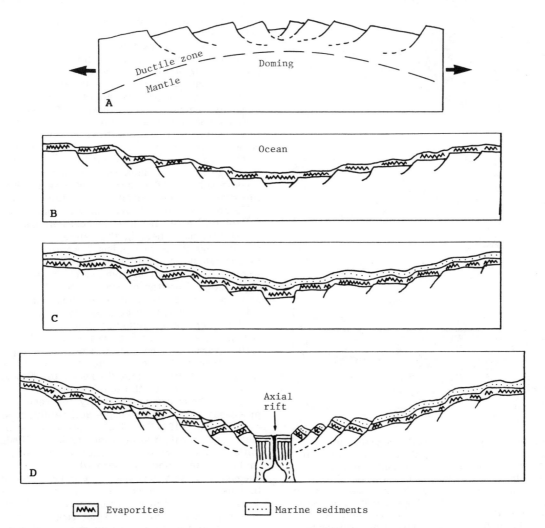

MWWW Evaporites   ⋯⋯ Marine sediments

Figure 24.2 Development stages of the Red Sea basin. (A) Pre-Miocene crustal spreading, thinning, and block faulting. (B) Middle-Late Miocene subsidence forming seaway; evaporites deposited. (C) Pliocene; deposition of marine sediments. (D) Late Pliocene to Recent; seafloor spreading starts, deep oceanic rift opens, and basaltic crust forms. (Based on data of D. A. Ross and J. Schlee, 1973, Bull., Geological Soc. of America, vol. 84, pp. 3827-48.)

The brine may have been derived from the earlier salt deposit by deeply circulating ground water.

A somewhat different situation is shown in Figure 24.3. As a new ocean basin widens, the passive continental margin develops a barrier coral reef, shutting off a broad lagoon between reef and mainland. The situation is analogous to that of the lagoon system shown in Figure 24.1. Salt beds are deposited in the lagoon, behind the reef. As subsidence continues, the reef grows upward and the salt beds thicken. Later, reef and salt beds are deeply buried under geoclinal sediments of terrestrial origin. The geologic section shown in the figure has been reconstructed from drill cores and seismic exploration of the continental margin of eastern North America.

Yet another geologic situation in which evaporites have accumulated is that of a small but deep ocean basin, cut off from interchange with seawater of a large ocean. The Mediterranean Sea occupies such a basin. It is what remains from the closing of the Tethys Sea as the African plate moved toward the Eurasian plate. The Tethys Sea was finally pinched off as the Arabian subplate collided with the Persian subplate. (See Figure 20.10.) At its western end, the Mediterranean basin had already been closed off by collision of the African plate with the Iberian Peninsula. Thus sealed off, the newly formed Mediterranean Sea underwent evaporation. Within about 1,000 years the entire 4 million cubic kilometers of seawater had evaporated, leaving the ocean basin almost completely dry (Hsu, 1972; 1978). Evaporites were left on the deeper parts of the basin floor. Then, the Gibraltar Strait was opened up, flooding the basin. The strait was then closed, and another desiccation occurred, with more evaporite deposits. Alternate flooding and evaporation occurred several times in the late Miocene time between -8 and -6 m.y. The evaporites total several hundred meters in thickness. During the ensuing period, terrestrial sediments from the surrounding continental margins buried the evaporites, causing the salt to form salt plugs (diapirs) that forced their way up through the covering sediments.

It appears, then, that plate tectonics has brought new knowledge of major situations favorable to accumulation of evaporites on a large scale. An enormous increase in information about such deposits has come about through deep-sea drilling of the ocean floors, both in deep ocean basins and along the passive continental margins. Seismic reflection profiling has revealed details of the layered structure of these sedimentary deposits.

Creationists' Arguments for Recent Evaporites

The scientific creationists' arguments in favor of Flood geology as an explanation of evaporites follow along familiar lines (Whitcomb and Morris, 1961, pp. 412-17). First, as standard operating procedure, they exhibit a straw-person in the form of a long-abandoned view of

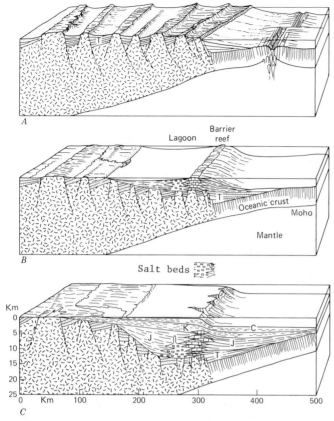

Figure 24.3 Accumulation of salt beds beneath the Atlantic coastal plain of the United States. (A) Early in the opening of the Atlantic basin, the continental crust was faulted into blocks, with basins of Triassic (T) sediment accumulation. (B) In Jurassic (J) time, a barrier reef of limestone was formed. In a lagoon behind it, salt beds were precipitated. (C) Today, the reef and salt deposits are deeply buried beneath strata of Cretaceous (K) and Cenozoic (C) age. (Copyright © 1981 by Arthur N. Strahler.)

uniformitarianism. They observe that present environments of salt deposition in desert regions, such as the Dead Sea basin, or in shallow coastal lagoons, salt pans, salinas, and sabkhas are producing salt at extremely slow rates and in small quantities that cannot account for the great evaporite deposits of the geologic record. Thus, they say, the present is not a key to the past. Nor is enlightenment to be found in observing present relict seas, such as the Caspian Sea and Salton Sea, which produce little evaporite today. Whitcomb and Morris find no published explanation (to 1960) adequate for salt domes that are the surface expressions of deep, narrow plugs of salt, or diapirs. All this commentary is negative, but the charge is implicit that mainstream science has no adequate explanation of the phenomena, hence, that "evolution science" has failed, and must be replaced.

Returning to modern mainstream science, the sequence of salts to be expected through evaporation of seawater is illustrated in modern-day shallow lagoons. A classic example is the Kara-Bogaz Gulf on the eastern shore of the Caspian Sea. There, the experimentally derived sequence of salts is matched by the natural occurrence. For each step in the sequence there occurs a zone of salt deposit of a particular kind (called a "facies" in geologists' language). Although ancient salt deposits of the stratigraphic record occur on a much larger scale, the successive occurrences of each salt type, or facies, exhibit the same sequences (Briggs, 1978, p. 302). The point is that the present may in some instances be a key to the past in a qualitative sense, but not in a quantitative sense. With this qualification, the alleged

uniformitarian principle has not failed, as the creationists claim it has. In any case, the alleged requirement that all phenomena of the past must be represented by modern examples is not really a recognized requirement of the modern view of uniformitarianism.

The final sequence of precipitated salts consists of the potash salts--compounds of potassium (polyhalite, sylvite, kieserite, carnallite, and others). Through direct evaporation of seawater, there occur two sequences of these salts. In some localities, however, changes in the chemical composition and sequential arrangement of the salts have occurred following initial precipitation. In the language of geochemists these changes are described as "metamorphism" (Braitsch and Kinsman, 1978). Chemically, the metamorphism consists of reactions between two different salts or between one salt and water to produce a salt of different chemical composition. One cause of such metamorphism is heat in the range of 70 to 90 C; it is called "thermo-metamorphism." Another form of salt metamorphism is solution metamorphism, which takes place when a salt deposit is invaded by brines that differ in composition from the original brine. Thermo-metamorphism acts on a large scale, affecting large salt deposits; solution metamorphism acts locally in limited zones.

The point of introducing this discussion of salt metamorphism is that the creationists allude to it in a rather vague and confusing way, distorting the meaning of the term and then relating it to cataclysmic events of the Flood. Whitcomb and Morris state: "Modern writers are gradually coming to the opinion that even the stratified evaporite beds are very largely the result of metamorphic processes rather than simple sedimentation and evaporation" (1961, p. 416). Note two features of this statement: First, the creationists imply elimination of the initial evaporative process entirely, something that is not implied or intended by those who investigate the chemistry of salt deposits. The correct statement would be that metamorphic processes have acted to modify those salts initially produced by evaporation of seawater. Second, the creationists do not make clear to the unwary reader that the "metamorphic processes" referred to are in no way to be connected with rock metamorphism as that process usually occurs, i.e., at much higher temperatures and under great confining pressures. This leads Whitcomb and Morris to their concluding paragraph, which reads as follows:

> In view of the difficulties encountered by uniformitarianism in attempting to explain the great evaporite beds and the need to postulate either some special kind of brine which does not now exist or else some special conditions of evaporation and metamorphism that are not known to exist at present, perhaps it is not too presumptuous to suggest that these unusual brines may have been generated during the volcanic upheavals accompanying the Deluge and that unusual conditions of vaporization and separation of precipitates may likewise have been caused by the locally high temperatures accompanying these same upheavals. The details of such reactions may be difficult to decipher at present, at least without considerable further study, but it does appear that the catastrophic environmental factors associated with the Flood provide a more satisfactory framework within which to develop a satisfactory hypothesis than does the alternate procedure of pure speculation! (1961, p. 417)

Now, do you realize what the creationists have offered us? They have made a connection between low-temperature/low-pressure chemical change in highly soluble salts and the high-temperature/high-pressure metamorphism that accompanies igneous activity (recrystallization of silicate minerals to produce gneiss, schist, eclogite, and other such regional metamorphic rocks). The connection is made through a single word-- "metamorphism"--by switching from one meaning to another, and from one geologic context to another.

In any case, the play on the term "metamorphism" has nothing to do with whether evaporites were deposited rapidly during the one year of the Flood. Whitcomb and Morris offer no suggestions or hints in answer to this question. What is the evidence that the Castile evaporite formation of Permian age was deposited in days or weeks, along with all fossiliferous strata of younger and older ages that enclose it?

Having nothing constructive to say about recency of deposition and, perhaps realizing the impotence of their position, the creationists have used an escape hatch, one that is resorted to only in case of extreme difficulty. Whitcomb and Morris say: "But there is always the possibility that the evaporite bed was formed by transportation from some previous location, where it may have existed since the Creation" (1969, p. 412). We should remind the creationists that, because God's intervention in the Flood consisted only of hydraulic control (turning on the water and turning it off), Flood geology must rely on naturalistic processes. These can include erosion of older rocks formed during the Creation and transportation of the sediment by Flood waters to a new location. The creationist motto for the transfer of responsibility should read: "When in doubt, let God do it!" Science is not so fortunate as to have a guaranteed, foolproof cop-out. But fortunately, mainstream science has sound hypotheses and a considerable store of information with which to explain evaporites of the geologic record. There are information gaps and inconsistent inferences to be sure, but these invite further study and testing; they invite the possibility of change in the form of restructured hypotheses.

Is there, on the other hand, scientific information that points in a positive way to the conclusion that evaporite sequences must have required long spans of geologic time to complete? There is such information. First, consider that physical laws, corroborated to the point of almost-certainty, would make virtually impossible the precip-itation of salts from a huge mass of seawater in a matter of only days or weeks. The rate of evaporation from a free water surface depends upon environmental conditions that, on our planet's surface, are constrained by climatic limits. One is the temperature of the air layer in contact with the water surface, another is the surface-water temperature. Still another is the relative humidity of the overlying air layer; then there is also the factor of wind. A great deal is known about the evaporative process and the factors that determine its rate. Every owner of a swimming pool has a general idea of the controlling factors, because in warm, rainless weather pool water evaporates rapidly and more must be frequently added.

Desiccation of the Mediterranean Basin

The evaporation rate is reduced about in proportion to the salinity of the water. For sea water, with salinity of about 3 percent, the rate is roughly 3 percent less than for fresh water. For the ocean surface, calculations of the mean annual rate of evaporation yield values of from 100 to 140 cm/y. in latitudes below 30° (Defant, 1961, pp. 223, 226). That means that a layer of water ranging in thickness from 100 to 140 cm is removed by evaporation each year. In the tropical desert zone, latitude range 15° to 35° N and S, precipitation is less than evaporation, and the net loss runs from 30 to 80 cm/y. The latter figures can be applied to estimates of the time required to desiccate a completely closed deep ocean basin, such as the closed Mediterranean basin; it is about 5 m.y. If such a basin lay in the tropical desert zone, with a net loss of 50 cm/y., the rate of lowering of the water surface would amount to 2,000 y./km. The deepest parts

of the Mediterranan basin are below -3 km, so 6,000 y. would be required for the desiccation. After allowing for the effect of increasing salinity, this figure might be increased to 6,500 y. Kenneth Hsu gives "1,000 years or so" for complete desiccation of the Mediterranean following its total closure (1978, p. 54). This is faster by a factor of 6, compared with my rough calculation. Hsu does not explain how he arrived at his value, but elsewhere he states that the Mediterranean basin holds almost one million cubic miles of water and that the annual evaporation loss is approximately 1,000 cubic miles (Hsu, 1972, p. 29). Dividing, we get 1,000 y. Actually, in the present Mediterranean region, the net deficit runs only 30 to 40 cm/y. in the drier parts, so my estimated rate of desiccation on modern terms would need to be reduced by perhaps half or less, for a total desiccation time of 13,000 y. Taking into account the influx of river water from surrounding lands would make a further increase in desiccation time, perhaps by 10 percent.

But suppose we adopt Hsu's 1,000-y. schedule for desiccation of the Mediterranean basin. There is no way it can be fitted into a one-year Flood. It might, however, be accommodated in post-Flood time (4,300 y.), but then we run into the difficulty that the Near East was populated by Noah and his descendents who could scarcely have failed to notice that their ocean had disappeared, leaving a huge desert basin. If the Egyptians failed to notice and record the phenomena of their Nile trenching its lower course below the first rapids and leaving the delta high and dry, they must have been, to say the least, quite blasé.

Let us pursue this subject on a somewhat different tack. Seawater of present salinity (about 3.5 percent), upon complete evaporation, yields slightly less than 2 percent thickness of salt with respect to the original depth of seawater (Kay and Colbert, 1965, p. 296). Or, in other words, the depth of seawater required to be evaporated is about 50 times the residual salt thickness. Suppose, then, that we take one of the great salt accumulations of the geologic record, such as the salt beds of Permian age in northwestern Texas and southwestern Oklahoma, where the total salt thickness exceeds 1 km. One kilometer of salt would have required evaporation of 50 km of seawater. Using a net evaporation rate of 50 cm/y., we come up with 100,000 years. There is no way the production of the Permian salt can be fitted into a one-year Flood, a 4,350-y. post-Flood period, a 1,650-y. pre-Flood period, or the entire time from Creation to the present (6,000 y.). And we should not forget the time requirement for nearly 1 km of Silurian salt in the Michigan basin, and each of the other salt formations of the fossiliferous strata in Paleozoic and later time.

Those among my readers who always want to check out the figures will think back to the paragraph on the desiccation of the Mediterranean basin. Using the 50-to-1 ratio of seawater depth to thickness of salt, how thick a layer of salt should we expect on the floor of the Mediterranean basin in a single desiccation? Answer: For 3 km of seawater, the salt residue would be 60 m thick. Actually, seismic refraction has revealed 1,600 to 2,000 m of salt in parts of the Mediterranean floor. For 2,000 m of salt to form would require evaporation of 100 km of water, which is about 33 times the depth of the Mediterranean basin. This means 33 desiccations and 33 refills. Hsu reports that sediments above and below the salt (anhydrite) layer proved to be marine sediments of deep-water origin, based on the microfossils they contain (1972, p.32). But he also says: "Because we discovered oceanic sediments interbedded with the evaporites, we concluded that the floodgate swung open and shut repeatedly during an interval of about a million years." Using my calculation of 12,000 y. per desiccation, 33 desiccations would require almost 400,000 y. Doubling that time to accommodate the episodes of refilling still keeps the total under one million years. But one cannot help but wonder what mechanism would close and open the Gibraltar Strait repeatedly. Did the African plate give a little push against the Iberian massif to close the gap each time? And did this happen 33 times at just the right intervals? Asking annoying questions is standard operating procedure within the community of mainstream scientists. In this case, the basic data--a known thickness of salt and a known ratio of dissolved solids to seawater-- requires an evaporating period enormously longer than Flood geology can supply.

In summary, there is good evidence of at least three geologic situations in which thick evaporites may have accumulated: (1) in young narrow basins produced in early stages of continental rifting and opening of an ocean basin; (2) in small but deep basins remaining in penultimate stages of continental collision; and (3) in shallow basins behind reef barriers, both on subsiding passive continental margins and marginal to epeiric seas on cratons. In all three situations time requirements for salt accumulation by evaporation of seawater enormously exceed the few thousand years available to recent Creation and Flood geology.

Juvenile Origin of Salt

The ICR textbook <u>Scientific Creationism</u> updates material presented by Whitcomb and Morris in <u>The Genesis Flood</u> by reference to a paper by V. I. Sozansky, a Russian geophysicist, published in 1973 in a reputable journal of mainstream science (Morris, 1974a, pp. 106-7). Sozansky is quoted as discarding the seawater-evaporation hypothesis in favor of one in which the salts emerge from faults in ocean basins where magma extrusion is also occurring (1973, p. 590). The salts, according to Sozansky, are "juvenile," meaning salts contained in rising magmas.

Sozansky's hypothesis predates Project FAMOUS, the 1974 voyage of discovery to the floor of the Mid-Atlantic spreading rift and subsequent deep dives to the Galapogos rift and other active spreading rifts of the Pacific basin. We now have evidence of the great intensity of hydrothermal activity in these zones, where basalt of the crust is being chemically altered by circulating hot solutions, releasing such metals as iron, manganese, copper, and zinc from the parent mafic rock and carrying them up to the ocean floor. In combination with sulfur (as sulfate) from the circulating seawater, they are precipitated as sulfides, or they may be deposited as oxides and silicates (Edmond and Von Damm, 1983). Conspicuously lacking in these mid-oceanic hydrothermal zones are fresh deposits of sodium chloride, but anhydrite is deposited in the walls of smoker chimneys. The hot brines of deep spots in the Red Sea are atypical when compared with conditions in the Atlantic and Pacific spreading axes, perhaps because the Red Sea is a very young, narrow ocean basin. That the Red Sea brines are concentrations of seawater, as distinct from juvenile brines, seems to be indicated by oxygen and deuterium studies (Ross, 1972, p. 1456). Thus, Sozansky's rather vague references to a supposed environment of submarine tectonic and volcanic activity are unacceptable in the light of a great deal of firm information as to what actually goes on at submarine spreading plate boundaries.

Sozansky derives support for this hypothesis of juvenile origin of salt from the asserted absence of marine organic remains. He states:

> It is well known that salts are chemically pure formations which are void of the remains of marine organisms. At the same time they contain numerous plant remains in the form of spores, pollen, branches, twigs, etc. If salt-bearing sections were formed in lagoons or marginal seas by the evaporation of seawater, then organic matter,

chiefly plankton, would have to enter the salt-forming basin together with the waters. As a result the bottom sediments would be rich in organic matter. (P. 589)

His contention that fossils are never present was discredited in a research report published in 1972, one year earlier than the date on Sozansky's paper. The evaporites of the Castile formation of Permian age in Texas, adjacent to the Capitan reef limestones, have been found to contain abundant fossils of plankton and other organic matter. These materials were discovered and identified by Roger Y. Anderson and colleagues, whose study of the Castilian evaporites we will refer to on later pages (Anderson et al., 1972). Fossils include fusilinids-- calcareous planktonic foraminifera--that must have been brought into the evaporating basin from the open ocean with entering currents, and that settled to the bottom to be enclosed in bands of calcium carbonate that are interbedded with the anhydrite of the formation. Thus Sozansky's strongest point is clearly eliminated.

Of course, those recent creationists who are also Flood geologists have a neat way out of their time dilemma. Heating of Flood waters by submarine volcanic eruptions could have set the entire ocean to boiling violently, quickly disposing of the entire ocean into a vapor canopy, and leaving behind thick layers of salt. With punctuated volcanism, the intervolcanic periods would have seen torrential precipitation of the vapor canopy to restore the world ocean, allowing partial solution of the crystallized salt. The whole process would have been repeated as many times as the known stratigraphy requires. But poor Noah, his family, and all those beautiful animals! They would have been repeatedly cooked in live steam, dumped on dry land, refloated, and cooked again. Life was really hard in those days! Talk about survival of the fittest-- that was real survival!

Mainstream geology has another trump card to be played in the evaporite game before the hand is finally played out. It concerns a remarkable banding structure found in some thick evaporites of the geologic record. The subject is best combined with other sediment banding phenomena that, like tree rings, seem to be records of vast numbers of annual cycles. This is our next topic.

Rhythmites and Flood Chronology

Fine sediment deposited in standing water in many cases exhibits <u>lamination</u>, consisting of thin layers (laminae) set apart, one from the next, by change of color, texture, or composition. In one kind of lamination a single lamina consists of particles size-graded from coarsest at the bottom to finest at the top. This is described as graded bedding; it is also found in turbidites. If the particles grade into fine clay at the top, that part of the lamina will usually be dark in color, sharply contrasting with the lighter color of the base of the overlying lamina. A second kind of lamination consists of an alternation of two compositionally different laminae. Two mineral varieties may alternate, or a mineral lamina may alternate with a lamina composed of organic particles. Two adjacent laminae constitute a <u>couplet</u>.

Laminated sediments, whether in a soft condition because of recent deposition, or fully lithified in older formations, are of great interest in the controversy over Flood geology. The lithified laminations, according to proponents of recent creation, were all deposited within the one-year period of the Flood, along with all other fossiliferous strata of the geologic record. Controversy arises when mainstream geologists assert that each lamina or each couplet represents the sediment accumulation of one year, so that a formation consisting of hundreds or thousands of laminae could not possibly have been deposited in a one-year Flood.

A single lamina or a couplet deposited in a one-year

cycle of environmental change is a varve. The term is of Swedish origin and was originally applied to laminated clays deposited in temporary freshwater lakes at the margin of the retreating ice sheet that covered northern Europe in the Pleistocene Epoch, or Ice Age. Now the adjective "varved" applies to any sequence of laminations in which each lamina or each couplet is known to be (or asserted to be) the deposit of one year. But a consistent repetition of a certain succession of sediments--as in the cyclothems of coal measures, described in an earlier section--need not be a result of annual cycling; it might represent much longer cycles, or perhaps even shorter cycles than one year. For such forms of lamination the term rhythmic layering is used, and the collection of laminae can be called a rhythmite. The two terms--varves and rhythmites--summarize the two sides in the dispute we are investigating: for mainstream geologists, the laminations are varves; for the creation scientists, they are rhythmites. If large sequence of laminations are demonstrated beyond all reasonable doubt to be varves, Flood geology lies demolished and totally discredited. As you can see, much is at stake in the argument over the meaning of laminated sediments.

Actually, the debate over the significance of laminated clays of glacial-lake (glaciolacustrine) origin has raged for decades within mainstream geology. The hassle began when a Swedish geologist, Baron Gerard De Geer, began as early as 1884 to use sequences of varves to establish the chronology of glacial retreat (Mörner, 1978, p. 84). The process is quite closely analogous to the use of tree rings to establish a chronology. Varves vary in thickness from one to the next, and a long sequence of varves is thus unique in time. Professor Richard Foster Flint, a distinguished scholar (now deceased) in matters of Pleistocene history, explains how varves are thought to have been formed:

> The environment of deposition is inferred from field data and laboratory experiments. Suspended sediment consisting chiefly of silt and clay was brought into a temporary glacial lake by streams from the melting glacier and from surrounding land during spring and summer. The silt settled early, but the clay particles remained in suspension, settling gradually during autumn and winter after melting had ceased. This separation caused the observed gradation from the coarse member to the fine member of the pair. The low temperature and hence relatively great density of the water was an

important factor in delaying settlement of the clay particles. With resumption of melting the next spring, new sediment entered the lake, and the coarse fraction, settling out rapidly, produced the sharp contact between the top of the older pair and the base of the younger one. Thus the couplet is a varve, the sediment year commencing with the spring. (Flint, 1971, p. 402)[1]

De Geer began to measure the thicknesses of successive varves in glacial clays of Sweden. When thickness is plotted against time for each year and the points are connected by straight lines, a toothed graph results (Figure 24.4). Sequences of separated lake deposits were correlated by moving one graph along the other until the two profiles showed close correspondence. As in the case of tree rings, overlap allows sequences to be added one to the next. By the early 1900s a chronology of glacial retreat was established. Flint notes: "From the resulting curves was compiled a Swedish varve chronology extending back, with one gap and one extrapolation, from A.D. 1900 over a period of nearly 17,000 yr., on the assumption that the rhythmites are varves" (1971, p. 403)[1]. Danish geologists who were critical of De Geer's interpretation claimed that on occasion a severe storm had disturbed the bottom sediment to form laminations of shorter than annual period. Mörner states that De Geer's chronology can be reliably brought down to -10,000 yr. (1978, p. 842; see also Flint, 1971, p. 404, Table 15-A).

De Geer's work was carried on in New England by Ernst Antevs in the 1920s and 1930s. Antevs, also assuming he was dealing with true varves, compiled a sequence for eastern North America spanning a north-south distance of over 100 km and yielding a chronology of 28,000 yr. There are three long gaps in this stretch, and much of the chronology rests on interpolation. Flint states that carbon-14 dates within the sequence "suggest a period of little more than 10,000 years" (1971, p. 406)[1].

Small wonder that creation scientists are highly critical of the varve studies of De Geer and Antevs. They need only to borrow and repeat the criticisms expressed by mainstream scientists. On the other hand, numbers of varves in these chronologies are larger than can be accommodated in post-Flood time. In fairness to both parties, it would be wise to simply judge the debate a draw on the basis of glacial varve evidence presented by both sides.

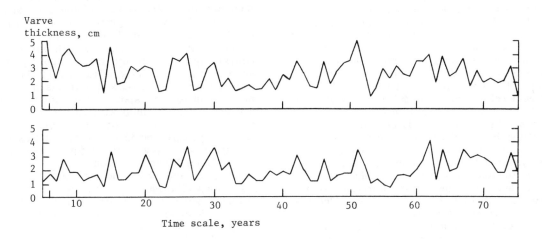

Figure 24.4 Graphs of thicknesses of seventy consecutive varves at New Haven, Connecticut (A), and seventy at Haverstraw, New York (B). The two graphs are placed in a position judged to correspond in time. (After E. C. Reeds, in R. F. Flint, 1957, Glacial and Pleistocene Geology, John Wiley & Sons, New York, p. 296, Figure 17-2.)

Varves in Modern Lakes

Well, that is not all there is to the subject of varves. Flint reviews an entirely separate and independent study of laminated sediments in which biological evidence seems to require the conclusion that in certain circumstances rhythmites are indeed varves, and that their chronology is of far too long duration to fit post-Flood time (1971, p. 400). The locale is a lake near Interlaken, Switzerland, in which sediments have accumulated. The rhythmites consist of couplets in which one lamina is light-colored and rich in calcium carbonate while the other is dark and rich in organic matter. Flint gives the following description of the deposit:

> Proof that the couplets are annual and are therefore varves is established on organic evidence. . . . The sediment contains pollen grains, whose number per unit volume of sediment varies cyclically, being greatest in the upper parts of the dark layers. The pollen grains of various genera are stratified systematically according to the season of blooming. Finally, diatoms are twice as abundant in the light-colored layers as in the dark. From this evidence it is concluded that the light layers represent summer seasons and the dark ones fall, winter, and spring. Counts of the layers indicate a record that is valid through at least 7000 years BP. Despite some obscurities the sequence agrees in many respects with that inferred from other kinds of evidence, so that the general inference that the couplets are varves is supported. (1971, p. 400)[1]

The sequence described above requires just about double the time available in the post-Flood period. Again, we find the severe time constraint that the environment imposes through the needs of organic activity.

My attention was caught by an obscure reference to studies of the sediments in Lake Biwa-Ko, Japan. The present sedimentation rate in that lake as "directly measured at 70 m depth, is approximately 1 mm/y." (T. Koyama et al., 1973, cited in Horie, 1978, p. 424). A 200-m core from the lake provides a record going back about 500,000 y., for an average sedimentation rate of 0.4 mm/y. At the current sedimentation rate, 200 m of sediment would require 200,000 y. Even allowing an average sedimentation rate faster by a factor of two, suggests a time requirement far in excess of the 4350 y. post-Flood allowance.

Rhythmites of the Green River Formation

Whitcomb and Morris treat at some length the interpretation of a celebrated mass of laminated sediments --the Green River formation of Eocene age that is widely exposed in part of Colorado, Utah, and Wyoming (1961, pp. 424-28). The formation averages about 600 m in thickness and consists of sediments deposited in inland lakes and around the shores of those lakes. A calcium-carbonate sediment called marlstone makes up much of the lake deposit. It contains organic matter in the form of kerogen, a hydrocarbon compound closely related to hydrocarbons found in petroleum. The hydrocarbon material occurs as thin laminae between laminae of marlstone. Also present in the formation is a fine-grained sandstone, interlaminated with kerogen. Laminae of the sandstone are thicker than those of the marlstone; the latter averaging only about 0.2 mm per couplet. The point at issue between mainstream scientists and creation scientists is whether the couplets of the Green River formation represent true varves (one couplet per year) or some other cycle of rhythmic sedimentation, or perhaps a secondary effect of chemical change (diagenesis). If varves, their great number clearly demolishes Flood

geology, which requires all fossiliferous strata of the world to be formed within one year.

Major field and laboratory research on the Green River formation and its couplets was carried out by Wilmot H. Bradley of the U.S. Geological Survey. His major report was published in 1929, followed by a brief review of his conclusions in 1948 (Bradley, 1929 and 1948). His investigation was extremely thorough and his report gave attention to all aspects of the problem of the origin and significance of the couplets. Although he referred to them as varves, Bradley was by no means positive that they represented annual phenomena. He noted that the separation of the hydrocarbon substance (kerogen) from the carbonate of the marlstone in the laminae was not perfect. His favored hypothesis of origin of the couplets was based on seasonal rates of production of kerogen through decay of living matter and of calcium carbonate through production by planktonic algae. To this process he added the possible seasonal influx of calcium carbonate and other mineral particles from streams draining the surrounding land surfaces. He used principles of limnology of middle-latitude lakes to visualize an annual winter overturning of the lake waters and a long summer period of stagnation of the colder, denser lower portion of the lake--the hypolimnion. He postulated that, while production of both hydrocarbon particles and calcium carbonate particles was at a maximum during the summer, the two kinds of materials settled through the hypolimnion at different rates--faster for the calcium carbonate and slower for the hydrocarbon. This separation by density resulted in the formation of a couplet. The long seasonal period of stagnation was ended by an influx of fine detrital material, which accounts for the sandy couplets, and these would have accumulated near the outer margins of the lake.

Bradley compiled evidence that seemed to him to indicate three cycles of thickness variation on longer periods than the yearly cycle assumed for one couplet (1929, pp. 103-5). One is a cycle of varve thickness in which the peaks are separated by an average of a little less than 12 y., but the length of individual intervals ranges from about 7 to about 18 y. This interval corresponds rather well with the established sunspot cycle of a little over 11 y., but with individual periods ranging from about 7 to more than 16 y. Bradley refers to work of climatologist C. E. P. Brooks indicating that a close correlation exists in Africa between changes of level in certain large lakes and the sunspot cycle. Bradley also found what appeared to be a thickness cycle in the alternation of oil shale with marlstone, lithologies that come in beds much thicker than the laminae couplets we have been referring to. The mean length of the cycle was calculated to be 21,630 y. and the suggestion is made that it reflects the cycle of precession of the equinoxes, which is between 25,000 and 26,000 y. The third cycle, about 50 y., does not seem to correlate with any well-established natural rhythm. Although the sunspot cycle correlation is potentially a source of confirmation of the annual cycle of couplet formation, it remains dubious because the actual interval between sunspot maxima cannot be independently discovered for matching with lamina thickness. In modern lakes a correlation of depositional cycles with the recorded sunspot cycle record is possible, but this in turn might be difficult to apply to the unique compositional sequence of the Green River formation.

Whitcomb and Morris do not refer to Bradley's description of these longer cycles. They consider Bradley's reasons for interpreting the couplets as varves to be inadequate. They criticize his attempt to equate the volume of sediment in the couplets with the quantity to be expected to be produced by the land surface contributing runoff to the lakes. The creationists state certain factors that, they assert, make it highly doubtful that the couplets are annual layers (Whitcomb and Morris, 1969,

p. 425). One factor is that the couplets are too thin and too wide in extent to have been deposited in a "normal" lake bed. They say that there would be occasional storms to stir up the bottom sediments, and occasional stream floods to pour large quantities of sediment into the lakes. How can such objections be answered?

The Green River couplets are indeed a remarkable accumulation; their regularity and vast numbers are mind-boggling. How could such uniform deposition continue for 5 to 8 million years? Bradley's hypothesis calls for exceptional uniformity of climate, but that requirement does not necessarily vitiate his hypothesis. As to bottom disturbance by storms and influxes of sediment, there is reason to infer that for large inland lakes of moderate depth, these effects would not be felt in the central areas, far from shore (Bradley, 1929, p. 103). Modern studies of deep lakes reveal prolonged accumulation of couplets undisturbed by such effects, as in examples already described (Switzerland and Japan).

Whitcomb and Morris call attention to the remarkable preservation of remains of fishes in one part of the Green River formation (1961, p. 427). These fossils are highly prized as conversation pieces and exhibits; they are widely sold to collectors. The creationists ask: How could a dead fish remain perfectly intact during the many years required for its burial under thin laminae? A reasonable answer is that the bottom zone to which the fish sank following its death was a stagnant water zone largely devoid of free oxygen. Bradley postulates that within the hypolimnion there prevailed a high content of carbon dioxide and hydrogen sulfide in which bottom scavengers would not live, and thus, the fish would remain intact through burial (1946, p. 646).

The creationists point to the absence of graded bedding in the Green River couplets. While this is true of the carbonate/hydrocarbon laminae, it is not so for the sandy couplets, which clearly show the graded bedding one would expect of a suspension of detrital sediment projected into the lake by streams in flood (Bradley, 1929, p. 102).

After stating these objections, Whitcomb and Morris turn to presentation of the Flood version of the Green River couplet sequence:

> The detailed manner of deposition may be hard to deduce at this time, owing to the catastrophic nature of the environmental factors during the Deluge. The only certain conclusion, from the very nature of the deposits, would seem to be that they could not have been formed as cyclic varves as claimed. A possible plausible explanation might be in terms of a vast sedimentary basin formed by the gradual uplift of the land surrounding it, in the later stages of the Deluge period. A complex of shallow turbidity currents, carrying the still soft surface sediments and organic slime from the surface of the rising lands would then enter the basin, mingle, and deposit their loads. Slight changes in velocities of compositions of the turbidity currents would account for much of the laminated appearance of the central deposits, although it is possible that the accumulation of the organic matter into a succession of thin seams was also partly caused by later physico-chemical factors affecting the sedimentary mass. The general appearance of the Green River formation as a whole seems consistent with this sort of concept. (1961, p. 427)

Vague, inadequate, and incomplete as this scenario is, it says enough to reveal an enormous difficulty in accepting the Flood explanation. Bradley states: "From measurements of the varves the Green River epoch is estimated to have lasted between 5,000,000 and 8,000,000 years" (1929, p. 87). If two turbidity curents are

required for each couplet, the Flood version calls for 10 to 16 million separate currents, one following the other within a small fraction of a year. Suppose that we allow 100 days for the total deposition, consisting of 10 million separate turbidity currents. This comes out (very roughly) to one event per second. One second must see a turbidity current spread over an area of several thousand square miles. The turbidity current would need to traverse a surface distance of not less than 100 km/sec. (360,000 km/hr.). A similar analysis has been done by Robert J. Schadewald, to whom I am indebted for the idea (1982, p. 14). He calculates layers forming at the rate of 3 layers every 2 seconds, so we are in fairly good agreement. Schadewald relates what happened on two occasions in which this kind of calculation was presented to creationists in debate:

> Henry Morris apparently has no answer to this. Biologist Kenneth Miller of Brown University dropped this bombshell on Morris during a debate in Tampa, Florida, on September 19, 1981, and Morris didn't attempt a reply. Fred Edwords used essentially the same argument against Duane Gish in a debate on February 2, 1982, at the University of Geulph, Ontario. In rebuttal, Gish claimed that some of the fossilized fishes project through several layers of sediment and that therefore the layers can't be semi-annual.

Whitcomb and Morris conclude their discussion with suggestions for other causes of lamination (1961, p. 428). The general idea is that chemical reactions often progress through a permeable rock mass in waves, producing a set of color bands that often represent varying amounts of iron or manganese oxides in spherical zones. Publications cited by the creationists in reference to such banding phenomena actually deal with banding in soils and in lake sediments rich in iron. Neither situation seems to relate to the Green River rhythmites.

Laminations of the Castile Formation

We referred earlier to the evaporites of the Castile formation of Permian age in West Texas and New Mexico. These evaporites display a remarkable system of laminations in which couplets consist of one lamina of calcite and one of anhydrite; in another part, the couplet consists of one lamina of anhydrite and one of halite. Roger Y. Anderson and his team of researchers completed a study of the Castile evaporites in which 260,000 couplets were counted in a total thickness of approximately 450 m (Anderson et al., 1972). Laminae extend for distances as great as 113 km. The couplets are interpreted as varves, but the authors note that this interpretation has never been conclusively demonstrated. The hypothesis advanced for formation of the couplets requires an annual cycle in which conditions of temperature or salinity resulted in the alternation of the respective minerals.

If, as the creationists have claimed in the case of the Green River couplets, the lamination is a result of catastrophic conditions during the year of the Flood, the problem arises as to the mechanics of catastrophic events that might form laminae. If, for evaporites, turbidity currents are ruled out, resort must be made to volcanic activity, as postulated by the Russian geochemist, V. I. Sozansky. (We reviewed this hypothesis in the discussion of origin of evaporites.) If each lamina represents the deposit of a single emission event in which juvenile solutions originating in magmas are being extruded from the ocean floor, enormous numbers of such events are required to complete the Castile sequence of 260,000 couplets. Each lamina requires one event, for a total of 520,000 events, and under Flood geology these must have occurred within one year, but more likely only a part of one year. If the emission were to take the form of a bottom-hugging current and 100 days are allotted to the total deposition, each event must be accomplished in about 17 seconds, on the average. While the demands for speed are not quite as drastic as for the supposed Green River turbidity currents, they are outrageous. To traverse 100 km of seafloor in 17 seconds requires a speed of 6 km/sec. (21,000 km/hr.). Surely this rate of flow of a thin layer of dense fluid on the floor of the ocean, without any mixing with the overlying water layer, would be impossible in view of the demands of boundary shear and viscosity.

The creationist's insistence that all events of the sedimentary record of more than 1.5 b.y. must have been completed in no more than a single year has brought upon itself a patent absurdity through the sheer number of events to be explained. Thus, the rejection of Flood geology is clearly rendered without recourse to methods of absolute age determination. Natural processes that are purely physical are constrained by laws of physics as to the rates at which those processes can act. Transport processes such as turbidity currents and bottom-hugging density currents involve masses in motion, and the forces of propulsion and resistance to which those masses are subjected. Conditions of mass, density, velocity, inertia, and viscosity are calculable for the situations in which those transport processes are postulated. Gross violations of the constraining conditions and parameters are simply not permitted by science. When such violations are nevertheless required, those postulates become statements of pseudoscience.

Credit

1. From <u>Glacial and Quaternary Geology</u>, by Richard F. Flint. Copyright © 1971 by John Wiley & Sons. Reprinted by permission of John Wiley & Sons.

Chapter 25

Petroleum and Ore Deposits of the Flood

Whitcomb and Morris devote several pages to their claim that the occurrences of petroleum and metallic ore deposits are expected consequences of the Flood scenario (1961, pp. 429-38). Their discussion begins with the familiar distortion of the modern concept of uniformitarianism, but with a new twist. Introducing their section on origin of oil and mineral deposits, they write:

> Uniformitarian geology is frequently defended on the ground that it has worked so well in leading to the discovery of economically important deposits of petroleum and metals. It is maintained that it must be basically correct, or else it could not have served so well as a guiding philosophy in economic geology.

> But two replies can quickly be given to this sort of statement. In the first place, it has apparently not worked very well, as the discovery of valuable deposits of any kind is hardly on anything approaching a fully scientific basis as yet. In the second place, such techniques as have actually been found helpful in exploration do not really depend on the historical aspects of geology at all but only on recognition of the structural and sedimentary markers that experience has shown are associated with such deposits. (P. 429)

The discussion that follows seems to center on the claim that fossils have not proven a useful or even accurate means of locating oil "pools" and, as a consequence, the petroleum exploration industry is looking increasingly to geophysical methods. The creationist authors are reluctant to offer Flood geology as a better avenue to petroleum exploration, but, on the other hand, they doubt that it would be any less effective than evolutionary stratigraphy.

Whitcomb and Morris' book, The Genesis Flood, was published several years before the full outlines of plate tectonics were drawn in several seminal papers published in 1967 and 1968. In years that followed, the recognition and interpretation of great sedimentary accumulations along both active and passive continental margins increased by an order of magnitude our understanding of the occurrences of petroleum.

Plate tectonics also led to new insights into the occurrences of metallic ore deposits. It would be unfair to criticize Whitcomb and Morris in the context of a style of geology (geosynclines, vertical tectonics) that no longer prevails. That would be a reverse application of the straw-person argument. On the other hand, by 1982 The Genesis Flood had gone through 26 printings, unchanged in 21 years! The ICR textbook Scientific Creationism (Morris, 1974a) in its twelfth printing in 1985, scarcely makes mention of plate tectonics and relies almost entirely on material drawn from its predecessor, The Genesis Flood. How can teachers and students take seriously the content of books purported to be basic text material for creation science, when those books ignore decades of

newer scientific discovery that also includes massive corroboration of previously established methods and findings (such as radiometric ages of rocks and magnetic polarity reversals)?

Creation science is, of course, frozen into a final and enduring view of the universe; it will never need to be revised because the concept of up-dating is contradictory of its premise. On the other hand, creation scientists should continually revise their criticisms of mainstream science to include its new findings and new or revised hypotheses. Having failed to do so, they have no valid case for introducing their arguments into the schools under the disguise of a debate about current science. But, keeping in mind that the creationists' books must be enjoying large sales, and that untold tens of thousands of copies are already in the possession of readers, we need to address ourselves to such arguments as they contain, taking opportunity to call attention to new evidence and new explanations that bear on those arguments.

Petroleum Occurrences--The Mainstream View

As used in science and technology, the term petroleum refers to a group of naturally occurring hydrocarbon compounds that spans the range from crude oil to natural gas in the one direction and to solid asphalt and bitumen on the other. Crude oil in its natural state is a mixture of a large number of hydrocarbon compounds (compounds of carbon and hydrogen) of which more than 200 individual compounds have been isolated and analyzed. Natural gas, which usually occurs in close association with crude oil, is a mixture of gases, the principal one being marsh gas (methane, CH_4).

There is general agreement in mainstream science that petroleum had its origin in organic matter that became incorporated into marine muds and clays accumulating in subsiding basins (Hunt, 1979, Chapter 4). Two sources of organic matter are recognized. One is a terrestrial source and consists of fragments of land plants carrying waxes that are hydrocarbon compounds of distinctive chemical formulation. These are carried to the sea and deposited mostly in nearshore environments. A second source, and by far the more important in terms of quantity, are hydrocarbons (oils, fats) produced by plankton (floating organisms). Upon death, the organism sinks to the ocean floor and what remains of it is incorporated into mud or clay, that may be calcium carbonate, silicate mineral matter, or a chemical precipitate (salt, anhydrite, chert).

An important point on which agreement is general is that the various kinds of hydrocarbon compounds described above account for no more than 10 to 15 percent of the hydrocarbons actually found in petroleum. Thus the bulk of petroleum is a product of chemical change, thought to be a result of prolonged heating of the compacted sediment.

Downward into the earth's crust, rock temperature rises rather steadily at an average rate of about 3 C degrees for each 100 m of descent. This phenomenon is called the geothermal gradient. It results from the steady production of heat in the crust and mantle as a result of radioactive decay. Upward heat flow is higher (or lower) in some regions than others. For example, the rate is high in the vicinity of the mid-oceanic ridge where plate spreading is in progress and hot mantle rock is rising at depth. Consider, now, that as sediment accumulates in a natural basin--such as a backarc trough, a forearc trough, or in the sediment wedge (geocline) along a passive continental margin--a given set of strata are taken into an environment of increasing confining pressure and increasing temperature. The effect is quite like that of putting a mass of dough or a meat roast into an oven. At a sufficiently high temperature, long-sustained, chemical changes take place. In the case of the parent hydrocarbon compounds, the molecules are reformed into new hydrocarbon compounds, which have different numbers of carbon atoms in the molecule. The end product depends in large part on the temperature zone that the strata occupy. In the deepest zone, where temperatures are around 150-200 C, natural gas is the main product. In an intermediate zone, 60-150 C, gasoline and oils are the main product. Duration of the "cooking" process is a factor directly affecting the total quantity of petroleum produced (Hunt, 1979, p. 131).

Petroleum produced by such chemical changes remains disseminated in a large mass of shale, but under favorable conditions begins to migrate out of that parent rock into adjacent rocks and structures. The process of petroleum migration is a subject of considerable debate. Recent reviews state clearly that the nature of petroleum migration is not known with certainty and that it may occur in several different ways (Wszolek and Burlingame, 1978, pp. 571-72). Moreover, the rate of migration is not known. It is, however, generally agreed that only a small percentage of the total petroleum in the parent rock is involved in migration. The end result is certain beyond reasonable question: petroleum occupies pore spaces or cavities in rocks quite different from those in which it was formed. These are <u>reservoir</u> <u>rocks</u>; most commonly they are sandstone or limestone. The former may contain numerous interconnected pore spaces between mineral grains; the latter may contain solution cavities or passageways, as in limestone cavern systems. Normally such reservoir rocks are filled with ground water in the near-surface zone, but this water may be displaced by migrating crude oil and/or natural gas. A substantial accumulation of extractable petroleum in reservoir rock is commonly called a "pool," but the term is highly misleading, incorrectly implying a large tank or container of liquid.

The reservoir rock, in order to retain petroleum, must have a special configuration and relationship to overlying or surrounding rock masses. The reservoir must be capped or otherwise sealed on the top to prevent the petroleum from rising to the surface and escaping. Favorable structures are called <u>traps</u>. One of the simplest is a dome or arch structure, in which an impervious caprock of shale overlies the sandstone reservoir rock (Figure 25.1). Very different is the salt-dome trap, in which the sandstone reservoirs are sealed in contact with the salt column; or the crude oil occupies cavities in a limestone caprock at the top of the salt column (Figure 25.2). Certain of the petroleum traps are produced by earlier tectonic activity--faulting or folding of strata-- while others arise in the course of changing sediment deposition.

In summary, then, both mainstream petroleum geology and Flood geology are faced with explaining a rather complex chain of events: accumulation of the raw hydrocarbon ingredients, chemical alteration to produce petroleum, migration into reservoir rocks, and coming into existence of suitable traps to hold the oil. Let us see what the creation scientists have to say about their explanation. How can all this be accomplished in one year, plus some 4,000 years of post-Flood time?

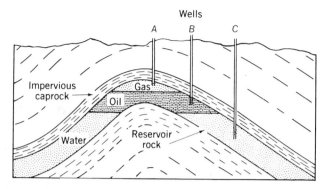

Figure 25.1 Idealized cross section of an oil pool on a dome structure in sedimentary strata. Well <u>A</u> will draw gas; well <u>B</u> will draw oil; well <u>C</u> will draw water. The caprock is shale; the reservoir rock is sandstone. (Copyright © 1972 by Arthur N. Strahler.)

The Creationists' View of Petroleum Occurrences

Whitcomb and Morris observe that (a) petroleum has been found in rocks of practically all geologic ages, and (b) an exception is Pleistocene and Holocene rocks, which do not contain petroleum (1961, p. 432). From (a) they reason that all petroleum must have a "universal explanation," i.e., "The conditions of its formation must have been the same everywhere." For them, that means one great global event--the Flood. These authors confuse the distinction between (1) strata in which petroleum originates, and (2) the reservoir rocks in which it is now found. Because petroleum can occupy any reservoir system of interconnected openings in any kind or age of rock, the second implication has no meaning in terms of

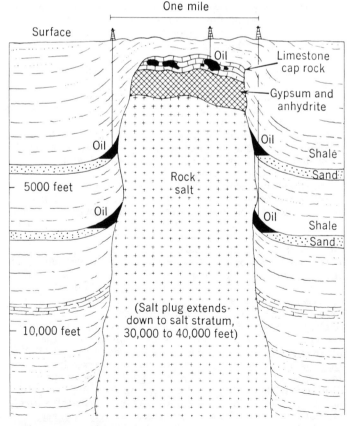

Figure 25.2 Structure of a salt dome showing locations of oil accumulations. (Copyright © 1971 by Arthur N. Strahler.)

origin of the substance itself. Perhaps they refer to the shale formations in which the disseminated hydrocarbon was originally enclosed. In that case, the competing mainstream hypothesis, outlined above, would also apply consistently throughout all of that part of geologic time in which primary hydrocarbon compounds were available from terrestrial or marine environments, and marine sedimentary basins were formed to receive the sediment.

Land plants appeared in abundance only in the Devonian Period, limiting that source of hydrocarbons to Devonian and later time. Whether marine plankton were available in Precambrian time in sufficient quantity is questionable. Nevertheless, some late Precambrian shale formations have a hydrocarbon content comparable to that of Paleozoic and younger rocks. Precambrian strata of Proterozoic age in the Irkutsk, USSR, area are producers of natural gas (Hunt, 1979, pp. 17-18). Generally, however, petroleum genesis is limited to Paleozoic and later time.

Although reservoir rocks of Precambrian age containing petroleum are rare (Levorsen, 1967, p. 38), it is conceivable that overthrusting of Precambrian rocks over younger rocks could provide the necessary reservoir rocks and traps. Pleistocene "rocks" have produced practically no petroleum, but tar and asphalt seepages from alluvium or other materials of Pleistocene and Holocene age might be counted in as conspicuous, if not important, occurrences. There are good and sufficient reasons that terrestrial Pleistocene rocks, which may be unconsolidated forms of transported regolith (and volcanic rocks, as well), should have very little potential as reservoirs. There is usually no impervious capping material of requisite thickness to form a trap. These young materials have also not been present long enough to be invaded by migrating petroleum.

The creation scientists associate original hydrocarbon accumulation in strata with the Flood. Their version of this process is as follows:

> Although the details are not clear, the Deluge once again appears to offer a satisfactory explanation for the origin of oil, as well as the other stratigraphic phenomena. The great sedimentary basins being filled rapidly and more or less continuously during the Flood would provide a prolific source of organic material, together with whatever heat and presssure might have been needed to initiate the chemical reactions necessary to begin the transformation into petroleum hydrocarbons. Of course, not all organic debris deposited during the Flood was converted into oil; apparently certain catalysts or other chemicals were also necessary, and where these were present, it was possible for oil to form. (Whitcomb and Morris, 1961, p. 434)

Migration of the petroleum into reservoir rocks is considered by the creationists as basically a hydraulic problem that, they say, has nothing to do with the fossil contents of the strata and very little to do with the regional tectonic history (p. 436). Migration and emplacement in pools is relegated to the post-Flood period (p. 436).

The time constraint upon the creationists is, as always, most severe. The only comment they offer here is that there is no reason to think the process required long periods of time. Actually, there are very good reasons to think that 4,000 years is hopelessly inadequate.

Migration of Petroleum

Primary migration is the movement of petroleum from strata of origin (shales at depths of 1 to 5 km) into surrounding or adjacent porous rocks; it is thought to take place along with migration of water out of the compacting sediment (Biederman, 1978, p. 216). This water is thought to be released from clay minerals, possibly as a chemical change in the clay mineral itself, because of application of heat and pressure. The theoretical problem is then essentially one of ground water movement where a pressure gradient exists. Some idea of the extremely slow speed of fluid motion to be expected can be gained by considering the movement of ground water at shallow depths in dense clays, classed as "impermeable." Under a moderate hydraulic gradient and a reasonable value of permeability for clay, we come up with flow speeds of ground water on the order of 2 to 3 million years per kilometer. Yet the permeability of source shales of petroleum is rated at only one-thousandth as great as for clays tested in the surface environment (Wszolek and Burlingame, 1987, p. 573). The time required for primary migration to occur thus appears to be many orders of magnitude longer than can be allowed in the Flood hypothesis. These estimates are based on principles of physics of fluid flow and not on absolute dating of rocks by radiometric methods.

Besides the primary migration of petroleum, geologists recognize secondary migration, in which petroleum within reservoir rocks is segregated into oil and gas or moves from one reservoir rock into another (Biederman, 1978, p. 218; Hunt, 1979, p. 251). In special cases, secondary migration can occur rather rapidly, geologically speaking. Whitcomb and Morris (1961, p. 436) make use of examples cited by petroleum geologist A. I. Levorsen in his authoritative treatise Petroleum Geology, first published in 1954. (My citation of Levorsen is to the second edition, 1967, which is identical in text.) Levorsen describes the Kettleman Hills pool in California, in which the stratigraphic evidence is quite clear that the petroleum trap (a low arch) cannot be older than Pleistocene in age (1967, p. 540). The creationists use this statement to show that migration of oil into a pool can occur within their post-Flood time allotment of about 4,000 years. In quoting from Levorsen, however, they fail to include a significant modifying phrase from the crucial sentence. To get it all and get it right, I quote Levorsen's full sentence, underlining the portion omitted by Whitcomb and Morris:

> This places the accumulation in late Pleistocene or post-Pleistocene time, possibly within the last 100,000 years and certainly within a million years.

For a mainstream geologist, Pleistocene means an epoch that began -1.5 m.y. (or perhaps as much as -2.0 m.y.) and ended about -12,000 y., when the Holocene Epoch began. The creationists' deletion obviously slants Levorsen's words in favor of post-Pleistocene migration, and that meaning is not excluded by Levorsen's statement. However, it should be pointed out that Levorsen was attempting to set the shortest possible time limit on secondary migration. In this case, the arching of the strata that formed the trap allowed crude oil to migrate rapidly through highly permeable sand, possibly from a nearby location in that same sand formation.

A second example of rapid movement of crude oil cited by the creationists is Levorsen's reference to something that happened in the Cairo pool in Arkansas. The pool tilted measurably within a period of 10 to 12 years. "Tilting" refers to the base of the oil-saturated zone, that is, the oil-water contact. Under normal conditions, this surface is horizontal, but it can become tilted because of oil withdrawal in nearby pools. Levorsen explains "The Cairo pool . . . has a tilted oil-water contact that is of special interest because it is probably the result of man-made fluid potential gradient formed by withdrawal of reservoir fluids in the nearby Schuler field within a period of ten or twelve years" (1967, p. 562). On this last example, which consists of an event totally irrelevant to the migration of oil into a trap, the creationists rest their case. Their closing statement reads:

There is thus no reason to reject the Deluge as a possible framework for formation of the great oil deposits of the world. Especially is this so since the uniformitarian hypothesis and the evolutionary framework of gological ages have been shown to be largely irrelevant to the actual practice of petroleum exploration. The character of petroleum deposits, and such information as has been accumulated regarding the origin and migration of oil, harmonize quite well with the Deluge hypothesis. (Whitcomb and Morris, 1961, p. 437)

The creationists have taken us from an initial assertion to an identical final reassertion, with nothing of substance between. Not one shred of evidence has been presented that positively limits the formation of all petroleum source rocks to a single year (2350 B.C.) and all petroleum migration and pool formation to four subsequent millennia.

Petroleum enclosed in rocks is briefly mentioned by the creationists in their ICR text for school use as a phenomenon suggesting rapid accumulation and, hence, by implication, a product of the Flood (Morris, 1974a, pp. 109-10). The argument scarcely deserves mention, let alone refutation. It is a quotation from an article in Science Digest to the effect that industrial experiments give promise that conversion of organic wastes (garbage) could yield large quantities of "oil" (reference cited by Morris, p. 110). What that reference has to do with the origin of petroleum and natural gas in natural sedimentary basins is anybody's guess.

In 1982, Australian creation scientist Andrew Snelling (Ph.D., Geology) published in Ex Nihilo a technical report on oil and gas deposits in southeastern Victoria (Snelling, 1982). His is outwardly a more sophisticated effort than creationists have previously made to press the case for rapid and recent origin of petroleum. The oil and gas accumulations described are in the Gippsland Basin in the shallow offshore zone under Bass Strait. The gas and oil have been generated in strata of the Latrobe Group of Upper Cretaceous and Paleocene age (roughly -100 to -50 m.y.) and have migrated into reservoirs in overlying strata of Oligocene age (-22 to -30 m.y.). Coal seams are also present in the Latrobe Group. Most of Snelling's text is given over to a detailed description of the geology of the petroleum-bearing strata, all of which conforms with the mainstream principles of origin and migration of petroleum as we have presented it on earlier pages.

Snelling quotes one geological report in which an observed "disequilibrium" in the existing hydrocarbon trends with depth is attributed to "recent rapid subsidence of the sediments" (Shibaoka et al., 1978). That subsidence is thought to be continuing. Snelling quotes Shibaoka as concluding that "hydrocarbon generation must be occurring strongly at the present time with the products migrating relatively rapidly either into traps or to the surface." The idea behind these suggestions is reasonable enough: as a mass of hydrocarbon-bearing strata sinks steadily to lower levels each stratigraphic unit passes into zones of higher temperature in which oil and gas can be generated. Snelling now makes his move, which is contained in the final five lines of type in his paper:

> The above evidence alone indicates recent rapid burial of the coal-bearing sediments, followed by rapid generation of hydrocarbons and very rapid migration of oil and gas into traps. This is clearly contrary to the popular concepts of slowly forming coal bearing sediments and oil formations. (P. 46)

What we find here is a simple case of switching adjectives and adverbs from one context to an entirely different context. "Recent rapid" modern subsidence of the sediments and "relatively rapidly" in the scientists' description of petroleum migration rate have been switched to "rapid burial" in Snelling's version of the accumulation of the hydrocarbon-bearing sedimentary sequence. The scientists never implied that the sedimentation period producing the Latrobe Group was anything other than the time span from late Cretaceous through early Miocene-- about 95 m.y. in conventional geologic time. Snelling has not given one iota of evidence that those 1000-5000 meters of sediments were deposited in one Flood year.

Snelling's abstract insists: "The evidence indicates that the oil and gas are still being formed, a factor which strongly supports the conclusion that Bass Strait oil and gas are of recent origin" (p.43). The scientists' report does indeed suggest that oil and gas are still being formed, but this conclusion has nothing whatsoever to do with the age of deposition of the sedimentary sequence with its original hydrocarbon contents. Snelling's attempt to force an irrelevant conclusion from a set of observational data is completely transparent as an act of desperation.

Ore Deposits and the Flood

Few areas of geology are spared from probings by the purveyors of Flood geology. Besides coal and petroleum, mineral resources of economic value coming under creationist scrutiny include the metalliferous ores that, since biblical times, have been extracted by humans for refinement or smelting to yield gold, silver, lead, zinc, tin, copper, iron, and mercury.

Whitcomb and Morris give two pages to the origin of ore and mineral deposits (1961, pp. 437-38). The ICR textbook has even less (Morris, 1974a, pp. 110-11). The tune in each is much the same: geologists know very little about the origins of ore deposits; the uniformitarian approach has been unsuccessful in explaining such deposits and is not successful in locating them. These claims are offensive to the large number of mainstream economic geologists, who know a great deal about the origins of the many kinds of ore deposits and who have been successful in locating and developing them. The creationists claim that Flood geology, with its special brand of volcanism and tectonic activity, is better suited to explain ore deposits.

Mainstream geology recognizes that the primary sources of metalliferous ores are in magmas that produce igneous rocks, both intrusive and extrusive. The advent of plate tectonics has allowed the origins of primary ore deposits to be unified under one general hypothesis. New knowledge of plate motions and plate boundary activities has brought new insights into differences among ore bodies of highly varied types. The first of the two creationist references cited above was published before plate tectonics was available, but there is no excuse for continuing to reprint the books unrevised. The new information was available in the year the ICR textbook was published, but was apparently ignored.

If that were the sum of creationist delvings into ore deposits, we would not have needed to mention the subject, but there is more. A recent technical article, published in the Australian creationist journal Ex Nihilo, claims to have evidence that formation of a major ore deposit in Queensland was both recent and rapid, in compliance with the conditions of Flood geology. The author, Andrew A. Snelling, cited earlier in this chapter, was formerly Chief Geologist with the company that developed the Koongarra uranium field. His article is written in a thoroughly scientific style and is complete with references to the scientific literature and detailed cross sections of the ore bodies (Snelling, 1984). The claims he makes warrant an answer. An effective answer requires use of principles of igneous activity and ore deposition in the framework of plate tectonics. A brief overview of this subject will help you to understand the arguments.

Metalliferous Ores and Plate Tectonics

The metalliferous ores (ores of metals) are highly diverse in origin, just as for all minerals and rocks. Limiting ourselves for the moment to the primary ores that are of direct igneous origin, three broad classes are easily recognized. One consists of grains of a particular mineral variety that have crystallized first in a cooling magma and, by reason of their being of greater density than the remaining magma, have settled by gravity to the bottom or floor of the magma chamber. The process is called <u>magmatic</u> <u>segregation</u>. It has given rise to some important ore bodies, for example, nickel ores forming a basal layer in a saucer-shaped pluton of gabbroic igneous rock. The second major class consists of ores precipitated from watery solutions (<u>hydrothermal solutions</u>) that are exuded by a large magma body in its final stages of crystallization (Figure 25.3). These minerals are deposited in fractures that often penetrate into the surrounding country rock. The filled cracks are mineral veins of various shapes and forms; where exceptionally numerous and thick, they form a lode. Hydrothermal solutions can also penetrate older igneous rock that was previously fractured, resulting in a <u>disseminated</u> <u>deposit</u>. Hydrothermal solutions may rise close to the surface to deposit ores in a shallow zone. An example is mercury ore, consisting of the mineral cinnabar.

Third, a form of rock metamorphism can take place in the country rock close to an invading magma body. Hydrothermal solutions may soak into the country rock, introducing ore minerals in exchange for mineral components of the rock. An example of a contact metamorphic ore deposit might be an oxide of iron replacing parts of a layer of limestone to form an iron-ore body (Figure 25.3).

Because primary metallic ores arise directly from magmas that invade the crust, we can turn to plate tectonics for a coherent explanation and classification of such deposits. We realize that igneous activity is largely concentrated in three types of environments or locations: (1) spreading (accreting) plate boundaries, where new oceanic crust is being created; (2) subduction zones, where magma is being produced by deep melting and is giving rise to intrusion and extrusion; and (3) hot spots of rising mantle rock that may be found at almost any location within a lithospheric plate, far from its boundaries. One of the most recent developments in plate tectonics has been the recognition of occurrences of important ore deposits associated with each of these zones of igneous activity.

Our new classification of metallic ore deposits recognizes five basic types, shown schematically in Figure 25.4. We shall give each a tentative name, keeping in mind that we are dealing with an area of scientific exploration in its early, formative stages.

1. Cyprus-type Ore Deposits. On the island of Cyprus in the eastern Mediterranean Sea, a huge deposit of copper sulfide ore occurs in the Troodos Massif, a broad

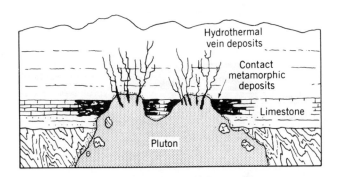

Figure 25.3 Schematic cross section of vein deposits and contact metamorphic deposits adjacent to an intrusive igneous body. (Copyright © 1972 by Arthur N. Strahler.)

band of mafic igneous rock cutting across the central part of the island. The ore bodies occur within a zone of pillow lavas of basalt composition, near the northern margin of the igneous rock zone. Adjacent to the pillow lava and grading into it is a zone of sheeted basaltic dikes, which in turn gives way to an adjacent zone of coarsely crystalline mafic rock of gabbro composition. Geologists now realize that what they are seeing on Cyprus is an <u>ophiolitic</u> <u>suite</u>.

Recall that ophiolites are masses of oceanic crust produced along accreting plate margins in the spreading zone of an ancient mid-oceanic rift (See Figure 20.16). The relationship of ore deposits to the igneous rocks of the oceanic crust is shown in Figure 25.5.

Cyprus-type ore bodies, which are usually sulfides of iron, copper, or nickel, seem to have been deposited in the pillow lava layer by hydrothermal solutions heated in contact with magma in the mantle beneath. Sediments deposited upon the pillow lavas also receive hydrothermal deposits, and these are commonly enriched with compounds of iron and manganese. Bodies of chromite (chromium ore) are sometimes found within the ultramafic rock at the base of the ophiolitic suite, as shown in Figure 25.5. These bodies may have formed by magmatic segregation in pockets of ultramafic magma at the same time that the seafloor spreading was in progress.

If the Cyprus ophiolite rocks were indeed formed at accreting plate margins in a spreading zone, how did they come to be elevated and exposed as a mountain range? During continental collision, masses of oceanic crust are broken away from the descending plate and slide up along overthrust faults to take an elevated position among the highly deformed strata of the tectonic belt, as we showed in Figure 20.29. Later erosion exposes the fragment of oceanic crust, as in the Cyprus example.

During the early 1980s many new observations were made of hydrothermal activity at the spreading axis of the mid-oceanic ridge. Particularly striking has been the sampling of hot water of the black "smokers" from which

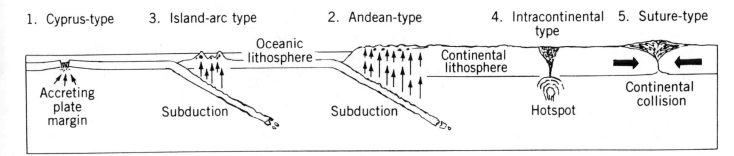

Figure 25.4 Schematic diagram showing five types of ore deposits in the frame of reference of plate tectonics. (Copyright © 1977 by Arthur N. Strahler.)

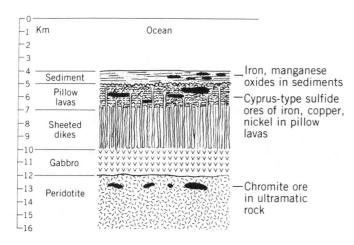

Figure 25.5 Cyprus-type ore deposits occur within an ophiolitic suite of rocks formed at accreting plate margins. The arrangement of rocks is the same as in Figure 20.16. (Copyright © 1977 by Arthur N. Strahler.)

water at 350 C issues in a jet, carrying iron sulfide and manganese. The chimney surrounding the jet is formed of sulfides of iron, copper, and zinc. Iron and manganese oxides produced from the "smoke" settle on the nearby rift floor. (For an excellent review of the entire process and its chemistry, see Edmond and Von Damm, 1983.)

In only a few years of deep-seafloor research it became evident that the hydrothermal process in and adjacent to active oceanic plate-spreading boundaries is an enormous hot-water recirculation system. Probably only a small fraction of the water is "juvenile" water, brought up from the mantle with the rising magma. Nearly all is seawater that penetrates the sediments and basalts of the oceanic crust, reaching a zone of extremely hot rock in the lower crust and upper mantle, then returning to the seafloor along major faults. Figure 25.6 is a highly pictorial schematic drawing of this system, showing the crustal structure (right), the magma chamber and conduit (center), and the ground-water flow system (left). Recent estimates give the total annual flow of seawater through this world system as about 10^{17} grams, at which rate the entire volume of the world ocean could be recycled every 10 million years (EOS, 1982, vol. 63, no. 42.).

Heat-flow measurements confirm the flow system: high heat flow where the hot water is rising; low heat flow where the seawater is sinking into the crust. The flow is driven by the residual heat from the solidified magma. The hot water is chemically active and reacts with the basalt, releasing the metallic ions, which combine with the

sulfur already present in seawater as sulfates. Water combining with the mafic silicate minerals of the basalt and gabbro forms a greenish mineral called serpentine that remains in place.

As an oceanic plate nears a subduction zone it begins to bend down, entering the trench zone. Now, consider that the spreading plate boundary producing that plate may also be drifting toward the adjoining continental plate, and, if this continues, the entire spreading plate boundary will be subducted (Figure 25.7). This event seems to have occurred repeatedly in the Pacific basin in Mesozoic and Cenozoic time, so that entire oceanic plates have simply disappeared beneath volcanic arcs of continental margins. (The Juan de Fuca plate is in the process of disappearing in this manner.) Our next thought is that a forearc basin at the subduction line may be receiving large quantities of terrestrial sediment--clays and muds--comprising a thick lens of sediment. Rising basaltic magma moves up into the sediment layer, spreading out in horizontal, platelike plutons called sills. The hydrothermal solutions of the spreading plate boundary can now move upward into the great mass of clay-rich sediment, where they can deposit large bodies of sulfide ores. This scenario seems to explain an important type of sulfide ore deposit, called a Besshi deposit, after a well known sulfide ore body in Japan. More recently, bodies of sulfides within thick river sediment overlying an active spreading plate boundary have been discovered in the Gulf of California (Edmond and Von Damm, 1983, pp. 92-93).

Here we have a remarkable example of the revised and modernized principle of uniformitarianism in geologic history. The hydrothermal process operating today in numerous places at plate spreading boundaries may be duplicating the process by which many ore deposits came into existence in the geologic past, even as far back as Precambrian time, in countless spreading zones long since vanished.

2. Andean-type Ore Deposits. A second class of ore deposits related to plate tectonics has been recognized by geologists in continental crust lying above a descending plate adjacent to a subduction zone. As shown in Figure 25.8, magma bodies rising from the descending plate reach the upper crust to solidify as felsic plutons or to emerge as volcanic rocks such as andesite and rhyolite. The rising magmas bring up a supply of metals that occur now as ores in the igneous rocks. Because this class of deposits occurs in the Andes Mountains of South America as important tin and copper ores, it is referred to as the Andean-type of ore deposit.

Magmas produced at various depths along the descending plate differ in dominant metallic element content. Consequently, the ore deposits show a distinct zoning by dominant metallic elements from continental

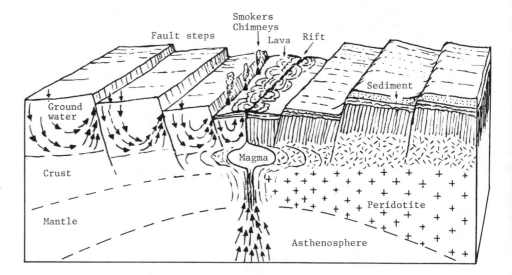

Figure 25.6 Schematic diagram of a spreading plate boundary with axial rift, rising magma, and block faulting. Geology is shown at right; ground-water circulation system at left. Features are not drawn to scale. (A. N. Strahler.)

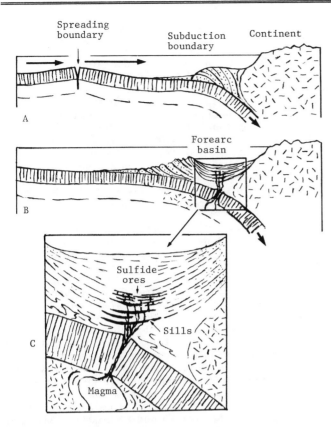

Figure 25.7 Conditions favorable to the accumulation of sulfide ores in forearc sediments above a subducted spreading plate boundary. (A. N. Strahler.)

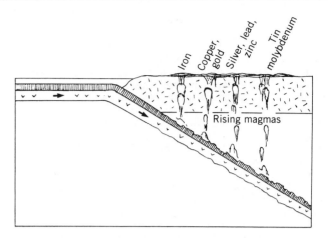

Figure 25.8 Andean-type metallic ore deposits are formed by magmas rising from an oceanic lithospheric plate undergoing subduction. (Copyright © 1977 by Arthur N. Strahler.)

the asthenosphere. The rock was probably blown upward through the crust in a powerful jetlike blast in a very short time, propelled, it is thought, by the pressure of expanding water or carbon dioxide. Geologists call the pipelike structure a <u>diatreme</u>, and, though many diatremes are known, only a few have yielded diamonds. Diatremes at Kimberly, South Africa, are certainly the most famous diamond producers of the world.

5. Suture-type Ore Deposits. Any one of the four ore deposit types previously named could, in the geologic past, have become involved in continental collisions. Today the rock types that bear these ores are found crumpled, overthrust, and metamorphosed in sutures within the continental shield. Take the case of the ore deposits of Newfoundland. Throughout the Palezoic Era, the Appalachian belt, in which Newfoundland lies, suffered arc-continent and continent-continent collisions interspersed with episodes of continental rifting, as the primitive Atlantic Ocean basin alternately opened and closed. The final suturing occurred at the close of the Paleozoic Era. Geologists can recognize in the rocks of Newfoundland the typical ore deposits of the Cyprus type (ophiolitic suites), island-arc ore deposits in great volcanic rock masses, and even disseminated ores of the Andean type that occur above subduction zones.

Although still in its early stages, the tie-in of ore deposits with plate tectonics has produced a great sweep of optimism within the ranks of economic geologists engaged in exploration for new mineral deposits. Now, for the first time, they have a frame of reference in which to place the major types of ore deposits. Interpretation of a particular rock sequence in terms of plate tectonics allows the geologist to predict what type of ore deposit may occur there. As with any search program, knowing what to look for and where to look for it gives the searcher a great advantage.

The lesson to be learned from this review of a modern synthesis of ore-forming processes is that the hypothesis of plate tectonics shows its strength by its ability to throw light on subject areas that might otherwise seem to be remote. Fecundity is the mark of a successful scientific hypothesis; fecundity is cetainly a characteristic of plate tectonics.

As for the creationists' Flood geology, it offers no dynamic framework for the origin of ore bodies of widely different types. The repeated claim is simply that the ores and the enclosing and related rocks were formed recently and rapidly, but that is merely a repetition of the Flood hypothesis itself. Is there independent evidence that the one-year time frame is required by compelling evidence? Let us now see if such evidence exists for the Queensland ore deposits.

margin to continental interior, as shown in Figure 25.8. Ores of iron are typical of a belt closest to the plate margin. Then come successive zones: copper and gold; silver, lead, and zinc; tin and molybdenum. This sequence has been recognized in the western United States and has been related to past episodes of subduction of the Farallon plate beneath the North American plate. Disseminated copper deposits are thought to be one of the forms of the Andean-type ore deposits. Similar copper deposits also occur in the Andes and beneath ancient subduction zones in Eurasia. The concept has guided recent searches for new ore deposits.

3. Island-arc-type Ore Deposits. A third group of metallic ore deposits occurs within massive accumulations of felsic lavas--andesite and rhyolite--within volcanic island arcs. An example is seen in important copper sulfide ores of Japan, one of the few rich mineral resources with which that island nation is endowed. The ore occurs in massive bodies and is usually a mixture of sulfides of copper, zinc, and lead, along with lesser amounts of silver and gold.

4. Intracontinental Ore Deposits. Some important ore deposits occur far from plate boundaries and lie in the heart of large areas of continental crust. An example is the lead and zinc ore deposits of the Mississippi Valley region. These ores are found in sedimentary strata of the continental platform. Although the explanation is highly speculative, it seems possible that ores of this type were formed by hydrothermal solutions rising from a hot spot in the mantle below.

The same concept of a hot spot beneath continental lithosphere has been called upon to explain the occurrence of diamonds within narrow, vertical, pipelike bodies of an ultramafic rock known as <u>kimberlite</u>. Fragments of garnet crystals within this rock provide evidence, through chemical analysis, of having been crystallized at a temperature of about 1360 C. These conditions might be expected at a depth of some 150 to 200 km, well within

Flood Geology and Ore Deposits

At Mount Isa, Queensland, is one of the world's largest deposits of base metals (Snelling, 1984b, p. 40). Within the same sequence of metamorphosed strata are contained both silver-lead-zinc ore bodies and copper ore bodies. The strata as a group (Mount Isa Group) are of middle Proterozoic age (about -1.65 b.y.), according to the conventional geologic time scale. The two distinct kinds of ore bodies are physically separated from each other; each kind of ore is enclosed in a different kind of sedimentary rock.

Snelling (1984b, p. 41) makes clear that in recent years the consensus of economic geologists is that the silver-lead-zinc ore bodies, which are sulfides (compounds of sulfur), originated by hydrothermal activity taking place on and near the axial rift of a spreading plate boundary. As we have seen, this is a deep-ocean environment in which basaltic crust is being formed. Deposition of thick black shales, consisting originally of muds of terrestrial origin, is not, however, typical of the axial-rift environment. Although not typical, such deposition could occur if the rift (plate-spreading boundary) were situated close to a continental margin with an active subduction zone, as in the case of the Besshi-types ore deposits or the terrestrial sediments in the Gulf of California.

Snelling (1984b, pp. 42-46) devotes the greater part of his paper to two matters: (a) evidence for recent formation, and (b) evidence for rapid formation. Evidence for recent formation hinges on whether or not there is substantial evidence of fossils or organic matter representing fossils in the shale formation. Snelling cites published reports showing that remains of blue-green algae cells are present in the shale. These findings of mainstream geologists are not now subject to serious doubt. Thus, both creation scientists and mainstream scientists agree that fossils are present; so how does this finding prove recent formation? Snelling states: "Wherever fossils or organic matter representing fossil remains are found in the geological column the rocks containing the fossils were deposited either by or after Noah's Flood regardless of their assumed evolutionary age" (p. 42). Snelling has not given us one shred of scientific evidence that all fossils are no older than a few thousands of years. He has, instead, stated his initial postulate, which is religious doctrine based on belief in the literal meaning and inerrancy of the Bible. His initial postulate is now being passed off as its own proof; he has taken no step toward scientific confirmation of his postulate.

Evidence for rapid formation of the Mount Isa ore bodies would, if acceptable, be only permissive of the postulate of recent formation. No matter how good the evidence for a very short period of deposition--even if a period of only one year--it would not give exclusive support to the Flood hypothesis. The rapid deposition (if it had been rapid) could have occurred in a time period lying within the middle Proterozoic, which radiometric age places around -1.65 b.y. Recent formation has not been suggested by even a shred of independent evidence; therefore, no amount of concentration upon evidence of rapid formation is capable of producing corroborative value. Therein lies the basic flaw in all of the creation-ists' claims for Flood geology. Evidence for rapid formation of a geologic structure, unit, or other entity does not substantiate recent formation. In even plainer language, "rapid" does not substantiate "recent."

Snelling's strategy is the same one followed by other creation scientists in arguments for rapid occurrence of natural phenomena. He cites a report by T. Finlow-Bates (1979) in Economic Geology, an established journal of mainstream geology. Finlow-Bates calculates that, in the case of Red Sea mineral deposits, a layer of lead sulfide (galena) 1 cm thick could be formed in just under 5 weeks from a dense, sluggish bottom layer of ore-bearing solution one meter in depth, carrying 50 parts per million of lead and flowing at a rate of one meter per minute (p. 1417).

If we were to accept Finlow-Bates's carefully calculated estimate of 1 cm in 5 weeks and apply it to the total thickness of the Mount Isa ore bodies, as shown on the detailed cross-section, we must account for thousands of 1-cm layers. One hundred meters of ore thickness (less than the cross-section shows) gives a total deposition time of 50,000 weeks, or nearly 1,000 years. This does not take into account time required for deposition of the intervening shale laminae, which with the ore totals some 1000 m. This total accumulation time is far in excess of the one-year limit of the Flood. Obviously, Snelling is unhappy with this discrepancy. He goes on to state:

> It is not difficult to see the implications of these calculations. If we make some appropriate and reasonable changes to Finlow-Bates' parameters and then recalculate the deposition rate the result is even more startling. Consider, then, a layer of dense ore solution, 15 metres deep, flowing on the sea floor at a rate of 500 metres/minute (30 km/hour, still relatively slow) carrying 1000 ppm lead all of which is to be deposited within a distance of 1,000 m. (It should be noted that these figures are reasonable even in present day terms: the Red Sea brine pools are up to 250 metres deep [32]; dense turbidity currents are known to have travelled thousands of kilometres down the continental slope and across the deep ocean floor at speeds up to between 65 and 80 km/hour [33]; and concentrations of metals such as lead carried by ore-forming solutions are by consensus stated to be in the range XO-X,000 ppm, where X = 1, 2, . . . [34], and by analysis of residual fluid inclusions in ore and ore-related minerals have been measured as up to 10,000 ppm [35]. A galena bed carrying 25% lead with an average thickness of 1 cm would then form in only about 20 seconds, a rate of about 1 metre/30 minutes!! (Pp. 43-44)

(Note: The bracketed numbers indicate references cited by Snelling. The authors cited are mainstream geologists.)

Snelling has increased each of three rates as follows:

    Layer depth: x 15
    Flow speed: x 500
    Lead concentration: x 20
    Total increase: x 150,000

Actually, 1 m/30 minutes represents an increase factor of about x 175,000. Justification for these increases is stretched to the breaking point. Observation of a brine pool 250 m deep in the Red Sea rift zone is of no value in the argument, because those brine pools are stagnant, stratified bodies of hot brine situated in closed depressions (Degens and Ross, 1976, p. 194, Figure 18). Finlow-Bates is discussing "sea-floor-hugging currents" (p. 1417), for which the fluid "is supplied in relatively rapid bursts and deposition occurs relatively rapidly" (p. 1418). Flow rate increase by a factor of 500 is in no way to be implied from the flow rates of turbidity currents, which are gravity phenomena on the continental slope and rise--far from any axial rift or fault where the ore solution is emerging. Lead concentrations 20 times greater than the initial value may have been inferred in some hydrothermal solutions under entirely different geologic conditions (e.g., above an intrusive magma body in continental crust); such high values are not necessarily reasonable for hot brines on the sea floor. Every step of the recalculation has been contrived from unrelated observations to achieve a truly astounding result. The futility of it all is that, if this enormous extrapolation of rates could be declared reasonable, it could also have been equally reasonable in the year 1,653,281 B.C.

Snelling continues with a lengthy discussion of details of the supposed processes of deposition of the ores (1984b, pp. 44-45). He makes a strong point of lead isotope evidence. He quotes from a work of J. R. Richards (1975) that lead isotopic ratios are remarkably uniform within the ore deposit. This is taken by Snelling as indication of the rapid rate of ore deposition. It is more likely, instead, that a reasonable total period of deposition--say one million years--would not appear as a significant difference in lead isotope ratios, considering the standard error of such calculations. Ratios of Pb-206/Pb-204 and Pb-207/Pb-204, measured by Richards (1975), are about 16.12 and 15.45, respectively. On a plot of the latter ratio versus the former, these values lie on an isochron about halfway between the -1.0 and -2.0 b.y. isochrons, which is in rough agreement with the radiometric age of the Mount Isa stratiform ores, i.e., -1.65 b.y. Just imagine, a creation scientist quoting as evidence lead-isotope ratios when it is convenient to do

so, but elsewhere labeling the entire method as fallacious!

Snelling concludes his paper with: "A creationist interpretation" (1984b, pp. 45-46). He concludes that all the silver-lead-zinc ore bodies of Mount Isa and their enclosing metasedimentary rocks could have been deposited in less than 20 days. He states flatly that because "Noah's Flood occurred approximately 4,300 years ago according to Biblical chronology, evolutionary ages for the rocks and ores at Mount Isa have to be discarded." In support of that assertion he cites work of creation scientists Slusher, Setterfield, Mathews, and others "who have shown that radioactivity is unreliable as a means for dating rocks." Well, then, the whole thing comes down to a question not of rapid deposition, but of recent deposition. That being the case, we are back to the basic issue of radiometric methods of age determination. This is a subject already covered in depth. Other creationist arguments for a recent age of creation have been explored in depth on earlier pages--speed of light, decaying magnetic field, and the lot. Have the creation scientists "severed the Achilles' tendon of evolution"? You be the judge.

PART V

Two Views of the Origin of Landscapes

Introduction

In this, Part Five, we turn to geologic events and processes of the comparatively recent past, geologically speaking. These have resulted in the detailed features of the continental surfaces that give the terrain, or landscape, its distinctive character, that varies greatly from place to place. The study of landscape features, or landforms, is a research field within geology known as geomorphology.

For mainstream geologists, landforms fall into the last two million years or so, being largely features of the Pleistocene and Holocene epochs of the Cenozoic Era, although many basic features of the landscape date much farther back into Cenozoic time.

Although tectonic and volcanic activity have continued to play an important part in landscape development in this final bit of geologic time, the major role in shaping the fine details of the land surface has been played by solar-powered processes that include rock weathering and spontaneous sliding and flowage (mass wasting), combined with processes of erosion, transportation, and deposition carried out by running water of streams, by waves and currents along the shores of oceans and lakes, by wind, and by glacial ice. These activities are encompassed under the term denudation. The landforms produced by denudation are classed as secondary landforms, distinguishing them from the class of primary landforms created by tectonic and volcanic activity.

One group of primary landforms is produced directly by faulting, a tectonic process. Fault landforms commonly consist of sharp, low cliffs, and these succumb rapidly to weathering, mass wasting, and the action of runoff. Active volcanoes are another primary landform type, being built of fresh lava and ash accumulating in successive layers. Volcanoes, too, are continually under attack by denudation processes, and these dissect the cone or flow into erosional landforms during long intervals between eruptions and in the period following extinction.

For the creationists, landforms belong in the post-Flood (postdiluvial) period, and its duration is limited to about 4,000 years, although some landforms may have originated in receding stages of the floodwaters as the land surfaces emerged. Creation scientists place much stress on the claim that geomorphic processes must have acted with much greater effectiveness immediately after the Flood than they do today. Only by making this assumption can they compress vast erosional and depositional events, like the carving of the Grand Canyon or the deposition of the Mississippi River delta, into the 4,350 years available since the year of the Flood. Justifying this forced speed-up of nature's activities cannot begin to compare in difficulty with the task of explaining all fossil-bearing rocks of the earth in a one-year Flood, but it is nevertheless a challenging assignment--one that the creation scientists undertake with remarkable optimism and persistence.

A major topic of post-Flood geomorphology is the action of glaciers during the last ice age and the sedimentary deposits associated with the spread and disappearance of great continental ice sheets. This is the subject of the following chapter.

Chapter 26

Glaciers and the Ice Age

Because most of us have dealt with ice only in small pieces or thin layers, we think of it as a brittle solid. Ice fractures easily and shows crystalline structure, for it is a true mineral. But where an ice layer has accumulated on land to a depth of about 100 m and rests on a sloping surface, the basal part of the ice becomes plastic, yielding by slow internal flowage. The entire ice mass is carried downslope and becomes a glacier. A glacier is defined quite broadly as any large natural accumulation of land ice affected by motion that may be occurring today or may have ceased.[1]

Glacier Ice

For glaciers to form, the quantity of incoming snowfall --on the average, year in and year out--must exceed the average quantity lost yearly by melting and evaporation, two processes that geologists combine under the single term ablation.

Freshly fallen snow has a very low density, with much of its volume consisting of air-filled openings. Changes quickly set it, however, and the elaborate hexagonal snow crystals change into more rounded, smaller particles. The mass becomes greatly compacted into old granular snow, in which the air spaces make up less than half of the volume. Under the load of newer snow layers and with the aid of some melting and refreezing, the old snow compacts further. As the older snow is buried still more deeply under new layers, it recrystallizes, finally becoming glacier ice. Although most of the air has by this time been expelled, some is enclosed in bubble holes and the ice density may not be greater than about five-sixths that of water.

Glaciers formed in high, steep-walled mountain ranges and occupying previously carved stream valleys are shaped into long, narrow ice streams. They have many of the basic elements of the fluvial drainage system, such as tributary channels leading downgrade to a trunk stream, and are thus classified as alpine glaciers. In the high interior area of a large landmass in the high latitudes, glacier ice accumulates in platelike bodies called icecaps. Small icecaps a few tens of kilometers across and roughly elliptical or circular in outline are found today on summit areas of arctic islands such as Iceland, Baffin Island, and Ellesmere Island. Alpine glaciers and small icecaps are sustained on high mountains and plateaus, where air temperatures are low and there is abundant snowfall.

Ice Sheets of Today

Vastly greater ice masses of continental proportions, such as those of Greenland and Antarctica, are continental glaciers, or simply ice sheets. A great ice sheet may reach a thickness of several thousand meters and may cover an area of several million square kilometers, spreading far beyond the limits of the highlands where it originated. Many ice sheets of the past covered low plains that otherwise would not have sustained glaciers.

The Greenland Ice Sheet occupies some $1,700,000$ km^2. This is 80 percent of the area of the island of Greenland, where ice covers all but narrow land fringes (Figure 26.1). The ice sheet contains almost 3 million km^3 of ice. In a general way, the ice forms a single, broadly arched, doubly convex lens. The ice thickness measures close to 3 km at its greatest. The center of Greenland is actually depressed under the great ice load, in conformity with the principle of isostasy, because 3 km of glacial ice is roughly equivalent to a rock layer at least 1 km thick. Because the Greenland ice surface slopes seaward, the ice creeps slowly downward and outward toward the margins, where it discharges by glacier tongues.

Like Greenland, the Antarctic continent is almost entirely buried beneath glacial ice, in an area of about 12.5 million km^3, or about 1.5 times the total area of the contiguous 48 United States (Figure 26.2). Ice volume is about 25 million km^3, which is over 90 percent of the total volume of the earth's glacier ice. (In comparison, the Greenland Ice Sheet has about 8 percent.) The Antarctic Ice Sheet reaches its highest elevations, just over 4 km, in a broadly rounded summit. Surface slope is gradual to within 300 km of the edge of the continent, where a marked steepening occurs. In general, the ice is thickest (4 km) where surface elevation is highest. A characteristic feature of the Antarctic coast is the presence of numerous ice shelves that are great plates of partially floating ice attached to the land.

The Antarctic Ice Sheet moves outward near its margin at a rate estimated between 25 and 50 m per year. Rate of movement of outlet glaciers is much faster, estimated at 400 m per year on the average. For the ice shelves, rate of outward movement is even more rapid, on the order of 1000 to 1200 m per year for the Ross and Filchner shelves. From these estimated flowage rates, scientists infer that a particle of ice following the longest possible path of flow might remain in the ice sheet for as long as 100,000 years.

The Antarctic Ice Sheet seems to be very closely balanced in terms of the net rate of accumulation and the rate of loss of ice to the atmosphere and surrounding ocean. Actually, the figures show a small rate of growth to be in progress, but this may be a short-term effect.

Minimum Ages of the Greenland and Antarctic Ice Sheets

Using drilling equipment similar to that used for drilling oil wells, scientists have secured cores of ice from the Greenland and Antarctic ice sheets. In 1966, at Camp Century in Greenland, a core nearly 1,400 m long was obtained, extending through the entire ice sheet to bedrock beneath. Age of the ice was determined at many sample depths using the oxygen-isotope method (described in later paragraphs). The age of ice at the bottom of the core was estimated to be just over 100,000 y. In 1981, another ice core (Dye 3) was taken in Greenland to a bottom depth of just over 2,000 m, with a bottom ice age of about 125,000 y. In 1968 a core taken at Byrd Station,

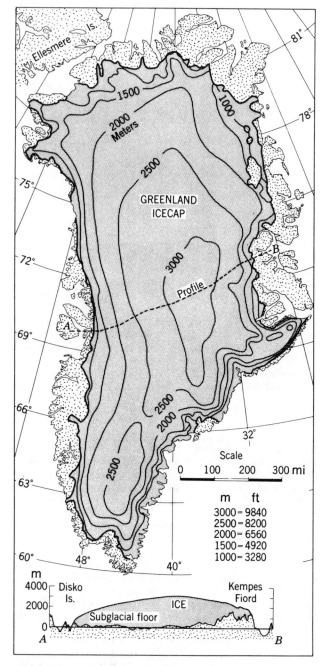

Figure 26.1 Greenland and its ice sheet. (Based on data of H. Bader, 1961, U.S. Army Corps of Engineers, Cold Regions Research and Engineering Laboratory.)

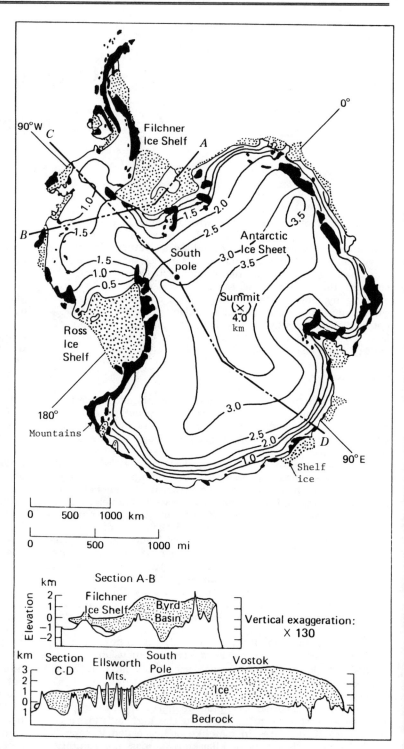

Figure 26.2 The Antarctic Ice Sheet and its ice shelves. (Compiled from data of the American Geophysical Union and American Geographical Society. Copyright © 1981 by Arthur N. Strahler.)

Antarctica, penetrated to a bottom depth of about 2,200 m, where the basal ice was estimated to have an age of about 75,000 y.

Ice of these cores consists of annual layers, easily recognizable in the upper part of the core by a seasonal cycle of variation in oxygen-isotope composition and various kinds of atmospheric fallout particles that include airborne continental dust, biological material, volcanic debris, sea salts, and cosmic particles containing isotopes produced by cosmic radiation (Dansgaard et al., 1982, p. 1273). Concentrations of particles occur in inverse proportion to atmospheric temperature as determined from oxygen-isotope ratios and are clearly cyclic on an annual basis, at least down through several thousand cycles. There also occur layers of volcanic dust and ash that represent distant sporadic volcanic eruptions (Thompson, et al., 1975). Once it was established that summer peaks

of oxygen isotope ratios correlate with other parameters such as acidity and concentrations of dust or of chemical trace elements, it was possible to count several thousand summer peaks downward in the core. "In this way it is possible to obtain an absolute time scale along the core with an accuracy that compares with that of tree ring chronology" (Dansgaard et al., 1982, p. 1275). For Camp Century, Greenland, the annual accumulation of snow, reduced to solid glacial ice, amounts to about 35 cm/y. (Dansgaard et al., 1969, p. 377). Annual ice layers can be counted directly from the surface down to nearly 1,000 m, while annual cycles are identifiable down to much

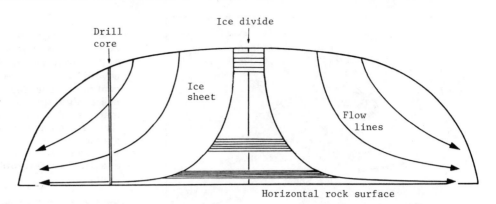

Figure 26.3 Schematic vertical cross section of an ice sheet resting on a horizontal rock floor, showing the travel paths of ice particles (arrows). Horizontal lines represent ice layers plastically deformed with increasing depth. (From W. Dansgaard et al., Science, vol. 166, p. 378, Figure 1. Copyright © 1969 by the American Association for the Advancement of Science. Used by permission.)

greater depths. The thickness of the annual ice layer diminishes with depth, because the ice is engaged in slow flowage outward and downward from the ice sheet divide to the margin, and is thus continually stretched and thinned (Figure 26.3). Figure 26.4 is a nomograph relating age of ice layers to depth in core, allowing for thinning. The age would tend to approach infinity at the bedrock floor as ice layers were stretched to approach infinite thinness.

If we were to use the surface value of 35 cm/y. throughout the entire core, the Camp Century core, 1,390 m long, would have required only about 4,000 y. to accumulate. But allowing for thinning, as shown in the nomograph (Figure 26.4), the age of ice near the bottom is well over 100,000 y. Clearly, this minimum estimated age is vastly greater than the 4,000 y. available in post-Flood chronology or even the entire 6,000 y. since Creation.

The creationists might wish to argue that snowfall was much heavier early in the post-Flood time than today. Granting a net annual accumulation rate of 3 or even 5 times larger than that of today, the minimum accumulation

time is still much greater than the post-Flood period can accommodate.

Notice that in this calculation we have not used any method of absolute age determination based on radioactivity. The only ingredients are measured rates of snow accumulation and measured depths of the ice sheet. Annual layers are identified by their contents, which are observed to vary through a yearly cycle. These cycles can be counted. Again, we have an example of environmental controls that must be recognized in the workings of nature. Precipitation is part of the hydrologic cycle; precipitation varies according to latitude, altitude, and relative position with respect to ocean bodies and continents. While the intensity of precipitation has certainly varied from glacial to interglacial periods, along with the net accumulation in the form of snow, there are reasonable limits to such swings. Creation scientists are not free to postulate outrageous rates of precipitation and ice accumulation. When they choose to do so, they deal in pseudoscience.

The Ice Age

A period of growth and outward spreading of great ice sheets is known as a glaciation. We can safely assume that a glaciation is associated with a general cooling of average air temperatures over the regions where the ice sheets originated. At the same time, ample snowfall must have persisted over the growth areas to allow the ice masses to grow in volume. The opposite kind of change--a shrinkage of ice sheets in depth and volume--would result in the receding of the ice margins toward the central highland areas and eventual disappearance of the ice sheets. This period is called a deglaciation. Following a deglaciation, but preceding the next glaciation, is a period in which a mild climate prevails; it is called an interglaciation.

A succession of alternating glaciations and interglaciations, spanning a total period on the order of 1 to 10 million y. (m.y.) or more, constitutes an ice age. Throughout the past 2½ to 3 m.y., the earth has been experiencing an ice age that has long been known simply as The Ice Age--a title adopted by Louis Agassiz. Present time is within an interglaciation, following a deglaciation that set in about 15,000 years ago (-15,000 y.). In the preceding Wisconsinan glaciation, ice sheets covered much of North America and Europe and parts of northern Asia and southern South America. The maximum ice advance of the Wisconsinan Glaciation was reached about -18,000 y.

Figure 26.5 shows the approximate extent to which North America and Europe were covered at the maximum known spread of the last advance of the ice. Canada was engulfed by the vast Laurentide Ice Sheet. It spread south into the United States, covering most of the land lying north of the Missouri and Ohio rivers, as well as northern Pennsylvania and all of New York and New England. Alpine glaciers of the Cordilleran ranges coalesced into a single ice sheet that spread to the Pacific shores and met the Laurentide sheet on the east.

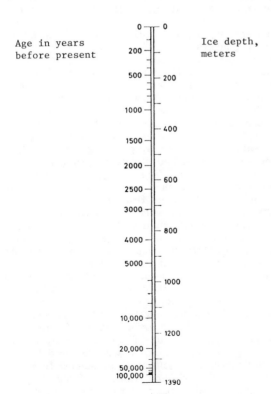

Figure 26.4 Nomograph for relating age of ice to depth. (Same data source as Figure 26.3. Copyright © 1969 by the American Association for the Advancement of Science. Used by permission.)

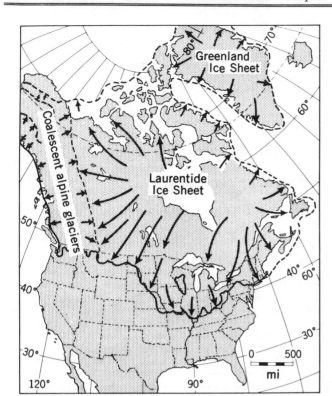

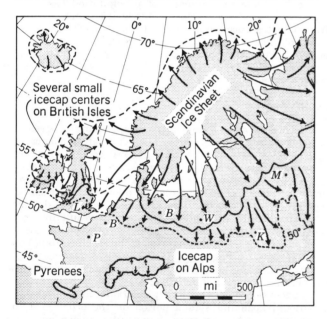

Figure 26.5 Maximum extent of Pleistocene ice sheets of North America and Europe. On map of Europe, the Wisconsinan maximum limit is shown by a solid line; limit for entire Pleistocene Epoch by a dashed line. (Based on data from many sources, as given by R. F. Flint, 1971, *Glacial and Quaternary Geology*, John Wiley & Sons, Inc., See chapters 18 and 23.)

In Europe, the Scandinavian Ice Sheet centered on the Baltic Sea, covering the Scandinavian countries. It spread south into central Germany and far eastward to cover much of Russia. In north-central Siberia, large icecaps formed over the northern Ural Mountains and highland areas farther east. At one time ice from these centers grew into a large ice sheet covering much of central Siberia. The British Isles were almost covered by a small ice sheet that had several centers on highland areas and spread outward to coalesce with the Scandinavian Ice Sheet. The Alps at the same time were heavily inundated by enlarged alpine glaciers.

South America, too, had an ice sheet. This grew from icecaps on the southern Andes Range south of about latitude 40°S, spreading westward to the Pacific shore, as well as eastward to cover a broad belt of Patagonia. It covered all of Tierra del Fuego, the southern tip of the continent. The South Island of New Zealand, which today has a high spine of alpine mountains with small relict glaciers, developed a massive icecap in late Pleistocene time. All high mountain areas of the world underwent greatly intensified alpine glaciation at the time of maximum ice-sheet advance. Today only small alpine glaciers remain and, in less favorable locations, the glaciers are entirely gone.

Evidences of Continental Glaciation

Evidence that such great ice sheets existed takes the form of deposits that the ice left behind and the marks and landscape forms that it carved into the bedrock over which it passed. All these evidences were available to field geologists of the early 1800s in the British Isles and northern Europe, but the process of recognition was painfully slow and was marked by many false steps. This early history of glaciology is recounted in detail by Professor Albert V. Carozzi, a geologist who has written extensively on the history of geology (Carozzi, 1984). The early investigators were intrigued by great boulders that seemed to have been transported from distant regions, but for which no adequate transportational medium was immediately evident. Through careful studies of the glaciers of the European Alps, the relationship between boulder deposits and living glaciers was finally established. Strange deposits of boulders and other surficial deposits of sand, gravel, and clay in low-lying areas, far from any mountainous sources, remained for long objects of extreme controversy. Flood geology offered an explanation that was widely favored by some. The concept of a great ice sheet of continental dimensions was extremely difficult to comprehend, as information on the great Greenland and Antarctic ice sheets was not then available.

Erosional activity of glacial ice, whether it be in the form of valley glaciers or ice sheets, is very much the same in terms of the minor features carved in bedrock. Beneath the glacier, ice flows plastically around joint blocks, then pulls them loose when a sudden forward movement of the ice occurs; this activity is glacial plucking. The blocks are then scraped and dragged along the rock floor, gouging and grooving the bedrock and chipping out fragments in an abrasive process called grinding. Features produced by plucking and grinding include fine scratches, called striations (Figure 26.6). At the same time, the rock surface is highly polished. Application of strong pressure from the sharp corner of a boulder creates nested curved fractures, of which a common type is the chatter mark. In particularly susceptible types of rock, such as limestones, glacial abrasion produces long, deep glacial grooves. Hills of bedrock are shaped into rounded knobs from which joint blocks have been plucked on the lee side. These features of abrasion are unique to the action of glacial ice and are important in mapping the extent of former ice sheets.

The term glacial drift applies to all varieties of rock debris deposited in close association with Pleistocene ice sheets. Drift consists of two major classes of materials: (1) Stratified drift is made up of sorted and layered clay, silt, sand, or gravel deposited as bed load or delta sediment from streams of meltwater, or settled from suspension into bodies of quiet water adjoining the ice. (2) Till is an unsorted mixture of rock and mineral fragments of a wide range of sizes--from clay to large boulders--deposited directly from glacial ice. One form of till consists of material dragged along beneath the moving ice and plastered on the bedrock or on other glacial deposits. Another form consists of debris held within the

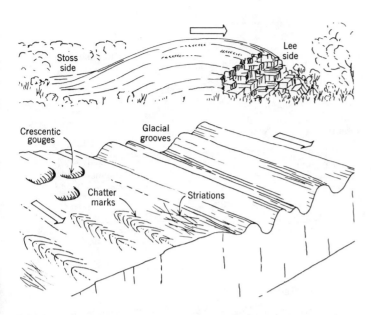

Figure 26.6 Sketch of glacially rounded and plucked rock knob. Arrows indicate direction of ice flow.(Copyright © 1981 by Arthur N. Strahler.)

ice but dropped in place as the ice wastes away. Large rock fragments within glacial till, some of which have been transported tens or even hundreds of kilometers, are called erratic boulders. Figures 26.7 and 26.8 show how features of erosion and deposition are related in the terminal region of an ice sheet. Figure 26.7 shows the moving ice wasting continually at the margin, so as to hold the ice limit approximately in equilibrium. Figure 26.8 shows the final stages of ice wastage and the deposits and landforms that remain. The major features are the hummocky terminal moraine of till, fronted by a gently sloping outwash plain of sand. In the relatively brief period since the last ice sheets disappeared from North America and Europe, these glacial landforms show very little modification by running water and mass wastage. Consequently, the limits of the ice sheet and successive halts in the retreatal stage are usually clearly defined.

Glaciations and Flood Geology

Creation scientists are interested in all aspects of continental glaciations, not only the latest of these--the Pleistocene glaciation--but also those of which there is a stratigraphic record in the geologic past, well back into the Precambrian. Pleistocene glaciation falls in the post-Flood (post-Diluvial) period, whereas glaciations indicated

by stratigraphic features of early Cenozoic and older ages would fall within the one-year Flood. Glaciations recorded in Flood strata present an insurmountable problem, with the entire earth inundated by floodwaters, so any such events must be rejected out of hand. Even the latest of the Pleistocene glacial stages--the Wisconsinan--poses some problems, but they are not so severe. Whereas the Wisconsinan glaciation ended about -11,000 y. (according to mainstream geology) and began perhaps as early as -75,000 y., that time schedule need only be reduced to about 20 percent to be contained in post-Flood time.

Acceptance of at least one glaciation in post-Flood chronology is almost obligatory for the creation scientists. First, the evidence is clear and fresh for the comparatively recent spread of ice sheets well into middle latitudes. Second, Louis Agassiz, the distinguished nineteenth century zoologist/geologist, almost single-handedly at first, proclaimed and defended the theory of a great Ice Age (Étude sur les glaciers, 1840). At the same time, he was stoutly anti-Darwinist and held out for repeated special creation in the multiple catastrophism of Baron Cuvier, under whom he had studied. Modern-day creationists cannot let Agassiz down on his Ice Age without impugning his reputation as a scientist; it certainly would not do to pull the rug of catastrophism out from under him as well.

Creation scientists accept the overwhelming evidence of ice sheets of the last glaciation--that of the Wisconsinan stage. This acceptance of the obvious applies to all processes of landscape evolution--geomorphic processes-- that can be observed in action today, and to volcanic and tectonic activity of recent record. They differ in philosophical position from mainstream geology by their insistence that the enormous cataclysmic forces of the Flood rapidly diminished in intensity after the flood waters vanished, but some catastrophic effects persisted. In this respect, the last, or Wisconsinan glaciation (there was only one glaciation), is viewed as catastrophic in nature, as opposed to a uniformitarian view in the sense of having happened repeatedly and at various intervals through geologic time. The single glaciation is a delayed catastrophe triggered by the sudden environmental changes that immediately followed the Flood. Whitcomb and Morris deduce from the Genesis description of the Flood that a cold period is called for (1961, p. 292). They refer to the dozens of hypotheses that have been proposed to explain an ice age and find none of them satisfactory. The implication is that disagreement among mainstream scientists is prima facie evidence that all those hypotheses are unsatisfactory and, therefore, probably false. Instead, they offer the following picture:

The Biblical Deluge, however, obviously offers an eminently satisfactory explanation. The combined effect of the continents and mountain-chains and the removal of the protective vapor blanket around the

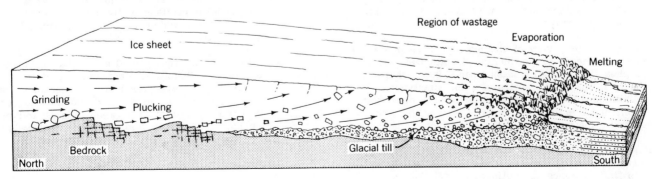

Figure 26.7 Glacial erosion and transportation in the marginal zone of an ice sheet. (From A. N. Strahler, A Geologist's View of Cape Cod, p. 10, Figure 5.

Copyright © 1966 by Arthur N. Strahler. Reproduced by permission of Doubleday & Company, Inc.)

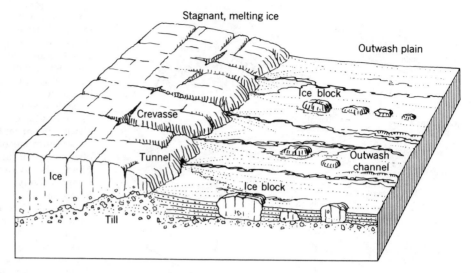

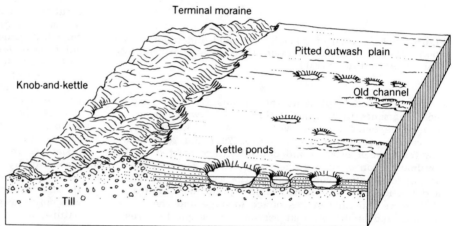

Figure 26.8 (Above) Building of outwash deposits along the margin of a stagnant ice sheet. (Below) Landforms remaining after disappearance of ice. (From A. N. Strahler, A Geologist's View of Cape Cod, p. 15, Figure 10. Copyright © 1966 by Arthur N. Strahler. Reproduced by permission of Doubleday & Company, Inc.)

earth could hardly have failed to induce great snow and ice accumulations in the mountains and on the land areas near the poles. And these glaciers and ice caps must have continued to accumulate and spread until they reached latitudes and altitudes at which the marginal temperatures caused melting rates in the summers adequate to offset accumulation rates in the winters.

The total amount of water locked up in these great glaciers during their greatest extent is not known as yet, but it may have been very great. The main evidence of this fact is in the greatly lowered sea levels of the Ice Age. In the past decade a large amount of evidence has been amassed to show that ocean levels were at least 400 feet lower than at present, possibly much more, as shown by such features as the continental shelves, sea-mounts, submerged canyons and terraces, etc. (Pp. 293-94).

This catastrophic view of the Ice Age is clearly a relict of Agassiz's concept of continental glaciation as one of the catastrophes that caused a mass extinction--perhaps fitted into the final one of the series of such catastrophes that Cuvier envisioned. Whereas previous catastrophes may have been in the nature of floods of water, the final extinction by ice was unique. Agassiz strongly believed that special creation followed the ice extinction, and this event is strongly divergent from that of Flood geology. Agassiz even went so far as to travel to Amazonia in order to find evidence of continental glaciation in even the warmest equatorial regions. If found, such evidence would show the Ice Age to have been universal, and thus the extinction would have been worldwide and total.

Albert Carozzi describes Agassiz's bizarre last fling in these words:

This idea, which now seems to have been one of the most extraordinary aberrations in the history of geology, was the reason for the Thayer expedition to Brazil (1865-1866). With boundless enthusiasm and many preconceived ideas he was able to discover "proofs" of a Pleistocene glaciation which extended from Rio de Janeiro to Amazonia. In the Bay of Rio, Agassiz interpreted the red lateritic soils as tills, the exfoliating granitic domes as "roches moutonnées," and the spheroidal boulders of granite as erratic blocks. His fame was such that even today tourist pamphlets describe such erratic blocks in the vicinity of Rio. In Amazonia, Agassiz interpreted the fluvial sediments of the great river as subglacial deposits held by a huge frontal moraine at Bélem which was later destroyed by longshore currents.

Agassiz reported that glacial action had covered great areas of Brazil, and concluded that the flora and fauna of the Americas had been created anew after the glacial ice had receded, so that no genetic relationship was possible between animals and plants which had lived before and after the worldwide glacial epoch. Here again, he stated, was undeniable evidence of the erroneous concepts of evolutionists. (1984, p. 167)

Today, there seems to be no question that most plants and animals survived the Ice age in North America and Europe, thriving even close to the ice margin, but forced to migrate to lower latitudes as the ice advanced. Radiocarbon ages show beyond reasonable doubt that humans were inhabiting central and southern Europe during the time of maximum advance of the Wisconsinan ice, -18,000 y. Their numerous cave paintings and drawings show the richness of the animal life that flourished around them. Thus Agassiz's last crusade lies repudiated and disowned by both mainstream science and fundamentalist creationism.

Multiple Glaciations in the Ice Age

From the middle 1800s until about 1950, the record of glaciations, deglaciations, and interglaciations was almost entirely interpreted from continental deposits. During this early period of research, there emerged a history of four distinct North American glaciations in the Pleistocene Epoch. A similar and possibly equivalent four-glaciation history was also established for Europe on the basis of studies in the Alps. Names of the four "classic" glaciations and interglaciations of North America are given in Table 26.1.

Table 26.1 Classical Names of North American Glaciations, based on Continental Evidence Before 1950.

Glaciations:	Interglaciations:
Wisconsinan	
	Sangamonian
Illinoian	
	Yarmouthian
Kansan	
	Aftonian
Nebraskan	

Unraveling the history of glaciations and interglaciations was first attempted solely on the basis of evidence from glacial and related deposits and landforms. Because ice-sheet advance and recession were evidently not synchronous over all parts of even one continent, the correlation of events proved extremely difficult. One form of evidence has come from stratigraphic interpretation of layered deposits. The till of one glaciation is typically followed by a layer of wind-transported dust (loess) associated with deglaciation, then by formation of an ancient soil (paleosol) and by deposition of organic matter such as a bog peat indicative of an interglaciation with its mild climate. Older tills show varying degrees of chemical alteration by weathering. Landforms of earlier glaciations, where not buried under new glacial deposits, show increasing degrees of modification by mass wasting and running-water erosion with increasing age.

Creation scientists admit that it is difficult to account for four or more glaciations under the Flood hypothesis, and they find it equally difficult for mainstream science to account for them under any current hypothesis (Whitcomb and Morris, 1961, pp. 295-303). Perhaps, they argue, there never were multiple glaciations. They do not accept the criteria put forward for the earlier stages, namely, the oxidation of iron and other chemical forms of alteration and the depth of leaching of calcium carbonate. They assert that there has been ample time for such observed effects to have occurred since the deposition of Wisconsinan drift. Whitcomb and Morris state:

> The length of time required to weather fresh material and develop a soil profile is quite unknown. Seldom if ever is there found in any one vertical sequence more than one apparently mature soil other than that on the surface, and there is no reason to insist that it took a long time to form. (1961, p. 298)

The authors continue with quotations from the mainstream literature pointing out the many uncertainties involved in judging the age of pre-Wisconsinan deposits. Frequently quoted is the late Professor Richard Foster Flint, widely recognized as the American dean of Pleistocene (Quaternary) geology. What they failed to include is his statement that horizons of maturely weathered till, called "gumbotill," required substantial amounts of time to form, spans of time that were more than the duration of the Wisconsinan Stage (Flint, 1957, p. 213). Flint, in his updated volume, comments that "very thick mature soils rich in clay or iron or both, represent far longer periods (than 2000 to 4000 years), possibly running into tens of thousands of years" (1971, pp. 291-92)[2].

Recall from our earlier section on radiocarbon dating that in the early 1950s the method began to be applied to various materials contained in deposits of late Wisconsinan age. A landmark date of this early work was obtained from wood of trees overwhelmed by the last ice advance at Two Creeks, Wisconsin: -11,850 ± 100 y. The age proved to be about half that previously guessed. Figure 26.9 shows the stratigraphic relationships. The tree trunks are enclosed in the red Valders till. Creationists reject this and all other radiocarbon dates, along with all radiometric ages. Throughout the 1960s and 1970s the limit of radiocarbon dating stood at about 50,000 y., but by 1980 the limit had been extended to about 75,000 y. by thermal diffusion isotopic enrichment of carbon-14 (Grootes, 1978). This limit corresponds approximately with the start of the Wisconsinan glaciation. And so, the impasse between mainstream science and creation science reappears in a familiar form that cannot be resolved by reason. But mainstream scientists have forged ahead, using every new investigative tool available with which to delve into the record of earlier glaciations and interglaciations. Let us review these developments.

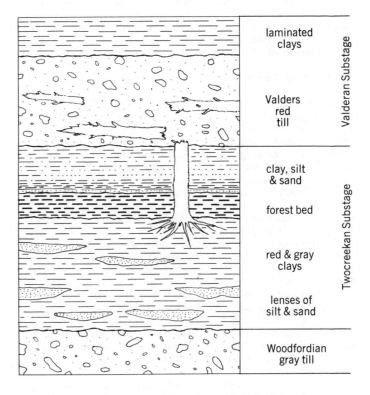

Figure 26.9 The Two Creeks forest bed, exposed near Manitowoc, Wisconsin. (Based on data of J. L. Hough, 1958, _Geology of the Great Lakes_, p. 102, Figure 31, Univ. of Illinois Press.)

The Ice Age Record on the Ocean Floors

We have seen that the record of Pleistocene events as read from glacial deposits on the continents is, at best, fragmentary. So we turn to the one environment in which a continuous and undisturbed depositional record of the Pleistocene can be found--the floors of the deep oceans. It is from the evidence of deep-sea sediment cores that a series of cold and warm episodes can be identified, dated, and perhaps correlated with glaciations and inter-glaciations.

To be successful, a program of Pleistocene research based on deep-sea sediments must develop two systems of information: (1) a system of establishing the absolute age of the sediment, and (2) a system of climate indicators within the sediment. Let us look into each of the information systems.

Direct radiometric methods, beside that of radiocarbon, have included thorium-protactinium (Th-230/Pa-231) and Th-230/Th-232 ratios. These methods carry back the deep-sea record over a range of several hundred thousand years, but errors of unknown magnitude may be introduced by contamination of the sediment samples.

Paleomagnetic age determination has been a mainstay of research on ocean sediment cores. Looking back at Figure 19.3, you can see that two magnetic epochs--the Brunhes normal and the Matuyama reversed--make up the record back to -2.4 m.y. Within the Matuyama epoch are two normal events that also serve as time markers. And, of course, the paleomagnetic record continues all the way back through Cenozoic time to date the earliest records of cold-climate periods that might indicate a glaciation. For a given deep-sea core, the points at which magnetic polarity reversals occur serve as rather precise fixes on age. Between the reversal points, age is estimated in proportion to depth in the core and is subject to some error.

As explained in Chapter 19, creation scientists reject summarily the paleomagnetic ages. Mainstream scientists can point out that the polarity reversals have been identified in lava sequences where radiometric ages can also be determined, so that the paleomagnetic ages have been locked in with radiometric ages. But, of course, creation scientists also summarily reject radiometric ages, so we arrive at the same impasse as before!

Interpreting Glacial Cycles from Foraminifera

Cores that are suitable for determining Pleistocene chronology consist of sediments that have accumulated very slowly in locations far from the continents and free from disturbance by bottom currents. These are pelagic sediments consisting of suspended particles that have settled from the surface layer to the bottom. The materials in these cores consist partly of terrigenous matter (fine clay) that has settled out from the overlying water body, and partly of the tests (hard parts) of planktonic microorganisms that lived in the near-surface zone of the overlying ocean.

In the cases of cores taken in low and middle latitudes, the microfossils are almost entirely foraminifera with calcareous tests. A small sample of the core can be washed on a sieve to remove the fine mineral particles, leaving only the tests for study. Two methods of using foraminifera have been followed in attempts to interpret water temperatures in the near-surface zone. The inferred seawater temperatures, in turn, are correlated with periods of colder and warmer atmospheric temperatures thought to be associated with glaciations and inter-glaciations.

Under the first method, percentages of the various species of foraminifera present in a core sample are determined by counting. From this information, it can be decided whether the plankton that lived in the surface water over the site of the core belonged to a cold-water

or a warm-water fauna. A cold-water fauna is assumed to be associated with a glaciation; a warm-water fauna, with an interglaciation. For the waters in which these organisms lived, a sea-surface temperature of 21 C is considered "cold," while 28 C is considered "warm." A total range of about 5 to 6 C is a reasonable maximum sea-surface temperature range from the coldest to the warmest periods during the Pleistocene Epoch.

Data of several cores must be averaged, because we expect that there were many local variations due to controls other than sea-surface temperature. Figure 26.10 shows a generalized curve of water temperature interpreted in 1968 by scientists of the Lamont-Doherty Geological Observatory of Columbia University, using deep-sea cores from the equatorial and tropical zones of the Atlantic Ocean (Ericson and Wollin, 1968). In this case, three foraminifera subspecies of the species Globorotalia menardii were counted as a group in relation to the total plankton content of the sample. A low count was interpreted as indicating cold water and possible glaciation, a high count as indicating warm water and a possible interglaciation. Also shown are the magnetic polarity scale and the corresponding age scale in millions of years before present.

A second method makes use of a single species of foraminifera in a rather remarkable way. The test of this species is coiled about an axis. Some of the tests show a left-hand direction of coiling and others a right-hand direction. It has been established that left-coiling tests are dominant in periods of cold water, while right-coiling tests are dominant in periods of warm water. Data for use in the two methods of foraminifera analysis can be gathered from the same deep-sea core and should show a good degree of correspondence within the same core.

In other core studies, the abundance of assemblages of radiolarian species or of a single radiolarian species have

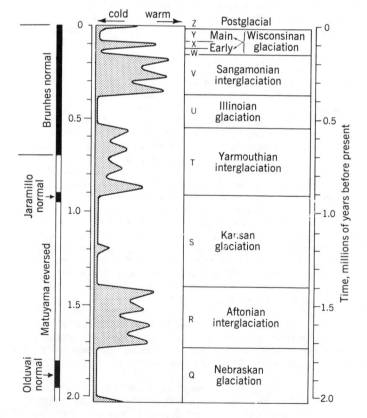

Figure 26.10 Generalized curve of ocean-water temperatures interpreted from data of foraminifera in deep-sea cores. (Based on data of D. B. Ericson and G. Wollin, 1968, Science, vol. 162, p. 233, Figure 7.)

been used to produce a curve of sea-surface temperature. The lower curve shown in Figure 26.11 is an interpretation of summer sea-surface temperature based on a statistical analysis of assemblages of radiolarians. The data come from two cores taken at about latitude 45°S at points located about intermediate between Africa, Australia, and Antarctica.

Interpreting Glacial Cycles from Oxygen Isotopes

Independent evidence of cycles of glaciation comes from an analysis of the ratio of abundances of isotopes of oxygen. In addition to the common form, oxygen-16, there exist two heavier oxygen isotopes, oxygen-17 and oxygen-18. In 1947, Nobel Prize chemist Harold C. Urey noted that the ratio of oxygen-18 to oxygen-16 (O-18/O-16) in ocean water depends partly on water temperature. He then reasoned that the ratio of those isotopes in the carbonate shell matter of marine organisms should reflect the surrounding water temperature at the time that matter was secreted. Thus, changes in water temperature should be reflected in changes in the oxygen-isotope ratio. Through inprovement in laboratory techniques, it became possible in ensuing years to measure very small differences in oxygen-isotope ratios and to interpret these differences.

The oxygen-isotope method was applied in the late 1950s to foraminifera tests by Professor Cesare Emiliani, who had begun his research under Professor Urey's direction. Emiliani translated the scale of oxygen-isotope ratios directly into a scale of water temperatures, called paleotemperatures. The result of his study was to reveal a paleotemperature curve with a large number of peaks and valleys--far more than would fit the classical four established continental glaciations. Emiliani first estimated that about eight climatic cycles, each representing a glaciation and an interglaciation, occurred following the last reversal of magnetic polarity, i.e., back to about -700,000 y.

Emiliani's paleotemperature analysis came under criticism of other scientists who pointed out that water temperature is only a small factor in determining the oxygen-isotope ratio of seawater. The true significance of variations of the isotope ratio in the tests of plankton, such as the foraminifera, lies in a more complex but nevertheless quite straightforward chain of events involving the hydrologic cycle. As Professor Urey had pointed out in 1946, when water evaporates from the sea surface, there is a tendency for a larger proportion of those water molecules containing the light isotope of oxygen, O-16, to enter the vapor state than of the molecules containing the heavier isotope, O-18. This selective process tends to leave the remaining seawater richer in O-18; in other words, the ratio O-18/O-16 tends to become larger. On the other hand, when the global hydrologic cycle functions in a state of balance, the precipitation of water on the continents and oceans and the runoff of water from continents to oceans assures that the rate of return of lighter isotopes to the oceans as liquid water will balance the quantity leaving by evaporation. Under this state of global water balance the quantity of water stored on the lands in both liquid and ice forms will hold constant, and so will the oxygen-isotope ratio of the seawater.

Suppose, now, that the total quantity of glacier ice begins to increase because of a climate change that favors the expansion of glaciers. As ice volume grows, more of the lighter oxygen isotopes are withheld in the ice and prevented from entering the return flow of the hydrologic cycle, resulting in an increase in the value of the ratio O-18/O-16 of seawater. On the other hand, when a climate change results in a decrease in the total quantity of glacier ice, more of the lighter isotopes are released as runoff, and the isotope ratio of seawater becomes smaller.

The oxygen-isotope curve determined from carbonate matter in deep-ocean cores is now regarded as a reliable indicator of the total volume of glacier ice present on the earth at the time the plankton secreted their tests. For this reason, the isotope-ratio curve can be called a paleoglaciation curve. Low points on the curve are associated with glaciations, while the high points are associated with interglaciations.

The oxygen-isotope method has been applied to many deep-sea cores, and the result has been a recognition of a succession of isotope-ratio stages extending back as far as about -2 m.y., and perhaps even to -3 m.y. Stages within the 2-m.y. time span are numbered from 1 to 41. An example is an isotope-ratio curve, or paleoglaciation curve, published in 1976 by a team of scientists representing several universities (Figure 26.11). It covers 500,000 years and is based on two deep-sea cores from the subantarctic ocean in a location between Africa, Australia, and Antarctica (Hays, Imbrie, and Shackleton, 1976). The record shows the first 13 isotope stages. If you smooth out the smaller wiggles in both curves, perhaps you will conclude that there were five glaciations in the 450,000 y. of record, for an average of about one glaciation every 90,000 y.

From this evidence it appears that glaciations may have been occurring as far back as -3 m.y., or well into the Pliocene Epoch, and that the total number of Cenozoic glaciations may have numbered more than 30, spaced at intervals of about 90,000 y.

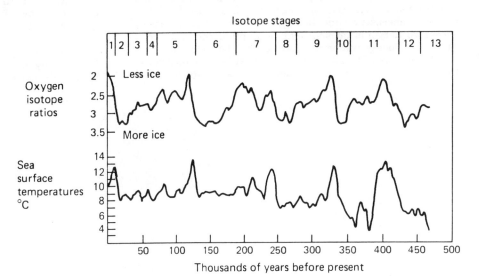

Figure 26.11 Curves of oxygen-isotope ratios (above) and inferred summer sea-surface temperatures (below) from two deep-sea cores in the subantarctic ocean. (Based on data of J. D. Hayes, J. Imbrie, and N. J. Shackleton, 1976, Science, vol. 194, p. 1130, Figure 9.)

Evidence of Earlier Cenozoic Glacial Ice

Deep-sea cores carry yet another kind of evidence of the presence or absence of great ice sheets on the continents. When a large ice sheet is present, sediment is carried out over the adjacent deep ocean by icebergs that melt and allow the mineral fragments they carry to be dropped to the ocean floor. In the 1960s, cores taken in the southern Indian Ocean showed the presence of glacial-marine sediments as old as -4 m.y., and it was then realized that the Antarctic continent bore an ice sheet long before the Pleistocene Epoch began. In 1973, drill cores taken near the Antarctic coast were found to contain ice-rafted materials dated at about -20 m.y., which is early in the Miocene Epoch. More recently, ice-rafted sediments in cores in that area have been dated at about -38 m.y., which is at the start of the Oligocene Epoch. These earliest ice-rafted materials may have come from alpine glaciers on Antarctica, rather than from an ice sheet. It has been guessed that the Antarctic Ice Sheet began to build up between -11 and -14 m.y.

Oxygen-isotope ratios from deep-ocean sediment cores show a rather definite shift from generally nonglacial climate to strongly and persistently glacial climate between -13 and -17 m.y., with the most rapid part of the shift occurring between -14 and -15 m.y. (Woodruff, Savin, and Douglas, 1981; Kerr, 1982, p. 974). Thus, for at least one group of research scientists, the tentative conclusion is that the late-Cenozoic Ice Age began around -14 to -15 m.y., which is in the middle of the Miocene Epoch. In any case, association of the Ice Age with the Pleistocene Epoch, starting at about -1.5 m.y., has long been abandoned by mainstream science. Increasing the duration of the Ice Age by a factor of about 10 greatly increases the stress upon the creation scientists, who must compress the events of 15 m.y. into 4,000 y. of post-Flood time. During the Miocene Epoch, the Antarctic Ice Sheet apparently underwent many changes in volume and may even have largely disappeared in interglaciations of exceptionally mild climate. Drastic changes in the volume of ice held on Antarctica and Greenland would have caused large swings in worldwide sea level, alternately inundating and exposing low coastal areas of all continents.

Deposition Rates of Deep-sea Oozes

We have already pointed out the near-impossibility of explaining certain sedimentary accumulations within the one-year limit of the Flood or within the 4,000-y. limit of post-Flood time. Without resorting to absolute-age determinations, which the creation scientists simply declare to be false, a basic argument can be made that certain physical and biological processes proceed at rates constrained by environmental conditions and by laws of physics, chemistry, and biology. When creationists make excessive demands on sediment production rates, their case is seriously undermined, often to the point of requiring outright rejection. One example we covered is the limitation on rate of production of salt beds by

evaporation of seawater; another is the limit upon rate of biomass production in the case of reef corals. Let us use this line of reasoning with respect to the rate of deposition of deep-sea cores.

In a study published in 1975, highly detailed analyses were made of three deep-sea cores obtained in the southeast Indian Ocean between Australia and Antarctica (Blank and Margolis, 1975). The cores, on the order of 5 m in length, consist primarily of siliceous ooze, made up largely of the skeletons of radiolarians. The sample area lies between the Antarctic divergence on the south and the Antarctic convergence on the north at latitude 55° to 60°S. This is a belt of high-level productivity of radiolarians. Oceanic divergence, accompanied by rise of deep water to shallow, sunlit depths, provides the nutrients required for silicious plankton to thrive. One such belt is a latitudinal zone straddling the 60th parallel south.

One of the cores in the study referred to above is about 5 m in length. Dated by paleomagnetic methods, it covers a time span of 2 m.y. The average rate of deposition is thus approximately 0.25 cm per 1,000 y. This rate compares favorably with those of Arctic Ocean cores, which have been measured as 0.1 to 0.3 cm per 1,000 y. (Herman and Hopkins, 1980, p. 557). If the same 5-m core were deposited in a post-Flood period of roughly 4,000 y., the average rate would need to have been 125 cm per 1,000 y., which is 500 times faster than the rate determined by mainstream geologists! A production rate 500 times greater than the rate measured by mainstream geology is required. Is such a production rate reasonable, or even possible in view of the obvious limits to the rate of reproduction and growth of the radiolarians? The rate of nutrient supply would need to be increased by a factor of 500, but this would require a similar increase in rate of upwelling, and with it the overall rate of mixing of water masses. The post-Flood scenario makes truly outrageous demands on environmental factors or organic growth rates, or both--demands that are unacceptable to science. The creation scientists would be calling upon catastrophism to accomplish the accumulation of radiolarian oozes found in such cores. Undoubtedly, at the same localities, radiolarian sediments extend down to much greater depths and older ages. All such fossiliferous deep-ocean sediments, according to the creationists, belong to either the one-year Flood or the 4,000 y. post-Flood period. Their collective occurrence requires a level of catastrophism totally outside the realm of natural science.

Credits

1. Much of the descriptive text of this chapter is taken from A. N. Strahler, Physical Geology, Chapter 18, Harper & Row, Publishers, Inc., New York. Copyright © 1981 by Arthur N. Strahler.

2. From Glacial and Quaternary Geology, by Richard F. Flint. Copyright © 1971 by John Wiley & Sons. Reprinted by permission of John Wiley & Sons.

Chapter 27

More Evidence of a Lengthy Ice Age

New evidence gathered in the 1970s and 1980s from terrestrial localities has provided corroborative evidence of multiple glaciations. These events correlate rather well with glaciations or cold periods inferred from deep-sea cores.

Evidence from Mauna Kea Volcano

One such line of evidence comes from the island of Hawaii, on which great basaltic lava domes, or shield volcanoes, have been built. Of these, Mauna Kea is the highest, 4,200 m summit elevation; it has produced no volcanic activity within the past 2,000 y. During the Wisconsinan glaciation Mauna Kea bore an icecap from which ice tongues extended down the slopes as short valley glaciers, leaving moraines and the usual evidences of glacial abrasion in the form of striations and grooves on bedrock surfaces (Macdonald and Abbott, 1970, pp. 234-37). It is the only known glaciated mountain summit in the central Pacific Ocean basin. Good field evidence of multiple glaciation of Mauna Kea has been uncovered in the form of glacial deposits (till with striated boulders and marginal lake sediments) overlain by volcanic rocks (tephra and lava flows). Radiometric ages have been obtained from the volcanic materials above and below the glacial materials, bracketing the latter in time (Porter, Stuiver, and Yang, 1977). Both potassium/argon and carbon-14 methods were used, the former on volcanic rocks, the latter on buried soils. These show glaciations culminating in icecap formation at about minus 20,000, 55,000, 135,000, and 250,000 y. These glacial events are correlated with isotope stages 2, 4, 6, and 8 from deep-sea cores. Because an icecap can form only when the snow line is lowered many hundreds of meters below its present level, the glacial deposits probably coincide with the maximum spread of continental ice sheets.

Needless to say, the compression of this history of great volcanic accumulation and its interspersed episodes of glaciation into 4,000 y. of post-Flood time stresses the creationists' theory beyond acceptability.

Evidence from Pluvial Lakes

Another line of corroborative evidence of multiple glaciation comes from a terrestrial locality far removed from the regions covered by ice sheets. Great Salt Lake in Utah lies at the western foot of the Wasatch range, a block mountain uplifted in Cenozoic time. During glaciations, the level of the lake rose greatly because of generally increased precipitation and lowered air temperatures. In reference to these enlarged stages, the water body bears the name of Lake Bonneville. It was only one, albeit the largest, of about 120 pluvial lakes that occupied the floors of tectonic basins in the Basin and Range Province. The adjective "pluvial" suggests the increase in depth and areal extent of the lakes as a result of a net gain in water input. Rise of the pluvial lakes was synchronous with the growth of alpine glaciers and icecaps on the Wasatch range and other mountains of the

region, such as the Sierra Nevada. As the level of Lake Bonneville rose, glaciers occupying troughs on the western face of the Wasatch range thickened and extended to lower elevations, where they emerged upon the sloping piedmont plain. At its high stages, the lake shoreline stood against the lower mountain slopes, at which time wave action carved a notch into the bedrock of the mountainside. Wave action also constructed deposits of sand and gravel at the water level in the form of shoreface embankments, deltas, bars, and spits. Glacial ice advanced over the lake sediments, leaving superposed morainal material as a record.

Extensive research on these deposits was carried out in the 1960s and 1970s by Roger B. Morrison of the United States Geological Survey. From field examination of exposures and from drill cores the sequence of synchronous rises in lake level and advances of glaciers was deciphered. Figure 27.1 is a schematic diagram in which time is scaled on the vertical axis; elevation on the horizontal axis (Morrison, 1965). It shows six major glacial advances associated with six rises in lake level, all within Wisconsinan and Holocene time. Morrison was then able to extend the record far back in time through examination of drill cores penetrating older sediments (1975). Altogether, 28 major cycles of rise in lake water with glacial advance were documented, going back as far as -800,000 y. Ten of these cycles are older than -500,000 y. Strong, well-defined glaciations were found at approximately minus 125,000, 200,000, 300,000, 400,000, and 440,000 y. (See also Science News, 1975, vol. 108, p. 296.)

Creation scientists are forced to compress the entire 800,000-y. Lake Bonneville scenario into 4,000 y. of post-Flood time. This speeds up the action by a factor of 200. Whereas mainstream science allows an average of over 28,000 y. for each cycle of glaciation and interglaciation, creation science allows about 140 y. for each cycle. In a cycle of such short duration, an enormous rate of net snow accumulation must have prevailed for several consecutive decades, and this would have been followed by some decades of extremely intense evaporation to dispose of both the glacial ice and the lake water. We are asked to believe that such enormously accelerated rates of net gain and net loss could have occurred alternately in conformity with known principles of meteorology and hydrology. Rates of glacial advance and recession vastly greater than any documented in historic time would have been required. Moreover, we are asked to believe that 28 such cycles occurred! The creationists have offered us a scenario of catastrophism far removed from natural science; indeed, they have offered us pseudoscience.

Isostatic Rebound and Deglaciation

Evidence of multiple glaciation, with the possibility of as many as 30 glaciations suggested by the oxygen isotope ratios in deep-sea cores and cores of the Lake Bonneville deposits, is so well corroborated by mainstream scientists that it cannot be brushed off cursorily by the creation

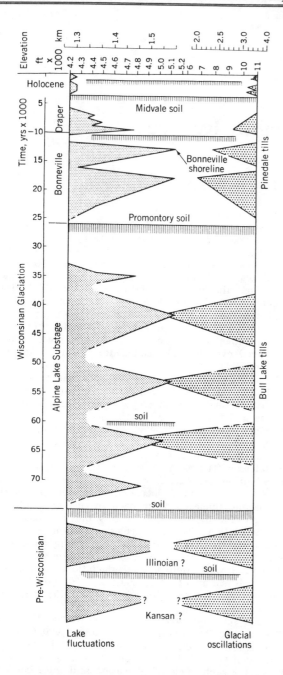

Figure 27.1 Fluctuations in Lake Bonneville levels (left) and corresponding episodes of glacial till deposition in the Wasatch Mountains (right). Note that two elevation scales are used. (Based on data of R. B. Morrison and J. C. Frye, 1965, Nevada Bureau of Mines, Report 9, Figure 2.)

this principle applies to the growth and disappearance of an ice sheet or a large inland lake, such as Lake Bonneville.

American geologist Grove Karl Gilbert made the first major field survey of the relict features of Lake Bonneville; the results were published in 1890 in a monograph of the U.S. Geological Survey. Gilbert's survey party determined the elevation of two of the high shorelines. It might be expected that the the height of each shoreline above the present lake surface elevation would be the same at all points throughout the area; but this proved not to be the case. The highest shoreline--called the Bonneville shoreline--is higher in elevation near the former center of the ancient lake than in the peripheral zone, the difference being on the order of 60 m. (The shoreline can be found carved into islands that occupied the central part of the lake in its highest stages.) Clearly, the surface that was formerly a high water-level plane (disregarding the earth's surface curvature) has been warped into a domelike surface. Gilbert ruled out the possibility that the up-doming could have been caused by tectonic processes. He proposed, instead, the hypothesis that is still favored today and strongly supported by physical theory. The weight of the water mass of Lake Bonneville was sufficient to cause the crust beneath the lake floor to subside, as called for by the principle of isostasy (explained in Chapter 20). This subsidence had been largely accomplished by the time the Bonneville shoreline became established as a stable water level, during which time the wave-cut notch and its related features were formed. As the lake began to shrink, the crust began to rise. With the water layer entirely gone and its load removed, the updoming largely restored the previous downwarp, but as a result, the Bonneville shoreline had been deformed into a domelike configuration. Later, the shoreline elevations were carefully resurveyed by the Geological Survey; the data are shown in Figure 27.2, a map using elevation contours to depict the upwarping (Crittenden, 1963).

Isostatic downwarping and subsequent recovery has been identified on a much grander scale in two major centers of ice sheet accumulation. One of these is the former site of the Laurentide Ice Sheet, centered about on what is now Hudson Bay and the Laurentian Highland immediately to the east. The other is the former site of the Scandinavian Ice Sheet, centered on what is now the Gulf of Bothnia. In both cases the ice load depressed the crust into saucerlike basins that became flooded by the ocean as sea level rose. Along the shores of these epicontinental seas, wave action marked the new shoreline with a wavecut notch or a ridge of cobblestones (a beach ridge). Crustal rebound set in immediately following the rapid thinning and final disappearance of the ice, so that each shoreline marker (called a "strandline") was lifted above the limits of wave action, becoming an elevated strandline. Large numbers of strandlines lie abandoned in the coastal belt of Hudson Bay, and elevated strandlines can be identified in the lands surrounding the Gulf of Bothnia.

Postglacial uplift of the Baltic region had been recognized in the mid-1700s, but until the glacial theory was proposed in the mid-1800s no suitable explanation was forthcoming. By 1860 a Scottish geologist, T. F. Jamieson, had identified upraised shorelines in his country and applied the principle of isostasy (first proposed in 1855) to the problem. He suggested that the crust is weak and flexible, allowing it to subside under the ice load, and to rise again after the ice had disappeared. The explanation became universally adopted and, as we have seen, was tested by Gilbert in the analogous case of Lake Bonneville. Geophysicists came forward with evidence substantiating the hypothesis. One of these was the distinguished seismologist, Beno Gutenberg, who in 1941 offered the following lines of support, as restated by R. F. Flint:

scientists. As we have shown, a large number of glaciations and interglaciations cannot be crammed into the early part of post-Flood time, which totals no more than about 4,300 y. When the creationists appeal to catastrophism, perhaps lingering after the Flood, to speed up the growth and disappearance of ice sheets and alpine glaciers after the manner of a time-lapse movie, their argument loses credibility because meteorological and hydrological phenomena must operate within reasonable limits, conforming with laws of physics. But there is yet another physical constraint to be considered in cramming numerous glacial cycles--or even one--into the first one or two millenia of post-Flood time. The earth's crust moves up or down as required by the principle of isostasy when mass is superposed at the surface or removed. Let us see how

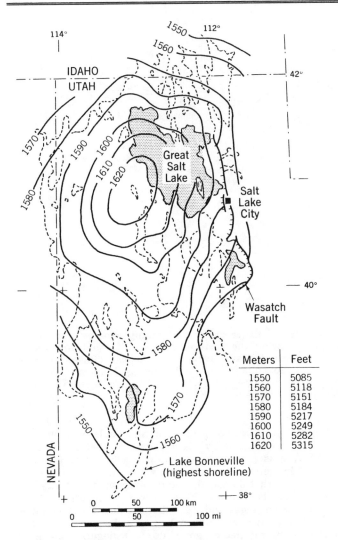

Figure 27.2 Deformation of the highest shoreline (dashed line) of Lake Bonneville, Utah, shown by elevation contours in meters. (After M. D. Crittenden, Jr., 1963, U.S. Geological Survey, Professional Paper 454-E.)

Meters	Feet
1550	5085
1560	5118
1570	5151
1580	5184
1590	5217
1600	5249
1610	5282
1620	5315

1. In both Fennoscandia and North America the outer limits of the upwarped region parallels the limit of the latest glaciation. 2. In both regions the isolines (lines connecting points on any strandline that have been equally uplifted) are concentric to the area in which, according to independent evidence, the former ice was thickest late in the process of deglaciation. 3. In both regions antecedent downwarping is suggested by the presence of marine deposits now bent up above sealevel, and in turn overlying subaerial deposits. 4. In both regions rate of uplift is of the same order of magnitude. During the last few thousand years the rate of uplift of Fennoscandia has slowed down from an earlier, much greater rate. 5. In both regions even the incomplete data obtained thus far show that the gravity anomalies are negative, and that they increase toward the central parts of the glaciated areas, indicating that crustal equilibrium in these regions has not yet been reached. 6. In other regions of former glaciation or former greater glaciation, such as Britain, Greenland, Svalbard, Novaya Zemlya, Siberia, Patagonia, and Antarctica, where postglacial upwarping is expectable, evidence of it has been observed. (1971, p. 343)[1]

Point 5 needs special explanation. Although the crustal downwarping is in part elastic in nature, like the flexing

of a steel spring, the later deformation is accomplished by plastic flowage in the soft asthenosphere. This flowage is directed radially outward from the area of loading into the unloaded peripheral zone, where the crust is actually forced to rise very slightly, as shown in the deformation curve in Figure 27.3. Geophysicists W. A. Heiskanen and F. A. Vening Meinesz describe what happened after the ice sheets were gone:

At the present time the flow of the subcrustal material should be directed toward the center. Since all of the mass that flowed outward during the glacial period has not had time to move back, there is too little mass under the uplift, and the gravity anomalies are therefore negative, being as high as -50 milligals in the center of the area. The existence of negative anomalies in the uplift area is of great significance. If the subcrustal masses had moved only vertically downward during the glacial period and upward later on, the mass under the glacial areas would not have changed, so that there would be no reason for the systematically negative gravity anomalies. The existence of the negative gravity field in this area is evidence of the horizontal movement of the subcrustal masses. (1958, p. 212)

The same authors go on to cite the opinion of another geophysicist, E. Niskanen, that in the Baltic region the land has yet to rise about 200 m before complete isostatic equilibrium prevails (Niskanen, 1939). It is not expected that the entire amount of recovery will be achieved, but recovery will continue through perhaps as long as 15,000 y. (Flint, 1971, p. 349).

Figure 27.4 is my rather crude attempt to diagram the program of glacio-isostatic crustal movement during a cycle of glaciation followed by a rapid deglaciation. (Purely elastic deformation is not shown). Both downwarping and upwarping are taken to be represented by exponential or logarithmic curves--rapid change at first, then slowing continuously to a much lower rate. This type of curve is indicated by the field evidence consisting of elevation measurements of strandlines and carbon-14 age fixes (Farrand, 1962; Flint, 1971, p. 345, Figure 13-1). Modern rates of uplift have been obtained by two programs of precision geodetic leveling in Finland, separated by a period of 40 to 50 y. (Heiskanen and Vening Meinesz, 1958, p.212). They show a maximum rate of 0.9 m per 100 y. On the Swedish side, the rate may be as high as 1.0 m/100 y.

Closely related in principle is another geodetic property clearly showing that isostatic recovery is far from being achieved in the North American and Fennoscandian areas: the configuration of the geoid. A few words of explanation are in order here. Although to a first approximation, the earth is a true sphere, a much better and second approximation is that the earth's figure is that of an ellipsoid that we can envision as a sphere deformed by shortening the polar axis and increasing the

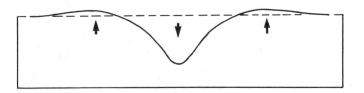

Figure 27.3 Idealized diagram of deformation of the earth's crust under a concentrated vertical load. The vertical scale is greatly exaggerated. (Based on data of W. A. Heiskanen and R. A. Vening Meinesz, 1958, The Earth and Its Gravity Field, p. 332, Figure 10A-3, McGraw-Hill Book Co., New York.)

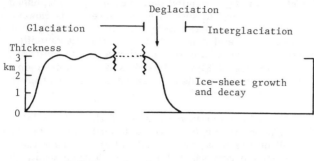

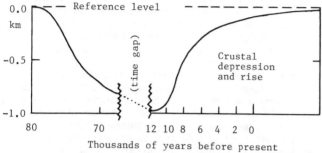

Figure 27.4 Schematic graph of the relationship of isostatic response to growth and disappearance of an ice sheet. Elastic deformation and recovery are not shown. (A. N. Strahler.)

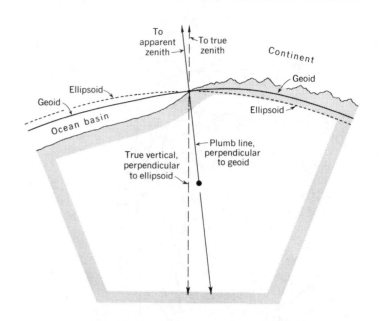

Figure 27.5 Diagrammatic and highly exaggerated cross section showing the relation between ellipsoid and geoid under continents and over ocean basins. (Based on data of W. A. Heiskanen and F. A. Vening Meinesz, 1958, The Earth and Its Gravity Field, p. 237, Figure 8-5, McGraw-Hill Book Co., New York.)

equatorial circumference. Exact dimensions can be set down for the earth ellipsoid, which is merely an abstraction in any case. Going one step further, we can observe the level of the surface of the ocean as it really is, and somehow do away with the effects of winds, tides, currents, and other disturbing effects. That leaves us with a surface realistic as a description of the shape of the earth--the geoid. We then imagine a network of sea-level canals (or better yet, sea-level tunnels), crisscrossing the continents. With this idealized device in mind, the geoid can be ascertained and mapped for the entire globe. Figure 27.5 shows the typical relationship between geoid and ellipsoid. Over the deep oceans the geoid usually lies beneath the ellipsoid, because of the deficiency of crustal mass. Over the continents the geoid usually lies above the ellipsoid, because the rock mass above the level of the ellipsoid exerts an upward attraction that raises the geoid. (For background reading in gravity and geodesy see Strahler, 1971, Chapter 10, The Earth's Figure and Gravity.)

Today, the complete geoid can be calculated from gravity data obtained by highly sophisticated orbiting satellites; it is presented as a map consisting of geoid contours in meters above or below the chosen ellipsoid of reference. With this map in hand, examine the geoid contours for the two ice sheet centers in question: Laurentide Ice Sheet and Scandinavian Ice Sheet (Figure 27.6). Centered over Hudson Bay is a deep closed depression of the geoid, going down to negative elevations of more than -48 m. Centered about over Finland is a deep trough, opening out to the northeast, in which the geoid surface plunges deeply. Although not a closed depression, it is nevertheless a true depression, shaped like a hand-scoop used to dispense candies or nuts. Returning to the North American case, consider the meaning of the deep geoidal depression centered where the ice sheet was formerly located. Like the negative gravity anomaly, it means that a mass deficiency exists. How could this deficiency be restored and the geoid surface become raised to zero elevation? The answer is by

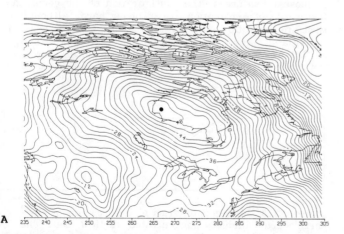

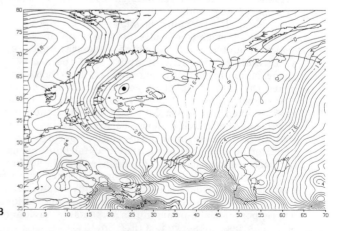

Figure 27.6 Portions of a world map of the global gravity geoid, based on satellite data, and showing depressions centered over the former sites of the Laurentide (A) and

Scandinavian (B) ice sheets. (NASA Goddard Space Flight Center.)

completion of the full process of crustal rebound, and this will require many thousands of years.

The whole point of this lengthy discussion of elevated strandlines, gravity anomalies, and geoidal depressions is to point out a time requirement far exceeding the narrow limits of post-Flood time. As Figure 27.4 shows in a highly schematic manner, the full isostatic recovery process must have been preceded by an equally lengthy process of isostatic downbending in which the dense, highly viscous material of the soft layer of the mantle was forced out from beneath the great ice sheets and then forced to return to its original location. To fit even a single cycle of glaciation into the first 1,000 or 2,000 y. of post-Flood time would require a drastic reduction of the viscosity of the moving mass, something completely removed from serious consideration because of the actual conditions of depth, rock density, and temperature that must have prevailed then, as now. Although the schedule worked out by mainstream geology is secured by carbon-14 age determinations, its long duration rests upon the laws of physics of deformation of matter and the constants that must be fed into computations based on those laws. Creation scientists cannot ignore such constraints and expect to be accepted as practitioners of natural science. For creationists to insist on their grossly inadequate schedule is to attempt to force a program of pseudoscience on students and teachers.

Elevated Reef Shorelines

Creation scientists point to features they consider evidence of subsiding Flood water continuing into early post-Flood time to the accompaniment of diminishing volcanic and tectonic activity. Whitcomb and Morris write:

> Glaciation was only one of the after-effects of the Deluge, though undoubtedly the most spectacular. Although the Pleistocene Epoch is generally thought of as the Glacial Period, there is much evidence of continuing catastrophic activity of other kinds.

> Evidently the tectonic and volcanic disturbances which played such a large part in the initiation of the Flood, as well as in the uplift of the land at its close, continued with only gradually-lessening intensity for many centuries thereafter. (1961, pp. 311-12)

The authors refer to post-Flood activity of this kind as residual catastrophism (p. 313). Elevated strandlines of Lake Bonneville and higher marine strandlines in regions occupied by the major ice sheet centers that we have already discussed are products of residual catastrophism. Creationists view these features as an expected consequence of either subsiding Flood water or tectonic uplift of the continents or of both processes acting simultaneously. To these kinds of strandlines they add evidences of numerous glacial lakes within the regions covered by the ice sheets or impounded between the ice sheet margins and adjacent rising ground. Former shorelines and outlets of the Great Lakes during and after late Wisconsinan time are important examples of such features.

An entirely different class of marine shoreline features produced during Pleistocene time is found on coasts far removed from former ice sheets. Examples abound on the coast of California and its offshore islands, where flattened marine benches are often stacked one above the other like flights of stairs. A particularly striking example is the succession of more then 20 marine terraces on the western side of San Clemente Island, the highest lying about 400 m above sea level. Whitcomb and Morris include a photograph of a remarkable succession of marine terraces on the north coast of New Guinea, on the Huon Peninsula, near Finschaven (1961, p. 316). This locality

lies in the equatorial zone (lat. 6°S.), where coral reefs form on shorelines exposed to vigorous wave action. Consequently, the elevated shorelines bear former coral reefs that consist of coral and algal carbonate rock and can thus be dated by carbon-14 and thorium/uranium methods.

Creation scientists regard this class of elevated marine shorelines as a feature of residual catastrophism, perhaps resulting from a combination of subsiding Flood waters and tectonic uplift. Mainstream science invokes a combination of two processes: cyclic rise and fall of sea level during deglaciations and glaciations, respectively, superposed on a more-or-less steady tectonic uplift to be expected on active continental margins as a consequence of subduction processes.

Glacial-Eustatic Changes in Sea Level

Alternate growth and disappearance of great continental ice sheets had profound effects on the level of the world ocean--changes that are described by the adjective eustatic. Eustatic changes can be understood by a simple model. Suppose we fill a deep tub or basin about half-full of water to represent the world ocean. We can change the level of the water surface in one of two ways. First, we can remove some water, using a dipper, and causing a drop in water level. This is analogous to what happens to sea level when ocean water is withdrawn to provide glacial ice. We can return the water from dipper to tub, representing the melting of the ice sheets. Second, we can gently place a brick on the bottom of the tub, which will raise the water level; removing the brick restores the water level to its former position. This second model of eustatic change of sea level illustrates a tectonic cause. Under plate tectonics, a doming of the ocean floor would cause a rise of sea level. Doming is to be expected when the rate of seafloor spreading is increased, causing a broad zone adjacent to the spreading plate boundary to become elevated, because the new lithosphere is comparatively hotter and less dense than before.

The creation scientists can argue that the ocean basins subsided following the Flood because the crust became cooler and denser as that great cataclysm subsided. Under that argument, eustatic change in sea level would be negative--a falling sea level, that is. It would leave a succession of marine terraces, giving the appearance of a rising coastline, even if no crustal rise affected the coastal zone.

The amount of sea-level change resulting from total melting of large ice sheets is calculated from estimates of the volume of those ice sheets. The Antarctic Ice Sheet alone holds sufficient water volume to provide a sea-level rise of about 60 m, should all of that ice be melted. Assuming that the added load of this water upon the oceanic crust would cause an isostatic downwarping of 20 m (about one-third the added water depth), the net sea level rise would still be about 40 m. The disastrous effect of a 40-m rise on the heavily populated lowlands of the Atlantic and Gulf coastal plains and along coasts the world over has been a favorite theme of journalists and fiction writers; it can scarcely have escaped your attention. For scientists, a more interesting topic is the lowering of sea level that must have accompanied each glaciation.

Sea-level lowering in the last major advance of ice sheets in the Wisconsinan Glaciation and its subsequent rapid rise in Holocene time are quite well documented. Evidence consists of radiocarbon dates of such materials as salt-marsh peat, oyster shells, coral rock, and lithified beachrock (carbonate-cemented beach materials). One recent study, summarizing a large amount of data collected by several scientists, estimates a maximum of lowering along the east coast of the United States of about 90 m, occurring about -17,000 y. (Dillon and Oldale, 1978, p. 59). An extremely interesting finding comes from an underwater cavern near Andros Island in the Bahamas. Here stalagmites were found in a "blue grotto" at a depth of 45 m. Radiometric analysis using uranium/thorium ratios yielded an age between -160,000 and -140,000 y. that is thought to correspond with the Illinoian Glaciation (Gascoyne et al., 1979). For a limestone cavern to have formed and subsequently to have accumulated speleothems, the sea level must have stood at a considerably lower level. Conservative opinion would perhaps be that maximum sea-level lowering throughout the Ice Age was probably on the order of 60 to 100 m. Assuming that the oxygen-isotope curve (Figure 26.11) gives a direct measurement of ice volume, that same curve should also indicate the glacio-eustatic fall and rise of sea level in terms of both timing and general magnitude.

And now, returning to the multiple raised coral reefs of the New Guinea coast, we can analyze the effects of a combination of cyclic glacio-eustatic change in sea level with a continuous uplift of the coastal region as a tectonic phenomenon.

Tectonic Uplift and Reef Shorelines

In a study published in 1970, H. H. Veeh and John Chappell made a detailed profile and geologic cross section of a succession of five New Guinea terraces, from sea level to a height of nearly 200 m (Figure 27.7). After each reef was constructed, sea level dropped, allowing marine erosion to act on the outer base of the reef, partially cutting it away. Sea level then rose and the rise of water level was accompanied by reef growth. The cycle was repeated five times, but because the crust was rising more or less steadily, each reef crest was preserved. The diagram shows by arrows the cyclic rise and fall of sea level. Figure 27.8 shows sea level plotted against time. Dots give the location of sample points at which age of the reef rock was determined by either carbon-14 or thorium/uranium ratio. The slanting dashed line is fitted to represent the crustal rise. In the lower diagram, the crustal rise has been removed, leaving only the cyclic glacio-eustatic curve, each low point representing a maximum of ice accumulation. The high stands represent interglaciations, of which five seem to be clearly defined, including the present condition.

A previous study of a succession of elevated coral reefs on the island of Barbados determined the ages of three reefs as -82,000, -103,000, and -122,000 y. (Broecker et al., 1968) These ages are shown by arrows (A, B, C) in the lower diagram. Notice that the correlation with the three oldest of the New Guinea reefs is quite close. In recent years similar studies of the Huon, New Guinea, reefs have extended the correlation with the Barbados reefs back to -170,000 and -230,000 y. (Chappell, 1974, p. 566 and Figure 16). Further corroboration was obtained from elevated reefs in Timor and a nearby island, Atauro (Chappell and Veeh, 1978). Good correlation exists there between the elevated reef cycles and the oxygen-isotope curve of paleotemperatures based on deep-sea cores (Shackleton and Opdyke, 1973). The highest reef on Atauro Island, now 7,300 m above sea level, is estimated by extrapolation to have an age of about -700,000 y.

Clearly, mainstream geology presents a consensus on an Ice Age schedule going back more than 200,000 y., but the creation scientists reject that schedule out of hand on the grounds that the radiometric ages are false. Again, as in our earlier analysis of coral reef deposits on sunken volcanic islands (atolls), the sheer bulk of the elevated reef deposits effectively rules out the possibility of their having been formed in post-Flood time. Using John Chappell's data on the Huon terraces, I counted 17 elevated reefs, each of which must contain on the average at least 20 m of coral rock in the vertical dimension (Chappell, 1974, figures 14 and 15). Thus we must reckon with the upward growth of a total of at least 340

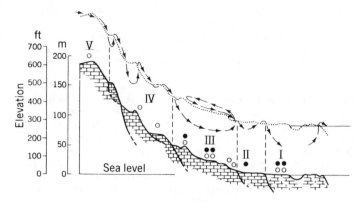

Figure 27.7 Cross section of coral reef deposits and terraces of the northeast New Guinea coast. Arrows on upper profile show inferred positions of shoreline during falling and rising of sea level. Solid circles indicate carbon-14 dates; open circles, thorium-230 dates. Compare with Figure 27.8. (Based on data of H. H. Veeh and J. Chappell, 1970, <u>Science</u>, vol. 167, p. 863, Figure 1.)

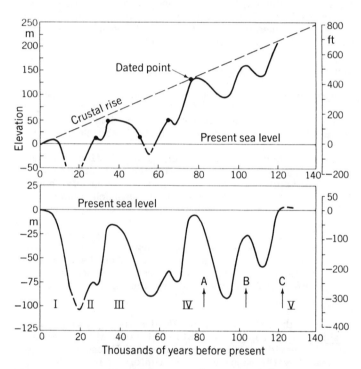

Figure 27.8 An interpretation of the data of Figure 27.7. Inferred sea level changes (above) are subtracted from estimated uniform crustal rise. Elevation difference (below) is attributed to glacio-eustatic sea-level change. (Data of upper diagram based on reference cited for Figure 27.7, p. 864, Figure 2.)

m of coral reef. Using a growth rate of 5 m per 1000 y. (five times faster than the rate of 1 cm/yr used in Chapter 23), 68,000 y. of coral growth is called for. This figure is larger by a factor of 15 than the 4,300 y. of post-Flood time. The creationists' schedule calls for a rate of upward coral growth 15 times faster than the "uniformitarian" rate. If reasonable time allowance is made for marine erosion that has obviously occurred in periods between reef building, the growth rate would need to be raised to a factor of perhaps 30 or more. Only a single layer of living coral polyps can occupy the reef surface. Their rate of calcium carbonate production is limited by environmental factors and nutrient supply. To postulate an enormously higher rate of biological activity goes beyond the limits of scientific hypothesis. Again, we find

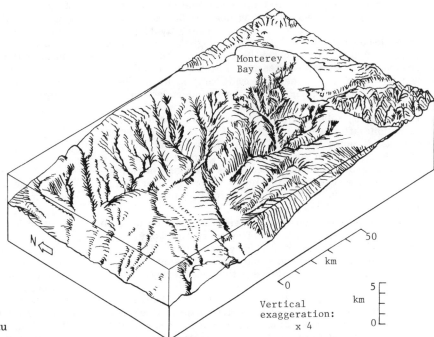

Figure 27.9 Block diagram of the Monterey and Carmel submarine canyons off the coast of central California. The vertical dimension is greatly exaggerated. (Redrawn by A. N. Strahler from an orthographic drawing by Tau Rho Alpha, U.S. Geological Survey.)

the creationists forced into an appeal to pseudoscience. Whereas the scientific method is open-ended and leads the investigator to whatever goal reason dictates, creationism pre-establishes an inflexible goal and forces all observations to conform with it, reason denied; or in plain English, "if it doesn't fit, it can't be right."

Submarine Canyons and Lowered Sea Level

Notching the continental slope of eastern North America are many narrow, steep-walled submarine valleys heading just below the brink of the continental shelf and debouching on the gentle slope of the continental rise. You might want to liken them superficially to gullies eroded by torrential rains in a steep road embankment made up of soft sediment. The Atlantic continental shelf is notched by the heads of a great number of submarine canyons, the shelf depth being on the order of 200 m. Atlantic submarine canyons can be traced down to depths of 1,500 m and more, which is far lower than any reasonable estimates of sea level in full glaciations. Deep, branching canyons are also found off the Pacific coast of the United States, an example being the Monterey Canyon, that heads in Monterey Bay and descends to a depth more than 1,700 m (Shepard, 1973, p. 315). Figure 27.9 is a three-dimensional drawing of the area based on numerous depth soundings.

As long ago as the early 1900s, a major submarine canyon had been charted offshore from the present mouth of the Hudson River (Veatch and Smith, 1939, Figure 2). It was traced landward across the shelf as a distinct channel indenting the surface of the shelf and even showing smoothly curving bends like those of a terrestrial river. A similar feature was mapped off the mouth of the Congo River. It is a huge trench up to several hundred meters deep, clearly traceable to a depth of over 4,000 m (Shepard, 1973, p. 321).

Extensions of river channels across an exposed continental shelf are easily understood for a sea level depressed as much as 100 m during glaciations, but how can we explain extensions of those channels to depths over 1,000 m? Geologists debated the origin of submarine canyons for several decades before modern methods of study were available (direct observation from submersibles, sidescan sonar, seismic reflection profiling, and other geophysical methods). The idea that a great eustatic

lowering of sea level had occurred was not very popular. (Where did all that ocean water go?) Seriously considered was the possibility that tectonic activity in the form of downwarping or downfaulting of the oceanic crust had carried subaerially formed stream valleys down to great depths. With the knowledge that isostatic equilibrium prevails generally along passive continental margins, the tectonic hypothesis was not attractive either.

The most serious attention was given to the possibility that processes of erosion act on the ocean floor in the form of strong currents flowing down grade on continental slopes and other sloping surfaces of the deep ocean. Specifically, the turbidity current, a spontaneous gravity flow of a mixture of water and mud, was singled out as the most important agent. A turbidity current might easily be set in motion by the detachment of a mass of soft sediment precariously perched high on the continental slope. A remarkable event gave impetus to the turbidity current hypothesis. In 1929, directly following the Grand Banks earthquake, several submarine cables in the vicinity of the earthquake epicenter were broken in rapid succession. In the early 1950s, Maurice Ewing and Bruce Heezen of the Lamont Geological Observatory reexamined the records of the cable breaks and found that the time lag of each break increased with distance down the continental slope, away from the epicenter. They postulated that a submarine landslide had been set off by the earthquake, initiating a turbidity flow capable of parting the cables over a total distance of over 200 km. They estimated the speed of the front of the flow as exceeding 80 km/hr. In ensuing decades the rapid accumulation of great volumes of data attesting to the potency of deep ocean bottom currents to scour the floor and transport sediment has reinforced the general hypothesis of a submarine origin of the canyons below depths of 100 to 200 m. Various processes of submarine erosion and deposition have been recognized as participating in the shaping of submarine canyons of several classes, including some that lie at abyssal depths.

For the creationists, submarine canyons constitute evidence that sea level was once much lower than today. This they attribute to either or both of two possibilities: (1) the amount of water in the oceans was formerly much less than today; (2) some parts of the ocean floors have dropped (Whitcomb and Morris, 1961, pp. 124-26). They find additional evidence of these possible explanations in the presence of guyots, flat-topped volcanic seamounts with their upper surfaces lying as deep as 500 m below sea level. The name "guyot" honors Arnold Guyot, a nineteenth-century Swiss-American geologist; it was given in 1946 by Princeton professor Harry Hess, who discovered the flat-topped seamounts while the naval vessel on which he was serving made continuous profiles of the Pacific Ocean floor, using the sound-reflection principle. The hypothesis advanced by Hess was that a volcanic island, following its formation, was beveled by wave action, after which a sinking of the seafloor carried the seamount down to depths below 200 m. Under the general hypothesis of plate tectonics, gradual sinking of the oceanic crust is to be expected as seafloor spreading takes place, because the lithosphere becomes cooler and denser as it moves away from the spreading plate boundary.

After citing some geological references, now obsolete in the light of modern knowledge of submarine geology, questioning the efficacy of turbidity currents to form submarine canyons, Whitcomb and Morris present the following scenario in accordance with Flood geology:

> It would seem, on the other hand, that Deluge conditions, as inferred from the Scriptural record, could give a reasonable explanation for their origin. As the lands were uplifted and the ocean basins depressed at the close of the Deluge period, the great currents streaming down into the ocean depths would quickly have eroded the still soft and unconsolidated sediments exposed by the sinking of the basins. Then, as these gorges were themselves submerged by the continuing influx of waters from the rising continental blocks, it may well have been that the turbidity currents entering the canyons may have deepened and extended them still further, a process which has continued on a smaller scale throughout the centuries since. (1961, p. 126)

In a later discussion of this same topic in the context of a Scriptural framework for historical geology, the hypothesis of downfaulting of the ocean basins is developed more fully (Whitcomb and Morris, 1961, pp. 324-26). Prevalent among geologists in the early 1900s was the hypothesis that movement along a major normal fault located at depth below the continental slope accounted for the difference in elevation of the continental crust and the oceanic crust. As seismic investigations accumulated in the 1940s and 1950s, revealing the differences between continental and oceanic crust, the need to postulate a fault contact was largely removed. Later, plate tectonics explained the passive continental

margins through continental rifting, which involves a great deal of block faulting in its initial stages, but explains the formation of new oceanic crust by upwelling of mafic magma. (See Chapter 20 for a description of continental rifting and opening of a new ocean basin.)

Whitcomb and Morris revive the discredited fault hypothesis of continental margins in the following paragraph:

> This of course accords quite well with the Biblical implication that the uplift of the lands, coincident with the subsidence of the ocean basins, marked the terminus of the universal inundation caused by the great Flood. This uplift (or fault slippage along the edge of the granite blocks of the continents) was intermittent, largely being completed during the Flood year but evidently continuing on a lesser scale for many centuries to come. (1961, p. 325)

In Chapter 22, in the discussion of diastrophism and volcanism during the Flood, we evaluated the creationists' tectonic hypothesis of sinking ocean basins and rising continents by normal faulting, reversing the previous opposite set of relative movements that assisted in causing a worldwide Flood. We showed that such imagined alternating crustal movements violate the principle of isostasy and would, in any case, be impossible within a one-year Flood because of the great viscosity of the asthenosphere. The creationists continue to reprint their obsolete 1961 text (its 26th printing is dated 1982), ignoring the growing mountain of geophysical data secured in the intervening decades by scientists working in independent groups and leading to a consensus that oceanic crust has originated by igneous processes acting more or less continuously at spreading plate boundaries.

Recently, ICR creation scientist Steven A. Austin revived the subject of submarine canyons and other deep-sea canyonlike features (1983). He insists, despite a vast literature on this subject showing great progress in observing deep-sea processes, that little is known about the origin of submarine canyons. For him the data simply indicate these features are relicts, presumably of catastrophic events that marked the end of the Flood. Surely, such a cursory dismissal of a large body of scientific information must be regarded as a nonscientific pronouncement.

Credit

1. From Glacial and Quaternary Geology, by Richard F. Flint. Copyright © 1971 by John Wiley & Sons, Inc. Reprinted by permission of John Wiley & Sons.

Chapter 28

Causes of Glaciations through Geologic History

One of the most remarkable geological displays to be seen anywhere is in the Dwyka formation of late Carboniferous age in South Africa. It is described as a <u>tillite</u>--a lithified glacial till containing pebbles and boulders that clearly show the effects of strong abrasion beneath moving ice. The boulders have facets (flattened surfaces) bearing scratches, called striations. In my collection of miscellaneous rocks and minerals, I have a large pebble from the Dwyka formation, and for all the world it looks like a striated, faceted pebble that one might find by the dozen in Pleistocene till in Connecticut or New Jersey. I also have a fragment of the bedrock surface that underlies the Dwyka tillite; it is a dark igneous rock, strongly abraded and polished, and bearing striations. In South Africa, the exposed basal rock surface shows great grooves and smoothly rounded small hills (rôches moutonées). The same topography can be seen today in heavily glaciated mountains--in the Klondike, for example. But the most remarkable specimen in my little collection is a finely banded sedimentary rock--evidently a siltstone or claystone. The perfection of the alternating light and dark bands leaves no doubt that it is the lithified equivalent of varved clays such as one finds widely in the Pleistocene glacial deposits of North America and Europe. That the Dwyka formation was left by a continental glacier that thrived in late Carboniferous time is an almost universally accepted interpretation of mainstream geology. In early decades of this century, Alfred Wegener and his followers accepted this interpretation as evidence that the ancient continent of Gondwanaland had occupied a polar position during the Carboniferous and early Permian periods, just prior to the time that Pangaea began to split apart and separate into the continental fragments we now know as Africa, South America, peninsular India, Madagascar, Australia, New Zealand, and Antarctica. Tillites of that geologic age can also be identified on each of those continental fragments. Moreover, when the fragments are reassembled and the orientations of bedrock striations are plotted, they reveal an outwardly radial pattern suggesting the flow pattern of a large ice sheet centered on the Antarctic landmass (Figure 28.1).

Creationists' Rejection of Ancient Glaciations

The Carboniferous tillites cannot be accepted by creationists as being of glacial origin for the obvious reason that the tillite formations are both overlain and underlain by fossiliferous strata, which are deposits of the Flood. During that great inundation, which lasted the better part of one year, there could have been no land ice formed by accumulation of snow. The difficulty is something like that which applies to the occurrence of ancient sand dunes in the stratigraphic record: terrestrial wind-deposited sediments also have no place on a globe totally inundated by Flood waters.

Creationists Whitcomb and Morris discuss this problem at some length in their book, <u>The</u> <u>Genesis</u> <u>Flood</u> (1961, pp. 245-49). They argue that the supposed tillites could have been formed by nonglacial processes; they cite statements by mainstream geologists that nonsorted aggregations of gravel, sand, and boulders in a clay matrix can result from flowage or slump movements of unconsolidated masses of rock fragments. They particularly stress the presence of coal seams interbedded with the Carboniferous and Permian tillites, and this, they say, must mean that the depositional environment was that of the warm, equatorial zone. What reply can be made to such seemingly rational arguments?

First, references to the scientific literature cited by Whitcomb and Morris are all dated in the 1950s or earlier, before plate tectonics came into full recognition (late 1960s) and while the paleomagnetic methods of latitude determination of rock formations were just beginning to make an impact through the plotting of polar-wandering paths. Because <u>The</u> <u>Genesis</u> <u>Flood</u> has remained unrevised through numerous printings, the 26th being in 1982, we must infer that its authors consider it a currently valid scientific presentation worthy of study by science students today. Thus the obsolescence of the work needs to be scored, and a modern, state-of-the-art review offered in its place.

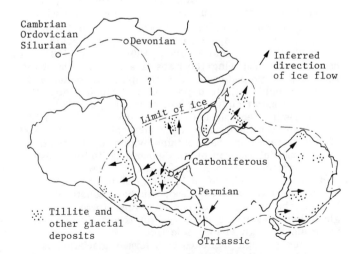

Figure 28.1 Hypothetical restoration of the nested continents of Gondwana, showing the location of tillites and other glacial deposits and striations of Late Carboniferous time, together with inferred directions of ice motion and the limit of a single great ice sheet. Pole positions are based on paleomagnetic data. (A. N. Strahler. Based on data from several sources.)

Paleomagnetism and Ancient Latitudes

How can the global latitude of a particular continent or continental fragment be determined for a given geologic period? Paleomagnetism has provided a tool for this purpose. In Chapter 19 in a section titled "Rock magnetism and polarity reversals" we reviewed the principles of paleomagnetism. Recall that for a basalt sample taken with great care so as to document its orientation in terms of geographic north and horizontality, a magnetometer can be used to determine the polarity of the specimen and the inclination (direction and inclination of the ancient lines of force). The principle of magnetic inclination is illustrated in Figure 19.1 by small arrows at the earth's surface. Over the magnetic equator the arrow is horizontal and inclination is zero. Over either magnetic pole the arrow is vertical and the inclination is 90 degrees. For intermediate latitudes, the inclination has a value proportional to the latitude. Compass direction of the lines of force in the sample tells the direction one would follow on a great circle to reach the former north (or south) magnetic pole (Figure 28.2A). This information allows the continent to be oriented correctly with respect to the former north direction. The amount of inclination read from the sample tells the latitudinal distance to the former magnetic pole (Figure 28.2B). But, as you can see from the figure, the sample point could have been located anywhere on the former geomagnetic parallel of latitude. In other words, while we know the former geomagnetic latitude of the sample, we do not know its former geomagnetic longitude.[1]

At the present time the geomagnetic axis does not quite coincide with the earth's axis of rotation, which determines the position of the geographic poles. This discrepancy (called the compass declination) follows a cyclic pattern of change, but it is thought that the geomagnetic axis, over a span of hundreds of millions of years, has coincided quite closely with the axis of rotation. This coincidence is required by the dynamo theory that we outlined earlier. Other causes of small errors exist, and thus every determination of a former position of the earth's pole of rotation is subject to an error on the order of five degrees of latitude and longitude. Even so, paleomagnetism provides us with some very strong evidence of past movements of the continents relative to one another.

Largely from the data of paleomagnetism, geologists have been able to draw maps reconstructing the positions of the continents in each period of geologic time within the Phanerozoic Eon (Paleozoic, Mesozoic, and Cenozoic eras). (For such maps refer to Seyfert and Serkin, 1979; Bambach, Scotese, and Ziegler, 1980; Scotese, 1984.) Also important in preparing the maps is evidence from climate-related geologic materials. For example, evaporites and redbeds would indicate deposition in a tropical desert zone; coal might suggest a warm, moist equatorial environment. Tectonic features are also useful in orienting an ancient continent. For the Proterozoic Eon (that part of the Precambrian that contains evidences of primitive life), some information on the general latitudinal position of the ancient cratons is also available. Now that we understand how Figure 28.1, a map of nested continents in the late Carboniferous time, can be reconstructed, we turn to the geologic evidence of several former glaciations, occurring in early Proterozoic time and at intervals through Phanerozoic time.

Strong geologic evidence of a former glaciation is exemplified in the Dwyka formation of South Africa. The tillite must contain pebbles and boulders having the characteristic shape and striations produced by glacial abrasion. Presence of lithified varved clays is valuable evidence. Perhaps essential is the finding of a glaciated bedrock floor lying beneath the tillite and showing the characteristic evidences of ice abrasion. Also present may be lithified deposits typical of marginal zones of ice

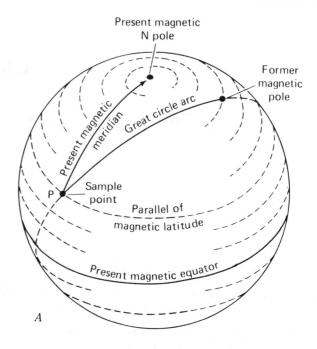

A

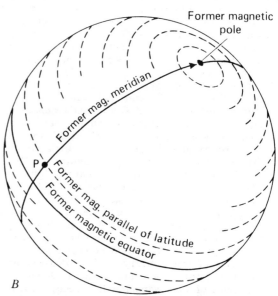

B

Figure 28.2 Present and past geomagnetic coordinates. (A) Present geomagnetic pole, equator, and meridian. Paleomagnetism of sample at Point P indicates where the former magnetic pole was located. (B) Former geomagnetic coordinates system based on the former magnetic pole. (Copyright © 1981 by Arthur N. Strahler.)

sheets: drumlins, eskers, and outwash accumulations. We should be skeptical of the finding of only a bed of supposed tillite without the bedrock floor. Mixtures of rock fragments in a mud matrix can form in other, nonglacial situations. There is also the possibility that a striated boulder was carried by an iceberg for a long distance across the ocean before being released to sink to the ocean floor and become incorporated into marine sediments. We must also be aware that glacial tills with striated boulders are produced by alpine glaciers flourishing in high mountain environments, and that these materials might become incorporated into sediments of nonglacial origin in adjacent lowlands. Creation scientists are well aware of the possibilities for misinterpretation of nonglacial sedimentary materials and tend to emphasize the negative aspects in evaluating published evidence of past glaciations. They also reject totally the value of paleomagnetism in reconstructing events that occurred more than a few thousand years ago. As we found earlier, creation scientists have their own hypothesis of a recent origin for the earth's magnetic field.

Seven Ancient Glaciations

Mainstream historical geology recognizes the possible occurrence of seven glaciations since the Proterozoic Eon began. For a detailed discussion of all of these glaciations, I recommend that you refer to a modern textbook of historical geology entitled Earth History and Plate Tectonics, by Carl K. Seyfert and Leslie A. Sirkin (1979). Their statements are accompanied by a full list of references to the scientific literature. In the following paragraphs, page numbers in parentheses refer to their text.

Of the seven glaciations, four are in Proterozoic time: one in the Early Proterozoic at -2.3 b.y., and three in the Late Proterozoic at approximately -975, -740, and -610 m.y. (p. 237). In Phanerozoic time three continental glaciations are identified. The first was in late Ordovician time, with a major ice sheet centered over Africa; the second was in late Carboniferous and early Permian times, centered over Antarctica; the third is our present Cenozoic Ice Age.

The evidence of early Proterozoic glaciation is found in several parts of the ancient North American craton (p. 199). It consists of striated pebbles in tillite that locally rests on a striated pavement. One remarkable exposure reveals a succession of varves containing a pebble that may have been dropped into the soft clays from floating ice (p. 199, Figure 9.18). Paleomagnetic data show that this part of what is now Canada lay within about 42 degrees of the north pole, which is within a latitudinal range comparable to ice sheets of the Cenozoic Ice Age. Evidence of glaciations in late Proterozoic time is perhaps less satisfactory, but striated pebbles are found in a tillite resting on a striated basement (p. 226). The main point of interest concerning these occurrences is that paleomagnetic data show that the African localities lay on or close to the equator and that tillites are associated with thick carbonate strata, including stromatolite reefs, that also suggest a warm climate (p. 226). Assuming that the latitude data are valid, a contradiction in evidence is indicated, something that should prove a source of delight to the creation scientists. However, we are dealing here with extremely old rocks that have had a long and complex geologic history. One suggestion that has been made to resolve the problem of glaciation in the equatorial zone is that worldwide climate at those glacial times was much colder than even the coldest periods of the Cenozoic Ice Age (p. 238). But this suggestion conflicts with the need to postulate a warm climate favorable for reef stromatolites to form. Bigger puzzles than this have beset mainstream geology throughout its history, but some of them have yielded to a reasonable solution when better evidence has been discovered. Human knowledge is imperfect, no matter what subject it deals with.

Fortunately, the geologic record improves greatly in quality in the Phanerozoic Eon, and the quality of evidence pointing to a major glaciation in Ordovician time in northern Africa is remarkably good. The paleomagnetic evidence shows the pole located off the West African coast in Ordovician time (see Figure 28.1). In 1961 geologists found in the Sahara Desert the distinctive markings of a glaciation in rocks of Upper Ordovician age. In 1970, an international expedition traveled to the central Sahara to evaluate this evidence. The visitors were convinced that a great glaciation had indeed occurred here, for the evidence includes tillites with striated and faceted boulders as well as grooved and striated pavements (p. 289). Even the distinctive "chatter marks," so common on recent glacial pavements, were identified. These features indicate that the ice moved in a direction that is

now toward the north, which is away from the more southerly position of the pole of that time.

The principal tillites of late Carboniferous to early Permian time are found in Antarctica, South Africa, South America, and India (Seyfert and Sirkin, 1979, p. 115). Typically, shales, sandstones, and coal beds overlie the tillites and are of Permian age. This is where we get into the question of how coal can accumulate in a cold climate favorable to glaciation. Actually, the coal seams formed after the tillites, suggesting an amelioration of climate following the glaciation. In South Africa, the coals occur in the Ecca formation; it contains continental deposits in which fossil wood is preserved (p. 341). This wood shows seasonal growth rings, indicating that the coals of this area were probably deposited in a moist, midlatitude climate (p. 341). From paleomagnetic data, a latitude of about 60°S is indicated for the Ecca strata; this is well within the latitude range of spread of ice sheets. That peat should accumulate in thicknesses necessary to produce coal seams in this latitude is not unreasonable, since cold inhibits the oxidation of peat. Peat bogs are typical of the glaciated regions of North America and Europe. Thus the creationists' objection that coal seams have been found in close association with tillites is deprived of its force when the circumstances and stratigraphic relationships are fully disclosed.

The remarkable displays of the work of great ice sheets in both Ordovician and Carboniferous times on the ancient continent of Gondwanaland makes a powerful case against the creationists' Flood geology. These glacial features and deposits lie squarely sandwiched within the total pile of fossiliferous Phanerozoic strata. At least two, and quite possibly seven or more, periods are recorded since the start of Proterozoic time, in which continental areas would have been exposed to the atmosphere at various times in the single year of the Flood--exposed long enough, that is, for great ice sheets to grow, spread, and disappear. The Genesis story of the Flood, written or revealed by God Himself, forbids such an interpretation.

Causes of Glaciations--The Mainstream Version

Few geological puzzles have proved so intriguing and so difficult to solve as the cause or causes of multiple glaciations and interglaciations. Despite scientific advances on many fronts in the earth sciences and an enormous gain in our knowledge of how the earth works through plate tectonics, the debate as to what causes a glaciation continues undiminished in its fervor and in the diversity of mechanisms proposed.

The puzzle must be approached on two quite different levels. First is an overall set of fundamental conditions that seems to have favored the repeated growth and disappearance of large ice sheets throughout an entire ice age lasting as long as three million years, or perhaps much longer (Beaty, 1978). In this broad view of the problem, we need to take into account ancient ice ages, particularly the Paleozoic glaciations so well documented on the Gondwana continent. The second level of inquiry is into the immediate or forcing causes that are responsible for actually precipitating a glaciation at a particular point in time, for causing a deglaciation to follow, and for this cycle to be repeated. The questions are these: Why are there cycles of glaciation and deglaciation? What controls the length of these cycles? What causes them to be initiated or triggered? In attempting to answer these questions, we may find ourselves considering several quite different favorable factors and triggering mechanisms acting together to cause a glaciation.

A meaningful inquiry into the causes and factors involved in glaciations and interglaciations requires application of the atmospheric and oceanic sciences, because climate change is the essence of glacial cycles.

Fundamental Conditions for an Ice Age

Four basic, or fundamental, conditions would seem to tend to favor or permit an entire ice age to occur: (1) A favorable positioning of the continents with respect to the polar regions. (2) A withdrawal of oceans from the continental cratons as a result of general relative lowering of sea level. (3) A sustained period of increased volcanic activity. (4) A sustained period of diminished intensity of solar energy reaching the earth.

Of the four fundamental factors, the first three are geologic in nature. Favorable positioning of continents, clustering around one or both poles, is evidently essential, as paleomagnetic determinations make clear. But the presence of a large landmass under or close to a pole does not seem to have guaranteed an ice age. It worked for the Ordovician and Carboniferous glaciations, but during the intervening Silurian and Devonian periods, as the pole path traversed North Africa, no continental glaciation appears in the stratigraphic record. Breakup of Gondwana ended the Carboniferous/Permian glaciation, as most of the fragments moved to lower latitudes. Only Antarctica ended up in the south polar position.

Look next at the north polar region in late Paleozoic time, as shown in Figure 20.20, Map A. (See also Seyfert and Sirkin, 1979, p. 312, Figure 11.1C.) By Permian time only the northeastern tip of the Asiatic continent penetrated the north polar region. As the Laurasian continent rifted apart and the Atlantic basin widened, North America moved westward and poleward to a position opposite Eurasia, while Greenland took up a position between North America and Europe. The effect of these plate motions was to bring an enormous landmass area to a high northerly latitude and to nearly surround a polar ocean with land. (See Donn and Shaw, 1977.)

The reasons why the present configuration of northern landmasses favors the onset of an ice age are fairly simple. First, the highlands on which ice sheets begin to grow are located in a cold climate zone where the rate of snowfall can greatly exceed ablation. Second and perhaps less obvious, the landmasses tend to block the poleward flow of warm ocean currents that might otherwise tend to make the polar climate relatively mild. At present the Arctic Ocean is connected with the Atlantic and Pacific oceans by narrow straits. The polar ocean with its year-around cover of sea ice maintains a cold climate that extends well into the fringes of the bordering lands.

The second factor on our list is a general relative lowering of sea level that increases the area of continents while at the same time tending to narrow or close entirely straits that separate land areas while also interconnecting ocean bodies. In Cretaceous time, sea level stood high relative to the continents, inundating continental shelves and large low-lying interior areas of the continents with shallow seaways. Global climate was remarkably mild, and land plants of subtropical affinities occupied latitudes as high as 70°; coal seams were produced in the full range of latitude from equator to poles. By late Cretaceous, the oceans had receded and the climate became cooler. This fall of sea level has been attributed to a general positive epeirogenic movement of the continental crust, but perhaps a more attractive explanation is an increase in the average depth of the ocean basins as the rate of seafloor spreading decreased following a high rate of spreading in Cretaceous time. Slow spreading increases the proportion of colder, denser oceanic lithosphere, which stands at a lower level than new lithosphere.

Reasons why relative lowering of sea level should favor an ice age are not easy to come by. Large landmasses in middle and high latitudes tend to have colder winters (but warmer summers) than adjacent oceans in the same latitude zone. While climate is cooler at higher land elevations, it seems unlikely that continental interiors underwent an amount of uplift sufficient to be of consequence.

Some geologists have argued that an increase in tectonic and volcanic activity in late Cenozoic time caused a rise of lofty mountain ranges capable of trapping large amounts of snowfall and thus made possible the initial growth of icecaps that would later coalesce into ice sheets. Others argue that with plate subduction going on more or less continuously throughout the Phanerozoic Eon, there were high mountain ranges on continental borders more or less continually at one place or another. The mountain-growth effect thus loses much of its effectiveness as a specific initiating cause of ice ages (Beaty, 1978, p. 453).

Our third factor, also geologic, is an increase in the intensity of volcanic activity on a global basis over a long time span. The link between volcanic activity and glaciation is through the emission of vast amounts of extremely fine volcanic dust during explosive volcanic eruptions, particularly those of composite volcanoes (stratovolcanoes) and caldera-forming explosions. The dust rapidly reaches the upper atmosphere and spreads throughout the entire global stratosphere within months or a year or two following an eruption. The dust particles, in turn, cause a climate change that may be global in scope.

Following the the eruption in 1883 of the volcano Krakatoa, lying between Sumatra and Java, a veil of volcanic dust formed in the stratosphere and spread into high latitudes. In Europe and North America, the dust veil brought on a large number of extremely brilliant sunsets and attracted a great deal of popular attention. Solar observatories recorded a 20 percent drop in the intensity of solar energy reaching the earth's surface in the first year following the explosion. For each of the next three years the reduction was about 10 percent. Eventually most of the dust settled out into the lower atmosphere and the blocking effect disappeared. Atmospheric scientists agree that the climatic effect of a stratospheric dust veil is to cause a measurable cooling of the average atmospheric temperature near the earth's surface. If we accept this conclusion, we can then examine the geologic record with regard to the intensity of volcanic activity before and during the late Cenozoic Ice Age.

A record of past volcanic activity can be obtained from deep-sea cores. A major volcanic eruption is recorded as a thin layer of volcanic ash consisting of minute shards of volcanic glass. Some indication of the general intensity of volcanic activity can be had by determining the number of ash layers per thousand years. Some research scientists claim the evidence points to greatly increased volcanic activity in the Pleistocene Epoch. One study of deep-sea cores led to the conclusion that volcanic activity in the Pleistocene was at a level four times higher than the average for the 20-m.y. span covered by the cores (Kennett and Thunnell, 1975). The data were seriously impugned by a study that followed shortly, suggesting that the apparent great increase in volcanic activity can be explained by the disappearance of older ash-bearing layers into active subduction zones near the volcanic arcs that supplied the ash (Ninkovich and Donn, 1976). The scientific debate continued, with new deep-sea core data being brought into the argument as evidence. A different and rather startling hypothesis put forward in 1979 proposed that climatic changes may have triggered the volcanic outbursts, and not vice versa, as most investigators had assumed. This suggestion was based on data showing that the cooling trend had in each case preceded the volcanic outburst (Rampino, Self, and Fairbridge, 1979). All this diversity of opinion is normal in the course of mainstream scientific investigation, which is open-ended and ever leading to surprises.

Our fourth possible basic cause of an ice age is in the realm of astronomy. Actually two very different mechanisms are recognized whereby the rate at which solar energy arrives at the outer limit of the earth's atmosphere: (1) The radiant output of the sun might be substantially reduced for long periods; (2) The sun's rays might for long periods of time be absorbed by cosmic dust clouds lying between the sun and the earth. As to the first hypothesis, recall from our discussion of the cyclic variation in the dipole magnetic field that reduced output of solar energy can definitely be correlated with sunspot activity. This information applies, however, to only the past few thousand years. A 1970 science news report described studies of the near-surface radionuclides on the moon's surface demonstrating "that during at least the last 1-2 million years solar flare activity has been comparable to that at present, and that there is no evidence of long-period fluctuations in solar activity that might relate to long-period climatic changes on earth" (NCAR Quarterly, no. 28, August, 1970). But then, in 1982, another science news report stated that studies of the solar nutrino flux carried out at the Los Alamos National Laboratory suggest that "terrestrial climatic changes during the past millenium, including the glacial epochs, may be related, even caused, by variations in the sun." (EOS, vol. 63, no. 37). The supposed variations are attributed to fluctuations in nuclear reactions in the sun's core. This is just an example of the on-and-off nature of scientific speculations.

The second hypothesis of solar radiation variation is based on new and more detailed knowledge of the structure of our galaxy. As our sun revolves around the center of the galaxy, once in 500 m.y., it crosses the oppositely placed spiral arms of the galaxy once each 250 m.y. In so doing, the solar system passes through a lane of dark cosmic dust. The dust absorbs a fraction of the sun's radiant output. According to English astronomer W. H. McCrea, who outlined the hypothesis in 1975, each passage of the solar system through the dust lane would cause an ice age, while fluctuations in density of the cloud would force glaciations and interglaciations within the ice age (Science News, 1975, vol. 108, p. 23). A news report that soon followed stated that soil samples brought back from the moon reveal increases in dust particles of the proper sizes and masses, dated at roughly the same time intervals as McCrea's hypothesis requires (Science News, 1975, vol. 108, pp. 309-10). The hypothesis received further support in 1978 by a suggested chemical process in which a cloud of molecular hydrogen in the spiral arm could stop the solar wind, ultimately leading to formation of a permanent mesospheric layer of ice clouds, starting an ice age (Science News, 1978, vol. 113, p. 148). Shades of the antediluvian vapor canopy! Just to keep you from jumping on this bandwagon, it should be noted that another version of the dust-lane hypothesis calls for the dust to be drawn into the sun's surface, where it is highly heated and increases the sun's brightness. Thus, the sun's energy output is increased for a time and could raise the earth's average temperature. Or would this warming, on the other hand, intensify precipitation processes that would nourish a new generation of ice sheets? A rich and varied menu from which to choose!

Causes of Ice Ages--The Creationists' Version

Now that we have reviewed the diverse evidence collected by mainstream scientists for numerous cycles of glaciation and deglaciation in a great Ice Age spanning at least the last 3 m.y., we can turn to consider possible causes of such cycles. First, however, the creationists' explanation deserves attention. Is their hypothesis worthy of serious consideration alongside those of mainstream science?

Whitcomb and Morris suggest a variety of circumstances favoring a single glaciation immediately following the Flood:

It has been argued that, once an ice sheet got started, it would probably grow rapidly and

extensively. This would perhaps be possible in the years immediately following the Deluge. An abundant supply of moisture, strong polar winds, lowered polar temperatures due both to removal of the thermal vapor blanket and probably dense accumulation of volcanic dust particles in the atmosphere, newly uplifted mountains, essentially barren topography of the denuded lands: all these and possibly other factors could have contributed to the rapid accumulation and growth of the ice sheets. These factors are all legitimately deduced from the record of the Flood and would be quite sufficient to explain the Ice Age. The catastrophic nature thereof, however, will of course be unacceptable to many geologists. (1961, p. 294)

The suggested influences favoring growth of ice sheets include some that are assumed in mainstream geology: lowered atmospheric temperatures in high latitudes, presence of highlands to trap precipitation, and abundance of moist air masses engaging in vigorous cyclonic circulation. Suspended volcanic dust at high altitude has long been regarded as effective in reducing temperatures in the lower atmosphere.

The unique feature of the creationists' hypothesis seems to be the sudden collapse of the vapor canopy, which had successfully preserved a mild, rainless global climate in the antediluvial period. The canopy included ozone and carbon dioxide as well as water vapor. During the Flood, most of the carbon dioxide was taken up in sediments to produce carbonate rocks and petroleum hydrocarbons (Whitcomb and Morris, 1961, pp. 306-9). With the greenhouse effect temporarily gone, atmospheric temperatures would have dropped dramatically in the high-latitude zones, where the solar radiation balance would have been strongly negative. Intense volcanic activity immediately following the Flood would have introduced much carbon dioxide into the atmosphere, eventually restoring the greenhouse effect, but the great quantity of volcanic dust would have initially had a salutary cooling effect.

The difficulty in the creationist's hypothesis lies not in the set of factors that would have favored the onset of a glaciation, but in the postulate of a vapor canopy that previously existed to prevent glaciation. We have already gone over in detail the reasons for rejecting the postulate of an antediluvial vapor canopy.

The rather vague and general hypothesis outlined above has more recently been fleshed out in terms of atmospheric and oceanic processes by a creation scientist, Reginald Daly, whose paper was published in the Creation Research Society Quarterly under the title "The Cause of the Ice Age" (Daly, 1972). He begins by noting that immediately following the Flood, areas that are now deserts would have been "soaking wet," as evidenced by pluvial lakes. Daly then infers that the global atmosphere would have maintained a relative humidity of 100 percent and the continued evaporation would have had a strong cooling effect, chilling the global climate to the freezing point. In his own words: "The earth was thus a new-style refrigerator--a modern evaporator type. For evaporation is a cooling process and if there is enough evaporation, and it is rapid enough, then it is a freezing process" (p. 210). It is hard to believe that such a scenario could be written by an individual who holds a master's degree in science and has served as an instructor in physics in several colleges (see Daly's footnote on p. 210). If the relative humidity were constantly maintained at 100 percent, how could rapid evaporation from water surfaces continue? In any case, evaporation from a free water surface results in a cooling of the remaining liquid, while at the same time the molecular kinetic energy thus removed from the water goes into storage in the latent form in the water vapor. This latent heat is then available for release during condensation as air masses rise, experiencing adiabatic cooling to the dew-point temperature. Daly goes on: "Freezing cold evaporation clouds were rising everywhere, resulting in rainfall in the tropical areas and snowfall in the temperate zones, and cold winds were flowing across the resulting snow and ice fields pouring continuous snowfall on the adjoining shore lands" (p. 210).

In comment on this most confused and confusing statement we can overlook the self-contradiction in the words, "evaporation clouds" (clouds result from condensation) and note that the missing activity in the total process is the loss of heat from atmosphere to space by longwave radiation, which, at high latitudes, exceeds the energy absorbed from incoming solar radiation. With a sustained average energy deficit, the production of snow and its accumulation as glacial ice is assured. Meantime, the global energy balance is being maintained by an energy surplus in low latitudes and poleward transport of that surplus energy by atmospheric circulation. On our planet this open energy system has functioned smoothly for hundreds of millions of years. Merely to describe its general functioning does not answer the question of why ice ages occur interspersed with mild nonglacial periods. Obviously, the energy system undergoes fluctuations, but what forces those fluctuations? Ever since the earth developed its expansive ocean, there has been no lack of evaporating surface to provide atmospheric water vapor, so the mere presence of the world ocean, whether it has endured for 3 billion y. or a mere 4,300 y., cannot be the crux of the problem of cause of an ice age.

Daly continues his narrative by stating that "moisture laden supersaturated clouds were carried by wind currents to northern Canada, Scotland, Norway, and Sweden where snow fell daily from November to April, accumulating to depths of 500 to 1,000 feet (150-300 m) during the first winter" (1972, p. 210). At the lower rate, in one century the total snowfall would be 50,000 ft (15,000 m; 15 km). Initially at a density one-tenth that of ice, that much snow would yield 5,000 ft (1,500 m; 1.5 km) of glacial ice, disregarding thinning by outward flowage.

Daly's snowfall rate is indeed catastrophic in comparison with observed rates today. In favorable mid-latitude mountain locations, such as the Sierra Nevada and Colorado Rockies, average snowfall somewhat exceeds 5 m/y.--only 1/30 of that which Daly imagines. As stated in an earlier section, annual snow accumulation at Camp Century on the Greenland Ice Sheet is on the order of 3.5 m/y., enough to keep that ice sheet approximately in dynamic equilibrium. Daly has concocted his snowfall rate purely from imagination. Would the natural hydrologic system of the globe be physically capable of functioning that rapidly? To do so would require a meridional energy gradient of enormous slope--far greater than can be accounted for by the distribution of solar energy per unit of horizontal surface on a sphere exposed to parallel solar rays. So again we have an example of the creation scientists' willingness to postulate outrageous rates of natural processes in order to squeeze certain events into an inflexible time span that derives not from observation of nature, but from rigid adherence to the literal interpretation of an ancient religious document. So let us turn next to an appraisal of current hypotheses of glaciation under debate within the community of mainstream scientists.

The Astronomical Hypothesis of Glacial Cycles

Turning now to our second level or scale of inquiry, we ask: What forces or factors cause alternating glaciations and interglaciations within an ice age? Although a number of hypotheses might warrant our attention, this review is limited to one major hypothesis that now seems to be the leading contender. It is based on astronomical information that is fully accepted by the entire scientific community: the orbiting of the earth about the sun and the spin of the earth on its axis of rotation. This information has been acquired and refined over many decades.

Most of my readers have at least a general idea about what causes the summer and winter seasons in middle and high latitudes but, unfortunately, the details we will need in order to understand today's leading hypothesis of glacial cycles are not rigorously taught in the schools. That the "year" is the period of earth revolution around the sun is clear enough, but we need to specify that our interest lies in the tropical year, which is the time elapsed between two successive crossings by the sun of the celestial equator at a point known as the vernal equinox. The tropical year has a length of 365 days, 5 hours, 48 minutes, and 46 seconds of mean solar time. This may seem like unnecessary detail, but the earth's year can be defined in other ways with different length values. The earth's seasonal climate rhythms are dictated by rhythmic changes within the tropical year, and these need to be understood.

Figure 28.3 shows the earth at its four key points in the orbit. At all times the earth's surface lies half in the sun's rays, half in shadow, giving us night and day alternately as the earth rotates on its axis. The dividing circle between night and day (the circle of illumination, or terminator) always bisects the equator. The plane of that circle is always at right angles to the sun's rays. The next point to notice is that the earth's axis of rotation is inclined, or tilted, with respect to the plane of the orbit (also called the plane of the ecliptic). The angle of tilt remains constant (speaking in terms of a few decades or so), a condition we can appreciate by observing that Polaris, the North Star, is always found in direct line with the earth's rotational axis, serving very well as a compass. The next step is to notice that on the instant of the vernal equinox, which opens the tropical year, the tilt of the axis is neither toward nor away from the sun, that the circle of illumination passes through the two poles, and that day and night are of equal duration over the entire globe. If equinox conditions endured continually (i.e., if the axis had no tilt), no climatic seasons would be possible. Looking next at the summer solstice position, we find the full amount of the axial tilt directed toward the sun in the northern hemisphere and away from the sun in the southern hemisphere. Through the remainder of the year the configuration passes through the autumnal equinox, and then through winter solstice, at which time the geometry is the reverse of that of summer solstice.

What effect has the amount of axial tilt that is presently $23\frac{1}{2}°$ from a perpendicular to the orbital plane on the distribution of incoming solar radiation (insolation, for short) over the earth's spherical surface? Figure 28.4 shows the answer in the form of a graph on which total annual insolation is plotted against latitude. The dashed line shows what that distribution would be if the axis were perpendicular (no tilt); it would decrease as the cosine of the latitude and would fall to zero at the pole. The solid line shows the actual energy distribution, which adds greatly to the total insolation in high latitudes and detracts somewhat for low latitudes. We can deduce that the intensity of the climate seasons in high latitudes might be strongly increased if the axial tilt were to increase, and would be weakened if the tilt angle were reduced. Now we are better prepared to understand the hypothesis of glacial cause, but another astronomical concept needs to be introduced.

As the great German astronomer Johannes Kepler (1571-1630) discovered, the orbit of each planet is an ellipse, with the sun located at one focus of the ellipse, as shown in Figure 28.5. The form of an ellipse is described in terms of eccentricity, defined as the distance, a, from the center of the ellipse to the occupied focus, divided by the length, b, of the semimajor axis.

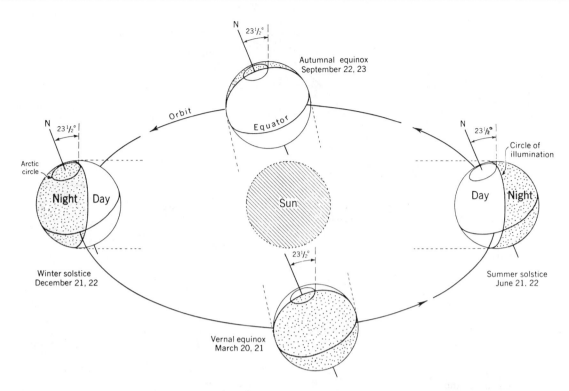

Figure 28.3 Orientation of the earth's axis remains fixed in space as the earth revolves about the sun, producing the astronomical seasons. (Copyright © 1971 by Arthur N. Strahler.)

(The semimajor axis is one-half of the long axis of the ellipse.) At present the eccentricity of the earth's orbit is 0.017. So what does eccentricity of orbit have to do with seasons and glaciations?

Figure 28.6 gives some details of the earth's elliptical orbit. The major axis lies in such a position that its ends coincide with the dates January 3 and July 4; the shortest distance between earth and sun occurs on January 3, when the earth is in perihelion; the longest distance is on July 4, when the earth is in aphelion. The difference in the two distances is about 5 million km, or about 3.3 percent more (or less) than the mean distance of about 149.5 million km. Because the intensity of solar radiation falling on a unit area of surface at right angles to the sun's rays diminishes inversely as the square of the distance from the source, the total insolation that the earth intercepts will fluctuate in a yearly cycle--maximum

at perihelion and minimum at aphelion. Under the existing situation shown in Figure 28.6, perihelion occurs close to the date of winter solstice, which is the winter of the northern hemisphere. Greater-than-average insolation at this season will tend to make northern hemisphere winters less severe, as compared with winters in the southern hemisphere, when insolation is less than average, and will tend to make those winters more severe. Actually, this effect is not measurable in terms of climate statistics because the two hemispheres have greatly unlike distributions and proportions of land areas and ocean areas. Nevertheless, it is a difference that could conceivably influence the potential for growth of ice sheets. But there is more to be said on this subject.

The elliptical earth orbit is subject to two cyclic changes that affect earth-sun relationships. First is a

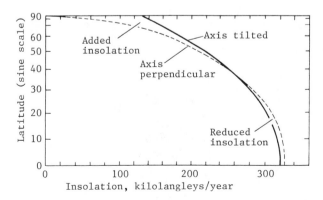

Figure 28.4 Meridional profile of annual total incoming solar radiation (solid line) compared with that for an imagined case of an axis always perpendicular to the ecliptic plane (dashed line). (A. N. Strahler.)

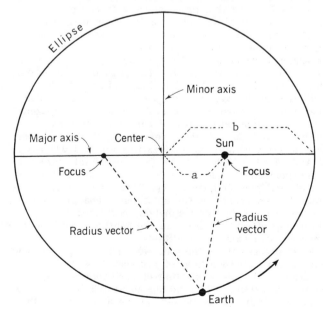

Figure 28.5 The orbit of every planet is an ellipse in which the sun occupies one focus. (Copyright © 1971 by Arthur N. Strahler.)

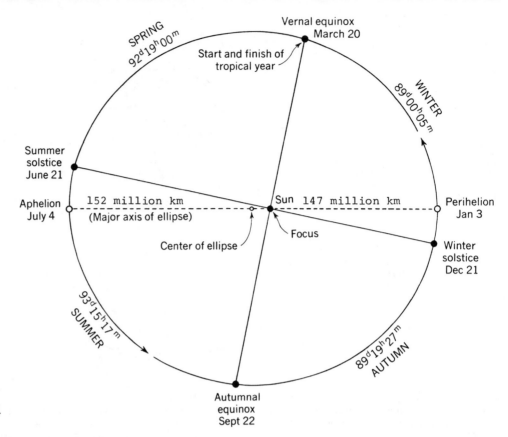

Figure 28.6 Dates of equinox and solstice and duration of the astronomical seasons. (Copyright © 1971 by Arthur N. Strahler.)

phenomenon known as <u>precession of the equinoxes</u>, in which the astronomical seasons slowly shift their positions in a direction opposite to that of the earth's revolution. This shift requires that each vernal equinox occur about 20 minutes sooner than the one that preceded it, measured against the fixed stars of the constellations. Precession of the equinoxes is an apparent effect of precession of the earth's axis of rotation. Think of the earth as a spinning top. As you probably know from experience, as a top loses its speed of rotation because of friction, it begins to wobble; its axis describes a cone in which the axial motion is opposite to the direction of spin. Figure 28.7 shows this precessional effect, which is caused by the tidal attraction of the moon upon the earth's equatorial bulge. Precession is extremely slow as measured against a human life span. A complete precessional revolution would take about 25,800 y. Over many centuries, however, the change in position of the celestial pole is easily noticed. Hipparchus, who is credited with discovery of the earth's axial precession about 120 B.C., became aware of it when he compared his own observations with those of earlier astonomers. Whereas Polaris is now the polestar, Alpha Draconis was the polestar for Egyptian civilizations about 3,000 B.C. By A.D. 7,500, Alpha Cephus (a rather faint star) will be the polestar. By A.D. 14,000, Vega will be close to the celestial pole (Figure 28.7). Precession, then, will cause the occurrence of the solstices to change cyclically relative to perihelion and aphelion, reversing the reinforcing effect on winter in the southern hemisphere as compared with existing conditions.

As if precession of the axis were not enough for you to cope with, another kind of precession is also in operation; it is a slow shifting of perihelion and aphelion around the earth's orbit, taking about 108,000 y. for a complete circuit. (It is caused by the attraction of the other planets.) This form of precession must be combined with that of precession of the earth's axis among the stars in order to arrive at the actual cycle of precession of perihelion and aphelion with respect to the vernal

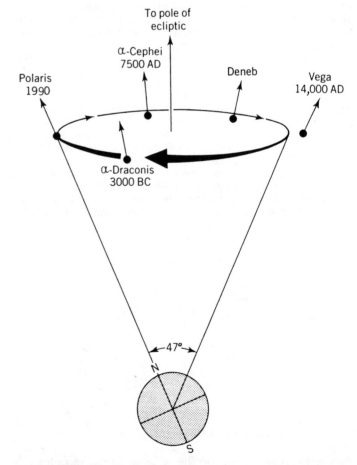

Figure 28.7 Position of the celestial pole among the stars during the cycle of precession. (Copyright © 1971 by Arthur N. Strahler.)

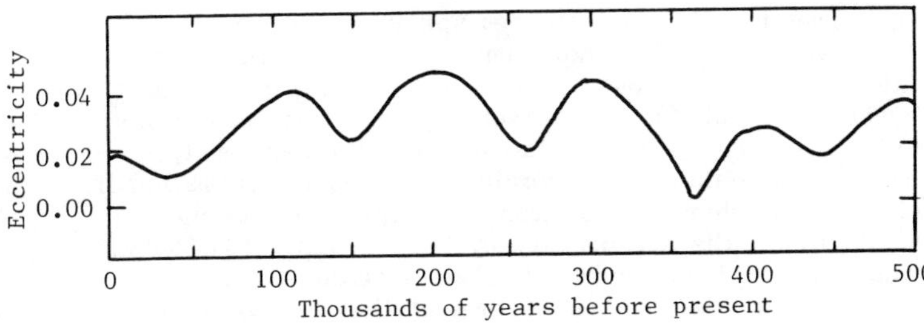

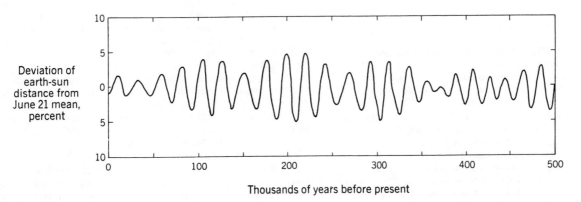

Figure 28.8 Curve of eccentricity of the earth's orbit, as computed by A. D. Vernaker in 1972. (Based on data of J. D. Hays, J. Imbrie, and N. J. Shackleton, 1976, Science, vol. 194, p. 1130, Figure 9.)

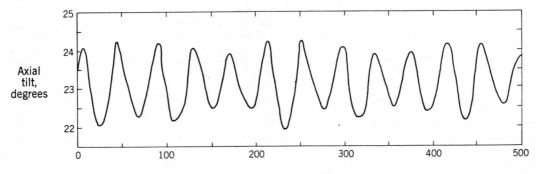

Thousands of years before present

Figure 28.9 Changes in earth-sun distances due to precessional and eccentricity cycles over the past 500,000 y. (Based on data of W. Broecker and J. van Donk, 1970, Reviews of Geophysics, vol. 8, p. 188, Figure 9.)

Thousands of years before present

Figure 28.10 Changes in tilt of the earth's axis with respect to the ecliptic plane during the past 500,000 y. (Based on data of same source cited for Figure 28.9.)

equinox. The actual precession takes about 21,000 y. to complete; a half-cycle (switch in positions of perihelion and aphelion) takes about 10,500 y.

We're not through yet, so hang in there a bit longer! The eccentricity of the earth's orbit does not hold steady, but changes in a cycle that requires nearly 100,000 y. to complete (actual value close to 93,000 y.). The cycle runs from a near-zero value of eccentricity to a maximum of about 0.06 (compared with the present value of 0.017). Figure 28.8 shows the pure eccentricity curve for the past 500,000 y. As eccentricity increases, the difference in solar distances in perihelion and aphelion increases, and that change has to be taken into account. We will need to combine the precessional and eccentricity cycles to get the curve of actual changing distance between sun and earth. Figure 28.9 shows the curve of distance change calculated back over the past 500,000 y. Distance change is given as percentage of the average distance at summer solstice, June 21. But hold on, another cycle needs to be considered before we can go to work on the hypothesis!

The inclination of the earth's axis, or tilt angle--also called the obliquity of the ecliptic--is known to be in change, the annual change being measurable as on the order of a half-second of arc. The change is thought to be cyclic with a period of about 41,000 y., and with a maximum range on the order of about 2.5°. Figure 28.4 is a graph of the change in tilt calculated for the past 500,000 y. Although the amount of tilt does not affect the total annual insolation received by the earth, it affects the latitudinal distribution of that insolation, as we have already seen (Figure 28.10). Increased tilt allots a greater proportion of the energy to the northern-hemisphere high-latitude zone at time of summer (June) solstice. Greater tilt means greater seasonal contrast in both hemispheres.

The suggestion that orbital changes that influence the earth-sun distances in relation to season could affect global climate was first put forward in 1830 by English astronomer John Herschel (Kerr, 1978, p. 1144). Specific application of the principle to the cause of glaciations seems to have gained serious attention of science in the 1860s with the work of James Croll, a Scot, who was an active participant in the great debate over Agassiz's glacial hypothesis. Croll's paper, published in the Philosophical Magazine in 1864 and subsequent works (1875), made use of the idea that when eccentricity was greatest, the hemisphere that enjoyed the occurrence of aphelion in its winter season should experience a long, cold winter of such severity that ice sheets would develop, assuming a favorable situation of land areas. At the same time the opposite hemisphere would be experiencing a short, hot summer and would be in an interglaciation (Brooks, 1949, p. 103). The precession cycle would call for a reversal in hemispheric conditions every 10,500 y. When orbital eccentricity was small, conditions would be unfavorable for ice-sheet formation in either hemisphere. This means that the major ice ages would follow the eccentricity cycle of roughly 100,000 y. What Croll did not know was that glaciations are experienced by both hemispheres at the same time, and not in alternation. Thus, abandonment of his hypothesis was assured.

The Milankovitch Curve Revived

We skip now to 1920, when a Yugoslavian mathematician, Milutin Milankovitch, entered the scene with a series of elaborate calculations taking into account two variable factors (shown in Figures 28.9 and 28.10): (1) Changing distance between earth and sun; (2) changing tilt of the earth's axis. Milankovitch did not consider the eccentricity cycle to be important, so he limited his analysis to combining the axial precession cycle with the tilt cycle. It was necessary to select a particular latitude zone for each curve he plotted and to calculate the intensity of solar radiation for that zone. His 1930 curves revealed conditions favorable to glaciations in a period between -175,000 and -250,000 y., but favorable to mild, nonglacial conditions between -300,000 and -400,000 y. Milankovitch's scientific career ended in 1941, long before modern radiometric and paleomagnetic methods were introduced to establish an absolute chronology of the climate cycles revealed in deep-sea cores.

New and greatly improved calculations of the orbital and tilt cycles were made in 1953 by Van Woerkom and in 1968 and 1972 by A. D. Vernekar. Thus, by the mid-1960s the new curves were available to fuel a revival of the Milankovitch hypothesis. Figure 28.11 is one such Milankovitch-type curve, calculated for latitude 65°N. Professor Wallace S. Broecker, a geochemist, and his collaborators at the Lamont-Doherty Geological Observatory of Columbia University now began to correlate the oxygen-isotope data of deep-sea cores with Van Woerkom's curves for summer insolation in high northern latitudes (Broecker, 1966). Dating of the isotope cycles was made possible by use of the protactinium-231/thorium-230 method. Broecker claimed to have found the necessary

support for the astronomical hypothesis of glacial cycles, as this and later successors of the Milankovitch hypothesis are generally called. Correlations were found between the insolation cycle and both sea-level changes and oxygen-isotope ratios over the past 140,000 y. As you might expect, Broecker's interpretations had both supporters and detractors.

About ten years later, the astronomical hypothesis was given new impetus by a group of scientists who had organized their research to concentrate upon the full set of global climatic and oceanographic conditions that prevailed about -18,000 y., when the last, or Wisconsinan, glaciation was in its final peak stage. Their project was given the name CLIMAP (Climate: Long-range Investigation Mapping and Prediction). Three scientists of this project, James D. Hays of the Lamont-Doherty Observatory staff, John Imbrie of Brown University, and N. J. Shackleton of Cambridge University, published in 1976 a major article with the bold title: "Variations in the Earth's Orbit: Pacemaker of the Ice Ages (Hays, Imbrie, and Shackleton, 1976). They reworked the available deep-sea core data on oxygen-isotope ratios and on sea-surface temperatures revealed by abundances of radiolarian species. New mathematical methods, including spectral analysis, were used to find peaks in the core data. These peaks were compared with spectral peaks in the astronomical data. The authors isolated peaks in the climatic record of the cores corresponding with the astronomical tilt period, the equinox precession cycle, and the eccentricity cycle.

Figure 28.12 shows their curves of oxygen-isotope ratios (showing ice-volume fluctuations) and estimated sea-surface temperatures (inferred from radiolarian assemblages). Superposed on these curves is the curve of the astronomical eccentricity cycle. You can make your own visual appraisal of the degree of correlation (or lack of it) between the climate curves and the eccentricity curve. It looks to me as if the correspondence in curve peaks is good (one-on-one relationship), but there is a distinct lag in the climate peaks behind the eccentricity peaks. This lag relationship might seem reasonable if the eccentricity is the forcing cause. Surprisingly, the 100,000-y. eccentricity component is dominant over the 42,000-y. and 23,000-y. components--surprising, because the eccentricity effect is the weakest of the three as an astronomical phenomenon. In a summary, the three scientists flatly stated: "It is concluded that changes in

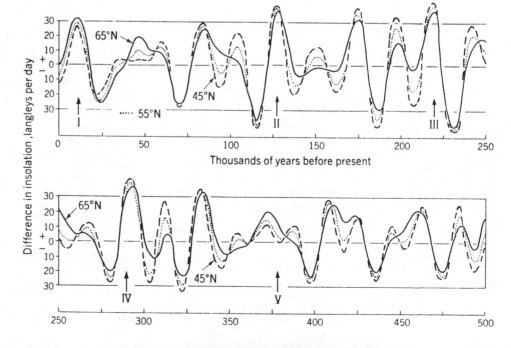

Figure 28.11 Fluctuations in daily insolation in summer at three northern latitudes over the past 500,000 y., based on calculations made by A. D. Vernaker, 1968. Zero value represents the present solstice insolation for latitude 65° N. Roman numerals inidcate inferred terminations of glaciations. (Based on data from Figure 10 in same source cited for Figure 28.9.)

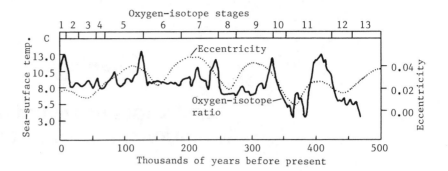

Figure 28.12 Oxygen-isotope ratio curve of Figure 26.11 superimposed on the eccentricity curve of Figure 28.8. (Based on same data sources as cited for figures 26.11 and 28.8.)

the earth's orbital geometry are the fundamental cause of the succession of Quaternary ice ages" (p. 1131). (They are referring to cycles of glaciation and interglaciation within the late Cenozoic Ice Age.)

Correlation between two different categories of data--in this case insolation data versus deep-sea core data--leaves a great gap in explanation. Insolation feeds into climatic variables, such as air temperatures, seawater temperatures, atmospheric carbon dioxide content, cyclonic activity, precipitation in the form of snow, and rates of evaporation and melting of snow and ice. The atmospheric and oceanic processes work in complex system-cycles, in which various forms of feedback are active. In the years that followed, the researchers of the CLIMAP group began to give increased attention to the links between insolation and the growth and disappearance of ice sheets. In this endeavor, the group was joined by William F. Ruddiman and Andrew McIntyre, both of the Lamont-Doherty Geological Observatory research staff. (McIntyre is a professor in Queens College of the City University of New York.) In the late 1970s and early 1980s they investigated the oceanic mechanisms that might amplify a 23,000-y. ice-volume cycle (Ruddiman and McIntyre, 1981; 1984). They found various time lags and time leads among the interacting parts of the earth's climate system. Because the Northern Hemisphere ice sheets have dominated during glaciations, the investigators looked at the sequence of changes leading to ice-sheet growth. Their findings were summarized as follows:

> Based on the record of the deep-sea cores, the steps that led into the most recent glacial period were: cooling of the circum-Antarctic Ocean surface, with an expansion of Antarctic sea ice; then several thousand years later, the beginning of accumulation of a significant volume of ice, mainly on continents of the Northern Hemisphere, and probably around the margins of the Arctic Ocean and northeastern Canada; this was followed several thousand years later by the beginning of a large amplitude cooling of the subpolar gyre in the North Atlantic Ocean, much of which had remained at or near full interglacial warmth for the first several thousand years of ice-sheet growth. (Ruddiman, McIntyre, Hays, 1979, p. 29)

They found the circum-Antarctic sea to be highly responsive to the astronomical cycles, with little or no lag in the response. The Northern Hemisphere lag in its sequence of responses seems capable of explanations that involve oceanic and atmospheric systems with feedback mechanisms.

The work of this research group shows promise of elevating the astronomical hypothesis to full status as a major scientific hypothesis complete with numerous, diverse interacting natural systems. Vigorous criticisms have been voiced by scientific colleagues, and these are aired at periodic conferences. Control of glaciations by volcanism continues to be supported by some scientists as an alternative hypothesis. Regular encounter of our solar system with dust lanes or molecular clouds in the spiral arms of the galaxy remains an attractive hypothesis. Nevertheless, we may be seeing in the revival of the Milankovitch hypothesis the makings of what might be called a paradigm, or ruling general hypothesis, of glacial/interglacial cycles.

Creation scientists have little or nothing to say about the updated Milankovitch hypothesis, so far as I can tell. One might, however, venture to predict their response. Perhaps their best argument might be that the astronomical cycles represent unwarranted extrapolations based on scarcely more than a few decades of modern, precision astronomical observations. Creation scientists have been sharply criticized by mainstream scientists for extrapolating the history of the earth's magnetic field on the basis of observations of less than two centuries, and for extrapolating a history of the speed of light from an equally brief observational record. In support of the extrapolation of astronomical cycles, it might help to point out that they are plotted in strict conformance with the principles of solar-system mechanics, greatly refined since the time of Sir Isaac Newton. Nevertheless, it is well to keep in mind that the earth's motions before human history was recorded have left no direct trace that can be found and examined. Only the indirect influences through climate have generated the geologic record we are reading. The proposition that the geologic record derives indirectly from astronomical controls is the hypothesis itself, and no hypothesis can contain validating evidence within the bounds of its statement as a proposition. Perhaps we will some day find independent confirmatory evidence of the Milankovitch curve in surficial materials on the moon, where, in the absence of atmosphere and hydrosphere, intensity of the solar beam has directly generated a continuous record unbroken over half a million years, subject to dating by radiometric methods.

Already, geologists have begun to discover what they interpret as convincing manifestations of the Milankovitch cycles in rhythmites of Phanerozoic eras. Rhythmic bedding sequences have now been fitted to Milankovitch variations in strata of Early and Late Cretaceous, Late Triassic, and Early Jurassic ages (Olsen, 1986b; Kerr, 1987; Laferriere, Hattin, and Archer, 1987). These are exciting discoveries, and we may expect many more to be disclosed in coming years.

Credit

1. Much of the descriptive text of this chapter is taken from A. N. Strahler, The Earth Sciences, 2d ed., and Physical Geology, Harper & Row, Publishers, New York. Copyright © 1971 and 1981 by Arthur N. Strahler.

Chapter 29

Landscapes of Denudation—
Evolutionary or Catastrophic?

About 1890, a Harvard professor named William Morris Davis put forward a conceptual model of the evolution of an entire landscape through a life cycle analogous to that of a living organism--he called it a geographical cycle (Davis, 1899). Because the cycle focuses on the denudation (down-wearing) of a continental surface, it is now usually referred to as a denudation cycle. There is no doubt that Davis had in mind the life cycle of an individual animal or plant, for he wrote:

> The larva, the pupa, and the imago of an insect, or the acorn, the full-grown oak, and the fallen old trunk, are no more naturally associated as representing the different phases in the life history of a single organic species than are the young mountain block, the maturely carved mountain peaks and valleys, and the old mountain peneplain, as representing the different stages in the life history of a single geographical group. . . . A young land form has young streams of torrential activity, while an old land form would have old streams of deliberate, or even of feeble, current, as will be more fully set forth below. (P. 254)

Actually, Davis's colorful adjectives, used profusely throughout his writing, suggest the image of a higher mammal, and perhaps specifically a human, as his model of the life cycle of a landform. The idea caught on rather well and for some decades Davis's model of landscape evolution was quite popular. But obviously, he was not referring to the Darwinian model of evolution of species that was then enjoying considerable popularity.

Did Landscapes Evolve?

Creation scientist Steven A. Austin, writing in ICR's Impact, titles his essay "Did landscapes evolve?" (Austin, 1983) In so doing, he makes the error referred to in the previous paragraph. He says: "Davis' system follows the concepts of organic development which also swept the scientific community in the late nineteenth century (even the stages 'youth', 'maturity', and 'old age' correspond nicely with organic evolution!)" (p. i). The statement in parentheses is in error; the stages mentioned correspond not with evolution of species, but with the developmental life cycle of a single individual of a species. Davis's life-cycle concept applied to individuals would also be valid in the framework of the creationists' model of special creation of each species.

Austin goes on to describe Davis's cycle of landscape development as uniformitarian in concept, involving slow and continuous change of a landscape. Austin contrasts this model of change with that of the creationists, who insist that the processes that first formed the landscape were catastrophic:

> The ancient processes which formed the landscape would be discordant with modern processes acting on the landscape. Modern erosion processes would

be viewed as entirely destroying an ancient landscape, not transforming it from one equilibrium stage to another. Such a landscape would contain relict landforms, surface features which were created by erosional or depositional processes no longer acting. Relict features on the earth's surface make the landscape appear as a "museum," and such features, in contrast to the Davisian system, would have a great degree of permanence. (P. ii)

Austin has thrown down a smokescreen of confusion upon the issues involved here. We need to get things straightened out!

The creationist view of landscape evolution follows their "party line" closely in postulating residual catastrophic activity persisting after the Flood, but subsiding rapidly and leading into the decay phase of irreversible increase in entropy. This means that the potential energy of position above a datum is not now being restored. Translated into terms of landform development, the continental crust no longer experiences significant amounts of uplift by tectonic and epeirogenic (broad upwarping) processes, capable of restoring the potential energy of elevation that allows gravity-powered erosion processes to operate. Neither is there more than trivial upbuilding of the surface by volcanic extrusion. True, the creationists are prepared to admit that following a major earthquake the land level is sometimes found to be raised and that occasionally a volcano spews forth ash and pumice or some lava flows. They expect a little catastrophe to persist, but only as the earth's last convulsive twitchings. Quite logically, under these postulates, landscapes can only undergo decay, i.e., downwasting of the surface accompanied by transport of sediment to the ocean shorelines and thence into residential sites on the ocean floor. Isostatic compensation will accompany this process, partially restoring the mass that is removed, but not changing the trend. This creationist scenario is described by Austin as "non-evolutionary or catastrophic," whereas Davis's model of landscape evolution and other models proposed by mainstream geomorphologists to improve on it or replace it are expressions of "evolutionary-uniformitarianism," tainted by Darwinism, and thus inherently false (1983, p. ii).

A parenthetical note is in order here. Although I have not seen reference to it in publications of the Institute for Creation Research, some sects of Christian funda-mentalists preach that catastrophic or cataclysmic events are presently on the increase, in apocalyptic anticipation of the coming Armageddon, or final battle, between forces of good and evil (Rev. 16:14-16).

Of course, "evolutionary-uniformitarianism" is a pure fabrication on the part of the creationists; it simply does not exist in the thought and writings of mainstream geologists, nor did it have a place in the thinking of such pioneers as G. K. Gilbert, J. W. Powell, C. E. Dutton, or even Davis, all of whom recognized the importance of tectonic and volcanic activity in producing landscape features and periodically restoring landmasses to elevated positions and thus bringing about intensifications of denudation rates. Davis made this concept an integral part of his denudation system--he introduced the term rejuvenation to describe what repeatedly takes place through geologic time. If it were not for rejuvenation, the continents would long ago have been reduced to low plains and submarine platforms.

Today, mainstream geologists have found in plate tectonics a model of uplift mechanisms by which denudation processes can be rejuvenated. Crustal uplift can be envisioned as persistent and continuous over hundreds of millions of years in mountain and island arcs adjacent to subduction boundaries. Uplift of mountain ranges in these active zones can equal or exceed the rate of denudation. Whereas Davis's ideal denudation cycle postulated a single initial uplift, followed by crustal stability, we must now recognize that his is not the only model. Perhaps Davis's cycle can work in the case of continental collision, where a sudden and catastrophic thickening of the crust by underthrusting causes a rapid rise of a great mountain mass, such as the Himalayan range and its adjacent Tibetan plateau, an event that could be followed by millions of years of relative quiescence and crustal stability.

Do you realize what I have written into the last sentence? I have given you the creationists' own model of initial catastrophism followed by decay and entropy increase! Yes, Davis wrote their very own scenario in 1899. It contains nothing that can be construed as "evolutionary" in the sense of organic evolution of species. It starts with catastrophe, that the creationists insist upon; it continues with destruction of the ancient mountain landscape (the collision suture). True to the creationist model, Davis calls for maximum rate, scale, and intensity of both tectonic and erosional processes at the outset of the entire history of that landscape. The only difference seems to be that Davis makes no mention of a biblical Flood; we can safely assume that although he was of the same Quaker faith as some of the early creationist writers, his geomorphology was fully naturalistic. Usually, the creationists beat the straw person of an abandoned uniformitarian concept, but in this case they have turned to self-inflicted torture.

Post-Flood Denudation--Was There Enough Time?

Creation scientists find the amount of removal of rock by denudation following the Flood to be so great as to require catastrophic erosion rates, far greater than those observable today. They also find evidence of post-Flood accumulations of alluvium (stream-deposited sediments) so thick and widespread as to require catastrophic rates of sediment transport and deposition. Considering their inflexible premise--a one-year Flood occurring in 2,350 B.C., give or take a few decades--their inference of catastrophism continuing after the Flood is not only reasonable, but inescapable.

Let us see what Whitcomb and Morris have to say about this matter in their major creationist work, The Genesis Flood (1961, p. 150). Referring to the Uinta Mountains, one of the great uparched or updomed mountain ranges characteristic of the central and southern Rocky Mountains, they quote a paragraph from Nevin M. Fenneman (1931, p. 147) in which he explains how the Uinta range (an east-west trending anticlinal structure) was buried in alluvium, from which cover a major stream became superimposed across the anticlinal structure. He is referring to the Green River that cuts through the eastern end of the range in a great gorge, the Ladore Canyon. Whitcomb and Morris comment as follows:

These mountain-burying sediments are believed to have been derived from the wearing down of more then 7,000 cubic miles from the summit of a great

fold, filling the surrounding area to a depth of at least several thousand feet, up almost to the summits of the remaining mountains themselves! After the formation of the now-anomalous rivers on these tremendous alluvial deposits, another uplift is postulated, permitting a new cycle of dissection to begin. This sort of phenomenon is frequently encountered in the study of geomorphology and provides still another evidence that present-day rates of erosion and deposition cannot account for the ancient deposits as they are found. (Pp. 150-51)

The paragraph is fully acceptable in mainstream geology, except for the last sentence. Contrary to that statement, present-day rates of erosion and deposition easily account for all waterlaid sedimentary deposits in all parts of the geologic record. All that is needed to complete the accounting is ample time; mainstream geology has ample time. In the case of the Uinta Mountains, mainstream geology has a full 65 million years allotted to the erosion, burial, and exhumation of the range. Can the creation scientists cram these events into 4,350 years by invoking catastrophism? Let us do some rough figuring based on geologic field observations all parties can accept as reasonably sound.

Our plan of study is to look at archlike and domelike crustal uplifts from which a thick succession of conformable platform strata has been removed. That the strata formerly extended over what is now the central uplifted area is inescapable from examination and measurement of the exposed edges of the strata, upturned in a belt that surrounds the uplift. Where the strata are unquestionably of marine origin and are continuously present in a large area surrounding the uplift, they can safely be assumed to have once been continuous over the area of the uplift. Geologists have measured the thicknesses of the individual formations in the peripheral zone of the uplift, and these thicknesses can be totaled for a reasonably reliable estimate of the minimum thickness of rock removed from the summit zone of the uplift. To this figure may be added an estimate of the thickness of older basement rocks (igneous and metamorphic) that were later removed below the level of the conformable sedimentary sequence. We will, however, make no attempt to include the latter kind of estimate. Two regions, widely separated in location and differing greatly in geologic history serve as sources of examples. One is that province of the Appalachian Mountains known as the Newer Appalachians (or Folded Appalachians); the other is the combined region of the Middle and Southern Rocky Mountains and the Colorado Plateau.

In south-central Pennsylavania, strata of Paleozoic age were deformed into large open folds during the Permian collision that marked the final closing of the Iapetus Ocean, an event called the Alleghenian (Hercynian) Revolution. (See Strahler, 1981, pp. 355-57.) The folds show moderate overturning and some high-angle overthrust faulting, but are otherwise quite simple in structure. Figure 29.1 is an idealized diagram of a symmetrical fold showing the present land surface, which bevels the folds. Total minimum erosion since folding is represented by the stratigraphic thickness of beds between points A and B, the axes of an anticline and syncline, respectively. That measured thickness would also represent the vertical distance from point A to point C. Table 29.1 shows that the key points are taken at the base of the Bloomsburg redbeds and the top of the Mauch Chunk redbeds, with a total thickness of approximately 3,700 m. Additional younger strata may have been present here, but we do not need to speculate about that possibility.

A reasonable estimate of the highest rate of fluvial denudation in high, small mountain watersheds is from 1.0 to 1.5 m/1000 y. (Schumm, 1963, p. 3). This is to say

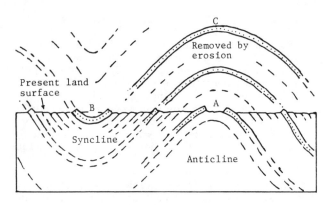

Figure 29.1 Folds beveled by erosion. The minimum amount of removal is represented by the stratigraphic column between A and B, equivalent to that between A and C. (A. N. Strahler.)

that on the average the surface is lowered by that amount each year. Multiplying the stratigraphic thickness, 3,700 m, by the lesser value (1.0 m/1,000 y.) gives a total elapsed time of 3.7 m.y. For the creationists' post-Flood version (3,700 m in 4,350 yr) the rate must averge 0.84 m/y., which is roughly 1,000 times faster than the mainstream estimate.

Actually, our estimate is unrealistic in two ways. First is that the rate we used is reasonable only for the very early stages in denudation, immediately following the cessation of orogenic uplift. As average surface elevation of the mountain mass is reduced, the rate itself diminishes--probably exponentially--and would rapidly

Table 29.1   Thicknesses of Stratigraphic Sections

A. Folded Appalachians, South-Central Pennsylvania.

Bloomsburg (Silurian) through
Mauch Chunk (Mississippian)   3,700 m

B. Front Range, Colorado Rockies.

Sawatch (Cambrian) through
Laramie (Cretaceous)   2,750 m

C. Black Hills Uplift, Wyoming/
South Dakota.

Deadwood (Cambrian) through
Lance (Cretaceous)   1,450 m

D. Grand Canyon/Kaiparowitz Region,
Colorado/Utah.

Moenkopi (Triassic) through
Kaiparowitz (Cretaceous)   2,650 m

Data sources: (A) Willard, B., and A. B. Cleaves, 1938, A Paleozoic Section in South-Central Pennsylvania, Bulletin G-8, Plate 10, Pennsylvania Topographic and Geologic Survey, (B) Henderson, C. W., 1933, Colorado. Guidebook 19, Plate 2, Internat. Geological Congress, XVI Session, Washington, D.C. (C) O'Hara, C. C., 1933, The Black Hills. Guidebook 25, Figure 2, Internat. Geological Congress, XVI Session, Washington, D.C. (D) Gregory, H. E., and R. C. Moore, 1931, The Kaiparowitz Region. Profes-sional Paper 164, Plate 5, U.S. Geological Survey, Government Printing Office, Washington, D.C.

drop to much lower values. For example, in the contiguous 48 United States, the present average denudation rate is estimated to be only 0.03 m/1000 y., which is less than 1/30 of the rate we have used (Judson, 1968, p. 367). What this diminishing rate means is that the creationists would need to postulate an enormously faster rate of denudation in the immediate post-Flood years to compensate for a greatly reduced rate in the past one or two millenia. On the other hand, mainstream geology will be forced to increase the time estimate substantially, but the hypothesis is open-ended in that direction and easily absorbs the increase.

The second reason why the initial estimate is inadequate is because of the need to accommodate the principle of isostasy. As rock is removed from a region by denudation, the crust must respond by rising to restore isostatic equilibrium. A rough working estimate of the amount of isostatic rise required is that for every 5 m removed by denudation, isostatic uplift restores 4 m. This means that our initial figure of 1.0 m/1000 y. nets only 0.2 m/y. of surface lowering. The creationists will need to inflate their denudation rate by a factor of five to take isostasy into account; the mainstream geologists will have to multiply their total time by a factor of five. The contrast between creationist catastrophism and mainstream uniformitarianism then assumes gigantic proportions.

A Model of Fluvial Denudation

Some years ago, in an attempt to produce a reasonable model of the denudation of a rapidly uplifted landmass, I made use of figures given by Stanley Schumm and Sheldon Judson (references cited above). I assumed rapid tectonic uplift to produce a mountain range with average surface elevation of 5 km. (That's the "catastrophism"!) Uplift was accompanied by a rapidly increasing rate of denudation that one would expect of any growing mountain range (Strahler, 1981, pp. 437-38). At an arbitrary "time-zero," with uplift having ceased, denudation assumes an initial rate of 0.2 m/1000 y. (200 m/m.y.), taking isostatic compensation into account. Figure 29.2 shows the graph of decreasing average elevation and diminishing denudation rate. Figure 29.3 is a graphic display of the changing profile of the mountain range. I had assumed an exponential decay with a half-life of 15 m.y. The situation at each half-life point is shown in these diagrams. What is important is that after four half-lives (60 m.y.) the average elevation has fallen to about 300 m and the denudation rate is reduced to 12 m/m.y. (0.012 m/1000 y.). This landscape might correspond with the concept of the peneplain, or penultimate plain of denudation. It is not likely that the crust could remain free of tectonic disturbances for 60 m.y. Instead, the model curve would be terminated much earlier by some tectonic event. Nevertheless, denudation might continue on another exponential decay schedule, and the system would tend toward achievement of a peneplain. Based on the model, the Appalachian region of south-central Pennsylvania might have attained its present erosion level in about 45 m.y. The Alleghenian Revolution occurred about -300 m.y., at the close of the Paleozoic Era. During the ensuing Triassic Period the region underwent great tectonic activity in the form of block-faulting, when crustal extension set in as a prelude to opening of the Atlantic ocean. Tectonic uplift may have compensated to a large extent for denudation during that time, but things quieted down in the Jurassic Period. By Cretaceous time the crust was stable in eastern North America and the denudation cycle probably followed the ideal curve for many tens of millions of years. At least one uparching (epeirogenic) occurred in early Cenozoic time, and perhaps several more occurred after that. However, the denudation process continued through to the present time as the marine geoclinal deposits of the stable Atlantic margin seem to suggest.

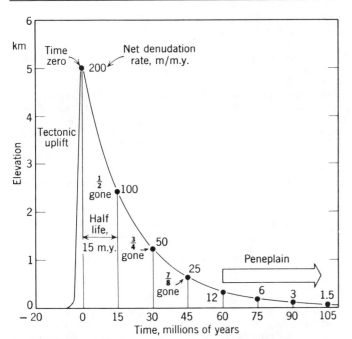

Figure 29.2 Graph of decrease of average surface elevation with time, as shown in Figure 29.3. (Copyright © 1981 by Arthur N. Strahler.)

Denudation of the Rocky Mountains

The second region where thickness of sedimentary rock removed by denudation can be studied combines two geologic provinces that share a common geologic history. One of these, the combined Middle and Southern Rocky Mountains, can be characterized as a collection of structural uplifts (domes, anticlines) with intervening sedimentary basins; these are outlined in Figure 29.4. The other is the adjacent Colorado Plateau, characterized by expanses of more-or-less horizontal platform strata separated by sharp flexures (monoclines) and normal faults that give a generally stepped structural pattern, resulting in plateaus of a wide range of altitudes. Uplifts of the Rockies show extensive removal of thick sequences of Paleozoic and Mesozoic strata, leaving exposed large core areas of Precambrian igneous and metamorphic rock. Of the numerous examples, data for two are shown in Table 29.1. The Front Range in Colorado and its northward extension into Wyoming as the Laramie Range is essentially a long arch, or anticline. Figure 29.5 is a restoration of the strata that would have arched over the Front Range in the Denver/Boulder area, if no denudation had occurred. Although formations of Paleozoic age occur at the bottom of the stratigraphic sequence, those of Mesozoic age are the most important. Triassic and Jurassic strata are present, but not in great thickness. It is the strata of Cretaceous age that make up most of the column. During Cretaceous time a great inland seaway existed here, accumulating marine strata up to 6,000 m thick in some areas. In the Front Range section, shown in Figure 29.5, strata of Cambrian through Cretaceous age total about 2,750 m (Table 29.1); this is our minimum denudation thickness. In the Black Hills uplift of Wyoming and South Dakota, the equivalent column is less: 1,450 m. These thicknesses are greatly exceeded by that of strata formerly covering the Uinta arch, where the Upper Cretaceous strata alone were more than 6,000 m thick (Kay and Colbert, 1965, p. 441, Figure 18-23A).

Uparching of the Front Range, the Black Hills uplift, the Uinta arch, and others like them occurred at the close of the Cretaceous Period, about -65 m.y., in an event known as the Laramide Orogeny. The denudation rate must have been extremely fast as the uplifts grew, and enormous quantities of coarse sediments accumulated in

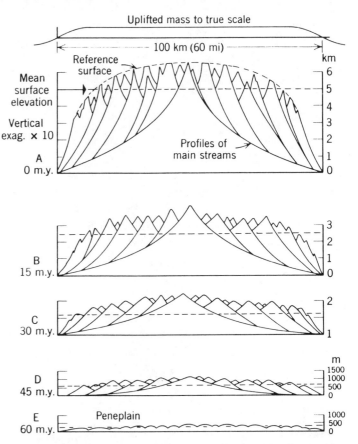

Figure 29.3 Schematic diagram of landmass denudation. In this model, the average surface elevation is reduced by one-half every 15 m.y. (Copyright © 1981 by Arthur N. Strahler.)

basins between the uplifts (Paleocene time). Mainstream geology assigns about 10 m.y. to this first denudational phase. Layers of sediment formed sloping plains (alluvial fans) that spread headward far up over the fringes of the receding mountains. Tectonic activity persisted into Eocene time, elevating the domes and arches by folding and faulting and reviving the rapid denudation. Throughout Eocene, Miocene, and Oligocene time stream-deposited sediment (alluvium) filled the intermontane basins, while rising sheets of alluvium buried large parts of the uplifts, allowing major streams to establish courses across buried core rocks of the uplifts. The sheets of alluvium also spread far eastward over what is today the High Plains belt. The creationists are interested in these great accumulations of sediment. While they accept the explanation of mainstream geology for the processes involved, they construe it in a different light, unique to the tenets of Flood geology:

> There is no reason to question the general correctness of the nature of the geomorphic origin of these plains, as attributable to widespread and overlapping alluvial fans formed by heavy-laden rivers coming down from the recently uplifted mountains to the west. The significant thing, however, is that here again one must visualize a phenomenon for which there is no parallel in the modern world except on a much smaller scale. The principle of uniformity is misnamed if, to interpret ancient phenomena on the basis of the present, the expedient of extrapolation must so continually be employed and to such a great degree. The example chosen is one taken almost at random from many similar deposits around the world. It seems that almost everywhere one looks, he can find evidence of widespread deposition, either alluvial or deltaic

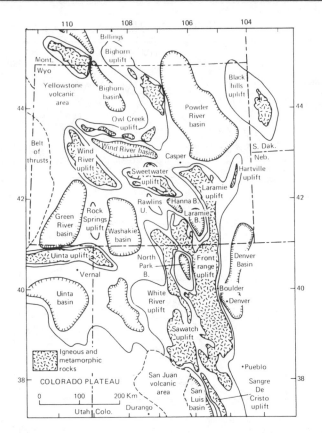

Figure 29.4 Outline map of the major structural uplifts and basins of the Middle and Southern Rocky Mountains. (Based on data of <u>Tectonic</u> <u>Map</u> <u>of</u> <u>the</u> <u>United</u> <u>States</u>, 1944, American Association of Petroleum Geologists and National Research Council.)

in nature, of magnitude quite beyond that of any deposits being formed in the present. (Whitcomb and Morris, 1961, p. 150)

I should add that the above paragraph is intended to include these authors' earlier reference to the huge troughs of marine deposition, formerly known as geosynclines, that we recognize today as the forearc and backarc troughs adjacent to active subduction boundaries. There is no question that contemporary sediment deposition is rapid in those troughs. As to the absence of

rapid deposition of alluvium throughout the world today, the creationists' claim is simply nonsense. One has only to travel to Death Valley and other similar tectonic basins to find sheets of gravel and sand deposited in recent years by raging streamflows emerging from the mouths of canyons flooded by thunderstorm downpours. These events have been witnesssed repeatedly by humans. Accounts of the destructive flows are seen at least a few times each year in the national news media. The creationists' statement in the final sentence of the earlier passage quoted above ("that present-day rates of erosion and deposition cannot account for the ancient deposits as they are found") is an assertion of no substance whatsoever. Such statements should be turned around to read: "Geologic evidence of vast amounts of erosion of bedrock and vast accumulations of terrestrial and marine sediment defy explanation by processes limited in time to a few thousand years since the Flood of Noah."

Denudation of the Colorado Plateau

And now, to the final case study, the Colorado Plateau of Utah, Arizona, Colorado, and New Mexico, where denudation on a vast scale is documented by the stratigraphic record. The Grand Canyon of the Colorado River in Arizona is trenched into a "plateau" underlain by strata ranging in age from Cambrian through Permian, i.e., Paleozoic strata. These total a good 950 m in thickness, while the bottom of the inner gorge lies another 300 m deeper, carved in Precambrian rocks (Figure 29.6). The canyon rim and its adjacent plateau surfaces are now at an elevation of 2,000 to 3,000 m above sea level, but there is good evidence that strata totalling at least 2,650 m in thickness were formerly present on top of what is now the plateau surface, as shown in Table 29.1. Strata ranging in age from Triassic through Cretaceous are exposed in lines of cliffs to the north of Grand Canyon. These are the Vermilion, White, and Pink cliffs shown in Figure 29.6. The structure through which the Grand Canyon is cut at its deepest point is best described as a broad arch, or anticline, bounded on the eastern side by a sharp downflexing of strata (a monocline) and on the west by major normal faults; it is named the Kaibab Plateau. The crest of the arch diminishes in elevation to the north, as shown in Figure 29.6. This arched structure was produced in the Laramide orogeny at the close of the Cretaceous Period. As in the case of the Rocky Mountain uplifts, denudation was intense in early Cenozoic time, rapidly stripping the Mesozoic strata from the summit of the arch. Repeated minor regional uplifts doubtless occurred throughout

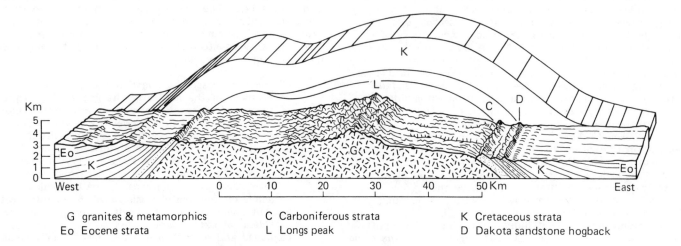

G granites & metamorphics   C Carboniferous strata   K Cretaceous strata
Eo Eocene strata   L Longs peak   D Dakota sandstone hogback

Figure 29.5 Stylized and generalized cross section of the Colorado Front Range, north of Denver. The restored arch structure is imaginary, because erosion would have

removed the strata as the uparching occurred. (Based on data of W. T. Lee, U.S. Geological Survey.)

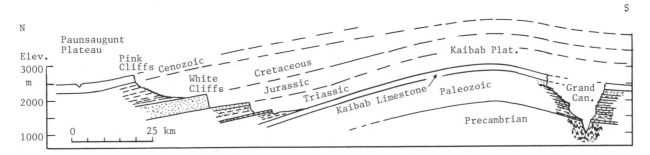

Figure 29.6 Generalized cross section from the Pink Cliffs in Utah to the Grand Canyon in Arizona. Mesozoic strata formerly over the region of the Kaibab Plateau are indicated as restored. (Data of G. K. Gregory, 1933, Guidebook 18, p. 19, Internat. Geological Congress, XVI Session, Washington, D.C.)

Cenozoic time, but the major part of the regional elevation of the crust that brought the arched Paleozoic strata to their present high position may have occurred as recently as -10 m.y., or late Miocene time (McKee and McKee, 1972, p. 1924). The Colorado River probably did not come into existence in its present course until after the final uplift began (McKee et al., 1967, p.61). This event challenged (so to speak) the mighty new Colorado to carve its great gorge. So the creationists are faced with a double feature to cram into post-Flood time: removal by denudation of over 2,600 m of Mesozoic strata, followed by river trenching through an additional 1,400 m of solid rock together with the opening out of the canyon to an enormous chasm 10 km wide from rim to rim.

As a closing note, there is some remarkable evidence, based on radiometric dating, that the Grand Canyon had already been cut to its present depth about one million years ago. At a point along the canyon known as the Toroweap, basaltic lavas erupted from the side of the gorge, forming a lava dam in the bottom of the gorge. That dam has since been completely cut through and the present level of the river is now only a few meters below the base of the lava dam. The lava has been dated as having solidified at -1.2 m.y., plus or minus 0.6 m.y. (Damon, 1965, p. 42). This evidence establishes the presence of the river at a position close to its present level at least a half million years ago. For mainstream geologists this one bit of evidence is fully sufficient to rule out the creationists' entire schedule of Creation, the Flood, and post-Flood events. But, of course, the creationists summarily reject the validity of all radiometric age determinations.

To support their scenario of enormously rapid rates of erosion immediately following the Flood, creationists emphasize that the strata--all of which were deposited in a one-year Flood--must have been soft and unconsolidated, and could thus be easily eroded and transported (Whitcomb and Morris, 1961, p. 153). One can visualize huge masses of soft, water-saturated clays, muds, and sands sloughing off the flanks of the great anticlines and arches in swift mudflows of enormous length and depth, flowing unchecked to the ocean. There is no geologic evidence of the catastrophic denudation proposed by the creationists, but the scenario can be challenged for other good reasons.

If all fossiliferous strata of the globe were in a soft and unconsolidated condition following the single year of deposition some 4,350 years ago, how could the great bulk of that sediment have become consolidated (lithified) into the dense, hard condition in which it exists today? Lithification, the process of transformation of soft, water-saturated sediment into rock, involves certain physical and chemical changes--changes that could not possibly have taken place in only a few thousand years. We are talking about strata occurring in thicknesses on the order of several thousand meters. Thick shale formations that were deposited as muds and clays must have undergone

the expulsion of water (dewatering) in large amounts. Sufficient dewatering to yield a dense, fissile shale could not occur in such a short time as 4,000 years in thick clay-rich formations of extremely low permeability. Carbonate sediments pose a particularly difficult problem of lithification (Friedman and Sanders, 1978, pp. 147-58). Whereas modern carbonate sediments have high porosity (60-70 percent), ancient limestones found in strata of Paleozoic and Mesozoic age have porosities of less than 2 percent or even than 1 percent. If compaction under load were the answer, it would involve an enormous amount of dewatering. If the addition of new calcium carbonate to fill the pore spaces is the answer, that process would take vastly greater spans of time than the creationists have available. Then there are sandstone formations, now strongly cemented by calcium carbonate or silica, that were formerly loose sand--delta sands or dune sands, for example. (Remember the ancient dune formations in the walls of Zion Canyon and Grand Canyon, formed when the entire earth was covered with waters of the Flood?) Cementation of the great sand layers must have required the slow circulation of ion-bearing solutions over spans of time vastly longer than the few millenia of post-Flood time. These and other processes of lithification of sediment, including many kinds of physical and chemical change, cannot simply be asserted to have operated at the incredible rates demanded by Flood geology. To insist upon such changes having occurred in less than 4 millenia puts the creationists in the position of invoking pseduoscience. Divine intervention in the form of miracles is ruled out in post-Flood time, because God no longer interfered or intervened in natural processes after the Flood waters subsided. It is precisely this interpretation of the Scriptures that puts the creationists "between a rock and a hard place" in creating their scenario of post-Flood events; they can choose only between science and pseudoscience. A difficult choice, indeed!

Underground Denudation in Caverns

Creationists find in limestone caverns and their deposits phenomena that indicate rapid excavation of rock cavities by solution processes and rapid partial filling of those same cavities by mineral deposits (Whitcomb and Morris, 1961, pp. 417-18). They do not spell out the time position of cavern formation in Flood geology, but it is regarded as a recent phenomenon and can be assumed to be of post-Flood time.

The subject was revived by creation scientists of the ICR through an article in Impact by Dr. Steven A. Austin, a geoscience research associate with ICR and Professor of Geology at Christian Heritage College (Austin, 1980). Much of his article is a review of the general subject of limestone caverns and current scientific hypotheses for their origin. In general, the creation scientists accept most of the published literature of mainstream science on processes of limestone solution and

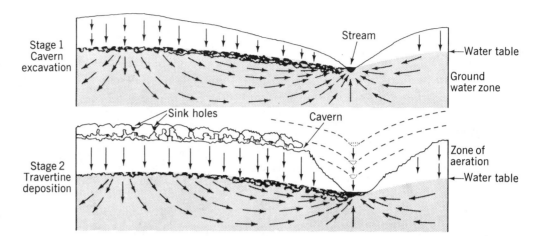

Figure 29.7 Stages in cavern development (Copyright © 1971 by Arthur N. Strahler.)

cavern formation, but suggest that the processes acted much more rapidly in the past than today.

Mainstream geologists favor a hypothesis that requires excavation of the caverns to take place within the ground-water zone, beneath the water table (Figure 29.7). Carbonic acid in solution in rainwater enters the ground surface, is enriched in the soil zone, and percolates down to the water table. Moving through various kinds of passageways in limestone strata--joints (rock fractures) and more permeable bedding layers--the acid reacts with the rock surfaces, dissolving the limestone and carrying along the solution products (calcium ions and bicarbonate ions). Flow paths of the ground water lead to the beds of streams occupying valleys cut slightly below the general level of the water table. In time, a system of open, waterfilled passageways is formed. This system could not have operated under Flood waters, which would have been highly turbid and rapidly depositing sediment on the floor of the world ocean. Only when the Flood waters had receded to expose land surfaces could the process begin. Hence, creationists must limit cavern formation to post-Flood history, spanning less than 5000 years.

In uplands on the continent, major streams deepen their valleys, whether steadily or in sporadic episodes of trenching. When this happens, the surface outlet for the ground-water system is also lowered, and with it the entire water table (Figure 29.7). New caverns are opened up at lower levels in the limestone formation, while the earlier formed caverns come to occupy the zone of aeration. The cavities are air-filled, but water percolates down from the land surface to emerge as drip from cavern ceilings and to flow in underground streams on the floors of tunnels. Water charged with calcium and bicarbonate ions, upon exposure to the air, loses carbon dioxide as gas and this triggers the deposition of calcium carbonate on ceilings and floors as <u>travertine</u>, a form of calcite. The result is growth of a large variety of depositional forms known collectively as <u>speleothems</u>.

If it can be shown that either the excavation of caverns or their subsequent filling must require a vastly longer time to accomplish than the post-Flood limit, literal acceptance of the Genesis chronology is untenable. We turn first to rates of removal of limestone by the process of carbonic-acid reaction.

Austin discusses the problem of estimating the rate of removal of limestone by solution (1980, pp. iv-v). He refers to publications of mainstream geohydrologists, who base their estimates on the volume of water that infiltrates the surface as rainfall or that emerges from the area as stream discharge. Concentration of calcium ions in the discharge flow enables the mass rate of removal per unit area to be determined. Austin uses the data of Thrailkill (1972) for the Sinkhole Plain/Mammoth Cave area in

Kentucky, where solution is presently occurring in ground water moving through an aquifer beneath the plain. Austin comes up with a value of about 60 mm of limestone per 1000 y. as the average equivalent rate of denudation (lowering of the surface). This figure is within the range of values given by J. N. Jennings for several localities in moist middle-latitude climates (Ireland, England, Yugoslavia) (Jennings, 1971, p. 181, Table 6). Austin shows that this rate is equivalent to the annual removal of enough limestone to produce a new section of cavern 1 m in cross section and nearly 60 m long. All well and good, but Austin then remarks: "The high rate of solution of limestone and dolostone should be a matter of alarm to uniformitarian geologists" (p. v). At the rate calculated for this limestone area, he says, a layer of limestone well over 100 m could be completely dissolved off of Kentucky in 2 m.y., assuming that present rates and conditions prevailed through that entire time. The calculation is correct as to thickness, but it can only apply to the Sinkhole Plain/Mammoth Cave region, not to the entire state, because sandstones and shales underlie much of the state. He assumes that the amount of limestone actually removed from Kentucky caverns and sinkhole areas (karst) is only an insignificant fraction of the estimate given above for 2 m.y. The inference, not stated, is that the quantity of limestone actually removed from Kentucky caverns could have been dissolved in a few thousand years. He suggests that the removal rate could have been much faster in cooler, more humid climates that may have prevailed earlier in the history of the area.

Actually, mainstream geologists, referred to by Austin as "uniformitarian geologists," are by no means alarmed about solution removal of the equivalent of 100 m of limestone in 2 m.y. Measurements of the sediment load-- solid matter and dissolved matter combined--carried to the sea by major rivers give comparable rates of denudation for areas of subcontinental size (Judson, 1968, p. 369, Table 3). Measured rates of denudation by running water in steep mountain watersheds runs very much higher-- 1 m/1000 y. and more (Schumm, 1963, p. 3).

A significant study of limestone denudation rates made in Puerto Rico by Ennio V. Giusti (1978, pp. 48-55) of the U.S. Geological Survey provides firm evidence that limestone solution acting over at least 3 m.y. has removed the equivalent of a layer of solid calcite 210 m thick. The original thickness of the limestone formation can be directly measured where both the base and top are exposed. Chemical analyses of water from streams, springs, and wells yields a current rate of denudation in the limestone area of about 55 mm/1000 y. At that rate, nearly 4 m.y. would be required to remove the 210 m of solid limestone. Taking into account that abrasion by running water has also been at work, the figure is reduced to 3 to 3.6 m.y. Thus all the required information is directly measurable and reliance need not be placed on radiometric ages. Even allowing for solution rates larger by a factor of five in the past, the time requirement exceeds by an enormous margin the post-Flood allowance.

Turning now to travertine deposits in caverns, the simplest speleothem is a narrow, straight tube of calcite that grows down from cave ceilings; it is called a straw stalactite, because it is a hollow structure, like an ordinary drinking straw. Straw stalactites have been observed to lengthen at a rate of about 0.2 mm/y. (Moore, 1968, p. 1040). Straw stalactites can reach lengths of several meters, but usually undergo rupture of the fragile wall, leading to the growth of a larger, more massive structure, the stalactite, constructed of layer upon layer of calcite. Directly below the stalactite there accumulates a blunt, massive stalagmite as drops fall from the tip of the stalactite and precipitate their calcite at the point of impact. Eventually the two forms meet to create a single column of travertine, and by similar processes curtainlike and wall-like forms make partitions within the cavern. Travertine also accumulates on the open floors of caverns as dripstone, and at the rims of water pools, forming travertine terraces.

Whitcomb and Morris quote from standard geologic publications statements to the effect that (a) the rate of production of cave travertine is probably highly variable and in any case the rates of accumulation are not known, and (b) stalactite growth rates are extremely rapid, as witness their formation in human-made tunnels (1961, pp. 417-18). In regard to point (a), the creationists say that in post-Flood time, the travertine deposition rate "would be rapid at first, gradually leveling off to the present rates" (p. 418). As to point (b), they say the evidence shows that it is both unnecessary and unreasonable to attribute great lengths of time to the accumulation of cave deposits.

Published studies in mainstream geology do not offer hard figures on growth rates of speleothems--the fragile straws excepted. There are, however, certain constraints on the rate of accumulation of travertine. Calcite deposition in open caves in the zone of aeration depends on a supply of calcium and bicarbonate ions from overlying rock, where they must be released by carbonic-acid activity at the contact between limestone bedrock and the soil cover, or from exposed limestone outcrops. Concentration of atmospheric carbon dioxide being relatively uniform through time, the intensity of carbonic-acid activity depends largely on the annual precipitation and, secondarily, on temperature. Studies of ion concentrations in stream water show that limestone solution is most rapid in moist climates, roughly in proportion to the annual rainfall. Rate of production of travertine can therefore not exceed the rate of production of the calcium and bicarbonate ions by the solution process. Creationists cannot take the liberty of postulating greatly increased rates of this total process in early post-Flood time and later, because it is climate-dependent.

Observation of the growth of stalactites and carbonate coatings from the ceilings of human-made tunnels has little bearing on the problem, because these are extremely fragile, thin-walled structures and represent very small mass accumulations.

Austin refers to a paper by Lloyd W. Fisher (1934) in which observations of rates of stalactite formation are collected from various sources (Austin, 1980, pp. v-vii). Fisher himself observed stalactites growing from concrete arches in an inspection tunnel of the Gulf Island Dam on the Androscoggin River in Maine. Over a five-year period he measured their average rate of length increase as about 1.4 cm/y. All of the comparison data deal with occurrences in artificial locations, such as mines, tunnels, and buildings. However, in reviewing this report, Austin fails to mention that the stalactites are not forming in

limestone caverns. Instead, Austin picks out two numbers for further use: "Stalagmites observed by Fisher grew 0.6 cm in height and 0.9 cm in diameter at the base each year." He then applies these growth rates to a large stalagmite in the Big Room of Carlsbad Caverns. Known as the Great Dome, it has a height of 19 m. Using the growth rate observed by Fisher (0.6 cm/y.), he concludes that the Great Dome could have grown to its present height in less than 4,000 y. Not only is the great extrapolation into time unwarranted, but the rate is based on an unrelated situation. Austin then goes on to calculate the flow of drip water what would be needed to deposit the mass of the Great Dome in less than 4,000 y. He finds the flow requirement to be extremely great, and seems to indicate his own uncertainty that it could have happened at such a fast flow rate. Perhaps we are to read between his lines the inference that catastrophic conditions, present immediately after the Flood, could have accounted for the enormous flow of water drops from which the stalagmite was precipitated.

Mainstream science does, however, have powerful tools for investigating the absolute ages of speleothems and the stages in excavation of caverns. Speleothems can be dated by uranium-isotope ratios. In a report by R. S. Lively (1983) this method was applied to speleothems in the Driftless Area of southeastern Minnesota. Calcium carbonate samples of stalagmites and flowstone were analyzed for the ratios U-234/U-238 and Th-230/Th-232. From a large number of samples there emerged four distinct periods of speleothem deposition. Individual samples ranged in age from as recent as a few hundred years to as old as 285,000 y.

A second tool is the established chronology of magnetic polarity reversals, which in turn is tied to the geologic time scale through associated radiometric ages. In a study of the paleomagnetism of sedimentary accumulations in Mammoth Cave, Kentucky, absolute ages were established for the excavation of the cavern levels (Schmidt, 1982). The method is called magnetostratigraphy, because it makes use of water-transported sediment that has been trapped in cave passages and left stranded as the removal activity progressed to lower levels. Polarity of sediment samples taken at a succession of levels allowed the samples to be identified with the established polarity time scale. Cave sediment at the highest level in Mammoth Cave yielded an age of about 2 m.y., while ages decreased with samples at progressively lower levels. This progression is what would be expected of a cave system opening to deeper depths as the downcutting of the nearby Green River and its subterranean tributary, Echo River, allowed the water table to fall.

Although the paleomagnetic and radioisotope methods yield an age of Mammoth Cave and other cavern deposits 400 times greater than Flood geology will allow, that evidence is not admitted into the argument by the creation scientists. They do not accept such methods, which they say are erroneous and worthless. This leaves only independent approaches based on natural environmental processes and their time constraints. So far such an approach has not been presented for the rate of accumulation of speleothems, but the possibility exists for a systems analysis of the total process of calcium carbonate solution and precipitation, beginning with rainfall and ending with cavern deposition. A calculated limit of mass solution of limestone rock overlying the deposits would place a limit on the rate of accumulation of travertine. So far, this calculation has not (to the best of my knowledge) been done.

Desert Denudation and Desert Varnish

Visitors to the great western American desert of the Basin and Range Province notice at once the black, shiny coating on surfaces of boulders and outcroppings of bare rock. Called desert varnish, it is a remarkably durable coating containing oxides and hydroxides of manganese and iron, that are extremely stable compounds under atmospheric conditions. Archaeologists are greatly interested in the primitive pictures and drawings, called petroglyphs, etched into desert varnish surfaces. Early inhabitants of pre-Columbian times chipped away the varnish coating to expose the lighter-colored native rock beneath, creating their geometrical patterns and animal images.

Because desert varnish appears to darken and thicken with time, judged from the apparent relative ages of the surfaces it occupies, it has long been a hope of geologists and archaeologists that some means could be devised to determine the absolute age of a varnish coating at a particular location. (For a comprehensive review of the subject of rock varnish see Dorn and Oberlander, 1982.)

Highly sophisticated laboratory techniques now enable the structure and chemical composition of desert varnish to be investigated in great detail. Scanning electron micrographs reveal many details of the structure, while infrared spectroscopy and energy dispersive X-ray analyses identify the elements present. The conclusion of one such study is that clay minerals constitute 70 percent or more of the varnish, with iron and manganese oxides making up most of the remainder (Potter and Rossman, 1977). The clay minerals present are illite and mont-morillonite. The contact between the varnish layer and the native rock is very sharply defined. Another study revealed that certain bacteria are present and serve to concentrate manganese, greatly increasing its proportion relative to iron (Dorn and Oberlander, 1981). From this new information has come strong support of the long-held hypothesis that the components of the varnish are imported to the rock surface in the form of dusts consisting of a variety of constituents, including the clay minerals, that are found in abundance in many kinds of desert soils (Dorn and Whitley, 1984). This hypothesis of transported ingredients acted upon by microorganisms seems highly favored, although some scientists continue to attribute the varnish to chemical ions that are exuded from the native rock. Because various kinds of surface coatings and crusts are identifiable in other places and environments, origin of constituents from within the native rock may apply to situations other than that in which the typical desert varnish of the American Southwest is found.

The problem of absolute age of desert varnish received a decisive push toward resolution through application of a new chemical method developed by Ronald I. Dorn and David S. Whitley of the University of California at Los Angeles (Dorn, 1983; Dorn and Whitley, 1984). It seems that the mobile cations, such as sodium, potassium, calcium, and magnesium, are more easily leached from the accumulating varnish than certain less mobile cations, for example, titanium. Thus, as the varnish ages, the ratio of the more mobile cations to the less mobile ones decreases. The ratio studied is that of the sum of potassium and calcium ions to the titanium ion: $(K + Ca)/Ti$. Measurements of the amounts of these cations in a sample of the varnish are made by the particle-induced X-ray emission method. The ratio for each sample is plotted against absolute age of the native rock beneath it. The study area is the Coso volcanic field in Inyo County, eastern California, where extrusive rocks on which the varnish is found can be dated by the potassium-argon method. Ages of the volcanic rock beneath the varnish samples ranged from -30,000 y. to -5 m.y. A much younger carbon-14 date, setting the age of a varnish sample on an abandoned shoreline of Searles Lake, gave a fix at approximately -10,000 y. The correlation between potassium-calcium/titanium ratios and native rock age proved to be remarkably consistent, even allowing for errors in the methods used (Figure 29.8).

Petroglyphs in the Cosos area are too young to be dated by the potassium-argon method, but when the

potassium-calcium/titanium ratios for the varnish into which the figures are carved are located on the fitted regression line, petroglyph ages are estimated as being from as young as -600 y. to as old as -6,400 y., with a considerable range of error present. A reasonable statement would be that none of the petroglyphs of this area has an age more than about 10,000 y.

The maximum age of the petrolgyphs is not entirely out of range of the allowance of post-Flood time, considering the probable errors involved. Creation scientists might find the data consistent with the time of arrival of humans who reached this part of North America after the dispersal from Babel. They have argued that radiocarbon ages on the order of -10,000 to -11,000 y. are corrected by their "non-equilibrium model" to reduce to less than -7,000 y. (Morris, 1974, pp. 192-93). What would the creation scientists say about cation ratios that, by the potassium-argon fixes, indicate ages greatly over 10,000 y.? This is a question not covered in their published works, to the best of my knowledge. I suspect that they would dismiss such ages as false, along with all radiometric ages. But can so cavalier a dismissal of evidence be allowed?

Creation scientists are willing to accept tree-ring chronology, with some reservations. The creationist textbook seems to concede that the oldest living thing on earth--a bristlecone pine not far distant from the Coso volcanic region--has an age less than 4,900 y. old and probably less than 4000 y. (Morris, 1974a, p. 193). The text concedes that radiocarbon ages track reasonably closely with the tree-ring ages and are in agreement with historical dates back to -3,000 y. (Morris, 1974, p. 167). Going to the graph of correlation between cation ratio and age in years, let us locate a point at -4,000 y. on the straight line connecting the modern ratio (about 9.2) with the Searles Lake fix. For -4,000 y. the cation ratio reads about 6.5, a decrease of 2.7, or about one-third of its initial value. The lowest measured cation ratios are in the range of 1.0 to 2.0. Disregarding all potassium-argon

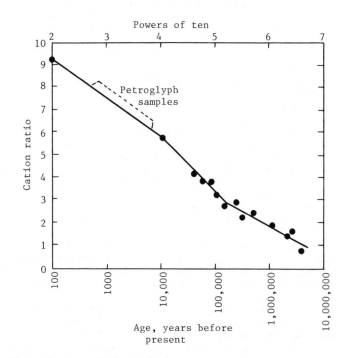

Figure 29.8 Cation-leaching ratios plotted against potassium-argon dates for samples of desert varnish in the Coso region, California. (From R. I. Dorn and D. S. Whitley, Annals of the Assoc. of American Geographers, vol. 74, p. 310, Figure 1. Copyright © 1984 by The Association of American Geographers. Used by permission.)

dates and extrapolating in an arithmetically linear rate, a decrease in cation ratio of 2.7 per 4,000 y., from the initial value of 9.2 to a terminal value of 1.0, requires a total time of 11,000 y., a figure within the range available to Flood geology. A much more reasonable assumption, inferred from the distribution of plotted points in Figure 29.8, is that the rate of decrease in cation ratio declines exponentially with time, asymptotically approaching 1.0 as a limit. This model is reasonable because, as the concentration of the more mobile cations diminishes, the rate of their removal must also diminish. On that assumption the region close to a ratio of 1.0 is reached only after 1 m.y. or longer.

Independent reinforcement of the great age of American desert varnish was reported in 1986 by Dorn and eleven associates. The study involved the sampling of rock varnish on artifacts and landforms of the Mojave Desert in eastern California (Dorn et al., 1986). Carbon-14 ages of the varnish were determined by tandem accelerator mass spectroscopy (TAMS). These ages, extending back in time to over -16,000 y., correlated well with cation ratios. The cation-leaching ratio for the Mojave River area showed a faster leaching rate than that of the Coso Range.

The creationists may wish to propose that in centuries immediately after the Flood rainfall was much higher than in the same regions today, and that most of the leaching of the older varnish samples was accomplished during that brief pluvial time. The trouble with that argument is that it calls for conditions of highly moist climate under which desert varnish does not form. Thus, the great antiquity of desert varnish can be derived at least qualitatively from the cation-ratio information as secured in time by tree-ring and historical data. Fortunately, that conclusion does not depend on an unsupported extrapolation into past time. Mainstream science has a strong case for the validity of its radiometric age-determination methods, which, in this case, point firmly to a great age for many of the samples of desert varnish.

Chapter 30

Landforms of Alluvial Rivers

Denudation of the continental surfaces produces great quantities of sediment. Eroded from land slopes by rainbeat and runoff, or dissolved by weak acids from the soil and bedrock, sediment is carried to lower levels and enters the converging system of streams and rivers that lead to the sea. The dissolved solids may go the whole distance to the sea, but the solid mineral particles and bits of organic matter are commonly detained, at least temporarily, in the lower course of the main river channels. These deposits of sediment in floors of river channels consist of alluvium; the river channels of low gradient in which they form are called alluvium rivers. The same rivers may, if conditions are right, abandon the deposition process in favor of erosion, in which the river bed is scoured and the river becomes deeply entrenched into its own alluvial deposits and even into the solid bedrock beneath. Thus many interesting varieties of stream-formed landscape features are possible, both depositional and erosional. At the terminus of the transport system, the ocean shore, deposition produces deltas that show the combined effects of stream action, wave action, and tidal action.

Creationists are interested in alluvial landforms and deltas. They see in these features indications of recent, rapid deposition occurring in the final stages of the Flood and early in the post-Flood period.

Alluvial Terraces as Flood Relics

Creationists Whitcomb and Morris have focused attention on a class of distinctive landforms found in river valleys (1961, pp. 318-24). Called stream terraces (or river terraces), they are steplike or benchlike features that occupy the sides of the valley at locations well above the existing floodplain. A single terrace consists of a more-or-less flat tread, bounded by a steeply descending slope, the scarp. Stream terraces often occur in a sequence of two or more terraces, as in a flight of stairs. Typically, terraces are found on both sides of the valley, but they are not necessarily paired off in matching levels. For stream terraces, the flat tread represents a former floodplain. (Stream terraces must not be confused with terraces controlled by alternations of hard and soft sedimentary formations, such as one finds in the walls of Grand Canyon and other gorges of the Colorado Plateau.)

To read the creationist text is to encounter an exercise in confusion following a familiar pattern or stratagem, which is to attack an outmoded and discredited hypothesis (i.e., to beat a dead horse), while leaving unstated a strongly supported current hypothesis of mainstream science. Instead, an explanation in the context of Flood geology is offered, but in rather vague form and without hard evidence to support it. In carrying out this stratagem, statements describing the discredited older hypothesis are quoted from publications of mainstream geomorphology.

Terraces formed by stream erosion are of two quite different basic types. A stream flowing for a long period of time in a delicate state of balance at an approximately fixed level erodes laterally (sidewise) into the bedrock that forms the base of enclosing valley walls. Rock weathering and gravity movements of rock debris cause the valley wall to retreat. This cutting activity is generally called lateral corrasion, a term correctly used by Whitcomb and Morris (1961, p. 321). Lateral corrasion can, in theory at least, result in a wide floodplain on which only a thin veneer of alluvium (well-sorted sand and gravel) lies upon an even rock floor. Later, crustal uplift (or some other cause) may result in rapid downcutting of the stream channel, producing a narrow inner rock gorge. The former floodplain is now abandoned and becomes an elevated rock terrace on one or both sides of the gorge (Figure 30.1). The rock terrace is thus analogous to the elevated marine terrace, a steplike feature discussed in Chapter 27. To see a small remnant of a rock terrace, visit the U.S. Military Academy at West Point, New York. Standing on the parade ground, you will be on that rock terrace, about 50 m above the Hudson River. Other patches of this same terrace can be found on both sides of the river gorge through the Hudson Highlands. According to mainstream geomorphology, there is good evidence to show that most such rock terraces are much older than the latest (Wisconsinan) glacial stage, and may date back some hundreds of thousands of years--far out of range of the post-Flood history of creationists.

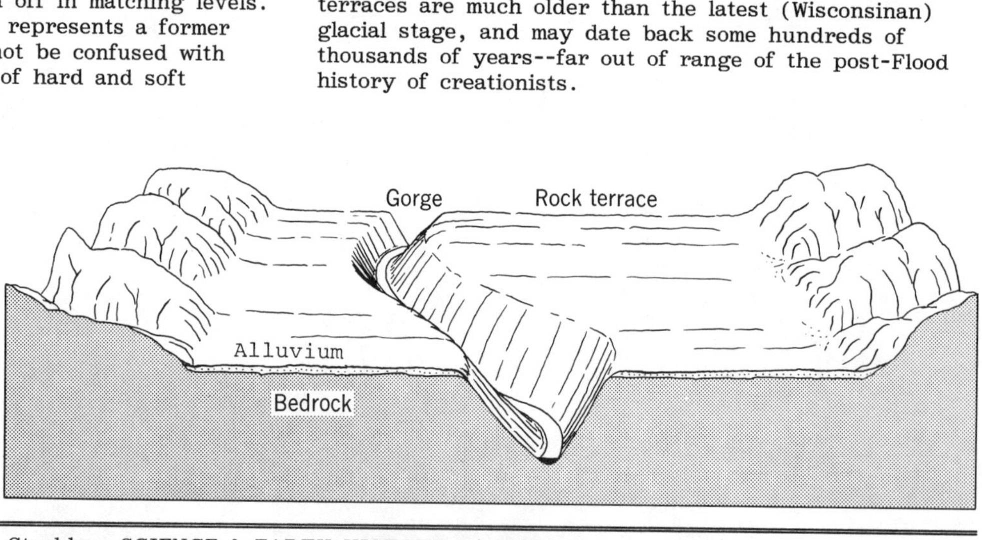

Figure 30.1 A rock terrace, formerly a floodplain belt, borders a steep-walled inner gorge resulting from stream rejuvenation. (Copyright © 1971 by Arthur N. Strahler.)

Whitcomb and Morris do not consider lateral corrasion by streams to be capable of creating a wide floodplain, such as that occupied by the Mississippi River (1961, p. 321). Large alluvial rivers such as the Mississippi, they say, are cutting into thick alluvial fill previously deposited when conditions were very different from today. That deposition occurred soon after the Flood subsided, when river flows were much greater than now.

In support of their view that such a broad valley as the Mississippi occupies between its bluffs could not have been caused by lateral corrasion, they draw upon statements by mainstream geomorphologists. One of these is Richard J. Russell, formerly a professor at Louisiana State University and a recognized authority on river floodplains and deltas. He is quoted as writing: "Broad flood plains are characteristic of most rivers leading to the sea. For many years these were explained on an erosional basis. The rivers were pictured as having cut down their valleys to a base-level established by the sea, after which their energies were directed toward lateral corrasion, or valley widening. The alluvium of flood plains was thought of as a thin veneer, resting on laterally planed bedrock" (Russell, 1957, p. 417). Russell then goes on to show that this early hypothesis had to be totally rejected when numerous borings revealed the presence of thick alluvium beneath the Mississippi floodplain in its middle and lower valley. Between Baton Rouge, Louisiana, and Cape Girardeau, Missouri, alluvium beneath the floodplain is about 60 m thick. The lower part of the deposit is sand and gravel deposited by the river during the last glacial stage (Wisconsinan); the upper part is dark clay, silt, and fine sand laid down by the modern meandering Mississippi in Holocene time (Fisk and McFarlan, 1955, p. 288).

The hypothesis of a thin alluvial cover resting on a shallow rock surface beneath doubtless derives from the writings of W. M. Davis (1899) on the subject of the denudation cycle. To describe an ideal cycle, it was necessary for Davis to assume absolute stability of sea level with respect to the continents in order to provide a fixed "base level" toward which denudation could trend without interruptions. Reference by Russell to this abstract concept provides the creationists with yet another opportunity to display and abuse the straw person of their incorrect version of Lyellian uniformitarianism. They take the opportunity to describe lateral corrasion as "mainly a uniformitarian assumption rather than an actual geomorphic process" (Whitcomb and Morris, 1961, p. 321). They wish to substitute their catastrophic version of "great swollen rivers pouring rapidly to the sea" (p. 323), and providing the enormous mass of alluvium now found beneath the Mississippi floodplain. Never mind that mainstream science has its own version of how that alluvium was deposited and how it fits into the chronology of alternate glaciations and interglaciations in the Ice Age. To present that mainstream hypothesis of changing activity of streams in response to the glacial cycle, we must resume the discussion of stream terraces.

The second class of stream terrace is the alluvial terrace, carved out of a thick layer of alluvium extending to a depth far below the normal range of channel scour and fill experienced by the stream in flood stages. This thick alluvium is explained by an upbuilding process, in which the stream deposits layer upon layer of alluvium on the valley floor. The process is called aggradation; it takes place when the stream channel is supplied with more (or larger) rock particles than it is capable of dragging along its bed. Aggrading streams can be recognized by their braided channel pattern, in which the water stream repeatedly divides, subdivides, and rejoins in a pattern resembling a braided cord. A good place to look for a braided channel is in a modern stream fed by meltwater from a wasting glacier. A classic example of aggradation caused by humans is the choking of stream channels of the Sierra Nevada foothills and Great Valley in California

when gold miners, using powerful water jets, washed large amounts of gravel out of river banks and terraces, sweeping it into the nearby stream channels.

Aggradation has occurred as a natural process on a large scale in middle latitudes as a result of the changing conditions of debris supply during alternate glaciations and interglaciations. Aggradation is the typical pattern in the upper and middle courses of streams during a glaciation and the immediately following deglaciation. During a glaciation, streams fed in summer by meltwater issuing from the ice margin became heavily charged with rock fragments released from the ice. In addition, sediment was swept into the streams from barren land surfaces in a region marginal to the ice, where conditions resembled the rocky tundra of today's arctic landscape. Soil-ice melting in summer produced saturated mud, which in turn yielded sediment for overland transport to streams during rains. Similar conditions applied during deglaciation as the ice thinned and disappeared, exposing more barren ground as a sediment source. Rapid aggradation at such times filled many preglacial river valleys almost to the upper limits of their walls (Figure 30.2A). What followed then? Sea level became stable during a long interglaciation of mild climate; in middle latitudes, forests and grasslands took hold on the watersheds, holding back most of the coarser debris. With

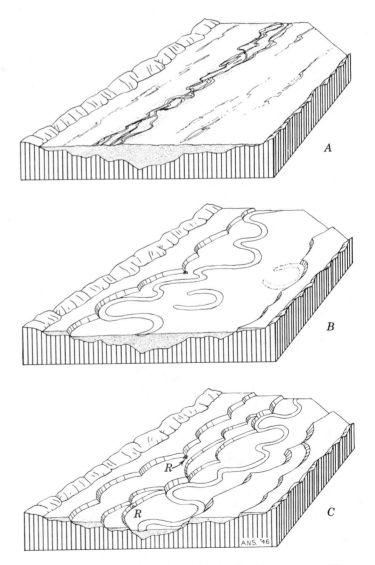

Figure 30.2 Formation of unpaired alluvial terraces. The letter R in Block C refers to a point where a terrace is defended by a rock outcrop. (Copyright © 1975 by Arthur N. Strahler.)

1. Pre-ice-age valley: thin alluvium on rock floor.
2. First glaciation: aggradation.
3. First interglaciation: trenching, excavation.
4. Second glaciation: aggradation.
5. Second interglaciation: trenching, excavation.
6. Third glaciation: aggradation.
7. Third interglaciation: trenching.

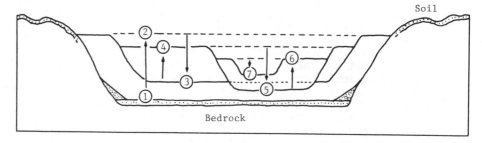

Figure 30.3 Schematic diagram of nested alluvial fills produced by alternate aggradation and trenching in upper and middle stream courses. (A. N. Strahler.)

less debris to transport, and most of that in fine particles such as silt and clay that are easily carried long distances in suspension, the streams were able to scour the floors of their channels. The result was <u>degradation</u>, the trenching and removal of the thick alluvial fill as the meandering channels shifted back and forth, undermining steep alluvial banks (Figure 30.2B). There remained high alluvial terraces on both sides of the valley, their tread elevations matched across the intervening gap.

As the stream continued to excavate the alluvium beneath its floodplain, it carved more terraces at lower levels. Meandering allowed for a great amount of lateral cutting of the steep valley walls as meander bends enlarged and shifted in a down-valley direction. The combination of downward and sideward cutting left behind crescentic patches of floodplain stranded above the level of the channel, as shown in Figure 30.2C. These terraces do not match in level across the valley. Terrace systems of this kind, formed in a single long episode of denudation, are common in valleys that lay close to the limit of Wisconsinan ice-sheet advance. Excellent examples are found in New England, where they were first studied by W. M. Davis (1902). In later decades, as terraces were intensively studied in many other parts of the midlatitude zone, it became obvious that a much more complex scenario is usually required--one that the creationists do not discuss.

Suppose, next, that the glaciation referred to above was only the first of a series of glaciations in an ice age (Figure 30.3). After excavation of alluvium had occurred in the first interglaciation, a second glaciation set in. The aggradation cycle was repeated and alluvium largely filled the trench between the two parts of the uppermost alluvial terrace, as shown in Figure 30.3, stage 4. During the ensuing interglaciation trenching again occurred (stage 5). The diagrams have been simplified by using a program of diminished amounts of both aggradation and degradation in successive glacial cycles. Other programs could be postulated. Actual relationships among alluvial fills are often complex. As the glacial cycle was repeated, alternate excavation and aggradation resulted in newer alluvial fills, each one nested within the next older one. Figure 30.3, stage 7, shows the final result: a succession of paired alluvial terraces. Geologists have worked out the details of nested alluvial fills in many parts of the midlatitude zone lying between former ice sheets and continental coasts. Using excavations, core drilling, and other probing methods, they have been able to reconstruct the former trenches and sets of alluvial fills. A layer of ancient soil is often present at the surface of a buried terrace. Dating by carbon-14 methods enables the deposits to be placed in an absolute age sequence and fitted into the glacial chronology based on deep-sea cores.

The rather complex hypothesis of "cut-and-fill" in response to glacial cycles of climate and sea-level change provides an explanation fully compatible with what is known from hydraulic engineering studies about the action of streams in eroding, transporting, and depositing

sediment. The creationists offer in its place a one-step scenario requiring catastrophic stream activity immediately following the Flood. They say that the alluvial fill now found in stream valleys was carried by great stream floods that washed down enormous quantities of sediment from land surfaces recently emerged from beneath the Flood waters. Whitcomb and Morris make no mention of the evidence of cycles of cut and fill clearly associated with glacial cycles (1961, pp. 318-28). They would prefer not to recognize multiple glaciations of an Ice Age because the time constraints of Flood geology allow for only one major glaciation, and that leaves unexplained the nested alluvial fills and the ancient soil layers (paleosols) on buried terrace surfaces.

Deltas--How Old Are They?

Creation scientists have long been interested in the great delta of the Mississippi River. Can it be accommodated within the 4,350-year post-Flood period? Do estimates of its age actually fall close to that mark? This question does not appear in Whitcomb and Morris's <u>The Genesis Flood</u> (1961) or in recent publications of the Institute for Creation Research, but is in Randy Wysong's comprehensive textbook of creation science, <u>The Creation-Evolution Controversy</u>. The following paragraph appears as one of his 33 "methods of showing youth" of the universe:

> 6. Delta Filling. The Mississippi River dumps about 300 million cubic yards of sediment into the Gulf of Mexico each year. If that river were millions of years old, the Gulf would have been long since filled. By measuring the rate of growth of the delta (about 250 feet per year), its age calculates to about 4,000 years. (1976, p. 163)

There is no clue in Wysong's text as to where his information came from, but I have found a likely source-- an article in a 1972 issue of <u>Creation Research Society Quarterly</u> written by Benjamin Franklin Allen with the title "The age of the Mississippi River" (1942; 1972). The article was reprinted from a 1942 issue of the <u>Bulletin of Deluge Geology and Related Sciences</u>, an organ of the Deluge Society of Los Angeles, of which George McCready Price was a leading light. Recall that Price laid the foundations of Flood geology in his 1923 textbook, <u>The New Geology</u>. Allen's paper is particularly interesting because he refers to Sir Charles Lyell's estimate of 60,910 y. as the age of the Mississippi River. The reference given is Lyell's ninth edition of <u>Principles of Geology</u> (1853, pp. 271-75). Allen quotes in full Lyell's statement as to how he reached this figure--a bold venture indeed, considering that few borings had been made to reveal what lay beneath the delta. It seems that in 1846 Lyell visited New Orleans and consulted with local persons knowledgeable on the subject of delta geology and river flow. Lyell learned what proportion of solid matter was carried in relation to water volume. Using the figures of measured discharge of the river, he calculated the amount of solid matter annually carried down by the river. He then estimated the volume of the delta sediment mass using an area of 13,600 sq mi and a thickness of 528 ft (one-tenth of a mile). Taking into account some additional sediment present above the delta, he came up with a figure of about 100,000 y. as the time required for the sediment to be deposited, but this was later reduced by him to 60,910 years (Allen, 1942, p. 109).

Needless to say, Allen was not happy with the arch-uniformitarianist's estimate--about 15 times longer than a Flood geologist would like to see. So Allen set about finding fault with the calculations, something very easy to do because of the near-impossibility of making even a rough guess about the actual sediment volume. Allen found the comfort he needed in volume estimates made independently by two Americans: Professor E. W. Hilgard, State Geologist of Louisiana, and General A. A. Humphreys of the U.S. Army Corps of Engineers (the organization then in charge of regulating the river). Hilgard (1869-1870) used a thickness of only 41 ft for the delta sediments, and in this opinion he was supported much later by A. C. Trowbridge (1930). Naturally, reducing the thickness of delta sediments from 528 ft to 41 ft would cut Lyell's age calculation down to about 4,800 y. That's serendipity for you! But an even nicer surprise was in store for Allen. General Humphreys used a thickness of 40 ft for the area within the 10-fathom depth curve, giving ages of 4,900 and 5,400 y., depending upon which set of sediment transport data he chose from those at his disposal (Humphreys and Abbott, 1876).

Humphreys then tried an entirely different approach, which was to average the annual rate of growth of all the distributary mouths into the Gulf of Mexico and divide that rate into the total distance in miles from the head of the delta to its present extremity. He divided his distance value, 1,152,800 ft, by an annual advance of 262 ft, yielding an age of 4,400 y. Marvelous! To add to the largess he unwittingly bestowed on the Flood geologists, Humphreys published a statement in which he implied a catastrophic beginning for the Mississippi River. As quoted by Allen (p. 111), the statement reads as follows:

> The age of the delta has been estimated at 4,000 years, upon the assumption that the river was of equal magnitude during the whole period of its delta-forming condition. This assumption implies that the river was suddenly brought into existence with its present condition, or was suddenly converted into that condition. The rapid, simultaneous upheaval of the whole basin of the river would have brought that river suddenly into existence with very much the same characteristics it now possesses; but geologists do not admit the possibility of such a rapid upheaval. (Humphreys, 1869-1870, p. 376)

Savoring this catastrophic scenario, Allen then paid tribute to his benefactors:

> Sir Charles Lyell had estimated its age at 60,900 years. But General Humphreys, who, with his staff of engineers and geologists, had charge of the delta for twenty-five years, and Hilgard, studying it until thirty-five years later, accumulated vast data which have apparently never been superseded by subsequent discoveries. Their basic evidence appears to prove that the river has been in existence only 4,400 to 5,000 years. Facts developed since then seem only to strengthen this conclusion. (P. 111)

In his concluding paragraph, Allen voiced the creationists' aspirations, unchanged to the present time,

in the writings of the creation scientists of the Institute for Creation Research:

> Labors on several other natural chronometers involving geological processes, besides the growth of river deltas, are apparently developing, each with considerable capacity for accuracy, and they will be of utmost value in correlating and counterchecking. Altogether, there appears to be some promise of satisfying all reasonable doubts not only that the Flood of Noah was universal, but that it occurred well within the range of dates set by sacred writings and archaeological evidences. (P. 113)

The year 1942, in which Allen's paper first appeared, concluded an era of unfounded assumptions compounded by ignorance of what lay beneath the deltaic plain and offshore zone of the Mississipi. The modern era of mainstream scientific study of the same region was ushered in by the following letter, dated December 1, 1944:

> Subject: Geological Investigation of the Alluvial Valley of the Lower Mississippi River
>
> To: The President, Mississippi River Commission, Vicksburg, Mississippi
>
> Transmitted herewith is the final report of the Geological Investigation of the Alluvial Valley of the Lower Mississippi River.
>
> Respectfully submitted, Harold N. Fisk, Ph.D., Consultant

Fisk, a geologist then holding the position of Associate Professor in Louisiana State University, had produced a report of monumental proportions (Fisk, 1944). Its text of 78 pages was enriched by 80 figures, 11 tables, and 33 plates, displaying the subsurface geology in remarkable detail. Studies of the alluvial deposits used data from approximately 16,000 borings, of which more than 3,000 penetrated the entire depth of the alluvial and delta deposits. Many oil companies contributed well data, as did water-drilling companies and individual drillers. The alluvial studies extended from Cape Girardeau, Missouri, where the Ohio River joins the Mississippi, to the Gulf of Mexico.

One important contribution of Fisk's report was to show a succession of nested alluvial terraces along the alluvial valley, between Cape Girardeau and Natchez (Fisk, 1944, Plate 26; reproduced here as Figure 30.4). These represent four earlier cycles of glacial alluviation and interglacial degradation, preceding the latest cycle of valley trenching and alluvial filling. Altogether, then, five glaciations are documented and correlated with the Nebraksan, Kansan, Illinoian, Iowan, and Wisconsinan glaciations. This history goes back a full one million years, according to mainstream geology. Creationists recognize only the last cycle, since it would be extremely difficult to fit the previous four cycles into the past four thousand years. Using Fisk's cross section at the latitude of Natchez, I measured both the degradation (depth of trenching) and aggradation (alluvium thickness), totalling the vertical dimension of each activity. In the five glacial cycles, degradation totalled about 310 m, the aggradation about 290 m. Creation scientists are required to compress this one-million-year history into 4,300 y.

To learn whether this degree of time compression is reasonable, let us turn to evidence of the rate of accumulation of the Mississippi delta, using the data collected by Fisk and his coworkers, and summarized in a more recent work (Fisk and McFarlan, 1955). Their report is limited to deltaic deposits younger then -60,000 y.,

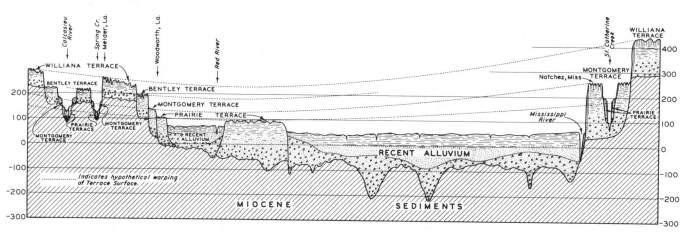

Figure 30.4 Cross section of alluvial deposits and terraces of the Mississippi River alluvial valley in the vicinity of Natchez. Elevations in feet. (From H. N. Fisk, 1944, Geological Investigations of the Alluvial Valley of the Lower Mississippi River, Plate 26, Section B-B, War Dept., Corps of Engineers, U.S. Army, Mississippi River Commission, Vicksburg, Miss.)

which are late-Wisconsinan and Holocene in age.

The deposition period is subdivided into two periods of 30,000 y. each. In the earlier of the two, the delta sediment was deposited in deep water of the continental slope. This deposition accompanied the lowering of sea level as ice sheets grew. As sea level fell, the Mississippi began to carve a deep trench in its lower part across the earlier deltaic plain. The trench extended far up the middle Mississippi valley and served to facilitate the down-river transport of coarse sediment derived from the glacial border. However, as sea level became steady at the low level, delta sediments began to fill the trench and to extend progressively farther northward. Deglaciation with rapidly rising sea level was accompanied by rapid deposition of sandy sediment, filling the trench far upstream and aggrading the entire valley bottom. Upbuilding of the delta continued over the region of the present deltaic plain. As delta sediments accumulated, crustal subsidence occurred, affecting the entire deltaic plain and the continental shelf. In a late phase, the river assumed its present meandering pattern, spreading fine-grained alluvium over its valley during floods. The delta developed branching distributaries and shifted from one location to another, forming altogether seven successive subdeltas within the past 5,000 y. (Kolb and Van Lopik, 1966). Figure 30.5 shows the outlines of these late-stage deltas and the locations of their former distributary channels. The birdfoot delta in existence today was built in only the past 450 y.

It seems clear that the estimates of delta volume made by Hilgard and Humphreys apply only to the uppermost layers of an enormous mass of deltaic sediment that subsided to great depth and is completely concealed from view by late-phase deltas. Hilgard and Humphreys, using a thickness of 12 m, were limiting themselves to late-phase deposits, and it is not surprising that their estimates of 4,000 to 5,000 y. are in close agreement with the time assigned to the seven deltas making up the present deltaic plain.

Fisk and McFarlan give 33,000 km³ as the total volume of the late Wisconsinan deltaic sediment, deposited in 60,000 y., for an average rate of accumulation of 0.56 km³/y.(1955, p. 299, Figure 10) Creationists must account for the same volume of sediment in only 4,350 y., for an average of about 7.7 km³/y., which is almost 14 times faster than the mainstream estimate. But we cannot leave the figures at this level of comparison. Let us extrapolate by direct proportion back through the Ice Age as far as five glacial cycles (the Nebraskan is the oldest). Now, five glaciations will have produced 167,000 km³ of sediment (5 x 33,300 = 167,000). Using -1.2 m.y. as the

starting time, and 167,000 km³ as the volume, the average rate is about 0.14 km³/y. The creationists will need to account for 167,000 km³ in only 4,350 y., for a rate of just over 38 km³/y., which is about 275 times faster than the mainstream rate. Can you imagine a Mississippi River that could have transported sediment at the rate of 38

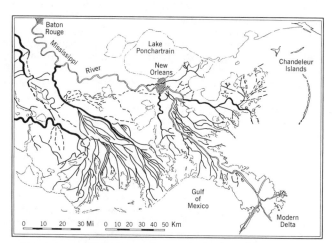

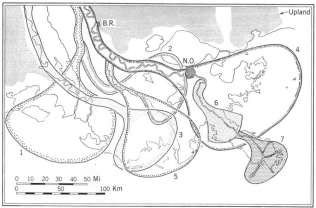

Figure 30.5 The deltaic plain of the Mississippi River. (Upper map) Abandoned river courses and distributaries are shown by bold lines. (Lower map) Seven deltas, numbered in order from oldest to youngest. The lower map covers a larger area than the upper map. (Redrawn and simplified from maps by C. R. Kolb and J. R. Van Lopik, p. 22, Figure 2 and p. 31, Figure 8, in M. L. Shirley, ed., 1966, Deltas and Their Geologic Framework, Houston Geological Society, Houston, Texas.)

km³/y.? Firm data for the past 450 y. gives the Mississippi's transport rate at only 0.25 km³/y. (Wysong's figure is in agreement here.) You would need to envision a river capable of carrying 150 times as much sediment as the present river. Geologists have found that the capacity of a stream to transport particles in suspension increases about as the square of the stream's discharge (m³ per second). This means that the creationists' Mississippi would have needed an average discharge about 12 times that of the present river to carry out its work in the limited time since the Flood. The creationists will need to give good reasons for postulating such a mighty river. The watershed area of the Mississippi in the Ice Age was probably no larger, even during the interglacials. That being the case, it would be necessary to postulate watershed runoff greater by a factor of 12. It is up to the creationists to show how this could be so. One possibility would be enormously increased precipitation; another would be extremely rapid melting of the glacial ice during deglaciations. Either explanation would involve recourse to completely unreasonable parameters for energy conversions and mass transports.

Entrenched Meanders--How Do They Form?

Creationists Whitcomb and Morris look for evidences of distinctive post-Flood geologic activities revealed in landforms made by rivers (1961). One kind of feature they have focused attention on is the phenomenon of meandering of river channels. The sinuous form of river channels on flat floodplains is particularly striking when seen from the air, as sunlight reflects brilliantly from the water surface. Meander loops can be seen in various stages of growth, ending in the cutoff that isolates a crescentic portion of channel; it persists as an oxbow lake or marsh.

Figure 30.6 shows sketches of meanders on three different scales. The transition from large loops to small ones might represent the same river seen from progressively higher altitudes or on maps of progressively smaller scale. On the other hand, the three examples might represent rivers of three different orders of magnitude shown to the same map scale. We are depicting by parallel lines the channel that the river or stream occupies, i.e., we show the two river banks. (Hereinafter, we use the word "stream" for large rivers as well as for streams and brooks, irrespective of their actual size.) Streams can be compared in magnitude in the bank-full stage, in which the water fills the channel from bank to bank, but does not spill over the banks, as it would in a major flood.

Numerous field and map measurements have shown that stream meanders are remarkably similar in their form characteristics, despite wide ranges in the actual channel size. Meander length, L, is analogous to wavelength of water waves or light waves; it includes two successive bends, one to the right and one to the left. Length ranges from 7 to 10 times the channel width, W, over a great range of stream sizes (Leopold and Wolman, 1960, p. 772). A meander bend can be fitted with an arc of a circle with a radius, R, that is typically about one-fifth of the meander length. Going further, it is possible to relate the flow-rate, or <u>discharge</u>, of a stream to the dimensions of its meanders. For a wide range in meander dimensions it can be shown that, on the average, meander length, L, varies about as the square-root of the discharge when the stream is in bank-full stage or some other measure of its discharge in a high stage (Carlston, 1965).

What we are leading up to is the concept that the meander dimensions give us a fairly good measure of the size of the stream--size being measured in terms of the quantity of water that the stream carries past a given crosssection in a given unit of time. If, then, we discover an ancient channel, now buried under later stream deposits, and compare its width with the present stream

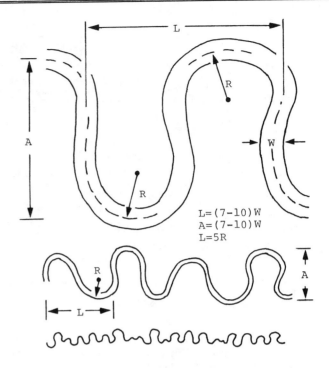

Figure 30.6 Outlines of meandering channels of a wide range of dimensions. L--length, W--width, A--amplitude, R--radius. (A. N. Strahler.)

channel in the same valley, we can make a good guess as to whether the ancient stream was of larger discharge than the present stream, and if so, about how much larger. This information might, in turn, allow us to make some inferences as to the climate that prevailed when the ancient channel was in use. Before proceeding further with this kind of analysis of past climate, it will be necessary to know about another kind of stream meander and its significance.

Vacationers taking a boat trip down the Mosel River in the upland of western Germany (between the Luxembourg border and the city of Koblenz) will find themselves at the bottom of a winding gorge enclosed by high, steep valley walls. Seen from above or on a detailed map, these bends are meander loops with only a narrow strip of floodplain next to the channel, and usually no floodplain at all on the outsides of the river bends. Actually, the entire gorge can be described as meandering--as a series of <u>entrenched</u> <u>meanders</u> (Figure 30.7). Entrenched meanders are a landform type distinctly different from <u>alluvial</u> <u>meanders</u>, which shift location and change form freely on broad, flat floodplains underlain by alluvium that the stream itself has put in place. Entrenched meanders are constricted within the winding rock gorge that was carved into bedrock by the stream itself in an earlier phase of erosional activity.

Whitcomb and Morris call attention to striking sequences of entrenched meanders in the Colorado Plateau region (1961, pp. 153-54). They include a photograph of the "goosenecks" of the San Juan River, one that appears in many geology textbooks. Strata here are almost horizontal and are etched out into sharp lines, like a set of natural contour lines. The authors call forth an explanation advocated several decades ago by some geologists to explain this and other occurrences of entrenched meanders. At one time, perhaps several million years ago, this region was a nearly flat plain lying close to sea level. (A surface of this kind is commonly identified as being a peneplain.) It was covered by a thin layer of alluvium, that is, sediment spread by freely shifting streams. Streams on that plain probably had alluvial meanders. Tectonic uplift of the entire region set

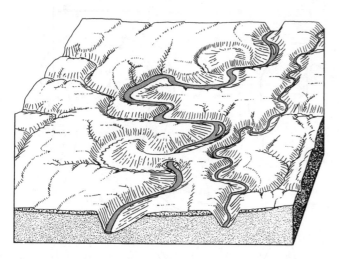

Figure 30.7 Block diagram of entrenched meanders with a natural bridge formed by cutoff. (Drawn by Erwin Raisz. Copyright © 1975 by Arthur N. Strahler.)

in, perhaps as a broad arching, covering a width of several hundred kilometers. The meandering alluvial rivers responded to the uplift, which steepened their gradients by quickly eroding down through the alluvial cover and coming into direct contact with underlying hard bedrock. Slowly, but surely, the meanders were worn into the bedrock and a meandering rock gorge came into being, preserving the original form and size of the alluvial meanders. The entrenched meanders are thus inherited from alluvial predecessors (Mahard, 1942, pp. 35-36).

Whitcomb and Morris discredit the hypothesis of inheritance on grounds that a degrading (down-cutting) stream tends to straighten its course, rather than to preserve alluvial meanders (1961, p. 154). This conclusion, they point out, was verified by model experiments performed by a hydraulic engineer, Joseph F. Friedkin, at the U.S. Waterways Experiment Station at Vicksburg, Mississippi. Friedkin's study, published in 1945, has been superseded by numerous later laboratory studies showing that the form a stream channel takes depends on factors of gradient and bed load (coarse sediment dragged along the bottom of the channel). Experiments show that on low gradients, the channel remains straight but will change into a meandering channel when the gradient is increased above a critical value (Schumm, 1977, pp. 121-31). Moreover, the changes are reversible by appropriate changes in gradient.

Laboratory experiments have also included simulated incision (entrenchment) of model streams into simulated "bedrock." The experiments were begun with a manually excavated sinuous channel, and from these initial bends it was possible to cause outward growth of the bends by erosion at the outsides of the bends (Shepherd, 1972). On the other hand, attempts to produce entrenched meanders inherited from an alluvial cover have not been successful. Professor Stanley A. Schumm, who has done extensive experimental study of stream channels in laboratory flumes, writes as follows:

During the experimental investigation of bedrock incision, alluvial meanders developed in a sand cover placed over the simulated bedrock, but in every case when the slope of the flume was increased to induce incision, the alluvial meanders were destroyed and a relatively straight channel incised into bedrock. From these observations it is difficult to accept the hypothesis that symmetrical incised meanders in nature inherited their patterns

from an alluvial channel superimposed from a warped peneplain. (1977, p. 199)

Schumm's results seem definitely in support of the claim made by Whitcomb and Morris, based on Friedkin's earlier findings. But Schumm has more to say:

However, another way in which incised meanders can form is by baselevel lowering or essentially vertical uplift. Gardner (1975) showed by additional experimentation that when a bedrock surface is horizontal, a lowering of baselevel causes headward incision up the meander pattern. (P. 200)

The term "baselevel" refers to a control point downstream on the channel at which the channel cannot be lowered. In a natural stream, baselevel might be the level of the lake or ocean into which the stream empties. In the experiment described above, incision took the form of the upstream migration of a steep rock face (a sort of waterfall) or a zone of rapids, known as a <u>nickpoint</u>. The migrating nickpoint closely followed the shallow channel in the alluvial cover and thus the meanders were successfully inherited.

An interesting consequence of inheritance by nickpoint migration was seen in the experiment. The entrenched meanders became deformed so as to show narrow necks, closely resembling natural entrenched meanders, such as the Goosenecks of the San Juan River. (Details of the process of shape deformation are given in the reference cited.) So it turns out that mainstream science has, after all, a reasonable hypothesis for the inheritance of entrenched meanders from alluvial meanders.

The creation scientists, after rejecting an outmoded hypothesis, propose in its stead an outrageous hypothesis that makes use of catastrophic events occurring at the close of the Flood (Whitcomb and Morris, 1961, p. 154). They imagine great systems of vertical fissures, presumably of tectonic origin, that were occupied and enlarged by streams. Those early streams were of great discharge and flowed at high velocities on steep gradients, enabling them to cut laterally at abrupt bends and thus form serpentine gorges. This scenario requires the creationists to postulate much heavier precipitation and runoff in the period just following the Flood--a favorite theme expressed in connection with other phenomena, such as pluvial lakes. The postulate of great systems of vertical fissures has no validity in terms of known geologic structures, past or present. Extensional (pulling-apart) stresses acting on the brittle crust do not produce deep, gaping fissures in intersecting sets on a scale that would allow them to be transformed into entrenched meanders.

Underfit Streams--Relics of Catastrophe?

Is there independent evidence that all streams were much greater in discharge in a pluvial period--evidence that might be interpreted as requiring the immediate postFlood period to which the creationists assign many geomorphic features? Perhaps that evidence exists. It, too, has to do with meanders.

Figure 30.8 is a sketch-map of a winding rock gorge-part of a system of entrenched meanders occupied by a small stream that has its own meandering course. The valley floor is underlain by alluvium, and the small stream forms alluvial meanders appropriate in size to its small discharge. Surely, the gorge was formerly occupied by a stream of much greater discharge, as required by the large dimensions of the bends. Geomorphologists describe the small stream as a <u>misfit stream</u>. The term was coined around 1900 by William Morris Davis to describe streams that he observed in Europe. He supposed that the small stream was a shrunken remnant of the former occupant, the upper portion of which had been diverted into another

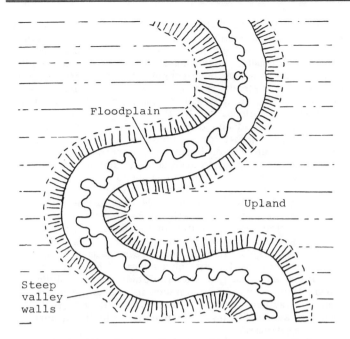

Figure 30.8 Sketch map of a meandering entrenched valley occupied by a small meandering stream. (A. N. Strahler.)

valley by a predatory stream in a dreadful maneuver known as "stream capture" (or "stream piracy").

Creationists find in underfit streams the evidence they seek of effects of the closing events of the Flood. They cite a passage from a once-popular but now obsolete geology textbook to the effect that the idea of a formerly much larger stream discharge is unreasonable, because no evidence exists for previously larger drainage basins (watersheds) to supply that additional water. This gives Whitcomb and Morris the opportunity to come to the rescue of science with the following explanation:

> If, as indicated, the reason for rejecting the plain indication of a former much greater stream flow is merely the lack of a source of the required waters, we would suggest for consideration once again the waters of the Flood, which in response to the uplift of the lands and subsidence of the ocean beds, required to be rapidly and powerfully transported to the sea. Furthermore, the rainfall of the early post-diluvian times must have been much greater in most places than it is now. (1961, p. 320)

Let us overlook the part about uplift of the lands and subsidence; that has been fully discussed on earlier pages. Does underfitness require the postulate of greatly increased precipitation while the large meandering gorge was being produced?

For years, Professor George H. Dury studied underfit streams in many areas. They occur in middle latitudes in such places as the northcentral United States, England, Europe, the USSR, and New Zealand. As his studies progressed, he pushed more strongly the hypothesis that misfitness on such a geographically widespread scale requires that the former streams that shaped the large meanders must have had discharges from 80 to 100 times greater than that of the present occupants of the valley floor (Dury, 1960, p. 235). He claimed to have been able to find, in some instances, buried beneath the present valley surface, filled channels of much greater width than the present channel. He supposed that these buried channels could be outlined by using a soil auger to bring up samples of the alluvial material at depth (Dury, 1958, pp. 105-6). By 1965, Dury had lowered his estimate of

the former discharges to an an increase of 20 times for most underfit streams, but as great as 60 for some (Dury, 1965).

If the watersheds of the rivers studied by Dury were essentially of the same size when the large streams were flowing in them as they are today--a condition that seems inescapable for most of the examples--the presumption of formerly greater stream discharges would be difficult not to accept. Dury was inclined to associate the carving of the winding valleys with the high-rainfall pluvial Atlantic climate stage of the Holocene Epoch (1960, pp. 235-37). That stage set in about -8,000 y. and lasted for about 3,000 y. It was a warmer climate than now and is referred to as a "climatic optimum." The question on which one must speculate is how a greatly increased stream flow was brought about in that climatic optimum. Stream erosion is largely accomplished at high-water stages. One possibility is that the climate favored rapid snowmelt, yielding large volumes of runoff in sudden bursts. Another possibility is that cyclonic storms of that period tended to produce rainfalls of much higher intensity than occur today. Nevertheless, the requirement of former stream discharges as great as 20 to 60 times those occuring today presents a difficult problem for mainstream science, which insists that principles of atmospheric science and hydrology be applied in a reasonable manner. That the explanation involved some uncertainties would not necessarily be a deterrent to a science that has faced many long delays in finding the evidence needed to shape successful hypotheses.

In this case, however, a rational explanation supported by field evidence appeared in 1967, when Stanley A. Schumm, mentioned earlier as a leading research scientist specializing in stream channels, presented solid evidence that a second factor is of major importance in determining the relationship between the width of a stream channel and the size of its meander bends (Schumm, 1967). Given two streams of equal mean annual discharge (or equal mean annual flood discharge), the one transporting a high proportion of its load as bedload of mixed sand and gravel will have a meander size (wavelength) larger by a factor of approximately 10 than the stream carrying mostly clay, silt, and fine sand held in suspension.

Carrying his research to Australia, Schumm studied both the present channel and ancient channel (or paleochannel) of the Murrumbidgee River on its floodplain in New South Wales. This was one of the examples used by Dury to illustrate a modern stream with small meanders occupying a valley showing clear evidence of much larger meanders of a former stream. Indeed, that relationship is established beyond doubt and shows clearly in air photos (Schumm, 1977, p. 168, Figure 5-34). Dury had interpreted the large paleomeanders as the work of a former stream of vastly greater discharge than the present stream, but this interpretation was in conflict with independent evidence from geomorphology and soil science that the oldest of the paleochannels had been formed during a period of climate drier than that of the present.

Schumm examined the channel deposits preserved in the paleochannel, comparing them with channel deposits of the present stream; the former are predominantly composed of coarse bedload particles; those of the former, predominantly of silt and clay. Schumm's measurements showed that the channel of the former bedload-carrying stream was also much wider than the channel of the present stream. The data are summarized in Table 30.1. The paleochannel was much wider, not only because of its former greater discharge, but also because of its dominant bedload. Table 30.1 shows that the estimated discharge of the paleochannel was only about twice that of the present channel--not the 20 or more times greater inferred by Dury. Greater bank-full discharge is to be expected of the paleochannel even though the climate was arid, because in a dry climate flood peaks tend to be higher,

Table 30.1 Comparison of the Murrumbidgee River and Its Paleochannel

	Modern river	Paleochannel
Channel form:	Narrow, deep; low gradient	Broad, shallow; steep gradient
Width (ft)	220	600
Depth (ft)	21	9
Gradient (ft/mi)	0.7	2.0
Meander size:	Small	Large
Wavelength (ft)	2,800	18,000
Load:	Suspended; fine	Bedload; coarse
Silt-clay (%)	25	1.6
Sand flow (tons/day)	2,000	54,000
Flood flow:	Small	Great
Bank-full flow (ft^3/sec)	11,000	23,000

Data of S. A. Schumm, 1977, *The Fluvial System*, p. 168, Figure 5-34, John Wiley & Sons, Inc., New York.

all other conditions being equal. The explanation lies in the density of vegetation that protects the ground. Dense ground cover under the humid regime holds back runoff and sediment at the same time reducing the flood peak; sparse vegetation of the dry regime allows runoff to move rapidly, gathering a heavy load of coarse particles and forming a high flood peak. The lesson here is that when all relevant factors are taken into account there is no need to resort to catastrophic events, such as enormous floods of magnitudes unheard of anywhere in the world today.

The subject of underfit streams was raised in 1983 by ICR staff scientist Steven A. Austin, who holds the Ph.D. in geology. Writing in Impact, Austin (1983) calls attention to Dury's studies and uses that author's hypothesis of former great stream discharges as part of his attack on "uniformitarianism." The context of Austin's mention of Dury's proposal is the familiar parading and chastising of the straw person that is the creationist version of uniformitarianism, a version that mainstream geology has for decades thoroughly repudiated. Geomorphic processes have varied in intensity from time to time in the past, following the fluctuations of climate that control those intensities. Austin repeatedly stresses the false claim that mainstream geomorphologists advocate slow, continuous changes in the landscape through long periods of the geologic past. The false claim includes Austin's statement: "Evolutionary theories for the origin of landscapes assume near constancy of discharge of streams and a steady rate of erosion as a landscape evolved" (1983, p. iii). Mainstream geomorphology makes no such assumptions, for its tentative conclusions are always open-ended and derive from observations. When observations suggest former faster or slower rates in processes of erosion or deposition, hypotheses are formulated or adapted to accommodate that information.

Austin goes on to make reference to statements by H. F. Garner (1974) to the effect that there are examples from all continents of dry channels that formerly carried surges of flood waters (Austin, 1983, p. iii). These features, strikingly exhibited in the Sahara Desert, are well known and are assigned to pluvial periods. They do not lack for explanation by mainstream geomorphology and are fitted into the Ice Age chronology of alternating glaciations and interglaciations. Catastrophism, biblical-Flood style, offers no advantages in explaining relict drainage features, but it carries, instead, the great disadvantage of forcing all explanations to fit an absurdly short time-frame.

Austin cites the Channeled Scablands of eastern Washington State as evidence of a catastrophic flood that caused channel erosion in bedrock on a fantastic scale within a brief time (1983, p. iii). That is, of course, the accepted and inescapable explanation put forward by mainstream geologists, the pioneering study being that of Professor J. Harlen Bretz of the University of Chicago in 1923. The flood (or floods) seems best explained by the breaching of a dam of glacial ice, behind which a large volume of water had been temporarily impounded. Relict dry waterfalls—one over 100 m high and 5 km wide—attest to an enormous discharge of meltwater with tremendous erosional capability. Most certainly, as Austin points out, these relict landforms are not being shaped today by the Columbia River that flows through the region (p. iii). The catastrophic hypothesis of an enormous meltwater flood generated at the margin of a wasting ice sheet conforms with the modern actualistic view of process.

That geomorphologists are well aware of the magnitude and importance of catastrophic geomorphic events is evident in Professor Dury's tabulation of many such events (Dury, 1980, pp.401-3, Table 2). The title of his paper is: "Neocatastrophism? A Further Look." This significant work was cited by Austin, who comments: "With the recent rebirth of interest in catastrophe as an important element of geomorphology the alternate landscape theory needs to be considered" (1983, p. ii). Most certainly, the biblical Flood catastrophe was never in the mind of Professor Dury as he wrote his paper, yet Austin has insinuated the absurd suggestion that mainstream geomorphologists may be turning to the Flood of Noah as a scientific hypothesis.

PART VI

Two Views of Stratigraphy and the Fossil Record

Introduction

In Chapter 22 we described the creationists' picture of Flood geology in terms of the way fossil-bearing strata were deposited and lithified, invaded by magma, broken and even overturned by tectonic processes, and locally converted into metamorphic rock--all of these events occurring within the single year of the Flood. We turn now to examine in detail the creationists' view of stratigraphy in terms of fossil content of the strata and the possible significance of the distribution of fossil types from bottom to top of the world stratigraphic section.

First, however, we must review in some detail principles of mainstream science relating to stratigraphy and paleontology. The fossil record through the Phanerozoic Eon needs to be outlined in sufficient depth and detail to place the beginnings of major taxonomic groups of animals and plants in the table of geologic time. The next question is: How is evolution, defined as descent with modification, inferred from the fossil record?

The first three chapters of Part 6 might be characterized as "classical" stratigraphy and paleontology. The biology it entails deals with whole individuals and, as such, may be called macrobiology. Since these principles were devised and applied, a whole new sphere of biology has emerged: molecular biology, the explanation of heredity in terms of complex molecules that carry the genetic code. Chapters 34, 35, and 36 cover essentials of molecular biology and the way it is applied to phylogeny and confirmation of descent with modification.

With these fundamentals of mainstream science firmly in place as a background, we can turn to the creationists' view of stratigraphy and the fossil record. Flood statigraphy and Flood paleontology are examined in terms of credibility when viewed against the stratigraphic record exposed to view for all humans to examine and interpret. How well do the field data fit the creationists' hypothesis?

Chapter 31

Stratigraphy and the Fossil Record

In Chapter 9, we outlined some basic concepts of stratigraphy, including the principle of superposition of strata in order of decreasing age (Steno, 1669) and the principle of faunal succession arrived at by direct observation of the fossil content of strata (William Smith, 1799). Recall that the faunal succession for periods of the Phanerozoic Eon (Paleozoic, Mesozoic, and Cenozoic eras) was established by field observations largely during the early decades of the 1800s; the broad outlines were almost complete before Charles Darwin and Alfred Wallace simultaneously published in 1858 their versions of evolution by natural selection. Recall, also, that the succession of fossil faunas was believed by many of those geologists who worked out the field details to be explained by a succession of catastrophic annhilations, each one followed by Divine Creation of a new fauna. This creationist view, largely attributed to Baron Cuvier, was thus the ruling hypothesis, or paradigm, under which the succession of faunas and its use in transatlantic correlation of strata emerged. True, Lamarck and his collaborator, Saint-Hilaire, had their alternative theory of continuous organic evolution going at the same time as Cuvier's, but this alternative seems not to have gained a following among the British geologists who were carrying forward the work of identifying and naming the Paleozoic periods and describing their fossil content (Faul and Faul, 1983, pp. 137-43). Table 31.1 presents thumb-box historical sketches of the naming of the Phanerozoic systems in Europe and the British Isles. Note that all systems were identified and ordered prior to 1860. A seeming exception is the Ordovician system, but that is because it was devised much later from overlapping parts of the Cambrian and Silurian systems.

Is the Fossil Sequence Arbitrary?

If, at the outset, we can divorce completely the establishment of faunal succession from any necessary connection with organic evolution--Darwinian or Lamarckian--we will have made a positive step in exposing an egregious error perpetrated by the creation scientists. Their claim is that "evolution scientists" are guilty of circular reasoning. This subject is discussed at some length by Henry M. Morris in an issue of the ICR's Impact Series titled "Circular Reasoning in Evolutionary Geology" (1977). He states:

> Creationists have long insisted that the main evidence of evolution--the fossil record--involves a serious case of circular reasoning. That is, the fossil evidence that life has evolved from simple to complex forms over the geological ages depends on the geological ages of the specific rocks in which these fossils are found. The rocks, however, are assigned ages on the fossil assemblages which they contain. The fossils in turn are arranged on the basis of their assumed evolutionary relationships. Thus the main evidence for evolution is based on the assumption of evolution. (1977, p. i)

In reply to Dr. Morris's statement, I need only suggest that he and his colleagues read my historical note that opens this chapter. Contrary to what Morris says, the strata were not in the first instance assigned relative geologic ages on the basis of either the fossils they contained or any possible evolutionary relationships that might have been assumed. They were, in the first instance, assigned relative ages on the basis of the principle of superposition, a principle based on elementary physics and nothing else, i.e., that gravity tends to pull sediment particles down through whatever fluid (air, water) encloses them, so that they accumulate in layers-- each layer younger than the layer below it. Once application of the principle of superposition told the relative ages of strata, the fossil faunas contained in those strata could in turn be arranged in order of relative age. Nothing in that arrangement was dictated by considerations of complexity of the organisms.

Paleontologist David M. Raup has commented on the significance of the historical development of stratigraphy, as I have reviewed it (Raup, 1983, pp. 153-54). If my explanation fails to refute the creationists' charge of circular reasoning, let's let Dr. Raup have a go at it. He states firmly that "the use of fossils in geologic dating is in no way dependent upon biological theories of evolution." He continues:

> The best evidence for this is that the geologic column as we know it was quite fully developed by about 1815, nearly a half century before Darwin published The Origin of Species. In other words, the geologic chronology was developed on the basis of fossils before we had any Darwinian theory, and it was developed by people who subscribed largely to a creationist view of life. Thus, geologists using a creationist paradigm developed the geologic column, and only later was evolutionary theory added as a means of understanding or interpreting the sequence of fossils found in the rocks. It is in this context that I would note that fossils would work just as well in geologic chronology if they were only funny marks on rocks. Geochronology depends upon the existence of a virtually exceptionless sequence of distinctive objects in rocks; that sequence just happens to exist in the fossils.[1]

Thank you Dr. Raup! That exposition surely drives home the last nail in the coffin lid of the creationists' erroneous charges about circuitous reasoning; it makes the historical record crystal clear. (For another excellent review of the alleged circuitry of reasoning see Steven D. Schafersman, 1983, pp. 221-32.)

Figure 31.1 is a highly schematic diagram to illustrate the initial application of the principles of superposition and faunal succession. At Locality 1 strata units (or groups) A, B, C, and D form a continuous rock exposure. (Unit G actually rests directly on unit C.) Each unit contains a distinctive fauna, or collection of

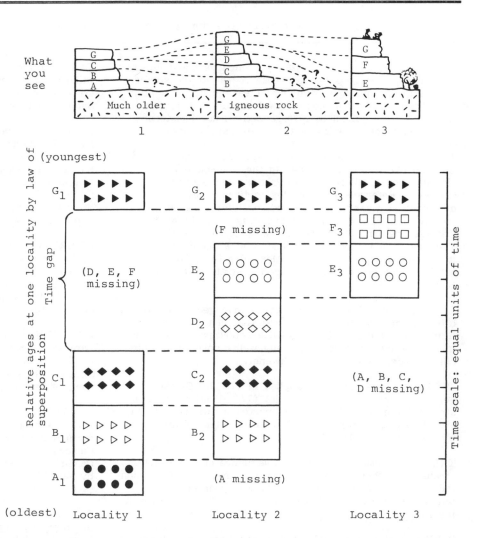

Figure 31.1 Schematic diagram of correlation between localities by the law of faunal succession. Horizontal lines are time-lines. A given fauna lies on a single time-line. (A. N. Strahler.)

taxa (species, genera, orders, etc.); that of each unit strikingly different from the fauna of units above or below it. We have used nonsense symbols to depict the faunas. Because of superposition, the order of faunal succession is established. At Locality 2, unit A cannot be found, but B and C are present in the same order as at Locality 1. On the assumption that a given fauna has the same age no matter where it is found, units B and C are correlated in time across the intervening geographic gap between the two exposures. The assumption that the age of a given fauna is the same everywhere is reasonable, whether faunas change through organic evolution or are created as new faunas following extinction. At Locality 2, unit C is overlain in succession by units D, E, and G. At Locality 3, we find unit E at the base and above it units F and G. Unit F is a new find, and clearly it is intermediate in age between E and G. Only now do we realize that three units (D, E, and F) are missing at Locality 1. Other, as yet undiscovered units, may exist elsewhere; when later found, they must be inserted into the complete stratigraphic column. On this diagram, any horizontal line is a <u>time line</u> and the rock units are <u>time-stratigraphic</u> units. (More about this subject later.)

Dr. Morris continues his article following a familiar creationist tactic: selection of passages from mainstream science publications, interspersed with comments that misinterpret the writer's intended meaning (1977). Morris cites as an important paper one by J. T. O'Rourke with the title "Pragmatism Versus Materialism in Stratigraphy" (O'Rourke, 1976). It is obvious that O'Rourke is addressing himself to colleagues in his own field of specialization, as is the practice of contributors to the <u>American Journal of Science</u>. Papers of this kind can be a ripe source for statements easily taken out of context.

A paragraph quoted from O'Rourke concerns the use of index fossils, which are individual species (or other taxa) already established as reliable indicators of relative age-- established, that is, as a consequence of geologists having first worked out relative ages of strata using the principle of superposition. O'Rourke remarks: "Each taxon represents a definite time unit and so provides an accurate, even 'infallible' date. If you doubt it, bring in a suite of good index fossils, and the specialist without asking where or in what order they were collected, will lay them out on the table in chronological order" (1976, p. 51). (How well I recall taking lab tests in paleontology in just that way. In graduate school we had to commit to memory hundreds of index fossils because, if we became exploration geologists, we would need that information.) Morris now proceeds to twist O'Rourke's statement into an entirely different meaning:

> That is, since evolution always proceeds in the same way all over the world at the same time, index fossils representing a given stage of evolution are assumed to constitute infallible indicators of the geologic age in which they are found. This makes good sense and would obviously be the best way to determine relative age--if, that is, we knew infallibly that evolution were true! (1977, p. i)

No, Dr. Morris, that doesn't make good sense, and it could not provide a viable means of determining relative geologic age. Dr. Raup again supplies us with clarification:

> The idea that geologists date a rock by the stage of evolution of its fossils is so deeply ingrained in

Table 31.1. Founding of the Phanerozoic Geologic Systems in Europe

QUATERNARY Named in 1829 by French geologist Jules Desnoyers, it was applied to "loose rocks" at the surface, including alluvium and "diluvium" (Ice Age deposits then thought to be Flood deposits), widespread in northern Europe. Charles Lyell in 1839 gave the name Pleistocene to a fossiliferous portion of the Quaternary deposits.

TERTIARY Named in 1760 by an Italian, Giovanni Arduino (1714-1795), it added to the then-popular classification of all rocks into Primary and Secondary divisions. Referred to as Montes tertiarii, they are described as weakly consolidated fossiliferous strata, largely of marine origin, but including volcanic layers. Charles Lyell in 1838 named three epochs within the Tertiary Period: Eocene, Miocene, and Pliocene. French geologists George Cuvier and Alexandre Brogniart in 1811 established the sequence of Tertiary strata of the Paris Basin.

CRETACEOUS Named in 1822 by Belgian geologist Omalius d'Halloy for a system of rocks exposed in the Paris Basin. Abundance of chalk (L. creta) characterizes this system in France and England, especially well displayed in white cliffs bordering the English Channel.

JURASSIC The name was applied in 1795 by German geologist Alexander von Humboldt (1769-1859) to limestone strata of the Jura Mountains of northern Switzerland. In 1829 these rocks were identified as a system of that name by Ami Boue (1794-1881).

TRIASSIC Named in 1834 for a sequence of salt-bearing strata in southern Germany, where a threefold subdivision was striking (Trias). This locality was poor in fossils, but was later correlated with richly fossiliferous strata of the Alps.

PERMIAN Named in 1841 by Roderick Murchison, then under the employ of Czar Nicholas I, for a sequence of fossiliferous limestone strata on the western flank of the Ural Mountains, where the city of Perm is located. In that area the Permian strata clearly overlie Carboniferous strata.

CARBONIFEROUS Named in 1822 by English geologists William Conybeare and William Phillips for coal-bearing strata of north-central England overlying the Old Red Sandstone (Devonian).

DEVONIAN Named in 1840 jointly by Roderick Murchison and Adam Sedgwick for marine carbonate strata on the coast of Devon and correlated with the Old Red Sandstone, a sequence of shales, sandstones, and conglomerates prominently displayed in Scotland and rich in fossil fishes. The Old Red unconformably overlies tilted Silurian strata in a striking exposure demonstrating a time gap with orogeny and erosion separating the two periods.

SILURIAN Named in 1835 by Roderick Murchison (1792-1871) for a sequence of richly fossiliferous strata in western Wales. Murchison began his studies at the base of the overlying Old Red Sandstone (Devonian) and worked down in the sequence, to which he gave the name of an ancient Welsh tribe.

ORDOVICIAN Named in 1879 by English geologist Charles Lapworth (1842-1920) to include strata previously claimed by Sedgwick to be Cambrian and by Murchison to be Silurian. In settling that controversy, Lapworth used the name of another ancient Welsh tribe.

CAMBRIAN Named in 1835 by Adam Sedgwick (1785-1873), a professor of geology at Cambridge University, for strata exposed in Wales in a region that was called Cambria in Roman times. Originally, Sedgwick carried his Cambrian system upward in age to include strata included by Murchison in the lower part of his Silurian system (see above).

Data sources: M. Kay and E. H. Colbert, 1965, Stratigraphy and Life History, John Wiley & Sons, New York; D. Eicher, 1976, Geologic Time, 2d ed., Prentice-Hall, Englewood Cliffs, N.J.; P. L. Abbott, 1984, Evolutionists Confront Creationists, Proc. 63rd Ann. Meeting, Pacific Div., American Assoc. Advancement Science, vol. 1, Part 3, p. 167.

creationist thought that it needs more discussion. In describing how the geologic column was developed, creationists have written that "the standard column was developed on the basis of the assumption of evolution. The fossils of 'early' ages are characterized by simplicity, of 'later' ages by complexity, because evolution must theoretically have proceeded generally in this manner" (Boardman, Koontz, Morris, 1973, p. 33); and in similar vein "that fossils are gathered from around the world . . . and assembled in a progressive order from simple to complex on a chart" (Wysong, 1976, p. 353). Nothing could be farther from the truth. (1983, p. 154)[1]

Raup goes on to say that the method described by the authors quoted above would not work even if it were tried; "there is no recognizable trend toward increasing complexity that is clear enough to use for dating purposes" (Raup, p. 154).[1] The subject of trends in evolution, whether toward increasing complexity, or diversity, or whatever, is best left for later discussion under the heading of nature of the evolutionary process. At this point, we have commented only on its relevance to the question of how the stratigraphic column became established.

Strata and Time

We have seen how the geologic systems were identified and named in early decades of the 1800s, based on direct observation of exposed strata in widely separated localities in Europe. Perhaps you are puzzled by the distinction between the words "system" and "period." For example, in earlier sections we may have referred to the Permian Period, whereas in the past few pages it would be called the Permian System. There is a distinction, and it is important in developing an understanding of the creationists' arguments about fossils and geologic time.

In a nutshell, a geologic system is a certain limited sequence of strata set apart from systems of younger age that lie above it and those of older age that lie below it. In the first instance, a system was recognized in a particular locality. Here, a record of all layers in the exposure, including a description of their thicknesses and rock type (lithology), and of the fossils they contain (if any) constituted the type section. For example, the Devonian System is represented by a type section described by Murchison and Sedgwick in Scotland, where it is represented by the Old Red Sandstone, a thick succession of shales, sandstones, and conglomerates of which certain beds contain a distinctive fauna of fossil fishes. In order to recognize strata of the Devonian

System anywhere else in the world, it had to be shown that those strata are of the same age as rocks of the type section, and this had to be done on the basis of the presence of the same distinctive index fossils or faunas as characterize the type section. To summarize, the geologic system includes all strata recognized through correlation methods as being of the same age, whatever their global location.

The concept of geologic time is an abstraction to be considered apart from actual strata or rocks of any kind. Time is a continuum, thought of as having no interruptions, but it can be marked off arbitrarily into segments--years, decades, centuries, millenia, etc. For each geologic system there is a corresponding time unit, called the period. For the Devonian System there is a Devonian Period. All strata of the Devonian System were formed within the time limits of the Devonian Period, but the elapsed time required for the actual deposition of the strata does not necessarily have to fill completely the entire period. Consider that in the type section of a system, the strata actually present may have required only one-half or one-third of the time allotted to the period. The Devonian Period, according to the latest information based on absolute age determinations, began at -408 m.y. and ended at -360 m.y., for a duration of 48 m.y. (There is an uncertainty of approximately 10 to 12 m.y. in the limiting ages.) We would not expect that the type section of the Devonian System in Scotland would fill all of the time allotted to the Devonian Period. Consider, for example, that the uppermost bed of the Devonian type section may be separated in time by many millions of years from the bottommost bed of the Carboniferous System, now directly in contact with it. We may have here a major time gap, possibly accounted for in one of two ways: (a) additional younger strata that were deposited in this locality were removed by erosion before the close of the period, or (b) there was simply no more deposition than the type section shows. Perhaps in other parts of the world, strata of Devonian age can be found that are younger than the youngest strata of the type section. That finding would partly fill the time gap. Actually, there are places in the world where deposition seems to have been continuous from one system into the next one above it. In such places, the fossils may show no abrupt changes and a boundary line would be difficult to establish.

The idea that a major time gap separates the strata of one system from the next seems to have been firmly in the minds of the geologists who first worked out the system limits, and for good reason, for they observed that the fossil assemblage in one system was quite different from that in the systems immediately below and above it. For them, a cataclysm closed the history of a system, while a new creation began the history of the next. They also noticed that in some cases, strata of one system showed evidence of having been seriously dislocated and then partly removed by erosion prior to onset of deposition of strata of the next younger system. The idea of a worldwide cataclysm marking the close of deposition of each system persisted well into the era of modern geology, but under plate tectonics has lost much of its credibility.

Geologic time is divided into periods, each one corresponding to one of the systems, as shown in the table of geologic time (Figure 31.2). By 1841, the periods had been lumped into larger time units, following the proposal of John Phillip, who suggested these be called eras, with the names Paleozoic, Mesozoic, and Cenozoic. (He spelled the last one "Kainozoic.") Translating from the Greek roots of these three titles, they can be pharaphrased as the eras of ancient (paleos), middle (mesos), and recent (kainos) life, respectively.

Rocks older than Cambrian in age can conveniently be designated as of Precambrian age, but that seems a cavalier dismissal of the three billion years of geologic

history it includes and, moreover, carries little information content. Ancient rocks that lay below the type section of the Cambrian System were visible in small patches in the northwesternmost tip of Wales, but these were simply designated as "granite" at the bottom of the stratigraphic column.

In 1930, G. H. Chadwick proposed that all geologic time for which there is a record in rock be divided into two eons: for Precambrian time, the Cryptozoic Eon, from Greek roots meaning "hidden life"; for all time since the start of the Cambrian, the Phanerozoic Eon, meaning "evident life." The latter name has become widely used out of necessity, for we cannot say "Postcambrian" without excluding the Cambrian Period.

As exploration of the great Precambrian shields of the continents progressed, a complex history emerged. It was soon recognized that at least a twofold subdivision was required, and so an older Archean time was differentiated from a younger Proterozoic time, the latter recognizing the presence of fossils interpreted as primitive one-celled organisms. (An alternative term for the younger time is Algonkian.) Figure 31.2 shows Precambrian time to consist of an Archean Eon and a Proterozic Eon: each is subdivided into early, middle, and late eras.

Both systems and periods are divided into smaller units. Systems are divided into series; periods into epochs. Beyond this level of classification we need not go at this time.

Time Gaps in the Stratigraphic Record

As we shall find out in Chapter 41, creationists consider one of their strongest arguments against the Darwinian theory of evolution by natural selection to be numerous gaps in the fossil record. They claim that the evolutionary descendency of one fossil organism from an earlier one (i.e., from a fossil organism in strata claimed to be of older age) is disproved because transitional forms of intermediate age are missing. To approach this argument, we need to investigate in more detail the nature of time gaps in the stratigraphic record, but to do so at this time quite independently of the fossil content of the strata.

As already noted, the existence of a time gap in a sequence of strata can mean one of two things: first, no deposition occurred during the time interval in question; second, some strata were deposited during the time interval, but were subsequently removed by erosion before the end of that interval. The very existence of a time gap is extremely elusive. There are no written instructions in the rocks, as there are in a theater program telling us that the next scene occurs "two weeks later."

Time gaps, or breaks, occur on a wide range of the temporal scale. We start with the smallest gaps, and these occur between individual beds. What then is a "bed"? Why and how is one bed separated from the bed below it and the bed above it? A bed is generally thought of as a layer of sedimentary rock that is uniform in its properties (homogeneous) within itself. A single bed may range in thickness from about a centimeter to several meters, or even more. Uniformity may be with respect to grain size, chemical (mineral) composition, color, or some other recognizable property. The idea of uniformity can run into trouble when, for example, we are looking at a single bed in a sequence that shows graded bedding. In one kind of graded bedding, within a single bed particle size grades upward from coarse at the base to fine at the top; but then so does the next bed above it, so each bed is like the next and they can easily be distinguished.

Perhaps we should be focusing attention on the surface that separates one bed from another--a bedding plane, that is. Very commonly sedimentary rock breaks apart easily along the bedding planes. Beds of sandstone and

Eon	ERA	PERIOD		Duration, m.y.	Age, m.y.	Orogenies
Phanerozoic Eon	CENOZOIC 66 m.y.	Quaternary		2	2	Cascadian
		Tertiary	Neogene 22, Paleogene 42	64	66	
	MESOZOIC 179 m.y.	Cretaceous		78	144	Laramian
		Jurassic		64	208	Nevadian
		Triassic		37	245	
	PALEOZOIC 325 m.y.	Permian		41	286	Alleghenian
		Carboniferous	Pennsylvanian	34	320	(Hercynian)
			Mississippian	40	360	
		Devonian		48	408	Acadian
		Silurian		30	438	(Caledonian)
		Ordovician		67	505	
		Cambrian		65	570	
Cryptozoic Eon	Precambrian time	Proterozoic Eon	(Era) Late	(b.y.) 0.3	(b.y.) 0.9	Grenville
			Middle	0.7	1.6	Hudsonian
			Early	0.9	2.5	Kenoran
		Archean Eon		1.3		
		Oldest dated rocks		3.8?		
		Earth accretion completed		4.6-4.7		
		Age of universe		17-20?		

Figure 31.2 The table of geologic time. (Based on data of The 1983 Geologic Time Scale, Decade of North American Geology, Geological Society of America, 1983, Geology, vol. 11, p. 504.)

limestone are often separated by an extremely thin clay lamina that facilitates breakup. One standard textbook of stratigraphy has this to say: "To many geologists . . . the word bed (in such compounds as well-bedded, for instance) carries the additional connotation that the strata are visibly distinct units, separated by bedding planes that are actual partings or else planes along which parting tends to develop" (Dunbar and Rodgers, 1957, p. 98).

The idea that beds are horizontal in attitude when deposited has to be used with some caution. For example, in sandy deltas being built out into standing water, the sand is spread in rather steeply sloping layers on the delta front; these are the "foreset beds." Actively building dunes accumulate in steeply sloping layers on the lee face (slip-face) of the dune. In Zion National Park you will see great cliffs of sandstone in which the ancient dune layering is rarely horizontal. In an entirely different situation, if a rugged land surface should be rapidly submerged beneath the ocean, the first sediment layers to form will slope in conformity with the former land surface. Ultimately, as depressions are filled and the ancient hilltops buried, the surface of deposition approaches horizontality. Geologists are well aware of all kinds of situations in which bedding planes are not originally formed in a horizontal position, but the point is that whatever the angle of inclination may be, the younger layer is always deposited on the upper side of the one below it (law of superposition). This observation holds true even on the wrinkled surface of an accretionary prism at an active subduction boundary. Here, the sediment layers are being crumpled, overturned, and overthrust even as they are being deposited. This would surely be a terrible place to try to apply William Smith's principle of faunal succession!

What meaning can we find in a single bedding plane? It is generally assumed, with good reason, that a single bed is a time-stratigraphic unit, and therefore that the bedding planes that bound it are time lines. However, this interepretation is justified in principle only for a small area, say on the order of a kilometer or so across, unless there is good reason to think otherwise. We found in the case of certain thick accumulations of rhythmites--laminated anhydrites of the Castile formation and lake sediments of the Green River formation--that there is reason to correlate individual layers over vast areas on the grounds that they are formed in seasonal rhythms.

Diastems

Generally, a bedding plane represents a time gap of indeterminate duration. The bedding plane between a sandstone bed and a shale bed might represent an extremely short time lapse; a current that dragged sand grains along the bottom and deposited them in a sand layer might have been followed in only hours by the raining down of clay particles from overlying turbid water. More likely in the case of sediments deposited on an ocean floor, the beds are separated by a time interval on the order of a few years, a century, or even several centuries. These time breaks are called diastems. It is almost impossible to make a reliable estimate of the

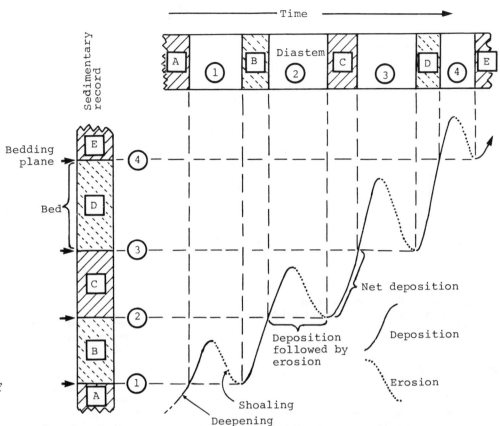

Figure 31.3 Schematic diagram of the Barrell model of diastems produced by cyclic changes in water depth. (A. N. Strahler.)

percentage of the time of actual deposition in ratio to the total elapsed time, including the diastems. In the early 1900s the use of radiometric dating made possible the estimation of the total time span elapsing for some thick sequences of conformable strata within a single system, but usually this estimate proved vastly greater than one would arrive at by summing the probable deposition times of the beds themselves. Don L. Eicher describes how this conclusion was reached:

> Joseph Barrell in 1917 pointed out that these estimates posed a problem because they indicated that strata deposited since the beginning of the Paleozoic Era, even where they are thickest, had accumulated much more slowly than anyone had previously inferred--thousands, rather than hundreds, of years per foot. Yet there is good evidence that many individual strata accumulate rapidly. Barrell concluded that innumerable diastems, most of which probably represent years or hundreds of years, must permeate stratigraphic sequences and must actually account for the bulk of elapsed time. (1976, p. 45)

We can conclude that the record of the stratigraphic column is actually highly "diluted" with lost time by innumerable diastems. Joseph Barrell introduced the idea that diastems in marine strata of continental shelves and shallow inland seas are under the control of changes of sea level relative to the land, which in turn control the water depth in the offshore zone. When water depth is increasing, sediment can accumulate to form a bed. When water depth is decreasing, the seafloor may be experiencing more or less continual erosive action by waves and currents that sweep sediment seaward to zones of greater depth, causing the upper part of a bed, or series of beds to be truncated. In the next cycle of depth increase a new bed, or series of beds, is deposited. Figure 31.3 is a graphlike presentation of Barrell's concept of cyclic depth changes leading to production of

diastems. Preservation of a succession of beds requires that there be an overall trend to increasing water depth through subsidence of the ocean floor. That trend is typical of passive continental margins because of long-term cooling of the lithosphere as the plate moves away from an oceanic spreading boundary.

The Barrell model we have outlined provides for a shifting of the site of sediment accumulation into another location in the direction toward deeper water during the erosional phase of each cycle, but one may think of other ways in which diastems would be formed. These are included by Charles W. Byers in a summary statement of the modern view of deposition of strata:

> Strata are still being stacked on top of one another, younging upwards, but we now recognize that this stacking is accomplished mostly by lateral sedimentation in shifting sedimentary environments. Each stratum is the record of some sedimentation-machine sweeping past, and the breaks between strata may encompass as much or more time than the sweeps. On a small scale, as bedforms migrate across the bottom, sedimentation is by pulsation, involving minutes to days to weeks. Episodic scour and deposition on a scale of years to centuries is seen in storm deposits, inlet migration, and individual turbidites. Punctuated aggradational cycles and lateral sweeping of delta and fan lobes may take thousands of years, but the principle is the same. Sediments accumulate rapidly in a local area for a short time, and then the influx ceases and the sediment surface is static for a long time. (1982, pp. 217-18)

(Note: Several literature references have been omitted from the above quotation.)

Professor Derek V. Ager of the University College of Swansea, England, summarized our conclusion succinctly in this excellent little volume titled <u>The Nature of the</u>

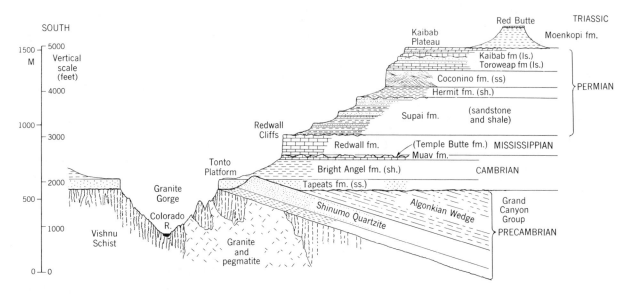

Figure 31.4 Stylized cross section of the Grand Canyon of the Colorado River. (Copyright © 1981 by Arthur N. Strahler.)

<u>Stratigraphical</u> <u>Record</u> (1981). He has a chapter titled "More Gaps Than Record," and it closes with a proposition reading as follows:

> The sedimentary pile at any one place on the earth's surface is nothing more than a tiny and fragmentary record of vast periods of earth history. This may be called the Phenomenon of the Gap Being More Important· than the Record. (P. 35)

Could diastems provide an explanation for gaps in evolutionary development so strongly emphasized by the creationists? Could intermediate forms have evolved and been destroyed during a diastem? A negative answer seems the more reasonable, considering the short duration of diastems.

And so we turn to large time gaps, in many cases involving many millions of years of missing time. These are time gaps that separate one geologic system from another and are associated with major evolutionary gaps as well.

Unconformities

To understand the kinds of major time gaps that occur in the geologic record, we will draw examples from strata of the Grand Canyon of the Colorado River. Here, the evidences of major time gaps are clear for all to read. We first need to know about the identification of lesser stratigraphic units within systems.[2]

Time-stratigraphic subdivisions within the system are <u>series</u> and <u>stage</u>, shown in descending order in the accompanying table. Their counterparts in terms of geologic time units within the period are <u>epoch</u> and <u>age</u>:

TIME-STRATIGRAPHIC UNITS	TIME UNITS
System	Period
Series (Lower, Middle, Upper)	Epoch (Early, Middle, Late)
Stage	Age

Within even a single stage at a particular geographic location, we may find confronting us a succession of varied strata totalling hundreds of meters in thickness. Our problem, as geologists, is to describe and give names to distinctive beds or groups of beds. From our limited perspective we do not know that these beds, if traced

across country for any considerable distance--say 50 or 100 km--would represent true time-stratigraphic units. We might even have difficulty at first in assigning these strata to a particular stage or series established in another part of the continent or on another continent. What we must do as a matter of practicality is simply to abandon the concept of time-stratigraphic units. Perhaps a geologic report with a geologic map is needed right away to describe a limestone layer that would be valuable in the making of portland cement, or a bed of sedimentary iron ore that a steel company would like to begin to mine, and perhaps coal seams that might be needed to fuel their blast furnaces. Description of the strata then proceeds on the basis of <u>rock</u> <u>units</u> (see Weller, 1960, pp. 420-26).

In the description of strata in terms of rock units, the working unit is the <u>formation</u>, recognized and named in terms of its rock composition, or lithology. A single formation might consist entirely of beds of limestone, or entirely of layers of shale, or entirely of sandstone beds. For examples, we use three Grand Canyon formations of Cambrian age, exposed in the lower part of the canyon, just above the Inner Gorge. As the cross section of Grand Canyon rocks shows (Figure 31.4), there is a conspicuous ledge called the Tonto Platform; it is underlain by massive sandstone beds that make up the Tapeats Formation. The cliff that represents the natural cross section of the formation sharply marks the base of the entire Paleozoic column in this region (Figure 31.5). Above the Tapeats sandstone is the Bright Angel Formation, made up largely of shale that is quite fissile and forms a long slope. This formation also contains some thin beds of dolomite and sandstone, illustrating the point that a formation may consist of more than one kind of rock. Clearly recognizable but relatively minor beds or groups of beds can be recognized as <u>members</u>, which are named subdivisions of the formation. Above the Bright Angel shale slope we arrive at a steplike succession of massive limestone beds making up the Muav Formation. Above the Muav rises a great limestone cliff, representing the Redwall Formation of Mississippian age. (Mississippian and Pennsylvanian are recognized in the United States as separate systems and periods, equivalent in Europe and elsewhere to the Carboniferous.)

Major time breaks below or above major units of strata are called <u>unconformities</u>. Four kinds of unconformities are recognized, each with its special name. We will arrive at an understanding of these four types by reconstructing the actual histories of representative examples in the Grand Canyon.

We begin with the depositional history of the Tapeats,

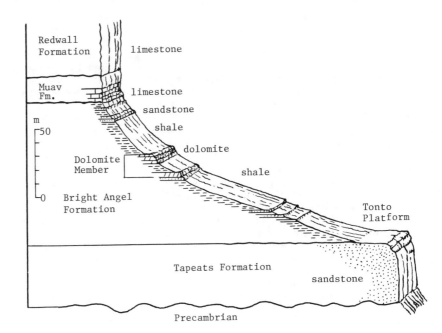

Figure 31.5 Stratigraphic profile of the Bright Angel formation on the north side of the Colorado River in Grand Canyon. (A. N. Strahler.)

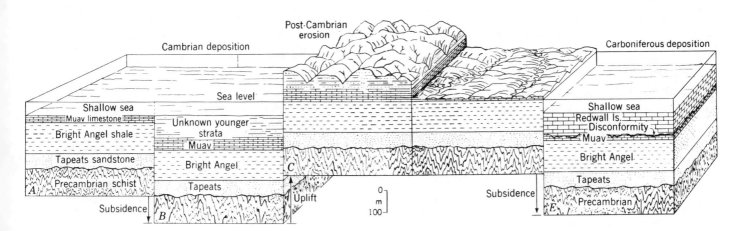

Figure 31.6 Sequence of events leading to the development of a disconformity in the walls of the Grand Canyon. (Copyright 1976 by Arthur N. Strahler.)

Bright Angel, and Muav formations, already described. Disregarding for the moment the surface separating the base of the Tapeats sandstone from Precambrian rocks that underlie it (a major unconformity), we begin with the submergence of the margin of a continent beneath the sea, to be come a continental shelf. As shown in Block A of Figure 31.6, the Tapeats, Bright Angel, and Muav formations were deposited in comparatively rapid succession. Above what is now the upper limit of the Muav Formation additional strata of which no record now exists may have been deposited, as shown in Block B. All of these Cambrian strata are said to be conformable, because they appear to be an unbroken succession, despite the vast number of hiatuses that are probably present between beds. Following deposition of the conformable sequence a broad, regional upward movement of the crust--an epeirogenic uplift, that is--brought the Cambrian strata above sea level, allowing erosion to remove the upper part of the sequence, as shown in Blocks C and D. The hilly land surface then consisted of shallow stream valleys and low divides. With only a small part of the Muav Formation remaining, a downward epeirogenic movement of the crust occurred, submerging the area as a continental shelf. Here, deposition of carbonate sediment occurred, forming the Redwall

limestone of Mississippian age. (We have simplified the story by omitting deposition of the Temple Butte Formation, also a limestone, directly on the eroded surface of the Muav. An explanation will follow.) As seen today in the canyon wall, the surface of separation between the Muav and the Redwall formations is an undulating line that is a profile of the former erosion surface, perhaps somewhat modified by wave action of the shoreline that advanced over it during submergence. This line of separation is a disconformity, a kind of unconformity in which the bedding planes above and below the break are essentially parallel. In this case, the disconformity represents an enormous time gap. Between the Cambrian and Mississippian periods lie the Ordovician, Silurian, and Devonian periods, which total about 245 m.y. It seems hard to imagine, but the parallelism between strata above and below the disconformity requires us to conclude that throughout that vast span of time the earth's crust in this region was not subjected to tectonic activity that elsewhere repeatedly deformed crustal rock by faulting and folding. We can only conclude that this region was part of a stable continental margin, far from any subduction boundaries where collision activity might occur and from areas of the continental lithosphere where rifting was in progress.

Although this long interval of freedom from tectonic activity is remarkable, it poses no special problem to mainstream geology. The creationists, however, would like to seize upon it as something impossible during the year of the Flood, when both tectonic and volcanic activity were at high intensity; they might wish to interpret the disconformity as a hiatus attributable to one of the frequent tsunamis that raged back and forth in the Flood waters.

For those concerned about the stratigraphic unit we have intentionally omitted, a note of expanation is due. The Temple Butte Formation, a limestone, is found at a number of places in Grand Canyon, where it occurs as lenselike bodies sandwiched between the Muav and the Redwall. Where these lenses occur, we are dealing with two disconformities. The Temple Butte beds are of Devonian age, as shown by the fossils they contain. This means that two, rather than one, erosion intervals must be recognized; one between Cambrian and Devonian (Ordovician and Silurian missing) and another between Devonian and Mississippian. The presence of the Temple Butte Formation requires that we insert a second cycle of epeirogenic movements causing erosion and deposition. Perhaps other such cycles occurred, but left no record.

Looking higher in the Grand Canyon walls, another disconformity is to be found at the top of the Redwall formation, where it is overlain by the Supai Formation of Permian age (Figure 31.4). Here the Pennsylvanian Period is not represented, for a time gap of over 30 m.y. Above, we encounter a sequence of about 750 m of Permian strata, represented by five formations. Two of these are interpreted as deltaic deposits (Supai, Hermit), one as a desert dune deposit (Coconino), and two as marine deposits (Toroweap, Kaibab). Minor disconformities appear to mark the formation contacts. When we add these conformable strata to the others, to give a total time span from Cambrian through Permian (the entire Paleozoic Era) of 325 m.y., the freedom that this region enjoyed from important tectonic activity is all the more remarkable. The total thickness of the section is about 1,000 m; so that the average rate of deposition, had it been uniform, comes to 1 cm per 3,250 yr. Excluding the missing periods reduces the time to about 145 m.y., for an average rate of 1 cm per 1,450 yr. The latter rate seems outrageously slow on the basis of what is known of processes of sediment deposition in similar environments of today. From what we have said about the importance of diastems in "diluting the record," one might hazard a guess that the existing Paleozoic strata of Grand Canyon account for only about 1 percent of the total time of the periods they represent.

This brings us to another kind of unconformity, recognized in a few places, but otherwise often passing unnoticed. Using fossils as evidence, geologists have found exposures in which what appears to be a smooth bedding line within a seqence of conformable strata must actually represent a major time gap. This relationship is called a paraconformity. Professor Charles Schuchert, a distinguished stratigrapher working in earlier decades of this century, took a particularly striking photograph of a paraconformity in Kentucky separating limestone beds of Middle Silurian age from those of Middle Devonian age; it shows the paraconformity as a straight, horizontal line, almost perfectly parallel with the bedding planes (see Dunbar and Rodgers, 1957, p. 121, Figure 61). Elsewhere, in the Appalachian region, the same time gap is occupied by strata measuring over 1000 m in thickness. The lesson of this example is that numerous lesser paraconformities may have thus far been overlooked.

Precambrian rocks in Grand Canyon lie beneath the Cambrian Tapeats sandstone, which forms the brink of the Inner Gorge (see Figure 31.4). Looking down into the narrow gorge, you notice that the rough, dark walls are completely lacking in horizontal bedding planes. Instead,

they show sets of nearly vertical partings, giving a grooved appearance to the rock. This rock, the Vishnu schist, is a metamorphic rock rich in quartz, mica, and hornblende. A granite that intrudes the schist has been dated by radiometric methods as having an age of -1.7 b.y. (Seyfert and Sirkin, 1979, p. 201). The orogeny that produced the schist from older sedimentary rock may have occurred between -1.75 and -1.80 b.y. and can perhaps be correlated with the Hudsonian Orogeny (Figure 31.2).

The surface of separation between the Cambrian strata and the underlying Vishnu schist is a nonconformity. It represents a long period of erosion in which a great but undetermined thickness of the older metamorphic and igneous rock was removed from the area. This erosion period saw great mountain ranges reduced by denudation to a peneplain surface of low relief and low elevation above sea level; this part of the history is pictured in blocks A and B of Figure 31.7.

In terms of creationist Flood geology, the Vishnu schist and its intrusive granites were formed during the six-day Creation, because these rocks contain no fossils. Assuming that the Cambrian strata are the first beds laid down in the Flood, this puts the long erosion interval into the 1,650-yr time gap between Creation and the Flood. But this scenario poses a grave dilemma. The Golden Age between Creation and Flood experienced no rain, so that erosion by running water would have been totally absent. To avoid this difficulty, it is necessary to assign the erosion interval to the Flood, perhaps as a cataclysmic event associated with enormous water surges produced by sudden cloudbursts from the collapsing vapor canopy. Perhaps we should avoid such gratuitous speculations on behalf of the creation scientists.

Thus far, you have been viewing the north wall of the Inner Gorge from a vantage point on the south side, near the Suspension Bridge. Suppose, now, you travel east along the canyon rim, keeping your eyes glued to the Inner Gorge. A new geologic feature soon enters the picture—a slanting wedge of sedimentary strata that appears between the Tapeats sandstone and the Vishnu schist (Figure 31.8). The wedge continues to thicken until several thousand meters of strata are exposed. This tilted sedimentary series consists of shales and sandstones belonging to the Grand Canyon Group. Referred to as the Algonkian Wedge, it is shown in the cross section in Figure 31.4. Rocks of this group are of Middle Proterozoic age, perhaps about -1.2 b.y.

From the principle of superposition, it is evident that the Algonkian sedimentary rocks are younger than the Vishnu schist, upon which they rest in a nonconformity, but are older than the Cambrian Tapeats sandstone, that lies across the beveled edges of the wedge strata. The line of separation between Cambrian strata and Algonkian Wedge strata is an angular unconformity, so designated because of the angle formed between the two sets of strata. An angular unconformity is evidence not only of a vast erosion period, but also of an orogeny that followed the deposition of the older strata and caused them to be tilted. As shown in blocks D, E, and F of Figure 31.7, the orogeny seems to have consisted of block-faulting on a large scale, so that blocklike masses of Algonkian strata were tilted between normal faults. Perhaps this event was the formation of a major continental spreading rift, associated with the opening up of a new ocean basin. The fault-block mountains were reduced by erosion to a peneplain (block E), after which crustal sinking took place and the region was gradually engulfed by the shallow Cambrian ocean. Only wedgelike masses of Algonkian sediment remained intact below the peneplain. The Cambrian strata rest in some places on the Vishnu schist and elsewhere on the Algonkian strata.

Figure 31.9 is a schematic representation of three of the four kinds of unconformities we have recognized.

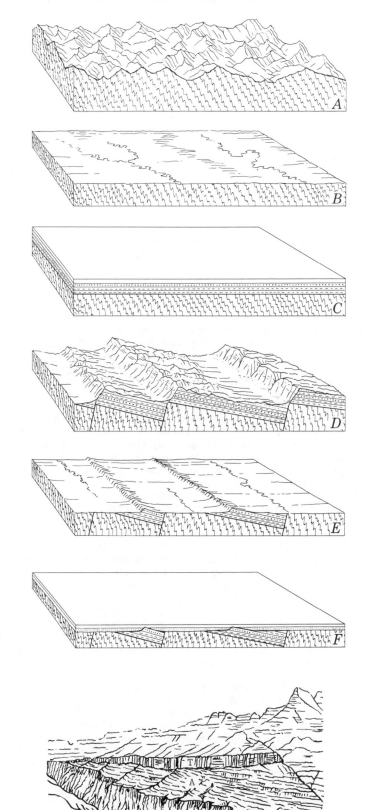

A. Precambrian mountains of Vishnu schist. —2 b.y.

B. Middle Precambrian peneplain. —1½ b.y.

D. Upper Precambrian. Block faulting during orogeny. —½ b.y.

C. Algonkian time. Sediment deposition. First unconformity. —1 b.y.

E. Late Precambrian peneplain. —¼ b.y.

F. Cambrian Period. Sediment deposition. Second unconformity. —500 m.y.

Figure 31.7 This set of block diagrams shows the manner in which the great wedges of Algonkian strata came into existence in the inner Grand Canyon. (Based on data of L. F. Noble, 1914, The Shinumo Quadrangle, U.S. Geological Survey, Bull. 594, p. 30, Plate 9.)

They are shown individually and combined in a single cross section. (Obviously, a paraconformity would not show on these diagrams.)

A special feature of the angular unconformity above the Algonkian Wedge deserves close attention, because it shows beyond question that the creationists' one-year Flood scenario is completely untenable. First, however, we note that the Algonkian strata (Grand Canyon Group) are (in creationists' terms) Flood deposits beyond doubt, because they contain clearly recognizable fossils, such as chitinozoans, and impressions of jellyfish, along with blue-green algae that were builders of stromatolites (Seyfert and Sirkin, 1979, pp. 232-33). Therefore, the tectonic events and later erosion involved in the unconformity occurred within the year of the Flood.

Referring to the schematic cross section of Grand Canyon, Figure 31.4, note that within the Grand Canyon Group is a massive bed called the Shinumo quartzite. Quartzite is an extremely hard rock, formed of quartz grains solidly cemented with silica: its resistance to erosion by running water, waves, wind, and ice is extremely great--perhaps greater than for any other

Figure 31.8 A wedge of Algonkian strata (A) lies beneath the Tapeats sandstone (T) of the Tonto Platform, but rests on the Vishnu schist (V) of Archean age. (Copyright © 1976 by Arthur N. Strahler.)

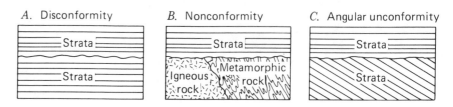

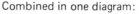

Combined in one diagram:

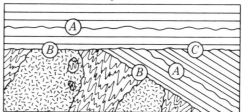

Figure 31.9 Three kinds of unconformities.
(Copyright © 1981 by Arthur N. Strahler.)

common kind of rock. As the cross section shows, the
quartzite bed projects up through the Tapeats sandstone.
Because of its great superiority in resistance to
denudation, the Shinumo quartzite rose above the
peneplain as a conspicuous ridge; it was a <u>monadnock</u>. As
the peneplain and its quartzite monadnocks were gradually
submerged beneath the Cambrian ocean, the monadnocks
formed islands, under attack by breaking waves. The
waves made little progress in getting rid of the
monadnocks, and instead, sands of the Tapeats Formation
were spread in layers around them. In some places, where
the quartzite monadnocks were low, they were completely
buried by the Tapeats sands, but in a few places the
higher monadnocks were finally buried by muds of the
overlying Bright Angel Formation.

Details of the two great unconformities of the inner
Grand Canyon were studied with great care by California
Institute of Technology geologist Robert P. Sharp (1940).
He documented some features of the monadnocks that have
great significance. First, the Shinumo quartzite was in a
fully hardened state at the time it was attacked by waves.
This means that the diagenesis of the original sand
formation was accomplished at a much earlier date,
probably before the orogeny that faulted the Algonkian
strata, and probably by slowly circulating ground waters.
That the Shinumo quartzite was hard at the time it was
attacked by Cambrian waves is specifically shown by its
relationship to the surrounding Tapeats sandstone. Sharp
describes details of the southwest flank of a prominent
monadnock of Shinumo quartzite: "Wave erosion steepened
the slopes of the monadnock to such an extent that a
great mass of material broke loose, slid down into the
sea, and moved southwestward over the fine beds,
deforming them locally. Waves and currents smoothed off
the top of the slide and rounded some of the boulders on
its upper surface; continued deposition of finer material
buried the mass completely" (1940, p. 1254). Figure 31.10
shows details of one of the Shinumo monadnocks. The
flank of the monadnock shows steepening by wave action.
Bouldery deposits were produced by wave attack as sea
level rose, but eventually, as the water became deeper
and wave energy greater, the top of the monadnock was
beveled off as a marine terrace and buried by horizontal
Tapeats strata. Sharp and others also called attention to
the presence of a zone of weathered rock immediately
below the surface of unconformity. For the Archean
Vishnu schist and its intruding granites, the weathered
zone is typically 3-4 m thick, but the original depth of
the weathered zone may have been much thicker just prior
to submergence of the erosion surface, because wave
action would easily remove soft regolith (Sharp, 1940,
pp. 1240-42, 1245-49).

Details of the great Algonkian-Cambrian unconformity

cannot be fitted into a Flood scenario that saw rapidly
rising flood waters cover the entire earth and all its
mountains in only a few weeks. Clear evidence of wave
action on the Shinumo monadnocks requires that after the
deposition of thousands of meters of fossiliferous
Algonkian strata in a marine environment, the Flood
waters must have receded, exposing the faulted strata to
subaerial erosion, after which the Flood water again began
to rise and inundate that land surface. A weathered rock
zone beneath the unconformity is clear evidence of long
exposure to the atmospheric environment. Wave action
against a sea cliff is clear evidence of an adjacent
exposed land surface, a condition not permitted by the
biblical account of the Flood. The circumstance is much
like that which we detailed earlier in the case of great
desert dune deposits, and also of deltaic sediments,
sandwiched between great thicknesses of marine strata. A
complex Flood history with numerous great water
withdrawals interspersed by full restorations of the Flood
waters is required to replace the straightforward and
clear biblical account, in which Noah's Ark experienced a
single rise of Flood water, followed by a single, long
episode of floating above the mountaintops, and terminated
in a single withdrawal of the waters. Wave action
producing marine cliffs, wind action producing great dune
layers, and river deposition on deltaic plains simply do
not take place on the bottom of a world ocean several
kilometers deep. Neither could fluvial denudation of
thousands of meters of fossiliferous rocks occur on such
an ocean floor; nor could the induration of a great bed of
quartzite occur in only a few weeks beneath that floor.
All these grossly incongruous events of an imagined

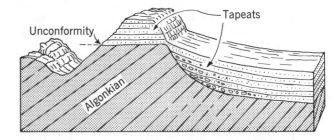

Figure 31.10 Diagram of a detail of the unconformity
between the Tapeats formation of Cambrian age and the
Algonkian strata of Precambrian age, Grand Canyon,
Arizona. A monadnock of dipping Algonkian beds (center)
resisted attack by waves of a rising sea approaching from
the right. The section shown here is about 180 m long.
(After R. P. Sharp, 1940, <u>Bull</u>., <u>Geological Soc. of
America</u>, vol. 51, p. 1263, Figure 8.)

universal Flood could, of course, be explained as miracles worked by God. Mainstream science would not then attempt to refute the miracles, since they would be part of a religious belief system. But remember that all details of Flood geology are considered by creationists to be fully naturalistic phenomena, except for the initial influx and final draining of the water itself.

Crosscutting Relationships

Closely related to the interpretation of unconformities between groups of strata is another direct method of recognizing the relative ages of masses of rocks, whether sedimentary, igneous, or metamorphic. It uses the principle of <u>crosscutting</u> <u>relationships</u>: a given rock unit is younger than any adjacent unit whose structure it cuts across. Nonconformities and angular unconformities are, in themselves, one class of crosscutting relationships. Another class consists of intrusive igneous bodies. For example, batholiths, dikes, and sills are obviously younger than the country rocks they intrude. The outer surface of a batholith cuts across structures of the country rock. Dikes often cut sharply across strata and their unconformities at a large angle.[2]

Figure 31.11 is a block diagram showing several kinds of crosscutting relationships as well as the two kinds of crosscutting unconformities. The rock bodies are arranged in alphabetical order from youngest to oldest. We must assume that these relationships are visible through exposures in canyon walls or in mine shafts. Interpretation combines the principle of superposition with the crosscutting principle. The volcano and lava flow (A) are youngest, by superposition. Igneous dike E is an interesting feature, because it intrudes sedimentary sequence F, but was leveled by erosion along the disconformity before sequence D was deposited. It reinforces the conclusion that a major time gap is present. Igneous sill H is younger than stratigraphic unit I, which it has separated into a lower and upper sequence. Although a sill is a concordant type of intrusion, most sills are not perfect tabular bodies and will be found to cut across strata at certain points. A dike (H) feeding the sill cuts across the lower sequence of strata. Batholith J is younger than the metamorphic rock, K, which it has intruded, so that the metamorphic rock body is the oldest rock shown.

Crosscutting relationships can, in some cases, reinforce the relative age relationships between fossil assemblages derived in the first instance from the principle of superposition. Crosscutting igneous rocks often give excellent opportunities to use radiometric age dating to provide limiting ages for stratigraphic sequences.

With this review of physical geological evidence capable of establishing relative ages of strata without use of fossils and without any assumptions as to the meaning of fossils, we turn to the significance of fossils in the stratigraphic column. Let us next introduce fossils into the stratigraphic record as indicators of relative time.

Time Lines Crossing Formation Boundaries

Recall that in an earlier section we dealt with the problem of explaining single formations of sandstone or conglomerate extending over a large part of a continent. The creationists asserted that formations of vast extent would be impossible to explain under a "uniformitarian" hypothesis. How, they asked, could a single sheet of sand be laid down rapidly in a shallow sea hundreds of kilometers wide? Our answer was that the formation in question is not everywhere of the same age, but represents instead a narrow zone of sand deposition—a sandy beach, for example—that made its way across a continent through time, keeping pace with a migrating shoreline. We used as an analogy the case of roofing shingles being laid from the eaves to the ridge, so the roof is younger at the ridge than at the eaves. We might also use the analogy of a transcontinental railroad that began to be constructed at Omaha, Nebraska, and ended up complete several years later at Promontory, near Great Salt Lake in Utah. The track is continuous and exactly uniform in physical construction along its entire length, but its age decreases from one end to the other. You might also like to use the example of a long scarf you began to knit in 1970 but completed in 1980; one end is 10 years older than the other, but you would never know it to look at the scarf.

The Tapeats, Bright Angel, and Muav formations that we examined in earlier paragraphs are all of the Cambrian system. They can be followed continuously along the walls of Grand Canyon from a point near Hoover Dam, on the west, to the eastern end of Grand Canyon National Park, an airline distance of about 350 km. This means that you could follow each formation from a helicopter and never loose sight of it in the winding walls of the canyon; no question whatsoever about their continuity. Your visual image would be that of three parallel formations lying almost perfectly horizontal most of that distance. Your first conclusion might be that the deposition of each formation began at exactly the same time and ended at exactly the same time, but there you would prove to be mistaken.

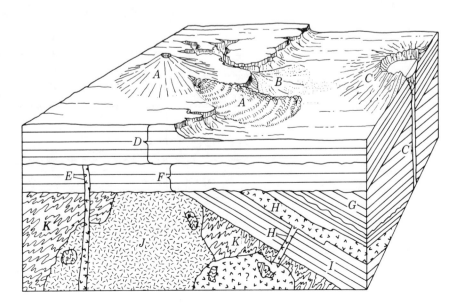

Figure 31.11 An imaginary set of igneous and sedimentary rock units arranged to illustrate the principle of crosscutting relationships. (Copyright © 1981 by Arthur N. Strahler.)

Creationists would have a fine opportunity here to explain the three formations as Flood deposits. First, sand would have settled out from bottom currents to form the Tapeats sandstone; then the mud would have settled down from a thick, turbid water layer to form the Bright Angel shale; finally, dissolved calcium carbonate would have precipitated to form the Muav limestone. Perhaps all the deposition would have taken only a few hours. And how about fossils that we find in these formations? They are marine invertebrate fossils, so they may have been living on the ocean bottom before the Flood began. Swept up by tumultuous currents, they were carried along with the turbid water and incorporated in the formations. Simple as this scenario may seem, it is totally at odds with what geologists have found to be the distribution of fossils in these formations.

We turn to information collected in large part by a distinguished geologist, Edwin D. McKee, who spent many years studying the strata of the Grand Canyon. He scaled the cliffs and slopes at many points throughout the canyon, examining the beds in minute detail and collecting fossils wherever they could be found. One of his monumental reports deals with the Cambrian System in Grand Canyon; it was published by the Carnegie Institution of Washington (McKee, 1945). What follows is based on that report, but is a simplification and generalization, concentrating on the major concept involved in the observed relationship between the strata and the fossils.

Cambrian fossils of the Grand Canyon strata occur in extremely thin key beds that extend over many tens of kilometers, making them excellent markers of time horizons. In the Tapeats formation a key fossil bed containing two distinctive species of trilobites can be traced for 55 km. Another key bed is in the Bright Angel formation and is continuous in horizontal extent for 150 km. It is unique in having a combination of two species of trilobites, a crinoid, a gastropod, and a brachiopod. The combination is easily recognized and is not found elsewhere in the Cambrian sequence. These fossils identify a <u>faunal zone</u>, which is a time line. Creationists have no reasonable explanation for the occurrence of several such faunal zones in the Cambrian sequence. Each zone is what remains of a continuous seafloor ecosystem. No such ecosystem could have been indigenous to the floor of the Flood ocean, since a single faunal bed would have needed a vastly longer time of habitation than one year. Even the individuals that are now fossils must have required several years to attain their sizes. Creationists say instead that the fossils were killed in the Flood and lie in zones of burial. But then we must ask: how could these fossils have been sorted out by turbulent Flood currents into a number of distinctive faunas, each deposited in a single thin sediment layer separated by thick accumulations of sands or muds? No suitable explanation is possible in the brief Flood scenario. Instead, we should expect random or irregular distribution of the fossils throughout the entire Cambrian sequence as dead individuals were flung about in turbulent currents, coming to rest in chance levels and locations along with the enclosing sediment. But our argument remains to be supplemented by a remarkable

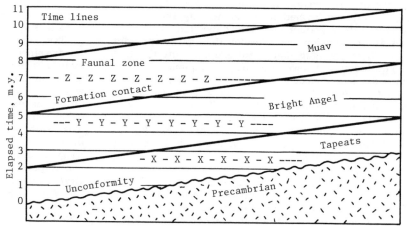

B. Time lines horizontal

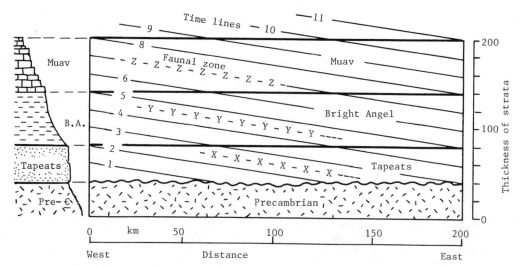

Figure 31.12 Schematic diagram of relationship between stratigraphic contacts and time lines. (A. N. Strahler.)

A. Formation contacts horizontal

finding that must surely deal the coup de grace to the moribund creationist hypothesis.

What we are going to tackle next is a bit complicated, but well worth following through with the aid of diagrams. Figure 31.12 makes use of two sets of parallel lines: time lines and formation boundaries. In the lower diagram (A), the formation boundaries are shown as horizontal and the vertical scale is marked off in meters of rock thickness. In the upper diagram (B) we have made the time lines horizontal and the vertical scale is marked off in time units (millions of years, for example). Three faunal zones are shown by rows of Xs, Ys, and Zs paralleling time lines. When McKee plotted the locations of points where he encountered the same faunal zone, he discovered that from west to east those points moved down through the formation to relatively lower levels. This phenomenon shows in diagram A by down-slanting faunal lines that descend as if to approach the base of the formation. The same thing holds true in diagram A but, by having made the time lines horizontal, we are required to show the formation boundaries as slanting up from left to right.

Well, now, what does this diagonal relationship between time lines and formation boundaries mean? The interpretation is given by means of two other kinds of diagrams, shown in Figure 31.13. At the left are schematic maps of the region in which the stratigraphic section is located. At the right are profiles (vertical slices) showing the relationship of the ocean floor to sea level. Both kinds of diagrams show the kind of sediment being deposited on the ocean floor at the time represented by the faunal zone. The diagrams are to be read from bottom to top. Taking time X first, the shoreline lay in the right portion of the map; immediately offshore was a zone of deposition of sand (Tapeats); farther left (west) in deeper water was a zone of deposition of clay-mud (Bright Angel). During all of this part of Cambrian time the sea was encroaching on the land, a process called transgression. The shoreline was shifting eastward and with it the zones of sediment deposition. Thus, at time Y the map shows three zones: sand, clay-mud, and lime-mud, the latter

representing the Muav limestone. By time Z, only the clay-mud and lime-mud zones were within the limits of the map.

For those of you who like to acquire strange new words, we note that geologists use the term facies to mean the general appearance, composition, and depositional environment of a unit of strata, such as a formation. As defined in the dictionary, "facies" means "general appearance"; it derives from the Latin word for "face." Specifically, we have been discussing lithofacies, referring to the physical properties of the strata. We have recognized and mapped three distinct lithofacies of the Cambrian System in the Grand Canyon region. The lesson has been that each lithofacies has its own environment of deposition--in this case, determined by water depth and distance from shore--and that lithofacies shift laterally with time. In this case they shifted toward the continental interior during trangression; at other times and places lithofacies have shifted oceanward during regression, as the ocean level has fallen relative to the continents.

Now, you ask, where is that coup de grace I promised? Creationists are asking us to believe that during a one-year Flood conditions were so highly uniformitarian that not only could narrow faunal zones just happen to form, but those faunal zones all run obliquely to the formation boundaries, all slanted down in the same direction. If the bodies of dead organisms were dropping out of suspension from highly turbid, swiftly moving flood waters, what would be the probability that organisms of only certain selected species would form a continuous thin layer cutting obliquely across the formation? May we suggest one chance in ten raised to the one-hundredth power (10^{100})? Keep in mind that we are concerned with a totally naturalistic geologic process, as the creationists agree must have been the case in all of Flood geology; we are not talking about Divine Creation by a Creator.

Examples like that from the Cambrian System of the Grand Canyon are numerous throughout all of the geologic

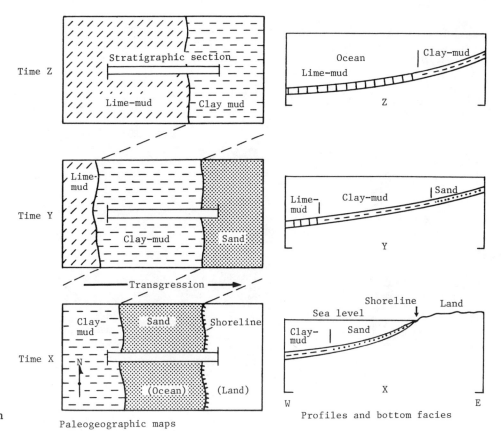

Figure 31.13 Paleogeographic maps showing depositional facies prevailing at times of the faunal zones X, Y, and Z of Figure 31.12. Not drawn to true scale. (A. N. Strahler.)

systems in all continents of the globe. One might go so far as to say that the crossing of time lines (faunal zones) by formation boundaries (i.e., lithofacies boundaries) is the typical or characteristic feature of stratigraphy, rather than the exception. It has been verified by an entirely unique natural process--the formation of thin layers of volcanic ash during major eruptions. A single ash layer appears as a natural time line over a vast area, because the ash settles upon whatever facies happens to be accumulating on the ocean floor. And, of course, absolute methods of age determination have confirmed the interpretation we have offered. Our overview of stratigraphic principles has provided a solid basis on which to examine the fossil record itself. We are ready to ponder a major question: Does the fossil record indeed suggest the kind of organic evolution proposed by Charles Darwin and Alfred Wallace?

First, however, a review of the fossil record of the Phanerozoic Eon can be most helpful for those readers who have not taken a course in historical geology.

Credits

1. From Scientists Confront Creationism, Laurie R. Godfrey, ed., W. W. Norton & Company, New York. Copyright © 1983 by Laurie R. Godfrey. Used by permission.

2. Parts of the descriptive text of this chapter are taken from A. N. Strahler, The Earth Sciences, 2d ed., and Physical Geology, Harper & Row, Publishers, New York. Copyright © 1971 and 1981 by Arthur N. Strahler.

Chapter 32

The Fossil Record of the Phanerozoic Eon

To understand the arguments used by creationists against organic evolution through geologic time, you will need a sound but general picture of the fossil record and its interpretation according to mainstream science. After the systems and periods were established by the principle of stratigraphic succession, the fossil faunas of each system and of subdivisions within each system subdivision could be plotted on a time chart. This endeavor produced the data base from which evolutionary interpretation was made. Let us now fill in some details of that chart. First, however, we need a general overview of the broad classification of life forms.

A Classification of Life Forms

A modern classification system, or taxonomy, of life forms appears in the creation science textbook, Biology: A Search for Order in Complexity (Moore and Slusher, 1970, pp. 141-265). Looking it over, you would probably accept it as a satisfactory presentation of the world of living things. To debate about the origin of the different groups of life forms, the participants must agree on a system of classification and nomenclature, for otherwise they cannot communicate. That the creationist textbook authors accept the Linnaean system is evident from their description of the life and work of Linnaeus, who believed in Divine Creation and who lectured passionately on Nature as witness to the skill and providence of the Creator (pp. 142-43).

The highest category in the the taxonomy is the kingdom. Depending on which modern taxonomic system is selected, there are either four or five kingdoms. Our concern in this chapter is with the fossil record of the animal and plant kingdoms in the Phanerozoic Eon. Taxonomic categories of descending rank within each of these kingdoms are the phylum (plural, phyla), class, order, family, genus, and species. Examples of taxa (singular, taxon) of an animal and a plant are shown in Table 32.1.

Animals are characterized by the need to ingest and digest food manufactured by other organisms. Animals have a mouth and a digestive system, a circulatory system for oxygen distribution, and a nervous system for control. Animals engage in respiration, in which oxygen is taken into the cell to burn food and release energy. Oxidation is accompanied by the release of carbon dioxide. Many forms of animals are capable of locomotion, but others are fixed in place and require that food be brought to them. In an older terminology, the animals belong to the Metazoa. The term is complimentary to Protozoa, applied to one-celled organisms classed as animals; both were widely used throughout the literature of geology and paleontology.

Members of the plant kingdom are characterized by growth through the process of photosynthesis, in which atmospheric carbon dioxide and water are combined with solar energy in the presence of chlorophyll to produce carbohydrate, at the same time liberating free oxygen. Plants lack mechanisms of locomotion and either remain stationary and anchored in place or are transported by other mechanisms.

Of the plant kingdom, the fossil record consists mostly of remains of the vascular plants. The word "vascular" refers to the presence of a vascular system of roots and connecting tubes by means of which water and nutrients are carried from one part of the plant to another. The vascular plants include all terrestrial herbs, shrubs, and trees; these will be the main objects of our review of plant evolution. (Not included in the vascular plants are the bryophytes--the mosses and liverworts.)

Nine animal phyla important in the Phanerozoic fossil record are named in Table 32.2. The Latin name is given along with the common name of the phylum; for each phylum, some common representative groups are listed. All animals except the vertebrates (a subphylum of the Chordata) are referred to as invertebrates. Whereas the vertebrates have an internal bony skeleton, the invertebrates have no internal skeleton but can have shells of mineral matter or external skeletons (exoskeletons) of hardened organic matter (chitin).

The taxonomic system devised by Linnaeus was based on his study of living species of plants and animals. It was completed by him long before organic evolution was envisioned by either Lamarck or Darwin. But, as fossils were collected and arranged in the stratigraphic order in which they were found, the evolutionary relationships began to become inescapable. A particular taxon--phylum, class, or order--could be recognized as persisting from one geologic system to the next younger system, albeit with changes in form. In recognition of this observation, biologists and paleontologists began to adapt the taxonomic system to reflect the continuity of taxa in the fossil record. A particular taxon, therefore, came to imply an evolutionary pathway, in which all members of a taxon are descended from a common ancestor (Figure 32.1). A

Table 32.1 Taxonomic Categories in Order of Descending Rank

Categories	Taxa	
KINGDOM	Animal	Plant
PHYLUM	Chordata	Tracheophyta
CLASS	Mammalia	Gymnospermae
ORDER	Primates	Pinales
FAMILY	Hominidae	Pinus
GENUS	Homo	Pinus
SPECIES	H. sapiens	P. ponderosa
Common name	Human	Pine

classification system that is based on this interpretation is described as a <u>phylogenetic classification</u>.

Phylogenetic (evolutionary) relationships are often difficult to establish, and taxonomists will differ among themselves as to the phylogeny of a particular taxonomic group. Not surprisingly, the creationists are quick to seize upon such differences within the mainstream science community. Creationists deny evolution in the first place and find support in gaps in the fossil record. Where phylogeny seems obscure, the obvious reason (they say) is that no evolutionary relationships exist in the first place. Where evolutionary relationships are strongly evident, the creationists simply say that an omniscient Creator made things that way, and that variation exists within created kinds.

The fossil record of Phanerozoic time for the major groups of plants and animals is shown in a summary time chart, Figure 32.2. Dashed lines show the probable (or possible) origin of each group by branching from another group. Where a band terminates in a dot at the top, the group has become extinct. More detailed charts of particular groups--vascular plants, invertebrates, reptiles, dinosaurs, and mammals--will follow in the course of our brief review of evolution through Phanerozoic time.[1]

Life of the Cambrian Period

In early decades of the 1900s, one of the most profound and puzzling observations about the fossil record was the faunal contrast between Precambrian and Cambrian strata. The Cryptozoic Eon, with which Precambrian time coincides, was so named--Eon of Hidden Life--because so few traces of life could be found. As we noted in describing the Algonkian Wedge of Grand Canyon, that late Precambrian sequence was found to contain stromatolite structures produced by calcareous algae, some fossils that appeared to be sponges or siliceous spicules of sponges, and what seemed to be worm trails. The Belt system of the Northern Rockies, also of late Precambrian age, revealed similar fossils, and other Proterozoic rocks in Brittany had produced what seemed to be the siliceous tests of radiolaria. My college historical geology text summarized the extent of the Precambrian fossil record in these words:

> This is certainly not an impressive array to represent the life record of more than half the history of the Earth! On the other hand, it is of the greatest import. Primitive miscrosocopic plants, a few single-celled animals, a trace of sponges, and a suggestion of some type of worm--these are all that a century of search has brought to light. . . . We may infer, therefore, that life was probably abundant in the seas of Cryptozoic time and especially during the Proterozoic, but was of a low order and doubtless small and soft-tissued, so that there was little chance for actual preservation of fossils. (Schuchert and Dunbar, 1933, pp. 115-16)

That textbook statement remained essentially valid until the early 1960s. (How it was changed by a series of remarkable fossil discoveries is a story we will relate in Chapter 42.)

The earliest occurrences of Cambrian strata, in striking contrast with the Proterozoic strata, show an abundance of invertebrate life forms, including representatives of several of the important animal phyla. Figure 32.3 shows the evolution of the major groups of invertebrate marine animals. The sponges, along with the radiolarians (shown at the left), can be considered to have persisted from Precambrian time. There appeared abruptly at the opening of Cambrian time corals (Coelenterata), brachiopods (Brachiopoda), trilobites (Arthropoda), and archaeocyathids (a spongelike group

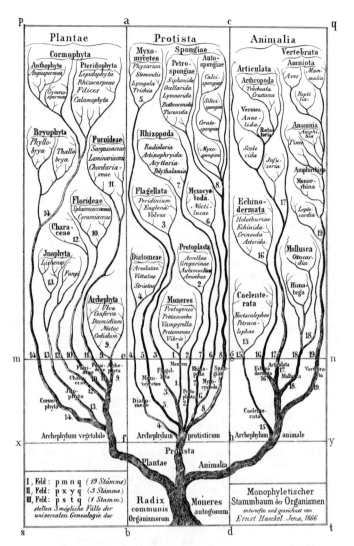

Figure 32.1 This phylogenetic tree, the first of its kind, was published in 1866 by German biologist Ernst Haeckel, a staunch supporter of Charles Darwin. Three main stems at level x-y represent the three great kingdoms ("Archephylum"). Nineteen stems crossing level m-n represent the phyla or equivalent taxa. (From E. Haeckel, 1866, <u>Generelle Morphologie</u>.)

not related to modern phyla). Also appearing early were crustaceans (Arthropoda).

The large increase in numbers and diversification of new groups within these new phyla following their introduction is known as <u>evolutionary radiation</u>. A period of rapid radiation is followed by a much longer span of time in which the new groups persist with little change. The causes of evolutionary radiation are complex and not well understood, but it can be reasoned that the onset of a particularly favorable set of environmental conditions is partly responsible. In the case of the almost explosive radiation of the earliest Cambrian faunas, a rapid increase in atmospheric oxygen was possibly a major factor, along with crustal stability of extensive passive continental margins with broad continental shelves and large expanses of shallow inland (epicontinental) seas. Whatever the reasons, early Cambrian time was especially favorable to the diversification of trilobites, brachiopods, and archaeocyathids (elongate conical fossils resembling small cornucopias). An estimate of the contents of the Cambrian fossil assemblage might give: trilobites, 60 percent; brachiopods, 30 percent; all others 10 percent (McAlester, 1968, p. 54). The trilobites are now extinct, while the brachiopods are now rare. Among the other phyla of

Table 32.2   Principal Animal Phyla Important as Fossils

Phylum	Common name	Representative groups
Porifera	Sponges	
Coelenterata	Coelenterates	Corals, jellyfish, sea anemone
Bryozoa	Bryozoans	"Moss animals," "sea mats"
Brachiopoda	Brachiopods	
Mollusca	Molluscs	Pelecypods, or bivalves (clam, mussel, oyster, scallop) Gastropods (snail, slug, conch) Cephalopods (squid, octopus, nautilus, ammonite)
Annelida	Annelids	Segmented worms, earthworms, leeches
Arthropoda	Arthropods	Crustaceans (crab, lobster, shrimp) Trilobites* Ostracodes Insects Arachnida (spiders)
Echinodermata	Echinoderms	Crinoids (sea lilies) Sea cucumbers, sea urchins, starfish, serpent stars
Chordata	Chordates	
Vertebrata (a subphylum)	Vertebrates	Fishes Amphibians Reptiles Birds Mammals

*Now recognized as a separate phylum.

Based on data of A. L. McAlester, 1968, The History of Life, Prentice-Hall, Inc., pp. 146-47.

Cambrian time, many of the representative groups are quite unlike their modern equivalents.

The richness of Cambrian life was strikingly demonstrated by a most remarkable fossil find in 1910 by paleontologist and stratigrapher Charles D. Walcott of the U.S. Geological Survey. In a remote mountain locality in British Columbia he came upon outcrops of black shale in which the presence of certain trilobite fossils showed the strata to be of Middle Cambrian age. Impressions on bedding surfaces of the shale preserve minute details of many species of soft-bodied invertebrates, the existence of which was previously unknown. These include jellyfish, anemone-like coelenterates, annelid worms, and a holothurian (sea cucumber). The Burgess shale, as the horizon came to be known, yielded some 35,000 fossil specimens. The lesson of the Burgess shale fauna is that the fossil record is dominated by animals with hard parts, giving a strongly biased set of data. It also raises the possibility that soft-bodied ancestors of the Cambrian invertebrates may have evolved in late Precambrian time, in which case the "Cambrian explosion," was perhaps not explosive at all. (See Chapter 42.)

Evolution has tended to proceed on a series of one-way branching tracks. New groups form readily and proceed on parallel or diverging tracks, but the reverse trend--to have those tracks rejoin so as to unite two groups in a common line--is forbidden by the inability of different species to interbreed and thus to exchange genetic materials. Thus, for many groups, the evolutionary track simply ended by failure to propagate, and the result was extinction.

Extinctions of large numbers of groups seem to have

occurred rapidly at certain points in the geologic column. Again, as in the case of rapid radiations, environmental changes of one sort or another are assumed to have been responsible for large-scale extinctions. Changes in atmospheric and oceanic temperatures can be expected to place severe stresses on faunas and floras. Widespread emergence of continental margins at certain times greatly reduced the extent of shallow seas, also isolating one sea from another. Major extinctions have been attributed to such constrictions of marine environments. Other, more catastrophic causes of mass extinctions are attributed to the physical effects of enormous impacts by asteroids or comets. (See Chapter 49.)

Rapid extinctions of large groups of organisms lead to vacant environments, and these are rapidly filled by adaptive radiation of other groups. Major episodes of extinction followed by rapid evolutionary radiation seem to have marked the transitions from one geologic era to the next.

By the end of the Cambrian Period, trilobites had been reduced somewhat by extinctions, while the archeocyathids had become completely extinct. Extinctions of minor classes of molluscs, arthropods, and echinoderms also occurred. Very late in Cambrian time two new orders of molluscs appeared: gastropods and cephalopods. On the other hand, the brachiopods, corals, sponges, and crustaceans held their own through the Cambrian and into the Ordovician. But when we look at the total picture of the invertebrates, the Cambrian seems to have "gotten off to a false start" with respect to life of the Paleozoic Era as a whole.

The opening of the Ordovician Period appears to have

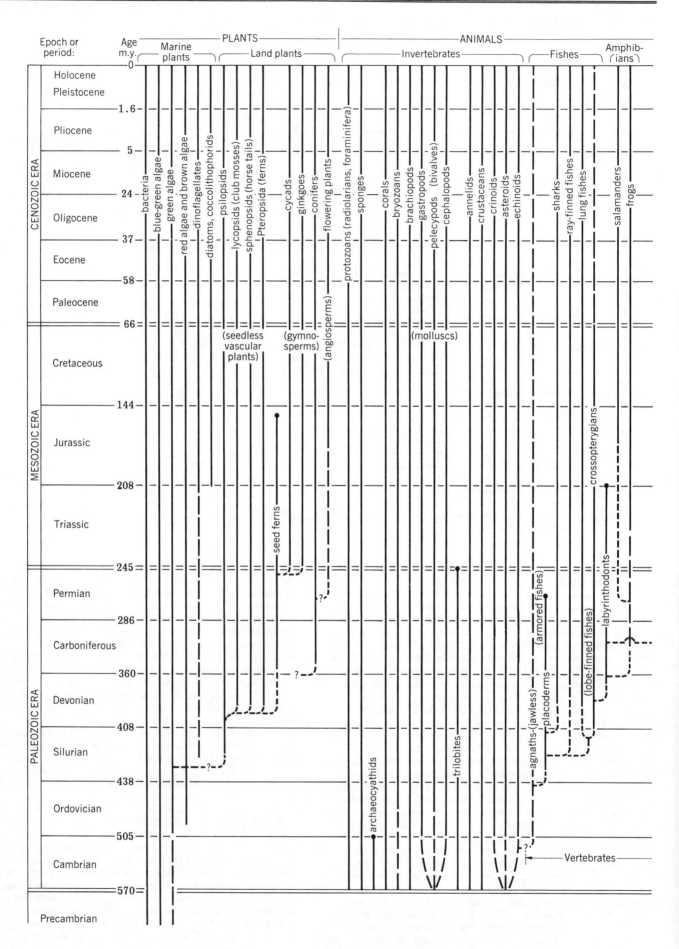

Figure 32.2 Summary chart of the evolution of major plant and animal groups from Cambrian time to the present. (Based on data of A. L. McAlester, The History of Life, Prentice-Hall, Inc., Englewood Cliffs, N.J.)

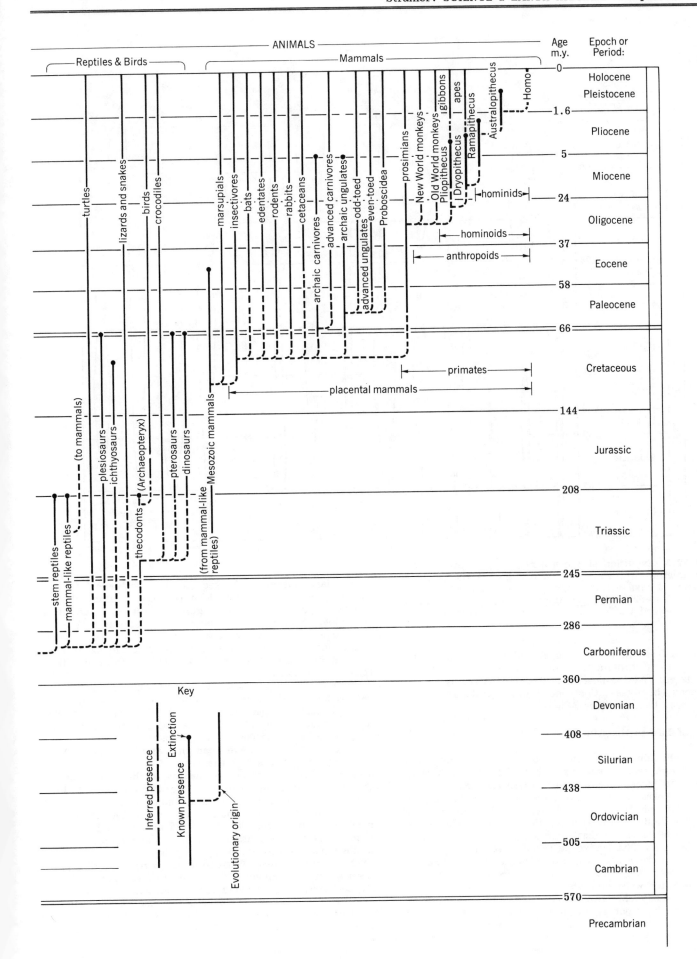

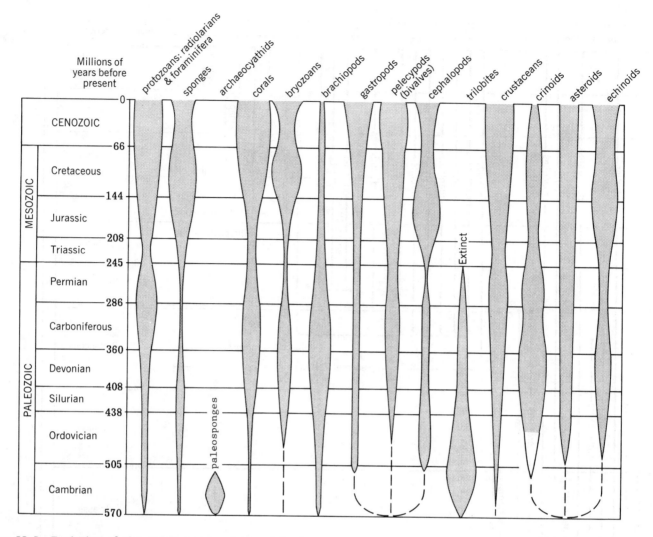

Figure 32.3 Evolution of the major groups of invertebrate marine animals. Width of each band suggests the relative abundance of taxa within a group. (Based on data of A. L. McAlester, 1968, The History of Life, Prentice-Hall, Inc., Englewood Cliffs, N.J., pp. 60-61, Figure 3-7.)

signalled a fresh start, setting evolutionary trends that were to persist through the Paleozoic. Asteroids and echinoids entered the scene, as did the bryozoans and pelecypods (bivalves). Gastropods and cephalopods rapidly rose to importance (Figure 32.3).

Spread of Life to the Lands

Until late Silurian time the lands were totally devoid of the tracheophytes, or vascular plants (familiar as herbs, shrubs, and trees). Neither were there any true terrestrial animals, which today are are in the arthropod and chordate phyla, and would also include land snails of the phylum Mollusca (McAlester, 1968, p. 64). The oldest known fossils of land animals are scorpion-like arachnids of Silurian age. The landscape would have presented everywhere a barren appearance, despite the presence of extensive microbial mats and scums where conditions of heat and water were favorable.

Lack of sufficient atmospheric oxygen has been cited as possibly the major single deterrent to occupation of the terrestrial environment by plants and air-breathing animals. Two factors are involved here: (a) effect of the concentration of ordinary molecular oxygen (O_2); (b) effect of the concentration of ozone (O_3). The first effect would be on the metabolism of oxygen-using organisms

(e.g., land animals); the second would influence the intensity of incoming ultraviolet radiation. Taking the second effect first, for all organisms, the lethal effect of ultraviolet radiation would have been a barrier to life in the surface layer of the oceans and on land. That barrier could only have been removed when sufficient ozone was present in a layer high in the atmosphere to absorb this form of radiation.

In 1964, L. V. Berkner and L. C. Marshall made calculations of the oxygen content of the atmosphere through geologic time (Figure 32.4). They estimated that, at the start of the Paleozoic, the atmospheric oxygen level stood at about 1 percent of its present level. The assumption seems to have been that at this level of ordinary oxygen (as O_2) the level of ozone (O_3) would also have been very low. According to their model of oxygen buildup, the value rose steadily until, by Middle Silurian (-240 m.y.), it stood at about 10 percent. Presumably, this level would have been sufficient to produce enough ozone to shield the ground surface from lethal ultraviolet radiation. Vascular land plants appeared in the fossil record at about this time.

The Berkner/Marshall schedule of oxygen buildup has been severely challenged. In 1970, Helen Tappan and A. R. Loeblich, Jr., asserted that an atmospheric oxygen level near the present value was attained as far back as -2.0 b.y. (1970, p. 283). Biologist Lynn Margulis of Boston University, distinguished for her investigations of the evolution of microscopic organisms (prokaryotes, eukaryotes) through Precambrian time, postulates high values of atmospheric oxygen (i.e., "about the present concentration") by -1.8 b.y. (Margulis, 1984, pp. 62-63).

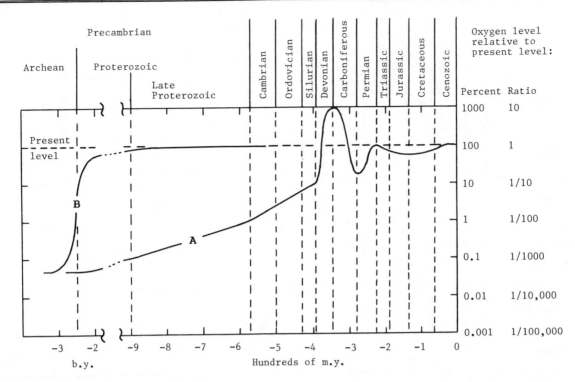

Figure 32.4 Postulated changes in the level of atmospheric oxygen (O_2) throughout geologic time. (A) Berkner/Marshall model. (B) Margulis/Walker model. (Based on data of L. V. Berkner and L. C. Marshall, 1964, p. 1223, Figure 10, and L. Margulis, 1984, pp. 62-63.)

The significance of the Berkner/Marshall schedule of oxygen buildup in terms of associated ozone levels was restructured in the late 1970s by a trio of scientists-- Joel S. Levine, Paul B. Hays, and James C. G. Walker. They reported that an ample supply of ozone was present in late Precambrian time to protect the surface layer from damaging ultraviolet radiation (Science News, vol. 111, p. 378, 1977). Perhaps, then, we must look to explanations other than protection from lethal ozone for the delayed appearance of land plants and air-breathing animals.

Transition from the water environment to the land environment posed serious difficulties to both plants and animals, requiring profound changes in structure and function. One difficulty for animals was to emerge from a salt solution to which the body fluids were matched and enter fresh water that could rapidly dilute those fluids. A second difficulty for animals was posed by leaving the aqueous environment, from which oxygen in solution is absorbed by specialized tissues, to enter the atmospheric environment in which oxygen must be absorbed from a gaseous medium.

For all organisms adapted to an aqueous environment, prolonged exposure to air would result in fatal desiccation (drying out). Survival in a surrounding of air required development of a protective coating to retain vital fluids or the means to replace them as fast as they were lost. Desiccation also constituted a barrier to sexual reproduction, for water-dwelling organisms release their reproductive cells (gametes) into fluid surroundings where fertilization takes place. New means of protecting these cells had to be developed.

The transition from salt water to land seems logically to have been a two-stage process: salt water to fresh water, then fresh water to atmosphere. Difficult as the transition was, an environment with abundant oxygen for animals and intense sunlight for plant photosynthesis was awaiting occupation. An almost explosive evolutionary radiation took place, first among the plants.

The Earliest Land Plants

The vascular land plants are thought to have evolved from the green algae, which use chlorophyll in photosynthesis in the same way as do the plants. The green algae were successful in making the transition to fresh water, but there is no fossil record of the evolution from green algae to the vascular plants. However, some measure of support for this relationship is found in the bryophytes, a plant phylum that includes the mosses and liverworts. These plants now live in wet, well-shaded environments on land but are limited to small size because they lack structures for transferring fluids from one part to another within the individual and, moreover, have no means of protecting the reproductive cells from drying out after they are separated from the parent plant. It thus seems possible that the vascular plants evolved through plants related to the bryophytes, but there is no fossil evidence of such an evolutionary chain.

The vascular plants met the restraints and opportunities of the land environment by developing roots for intake of soil moisture, leaves for photosynthesis, and stems containing a vascular system of specialized conductive tissue for transporting fluids between roots and leaves. In addition, plants have developed on their leaves a specialized outer cell layer covered by a protective layer, the cuticle, capable of preventing evaporative water loss, but also containing openings (stomata) through which the release of transpired water can be regulated.

The vascular plants are subdivided into two groups: those that are seedless, designated as pteridophytes, and those that are seed-bearing, the spermatophytes. The pteridophytes are ferns or fernlike plants and include the modern club mosses and horse-tails. The spermatophytes include the gymnosperms, with naked seeds (seed ferns, cycads, ginkgoes, and conifers) and the angiosperms with covered seeds (flowering plants).

Evolution of the vascular plants is shown in Figure 32.5. First to evolve were the seedless plants. One group appears first in rocks of upper Silurian age; these were the psilopsids, simple plants lacking in roots or leaves. They diversified rapidly in the Devonian Period but then rapidly subsided to minor importance. In Devonian time,

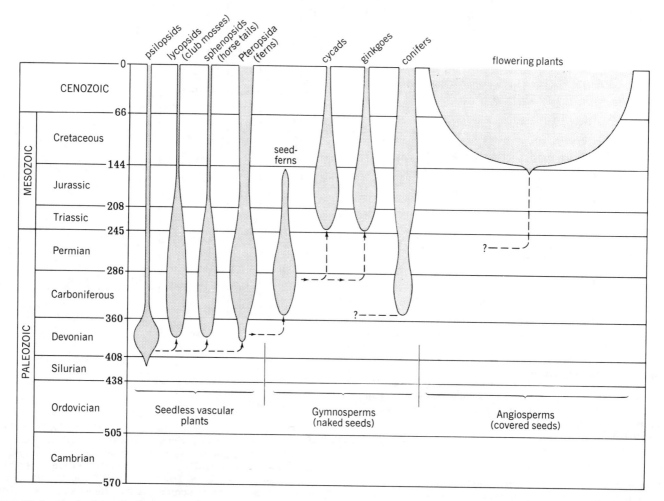

Figure 32.5 Evolution of the vascular plants. (Based on data of A. L. McAlester, 1968, The History of Life, Prentice Hall, Inc., Englewood Cliffs, N.J., pp. 86-87, Figure 5-2.)

the remaining groups evolved rapidly. These are the lycopsids, represented today by the club-mosses; the sphenopsids, seen today in the horse-tails; and the pteropsids, or ferns. All three plant groups included species of tall trees, although today only the ferns attain tree size. By the middle of Devonian time, rich forests of seedless vascular plants covered the lands.

Fishes to Amphibians

Origin of the vertebrates is obscure--there is no fossil record preceding the occurrence of fishes in the late Ordovician time. From that point on, the record is clear for the evolutionary succession from fishes to amphibians, then to reptiles, birds, and finally mammals. Figure 32.2 shows the evolutionary sequence. (Inferred evolution of the vertebrates, a subphylum of the phylum Chordata, from more primitive chordates is discussed in Chapter 42.) By Silurian time fishes with thick bony armor were well developed and included both bottom scavengers and carnivorous types. Both the sharks and the bony fishes evolved from these early armored fishes before the close of the Devonian Period.

Transition from fishes to land animals began with fishes having lobelike fins containing muscles and articulated bones outside the body. These fishes must have entered the freshwater environment and developed the capacity to breathe air by means of lungs. The modern lungfish retains this capacity, and by resorting to air breathing survives the annual drying up of streams

and lakes in the wet-dry tropical (savanna) climate.

The fossil record shows that a group of lobe-finned fishes, the crossopterygians, developed the capacity to move about on land through the modification of fins to stubby legs. Thus there emerged in late Devonian time a group of amphibians, the labyrinthodonts, that became the dominant land vertebrates in the ensuing Carboniferous Period. These animals, which resembled a modern alligator, had representatives up to two or three meters in length. They probably spent much of their time in or near water and were inhabitants of the great swamp forests. The labyrinthodonts declined rapidly in the Permian Period and became extinct in the Mesozoic Era.

Life of the Carboniferous Period

The Carboniferous Period includes two subperiods recognized in North America--Mississippian and Pennsylvanian--and usually thought of as having full status as periods. The Mississippian Period, which we encountered in the Grand Canyon (Redwall Formation), was a time of extensive shallow inland seas providing a favorable environment for marine life. For the most part, the marine invertebrates held their own in terms of abundances, except for the trilobites, which were steadily declining (see Figure 23.3). The Pennsylvanian Period, or upper Carboniferous, saw a worldwide trend toward extensive areas of low-lying freshwater swamps repeatedly inundated by shallow seas. The present consensus is that these were deltaic environments, as discussed in Chapter 23 in our section on the origin of coal measures. The terrestrial environment, with its great swamp forest and rich insect life, is thus an outstanding feature of the upper Carboniferous.

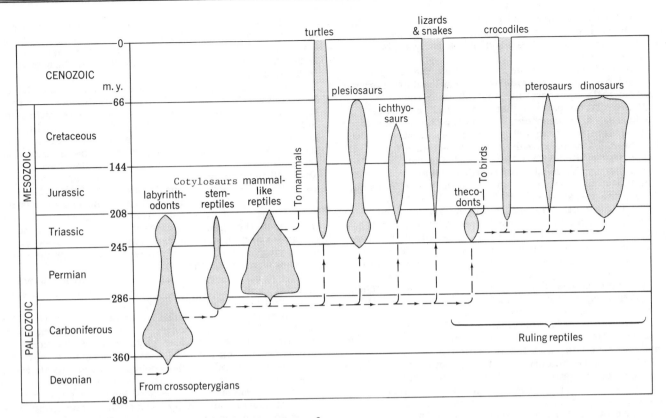

Figure 32.6 Evolution of the reptiles. (Based on data of A. L. McAlester, 1968, The History of Life, Prentice-Hall, Inc., Englewood Cliffs, N.J., pp. 104-5, Figure 6-1.)

In addition to the three groups of seedless vascular plants that had arisen in Devonian time, there evolved in the Carboniferous Period the first of the seed-bearing plants. These were the seed ferns, one of the four classes of seed-bearing plants comprising the gymnosperms. As Figure 32.5 shows, the seed ferns later gave rise to two of the four gymnosperm classes--the cycads and ginkgoes--but these became prominent only in the Mesozoic Era. The fourth class consisted of the conifers, or cone-bearing, needle-leaf gymnosperms. These also arose in lower Carboniferous time and had many representatives throughout the period, but they were primitive forms now largely extinct. Carboniferous forests must have grown in great luxuriance over long periods of time, judging from the thickness of coal seams derived from compaction of the partially decayed vegetative mass that was produced. A decline of atmospheric oxygen to levels near that of the present has been interpreted on the supposition that the decay of great quantities of plant matter would have used substantial amounts of oxygen.

Among the invertebrates, two phyla invaded the terrestrial environment to become abundant in Carboniferous time. One of these is the gastropod, which was able to develop air-breathing apparatus and to emerge from the water as a land snail. These animals were not abundant, however, until much later in the geologic column. By far the more striking evolutionary development was that of the insects and arachnids (spiders and scorpions), which became very abundant in the Carboniferous Period.

Insects are of the phylum Arthropoda, which includes the crustaceans. Like the crustaceans, the insects have a tough, waterproof outer covering of organic matter. This covering was very likely an important factor in successful emergence from a water environment into the air. The development of wings for flight was an enormous advantage to the insects both in obtaining food and in evading predators. As a result the evolutionary radiation

of the insects was spectacular. Some 400 kinds of insects are known from the Pennsylvanian Period. Although most were unlike modern insects, an exception was the cockroach, some species of which attained a length of a dozen or more centimeters. A dragonfly-like insect was perhaps the largest--a fossil specimen with wingspread of 75 cm has been found. Evolutionary radiation of the insects continued throughout geologic time, and today about 500,000 insect species are known.

The first reptiles had evolved from amphibians during the Carboniferous Period, and by late in that period were competing strongly with the amphibians for the terrestrial environment. Reptiles had developed the capacity to lay eggs with protective shells that could be hatched on dry land. Thus freed from dependence on water bodies, the reptiles enjoyed an expanded environment and were soon to undergo a great evolutionary radiation.

The Permian Period--End of an Era

An almost explosive evolutionary radiation of early forms of reptiles is a particularly outstanding feature of the Permian Period. It was preceded in the Carboniferous by the emergence of labyrinthodonts, amphibians that were the dominant land vertebrates. Their progress is shown at the extreme left in Figure 32.6, a chart of reptilian evolution. From them descended the earliest reptiles, the cotylosaurs. Because all other reptiles descended from them, the cotylosaurs have been designated as the stem reptiles. On the chart, they are shown as entering the picture in late Carboniferous time. Cotylosaurs were heavy-bodied alligator-like creatures and attained lengths of a meter or two. Early in their history, at the opening of Permian time, they gave rise to a most important group, the mammal-like reptiles, that became dominant in Permian time and continued into the Triassic Period, when they declined rapidly and finally disappeared. Their importance lies in giving rise to the mammals in Triassic time. We shall have more to say about this subject in later paragraphs.

The stem reptiles are shown on the chart as giving rise to five major groups of Mesozoic reptiles, but these did not appear until after the close of the Permian. (The chain of horizontal arrows is misleading if interpreted to mean that those reptiles branched off as early as late Carboniferous time.)

The general impression we have is that the labyrinthodonts and early reptiles made the transition from Permian to Triassic rather successfully, not showing the extinction-radiation pattern that one usually associates with a boundary between eras. Actually, the number of reptilian genera fell from 199 in the upper Permian to 56 in the lower Triassic, revealing the extinction effect, but this drop was mostly within the mammal-like reptiles. On the other hand, the number of reptilian orders rose from four in the upper Permian to seven in the lower Triassic, exhibiting the radiation effect (Kay and Colbert, 1965, p. 351). The radiation in orders shows that the land environment continued to be especially favorable for the reptiles. Vascular plants, too, show a transition into the Mesozoic Era with little disruption (see Figure 32.5).

For the invertebrates, the close of the Permian Period was, in contrast, a time of evolutionary crisis. The decline in abundances of invertebrate groups can be seen in the narrowing widths of bars in Figure 32.3, but this device does not show the extent to which extinctions and replacements occurred. Approximately half of the invertebrate families present in the Permian Period became extinct and failed to make the transition into Triassic time. The trilobites underwent total extinction along with several orders of corals, crinoids, bryozans, and brachiopods. New groups that expanded in the early Mesozoic make up most of the modern invertebrates.

The case of the ammonite genera is quite instructive on the meaning of extinction followed by radiation. Figure 32.7 is a graph of the abundances of ammonite genera from the Devonian through the Cretaceous. It shows (a) the numbers of genera retained from the preceding epoch and (b) the numbers of genera introduced in each epoch. Across the Permian/Triassic time boundary, the numbers of genera retained falls to a very low number, while the number introduced rises sharply from under 20 to a full 120, for an increase more than sixfold.

In the oceans, vertebrate groups were differently affected. The armored fishes became extinct, while there was a sharp decline in the sharks. In contrast, the bony fishes underwent an expansion as Permian time gave way to Triassic time.

Late Permian extinction of marine animals may have been related to a major withdrawal of shallow seas from the lands. The final collision that brought together the North American and Eurasian/African plates during the late Permian Alleghenian-Hercynian orogeny destroyed major seaways of sediment deposition. One estimate places the reduction in global coverage by shallow seas as from about 40 percent in early and middle Permian to only 15 percent at the close of the Permian (Seyfert and Sirkin, 1979, p. 363). Change in salinity of the oceans has also been suggested as a cause. Another suggestion is that the end of the great Carboniferous/Permian glaciation of Gondwanaland brought a rapid warming of the oceans, destroying cold-water environments and reducing the nutrient content of the seawater.

The word "Paleozoic" comes from the Greek words for "ancient" and "animal life." Ancient as this life may seem, it is significant that most of the animal phyla of the modern world had developed at or near the start of the Paleozoic Era. In that sense, ancient and modern life are one. However, within those phyla most of the early groups became extinct and were replaced with new ones at one time or another in the geologic record. In that sense, Paleozoic life is truly ancient. The evolutionary process led to many mistakes in terms of longterm survival; yet an environment vacated by extinction of one group seems never to have remained vacant for long, and evolutionary radiation has never lagged far behind.

Mesozoic Life

The era of middle life, the Mesozoic Era, offers many lessons of life adaptation to environmental opportunities as well as of survival failures. Because the Mesozoic Era saw both the rise and fall of several reptilian groups, including species of enormous individuals with almost unbelievable conformations, this era has been aptly subtitled the Age of Reptiles. Yet, almost equally dramatic evolutionary radiations and extinctions are to tbe found among the land plants and the marine invertebrates. In terms of plate tectonics, the Mesozoic Era represents the time of fragmentation of the supercontinent of Pangaea, with widening ocean basins separating the fragments. We must therefore look for effects of that event on organisms incapable of crossing broad, deep ocean basins.

Evolution during the Triassic Period

During Triassic time Pangaea was largely intact, and it is inferred that the continental surfaces stood relatively high. Sedimentary accumulations on the cratons were in large part continental types--redbeds, evaporites, desert sands, and coal measures. The strata contain abundant remains of reptiles and land plants. Absence of glacial deposits within the Triassic column suggests a prevailingly mild-to-warm climate.

The mammal-like reptiles persisted into the Triassic, although greatly depleted in numbers of groups. Among the most interesting of the mammal-like reptiles was the genus Lystrosaurus, a small animal, somewhat resembling a hippopotamus, with massive, wide-set legs (Figure 32.8). Fossil remains of Lystrosaurus are abundant in Triassic strata of southern Africa and are also found in India, Russia, and China. Search for fossils of the genus in Triassic rocks of Antarctica was intensified as plate tectonics provided documentation of the breakup of Pangaea along with a timetable for the opening of ocean basins revealed by magnetic anomaly zones on the seafloor. It is considered a near-impossibility that Lystrosaurus could have migrated across the broad, deep

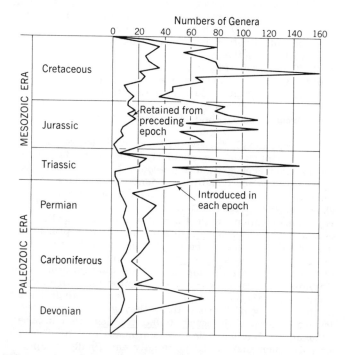

Figure 32.7 Graph showing abundances of ammonite genera from Devonian through Cretaceous periods. (Based on data of B. Kummel, as presented in M. Kay and E. H. Colbert, 1965, Stratigraphy and Life History, John Wiley & Sons, Inc., New York, p. 380, Figure 16-25.)

Figure 32.8 Mounted partial skeleton of <u>Lystrosaurus</u>, a mammal-like reptile of Triassic age, about 1 m long. (Sketched from a photograph by A. W. Crompton.)

ocean basin that now separates Antarctica from all other continents. The search met with success in 1969, when remains of <u>Lystrosaurus</u> were found in the Transantarctic Mountains, about 650 km from the South Pole. The fossil find was hailed as one of the most significant in modern times, for it threw paleontologic evidence strongly in favor of a unified Pangaea as late as the Triassic Period.

More important from the standpoint of evolutionary radiation was the emergence in Triassic time of several important reptile groups that were to dominate the Mesozoic Era. Figure 32.6 shows the evolution of the reptiles. Evolving from the stem-reptiles were turtles, lizards, snakes, and two groups of marine reptiles now extinct, the <u>plesiosaurs</u> and <u>ichthyosaurs</u>. Turtles, quite similar to those of today, entered the scene in the Triassic Period and persisted steadily through all succeeding time. Their success is attributed to an effective protective shell, to aquatic habits, and to being omnivorous. Lizards do not appear until the upper Triassic, and they are rarely preserved as fossils. The snakes are thought to have evolved from the lizards but are not found as fossils until late in Mesozoic time.

The large, carnivorous marine reptiles are particularly interesting, because they bear somewhat superficial resemblance to whales (which are mammals) and can be thought of as having occupied an equivalent niche. Like the whales of much later time, the plesiosaurs and ichthyosaurs returned from the land to the sea. Both were predatory animals living mostly on fish.

Lastly, of the reptile groups that arose from the stem-reptiles we come to the <u>thecodonts</u>, that had evolved as small animals in the Triassic Period. They had achieved a major evolutionary advantage by acquiring good running legs. Their limbs, instead of spreading outward, projected down, directly beneath the body. In some thecodonts the hind legs took over the running function, becoming greatly enlarged. Simultaneously, the forelegs became smaller and could easily be held off the ground, developing new functions such as grasping, clawing, and holding. These bipeds were to become the ancestors of the <u>ruling reptiles</u>--dinosaurs, flying reptiles, crocodiles --and the birds. By late Triassic time, there had evolved small predatory bipedal dinosaurs whose agility enabled them to prey on more sluggish reptiles and other animals.

Plant life underwent major evolutionary changes in the Triassic Period, setting the pattern for all of the remaining Mesozoic Era (see Figure 32.5). Among the gymnosperms, the <u>cycads</u> and <u>ginkgoes</u> had risen from the seed ferns and showed rapid evolutionary radiation in Triassic time. Representatives of both cycads and ginkgoes survived through to the present, although most became extinct at the close of the Mesozoic. The conifers persisted through the entire Mesozoic.

Turning to the marine invertebrate animals, we find that most groups showed evolutionary expansions

throughout the Mesozoic Era. An exception was the brachiopod phylum, which never regained importance after the extinctions of the Permian crisis. A particularly interesting group is the subclass of ammonites, within the cephalopod phylum. We have already noted that they came close to extinction in earliest Triassic time. Figure 32.7 shows that the great radiation of new ammonite genera in the Triassic was followed by another near-extinction in the transition from Triassic to Jurassic. Again a great radiation followed in the Jurassic, a severe depletion at the Jurassic-Cretaceous boundary, another great radiation in Cretaceous time, and a final collapse to almost total extinction at the end of Cretaceous time. It is easy to see why the many distinct varieties of fossil ammonites provide an excellent set of guide fossils for distinguishing among age zones of the Mesozoic Era.

The Dinosaurs

Nearly everyone is familiar with the appearance of several varieties of the great dinosaurs of Jurassic and Cretaceous time. The word "dinosaur," a popular term, means "terrible lizard." Even such names as <u>Tyrannosaurus</u> and <u>Brontosaurus</u> are almost household words. The dinosaurs include two orders distinguished according to an anatomical difference that is not readily apparent to anyone but a specialist. One order, the Ornithischia, had a birdlike pelvis, while a second, the Saurischia, had a pelvis of a type normal to reptiles. Both orders arose in the Triassic from common ancestors, the thecodonts, and gave rise to five groups of herbivorous dinosaurs and one carnivorous group (Figure 32.9).

Of those dinosaurs with a reptilelike pelvis, there developed one carnivorous branch, the <u>theropods</u>, and one herbivorous branch, the <u>sauropods</u>. The theropods were three-toed bipedal animals with a high degree of mobility. They had huge heads and fiercely armed jaws. <u>Tyrannosaurus</u>, the largest terrestrial predator of all time, reached a length of 15 m and stood over 5 m high. The sauropods reverted to the quadripedal stance and herbivorous diet. They became huge animals, perhaps sluggish in moving about, and may have had to seek safety in rivers and lakes. Largest of the sauropods, and also largest of all dinosaurs, were <u>Apatosaurus</u> (<u>Brontosaurus</u>) and <u>Brachiosaurus</u> of Jurassic time, reaching a length of 20 m. Longer, but not as heavy, was <u>Diplodocus</u> with a length of 26 m.

Of the four groups of dinosaurs with a birdlike pelvis, two became important in the Jurassic Period. The <u>ornithopods</u> were bipedal herbivores. They culminated in late Cretaceous time in the familiar "duck-billed" dinosaurs, the <u>trachodonts</u> (or hadrosaurs). Also from the ornithopods there developed in Cretaceous time the <u>Ceratopsia</u>, or horned dinosaurs. The <u>stegosaurs</u>, protected with a bony armor, arose in later Triassic time and were important dinosaurs throughout the Jurassic Period. Bony plates running down the spine and a spiked tail, used as a weapon, characterize <u>Stegosaurus</u>. From the stegosaurs there evolved in the Cretaceous Period the <u>ankylosaurs</u>, heavily armored with body plates.

While the dinosaurs have drawn first attention among the Mesozoic reptiles, other groups, such as the marine reptiles, are equally interesting. Those reptiles that took to the air include the <u>pterosaurs</u>, winged batlike animals of the Jurassic and Cretaceous periods. With wingspreads of up to 6 m, they were probably gliding and soaring animals. The wings were formed of a thin membrane of skin supported by bones of a single elongated finger.

Probably the most celebrated of all fossil finds is the remarkably preserved skeleton and feathers of the first known bird, <u>Archaeopteryx</u> (see Figure 44.1). This animal was found in Bavaria enclosed in fine-grained lithographic limestone of Upper Jurassic age. Until recently, predecessors of this highly developed animal were unknown, perhaps because the preservation of a bird as a

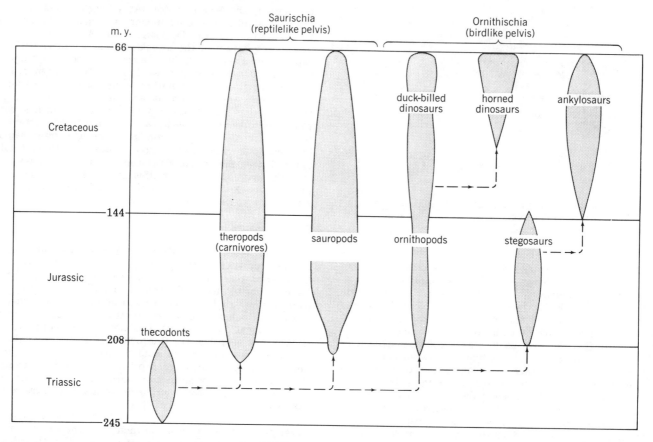

Figure 32.9 Evolution of the dinosaurs. (Based on data of A. L. McAlester, 1968, The History of Life, Prentice-Hall, Inc., Englewood Cliffs, N.J., p. 112, Figure 6-9.)

fossil would have been an extremely unlikely event. In 1986, what has been identified as an older bird fossil was found in Texas and given the name Protoavis ("ancestral bird") (Science News, vol. 130, p. 103, 1986). The strata in which it was found are 75 m.y. older than those enclosing Archaeopteryx, placing Protoavis in the Upper Triassic. Protoavis, too, shows distinctly dinosaurian features, including clawed fingers and the avian teeth and tail, but no feather impressions were found. The Mesozoic birds are thought to have evolved from the thecodonts and retain many characteristics of those reptiles, i.e., the primitive birds have teeth, and their skeletal structure is like that of the small reptiles (see Chapter 44). Unlike reptiles, however, birds today are warm-blooded animals. When the transition to the warm-blooded state was made is not known, but it must have had great survival value in adapting to life in cold climates.

The Cretaceous Period--Close of an Era

As the continents continued to separate throughout Jurassic and Cretaceous time, the North and South Atlantic basins grew steadily wider. Passive continental margins on both sides of the Atlantic subsided steadily as the lithosphere beneath became cooler and denser. We have already explained how the great geoclines--sediment wedges beneath the continental shelves and rises--accumulated during this subsidence. Although the pulling apart of the Gondwanaland fragments from southern and eastern Africa was delayed well into Cretaceous time, the Indian Ocean basin was broadly developed in late Cretaceous time and enormous extensions of passive margins were formed around Africa and Antarctica. Shallow seas lapped far over the passive margins and invaded the stable cratons to form large inland seas.

Under the concept of the Wilson cycle, when continents spread apart, opening up new ocean basins, the distal margins of those continents lie on subducting plate margins. Consequently, the western borders of North and South America were experiencing almost continuous subduction of oceanic plates underlying what is now the general region of the Pacific Ocean. These active margins were replete with trenches, accretionary prisms, forearc troughs, volcanic arcs, and backarc basins. Occasionally, a volcanic arc carried along on a subducting plate slammed into the large continental lithospheric plate, resulting in an orogeny and accretion of a microplate. Despite all the tectonic and volcanic activity, narrow zones of shallow ocean water were continuously present and provided environments for marine animals, just as they do today.

Throughout much of the Cretaceous Period climate was warm and moist over a wide latitude range. Subtropical floras spread to latitudes as high as 70 degrees from the equator of that time. Coal seams were forming in almost the full range of latitudes. A cooling of climate seems to have set in during late Cretaceous time, perhaps presaging the great extinction that was to terminate the period and the Mesozoic Era.

Although the Age of Mammals (Cenozoic Era) was yet to come, mammals actually made their appearance in Mesozoic time, but kept a low profile, presaging a great radiation at the start of the Cenozoic Era. An evolutionary event of great significance was the appearance in Jurassic time of primitive mammals, arising from the mammal-like reptiles that were abundant in the Triassic Period.

Mammals are distinct from reptiles in several respects. First, the young of mammals are born alive; the mammalian egg is fertilized within the female body and (in the placental mammals) attains there an advanced state of development before entering a hostile environment. Along with this form of reproduction, the female mammal

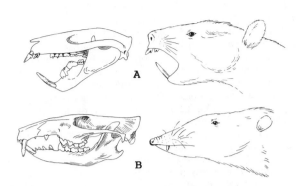

Figure 32.10 Mesozoic mammals--skulls and restorations of heads, about natural size. (A) A multituberculate mammal of the Jurassic Period. (B) A shrewlike insectivore of Cretaceous age. (After C. O. Dunbar and K. M. Waage, <u>Historical</u> <u>Geology</u>, 3d ed., p. 365, Figure 15-20, and p. 395, Figure 16-24. Copyright © 1969 by John Wiley & Sons, Inc. Used by permission.)

possesses milk glands to provide food for the young after birth. Second, the mammals are warm-blooded. They can not only generate heat, but can also reduce body temperatures below that of the surrounding air by means of evaporation. When provided with an insulating coat of hair, mammals can adapt to life under extremely cold conditions (So can the birds, also warm-blooded and often well insulated, but they are not mammals.) Reptiles--the modern ones, at least--are cold-blooded animals, which is to say they lack a body-heating mechanism and tend to take on the temperature of the surrounding air or water. Consequently, reptiles require warm climates, or climates

with a warm season. Reptile growth continues throughout the life of the individual, whereas a mammal quickly matures to a limiting skeletal length that it maintains.

Paleontologists distinguish the early mammals from the mammal-like reptiles on the basis of jaw and tooth structures. Mammal teeth developed specialized functions, including sharp incisors at the front and massive molars for grinding at the back (Figure 32.10). On the basis of teeth, the primitive mammals of the Jurassic Period are grouped into four orders, but only one of these, the Pantotheres, gave rise to all later mammals. By late Cretaceous time two new orders of mammals had arisen; the marsupials and the insectivores (Figure 32.11). <u>Marsupials</u> are pouch-bearing mammals, of which the kangaroo and opossum are modern survivors. <u>Insectivores</u>, represented today by the moles and shrews, were later to give rise to the placental mammals, which underwent a phenomenal radiation in the Cenozoic Era. The Cretaceous Period was, however, the initial period of "experimentation" for the mammals, just as the Cambrian was the period of "experimentation" for the marine invertebrates. (Do not infer an "experimenter" in control.)

Equally significant to the development of early mammals in late Cretaceous time was the sudden and explosive evolutionary radiation of the <u>angiosperms</u>, those flowering plants with covered seeds (see Figure 32.5). They succeeded the <u>gymnosperms</u>, or plants with naked seeds, that were the dominant vascular plants of the Mesozoic Era.

The reproduction of the gymnosperms takes place through the dispersion by wind of embryonic seeds and pollen grains carrying the sperm. Union of a pollen grain with an embryonic seed allows fertilization to take place at some point on the ground far removed from the parent plant. Obviously, this mechanism has the disadvantage

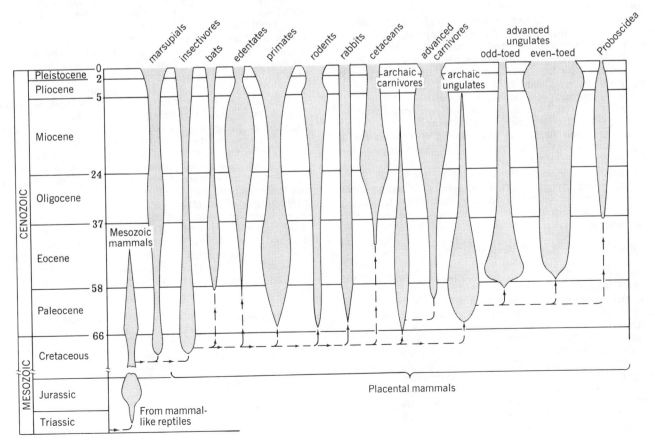

Figure 32.11 Evolution of the mammals. (Based on data of A. L. McAlester, 1968, <u>The</u> <u>History</u> <u>of</u> <u>Life</u>, Prentice-Hall, Inc., Englewood Cliffs, N.J., pp. 122-23, Figure 6-19.)

that the fertilization process takes place in unprotected surroundings, while the dispersal and union of spores and embryonic seeds depend upon the random actions of the wind.

In the angiosperms, pollen fertilizes an embryonic seed held within the base of the flower, producing a mature seed attached to the parent plant until it is ready to be dropped to the ground. Birds or insects, attracted to the flower, carry out the pollination process, which thus has a higher degree of reliability than in the case of the wind-transport of pollen. Enclosure of the angiosperm seed in a fleshy fruit, attractive to birds as food, further aids in the dispersal of the seeds over long distances.

Perhaps as a result of these advances in seed fertilization and dispersal, the angiosperms underwent a phenomenal evolutionary radiation at the very end of the Mesozoic Era so that, when the Cenozoic Era began, the angiosperms had already replaced the gymnosperms as the dominant vascular land plants. As the Cretaceous Period drew to a close, the cycads and ginkgoes suffered a great decline. Of the gymnosperms, only the conifers persisted abundantly into the Cenozoic Era. The ferns made this transition in abundance.

The close of the Cretaceous Period is marked by great extinctions, of which the most celebrated is the total disappearance of the dinosaurs. The large marine reptiles (turtles excluded) and flying reptiles also suffered total extinction. Recall, also, that the ammonites became almost extinct at this time (Figure 32.7). Extinctions also affected certain groups of pelecypods and cephalopods other than the ammonites. Yet other animal groups successfully made the transition to the Cenozoic Era and some groups reflect no special evolutionary crisis at this time.

The cause or causes of late-Cretaceous extinctions are currently a subject of major scientific debate--one of the most heated in many a decade. Some paleontologists strongly favor a terrestrial environmental factor, for example, lowering of seawater temperature. But much greater excitement attaches to a catastrophic astronomical hypothesis requiring impact of a large extraterrestrial body--a comet or an asteroid--that created a short period of extremely adverse or lethal atmospheric and oceanic environments. The new geochemical findings in support of such a great impact near the Cretaceous/Cenozoic boundary seem very convincing as presented by the research teams that have made the findings and analyses. On the other hand, experienced stratigraphers and paleontologists can point out cases in which extinctions of certain groups occurred before or after that postulated impact event. Conservative scientists suggest that hypotheses of mass extinctions should be tested by application over the entire geologic record. Keeping this advice in mind, we postpone further discussion to Chapter 49 as part of a more general and inclusive treatment of the problem of extinctions.

The Cenozoic Era--Age of Mammals

The Cenozoic Era, or era of "recent life," consists of 66 million years of geologic time, whereas the Cretaceous Period that preceded it endured 78 million years. The nine periods of the Paleozoic and Mesozoic combined average about 56 years each, which is only a little less than the entire Cenozoic Era. Why, then, does the Cenozoic deserve the status of an era? The answer lies in the history of geology, which resembles human cultural history in the density distribution of its documentation. The farther back we go in history, the sparser is the record. The older the strata, the smaller is the proportion of the original amount that has survived until today. The richness of the fossil record in Cenozoic strata appeared to the early geologists of Europe to be worthy of era status, and there was no scale of absolute time to suggest otherwise.

Figure 32.12 gives the subdivisions of the Cenozoic Era. For most practical purposes, geologists can skip across the column headed "Period," focusing on the epochs as working subdivisions of the Cenozoic. Of the epochs there are seven: Paleocene, Eocene, Oligocene, Miocene, Pliocene, Pleistocene, and Holocene. The syllable "cene," found in all epoch names and also in the first syllable of Cenozoic, comes from the Greek kainos, meaning "recent." The table gives the ages of the time boundaries and the durations of the epochs.

Now, getting back to the periods of the Cenozoic Era, we find a dichotomy of usage by geologists, leading to agonizing confusion among students. In Figure 32.12, the two alternatives are labeled A and B; of these A is the older, and indeed grossly archaic. The terms "Tertiary" and "Quaternary" are relics of the early nineteenth century when they enjoyed era status within an age sequence reading: Primary (or Primitive), Secondary, Tertiary, Quaternary (Bates and Jackson, 1980). "Primary" and "Secondary" were relegated to obscurity as the Paleozoic and Mesozoic eras were instituted, but Tertiary and Quaternary persisted as younger periods within the Cenozoic. The Tertiary Period, lasting 64 m.y., is vastly longer than the Quaternary Period, enduring less then 2 m.y. You should know about the Tertiary and Quaternary periods because American geologists refer to them repeatedly in modern writing, both technical and popular. There is a worldwide organization, the International Quaternary Association (INQUA), whose members will treat you very harshly if you fail to pay homage to their title. In the prestigious North American geologists' organization, The Geological Society of America, there is a Division of Quaternary Geology and Geomorphology (to which I belong); many of its members are also connected with INQUA, so watch your step!

Alternative B in Figure 32.12 divides the Tertiary into two periods: Paleogene and Neogene; their durations are about 42 and 22 m.y. respectively. This alternative has long been the choice of Europeans but is now standard in North American research publications that deal with plate tectonics and its aura of related topics. Even so, that sacred cow, the Quaternary Period, enjoys immunity from change, for as you see in the table, it remains a 1.6-m.y. period following the Neogene.

The history of the Cenozoic Era was first deciphered from marine fossil assemblages within sequences of strata exposed in areas surrounding London and Paris. The southern area we have referred to earlier as the Paris Basin. The Cenozoic strata have been downwarped in these areas, escaping erosional removal and surrounded by Cretaceous strata. It was soon recognized that among the invertebrate faunas of this stratigraphic column the percentages of species living today increased from a low value near the base to a high value near the top. French paleontologist Alexandre Brogniart had made detailed studies of the faunas and had identified about 5,000 species in the column, of which about 3,000 are now extinct. Based on this information, Charles Lyell proposed in 1839 the following names and definitions for three epochs:

Epoch:	Percentage of species now living:
Pliocene ("more recent")	30-50
Miocene ("less recent")	about 16
Eocene ("dawn of recent")	about 3.5

Subsequent study led to the insertion of the Oligocene Epoch into this sequence and the addition of the Paleocene Epoch at the base to replace a part of the original Eocene.

Period		Epoch	Duration	Age
A	B			
Quaternary		Holocene (Recent)	10,000 y.	
Quaternary		Pleistocene	m.y. 1.6	m.y.
				— 1.6 —
Tertiary	Neogene 22 m.y.	Pliocene	3.4	
				— 5 —
Tertiary	Neogene 22 m.y.	Miocene	19	
				— 24 —
Tertiary	Paleogene 42 m.y.	Oligocene	13	
				— 37 —
Tertiary	Paleogene 42 m.y.	Eocene	21	
				— 58 —
Tertiary	Paleogene 42 m.y.	Paleocene	8	
				— 66 —

(CENOZOIC ERA spans the entire table)

Figure 32.12 Subdivisions of the Cenozoic Era. (Based on data of A. R. Palmer, 1983, The Decade of North American Geology 1983 time scale, Geology, vol. 11, p. 504.).

Evolution of the Mammals

While the marine invertebrates have provided us with the means of subdividing the Cenozoic Era into epochs and lesser units of time, it is the phenomenal rise of the mammals on land that makes the era unique.

The mammals that began this spectacular evolutionary radiation were not particularly prepossessing in any sense. They were small, long-tailed creatures with short legs and five-toed feet on which they walked flat-footed. The modern hedgehog has been cited as giving us a good example of the body shape and walking gait of these early mammals. Their brains were correspondingly small. Like the primitive mammals of the Jurassic and Cretaceous periods, from which they evolved, the early Cenozoic mammals had long, pointed jaws.

Evolution among the groups of mammals has certain trends in common. Increase of size is very marked, along with the disproportionately large increase in brain size and hence in intelligence of several of the groups. Teeth show a high degree of specialization, the changes depending on function. The principal divergence in tooth development has been into the high-crowned, deeply involuted teeth of the grazing herbivores, as contrasted to the sharply pointed or sharpened teeth of the carnivores, adapted to tearing and cutting of flesh.

One of the most striking changes setting apart one mammal group from another has been in the feet and limbs. Grazing animals of the plains regions developed the capacity to rise up on one or two toes, which became elongated and strengthened, while the remaining digits disappeared. Thus equipped, these animals would run swiftly to escape from the predators. Carnivores developed powerful, sharp claws for slashing and holding their prey, while another mammal group developed long digits suited to tree climbing.

In this very limited review of the evolution of life forms we cannot begin to give a full picture of the Cenozoic history of mammals. Even the classification of mammals poses a formidable problem in itself. But by focusing attention upon a few evolutionary principles and trends, we can at least gain a perspective from which to view the place of the human species among the mammals.

First, it is important to make a distinction between two groups of mammals, the marsupials and the placental mammals. Marsupials, or pouch-bearing mammals, give birth to very small immature young that must enter a pouch on the mother's abdomen. Here the offspring are suckled and grow to sufficient size to survive in the outside environment. Placental mammals retain the embryonic young in a fluid-filled membrane within the uterus where, until it attains an advanced stage of development, it can be nourished through an umbilical tube receiving nutrients from a highly developed part of the membrane, the placenta. This mechanism would seem to offer advantages in better protection of the young, although the marsupials have managed very well, as witness the kangaroo and opossum. Under competition, the placental mammals were the more successful in terms of evolutionary radiation and came to dominate the mammal world on all continents except Australia.

The placental mammals show an extreme diversification in terms of anatomical structure and adaptation to varied environments. Altogether, 28 orders evolved during the Cenozoic Era, but 12 are now extinct. Table 32.3 lists eleven of the living orders with common names and representative animals. The remaining five orders, not listed, include rather rare and bizarre creatures, such as the sea cows, aardvarks, pangolins, and conies. Several extinct orders are grouped together under the title of the ungulates. The carnivores have been broken into two groups: the advanced carnivores, which include all living groups, and the archaic carnivores, all of which are extinct.

The Archaic Mammals

As the Paleocene opened, the marsupials and insectivores, both of which were important in the Cretaceous Period, were the two most important orders on the scene (Figure 32.11). Their predecessors, the pantotheres, survived through the Paleocene but became extinct shortly thereafter. The insectivores, which were the original order of placental mammals, gave rise to all other placental mammals. Among the first to appear were two now-extinct groups, the archaic carnivores and the archaic ungulates.

The archaic carnivores, known as creodonts, were at first small animals with slender bodies, long tails, and short legs. Many remained small, but some became large animals, outwardly resembling bears, during the Paleocene Epoch. Most of the creodonts were extinct by Oligocene time, and by Pliocene time none remained. However, in Paleocene time the creodonts had given rise to the advanced carnivores, which subsequently underwent a rapid evolutionary radiation (Figure 32.11).

The archaic ungulates were hoofed animals that quickly became the dominant herbivores in Paleocene time. No single description will fit these animals. They ranged in size from animals as small as a modern sheep to huge creatures as large as a rhinoceros. An important characteristic of these early ungulates was specialization

Table 32.3 Principal Orders of Living Placental Mammals

Order	Common name	Representatives
Insectivora	Insectivores	Shrews, moles, hedgehogs
Chiroptera		Bats
Edentata	Edentates	Anteaters, sloths, armadillos
Primates	Primates	Lemurs, tarsiers, monkeys, apes, humans
Rodentia	Rodents	Squirrels, beavers, mice, rats, porcupines
Lagomorpha	Lagomorphs	Rabbits, hares
Cetacea	Cetaceans	Whales, porpoises, dolphins
Carnivora	Carnivores	Dogs, foxes, bears, cats, mink, sea lions
Perissodactyla	Perissodactyls, or odd-toed hoofed ungulates	Horses, tapirs, rhinoceroses
Artiodactyla	Artiodactyls, or even-toed hoofed ungulates	Pigs, camels, deer, giraffes, antelopes, goats, sheep, cattle
Probiscidea	Proboscideans	Elephants (extinct mastodonts, mammoths)

Based on data of E. H. Colbert, 1980, Evolution of the Vertebrates, 3d ed., John Wiley & Sons, Inc. pp. 274-75; and A. L. McAlester, 1968, The History of Life, Prentice-Hall, Inc., p. 147.

of the teeth. The front teeth were set chisel-like across the front of the mouth, facilitating the cropping of vegetation. The rear teeth were massive, high-crowned, and with more or less square or rectangular upper surfaces adapted to grinding grasses and leaves. These dentition characteristics persisted through all later ungulates and are seen today in the modern horse, camel, and cattle. In the larger ungulates the feet were broad and short, somewhat like those of the modern elephant.

An instructive story lies in the occupation of South America by archaic ungulates early in the Cenozoic Era. Evidently a land connection existed between North and South America in Paleocene time, enabling these animals to migrate from North America, Eurasia, and Africa, which retained connections generally throughout the entire Cenozoic Era. From the standpoint of the evolution of land animals, these connected continents can be designated as the World Continent (not to be confused with super-continents such as Pangaea in the context of plate tectonics). The connection between North and South America was by a narrow isthmus, possibly a volcanic arc rising adjacent to a subduction zone. The connection between North America and Eurasia was probably a narrow land bridge between what are today Alaska and Siberia. A connection between Africa and Europe may have been via one or more islands in the ancestral Mediterranean, corresponding to what is now Italy, the Adriatic coast, and Sicily. These connections are suggested by paleogeographic maps (Seyfert and Sirkin, 1979, Figure 13.1, pp. 448-50).

In Eocene time the connection with South America was severed and that continent became isolated. Evolution of the ungulates continued independently thereafter in South America and led to the development of five new orders. A number of marsupials had also reached South America in Paleocene time and continued their evolution independently.

Late in the Cenozoic Era, toward the end of Pliocene time, connection was reestablished with the World Continent and migrations into South America were

resumed. The mixing of very diverse faunas resulted in the extinction of many of the earlier South American mammal groups, including all five primitive ungulate orders. Extinction is attributed in part to the onslaught of intelligent, cunning carnivores, and to competition from more efficient advanced ungulates, all of which invaded from the World Continent.

In a similar vein is the even stranger case of Australia, which was populated largely by marsupials in early Cenozoic time, when a land connection existed with the World Continent. The best guess as to the location of the connection might be through Antarctica to the southern tip of South America, then to North America and finally Eurasia. This route seems to have been open in late Cretaceous time (Sclater and Fisher, 1974, p. 700, Figure 15). The situation may have persisted into Paleocene time (Seyfert and Sirkin, 1979, p. 448, Figure 13.1). Placental mammals, on the other hand, did not establish a foothold in Australia, with the exception of bats and rodents. The connecting bridge was then broken and Australia remained isolated thereafter. In isolation and free of competition from the placental mammals, the marsupials of Australia evolved into a bizarre mammalian fauna seen today in the kangaroo, wombat, koala, and Tasmanian wolf.

Evolution in isolation, seen in the examples of South America and Australia, illustrates the principle of parallelism in evolution. The primitive ungulates performed the functions of herbivores in South America in much the same manner as the advanced ungulates served as herbivores in the World Continent. The carnivorous marsupials of South America functioned in the same role as the advanced carnivores of the World Continent. Under parallelism of evolution, unlike animal groups perform like functions, adapting to similar environments and developing analogous structures (such as grinding teeth) of quite similar form. (Note: Parallelism in evolution is also designated by the term convergence, although as strictly defined the two terms have somewhat different meanings.)

Evolution of the Advanced Ungulates

The advanced ungulates consist of two orders, one of which we can simply designate as odd-toed, the other as even-toed (see Table 32.3). These herbivores were browsing or grazing animals, and many of the groups were forced to develop unusual leg and foot structures to secure the speed needed to escape from the carnivores. The ability to stand on the toes favored strengthening of the middle digits at the expense of those on the sides. Thus the five toes of the ancestral ungulates were reduced to three, two, or one. The odd-toed ungulates may have three toes, with the center line of the foot running down the center toe; alternatively, there remains a single toe equipped with a solid hoof. The three-toed foot is seen in the tapirs and rhinoceroses, and the single-toed foot in the modern horse.

Evolution of the horse throughout Cenozoic time is a particularly fine example of adaptation to the need to run swiftly (see Figure 45.9). As the animal increased in size, the legs and feet were lengthened and the lateral toes were lost. Ultimately there remained only a single toe equipped with a solid hoof. At the same time important changes took place in the skull. The sharp incisor teeth were moved forward and separated from the molars, which moved to the sides and developed higher, flatter crowns. The jaws became heavier and deeper to accomodate the larger teeth, causing the face to project farther forward beyond the eyes.

In the even-toed ungulates three digits were lost, while the remaining two developed hooves. These cloven-hoofed animals are exemplified today by the camel, giraffe, goat, sheep, and cattle.

While some ungulates were developing improved facilities for running to escape predators, others were developing enormous bulk and tough hides. An example is seen in the formidable African rhinoceros, which has a sharp horn and an armorlike hide. In Oligocene time there lived a member of the rhinoceros family known as Baluchitherium, which was over 7 m long and had a shoulder height of over 5 m. This animal was the largest land mammal ever to have lived. Interesting also were the titanotheres, a now-extinct group of the odd-toed ungulates. In Oligocene time a representative of this group reached a height of 2.5 m.

The elephants, constituting a separate order of mammals (the proboscideans), arose early in the Cenozoic Era from the primitive ungulates. Evolution of the long trunk and tusks seen in the modern elephant and in the extinct mastodons and mammoths of the Pleistocene Epoch was slow in coming.

Evolutionary development of certain other important mammalian orders is covered in later sections. In evaluating the creationists' arguments against evolution, we will review the rise of the whales (Cetacea), bats (Chiroptera), rodents (Rodentia), and primates (Primates). Our review of the primates will concentrate on the evolution of the anthropoids (monkeys, apes, humans), a topic of major importance in the creation/evolution debate.

In one chapter we have taken a rapid tour through Phanerozoic evolution, following the evolution of various plant and animal groups at significant points where extinction was followed by evolutionary radiation. This material has been presented as a consensus of mainstream geology and biology. The evolutionary changes described have been inferred from the fossils themselves, under the assumption that all life has streamed continuously from previous life and, in the beginning, from a single life form. Creationists challenge this ordering of life forms in an evolutionary sequence; they claim that the "tree of life" and its entire phylogenetic basis is an artifact. We will move on to consider their arguments and how those arguments can be answered.

Credit

1. In this chapter, numerous paragraphs and most of the illustrations are taken from The Earth Sciences, Harper & Row, Publishers, Inc., Chapters 29 and 30. Copyright © 1971 by Arthur N. Strahler.

Chapter 33

Evolution Inferred from the Fossil Record

In our overview of the growth of the tree of life through the Phanerozoic Eon, the evolution of several major groups of plants and animals was traced from period to period and from era to era. Just how are these phylogenetic relationships established? Prior to the rise of molecular biology, evolutionary descent (phylogeny) was inferred from resemblance and dissimilarity of fossils with one another and with living organisms. Closeness of resemblance was equated to closeness of descent, and thus hierarchies of succession were built.

Evolution and Comparative Anatomy

The methods of comparative anatomy include ordering of two types of information. One is homology, which ascertains the degree of similarity or difference in corresponding features of different organisms. Another is ontogeny, which examines successive growth stages in an individual of a species. Stages in this life history can be compared with adult forms in earlier geologic time in related species or other taxa.

The principle of homology is nicely illustrated by comparison of the bones of the forelimb of four kinds of modern mammals: human, seal, bat, and dog, as shown in Figure 33.1. In each skeleton there is found a single upper-arm bone (humerus), paired lower-arm bones (ulna and radius), several wrist bones, and five digits. All four mammals shown have the same complement of arm bones, but the bones differ in proportions and sizes. The same statement would apply if we were to compare the same forelimbs with those of certain lower mammals, such as the lemur and opossum, and with those of birds and of certain reptiles, such as the theriodonts (Figure 33.2).

Corresponding organs that agree in their basic structures are said to be homologous. Of course, the same homologous relationships would also extend to all parts of the skeleton and to the soft parts, such as muscles, nerves, and internal organs. When the vast number of homologous relationships is taken into account, the

evidence for descent from a common ancestor becomes overwhelming. A striking example of homology in the fossil record is the evolution of the forelimb and skull of the horse, mentioned in Chapter 32 and described in more detail in Chapter 45.

Ontogeny describes the entire sequence of changes in an organism, from its inception in a seed or egg, through its growth stages, to the final mature form. Ernst Haeckel, whom we mentioned earlier, is credited with suggesting that the ontogeny of a particular species or genus repeats the evolutionary stages of the time sequence of forms that preceded it; hence the saying "Ontogeny recapitulates phylogeny."

In the fossil record are found some good examples of an ontology that recapitulates phylogeny. A particularly striking example is illustrated in Figure 33.3. Growth stages of a Jurassic brachiopod in the genus Pygope are shown from left to right in the upper row. In early stages the shell has a deep notch in the front edge between the two lobes. Growth lines on the shell clearly show the development. As the individual grows, the notch deepens and finally closes in to give a doughnutlike shell, sometimes called a "keyhole" shell. Illustrations in the lower row are of four distinct species of the same genus in older and younger strata, the oldest at the left and the youngest at the right. The first three correspond with stages in the upper series, while the fourth, of early Cretaceous age, has evolved further, as seen in the near-closing of the keyhole. In this example, descent with modification seems an almost inescapable conclusion.

When ontology is applied to development of embryos of living forms, the results are often unsatisfactory because the embryo does not necessarily recapitulate all stages of its evolutionary past. For example, the embryo of the horse does not go through the five-toed stage of the early Cenozoic predecessor (shown in Figure 45.9). However, the degree of similarity of two embryos is thought to be a general indicator of closeness of phylogenetic relationship.

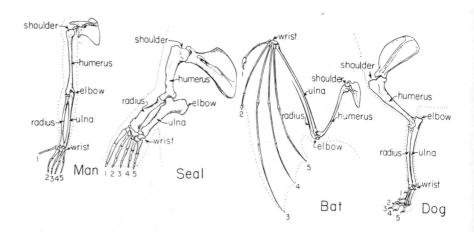

Figure 33.1 Sketches showing the forelimb bones of human, seal, bat, and dog. (From C. O Dunbar and K. M. Waage, Historical Geology, 3d ed., p. 112, Figure 6-4. Copyright © 1969 by John Wiley & Sons, Inc. Used by permission.)

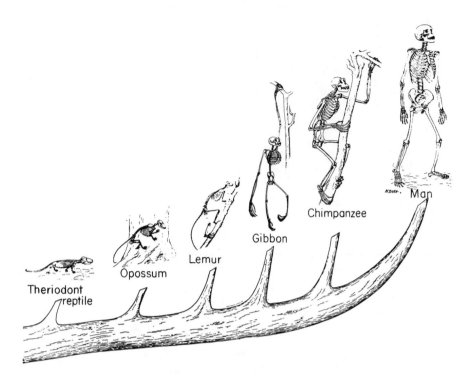

Figure 33.2 A succession of skeletons from reptile to human. (From C. O. Dunbar and K. M. Waage, <u>Historical Geology</u>, 3d ed., p. 111, Figure 6-3. Copyright © 1969 by John Wiley & Sons, Inc. Used by permission.)

The Hierarchical Structure

Studies in comparative anatomy, such as we have described in homology, reveal <u>hierarchical patterns</u> that are seen as trees--branching systems in which the fingertip stems bear different levels of relatedness among themselves. A hierarchical tree can be constructed from the data of comparative anatomy. The principle is that of shared similarities: the more homologous anatomical features that two groups of organisms share in common, the closer is their evolutionary relationship. In terms of a branching tree, the greater the number of shared homologous anatomical features, the closer is the common fork, or junction, of the single branch that supports them both. Homologous similarities, or <u>characters</u>, are of two kinds: primitive and derived. Derived characters are new evolutionary features derived from characters previously present. For example, feathers of birds are derived from reptilian scales. Derived characters (but not primitive characters) are used to set up branching trees (Cracraft, 1983, pp. 171-72).

The hierarchical structure is best understood by means of a tree diagram, called a <u>cladogram</u> (Cracraft, 1983, p. 170). A set of four cladograms is shown in Figure 33.4. The black box in each represents a list of shared derived similarities, shared, that is, by all of the organisms represented by fingertip branches emanating from the common junction above the box. Five different organisms are involved in the classification: lamprey, perch, lizard, mouse, and cat. In cladogram (1) the shared anatomical features are a dorsal nerve cord, a notochord, and a chambered heart. Thus the five organisms are all chordates (phylum Chordata). (The notochord is found only in the early embryo stage of vertebrates but persists through life in the lamprey.) In cladogram (2) the shared anatomical features are jaws, paired appendages, and a vertebral column. This excludes the lamprey, which is a jawless fish (one of the agnaths). The lamprey is given its own branch and the black box is placed between that junction and the common junction of the four remaining organisms. Cladogram (3) uses the amniote egg as its shared anatomical feature. (The amniote egg has internal membranes and is adapted to life on land.) This excludes the perch (fishes, superclass Pisces), which now is assigned its own branch with a

junction below the box and above that of the lamprey. In cladogram (4) the shared similarities are three ear ossicles, hair, and mammary glands. Lacking these characteristics, the lizard (a reptile) is excluded and is assigned its own branch, above that of the perch. We end up with four groups, one for each junction. Closeness of evolutionary descent is indicated by the number of junctions separating any two groups: the fewer the junctions, the closer the relationship.

Our cladistic analysis has produced a natural hierarchy. The cat and mouse, as mammals, are at the top because they possess all of the anatomical features we used as shared similarities--ten in all (Figure 33.5). At the base is the lamprey, possessing only three of the ten anatomical features. Note especially that the cladistic tree has been drawn from living species, entirely independently of the fossil record. The structure of the resulting

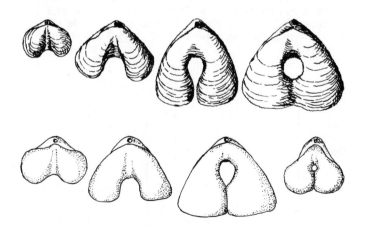

Figure 33.3 (Above) Growth stages of a single species of a Jurassic keyhole-shelled genus, <u>Pygope</u>. (Below) Adult shells of four distinct species of keyhole-shelled brachiopods found in successive horizons in strata of Jurassic and Early Cretaceous age. (From C. O. Dunbar and K. M. Waage, <u>Historical Geology</u>, 3d ed., p. 113, Figure 6-5. Copyright © 1969 by John Wiley & Sons, Inc. Used by permission.)

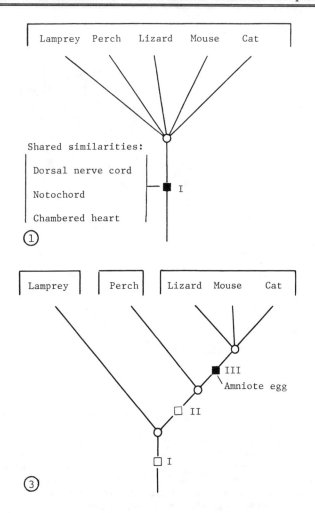

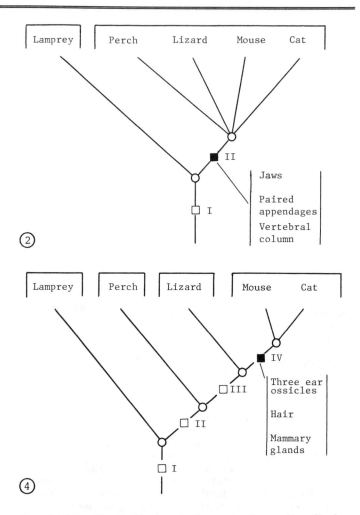

Figure 33.4 Cladograms based on shared anatomical characteristics. (Modified from Joel Cracraft, p. 171, Figure 1, in L. R. Godfrey, ed., <u>Scientists</u> <u>Confront</u> <u>Creationism</u>, W. W. Norton & Company, Inc., New York. Copyright © 1983 by Laurie R. Godfrey. Used by permission.)

tree is arrived at in an objective manner, based on anatomical observation.

Now, if descent with modification has occurred through geologic time, as the total hypothesis of evolution requires, we should expect to find corroboration in a correspondence between the phylogeny inferred from the hierarchal analysis and the fossil record. Figure 33.6 fits the final cladogram of Figure 33.4 to the geologic time scale, placing the branches (called <u>clades</u>) where the

Anatomical features present:

Rank:	I			II			III	IV			
Highest	x	x	x	x	x	x	x	x	x	x	Mouse, cat
	x	x	x	x	x	x	x				Lizard
	x	x	x	x	x	x					Perch
Lowest	x	x	x								Lamprey

Figure 33.5 Hierarchical structure derived from cladogram analysis of shared similarities shown in Figure 33.4. (A. N. Strahler.)

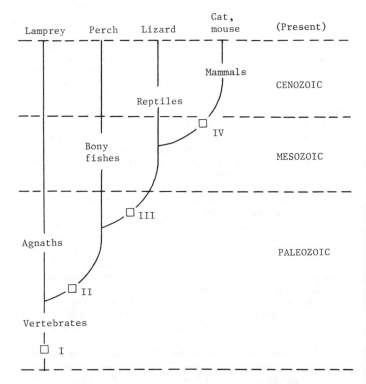

Figure 33.6 Cladogram number four of Figure 33.4, adjusted to the fossil record of Phanerozoic time. (A. N. Strahler.)

fossil record requires them to be. The fossil record shows that the agnaths appeared first, followed in sequence by the bony fishes, the reptiles, and the mammals. According to biologists and paleontologists, the total natural hierarchy discoverable by the use of comparative anatomy is best explained as representing the phylogenetic history of life (Cracraft, 1983, p. 177). As we have shown by a simple example, that conclusion can be tested. Success of those tests on a large scale and in great detail provides extremely strong independent support for that part of the hypothesis (or theory) of evolution embodied in the concept of descent with modification through geologic time.

Is Evolution a Fact?

All too often these days we read or hear these categorical assertions by leading mainstream scientists: Evolution is a fact! Scientists believe in evolution! Let us examine some of these public utterances.

At the 147th annual national meeting of the American Association for the Advancement of Science, held in January 1981, a special session of papers dealt with "science versus tradition." At that time, mainstream science was taking little or no notice of the creationists' attempts to introduce their views into public education. Jolting the scientists out of their complacency was a paper delivered by Porter M. Kier, a senior scientist of the Smithsonian Institution and former Director of the National Museum of Natural History. Kier flatly asserted: "Evolution is a fact." He went on: "Over the eons, we know beyond doubt that life has changed dramatically, from simple unicellular organisms into animals as complex as man. Our knowledge of this past life and how it has changed--the process of evolution--is extraordinarily detailed. This knowledge is based on unassailable evidence, the fossil remains of thousands and thousands of species of plants and animals which no longer exist. . . . Scientists may argue over the details of evolution, but they agree that evolution is a fact and should be so labeled." (Quoted statements appear in news reports in Geotimes, vol, 24, no. 4, p. 18, and Science News, vol. 119, p. 19.)

In discussing the nature of science (see Part One), I took the position that no theory or important scientific hypothesis should be described as "a fact." Quite apart from the reasons I gave for this preference is a reason applying to public relations between mainstream scientists and nonscientists. The arrogance displayed by the claim of fact--absolute truth, that is--incites resistance in a substantial sector of the public and can easily generate hostility toward the scientific community. This effect was seen in the case of a pseudoscience cultism rising in response (in part, at least) to denunciation by scientists of Velikovsky's Worlds in Collision, immediately following its publication in 1950. Harlow Shapley's imperious attack on the work seems to have polarized a substantial sector of the public against mainstream science.

Equally unfortunate, in my opinion, is the widespread public attestation of some distinguished mainstream scientists to "belief in evolution." In September 1984, the American Geological Institute announced publication of a pamphlet titled Why Scientists Believe in Evolution, written by Norman D. Newell, a distinguished paleontologist, Curator Emeritus of the American Museum of Natural History, and Professor Emeritus of Columbia University. My reasons for avoiding the expression "scientists believe" were set forth fully in Chapter 1, where I proposed that expressions of belief be left to the realm of religion.

As an anticreationist writer, Newell has written a book titled Creation and Evolution; Myth or Reality? That he, too, considers evolution to be a fact is evident in two sentences in his preface: "In making much of the idea that scientists do not agree even among themselves about

evolution, the creationists fail to distinguish between the fact of evolution and the theories of evolution." (He quotes Webster's definition of "fact.") "From many converging lines of evidence biologists and paleontologists long ago accepted organic evolution as a fact" (Newell, 1982, p. xxx). Stephen Jay Gould, in a 1987 essay in Discover, closely follows Newell's usage of "fact" and "theory." The "fact of evolution" referred to by both Newell and Gould is, of course, the proposition that descent with modification has occurred through geologic time. What they refer to as "theory" is the collection of hypotheses dealing with the causes and mechanisms of that modification process. Gould's essay is devoted to explaining the distinction between "fact of evolution" and "theory of evolution," a task he accomplishes with great skill and effectiveness. But will the public follow through lengthy expositions of this kind? I suspect not. For the general public and the news media, the "fact of evolution" will continue to be assumed to refer to the whole evolutionary package.

Here's another example--one calculated to compound confusion. In a review of several recent anticreationist books, geologist Richard K. Bambach commented as follows:

> Lay persons sometimes view science as a body of demonstrated truths, but this is a misimpression. Philosophers of science tell us that certain proof is virtually impossible. As Kitcher says (Philip Kitcher, 1982, p. 32), "Complete certainty is best seen as an ideal toward which we strive and that is rarely, if ever, attained." (Bambach, 1983, p. 851)

Referring to discussions of two of the authors he reviews, Bambach continues:

> Therefore I agree with Futuyama that the ideas of Dorothy Nelkin. . . . expose the central difference between evolutionary science and creationism: namely, whether we require evidence for beliefs or whether we accept beliefs without evidence. Beliefs, not proofs, are the issue. Both tenets of religion and widely accepted scientific theories are beliefs. However, scientific beliefs require evidence of some sort before they can claim to be convincing, whereas religious beliefs are frequently held without substantiating evidence. (1983, p. 851)

I find the statement "beliefs require evidence" to be self-contradictory. The nature of "belief" is such that it does not require evidence. Just look in Webster's Ninth New Collegiate Dictionary and you will read that "to believe" is to "accept trustfully and on faith." Truth and belief are closely linked in concept, and I am inclined to associate their meaning with an expression of faith in something transempirical, thereby automatically excluding those concepts from empirical science. The trouble with using words such as "truth" and "belief" in connection with any scientific theory or hypothesis is that it puts the user in the same church and pew as the fundamentalist creationist. Of course, what happens then is that the creationist says to the scientist seated alongside: "You stay here and worship your doctrine of evolution, which is indeed a religion, but I'm going to go sit in your office at the university and practice my creation science, which is genuine science." How this juxtaposition works is nicely shown in a letter by Duane Gish to the editor of Science 82 (vol.3, no.1, p. 16). Dr. Gish speaks for the Institute for Creation Research on matters of evolution and the fossil record. He says:

> Evolutionists, because of their arrogance (such as your article's reference to my response during an interview as a "performance") and the often frivolous or shallow dismissal of creationists'

arguments, are their own worst enemy. Our students, and the public in general, are not as ignorant nor gullible as some evolutionists seem to believe, and the logically coherent scientific evidence for creation appears compelling to those who have an opportunity to hear it. We do not demand that "religious stories of creation be given equal time." We request a hearing for the <u>scientific evidence</u> supporting creation, in perfect accord with academic freedom.

But not all evolutionists "are their own worst enemy." A somewhat more tempered statement of the status of the theory of evolution is given by anthropologist Bernard G. Campbell in his book titled <u>Human Evolution</u>:

> Science does not claim to discover the final truth but only to put forward hypotheses based on evidence that is available at the time of their presentation. Well-corroborated hypotheses are often treated as facts, and such a fact is that of organic evolution. If a hypothesis is fairly general in its presentation, it is difficult to test, but a detailed hypothesis like that of organic evolution is readily susceptible to disproof. The evidence for evolution is overwhelming, and there is no known fact that either weakens the hypothesis or disproves it. (1966, p. 1)[1]

The above statement would satisfy me if we agreed to modify it to say that "the evidence for evolution is so overwhelming that we can treat it <u>as if it were</u> a fact." Campbell's reference to "fact" in the last sentence seems to be what I referred to in Part One as "a statement of fact," a simple declarative statement describing a singular observation that is not subject to meaningful challenge and for which no conceivable alternative is reasonable.

I would not recommend that the descriptive statement of evolution as descent with modification be considered a simple statement of fact because, although it draws massive support from a vast and complex history of events inferred from fossil and stratigraphic evidence, that corroborative evidence comes only from a sample of the total stratigraphic record. Scientists have not even begun to sift through all fossiliferous strata of the continental crust. Until all this unstudied rock has been fragmented and examined particle by particle, we cannot be sure (absolutely, positively sure) that the biostratigraphic record is perfectly consistent. But much of the former biostratigraphic record has been lost forever by weathering and erosion and recycling of sediment. Can we really say with complete assurance that nothing in that lost record could possibly have proved contradictory or anomalous in the context of our stated theory of evolution? Surely, some expression of reservation is in order in keeping with a scientific tradition that is summed up in the single word "skepticism." Stephen Gould, in his essay mentioned earlier, seems to include this requisite expression of reservation when he writes: "The fact of evolution is as well established as anything in science (as secure as the revolution of the earth about the sun),

though absolute certainty has no place in our lexicon" (1987, p. 64).

In reading a recent publication of the National Academy of Sciences titled <u>Science and Creationism</u>, I was favorably impressed by the absence of dogmatic statements of belief in evolution or of evolution as truth (Ebert, 1984). The report states: "Evidence that the evolution of the universe has taken place over at least several billion years is overwhelming" (p. 12.) On biological evolution we read:

> As applied to biology, a distinction is to be drawn between the questions (1) <u>whether</u> and (2) <u>how</u> biological evolution happened. The first refers to the finding, now supported by an overwhelming body of evidence, that descent with modification occurred during more than 2.7 billion years of earth's history. The second refers to the theory explaining how those changes developed along the observed lineages. The mechanisms are still undergoing investigation; the currently favored theory is an extensively modified version of Darwinian natural selection. (P. 15)

And, finally this appraisal of the theory of evolution as descent with modification:

> Evolution pervades all biological phenomena. To ignore that it occurred or to classify it as a form of dogma is to deprive the student of the most fundamental organizational concept in the biological sciences. No other biological concept has been more extensively tested and more thoroughly corroborated than the evolutionary history of organisms. (P. 22)

I find the academy committee's wording adequately supportive of the description of organic evolution as descent with modification through geologic time. For me, their statements carry much greater power of persuasion than flat assertions of truth or belief. They do not categorically rule out the possibility that negative evidence may yet appear--that a new paradigm will arise out of the ashes of a now unforeseen scientific revolution --but they give us confidence that the probability of that event occurring is extremely small.

We have by no means presented all of the lines of evidence in favor of organic evolution having occurred as a part of geologic history. Actually, we have only begun by summarizing what was largely established as massive corroborative evidence as early as a century ago. We have supplemented biostratigraphic evidence given by the sequence of fossils with independent evidence from comparative anatomy of living organisms. Beyond that, modern biochemistry has much to tell about the theory of descent with modification, providing strongly corroborative evidence in an entirely independent area of science.

---

[1]From: Bernard Campbell, HUMAN EVOLUTION, Second Edition. Copyright © 1974 by Bernard Campbell. (Aldine Publishing Company, New York). Adapted by permission.

Chapter 34

The Genetic Code—A Review of Mainstream Molecular Biology

Up to this point we have treated evolution as Darwinists defined it: descent with modification. Darwin offered a description of the process of natural selection forced by the environment, but he did not have any explanation to offer in terms of biological process, beyond the assumption that individuals of a species show differences among themselves and that such differences are inheritable. True, the Austrian monk, Gregor Mendel, had discovered his laws of heredity and published his findings in 1866, but they were ignored. It was not until about 1900 that others rediscovered Mendel's laws of heredity, allowing individual variation to be conceived of in terms of genes.

Surprisingly, Mendel's principles were used by geneticists of that time to discredit Darwinian natural selection. For the first two decades of the 1900s, biological theory favored the idea of mutations (sudden and permanent changes in genes) as guiding evolution. Natural selection, these biologists claimed, performed only a minor and negative function in the evolutionary process. All that changed in ensuing decades as new findings reinforced Darwinian natural selection. Gradually there emerged a new synthesis of Darwinian theory with the new observations of biology. The turning point is often keyed to the publication in 1937 of a work titled Genetics and the Origin of Species by Columbia University geneticist Theodosius Dobzhansky.

We can go no further in assessing the differences in views of mainstream biologists and creation scientists without arming ourselves with some basic principles of modern genetics and evolution. Going one step further, we must become armed with at least a descriptive understanding of molecular biology that lies at the very heart of genetics and evolution. We must carry out the reductionism discussed in Part One as a controversial principle in the life sciences and historical sciences. Reversing the historical order of discovery, we start with molecular biology and the remarkable double helix that carries the code for reproduction of life. There follow chapters on mode and tempo of evolution and the application of molecular biology to phylogeny.

To establish this broad informational base requires a substantial digression. For those familiar with this material, it is easy enough to skip directly to later chapters, where we resume the debate by presenting the creationists' view of the fossil record as a product of the Flood of Noah.

The Genetic Code

We turn now to the question: How does evolution take place? What is the biological system behind evolution and how does that system function? If evolution is a naturalistic phenomenon, as mainstream scientists assume, every organism must carry within itself a complete set of instructions for (a) reproducing itself, and (b) carrying out its own growth and development as an individual, from a single complete cell to an adult. This information is described by the adjective genetic. Every cell of an organism contains the essential store of genetic information and, as everyone knows, it is held in chromosomes and genes. If we think of a cell as resembling a manufacturing plant run by humans, there must be a central authority with the knowledge necessary to operate the factory effectively. This knowledge includes plans and specifications showing exactly what is required and how every part is to be made and assembled.

In the organic cell, the plans and specifications are contained in extremely large, compound molecules of a class known as nucleic acids. There are two kinds, designated in shorthand as DNA and RNA. The nucleic acids are quite unique in chemical composition; they form long, twisted chains of molecules, containing thousands of atoms. Almost everyone who gets input from the news media has learned of DNA and RNA in statements to the effect that these substances "carry the genetic code."

DNA and RNA work together to assemble proteins that are mostly enzymes, and thus they direct the manufacturing operations of the whole organic factory. To carry the technology analog further, DNA includes complete plans and specifications for setting up new factories essentially identical to the existing ones. To get at the heart of the genetic system, we need to focus special attention on these all-important control molecules that seem to know exactly what to do at all times and are able to do it exactly right (most of the time at least). These are molecules that can assure almost unerring replication, not only of single cells, but also of entire organisms of extraordinary complexity.

Logic might seem to suggest that we start with the control molecules, DNA and RNA, then go to their products, the proteins. In terms of levels of complexity, however, it would be better to start with the relatively simpler protein molecules. Some basic principles of organic chemistry will also be useful for those who (like myself) strenuously avoided that malodorous subject while satisfying the science requirements of a liberal arts degree.

Composition of the Biosphere

First, consider the elements that make up the substance of the global stock of living matter, or biosphere. Table 34.1 gives the list of those elements in order of abundance. Percentages are relative to the total quantity of the fifteen most abundant elements in the biosphere, named in the table. All but one-half percent consist of three elements: hydrogen, just under one-half of the total; carbon and oxygen, each just under one-quarter of the total. These are the major macronutrients. The remaining one-half percent is divided among twelve elements. Six of these are also macronutrients: nitrogen, calcium, potassium, magnesium, sulfur, and phosphorus. The macronutrients are all required in substantial quantities for organic life to thrive.

Quite a number of additional elements, not among the nine macronutrients, are also vital to life processes. Their

Table 34.1 Elements Comprising Living Matter

Basic carbohydrate	Percent[*]
Hydrogen (M)	49.74
Carbon (M)	24.90
Oxygen (M)	24.83
Subtotal:	99.47

Other nutrients	
Nitrogen (M)	0.272
Calcium (M)	0.072
Potassium (M)	0.044
Silicon (M)	0.033
Magnesium (M)	0.031
Sulfur (M)	0.017
Aluminum	0.016
Phosphorus (M)	0.013
Chlorine	0.011
Sodium	0.006
Iron	0.005
Manganese	0.003

M - macronutrient

[*]Relative to total of the 15 most abundant elements in living matter.

Data source: E. S. Deevey, Jr., Scientific American, vol. 223, September, 1970.

presence is needed in mere traces and, for that reason, they are called micronutrients. The micronutrient list includes iron, copper, zinc, boron, molybdenum, manganese, and chlorine.

Organic compounds, naturally produced by living organisms, are molecules consisting of atoms of carbon (C), nearly always with bound atoms of hydrogen (H), and in most instances also including bound atoms of oxygen (O). In the important molecules that direct life processes and reproduction, nitrogen (N) must also be present, while phosphorus (P) plays an important role. Understanding nucleic acids, proteins, and the "genetic code" is at least to some extent made much simpler than it might otherwise be, involving as it does (for the most part) only five elements: C, H, O, N, and P. Does this simplicity of chemical ingredients carry a message about the origin of life--that all life derives from the same set of primitive control molecules?

Proteins and Amino Acids

Almost everyone is familiar with the essential classes of organic compounds making up most of the substance of plants and animals: carbohydrates, fats, and proteins. Carbohydrates are composed entirely of oxygen, carbon, and hydrogen. We are all familiar with such carbohydrates as the sugars, starch, and cellulose. The various kinds of fats belong to a chemical group called the lipids. A distinguishing feature of the lipids is that they do not dissolve in water. Like the carbohydrates, they are composed of carbon, hydrogen, and oxygen. Although carbohydrates and lipids are essential to the life processes and structures of plants and animals, we need not be concerned with them as being directly involved in controlling the growth processes or reproduction.

We must give our attention to the proteins, if we are to understand how growth is directed and how the genetic code works to pass on the same life form from one generation to the next. All proteins consist of chemical building blocks known as amino acids, of which there are twenty common kinds in nature. Each amino acid shares in common with the others a certain basic, or general,

structure, shown in Figure 34.1. The straight lines connecting the letter groups are chemical bonds; they hold the atoms or groups of atoms together in a molecular group. As shown here, the amino acid is a monomer, i.e., a single group of atoms acting as a whole molecule. The letter "R" stands for an organic group attached as a side chain; it is a different group for each of the twenty amino acids. The amine group (NH_2) at the left is free to attach to the OH group, as shown in Figure 34.2. We have written H---NH for NH_2 for the monomer on the right. The dotted oval line in (A) isolates the makings of a molecule of water (H_2O), which drops out, leaving the NH group bonded directly to the carbon atom, C. Thus, as shown in (B), two monomers have been joined into one molecule. This kind of bond is called a peptide bond; the molecule itself is called a dipeptide.

Going one step further, try making the same linking with two different amino acids, for example, alanine (ala) and glutamic acid (glu). (Chemists shorten the name of each amino acid to a three-letter combination, or code, making it easy to write down a long string of linked units.) In Figure 34.3, (A) shows the two acids as monomers, each with a different organic group as a side chain (equivalent to "R" in Figure 34.1). The resulting dipeptide (B) is called alanylglutamic acid (ala•glu).

The important point is that different amino acids can be joined together in a chain in almost any conceivable order. In the above example, alanine could just as well be on the right with glutamic acid acid on the left, in which case we can call the dipeptide "glutamylalanine" (glu•ala). The "backbone" of the chain, shown in Figure 34.3 by the single horizontal line of linked groups, is always the same regardless of the order of arrangement. You might think of the way in which beads of various colors and sizes can be threaded on a string, using any order you wish and any number of beads. With three or more beads, you have a polypeptide. Chemists refer to repeating molecules linked together in a chain as polymers. Table 34.2 lists the twenty amino acids and their code letters. The order in which they are given is preferred by

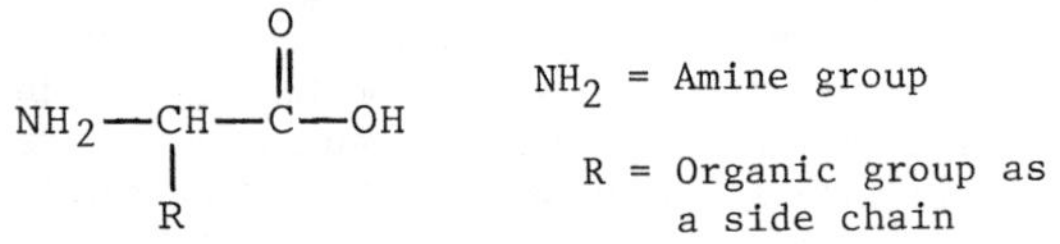

Figure 34.1 The basic structure of amino acids.

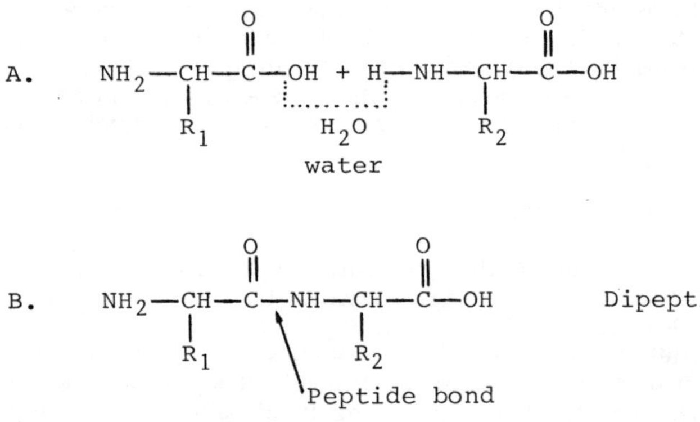

Figure 34.2 A dipeptide molecule, formed from two monomers joined by a peptide bond.

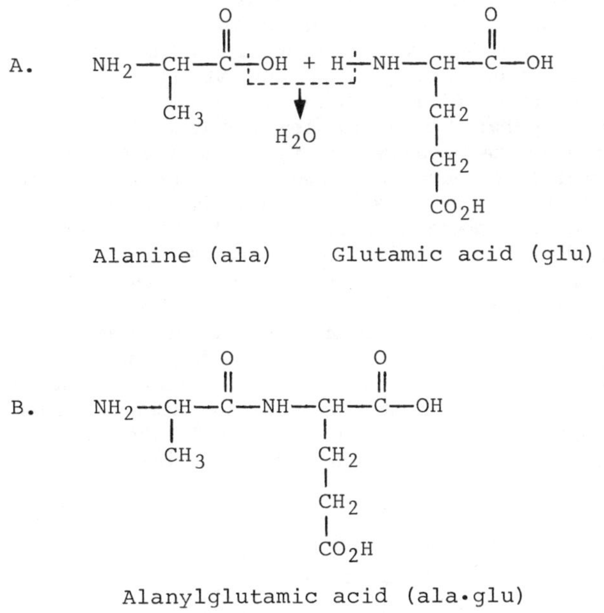

A.

$$NH_2-CH-C-OH + H-NH-CH-C-OH$$

Alanine (ala)   Glutamic acid (glu)

B.

$$NH_2-CH-C-NH-CH-C-OH$$

Alanylglutamic acid (ala·glu)

Figure 34.3 Peptide bonding of alanine and glutamic acid to form alanylglutamic acid. (Based on data of K. R. Atkins, J. R. Holum, and A. N. Strahler, 1978, <u>Essentials</u> of <u>Physical</u> <u>Science</u>, John Wiley & Sons, New York, p. 297.)

chemists and is one of increasing complexity of the "R" side chain.

Protein molecules are large polypeptides, often with over 100 amino acids in a single chain. It is also possible for two or more chains, as polypeptide subunits, to be joined together at certain points, called "bridges," or simply held together by weak chemical attraction. An example is the protein molecule of beef cattle insulin, shown in Figure 34.4. It has two strands connected by bridges. Note that at one end of each strand is an NH$_2$ group; at the other end a carbohydrate group. The insulin molecule is small compared with many other natural proteins. For example, the gamma globulin molecule contains 1,320 amino acids and consists of a total of almost 20,000 atoms.

Table 34.2 Amino Acids Specified by the Genetic Code

Name	Symbol		
Glycine	Gly	Asparagine	Asn
Alanine	Ala	Glutamic acid	Glu
Valine	Val	Glutamine	Gln
Leucine	Leu	Lysine	Lys
Isoleucine	Ile	Arginine	Arg
Serine	Ser	Histidine	His
Threonine	Thr	Phenylalanine	Phe
Cysteine	Cys	Tyrosine	Tyr
Methionine	Met	Tryptophan	Trp
Aspartic acid	Asp	Proline	Pro

Notice in Figure 34.4 that insulin of certain other animals--sheep, horse, and whale and pig--differs slightly in structure and in sequence of amino acids from that of beef cattle. Not only is a bridge missing, but ile substitutes for val; gly for ser; and thr for ala. This may seem like just so much trivia to take note of, but such differences in protein molecules from one species to another and from one taxonomic group to another have great significance to the study of evolution. The differences allow phylogenies to be constructed, thus confirming phylogenies based on comparative anatomy and the fossil record. But we must defer this fascinating subject to Chapter 36.

A complex protein chain tends to twist naturally into a coil, or <u>helix</u>, as if it were a twisted ribbon. At the continual line of contact (or nearest approach) of the coil on itself, hydrogen atoms of the amino acid segments form weak bonds with the opposing atoms, tending to hold the coil together. The direction of coiling, for most natural proteins, is into a right-handed helix.

When you think of the large number of different amino acids that must be brought together in exactly the right order from out of a collection of free, individual amino acid molecules floating around in the cell medium (cytoplasm), you realize that for this to happen by a purely random bumping together of the ends of the backbones of the molecules is so highly improbable as to be dismissed as, for all practical purposes, impossible. So the question really is: What directive mechanism in the cell gets these chains organized in exactly the right way, does it rapidly, and rarely makes an error?

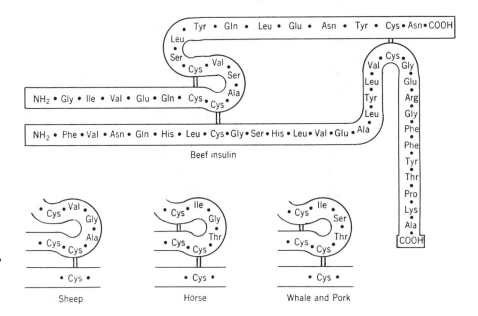

Figure 34.4 The structure of insulin. See Table 34.2 for names of the amino acids. (From K. R. Atkins, J. R. Holum, and A. N. Strahler, <u>Essentials</u> of <u>Physical</u> <u>Science</u>, p. 298, Figure 19-8. Copyright © 1978 by John Wiley & Sons, Inc. Used by permission.)

Enzymes

Protein molecules of certain types play an important role in chemical reactions within a cell by acting as enzymes. The normal activities within a cell, collectively called <u>metabolism</u>, consist of breaking apart or putting together organic molecules, such as carbohydrates, lipids, and proteins. To activate or speed up such chemical reactions, a <u>catalyst</u> is usually required. Protein molecules can act as catalysts in living systems, playing an essential role in cell metabolism. Protein molecules that perform this function are called <u>enzymes</u>. Molecules that must be joined together or broken apart in chemical reactions approach a protein molecule that is an enzyme and are temporarily held fast to it by electrostatic forces. The enzyme then exerts other forces that assist in the formation of the required chemical bonds, or act to break such bonds. In order for the enzyme to do its work, the beneficiary of this attention (called the "substrate") must fit physically into the surface configurations of the enzyme for close contact. Think of the enzyme as a polypeptide made of various amino acids, their side chains sticking out in various lengths and making "bays" and "promontories." Chemists think of this geometry as resembling that of a toothed key in a tumbler lock. If the key fits all the tumblers just right, the key will turn. Under this "lock-and-key" theory, a particular enzyme is limited to doing its work on just the right beneficiary molecules and doing it in just the right way. After the job is done, the beneficiaries are repelled by the enzyme, which is then free to perform the same service for other substrate molecules.

Understanding the role of certain proteins as enzymes brings us a step nearer to understanding how the activities of a cell are directed, but we still have to deal with the question: What directs the formation of complex polypeptide molecules in the first place? "Who has the blueprint?" so to speak, and "Who gives the orders?"

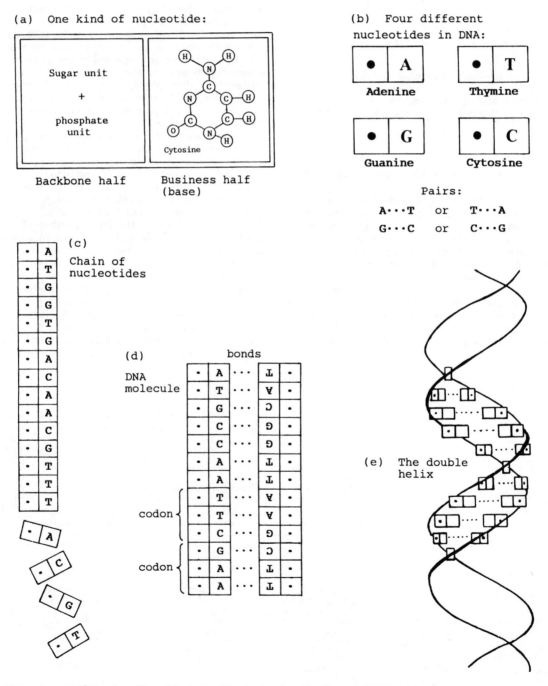

Figure 34.5 Using a set of domino-like objects to illustrate the structure of DNA. (A. N. Strahler.)

DNA and Nucleotides

Molecular biology, emerging in only the past three decades or so, has finally revealed the nature and workings of the DNA and RNA molecules. Undoubtedly the best-known scientists in this connection are James Watson and Francis Crick, who in 1953 first suggested a structure for DNA that could explain how it served as a genetic code. In so doing, they established the molecular nature of what had been, until that time, called simply a gene. The <u>Crick-Watson</u> <u>theory</u> won its authors a Nobel prize, which seems richly deserved. Although the components of the DNA molecule are comparatively simple as units, they are assembled in chains of enormous length.

Although we will go into some details of the chemical structure of DNA and RNA and the way in which they carry and make use of information necessary for the synthesis of proteins, the essential concepts can be grasped in a purely descriptive and pictorial way.

First, what is DNA? Think of a set of special dominoes. Each domino is a rectangular tablet containing two squares bearing information. Let one domino represent one molecule, called a <u>nucleotide</u> (Figure 34.5). One half of the nucleotide domino is the same in all dominoes of the set. We can call this half the "backbone half," for it serves only two purposes: first, to carry the other or "business half" of the nucleotide, and, second, to form a strong attachment to the backbone halves of other nucleotides. Imagine that there are little magnets in the backbone half of the domino, so that when two are brought close together, they are strongly attracted and latch onto each other.

Details of a single nucleotide are shown in Figure 34.6. There are three distinct parts. Two of these parts form the backbone of a chain: (1) a <u>phosphate</u> <u>unit</u>; (2) a <u>sugar</u> <u>unit</u>. Phosphate and sugar units are connected by strong bonds. The third part of the nucleotide is an <u>amine</u>. An amine is the <u>nitrogen</u> <u>base</u> of an amino acid; it

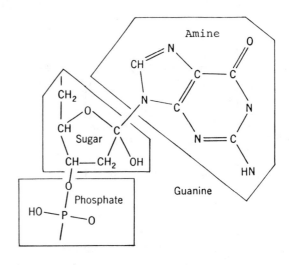

Figure 34.6 Details of a single nucleotide, guanine. (A. N. and A. H. Strahler.)

is a ring structure synthesized from an amino acid and is identified by a distinctive name. For this reason, the unit is usually referred to simply as a <u>base</u>. In Figure 34.6 the base is named <u>guanine</u>.

There are two kinds of nucleic acids: <u>RNA</u> and <u>DNA</u>. The first initials "R" and "D" refer to the sugar unit of the nucleotide. "R" stands for <u>ribose</u>, a simple sugar with the formula $C_5H_{10}O_5$. It differs little in formula from glucose, the familiar natural sugar of ripe fruits, with the formula $C_6H_{12}O_6$. "D" stands for <u>de-oxy-ribose</u>. Its formula would be that of ribose if we removed one oxygen from ribose, leaving $C_5H_{10}O_4$. The implied word, "deoxygenate," seems to indicate quite clearly that we recognize removal of some of the oxygen, which is the case. Thus we have RNA (<u>ribonucleic</u> <u>acid</u>) and DNA

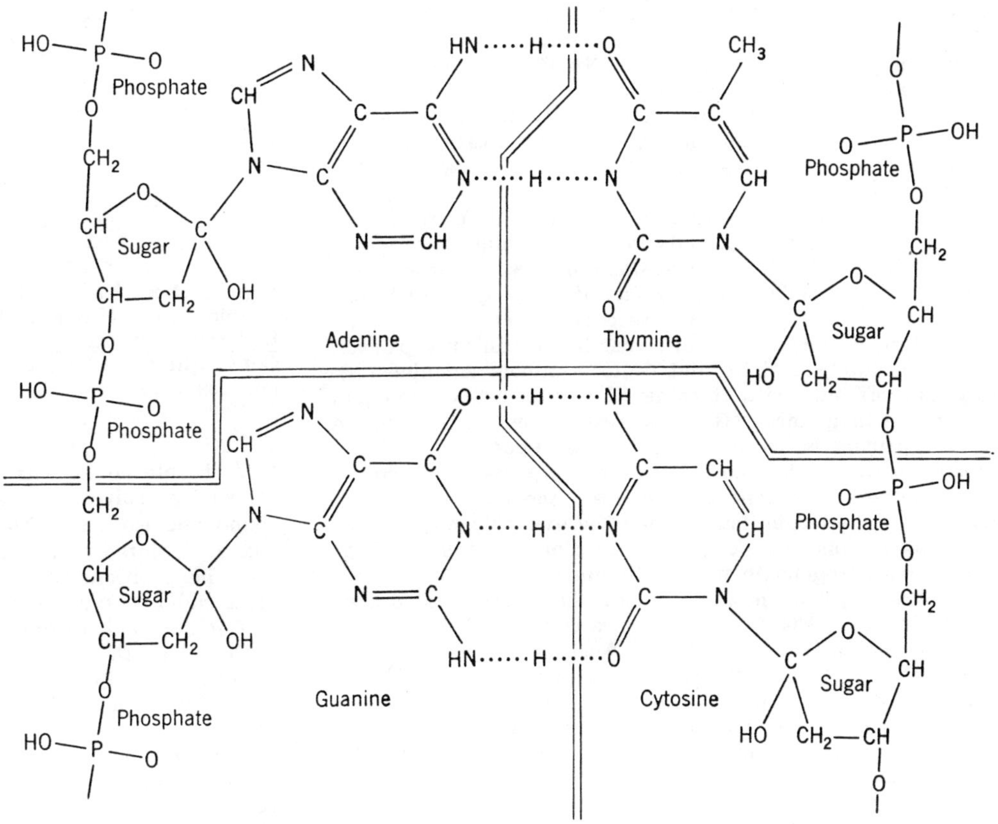

Figure 34.7 Bonding of nucleotides in the DNA molecule. (A. N. and A. H. Strahler.)

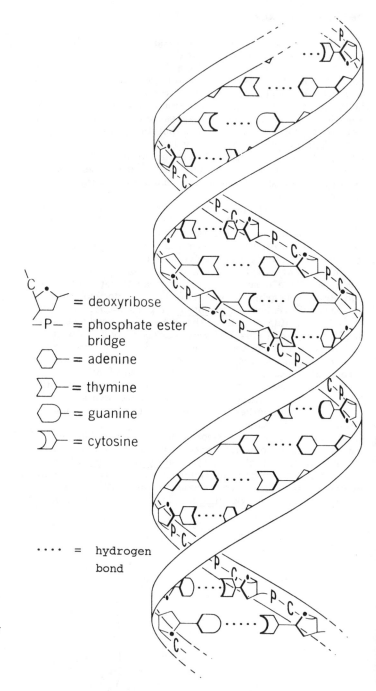

Adenine ... Thymine

Guanine ... Cytosine

= sugar (deoxyribose)

= base

P = phosphate ester bridge

——— = hydrogen bond

Figure 34.8 Pairing and bonding of nucleotides in DNA. (Adapted from K. R. Atkins, J. R. Holum, and A. N. Strahler, Essentials of Physical Science, p. 302, Figure 19-13. Copyright © 1978 by John Wiley & Sons, Inc. Used by permission.)

= deoxyribose

—P— = phosphate ester bridge

= adenine

= thymine

= guanine

= cytosine

.... = hydrogen bond

Figure 34.9 Simplified schematic diagram of the DNA double helix. (From J. R. Holum, Elements of General and Biological Chemistry, 7th ed., p. 448, Figure 21.7. Copyright © 1987 by John Wiley & Sons, Inc. Used by permission.)

(deoxyribonucleic acid). We must now turn to another, and very important difference between the two nucleic acids.

Both RNA and DNA have four different bases that can fill the nitrogen-base position. Three of these four bases are the same for both RNA and DNA, namely, guanine (G), cytosine (C), and adenine (A). The fourth amine in RNA is uracil (U). The fourth amine in DNA is thymine (T).

In going further, we take up DNA first, followed by RNA. This is because DNA is the prime controller of the whole biochemical molecular system. Only copies of DNA are taken out of the cell nucleus. The information held by DNA is copied onto RNA, and those copies move out into the body of the cell.

DNA is particularly remarkable because it forms into a double helix. Each helix coils in the same direction-- toward the right. This arrangement brings the amines (bases) of one helix into positions directly opposite bases of the other helix. But now, we come to the most fascinating observation of all, that bases can only pair off in an absolutely definite and limited manner. Adenine (A) may pair off only with thymine (T); cytosine (C) can pair off only with guanine (G). The reason for this restriction is that pairing is accomplished by the opposition of a hydrogen with either an oxygen or a nitrogen, but not with another hydrogen; besides that, the number of pairings of hydrogen must match the available number of opposing nitrogens or oxygens. Adenine and thymine have only two hydrogens to pair with; guanine and cytosine have three hydrogens to pair with. The combinations are shown in Figure 34.7. (These pairs can be called base pairs.) The hydrogen bonds are weak, but they enable the double helix to keep its identity, until greater forces are applied to split the two helices apart. The backbones of the two helices run in opposite directions, and thus the base of one pair can be visualized as "upside-down" with respect to the other, as is obvious from Figure 34.8.

Figure 34.9 is a simplified schematic diagram of the

DNA double helix. The A--T and G--C base pairs are shown by arbitrary geometrical symbols, devised so that a round end must fit into a semicircular cavity and a sharply pointed end into an angled embayment. But notice that the positions of the members of pairs in the double helix can be reversed from left-right to right-left.

Getting back to our set of dominoes as models of nucleotides, DNA consists entirely of four kinds of dominoes: A, T, G, and C. With an unlimited number of each of the four kinds of dominoes (nucleotides), we can arrange them in a chain, backbone halves touching each other; base halves touching each other. You would want to select the dominoes at random, since, at this point, you have no reason to put them in any particular order.

SECOND LETTER

FIRST LETTER		A	T	G	C
	A	A A	A T	A G	A C
	T	T A	T T	T G	T C
	G	G A	G T	G G	G C
	C	C A	C T	C G	C C

In groups of two: 16 codons

SECOND LETTER

FIRST LETTER		A	T	G	C		THIRD LETTER
A		A A A	A T A	A G A	A C A	A	
		A A T	A T T	A G T	A C T	T	
		A A G	A T G	A G G	A C G	G	
		A A C	A T C	A G C	A C C	C	
T		T A A	T T A	T G A	T C A	A	
		T A T	T T T	T G T	T C T	T	
		T A G	T T G	T G G	T C G	G	
		T A C	T T C	T G C	T C C	C	
C		C A A	C T A	C G A	C C A	A	
		C A T	C T T	C G T	C C T	T	
		C A G	C T G	C G G	C C G	G	
		C A C	C T C	C G C	C C C	C	
G		G A A	G T A	G G A	G C A	A	
		G A T	G T T	G G T	G C T	T	
		G A G	G T G	G G G	G C G	G	
		G A C	G T C	G G C	G C C	C	

In groups of three: 64 codons

Figure 34.10 Codons for A, T, G, and C in groups of two and three. (A. N. Strahler.)

The same base can repeat one or more times, if chance dictates. The chain of nucleotides thus formed is only one-half of the DNA structure. The next step is to form a second chain of dominoes and place it opposite the first, with bases opposed to each other. This will necessitate rotating each domino through 180 degrees. The sequence in each strand is thus the complement of the other strand, and each strand can serve as a model, or template, for the formation of the other strand. In building the complementary chain of dominoes, a strict rule must be followed. As explained earlier, only A and T can become opposed as a pair; only C and G can be a pair. However, the pairs can be A-T or T-A; C-G or G-C. Notice that this rule has been followed in Figure 34.5(d).

Stop and consider, now, how the succession of bases might serve as a code to convey information, four kinds being available in any order and with repetitions of the same base being allowed. Suppose you are thinking about sending English language words by such a code. Four single letters to work with is not enough for an alphabet of 26 letters. Taken in sequences of two you have 16 different groups, allowing for repetition, as shown in Figure 34.10. Since this is not enough, try sequences of three. Now you have 64 groups, more than enough, and you can assign the same meaning to two or more different groups. But you can also use some of the surplus groups for various punctuation marks, or for numerals 0 through 9.

Remarkably enough, research by molecular biologists led to the conclusion that groups of three successive bases, with bases taken in any order or repeated three times, comprise the genetic code units, called <u>codons</u>. Biochemists already knew that there are only twenty different amino acids with which the cell's protein molecules are put together. The amino acid units are strung together in chains, but the order and number of the amino acids must be exactly the same to produce the same protein. I think by now you can leap ahead to the heart of the molecule manufacturing process in the cell factory. DNA carries in code form the names of the amino acids that are required and the order in which they must appear. But just how does the code of instructions lead to the finished product? How can the DNA molecule be replicated?

The observation that the DNA molecule is coiled into a double helix is of no special consequence in understanding how replication works; a straight flat chain will do just as

well. So we continue with the domino analogy in Figure 34.11.

Now it's time to mix metaphors and look at the double chain of nucleotides as an ordinary zipper or, at any rate, a slide fastener of sorts. Because the bonds of the two opposing halves of the DNA molecule are weak, they can be pulled apart rather easily. If you take an ordinary zipper, unattached to a garment or bag, you can unzip it completely into two separate parts. Suppose that you send one half to a zipper factory and request that they make

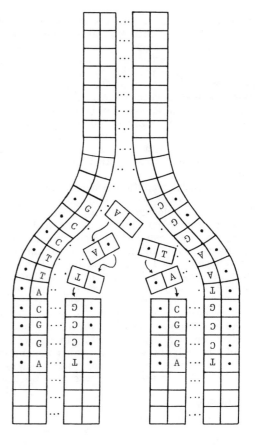

Figure 34.11 Schematic diagram of replication of DNA. (A. N. Strahler.)

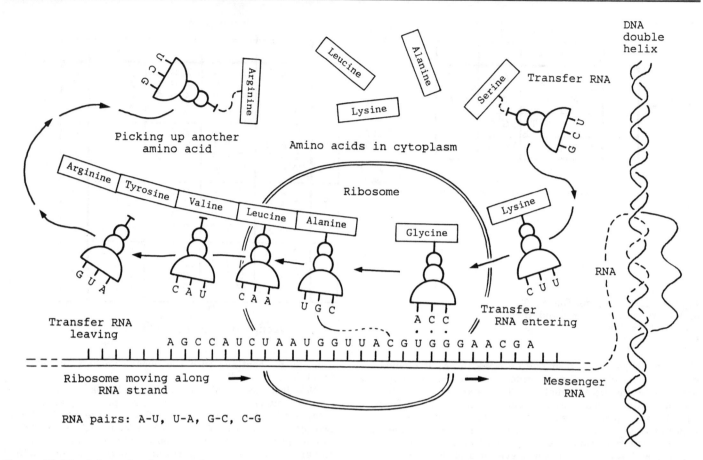

Figure 34.12 Schematic diagram of synthesis of protein molecules by messenger RNA and transfer RNA. (A. N. Strahler.)

up a matching half to fit. Suppose you send the other half to another factory with the same request. You will get back two complete zippers, exactly alike. DNA can replicate itself by an analogous process. It splits apart, allowing access to free nucleotides that can be pulled in from the surrounding liquid medium within the cell, i.e., from the cytoplasm (Figure 34.11). A nucleotide with the correct base will form a new pair just like that which was pulled away. Nucleotides are added in succession to replicate the missing half. Meantime, the same thing is happening on the other half of the DNA molecule. Each half is serving as a template with which to build the missing half.

From this greatly simplified model of the DNA double helix, it is easy enough to see that exact instructions for assembling protein molecules of many kinds and many lengths can be replicated, just as an original set of engineering drawings can be duplicated innumerable times by a copying machine. In the cell of an organism this ability to replicate has two kinds of uses. First, all of the information relating to building the cell can be transmitted to daughter cells, formed by fission (splitting and resplitting) of the original cell. Transmitting the genetic code intact from one generation to the next is, after all, an essential and unique characteristic of life, as distinct from nonlife. Second, copies of DNA can be taken out into the body of the cell where they serve as instructions essential to the process of assembling new protein molecules. It is to this second process that we turn next.

Protein Assembly

Here is where RNA comes into the picture. Everything said here about a DNA molecule as to its chemical makeup and structure also applies to RNA, with the important difference already noted. While RNA also has only four

different bases, they are not the same four. Three of them are the same: G, C, and A. RNA does not have T, but instead has uracil (U). The two paired bases in RNA are thus A--U and G--C. Keep this in mind as we continue. When a double helix of DNA is split open, it can serve as a template with which to construct a single strand of RNA, composed of G, C, A, and U (Figure 34.12). Where A occurs in DNA, it will be paired with a nucleotide of RNA that is U (rather than T). Despite this difference in opposing bases, the message is not changed. What happens is that the new RNA strand pulls away from its template (the DNA strand), carrying the message out into the body of the cell and transporting it to little "factories" called ribosomes.

Now, things get a bit complicated, but please continue to follow me. The RNA strand formed from the DNA template is called messenger RNA (Figure 34.12). It moves out into the cytoplasm and makes contact with a ribosome (pictured as a spherical enclosure), where the protein molecule is assembled. We now recognize a different form of RNA molecule, called transfer RNA; it consists of a relatively short strand of RNA, formed from the DNA template. The strand has doubled over on itself to form a compact unit. We depict it as a sort of "robot" with a flat base exposing a sequence of three bases, representing the code for a particular amino acid. Notice, however, that this codon carries the opposing bases of the four possible pairs of bases, with respect to the messenger RNA.

A particular transfer RNA molecule finds the amino acid for which it is coded, attaches to it, and brings it into the ribosome, where it temporarily attaches to the messenger codon for the same amino acid. As successive transfer molecules arrive, the amino acids are firmly joined in a chain, called a polypeptide chain. When attachment to the chain is completed, the transfer molecule separates itself and goes back into the cytoplasm to look for another amino acid molecule of its designated kind. When the codon for "stop" is reached, the synthesis ends and the protein molecule is complete. Other ribosomes can use the same messenger chain to produce more of the same kind of protein molecule. The proteins produced in this way are often enzymes, essential to the chemical synthesis and breakdown of many other kinds of molecules in the cell. Much remains to be learned about details of the entire biochemical process, but with the knowledge that is presently available, the role of DNA and RNA in evolution can be quite specifically described.

We have omitted mention of the presence of a whole list of special proteins that are essential in polypeptide synthesis. Fortunately, a detailed description of these is not essential for a general understanding of the process of protein synthesis. Also required is a source of energy; much of this is adenosine triphosphate (ATP).

Some details of the coding for specific amino acids are given in Figure 34.13. Part A gives a typical example of the codons as they are transcribed from the DNA bases to the messenger RNA. Names of amino acids are given for each codon. Part B shows the full, three-letter RNA code for all the amino acids, of which there are twenty different kinds. Individual amino acids are identified by a three-letter symbol (see Table 34.2). Notice first that a particular amino acid may be coded for by more than one codon. Leucine (leu) and arginine (arg) can be specified by six different codons; glycine (gly), valine (val), and four others by four codons; all others by two, except methionine (met), which has only one codon. Notice also that three different codons indicate "stop," which will terminate the assembly of the polypeptide chain.

This description of the way in which DNA and RNA direct and control the construction of the proteins needed by a cell to grow and to multiply by subdivision is, of course, greatly simplified and omits important details of the chemistry of the actual formation of proteins.

DNA in the Eukaryotic Cell

Most organisms large enough to be seen without a microscope are eukaryotes; their cells have a true nucleus. Bacteria, on the other hand, are in many ways less complex than the eukaryotes. In bacteria the DNA is massed in a central part of the cell, forming the nucleoid, which is surrounded by cytoplasm containing many small ribosomes. When the bacterial cell subdivides, the DNA simply divides into two masses, one going to each of the new cells. For the eukaryotes, the cell is much more complex and contains many structures not found in the bacterial cell. In the eukaryotic cell the DNA is contained in the nucleus, which is enclosed by a nuclear membrane, as shown in Figure 34.14. When the eukaryotic cell is ready to subdivide, by a process called mitosis, the genetic material can be seen to take the form of dark, slender rodlike bodies, called chromosomes (see Figure 35.1). The chromosomes are visible under a microscope and have for many decades been known to hold genetic information. It was known that chromosomes come in pairs and that certain sites on individual chromosome pairs serve as genes. Each gene site was known to control a particular structure, or organ, or characteristic of the individual during its growth. Beyond that, the chemical nature of the gene was a complete mystery until DNA was understood. Now, it is clear that the chromosomes contain DNA molecules and that a gene is a particular sequence of codons. The typical gene contains over 1000 nucleotide pairs. But a single DNA molecule may contain as many as 10 million nucleotide pairs. Using that ratio, a single DNA molecule could carry 10,000 genes!

Nearly all the DNA in the typical eukaryotic cell is found in the chromosomes within the cell nucleus. Therefore, the quantity of DNA in each eukaryotic cell of an individual plant or animal of the same species is a constant quantity, the same for all its cells. As a somatic cell prepares to divide by mitosis into two cells, the DNA is replicated once, supplying an equal quantity of DNA to each of the daughter cells. The situation is quite different for those specialized sex cells--the ovum and the sperm (gametes, they are called)--that must unite and pool their DNA to create the first complete cell of an individual organism. The ovum and sperm each have half the quantity of DNA of the complete cell, because in previous production of an egg or sperm (by meiosis) the total DNA was divided equally among two cells.

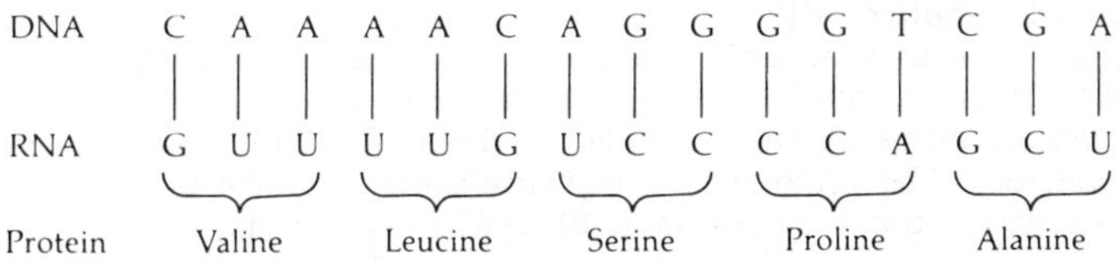

DNA   C A A A A C A G G G G T C G A

RNA   G U U U U G U C C C C A G C U

Protein   Valine   Leucine   Serine   Proline   Alanine

A

Figure 34.13 The RNA code for the amino acids. (A) An example of the transcription of DNA to RNA. (B) The complete three-letter messenger RNA code for all the amino acids. (From G. Ledyard Stebbins, DARWIN TO DNA, MOLECULES TO HUMANITY, p. 35, Figure 2-3. Copyright © 1982 W. H. FREEMAN AND COMPANY. Used by permission.

Second letter

First letter		U	C	A	G	Third letter
U		UUU } Phe UUC UUA } Leu UUG	UCU UCC } Ser UCA UCG	UAU } Tyr UAC UAA Stop UAG Stop	UGU } Cys UGC UGA Stop UGG Trp	U C A G
C		CUU CUC } Leu CUA CUG	CCU CCC } Pro CCA CCG	CAU } His CAC CAA } Gln CAG	CGU CGC } Arg CGA CGG	U C A G
A		AUU AUC } Ile AUA AUG Met	ACU ACC } Thr ACA ACG	AAU } Asn AAC AAA } Lys AAG	AGU } Ser AGC AGA } Arg AGG	U C A G
G		GUU GUC } Val GUA GUG	GCU GCC } Ala GCA GCG	GAU } Asp GAC GAA } Glu GAG	GGU GGC } Gly GGA GGG	U C A G

B

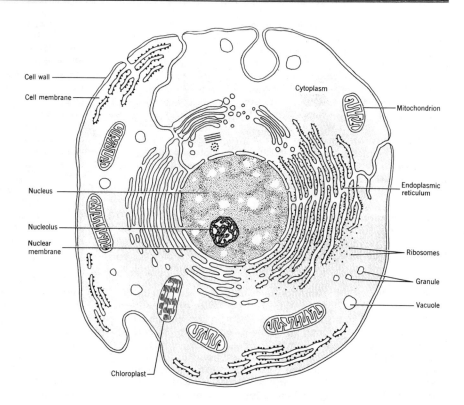

Figure 34.14 Parts of the eukaryotic cell, drawn by John Balbalis. The composite diagram shows some of the more important cellular structures of different types of cells. For clarity, structures are shown only on a portion of the cell diagram. Most structures, for example, the endoplasmic reticulum and ribosomes, are actually present throughout the cytoplasm. (A. N. and A. H. Strahler.)

The Creationist View of the Genetic Code

As a general rule, creationists accept the findings of mainstream science concerning ongoing processes and observable phenomena of physics, chemistry, biology, and geology. They have no reason to challenge that which can be observed and replicated by scientists in the field and laboratory. Thus, they accept the findings of molecular biology as to the structure and function of DNA and RNA.

The creationist biology textbook, Biology: A Search for Order in Complexity, closes its chapter on the science of genetics with a supplement on nucleotides, DNA, and the genetic code (Moore and Slusher, 1970, pp. 105-14). It is conventional in treatment and quite thorough in terms of the secondary school level for which it is written. The opportunity to imply some measure of uncertainty in the Crick-Watson DNA-RNA system is, however, seized upon through the scientists' own in-house debating. Crick's "Central Dogma" is singled out for attack:

> According to the Dogma, information transfer is in only one direction; information from protein never specifies RNA or DNA sequences, and RNA never specifies DNA. In 1970 the Dogma was overthrown by the dramatic discovery of several cases where RNA sequences are used to specify DNA nucleotide sequences. This does not mean that the Central Dogma is a useless concept, but it should not be taken "dogmatically." (Moore and Slusher, 1970, p. 114)[1]

The theme of the above criticism is a familiar one. It is a play on the word "dogma," which, in the minds of most readers, is associated with a hard-nosed belief in religious dogma of a church. In a subtle way, the creationist authors are throwing in the implication that mainstream science deals in religion.

The creationist textbook continues with a critique of the DNA concept (pp. 115-17). Here, a difference of viewpoint between mainstream biology and creationism surfaces. Molecular biology is accused of excessive materialism:

> Some molecular biologists in their enthusiasm would have us believe that all activity of organisms, and this includes human beings, is directed by the order of the nucleotides in the DNA. Thus we are presumably merely the sum total of expression of the information carried by the sequence of the four nucleotides made up of purines and pyrimidines. In fact, it could be said that individuals exist merely as custodians of, and to maintain the continuity of, the complex DNA molecule. We are in effect reduced to a sort of complex automaton, differing from other automatons only by virtue of having certain DNA nucleotides arranged in a slightly different order of sequence. (1970, p. 115)[1]

Clearly, this excursion into philosophy is unbefitting a science textbook, because it introduces vague yearnings for signs of design and a designer. Creationists do not want DNA to be the master of human destiny. They would prefer, instead, to see the role of DNA as that of an agent in the service of some higher power (p. 115). The text quotes Barry Commoner (no reference cited) with remarks to the effect that DNA is not completely in charge, that "certain traits are inherited by way of factors present in the cytoplasm, completely independent of the nucleus with its DNA genes." Two other scientists are said to "have found evidence of a pattern in insect development that is in control regardless of the DNA codon" (p. 116). The creationists do not wish to grant "god-like generalized planning powers" to the master codons within the DNA molecule (pp. 116-17).

[1]Taken from BIOLOGY: A SEARCH FOR ORDER IN COMPLEXITY by John N. Moore and Harold Slusher. Copyright © 1970, 1974 by the Zondervan Corporation. Used by permission.

Thus, creationists' acceptance of the genetic code is tinged with reluctance, despite their view (unstated here) that the genetic apparatus is the work of the Creator, completed within the six-day Creation some 6,000 years ago. Are they, in effect, casting aspersions of inadequacy on the work of the Creator? That work was perfect in all respects and is not to be questioned--only revealed to humans through scientific research.

I found in the Australian journal, Ex Nihilo, a more recent creationist view of the genetic code written by J. G. Leslie, holder of the Ph.D. in experimental pathology from the University of Utah (Leslie, 1984, pp. 38-45). His article reviews genome structure and function, including the replication of DNA. The article is highly technical and has fifty-nine citations from the scientific literature. Much of the text deals with mutations, a subject we treat in later chapters. (Creationists consider all mutations to be harmful.) The idea that DNA might be capable of evolution is questioned, of course. Only in the final paragraph does Leslie come right out with the creationist thesis, which is that the entire genetic apparatus was created as a perfect structure, but that through mutations, it has undergone some deleterious changes. This view is, of course, in line with the strict creationist dogma of inexorable decay of the universe that has set in since the Creation.

Chapter 35

Mode and Tempo of Evolution—
The Mainstream View

At this point we need to review briefly some essential principles of genetics, the study of heredity, or how the reproductive process determines the characteristics of offspring. Much of this information was available long before DNA and RNA were discovered; it was based on inferences about the nature and function of genes and chromosomes. Under the microscope chromosomes could be seen individually whenever the cell prepared to undergo division; they appeared as dark Xs of various shapes and could be numbered and counted. The locations of genes could be identified, but the gene was just "something" that controlled the heredity of a character or trait of the organism, such as a structural feature or metabolic function.

Chromosomes, Genes, and Alleles

Chromosomes come in pairs, identifiable through their distinctive shapes and sizes (Figure 35.1). One chromosome of the pair comes from one parent; the other chromosome from the other parent. This duplication of chromosomes means that each member of the pair contains a gene for the same character (Figure 35.2). Two (or more) genes coding for the same character or trait are called alleles (or allelic genes), but only one allele can occupy the assigned space on the chromosome. This relationship has been likened to the position of pitcher on a baseball team. Only one pitcher can occupy the mound at a given time, but two or three different pitchers may be available for assignment to the duty.

Alleles often differ rather strongly in effect and capability. One allele may code for a trait to appear, while the other codes for that trait not to appear. For example, in a corn plant one allele may code for the production of chlorophyll, essential for plant growth; the other may code for no chlorophyll production and the resulting plant would be an albino and would soon die. In such cases, the allele that performs the useful function is usually dominant, the defective allele is usually recessive. The gene coding for chlorophyll to be produced is dominant; the other is recessive. The dominant gene for normal chlorophyll production has a high probability of appearing in successive generations; the recessive gene for defective chlorophyll production, a low probability of appearing. However, where both dominant and recessive genes are paired in an individual, the dominant gene determines the characteristic. Each gene has an equal probability of being placed in a gamete, or sex cell.

The rules governing heredity are straightforward mathematically for simple cases of dominant and recessive genes for single characters, but there are complications in nature. For example there may be multiple alleles, as in the case of those coding for blood type, for which there are three alleles. A single trait may be controlled by several pairs of genes located at different sites on the chromosome. These multiple genes can produce a wide range of variations in the degree or magnitude of a character, such as length of an individual or organ. In other cases, genes interact to produce the character or trait. A further complication is introduced when we consider that genes are confined to a given chromosome and are passed on as a group; they are said to be linked.

Geneticists recognize within a given organism two physical aspects or entities that are involved in the study of heredity. One is the gene makeup for a particular characteristic; it is called the genotype and obviously refers to the genetic code held by the DNA that makes up the gene. The other is the actual physical expression of a particular characteristic or trait in the organism; it is the phenotype. For example, a particular hair color in an animal is the phenotype; the gene that codes for that hair color is the genotype. The phenotype is also influenced by environmental factors, so that the full potential of the phenotype may not always be realized. For example, the synthesis of chlorophyll in a plant depends upon the presence of sunlight, which is an environmental factor. In another example, the gene might code for large size in the mature individual, but lack of adequate nutrition would result in a stunted individual, quite different from the required phenotype. The meaning of "genotype" also extends to the sum total of the hereditary materials in an organism, the meaning of "phenotype" to the appearance of the entire organism, including its form and life processes.

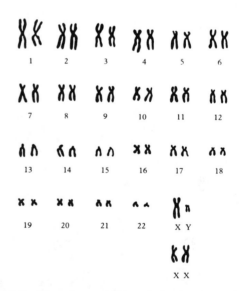

Figure 35.1 The chromosomes of the human male, outlined and enlarged from a photograph, and arranged in a standard numbered sequence. Pairs 1 through 22 are known as autosomes; the X and Y pair are the sex chromosomes. Shown below are the corresponding sex chromosomes of the human female; both are X chromosomes. (From HUMAN GENETICS: AN INTRODUCTION TO THE PRINCIPLES OF HEREDITY, 2d ed., by Sam Singer. Copyright © 1985 by W. H. FREEMAN AND COMPANY. Used by permission.)

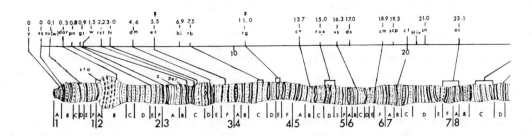

Figure 35.2 Drawing of a portion of the chromosome of the fruit fly, D. melanogaster, showing bands seen under microscopic examination. Below the drawing is the cytological map, numbering the sections and assigning letters to the bands. Above the drawing is the genetic map, showing the loctions of specific genes. For example,

"y" shows the location of the gene for "body yellow, bristles brown"; "rb" refers to "eyes ruby." (From R. C. King, Handbook of Genetics, vol. 3, Chapter 22, Figure 1. Copyright © 1975 Plenum Press, New York. Used by permission of Plenum Publishing Company and Robert C. King.)

Consider next the enormous variability to be expected among individuals of the same species when large numbers of gene pairs are taken into consideration. The figures in Table 35.1 show that, as the number of gene pairs increases, the number of different phenotypes that are possible in offspring from the union of two gametes (sex cells) increases very rapidly.

The number of specific traits, each controlled by a gene pair, has been estimated for the human species. There is sufficient DNA in the human cell nucleus for one and a half million pairs of genes (Ayala, 1977, p. 71). However, it is thought that all of the necessary genetic information is contained in only a small fraction of that quantity, and that the actual number of pairs of genes may be on the order of 100,000. For only 100 gene pairs, the number of possible phenotypes contains 31 digits (about 2.5×10^{30})! Thus, the number of varieties of genotypes within a single species is truly astronomical, and the probability of exact duplication of individuals by random combination is correspondingly minute.

Mutations

Suppose that a particular individual of a species shows a characteristic that is not explained by examining the characteristics of the ancestors. It is possible that this new characteristic is the result of a mutation, which is a change in the hereditary materials. In terms of molecular biology, mutation may consist of a localized change in the DNA nucleotide sequence of a gene, caused by an error in replication of the DNA molecule. These errors are called gene mutations (or point mutations). Errors or changes in the nucleotide sequence include substitutions of one or more nucleotides for others or simply the deletion or addition of one or more nucleotides. Following

the mutation, the revised genotype then codes for a revised phenotype, and thus the offspring may differ markedly in some respect from both parents. Another category of mutations affects the number of chromosomes or the number and placement of genes in a chromosome.

Agents that can cause mutations--mutagenic agents, that is--include application of extreme temperature, exposure to various chemical compounds, certain forms of electromagnetic radiation, such as ultraviolet rays, X rays, and gamma rays, and high-energy particle radiation, such as neutrons, beta rays, and cosmic rays.

When a mutation occurs within cells that make up the body structure--somatic cells, they are called--it will persist throughout the multiplication of that kind of cell during the life of the organism but will not be passed on to ensuing generations. On the other hand, when the mutation is in the DNA of a gamete (a germ cell, or sex cell), it can be passed on to succeeding generations. Most such mutations reduce the probability of survival of the individual so affected, because presumably the DNA code prior to the mutation represented a successful individual in terms of adaptation to the environment. Because accidental change in the code could not be purposefully directed to improved function it would most likely be detrimental. An example of a harmful mutation is the occurrence of hemophilia, a disease in which the clotting of blood does not occur. Hemophilia occurs by mutation, about thirty such events for every one million gametes produced. Another example is a mutation affecting hemoglobin and causing sickle-cell anemia.

But not all mutations are undesirable or harmful. A few, by sheer blind chance, increase the ability of the individual to survive in the existing environment, or in a changing environment, or in a new environment. That individual has a higher probability of survival and of producing more offspring than other individuals of the population; the genotype is thus subject to being "naturally selected" for eventual dominance over the others. The frequency of useful mutations in a species has been estimated conservatively at one in a thousand mutations (Stebbins, 1971, p. 30). The important concept is that there is no natural force or influence directed toward the preferred occurrence of mutations favorable to adaptation of an organism to its environment. Whether we judge a mutation to be "harmful" or "beneficial" implies no such intrinsic value in the mutation itself, but only in terms of the environment of the organism. A given mutation might be harmful in one environment but beneficial in another. Francisco Ayala gives the following examples:

A mutation increasing the density of hair in a mammal may be adaptive in a population living in Alaska, but it is likely to be selected against in a population living in the tropics. Increased melanin

Table 35.1   Numbers of Phenotypes Possible from Union of Two Gametes.

Gene pairs	Number of possible phenotypes in offspring
1	1
2	4
4	16
10	1,024
20	1,048,576
n	2 raised to nth power

pigmentation may be beneficial to men living in tropical Africa, where dark skin protects from the sun's ultraviolet radiation, but not in Scandinavia, where the intensity of sunlight is low, and light skin favors the synthesis of vitamin D. (1977b, p. 66)

Actually, individual beneficial mutations do not by themselves direct evolutionary change, according to the modern view of evolution (Stebbins, 1982, p. 79). The process is not that simple, and we must look for new insights into the genetic makeup (genotype) of entire species populations.

Species Populations and Gene Pools

We next begin to think of heredity in terms of large numbers of individuals of a species making up a statistical population of variates. First, the meaning of "species" needs to be established. Thus far, we have considered species to be the bottom rung of the taxonomic ladder, below the genus and indicated by a Latin name with lowercase first letter. What is essentially the modern meaning of "species" was proposed by an English biologist, John Ray (1627-1705), who lived before Linnaeus. Ray described species as groups "with mutual fertility," meaning that a species is composed of individuals capable of interbreeding freely. Groups of organisms biologically incapable of interbreeding belong to different species. For our present purposes, this definition is satisfactory, but we need to realize that it is not always a practical one for taxonomy. Sometimes a population of one species is physically isolated from another that it closely resembles, so there would be no way to show that the two populations are, in fact, capable of interbreeding. In such cases, decision is based on comparison of the phenotypes, using such criteria as anatomy, metabolic processes, and behavior. In the case of fossils of extinct groups, decision as to which belong to the same species must be made entirely on the basis of body parts that have survived, or impressions of those parts. Naturally, there are differences of opinion between scientists on such matters of classification.

For the purpose of discussing evolution, we can define the population of a species as all individuals of that species actually capable of engaging in sexual reproduction, i.e., they form a reproductive community. No barriers exist within that community to prevent interbreeding.

The modern view on evolution is expressed by Francisco Ayala as follows:

Evolutionary change occurs in populations, not in individuals. Although individual organisms change more or less throughout their lifetimes, their genetic constitution remains constant. The genetic constitution of a population, on the other hand, may change from generation to generation by the processes of genetic mutation, migration, drift, and natural selection. . . . From the evolutionary point of view the individual is ephemeral; only populations persist through time. The continuity derives from the mechanism of biological heredity. (1977a, p. 31)

Within a reproductive community the gene pool is the sum total of the genotypes of all individuals it contains. Thus, if we agree that all evolutionary change occurs in populations, rather than in individuals, we must also say that evolution occurs in gene pools rather than in the individual genotypes. That this point needs to be reiterated is in consideration of a now-outmoded hypothesis of evolution to the effect that a single, massive mutation in one individual can immediately create a new species. Let us now examine the ways in which populations evolve.

Evolution in Gene Pools

Mutation is a random process, constantly going on regardless of the rate at which the population is evolving. Free recombination between genes tends to keep the gene pool well mixed. As mutations occur, they are included in the recombination process and tend to increase the number of adaptive combinations available for selection. Thus mutations enrich the diversity of the gene pool; they can be said to replenish that pool.

Natural selection will be tending to eliminate large and/or unfavorable mutations and to retain favorable ones, the latter spreading through the population. Those mutations that are neutral--neither beneficial nor harmful --can be retained for long periods of time while environment holds steady. Occurring at the normal rates, mutations can keep up with the depletion of the gene pool brought about by natural selection. Under the circumstances, the gene pool may (according to this model) show a slow change in overall composition in response to gradual environmental change, but for many modern geneticists, this form of change is not considered in itself sufficient to produce a new species.

What, then, can cause rapid evolution--a spurt, that is--capable of producing a new species? Geneticist G. Ledyard Stebbins puts it this way: "Rapid evolution results from a strong challenge generated by a rapidly changing environment and the presence of organisms with gene pools capable of meeting that challenge" (1982, p. 79).[1]

In terms of evolution, environment includes a wide range of influences. We tend to think of environment in terms of climate--warm or cold, wet or dry--or of soil quality, or of geologic processes of erosion and deposition that change the substrate or habitat. But environment also includes biologic factors, perhaps the most important of which is the availability of nutrients to sustain life. For a predatory animal, the environment includes the availability of prey in satisfactory quantity; for the prey, the presence and numbers of predators are part of the environment. Disease can be included as a part of the biological environment.

During times of rapidly changing environment the balance of an entire ecosystem can be seriously upset, placing severe stresses on many kinds of organisms. In such stressful times, genotypes that may have previously been neutral or nonadaptive assume the role of adaptive genotypes. When this happens, genetic change may be rapid and a new species can arise. Where the gene pool lacks this reservoir of varied genetic resources, the species may become extinct.

Punctuated Equilibrium or Gradualism?

The evolutionary model I have described has come to be described as punctuated equilibrium, implying that relatively long periods of genetic near-equilibrium in adjustment to a nearly constant environment are punctuated by relatively short periods of rapid genetic change. This concept was shaped and brought to public attention by two paleontologists, Niles Eldredge and Stephen J. Gould, who used the expression "punctuated equilibria" in the title of an article they published in 1971 (Eldredge and Gould, 1971). The idea was not entirely novel: it had been proposed nearly three decades earlier by paleontologist George G. Simpson; he called it quantum evolution (1944).

Punctuated equilibrium found itself in opposition to a longstanding theory or paradigm known as gradualism, which held that evolutionary change proceeds steadily through time. This view of evolution, which arose from Darwin's own thinking, is, of course, a prime target for attacks by creation scientists. They see gradualism as an essential part of uniformitarianism, as they mistakenly present that doctrine. Although Darwin supported gradual evolution, he argued that the geologic record is imperfect, with many gaps preventing the verification of gradual evolution. When the "modern synthesis" arrived in the early 1940s, wedding genetics with natural selection, it continued to include the concept of gradual evolution. The "gradualist-punctuationalist debate," as Gould calls it (1982, p. 383), resurfaces from time to time, and often with considerable fervor (Lewin, 1980).

Gould does not find punctuated equilibrium in any way inconsistent with Darwin's original version of evolution of species by natural selection, for he writes as follows:

> The modern theory of evolution does not require gradual change. In fact, the operation of Darwinian processes should yield exactly what we see in the fossil record. It is gradualism that we must reject, not Darwinism. The history of most fossil species includes two features particularly inconsistent with gradualism: 1. <u>Stasis</u>. Most species exhibit no directional change during their tenure on earth. They appear in the fossil record looking much the same as when they disappear; morphological change is usually limited and directionless. 2. <u>Sudden appearance</u>. In any local area, a species does not arise gradually by the steady transformation of its ancestors; it appears all at once and "fully formed."

> Eldredge and I believe that speciation is responsible for almost all evolutionary change. Moreover, the way in which it occurs virtually guarantees that sudden appearance and stasis shall dominate the fossil record. (1980, pp. 182-83)

The subject of rate (tempo) of evolution leading to the production of new species will come up again as we look into the ways in which rapid environmental change can be caused.

Let us continue on the theme of punctuated equilibrium. Just how is it supposed to work? First, how does the punctuation take place? The concept of <u>discontinuous</u> change seems to be involved, a hypothesis dating back to the early 1900s, when Darwinian natural selection was out of style and mutation was the evolutionary mechanism of choice. At that time a Dutch botanist, Hugo de Vries, advocated the appearance of a new species by sudden and drastic genetic change. In 1940, a leading zoologist, Richard Goldschmidt, proposed that for large evolutionary changes to take place, producing new species and genera, <u>macromutations</u> must occur (Stebbins, 1982, pp. 144-45). A macromutation, Goldschmidt supposed, involves three kinds of changes occurring at one time. First, parts of chromosomes are rearranged, causing drastic reordering of the genes. Second, these chromosomal changes largely affect the earliest, or embryonic, stages in the development of the organism. Third, there results a "monster" that can be recognized as a new species. While nearly all monsters are incapable of evolutionary survival, on very rare occasions a "hopeful monster" appears, is highly successful in adapting to the environment, and initiates a new species. Goldschmidt's hopeful monster was soon put to rest, as natural selection was put back into the driver's seat, but one sees vestiges of it in the new idea of punctuation.

Stephen J. Gould's picture of the punctuation process includes the following steps: A single important mutation occurs and spreads rapidly through the population; it "serves as a 'key' adaptation to shift its possessor toward a new mode of life" (1980, p. 191). (Does this sound like mutation is in the driver's seat?) Gould continues: "Continued success in this new mode may require a large set of collateral alterations, morphological and behavioral; these may arise by a more traditional, gradual route once the key adaptation forces a profound shift in selective pressures" (p. 191). Gould repudiates Goldschmidt's hypothesis but nevertheless retains the idea that small

genetic changes affecting early embryonic stages can greatly alter the rates of later development in the individual. (The genes involved are called "rate genes.") Thus, the adults will be profoundly different from the ancestors and the differences can have great adaptive value. But that greatly improved adaptive value can be fully realized only if it occurs against a background of environmental change.

There seems to be general agreement among modern evolutionists that "budding" or "splitting" to produce a new species takes place quickly and in small populations. But how is a small population formed from a very large one, such as would be typical during equilibrium periods? One answer lies in actual geographic migration of a small group of individuals of a species to an isolated region. For example, a herd of individuals of a herbivorous animal species might succeed in migrating across a barren desert to reach a distant, well-watered habitat capable of providing ample nourishment. The new group is now physically isolated from the parent population, preventing interbreeding; the situation is one form of <u>reproductive isolation</u>. (Reproductive isolation can also be achieved through development of chromosomal differences or behavioral differences.)

Perhaps the new geographical environment offers some important environmental differences over the previous environment. For example, the food plants may be of different textural or nutritive properties, or the habits and effectiveness of the indigenous predators may be different. At this point, genotypes that were previously neutral would become highly adaptive, spreading rapidly through the small new population and quickly changing its genetic character. Thus a new species would arise but, because of the rapidity of its formation and the small size of its population, would not be likely to leave a readily recoverable fossil record.

Modern geneticists recognize other forms of speciation that do not require geographic isolation by migration, but these are beyond the scope of our treatment, which has to be extremely brief.

What we have described is a hypothetical process of origin of a new species, better adapted to new or changed environments, that has branched from the line of a preexisting species. Typically, the ancestral species became extinct a short time thereafter, but often this was also the fate of the new species. Surviving species thus replace the old to carry on the lineage of the genus or higher taxa to which they belong.

This replacement program differs from an idealized model in which every species branches (bifurcates) into two species, both of which survive, so that the total number of species increases by simple doubling.

If species replacement had not been the dominant activity through geologic time, the number of different species surviving today would be astronomical. Stebbins makes a calculation based on the assumption that splitting occurred at the rate of one per five million years, and that all branches survived to the present:

> After the first 5 million years, there would have been two species; after 10 million years, four; after 15 million years, eight. If none of these species became extinct, and if each split to form two species once every 5 million years, the number of species among their descendants, after 1 billion years, would be 10 to the 60th power, or ten followed by 59 zeroes! (1982, p. 23)[1]

Stebbins goes on to say that the number of species existing today on earth is estimated as between one and 10 million. Using the latter figure, the ratio of potential living species that have either become extinct or have never been formed exceeds the number of living species by a factor of 10 to the fifty-third power. Stebbins continues:

We can only conclude that by far the commonest fate of every species is either to persist through time without evolving further or to become extinct. Based on his knowledge of fossil animals, Stephen J. Gould estimates that "more than 99.9 percent of species are not sources of great future diversity." Evolution is not a universal property of life, like self-reproduction, growth, and individual response to the environment. Its most significant changes result from unusual combinations of events. (1982, p. 23)[1]

But species replacement has not been on a one-to-one basis through time (we would now have only one species). Diversity--seen in the increase in number of branches in the tree of life--has been an important characteristic of the course of evolution. With it has come the kind of multiple branching described in an earlier section, in which hierarchies are identifiable on the basis of relative evolutionary distance of separation. We recognize these degrees of relatedness or unrelatedness in the various levels of the phylogenetic classification.

The evolutionary trend in the earth's biota has been one of enrichment by diversity, in which the budding, or branching of species has led to an increase in the numbers of species and, in turn, to an increase of genera and higher taxa. But the enrichment has not proceeded at a steady pace. As we saw in tracing the evolution of groups through the fossil record, the pace of enrichment has been punctuated by episodes of displacement and replacement, which we referred to earlier as "extinctions followed by radiations." Perhaps we need to recognize a hierarchy of levels of punctuated equilibrium--perhaps a level for each of the taxa.

Looking a bit more closely at the process of displacement and replacement, we find an intensifying effect that is a result of environmental interactions--a kind of feedback effect. The rise of a new kind of dominant plant or animal has often stimulated the rapid evolution and diversification of other kinds of organisms that are a part of its environment. This interaction has been called coevolution. We have described examples of coevolution in reviewing the fossil record. For example, the rise of extensive prairie grasslands in Cenozoic time must have stimulated the radiation of grazing ungulates, which in turn stimulated radiation in their associated predator groups. We see this interaction particularly clearly in the case of the changes in dentition and leg structure of the horse, which adapted to the need for cropping coarse stems of grass and for running swiftly to escape predacious carnivores.

During rapid and stressful coevolution, many extinctions occurred because a species had become too narrowly specialized to permit rapid adaptation to changing environmental conditions. Often cited as an example is the group of saber-toothed cats that lived from Oligocene through Pliocene time, successfuly adapted for about 35 million years in the role of dominant carnivores (Colbert, 1955, pp. 324-25). The culmination of their specialized evolution was achieved during Pleistocene time in Smilodon, an animal as large as a modern lion. Its long upper canine teeth served as curved sabers that were bared by a remarkable dropping back of the jaw, allowing the sabers to be brought straight down with full force on the prey. Thus armed, it could successfully attack large, slow-moving elephantlike animals and giant sloths, easily penetrating their thick hides. But with the coming of the Ice Age, those animals on which Smilodon depended for food became extinct. Other predatory cats, swifter and smaller, were better able to kill the remaining prey. As a result Smilodon, too, became extinct.

The idea of coevolution was strongly presented by Darwin in his description of the process of natural selection. Stebbins summarizes the importance of coevolution in these words: "Most, if not all, examples of continued evolution in a particular direction are based on such coevolutionary, reciprocal challenges between unrelated kinds of animals or plants that inhabit the same ecosystem" (1982, p. 116).

As to the creationists' position on the topics we have covered in this chapter, it ranges from approval to rejection, depending on which topic is involved. The creationist textbook, Biology; A Search for Order in Complexity, has a full chapter on the science of genetics, which is conventional and thorough in the area of principles of heredity (Moore and Slusher, 1970, pp. 83-104). Emphasis is given to selective breeding (artificial selection), perhaps because it throws light on their acceptance of a limited amount of genetic change occurring naturally in species. They call this form of minor change in the genotype "microevolution within kinds," a subject we cover in Chapter 37. The creationists' text strongly maintains that all mutations are harmful, even those that may seem to be in some respect beneficial. Most important, perhaps, is their stated position that success in selective breeding of plants and animals does not in any way give evidence favoring the theory of evolution (p. 93).

The evolutionists' hypothesis of punctuated equilibrium is strongly attacked by the creationists as a flimsy excuse for the presence of major gaps in the fossil record. This is a topic we discuss at some length in Chapter 41.

Credit

1. From G. Ledyard Stebbins, Darwin to DNA, Molecules to Humanity. Copyright © 1982 by W. H. Freeman and Company. Used by permission.

Chapter 36

Molecular Biology and Phylogeny

The discovery of DNA and its role in coding for the synthesis of proteins brought with it a new and powerful tool for the study of phylogeny. Through biochemistry there has emerged in only the last two decades a remarkable means for corroborating the pattern of organic evolution through geologic time, already well established through the study of comparative anatomy and embryology in living organisms and fossils. The traditional approach can be referred to as <u>macrobiology</u>, the newer as <u>molecular</u> biology, and the corroboration of one by the other has been hailed as "the beginning of a grand new synthesis in evolutionary biology" (Lewin, 1982, p. 1091).

Homology and DNA

It is easy to understand how homology can be applied in comparative anatomy--matching corresponding bones or muscles in two or more related groups and arranging them in a developmental sequence. But how does one go about practicing homology with molecules of DNA or protein? The protein molecule consists of a chain of amino acids. There are 20 different amino acids, so the probability that a given amino acid will appear at a particular position in the chain is one in 20. For two adjacent positions to acquire the same amino acids in each, by chance alone, is one in 400 (multiply 20 by 20). Commonly, protein chains consist of from 100 to 160 amino acids. The probability that by chance alone two identical chains of 100 amino acids would originate independently is extremely minute. Thus homologous relationships can be found in comparing the amino acid sequences of proteins in different groups of organisms.

Because the assembly of protein molecules in the cell is coded for by the DNA and RNA molecules, comparisons of the similarities and differences between DNA molecules in two groups of organisms can also serve to establish homologous relationships. Before getting into methods of establishing phylogenies that make direct use of protein amino acid sequences and DNA, we will review an earlier method of practicing molecular biology for that purpose.

Immunological Differences

When the protein albumen from an animal--the human, for example--is purified and injected into another mammal --a rabbit, for example--the recipient shows an immunological reaction that causes it to produce antibodies against the foreign protein, i.e., it produces an antigen. These antibodies, in turn, will react against antibodies from other mammals. This principle was known as early as 1904, when George H. F. Nuttall was able to demonstrate that immunological differences might be used to classify animals on a biochemical basis (Washburn, 1978, p. 202).

Without going into further detail, we can simply say that there are various degrees of dissimilarity between human albumin and that of other species, and these dissimilarities can be expressed in terms of "immunological distances." This information can, in turn, be used to construct a phylogenetic tree.

Morris Goodman of Wayne State University set a chain of studies in motion in 1962 by his announcement that protein albumin studies showed humans, gorillas, and chimpanzees to be closely related to one another. A study carried out in the mid-1960s by Vincent Sarich and Allan Wilson of the University of California at Berkeley yielded the results shown in Figure 36.1. Their immunological data indicated that human, chimp, and gorilla are more closely related to one another than any one of the three is to the orangutan. Another study reported in 1974, using the enzyme lysozyme, shows the immunological distance to be particularly small between human and chimp (Wilson and Prager, 1974, as presented in Ayala, 1977c, pp. 292-93). Other methods of arriving at a phylogeny of the same primate groups generally support the interpretation based on comparison of immunological distances, although with some important differences in the details of the phyletic trees that are constructed from the data.

Amino Acid Sequences

Amino acid sequences in proteins have proven extremely useful in establishing phylogenies, extending them over almost the entire spectrum of life. In our earlier review of protein chemistry, we found that the same twenty kinds of amino acids are found in all organisms.

Amino acid sequences have now been determined for a large number of proteins. The sequence for insulin was the first to be established for a large, complex protein molecule (see Figure 34.4.). We took note of the differences in the insulin sequence for different animals.

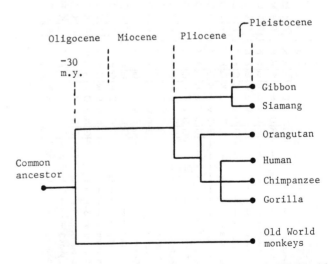

Figure 36.1 Phylogeny of humans, apes, and Old World monkeys, based on immunological differences between their albumin proteins. (From V. M. Sarich and A. C. Wilson, <u>Science</u>, vol. 158, p. 1202, Figure 1. Copyright © 1967 by the American Association for the Advancement of Science. Used by permission.)

Figure 36.2 Amino acid sequences of cytochrome c proteins from twenty different species of organisms. (From T. Dobzhansky, F. J. Ayala, G. L. Stebbins, and J. W. Valentine, EVOLUTION, pp. 296-97, Figure 9-18. Copyright © 1977 by W. H. FREEMAN AND COMPANY. Used by permission.)

These differences immediately suggest a relationship to evolutionary distances of separation in their phylogeny. By 1972, about 500 amino acid sequences or partial sequences had been established; many more have been completed in every year since then (Ayala, 1977c, p. 294).

Although the same basic molecular construction of a given protein is recognized in many different organisms, the molecules can differ in one of two ways: (a) they show replacement of one or more amino acids; (b) they show additions and/or deletions of amino acids. The latter class of differences causes the total sequence to differ in length. Using sophisticated statistical techniques, amino acid sequences can be matched in an optimum manner, revealing the replacements, additions, and deletions.

We will use a single example to illustrate the method and its use in homology on the molecular scale. Cytochrome c is a protein molecule involved in cell metabolism. Its acid sequence has been determined for 20 organisms, a list so diverse as to include fungus, wheat, and various insects, fishes, reptiles, birds, and mammals. These are named in Figure 36.2. The number of acids in the sequence ranges from 103 in the frog to 112 in wheat, while most vertebrates have 104. The amino acid sites are numbered from 1 through 112. When the sequences for all 20 organisms are matched up, they prove to have the same amino acids in 37 sites; this is the total fully shared identity. Figure 36.2 shows the complete data chart, allowing you to see how the sequences differ. Comparing the sequence for humans with that for the rhesus monkey, there is complete identity in 104 sites; the only difference is in the 66th position, where the human has isoleucine, whereas the rhesus has threonine. Although the degree of difference increases downward in the list, the number of shared positions is remarkably large throughout. Humans and baker's yeast--about as far apart as you can get in the spectrum of life forms that have a nucleus in the cell--share identical amino acids in 64 positions.

One way to measure the level of difference, or evolutionary distance, shown by the amino acid sequences is simply to count the number of positions in which differences exist between any two sequences, including the presence or absence of a given amino acid. While this would allow a phylogeny to be constructed, it is not the best way to infer the corresponding differences in the DNA of each organism. The reason is that a given amino acid can be coded for by more than one DNA or RNA codon. For example, the replacement in position 66 of threonine in the rhesus monkey by isoleucine in the human involves the substitution of one of three RNA codons in isoleucine for one of four codons in threonine, thus:

Isoleucine	Threonine
AUU	ACU
AUC	ACC
AUA	ACA
	ACG

The minimum number of nucleotide differences between the two sets of codons is one. Compare this case with the substitution in position 20 of methionine in the human for glutamine in the horse:

Methionine	Glutamine
AUG	GAA
	GAG

Here, the minimum number of nucleotide differences is two. A better estimate of the substitutions that actually occurred in the DNA of the two organisms being compared is obtained by noting the total minimum nucleotide differences between two organisms. This number is called the mutation distance. Actually, we have no way of knowing that the differences we have counted represent the actual number that occurred during evolution; there may have been a sequence of replacements through time. Thus the method tends to underestimate the actual mutation distance.

Figure 36.3 is a two-way table of minimum numbers of nucleotide susbtitutions in genes coding for cytochrome c. It represents the research effort of Walter M. Fitch and Emanuel Margoliash of Northwestern University. The table gives all possible mutation distances between any two organisms. Remember, the mutation distance sums the distances of the two tree branches to their common fork. Figure 36.4 is a phylogenetic tree constructed by Fitch and Margoliash from the distances given in the table (1967). The distances assigned to each segment of the tree are not drawn to true length scale, but if you sum the segment distances from, say, the human to Neurospora, they should total 63. (Actually, the total is 62.7.) Minor adjustments have been required in assigning lengths to individual segments of the tree, including negative values in two cases. (Did evolution back up a couple of times?)

Overall agreement with phylogenies prepared from comparative anatomy and the fossil record is judged to be good (Ayala, 1977c, p. 302). On the other hand, there are certain peculiar disagreements with the fossil record, described as follows by Francisco Ayala:

> Chickens appear more closely related to penguins than to ducks and pigeons; the turtle, a reptile, appears more closely related to the birds than to the rattlesnake; men and monkeys diverge from the other mammals before the marsupial kangaroo separates from some placental mammals. (1977c, p. 299)[1]

As we will see in a later section, creation scientists are delighted with this list of discrepancies, which they cite in proof of the total inadequacy of the method to support evolution. But Ayala concludes his paragraph on an up-beat note:

> In spite of these erroneous relationships, it is remarkable that the study of a single protein yields a fairly accurate representation of the phylogeny of twenty organisms as diverse as those in the figure. (1977c, p. 299)[1]

Whereas the cytochrome c molecules have evolved very slowly and are excellent for phylogenies going far back into geologic time, the same slowness of mutational change makes them of little value for phylogenies of closely related groups. For the latter, we can turn to rapidly evolving proteins. One of these is carbonic anhydrase I (CA I), for which the amino acids have been determined for 115 of the 260 positions for the human and five other primates. Using the estimated numbers of nucleotide substitutions, a phylogenetic tree has been drawn, as shown in Figure 36.5.

The phylogenies based on amino acids of cytochrome c and CA I agree rather well with phylogenies inferred from the fossil record. We must regard this success as a remarkable achievement of molecular biology in seeking corroboration of the Darwinian theory of descent with modification. Keep in mind that the molecular data are totally independent as a class of scientific information from the data of both comparative anatomy and the fossil record. You can think of the molecular data as making possible a test of a scientific hypothesis. Once the determination of amino acid sequences became a possibility, it could be predicted that numbers of nucleotide differences of two organisms should confirm the phylogeny derived from the fossil record. The test was passed with flying colors.

	1. Human	2. Monkey	3. Dog	4. Horse	5. Donkey	6. Pig	7. Rabbit	8. Kangaroo	9. Duck	10. Pigeon	11. Chicken	12. Penguin	13. Turtle	14. Rattlesnake	15. Tuna	16. Screwworm fly	17. Moth	18. Neurospora	19. Saccharomyces	20. Candida
1. Human																				
2. Monkey	1																			
3. Dog	13	12																		
4. Horse	17	16	10																	
5. Donkey	16	15	8	1																
6. Pig	13	12	4	5	4															
7. Rabbit	12	11	6	11	10	6														
8. Kangaroo	12	13	7	11	12	7	7													
9. Duck	17	16	12	16	15	13	10	14												
10. Pigeon	16	15	12	16	15	13	8	14	3											
11. Chicken	18	17	14	16	15	13	11	15	3	4										
12. Penguin	18	17	14	17	16	14	11	13	3	4	2									
13. Turtle	19	18	13	16	15	13	11	14	7	8	8	8								
14. Rattlesnake	20	21	30	32	31	30	25	30	24	24	28	28	30							
15. Tuna	31	32	29	27	26	25	26	27	26	27	26	27	27	38						
16. Screwworm fly	33	32	24	24	25	26	23	26	25	26	26	28	30	40	34					
17. Moth	36	35	28	33	32	31	29	31	29	30	31	30	33	41	41	16				
18. Neurospora	63	62	64	64	64	64	62	66	61	59	61	62	65	61	72	58	59			
19. Saccharomyces	56	57	61	60	59	59	59	58	62	62	62	61	64	61	66	63	60	57		
20. Candida	66	65	66	68	67	67	67	68	66	66	66	65	67	69	69	65	61	61	41	

Figure 36.3 Minimum numbers of nucleotide differences in the genes coding for cytochrome c in organisms listed in Figure 36.2. (Data of W. M. Fitch and E. Margoliash, Science, vol. 155, p. 281, Table 3. Copyright © 1967 by the American Association for the Advancement of Science. Used by permission.)

Is There a Molecular Clock?

The possibility that protein evolution might serve as a "molecular time clock" of evolution has been given a great deal of attention by molecular biologists (Ayala, 1977, pp. 308-13). The idea seems simple enough. If mutations occur at a constant rate, one has only to tie in a certain number of nucleotide substitutions with a point in time identified by a radiometric age, where an evolutionary event is recorded in the fossil record, for example, a point of branching of two distinct lines of organisms. This age allows the evolutionary clock to be calibrated, after which it becomes an independent timekeeper.

In the absence of evidence to the contrary, the assumption of constant rates of gene mutations over geologic time is reasonable as a point of departure. Under the neutrality theory of protein evolution, favored by several geneticists in the late 1960s and early 1970s, a large fraction of gene mutations is neutral in terms of natural selection: they do not affect the fitness of the carrier. Whereas mutants that are harmful are eliminated or kept to low values by natural selection, the neutral mutations are retained and make only very small, unimportant changes in the amino acid sequences. Favorable mutations, which are very rare, have little effect on the rate of amino acid substitutions. According to the neutrality theory, the rate of substitution of neutral alleles exactly matches the rate at which those alleles arise by mutation. A consequence of this conclusion is that the rate of evolutionary change in the amino acid sequence of a protein should hold constant. Thus, the number of nucleotide substitutions that must be assigned to each segment of the phylogenetic tree is in direct proportion to the elapsed geologic time. It should be made clear, however, that under this theory mutations expressed as amino acid substitutions do not occur at exactly equal intervals of time. The intervals are subject to random variations according to a stochastic model, but the rate would appear essentially constant over long periods of time. In this respect, the process is similar to radioactive decay, in which the decay "constant" is the expected rate over a long time period or for a large number of atoms.

Although early protein studies seemed to support the neutrality theory and its constant rates of mutations, newer studies seemed to show that, instead, molecular

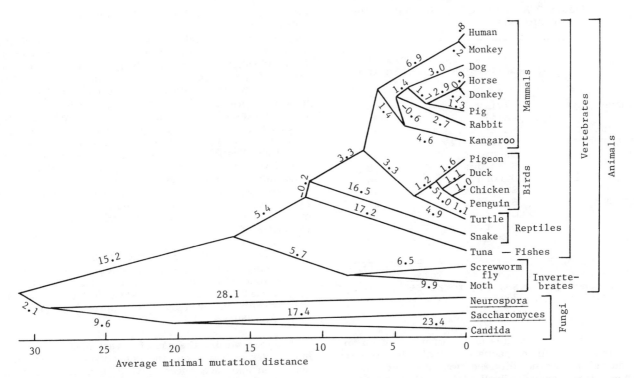

Figure 36.4 Phylogenetic tree for twenty organisms, based on cytochrome c matrix of Figure 36.3.(From W. M. Fitch and E. Margoliash, Science, vol. 155, p. 282, Figure 3. Copyright © 1967 by the American Association for the Advancement of Science. Used by permission.)

evolution is not constant. Deviations from a constant rate have been judged to be statistically significant (Ayala, 1977c, p. 311). A newer appraisal of the status of the protein "clock" by science writer Roger Lewin reads as follows:

> It was not a metronome but an approximate, sloppy clock. . . . Powerful though the various protein clocks were, their degree of resolution was inevitably limited because they are one step removed from the detailed structure of the gene. (1984, p. 1179)

Thus, evidence suggests that while changes in amino acid sequences are approximately correct in a relative sense and can be used to assist in construction of phylogenetic trees, firm decisions on the time-location of the points of branching should be made consistent with the fossil record and its fixation in time by radiometric ages. This caution notwithstanding, molecular phylogeny seems a genuinely independent method of testing and corroborating the theory of evolution by natural selection. That it cannot at this time be accepted as an accurate "molecular clock" does not reduce its value in lending qualitative support to phylogenies arrived at by comparative anatomy and biostratigraphy.

DNA Phylogenies

Turning finally to DNA and its genetic code, how can a scientist compare the DNA of one kind of organism with that of another? Recall the analogy used earlier of the DNA molecule as a slide fastener, or zipper. Suppose we wished to compare two different zippers. We would separate the halves of each and see if the left half of one pair could be made to mesh with the right half of the other. It is possible in the laboratory to cause the DNA of an organism to separate into single strands (the process is called "denaturing"). The single strands of one

organism can be mixed with those of another, allowing the unlike strands to lock together along those sections in which the codons are in the same order. After this has happened, the single strands that have not matched and joined are washed out, and the ratio of binding of the two kinds of DNA is determined. (Use of a radioactive tracer in the DNA of one organism allows the two materials to be distinguished from each other.) For two individuals of the same species, the ratio of binding will be close to 100 percent; for two different species, or other taxa, the percentage will be less than 100 percent and will be a smaller percentage in proportion to the evolutionary distance separating the two. The principle behind the observed difference in level of homology is that DNA undergoes change through mutations, so that the longer the elapsed time since the two groups split into two lines from a common ancestor, the greater will be the difference in their DNA.

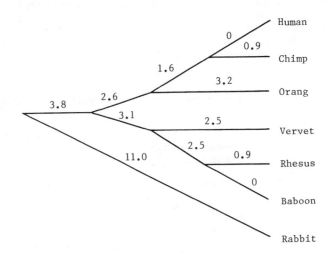

Figure 36.5 Phylogeny of six primates and rabbit, based on sequence of amino acids in carbonic anhydrase I. The numbers give nucleotide replacements (per 100 codons) that have occurred during evolution. (Based on data of R. E. Tashian, M. Goodman, R. E. Ferrell, and R. J. Tanis, in Molecular Anthropology, M. Goodman, R. E. Tashian, and J. H. Tashian, eds. Copyright © 1976 Plenum Press New York.)

Table 36.1   Human DNA Compared with That of Some Other Animals

	Taxonomic difference	Percent DNA binding
Man	--	100
Chimpanzee	Family	100
Gibbon	Family	94
Rhesus monkey	Superfamily	88
Capuchin monkey	Superfamily	83
Tarsier	Suborder	65
Slow loris	Suborder	58
Galago (a loris)	Suborder	58
Lemur	Suborder	47
Tree shrew	Suborder	28
Mouse	Order	21
Hedgehog	Order	19
Chicken	Class	10

Based on data of B. H. Hoyer and R. B. Roberts, pp. 425-79 in H. J. Taylor, ed., Molecular Genetics, Academic Press, New York. See Ayala, 1977c, Table 9-2, p. 279.

Table 36.1 shows the results of a comparison of human DNA with that of 12 other species of animals. Between the human and chimpanzee no difference was observed, but in the gibbon, also of superfamily Hominidae, the percentage of binding was 94. It dropped progressively through monkeys of the same superfamily to smaller values in the suborder of prosimians, to much smaller values in the lower mammal orders, and finally to only 10 percent in the chicken (class Aves).

Another technique based on binding of DNA of two groups examines rates of increase in the percentage of bound strands released as the material is heated through a temperature range from 60 to 100 C. Data for humans and six other primates obtained by this technique are shown in Table 36.2. The percentages of nucleotide changes allow the rates of changes to be estimated, using paleontological data of the times at which evolutionary branching occurred, as shown in the two columns at the right. Note that the rate at which changes have occurred slows down considerably from oldest to youngest age of divergence. Actually, this apparent decrease is partly related to the longer generation time in progressively higher groups shown, the longer generation time being associated with greater size of the individual. Figure 36.6 is a phylogenetic tree constructed from the data of Table 36.2. The numbers tell the estimated percentage of

nucleotide-pair substitutions that have occurred in each link of the tree.

Whereas the protein phylogenies have remained suspect as to their reliability as molecular clocks, the method of DNA hybridization has recently gained in strength in that area of concern. In 1983, Charles Sibley and Jon Ahlquist of Yale University's biology department completed a study in which more than 20,000 hybridization tests were run on 1,600 species of birds in an attempt to clarify the phyletic relationships. Their quest seems to have been highly successful. Sibley and Ahlquist claim that the method of DNA hybridization overcomes the major criticism of the use of DNA mapping or sequencing events--that mutation rates differ among the various classes of genes and that the rates within a single class are not time-uniform. The hybridization method uses a large proportion of the DNA of the entire genome, which in mammals may contain on the order of two billion nucleotides. Lewin states:

> Because of the very large number of nucleotides being compared, each of which is effectively a single data point, the DNA hybridization technique immediately commands a statistical robustness not readily achieved by other approaches. (1984b, p. 1180)

In other words, the use of a vast number of nucleotides averages out the departures in rates shown by individual gene sequences, giving "a uniform average rate of change, which is likely to be the same in all classes of animals" (Lewin, 1984b, pp. 1180-81).

Sibley and Ahlquist calibrated their DNA clock by means of a geological event--the opening up of the South Atlantic Ocean 80 million years ago, forming an impassable barrier between the ostrich in Africa and the rhea in South America (both are flightless birds). Among the hominoids, calibration was based on fossil evidence of the origin of the orangutan, 13 to 16 million years ago. It is tempting to present here this latest hominoid phylogeny, but it is best held back for use in our later discussion of the evolution of the hominoids in Chapter 49.

The Creationists' Attack on Molecular Phylogeny

Perhaps the first major attack mounted by the creationists on molecular biology and its evolutionary implications was that of Robert E. Kofahl and Kelley L. Segraves in their 1975 book titled The Creation Explanation. In it appears a direct challenge to a 1972 Scientific American article by Richard E. Dickerson explaining how cytochrome c reveals the progress of evolution. For an answer to that challenge, I turn to an

Table 36.2   Rates of Nucleotide Change in the Evolution of Primate DNA

Comparison of human and	Millions of years since divergence	Percent nucleotide changes	Percent change per m.y.	Nucleotide changes per y.
Human	0	0	-	-
Chimpanzee	15	2.4	0.08	1.6
Gibbon	30	5.3	0.09	1.8
Green monkey	45	9.5	0.11	2.1
Capuchin	65	15.8	0.12	2.4
Galago	80	42.0	0.26	5.2

Based on data of D. E. Kohne, J. A. Chiscon, and B. H. Hoyer, 1972, Jour. of Human Evolution, vol. 1, pp. 627-44. See Ayala, 1977c, Table 9-3, p. 283.

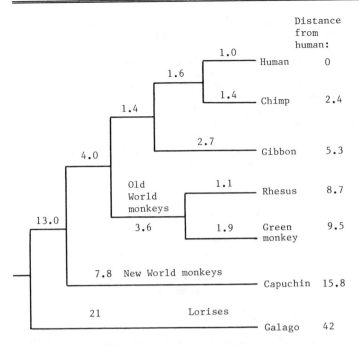

Distance from human:

		1.0	Human	0
	1.6			
		1.4	Chimp	2.4
1.4				
	2.7		Gibbon	5.3
4.0				
	Old World monkeys	1.1	Rhesus	8.7
	3.6	1.9	Green monkey	9.5
13.0				
7.8	New World monkeys		Capuchin	15.8
21	Lorises			
			Galago	42

Figure 36.6 Phylogeny of several primate species, based on thermal studies of hybrid DNA. Figures give percent change in nucleotides. (Data of D. E. Kohne, J. A. Chiscon, and B. H. Hoyer, 1972, Evolution of Primate DNA Sequences, _Jour. of Human Evolution_, vol. 1, p. 635, Table 4 and Figure 4.)

article by Professor Thomas H. Jukes, biochemist on the faculty of the University of California at Berkeley (Jukes, 1983). In reviewing the subject of cytochrome c, Jukes states: "Given the slow rate of change of cytochrome c, it should come as no surprise that the cytochrome c molecules in humans are identical with those of chimpanzees" (pp. 130-31). That particular statement by itself is enough to enrage any fundamentalist creationist. Small wonder that all molecular biology evokes anathema from the creationists!

The clearest way to present the creationists' argument and an evolutionist's reply is to quote an entire long paragraph by Professor Jukes:

On pages 167 and 168 of their book, Kofahl and Segraves document the structure of cytochrome c by reproducing Tables I and II from Dickerson's article. They state that "the segment from position 70 to 90 is invariant, for it is the heart of the active center of the enzyme molecule," and that "there is no evidence for the evolutionary development of the overall structure nor of the active center. . . . Thus the evidence supports the view that cytochrome c was designed, not evolved." The constancy of residues 70 and 80 for most species is well known, and the evolutionary interpretation is, of course, that mutations in this region are lethal and hence do not persist. Kofahl and Segraves explain the constancy of this region by saying that "cytochrome c was designed, not evolved," implying that the constancy of this polypeptide sequence of eleven amino acids was a decision specifically made by God and that "the invariance of all the cytochrome c's in their parts is evidence for intelligent, purposeful design." Why constancy of structure should be evidence of design rather than stability mediated by natural selection is not clear, but if Kofahl and Segraves mean to imply that variability of this segment would constitute falsification of their hypothesis of design, then their hypothesis must be taken as having been falsified. (P. 132)[2]

How falsified? All we need to do, says Jukes, "is to go back further in evolutionary time, to protozoa, and constancy of residues 70 to 80 disappears" (p. 132). Jukes does just that by giving complete data to back up his statement (pp. 132-33). Clearly Kofahl and Segraves have been caught in a technical error. The record, says Jukes, shows that evolution of the cytochrome c sequence takes place very slowly. In a last parting shot, Jukes remarks: "Why not conclude that the Creator planned that evolution could take place, rather than concerning Himself with the invariance of amino acids 70 to 80 in the cytochromes c of some, but not all, eukaryotes?" (P. 133) A good question, but it falls on deaf ears. I strongly recommend that anyone interested in the general subject of molecular evidence for evolution read all of Professor Jukes's article. It is an excellent example of how a busy research scientist has taken time to present to the public the latest knowledge in a rapidly changing field and to use the data to refute the claims of the creation scientists.

To the creation scientists of the Institute for Creation Research, molecular phylogeny is a red flag waving over the evolutionists' camp, inviting yet another joust with that enemy. In a 1982 issue of _ICR Impact Series_ titled "Molecular evolution?" Clete Knaub and Gary Parker rise to the challenge. Gary Parker is a familiar figure as Professor of Biology in the ICR Graduate School; Clete Knaub was then a graduate student in biology in that school. Their opening paragraph reads as follows:

Molecular homology has been acclaimed as the field of study that saved the house of evolution from collapsing by serving as an independent check that confirms evolution to be a fact.* What is molecular homology? Is it an independent confirmation of evolution? Can it "clock" the course of evolution? (1982, p. i)

(*Reference is to microbiologist-creationist Glen Klassen's article titled "Scientific Creationism vs. Evolution" in the October, 1981, issue of _Mennonite Mirror_.)

The _Impact_ article gives a good summary of the methods of molecular phylogeny we have reviewed on earlier pages of this chapter. The authors have no difficulty in finding fault with phylogenies based on the amino acid sequences; they quote almost verbatim Francisco Ayala's complaints about the cytochrome-c phylogeny as derived by Fitch and Margouliash (1967) and shown in our Figure 36.4. Recall that the turtle is shown as more closely related to the birds than to the snake, the chicken is grouped with the penguin rather than the duck, and so forth (see earlier Ayala quote). To this we can add a statement by Jukes that "it is well known that the rate of evolution of cytochrome c does vary in different species of vertebrates, as, for example, turtles and rattlesnakes (Jukes and Holmquist, 1972)" (1983, p. 133).

To the extent that I can judge the situation, there is nothing in Knaub and Parker's list of flaws in molecular phylogeny that has not been fully and openly presented by the mainstream molecular biologists, for example, Ayala's review of the methodology from which we have drawn excerpts (Ayala, 1977c). What is completely missing from the creationists' presentation is the strong vote of confidence given by the mainstream scientists to the general results of the molecular methods as strong corroboration of evolution. There is not even an attempt by the creationists to explain away the remarkable correlation that the great bulk of the molecular data shows with the classical evolutionary interpretation. Instead, the creationist authors pick and choose their quotations to maximum advantage, as in the following example by Knaub and Parker:

Reviewing recent data before a prestigious group at the American Museum, Nov. 5, 1981, Colin Patterson, himself an evolutionist, stated that if only the data of molecular homology were considered, then descent from common ancestry, the foundational concept of evolution, was "precisely falsified." (1982, pp. ii-iii)

Only the last two words of the above statement are Patterson's own words, and their context is suspect because of the circumstances surrounding his presentation. The alleged views and opinions of Colin Patterson, senior paleontologist of the British Museum, were recapitulated by Luther Sunderland and Gary Parker in a previous issue of ICR Impact Series (no. 108, June, 1982), titled "Evolution? Prominent Scientist Reconsiders." According to this ICR report, no official transcript of the speech was made available to the public, but the authors (Sunderland and Parker) had access to a nonofficial tape recording (their own?) and were able to cross-check it against several other nonofficial (personal?) tapes. Patterson, they say, declined to check their edited tape. Thus, Sunderland and Parker have had pretty much of a free editorial hand in relating to us what Patterson actually said. They say of Patterson: "He had been invited to speak on evolution and creation because, as he put it, he had been kicking around non-evolutionary and even anti-evolutionary ideas for about eighteen months" (p. i). They quote Patterson as having made the following remark in an earlier meeting of evolutionists in Chicago in 1980: "Can you tell me anything you know about evolution, any one thing, any one thing that is true?" (Sunderland and Parker, 1982, p. i).

Our topic here is molecular phylogeny, and the question is: Did Patterson, a mainstream paleontologist, actually say that the theory of evolution--descent with modification--stands falsified by molecular biology? According to Sunderland and Parker, their tape contains the following concluding statement by Patterson: "The theory makes a prediction, we've tested it, and the prediction is falsified precisely" (p. iii). The "theory" in this quote is clearly the theory of evolution. The "prediction" seems to be that under evolutionary theory "those forms more recently descended from a common ancestor have a greater similarity among their genes and gene products than those more distantly related." (Words in quotes are those of Sunderland and Parker, 1982, p. iii.) The test to which Patterson refers is obviously the output of the molecular biologists in establishing phylogenetic trees based on protein amino acid sequences, dissociation of hybridized DNA, and the other methods we have covered. The alleged falsification of the theory of evolution arises because the molecular phylogenies show some rather glaring discrepancies when compared with the standard phylogenies based on homology of living and fossil forms. Let us get the details as reported in the following paragraph by Sunderland and Parker:

Common ancestry falsified!? The first example Patterson used in leading up to his dramatic conclusion involved data obtained in Ann Arbor only a month earlier on the amino acid sequences for the alpha hemoglobins of a viper, crocodile, and chicken. On the basis of evolutionary theory we "know" that vipers and crocodiles, two reptiles, are much more closely related to each other than either is to a bird which is presumably a much more distant relative. But this evolutionary knowledge turns out to be "anti-knowledge" and the theory an "anti-theory." (P. iii)

The text goes on to say that the crocodile and chicken showed the greatest similarity, the two reptiles the least. The text continues: "In this particular example, the

evolutionary 'prediction is falsified precisely,' insisted Patterson." There follow several more examples from Patterson; these in turn are followed by Patterson's disclosure of manipulation of data in studies of mitochondrial DNA. Patterson is said to have followed with more examples of discrepancies in molecular phylogeny. Patterson is then quoted as saying: "Something is wrong with the theory." Is he referring to theory of molecular phylogeny or to the general theory of evolution? Sunderland and Parker cannot resist substituting their own words for Patterson's: "It certainly would appear that if evolution is a scientific theory that can be falsified, then it has indeed been falsified" (p. iv). The Impact text ends with praise for Dr. Patterson:

Dr. Patterson's open and objective approach to the evidence seems to embody the scientific spirit we encourage in students, but so rarely see in practice among professionals. It's a sad cliche in science to say that old theories never die, only their proponents do. So it is easy to understand the growing interest in creation science among science students still formulating their personal and professional paradigms. But science has precious few examples of scientists who, because of the evidence at hand have made paradigm shifts at the peak of their professional careers. Dr. Patterson is thus to be highly commended for his intellectual honesty and the courage to face possible ridicule from his scholarly peers. (P. iv)

Not being privy to what actually went on at the meeting described, and not being able to read the full and official transcript of Patterson's speech, I could only guess what the thrust of his argument was. My uncertainty was removed and my confidence in Patterson fully vindicated by an article in Creation/Evolution Newsletter (vol. 4, no. 4, pp. 4-5) written by its editor, Karl D. Fezer of Concord College, Athens, West Virginia. It seems that following the appearance of the Sunderland/Parker article, several scientists wrote to Patterson asking for clarification of his position. Patterson's reply sent to Steven Binkley of Burlington, North Carolina, contained these comments:

Obviously I have not helped you fight your local creationists--sorry. The story behind the 'Impact' article is that last November I gave a talk to the systematics discussion group in the American Museum of Natural History. I was asked to talk on 'evolutionism and creationism,' and knowing the meetings of the group as informal sessions where ideas could be kicked around among specialists, I put a case for difficulties and problems with evolution, specifically in the field of systematics. I was too naive and foolish to guess what might happen: the talk was taped by a creationist who passed the tape to Luther Sunderland. Sunderland made a transcript, which I refused to edit, since it was pretty garbled, and since I had no exact record of what I did say. Since, in my view, the tape was obtained unethically, I asked Sunderland to stop circulating the transcript, but of course to no effect.

There is not much point in my going through the article point by point. I was putting a case for discussion, as I thought off the record, and was speaking only about systematics, a specialized field. I do not support the creationist movement in any way, and in particular I am opposed to their efforts to modify school curricula. In short, the article does not fairly represent my views. But even if it did, so what? The issue should been resolved by rational discussion, and not by quoting 'authorities,' which seems to be the creationists' principal mode of argument. (Fezer, 1984, pp. 4-5; Patterson, C., 1984)

The news article goes on to report a dialog between Patterson and contributing editor Tom Bethell of Harper's magazine, who had included the Patterson incident in a February 1985 anti-evolution article. According to Bethell, Patterson said: "I don't think we shall ever have access to any form of [phyletic] tree which we can call factual." Bethell then asked: "Do you believe it to be, then, no reality?" Bethell states that Patterson replied: "Well, isn't it strange that this is what it comes to, that you have to ask me whether I believe it or not, as if it mattered whether I believe it or not. Yes, I do believe it. But in saying that, it is obvious it is faith."

Fezer goes on to evaluate a 1982 letter he received from Patterson that "indicates that Patterson (like your editor) is among those who prefer to restrict the use of the word 'fact' to direct observations and perhaps the most directly obvious generalizations based on them" (1984, p. 5). Fezer explains:

A system of explanatory concepts inferred from facts is called a theory. On this view, "theories never become facts, theories explain facts." But it does seem that tentative acceptance of a well-supported theory, even though it is based in part on "faith" in standard principles of reasoning, had better be called something other than "faith," a word that ordinarily connotes a very different kind of intellectual and emotional commitment. When a jury decides that a man is guilty as charged "beyond reasonable doubt," we do not call the verdict an act of "faith." (1984, p. 5)

It may seem that I have devoted far too much attention to an informal debate and its misuse by opportunistic creation scientists to suggest that a leading mainstream evolutionary scientist has succeeded in falsifying the theory of molecular phylogeny. I suggest that it is only through a scrutiny of such details that we gain understanding of how creationists work, using small and carefully selected pieces of an evolutionist's text to form an entirely misleading facade that hides the evolutionists' true position.

Falsification of a strongly supported major theory of broad application is rarely, if ever, completely achieved. Falsification in the sense of Karl Popper's definition is certainly not indicated by the kind of evidence Patterson is alleged to have presented. Those who actually conduct the experimental work in molecular biology are the first to recognize and report inconsistencies and internal contradictions within their findings. Considering the enormous complexities of biochemistry that are involved, the wonder is that molecular phylogenies agree as well as they do with those from macrobiology and paleontology.

Knaub and Parker sharply criticize the molecular biologists for assuming that molecular clocks keep good time (1982, iii). This observation puts them in the company of some leading molecular biologists. In a 1986 Science News report, Roger Lewin cited two papers pointing to "potential problems for would-be users of the clock" (Lewin 1985a, p. 571). One, by Professor Ayala and colleagues, reported on their studies of the amino acid sequence of a particular enzyme (superoxide dismutase) in the fruit fly compared with the same enzyme in human, horse, cow, and yeast. They found a five-fold variation in the number of amino acid substitutions per million years. Lewin quotes Ayala's conclusion that "using the primary structure of a single gene or protein to time evolutionary events or to reconstruct phylogenetic relationships is potentially fraught with error" (p. 571).

Lewin goes on to say: "Vincent Sarich of the University of California, Berkeley, has always conceded that some proteins change in a distinctly un-clockwise manner, and practitioners must be sure to demonstrate metronomic change (using the relative rate test) before drawing evolutionary inferences from protein data" (p. 571).

Mutation rates of DNA sequences have recently been found to differ substantially between taxonomic groups. Researchers, Chung-I Wu and Wen-Hsiung Li of the University of Texas, Houston, reported in 1985 that mutational changes in DNA rodent genes occur faster than in human genes (Lewin, 1985a, p. 571). Reporting in Science, Roy J. Britten of the California Institute of Technology, Biology Division, stated:

Examination of available measurements shows that rates of DNA change of different phylogenetic groups differ by a factor of 5. The slowest rates are observed for higher primates and some bird lineages, while faster rates are seen in rodents, sea urchins, and drosophila. The rate of DNA sequence change has decreased markedly during primate evolution. (Britten, 1986, p. 1393)

A 1987 article by science reporter Ivan Amato casts increasing doubt on the validity of the neutrality theory. He states: "Research from a number of labs is showing that clocks based on different molecules tick at different, and often varying, rates. For instance, clocks based on a particular molecule such as ribosomal RNA sometimes run at different rates for different species. Moreover, the rates of DNA clocks based on different cellular sources of DNA can differ within the same organisms" (Amato, 1987, p. 74). Two forms of DNA--nuclear and mitochondrial-- were studied by Lisa Vawter and Wesley M. Brown of the University of Michigan. They found that animal mitochondrial DNA (mtDNA) evolves at a rate 5 to 10 times that of vertebrate single copy nuclear DNA (scnDNA). Evidently, these two genomes are evolving independently (Vawter and Brown, 1986, p. 194). In contrast to this rather gloomy report is an optimistic one in which ribosomal RNA (rRNA) seems capable of allowing the construction of a master phylogenetic tree relating the three basic forms of earthly life--archaebacteria, eukaryotes, and eubacteria (Amato, 1987, p. 75). Amato concludes:

Most evolution biologists now agree that no single molecular clock is going to answer all of their questions. Ayala argues that biologists first must learn more about how and why molecular clocks vary in order to build a theory about their molecular clockworks. Such a theory might allow scientists to synchronize their many molecular clocks, each one fit to answer a limited set of evolutionary questions. And this could lead scientists now wandering in the complex molecular forest to the one tree they seek-- the tree of life that shows even how bacteria are related to giant redwoods and humans.

The current scene in molecular phylogeny research gives us a fine display of a basic attribute of empirical science--its tentativeness. Seemingly conflicting sets of data, so evident here, are accepted by mainstream science as an essential byproduct of its investigative process. Let the creation scientists make of it what they will. The intellectual richness of mainstream science will stand in sharp contrast to the stark poverty of the inflexible religious premise reiterated endlessly by the creationists without one iota of change.

This review of the contribution of molecular biology to phylogeny concludes our section on the general principles of mode and tempo of organic evolution. With an adequate background of evolution and the fossil record as put together by mainstream geology and biology, we are ready to turn to the creationists' nonevolutionary view of the origin of the earth's biota, fossil and living, and to their catastrophic picture of the origin of the strata in which fossils are enclosed.

Credits

1. From Francisco J. Ayala, Phylogenies and Macro-molecules, Chapter 9, pp. 262-313, in Evolution, by T. Dobzhansky, F. J. Ayala, G. L. Stebbins, and J. W. Valentine. Copyright © 1977 by W. H. Freeman & Company.

2. From Thomas H. Jukes, Molecular Evidence for Evolution, Chapter 7, pp. 117-46 in L. R. Godfrey, ed., Scientists Confront Creationism, W. W. Norton & Company, New York. Copyright © 1983 by Laurie R. Godfrey.

Chapter 37

The Creationist View of Life Forms and Fossils

In earlier chapters dealing with stratigraphy, we presented the creationists' catastrophic scenario of the Noachian Flood, in which raging flood waters--perhaps repeatedly made chaotic by tsunamis--rapidly transported great quantities of sediment and organic debris from place to place and eventually to sites of permanent deposition to form what we now recognize as all of the fossiliferous sedimentary rocks of the earth. The creationists' claims of extremely rapid deposition of all varieties of sediments, including coal, petroleum, and sedimentary ores, were carefully analyzed. We did not, however, consider the creationists' explanation of the occurrence and ordering of fossils in the stratigraphic column. How do they explain the biotic succession that follows unmistakably from the application of the principle of superposition?

The Early Creationists

Among those early geologists who worked out biotic succession were William Smith, Adam Sedgwick, and Roderick Murchison, all devout creationists. We have not previously mentioned William Buckland (1784-1856), a leading British paleontologist who led a group of staunch creationists. Faul and Faul refer to Buckland as "the last great diluvialist. . . . He was an excellent field observer and a perspicacious interpreter of geological data with an unusually broad grasp of geological relationships" (1983, p. 120). And, of course, Baron Cuvier, using Brogniart's biostratigraphy of the Paris basin, explained the succession of fossil forms as products of repeated extinctions followed each time by divine re-creation. (Brogniart, too, was a creationist.) These early creationists, who were actually compilers of the biostratigraphic column, seem not to have had a problem of reconciling what they found with the biblical account of the Creation and the Deluge. Cuvier voiced with great conviction what is now a fundamental tenet of modern creation science, "species were fixed, immutable, and independent," that "the only connection between fish and mammals was that they both had come from the hands of the same Creator" (Faul and Faul, 1983, p. 139).

Not only were these founders of the Phanerozoic systems creationists, but so was textbook-writer Sir Charles Lyell, now regarded by creationists an an archenemy because, they assert, he campaigned to replace catastrophism with uniformitarianism. Steven D. Schafersman summarizes the contributions of the persons I have mentioned above in these words:

> These men, by their efforts in the first half of the nineteenth century, built the geologic time scale by correlating strata with fossils. These men were the authentic scientific creationists; they were also deists or providentialists who believed in a First Cause, but thought that the province of science was to study the secondary causes of nature. (1983, p. 242)

So now I must ask, why does modern creation science

strive to undermine and repudiate a biostratigraphic record established by kinfolk in religious belief at a great cost in scholarly investigation? After all, their own findings caused the early creationists no inner conflict. The answer seems to lie in the successful rise of a new paradigm, voiced first by Lamarck (unsuccessfully), then with resounding success by Russell and Darwin: species can and do change; by transmutation one species can give rise to another, and it happens in a completely naturalistic manner. In a desperate attempt to crush this heresy, modern creation scientists have set as their goal the total destruction of everything their creationist predecessors established. Modern creationists attack even the very existence of geologic systems of strata in the order of relative age established by the early creationists. In so doing, the modern creationists denigrate the findings of their own religious forebears.

Position of the Modern Creationists

That the repudiation of classical biostratigraphy is the position of the creation scientists is evident in the ICR textbook, Scientific Creation (General Edition, Dr. Henry M. Morris, editor). On the subject of contradictions between Genesis and the geological ages Morris writes:

> The vague general concordance between the order of creation in Genesis and the order of evolutionary development in geology (and as noted earlier such a vague concordance is to the expected in the nature of the case and thus proves nothing) becomes a veritable morass of contradictions when we descend to an examination of details. (1974a, p. 227)

For example, the Bible states that land plants were the first life forms created, whereas biostratigraphy shows that marine organisms were the first life forms; that fruit trees appeared before fishes and birds before insects, whereas biostratigraphy shows the reverse ordering. Referring to his comparison table of such discrepancies, Morris writes:

> The above very sketchy tabulation shows conclusively that it is impossible to speak convincingly of a concordance between the geological ages and Genesis. Apart from the question of evolution or creation, the Genesis record is stubbornly intransigent and will not accommodate the standard system of geologic ages. A decision must be made for one or the other--one cannot logically accept both. (P. 228)

Despite Dr. Morris's absolute rejection of classical biostratigraphy, he seems willing to tolerate it in setting up a chronology of Flood stratigraphy (Morris, 1974a, p. 129). He has shown a table in which a succession of "phases" of the Flood corresponds with Proterozoic, Paleozoic, Mesozoic, and Cenozoic eras (see our Table 23.1).

Reviewing the position of creation science on the origin and history of life on earth, it is seen to consist of the following steps: (1) Divine Creation in a six-day period about -6,000 y. (4,000 B.C.), as described in Genesis 1, of all life forms (known as "kinds") that have ever existed. (2) Perpetuation of the full range of life forms (kinds) by reproduction without significant change during the antediluvian Golden Age of benign, rainless climate under a vapor canopy, a period lasting from -6,000 y. to -4,350 y. Populations of kinds may have experienced increases in numbers of individuals during that period. (3) The one-year Noachian Deluge, or Flood, that saw the extinction of many of the original kinds and their preservation as fossils in sediments deposited in the Flood year. (4) The post-Flood period of 4,350-y. duration in which (a) surviving kinds underwent a worldwide dispersal by radial migration from the Ark at Mt. Ararat, (b) some of the surviving kinds became extinct from natural causes, and (c) some or all of the surviving kinds have undergone relatively minor genotypic and phenotypic changes described by the term "microevolution," but otherwise, kinds are and always have been immutable.

The Edenic Curse and Structural Changes

We shall recognize as religious doctrine, beyond the scope of analysis by science, the divine initial creation of kinds. There is, however, a somewhat puzzling matter of interpretation of a possible secondary event of creation greatly affecting both animals and plants. It occurred at the time of the Fall of Man, when God pronounced the Edenic curse. Whitcomb and Morris discuss this subject at some length in a special appendix titled "Paleontology and the Edenic Curse" (1961, pp. 454-73).

In Gen. 1:29-30, God states that he has given man and all the animals seed-bearing herbs and trees to serve as their food. The creationists accept this passage as a clear indication that God ordered a vegetarian diet for all animals. Whitcomb and Morris conclude: "Under such conditions, there could have been no carnivorous beasts on earth before the Fall" (1961, p. 461). No beasts preyed on others and no rapacious or ferocious wild beasts existed at that time. Small wonder the earth was a paradise before the Fall! But, of course, there are carnivores in the fossil record and certainly many carnivores living today as survivors of the Flood. The only way to get carnivores on the antediluvian scene is to fit that "evolutionary" changeover into God's curse of Adam and Eve and their expulsion from the Garden of Eden into a harsh environment. Actually, I don't find any hint of a dietary shift to meat in the account of the Fall in Genesis 3, but Whitcomb and Morris quote a theologian (Oswald T. Allis) as saying that the eating of flesh was permitted to man after the Fall (p. 464).

Well, now, I think you can see what the problem is here. For certain of the animal kinds to go carnivorous would require structural changes. Teeth and jaws designed by God for cropping and grinding grasses and leaves must be supplemented by a set of long, sharp canines capable of piercing and tearing flesh, while hooves must be redesigned as sharp claws. Part of the animal's alimentary system would need redoing to assimilate raw meat. Whitcomb and Morris are well aware of these requirements, for they write:

> Some have objected that vast structural changes would have been involved in making an herbivore into a carnivore and such a transformation would have been tantamount to a creation of new Genesis "kinds" after the termination of the Creation Week. (P. 464)

But these authors think that such structural and organic changes did not necessarily involve a completely new creation or a loss of the creature's identity. They discuss God's punishment of the serpent by loss of its legs and conclude that it required far more drastic structural changes than making herbivores into carnivores. They also make reference to the structural consequences of God's punishment of Eve by making childbirth severely painful for her, presumably by altering the structure and dimensions of the birth canal or by increasing the diameter of the fetal head. Whitcomb and Morris conclude:

> While it is true that this case does not prove a similarly drastic change in the animal kingdom at the time of the Fall, it serves as an important illustration of how God could have introduced significant changes in the physical make-up of His creatures without at the same time eradicating their identity and producing thereby newly created "kinds." (P. 466)

Well, that was a close brush with repeated creation, wasn't it? But all's well that ends well, so we can go ahead with questions of the environment of the antediluvial period.

The Antediluvial Environment

As to the antediluvian Golden Age under the vapor canopy, many details about the canopy itself and the global environment are simply speculations unsupported by the scripture itself. This topic has been discussed in Chapter 21, in which the existence of a stable vapor canopy capable of producing a worldwide flood has been shown to be unrealistic under the constraints of principles of physics. Be that as it may, we must abide by the creationists' insistence on the antediluvial vapor canopy and its climatic influence; that is part of their hypothesis. The ICR textbook (General Edition) has described this environment in detail and its effects on life forms (Morris, 1974a, p.210). Because the vapor canopy would have efficiently filtered out harmful (ionizing) radiation from outer space, somatic mutations would have been fewer than today, and this would have slowed the aging of individuals and increased their life spans (see Wysong, 1976, pp. 390-91). Higher barometric pressure would have facilitated oxygenation in animal systems, also tending to promote longevity. Lands were more extensive then than now, with smaller expanses of ocean surface. Morris states that the mild climates and fertile soils would have supported much greater numbers of plants and animals the world over than exist today (p. 211). Keep in mind that these assertions about the character of the environment are human inventions, not derivable from statements in the scriptures.

Of course, we recognize this version of the antediluvial environment as an expression of religious belief. Nevertheless, it is an essential part of the hypothesis of creationism--a statement that all created kinds persisted without extinctions through the antediluvial period. Ignoring the question of cause, that statement can be subjected to scientific analysis, deducing the consequences of the premise and devising tests based on observation.

Biota of the Antediluvial Period

Consider the consequences of the creationists' premise that every kind of living thing was made by God during the six-day creation, and that no new kinds were created thereafter. This premise requires us to believe that all the kinds that were fossilized during the Flood were living on earth immediately prior to the onset of the Flood. There is a small possibility that a few of the fossils represent individuals that had died during the antediluvial period and whose remains had escaped destruction, but in the warm, oxygen-rich environment that prevailed this seems of little importance. After all, there were present all of the decomposers ever created, and they must have been busy oxidizing the organic matter of dead individuals, while chemical weathering would have quickly disposed of hard parts lying on the surface. So the biotic situation that prevailed was a coexistence of all the taxa of creation, pending their annihilation. Would this coexistence have been possible?

The question can be illuminated by considering the numbers of living taxa--genera and species in particular--existing on earth today and comparing these data with the numbers contained in the total fossil record. For guidance, I have turned to a standard textbook, Principles of Paleontology, by David M. Raup and Steven M. Stanley who have summarized the findings and estimates of other taxonomists and paleontologists (1971, pp. 6-13). They state that about 1.5 million different kinds of plants and animals are known to be living today, a figure based on species that have actually been found, described, and classified. Because new species are being discovered in large numbers each year, the actual number of extant species must be much greater than 1.5 million. One estimate (by V. Grant) is that a complete list of extant species would number as many as 4.5 million. G. Ledyard Stebbins suggests that the maximum number might be as large as 10 million (1982, p. 23).

The number of fossil species now known and named is only about 130,000, less than 3 percent of the probable total number of living species (taken as 4.5 million). Yet there is good reason to suppose that the actual total number of fossil species, if it could be determined, would greatly exceed the number of living species. I will not go into the reasons for the disparity, covered at some length by Raup and Stanley (1971, pp. 6-11). There is, however, a means to estimate indirectly the number of species that have lived in Phanerozoic time (from the start of the Cambrian Period). Data are available on the average duration of species of certain taxa. George G. Simpson's best guess was that an average of 2.75 m.y. elapses between the origin and extinction of a species, but with the actual average being as low as 0.5 m.y., or as high as 5.0. Using Simpson's average rate of 2.75 m.y. and the figure of 4.5 million living species, it is simple to calculate that nearly one billion species have lived since the start of the Cambrian. Raup and Stanley also report an estimate made by Grant in 1963 of at least 1.6 billion species since the start of the Cambrian (Raup and Stanley, 1971, p 11).

Using the estimate of one billion fossil species in the Phanerozoic record and dividing by 4.5 million, we estimate that the fossil species outnumber the living species by a factor of about 220. Now, according to the doctrine of recent creation, all of these billion-plus species, fossil as well as now living, would have had to cohabit the planet throughout antediluvian time. Would this have been reasonable? Indeed, would it have been possible? All organisms function within natural ecosystems, in which the producers (photosynthesizing plants) support the consumers through several trophic levels, while the decomposers subsist on dead individuals of both producers and consumers. If species populations were then of the sizes present today, it seems most unlikely that the earth could have supported this enormous population load, even with a worldwide equable climate.

As an alternative suggestion, we might postulate that populations of species were much smaller than today. But this condition would make species survival extremely difficult on an earth of homogeneous climate, where many species and genera of the same order were strongly competing for the same environmental niche. Consider a specific example, namely, the ammonites that entered the scene in Devonian time and persisted through the Cretaceous. In an earlier section on the fossil record, specifically dealing with extinctions of genera in Permian

time, we presented a graph of numbers of ammonite genera retained from the preceding epoch and the numbers introduced in each epoch (see Figure 32.6). My estimate of the total number of new genera, based on the values indicated by the "joints" or "doglegs" in the graph, is just over 2,000. Creationist hypothesis would require all 2,000 genera to have lived simultaneously throughout the 1,650 years of the antediluvial period. All 2,000 genera would have occupied the same marine environment, competing for survival in the same niche. They could not have been geographically separated and sequestered by physical barriers. Even supposing that their food supply was exceptionally abundant, there would be a limit to the numbers supportable, so that the population within each genus must have been very small, and each must have been on a nearly equal footing with all the others in ability to compete for food. What we come up with is a set of circumstances that is highly unlikely, to say the least. On the other hand, according to mainstream science, the ammonite genera came and went, one replacing another following an extinction. The record shows that the maximum number of genera present at one time never exceeded 140 and much of the time was fewer than 20. The evolutionary principle of replacement is totally lacking in the creationist view of initial creation of all the kinds, followed by coexistence of all without change through the antediluvial period.

So we must turn next to the Flood itself, which is described in Genesis. Many added details and embellishments of the Flood scenario as written by the creationists--but totally lacking of statement in Genesis--fail to stand up under close scrutiny; we covered these points in Chapter 21. But we did not give full attention to the question of the biota taken into the Ark and the fate of those not so included, nor have we considered the disposition of those taxa that survived on the Ark and were released into the strange new global environment resulting from the destruction of the vapor canopy.

What Life Forms Were Passengers on the Ark?

Assuming that all created kinds of all life kingdoms were extant up to the first day of the rains that began the Deluge, we want to know (a) whether all of them were taken by Noah into the Ark, or (b) whether some of them were excluded and must have had to fend for themselves during the great calamity. If the answer to (b) is "yes," what was the basis or criterion of exclusion? We look to the scriptures for guidance. This was God's command to Noah:

> Gen. 6:19-20: And of every living thing of all flesh, two of every sort shalt thou bring into the ark, to keep them alive with thee; they shall be male and female. Of fowls after their kind, and of cattle after their kind, of every creeping thing of the earth after his kind, two of every sort shall come unto thee, to keep them alive.

Genesis 7:21-23 tells us that, after the floodwaters rose high enough to cover the mountaintops, all of the above-mentioned life forms died, excepting only those that were with Noah in the Ark. In Genesis 8 and 9, the same life forms are described as leaving the Ark after the floodwaters subsided, so that they might "breed abundantly in the earth, and be fruitful, and multiply on the earth."

Although the creationists use only the King James Version of the Bible as their source of God's word, I was interested to see if the fruits of scholarly research over many subsequent decades, based on numerous more recent archaeological discoveries, might present the Genesis verses in more clearly understandable form. Accordingly, I consulted The Jerusalem Bible in English (1966). Genesis 7:13-16, which relates the Ark's passenger list, tells us that in addition to members of Noah's family, the boarders consisted of "wild beasts of every kind, cattle of every kind, reptiles of every kind that crawls on the earth, birds of every kind, all that flies, everything with wings. One pair of all that is flesh and has the breath of life boarded the ark with Noah."

Using The Genesis Flood by Morris and Whitcomb as a source of information, I attempted to pin down in terms of modern taxonomy the composition of the Ark's living cargo and sort it out from the hapless plants and animals that had to fend for themselves in the raging floodwaters. We read that God's original intention (before selecting Noah and his Ark passengers for survival) in sending a Flood was to destroy all of the kinds of land animals He had originally created, i.e., "the totality of air-breathing animals in the world" (1961, p. 13). It seems clear enough, both here and in other parts of the creationist text, that all animals normally living in an aqueous habitat were not intended victims of God's wrath, and were to be left to survive or perish in the floodwaters. That excludes from the Ark the marine and freshwater invertebrates, the fishes, and the aquatic mammals.

God made no mention of "microbes"--one-celled organisms not visible to the unaided human eye. God must have created these invisible life forms, but perhaps He reasoned that there was no point in mentioning it to a people who had no awareness of their existence. Perhaps microbes were ignored because they could not respond to God's command to enter the Ark, but many kinds of viruses, bacteria, tiny parasites, and spores must have come aboard on animals and their food stores and bedding materials. As to plants, they got on the Ark only incidentally, as animal food and bedding materials, so plant seeds of many varieties must have been given sanctuary. Land plants survived the Flood "either as floating vegetation rafts or by chance burial near enough to the surface of the ground for asexual sprouting of new shoots" (Walter E. Lammerts, quoted in Whitcomb and Morris, 1961, p. 70). The probable lethal effect of immersion of plants in salt water is dismissed with the explanation that dilution of the oceans by fresh canopy water would allow the plants to survive. (Would the same dilution have had an adverse effect on marine animals adapted to normal ocean salinity?)

Now that we have before us a fairly firm identification of the taxa that rode out the raging floodwaters on the Ark and of those that took their chances in the melee, we can turn to consideration of the relation of survivors and nonsurvivors in terms of the fossil record. We are assured that all taxa of the entire fossil record, from Proterozoic time to the present, were living immediately prior to the Flood. This means that Noah would have been under God's orders to load on male and female representatives of all air-breathing animals then alive, including, of course, all reptiles of the fossil record arising from the stem reptiles in late Carboniferous time and representatives of all land reptiles of the Mesozoic Era (thecodonts, pterosaurs, and dinosaurs).

Whitcomb and Morris specifically refer to the question of dinosaurs as Ark passengers; they say: "Extinct animals such as the dinosaurs may also have been represented on the Ark, probably by very young animals, only to die out because of hostile environmental conditions after the Flood; it seems more likely, however, that animals of this sort were not taken on the Ark at all, for the very reason of their intended extinction" (1961, p. 69, footnote). Elsewhere, in a footnote, Whitcomb and Morris have this to add: "If representative dinosaurs were taken on the Ark (presumably young ones), then it is likely that their final extinction is accounted for by the sharp changes in climate after the Flood. On the other hand, some may have persisted for a long time, possibly accounting for the universal occurrence of 'dragons' in ancient mythologies" (1961, p. 280).

The dragon legend as a possible reference to dinosaurs that survived the Flood is presented by Duane T. Gish, the ICR's biology expert, in a childrens' book, Dinosaurs: Those Terrible Lizards (1977, p. 51). In speculating as to the fate of the dinosaurs, Gish favors the hypothesis of extinction by the oncoming Flood (pp. 56-60). He says specifically that, as a result of the Ice Age, the dinosaurs along with the flying reptiles and the marine reptiles all died out after the Flood (p. 60). Clearly, Noah had to put up with these horrible creatures as passengers on the Ark. Could his gangplank have stood up under the weight of these huge creatures? Because they were all young individuals of the species, this would not have been a problem.

Actually, these creationists have found themselves facing a serious dilemma over the question of the required presence of now-extinct animals on the Ark. Genesis is clear on this point: all animals then living ("every living thing of all flesh") were to be brought to the Ark "to keep them alive with thee." Clearly, God's order was "Get them all on board and keep them alive." Surely, Noah would not have disobeyed God's order, nor would he have presumed to know which animals should be excluded in order to assure their extinction.

The suggestion by Whitcomb and Morris that dinosaurs (along with hundreds or thousands of species of long-extinct animals) emerged alive from the Ark, but perished in the unfavorable post-Flood climate is completely contrived to exonerate Noah. It does, however, give rise to a remarkable deduced consequence. With thousands of mating pairs of ancient reptiles, birds, mammals, and insects dumped off the Ark to fend for themselves, they could have become fossilized and a part of the record of the early Pleistocene. For example, primitive hominids-- also disembArking and searching for food--might well have preyed on the animal pairs, devouring their flesh and leaving their bones to accumulate around campfires on the floors of caves. Perhaps creationists should be looking for dinosaur bones in the caves of Olduvai Gorge and in the Chinese cave that bears the remains and leavings of Peking Man.

Facetious musings aside, we can legitimately ask: If the Ark took aboard representatives of all of the air-breathing animals now found as fossils in the stratigraphic record from Carboniferous time through the Cenozoic, why would all but the relatively few kinds that populate the earth today have become extinct almost immediately after walking off the gangplank of the Ark? The creationists have an answer. We have already covered the details in our section on the late-Cenozoic Ice Age. Creationists claim that the subsidence of the floodwaters was immediately followed by the onset of a highly varied global climate, extremely cold in high latitudes, and giving rise to large ice sheets. Lingering tectonism and volcanism would have assured much rugged, compartmented terrain with barriers separating lowlands. Plant cover would have been sparse or absent over large expanses of arid lands. Torrential rains, high winds, flash floods, and overbank floods of major rivers would have posed new hazards the animals were not prepared to endure. Species adapted to a benign, moist, but rainless climate with abundant supplies of plant matter for food would have quickly succumbed in the hostile environment. The small numbers of individuals of each species, descended from a single mating pair, would have had only the barest chance to perpetuate the species. Those that did survive and today constitute our global biota were able to adapt quickly to the harsh environment, expanding into large populations, and annihilating the poorly adapted kinds in short order.

The chances for fossilization of these post-Flood victims would be extremely slight, both because of the small sizes of the populations and the short spans of time that those populations endured. In other words, the reasons for the lack of such fossil remains are exactly the same reasons that the evolutionists propose to explain gaps in the fossil record of the stratigraphic column. I hope creation scientists will thank me for getting them out of the corner into which they seemed to have painted themselves.

There remains, however, a painful and perhaps insurmountable difficulty for the Ark hypothesis. By the requirement of taking aboard representatives of all air-breathing land animals for which fossils have been found, the logistical problem of Arkeology is greatly increased, perhaps by a factor of 100 to 200. The problem is not how to get rid of the beasts once they disembArked; it is how to accommodate them and keep them alive for a full year aboard the Ark.

Problems and more problems! Here's another one. As already noted, life in the halcyon antediluvian period was enormously diverse--it contained populations of all taxa known from the fossil record plus all those of taxa living today. Because only mating pairs of each kind of air-breathing land animal were taken into the Ark, millions upon millions of individuals of each kind were doomed to death and burial. This fate also awaited the numerous unchosen individuals of kinds that are alive today, i.e., those that could not be preserved on the Ark. But where are their fossils? Paleontologists tell us that very few of the mammalian fossil species or genera found in the geologic record are living today. The reverse is also the case. Very few living mammalian species and genera are found in the fossil record older than, say, the beginning of the Pleistocene Epoch (-1.6 m.y.). Yet all kinds, extinct and living, would have an equal chance to be in the fossil record because the Flood caught them all at the same time and wiped them out more or less simultaneously (within one year, that is). Something is very wrong with the Flood hypothesis as told by the creationists. Surely, the turbulent floodwaters would have swept dinosaurs to their watery doom along with modern ungulates and carnivores, carrying them all to a common grave along with many other kinds of animals and plants. But of course, we find no such heterogeneous mixing of plant and animal kinds throughout the stratigraphic column.

Biostratigraphy, if it shows anything at all with crystal clarity, shows that various kinds of plants and animals occupy unique time zones. The salient message of stratigraphy is that sorting of kinds--species, genera, families, etc.--into timebound layers has been accomplished in some way. Time sorting is just the antithesis of what we would expect of a raging, turbulent flood, overwhelming in only days or weeks the entire biota of a rapidly submerging land surface. But, as we will see in Chapter 39, the creationists have contrived a bizarre explanation of the time order of fossil zones; it is based on what they call "hydraulic selection."

What Are Kinds? Creationist Systematics

In Chapter 33 we covered some of the principles of systematics, which is that area of biology looking for order in taxonomic diversity and structural complexity. The classification of life forms into a hierarchic structure seems to have posed no problem for the early creationists, among them Linnaeus. The order they discovered was simply a manifestation of God's Divine Creation. Joel Cracraft, a biologist specializing in systematics, has recently written at some length on the differing views of mainstream science and creationism on systematics (1983, 1984). He places the differences in perspective in these words:

> Of course, if we are honest, we must admit that many present-day systematists also do not think much about the implications of their work, and prefer to accept, rather uncritically, their findings as evidence of evolution. The moral, then, is not that science is any less susceptible to a priori assumptions than is creationism. It is, indeed, that we must have these assumptions if we are to

interpret the data of science at all. The solution to this apparent dilemma is quite simple: The assumptions themselves must be scrutinized when we examine the empirical structure of nature. As I will discuss below, the assumptions of evolution and creationism imply very different predictions about the way nature is structured. However, in science no a priori assumption is immune to criticism and possible rejection, whereas in creationism all these assumptions are taken as religious truths, immune to rejection. (1984, p. 191)[1]

Creation science takes as its basic taxon of classification the kind, using the language of God's command in the Flood account in Genesis (Whitcomb and Morris, 1961, p. 66). Each kind is a unique creation, unrelated by descent to any other kind, and all of them were produced by God at the same time. Thus, there is no need for genera, families, orders, etc., in a hierarchic structure such as in mainstream systematics.

Why do the creationists speak only of kinds and not of species? The answer is obvious enough upon reading their discussion of nongeological arguments against the universal Flood (Whitcomb and Morris, 1961, p. 65-70). It is a matter of accommodations in the Ark--the problem of space. Earlier, we gave the count of identified living species as being about 1.5 million and an estimate of the total number of living species possibly as great as 10 million (Stebbins, 1982, p. 23). If it can be shown that there are far fewer kinds than there are species, the housing problem of the Ark can be made tractable. It seems that back in 1840, a proponent of the local-Flood theory, John Pye Smith, took pleasure in pointing out that if "species" are identified with "kinds" on a one-to-one basis, the Ark could not possibly have held them all. Whitcomb and Morris counter this argument with these words:

> But a hundred years of further study in the science of zoology has brought to light some interesting facts concerning the amazing potential for diversification which the Creator has placed within the Genesis kinds. These "kinds" have never evolved or merged into each other by crossing over the divinely-established lines of demarcation; but they have been diversified into so many varieties and subvarieties (like races and families of humanity) that even the greatest taxonomists have been staggered at the task of enumerating and classifying them. (P. 66)

Perhaps the reader can anticipate that a great trick of magic is about to be performed. A new taxonomy is about to be set up in order to decrease drastically the number of kinds but at the same time to preserve the great biotic diversity seen in nature, and even to allow for a modest degree of evolutionary descent!

Whitcomb and Morris turn to a 1947 work by Frank L. Marsh, letting him play the magician (1961, pp. 66-67). Marsh described eloquently the apparent great diversity among strains of pigeons, all descended from the wild rock pigeon of European coasts, Columbia livia. The wide differences in appearances of these strains, Marsh says, could easily result in their being assigned to different species and even to different genera. An observer who did not know that all of these pigeon strains lie within a single interbreeding population might fall for Marsh's ploy. He has built a "tree of life" with the "kind" as the common root.

Marsh's trees are named baramins (from the Hebrew: bara, "created"; min, "kind"). Figure 37.1, adapted from Marsh's work, is a diagram of the trees of three Genesis kinds, illustrating baramins of three types. Each kind originated at the Creation, but a single created kind might include more than one "variety." A baramin having

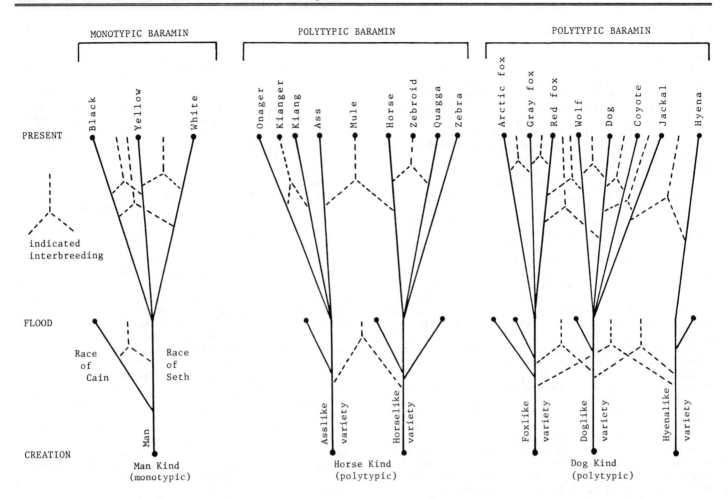

Figure 37.1 Diagram of three Genesis kinds. (Based on data of Frank L. Marsh, 1947, <u>Evolution</u>, <u>Creation</u>, <u>and</u> <u>Science</u>, Review and Herald Publ. Assoc., Washington, D.C., p. 179.)

created varieties is designated as "polytypic"; the man kind, with no varieties, is "monotypic." Between the Creation and the Flood, branching into subkinds and subvarieties occurred, an example being the bifurcation of the man kind into the races of Cain and Seth. Notice that these races can interbreed, leading to intermediate racial types. The diagram suggests that during the antediluvial period the kinds diversified into subkinds, the varieties into subvarieties, but that many of these became extinct during the Flood. Whitcomb and Morris do not endorse Marsh's category of created "varieties," and thus we must ignore Marsh's gratituous addition of that taxon. Only "kinds" were created.

Now comes the pièce de résistance: God needed to save only one representative of each kind, greatly reducing the numbers of mating pairs that needed to be accommodated on the Ark. The "horse kind" shown in Figure 37.1 consists of species of the genus <u>Equus</u>. The horses, zebras, and asses are recognized by mainstream taxonomy as three separate species of that genus (Colbert, 1955, p. 364). Thus, in mainstream biology the "horse kind" is at the taxonomic level of a genus. In mainstream biology the "dog kind" includes two superfamilies. Dogs, wolves, and foxes are within the superfamily Canidae; the hyenas are of the family Hyaenidae within the superfamily Feloidea (which also contains the cats, family Felida). From Marsh's taxonomy we can infer that a "kind" may be a taxon as large as a family or a superfamily. Confusion reigns! Is a "kind" equivalent to a genus, a family, or a superfamily? Will the real "kind" please stand up? If a kind may be represented by an entire superfamily, then it would not have been

necessary for both a dog pair and a cat pair to have been taken on the Ark. They are both of the same kind, and one pair of each kind would be enough. But it is hard to accept the interpretation that, for lack of cabin space, Fido rode the Ark, while Tabby had to swim for it (or vice versa). Surely Noah's family could never have made a decision like that! God would never have let the swimmer survive.

In a more serious vein, Joel Cracraft reviews critically the creationists' concept of the kind; he says:

> In science, the derivation of a definition is relatively unimportant, but how that definition is used is crucial. Science thus asks the question, if an entity is defined in a particular way, will that permit us to investigate nature effectively? If not, then the definition is of little use. Judged in these terms, is the "created kind" a useful scientific concept? The answer is decidedly no.
>
> In the first place, the creationists themselves cannot define "created kind" in such a way that it can be employed empirically. Obviously the Bible does not list the different "created kinds" nor does it specify how they might be recognized. This has led to immense difficulty for the creationists, so much so that even the leading creationist spokesmen are exceedingly confused as to exactly what a "created kind" might be. (1984, p. 193)[1]

Cracraft continues by pointing out specific examples from Duane Gish's work <u>Evolution? The Fossils Say No</u>. In one passage, Gish seems to equate "kind" to "species"; in another, to "genus." Gish is quoted as saying the creationists cannot always be sure what constitutes a separate kind (1978a, pp. 34-35). Examples cited by Gish suggest that it is a simple matter of eyeballing the animal

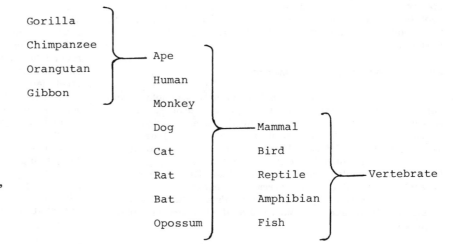

Figure 37.2 Hierarchical structure within kinds. (Based on data of J. Cracraft, 1984, Evolutionists Confront Creationists, <u>Proceedings</u>, 63rd Annual Meeting of the Pacific Division, Amer. Assoc. for the Advancement of Science, San Francisco, p. 194.)

and making a quick subjective decision--some animals obviously look alike and are of the same kind, others obviously don't look alike and are of different kinds. Cracraft concludes that Gish does not understand what a "created kind" is (p. 194).

I would suggest that the sloppy handling of "kinds" by the creationists renders the concept unworthy of serious consideration by the mainstream scientific community. Let the creationists come up with a rigorous definition or keep the term out of their "creation science."

Is there actually a hierarchical system within "kinds"? This question is raised by Joel Cracraft, whose answer is "Yes" (1984, pp. 194-95). Using Gish's own attempted explanation of the meaning of "kinds," Cracraft points out that creationists recognize that (a) gorillas, chimpanzees, orangutans, and gibbons are apes; that apes, along with humans, monkeys, dogs, cats, rats, bats, and opossums, are mammals; and that mammals, along with birds, reptiles amphibians, and fishes, are vertebrates. Presented graphically, this nested arrangement is clearly hierarchic (Figure 37.2). The "kind" that is called a "vertebrate" contains five other kinds, of which one, called "mammal," contains seven other kinds, of which one, called "ape," contains four other kinds. From this analysis it is clear that if God created each kind as separate and distinct from all others, He must also have created a hierarchy of kinds. That takes us back to precisely the position held by the early creationists, including the father of systematics--Linnaeus. There is no reason, as I see it, why, if God created kinds, He could not also have created a hierarchy of kinds. What is all the fuss about? Creationists, return to the fold of your forebears and ask them for forgiveness!

Microevolution--Creationist Style

Creationists recognize genetic change leading to the formation of races and varieties within kinds. Their textbook, <u>Biology; A Search for Order in Complexity</u>, presents accepted principles of genetics in considerable detail (Moore and Slusher, 1970). Their discussion of species seems conventional enough, emphasizing interfertility within a species (pp. 145-46). They make a strong point of the capability of two species of the same genus of being interbred in certain instances, their example being a cross between horse and donkey (ass) to produce the mule, which is usually sterile. They explain the sterility in terms of the chromosomal differences and differences in DNA nucleotide sequences.

As to the origin of variation within a species or a kind, the creationists reject the role of mutation, considering that to be an "evolutionary concept" (Moore

and Slusher, 1970, p. 147). Instead, they contend that the various created categories (kinds) were designed with a "variability potential" making it possible for them to survive in various environments. They say: "In other words there are latent recessive genes that later become expressed" (p. 147). They add: "Also, some variation [from this viewpoint] is simply an expression of the Creator's desire to show as much beauty of flower, variety of song in birds, or interesting types of behavior in animals as possible. It would be a monotonous world if all roses looked alike, or if all birds sang like the meadowlark, lovely as the song of this bird is" (p. 147).[2]

In summary, it seems that the creationists explain variation through recombination alone, using genetic differences purposefully introduced by the Creator to enable adaptation to be made to varied environments. Thus, Darwin's finches of the Galapagos Islands represent variations within a single kind, having made use of latent recessive genes supplied by the Creator and not requiring mutations. This view differs from a prevailing view of mainstream genetics, expressed on earlier pages, that mutations of neutral value accumulating in a stable population over a long period of time become advantageous in a new or changed environment, facilitating rapid emergence of a new species.

Moore and Slusher do not, however, deny the role of mutations in alterating the geneotype. They describe observed mutations in fruit flies. They show how mutations are used in selective breeding of plants and animals, citing as an illustration the production of the first hornless cow from a mutation (p. 90-91). They do, however, point out that because hornlessness is a dominant character, it was difficult to breed a pure hornless type. They stress that most mutations are harmful:

A few may be advantageous in some unusual environment. It is accepted by most geneticists that less than one mutation in a hundred could be considered "good" in any environment, and some say it is closer to one in a thousand. Some geneticists say the number of mutations that might be considered advantageous to the individual is so small, if indeed there are any at all, that all mutations may be considered deleterious. Many mutations are lethal. (1970. pp. 89-90)[2]

---

[2]Taken from BIOLOGY: A SEARCH FOR ORDER IN COMPLEXITY by John N. Moore and Harold Slusher. Copyright © 1970, 1974 by the Zondervan Corporation. Used by permission.

Clearly, the playing down of mutations as having any positive value is aimed at weakening the mechanistic evolutionary position vis-a-vis the creationist position that God endowed each kind with the required genetic variation.

How Creationists Look at Homology

In Chapter 33 we explained how mainstream biology and paleontology make use of the principle of homology to establish a taxonomy based on shared characteristics. One example compared the bones of the forelimbs of four modern mammals. All four have the same complement of arm bones, but they differ in proportions and sizes. Homologous structures are considered to offer strong evidence of descent from a common ancestor.

Creationists challenge the evolutionists' interpretation of structural similarities. Moore and Slusher present their argument:

> At the outset you should be aware of the fact that the entire line of reasoning by evolutionists is based on this single assumption: that the degree of similarity of organisms indicates the degree of supposed relationship of said organisms. It is argued that if animals look alike, then they are supposed to be related; if they do not look very much alike, then they are more distantly related. This is just an assumption; and none of these "evidences" from similarity can be used to demonstrate genetic relationships among organisms. (1970, p. 431)[2]

Their text then turns to morphology and anatomy, using the familiar set of drawings of the forelimbs of several animals (Moore and Slusher, 1970, p. 432, Figure 22.1). They are a somewhat different assemblage from those we used in Figure 33.1, but that is of no importance. The creationist text compares the limb bones of the salamander, crocodile, man, bird, bat, and horse. Whereas the evolutionists emphasize the correspondence in number and sequence of individual bones, which is remarkably complete, the creationists emphasize the differences in functions performed by the forelimb bones in different animals. For the salamander and crocodile, the forelimbs are used for walking; both the bird and bat use theirs for flying, but their manner of flight is quite different; humans use the hand for grasping. Creationists interpret these differences in function as evidence of independent origin, i.e., of creation especially for the function performed. They write:

> Creationists believe that when God created the vertebrates, He used a single blueprint for the body plan but varied the plan so that each "kind" would be perfectly equipped to take its place in the wonderful world He created for them. (1970, p. 432)[2]

As they do over and over again, the creationists retreat to the impregnable fortress of an omniscient and omnipotent Creator. The Creator could have done anything He chose to do. Animals and plants were created before Adam, and there were no human witnesses to record how creation took place. Evolutionists can point out that the bat's wing (i.e., a wing formed of skin stretched over a bone framework) could have been created with a very different skeletal structure, perhaps one much more efficient than the bat now enjoys. It might be argued as a principle that intelligent optimum design for function would result in far greater differences in structure than similarities. For example, the bird wing (a wing using long, stiff feathers) might be more efficient with many more bones, or perhaps a double wing as in a biplane. Nothing in such an argument would in the slightest degree detract from the religious belief that God did what He did for His own good and suffcent reasons. Homology, then, has no power to test the creationists' belief in independent creation of kinds. To the extent that similarities of structure can be shown, these can be attributed to God's ways of doing things--ways that cannot be questioned.

On the other hand, the existence of structural similarities can be deduced as a consequence of the hypothesis of descent with modification. A consistent naturalistic explanation can be formulated for the observed modifications. Evolutionists do not claim that structural adaptations have always been perfect or ideal for the required purpose. More realistically, it has been a case of making do with the inherited resources. For ancestors of the the bird and bat, the opportunity existed to adapt to the air, using only a set of bones previously suited to walking on solid ground. Given the circumstances, the naturalistic evolution process scored remarkable successes.

Creationists attack evolution on other matters related to structure and function, namely, ontogeny and the presence of vestigial organs. These we shall discuss in Part Seven.

Credits

1. From Joel Cracraft, The Significance of the Data of Systematics and Paleontology for the Evolution-Creationism Controversy, pages 189-205 in F. Awbrey and W. M. Thwaites, eds., <u>Evolutionists Confront Creationists</u>, Proc. of the 63rd Annual Meeting of the Pacific Division, A. A. A. S. Copyright © 1984 by the Pacific Division of the American Association for the Advancement of Science. Used by permission.

2. Taken from BIOLOGY: A SEARCH FOR ORDER IN COMPLEXITY by John N. Moore and Harold Slusher. Copyright © 1970, 1974 by the Zondervan Corporation. Used by permission.

Chapter 38

Biogeography and Demography—
The Creationist Version

Biogeography is the study of the geographical distribution of the earth's life forms, or biota. There is a biogeography of the living plants and animals--the flora and the fauna--as well as a biogeography of the geologic past. The kind of biogeography that interests us can be described as "floristic" and "faunistic." What this means is that the early botanists and zoologists traveled to various regions of the earth's land surface and compiled lists of the genera and species of the plants and animals living in each locality. When this was done for all areas of the globe, these scientists perused their lists and tried to define distinctive biogeographical regions, each characterized by the presence or absence of various plant or animal species. The task was exceedingly difficult, but despite many differences of opinion as to how these regions were to be defined and their boundaries drawn, a substantial area of agreement emerged about the general global pattern of floral and faunal distribution.

The Post-Flood Dispersion

Creationists have their own version of biogeography, designed to fit the biblical account of the Flood. It is an extremely simple version, because it can only be structured to agree exactly with the words of Genesis. First, it can ignore the plants, because they were not taken on the Ark, except by accident. Distribution of the earth's land plants as we find it today is simply a result of accidental grounding of rafts of floating vegetation or of haphazard shallow burial of seeds and roots of plants that sprouted new life as the waters of the Flood receded. Distribution of the animals is also simplified by merely disregarding the aquatic animals, which fended for themselves and must have been more or less randomly distributed by ocean currents. The creationist writings make no mention of the aquatic animals, but I can suppose that once the waters quieted down, those sessile invertebrates that happened to be in the right marine environment settled in without difficulty. Those capable of locomotion by swimming (nekton) or by floating (plankton) managed to reach suitable environments or quickly adapted by microevolution to new environments. All attention thus focuses on the land animals--the air-breathers--that disembArked on the slopes of Mount Ararat.

Whitcomb and Morris summarize the creationist view of biogeography in these words:

> The more we study the fascinating story of animal distribution around the earth, the more convinced we have become that this vast river of variegated life forms, moving ever outward from the Asiatic mainland, across the continents and seas, has not been a chance and haphazard phenomenon. Instead, we see the hand of God guiding and directing these creatures in ways that man, with all his ingenuity, has never been able to fathom, in order that the great commission to the postdiluvial animal kingdom

might be carried out, and "that they may breed abundantly in the earth, and be fruitful, and multiply upon the earth" (Genesis 8:17). (1961, p. 86)

In Chapter 37, we took up the vexing question of the Ark passengers that are represented in the fossil record, but that are not living today. We concluded that they quickly became extinct but left no record in post-Flood sediments of Pleistocene age. Land animals that are found widely distributed over the earth today moved out from the Ark in waves of migration. They were multiplying rapidly and thus were able to populate ever increasing areas of land encompassed by the widening circle of their frontier. Those kinds that were bound to terra firma moved out by walking and running, or by crawling and slithering, according to their locomotive endowments. We can imagine that the swiftly running herbivores would be in the forefront, accompanied by the equally swift carnivores that preyed on them. They would be channeled into favorable travel routes--stream valleys recently carved by receding floodwaters and newly created rivers, or open grabens such as those of the East African rift valley system. Others, like the mountain sheep, invaded and colonized high mountains and plateaus. Arriving at an oceanic shoreline, the emigrants would be halted, and the slower moving animals would soon catch up to complete the ecosystem. A few, however, would be carried as passengers across the ocean on large rafts of matted vegetation, emerging from the mouths of major rivers and drifting on the newly formed ocean currents, such as the equatorial current and the equatorial countercurrent. In this way some colonists could populate an island, or reach the shore of a completely isolated subcontinent. But there were many land bridges, such as that connecting Eurasia and North America where the Bering Strait is now located. There, the land bridge existed while large ice sheets were over the northern continents and the sea level was much lower than now. And so the animal horde swept into Alaska, moving southward into the heartland of North America, thence into Central America, and from there by way of the isthmus of Panama into South America. Wherever they went, these animals encountered established populations of birds, bats, and flying insects, which had little difficulty in traveling thousands of kilometers in only days or weeks, taking advantage of the new global system of prevailing winds unleashed after the collapse of the vapor canopy.

Simple as it all seems, there are some problems-- voiced, incidentally, by advocates of the local-Flood hypothesis, a vocal group of evangelical scientists who have added much of the heat to the entire Arkeological debate. One member of this group, Russell L. Mixter, a professor of zoology at Wheaton College (a fundamentalist school), is quoted by Whitcomb and Morris as follows:

> If kangaroos were in the ark and first touched land in Asia, one would expect fossils of them in Asia.

According to Romer, the only place where there are either fossil or living kangaroos is in Australia. What shall we conclude? If the fossil evidence means that there never have been kangaroos in Asia, then kangaroos were not in the ark or if they were, they hurried from Australia to meet Noah, and as rapidly returned to their native land. Is it not easier to believe that they were never in the ark, and hence were in an area untouched by the flood, and that the flood occurred only in the area inhabited by man? (From Russell L. Mixter, 1950, Creation and Evolution, American Scientific Affiliation, Monograph 2, p. 15, as cited in Whitcomb and Morris, 1961, p. 81.)

(Note: some of the harshest criticism of the creationists' universal Flood scenario has come, not from the materialistic evolutionists, but from the the creationists' nearest brethren, the "evangelical scientists" rooting for a tranquil flood or a limited flood.)

Whitcomb and Morris give a surprisingly wimpish reply to this criticism:

It should be be observed at the outset, however, that our purpose cannot be to prove that all modern animals have migrated from the Near East; for little is known about the movements of animals in the past from either science or Scripture. It is necessary to show only that a general migration of animals from the Near East since the Flood is reasonable and possible. (P. 81)

Whitcomb and Morris seem to have passed up any reply on the subject of kangaroos hurrying to meet Noah before the Flood, but I can clear that up very easily. The climatic environment of the antediluvian period was equable and homogeneous over the entire globe. Kangaroos lived everywhere, as did all the kinds of land animals. There would have been an acceptable mating pair of kangaroos close to where Noah was building his Ark. They would have come leaping at the slightest signal of encouragement. Besides, Whitcomb and Morris point out that, in the first place, it cannot be proven that the Ark was built in the same part of the earth as that at which it grounded (p. 83).

In Chapter 32, covering the major points of evolution through the Phanerozoic Eon, we presented several topics in biogeography (paleobiogeography), among them the strange case of the marsupial mammals in Australia. We explained their dominance, to the almost complete exclusion of placental mammals, by the events of plate tectonics. The breakup of Pangaea eventually severed the final link between Australia and the other fragments of Gondwanaland. The marsupials evolved in Australia without competition from placentals, remaining in isolation until the arrival of humans in modern time.

Creation scientists do not accept plate tectonics, but they are happy with land bridges. Today, Australia lies close to the Sumatra-Java-Timor island arc, New Guinea, Borneo, and many other islands of the East Indies. These could have served as stepping stones for post-Flood migration of marsupials from the Ark in Anatolia to Australia (Whitcomb and Morris, 1961, p. 84). But this possibility was open to the placental mammals, as well, and it is difficult for creationists to find an explanation for the presence of the marsupials in Australia to the exclusion of the placentals. I find no mention of this crucial problem in creationists' writings. We must presume the migration pattern to have been under direct control of God, who ordered different animal groups to different destinations, but this speculation is not, of course, within the aegis of science.

Joel Cracraft has attempted to compare the creationists' version of biogeography with that of mainstream biology (1984, p. 199). He has set down some deduced consequences of each version, and these can be tested by comparison with distributional information acquired by biologists through field observations over many decades and comprising a data base independent of any hypothesis of origins.

First, however, an important concept of biogeography needs to be introduced. The geographical distribution of a particular species is said to be cosmopolitan when the species is found widely distributed over the lands of the globe. The phenomenon of cosmopolitan occurrence is called cosmopolitanism. When the species occurs only in a few relatively small areas, or in only one small area, the distribution is said to be endemic. The phenomenon of endemic distribution is called endemism. For example, many species having endemic distribution are island-dwellers; their presence is explained by evolutionary biology in terms of physical isolation in a new environment leading to speciation.

Cracraft reasons that under the hypothesis of dispersal of all animals from Mount Ararat only a few thousand years ago, strong patterns of endemism should be lacking, because this style of dispersal instead promotes cosmopolitanism (1984, p. 199). He reasons that there would not have been enough time for endemism to have developed. Under the hypothesis of evolution through great spans of geologic time, during which land distributions were being radically altered by lithospheric plate motions, "narrow endemism" is to be expected (p. 199). That endemism is the dominant feature of global biogeography today strongly supports the evolutionary hypothesis but militates against the creationist hypothesis.

Cracraft makes other deductions that can be used as a test of each of the two hypotheses, finding in each case that observed distributions cannot be predicted from the post-Flood dispersion scenario but are to be expected under evolutionary biology through geologic time. The arguments are highly technical, requiring an extensive background in biogeography, so we shall close with Cracraft's own words summarizing the argument we have presented:

We can now ask which set of predictions is confirmed by the geographic data available to science. The answer is that the predictions of evolutionary biology are strongly supported by scientific data and those of the creationists are resoundingly rejected. Without question the vast majority of organisms are narrowly endemic, with only a small minority exhibiting cosmopolitanism or widespread distribution. Our familiarity with widespread species creates an intuitive bias. They seem to be relatively common, but such species actually are quite rare. If these distributions were the result of very recent dispersal from Mount Ararat, just the opposite would be expected, that is, cosmopolitanism would be the rule, not the exception. (Pp. 199-200)[1]

The Post-Flood Human Population Explosion

In our haste to disperse the Ark's animal cargo over the lands of the earth, we should not overlook that humans also took part in that post-Flood migration. Homo sapiens can be said to have a cosmopolitan distribution as a species, overlooking the contrasting endemism of the races or subspecies of which it is comprised. Did Noah's descendants fan out behind the other air-breathing animals, or did they keep up with the front-runners? Perhaps humans had a great advantage in crossing ocean barriers, because, as Noah's descendants, they were skilled at building boats and weatherproofing them with pitch.

Creationists John Whitcomb and Henry Morris have found food for thought in the statistics of human population growth, starting with the creationist postulate that the initial human population consisted of two people (1961, pp. 396-98). They place this point in time at approximately 3300 B.C. (-5,300 y.), which is midway between the Creation and the Flood. This, they say, is about the time of the birth of Noah's first son.

The figuring, as carried out by Whitcomb and Morris, consists of establishing a doubling time, which is the number of years it takes for a growing population to double in numbers of individuals. The doubling time varies according to the percentage rate of increase, as in compound interest on a savings bank account. The higher the rate of interest, the shorter the doubling time of the initial investment. For example, at 10 percent per year interest (reinvested), the doubling time is 7 years, whereas at 1 percent the doubling time is 70 years. What Whitcomb and Morris have done is to look at world population figures for as far back as they can be considered reasonably reliable and from these figures have derived a value for the doubling time. Their data on world population are essentially in agreement with figures shown in Table 38.1 for the years A.D. 1650 to the present.

Whitcomb and Morris are well aware that the doubling time has consistently decreased through this period of record. (As of 1984, it has been known that the world population growth rate has declined, going from 2.0 percent to 1.7 percent in the decade from 1973 to 1983, with a small corresponding increase in the doubling time.) Nevertheless, their strategy is to ignore the firm historical evidence of declining doubling time and pick out instead a single representative doubling rate. From the data they cite, they select a value of 175 years. They also calculate the number of doublings needed to yield a world population of 2.5 billion, which is about what it was in 1950. From a table of powers of two, we can quickly learn that 30 doublings of the number two yield the number 2.15 billion (approximate); so we settle on a scheme of 30 doublings, each requiring 175 years.

Table 38.1   Growth and Doubling Time of the Human Population

Year	Population	Doubling time (years)
1,000,000 B.C.	125,000	
		230,000
300,000	1,000,000	
		160,000
25,000	3,340,000	
		22,000
8,000	5,320,000	
		1,000
4,000	85,500,000	
		6,400
A.D. 0	133,000,000	
		830
1650	545,000,000	
		240
1750	728,000,000	
		160
1800	906,000,000	
		120
1900	1,610,000,000	
		87
1950	2,400,000,000	
		36
2000 (est.)	6,270,000,000	

Data source: E. S. Deevey, Jr., 1960, Scientific American, vol. 203, no.5, pp. 194-204.

Multiplying 175 by 30 gives 5,250 years, which takes us back to the year 3,300 B.C., calculated from 1950. After arriving at this figure, Whitcomb and Morris comment as follows:

> It could not be maintained, of course, that this calculation is completely rigorous, but it certainly is reasonable--far more so than to say that the population has been doubling itself since a hypothetical beginning several hundred thousand years ago. Added to all the other evidence for the beginning of the present order of things on the earth after the Deluge several thousand years ago, this further testimony is quite impressive. (1961, p. 398)

Actually, these creationists have testified against their own conclusion by disregarding the data they themselves cite, which is to the effect that the doubling time has decreased dramatically since 1650--from 240 years to around 35 years in 1930. Not only is the assumption that population growth rate (in percent per year) has remained constant for more than 5000 years completely unwarranted, but it is shown by the population data to be an invalid assumption.

Mainstream science has no really firm data on world population going back before about A.D. 1650, but there are ways to make estimates through various archaeological procedures. Figure 38.1 is a graph showing the growth of world population throughout the past 8000 years as estimated by demographers of the Population Reference Bureau of Washington, D.C. Using arithmetic scales on both axes of the graph emphasizes the extremely slow rate of growth until well into the Christian Era. At the same time, the extremely rapid growth rate in the past two or three centuries makes that part of the curve resemble a nearly vertical line. To present the data in a way that allows growth rates to be shown in true proportion over vast spans of time and a huge range in population numbers, we can use a graph on which both scales are logarithmic (i.e., a log-log plot), as shown in Figure 38.2. This graph shows an estimate made by Edward S. Deevey in 1960 of the human population/time curve over the past one million years. Deevey, who at the time was professor of biology in Yale University, shows three surges in population growth, each slackening off to near constancy to hold for a considerable period. Of course, much speculation is involved in any such reconstruction going so far back in time with no written record. But even if we insert into our simplified equation the necessary rate of change in the population growth rate seen over the past few centuries, we will end up with a starting point vastly older than that arrived at by the creationists.

The creationists' arguments for a young earth, based on population statistics, has been reviewed in depth by David A. Milne, a professor of evolutionary biology at the Evergreen State College in Olympia, Washington. The title of his paper is a real teaser: "Creationists, Population Growth, Bunnies, and the Great Pyramid" (Milne, 1984).

Milne finds a fatal flaw in the creationists' calculations based on a constant population growth rate. I cannot resist quoting two rather long paragraphs from Milne's paper, both because they will amuse you and because I could never paraphrase them half as nicely. He is referring to a somewhat different set of calculations given by Henry Morris (1974a, pp. 167-69), but with essentially the same assumptions as in the method I reviewed above. It shows that from a starting date of 4,300 B.C. and a uniform growth rate of one-third of one percent per year, the estimated world population figure for A.D. 1800 would be achieved. Figure 38.3 is a plot of the two creationist schedules we discuss here. Milne uses the Morris schedule of 1974 with devastating results to calculate the world population in the year 2500 B.C.:

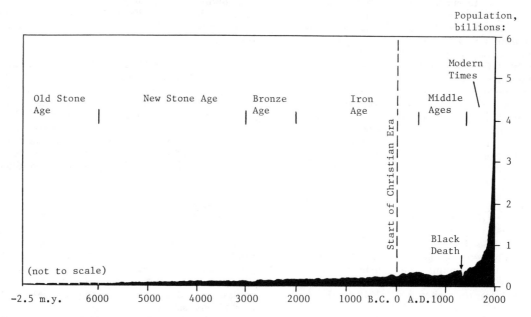

Figure 38.1 World population growth throughout human history. (From Population Reference Bureau, Washington, D.C., 1984, Population Bulletin, vol. 18, no. 1, Figure 1.)

By Morris' calculation, that number is 750 individuals. If Egypt, with about 1% of the earth's surface area also had 1% of its population, then about eight people must have lived in Egypt at that time. However, the Great Pyramid of the Egyptian king Cheops was built in about 2500 B.C. If the creationists are right, then the Pyramid was built by eight people. In fact, suppose that the entire population of the Earth lived in Egypt at that time. Half of the 750 souls were women (who I don't think worked on the Pyramid); half of the males were children (ditto) and a few exalted characters (Cheops himself and his assorted advisors) undoubtedly convinced the others that nobility should not have to haul heavy limestone blocks. That leaves about 150 able-bodied men to quarry 2,300,000 blocks (ranging from two and one-half to 50 tons in weight), haul them to the construction site and raise the 480-foot Pyramid. Does anyone who has seen this colossal monument believe that 150 men could have built it? Yet that is what Morris, through the magic of his calculation, must boldly assert. (1984, pp. 3-4)[2]

Having once gotten on this trail of destruction by mathematical logic, Milne pursues it to its ridiculous conclusion:

World history prior to 2500 B.C., in the Morris scenario, becomes even more remarkable. Six pyramids, some comparable in size to the Great Pyramid, were built at nearby sites within the previous 200-year period (as were numerous accessory causeways, temples, etc.). The parents and grandparents of the 750 people at the Great Pyramid site must have built them, at a rate of one every 33 years. Their numbers (which, recall, constituted the entire human population of the Earth) were fewer then--only about 300-400 souls-- and they were distracted by the need to perform a fast migratory quick-step over to Mesopotamia to build (and abandon) a number of fortified towns that appeared at about that time. The action was even more frenzied in earlier centuries. World population in 3600 B.C., as calculated by the Morris equation, was 20 people. A century earlier, in 3700 B.C., it was 14 people. So, in the Morris scenario, a world population of one or two dozen people must have rushed back and forth between Crete, Mesopotamia, the Indus River valley, and other sites of ancient civilization, energetically building

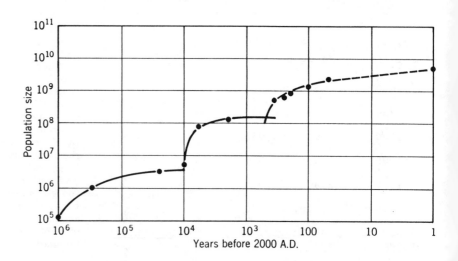

Figure 38.2 Human population size in past years, plotted on logarithmic scales. (Based on data of Edward S. Deevey, Jr., 1960, The Human Population, Scientific American, vol. 203, no. 5, table on p. 196.)

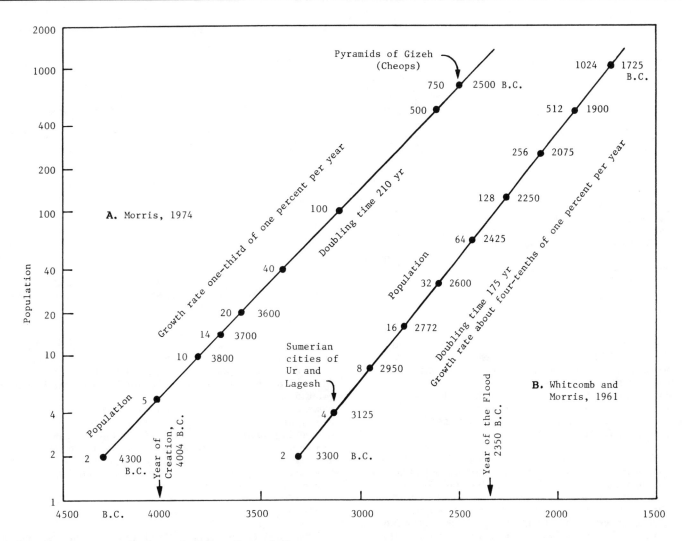

Figure 38.3 Two creationist schedules of population growth, starting with two individuals at two different points in time and following two different rates of growth. (A. N. Strahler.)

and abandoning enough cities, irrigation works, monuments and other artifacts to leave us with the mistaken impression that millions of people populated the ancient world. (1984, p. 4)[2]

Perhaps you are curious about the "bunnies" in Milne's title. He uses what he calls the "bunny blunder" as an even more extreme example of the erroneous conclusion one can come to by following the creationists' schedule of assumed constant population growth rate (or doubling time). The bunnies referred to are snowshoe hares somewhere in Canada in the early 1930s. Published studies show that at that period, the hares were multiplying at a rate of about 57 percent per year for a doubling time of under two years. Compare that rate with a human population increase of 2 percent per year in recent decades for a doubling time of 35 years. Extrapolated back to a single breeding pair, the snowshoe hare must have made its first appearance on earth in 1885, during the Cleveland administration, writes Milne. Did the Creator create the hares on that date? Obviously not, since God created all kinds once and only once, during the six-day creation. Field observations of many mammals such as the snowshoe hare show their population size to change radically in cycles of only a few seasons' duration. Explosive population growth is followed by population collapse. This pattern shows clearly in the records from 1790 to 1930 for snowshoe hare and lynx in Figure 38.4.

If the fluctuations were averaged out, Milne points out, their rate of increase would be essentially zero. This element of the "bunny blunder" carries over into the human population variation with time. Great plagues and famines have at times drastically reduced the human population over large areas. For example, a dramatic population drop was caused by the Black Death (bubonic plague) in Europe, starting in the year 1334 and raging for about 20 years (see Figure 38.1). Fluctuations such as these make even more hazardous any extrapolation to a beginning point based on a single assumed rate of growth.

Morris has used his mathematical model in an attempt to make the timetable of mainstream evolution seem absurd (1974a, p. 169). He states that with a population increase rate of only one-half percent per year extending over a million years, the number of individuals of the living population today would exceed 10 raised to the power 2,100. This number, he says, is "utterly impossible." He asks us to compare this body count with his estimate that the entire universe could contain only the number of electrons represented by 10 raised to the power 130. Morris goes on to say that even if the human growth rate were so small that the total population could reach 3.5 billion only after one million years, we would have to believe that at least a total of 3,000 billion individuals lived and died on earth during that period. He then makes the point that if such an enormous number of humans have lived on earth at one time or another, it is beyond belief that there would be so few human fossils or cultural evidence of ancient man (p. 169).

A conservative estimate, given by Paul and Anne Ehrlich, places as between 60 and 100 billion the number

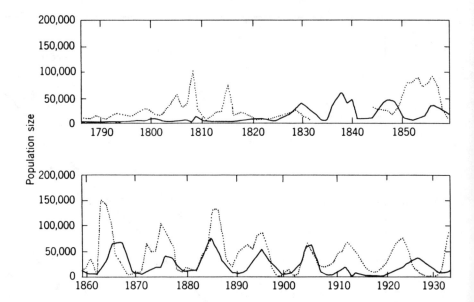

Figure 38.4 Cycles in the populations of the snowshoe hare (broken line) and the lynx (solid line). Data are incomplete for the hare from 1832 to 1844. (Based on data of D. A. MacLulich, 1937, Univ. of Toronto, <u>Biological Series</u>, no. 43.)

of individuals of <u>Homo</u> <u>sapiens</u> that have lived on earth (1970, p. 5). This is indeed an enormous number to contemplate, but it should not require us to expect a much larger number of fossil finds of their remains. Deevey makes some interesting comments relevant to this question. He estimates that "there were 36 billion Paleolithic hunters and gatherers, including the first tool-using hominids" (1960, p. 51). He continues:

> One begins to see why stone tools are among the commonest Pleistocene fossils. Another 30 billion may have walked the earth before the invention of agriculture. A cumulative total of about 110 billion individuals seem to have passed their days, and left their bones, if not their marks, on this crowded planet. Neither for our understanding of culture nor in terms of man's impact on the land is it a negligible consideration that the patch of ground allotted to every person now alive may have been the lifetime habitat of 40 predecessors. (P.51)

In closing, I might comment on another aspect of the creationists' belief that all surviving air-breathing animals entered upon their post-Flood lives in the number of only two individuals per created kind. What chance would a single pair of animals have to survive and multiply under the best of environmental circumstances? Precious little, I would think, especially when we consider that the post-

Flood climate was extremely harsh in comparison with that of the pre-Flood period. Experience today shows that when a breeding population of a given rare species is reduced to only two individuals, the chance that it can rebuild to a viable population in its natural habitat is close to nil. Even the most determined and inventive intervention by wildlife experts may not prove successful in keeping such a thin thread of descent from ending forever. Seen in this realistic light, the legend of the animals of the Ark and how they populated the lands of the earth with new life can scarcely be taken seriously as a scientific hypothesis.

Credits

1. From Joel Cracraft, The Significance of the Data of Systematics and Paleontology for the Evolution-Creationism Controversy, pages 189-205 in F. Awbrey and W. M. Thwaites, eds., <u>Evolutionists</u> <u>Confront</u> <u>Creationists</u>, Proc. of the 63rd Annual Meeting of the Pacific Division, A. A. A. S. Copyright © 1984 by the Pacific Division of the American Association for the Advancement of Science. Used by permission.

2. From David H. Milne, Creationists, Population Growth, Bunnies, and the Great Pyramid, <u>Creation</u>/<u>Evolution</u>, vol. 4, no. 4, pp. 1-5. Copyright © 1984 by David H. Milne. Used by permission.

Chapter 39

Fossils and Flood Strata—
A Hydraulic Stratigraphy

Creationists, as we have seen, do not accept the biostratigraphic column of mainstream geology and paleontology, nor do they accept the geologic time scale that goes with it. But then, having protested that the mainstream interpretation is erroneous, they make an attempt to explain its evolutionary sequence of life forms, revealed through fossils, by means of a purely mechanistic, cataclysmic Flood history that comes up with the same succession of fossil forms as mainstream biostratigraphy, i.e., the same succession they reject as nonexistent. (See Table 23.1.) It seems to me that by attempting to fit their Flood history to the mainstream biostratigraphy, they tacitly support the latter. On the one hand they assert that the succession of fossils worked out by evolutionists is to be expected as a result of the cataclysmic processes of the Flood; yet they also challenge the validity of that fossil succession. The absurdity of the contradiction reminds one of a person trying to pull the rug out from under the chair on which he or she is sitting.

The Creationists' View of Stratigraphy

The creationists' position on stratigraphy and the fossil record has been recently reviewed by Dr. Andrew Snelling, Ph.D. (geology), in a technical paper published in the creationist journal _Ex Nihilo_ (1983). After first stating the mainstream view of geology and evolution, Snelling continues:

> On the other hand, creationists interpret the majority of fossiliferous sedimentary rocks of the Earth's crust as testimony to Noah's flood. . . . Creationists do this because they regard the Genesis record as implying there was no rain before Noah's flood, therefore no major erosion, and hence no significant sedimentation or fossilization. However the flood was global, erosional, and its purpose was destruction. Therefore the first major fossilization commenced at this time, and the majority of the fossils are regarded as having been rapidly formed during this event. Creationists therefore regard sedimentary strata as needing to be classified into those formed during the time of the creation week, preflood, flood, (early, middle, and late) and post flood, and recent. (P. 42)

Snelling states that the creationists' principles differ so completely from those of the evolutionists that the creationist geologists must place a new interpretation on the geologic evidence.

Snelling then introduces a statement by Joseph C. Dillow, who has published creation science papers on the subject of the antediluvian vapor canopy. Dillow writes:

> It should be obvious that if the earth is only 6,000 years old, then all the geological designations are meaningless within that framework, and it is deceptive to continue to use them. If, as many

creationist geologists believe, the majority of the geological column represents flood sediments and post-flood geophysical activity, then the mammoth, the dinosaur, and all humans existed simultaneously. . . .Some limited attempts have been made by creationist geologists to reclassify the entire geologic column within this framework, but the task is immense. (1981, pp. 405-6)

Snelling turns next to the question of the extent to which the Precambrian rocks are Flood deposits. This will prove to be a crucial point in testing the validity of the creationists' flood scenario of the sequence of fossil deposition, so it is well worth putting straight at the outset. It seems that in recent years, Dillow and some others have been limiting the Flood strata to the Phanerozoic Eon (Cambrian and younger). Snelling will have none of this, for he says:

> It is my contention that those who do this have failed to study carefully the evidence for the flood deposition of many Precambian strata and have therefore unwittingly fallen into the trap of lumping together the Precambrian strata to the creation week. The usual reason for doing this is that the evolutionists regard the Precambrian as so different, so devoid of life in comparison with other rocks that creationists have simply borrowed their description. (P. 42)

Having suppressed the heresy, Snelling goes on to push the Flood strata farther back into Precambrian time than simply the time of first appearance of certain fossils, now interpreted with some confidence as blue-green algae. He would interpret graphite in Precambrian metamorphic rocks as clear evidence of early life:

> What I am contending here is that fossils, whether they be microscopic or macroscopic, plant or animal, and the fossil counterpart of organic matter, along with its metamorphosed equivalent graphite, are the primary evidence which should distinguish flood rocks from preflood rocks, regardless of the evolutionary age. (P. 45)

The significant point is that the fossil content of the lowest division of the Flood strata--those of "early" Flood time--must be limited to one-celled life forms (within the Monera and Protoctista). How does this requirement stack up against the creationists' mechanistic theory of deposition and fossilization during a catastrophic one-year flood? The answer will emerge in later paragraphs.

Physical Conditions of Flood Sedimentation

Let us review the physical conditions under which fossils became enclosed in sediments during the Flood. The cataclysmic nature of the Flood cannot be understated. Whitcomb and Morris describe it in such words as "a cataclysm of absolutely enormous scope and potency" (1961, p. 258). Actually, three different forms of physical activity were occurring simultaneously.

First, the bursting forth of great underground springs from the floor of the ocean generated enormous turbulence and tore out great masses of rock from the oceanic crust. Doubtless the emerging water was extremely hot--perhaps it was in the form of superheated steam--and would have been heavily charged with the volatiles exuded by extruding magma. Tectonic activity--faulting, that is--was intense and would have generated tsunamis. The conditions of extreme turbulence, high temperature, and high toxicity would have prevailed while the waters were rising. Marine animals would have all rapidly perished by cooking and poisoning, but their remains could not have come permanently to rest until the springs ceased flowing and volcanism subsided, i.e., until the first 40 days had ended.

Second, as the torrential rains descended, fed by the collapsing vapor canopy, the land surfaces were scoured by heavy overland flow, gullied deeply, and carved into great new systems of ravines and canyons. Huge quantities of sediment were transported to the coasts to become deltas, and these were continually undergoing submergence as the waters rose. Recall that in the antediluvial period there was no rain whatsoever, but in the warm, moist atmospheric environment, exposed bedrock was deeply weathered and decayed to form thick saprolite, which was the source of the fluviatile sediment. Land plants and animals would have been heavily impacted by the torrential rains and floods. Whitcomb and Morris describe the scene:

> On the land, the raging rivers would carry great quantitites of detritus seaward, occasionally entombing animals or reptiles [sic], together with great rafts of vegetation. These would normally be deposited finally in some more or less quiescent reach of stream or finally in the sea on top of other deposits or perhaps on the exposed bottom itself. (1961, p. 266)

Third, the great tsunamis (seismic sea waves, popularly miscalled "tidal waves") reaching the shores of the continents would have wreaked havoc on all coastlines. It is reasonable to suppose that these tsunamis were of much greater energy and dimensions than any occurring in recent time, because tectonic activity was tremendous during the Flood, with fault movements releasing energy in bursts larger, perhaps by a factor of several Richter-scale units, than the greatest known in recorded human history. Each tsunami, a train of water waves of great wavelength, upon arriving at a coastline, produced a sequence of alternating rises and drops of water level, and these may have been of an amplitude of several or many tens of meters. Storm surf superimposed on the water-level rise would have impacted the land surface over a zone extending many kilometers inland, uprooting trees and drowning large numbers of land animals. The plant debris and corpses would have been swept seaward during the falling water stages. Accompanying the tsunamis would have been numerous, great submarine turbidity currents, triggered by earthquake shocks. Although the incidence of tsunamis would have fallen off greatly after the floodwaters reached their maximum elevation, the onset of violent winds of superhurricane intensity, set in motion following the total collapse of the vapor canopy, would have generated powerful surf, continuing the coastal erosion during the falling stage of the Flood.

A Model of Flood Sedimentation

Putting these various forms of activity together, we can reconstruct a general plan for the successive phases of sediment and fossil deposition. The phase of rising floodwater would have prevented deposition of marine strata capable of permanency. In this phase, remains of dead organisms were circulating in a high-turbidity regime along with suspended sediment of all grade sizes, from the finest colloidal clays to the largest rock masses capable of being suspended in upward currents. No marine organisms could have remained alive; those terrestrial organisms introduced into the convulsing ocean would have all been dead on arrival, or shortly thereafter. These terrestrial organisms would have also been kept in suspension, thoroughly mixed with the suspended marine organisms.

A second phase would have been the raining down of sediment and dead organisms upon the ocean floor, permitted by the rapid cessation of submarine spring flow, and the drastically decreased volcanism and tectonic activity. During this phase, the principle of "hydro-dynamic selectivity" would have been in operation (Whitcomb and Morris, 1961, p. 273). Discrete solid particles of both mineral matter and organic matter of noncolloidal dimensions would have settled out in preferential order according to density (specific gravity), outward form (shape), and bulk (diameter or volume). We must assume that turbulent intensity diminished over a finite period of time, so that the layers of sediment capable of accumulating as the permanent record would have been graded in terms of texture of the inorganic particles and in terms of the the hydraulic characteristics of the organisms. Generally, then, the coarsest mineral particles and the organisms with the most rapid settling velocities would have come to rest first, followed by particles of increasing fineness and organisms of less rapid settling velocity. The result would have been a stratigraphic sequence in which not only the mineral particles were sorted by size, but the fossils would have been sorted strictly according to their physical parameters (but not according to phylogenetic affiliations).

One can allow for the possibility that continued cataclysmic events of diminishing scale and lessening intensity disturbed and reworked an upper portion of the sediment sequence but left a lower portion intact. This punctuated residual cataclysm could account for unconformities and disconformities of the geologic record. The end result would be much the same in any case, so far as sorting is concerned, namely, a well-graded deposit produced in less than one year. Keep in mind that all organisms were of essentially the same age; all were alive on the day before the Flood began, and nearly all were dead by the time the floodwaters submerged the highest mountaintop. All were dead (except those in the Ark) within one-year's time at the most, as God intended.

You might speculate as to whether some fossils, those last to be enclosed in sediment, were arranged, not by the process of hydrodynamic selectivity, but by other processes of transport and deposition. Perhaps as the floodwaters receded, exposing the remains of plants and animals on the land surface and along the beach and tidal zone of the coasts, these remains were swept into the ocean by streams, waves, and currents, coming to rest enclosed in continental shelf sediments. Perhaps turbidity currents carried some of this sediment to deeper water, where it came to rest on the sedimentary pile produced by the earlier phase of settlement of suspended sediment. If this late-Flood and early post-Flood activity added to the fossiliferous sedimentary sequence, as seems likely, those younger strata would have a very different appearance, being either shallow-water clastics spread by tidal currents or turbidites showing graded bedding. The fossil content of the younger sequence would, however, be distinctly different in biotic makeup from the older, graded sequence.

And now for the fossil sequence to be expected from changing Flood conditions as described above, Figure 39.1 shows the restored section and its fossil contents. At the

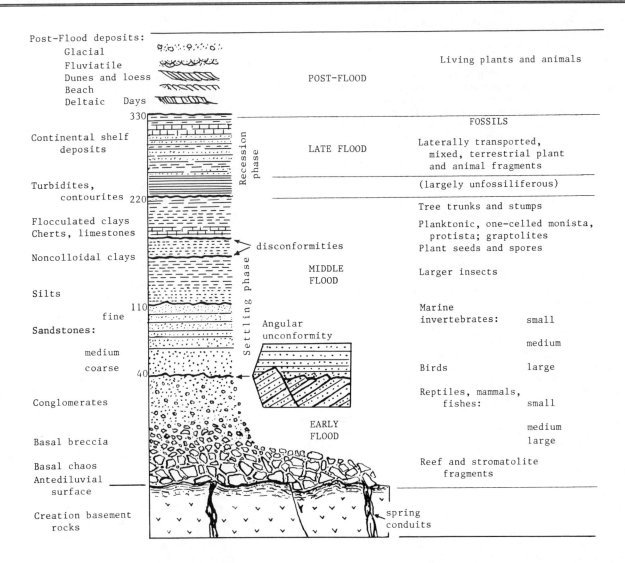

Figure 39.1 Schematic diagram of Flood stratigraphy as determined by hydraulic selectivity. (A. N. Strahler.)

base is the antediluvial land surface, with its gently rolling terrain and thick saprolite, overlying a basement of Creation rocks (nonfossiliferous). Recall that, as the Flood began, downfaulting on a grand scale dropped large areas of the Creation crust vertically downward along major normal faults to form the ocean basins that were required to receive the Flood waters. As underground waters burst forth from these basin floors, huge rock masses were broken out and lifted in the rising water jets. The angular rock masses could not have been lifted very high, and perhaps some were only overturned or rolled about, so that when the turbulence subsided they quickly formed the <u>basal chaos</u>. In the basal chaos we should expect to find large fragments of stromatolites and coral-algal reefs, which had thrived in the antediluvial period. The huge exotic blocks would grade upward into angular fragments of diminishing size, forming the <u>basal breccia</u>.

At about this point, we would find fossil remains of large reptiles and mammals, for example, large dinosaurs and large herbivorous mammals. Representative fossils of land animals would include <u>Brontosaurus</u>, <u>Tyrannosaurus</u>, <u>Stegosaurus</u>, and <u>Uintatherium</u>, along with mammoths and mastodonts, and also large marine animals, such as the plesiosaurs and whales. By the time their carcasses had circulated in the Flood melee for a couple of weeks, the expanding internal gases of putrification would have exploded their abdomens, preparing them for a rapid

plunge to the ocean floor. The basal breccia would have graded upward into conglomerate that contained pebbles rounded by stream and beach action; it would mark the first appearance of particles of terrestrial origin in the marine sequence. In the conglomerate beds we would find an upward gradation of reptiles and mammals from medium size to small size, along with large fishes and a few of the largest forms of invertebrates, such as the giant squid. The conglomerate would have graded upward into sandstone of diminishing grain size. In these beds we might find a heavy representation of marine invertebrates, grading upward from large to small individuals. Here, trilobites would abound, along with brachiopods, pelecypods, gastropods, crinoids, and small fishes as well.

Above the level of the sand and silt grades there would have been deposited noncolloidal clays and flocs of colloidal clay minerals. Here, too, would be found thinly bedded chemical precipitates, such as deepwater cherts and limestones. In this zone would be enclosed fossils of small, slowly settling organisms--insects, larval forms of various animals, and numerous one-celled organisms. An interesting fossil class of this zone would be the graptolites, whose large surface area and small mass would cause them to remain in suspension fully as long as such plant materials as seeds and spores. Leaves and flowers of the vascular plants would also be found here. Finally, at the top would be tree trunks and stumps, capable of floating for years (as we see in the case of the flotsam on Spirit Lake adjacent to Mount St. Helens). This would mark the termination of the marine settling phase, largely complete about 110 days after the Flood began, but with final stages of sedimentation persisting for decades after the Flood.

Looking more closely into the sedimentation process, we realize that the subterranean springs would emit upward jets of hot water, each powering a convection system consisting of a rising water column, spreading out radially in a near-surface layer and subsiding as a strong down-current in the zones between the vents. Thus the inter-spring zones of the ocean floor would be swept by strong currents completing the convection cells. Here, currents would prevent the coming to rest of all but the largest rock masses, but these would in turn act to trap and hold finer particles. As the jet-powered convection systems weakened and finally ceased (except for residual thermal convection), the effects of differential settling took over. However, sorting according to different settling velocities would not be perfect. At any level, from base to top, there would be included by entrapment of finer particles between coarser particles some particles of all grades finer than the dominant grade at that level. For example, the basal chaos would include representatives of all the fine grades; the coarse-sand grade would include some particles of medium and fine sand, silt, and clay. By this same process of entrapment of fines, there would be found within the entire stratigraphic section a repre-sentation of at least a few individuals or parts of individuals of all organisms living at the onset of the Flood. Parts and fragments of organisms would have abounded in the convection currents because of the continual crushing and grinding action of boulder-sized rock masses. Also, many of the organisms were abraded and fragmented in river floods and surf before entering the ocean. Because of their abundance, fragments should be found in all parts of the stratigraphic column. For example, hairs, feathers, teeth, claws, ears, eyes, bits of torn skin and muscle, and bone fragments of all species of animals should be universally present in the Flood strata.

During late-Flood time (the recession phase), turbidites would be laid over the deep ocean floors. These strata would be largely nonfossiliferous but might contain fragmental organic materials. Contemporaneously, continental shelf deposits would be forming, and these would contain fossils of all organisms that were alive before the Flood, their remains having been swept off the newly exposed land surfaces as the floodwaters receded. In other words, this accumulation of sediment and fossils represents a reworking by running water, waves, and tidal currents of materials that had settled out over higher surfaces at full Flood stage. This model makes no provision for terrestrial sedimentary accumulations in the fossiliferous record. Sand dunes, deltaic and fluviatile sediments, and glacial or eolian sediments could not have formed below the surface level of the floodwaters. As to occurrences such as the desert-dune sandstones of the walls of Zion Canyon and the Coconino cliff of Grand Canyon, both of which are overlain by fossiliferous strata, an explanation must be sought in later displace-ments and juxtapositions of sequences of strata, probably by tectonic activities (such as low-angle overthrust faulting) that resulted in interlayering of Flood strata with post-Flood deposits.

We are not home free just yet. Another problem looms, but it can be solved--like all the others--by a little ingenuity. The problem relates to tectonics and the modern continents. The stratigraphic section produced during the early and middle Flood stages is composed entirely of marine rocks and it lay on the floor of a deep world ocean. Today we see this section exposed in canyon and mountain walls, such as those of the Colorado Plateau, the Rockies, and the Appalachians. We must explain how ocean bottom became high-standing continent. Creation scientists have dealt with this question and we have presented their solution in Chapter 22.

Whitcomb and Morris state that, in the antediluvial

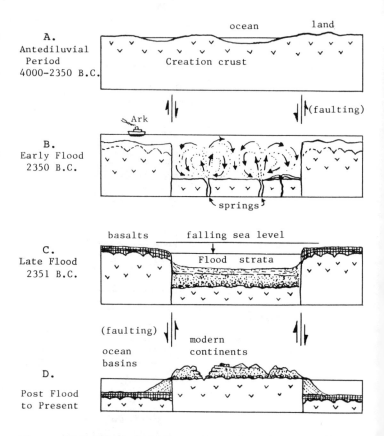

Figure 39.2 Flood tectonics and stratigraphy, according to the creationists' view. (A. N. Strahler.)

period, the earth's crust was in a condition of isostatic equilibrium, and that there were no very substantial regional density differences (1961, pp. 268-71). Mountains were relatively low and ocean floors were shallow. This condition is shown in Diagram A of Figure 39.2, in which I have attempted to show the various events by schematic diagrams. The crustal equilibrium state was a very precarious one, however, because of the great quantity of water locked up under high confining pressures deep in the crust. All this changed when the subterranean springs let loose, along with much volcanic activity. Crustal instability was such that the ocean basins deepened greatly by downfaulting (Diagram B), allowing the enormous quantities of floodwater to be accommodated, as well as providing a site for the enormous thickness of sediment that was to accumulate (Diagram C). I have shown great outpourings of flood basalts on the land surfaces, for reasons soon to become apparent. What happened next is not clear in detail from the Whitcomb-Morris text (see bottom paragraph on p. 268), but evidently a topographic reversal set in during the late-Flood period. Again, by faulting, but this time in reversed direction, the oceanic fault blocks rose to form continents, while the land blocks dropped down to become ocean basins. The terrestrial flood basalts came to form the oceanic crust as we know it today, and the ocean floors would have begun to received turbidites and other deep-ocean sediments. Continental shelves would have formed and receive terrigenous sediment, starting in late-Flood time.

I realize that you will use against this scenario my earlier argument (Chapter 22) that such drastic crustal displacements could not possibly have taken place in only a year's time because of the great viscosity of the asthenosphere. Creationists consider these geologic events to have been God's work, which did not need to conform with modern geophysical principles. The tectonics of the

Flood goes hand in hand with the starting and stopping of the Flood, these events being under God's direct control.

Please take note that Figure 39.1 and the foregoing paragraphs make no mention of evolutionary stratigraphic or time systems, such as "Proterozoic" or "Silurian" or "Eocene." Pure, unadulterated Flood stratigraphy makes no use of such evolutionary/uniformitarian terminology, because that brand of geology stands rejected out of hand. The biggest mistake for creation scientists to make (as Snelling has so forcefully pointed out) would be to attempt to make the biblical story of the Creation and Flood fit the evolutionary model. No such procrustean measures are permissible, nor are they necessary.

The depositional sequence of sediment and fossils I have outlined above is based on the description of the physical events of the Flood as described by the leading creationist writers: Whitcomb and Morris, 1961; Morris, 1974a; Wysong, 1976, and other creationists qualified in science. Using their descriptions, I have deduced a program of sediment deposition and fossil succession based on accepted hydrodynamic principles. I have not, however, arrived at the biostratigraphy observed and documented in the earth's strata by mainstream stratigraphy and paleontology--but that's no surprise at all, because the two scenarios are so unlike in their basic premises. What comes as a surprise is that my deductions of what happened during a cataclysmic one-year Flood differ in many important respects from the deductions made by the creation scientists themselves. Could it be that Whitcomb and Morris and their colleagues made an erroneous set of deductions to come up with a general sequence of fossils that seems to agree in many respects, but only generally, with the biostratigraphy of the evolutionists? To answer this question, we turn to a strict reporting of the story of fossils according to creation science.

The Creationists' Explanation of the Fossil Record

Creationists note that geologists have found a certain general ordering of fossil life forms, where a single exposure or well-log involves more than one geologic system. They say that in such cases "it is frequently found that the lowermost strata are those containing the simpler (and therefore supposedly more ancient) organisms, usually marine organisms" (Whitcomb and Morris, 1961, p. 273). In an earlier section, we showed that paleontologists reject the concept of "earlier being simpler" as necessarily descriptive of the full fosssil record. But it is fully accepted that (as Whitcomb and Morris point out) the fossil record of the Cambrian Period consists wholly of marine invertebrates of which about 60 percent are trilobites and 30 percent brachiopods. Marine invertebrates also dominate the Ordovician and Silurian periods, although the earliest vertebrates--jawless fishes --appeared in the Ordovician. Although creationists, as already pointed out, reject the Cambrian-Ordovician-Silurian stratigraphic sequence, they feel the necessity of explaining the dominance of marine invertebrates in those periods in terms of hydrodynamic selectivity.

Ignoring this puzzling and seemingly irrational ambivalence exhibited by the creationists, let us see how hydrodynamic selectivity would place the marine invertebrates at the bottom of the Phanerozoic column.

First, Whitcomb and Morris (p. 273) quote from a standard textbook of sedimentology by Krumbein and Sloss (1951, p. 156), stating the variable factors on which settling velocities of large particles depend. These are particle diameter, particle sphericity, and particle density relative to the fluid density. Whitcomb and Morris ignore the first variable factor, diameter, i.e., that large spheres settle faster than small ones. Actually, for size grades of medium sand and larger, settling velocity increases in direct proportion to the square root of the diameter of a sphere (the "impact law"). Whitcomb and

Morris make use only of the sphericity of the fossil individual and its mean density. Quite rightly, they point out that trilobites (when curled up tightly) and brachiopods with valves tightly closed are quite "streamlined" (smooth-surfaced and more-or-less spherical in outline, i.e., they are compact forms). They note that the density of shell materials--calcite and aragonite--is more than 2.6 greater than that of pure water. They conclude:

> These factors alone would exert a highly selective sorting action, not only tending to deposit the simpler (i.e., more nearly spherical and undifferentiated) organisms nearer the bottom of the sediments but also tending to segregate particles of similar sizes and shapes, forming distinct faunal stratigraphic "horizons" with the complexity of structure of the deposited organisms, even of similar kinds, increasing with increasing elevation in the sediments. (1961, p. 274)

On the same page they explain that the sorting powers of hydraulic action are valid statistically, rather than in an absolute sense, so that local variations are to be expected in organic assemblages. Nevertheless, they conclude:

> But, on the average, the sorting action is quite efficient and would definitely have separated the shells and other fossils in just such fashion as they are found, with certain fossils predominant in certain horizons, the complexity of such "index fossils" increasing with increasing elevation in the column, in at least a general way. (P. 274)

Let us ignore the part about increasing complexity, because it must be obvious that organic complexity can in no way be shown to have a significant positive correlation with the settling velocity of an individual of a species. Let us stick to the principle of segregation because of differences in settling velocities. That is, after all, the basis on which I deduced a fossil sequence in adherence to Flood conditions.

I now offer two arguments for complete rejection of the hypothesis that the lower Paleozoic strata are populated by fossils of marine invertebrates because those individuals had faster settling velocities than all other marine animals.

First, as Snelling has assured us, the fossilized life forms of the Proterozoic rocks must be included in the Flood record. The Proterozoic strata are known to be older than the Phanerozoic strata because of the stratigraphic order in which they always occur. This arrangement is vividly shown in the inner walls of Grand Canyon, as we explained earlier in describing the Algonkian wedge. The Proterozoic rocks contain no metazoan life forms (the Ediacaran fauna is actually within Phanerozoic time, according to the newer interpretations), only single-celled forms, whether living as independent individuals or in colonies. (The presence of what appear to be worm trails in Precambrian strata does, however, suggest the presence of metozons.) Some of the single-celled forms have been identified as members of the blue-green algae (Moneran Kingdom, Phylum Cyanophyta); others as bacteria (also monerans). Individual spherical algal spores and enigmatic forms (such as Eospheria) have been identified in microscopic sections; they have diameters on the order of 10 to 20 microns. These must have been free-floating (planktonic) organisms and, when dead, they must have had extremely slow settling velocities.

Putting these microscopic life forms into the perspective of Flood geology, we must accept their coexistence with all other life forms in the antediluvial period. They swarmed as plankton in the antediluvial ocean and built stromatolites in shallow tidal waters of that period. Under my reconstruction of the fossil record on a strictly hydrodynamic basis, the free-floating algal spores, bacteria, and other microscopic organisms coexisting in antediluvial time would have remained in suspension until the floodwaters had become almost perfectly tranquil; they would be incorporated into sediments of the late-middle and upper Flood period. They would thus be found above the marine invertebrates (metazoans) such as trilobites, brachiopods, and pelecypods; whereas it is obvious from field geology, open for all to inspect, that they are now found abundantly in strata older (and beneath) the lower Paleozoic strata.

On the other hand, masses of stromatolites, along with reef deposits and other bioherms, would have been broken into large blocks, which would have come to rest with exotic crustal blocks in the basal chaos below the zone of the marine invertebrates. While this position is correctly deduced from hydrodynamic principles, the stromatolites are actually found undisturbed in their places of origin in the Proterozoic strata; they were not fragmented and tossed about, as Flood geology demands.

Clearly, hydraulic selectivity has failed dismally as an explanation of the fossil sequence, as any alert person can verify by identifying fossils enclosed in the strata of the walls of Grand Canyon and many other such localities. Sequential sediment deposition in highly varied environments over hundreds of millions of years, entombing organisms that were alive at the time the enclosing layers were deposited, explains their distribution, both vertically and horizontally, quite apart from any consideration of whether evolution took place or not.

And now to a second point, that of the actual occurrence of fossils of large animals--reptiles and mammals--in strata of younger age (Mesozoic and Cenozoic) than those of older age (Paleozoic) dominated by invertebrates. Whitcomb and Morris seem to accept the conventional biostratigraphy but find the fossil arrangement to be easily explained by a number of causes acting in concert during the Flood:

> In general though, as a statistical average, beds would tend to be deposited in just the order that has been ascribed to them in terms of the standard geologic column. That is, on top of the beds of marine invertebrates would be found amphibians, then reptiles and finally birds and mammals. This is the order: (1) of increasing mobility and therefore increasing ability to postpone inundation; (2) of decreasing density and other hydrodynamic factors tending to promote earlier and deeper sedimentation, and (3) of increasing elevation of habitat and therefore time required for the Flood to attain stages sufficient to overtake them. The order is exactly what is to be expected in light of the Flood account and, therefore, gives further circumstantial evidence of the truthfulness of that account; in no sense is it necessary to say that this order is evidence of organic evolution from one stage into the next. (1961, p. 276)

Contrary to the last sentence, the order predicted and the reasons for it are not in keeping with the creationists' own account of the Flood, which I have already covered in detail and worked into a reasonable scenario based on the constraints of the flood dynamics. Point (1), increased escape potential through greater mobility, might be of value in a tranquil Flood, but the creationists describe great stream floods that swept animals into the sea. Surely, many of the land animals, of a wide range of sizes and orders--amphibians, reptiles, mammals--would have been rapidly swept into the raging, turbulent ocean, and their bodies there kept in suspension through the rising stages of the Flood. Tsunamis would have produced

a similar result. There is no reason at all to suppose that mammals would have greater mobility and escape potential than reptiles. Who is to say that _Tyrannosaurus_ was less able to take care of itself than a sabre-tooth cat during such trying conditions? Point (3) also implies a tranquil Flood but postulates an altitude zoning of animal habitats that is sheer conjecture and also sheer nonsense. Amphibians that might reasonably be expected to live in swampy coastal lowlands would have been among the first to be hit by the rising flood, but would not their amphibious habit have given them superior ability to resist submersion for long periods? For what reason would dinosaurs have been limited to a low-elevation habitat, segregated from mammals that required a high-altitude habitat? The creationists describe the antediluvial landscape as one of low relief, surely not likely to have produced strongly zoned habitats. Point (2) is that of hydrodynamic selectivity, and it is here that we can take issue with the hypothetical effects ascribed to it by the creationists.

If, as seems to follow from the effect of the Flood on land animals, both mammals and reptiles were swept to their deaths at the same time in the raging Flood, we must ask if there is any way in which they could have become so strongly segregated in the Flood strata. All dinosaurs are limited to the Mesozoic Era, none making the transition from Cretaceous to Paleocene. Most of the principal orders of the mammals do not appear until Paleocene or Eocene time.

Whitcomb and Morris go to some length to criticize the biostratigraphy of the Cenozoic Era as mainstream paleontologists have developed it since the time of Lyell (1961, pp. 281-83). They quote from conventional sources some criticisms of the use of the foraminifera as Cenozoic index fossils; they note that Lyell's original subdivision of the Cenozoic on the basis of marine invertebrates "is no longer considered definitive" (p. 283). But they seem to accept the mammals as bona fide guide fossils:

> Fossil mammals, however, are now considered the chief indicators of the various stages of the Tertiary, despite frequent popular textbook claims as to the worldwide provenance of marine index fossils. (P. 284)

At this point they quote supporting statements from a 1955 work of stratigrapher Maurice Gignoux. They continue:

> It must not be surmised from the above however, that these mammalian deposits are precisely identified and correlated on this worldwide basis. (P. 284)

They then recite their credo about the errant nature of mainstream biostratigraphy. Familiar as this statement is, it bears repeating, using their own words:

> The foregoing recital of past and present criteria for subdividing the Tertiary era seems to illustrate quite clearly our contention that the orthodox concepts of historical geology are almost entirely subjective in character, based squarely on the assumption of the fact of organic evolution. The variously correlated stages and even epochs are not at all based on the evidence of physiographic superposition, but rather on the paleontologic contents of the deposits, interpreted almost entirely in terms of assumed evolutionary development. (P. 284)

Having recited their credo, Whitcomb and Morris go back to the evolutionists' finding that the animal faunas of the Mesozoic are segregated stratigraphically from those of the Cenozoic, and furthermore, that within the Mesozoic

Era dinosaurs show evolutionary sequences within groups, and that within the Cenozoic Era mammals also show evolutionary sequences (p. 285). Their problem is to explain how events of the Flood would have segregated fossils at two levels: (1) at the level of the era--Mesozoic segregated from Cenozoic; (2) at the levels of periods and epochs within those eras.

At the first level, their explanation is that mammals had superior survival and escape capabilities, enabling them to hold fast to the land while the reptiles perished, but with the mammals eventually succumbing to the Flood. We have already protested that this explanation has no merit whatsoever and cannot be taken seriously.

At the second level, the creationists introduce a different principle of survival: the phenomenon of increase in size of mature individuals as each group undergoes evolution. In an unabashed acceptance of the evolution of taxa of dinosaurs and mammals, Whitcomb and Morris quote from paleontologists Edwin H. Colbert and George G. Simpson in support of the observed evolutionary size increase; it has become known as "Cope's law" (Whitcomb and Morris, 1961, pp. 280 and 285, respectively). Colbert pointed out:

> It is interesting to note that gigantism was achieved independently by various separate lines of dinosaurian evolution. Time and again in the collective history of these reptiles a phylogenetic line had its beginning with small animals and very quickly progressed to animals of large or even huge size. (1949, p. 71)

Simpson is quoted from Scientific Monthly (vol. 71, October 1950, p. 265) as pointing out that an increase in body size among Cenozoic mammals is very common, as in the case of the evolution of Eohippus (Hyracotherium) into the modern horse, but also adding that several evolving lines became smaller rather than larger.

And how do the creationists explain the increase in body size upward in the stratigraphic section? Not by evolution--most certainly not--but by the claim that the larger individuals had greater survival capability than the smaller! Large individuals could have outraced the rising floodwaters that drowned the smaller ones. Seen within the context of a single species, this claim might seem to have some merit, but it has to work perfectly or it fails totally.

Greater longevity equates to larger size. That being the case, consider that within even the largest species of dinosaur, the individual began life as a very small creature hatched from an egg. The life history of every individual reptile is a continuum of growth stages, growth ceasing only with death. If, as creationists believe, all species of dinosaurs were alive on the eve of the Flood, they would have perished strictly in order of body size. We would find the infantile individuals of all species of dinosaurs in a particular stratum; above it the next larger, and so forth, until the very biggest of all are in an uppermost layer.

Now, the same reasoning applies to the mammals, all of which were alive and cohabiting with the reptiles (and with all species of animals and plants) on the eve of the Flood. Thus, all mammal fossils will be matched in size with all reptile fossils in a stratigraphic sequence. That arrangement is most certainly a totally erroneous picture of biostratigraphy. We must reject the mechanism of segregation of sizes through graded survival and escape potential. That it is an absurd idea in the first place is patent when you realize that every animal faces the possibility of becoming weakened and incapacitated by disease or injury. In the population of animals living on the eve of the Flood there must have been many large animals in a weakened state, perhaps barely able to move and awaiting certain death by predation. These large

individuals would have perished with the smallest and would now be found in that relationship in the Flood strata.

Hydrodynamic Principles Applied

And so, back to the hydrodynamic principle of settling velocities proportional to particle size. All living forms of the lands were swept into the turbulent, toxic floodwaters within only a few weeks during the early rising phase of the Flood. Dead on arrival, if not sooner, the land animals were mixed by convection eddies with the dead marine biota. The mixing was thorough and sustained until the time that the fountains of the ocean floor were turned off. What followed is a simple exercise in pure democracy: each discrete particle in suspension, whether it was organic or inorganic in composition, settled to the bottom at a speed proportional to (a) its relative density, (b) its compactness or form (sphericity), and (c) its diameter or volume. A fourth factor that would affect the travel time to reach bottom would be the initial position of the particle when the water jets were turned off. Obviously, if the particle happened to be close to the bottom at the instant of turnoff, it would reach the bottom much sooner than an identical particle in an upper layer of the ocean.

It turns out that the factor of diameter or volume (size, that is), ignored by the creationists, is by far the strongest determinant of settling velocity within the class of biota with which we are most concerened, i.e., the animals. Bulk density of the body of a dead animal varies somewhat according to species. Creationists make a big point of this variation in allowing the marine invertebrates to settle out first, vis-a-vis land animals. But the differences are not consistent between, say, invertebrates and vertebrates. A bivalve, such as a particular species of pelecypod (a quahog, for example), may have a thick, dense mineral shell, but consider the turtle for comparison. The turtle has a thick, heavy shell and its shape, with head and feet pulled in, is just as streamlined as that of the quahog. The difference in settling velocity of the two, given equal body weights, may not be very great. Yet no turtle appears in the geologic record before mid-Triassic time. And consider the differences and similarities between certain reptiles and certain mammals. Would a dinosaur and a rhinoceros, both displacing the same volume of water, have markedly different settling velocities? I think not. And even if that were the case, we could find a dinosaur and a rhinoceros of nearly equal weights, both large animals, that would settle at the same speed. Yet with all the searching for fossils that has gone on, fossils of both a dinosaur and a rhinoceras have never been found side by side in the same stratum, or even in the same group of strata.

A little rough calculation, using the so-called "impact law" of settling of particles, will show that size is what counts most in settling velocities of dead animals. The impact law is stated as follows: Settling velocity, $\underline{v}$ (cm per second), is directly proportional to the square root of particle diameter, $\underline{d}$ (centimeters):

$$v \propto \sqrt{d}$$

The particle is assumed to be perfectly spherical, but we can reduce any particle, no matter how irregular its shape or rough its surface, to an ideal, imaginary equivalent perfect sphere. We can also introduce as a modifying parameter the difference in density between the particle and the fluid in which it is immersed. (Viscosity of the fluid is not a factor in the impact law.)

Figure 39.3 is a graph using logarithmic scales on both axes; it is often called "log-log paper." On this paper, any straight line rising from left to right in such a way as to travel two log cycles to the right as it travels one log cycle upward displays the square-root function in the

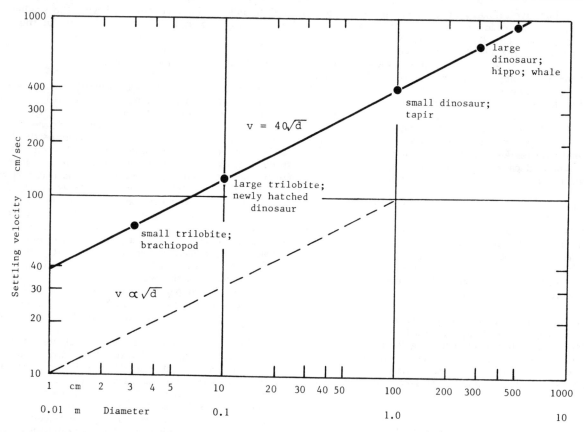

Figure 39.3 Relationship of settling velocity to particle diameter, with estimates of values for several kinds of animal bodies. (A. N. Strahler.)

impact law. In other words, a hundredfold increase in the particle diameter, $\underline{d}$, on the abscissa, is matched by a tenfold increase in settling velocity, $\underline{v}$, on the ordinate. The slanting line shown on the graph is drawn for perfect spheres of quartz, which has a mineral density of 2.65 g/cm^3. For spheres made of animal tissue--perhaps an assortment of eyeballs would be an example--the line would run a bit below the quartz line but parallel with it. We can use any square-root line to calculate the relative increase in settling velocity with increasing diameter. Table 39.1 gives examples useful for general comparisons.

Note that the large dinosaur would settle more than five times faster than the infant dinosaur, the latter being on the same order of size as a large trilobite There is no reason to think that a trilobite with a thin exoskeleton of chitin would have a higher density than the dinosaur. Even if we substitute a thick-shelled clam

Table 39.1   Settling Velocities of Selected Individual Animals

Representative individual	Equivalent sphere diameter, cm	Settling velocity, cm/sec
Small trilobite or brachiopod	3	70
Large trilobite; infant dinosaur	10	125
Small dinosaur; tapir	100 (1 m)	400
Large dinosaur; blue whale	300-500 (3-5 m)	700-900

for the trilobite, its density would probably not be increased to more than 150 percent that of the trilobite. This means that the large dinosaur would sink faster than any of the marine invertebrates by a factor of three or more. Thus, on purely hydrodynamic grounds, the creationists' explanation of the position of the early Paleozoic invertebrates below the large dinosaurs and large mammals must be rejected as erroneous. It seems, then, that no matter what selective mechanism the creationists choose to achieve the segregation of animal groups in the stratigraphic record, it fails to agree with what is logically deduced from their own scenario of the physical conditions and events of the Flood.

As if the stratigraphic segregation of higher taxa-- phylum and class--between geologic eras and periods does not pose enough problems for the creationists, we carry their hydrodynamic selectivity to lower taxa, such as order, family, and genus. I think of this level of selectivity as "fine tuning" and even "ultrafine tuning." They do this with the trilobites and brachiopods within the lower Paleozoic strata, with the ammonites in Mesozoic strata, and with the foraminifera in Cenozoic strata. With regard to the Mesozoic strata, Whitcomb and Morris state:

Proceeding higher in the geologic column (though not always, or even usually, higher in actual formational superposition), we come to the extensive Mesozoic strata, including the Triassic, Jurassic and Cretaceous systems. The "index fossils" for these strata are again marine organisms, especially the ammonites. Again there are many different kinds of these and of the other characteristic marine creatures of the period, and apparently they fall into large numbers of more or less distinct "horizons," which have been used as a basis for inter-regional and even inter-continental correlation. It is probable that these zones of similar assemblages can be explained on much the same basis as the zones of similar assemblages of trilobites and brachiopods in the Paleozoic strata. (1961, p. 279)

Their reference is, of course, to hydrodynamic selectivity as the process whereby the taxa of ammonites are arranged in order of complexity of sutures. The subclass of ammonoids consists of three orders-- goniatites, ceratites, and ammonites. The goniatites and ceratites range from Devonian through Triassic, the ammonites from Triassic through Cretaceous. Looking at comparative drawings of representatives of these orders, I judge them to be remarkably similar in general outward form, and there seems to be no valid reason to conclude that the older the form the greater was its settling velocity (Figure 39.4). The complexity of the sutures, which shows a steady increase from oldest to youngest, would surely not have made a significant difference in settling velocities. Remember that on the eve of the Flood the antediluvian ocean contained a full range of growth stages of individuals of all of the kinds of ammonoids, and that these were all to be thoroughly mixed in the floodwaters and sorted out by their settling velocities. If there were small but consistent differences in hydro- dynamic properties among the various orders, families, and genera, these would have been greatly overshadowed by the variations in size attributable to the growth cycle. The probability that, through hydrodynamic selection alone, the fossil sequence from Devonian through Cretaceous, with its remarkable sequence of changes in suture patterns from simple to highly complex, would happen to agree perfectly with the hydrodynamic properties of the individuals is infinitesimally small. Only "descent with modification" provides a fully logical and consistent explanation of the record as it is found in the strata.

We do not need to go over the same ground in connection with the foraminifera, which are one-celled organisms of the protistan kingdom. Nearly all are marine organisms of the deep-water zone and they have accumulated in deep-sea oozes. Whitcomb and Morris make a big point of the suggestion that the great variation in outward form of the foraminifera represents variation within a single species (1961, p. 282). That being the case, they say, the faunal zones of foraminifera are not related to evolution but are explained by hydrodynamic selectivity. There are, however, great differences in size between foraminiferal species. The Cretaceous and Cenozoic nummulites are disk-shaped or coin-shaped objects reaching a diameter as great as 2 to 3 cm, whereas most foraminifera are of microscopic size. These relative giants, which would have had the fastest settling velocities, are found in the Cenozoic record, strati- graphically well above the microscopic varieties that became abundant as early as the Carboniferous Period. Here the hypothesis of hydrodynamic selectivity fails utterly.

As a final note, consider the graptolites. They flourished in the Ordovician and Silurian periods, after which they had almost totally disappeared. Graptolites present themselves as delicate images on the bedding surfaces of black shale, probably deposited in very deep water. The word comes from the Latin graptos, for "painted" or "written." Graptolites come in an amazing variety, some being of branching patterns (dendroid, or dendritic) like the veins of a leaf; others are bladelike with sawtooth edges, like fragments of a jigsaw blade (Figure 39.5). The graptolites have been extinct since early Carboniferous time, and their taxonomic classifi- cation is uncertain, at best. They are interpreted as colonial animals equipped with floats to sustain them as plankton or attaching themselves to floating seaweeds. In other words, their settling velocity when alive would have been zero. Even without floats, their diaphanous form would have given them enormous surface area relative to mass, so they would have settled through the water with extreme slowness. Yet the graptolites are abundant in the lower Paleozoic strata, found at the same stratigraphic level as the trilobites and brachiopods (but usually in a

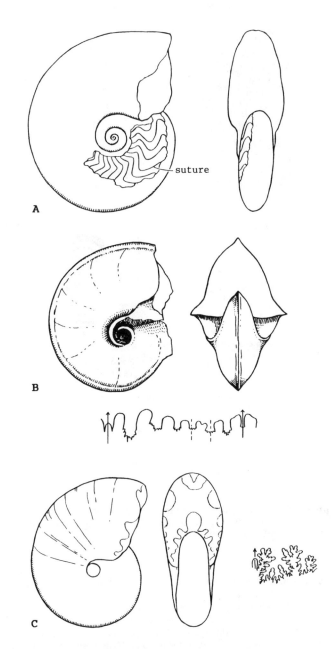

suture

A

B

C

Figure 39.4 Cephalopods (Ammonitoidea) of three different orders. A. Goniatite cephalopod from the Devonian. B. Ceratite cephalopod from the lower Triassic. C. Ammonite cephalopod from the Cretaceous. (From Stratigraphy and Life History by M. Kay and E. H. Colbert, p. 657, figures 26-15, 16, 17. Copyright © 1965 by John Wiley & Sons, Inc. Reprinted by permission of John Wiley & Sons, Inc.)

different rock type, or facies). Creationists: how do you explain that? Whatever happened to your hydrodynamic selectivity? Did it desert you when you needed it most?

Fossil Graveyards and Frozen Mammoths

Creationists are attracted to graveyards--the fossil kind, of course--because of their preoccupation with natural catastrophe. In The Genesis Flood, Whitcomb and Morris devote several pages to the subject (1961, pp. 154-69); the school textbook Scientific Creationism recapitulates this same material (Morris, 1974a, pp. 97- 101). Following their usual strategy, these authors draw from the literature of mainstream paleontology for numerous examples of the occurrence of "fossil grave- yards," where large numbers of individuals, sometimes of

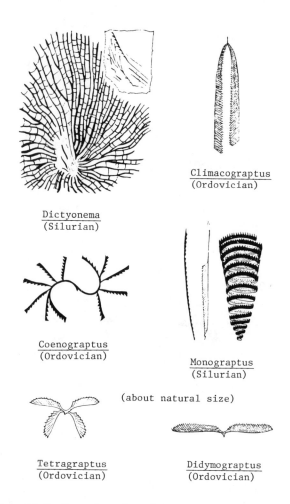

Dictyonema
(Silurian)

Climacograptus
(Ordovician)

Coenograptus
(Ordovician)

Monograptus
(Silurian)

(about natural size)

Tetragraptus
(Ordovician)

Didymograptus
(Ordovician)

Figure 39.5 Examples of varied types of graptolites. (After Karl von Zittel, 1900, <u>Textbook</u> of <u>Paleontology</u>, Macmillan and Co., New York.)

only a single species or kind, assembled or were brought to a single locality to perish together.

A good example of a fossil graveyard (good, because the way in which it came to be has been worked out in detail) is an accumulation of dinosaur bones found in 1877 in a Belgian coal mine at Bernissart. The discovery is recounted by Professor Edwin H. Colbert in his popular book <u>Men</u> <u>and</u> <u>Dinosaurs</u> (1968, pp. 55-59). Colbert is a leading authority on dinosaurs, and the creationists quote four such examples from his published works (Morris, 1974a, p. 98-99). At Bernissart, coal miners extending a horizontal tunnel at a depth of about 1,000 m below the surface penetrated a mass of fossil bones. Upon investigation by a noted Belgian paleontologist, it was ascertained that these were bones of the Cretaceous dinosaur <u>Iguanodon</u>, a huge herbivore often reaching a length of 10 meters. Three years of painstaking digging produced numerous complete and partial iguanodont skeletons. How did they arrive in one place in such great numbers? We will save the answer for later telling.

Colbert's accounts of dinosaur graveyards are perfectly written for quotation by the creation scientists. They include expressions such as these: "there were literally scores of skeletons one on top of another. It would appear that some local catastrophe had overtaken these dinosaurs, so that they all died together and were buried together" (p. 141); "in short, it was a veritable mine of dinosaur bones. . . . The concentration of the fossils was remarkable; they were piled in like logs in a jam" (p.151); "Innumerable bones. . . of dinosaurs and other associated reptiles . . . in . . . a stretch that is a veritable

dinosaurian graveyard" (p. 169); "the fossil boneyard was evidently one of gigantic proportions" (p. 58).

Also cited by Morris as an accumulation of vast numbers of fossils of a single kind in one stratum is that of fossil herring preserved on a single bedding plane in the Monterey shale of Miocene age in California (1974a, p. 97). Here more than one billion individuals died in an area of about four square miles (10 sq km).

From the work of Professor Norman D. Newell, a distinguished paleontologist specializing in the invertebrates, Whitcomb and Morris draw other examples (1961, pp. 159-60). One is the fossil graveyard in Eocene strata in central Germany where remains of thousands of vertebrate animals, insects, molluscs, and plants have been found preserved in lignite. Yet another example comes from the Karoo formation of Permo-Triassic age in South Africa. Newell cites the estimate of a South African paleontologist that "there are eight hundred thousand million skeletons of vertebrate animals in the Karroo formation" (Newell, 1959, p. 496). (Eight hundred billion skeletons--that's a lot! Think of that many individual vertebrate animals living on earth at one time--the night before the Flood began!)

The creationists move in fast and hard on this collection of graveyard stories, gathered from impeccable sources within mainstream geology. They say:

> To attempt to account for these vast graveyards in terms of present-day processes and events, except via the most extreme and unscientific extrapolation, is absolutely impossible! (Whitcomb and Morris, 1961, p. 161)

The creationists insist that fossil graveyards are proof of the Flood, which provided ideal conditions for entrapment, burial, and fossilization of vast numbers of individuals in a very short time. The fatal flaw in their explanation is one we have already developed at length in predicting the consequences of a chaotic flood dominated by turbulent mixing and subsequent settling according to the principle of hydrodynamic selectivity. Under that principle, there might well have been great fossil graveyards, but each would contain a heterogenous mixture of animals of many taxa, but of essentially the same hydrodynamic characteristics.

And now, back to that Belgian coal mine with its horde of fossil iguanodonts. Colbert explains what happened (1968, p. 58). Full excavation of the bone-rich material, an unstratified clay, showed that the deposit occupies a deep, narrow fissure or ravine that had been carved in Cretaceous time into layered shales and coals that comprised the indigenous bedrock of that landscape. In only a few years, numerous iguanodons had slipped by accident into the ravine, falling to their deaths and accumulating in disorder. Clay, washed into the bottom of the ravine by rains, continually buried the remains and allowed them to be preserved.

Similar examples have been reported of fossil accumulations in the floors of ravines or shafts in regions of cavernous limestone formations. Dunbar and Waage report one such occurrence near Cumberland, Maryland, where bones of 46 species of Pleistocene vertebrates have been found in the floor of a deep sinkhole in limestone (1969, p. 56). Whitcomb and Morris refer to this locality, describing the varied fauna it contains, but they ignore the obvious explanation of entrapment (1961, p. 158). They say that this kind of occurrence does not lend itself well to uniformitarian interpretation but strongly suggests, instead, some sort of very unusual catastrophe (1961, p. 158). I guess we might concede that for an individual taking that final fall to extermination, the mechanism could be regarded as "some sort of very unusual catastrophe."

For most assemblages of large numbers of fossils of a single species or group, the explanation seems to lie in a "catastrophic" incident such as a local storm or flood, which either swept large numbers of individuals to their death in a localized site of sedimentation, or buried them in place with sediment transported from an adjacent locality. Such incidents through geologic time would be predicted by mainstream geology because similar events are recorded in modern times. Another catastrophic cause of mass killing is by burial under volcanic ash. Examples are the ash beds of the John Day Basin in Oregon and Lake Florissant in Colorado, which contain excellent fossils of plants and insects.

Dunbar and Waage give a full discussion of the various ways in which quick burial of animals and plants can take place, leading to their fossilization (1969, pp. 48-58). Particularly interesting is the example of a slate formation of lower Jurassic age at Holzmaden, Germany, where fossils of hundreds of marine reptiles in an excellent state of preservation have been excavated. Here is the explanation of that fossil accumulation:

> The black muck of the sea bottom on which the ichthyosaur carcasses accumulated was obviously toxic. No scavengers were present to tear apart the bodies and scatter the bones; indeed the Holzmaden fauna lacks any kind of indigenous bottom-dwelling animals. Swimmers are dominant and include plesiosaurs, marine crocodiles, a few fish, squidlike cephalopods, and a few others in addition to the great numbers of ichthyosaurs. . . . Sulfur compounds in the shale indicate that the bottom was made toxic by hydrogen sulfide. Stagnant and toxic areas of black mud are known in areas of present-day seas where the bottom is in a depression or otherwise cut off from circulation of oxygen-bearing water. What attracted the ichthyosaurs and other animals into the stagnant bay that apparently existed at the Holzmaden site is not known, but their death was most likely due to toxic or poorly oxygenated water and their preservation was assured by the antiseptic mud that buried them. (Dunbar and Waage, 1969, p. 51)

This example is particularly valuable because the process of burial is not catastrophic in the sense of a single event causing mass killing; rather it is quite uniformitarian in principle. Other forms of continuous burial that might be characterized as "uniformitarian" are in lake sediments, where toxic conditions may also have existed (as in the case of the Green River shales, discussed earlier) and in asphalt seeps, such as the famed Rancho La Brea "tar pits," where hundreds of thousands of bones of trapped Pleistocene animals have been recovered. Similar traps for Pleistocene animals were provided by bogs and by quicksand in artesian springs.

The arctic tundra regions of North America and Siberia are characterized by vast areas of permanently frozen subsoil, a condition known as permafrost. Although thawing affects a thin surface layer during the summer, all soil water below that depth remains frozen as ground ice, which often takes the form of ice layers or ice wedges, the latter being vertical in orientation and narrowing at depth. In both Alaska and Siberia the remains of Pleistocene animals, now extinct, have been found in silts and clays in the permafrost areas. In some instances the continuous state of frost has resulted in preservation of skin and body tissue along with the bones.

Whitcomb and Morris (1961, pp. 288-89) quote at some length from Richard Foster Flint's major work, Glacial and Pleistocene Geology, descriptions of the faunas preserved in the frozen substrate (1957, pp. 471-72). Flint states that the Alaskan faunas are enclosed in silty alluvium that is older than the Wisconsinan glaciation and is probably from an interglacial stage.

Much more exciting to the average person is the finding in Siberia of a vast number of mammoth tusks-- estimated at 50,000 tusks--that have been collected and sold over the past two centuries to the ivory trade. In a few instances entire and partial mammoth bodies have been found preserved in the frozen ground. Flint goes on to give some details:

> These finds have fostered many tales of great catastrophes, for which there is no factual support. Being chemically durable, the tusks long outlast the rest of the fossil skeleton; there is little reason to doubt that they could have accumulated throughout many thousands of years. In the most fully documented case, that of the Berezovka mammoth, death occurred at the beginning of autumn and was caused by suffocation following a fall that broke many bones. The occurrence can be explained on the assumption that the elephant, feeding at the top of a river bluff, caused a slumping movement that precipitated him into the muck of the floodplain below, where his carcass was soon frozen in. (P. 470)

Flint places the Berezovka event near the end of the warm climatic stage (-8,000 to -5,000 y.) of the Holocene Epoch in which air temperatures were somewhat higher than today. Whitcomb and Morris quote only the first sentence of the above passage (1961, p. 289). They leave out Flint's simple and straightforward explanation of the mammoth's demise as based on an autopsy and a study of the enclosing sediment. Instead, they prefer to take off on a rather speculative excursion, emphasizing the theme of great catastrophe:

> And the Arctic Islands north of Siberia have been described as even more densely packed with the remains of elephants and other mammals, as well as dense tangles of fossil trees and other plants, so much that the entire islands seem to be composed of organic debris. No wonder these things have "fostered tales of great catastrophes"; the wonder is that uniformitarians could possibly offer any other explanation in any seriousness! There is most certainly no modern parallel entombment of elephants or any other kind of mammal taking place anywhere in the modern world. It may not be clear as yet whether these deposits were made directly during the Deluge period or soon after, or both, but it seems fairly evident that the extermination of such immense hordes of animals and their interment in what has ever since been frozen soil must somehow be explained in terms of the events accompanying just such a universal aqueous, catastrophe as the Bible describes. (P. 289)

A footnote explains that the sediments of the arctic tundra region settled out from Flood water and froze rapidly, forming the permafrost and at the same time entrapping the animals (p. 290). But nothing in the scientific accounts of occurrences of fossils in frozen ground in the arctic lands can be construed as representing "the extermination of such immense hordes of animals." The entrapment and burial of one or a few individuals at a time could have gone on over hundreds or thousands of years during interglacial stages.

Whitcomb and Morris go on to quote at some length from the writing of Ivan Sanderson, an individual described in a footnote as "a field zoologist and author of numerous volumes on wild life" (pp. 290-92). In this case, the material quoted is from Sanderson's article in the Saturday Evening Post (January, 1960, p. 83). But Sanderson has his own version of catastrophe and it is not the biblical Flood. It is a sudden extrusion of dust and gases that would have caused a catastrophic global climate disturbance, during the course of which great globs of volcanic gases, cooled to subzero temperatures during ascent to high altitudes, descended to the earth's surface and quick-froze the mangled mammoths. The creationists do not like Sanderson's hypothesis, for it conflicts with the Bible, but they do like his catastrophic point of view.

Under the more gradualistic explanation supported by Professor Flint and others of the mainstream science community, the frozen bodies would be expected to show signs of having been mangled and torn, perhaps by predators and scavengers, and by natural soil movements accompanying freezing and thawing of the active permafrost layer. Their remains might well have been piled together in a single wide, deep crevasse produced by slumping of alluvial silts of a caving river bank.

Claims have been made by creationists that the frozen mammoths, such as the Berezovka mammoth, showed from the frozen stomach contents a diet of "subtropical plants like buttercups," and that the flesh of some mammoths is so perfectly preserved that it has been cooked and eaten by humans in modern times (as reported by Christopher Weber, 1980, p. 15). These claims have been examined by Professor William R. Farrand, a geologist specializing in the Pleistocene Epoch (1961). He shows that plant materials found in the stomach of the Berezovka mammoth were all of arctic plants. The flesh of that animal was greatly putrified, with deeply penetrating chemical alteration, and would not have been edible by humans except at severe risk of poisoning. Christopher Weber, in reporting on Farrand's findings, presents the following counterattack on the creationists' catastrophic explanation:

But on top of all this, there is additional evidence that a literal Flood of Noah could <u>not</u> have deposited these mammoth remains. Farrand points out that we find no other species of frozen animals in Siberia <u>except</u> mammoths and woolly rhinoceri. Since these animals were so big and clumsy, they had trouble in crossing crevices in the earth's surface, just as modern elephants do. This evidence fits well with the theory that mammoths fell off cliffs and were buried in ways that more mobile animals like horses and bison were able to avoid. Yet, if the Flood of Noah were literal history, we would expect to find many <u>different</u> species of frozen animals, not just the mammoths and woolly rhinoceros. (1980b, pp. 15-16)

The argument seems valid for Siberia, but Flint gives a long faunal list of animals--24 in all, including many small, agile species, such as wolf, fox, jaguar, and lynx found frozen in central Alaska (1957, p. 471). But then, the list of animals found in the Rancho La Brea asphalt includes similarly agile creatures. Perhaps the answer is the same in both cases: the cries of trapped large animals attracted the agile predators, which also became mired in the sticky mud or tar. In any case, there is no compelling reason to turn to a hypothesis that requires mass extinction of a great horde of animals in a matter of days or hours.

PART VII

Integrity of the Evolutionary Record
under Attack by Creationists

Introduction

Creation scientists go to great lengths to argue against the biostratigraphy and phylogeny that are interpreted by mainstream biologists and geologists to be products of evolution through descent with modification. We can recognize three major categories of subject matter in this argument: (1) discontinuities of the fossil record; (2) inversions of the record; (3) anomalous coexistences within the record.

Inversions of the record include (a) reversal of the standard order of fossil progression, explained by overturning of strata in recumbent folds, and (b) juxtaposition of entire stratigraphic units in reverse order of the standard biostratigraphy, a condition explained by mainstream geology through overthrust faulting.

Discontinuities of the fossil record include (a) sudden appearances of new major taxa, without a fossil record of ancestral forms, (b) gaps in the fossil record of descent of evolutionary lines that must have been continuous through time according to the hypothesis of evolution, and (c) sudden extinctions of major taxonomic groups.

An anomalous coexistence consists of the finding, or alleged finding, in a single stratum or on a single bedding surface, of two or more fossil varieties that, in the standard biostratigraphic column, are widely separated in age. In simpler language, fossils are claimed to be found grossly out of place in time.

Both inversions of record and anomalous coexistences are regarded by creationists as capable of summarily falsifying the hypothesis of evolution--they are "sudden-death" situations, since the finding of only a single clear example, inexplicable by naturalistic causes, would vitiate the entire hypothesis of evolution. In the area of discontinuities of record, the potential for summary falsification is far weaker, because the argument is over something that is missing, or absent, i.e., something that is claimed not to exist. The difficulties of proving the existence of nonexistence are obviously monumental. In contrast, juxtaposed and inverted strata can be seen as such by anyone who cares to go and look at the outcrop, and any two observers can argue over the interpretation of what both of them accept as reality. A simple analogy might be the case of a crime you claim has been committed in your dwelling by an intruder: it is not difficult to convince the police that someone bashed your stereo set into junk, but not so easy to convince them that your stereo set has been stolen without a trace.

Chapter 40

Inversions of the Order of Strata

At various places on the continents, geologists observe that the order in which fossil faunas progress from bottom to top of a series of strata is the reverse of that shown in the standard table of geologic time. In one general case, a particular age sequence is repeated in reverse order. For example, the steep rock face of a mountainside or cliff may show from the base upward the normal, expected succession: Triassic-Jurassic-Cretaceous. But above the Cretaceous is more Cretaceous in which the order of the particular lithologic units (limestone, shale, and sandstone) and the faunal zones within them occurs in the exact reverse of the lower sequence. Moreover, this reversed upper sequence continues through Jurassic and Triassic formations. The conclusion is clear enough; the entire succession of strata has been overturned upon itself, much as if you took a pile of carpet samples and folded it back on itself. Overturning can be verified by examining the surfaces of bedding planes and observing any of a number of features that tell which side is "up." Ripple marks, cross laminations, mud cracks, the attitude of fossil shells, and the direction of size gradation within graded bedding are examples of such diagnostic features.

Overturning, Thrust Sheets, and Nappes

Overturning of strata is associated with intense folding in tectonic belts formed by continental collision. Figure 40.1 illustrates stages in the growth of a large fold that became asymmetrical, then overturned, and finally fully recumbent. The reversed sequence is within the lower limb of the recumbent anticline. Figure 40.2 is a cross-section of a recumbent fold exposed on a steep face of a mountain, the Grand Morgon, in the French Alps. The European names of the Triassic and Jurassic units reappear in reverse order in the higher part of the slope. The lower diagrams in Figure 40.1 show that in a more advanced stage of development the recumbent fold becomes torn by a low-angle overthrust fault. As the upper mass overrides the lower, a thrust sheet is formed. European geologists who first studied these complex structures in the Alps named the thrust sheets nappes. Figure 40.3 shows multiple nappes and recumbent folds in the Helvetian Alps.

The second general type of inversion of the normal stratigraphic order is the case in which two rock layers--call them A and B, with A being older than B--are found to be superimposed in reverse order of age, i.e., A lies upon B. In this case, however, neither layer has been overturned. The only rational geologic interpretation of such a situation is that A, the older layer, has been forced to slide up over layer B on an inclined fault surface--an overthrust fault, that is. In the simple case shown in Figure 40.4, older rock lies above younger rock at any given point on the fault plane. Most such occurrences are easily recognized by geologists and cause no problem of interpretation. The geometry of the overthrust fault is such that it suggests the action of compressional forces within the earth's crust and lithosphere, because any two reference points, one on

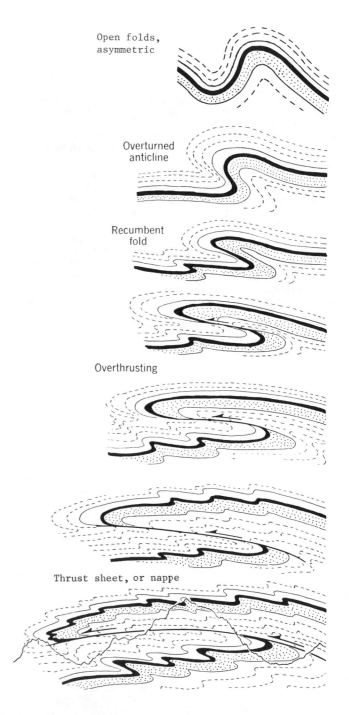

Open folds, asymmetric

Overturned anticline

Recumbent fold

Overthrusting

Thrust sheet, or nappe

Figure 40.1 Development of overturned and recumbent folds and an overthrust sheet, or nappe. (A. N. Strahler. Suggested by drawings of A. Heim, 1922, Geologie der Schweiz, vol. II-1, Tauschnitz, Leipzig.)

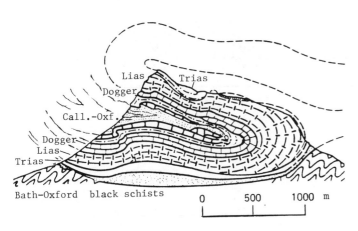

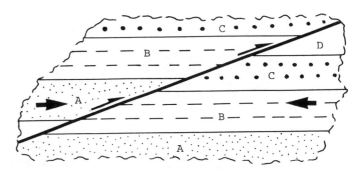

Figure 40.4 Schematic cross section of an overthrust fault, showing age relationships across the fault trace. (A. N. Strahler.)

Figure 40.2 Normal and reversed stratigraphic sequences in a recumbent fold of the Grand Morgon in the French Alps. Trias is Triassic; Lias, Lower Jurassic; Dogger, Middle Jurassic; Callovian (Call.) and Oxfordian (Oxf.), Upper Jurassic. (After Schneegans, 1938, in de Sitter, 1956, <u>Structural Geology</u>, McGraw-Hill, New York. From B. E. Hobbs, W. D. Means, and P. F. Williams, <u>An Outline of Structural Geology</u>, p. 416, Figure 9-23a. Copyright © 1976 by John Wiley & Sons, Inc. Reprinted by permission of John Wiley & Sons, Inc.)

each side of the fault, are moved closer together in horizontal distance of separation as fault motion continues.

Complications and difficulties of interpretation arise in tracing and interpreting overthrust faults as the attitude of the fault surface approaches the horizontal. The upper layer now assumes the nature of a vast thin sheet, often several tens of kilometers in horizontal extent. Such thrusts are definitely associated with the great orogens that form during arc-continent and continent-continent collisions. They seem to lie always on the continent side of the nappe zone in what is known as the <u>foreland belt</u>; it is characterized by deformation of a layer of horizontal sediments that was formerly a continental shelf wedge of a passive continental margin. Here we find that the thrust sheet characteristically coincides with a stratigraphic horizon, which may be (a) the upper surface of the basement of ancient crystalline rocks (metamorphic and igneous) or (b) a bedding surface between two sedimentary formations.

The first kind of foreland thrust--that in which a sedimentary layer moves over a crystalline basement--is referred to as a <u>décollement</u>, a French word meaning "detachment" in the sense of "coming unstuck." The rock sheet above the décollement zone may deform in one of two ways, as shown in Figure 40.5A. One is by wrinkling into a succession of open, wavelike folds. The other is by breaking up into a set of overlapping overthrust slices. The faults tend to steepen upward and the slices between them are stacked like shingles on a roof--"imbricate structure," it is called. Both thrust faulting and folding

may occur at the same time, giving an extremely complex structure to the moving rock sheet as a whole. Figure 40.6 shows the extremely complex combination of thrust faulting and folding seen in the Canadian Rockies. Geologists refer to the crustal shortening resulting from such faulting and folding as "telescoping." The overthrust layer at its outermost limit (away from the axis of the orogen) may ride out over the land surface, much as an advancing ice sheet spreads over the ground. In so doing, the thrust sheet may override very young layers of sediment that have been eroded from the sheet itself.

In the second case, the fault plane closely follows a bedding plane and is known as a <u>bedding thrust</u> or a <u>bedding slip</u>. Apparently, the resistance of shearing was least along that particular stratigraphic horizon, which may consist of a weak shale formation between two more massive formations of limestone or sandstone. Figure 40.5B is a schematic drawing of a bedding fault; it shows that a cross-cutting thrust fault (left) merges with the upper surface of formation B. Now sequence A-B is repeated. A bedding thrust may be difficult to identify in the field, but its presence must be inferred when an older group of strata overlies a younger group, as shown by their respective fossil contents.

The foregoing description of some of the strange features of foreland overthrusts may seem unnecessarily complicated but can be most helpful in interpreting and evaluating the creationists' arguments against the conventional or mainstream stratigraphy based on principles of superposition and succession of faunas. So let us turn to their allegations and the way in which they use published reports and statements by mainstream geologists to support their own version of Flood geology superimposed on the religious doctrine of recent creation.

The Creationist View of Bedding Thrusts

Bedding thrusts, in particular, draw the attention of the creationists because, as seen from a distance, the strata above and below the fault plane may seem to form a continuous depositional sequence, such as we find in the

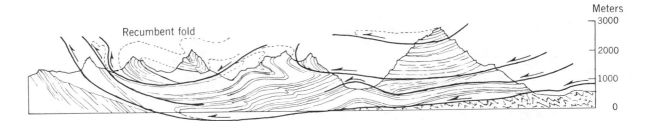

Figure 40.3 Structure section through a portion of the Helvetian Alps, Switzerland, showing nappes. Horizontal and vertical scales are the same. (Simplified from A.

Heim, 1922, <u>Geologie der Schweitz</u>, vol II-1, Tauschnitz, Leipzig.)

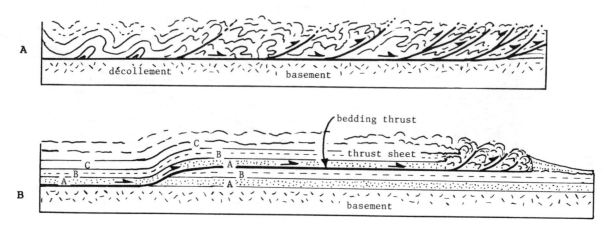

Figure 40.5 Schematic cross sections of (A) décollement with splay thrusts branching upward, and (B) bedding thrust in horzontal strata. (A. N. Strahler.)

upper walls of Grand Canyon, broken only by unseen disconformities. Creationists, looking at this scene, claim that the physical principle of superposition applies, and that if the fossil zones are actually found in what evolutionists claim is an inverted order, it is only because evolution is a false theory. Faunal succession, they say, is an erroneous concept to begin with. If, at a particular locality, we find so-called "Cambrian" fossils in strata overlying a formation that contains so-called "Jurassic" fossils, it is because the entire geologic column of eras, periods, and epochs based on faunal succession is pure fiction. This argument was strongly voiced by creationist writer George McCready Price whose seminal 1923 work, The New Geology, carried most of the standard geological topics picked up much later by the self-designated "creation-science" school, beginning with publication of The Genesis Flood by John C. Whitcomb and Henry M. Morris in 1961.

Whitcomb and Morris correctly review the standard explanations of age-inversions of strata as being low-angle overthrusts, nappes, or detachment thrusts (décollements) (1961, pp. 180-81). They note that subsequent erosion is considered to have removed a substantial layer of younger strata from the upper part of the original thrust sheet. They comment as follows on this conventional explanation:

> It is recognized that phenomena of this sort have taken place on a small scale, in certain localities

Figure 40.6 Telescoping of a great thrust sheet is seen in this structural cross section of the Canadian Rockies, between the Bow and Athabasca rivers in Alberta and British Columbia. Horizontal and vertical scales are the same. Notice that the dip of the thrust planes steepens toward the surface. (Prepared by R. A. Price from data of the Geological Survey of Canada, Geological Association of Canada, Special Paper, No. 6, 1970.)

where there is ample evidence of intense past faulting and folding. However, these visible confirmations of the concept are definitely on a small scale, usually in terms of a few hundreds of feet, whereas many of the great overthrust areas occupy hundreds or even thousands of square miles. It seems almost fantastic to conceive of such huge areas and masses of rocks really behaving in such a fashion, unless we are ready to accept catastrophism of an intensity that makes the Noachian Deluge seem quiescent by comparison! Certainly the principle of uniformity is inadequate to account for them. Nothing we know of present earth movements--of rock compressive and shearing strength, of the plastic flow of rock materials, or of other modern physical proceses--gives any observational basis for believing that such things are happening now or ever could have happened, except under extremely unusual conditions. As Hubbert and Rubey[*] admit: "Since their earliest recognition, the existence of large overthrusts has presented a mechanical paradox that has never been satisfactorily resolved." (Whitcomb and Morris, 1961, pp. 180-81)

[*](Hubbert and Rubey, 1959, p. 122.)

These creationist authors have completely thrown over their allegiance to catastrophism in the matter of thrust sheets. Catastrophism is the keystone of Flood geology. Catastrophism is the criterion by which creationists distinguish their view of the universe from that of mainstream geology, which they characterize as a uniformitarian system. Why then would they now espouse uniformitarianism and reject catastrophism? The answer is rather obvious: they are in this instance in the hot pursuit of a means whereby they can (they think) upset faunal succession and the theory of descent with modification. All they need to accomplish this aim is to show that extensive bedding thrusts do not exist, and that simple superposition was, instead, the mode of

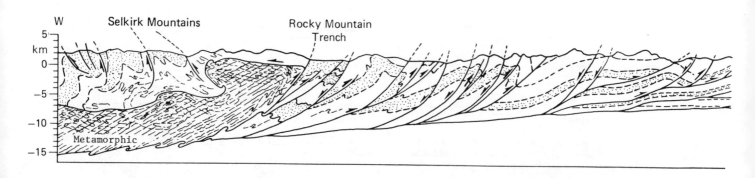

accumulation of the rock sequence in question. They seem to view this goal as worth any price that must be paid. In this case, the price is to repudiate their keystone doctrine of catastrophism. Never mind that in describing the Flood they invoked incredibly enormous rapid displacements of the crust in defiance of the principle of isostasy, that they appealed to volcanism on a titanic scale, along with fierce seismicity that generated truly incredible tsunamis, and to top it all, that greatest of cataclysms--the horrendous collapse of the global vapor canopy to dump a world ocean of water on the earth in only weeks. From unbridled imaginative excursions of unmatched scope, they have retreated to a sober-minded, uniformitarianist contemplation of the earth. Well, gentlemen, you cannot say it both ways in the pages of one book and expect anyone to take you seriously. Which is it to be--catastrophism or uniformitarianism?

Whitcomb and Morris devote a full twenty pages to thrust sheets, focusing upon two examples (1961, pp. 180-200). One is the great Lewis Overthrust seen in Glacier National Park in Montana; the other is the Heart Mountain Thrust of Wyoming. Although Whitcomb and Morris take up the Heart Mountain Thrust first, we shall place it second, for reasons that will become clear later.

The Lewis Overthrust

The Lewis Overthrust is exposed for some 200 km along the eastern front of the northern Rocky Mountains from northern Montana into Alberta. Here, thick, massive Precambrian sedimentary formations form the bulk of the mountain mass. Known as the Belt Series, its formations total from 3 to 7 km in thickness. Prominent as cliff makers in the area of the Lewis Overthrust are two great limestone formations, the lower one about 600 m thick; the upper one twice that thickness. Also present are quartzite and shale beds. A particularly noteworthy feature of the massive limestones is the presence in abundance of reeflike bodies of algal stromatolites, to which we shall make reference later. During the Laramide Orogeny that closed the Cretaceous Period, collision impacts from continental fragments and island arcs approaching from the west caused the Precambrian strata, which may have been little deformed up to that time, to be folded and thrust eastward over Cretaceous strata.

Figure 40.7, a diagram published about 80 years ago, is a greatly simplified version of the tectonic changes that took place. The Lewis Overthrust is shown as originating in a recumbent fold that broke along a low-angle thrust fault, carrying the Belt Series over the Cretaceous strata. Keep in mind that the upper part of the overthrust that seems to hang out over "thin air" would have been undergoing rapid erosion from the time of the first folding, for as long as the fault motion continued and, of course, long after the motion ceased. (Figure 40.6 gives a much more detailed rendition of the details of overthrusting and folding in this mountain belt, but at a point much farther north than the U.S.-Canadian boundary.) Horizontal displacement on the Lewis Overthrust is estimated to have been 55 to 65 km. The fault surface is gently inclined to the southwest at a dip

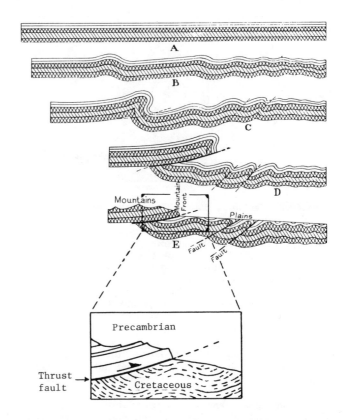

Figure 40.7 Simplified cross sections showing folding and overthrusting leading to production of the Lewis Overthrust in the region of Glacier National Park. (From National Park Service, 1937, <u>Glacier National Park</u>, U.S. Government Printing Office, Washington, D.C., p. 32.)

angle of about three degrees, more or less. Because the Cretaceous strata lying below the thrust plane are soft and easily eroded by running water, their rapid removal has tended to undermine the overlying massive limestone formations of the Belt Series, producing clifflike mountain walls and a rather sharp break in slope along the line of exposure of the fault plane.

A very large proportion of the eastern part of the thrust sheet disappeared entirely by erosion through Cenozoic time. At the same time the mountain front became ragged in plan (as seen from the air or on a map), with many promontories projecting eastward into the plains country. Here and there an isolated, islandlike mass of the Belt rocks can be seen as an erosional outlier, completely surrounded by Cretaceous rocks. One such outlier is Chief Mountain, shown in Figure 40.8. Its resemblance to a typical butte, such as those so splendidly displayed in the Monument Valley of Arizona far to the southwest, has established it as a tourist attraction, often a convenient object over which to raise the creationist controversy that surrounds the meaning of the Lewis Overthrust. With this background description of

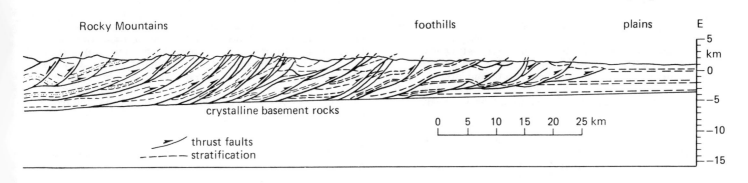

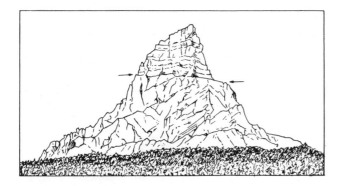

Figure 40.8 Geologist Bailey Willis made this sketch of Chief Mountain while doing pioneer field work in the region of Glacier Park. The upper part of the peak (above the line marked by arrows) is solid, flat-lying limestone, whereas the lower part is limestone broken by many oblique thrust faults. The Lewis Overthrust lies under the base of the mountain and is underlain by much younger shales. (B. Willis, 1902, Stratigraphy and Structure of the Lewis and Livingstone Ranges, Montana, Bull., Geological Soc. America, vol. 13, p. 307.)

the Lewis Overthrust in mind, let us turn to the creationists' critique of the accumulated data of several decades of field study of this rugged area by mainstream geologists.

Whitcomb and Morris are commendably cautious in their description of the Lewis Overthrust. It seems that Price had stressed the appearance of a normal stratigraphic conformity of the Belt strata on those below it, ignoring the numerous indications of compressional tectonic activity in the form of folding and cross-thrusting both above and below that contact. These conspicuous deformational structures had been described in earlier works of the U.S. Geological Survey (Willis, 1902; Campbell, 1914, pp. 10-11). A Geological Survey report by C. P. Ross and Richard Rezak emphasizes the deformation of the Belt strata; a photo shows "one of several zones of relatively intense crumpling" (1959, pp. 421-22, Plate 53).

Geochemist J. Laurence Kulp (mentioned earlier as a theistic scientist who urged repudiation of the extreme fundamentalist position), in his general article refuting many of Price's creationist statements, brought out the significance of structures that favored the overthrusting explanation (Kulp, 1950). Kulp was contending that there is ample evidence for the great extent of overthrusting quite apart from the reversed geologic ages of the strata as indicated by their fossil content. Whitcomb and Morris acquiesce to Kulp's point but consider the tectonic features to be on a small scale, not indicative of great overthrusting. They write:

> Such activity is to be expected in connection with mountain-uplift processes, whatever the nature or cause of those processes may be. On a small scale, it is evident that overthrusting has actually occurred in many places.

> Nevertheless, it requires a tremendous and entirely unwarranted extrapolation to infer from these small-scale folds and thrusts that over-thrusting can occur on the infinitely greater scale required to account for the Lewis "overthrust" and others like it. If such had occurred, it would seem that every part of the overriding block would be intensely deformed and that the fault plane especially would everywhere be brecciated, deformed and perhaps metamorphosed. But although there is evidence of disturbance at many points of the supposed fault-plane, and above, there are also many points where

there seems to be no physical evidence whatever of the tremendous sliding that is supposed to have taken place. (P. 185)

Whitcomb and Morris turn to the observations of Dr. Walter Lammerts, a creationist whose scientific qualifications were as a horticulturist and who studied geology as a hobby (pp. 189-91). Lammerts traveled to Glacier National Park to view the fault and reported his findings to Dr. Morris in a 1957 personal communication. Lammerts found very thin layers of shale, cemented both to the limestone above it (the Altyn Formation) and the Cretaceous shales below it at the actual contact of the Belt and underlying rocks. He found "no evidence of any grinding or sliding action or slicken-sides such as one would expect to find on the hypothesis of a vast overthrust."

One can discern in the foregoing creationist arguments a thread of weakness in the form of a nonsequitur. The first step of the argument is to set up some physical criteria on the basis of which an overthrust fault is to be identified. The criteria are sought in the field and fail to appear. Therefore, no overthrust fault could have occurred.

While finely ground rock (indurated as mylonite) and slickensides (glazed fault surfaces with striations) are considered good indicators of faulting on steeply inclined fault surfaces (normal and reverse faults), there is no basis for transferring that association to a thrust plane of nearly horizontal attitude, which may have formed under an entirely different set of physical environmental conditions.

Extensive low-angle thrust sheets seem a favorite topic for the creationists, one that bears repeating. Henry Morris, for one, did not forget his early geologic arguments on the Lewis and Heart thrusts, thinking well enough of them to do an update in the light of some more recent papers in the geologic literature. In an ICR Impact article titled "Those Remarkable Floating Rock Formations," Morris calls attention to a 1977 paper by geologist P. E. Gretener in which the common characteristics of great low-angle overthrust faults are reviewed and summarized (Morris, 1983, p. ii).

Gretener summarizes his conclusions as follows:

> The following observations seem to have universal validity: (1) The contact is usually sharp and unimpressive in view of the great amount of displacement. (2) Structures which have been named "tongues" seem to be common. (3) Secondary, or splay thrusts are common. (4) Coalescence of tongues may produce pseudo-boudins. (5) Minor folding and faulting can usually be observed in both the thrust plate and the underlying rocks. (1977, p.110)

Point 1 is welcomed by the creationists, who would turn it into an argument that the thrusts do not exist. However, geologists have good reasons to think the thrusts do exist quite apart from the sharpness and unobtrusiveness of the contacts. Point 2 is technical but highly significant, because it is a feature not found at conformable sedimentary contacts in regions unaffected by strong compressional stresses. A basal tongue, seen in cross section, is a toothlike intrusion of rock material from the strata immediately beneath the fault surface upward into the basal part of the thrust plate. Because Gretener reports that he found no such tongues in the Lewis Overthrust, we must pass on to point 3. Splay thrusts originate at the basal thrust and pass upward through the thrust sheet at a moderately steep angle. Many such splay thrusts are shown in Figure 40.6; they play an important role in telescoping of the thrust sheet. Gretener shows a photograph of "a large, well-defined secondary thrust in the Lewis plate at Marias Pass"

(p. 118, Figure 9). Here we should also note that Chief Mountain exhibits strong oblique thrust faulting in the lower part of the Belt limestone lying on the fault plane (Campbell, 1914, p. 33). Point 4 we shall pass by (as does Morris) because of its technical complexity. Point 5 is picked by Morris because Gretener adds: "The intensity of such deformations is normally comparatively weak, at least in view of the large displacements these thrust plates have undergone." Gretener's statement seems countermanded by R. A. Price's structure section of the Canadian Rockies, reproduced here as Figure 40.6, but both of those authors are specialists with a great deal of field experience, so I am reluctant to express an opinion. Dr. Morris shows no such reluctance, for he writes in commentary on Gretener's conclusions:

> But now suppose that all these physical phenomena --brecciation, rock powder, striations, basal tongues, splay thrusts, etc., that a real overthrust would produce, are actually present, does this finally prove that the rocks have really been moved out of their original depositional order?
>
> Of course not. Such phenomena merely prove that the upper block has moved somewhat with respect to the lower block. This is quite common, even with formations in the "correct" sedimentary order, due to the different physical properties and time of deposition of the two formations, and proves nothing whatever about overthrusting.
>
> Admittedly, such phenomena do not rule out the possibility of overthrusting, as their absence might do. They are necessary, but not sufficient, conditions for overthrusting. More evidence is needed--notably of the "roots" from which the alleged thrust block was derived, along with evidence that its incredible journey was physically possible. (1983, p. ii)

In his last sentence, Morris has thrown down a challenge that needs to be answered. He rightly says that the roots or sources of the thrust sheets are rarely discernible and invite much speculation. Figure 40.6 shows part of the root zone of the great décollement of the Canadian Rockies. Obviously, much of the deep structure shown is speculative. In general, however, such thrusts can be traced back to a root area of nappes involving metamorphic rocks, which are metasediments (gneisses and schists). Here the thrust fault mechanism probably gave way to flowage in ductile rock at greater depth.

All things considered, Morris's skepticism is no more intense than what is frequently encountered within the fold of mainstream geology. To answer Morris in a definitive way, we need to take a different perspective. The crux of the matter really is this question: What independent evidence can we find in the strata of the alleged thrust sheet and the strata that lie below, showing clearly that the former are older than the latter? We will leave out for the moment evidence from methods of absolute age determination--methods that the creationists do not consider valid. Neither can we base our argument on the fossils found in the strata, because the validity of the age significance of faunal zones is itself a question at issue. Instead, we must look to nonfaunal criteria.

Creationists argue that the contact between the Belt Series and the Cretaceous strata is one of in situ (in place) deposition of the former on the latter, with little or no subsequent dislocation by tectonic processes. The criteria involved in this argument are, of course, nonfaunal in substance. Creationists Whitcomb and Morris have fully addressed only one side of the nonfaunal evidence of the nature of this contact, arguing that features to be expected of a thrust dislocation are

lacking. This is arguing negatively, i.e., in favor of the absence of certain evidence they think should be present but is not. Positive evidence that the contact is either a disconformity or an unconformity represents the obverse side of the coin in the argument. We should also be asking this question: What evidence can be brought forward to show that the Belt sediments, which are agreed to have been deposited in the shallow-water marine environment, were actually laid down upon an erosional surface of preexisting Cretaceous strata? We specify an erosional surface, because in places the Cretaceous strata exhibit folds independent of the structure of the Belt strata.

Dr. Lammerts's statement, quoted by Whitcomb and Morris, clearly recognizes that Cretaceous strata are locally tilted and that deposition of the Altyn limestone occured after the tilting (1961, p. 189). This observation indicates that Lammerts recognizes the presence of an unconformity. Elsewhere on the contact line Lammerts finds a perfectly conformable relationship, i.e., a disconformity. In both situations, Lammerts claims that he observed (as did Price) the presence of thin layers of shale, and that these were cemented both to the Altyn limestone above and the Cretaceous shale below at places along the almost perfectly horizontal contact line extending for 0.8 km. He observed that where the two formations have split apart, the thin band of soft shale has stuck to the Altyn limestone. The shale layer is described as "wafer-like" and 1.5 to 3.0 mm thick. The concluding argument given by Whitcomb and Morris is contained in the caption of a photograph showing the contact: "It seems inconceivable that this very fine layer would have been left so intact if the limestone had actually been thrust over the shale as the Lewis 'Overthrust' interpretation demands" (p. 190, Figure 17). Alternatively, one might argue that the thin shale layer is actually a layer of rock flour derived by frictional pulverization of the shale on the sole of the thrust. In any case, it is a feature of dubious significance. More important, it seems to me, is the finding by Lammerts of "two four-inch layers of Altyn limestone intercalated with Cretaceous shale." He states further: "These always occurred below the general contact line of Altyn limestone and shale. Likewise careful study of these intercalations showed not the slightest evidence of abrasive action such as one would expect to find if these were shoved forward in between layers of shale as the overthrust theory demands" (pp. 190-91). He makes the point negatively (against thrusting) but completely ignores the positive interpretation, which is that the intercalation itself requires the thrust mechanism for its existence.

What is needed, it seems to me, is an instance in which a late-Precambrian sedimentary formation containing stromatolites, such as those in the Belt series, is found beneath a conformable pile of post-Precambrian (Phanerozoic) strata including Cretaceous formations, in a region where signs of severe compressional tectonics are completely absent. Here we might be able to build a case favoring the normal geologic sequence as found in the standard table of geologic time. Where to look? Deep down in the Grand Canyon, of course. We should examine the contact between the Grand Canyon Series and the Cambrian strata, of which the Tapeats sandstone is the lowest unit. (See our earlier description of this unconformity in Chapter 31.) Is there evidence that what we see in the Grand Canyon is the "normal" sequence of deposition? If so, by simple logic of mutual exclusion, the sequence found at the Lewis Overthrust is the reversal of the normal sequence. The Algonkian wedge, formed of tilted Proterozoic strata of the Grand Canyon Series, is in contact with the overlying Cambrian strata along an obvious erosional surface--definitely not an overthrust fault, or any kind of fault. That surface has sharp prominences of very hard quartzite (monadnocks), showing the effects of wave erosion by a rising Cambrian sea. Essentially conformable strata overlie the Cambrian formations, and these can be followed continuously, interrupted only by disconformities, right up to the Cretaceous strata of the Kaiparowits Plateau. Clearly, the evidence favoring the normal depositional stratigraphic sequence in the Grand Canyon region is strongly positive. It depends upon the principle of stratigraphic succession alone and is free of any possible implications of tectonic disturbance that could have reversed the sequence. Broad warping, monoclinal flexing, and minor high-angle faults are present in this region, but they do not indicate compressional tectonics. In contrast, the Lewis Overthrust sequence is so clearly and heavily implicated in compressional tectonic activity, seen in its telescoping folds and imbricate overthrusts, that any interpretation other than one of major overthrusting is suspect on the face of it.

While on the subject of comparison of the Lewis Overthrust stratigraphic sequence with that of Grand Canyon, we should be aware of a serious internal contradiction in the creationists' general theory of Flood geology. We must look back to the subject of Flood stratigraphy and, in particular, to a topic we have called "hydraulic stratigraphy." Recall from Chapter 39 that the creationists accept the general sequence of fossil forms as set forth by the evolutionists, but the ordering of the fossils is explained not by evolution, but by the rate of settling of the drowned animals in the floodwaters. We predicted that the basal layer, or basal chaos, would include large, dense masses of fractured stromatolites and coral-algal reefs. Later, the largest animals, such as dinosaurs, elephants, and whales would accumulate, followed by small animals, such as marine invertebrates. How does this creationist version of fossil succession stack up against the fossil succession in the Lewis Overthrust area? Instead of being at the bottom, the stromatolite masses are found above the Cretaceous strata, which (with Jurassic strata included) in Alberta and Montana contain some of the world's richest dinosaur quarries--the Red Deer River locality, for example. This arrangement requires that the stromatolite masses be held in suspension in the floodwaters while the remains of animals both large and small were allowed to settle out along with a great deal of very fine silt and clay that now encloses the fossils. Ross and Rezak describe the algal colonies of the Siyeh limestone, a member of the Belt Series, as being 30 cm to 1.8 m in diameter and 25 to 60 cm high (1959, p. 410.) The Belt strata show severe fracturing by secondary thrust faults. The larger of these fractured stromatolite masses, being of the density of mineral carbonate, would have plunged rapidly to the ocean floor, accumulating long before the animal bodies arrived. One might wish to argue that the stromatolite bioherms formed in quiet water near the close of the Flood year, but this explanation has to be ruled out because of the long time required for algal colonies to produce such massive limestone accumulations. They simply could not have accumulated in the final weeks of the Flood year as the waters were subsiding.

In contrast to the Lewis Overthrust relationship, Proterozoic rocks containing stromatolites occur in the Grand Canyon region and in many other localities, lying below the Cambrian and younger strata. From the standpoint of creationist hydraulic stratigraphy, this relationship is the correct one. Therefore, the Lewis Overthrust does indeed display an inversion of the normal sequence. In terms of Flood geology this inversion can be explained by low-angle overthrusting in late stages of the Flood, and even in early post-Flood time, since considerable residual catastrophic deformation is allowed. Indeed, a great deal of compressional tectonic activity can be invoked in bringing the strata deposited on the floor of the Flood ocean basin to the present elevated continental position, where all regions of nappes and thrust sheets are now located. As I pointed out earlier in

this section, the creationists' abandonment of their own doctrine of catastrophism has led them into serious contradictions, rendering their view of earth history untenable. Indeed, they have falsified their own theory in their desperate attempt to prove it at any cost.

Whitcomb and Morris turn next to an entirely different kind of argument; it deals at some length with the supposed physical impossibility of a thrust sheet of the dimensions of the Lewis Overthrust moving as a coherent layer without buckling and otherwise disintegrating (p. 191). In other words, these authors now challenge the mechanism of thrust sheet movement. What they are saying is: Never mind where the stratigraphic evidence leads; overthrusting is impossible in the first place; therefore, there is no Lewis Overthrust. The same argument for the physical impossibility of occurrence of large, low-angle thrust sheets has been used for decades by structural geologists and is familiar to all geology students. It has been presented in full mathematical detail by geophysicist M. King Hubbert and geologist W. W. Rubey in their now-classic paper on the possible mechanisms of motion of large, low-angle thrust sheets (Hubbert and Rubey, 1959). Whitcomb and Morris (p. 191) quote from the latter work the authors' conclusion: "Consequently, for the conditions assumed, the pushing of a thrust block, whose length is of the order of 30 km or more, along a horizontal surface, appears to be a mechanical impossibility" (Hubbert and Rubey, 1959, p. 126). Actually, it is a valid negative conclusion to draw based on assumptions of a typical shallow crustal situation. But Hubbert and Rubey then go on to set up a hypothetical environmental situation in which a mechanism could exist for relieving the enormous frictional force that would otherwise be present. Whitcomb and Morris have accepted a set of initial conditions unwarranted in the first place (a straw person), derived the deduced consequences of those conditions, and concluded that no overthrust could have occurred.

Gravity Gliding and Pore-Water Pressure

The basic problem of great thrust sheets, which is how lateral stress can be transmitted through an entire rock layer that is extremely thin in comparison with its horizontal extent, has invited speculation for almost a century. Looking at a great series of nappes, whether they are viewed in nature in alpine mountains or on true-scale cross sections, it is hard to visualize an individual nappe as a strong layer capable of being pushed uphill or even horizontally by compressive stresses applied at one margin of the thrust sheet. Folding within a nappe suggests that the rock was weak and behaved as a plastic solid. We might liken a nappe to a layer of soft pastry dough rolled out on a pastry board. Even if the dough is separated from the table by a layer of flour, so that it does not stick, it is impossible to force the entire sheet of dough to move as a unit merely by applying horizontal pressure at one edge. The dough simply does not have the internal strength to transmit the horizontal stress.

Reasoning on these lines, geologists suggested an alternative hypothesis to that of pressure applied horizontally at one source. Suppose, in the case of the sheet of dough, we tilt the pastry board. When the tilt becomes sufficiently steep, the dough begins to slide under the force of gravity. When the leading edge of the dough sheet reaches the horizontal table top it begins to buckle into folds, which pile up behind the leading edge. Our simple model illustrates the hypothesis of gravity gliding of overthrust sheets and nappes, proposed in the 1880s by E. Reyer, a geologist familiar with the structure of the European Alps (see Holmes, 1978, pp. 678-79). It postulates sufficient crustal uplift occurring along a collision zone to generate the downward gradient parallel with the base of a series of strata. Figure 40.9 shows the

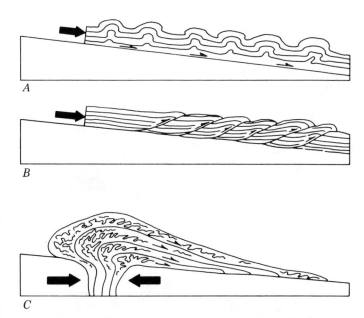

Figure 40.9 Schematic drawings of tectonic features of décollement produced by gravity gliding. A. Folds, such as those of the Jura Mountains. B. Imbricate thrust sheets. C. Nappe structure formed during continental collision. (Copyright © 1981 by Arthur N. Strahler.)

gravity-gliding explanation applied to décollements and nappes.

The problem of sufficiently reduced friction between the gliding layer and the base on which it moves remains to be solved and can be independently approached. We might imagine that the gliding is facilitated by some natural lubricant, such as a layer of saturated plastic clay. It remained for Karl Terzaghi, a specialist in the engineering mechanics of soils, to put forward a hypothesis based on the principle that <u>pore-water pressure</u> is capable of exerting a lifting force on a mass of soil or rock and thereby counteracting, and even nullifying, the frictional effect of the weight of the overlying mass (Terzaghi, 1950, pp. 92-94). The force is generated by water confined by the overlying rock layer and saturating the pore space beneath it. The mechanism is quite like that of a hydraulic jack, by means of which the power of the human arm can lift an automobile high off the ground. Although Terzaghi applied his hypothesis to the setting off of landslides on mountainsides, the principle can be applied to large overthrust sheets. This step was taken in 1959 by Hubbert and Rubey, who showed that in theory water trapped in interconnected pores in rock of the thrust sheet can experience a buildup of pressure to the point that the thrust sheet is actually "floated" off the thrust contact and made free to move under gravitation force.

Morris reviews the pore-water pressure hypothesis put forward by Hubbert and Rubey (1983, pp. ii-iii). His description and critique show an excellent command of principles of physics and engineering, as would be expected of one trained in engineering. Morris need not, however, undertake criticism of the hypothesis entirely upon his own resources, for there is plenty of expert criticism from within the fold of mainstream geology. Morris quotes from papers by a trio of Massachusetts Institute of Technology authors--Peter L. Guth, L. V. Hodges, and J. H. Willemin (Guth et al., 1982; Willemin et al., 1980). The main thrust (!) of their argument lies in imagining circumstances under which extremely high pore-water pressure can be held trapped in the rock of the thrust zone. These authors are addressing their criticism to papers published by two other geologists, Stephen

Ayrton (1980) and P. E. Gretener (1980), who currently support the Rubey/Hubbert hypothesis and press for its application to the interpretation of gliding of nappes and thrust sheets. Ayrton suggests that as metamorphism affects sedimentary rocks deep within nappes, mineralogical changes involving dehydration (dewatering) release free water that can enter the thrust zone and replenish water lost by upward leakage. His suggestion is that periodic renewal of the pore-water supply from this source can allow movement on the thrust plane to proceed intermittently. The M.I.T. authors maintain, however, that under pressures and temperatures high enough to cause the plastic flow seen in metamorphic rocks, open and interconnected pores could not be maintained (Willemin et al., 1980, pp. 405-06). They doubt that the overlying rock could maintain a sufficiently low permeability to hold in the great pore-water pressure needed for gravity gliding to occur. Furthermore, they argue, the overlying rock could not possess the tensile strength to hold in the confined water; it would break through along fractures and pressure would drop.

Morris capitalizes on this negative opinion from within the group of mainstream geology specialists. He writes:

> If fractures develop, of course, this increases the permeability and the water flows out, lowering the pressure and stalling any incipient flotation. Furthermore, it is simply inconceivable that these huge (often many miles long, wide, and thick) slabs of rock could traverse the long distances necessary without fractures developing from other causes as well. There seems no way to avoid escape of pore water through at least some fractures. By Pascal's law, if the pressure is lowered at any point in a continuous water body, it must drop by the same amount throughout the entire body. The whole scenario seems impossible, hydraulically, over any significant distances. (1983, p. iii)

Morris ends his paper with what could well be taken as the creationists' slogan in this particular skirmish: "Floating rock formations won't float!"

So much for the negative point of view in the matter of theory. The debate is in full force among geologists, and it may well be that the arguments against pore-water pressure flotation will put that hypothesis to rest; but it is perhaps premature to anticipate that verdict just now. What we can do, however, is to go back to the crustal rock itself to see if there is new evidence that low-angle overthrust sheets actually did move long distances. There is, indeed, some exciting new information to relate.

A New Look at the Deep Crust

For many years, an important geophysical tool in the search for petroleum accumulations has been the analysis of artificially generated seismic waves, sent down from the surface into the rocks beneath and picked up again after they have been reflected or refracted from various kinds of layerlike structures or abrupt rock interfaces. To apply these highly sophisticated methods to pure science, there was formed in 1973 the Consortium for Continental Reflection Profiling (COCORP). Its members have included geologists from various universities and from within the petroleum industry (Kerr, 1978). COCORP seeks to explore the continental crust to depths as great as 45 km, which is much deeper than petroleum exploration is normally carried. Equipment is mounted on a caravan of trucks that travels slowly across country. Seismic waves are generated by a hydraulically vibrated pad. The waves travel into the crust and are reflected back to the surface, to be picked up by a long chain of geophones. The data are presented as optical images of the structures in a vertical plane of section beneath the line of geophones. The reflection method is used in conjunction

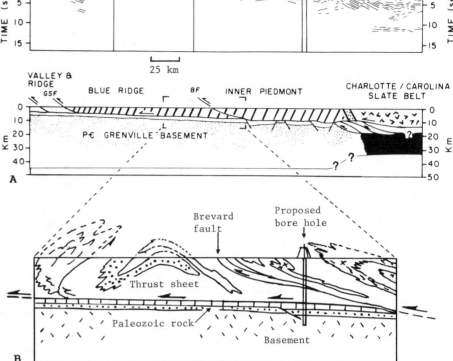

Figure 40.10 A nearly horizontal décollement carrying metamorphosed early Paleozoic rocks over an immobile layer of lower Paleozoic strata affected by metamorphism and lying on older basement rock. A. General geologic cross section, below, shows full breadth of the collision zone; section above shows COCORP reflection seismology data with principal reflectors. B. Detail of Brevard fault section showing loction of proposed deep bore hole. (A. From F. A. Cook, Geology, vol. 11, p. 89, Figure 4. Copyright © 1983 by the Geological Society of America. Used by permission. B. From R. D. Hatcher, Jr., reproduced in Geotimes, 1984, vol. 29, no. 11, p. 13. Used by permission of Robert D. Hatcher.)

with refraction studies, requiring a different kind of instrumentation. Refraction data are necessary to determine the speeds at which seismic waves travel through various rock formations, information that can be used to identify the kind of rock through which the waves travel.

Of the several COCORP crustal profiles thus far completed, one in particular relates to the problem of great overthrust sheets; it is a profile across the Southern Appalachians. This is the orogenic belt completed at the close of the Paleozoic Era by the collision of North America with Africa in Pennsylvanian time, finally eliminating the former ocean basin that lay between them. The deep profile extended across the Piedmont and Blue Ridge belts of crystalline metamorphic rock--schist, slate, and gneiss--and into the belt of foreland folds known as the Newer Appalachians (or Folded Appalachians), in which strata of Paleozoic age were thrown into open folds with some overturning and moved forward along a major overthrust layer as well. The crystalline metamorphic rocks of the Piedmont and Blue Ridge are comparatively ancient. They probably represent mostly early and middle Paleozoic strata metamorphosed in earlier arc-continent collisions such as one that occurred in Ordovician time (Taconian Orogeny) and another in Devonian time (Acadian Orogeny). They show complex nappe structures and some major low angle overthrust faults.

COCORP seismic reflection profiling has revealed a startling feature. Beneath the metamorphic rocks just described, there appears to be a flat-lying rock layer, interpreted as being composed of lower Paleozoic strata. These beds were deformed by the pressures and shearing forces of overthrusting and were altered to a metamorphic rock of low grade. This relatively immobile layer rests on the Precambrian basement on which the strata were originally deposited. The crystalline thrust sheet (or sheets) above this immobile zone apparently moved forward on a nearly horizontal décollement paralleling the underlying Precambrian surface. Figure 40.10 is a geological cross section based on the seismic reflection

data as well as on surface rock exposures. Overthrust sheets appear to have traveled toward the continental interior over distances of several tens of kilometers along a nearly horizontal thrust fault at a depth of several kilometers. Identification of the immobile basal layer of deformed strata seems reasonably certain, because the seismic methods used have been used repeatedly in the past and have often been verified by later drilling (although at much shallower depths) in searches for petroleum accumulations. It is now proposed that a ten-kilometer hole be drilled through the thrust sheet, at the location shown in Figure 40.10, in hopes of penetrating the strata beneath it (see Geotimes, 1984, vol. 29, no. 11, p. 14). When this is done, and if the strata are identified, there can be no further doubt about the existence of great thrust sheets and the tectonic inversion of order of rock ages they reveal. Perhaps then the creationists will return to the ranks of the catastrophists and include the phenomenon--known today as "thin-skinned tectonics"--as among the catastrophic happenings they ascribe to the Flood year and early post-Flood time.

The Heart Mountain Fault

The Heart Mountain fault is best described as bizarre and its cause enigmatic, using two adjectives that may seem to give overstatement, until one looks at the details of the phenomenon. In northwestern Wyoming, along the eastern side of Yellowstone National Park, is a rugged mountain mass, the Absaroka Range, built mostly of volcanic rocks of Eocene age. As these volcanic breccias and flows emerged, they nearly buried a landscape eroded into more-or-less horizontal strata of Paleozoic age. Figure 40.11A is a simplified schematic diagram of this pre-volcanic landscape in very early Eocene time. Two groups of strata overlie the Precambrian crystalline basement. The lower group is of Cambrian age, mostly shales and limestones. The upper group consists of Ordovician, Devonian, and Mississippian strata, the bulk of which is massive dolomite and limestone; it is this group that was to become detached and move as a unit, gliding over the

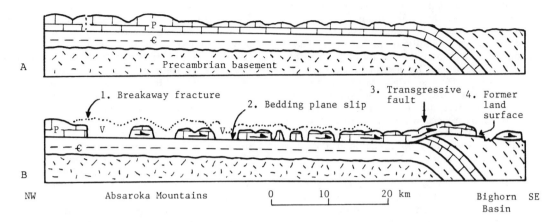

P = Post-Cambrian Paleozoic strata   Є = Cambrian strata   V = Eocene volcanics

NW   Absaroka Mountains   0   10   20 km   Bighorn   SE
Basin

Figure 40.11 Schematic cross section of the Heart Mountain bedding fault. Vertical scale is greatly exaggerated. (After W. G. Pierce, 1968, <u>Trans.</u>, <u>XXIII</u> <u>International</u> <u>Geological</u> <u>Congress</u>, vol. 3, p. 192, Figure 1.)

Cambrian formations. Notice that near the eastern end of the cross section the strata bend downward. This feature is the western margin of a great downwarp known as the Bighorn Basin.

Cross section B of Figure 40.11 shows the situation after the rapid occurrence of a great mass displacement. For reasons as yet undetermined, the entire layer of post-Cambrian strata simply began to glide as a unit southeastward over a bedding surface located immediately under the massive Bighorn dolomite formation of Ordovician age and above the topmost Cambrian formation. This layer detached itself along a vertical breakaway fracture shown at the left. Movement was evidently on a very low downgrade, declining some 650 meters in elevation from the breakaway fracture to the end of the bedding slip zone, a horizontal distance of some 50 km. As the rock sheet traveled, it broke up into blocks on a succession of vertical tension fractures. The blocks thus became separated by open gaps, in which the bedding plane of gliding (identified as the Heart Mountain fault) was exposed at the surface. Geologists have applied the term "tectonic erosion" to surface exposure of a fault plane by sliding away of the overlying mass. It is an unfortunate term at best, since no erosion by fluvial processes has occurred. A better term is "tectonic denudation," since "to denude" means "to lay bare" (Pierce, 1968, p. 191.). Where the sliding rock layer met the downbend of strata, the plane of movement turned upward, crossing the strata diagonally in a true thrust fault, labeled here as a "transgressive fault." Beyond that point, the thrust sheet moved out over the former land surface. Immediately after the rock sheet completed its movement, volcanic breccias and other extrusive forms of andesitic magma descended upon the landscape, burying the displaced blocks and the exposed sections of the glide plane. Stream erosion occurring later throughout Cenozoic time has cut deeply into the volcanic cover and the various rocks beneath it, revealing the fault structure in the steep walls of canyons.

The above description is based on published research of William G. Pierce of the U.S. Geological Survey (Pierce, 1957; 1968; 1979). It is a greatly simplified description of what is actually an extremely complex structure, but it must suffice as a basis on which to examine the creationists' commentary and critique.

Whitcomb and Morris consider the Heart Mountain fault (which they call the "Heart Mountain Thrust") as a valid example of the kind of low-angle overthrust fault treated by Hubbert and Rubey in the classic paper to which we have already referred (1961, pp. 181-84). In so doing, Whitcomb and Morris have at the outset of their critique completely misclassified the Heart Mountain structure, making it appear to be something it is not. Pierce, in his 1968 and 1979 reports, does not refer to the Heart Mountain structure as a "thrust," as he did in his 1957 paper, from which Whitcomb and Morris quote, and which could have been the only version of Pierce's work available to them. Reading of Pierce's two later papers reveals a much better understanding of the Heart Mountain structure. Pierce has identified the breakaway fracture, showing conclusively that we are not dealing with a compressional phenomenon in which a root zone would be required. Pierce now refers to the structure as an exhibit of "tectonic denudation," which must be distinguished from "deformation due to gliding from deformation coming from a root zone or energized by a push from the rear" (1968, p. 191).

In recently reviving the question of the Heart Mountain fault, Morris (1983) has no excuse for not recognizing it as a structure entirely different in category from the low-angle overthrusts found in nappes and their adjacent foreland belts. Oblivious to the distinction, Morris states that Pierce has been studying the Heart Mountain "overthrust" for many years (1983, p. iv).

Whitcomb and Morris attack Pierce's interpretation of the contact between the massive limestones of Devonian through Mississipian age and the Cambrian strata as being a nearly horizontal fault surface of substantial displacement (1961, pp. 181-84). To creationists, this feature is a normal depositional contact in a sequence in which the formations are in their correct stratigraphic order according to the conventional stratigraphic column. Considerable discussion centers upon Pierce's 1957 description of the presence of a "breccia" at the supposed bedding fault contact. Pierce writes:

> The fault contact of the bedding thrust may either be clean-cut and sharp, with essentially no brecciation of the beds above or below the fault, as observed at several places, or it may have a zone of broken limestone and limestone debris, such as observed at the northwest end of Sugarloaf Mountain. There the broken limestone bed is about 30 feet (10 m) thick; its lower contact with the Grove Creek formation is sharp, but its upper contact is indistinct. (P. 598)

Whitcomb and Morris quote the first of the above sentences, but not the second, which clearly favors the fault hypothesis (1961, p. 183). They reply to Pierce's description in these words:

Uniformitarians will say that these brecciated areas at the fault plane are evidence that movement has actually occurred and, therefore, that the "thrust-fault" concept of these Heart Mountain blocks is valid. However, it should be remembered that breccias occur widely, usually in places where no such phenomenon is in question at all. They might easily have been produced by means other than this hypothetical sliding. On the other hand, the really pertinent question is: Why is not the entire fault-plane heavily brecciated and distorted? The fact that there are many places where the contact line is clean-cut and sharp, looking very like a normal bedding plane, is seemingly inexplicable if the plane is in reality a thrust plane. (P. 184)

The answer to the creationists' point that sedimentary breccias are of many kinds and origins, and hence may not indicate fault grinding, is that it is acceptable only in a very general sense. In this case, however, the elucidation of the problem appears in Pierce's 1979 paper, dealing specifically with clastic dikes of fault breccia above the Heart bedding fault. Pierce writes:

> The Heart Mountain fault breccia is a cataclastic rock (cataclastite) composed of angular carbonate clasts in a microbreccia of smaller fragments without primary cohesion. The fault breccia is not mylonite, even in a broad sense, nor does it have a fluxion structure; it is not only found on the fault surface beneath the large blocks of Paleozoic rock composing the upper plate, but is irregularly distributed on the surface of tectonic denudation between these blocks. (P. 3)

On the basis of this description, the breccia is clearly distinguished from limestone breccias resulting from subaqueous deposition of angular particles. However, the conclusive association of this breccia with a large rock slab gliding under gravity and experiencing tensional fracturing comes in the presence of features called clastic dikes. The dikes are irregular fracture fillings, as much as 5 cm wide, that have been forced up into the overlying block (Pierce, 1979, p. 4). The clastic dikes never occur below the fault plane; instead, they pinch out upward and exhibit an upward branching trend. Pierce considers the possible mechanism of clastic dike emplacement (1979, pp. 21-22). He seems to favor the injection as a response to pressure of overlying rock as the volcanic breccias accumulated, burying the limestone blocks and the exposed fault plane surface.

On another tack, Whitcomb and Morris allude to the finding of the limestones of the fault slab resting on strata of Eocene age (1961, p. 184). This relationship occurs in the distal region of the fault slab, where it rode out upon much younger sediments underlying the land surface of that time. Here an age reversal exists, but the explanation in terms of a rock slab moving by gravity gliding is simple and consistent. In no way can it be used to vitiate the entire system of established age relationships in the stratigraphic column.

Finally, we refer again to Morris's 1983 paper, in which he refers to Pierce's 1979 paper on clastic dikes. Pierce provides an opening for the creationists to attack by expressing his opinion that "the Heart Mountain fault movement was a catyclismic event" (p. 24). Pierce is not happy with the Hubbert/Rubey hypothesis of pore-water pressure as the mechanism of initiating the sliding of the limestone slab. He argues that "the high permeability of the extensively brecciated rocks in the lower part of the upper plate and separation of the upper plate into numerous blocks and pieces indicate that the Heart Mountain fault movement was not aided by high pore pressure of water, air, or volcanic gas." Instead, he offers a most interesting suggestion--one that occurred to me independently as I was reading the earlier pages of his paper--that "the extensive brecciation of rock above the fault surface . . . may well have been caused by a catastrophic earthquake of vertical acceleration approaching 1 g." The idea had occurred to me in recalling my recent visit to the Hebgen Lake locality on the far side of Yellowstone Park, where a great earthquake in 1959 set in motion the huge Madison Slide. Why not a series of great earthquakes, each capable of momentarily lifting the rock slab and relieving the basal friction? Could such intense and prolonged seismic activity have caused a succession of gravity movements, ultimately achieving a cumulative displacement of several kilometers?

Morris concludes his paper with a summary of Pierce's evolving thought on the possible mechanism of the Heart Mountain fault:

> William Pierce has been studying the Heart Mountain "overthrust" (Paleozoic over Eocene) for many years. Originally he thought it was caused by simple gravity sliding, but this proved impossible mechanically. Then it was suggested that the mechanism might have involved the Hubbert and Rubey fluid-pressure concept. Now that he finds that won't work either, he has invoked catastrophism--and catastrophism with a vengeance, postulating an explosive transplantation of the thrust blocks! Catastrophic events, however, obviously can neither be observed in process nor modeled in the laboratory. (1983, p. iv)

My comment at this point is that I found no mention of "explosive transplantation" in Pierce's paper--unless, of course, a great earthquake can be described as "explosive." In Morris's final paragraph, he unwittingly reasserts his transient defection from catastrophism and momentary allegiance to uniformitarianism (adopted when convenience demands) in these words:

> One can believe in catastrophic overthrusting if his motivation to do so is sufficiently strong, of course. If evolutionists want to retain their cherished evolutionary sequences, therefore, they can do so only by faith in catastrophism. Floating rock formations won't float! (P. iv)

Chapter 41

Gaps in the Fossil Record—What do They Mean?

Darwin's model of organic evolution is generally assumed to require a unique time and place for the origin of life on earth (whether a naturalistic origin or by Divine Creation); it requires a continuum of life through all geologic time since that singularity. Descent with modification, under the principle of biogenesis, requires that every "new" life form be a modification of a previous life form. In that sense, evolution is a continuum without gaps. But the fossil record is another matter--another category of information entirely. The concept of continuous life is a deduced consequence of the premise of a single beginning; it is an ideal or theoretical concept. As such, the concept may be regarded as unverifiable in any practical sense. Observation of the strata fails to confirm this continuum of life forms; that continuum can only be inferred from the fragments that remain. This is not to say that the inference of continuity does not enjoy an extremely high level of probability of survival as a scientific conclusion.

Yes, the Fossil Record Has Gaps

Both parties--evolutionists and creationists--agree that the fossil record has gaps. The history of human civilization shares this same quality, for while we never think to question the continuity of Homo sapiens throughout human history, much of that history lacks a written record. Why do the creationists accept without any question whatsoever a historical record of the ancient peoples of the Fertile Crescent that is full of gaps? To creationists, gaps in the fossil record mean special creation of kinds. Why, then, doesn't a gap-ridden written record of the history of humans since the Flood also mean repeated special creation of new kinds, subkinds, and varieties of humans? Would it be asking too much of the creationists to suggest that they be consistent as to the meaning of information gaps for all time, from the instant of God's creation of the universe to the very instant of the present?

A good introduction to the problem of gaps in the fossil record is given by a leading paleontologist, Professor David M. Raup of the University of Chicago:

> Darwin predicted that the fossil record should show a reasonably smooth continuum of ancestor-descendant pairs with a satisfactory number of intermediates between major groups. Darwin even went so far as to say that if this were not found in the fossil record, his general theory of evolution would be in serious jeopardy. Such smooth transitions were not found in Darwin's time, and he explained this in part on the basis of an incomplete geologic record and in part on the lack of study of that record. We are now more than a hundred years after Darwin and the situation is little changed. Since Darwin a tremendous expansion of paleontological knowledge has taken place, and we know much more about the fossil record than was known in his time, but the basic situation is not much different. We actually may have fewer examples of

smooth transition than we had in Darwin's time because some of the old examples have turned out to be invalid when studied in more detail. To be sure, some new intermediate or transitional forms have been found, particularly among the land vertebrates. But if Darwin were writing today, he would probably still have to cite a disturbing lack of missing links or transitional forms between the major groups of organisms. (1983, p. 156)[1]

What Gaps Mean to the Creationists

I find no mention of gaps of the fossil record in Whitcomb and Morris's 1961 work, The Genesis Flood, but the subject appears in strength in the school textbook Scientific Creationism (Morris, 1974a, pp. 78-90). Taking their cue from straightforward evaluations of the fossil record, such as that quoted above, the creation scientists of the Institute for Creation Research now "accentuate the negative" for all it's worth. One of the ICR staff members most skilled in this activity is Dr. Duane T. Gish, Ph.D (biochemistry, University of California, Berkeley) and Professor of Natural Sciences at Christian Heritage College. Gish's major work, Evolution? The Fossils say NO! has been widely distributed for school use and has attracted the attention of mainstream biologists and paleontologists (1978a). The book has been followed by articles in the ICR Impact Series (Gish, 1980; 1981; 1983). A revised and enlarged edition of NO! was published in 1985 under the title Evolution: The Challenge of the Fossil Record.

A good sample of the characteristic opening attack by creation scientists on the fossil record is this paragraph by Gish:

> Ever since Darwin the fossil record has been an embarrassment to evolutionists. The predictions concerning what evolutionists expected to find in the fossil record have failed miserably. Not only have they failed to find the many tens of thousands of undoubted transitional forms that are demanded by evolutionary theory, but the number of arguable, let alone demonstrable, transitional forms that have been suggested are few indeed. This has placed evolutionists in a most difficult situation, made even more embarrassing by the fact that the fossil record is remarkably in accord with predictions based on special creation.
>
> An intense search spanning 120 years has produced an immensely rich fossil record but has failed to produce the expected transitional forms, and many geologists now realize the impossibility that a combination of geological processes would have miraculously eliminated all the billions of transitional forms while leaving billions of fossils of the terminal forms intact. (1983, p. i)

Those of my readers who are familiar with the newer views of mainstream paleontologists (presented in Chapter

35) will recognize in Gish's statements the standard creationist tactic of "beating a dead horse." Largely as a result of the persistence of evolutionary gaps of record, despite the steady growth of fossil collections, paleontologists began to replace the Darwinian paradigm of slow-but-steady evolution of species with one of spasmodic events of rapid evolution. These included George Simpson as a pioneer with the concept of "quantum evolution," expressed as early as 1944, and several other scientists with "rapid" or "quantum" speciation in the 1950s and 60s (see Godfrey, 1983, p. 205). The Eldredge/Gould/Stanley version of "punctuated equilibrium" followed in the early 1970s and has gained some enthusiastic supporters as a hypothesis of evolutionary tempo claimed to be better adapted to available knowledge than was Darwin's gradualism.

Although modern punctuated equilibrium seems to have been lost on Henry Morris, who ignored it in his 1975 school textbook, <u>Scientific Creationism</u>, Duane Gish decided to cope with the new view of evolutionary tempo. In <u>NO</u>! Gish briefly reviews Stephen J. Gould's 1977 articles in <u>Natural History</u>, in which punctuated equilibrium is described (1978, pp. 150-65; see Gould, 1980, p. 179, 186). In 1983, Gish comes to full grips with the idea in ICR <u>Impact Series</u>, No. 123. In both references Gish also reviews the story of Richard Goldschmidt and his "hopeful monster," a now-dead hypothesis, but one that gives Gish a chance to link Gould to Goldschmidt, and even to suggest that Gould looks for Goldschmidt's eventual vindication (Gish, 1983, p. ii). Anyone who has read Gould's articles knows that such a suggestion is completely unwarranted, but it is an unwarranted charge that not many readers will check out for themselves. Commenting on punctuated equilibrium, Gish writes as follows:

> This notion is merely a new scenario but certainly does not provide a mechanism. Left unexplained is how or why such rapid changes occur. Such rapid changes are in fact contrary to what we know about genetics. Furthermore, this notion does not solve the problem of the missing links. This suggestion is only an attempt to explain the absence of transitional species. That is not the serious problem--the real problem is the big gaps between the higher categories, the systematic absence of transitional forms between families, orders, classes and phyla. These gaps, for example, include those between invertebrates and fishes and between fish and amphibia, gaps of 100 to 30 million years or so, respectively, on the evolutionary time scale. The idea of punctuated equilibrium makes no pretense of addressing this problem, let alone providing a solution. (1983, p. ii)

Summary rejection of punctuated equilibrium allows Gish to reaffirm the creationists' imagined view of the "evolution model," described as follows in Chapter 2 of <u>NO</u>!

> Evolution Model: (1) By naturalistic mechanistic processes due to properties inherent in inanimate matter. (2) Origin of all living things from a single living source which itself arose from inanimate matter. Origin of each kind from ancestral form by slow gradual change. (3) Unlimited variation. All forms genetically related.

> Predictions: (4) Gradual change of simplest forms into more and more complex forms. (5) Transitional series linking all categories. No systematic gaps. (Gish, 1978a, pp. 50-51; 1985, pp. 44-45)

> (Note: Numbers were added by the writer to facilitate discussion.)

By now, you should be so well versed in the ways of creationism that you can easily pick out the errors in their statement of the evolution model. But, for the record, we need to review each point. Point (1): We can let it pass as applicable to all of mainstream science. Point (2): Organic evolution has nothing to say about biopoesis --the origin of the first life. Darwin allowed for Divine Creation of that first life, a deistic view, and many theistic evolutionists subscribe to that supernatural origin. Also important is an objection to limiting the tempo of evolution to slow gradual change. That view is now in competition with a hypothesis of stasis punctuated with bursts of relatively rapid change. Point (3): OK. Point (4): Evolutionary change always from simple to more complex is not an accurate description of the entire fossil record. In any case, change only from simpler to more complex forms cannot be predicted from the evolutionary hypothesis. Point (5): While transitional series must have existed throughout, gaps of the fossil record as it presently stands are to be expected for several good reasons. The creationists' prediction of "no gaps" is wrong--clearly, a non sequitur.

With this general introduction to the status of conflict between mainstream science and creation science on the subject of the nature of the fossil record, we are prepared to examine specific cases of sudden appearances, lack of intermediates, and extinctions.

Transitions at the Species Level

Creationists have claimed that if descent with modification from a common ancestor has occurred, "it seems inexplicable that there should be any distinct categories of organisms at all. One would certainly expect that nature would instead exhibit a continual series of organisms, with each grading into the other so imperceptibly that any kind of classification system would be impossible." (Boardman, Koontz, and Morris, 1973, p. 68, as quoted in Raup, 1983, p. 157). As David Raup explains, this statement does not follow from what is known of the genetics of species evolution:

> There is little or no gene flow between species because they do not normally interbreed. Thus each species is able to evolve on a course independent of all others, and there is no opportunity for blending once speciation has taken place. Given time, and perhaps subsequent speciation events, organisms become distinct. By the same reasoning, major groups such as molluscs and arthropods become increasingly distinct and separated by anatomical gaps. Thus, the presence of distinct kinds of organisms (especially when viewed at an instant in time) is a reasonable prediction of the evolutionary model. (1983, p. 157)[1]

Looking at it in another way (as I do from the viewpoint of a geologist interested in stream networks), think of evolution as individual salmon traveling upstream to spawn. As each fork, or junction with an entering tributary, is reached, a choice must be made--either go left or go right. As one branch is followed upstream, it diverges farther and farther from the other. There is no turning back and no way the salmon can cross from one stream segment to the other. Every one of the succession of choices is final and irreversible.

Why Should Intermediates Be Lacking?

Raup discusses our principal problem now under consideration: How does the evolutionist explain the lack of intermediates? (1983, pp. 156-158) Two of his three points are already familiar and need no further discussion here: (a) the fossil record is itself incomplete and (b) transitions between taxa may occur in short time intervals within small populations (punctuated equilibrium, or quantum evolution). His third point (stated as his first) is so simple that it is often overlooked. Our taxonomic system of classification is an artifact, so rigidly established that it cannot accommodate the discovery of an intermediate form. Raup explains:

> The practicing paleontologist is obliged to place any newly found fossil in the Linnean system of taxonomy. Thus, if one finds a birdlike reptile or a reptilelike bird (such as Archaeopteryx), there is no procedure in the taxonomic system for labeling and classifying this as an intermediate between the two classes Aves and Reptilia. Rather, the practicing paleontologist must decide to place his fossil in one category or the other. The impossibility of officially recognizing transitional forms produces an artificial dichotomy between biologic groups. It is conventional to classify Archaeopteryx as a bird. I have no doubt, however, that if it were permissible under the rules of taxonomy to put Archaeopteryx in some sort of category intermediate between birds and reptiles that we would indeed do that. Thus, because of the nature of classification, there appear to be many fewer intermediates than probably exist. (P. 157)

In chapters to follow, as we examine the creationists' claims of lack of intermediates, we shall be on the lookout for actual intermediates that, through taxonomic rules, fail to be recognized as such.

Is Speciation Occurring Now?

Let us now look into the question of whether the transition from one species to another has been adequately documented. That question divides itself into two parts. First, have biologists documented the contemporary process of speciation being carried to fulfillment in living organisms? Second, has speciation been documented in the fossil record?

Taking these two questions in sequence, we ask: Can one or more examples be cited of speciation occurring in living organisms, either in the laboratory or in the field under natural circumstances?

Note that our first question is not simply whether processes of biological evolution have been adequately established by study of living organisms. On that score, geneticist G. Ledyard Stebbins states:

> One of the most important discoveries of evolutionists during the past fifty years is that, at the level of populations and species, both the processes and the course of evolution can to some extent be duplicated by experiments conducted under strictly controlled conditions. (1982, p. 108)

There seems to be general agreement that genetic change by natural selection can be inferred to have occurred in populations of existing species. Perhaps the best known example is that of the melanic (black) variety of the peppered moth Biston betularia that appeared in abundance in the 1800s in industrial cities of England in response to the darkening of tree trunks by industrial soot. Now we observe that with a general cleanup of Britain's air and consequent lightening of the color of the tree trunks the original light-colored peppered moth, which had survived in the surrounding countryside, is returning to the cities while the melanic strain within the population is declining in numbers. In terms of the population of B. betularia, the process of protective color change can be seen as reversible by natural selection.

Many biologists prefer to define "evolutionary change" as an irreversible change in the gene pool. For example, Stebbins defines "biological evolution" as a "series of

irreversible transformations of the genetic composition of populations" (1982, p. 9). Using that definition, it can be argued that a reversible population change, such as that shown by the peppered moth, does not illustrate genuine evolutionary change. What we need to seek in order to answer our first question are examples of irreversible changes in genetic composition of species populations that have led to the rise of new species enjoying demonstrated reproductive isolation.

Writing in Creation/Evolution Newsletter (1986, vol. 6, no. 5, p.22), biologist Stanley L. Weinberg replied to a published statement by creationist Michael Denton that "No one has ever witnessed evolution of a new species." Weinberg's response: "Of course many such examples are known." He cited first the well-known examples of mutation and evolution of bacteria and viruses. Weinberg continued:

> Among eukaryotes also, many examples of speciation have been seen: a new marsh grass in Western Europe; a new copepod in the Salton Sea; five new cichlid fish species in Lake Nabugabo in Uganda; five new banana-eating moth species which have evolved in Hawaii since the Polynesians introduced bananas into the Islands. Herring gulls are clearly in the process of evolving into two new, separate and distinct species. So is the red deer-wapiti complex. New species are even created synthetically; triticale, a new, true species and a productive crop, resulted from a wheat-rye cross. And so on. (1986, p. 22)

I wrote to Weinberg, asking for more information on the above examples. He kindly sent several references, at the same time cautioning that not all biologists agree on the precise definitions of "species" and "speciation." Two major works contain several of the best examples. Ernst Mayr's 1970 work, Populations, Species, and Evolution, contains examples of speciation caused by geographical isolation over a period of several thousand years during the Holocene Epoch. These include the Red Deer in Britain and several mammal species on the island of Newfoundland (pp. 346-7). Speciation in small, freshwater lakes can occur much more rapidly, states Mayr, citing fish faunas in creeks and springs of the desert western United States (p. 348). He specifically cites the six species of cichlid fishes of the genus Haplochromis that live in an isolated lagoon, Lake Nagubago, detached from contact with Lake Victoria about 4,000 years ago, according to radiocarbon dating of the sandbar that forms the lagoon. Five of the six are new, endemic species. Note, however, that all of these cases, while qualifying as valid examples of speciation, were not actually documented by biologists through the process of speciation. Steven M. Stanley has also discussed evidence of rapid speciation (1979, pp. 40-47). He describes some of the same examples used by Mayr. Excellent as as they are as examples of speciation, nearly all involve time spans of 1000 years or much more. I did find one exception: the Faeroe Island house mouse, Mus musculus faeroensis, that was transported there not much more than 250 years ago. Some authors, says Stanley, have designated it a new species, but he observes: "It should be appreciated that taxonomists seldom actually test reproductive incompatibility of populations considered to be separate species" (p. 41).

We should not lose sight of the powerful evidence supporting the theory of evolution of new species by natural selection contained in several of the examples of Holocene speciation, such as the cichlid fish species of Lake Nabugabo, the desert pupfishes of Death valley, the mammals of Newfoundland, and the Red Deer of Britain. Although scientists were not on the scene, they have adequately documented the geomorphologic and hydrologic changes that forced the isolation of populations in islands or inland lakes. Isolation of Newfoundland and the British Isles from their respective mainlands resulted because of the post-Wisconsinan rise of sea level, an event firmly established by the radiocarbon dating of organic sediments laid down as the level of the ocean rose. Isolation of springs and small lakes in the American desert, as pluvial conditions gave way to the present arid regime, is documented independently by similar lines of solid evidence.

Turning briefly to speciation documented in plants, two cases are worthy of scrutiny. One I found problematical is a 1973 report by L. D. Gottlieb, a botanist, in the American Journal of Botany. In eastern Oregon, Gottlieb discovered 250 individuals of a new species of the genus Stephanomeria (wirelettuce), growing within a much larger population of a parent species. He states that the parent species, S. exigua, "has recently given rise by a process of sympatric speciation to a diploid species presently designated as "Malheurensis" (p. 545). He comments that this instance "appears to be an exception to the theory of geographical speciation because spatial isolation is not necessary at any time for the origin or establishment of its reproductive isolating barriers" (p. 545). Actually, the botanist did not directly document the speciation process, asserted to have occurred. The "new" species was discovered within the population of the "parent" species, but the evolution of the former from the latter is only assumed to have occurred. How long ago it occurred, if it did occur, is not known.

The emergence of triticale as a new genus engineered by humans seems clear beyond question. The history of this achievement is reviewed in a Scientific American article by Joseph H. Hulse and David Spurgeon (1974). The cross breeding of wheat (genus Triticum) with rye (genus Secale) to produce a fertile hybrid genus was not easily achieved and required special laboratory intervention. Because the first-generation seeds are usually sterile, colchicine treatment was required to double the number of chromosomes. For that reason, I would be inclined to rule out the engineering of triticale as an acceptable example of demonstrated speciation in natural ecosystems.

Can Speciation Be Seen in the Fossil Record?

We turn now to the fossil record for evidence of speciation having occurred through a succession of genetic changes leading from a parent species to a daughter species.

Although Niles Eldredge and Stephen Gould in 1972 argued for punctuated equilibrium as the rule in speciation and, hence, provided one good reason for the typical lack of intermediates in the fossil record, they were aware that in a few cases at least an excellent record with numerous intermediates might be found. Godfrey explains:

> Eldredge and Gould (1972) agreed that a perfect fossil record would document morphological intermediates between species, but they suggested that many of these would exhibit relatively brief and geographically limited existences. Indeed, Eldredge had such a near perfect record of the evolution of the Devonian trilobite Phacops. It was a record of stepwise evolutionary change in only two brief periods during a span of eight million years! One interval was recorded in a single easy-to-miss quarry in New York State. This quarry contained perfect intermediates between the geographically widespread mother and daughter species. In effect, due to the realities of an imperfect record, most such intermediates will simply not be sampled. (1983, p. 206)

In 1981, paleontologist Peter Williamson of the Harvard Museum of Comparative Zoology reported in the British journal Nature what is claimed to be the "first detailed documented evolution of one species into another, as revealed in the fossil record" (Lewin, 1981, p. 645). The locality is near Lake Turkana in northern Kenya; the strata consist of waterlaid floodplain and delta sediments deposited adjacent to an isolated ancient lake. The fossils are those of molluscs, representing 13 different species, and occuring through a total thickness of 400 m of sediments. Throughout most of this time--several million years--the species showed little change, being in a state of what is called "morphological stasis" (equivalent to "equilibrium" under punctuated equilibrium). However, the stasis was interrupted twice by a sharply dropping lake level, abruptly changing the environment of the molluscs and inducing speciation. Within a span of perhaps 5,000 to 50,000 y., new species appeared. The brief periods of speciation are documented with "intermediates along the path from the old species to the new ones." Equally important is "a demonstration that, during the transition, the morphological variation of the species increases markedly." When the lake level rose to its former position, the new species disappeared and the ancestral species returned to the same locality to take over. Williamson's work seems to have been greeted with enthusiasm by other paleontologists and geneticists, including Anthony Hallam of the University of Birmingham, and viewed as strong support of punctuated equilibrium.

Another case study in which evolutionary transition from one species into another has been documented in detail comes from fossils of early primates of Paleocene and Eocene time. These evolutionary sequences have been studied by Professor Philip D. Gingerich of the University of Michigan (1983). From two fully documented examples, I have chosen one that seems to put the point across rather well. It is the record of change of the tooth morphology of the Eocene primate genus Cantius upward through a 600-m sequence of strata in the Clark's Fork Basin of Wyoming. The research strategy was to collect fossils of the genus from beds between those in which species have been previously found and named. The objective was to establish the fine structure of morphological change from one established species to the next. If morphological change was more or less gradual throughout time, intermediates should conform to the overall pattern of change established from the named species.

The results of Gingerich's study are shown in a graph, Figure 41.1, which may seem highly technical, but which is actually quite straightforward in its message. The stratigraphic succession, equivalent to about 2 m.y. of time, is scaled on the vertical axis. On the horizontal axis is given information on the size of upper and lower molars of Cantius. The numbers are logarithms of measurements and need not be understood in detail. Horizontal lines show the arbitrary time boundaries between species, based on previously established information on the three species named in the diagram. Dashed lines running diagonally show the spread of the numerical data. The black dots represent individual fossils collected by Gingerich (where more than one specimen, a digit is shown instead of a dot). The remarkable feature of the graph is that at each stratigraphic level sampled, the measurements fell into a range transitional between sample levels above and below it, conforming with the trend in size change, whether toward increase or decrease. Gingerich summarizes the results in these words:

> As shown, each sample of Cantius from successive stratigraphic levels is itself intermediate and transitional between earlier and later samples. There is relatively little change from one sample to the next, yet the net change over two million years or so of geological time is profound. Change is so

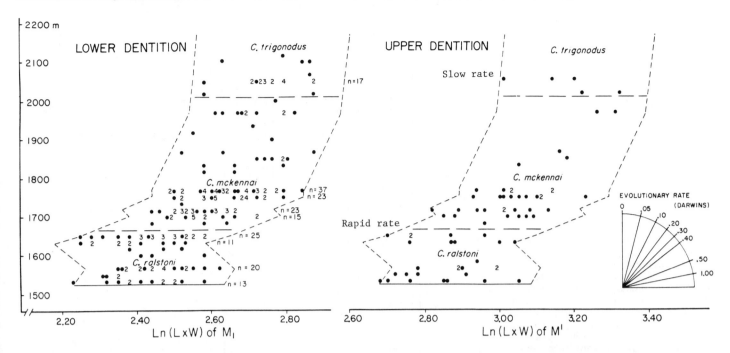

Figure 41.1 Carefully measured changes in size of lower and upper molars of an early Eocene primate, Cantius, show the course of transitions from one species to another upward in the succession of strata. (From P. D. Gingerich, 1982, Time Resolution in Mammalian Evolution; Sampling, Lineages, and Faunal Turnover. Proc., Third North American Paleontological Convention, Montreal, vol. 1, p. 209, Figure 4. Reproduced by permission of the author.)

gradual that the boundaries between successive species (heavier horizontal dashed lines) are necessarily arbitrary. Cantius ralstoni and its descendant C. trigonodus are linked by an insensibly graded sequence of intermediate transitional forms. (P. 142)

In this example, gradualism might seem to dominate, but this may be only because of the time scale involved. Punctuations, or spurts, may have occurred within time intervals too small to show up here. We can never know how the criterion of lack of interbreeding capability (genetic isolation) separates species in a case such as this. Yet the basic message seems clearly to be: "The closer you look the more intermediates you find." But by the same token, the more intermediates that are found, the more gaps that are established.

A letter to Geotimes, written by Caoimhin Mac Aoidh (1983) of the Geology Department, University College, Cork, Ireland, suggests a plentiful source for examples of speciation in the fossil record. The source is the micropaleontological record, in which fossils in one sample from a single horizon are abundant on a scale several orders greater than in conventional macropaleontology. The letter cites a published example in which a fossil form has been found to have changed "not only at the specific level but most certainly at the supragenic level." The writer concludes: "If more attention were given to the evolutionary patterns occurring in micropaleontological assemblages, an abundance of evidence for testing and explaining the mechanism for biological change could be gathered."

Research along these lines had already been discussed in a symposium titled "Tempo and Mode of Evolution from Micropaleontological Data," included in the program of the North American Paleontological Convention at Montreal in 1982. In 1985, an article in Science by Thomas M. Cronin of the U.S. Geological Survey offered examples of both rapid speciation (punctuation) and stasis in a genus of Cenozoic marine ostracodes (Cronin, 1985). The

ostracodes are microscopic crustaceans, many species of which reproduce sexually; their fossil remains can be studied in deep-sea cores and in shallow-water sediments exposed on land. A study of the phylogeny of four species of the ostracode genus Puriana from Neogene and Quaternary deposits of the Atlantic Coastal Plain and the Caribbean Ocean showed that three new species of the genus appeared suddenly between -3 and -4 m.y., having evolved from a fourth species during a period of sustained change in ocean temperature (Figure 41.2). The temperature change is attributed to oceanographic changes that resulted from a closing of the isthmus of Panama. The four species thereafter showed stasis lasting to the present time. An interesting observation is that stasis persisted despite cyclic changes in ocean temperature and sea level during nearly 1.6 m.y. of Pleistocene glacial/interglacial climate swings.

According to Cronin (p. 61), the speciation event shown in Figure 41.2 required a maximum of 300,000 to 500,000 years. (Speciation is labeled "cladogenesis.") This is about as long as the elapsed time from the appearance of genus Homo to the present. Let us say that the generation period of an ostracode is two years. That gives us 150,000 to 250,000 generations to work with. Perhaps if the speciation of these ostracodes were studied on a fine scale over 200,000 y. at, say, 1,000-y. intervals, the record would take on a distinctly gradualistic appearance. No hopeful monsters would be needed. Perhaps what looks like a tiny punctation mark on a printed book page of geologic history, if magnified by a factor of 10,000, would be revealed as a long narrative of fine historical details requiring many pages of readable print to describe fully.

Could there be such a program as stasis punctuated by periods of gradualism"? Would it then be called "punctuated gradualism" Here's your answer: A 1984 Science article with the title "Species Formation through Punctuated Gradualism in Planktonic Foraminifera" (Malmgren et al., 1984). The authors find speciation of foraminifera in deep-sea drill cores occurring at too slow

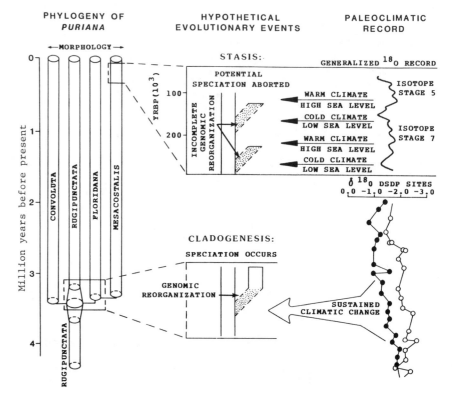

PHYLOGENY OF *PURIANA*

HYPOTHETICAL EVOLUTIONARY EVENTS

PALEOCLIMATIC RECORD

Figure 41.2 The phylogeny of four species of ostracodes of genus <u>Puriana</u> is shown at the left, against a time scale in millions of years before present. The rapid speciation (cladogenesis) shown here occurred during a period of sustained climate change, documented by oxygen-isotope ratios in the deep-sea core (right). Rapid speciation (punctuation) was followed by stasis persisting to the present time, despite Pleistocene cycles of climate change. (From Thomas M. Cronin, 1985, <u>Science</u>, vol. 227, p. 62, Figure 2. Courtesy of the U.S. Geological Survey; reproduced by permission of the U.S.G.S. and the author.)

a rate to qualify as punctuationism, Gould-style. The authors write:

> We propose the term "punctuated gradualism" for an evolutionary pattern of the type observed here in which populations are in stasis for long periods of time and in which these periods are interrupted by relatively rapid, gradual phyletic transformations (species formations) without lineage splitting. This model differs from the punctuational model in that new species are not assumed to evolve through lineage splitting and in that the transformations are not assumed to be geologically instantaneous. (pp. 318-19)

This looks to me a lot like the evolutionary program described by Gingerich for the chainlike succession of species of <u>Cantius</u>. Why shouldn't all combinations of evolutionary rates be entertained as possibilities? Surely, the modern view of uniformitarianism, which we presented in Chapter 21, can encompass whatever empirical science turns up through observation of the real world.

Gish's Law

To close this chapter, allow me to introduce a bit of comic relief contributed by geologist Robert S. Dietz of Arizona State University at Tempe (Dietz, 1983). In a letter to <u>Geotimes</u>, Dietz takes creationist Duane Gish of the ICR to task for not being aware of the full consequences of the creation scientists' attempt "to lose evolution down a gap in the fossil record." Dietz points out that geologists have had some good successes in closing the gaps in the fossil record, but perhaps Gish would see such success as simply the production of more gaps than ever before. "His reasoning is that if fossils A and C exist there is only one gap. But if B is discovered then there are two gaps--between A and B and between B and C." Dietz continues, relentlessly:

> We can apply this reasoning nicely to the Class Aves. Some 9,000 species are now extant and it is estimated that 1.2 million species (more than 99 percent extinct) have existed since the likes of <u>Archaeopteryx</u> in the Jurassic. If all 1,200,000 species were ever found by paleontologists we will wind up with 1,199,999 gaps. There is something about this which saddens. . . . In deference to Duane T. Gish I would like to offer Gish's Law: "As the fossil record becomes ever more complete, the number of gaps increases."

Credit

1. From David M. Raup, The Geological and Paleontological Arguments of Creationism, pp. 147-62 in <u>Scientists Confront Creationism</u>, Laurie R. Godfrey, ed., W. W. Norton & Company, New York. Copyright © 1983 by Laurie R. Godfrey. Used by permission.

Chapter 42

Transitions: I. Invertebrates to Vertebrates

Creation scientists claim that major gaps between the higher taxa exist at several points in the fossil record. Duane Gish devotes some eighty pages to this subject in Evolution? The Fossils Say NO! (1978a). Additional topics have been introduced by Morris (1974a) and Gish (1980; 1981; 1983). A summary list of the supposed unfilled major gaps and sudden beginnings runs about as follows:

Protozoans to metazoan invertebrates
   --the Cambrian explosion
Rise of insects
Invertebrates to vertebrates
Rise of the major fish classes
Fishes to amphibians
Amphibians to reptiles to mammals
      Rise of mammal-like reptiles
      Reptiles to birds
      Rise of bats
      Rise of rodents
      Rise of marine mammals
      Rise of horses
      Rise of man
Rise of plants

Throughout this entire section on gaps, chapters 42-45, I have relied heavily on published arguments of the following mainstream paleontologists and geneticists: Alfred S. Romer (1966), Edwin H. Colbert (1969; 1980), Laurie R. Godfrey (1983), David M. Raup (1983), Chris McGowan (1984), G. Ledyard Stebbins (1982), and James W. Valentine (1977). The total text by these and other authors devoted to the pros and cons of each gap or sudden appearance is far too great to be fully covered here, involving as it does many details of anatomical comparisons. I will offer only a brief summary of the paleontologists' findings.

Stating in full the creationists' case for the existence of gaps can easily be done in a single sentence: The gaps exist and no fossils have been found to fill them. Thus, the burden of proof is on the evolutionists, but this is as it should be throughout all of science. Bringing forward evidence--such as it is--of intermediate evolutionary forms is a most useful and instructive exercise, because it arouses us from the complacency we tend to fall into after repeatedly hearing from mainstream scientists that "evolution is a fact, proved beyond any doubt."

Sudden Appearances--The Cambrian Explosion

The Cambrian "explosion"--what a juicy morsel for the creation scientists! What deeper gloom and doom could there be for evolutionists than great faunal complexity and diversity, preceded by nothing, or almost nothing? I must admit that as a geology student, unreservedly committed to Darwinian evolution, I found the Cambrian explosion to be an enigma that really "tried one's faith." Our college textbook of historical geology expressed the problem in a forthright manner:

At the dawn of the Cambrian, life had already existed upon the Earth for probably a thousand million years, slowly evolving throughout the Proterozoic and much, if not all, of Archeozoic time. Small wonder, therefore, that the great branches of the animal kingdom were nearly all represented in the Lower Cambrian faunas, and that complex forms of Crustacea, like the trilobites, held the center of the stage! . . . Obviously the evidences of the truly primitive stages of life and of its differentiation into the great phyla must be sought far earlier than the Cambrian. It is a tantalizing thought that probably far more than half the drama of evolution had been enacted before the rising curtain gives us a clear glimpse into the Cambrian scene. (Schuchert and Dunbar, 1933, pp. 140-41:)[1]

This passage certainly expresses the concept of slow, gradual evolution that the creationists still repeat. The text continues with speculations then current as to why the hordes of predecessors of the Cambrian metazoans left no fossil record. The postulated reasons offered by the text for lack of fossilization are much the same as we have continued to hear until only very recently: lack of armor because of a lack of sufficient dissolved carbonates in seawater or because there were no carnivorous animals to stimulate the development of armor, and so forth.

Yet one cannot help but wonder how the lack-of-armor hypothesis could have had any credibility at all in the 1930s in view of the remarkable preservation of the soft-bodied Burgess fauna, brought to light back in 1910. Surely, there must have been similar fine-textured carbonaceous mud accumulations throughout late Proterozoic time capable of enclosing soft-bodied metazoans--precursors of the Burgess fauna--and giving us a record of their evolutionary development. On rereading this section of text, I was pleased to find that one W. K. Brooks in 1894 had published his suggestion that a rapid increase in size of individual animals, taking place about at the close of the Proterozoic, had caused crowding of the habitats and set off intense competition among species, which "felt the need of protective armor." (Professor Schuchert, your phrase does surprise me!) I hadn't realized the "punctuation" theme goes back so far, seemingly obscured and completely overshadowed by the Darwinian gradualism that must have been in full force in the 1890s.

Creationists capitalize on the Cambrian explosion and the total void of metazoans in the Precambrian that preceded it. Duane Gish makes full use of a consensus of mainstream paleontologists to bolster his creationist position:

What do we find in rocks older than the Cambrian? Not a single indisputable, metazoan fossil has ever been found in Precambrian rocks! Certainly it can be said without fear of contradiction that the evolutionary ancestors of the Cambrian fauna, if they ever existed, have never been found. (1978a, p. 63)

From all appearances, then, based on the known facts of the historical record, there occurred a sudden great outburst of life at a high level of complexity. The fossil record gives no evidence that these Cambrian animals were derived from preceding ancestral forms. Furthermore, not a single fossil has been found that can be considered to be a transitional form between the major groups, or phyla. At their earliest appearance, these major invertebrate types were just as clearly and distinctly set apart as they are today. (P. 64)

Gish then compares these "facts" with the predictions of the "evolution model" and finds that they clearly contradict such predictions. He finds, instead, that the facts are in full agreement with the predictions of the "creation model." It does look, on the face of it, as if the creationists have won this round. What do the evolutionists have to offer in their own defense?

The Ediacarian Fossils--Precursors of the Cambrian Metazoans?

The year was 1947. An Australian geologist, Martin F. Glaessner, searching for Cambrian fossils in strata in the Ediacara Hills of South Australia, discovered some strange rounded impressions that looked as if they might be fossil jellyfish. As these primitive coelenterate animals are abundant as fossils in lower Cambrian strata, Glaessner's find attracted little notice at the time. Later, however, other collectors found in the same area not only jellyfish impressions, but also what seemed to be the impressions of segmented worms. Besides these, there were some very strange geometrical forms not resembling anything known from the Cambrian Period. The patterns consisted of rounded outlines with radial lines or leaf-shaped outlines with bilateral subdivisions, as in a simple fern leaf (Figure 42.1).

Known as the Ediciarian fauna, Dr. Glaessner's collection eventually came to over 1,500 specimens, all consisting only of impressions on the sandstone bedding surfaces. Besides the jellyfish, representing the phylum Coelenterata, impressions of segmented worms strongly suggest members of the phylum. Particularly striking is an impression of a segmented worm with a distinct head, followed by about 40 identical body segments narrowing toward the tail end. A shield-shaped form suggests a primitive crustacean.

In the Ediacarian locality the strange fauna is found in a lower stratigraphic position than strata with Early Cambrian fossils, but the sequence of beds between the two is continuous and undeformed. Thus, there is little doubt that the Ediacarian fauna is older than Early Cambrian. On these grounds, the strange fauna was first assigned to the late Precambrian.

One of the Ediacarian forms, the one that resembles a ribbed leaf, soon turned up in other localities around the world. Given the name of Pteridinium, it was identified in South West Africa, England, northern Russia, and Siberia. In 1966, two geologists from the University of California at Los Angeles, Preston E. Cloud and C. A. Nelson, announced the finding of the same fossil near the base of the Paleozoic series in mountains of eastern California (Cloud and Nelson, 1966). In 1973, Cloud reviewed four regions of occurrence of Ediacarian fossils in North America and proposed that these strata be assigned to the "Ediacarian System," a new system to be placed directly below the Cambrian System. Cloud recommended that it be included in the Phanerozoic Era because it contained metazoans; as such, it would not be appropriate within the Proterozoic Era, for the obvious reason of the meaning of "Proterozoic." The Ediacarian

Period began around -670 m.y.; the Cambrian Period at about -570. In a jointly authored paper in 1982, Cloud and Glaessner reviewed the status of knowledge of the Ediacarian System. Its distinctive biota consists of "naked" metazoans that, in the type Australia region, are represented by 26 species in 18 genera and 4 or more phyla (Cloud and Glaessner, 1982, p. 783). The authors concluded that this 100-m.y. period, along with the Early Cambrian, "was the time during which metazoan life diversified into nearly all of the major phyla and most of the invertebrate classes and orders subsequently known."

That 100 million years of evolution immediately before the start of the Cambrian is fully adequate to accommodate the rise of the metazoans seems entirely reasonable. Holding this view is paleontologist Kenneth E. Caster, who commented as follows on the Ediacarian fauna of Australia:

Mine is the traditional view that the Ediacarian fauna is a backward projection of the Cambrian shallow benthos requiring no exceptional explanations of any sort. In the more than 130 million years between the Ediacarian time and the earliest Cambrian, hard parts (calcium phosphate, calcium carbonate, chitin), which may have been incipient in the Ediacarian organisms, were widely developed by invertebrate creatures. This made their basal Phanerozoic preservation not only abundant, but so much so by way of contrast as to seem to some paleontologists to have been the result of essentially instantaneous "explosive" evolution. (1984, p. 1128)

Caster goes on to say that the Ediacarian fauna "seems to me to fit comfortably into extant phyla." Caster's comments are in the form of a letter replying to a news report in Science in which Roger Lewin reviews the suggestion of a German paleontologist, Adolf Seilacher, that the organisms of the Ediacarian fauna were not precursory to the Cambrian explosion, "but instead represent a widespread, but ultimately failed, biological experiment" (Lewin, 1984, p. 39). Seilacher does, however, see in the wormlike forms within the Ediacarian fauna possible ancestors of the later Cambrian fauna. Despite this difference in interpretation of the evolutionary meaning of the Ediacarian fauna, there is broad agreement that at least 100 million years of evolutionary development of the metazoans have been established as part of the biostratigraphic record. Caster, in a summary statement, puts the whole matter in perspective:

When one considers the vastness of Precambrian time, during which life existed in oceanic waters, and the undoubtedly gradual change in the nature of those waters, it would be indeed strange if there were not a multitude of these ancients yet to be discovered as fossils: moreover, it would be surprising if some of these lineages were not unique to the Precambrian; thus an open mind is required in testing the oldest organisms with respect to their relevancy to the Phanerozoic biota. (P. 1131)

I found no references to the Ediacarian fauna in the creationist writings prior to 1985, despite the fact that the subject has been presented in leading journals since the late 1950s and is discussed in popular college textbooks published in the 1960s (e.g., McAlester, 1968, pp. 16-19; Dunbar and Waage, 1969, pp. 162-63).

More recently, Gish has introduced the Ediacarian topic briefly and makes these observations:

These discoveries do not alleviate the problem for evolution theory. These creatures are in no way intermediate between single-celled organisms and the

Figure 42.1 An artist's reconstruction of sea-bottom life (A) in Ediacarian time, and (B) in Early Cambrian time. (Drawings by Laszlo Meszoly, from <u>Early Life</u>, by Lynn Margulis. Copyright © 1984 by Jones and Bartlett, Publishers, Inc. Reproduced by permission of the author and publisher.)

complex invertebrates previously found in Cambrian rocks. They <u>are</u> complex invertebrates. Furthermore, it has been recently established that the creatures of the Ediacaran Fauna are not the same as the worms, coelenterates and echinoderms of the Cambrian. In fact, they are so basically different that it has been stated unequivocally that they could not possibly have been ancestral to any of the Cambrian animals.* It is asserted that a previously unrecognized mass extinction eliminated all of these creatures many millions of years before the Cambrian. (1985, p. 57)

(*Reference is made to an article by S. J. Gould in <u>Natural History</u>, vol. 93, no. 2, 1984.)

The problem of transition from one-celled organisms to the complex, multi-celled metazoans does, indeed, remain to be given our careful consideration.

A New Look at the Cambrian Explosion

Modern solutions to the problem of the seeming explosiveness of metazoan evolution in late Precambrian time have been reviewed by Stephen J. Gould in two delightful essays (they all are!) that first appeared in <u>Natural History</u> and are reprinted in <u>Ever Since Darwin</u> (Gould, 1977, pp. 119-33). He comments first on the decades-old suggestion that soft-bodied animals evolved

over a long time span in late Precambrian time without record, producing fossils only in the Early Cambrian, because they rather suddenly developed hard parts-- shells or outer skeletons--capable of widespread preservation. Gould's point is that these hard parts are not mere armor sheathing but are integral working parts of the organism. He writes:

> A clam without a shell is not a viable animal; you cannot clothe any simple soft-bodied organism to make one. The delicate gills and the complex musculature clearly evolved in association with a hard outer covering. Hard parts often require a simultaneous and complex modification of any conceivable soft-bodied ancestor; their sudden appearance in the Cambrian, therefore, implies a truly rapid evolution of the animal they cover. (1977, p. 122)

There seems little prospect of a viable alternative to "truly rapid evolution" of the metazoans. We must look for some possible change in the marine environment that could have triggered the rapid evolution. Environmental change could be a physical change, such as a rapid increase in dissolved oxygen in the ocean, or it could be a biological change. As we pointed out in an earlier section, within an ecosystem one aspect of the environment consists of the presence of organisms that interact with one another. For example, for a predatory animal the prey is part of its environment; for a herbivore the presence of certain plants on which it feeds is a part of its environment. Could it be that in late Precambrian time some significant change in the biological environment set off the rapid evolution of the metazoans?

Paleobiologist Steven M. Stanley of Johns Hopkins University suggested that the metazoan explosion may be explained by applying an ecological concept known as the "cropping principle" (1973). His hypothesis was reported in <u>Science</u> <u>News</u> (1973, vol. 103, p. 324). He pointed out that a pasture densely seeded with grass will prevent the appearance of weeds, but if grazing sheep pull up some of the grass by its roots, weeds move in and the ecosystem rapidly diversifies in species representation. Applying this principle to the late Precambrian marine environment, Stanley reasons that the complete dominance of the ecosystem by algae and bacteria for a couple of billion years effectively suppressed the growth of other life forms. New species that fed on the dominant inhabitants evolved very slowly at first and they were limited to very small populations. These initial croppers were single-celled protists that fed on algae and bacteria. Stanley is quoted as saying: "But once they started appearing, once this barrier was crossed, herbivorous and carnivorous organisms arose almost simultaneously, and the whole self-limiting character of the system changed." The new animal forms reduced the population of the algae, and this reduction in turn speeded up the diversification of the animals and the rise of the complex, multicelled metazoans. In Stanley's words, "The whole system suddenly took off."

Gould, reviewing Stanley's hypothesis, explains that communities of primary producers (blue-green algae, for example, that photosynthesize carbohydrates) consist of one or only a very few species (1977, p. 123). These achieve superiority and tend to monopolize the available space. When a cropper arrives to prey on the dominant species, it frees space for other species. The result is diversification of species but with fewer numbers of each. The cropping principle has been demonstrated in living ecosystems, including an algal community grazed by sea urchins.

In Proterozoic time the most abundant primary producers were the blue-green algae, which built widespread stromatolites, but today stromatolites are largely restricted to environments hostile to metazoans.

This observation gives some measure of support to Stanley's hypothesis. His is not an ad hoc hypothesis simply dreamed up to explain away the Cambrian "explosion."

In his second essay, cleverly titled "Is the Cambrian Explosion a Sigmoid Fraud?" Gould develops another concept relating to population explosions (1977, p. 126-33). Studies of bacterial colonies show a typical model of growth following a sigmoidal, or S-shaped, curve (Figure 42.2). When a colony is started in an uninhabited growth medium, the increase in numbers of individuals by doubling at regular time intervals is at first almost imperceptible on the graph; this is called the "lag phase." Then, the population curve starts a strong upturn and produces a steeply rising phase, called the "log phase," which would continue indefinitely were it not for counter factors that curb the rise. For example, the food supply would tend to run low, or toxins would be produced to slow the rate of increase. Now, the curve reduces in steepness and finally levels off.

Perhaps the same kind of growth curve was followed in the increase in numbers of species and individuals in the transition from late Precambrian through Early Cambrian. We can liken the conditions in the Precambrian oceans to those in an uncontaminated petrie dish of nutrient matter in the laboratory. There would be almost unlimited space and nutrient supply at the outset. The lag phase might pass unnoticed because of the paucity in numbers of the growing colony of consumer forms. The log phase might correspond with the Early Cambrian, dominated by rapid evolution of metazoans. Gould and his colleagues pursued this suggestion by making detailed counts of Cambrian animal species, plotting them as diagrams (spindle diagrams) to show the relative rates of change in diversity upward in time. The subject is highly technical and space does not permit me to try to explain it here, but the diagrams seem to show clearly a log phase of uninhibited growth in the Early Cambrian. Similar data plots for post-Cambrian time show changes in diversity with time that are to be expected in a general condition of equilibrium in which all facilities offered by the environment are fully occupied and one species rises to replace another that becomes extinct. Gould states that the log phase of the Cambrian explosion fully occupied the oceanic environments. He adds:

> The log phase of the Cambrian filled up the earth's oceans. Since then, evolution has produced endless variation on a limited set of basic designs. Marine life has been copious in its variety, ingenious in its adaptation, and (if I may be permitted an anthropocentric comment) wondrous in its beauty.

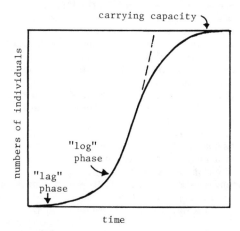

Figure 42.2 The typical sigmoidal (S-shaped) curve of population increase of bacteria with time in an uninhabited growth medium. (A. N. Strahler.)

Yet, in an important sense, evolution since the Cambrian has only recycled the basic products of its own explosive phase. (P. 133)

I find in this last statement a theme of catastrophism-- a one-time event of phenomenally rapid biotic change-- followed by a sort of uniformitarian regime persisting ever after. Creationists do not take kindly to the use of catastrophism by mainstream evolutionists. Catastrophism is claimed by creationists as their own exclusive property; its use by the archenemy is as abhorrent as copyright violation by blatant plagiarism is to an author. But catastrophism is free for the taking by all who need it to explain the workings of nature. Mainstream science has always invoked catastrophe when it has seemed an appropriate solution to a problem. The only restrictions have been that a catastrophic solution be totally naturalistic, and that it conform with the accepted body of universal laws and principles pertaining to the behavior of matter and energy. Creationists, on the other hand, seem to cultivate simultaneously two forms of catastrophism--the natural and the supernatural--passing off the one for the other as convenience dictates.

Evolutionary science has a long way to go to illuminate fully the nature of the Precambrian-Cambrian faunal transition, but real progress has been made in both the opening up of the fossil record and in developing viable hypotheses grounded in sound principles of ecosystem functioning and population dynamics. In its handling of this difficult problem, mainstream science has been functioning commendably well in making maximum use of interchanges between subdisciplines within paleontology and biology.

Invertebrates to Vertebrates

Duane Gish disposes of the transition from invertebrates to vertebrates in less than a page:

> The idea that the vertebrates were derived from the invertebrates is purely an assumption that cannot be documented from the fossil record. On the basis of comparative anatomy and embryology of living forms, almost every invertebrate group has been proposed at one time or another as the ancestor of the vertebrates. The transition from invertebrate to vertebrate supposedly passed through a simple chordate stage. Does the fossil record provide evidence for such a transition? Not at all. (1978a, p. 66; 1985, pp. 65-66)

At this point, Gish quotes a paragraph from F. D. Ommaney's popular book, The Fishes, one of the Life Nature Library series, published by Time-Life, Inc., in 1964. It states that a gap of 100 m.y. exists between Cambrian time, when the chordates first appeared, and the late Ordovician, when the first fishlike animals appeared, a gap that will probably never be filled. Gish continues:

> Incredible! One hundred million years of evolution and no fossilized transitional forms! All hypotheses combined, no matter how ingenious, could never pretend, on the basis of evolution theory, to account for a gap of such magnitude. Such facts, on the other hand, are in perfect accord with the predictions of the creation model.

Henry Morris is equally brusque in expressing his opinion:

> The evolutionary transition from invertebrates to vertebrates must have involved billions of animals, but no one has ever found a fossil of one of them. Invertebrates have soft inner parts and hard outer

shells; vertebrates have soft outer parts and hard inner parts--skeletons. How did the one evolve from the other? There is no evidence at all. (1974a, pp. 81-82)

Morris quotes a 1966 statement by Alfred Romer, a distinguished vertebrate paleontologist in his time, pleading ignorance to knowledge of how the transition to fishlike vertebrates occurred. "Which means, simply," says Morris, "that there are no fossils yet available of incipient forms leading up to these fish from their assumed ancestors. Surely it is more reasonable to believe that vertebrates and invertebrates were separate creations from the beginning" (p. 82).

If Gish and Morris are right about the total lack of intermediate forms, then theirs is indeed a difficult act to follow. One paleontologist who has moved into the argument with considerable vigor is Chris McGowan, Curator-in-Charge of the Department of Vertebrate Paleontology at Toronto's Royal Ontario Museum, and also a faculty member of the University of Toronto (1984, pp. 74-78). His argument, which originated with N. J. Bernal in the 1950s, comes not from fossils, but from a living organism called Amphioxus. This living animal is a lancelet, shaped like a long, narrow fish, almost transparent, able to swim, but not very well (Figure 42.3). Most of the time, Amphioxus lies buried in the sand of the ocean floor with only its front end exposed. It resembles a fish in that it has a row of gill slits on either side, but it has no distinct head and no jaws or eyes. The mouth, rimmed with small tentacles, sucks in seawater, which it passes through a sieve to collect microorganisms that are its food. It has body muscles and a tail, a liver, and an intestine. Lacking in mineralized hard parts, Amphioxus would not be likely to have left a fossil record, and indeed it "has no ancestors or counterparts in the fossil record" (Stebbins, 1982, p. 274).

What interests evolutionists about the anatomy of Amphioxus is that it contains two structures relating it closely to fishes. These are a spinal cord and notochord. The former, a hollow bundle of nerve fibers, runs the length of the body. The notochord is a firm, rodlike structure paralleling the spinal chord and providing firmness and rigidity to the body, preventing it from collapsing. Presence of a notochord in the embryonic stage is diagnostic of all vertebrates, which along with Amphioxus are within the phylum Chordata.

McGowan comments: "Because it lacks a vertebral column, amphioxus cannot be described as a vertebrate, but it is obviously closely related to vertebrates, having more things in common with them than with any of the invertebrate animals" (1984, p. 76). He continues:

> How should we interpret amphioxus? I regard it as a surviving member of a group of organisms from which the vertebrates evolved. I am not suggesting that amphioxus is the actual ancestor, of course, but only that the vertebrate ancestors were probably similar to amphioxus. Drs. Morris and Gish both discuss the transition from invertebrate to vertebrates, but they are so concerned with demonstrating the absence of fossil forms that they have nothing to say about living animals. Perhaps they would dismiss amphioxus as being merely an unusual vertebrate and thus maintain that we still had not found a bridge across the invertebrate-vertebrate gap. Aside from the fact that it is not a vertebrate as it lacks a vertebral column, this is a reasonable argument in itself, except that amphioxus is not the only primitive chordate animal. (1984, p.86)[2]

McGowan makes no mention of any fossil record of an early ancestor of Amphioxus. I could not help but wonder

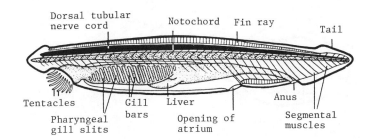

Figure 42.3 Amphioxus, shown in cross section. (From Ralph Buchsbaum, 1948, Animals Without Backbones, University of Chicago Press, p. 313. Copyright by Ralph Buchsbaum. Reproduced by permission.)

if there could be one in the Middle Cambrian Burgess Shale fauna, in which so many varieties of soft-bodied organisms were preserved in great anatomical detail. Accordingly, I turned to an update of that fauna in a 1979 article by Simon Conway Morris and H. B. Whittington in Scientific American. There I found exactly what I was looking for:

> Finally, we find among the Burgess Shale fauna one of the earliest-known invertebrate representatives of our own conspicuous corner of the animal kingdom: the chordate phylum. . . . The chordates are represented in the Burgess Shale by the genus Pikaia and the single species P. gracilens.
>
> What about Pikaia, formerly considered a poplychaete worm? Some 30 well preserved specimens show a prominent rod along the animal's back that appears to be a notochord, the cartilagelike stiffening organ that gives the chordate phylum its name. In addition to this key anatomical feature the blocks of muscle in Pikaia form a zigzag pattern that is comparable to the musculature of the primitive living chordate Amphioxus and of several fishes. Although Pikaia differs from Amphioxus in several important respects, the conclusion that it is not a worm but a chordate appears inescapable. The superb preservation of this Middle Cambrian organism makes it a landmark in the history of the phylum to which all vertebrates, including man, belong. There are possible instances of even earlier chordates from Lower Cambrian formations in California and Vermont but none is as rich in detail. (1979, pp. 122, 131)

So there is a fossil ancestor of the living chordate Amphioxus! A photograph of Pikaia clearly shows the probable notochord and reveals an animal strikingly like modern Amphioxus (Morris and Whittington, 1979, p. 133).

McGowan follows his winning ace lead with the king, hoping to take a second quick trick. This one is not quite so easy, however. He brings in the sea squirts, sessile sacklike animals firmly attached to rocks. They are purely invertebrate in anatomy with almost nothing to suggest a swimming lancelate or a fish. But the larva of the sea squirt offers the surprise. Looking like a small tadpole, it is remarkably like Amphioxus in its anatomical equipment. It has a a hollow spinal cord, a notochord, and a pharynx with a pair of gill slits (Figure 42.4). The implication is clear enough: Amphioxus and the sea squirts, and other coelenterates with similar larval forms, share a common ancestry.

McGowan's argument based on larval forms is, according to Stebbins, the prevailing opinion among zoologists. Stebbins explains:

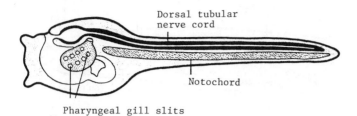

Figure 42.4 Schematic cross section of the free-swimming tunicate larva, with the three chordate characters labeled. (From Ralph Buchsbaum, 1948, <u>Animals Without Backbones</u>, University of Chicago Press, p. 318. Copyright by Ralph Buchsbaum. Used by permission.)

Animals like <u>Amphioxus</u> gave rise to primitive jawless fishes through changes in the head region, body covering, and internal skeleton. They developed eyes, nostrils, a rudimentary brain, and bony plates that formed the beginnings of an internal skull. Behind the head region, their bodies became covered with bony scales that in some forms were not very different from the modern fishes. The notochord was retained and became partly surrounded by the beginnings of vertebrae and a backbone. (1982, pp. 275-76)[3]

Worms to Insects

Although the origin of the arthropods (crustaceans, arachnids, and insects) lies within the invertebrate division, it is appropriate to discuss it here because the evidence is from anatomy of living forms, and the argument is much the same as for the transition from invertebrates to vertebrates by way of the lancelets.

Henry Morris gives brief attention to the problem:

If the evolutionary origin of the higher animals is obscure, the origin of insects is completely blank. Insects occur in fantastic number and variety, but there is no fossil clue to their development from some kind of evolutionary ancestor. (1974a, p. 86)

Gish expresses a similar opinion that transitional forms of the Insecta are absent (1985, pp. 61-62). Contrary to

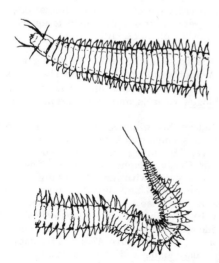

Figure 42.5 <u>Nereis</u>, a free-swimming polychete worm. Head and tail portions sketched from a photograph. (A. N. Strahler.)

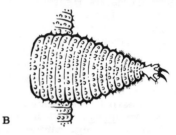

Figure 42.6 The peripatus, of the phylum Onychophora. A. Stylized drawing of the entire animal. B. Drawing of the leg, showing the arthropodlike claws. (From Pearse and Buchsbaum, <u>Living Invertebrates</u>, Blackwell/Boxwood, Publishers. Copyright © 1986 by Ralph Buchsbaum. Reproduced by permission.)

these statements, there is anatomical evidence of the evolutionary ancestor of the insects, which made their first appearance in Carboniferous time. Biologists suggest that the ancestor was one of the ringed, or segmented, worms of the phylum Annelida. A specific example is one of the more primitive annelid forms, <u>Nereis</u>, a living marine bottom dweller (Figure 42.5). Each segment of a nereis has projecting "side feet" from which project horny bristles. So equipped, a nereis can swim or burrow in sand. It is familiar to New Englanders as the "clam worm" used to bait fishhooks. The nereis is covered by a horny but flexible cuticle. The musculature of the annelids consists of longitudinal muscles; they have nothing in the form of the insects' articulated exterior skeleton operated by bundles of attached muscles.

Insects are a class (Insecta) within the phylum Arthropoda. Differences between the insects and the annelids, above and beyond those mentioned above, are profound. Arthropods have a rigid chitinous outer layer that serves as protective armor, divided into segments by a system of joints, i.e., an articulated chitinous exoskeleton. The articulated body segments and appendages of the insects are adapted to many different functions; the annelids show little such differentiation. Annelids breathe through the skin, which must be kept moist; insects have a breathing system (tracheal system) of air tubes opening through the exoskeleton and piping the air directly to the tissues.

The intermediate between annelids and insects (along with the other arthropods) is generally agreed by biologists to be a little-known animal called <u>Peripatus</u> (Figure 42.6). It resembles a worm in outward appearance, but most people would say it looks like a caterpillar with its velvety wrinkled skin and numerous pairs of "feet." Although usually assigned to a separate phylum (Onychophora), it has alternatively been designated as a class of the arthropods. It lives in moist tropical forests, usually hidden beneath bark or leaves.

<u>Peripatus</u> resembles annelids in having a thin cuticle, but like insects, it breathes through a tracheal system.

The feet of Peripatus resemble those of the arthropods in having claws. Its three-segmented head seems intermediate between that of an annelid and an arthropod. The circulatory system resembles that of an arthropod, but the excretory system is annelidlike.

McGowan gives his general impression of the intermediate onychophoran in these words:

> If an evolutionist had to sit down at a drawing-board and invent a hypothetical link between worms and arthropods, he could not do better than draw an onychophoran. What is more, there are beautiful fossil onychophorans which date back to the Cambrian and which look just like their living descendants. Dr. Gish overlooks the onychophorans when he tells us "not a single fossil has been found that can be considered to be a transitional form between the major groups, or phyla." (1984, pp. 73-74)[2]

Does the Middle Cambrian Burgess Shale fauna include a possible ancestor of Peripatus? Yes, indeed! Simon Morris and H. B. Whittington say the fauna includes "a peripatus-like animal that was aquatic rather than terrestrial" (1979, p. 122). The fossil genus is Aysheaia, shown in their photograph captioned "ancestral arthropod with a striking resemblance to the living onychophore Peripatus" (p. 130). It is an impressive photograph because it shows the tentacles and claw-equipped legs with remarkable clarity.

McGowan summarizes his case for these two supposed "gaps" with as much strength as he can muster:

This brief survey of living organisms has shown that the creationists are wrong when they say that there are no connections between the major groups of organisms.

The fact that we cannot draw a firm line between plants and animals, or between unicellular organisms and multicellular ones, is difficult to reconcile with the creation model. Taken with the evidence for a link between the two major invertebrate groups (insects and worms), a link that is also documented by fossils, and between invertebrates and vertebrates, we have an overwhelming case for evolution. (1982, p. 78)[2]

Credits

1. From A Textbook of Geology, Part II--Historical Geology, by Charles Schuchert and Carl O. Dunbar, John Wiley & Sons, New York. Copyright, 1933, by Charles Schuchert and Carl O. Dunbar. Used by permission of the publisher.

2. From IN THE BEGINNING © 1983 by Chris McGowan. Reprinted by permission of Macmillan of Canada, A Division of Canada Publishing Corporation.

3. From Darwin to DNA, Molecules to Humanity, by G. Ledyard Stebbins, W. H. Freeman and Company, New York. Copyright © 1982 by W. H. Freeman and Company. Used by permission.

Chapter 43

Transitions: II. Fishes to Reptiles
to Amphibians to Mammals

In this, our second chapter on transitions between major taxa, we sweep across an enormous sector of Phanerozoic time, during which waterbound, swimming vertebrates took to the lands, went through a transitional amphibious phase, in which they were comfortable in both water and air, then forsook the water for a totally air-breathing existence, and finally went on to exchange the hazardous, reptilian egg-laying habit for the mammalian device of sheltering the offspring in a placenta.

To defend this evolutionary sequence against the attacks of the creation scientists requires that we go into considerable detail on the anatomical similarities and dissimilarities between each of the contiguous pairs or classes of vertebrates mentioned above. This investigation will be a profitable exercise in recognition of mosaic evolution, a succession of morphological changes that affects certain characters while leaving others unchanged to provide little bridges across the time gaps in the fossil record.

The Creationists' View of the Origins
of the Classes of Fishes

Duane Gish finds from reading Alfred S. Romer's 1966 treatise, Vertebrate Paleontology, that mainstream paleontologists have found no fossil record of transitional chordates leading up to the appearance of the first class of fishes, the Agnatha, or of transitional forms between the primitive, jawless agnaths and the jaw-bearing class Placodermi, or of transition from the placoderms (which were poorly structured for swimming) to the class Chondrichthyes, or from those cartilaginous-skeleton sharklike fishes to the class Osteichthyes, or bony fishes (1978a, pp. 66-70; 1985, pp. 65-69). The evolution of these classes is shown in Figure 43.1. Neither, says Gish, is there any record of transitional forms leading to the rise of the lungfishes and the crossopterygians from the lobe-finned bony fishes, an evolutionary step that is supposed to have led to the rise of amphibians and ultimately to the conquest of the lands by air-breathing vertebrates.

In a series of quotations from Romer (1966), Gish finds all the confessions he needs from the evolutionists that each of these classes appears suddenly and with no trace of ancestors. The absence of the transitional fossils in the gaps between each group of fishes and its ancestor is repeated in standard treatises on vertebrate evolution. Even Chris McGowan's 1984 anticreationist work, purporting to show "why the creationists are wrong," makes no mention of Gish's four pages of text on the origin of the fish classes. Knowing that McGowan is an authority on vertebrate paleontology, keen on faulting the creationists at every opportunity, I must assume that I haven't missed anything important in this area. This is one count in the creationists' charge that can only evoke in unison from the paleontologists a plea of nolo contendere.

Fishes to Amphibians

Duane Gish turns next to "the assumed evolutionary sequence of life, fish gave rise to amphibia" (1978a, p. 73; 1985, p 72). With a drop of precious humor that the creation scientists dispense on rare occasions, Gish shortens this transition to "fin-to-feet," a segment in the total "fish-to-Gish" scenario. Gish says:

> The fossil record has been diligently searched for a transitional series linking fish to amphibian, but as yet no such series has been found. The closest link that has been proposed is that allegedly existing between rhipidistian crossopterygian fish and amphibians of the genus, Ichthyostega. There is a tremendous gap, however, between the crossopterygians and the ichthyostegids, a gap that would have spanned many millions of years and during which innumerable transitional forms should reveal a slow gradual change of the pectoral and pelvic fins of the crossopterygian fish into the feet and legs of the amphibian, along with loss of other fins, and the accomplishment of other transformations required for adaptation to a terrestrial habitat.

> What are the facts? Not a single transitional form has ever been found showing an intermediate stage between the fin of the crossopterygian and the foot of the ichthyostegid. The limb and the limb girdle of Ichthyostega were already of the basic amphibian type, showing no vestige of a fin ancestry. (Pp. 72-74)

To understand what Gish is talking about, we need to take a look at the anatomy of the two creatures in question, as shown in Figure 43.2. Whereas the fins of the crossopterygian (A) are attached to small bones loosely embedded in muscle (as Gish points out), the leg bones of the amphibian are securely anchored in massive pectoral and pelvic girdles. Whereas the lobe-finned fish probably used its fins against the bottom to push itself along in very shallow water, it did not need to support its body weight in air, as did the amphibian. As Gish rightly points out:

> In tetrapod amphibians, living or fossil, on the other hand, the pelvic bones are very large and firmly attached to the vertebral column. This is the type of anatomy an animal must have to walk. It is the type of anatomy found in all living or fossil tetrapod amphibians but which is absent in all living or fossil fishes. There are no transitional forms. (P. 75)

At this point, Gish turns to discuss Latimeria, a living coelocanth that has descended with very little change from the early crossopterygians. We will postpone discussion of

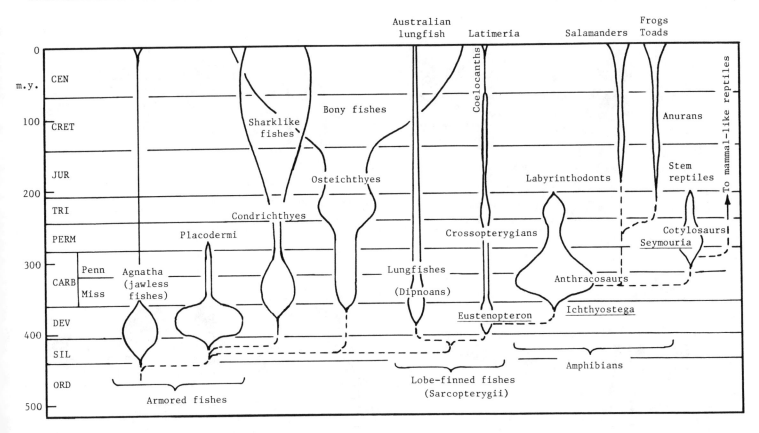

Figure 43.1 Chart of the evolution of the fishes, amphibians, and stem reptiles. (A. N. Strahler. Based on data of A. Lee McAlester, 1968, and E. H. Colbert, 1980.)

this remarkable animal to Chapter 47, devoted to "living fossils." Gish goes on to question why the lobe-finned fishes were chosen by paleontologists as the ancestors of the amphibians. For lack of an intermediate as a candidate, he says, the choice was made because of certain similarities in the skull patterns and vertebrae of the two forms.

Let us have a look, then, at the skulls of the two animals in question--fish and amphibian. Figure 43.3 shows the skulls, as seen from above and from the side.

The complement of individual skull bones corresponds closely in the two forms, but the sizes and shapes of the corresponding bones are different. In the lobe-finned fish (A), the portion of skull forward of the eyes is short, whereas in the amphibian (B), this region has become greatly lengthened and the eyes pushed back. This difference is very obvious in the side views.

The probability that two skulls with so many bones correspondingly present in the same relative positions could arise by pure chance in two totally unrelated animal forms is extremely small. The situation is essentially that of two nearly identical bridge hands dealt from two decks of cards in which the packs are of different sizes and proportions and in which the rendering of the symbols

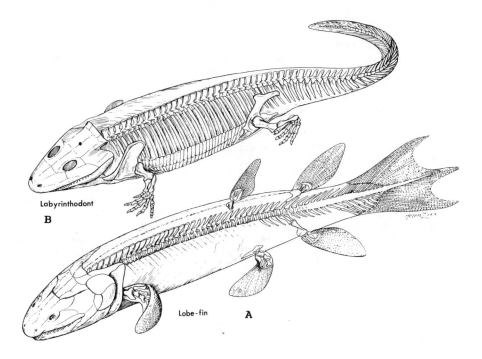

Figure 43.2 Reconstructions of (A) Eusthenopteron, a rhipidistian crossopterygian (lobe-finned fish) and (B) a labyrinthodont amphibian. (Courtesy of Department Library Services, American Museum of Natural History. Negative No. 321683.)

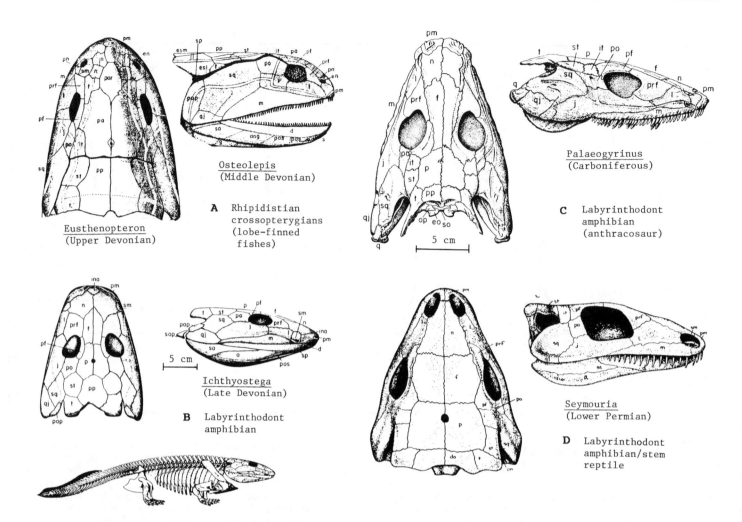

Figure 43.3 Reconstructions of skulls of (A) a lobe-finned fish, (B and C) two labyrinthodont amphibians, and (D) a stem reptile. (From <u>Vertebrate Paleontology</u>, 3d ed., by Alfred S. Romer, University of Chicago Press, p. 49, Figure 69; p. 81, Figure 109; p. 87, Figure 119; p. 96, Figure 133. Copyright © 1933, 1945, and 1966 by The University of Chicago. Reprinted by permission.)

follows two different artistic styles. The cards match closely in terms of suit and value, which is what counts in evaluating the probability of dealing two similar hands.

In the matter of the two skulls, no intermediates are needed to establish beyond reasonable doubt that the two animals are related through descent with modification. Either one has descended from the other, or both have descended from a common ancestor differing from both, but having the same complement of skull bones. If the former is the case, we might not know from this evidence alone which came first, but that would be established by the stratigraphic relationship.

The transition from lobe-finned fish to tetrapod amphibian offers an excellent opportunity to apply the principle of <u>mosaic evolution</u>, a concept that allows us to span fossil gaps with information capable of establishing the evolutionary sequence. As explained by G. Ledyard Stebbins, mosaic evolution describes "the way in which different characteristics evolve at different rates in a particular evolutionary line. An animal that is transitional between two major groups is much more likely to be a mixture, or <u>mosaic</u>, of old and new characteristics than to be intermediate between two groups with respect to all its characteristics" (1982, p. 283).[1]

Consider as a simple analog of mosaic evolution the development of the gasoline-powered automobile. As Table

43.1 shows, equipment change in each category of function proceded in steps. Each of the intermediates (1930 and 1950) shares two characteristics with the years before and after, but a third has undergone a radical change. Equipment items in the final year's model are all different from those of the first, but the last has evolved from the first.

Let us apply the principle of mosaic evolution to the transition, lobe-finned fishes to amphibians. Figure 43.4 shows transitions from the fossil lobe-fins represented by Eusthenopteron (A) to ancient amphibians, represented by Ichthyostega (B), to labyrinthodont amphibians of a group called the <u>anthracosaurs</u> (C). A fourth group, the stem reptiles represented by <u>Seymouria</u>, serves as a transition into the true reptiles. Six characteristics are selected; they are particularly important for the A-B-C transitions.

As we have already shown, changes in the skull are relatively small from A to B. From B to C to D the changes are also relatively minor with the same complement of bones being present in all (see Figure 43.3).

Changes in appendages are profound between the lobe-fins and the ancient amphibians. (These can be seen in Figure 43.2). The lobe-finned fish (A) has a long fin flap and only a few short, poorly articulated bones; the ancient amphibian, <u>Ichthyostega</u>, has a full set of well-articulated foot bones, with digits tapering to points. The digits may have been connected by a membrane of skin to give a webbed foot. From this point on, changes in the limb bones are relatively minor, even leading into the earliest reptiles, but the bones become more massive where required to support larger animals.

The development of strong pelvic and pectoral girdles in <u>Ichthyostega</u> (B), but absence of this equipment in the

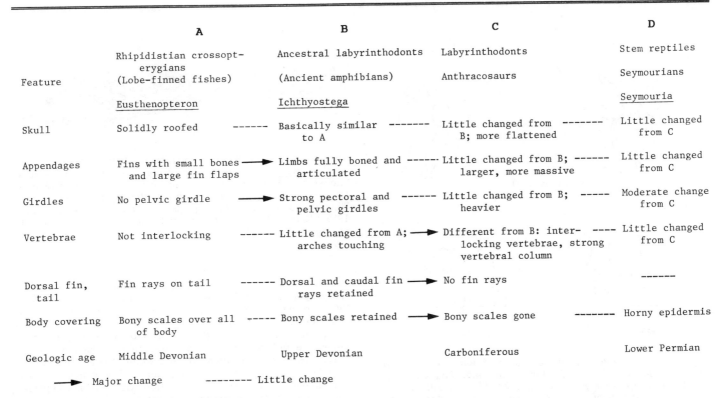

Feature	A Rhipidistian crossopterygians (Lobe-finned fishes) Eusthenopteron	B Ancestral labyrinthodonts (Ancient amphibians) Ichthyostega	C Labyrinthodonts Anthracosaurs	D Stem reptiles Seymourians Seymouria
Skull	Solidly roofed	------ Basically similar to A	------- Little changed from B; more flattened	------- Little changed from C
Appendages	Fins with small bones and large fin flaps	→ Limbs fully boned and articulated	------ Little changed from B; larger, more massive	------ Little changed from C
Girdles	No pelvic girdle	→ Strong pectoral and pelvic girdles	------ Little changed from B; heavier	----- Moderate change from C
Vertebrae	Not interlocking	------ Little changed from A; arches touching	→ Different from B: interlocking vertebrae, strong vertebral column	---- Little changed from C
Dorsal fin, tail	Fin rays on tail	------ Dorsal and caudal fin rays retained	→ No fin rays	-------
Body covering	Bony scales over all of body	----- Bony scales retained	→ Bony scales gone	------- Horny epidermis
Geologic age	Middle Devonian	Upper Devonian	Carboniferous	Lower Permian

→ Major change   -------- Little change

Figure 43.4 Mosaic evolution illustrated by transition from fishes to amphibians to stem reptiles. (A. N. Strahler. Based on data of A. S. Romer, 1966, and E. H. Colbert, 1980.)

lobe-finned fish (A) is a change we have already noted.

The vertebrae of both the lobe-fin (A) and the ancient amphibian (B) are quite similar. Both are of a primitive type not well adapted to supporting body weight, because the bones do not interlock. Figure 43.5 shows details. Here the major evolutionary step was delayed to the stage from B to stage C. The vertebrae of the labyrinthodonts (B to C) developed neural arches with facets that overlapped, giving strength to the backbone as required to support a large animal on land. There is little change from C to D, but the neural arches are expanded.

The tail section of Ichthyostega (B) is shown in Figure 43.6. The dorsal and caudal fin rays have been retained from the lobe-finned fishes (A). Body covering in the form of bony scales persisted from the lobe-finned fishes (A) to the ancient amphibians (B), but was subsequently lost.

The mosaic pattern is clear enough in this case. Ichthyostega is clearly transitional between the lobe-finned fishes that preceded it and the labyrinthodonts that followed. The creationists' premise that every character of the animal must show simultaneous gradual change from one form to the next is completely unrealistic. The strategy of proposing an unrealistic situation, and then finding that it does not exist is simply without scientific value.

Amphibians to Reptiles

The transition from amphibians to reptiles causes some difficulties for the creationists because the fossil record shows all the intermediates one could ask for. Indeed, there is a plethora of transitional forms. Romer writes: "The ties between primitive reptiles and anthracosaurian amphibians are so many and so clear that it is reasonable to believe that the reptiles stem from that group" (1966, p. 102).[2] Colbert reinforces this conclusion:

> The mixture of amphibian and reptilian characteristics seen in Seymouria is indicative of the gradual transition that took place between the two classes during the evolution of the vertebrates. Because the change was gradual rather than abrupt it is difficult to draw clear-cut distinctions between amphibians and reptiles when all fossil materials are taken into consideration. (1969, p. 114; 1980, p. 111)[3]

[2]From Vertebrate Paleontology, Third edition, by Alfred S. Romer, University of Chicago Press. Copyright © 1933, 1945, and 1966 by The University of Chicago. Used by permission.

[3]From Evolution of the Vertebrates, First edition 1955, Second edition 1969, Third edition 1980, by Edwin H. Colbert, John Wiley & Sons, Inc., New York. Copyright © 1955, 1969, 1980 by John Wiley & Sons, Inc. Reprinted by permission of the author and publisher.

Table 43.1   The Gasoline-Powered Automobile as an Example of Mosaic Evolution

1910	1930	1950	1980
Hand crank	→ Self starter	→ Self starter	→ Self starter
Manual gearshift (unsynchronized)	→ Manual gearshift (synchronized)	→ Automatic shift	→ Automatic shift
Carburetor	→ Carburetor	→ Carburetor	→ Fuel injection

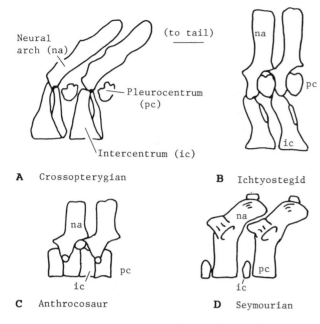

A Crossopterygian
B Ichtyostegid
C Anthrocosaur
D Seymourian

Figure 43.5 Comparison of vertebrae of lobe-finned fishes (A) with those of labyrinthodont amphibians (B, C, D). (A. From Vertebrate Paleontology, 3d ed., by Alfred S. Romer, University of Chicago Press, p. 80, Figure 108a. Copyright © 1933, 1945, and 1966 by The University of Chicago. Reprinted by permission. B. From Edwin H. Colbert, 1980, Evolution of the Vertebrates, 3d ed., John Wiley & Sons, New York, p. 90, Figure 30d; p. 102, Figure 37d; p. 94, Figure 32f. Copyright © 1955, 1969, 1980 by John Wiley & Sons, Inc. Reprinted by permission.)

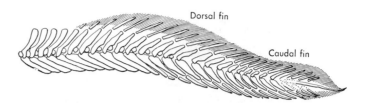

Figure 43.6 Tail of the labyrinthodont amphibian, Ichthyostega, showing the fin rays retained from its fish ancestors. About two-thirds natural size. (From Edwin H. Colbert, 1980, Evolution of the Vertebrates, 3d ed., John Wiley & Sons, New York, p. 90, Figure 30c. Copyright © 1955, 1969, 1980 by John Wiley & Sons, Inc. Reprinted by permission.)

So here the creationists have nothing to complain about on the matter of intermediates. But they must find something to complain about, so they turn to two features of the geologic record they don't like but are nevertheless pleased to find present. One is that soft parts are lacking to complete the documentation of transition. Missing, in particular, is that nice, soft reptilian egg. Second, the stratigraphic record does not show a satisfactory age sequence from amphibian to reptile and from reptile to mammal.

That soft parts are lacking in the record is fully expressed by both Romer and Colbert. Romer states that it is almost impossible to tell when we have crossed the boundary between primitive Paleozoic reptiles and some of the earliest amphibians. He says: "The true test, of course, is the type of egg the animal laid; but direct evidence for this is practically impossible of attainment in the case of extinct forms" (1966, p. 102).[2]

Colbert asks: "Was Seymouria an amphibian or a reptile?" He goes on to say:

> The ultimate answer to this question depends on whether Seymouria, like modern reptiles, laid an amniote egg on the land, or whether, like modern frogs, it returned to the water to deposit its eggs. Unfortunately, there is no direct paleontological evidence at the present time that gives us a clue about this important and diagnostic attribute. (1980, pp. 93-94)[3]

Colbert, however, cites some paleontological evidence that Seymouria laid amniote eggs, which must have been very large compared with amphibian eggs (pp. 94-95). Large eggs require a large cloaca out of which the eggs must pass, and this requires a favorable structure of the pelvic arch and other arches that project below the

vertebrae. It seems that Dr. T. E. White, who studied many specimens of Seymouria, found that individuals fall into two classes, one of which seems to represent the male and the other the female. In the latter group, the first of the arches (specifically the chevron bone) that projects below the tail vertebrae is situated farther back from the pelvic arch, an arrangement that would permit the exiting of a large amniotic egg. Colbert adds: "Of course, this is only speculation, but in conjunction with various reptilian characters of the postcranial skeleton, it is perhaps significant; it places Seymouria on the side of the reptiles" (1980, p. 95).[3]

I consider this a good example of the way in which science attempts to open doors that might seem forever closed. Persistence and ingenuity applied to the evidence presently on hand occasionally produces a new and interesting discovery bearing on a difficult question. Creation science is so fully committed to stasis of doctrine and negativism that it is incapable of making similar discoveries and inferences.

According to Colbert, the oldest known fossil reptile egg is from the lower Permian of Texas (1980, p. 110). This is the same approximate age and geographical region of occurrence as fossil remains of Seymouria. Could this oldest known egg be a seymourian egg? If the amniote egg developed as early as Carboniferous time, as paleontologists suspect it did, fossil reptilian eggs may yet turn up in Carboniferous strata.

Reptiles to Mammals

As we noted in Chapter 32, the earliest reptiles were the cotylosaurs, or stem reptiles, that appeared in Carboniferous time. Very shortly thereafter--well before the five major groups of Mesozoic reptiles appeared--the cotylosaurs gave rise to the mammal-like reptiles, a group dominant in Permian and Triassic times (see Figure 43.15). The mammal-like reptiles gave rise in Triassic time to the first true mammals, but this last group "sat on the sidelines," so to speak, while the reptiles had their reign of land supremacy through the Mesozoic Era. Insignificant creatures, some as small as a mouse and a few as large as a cat, these primitive Mesozoic mammals kept well out of sight (and out from under foot) of the ruling reptiles of Jurassic and Cretaceous time.

It is precisely this early origin and long-delayed rise to prominence of the mammals that attracts the attention of the creationists. Gish calls this Mesozoic phase a "great hiatus in mammalian evolution" (1985, p. 94). He describes it as follows:

> Since evolution is supposed to have involved natural selection, in which the more highly adapted creatures reproduce in larger numbers and thus gradually replace the less fit, we would now expect

the mammals, triumphant at last, to flourish in vast numbers and to dominate the world. A very strange thing happened, however. For all practical purposes, the mammals disappeared from the scene for the next 120 million years! During this supposed vast stretch of time, the "reptile-like" reptiles, including dinosaurs and many other land-dwelling creatures, the marine reptiles, and the flying reptiles, swarmed over the earth. As far as the mammals were concerned, however, the "fittest" that replaced the mammal-like reptiles, they were almost nowhere to be found. (P. 94)

Gish goes on to say that nearly all of the collected fossils of Jurassic and Cretaceous mammals are teeth, and all could be held in two cupped hands. From this, he supposes that there were very few individuals of the mammmal species living at any one time. Because they survived in very small numbers, Gish says, it was a case of "survival of the unfit" (p. 95). This last statement can be disregarded as sheer nonsense; these mammals must have had fitness for the niches they occupied or they would not have persisted through the Mesozoic. Moreover, their fossil remains are not so scarce as Gish implies. Referring to the upper Triassic mammals, Colbert states:

Originally, these forms were known from scattered teeth and a few jaw fragments but, within recent years, fossil bones in large numbers have been recovered from Triassic fissure fillings within Carboniferous limestones in South Wales. Skulls with associated skeletons are known from South Africa. In addition, some closely related materials have been found in southwestern China. (1980, p. 250)[3]

Professor Colbert has related to me that fossil remains of Mesozoic mammals "are constantly being discovered and studied on the various continents. There is indeed a flood of Mesozoic mammals coming to light these days" (personal communication, 1986).

Gish's statements are seriously flawed in terms of presenting an accurate picture of evolution as held by mainstream science. Unstated but implicit in his paragraph is the supposition that reptiles and mammals were competing for the same environmental niches. That supposition has no evidence to back it up. The very finding of mammal fossils throughout the Age of Reptiles is ample de facto evidence that they possessed fitness for survival. About one of the orders of Mesozoic mammals, the multituberculates, Colbert states that adaptations of their skull and dentition "were broadly similar to those seen in the later rodents, and it is reasonable to think that these early mammals lived a type of life that was imitated many millions of years later by rodents" (1980, pp. 252-53).[3] Colbert adds: "Indeed, since the multituberculates persisted into early Cenozoic time, as we shall see, it is quite possible that their extinction was brought about by competition from the early rodents" (p. 254).[3] That statement makes good evolutionary sense, namely, that the rodents entered into competition over the same environmental niche as the multituberculates, which lost out. Rodents today are preyed upon by reptiles (snakes) and birds, as well as by other mammals, yet they manage very well.

Gish persists in his theme of a "great hiatus in mammalian evolution":

Evolutionists would have us believe that mammalian evolution stood still for about 120 million years. For 120 million years, according to evolutionary theory, mammals, apparently existing for that vast, vast stretch of time in extremely few numbers, remained evolutionarily dormant as rather small, generalized forms. (1985, p. 95)[3]

Nothing could be more grossly in error than the above statement. Two phases of adaptive radiation of the primitive mammals in Mesozoic time are documented in the fossil record. Colbert describes these as follows:

The first embraced the late Triassic and Jurassic periods, during which five mammalian orders appeared and developed; these orders are the Docodonta, Triconodonta, Symmetrodonta, Eupantotheria, and Multituberculata.

The second phase of mammalian radiation occurred in the Cretaceous period, during which all of the Mesozoic orders except the docodonts continued their evolutionary development, but during which the triconodonts, symmetrodonts, and eupantotheres became extinct--their extinction taking place at the close of early Cretaceous history. (1980, pp. 256-57)[3]

Colbert adds that the eupantotheres gave rise in Cretaceous time to the marsupial and placental mammals (p. 257). That's a lot of evolutionary change over the 140 million years of Jurassic and Cretaceous time! The record shows Gish's claims of evolution standing still during that time period to be completely wrong; moreover his claim that it was the evolutionists who made that statement is completely false. We have already wasted too much time correcting an egregious and inexcuable error, so let us get on with the fossil record of transitions.

Jawbones and Ear Bones

Gish discusses the transition from reptiles to mammals, correctly pointing out that many diagnostic features of mammals are not seen in the skeletal parts we have as fossils, but rather are present in the soft tissues and physiology that escape fossilization (1978a, p. 79). The latter features include mode of reproduction, warm-bloodedness, mode of breathing, suckling of the young, and possession of hair. So Gish turns to differences in the jaw and ear bones of reptiles and mammals. (Later in this chapter we list and compare the major skeletal differences between reptiles and mammals.) Gish writes:

The two most distinguishable osteological differences between reptiles and mammals, however, have never been bridged by a transitional series. All mammals, living or fossil, have a single bone, the dentary, on each side of the lower jaw, and all mammals, living or fossil, have three auditory ossicles or ear bones, the malleus, incus, and stapes. In some fossil reptiles the number and size of the bones of the lower jaw are reduced compared to living reptiles. Every reptile, living or fossil, however, has at least four bones in the lower jaw and only one auditory ossicle, the stapes.

There are no transitional forms showing, for instance, three or two jaw bones, or two ear bones. No one has explained yet, for that matter, how the transitional form would have managed to chew while his jaw was being unhinged and rearticulated, or how he would hear while dragging two of his jaw bones up into his ear. (1978a, p. 80)

Gish repeats these charges in essentially the same words in 1981 (p. iv) and 1985 (pp. 100-101). I suspect that the average reader of Gish's book, whether creationist or evolutionist, would like a description of the tranformation in jaw and ear bones, particularly because everyone must feel sympathy for Gish's poor deaf half-caste--shunned by both the reptile and mammal congregations--and forced to subsist on a liquid diet. Explanations of the evolution of the lower jaw and the

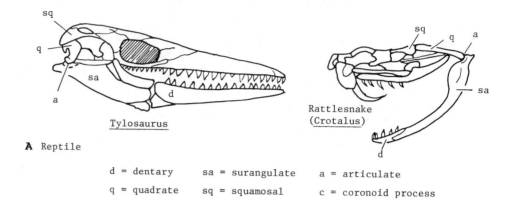

A Reptile

d = dentary   sa = surangulate   a = articulate

q = quadrate   sq = squamosal   c = coronoid process

Figure 43.7 Comparison of the jaw bones of reptiles and mammals. (A. From Edwin H. Colbert, 1980, Evolution of the Vertebrates, 3d ed., John Wiley & Sons, New York, p.241, Figure 90. Copyright © 1955, 1969, 1980 by John Wiley & Sons, Inc. Reprinted by permission. B. From Vertebrate Paleontology, 3d ed., by Alfred S. Romer, University of Chicago Press, p. 190, Figure 295. Copyright © 1933, 1945, and 1966 by The University of Chicago. Reprinted by permission.)

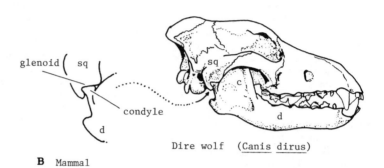

B Mammal

ear bones are available in standard works, accessible to creation scientists and anyone else who cares to use them. I recommend A. S. Romer (1966, pp. 85, 191), E. H. Colbert (1955, pp. 228-29; 1980, p. 247), A. W. Crompton and P. Parker (1978), T. S. Kemp (1982, pp. 208-15, 233-37), G. L. Stebbins (1982, p. 289-92), and C. McGowan (1984, pp. 128-39). The first two were available to Duane Gish before 1978 and contain most of the needed information. Gish recommends "a careful reading of Romer's book, Vertebrate Paleontology" (pp. 66-67). Heeding this admonition, I quickly found Romer's account of the evolution of the jaw and ear bones.

Consider first the bones and hinge structure of the reptilian lower jaw, illustrated by the order Squamata, which consists of the lizards and snakes. The example shown in Figure 43.7A is Tylosaurus, a mosasaur or giant marine lizard with a skull about one meter long. Notice first that the jaw consists of several bones. The dentary (d) bears the teeth, which are sharply pointed and undifferentiated (uniform) throughout. About midway on the jaw is a well-defined joint, not found in modern lizards and not typically present in reptiles either living or fossil, but nevertheless interesting and thrown in here for your amazement and amusement (like the rear-wheel steering system of a ladder fire truck, very handy in certain difficult situations). The principal jaw hinge at the rear consists of the quadrate bone of the skull (q) resting in a shallow socket in the articular bone (a) of the jaw. This is a diagnostic feature of the reptiles, fossil and living. In the modern rattlesnake, also one of the Squamata, the quadrate has an articulation at its upper end where it contacts the squamosal bone (sq), as shown in Figure 43.7A (right); this double-jointed feature allows the mouth to gape open and present its fangs for striking.

The mammalian jaw is illustrated in Figure 43.7B by the skull of the Pleistocene dire wolf. The jaw consists of a single bone, the dentary (d), which has an upward flangelike extension at the rear, the coronoid process (c), to which the muscles are anchored. The jaw joint consists of a projecting rounded process (the condyle) at the rear of the dentary, fitted into a groove (the glenoid) in the squamosal. In comparison with the reptile joint, it would

seem as if the point of articulation has moved from the lower end of the quadrate to the upper end of the quadrate, the quadrate having fused with the dentary.

I think, now, that you can figure out how the transition was made from reptile to mammal with no interruption in the ability to open and close the mouth. A transitional form must have had two joints in operation simultaneously (as in the modern rattlesnake), and this phase was followed by fusion of the lower joint. Do you suppose there ever was such an intermediate? McGowan has one for us: the Argentinian mammal-like reptile Probainognathus. Says McGowan: "if we examined its jaw joint we would see that there is a contact between the articular and the quadrate bone (reptilian condition) and between the dentary and the squamosal bone (mammalian condition)" (1984, p. 137).[4]

T. S. Kemp, a British vertebrate paleontologist in Oxford University, gives this description of the same features:

> The single most dramatic feature of Probainognathus is that the dentary may have become involved in the jaw articulation, for the first time amongst cynodonts (Romer, 1970). The facet of the squamosal, which in other advanced cynodonts receives the surangular, is further forwards in Probainognathus, and may be in contact with the posteriormost tip of the articular process of the dentary, as well as the surangular (Crompton, 1972b; Crompton and Jenkins, 1979). (Kemp, 1982, p. 212)

(See Kemp for references cited.)

Figure 43.8 shows details of the jaw joint of Probainognathus, viewing the joint sidewise from both outside and inside. The two kinds of joint are very close together, and it is not difficult to see how they could operate simultaneously.

From Romer comes the clincher, found thanks to Dr. Gish's advice to read Romer carefully. The animal is Diarthrognathus (translated freely as "old double-joints"),

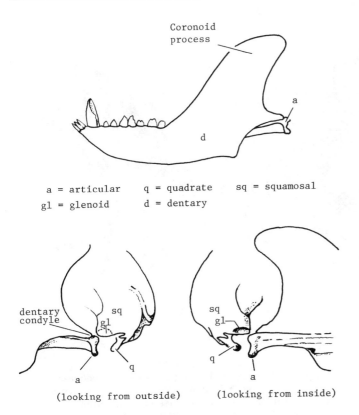

a = articular   q = quadrate   sq = squamosal

gl = glenoid   d = dentary

Figure 43.8 Jaw of <u>Probainognathus</u> with details of the joint. (From A. S. Romer, 1970, <u>Brevoria</u>, no. 344, p. 6, Figure 4. Museum of Comparative Zoology of Harvard University. Used by permission.)

a mammal-like reptile from the upper Triassic of South Africa. Romer states:

> In one regard it is a step more advanced than the tritylodonts (the last survivors of the therapsids). As described, the old quadrate-articular joint is still present; but close to and forming a minor part of the same joint is a squamosal-dentary contact. Thus <u>Diarthrognathus</u>. . . . appears to be almost exactly on the boundary between reptiles and mammals. (1966, p. 186)[2]

Colbert covers the same topic. Describing the tritylodonts, theriodont mammal-like reptiles of the upper Triassic, he notes that the skull was very mammalian in its features and there were other very advanced features (1980, pp. 135-36). He goes on:

> Yet, in spite of these advances, the tritylodonts still retained the reptilian joint between the quadrate bone of the skull and the articular bone of the lower jaw. It is true that these bones were very much reduced, so that the squamosal bone of the skull and the dentary bone of the lower jaw (the two bones involved in the mammalian jaw articulation) were on the point of touching each other. Nevertheless, the old reptilian bones were still participants in the articulation and, therefore, the tritylodonts technically may be regarded as reptiles. (Pp. 135-36)[3]

Colbert finds the double jaw articulation of <u>Diarthrognathus</u> fascinating:

> In this animal not only was the ancient reptilian joint between a reduced quadrate and articular still present but also the new mammalian joint between

the squamosal and dentary bones had come into functional being. Thus, <u>Diarthrognathus</u> was truly at the dividing line between reptile and mammal in so far as this diagnostic feature is concerned. (1969, pp. 143-44; 1980, p. 136)[3]

For a more recent discussion of the double jaw articulation of <u>Diarthrognathus</u>, see Kemp, 1982, p.213.

The evolution of the ear bones, or ossicles, is fully discussed by Romer (1966, p. 191). Romer's exposition was available to Gish a full decade before <u>NO!</u> was published, so it is hard to understand how Gish could claim that no one has explained the transition from one to three ossicles "or how he would hear while dragging two of his jaw bones up into his ear" (1987a, p. 80).

Romer's explanation makes use of a series of diagrams showing transverse sections (cross-cuts) through the skull to show the inner ear and the ossicles (Figure 43.9). Diagram C shows the arrangement for the mammal-like reptiles. Three ossicles are already in place, serving to transmit sound to the inner ear. The long slender stapes bone (s) receives auditory impulses from both the incus and the malleus, which are also serving as bones of the outer skull; they are the same quadrate and articular bones inherited from the amphibians, shown in Diagram B. We see that in the amphibian skull the stapes alone served to conduct sound from the tympanic membrane (tm) lying exposed on the outer surface of the skull. Gradually the stapes moved down and, quite by chance, came into contact with the quadrate and articular bones, which began to transmit sound to the stapes. The contact between quadrate and articular had been serving as the jaw joint of the amphibian, but this function was in the process of being abandoned (or had already been abandoned) for a new jaw joint between dentary and squamosal in the reptile. Romer summarizes this development as follows:

> The malleus is the old articular; the incus, the reptilian quadrate. Once important bones, they had become useless for their original purpose and, lying close to the auditory region, were salvaged and put to a new use. This change was presumably facilitated by the position of the eardrum, close to the jaw region, in mammal-like reptiles. Very likely quadrate and articular had begun to function in sound transmission before they had lost their function as jaw elements. (1966, p. 191)[2]

Going back one more step to the fishes (Diagram A), we find that the stapes was originally the hyomandibular bone, which served as a sort of strut within the skull. When no longer needed as a structural element, it rotated until it came to occupy the first gill slit, or spiracle (sp), which also came to form the Eustachian tube (eu) in the amphibian. Bone, wherever it is located in the vertebrate body, has the physical property to transmit sound waves efficiently, as does any dense, rigid solid. If you want to hear better what is going on in the adjacent room, you press your ear to the wall. But you could also hold the legbone of a cow or horse between your ear and the wall to enhance the sound. The point is that bone did not need to acquire a new physical property to take on a new function. All that was necessary was that a particular bone move into a favorable position, quite by accident, and become useful thereafter as a new part of a previously existing organ.

You may be skeptical of Romer's explanation, even if it sounds reasonable. If so, we can offer corroboration from the field of embryology. Romer mentions this topic, but I quote from McGowan for a more complete statement:

> By studying the embryonic development of mammals it was found that the outer bone, the malleus, actually started off as part of the lower jaw, and

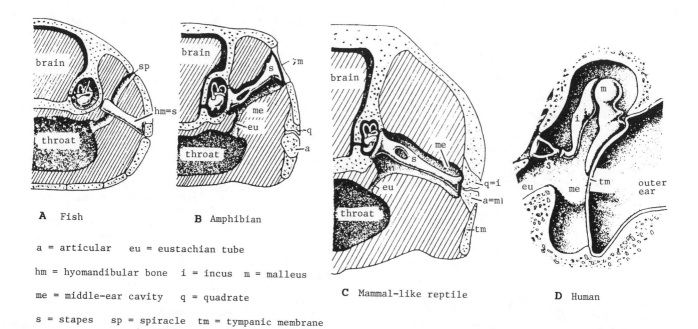

A Fish   **B** Amphibian

a = articular   eu = eustachian tube

hm = hyomandibular bone   i = incus   m = malleus

me = middle-ear cavity   q = quadrate

s = stapes   sp = spiracle   tm = tympanic membrane

C Mammal-like reptile   **D** Human

Figure 43.9 Stages in the evolution of the ear apparatus. (From Vertebrate Paleontology, 3d ed., by Alfred S. Romer, University of Chicago Press, p. 85, Figure 116. Copyright © 1933, 1945, and 1966 by The University of Chicago. Reprinted by permission.)

that this articulated with the incus, which is part of the skull. The embryonic jaw joint is therefore formed between the malleus and the incus, and this led to the conclusion that they represented the articular and quadrate bones of the reptilian ancestor. (1984, p. 139)[4]

McGowan returns to examination of the mammal-like reptile Probainognathus, referred to in an earlier paragraph as having both kinds of jaw joints:

If we examined Probainognathus, we would see that the quadrate is already partially free from the skull. Furthermore, the stapes, which has been found in an incomplete condition in some of the specimens, lies in line with the quadrate, and it is believed that it actually articulated with it, as it is known to do in other cynodonts. In Probainognathus, then, we see an early stage in the loss of the reptilian articular-quadrate jaw joint. It is not difficult to visualize the next step, when the articular and quadrate bones are freed to link up with the stapes to form the mammalian ear ossicles. Dr. Gish's contention that nobody has been able to explain how the transitional fossil could have chewed its food while its jaw was being unhinged and reconnected is without foundation. (1984, p. 139)[4]

For additional information on the evolution of the ear bones, see Crompton and Parker (1978) and Kemp (1982, pp. 233-37).

This has been a rather lengthy excursion into technical details of the evolution of the jaw and ear bones, but I think it has been worthwhile. Dr. Gish, in only a terse sentence or two, threw down a challenge to the evolutionists, at the same time failing to mention an answer long on the books, let alone commenting critically on that detailed information. Setting the matter straight takes hard work and concentration on details, but the process benefits teachers and students of evolutionary biology and geology. Challenges to established explanations are seldom raised in our schools, and even graduate students are rarely asked to question what they are given. Thank you, Dr. Gish, for prodding us into a most rewarding learning experience!

Mammal-like Reptiles to Mammals

We continue with the subject of an intermediate between reptiles and mammals, and more specifically between the mammal-like reptiles and the mammals. Of the differences between reptiles and mammals we have now covered the bones of the lower jaw, the jaw joint, and the ear bones. A number of other important differences in skeletal structure remain to be noted before we can examine a candidate for the intermediate position. An excellent review is given by McGowan (1984, pp. 128-33).

The teeth of modern reptiles, as in the case of the fossil mosasaur, Tylosaurus, whose skull is shown in Figure 43.7A, are shaped generally like simple conical pegs with sharp points. They are numerous and may vary in size, depending on the particular group, but are uniform in shape and construction, i.e., they are usually undifferentiated. Mammal teeth are specialized for different functions and follow a generally typical pattern. There are cheek teeth--molars, that is--possessing multiple cusps and used for breaking up or grinding food. At the front is a row of chisel-shaped incisors for nipping or tearing. On either side of the incisors is a long conical tooth, the canine, used by carnivores for stabbing prey. The dire wolf skull shown in Figure 43.7B illustrates the three kinds of teeth. Whereas reptiles continually replace their teeth as they drop out, mammals keep one set for life, once the milk teeth (deciduous teeth) have been replaced--as every human knows.

Turning next to the cranium, or "brain box," it is a very small, rather insignificant feature of the reptilian skull, located far to the rear and close to the neck. In contrast, the mammalian cranium is large and occupies a substantial part of the skull. The reptilian skull is attached to the first neck vertebra--the atlas--by a ball-and-socket arrangement in which a single ball, the condyle, fits into a socket in the atlas (Figure 43.10). In the true mammals there is a pair of condyles. In the mammals the second vertebra, called the axis, is jointed to the atlas by a specialized structure, described by McGowan: "The joint between these two vertebrae is a

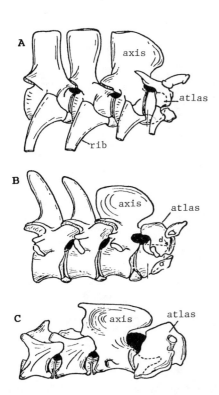

Figure 43.10 Neck vertebrae of (A) <u>Dimetrodon</u>, an early mammal-like reptile, (B) <u>Galesaurus</u>, a cynodont (later mammal-like reptile), and (C) a living mammal (cat). (From A. W. Crompton and F. A. Jenkins, Jr. Reproduced, with permission, from the <u>Annual Review of Earth and Planetary Sciences</u>, vol. 1, p. 146, Figure 6. Copyright © 1973 by Annual Reviews, Inc.)

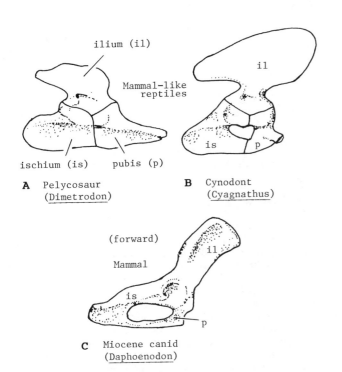

Figure 43.11 Comparison of pelvis of the mammal-like reptiles (A, B) with that of the mammals (C). (From <u>Vertebrate Paleontology</u>, 3d ed., by Alfred S. Romer, University of Chicago Press, p. 177, Figure 270. Copyright © 1933, 1945, and 1966 by The University of Chicago. Reprinted by permission.)

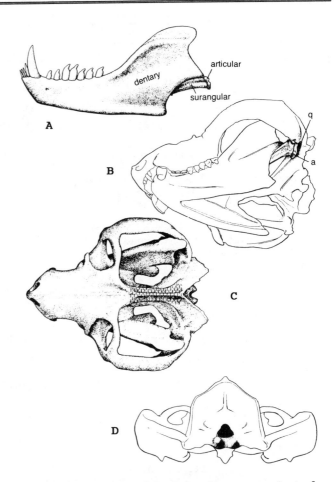

Figure 43.12 Features of <u>Probelesodon</u>, a cynodont of middle Triassic age. (A) Lower jaw. (B) Oblique underside view, showing the quadrate (q) and articular (a). (C) Underside of skull, looking up. (D) Rear view, looking forward, showing the double condyles. (Drawn by Anker Odum. From Chris McGowan, <u>In The Beginning</u>, Prometheus Books, Buffalo, N.Y., p. 136. Copyright © 1984 by Chris McGowan. Used by permission of Anker Odum.)

peg-and-socket arrangement, the atlas essentially hanging on the peg (called the odontoid process) of the axis like a hat on a coathook" (1984, p. 132).[4] Reptiles lack the odontoid process. Other differences exist in the vertebral columns. In reptiles ribs attach to the neck vertebrae, but in the mammals these ribs are very small and have become fused to the neck vertebrae. In reptiles the ribs continue along the vertebrae to the pelvis, but in mammals the ribs are limited to an upper section and missing in a lower, or lumbar, section.

In reptiles the legs are typically splayed outward from the body, inducing a sort of body-waddling motion to their walk, but in mammals the legs project directly downward beneath the body, making possible a smooth walking or running motion. Another difference in the limbs is in the number of finger and toe bones (phalanges); reptiles have more of them than do mammals. Finally, the hip or pelvis differs significantly. The upper pelvic bone, or ilium, of the reptile is generally small and is directed upward and sometimes also backwards, whereas in mammals the ilium projects forward on an upward slant (Figure 43.11).

With this list of differences in mind, we can turn to examine an intermediate example from among the mammal-like reptiles. McGowan has done an excellent job of singling out and comparing an intermediate from among the cynodonts, which are thought be the group of mammal-like reptiles that gave rise to the mammals (1984,

Table 43.2 Reptilian and Mammalian Features of Cynodonts

Reptilian condition	Intermediate condition (as in Probelesodon)	Mammalian condition
		1. Teeth specialized for different functions.
	2. Cheek teeth with cusps but these are not complex.	
3. Lower jaw comprises several bones (note, however, that the dentary is the largest, the other bones being small).		
4. Jaw joint formed between articular bone and quadrate bone (the quadrate bone, however, is very small).		
	5. Jaw joint formed between a hollow in the lower jaw and a flat surface in the skull.	
		6. Lower jaw with prominent coronoid process.
7. Small cranium.		
		8. Double condyle at back of skull for neck articulation.
		9. Axis with odontoid process.
10. Ribs in neck region.		
	11. Prominent ribs confined to chest region, but there are short ribs in front of the pelvis.	
	12. Legs not splayed, but not placed vertically beneath the body either.	
13. Number of bones in fingers and toes exceeds 2, 3, 3, 3.		
		14. Ilium slopes forward.

From IN THE BEGINNING © 1983 by Chris McGowan. Reprinted by permission of Macmillan of Canada, A Division of Canada Publishing Corporation.

pp. 135-238; see also Kemp, 1982, pp. 269-70). McGowan bases his comparison on two genera of cynodonts, both from the middle Triassic; one is Probelesodon, from Argentina, the other is Massetognathus. (Figure 43.12 shows some of the significant features of the skull of Probelesodon; compare them with the reptile and mammal skulls shown in Figure 43.7.) Taking the checklist of differences in skeletal features between reptiles and mammals, McGowan has set up a table in which each feature is examined in the cynodonts and scored as reptilian, mammalian, or intermediate (1984, p. 138). The checklist is shown here as Table 43.2. Totaling up the scores, we find: reptilian condition, 5; mammalian condition, 5; intermediate condition, 4. The scores lead McGowan to observe:

> Little wonder that we sometimes have difficulty in deciding whether these fossils are reptiles or mammals, and there is at least one case where a paleontologist first identified a particular species as being a reptile, only to change his mind a few years later and classify it as a mammal. (1984, p. 137)[4]

This looks like another excellent example of mosaic

evolution, a phenomenon the creation scientists seem not to grasp, or if they do, they simply ignore its significance.

A Second Look at the Jaw Joints

This would seem like a good place to close the subject of intermediates between reptiles and mammals, but Duane Gish has given us an addendum to his earlier coverage of the topic and in so doing has thrown further confusion into the question of intermediates (1981, pp. iv-vi; 1985, pp. 96-100). This material is technical but needs examination and clarification because it appears in the widely distributed ICR Impact Series that reaches many creationist teachers and students. If you have already had enough, please skip the next few paragraphs.

Gish takes a second look at the significance of jaw joints in distinguishing between mammal-like reptiles and true primitive mammals (1981, p. iv; 1985, p. 96). This time, he picks up references to two genera that are currently classified by most paleontologists as primitive mammals. These genera are Morganucodon and Kuehneotherium, of Upper Triassic age. Kuehneotherium is represented only by its teeth and lower jaw, and for that reason, as well as to keep things as simple as possible, we will refer only to Morganucodon, of which a very large number of fossil parts exist, including a complete skull from China, and parts with which to reconstruct the whole skeleton (Kemp, 1982, pp. 253-54). (Together with certain other closely related genera, these animals can be collectively called morganucodonts.) As we proceed, keep in mind that it is the creation scientists' strategy to argue that the morganucodonts are not mammals, but that they are really reptiles, albeit mammal-like reptiles with certain mammalian features. The strategy behind this claim rests in the problem of defining "kinds," something we will comment on in a later paragraph.

Gish's opening assertion is that "Morganucodon and Kuehneotherium each possessed a _full_ _complement_ _of_ _the_ _reptilian_ _bones_ _in_ _its_ _lower_ _jaw_" (1981, p. iv). Gish cites as a reference for this statement papers by vertebrate paleontologists who have described the morganucodont fossils, particularly the lower jaw (Kermack et al., 1973). Actually, I find no such statement about a full complement of reptilian bones in the Kermack paper cited, which deals exclusively with the lower jaw of Morganucodon; nor can I find any similar statement in any of the mainstream references cited in earlier paragraphs. My list of lower-jaw bones in amphibians and reptiles includes the following: dentary, predentary, angular, splenial, postsplenial, surangular, articular, prearticular, coronoid. Of these, I find that among the cynodonts (Cynognathus, for example) two are missing: predentary and postsplenial; presumably they became fused with the dentary and splenial bones, respectively. Are the remaining seven bones present and accounted for in Morganucodon? The Kermack paper states that the splenial has completely vanished (1973, p. 156). Thus Gish's assertion is refuted. Actually, the missing splenial is not involved in the distinction between reptiles and mammals. Gish's opening statement with its emphasized words is simply a device intended to suggest that Morganucodon can't possibly be a mammal, but alternatively it can also be taken to mean that there is really no difference between mammal-like reptiles and primitive mammals. From there, we can go on to conclude that reptiles and mammals belong to the same created kind (which could hardly have been Gish's intent).

Gish goes on to discuss the double system of jaw joints in Morganucodon, which has both the quadrate-articular joint of the reptiles and the squamosal-dentary joint of the mammals. The idea that Gish picks up from the 1973 Kermack paper is that Morganucodon had a very strong quadrate-articular joint. Gish states:

> The most striking characteristic of the accessory jaw bones of Moranucodon is their cynodont character. Compared with such a typical advanced cynodont as Cynognathus, the accessory bones present show no reduction either in size or complexity of structure. In particular, the actual reptilian jaw-joint itself, was relatively powerful in the mammal, Morganucodon, as it was in the reptile, Cynognathus. This is quite unexpected. (1981, p. v)

Gish continues to develop the jaw-joint theme by questioning whether the morganucodonts actually had a functioning dentary-squamosal joint (p. v). Were these bones actually in contact, as they should be in a working mammalian jaw joint? Gish says this contact is believed by Kermack and associates to have existed, but no fossils are available showing the dentary in actual contact with the squamosal of the skull. Paleontologists, says Gish, only infer the contact from the presence of a condyle on the dentary. Gish asks: "How then could a powerful, fully-functional reptilian jaw-joint be accommodated along with a mammalian jaw-joint" (p. vi). In Kemp's paper, I found a completely different version of the situation Gish

describes. Kemp describes the jaw joints of the morganucodonts as follows:

> The axes of the two jaw hinges, dentary-squamosal and articular-quadrate, coincide along a lateral-medial line, and therefore the double jaw articulation of the most advanced cynodonts is still present. . . . The secondary, dentary-squamosal jaw hinge had enlarged (in the morganucodonts) and took a greater proportion if not all of the stresses at the jaw articulation. The articular-quadrate hinge was free to function solely in sound conduction. (1982, p. 256)

So there you have a quite different picture, based on close scrutiny of the morganucodont skull. It's up to you to decide: Were the morganucodontids mammals or were they reptiles? In nearly all of the standard works, they are classed as mammals, and that decision is based on dentition, not on jaw joints.

Intermediates Lurking in Hierarchical Trees

Our study of intermediates has now bridged the so-called "gaps" between fishes, amphibians, reptiles, and mammals. Figure 43.13 shows seven skulls in which the progression in jaws and teeth from stem reptiles to mammals is particularly striking. Well-documented intermediates are far more than adequate to illustrate the descent by mosaic evolution. There is, however, one more topic raised by the creationists that invites discussion because it illustrates an important principle in setting up a phylogenetic tree.

Creationists seem to view evolutionary descent as linear, a single chain, which I would liken to the growth pattern of a stalk of bamboo. The unbranched bamboo stalk is segmented with solid joints that separate the hollow growth sections. Under this plan, every fossil of a transitional sequence must lie on an unbranched chain, and thus a transitional fossil must always be intermediate in geologic age between the older fossil from which it evolved and the younger fossil into which it evolved. Creationists cannot understand that intermediates may occupy positions at the ends of branches and can thus be found in strata of the same age, or younger age, than the descendant form. The descendants of an evolutionary branch can carry on the intermediate features through several geologic periods. We must substitute for the bamboo-stalk model of evolution a shrub-form model of numerous branches, their forks occurring at various levels above the base.

Laurie R. Godfrey, an anthropologist on the faculty of the University of Massachusetts at Amherst, has commented on the creationists' debating technique in which a laugh can be drawn from the audience by a slide showing a modern cow evolving into a whale (see Figure 45.1). Godfrey writes:

> Evolution does not demand that the "series" connecting any two extant forms will follow the shortest morphological line between them; it demands that any two extant organisms share a common ancestor--an organism that may be very far indeed from the morphological "mean" of the extant pair. In the case of the shark and whale, that common ancestor was a primitive fish! Given numerous speciation events subsequently separating the pair, most members of intermediate taxa will not be in the direct line of ancestry of _either_ modern form. They will be "cousins." (1983, p. 204)

In Chapter 33, on phylogeny (evolution and comparative anatomy), we showed how hierarchical trees are constructed from the data of comparative anatomy. Using a set of anatomical features, or characters, a

A

Stem reptile

Captorhinus,
a cotylosaur.
Lower Permian

Length 7 cm

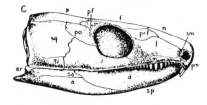

B

Mammal-like reptile

Ophiacodon,
a pelycosaur.
Permian

Length 38 cm

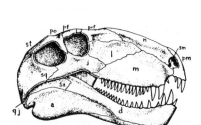

C

Mammal-like reptile

Dimetrodon,
a pelycosaur.
Lower Permian

D

Mammal-like reptile

Cynognathus,
an advanced cynodont.
Lower Triassic

Length 46 cm

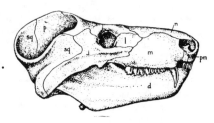

E

Mammal-like reptile

Diarthrognathus, an
advanced therapsid.
Middle-Upper Triassic

Length 4 cm

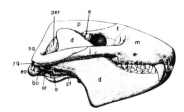

F

Primitive mammal

Sinocodon,
a triconodont.
Late Triassic

Length 4 cm

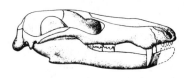

G

Mammal

Sinopa, a
a hyaenodont.
Middle-Late Eocene

Length 15 cm

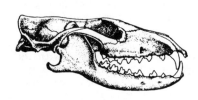

Figure 43.13 Reconstructions of skulls to show the progression from stem reptiles, through mammal-like reptiles and primitive mammals, to mammals. (From Vertebrate Paleontology, 3d-ed., by Alfred S. Romer, University of Chicago Press, pp. 105, 175, 176, 179, 186, 198, 231. Copyright © 1933, 1945, and 1966 by The University of Chicago. Reprinted by permission.)

cladogram was constructed step by step, based on shared characters (see Figures 33.4, 33.5, and 33.6). That construction was based on anatomical features present in living forms, but, of course, the same procedure is also carried out using the anatomical features of fossil forms, which is just what we have been doing in examining the transitional forms between reptiles and mammals.

Joel Cracraft has discussed the cladistic method as it applies to the question of transitional or intermediate forms in the fossil record (1984, pp. 202-3). Being based strictly on the observed morphological characters of the fossil forms, the method is (in concept, at least) free of the stratigraphic positions of the individuals studied. Cracraft explains:

At the outset, the question of transitional forms is primarily an issue of systematic analysis. Most paleontologists have recognized this to one degree or another, whereas creationists have been less interested in the scientific investigation of the problem than they have in manipulating the data to support their theological worldview. The identification of transitional forms is a systematic issue because it entails the comparative analysis of morphological data, the interpretation of genealogical relationships among all the taxa involved, and a comparison of these results with detailed stratigraphic investigation. One of the factors contributing to the controversy within contemporary paleontology over the identification of transitional forms has been the historical predisposition of

paleontologists to lend more weight to the stratigraphic sequences of their fossils than to their comparative systematics. Thus, some paleontologists have used stratigraphic position as an important criterion for identifying ancestral taxa. (1984, p. 202)[5]

A transitional form, then, is judged to be an "intermediate" when its morphological features, or characters, are a combination of those of two distinct taxa. (Actually, only species evolve, so that the transitions are between species.) We saw this in the case of the checklist of the reptilian and mammalian features of the cynodonts (Table 43.2), but this does not require that all reptiles be of older geologic age than all mammals, and that all intermediate mammal-like reptiles be older than all reptiles but younger than all mammals. Such time-stratigraphic strictures are patently absurd when we consider that representatives of both the reptiles and the mammals have persisted to the present. Mammals share the modern world not only with reptiles, but also with the fishes and the invertebrates.

To clarify this concept of intermediates, we can use a simple, idealized cladogram, shown in Figure 43.14. Each of three species (or genera)--X, Y, and Z--has four characters, which in a primitive state are a, b, c, and d. A new, derived state of each character is indicated by a horizontal arrow. First, a evolves into a', giving rise to X (a'bcd); then b evolves into b', giving rise to a species not as yet represented by a known fossil (a'b'cd). Then c evolves to c', giving rise to Y (a'b'c'd); and d

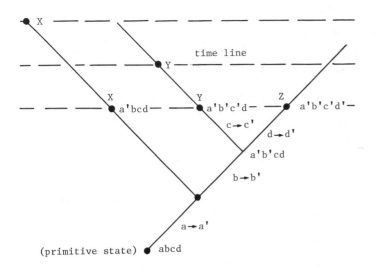

Figure 43.14 An idealized cladogram to illustrate the concept of intermediates. (Based on data of T. S. Kemp, 1982, p. 11, and Joel Cracraft, 1984, p. 302.)

evolves into d', giving rise to Z (a'b'c'd'). Clearly, Y is a true intermediate between X and Z. Both Y and Z share a common ancestor, inferred because they share three of the characters in common, differing only in the fourth. But, as the diagram shows, the fossil representatives of Y may actually have been found in strata of younger geologic age than either X or Z. It is also conceivable that fossils of X are found in strata younger than any in which both Y and Z are found.

That the creationists do not understand the nature of branching hierarchical systems, explained above, is evident in Duane Gish's contention that "in order for the

facts of the fossil record to fit the predictions of the evolution model, a true time-sequence must be established that accords with these predictions. This has not been possible with the amphibian-reptilian-mammalian sequence on the basis of fossil material so far discovered" (1978a, p. 80; 1985, p. 77). Gish continues:

> The known forms of Seymouria and Diadectes, which are said to stand on the dividing line between amphibians and reptiles, are from the early Permian. This is at least 20 million years too late, according to the evolutionary time scale, to be the ancestors of the reptiles. The so-called "stem reptiles," from the order Cotylosauria, are found, not in the Permian or later, but in the preceding period, the Pennsylvanian.
>
> In fact, the "mammal-like" reptiles of the suborder Synapsida, which are supposed to have given rise to mammals, are found in the Pennsylvanian. Thus, Seymouria and Diadectes, the "ancestors" of reptiles, would not only postdate the reptiles by tens of millions of years, but postdate even the "ancestors" of mammals by an equal length of time. (1978a, pp. 80-81; 1985, p. 77)

To try to straighten things out look at Figure 43.15, a rough sketch of the phylogenetic and stratigraphic relationships among the vertebrate groups we have reviewed thus far. Various representative genera have been plotted at the appropriate time positions in which their fossils have been found. Although stem reptiles Seymouria and Diadectes are much younger than the earlier mammal-like reptile Varanosaurus, both of the former are intermediates between the amphibians and reptiles in terms of their morphological characters. Where Gish has erred is in calling the Permian representatives of cotylosaurs "ancestors" of the reptiles. He has substituted the concept of ancestry determined strictly by geologic

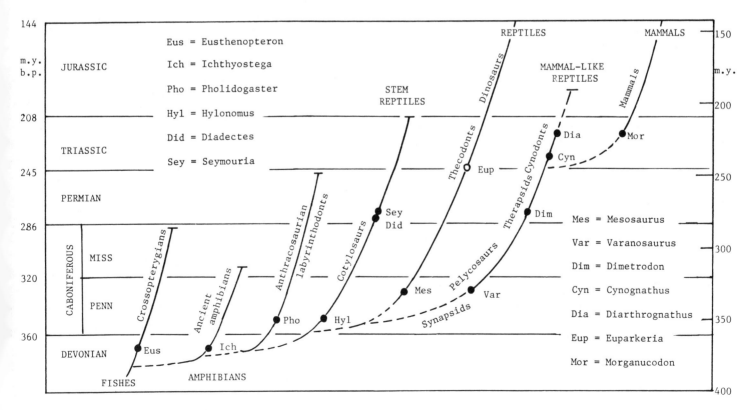

Figure 43.15 Rough sketch of the phylogenetic and stratigraphic relationships of vertebrate groups referred to in Chapter 43. (A. N. Strahler. Data compiled from several sources.)

age relationship for that of common ancestry by morphological relationship.

The error is fully as egregious as if one were to claim that evolutionists consider the living salamanders to be the ancestors of the Permian cotylosaurs. That statement in turn would be fully as outrageous as if I were to claim that I, living today, am the ancestor of a man named Richard Newton, who came From England to Sudbury, Massachusetts, in 1683. (Actually, he is certified as one of my ancestors.) The true ancestor of both the reptiles and the cotylosaurs must (in line with evolutionary theory) have been positioned at the point of branching off of the earliest reptile species from an early stem-reptile species, as indicated in Figure 43.15. To date, that common ancestor has not been found, but it was probably older than the oldest known stem reptile, Hyolonomus, of early Pennsylvanian age (Colbert, 1980, p. 112).

Are you now completely and hopelessly confused? If so, keep pondering this question: Am I the ancestor of my grandfather and grandmother, or are they two of my ancestors? If this is difficult to decide, keep thinking: How could I be the ancestor of my grandparents if they were all born before I was? In a more practical vein it may also help to substitute the words "descendants of a common ancestor" at every point where a creationist uses "ancestor." "Ancestor" and "descendant" are antithetical terms. In a roundabout way the creationists have switched the meanings of these terms in presenting their peculiar version of mainstream evolutionary theory. Again, we see the familiar straw person, fabricated only to be flayed.

Credits

1. From Darwin to DNA, Molecules to Humanity, by G. Ledyard Stebbins, W. H. Freeman and Company, New York. Copyright © 1982 by W. H. Freeman and Company. Used by permission.

2. From Vertebrate Paleontology, Third edition, by Alfred S. Romer, University of Chicago Press. Copyright © 1933, 1945, and 1966 by The University of Chicago. Used by permission.

3. From Evolution of the Vertebrates, First edition 1955, Second edition 1969, Third edition 1980, by Edwin H. Colbert, John Wiley & Sons, Inc. Copyright © 1955, 1969, 1980 by John Wiley & Sons, Inc. Reprinted by permission of the author and publisher.

4. From IN THE BEGINNING © 1983 by Chris McGowan. Reprinted by permission of Macmillan of Canada, A Division of Canada Publishing Corporation.

5. From Joel Cracraft, The significance of the data of systematics and paleontology for the evolution-creationism controversy, pp. 189-205 in Frank Awbrey and William M. Thwaites, Eds., 1984, Evolutionists Confront Creationists, Proceedings of the 63rd Annual Meeting of the Pacific Division, American Assoc. for the Advancement of Science, vol. 1, part 3, San Francisco, 213 pp. Used by permission of the author and publisher.

Chapter 44

Reptilian Transitions

Reptiles--especially the great dinosaurs and those that glided or flew through the air on batlike wings--fascinate the public to a degree scarcely matched by the appeal of any other animal group. Perhaps this is one reason why the creationists go to great lengths to argue that several of the reptile groups appeared from nowhere in the fossil record, their ancestry obscure and doubtful. This chapter is devoted to quite diverse reptilian types, both in terms of their appearance and size and in the environments they came to occupy and dominate. We start, however, with the question of the origin of birds, which most persons don't associate with reptiles.

Reptiles to Birds

The origin of birds--class Aves--is a major topic with the creationists, both in their writings and in public debates. They have adopted for purposes of argument a longstanding conclusion of the mainstream evolutionists, namely, that the birds arose from the reptiles, making it into a key target. What is this big "flap" all about? Basically, the argument is over evidence, or lack of evidence, of an evolutionary transition between dinosaurs and birds; it is a question of whether intermediates exist, and in particular whether the first birdlike fossil, Archaeopteryx lithographica, is that intermediate (Figure 44.1).

Henry Morris takes the initiative: "Evolutionists universally maintain that reptiles are the evolutionary ancestors of birds. Again, however, there is no fossil evidence of this, despite the famous Archaeopteryx" (1974a, pp. 84-86). Morris then introduces a 1960 statement from biologist W. E. Swinton: "The origin of birds is largely a matter of deduction. There is no fossil evidence of the stages through which the remarkable change from reptile to bird was achieved." That statement seems also to apply in the mid-1980s. The number of specimens of skeletons of Archaeopteryx has grown to six--two more than in 1960--but no intermediate between that genus and dinosaurs has yet been found. Thus, as Morris states, "If there is any transitional form at all, Archaeopteryx is the one" (p. 85). Morris says that paleontologists classify Archeopteryx as a bird--100 percent bird, and not part reptile at all:

The fossilized impressions of the features on the wings of Archaeopteryx have been found and this shows it was warm-blooded, not a reptile with scales and cold blood.

Thus, Archaeopteryx is a bird, not a reptile-bird transition. It is an extinct bird that had teeth. Most birds don't have teeth, but there is no reason why the Creator could not have created some birds with teeth. Not all reptiles have teeth, though some do. The same is true of fishes, amphibians and mammals. Some have teeth and some don't. The same is evidently true of the original birds. For some reason, those that were created with teeth have since become extinct. (1974a, p. 85)

At this point I need to comment on Morris's statement about extinction. Flood geology requires that Archaeopteryx was living on the day before the Flood started. Under creationism, the "extinction" mentioned by Morris was the death by drowning of all living members of that genus. Only toothless birds escaped drowning and fossilization. To continue, Morris states:

At the very least, there must have been a tremendous number of transitional forms between Archaeopteryx and its imaginary reptilian ancestor. Why does no one ever find a fossil animal with half-scales turning into feathers, or half-forelimbs turning into wings? Such animals must have lived in great numbers over long periods of time, but no fossils of them have ever been found. (P. 85)

Duane Gish repeats the same material covered by Morris but broadens the argument to include all the cases in which flight is said to have evolved independently in different animal groups: insects, birds, mammals (bats), and reptiles (pterosaurs) (1978a, pp. 82-89; 1985, p. 103). He emphasizes that not a single case of a transitional form exists for any of these groups. Gish stresses the affinity of Archaeopteryx with the birds, saying that it was not a half-way bird but a true bird.

Gish refers to the presence of clawlike appendages on the leading edges of the wings of Archaeopteryx (p. 84). These might seem to be reptilian features, he says, but a living South American bird, the hoatzin, and an African bird, the touraco, have claws on their wings in the juvenile stage. The ostrich has three claws on its wings. So, Gish reasons, claws don't make Archaeopteryx a reptile or part-reptile. Gish also rejects the presence or absence of teeth as a criterion of class affiliation.

Gish refers to a 1977 note in Science News (vol. 112, p. 198) reporting the announcement by Professor John Ostrom of Yale University (a paleontologist who has made a special study of all available fossil remains of Archaeopteryx) of the finding of an "undoubted true bird" in Jurassic strata, which is of the same vintage as Archaeopteryx. Gish says: "This means . . . that formations, much older than those containing Archaeopteryx (the Jurassic) will have to be searched for the ancestor of the birds. Obviously, Archaeopteryx cannot be the ancestor of birds if true birds existed at the same time" (1978a, p. 87). Now that you are familiar with the definition of an intermediate, as explained in our preceding chapter on transition from reptiles to mammals, you will recognize at once the error in Gish's last statement. In a branching hierarchical system an intermediate and one of its descendants can coexist in the same time plane, along with descendants of the common ancestor of both of them. So we need say no more about the reported discovery of a true bird in Jurassic strata.

Ever since 1861, when a single bird feather was found remarkably well preserved in the Jurassic Solnhofen limestone of southern Germany, vertebrate paleontologists have looked for suitable ancestors for the birds. They have looked back as far as the early ruling reptiles,

1 = furcula

2 = pubis

3 = pubic peduncle

Figure 44.1 Drawing of a remarkably fine fossil specimen of Archaeopteryx, showing the skeleton and feather impressions. The drawing is somewhat less than one-third natural size. (From Edwin H. Colbert, 1980, Evolution of the Vertebrates, 3d ed., John Wiley & Sons, New York, p. 184, Figure 70. Copyright © 1955, 1969, 1980 by John Wiley & Sons, Inc. Reprinted by permission.)

particularly the thecodonts of Triassic time. One of these was a small bipedal Triassic thecodont, Saltoposuchus; lightly built, it possessed long hind legs, but very short front legs (Figure 44.2). Alfred S. Romer writes: "It is obvious that it was forms of this sort from which rose the pterosaurs, birds, and dinosaurs" (1966, p. 140).[1] Romer adds, however, that there are no known thecodonts that can positively be identified as ancestral to the above-named groups.

The dinosaurs consist of two orders: Saurischia and Ornithischia. (At this point you may wish to review the earlier section in Chapter 32 titled "The dinosaurs" and examine the evolutionary diagram of the dinosaurs, Figure 32.9.) Within the Saurischia are two suborders:

Theropoda and Sauropoda. The theropods were mostly carnivores; the sauropods were herbivores. Ancestors of the birds are usually placed among the theropods, which are familiar to everyone through the huge flesh-eating creature, Tyrannosaurus, of Cretaceous time. In Upper Triassic time, however, the theropods included smaller animals, such as Coelophysis, lightly built, with hollow bones, and less than 3 m in length (see Colbert, 1980, pp. 156-59). Its hind legs were strong and birdlike, while the front limbs were short, with hands well adapted for grasping. It seems likely that theropods such as Coelophysis were ancestral to Archaeopteryx.

McGowan compares the skeletal features of the theropod dinosaurs with those of birds, showing many similarities, but also significant differences (1984, pp. 111-15). Both the dinosaur foot and the bird foot are three-toed. Each toe ends in a sharp claw. In the dinosaur foot there is usually a backward projecting big toe, corresponding with the spur of the bird foot. Both have three rodlike metatarsal bones; in the bird these are fused together, but can be seen separately in the immature bird skeleton. At the back of this single fused bone is a protuberance called the hypotarsus; it is missing in the dinosaurs. Other similarities, which we skip over here, are found in the ankle bones, shinbone, and thigh bone. In both dinosaur and bird, the pelvis is generally similar and the legs project directly down.

Differences between theropod dinosaurs (Saurischia) and birds can be seen in the form of the pelvis: the dinosaur pubis projects forward; that of the bird projects to the rear. Moreover, the dinosaur pubis has at its end a curious knob called a peduncle. Dinosaurs had long, heavy tails; birds have almost no skeletal tail, its place taken by a flattened wedge of bone, the pygostyle. There are also important differences in vertebrae of the neck. In birds, the contact surfaces of the vertebrae have a complex double curvature, i.e., a saddleshape. Birds have a breastbone, or sternum; dinosaurs do not. Birds have a wishbone, called the furcula, formed by a fusion of the collarbones; dinosaurs have no furcula. Dinosaurs have abdominal ribs; birds do not. Bones of the hand are closely similar, except that the birds have two instead of three metacarpal bones and these are fused. Finally, birds have feathers; the dinosaurs apparently did not.

McGowan summarizes the differences and prepares a checklist to separate those that are reptilian features from those that are avian features. Abstracting from his tables the features present in one group but not the other, I come up with this list:

A. Present in dinosaurs but not in birds: pubic peduncle, long bony tail, abdominal ribs.

B. Present in birds but not in dinosaurs: pygostyle, bony sternum, furcula (wishbone), hypotarsus, feathers.

If we were to take into consideration the pubic peduncle and long bony tail that Archaeopteryx possessed and the development of the finger, metacarpal and metatarsal bones, and the lack of a bony sternum, the majority of the features of Archaeopteryx would be reptilian, according to McGowan, and this contradicts Gish's contention that Archaeopteryx was in fact a bird. Colbert comments: "If the indications of feathers had not been preserved in association with Archaeopteryx it is likely that these fossils would have been classified among the reptiles, for they show numerous reptilian characters" (1980, p. 183).[2]

---

[1]From Vertebrate Paleontology, Third edition, by Alfred S. Romer, University of Chicago Press. Copyright © 1933, 1945, and 1966 by The University of Chicago. Used by Permission.

[2]From Evolution of the Vertebrates, First edition 1955, Second Edition 1969, Third edition 1980, by Edwin H. Colbert, John Wiley & Sons, Inc., New York. Copyright © 1955, 1969, 1980 by John Wiley & Sons, Inc. Reprinted by permission of the author and publisher.

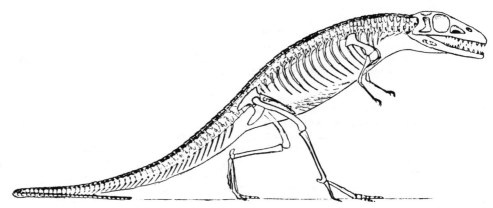

Figure 44.2 Reconstruction of the skeleton of <u>Saltoposochus</u>, a lightly built thecodont, about 1.2 m in length. (From <u>Vertebrate Paleontology</u>, 3d ed., by Alfred S. Romer, University of Chicago Press, p. 140, Figure 214. Copyright © 1933, 1945, and 1966 by The University of Chicago. Reprinted by permission.)

Viewing the evidence as an outsider not versed in the intricacies of vertebrate anatomy, I judge <u>Archaeopteryx</u> to be a genuine intermediate between dinosaur and bird. Compare in Figure 44.3 the skeleton of <u>Archaeopteryx</u> with that of the modern pigeon, as Colbert has done, and come to your own decision (1980, p. 186, Figure 58).

Gish comments that no intermediate forms of feather development have been found (1978a, p. 84). McGowan points out that of the only two specimens of <u>Archae-opteryx</u> on which imprints of feathers can be clearly seen, only the larger feathers are represented (1984, p. 119). Transition from reptilian scales to bird feathers are indeed lacking from the fossil record, but this transition is suggested by a study of the various kinds of feathers present on the wings of the modern penguin. There, we find cover structures that look more like scales than feathers, and we can also find transitional forms grading into the typical bird feather. Moreover, penguins have true scales on their legs. Both scales and feathers are made of a horny protein called keratin. McGowan assures us that he does not regard the penguin as an intermediate between reptiles and birds, but the range of covering forms on the penguin suggests that functional transitional forms could have existed during the transition from dinosaurs to birds.

McGowan also discusses the creationists' claim that claws on the wings of juvenile forms of the hoatzin, a modern bird, proves that the presence of wing claws is a natural feature of birds, not indicating reptilian ancestry (pp. 123-24). To the contrary, McGowan considers these wing claws in juvenile specimens to be a point of positive evidence of their reptilian ancestry.

Gish emphasizes that because <u>Archaeopteryx</u> flew, it was indeed a bird and not a half bird (1978a, p. 84). As McGowan says, we really don't know that this creature actually flew (1984, p. 125). Flight would not be the criterion of an animal being a bird, as witness the various species of true birds living today or in the recent past that are flightless: ostriches, cassowaries, emus, rheas, kiwis, and others.

Among mainstream evolutionists a fascinating subject of scientific debate, for which we have little space here, is the manner in which <u>Archaeopteryx</u> developed its capability for flight (if it had that capability). A traditional view is that of the arborealists, who imagine that the feathered creature developed the habit of climbing into trees and taking off in a glide, much as gliding tree-squirrels do today. This practice led to improved wings capable of flapping vigorously, and eventually, to free sustained flight. More recently, under the leadership of John Ostrom (1979), the possibility under consideration is that the feathered creature, which could run swiftly on its two strong hind legs, was able to increase its speed by wing flapping to the point that it could become airborne--this is the cursorial hypothesis (see Lewin, 1983b).

There is in this debate something we should think about that relates to the evolutionary process and the differences between creationism and naturalistic evolution. It is quite easy to imagine that birds were created by God magnificently and wonderfully equipped to fly. It is not so easy to understand how a reptile could develop flight-worthy feathers and wings with what might seem like the "intent" or "desire" to take to the air. The answer to this problem can be found in an evolutionary principle known as <u>preadaptation</u>; it operates without the need for design or purpose. The solution is really rather simple: feathers evolved from scales as the adaptation of a warm-blooded animal to a cold environment--a means of keeping warm. Small warm-blooded animals, whether birds or mammals, find it very difficult to retain body heat, so that efficient insulation by a body cover, whether of feathers or fur, is

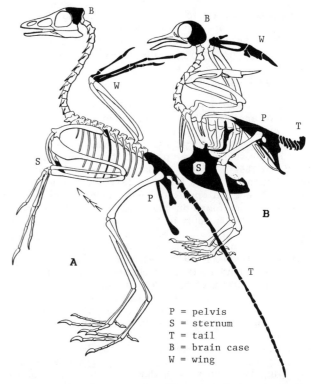

P = pelvis
S = sternum
T = tail
B = brain case
W = wing

Figure 44.3 Skeletons of <u>Archaeopteryx</u> (A) and modern pigeon (B) compared, with emphasis on the pelvis, tail, sternum, brain case, and wing. (From Edwin H. Colbert, 1980, <u>Evolution of the Vertebrates</u>, 3d ed., John Wiley & Sons, New York, p. 186, Figure 71. Copyright © 1955, 1969, 1980 by John Wiley & Sons, Inc. Reprinted by permission.)

an absolute necessity in cold surroundings. We need to propose as a hypothesis that small dinosaurs had developed warm-bloodedness and that some of them occupied cold climates (see Gould, 1980, pp. 267-77). Adaptation to this environmental situation took the form of transition of body scales to feathers. This change represents a preadaptation with respect to the next step, which was gliding or flying.

This explanation of the development of feathers does not, however, explain why wings should evolve from the bones of forelimbs equipped with hands suited for grasping. This is where John Ostrom's ingenuity came into play. He suggested in 1979 that the feathered reptile found its heavily feathered forelimbs useful in entrapping flying insects. Exercising this function could have led by natural selection to enlarged wings with longer bones and longer feathers and a stronger set of muscles for flapping those appendages. Then, it began to happen--flapping those wings while running into a headwind momentarily lifted the animal off the ground. Like the first short powered glide by the Wright brothers at Kittyhawk, it was a small first step to flight, but an essential one. Does this sound to you like just a string of improvisations thought up to keep evolution in a purely mechanistic mode? For myself, I find it a fascinating explanation, vastly more interesting than imagining a divine magician pulling a perfect bird out of a hat. Ostrom has since abandoned his insect-trapping hypothesis, but the controversy between arborialists and cursorialists continues unabated (Lewin, 1985, p. 530).

1986 saw Archaeopteryx joined by a relative. The discovery of another bird fossil, named Protoavis and touted as being even more birdlike than Archaeopteryx, was made in western Texas by paleontologists from Texas Tech University at Lubbock (S. Weisburd, 1986, p. 103). Protoavis, too, has dinosaurian features, such as claws, tail, and teeth; its skull is judged to be more advanced and very similar to that of modern birds. No feather impressions were found with the new fossil specimen, but the wing bones seem to leave little doubt that Protoavis could fly. As to its age, the specimen comes from strata of the Late Triassic (-220 m.y.), which is some 75 m.y. older than Archaeopteryx (Late Jurassic, -150 m.y.)

Is Archaeopteryx a Fake?

We can't leave Archaeopteryx without at least a brief account of another "flap" over its feathered wings. It is a recent claim that the feathers were faked. Of the skeletal remains of Archaeopteryx that show what seem to be feather impressions, one specimen (the holotype) found in Bavaria in 1861 is in the British Museum and a second from the same locality, found in 1877, is in the Berlin museum. The third and fourth--the Eichstätt and Maxberg specimens--were found in 1951 and 1956. A fifth specimen is in the Teyler Museum in the Netherlands. Although collected prior to 1857, it was identified as Archaeopteryx only in 1970, in part from the faint presence of feather impressions (Ostrom, 1970). An isolated feather impression, found in 1861, is counted as the sixth "specimen" (Charig et al., 1986, p. 622).

The claim of forgery of the London and Berlin specimens seems to have been started by Dr. Lee Spetner at a meeting of orthodox Jewish scientists in Jerusalem in July 1980 (Trop, 1983, p. 121). Spetner reported that in 1978 he had spent several hours in the British Museum examining their specimen of Archaeopteryx, using both a magnifying lens and a binocular microscope. He got the impression that areas of the rock surface adjacent to the tail and the wings (or forelegs) had been deeply gouged out and replaced by cement, perhaps it was a paste made up of the original limestone. It seemed to him that feather markings had been carved into the cement to give the likeness of fully feathered tail and limbs (or that chicken feathers had been pressed into the soft cement). Now, it is important to note that Spetner is not a paleontologist, but a physicist, "an Israeli consultant in electronic systems in Rehovot" (Vines, 1985, p. 3). Perhaps significantly, others who later entered the controversy on the side of forgery are also scientifically competent in areas outside of geology, biology, and paleontology.

American creationists were introduced to the subject in 1983, when the Creation Research Society Quarterly carried a letter by Dr. Moshe Trop of Ben Gurion University of the Negev reporting in detail on Spetner's alleged findings of forgery (Trop, 1983). I gather that Dr. Trop is an antievolutionist--a Jewish creation scientist--because he begins his letter by saying that the lack of gradual transitions in the fossil record undermines the theory of evolution.

As reported by Trop, both the London and Berlin specimens of Archaeopteryx, collected in 1861 and 1877, respectively, once belonged to a Dr. Karl Häberlein, District Medical Officer of Pappenheim (p. 122). Both were in the hands of Häberlein and his son before being sold to the two museums that now house them. Spetner reported that the specimens were sold for a good price; moreover, that no specimens have since been found with feather impressions. These facts, says Trop, "are consistent with the hypothesis that the London and Berlin Archaeopteryx fossil specimens are fraudulent" (p. 122).

The next act in this drama is the coming on stage of Sir Fred Hoyle, distinguished British astronomomer and cosmologist, and his associate Chandra Wickramasinghe, an astrophysicist, both on the faculty of the University College, Cardiff. This pair managed to induce Robert Watkins of the Cardiff physics department to take high-resolution photographs of the London Museum specimen. The photographs were published, along with an article by the Cardiff group, in the British Journal of Photography, and so the charge of forgery came in for a new round of exposure to public and press (Watkins et al., 1985). The group reiterated the claims of Spetner that the feather impressions must have been added to an otherwise featherless reptilian skeleton (Hoyle et al., 1985).

At once, the British paleontologists reponded vigorously in protest to the fraud charges. Paleontologist Alan Charig of the British Natural History Museum is quoted as saying: "We think the suggestion that it's a fake is a load of codswallop. . . . I don't understand why people who are eminent in astronomy or physics think they can write papers about vertebrate paleontology. . . . Perhaps it's because we all get dinosaurs in our cornflakes" (Vines, 1985, p. 3). Another paleontologist who responded quickly was Michael E. Howgate of the Department of Zoology of University College, London. He pointed out that the incredible detail of feathers which could be seen in the physicists' photographs "should rule out the cement carving theory of Dr. Spetner as a creationist fantasy" (Howgate, 1985, p. 348). The Cardiff physicists would have liked to take some samples from around the wing areas of the specimen for chemical and physical analysis that might reveal the human origin of the "cement," but this idea was stoutly resisted by Museum authorities.

A British Museum team of paleontologists, headed by Alan Charig, Chief Curator of Fossil Amphibians, Reptiles, and Birds, undertook to present an in-depth answer to the claim of forgery put forward by Hoyle's group. Their justification of this response is interesting in the context of our study of creationism:

It may seem that we, in refuting the consortium's allegations, are using a sledgehammer to crack a rather trivial nut; yet, if we bear in mind the high esteem in which the general public holds Professor Hoyle, together with its lack of knowledge of the facts concerning Archaeopteryx, then it is important that such doubts be finally removed--especially where students of zoology are concerned. More important still, we must put the record straight because of the Creationists, who are interested in any new ideas that, implicitly or explicitly, appear to threaten the concept of organic evolution. (Charig et al., 1986, p. 623)

We cannot go over all facets of the museum team's refutation of the charges of forgery, which includes many and varied aspects of the problem. There is, however, one conclusive point of evidence we can use to settle the matter once and for all. Under ultraviolet photography, the surfaces of both limestone slabs show a network of fine hairline cracks, generally filled with mineral matter. Details of the geometry of the cracks match perfectly on the counterslab. Says the Charig report, "their exact correspondence has been demonstrated by superimposing a negative of the counterslab on to a print of the main slab, the two photographs having been enlarged to the same degree" (p. 624). Obviously the limestone block was cracked naturally by a set of vertical joint fractures long before it was split during quarrying. There is thus no way in which cement could have been added to the natural limestone surface and still retain the exposure of the vertical cracks. The Charig team also took silicone rubber "peels" of the feather impressions, which were then examined under the scanning electron microscope. The extraordinarily minute details they reveal would have been virtually impossible for a human to carve and, besides, no chisel marks were found.

An interesting historical note is the Cardiff group's suggestion that a key figure involved in perpetrating what they claim is a hoax was Richard Owen, then Superintendent of the Natural History Section of the museum and later to become director of the museum. Owen was one of the strongest of Darwin's critics from among the scientists of that time. The reasoning of the Cardiff group runs about as follows (Hoyle et al., 1985, pp. 694-95): Owen's lavish spending of museum and personal funds to purchase the specimens from Häberlein would be viewed by the public as an act strongly supportive of Darwin's theory of evolution--supportive, that is, if Owen endorsed the authenticity of the specimens as naturally occurring fossils. To the Cardiff group, such a simple, innocent act by Owen seems illogical and unthinkable. Fossil forgeries were being turned out in that part of Bavaria in which the Solnhofen quarries are located; these forgeries were being sold to museums. The Cardiff group implies that Owen must have anticipated that the specimen he was purchasing was a forgery. They point out that Owen made no attempt to examine the specimen in advance of its purchase and delivery--a strange position to take in view of the high cost of the specimen and the bad reputation of dealers in the source area. In deliberately letting himself be sold a forgery, sight unseen, Owen saw an opportunity to trap Darwin's leading supporter, Thomas Huxley, who had previously predicted that a transitional birdlike fossil would some day be found in support of evolution. The forgers were perhaps also aware of the importance of Huxley's prediction of birdlike fossils, and they could fill the void. Owen would need only to purchase the forged birdlike fossil for the London Museum, exhibit it to the public, then allow it to be recognized by others as a forgery. In this way Owen would have been able to score a first-class hit against his enemy, Huxley, and at the same time enhance his own reputation. In 1862, Owen introduced his specimen to the Royal Society, hopeful that Darwin's supporters would accept it as genuine and hail it as proof of evolution. Disclosure by Owen of evidence of forgery was scheduled to follow. But, says the Cardiff group, Huxley was too smart to fall into such a trap, instead keeping his silence for the next several years and waiting until 1868 to deliver a scathing attack on Owen (Hoyle et al., 1985, p. 700). Charles Darwin, in revised editions of The Origin of Species, made only passing reference to the

Archaeopteryx specimen; this suggests that he, too, suspected it might be a forgery. I gather that Owen never hinted that his purchase might be a forgery, and no one else made any suggestion to that effect at the time. Do we have here a trumped-up "explanation" of an unsupported claim of forgery? The Cardiff team concludes: "So it came about, we suspect, that Richard Owen hung a millstone around both his own neck and that of the British Museum, a millstone that has only grown heavier with the years" (p. 700). Has that millstone now been transferred to a new set of owners? You be the judge.

Other Flying Reptiles--The Pterosaurs

While on the subject of flying reptiles, we can dispose of another of them, the pterosaur, a true, undisputed reptile. Pterosaurs had wings consisting of membranes attached to greatly elongated finger bones. Mainstream paleontologists and creationists are in full agreement that fully developed pterosaurs, well equipped for flying, appear suddenly in the geologic record, seemingly without intermediates linking them to any ancestral group.

The pterosaurs, an order (Pterosauria) of the reptile class, appeared during the final stages of the Triassic Period and continued through the Cretaceous Period. They are represented by about 85 species. Wann Langston, Jr., a specialist in these strange animals, has given us a comprehensive and highly readable treatment of the pterosaurs and the problems related to the acquisition of their flight capabilities; he writes:

Perhaps the least controversial assertion about the pterosaurs is that they were reptiles. For one thing, their skull is reptilian, including the shape of the teeth. For another, the pelvis and hind feet are those of a reptile. It seem clear, however, that in their adaptations to flight the pterosaurs departed so far from the reptiles popularly known as dinosaurs that no investigator would now confuse them with either of the two dinosaurian orders. The pterosaurs and the dinosaurs appear to have evolved on divergent paths from earlier forms of reptilian life. (1981, p. 122)

Pterosaurs do, however, share the same subclass--the Archosauria, or ruling reptiles--with the two orders of dinosaurs and it seems agreed that all three orders arose from archosaurians of the Triassic Period (Romer, 1966, p. 144; Colbert, 1980, pp. 153, 179-80). Langston continues:

It also seems clear that the pterosaurs did not evolve into the birds. In this regard the telltale anatomy is that of the wing. In a pterosaur the fourth finger of each forelimb was greatly elongated. It supported the front edge of a membrane that stretched from the flank of the body to the farthest tip of that finger. The other fingers were short and reptilian, with a sharp claw at the end of each one. In a bird it is the second finger that is the principal strut of the wing, and in the bird much of the extent of the wing consists of course of feathers. (P. 122)

There were two suborders of pterosaurs, distinctly different one from the other. The more primitive suborder, represented by Rhamphorhynchus, lived in Jurassic time. It is characterized by a long tail, a long flexible neck, and a set of sharp, pointed teeth (Figure 44.4A). The more advanced forms, appearing in late Jurassic time and living through the Cretaceous Period, are the pterodactyloids, or as popularly known, pterodactyls. The genus Pteranodon, with a wingspread of more than 8 m, lacked a tail and had a huge skull with a

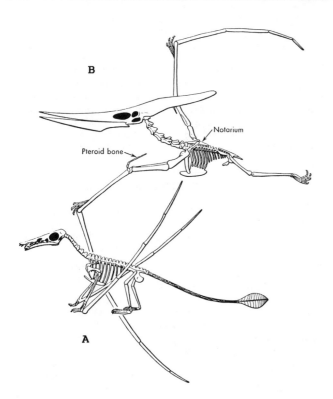

Figure 44.4 Restorations of skeletons of pterosaurs, or flying reptiles. (A) <u>Rhamphorhynchus</u>, of Jurassic age; wingspread of about 0.6 m. (B) <u>Pteranodon</u>, of Cretaceous age; wingspread about 6 m. (From Edwin H. Colbert, 1980, <u>Evolution of the Vertebrates</u>, 3d ed., John Wiley & Sons, New York, p. 181, Figure 69. Copyright © 1955, 1969, 1980 by John Wiley & Sons, Inc. Reprinted by permission.)

strange back-pointing crest (Figure 44.4B). Arm bones, vertebrae, and other parts discovered in 1975 in the Big Bend National Park in Texas indicate a pterodactyl with a wingspread of 11 m! Named <u>Quetzalcoatlus northropi</u>, this great creature lives again in the form of a flying model scaled to half size, powered by batteries, its electronically controlled wings flapping once each second.

It is suspected that the pterosaurs may have been warmblooded, and one fossil of a very small pterosaur shows what seem to be impressions of hairs that covered the body (Langston, p. 132). Although there seems little doubt that the pterosaurs were capable of sustained flight, they probably flew slowly, at best, and may have spent long periods soaring, taking advantage of strong updrafts that developed in front of high marine cliffs.

Creationists will take great satisfaction in Langston's comment on the ancestry of the pterosaurs:

> The sudden appearance of both suborders of the pterosaurs without any obvious antecedents is fairly typical of the fossil record. It emphasizes the random nature of discoveries in paleontology. (P. 123)

Gish capitalizes without mercy on the paleontologists' puzzlement:

> How could these strange creatures have evolved through innumerable intermediate forms over millions of years of time without leaving a single such intermediate in the fossil record? The answer is, they did not evolve--they were created! (1978a, p. 90)

Gish's argument is a familiar one and we cannot leave it

unchallenged. First, we do not know that the fossil record lacks intermediates between the pterosaurs and their ancestors; we can only say that intermediates have not yet been discovered. Second, the present lack of evidence of intermediates is totally irrelevant to the question of whether recent creation is a supportable hypothesis. Through the non sequitur, Gish has given us yet another example of the well-known fallacy of "proof from ignorance."

The Horned Dinosaurs

Everyone loves dinosaurs, perhaps because they are often found in the company of fur-clad cave-dwelling people, with whom they seem to have gotten along rather well, all things considered. That dinosaurs walked hand-in-hand with humans, leaving trails of footprints is common knowledge. Perhaps this is why Duane Gish of the Institute for Creation Research has found dinosaurs useful in creation/science debates. At least, this is what I learn from an article titled "The Dilemma of the Horned Dinosaurs," by Frederick Edwords, a seasoned participant in such debates and also editor of <u>Creation/Evolution</u>, the anticreationist journal in which his paper appears (Edwords, 1982). Again, the subject is gaps in the fossil record and the asserted lack of ancestors for an advanced form that may seem to have appeared suddenly.

According to Edwords, Dr. Gish shows a single slide of the dinosaur <u>Triceratops</u>, a huge animal with three formidable horns and a kind of bony shield rising up over the back of its neck. At this point, Gish tells his audience that <u>Triceratops</u> has no fossil ancestry--no similar dinosaur ancestor with fewer horns or other transitional features. Edwords quotes the following passage from Gish's book, <u>Dinosaurs</u>, <u>Those</u> <u>Terrible</u> <u>Lizards</u>:

> Nowhere do we find in-between forms with spikes starting out as little spikes which gradually got bigger and bigger and finally ending up as a <u>Triceratops</u> dinosaur. The first time you see a dinosaur with armor plate on its head and with three spikes, he is a full-fledged <u>Triceratops</u>, with a huge armor plate and with three big spikes. This is strong evidence for creation! (1977, p. 21)

Edwords strikes back in no uncertain terms:

> Every sentence of this is false. First, there definitely are in-between forms in the fossil record which have lesser and smaller "spikes" (horns); Dr. Gish denies that these exist. Second, <u>Triceratops</u> is not the only dinosaur with "armor plate (bony frill) on its head and with three spikes." He ignores <u>Pentaceratops</u> and <u>Torosaurus</u>, among others, which also fit this description. (1982, p. 2)[4]

To develop an opinion on who is right in this dispute, we need to review briefly the ceratopsians, or horned dinosaurs, which appeared late in Cretaceous time, shortly before all dinosaurs were to disappear from the earth. A possible ancestral form is, however, found somewhat earlier in the record; it is <u>Psittacosaurus</u>, a small genus of the primitive bipedal ornithischian dinosaurs of Lower Cretaceous time (Colbert, 1980, pp. 209-10). The clue to its possible ancestral position lies in the shape of the skull; the front portion of the skull was narrow and distinctly hooked downward, like a parrot's beak (see Colbert, 1965, p. 157, Figure 57).

The first of the true ceratopsians may have been <u>Leptoceratops</u>, a North American dinosaur of Upper Cretaceous age, with a pronounced beaklike skull and the beginnings of a collarlike ridge of armor at the back of the skull. This genus was followed by <u>Protoceratops</u>, found in Mongolia; it had a larger skull than its

Figure 44.5 Three horned dinosaurs, or ceratopsians, shown to the same scale. Triceratops was about 8 m long. (From Edwin H. Colbert, 1980, Evolution of the Vertebrates, 3d ed., John Wiley & Sons, New York, p.211, Figure 81. Drawn by Lois Darling. Copyright © by Lois M. Darling. Used by permission of the artist.)

predecessor and greater development of the beak form and the collarlike "frill" at the back of the skull (Figure 44.5A). Protoceratops was still a small animal, perhaps only 2 m in length, but its general appearance is strikingly similar to Triceratops. An important new feature of this animal was the transformation of its reptilian foot into a more hooflike form.

Perhaps in the direct line of descent from Protoceratops was Monoclonius, a much bigger creature than its predecessor and sporting a single large horn on its nose (Figure 44.5B). The descent from Monoclonius to Triceratops is broken by a substantial time gap, but there are time intermediates from two other branches having a common junction with that of Protoceratops (Figure 44.6). One of these branches, distinguished as the long-crested forms, gave rise to Torosaurus, a contemporary of Triceratops.

Edwords continues the narrative of his debate experiences with Gish:

> It should now be clear that the facts from the fossil record utterly destroy Dr. Gish's claim that Triceratops appears abruptly in the fossil record. It was certainly clear to me when I presented a small portion of this data to him in debate on February 2, 1982, at the University of Guelph in Ontario, Canada. But his response was interesting. He declared that, since all the fossils from Protoceratops to Triceratops were found in the Upper (or Late) Cretaceous strata, they couldn't be an evolutionary sequence. To be an evolutionary sequence, he claimed, these examples would have to stretch back to the Jurassic or Late Triassic. (1982, p. 8)[4]

Edwords counters by noting that the transition from Protoceratops to Triceratops spans some 25 million years, more than enough time for the evolution to take place. (Actually, the duration of the Late Cretaceous, within which ceratopsian evolution occurred, is now given as about 20 m.y.) Furthermore, Edwords says, the

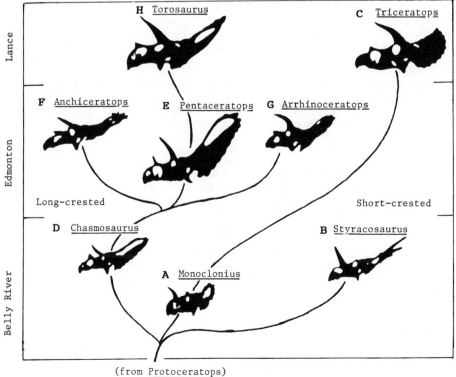

Figure 44.6 Evolution of the horned dinosaurs. Formations in which they occur are named at the left; they are Upper Cretaceous strata of North America. (From E. H. Colbert, The Age of Reptiles, p. 170, Figure 60, W. H. Norton & Company, New York. Copyright © 1965 by Edwin H. Colbert. Reproduced by permission of the author and publisher.)

progressive changes in morphology of the genera we have described are found in the correct time sequence in the strata. Other details of the continued debate follow, but we will skip them here.

Edwords analyzes the significance of the creationists' possible responses to fossil records of evolutionary intermediates such as that shown for the ceratopsians. Creationists cannot, of course, admit to evolution having occurred as mainstream paleontologists describe it, but they still have two options open. First is to claim that each of the genera cited was separately created, but this is not good strategy because, Edwords says, "it implies a creator who experiments with first this and then that until he comes up with something he likes" (1982, p. 9).[4] Second, the creationists can claim that all of the ceratopsian genera are of the same basic "kind," since they look so much alike.

It is this second alternative that seems to represent the current drift in the creationists' dogma. They are applying it to any number of evolutionary sequences in the fossil record--evolution of the horse, for example.

This reduction in numbers of kinds, while it may get the creationists out of one difficult situation, creates a different sort of difficulty. The "microevolution" that they say produced a sequence of "minor" morphological changes must have occurred in a period of only about 1,650 years, which is the time between the original creation of the kinds and the start of the Flood. Does it seem reasonable that the "microevolution" represented by the change from little Psittacosaurus to gigantic, horned Triceratops could have taken place in so little time by gene mutations and recombinations? I would judge that scenario to be preposterous on the face of it. Consider that the period between Creation and Flood was one of mild, rainless climate, uniform over the entire earth--clearly an unchanging and nonstressful environment, lacking in environmental changes that might induce genetic changes. Why would a herbivorous dinosaur need great sharp horns in a benign world ecosystem? The creationists' scriptural interpretation precludes the very evolutionary change they are willing to accept. My own suggestion is that the creationists should stick to Divine Creation of all taxa at the species level and higher, limiting microevolution to the formation of subspecies and race differentiation at the very most. Who is authorized to challenge God's wisdom and purpose in any of His acts of creation? If God chose to experiment with different genera, one after the next, improving each one in turn, what is wrong with that? Who are the creationists that they should judge and find fault with God's work?

The creationist's confused and ambiguous description of created "kinds" is a subject we have discussed at some length in Chapter 37. Recall that it can be shown that there is a hierarchy of "kinds." In his concluding paragraphs Edwords comments:

> Clearly, Dr. Gish has a loose enough definition of kind that, if people keep throwing the ceratopsians at him in debate, he can eventually fall back on the argument that they are all the same kind. It is no problem for evolution if creationists do this. It is rather a problem for creation. It means that creationists are retreating in the face of over-whelming evidence. It means that they are admitting to more and more evolution. It means that they are gradually giving their case away. (1982, p. 10)[4]

The Marine Reptiles

A number of reptile groups, some starting as early as Permian time, reverted to the aquatic environment from which their amphibian ancestors had earlier emerged. Instead of returning to fresh water of estuaries and marshes, however, some of these entered the saltwater environment. Some were poor swimmers, paddling about in shallow water in search of food. But two major groups, the ichthyosaurs and plesiosaurs, became excellent swimmers inhabiting deep water. None of these aquatic reptilian groups abandoned the air-breathing habit and apparatus.

Evidently the marine environment offered an abundant food supply, more than enough for the predaceous sharks and bony fishes. One can imagine that terrestrial reptiles, finding an abundance of fishes and invertebrates while wading in shallow tidal waters, were subject to strong adaptive pressures to occupy that habitat with increasing effectiveness, but competition would also have been intense. Structural modifications may have occurred very quickly, geologically speaking, explaining in part, at least, why the fossil record of the transitions is largely a blank.

Creation scientists seem not to have made an issue over the ancestry of the marine reptiles, but should they do so, they will find themselves forestalled by vertebrate paleontologist Chris McGowan, who has specialized in these creatures, particularly the ichthyosaurs (McGowan, 1984, pp. 158-60).

We are going to be discussing members of two different reptilian orders. The ichthyosaurs, or fishlike reptiles, belong to an order of the same name, Ichthyosauria. The plesiosaurs--and the nothosaurs, also marine reptiles--are of the order Sauropterygia. Members of the two orders looked very different and had quite different swimming structures. The two orders are entirely separate from those orders in which we find the reptiles already discussed, such as the lizards and snakes, crocodiles and alligators, flying reptiles, and dinosaurs. But neither are the marine reptiles primitive types; they flourished in the Jurassic and Cretaceous periods as contemporaries of the great ruling reptiles of the lands.

Ichthyosaurs certainly deserve the name of "fishlike reptiles." You could easily mistake a pictorial restoration of an ichthyosaur for a modern fishlike mammal, such as a dolphin or porpoise, or a killer whale performing in a marine zoo (Figure 44.7). They were sleek, streamlined animals, probably capable of very high speeds. Their resemblance to some of the fishes of today--the tuna, salmon, or mackerel--would also be striking. The lesson seems to be that all marine animals that swim efficiently must be shaped more or less alike, conforming to a common optimum hydrodynamic design.

In the ichthyosaurs, the reptilian limbs had evolved into finlike paddles, the tail into a caudal fin. There was also a dorsal fin. Fortunately, a large number of ichthyosaurian fossils have been found preserved in fine-grained shale, on which the outline of the body is

Figure 44.7 Reconstruction of an ichthyosaur. (From Edwin H. Colbert, 1980, Evolution of the Vertebrates, 3d ed., John Wiley & Sons, New York, p. 168, Figure 65. Drawn by Lois Darling. Copyright © by Lois M. Darling. Used by permission of the artist.)

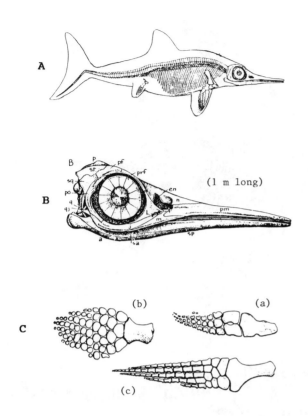

Figure 44.8 Anatomical details of the ichthyosaurs. A. Skeleton of a Jurassic ichthyosaur, showing body outline. B. Side view of the skull of Opthalmosaurus, showing the huge eye. C. Limb bones of ichthyosaurs (a, b) and plesiosaurs (c). (From Vertebrate Paleontology, 3d ed., by Alfred S. Romer, University of Chicago Press, p. 118, figures 172, 174; p. 119, Figure 175. Copyright © 1933, 1945, and 1966 by The University of Chicago. Reprinted by permission.)

preserved as a carbonaceous film (Figure 44.8A), revealing that the dorsal fin is lacking in any bony structure. Moreover, the upper half of the caudal fin was entirely fleshy, for the bony tail was bent down to enter the lower half. Very striking is the huge eye of the ichthyosaur, evidently a great asset in seeing its prey, which probably consisted of fast-moving fishes (Figure 44.8B). The paddles contained a large number of small bones, reduced from long, slender shapes to circular or hexagonal pieces (Figure 44.8C). All of these structures indicate that the ichthyosaurs propelled themselves by a rhythmic oscillation of the body that transmitted its thrust to the tail fin. The paddles served to steer the body, while the dorsal fin helped as a stabilizer.

One of the questions that long concerned paleontologists was the ichthyosaurian manner of reproduction. Reptiles lay shelled eggs, a great advantage on land where the eggs can be hidden in sand to hatch by themselves. This plan would seem unworkable in the ocean. Eggs dropped directly into the water would settle at random, subject to being moved around by bottom currents and eaten by scavengers. Some remarkable ichthyosaur skeletons have been found enclosing skeletons of very young individuals, seemingly hatched within the body. This happens in the modern snakes and lizards, so is not an outrageous idea. Romer writes:

> In agreement with the idea that the young were born alive are specimens which actually show skeletons of young ichthyosaurs inside the body of a large individual. It has been argued that these may have been youngsters which had been eaten by

mistake. But several specimens show the young partially emergent from what would have been the cloacal region in life. The mother here apparently died during childbirth, or (there are human parallels) labor may have taken place after the death of the mother. (1966, p. 119)[1]

Much more could be told about the structural features of the ichthyosaurs as adaptations of land reptiles to the marine environment, but we must turn next to the question of their ancestry. Can the creationists rightly claim, should they wish to do so, that no ancestors can be identified?

Colbert offers an interesting lead as to the ancestry of the ichthyosaurs: "The jaws were provided with numerous teeth, which had the labyrinthine structure that was so characteristic of the labyrinthodont amphibians and the primitive cotylosaurian reptiles, and this is one clue as to the ultimate ancestry of the ichthyosaurs" (1980, p. 168).[2] Colbert elsewhere states:

> It is only through an interpretation of the anatomical structures of these highly specialized reptiles that we are able to make some deductions as to their probable origin, which undoubtedly was from cotylosaurian ancestors. It may be that between the basic cotylosaurian stock and the first ichthyosaurs were connecting links related in a general way to some of the aquatic pelycosaurs, such as the ophiacodonts. (1980, p. 167)[2]

Such reasoned suggestions of early ancestry nothwithstanding, the prospect of identifying intermediates looks bleak. Romer minces no words:

> Although the Triassic ichthyosaurs were slightly more primitive than their Jurassic descendants, they were already very highly specialized marine types. No earlier forms are known. The peculiarities of ichthyosaur structure would seemingly have required a long time for their development and hence a very early origin for the group, but there are no known Permian reptiles antecedent to them. (1966, p. 120)[1]

McGowan, seems to have little in the way of new information and new hope on ichthyosaurian ancestry. He notes that the bestknown of the early representatives, Mixosaurus, lacks some of the typically ichthyosaurian features of the Jurassic and Cretaceous forms: "But Mixosaurus was already recognizable as an ichthyosaur; where, then did the ichthyosaurs come from? Do we have any fossils intermediate between Mixosaurus and an unspecialized land reptile?" (1984, pp. 158-59)[3] So, in a final quote, I must turn McGowan over to the tender mercies of the enemy:

> When Professor Robert Carroll of McGill University, a paleontologist who is particularly interested in the relationships among reptiles, last asked me that question, I believe I suggested that ichthyosaurs had just dropped out of the sky. The embarrassing fact is that we have not found the ancestor of the ichthyosaurs. (1984, p. 158)[3]

So now, on to the plesiosaurs. Were their ancestors more obliging and less determined to conceal their evolutionary tracks? Fortunately, some help comes from their earlier cousins, the nothosaurs. First, let us look at the anatomical configuration of the plesiosaurs.

I continue to be grateful to Dr. Duane Gish (1978a, pp. 66-67) for his admonition to read Romer's great treatise carefully. When I do read Romer carefully, I find here and there a little nugget of humor, which he inserts to relieve the inevitable pedantry into which his

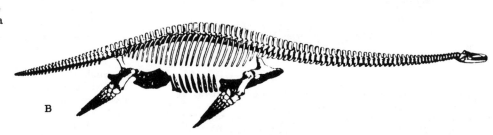

Figure 44.9 Two reconstructions of skeletons of pleisosaurs. A. Thaumatosaurus, a Jurassic plesiosaur about 3 m long. The underside is shown here. B. Muraenosaurus, a long-necked plesiosaur of Jurassic age, about 6 m long. (From Vertebrate Paleontology, 3d ed., by Alfred S. Romer, University of Chicago Press, p. 123, figures 183, 184. Copyright © 1933, 1945, and 1966 by The University of Chicago. Reprinted by permission.)

profession has forced him. Romer encapsulates the plesiosaurian anatomy in one short sentence:

> A plesiosaur has been compared by an old writer to "a snake strung through the body of a turtle." (1966, p. 124)[1]

That description certainly applies well to the Jurassic plesiosaur whose skeleton is shown in Figure 44.9A. Other genera had longer necks and a body less turtle-like (Figure 44.9B), but the main feature that set them apart from the ichthyosaurs was the way in which they were designed to swim. The limbs were mighty paddles, functioning as do the oars of a clumsy dory. The shoulder and pelvic girdles were large, shaped like great plates almost enclosing the abdominal region, and making a good anchor for the powerful muscles that drove the paddles. The body moved forward rigidly as a unit,

reminding us of the swimming habit of the sea turtles, but the plesiosaurs were huge animals reaching a maximum length of about 15 m.

An excellent candidate for an ancestor, or common ancestor of the plesiosaurs, is an older and more primitive group, the nothosaurs, found in strata of Middle Triassic age. Smaller than the plesiosaurs, they were not really very well adapted to an aquatic life. In particular, the limbs are intermediate between those of land reptiles and the plesiosaurian paddles, as you can see in Figure 44.10. The nostrils of the nothosaurs were in an intermediate position--farther back than in the land reptiles but not as far back as in the plesiosaurs. Nothosaurs, like plesiosaurs, McGowan writes, "have well-developed shoulder and hip girdles, and although these are plate-like, they are not as well developed as in the plesiosaurs. Indeed, the two groups merge into one another in a way that is most satisfactory to an evolutionist, but a

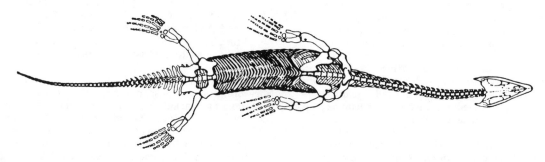

Figure 44.10 Skeleton of a Triassic nothosaur, Ceresiosaurus, about one meter long. View is of the underside. (From Vertebrate Paleontology, 3d ed., by Alfred S. Romer, University of Chicago Press, p. 124, Figure 188. Copyright © 1933, 1945, and 1966 by The University of Chicago. Reprinted by permission.)

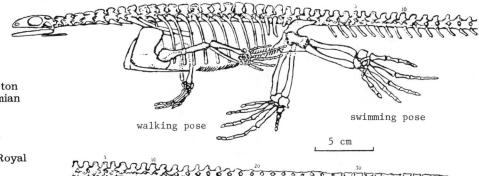

Figure 44.11 Restoration of the skeleton of _Claudiosaurus germaini_, a late Permian quadruped, possibly ancestral to the plesiosaurs. (From R. L. Carroll, _Philosophical Trans. of the Royal Soc. of London_, vol. 293, facing p. 368, Figure 29. Copyright © 1981 by The Royal Society of London and R. L. Carroll. Reproduced by permission of publisher and author.)

walking pose

swimming pose

5 cm

creationist would probably be unimpressed and would want to see a link between nothosaurs and unspecialized land reptiles" (p. 159).[3]

Romer is less happy with the relationship, saying that in certain respects "the typical nothosaurs were already a bit more specialized than the plesiosaurs and hence, despite similarities in other respects, cannot have been plesiosaur ancestors" (1966, p. 125).[1] I view the relationship as one of "cousins" sharing a common ancestor, but as we have seen, the role of the nothosaurs as a cousin is nonetheless that of a true intermediate.

McGowan now treats us to a case of serendipity in paleontology that illuminates the way the study of evolution actually works (1984, p. 160). Finding unusual fossils is something that can't be planned--it just happens. It seems that in 1950 some fossils of Late Permian age were unearthed in the Republic of Madagascar in the course of a petroleum search and were judged sufficiently interesting that they were sent to Paris for study. The French paleontologist, J. Piveteau, recognized them as possibly ancestral to the plesiosaurs, but he was unable to study the fossils further and they were passed on to Professor Robert Carroll (mentioned earlier) who gave them the full attention they deserved. Carroll published his findings in 1981, commenting in his introduction:

> In view of the previous lack of any definite information as to the origin of nothosaurs and plesiosaurs, it is a pleasure to be able to describe an extensive series of excellently preserved specimens that appear to be nearly ideal structural intermediates between ancestral terrestrial reptiles and the Mesozoic nothosaurs and plesiosaurs. (P. 317)

Figure 44.11 shows a reconstruction of the skeleton of one of Carroll's newly discovered reptiles, _Claudiosaurus germaini_ (named after Claude Germain, who had collected the fossils and presented them to Piveteau). The legs are shown in two modes: as if walking and as if swimming. Carroll notes that the hands of _Claudiosaurus_ are proportioned "in a manner similar to that seen in a variety of vertebrates that use the limbs as paddles" (1981, p. 369). He also says: "_Claudiosaurus_ shows some features that might be expected in any group beginning adaptation toward an aquatic way of life, but little that approaches the specifically nothosaurian pattern" (p. 380). The features he is referring to are details of the vertebral column, a rather technical subject.

McGowan, commenting on Carroll's findings, points out that the shoulder girdles were large and platelike, somewhat like those of a nothosaur, but not so extensive (p. 160). The skull seems to be intermediate between the eosuchian skull and that of the nothosaur. McGowan summarizes his discussion in general terms:

> We are therefore able to trace an evolutionary pathway from fully terrestrial reptiles to the fully aquatic, and highly specialized plesiosaurs. We obviously do not have a complete series of finely graded steps, nor would we expect to have all the intermediates, given the vagaries of the fossil record, but we do have compelling evidence that the plesiosaurs did evolve from a terrestrial ancestor. One day we may be able to do the same for the ichthyosaurs. (1984, p. 160)[3]

I suggested earlier that there were other marine reptiles besides those we have described in detail. One group, the geosaurs, were members of a family of crocodilians, order Crocodilia, that took to the sea and flourished in late Jurassic time. They persisted into the early Cretaceous. The limbs of the geosaurs were modified into paddles, while the tail developed into a fishlike fin. A second group, the mosasaurs, appeared in late Cretaceous time and are thought to have arisen from early lizards (Colbert, 1969, p. 179). A good example of one of these marine lizards was _Tylosaurus_, reaching a length of 10 m (Figure 44.12). This creature, too, had paddles modified from the reptilian limbs but had no dorsal fin or tail fin. (The skull of _Tylosaurus_ is shown in Figure 43.7A, where we used it to illustrate the bones of the reptilian lower jaw.)

Figure 44.12 A Cretaceous marine lizard, or mosasaur, six or more meters in length. (From Edwin H. Colbert, 1980, _Evolution of the Vertebrates_, 3d ed., John Wiley & Sons, New York, p. 174, Figure 67. Drawn by Lois Darling. Copyright © by Lois M. Darling. Used by permission of the artist.)

Origin of the Turtle's Shell

Where did the turtles get their shells? Creationists say that mainstream evolutionists have no satisfactory explanation, because (they say) the evolutionists would like to explain the upper (dorsal) bony shell as a fusion of the ribs. If that is so, how did the rib cage manage to get from the inside of the shoulder girdle (where it is located in all self-respecting reptiles) to a position outside of the shoulder girdle? Sounds impossible doesn't it? In analyzing such complexly worded statements, you will want to be suspicious of the creationists' premise itself. Is there a straw person lurking in this argument? Does any scientist really propose that the turtle's upper shell formed from its rib cage?

Fortunately, the order Chelonia has living representatives--many of them--and their anatomy can be studied along with their embryology. This information can shed a great deal of light on their phylogeny and possible ancestry, even if the fossil record is incomplete.

The Institute for Creation Research is evidently not into turtles, for I find no mention of the animal in either Henry Morris's school textbook (1974a). Nor is it mentioned in Duane Gish's NO! or the more recent version of the same work (1978a; 1985). For creationist sources mentioning the turtle we are indebted to Andrew J. Petto, a graduate student in the Department of Anthropology of the University of Massachusetts at Amherst, who covers the creationists' arguments and answers them in detail in Creation/Evolution (Petto, 1982). On page 20, Petto quotes from creationist Bolton Davidheiser:

> If the turtle evolved from animals of more "orthodox" structure, it is a mystery how they managed to get their shoulder bones inside their rib cages. If they (the shoulder bones) were outside the ribs, as in other animals, they would also be outside the shell. (1971, p. 246)

(Note: Words in parentheses added for clarity.)

A second quotation is from creationist J. W. Klotz (Petto, pp. 20-21):

> All reptiles are supposed to have developed from the stem reptiles, the cotylosaurs. A modified cotylosaur, Eunotosaurus, is sometimes postulated as the ancestor of all the turtles in that it comes from the proper time and it appears to be on the verge of developing a shell. But it has one serious drawback as a turtle ancestor. The carapace of modern turtles does not develop just from wide ribs but from independent plates of dermal bone which expand markedly and fuse with one another and with the underlying ribs and any shell or plastron. This unique armor and the contortions which the skeleton had to undergo to fit into it, combined with the toothless beak, have suggested to some that turtles are entirely different from any living reptile. (1979, p. 457)

The above statements are undocumented to begin with and are a mixed collection of both acceptable and unacceptable assertions.

Fortunately, when we follow Dr. Gish's admonition to make a careful reading of Alfred Romer's great treatise, Vertebrate Paleontology, the sources of Klotz's statements do not long remain concealed. In particular, Romer supplies what we need to know about the very strange turtlelike animal, Eunotosaurus, which has indeed been given consideration as a possible ancestor of the chelonians (turtles) (1966, p. 116). This creature is represented by a fossil from strata of Middle Permian age in South Africa. It shows eight pairs of very broad, flat ribs, one touching the next and extending out to the

limits of the body. Romer is very cautious in his appraisal of this fossil, for he says: "This may be compared with the condition in turtles in which there are eight ribs supporting the costal plates which make most of the carapace" (1966, p. 116).[1] As we will soon find, the "costal plates" of the turtle are structures very different from the ribs. Romer continues:

> This creature was not, of course, a true turtle; on the other hand, it was far from the typical cotylosaurs. Perhaps we may include it provisionally in the Chelonia in a broad sense but place it in a separate suborder, the Eunotosauria. (P. 116)[1]

Clearly, the creationists have attempted to set up a straw person in the form of a supposed evolutionary hypothesis of origin of the turtle shell directly from its ribs, by enlargement, broadening, and fusion of the latter into an armor plate. Actually, that explanation has been examined and rejected by mainstream science. Moreover, mainstream science has a reasonable evolutionary explanation of the shell, backed by the fossil record and by embryology of living turtles.

Creationists have hastened to forestall any viable alternative hypothesis of a reasonable evolutionary pathway. Bolton Davidheiser, as quoted by Petto on page 21, uses the words of a mainstream biologist to accomplish this mission:

> It was hoped that a study of the embyronic development of the turtles would clarify this. The problem is discussed by Archie Carr, professor of biological sciences at the University of Florida. He says, "It might accordingly have been hoped that the evolution of the relationship between the shell, ribs, and the girdles during embryology would shed some light on the original history of these events, but such is not the case." (1971, p. 246)

Petto suspected, as experienced creation-watchers have learned from experience, that Carr's statement is taken out of context. What Petto found is that Carr actually shows that "the turtle conforms to expectations for an animal maintaining a conservative reptilian embryology with an anatomical specialization for external armor" (p. 26). Petto continues:

> In this context, Carr's comments argue for the reptilian ancestry of the turtle, not against it, as Davidheiser would have us believe. Carr says that the embryology of the turtle confirms that it is a "good" reptile. No new or unusual--nonreptilian-- structures appear in its embryology. (Petto, 1982, p. 22)

No question about it, we have gotten ourselves deeper and deeper into the problem of the turtle's shell. For those of you who don't want to simply drop the problem, we follow through on lines of both embryology and paleontology.

First, we should examine the cross section of a living turtle. Petto gives us such a cross section drawn by his wife, Sarah Petto; it depicts the body structure of the snapping turtle, Chelydra (Figure 44.13). The bony shell comes in two halves: the top shell, or dorsal carapace, and the bottom shell, or ventral plastron. These shells consist of dermal bone, which forms in the skin of the reptile. Actually, the bony shell consists of tightly fitted individual plates. Figure 44.14 shows the entire turtle shell in generalized cross section.

Dermal bone differs from the other bones of the skeleton, which first appear in the embryo as a cartilage model that is hardened in later growth stages (see Zangerl, 1969, pp. 311-14). The dorsal carapace also has a horny shield added on top of the dermal bone. The

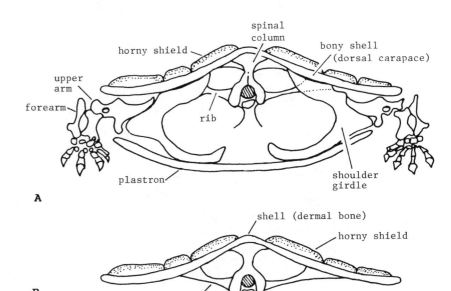

Figure 44.13 Generalized cross sections of the bone and shell structures of the modern turtle, <u>Chelydra</u>. A. Region of the shoulder girdle. B. The trunk region, showing the relationship of the ribs to the dorsal carapace. (Modified from A. J. Petto, <u>Creation/Evolution</u>, vol. 3, no. 4, p. 26, Figure 3. Drawn by Sarah Petto. Copyright © 1983 by Andrew J. Petto. Used by permission.)

horny material is keratin and is rarely preserved in fossils. The horny shield consists of individual plates, called scutes, and these fit closely in a geometrical pattern that overlaps the joints in the bony carapace. This arrangement thickens and strengthens the upper shell.

We turn next to the shoulder girdles that support the front legs. Each girdle is a curved bone to which the leg muscles are attached. The two girdles are attached at top and bottom to the dorsal carapace and ventral plastron, respectively, giving a very strong, solid anchorage. The rear legs are supported by a single strong pelvic girdle, also attached at top and bottom to the carapace and plastron. The ribs of the turtle have been displaced upward on either side so as to attach to the underside of the bony dorsal shell, as shown in Figure 44.13B and Figure 44.14. In no sense is the shoulder girdle "inside" of the rib cage, as the creationists claim; the ribs simply do not form an abdominal cage.

Turning to the embryology of the turtle, we find that in a near-term embryo the horny shield scutes are already laid out. Beneath them, the thick dermal layer has, as yet, no dermal bone in place, but the dermal plates develop later following preformed structures in the dermal connective tissue (Zangerl, 1969, p. 312). As the dermal plates grow, the outer ends of the ribs become attached to the carapace above them. The gaps between rib and carapace on either side of the spinal column form the vertebro-costal tunnel, shown in Figure 44.14, giving a supporting structure resembling the flying buttresses of a gothic cathedral, but upside down (Hoffstetter and Gasc, 1969, p. 224). The ribs, fused with the dermal bone, usually extend outwards to the limits of the bony carapace. Here, a marginal bony plate is formed and contacts the plastron below, forming a solid bony box in the trunk region of the turtle.

In the embryonic stage, the shoulder girdle gets involved with the dermal bone of the plastron, or lower shell. The shoulder girdle of the adult turtle and its ancestors--including the cotylosaurs and amphibians-- consists of several bones fused together. (These component parts are shown and labeled in Figure 44.17.) Three of the components are composed of dermal bone: cleithrum, clavicle, and interclavicle. In the turtle embryo, parts of the shoulder girdle that consist of dermal bone become incorporated into the plastron.

At this point you might raise a good question: is the growth of dermal bone to form armor plate a new feature entirely unique to the turtles, or is this a growth phenomenon found in other reptiles?

Romer, describing the plates of dermal bone of the turtle carapace writes: "Such plates are developed in many crocodiles and some lizards, as well as in other groups now extinct. The turtles differ from them in the consolidation of the plates into a complete, compact covering" (1966, p. 112).[1] In describing the primitive crocodiles (which are in a different subclass and order from the turtles), Romer states: "There is always a well-developed set of dermal armor plates down the back beneath the horny scales and sometimes down the ventral (belly) side as well--a feature inherited from primitive thecodonts" (p. 142).[1]

While perusing the content of a 1975 symposium on the morphology and biology of living and fossil reptiles, sponsored by the Linnean Society of London, I chanced upon an interesting paper by Frank Westphal on the dermal armor of some Triassic placodont reptiles (Westphal, 1976). Here I found dermal bone enclosing a group of fossil aquatic reptiles that, at first glance, would seem to be perfectly good turtles. They are, instead, placodonts, relatives of the nothosaurs and

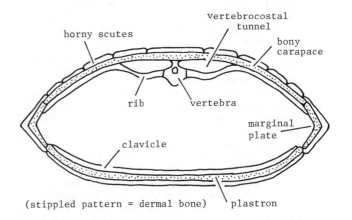

Figure 44.14 Generalized cross section through the trunk region of a turtle, showing the extent of dermal bone (stippled), ribs, and horny scutes. (Based on data of R. Zangerl, 1969, The Turtle Shell, p. 328, in Carl Gans, ed., <u>Biology</u> <u>of</u> <u>the</u> <u>Reptilia</u>, vol. 1, Academic Press, London.)

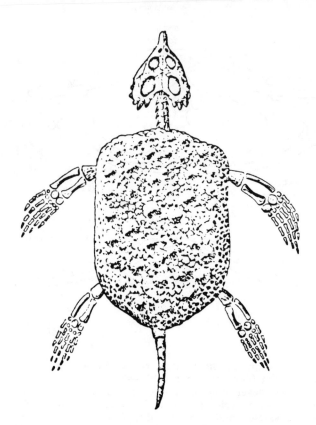

Figure 44.15 A placodont, <u>Placochelys</u>, superficially resembling a turtle. About one-half natural size. View is of the top of the shell. (From <u>Vertebrate</u> <u>Paleontology</u>, 3d ed., by Alfred S. Romer, University of Chicago Press, p. 126, Figure 192. Copyright © 1933, 1945, and 1966 by The University of Chicago. Reprinted by permission.)

plesiosaurs of the order Sauropterigya, within the subclass Euryapsida. True turtles belong not only to a different order (Chelonia) but also to a different subclass, the Anapsida. This may seem like a simple technicality of no consequence, but it puts the placodonts in a very different reptile group than the turtles. The basic difference lies in the construction of the skull-- whether or not there is a temporal opening. The anapsids have no temporal opening in the skull behind the eye; the eurapsids have a single temporal opening.

The placodonts, while turtle-like, were adapted to shell-crushing; they had large flattened teeth forming a pavement at the back of the mouth. Their similarity to turtles lies, then, in the body rather than the skull. The placodont <u>Placochelys</u>, shown in Figure 44.15, has an armor of small plates of dermal bone, but the individual plates--called ostoderms--did not contact each other. A better example is <u>Henodus</u>, a placodont of upper Triassic age. It had a turtle-like carapace, broad and short, with seven longitudinal rows of hexagonal ostoderms bordered by additional rows of ostoderms, all closely fused together in a solid shell (Westphal, 1976, p. 36-37). The shell also had a layer of horny scutes, which were rhombic in outline and overlapped the joints of the ostoderms beneath. The ribs did not make contact with the carapace, as they do in the turtles. <u>Henodus</u> also had a bony lower plate, analogous to the plastron of the turtles, but it was thin and weak. The development of the armor shell of these two very different reptile groups, while not similar in its details, is remarkable as a case of parallel evolution exploiting similar kinds of tissue in a similar manner.

Clearly, the development of armor plates of dermal bone is a common reptilian feature. In this respect, the

dermal bone shells of the turtle are "normal" reptilian features, their extreme development nothwithstanding. Petto emphasizes that "it is chiefly the <u>extent</u> of the development of dermal bone in the skin which distin- guishes the turtle from its reptilian relatives" (1982, p. 27). He goes on:

> The fact that this dermal bone forms in the skin without a cartilage model makes it precisely the sort of bony shield evolutionary biologists would predict for a reptile committed to enclosing itself in armor. The fact that it forms directly in the skin accounts for its location outside the limb bones. (P. 27).

Petto seems fully confident that the problem of the turtle's shells has been solved in a manner fully in accord with existing knowledge of biology:

> A plausible way to develop a turtle from a basic reptilian ancestor has been proposed. It is plausible because it relies on structures and developmental processes which we can observe in living animals and because it is based upon the natural laws which we have observed operating in so many other cases. No new or special mechanism is necessary to explain the result. (P. 28)

Despite what we have reviewed here, the creationists will continue to insist that the turtles appear suddenly in the geologic record without obvious ancestors and, as in the numerous other cases of seemingly sudden appear- ances, the evidence points to Divine Creation rather than to descent with modification from a common ancestor. Actually, quite aside from the unique development of armor shells, anatomical development of the turtles as just another variety of reptile is demonstrated with many intermediates when one looks at the homology of other parts of the skeleton--skull, vertebrae, shoulder girdles, pelvis, and legs. These establish its reptilian lineage beyond any reasonable question. The homology of the limb bones is in itself sufficient to make this determination.

Finally, Figure 44.16 shows the carapace of <u>Proganochelys</u>, a primitive turtle of Triassic age; it is among the oldest known of the true turtles. Romer

Figure 44.16 Reconstruction of the carapace of <u>Proganochelys</u>, a primitive turtle of Triassic age. Shown are the impressions of the horny scutes. Length about 69 cm. (From <u>Vertebrate</u> <u>Paleontology</u>, 3d ed., by Alfred S. Romer, University of Chicago Press, p. 115, Figure 167. Copyright © 1933, 1945, and 1966 by The University of Chicago. Reprinted by permission.)

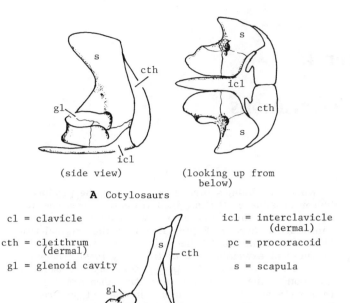

(side view)   (looking up from
                       below)

A Cotylosaurs

cl = clavicle

cth = cleithrum
          (dermal)

gl = glenoid cavity

icl = interclavicle
            (dermal)

pc = procoracoid

s = scapula

B Chelonians (turtles)

Figure 44.17 Comparison of the shoulder girdle of a cotylosaur, Labidosaurus (A), with that of a primitive turtle, Proganochelys (B). (From Vertebrate Paleontology, 3d ed., by Alfred S. Romer, University of Chicago Press, p. 107, Figure 154, p. 113, Figure 164. Copyright © 1933, 1945, and 1966 by The University of Chicago. Reprinted by permission.)

retraction into the body. He regards the shoulder girdle as "rather primitive" and reports: "The clavicles and reports that in these animals "all the shell elements of later turtles were already present, and their arrangement was essentially that of later forms" (1966, p. 113).[1] In describing the skull of this primitive turtle, Romer comments: "the most striking feature of all is that, although a horny beak is apparently already developed, there are still teeth on the palate" (p. 114). Surely these residual teeth are transitional from the earlier reptilian form. Romer thinks it improbable that the limbs could be withdrawn, nor were the head and tail capable of

interclavicle were already incorporated into the plastron (lower shell) but show their original character clearly, and there was possibly even a small cleithrum" (p. 114).[1] Figure 44.17 compares the shoulder girdle of Proganochelys with that of the cotylosaur Labidosaurus that, as a stem reptile, would have been ancestral to the turtles. Notice that the lowermost dermal bone of the cotylosaur shoulder girdle, the interclavicle, is extended backward as a long pointed keel-like feature. It is in a position to be greatly expanded into a ventral armor plate of dermal bone.

In keeping with the current and popular paradigm of punctuated evolution, we can find it a reasonable hypothesis that the development of the full turtle shells of dermal bone took place rapidly from small dermal elements in the shoulder girdle and elsewhere. Occurring in a small isolated population, this remarkable structural change would have given a tremendous advantage to the animal through its fortresslike defense against predators, making a rapid radiation possible in only a few millions of years. Although still partially water dwellers, like the primitive reptiles from which they arose, the turtles came to occupy dry terrestrial environments as well as the marine environment. In the marine environment, buoyancy allowed the turtles to grow to enormous size; those of Cretaceous time reached lengths of 4 m. In some aquatic terrestrial forms, the dermal bone armor seems to have been greatly reduced, along with disappearance of the horny shield, leaving only a leathery skin. Perhaps we can say in a final salute to the turtle's armor plate of dermal bone: "Easy come, easy go!"

Credits

1. From Vertebrate Paleontology, Third edition, by Alfred S. Romer, University of Chicago Press. Copyright © 1933, 1945, and 1966 by The University of Chicago. Used by permission.

2. From Evolution of the Vertebrates, First edition 1955, Second edition 1969, Third edition 1980, by Edwin H. Colbert, John Wiley & Sons, Inc., New York. Copyright © 1955, 1969, 1980 by John Wiley & Sons, Inc. Reprinted by permission.

3. From IN THE BEGINNING © 1983 by Chris McGowan. Reprinted by permission of Macmillian of Canada, A Division of Canada Publishing Corporation.

4. From Frederick Edwords, The Dilemma of the Horned Dinosaurs, Creation/Evolution, vol. 3, no. 3, pp. 1-11. Copyright © 1982 by Frederick Edwords. Used by permission.

Chapter 45

Mammalian Transitions

The placental mammals arouse in humans emotional responses of an intensity enormously greater than do the reptiles. Perhaps most of us are aware of an evolutionary kinship with the other mammals, so much so that we like to give them the same names we use for ourselves and often speak to them as if they were our children. Creationists play on these emotional reactions in their writings and debates against evolutionary scientists. In this chapter we detail the creationist/evolutionist debate over two of our favorite mammals, the whales and the horses. It is, as we shall see, an occasion for more than the usual display of humor attached to incongruous absurdities.

Origin of the Whales

Creationist interest in the origin of whales is comparatively recent. Duane Gish did not mention them in NO! (1978a), but moved into this area in a 1980 issue of ICR Impact (No. 87), a general article on the origin of mammals. Gish starts with the usual citations from leaders in mainstream paleontology, confessing their ignorance of ancestors of the marine mammals. Romer's 1966 treatise is, as usual, a rich source for such confessions.

What seems to have put Gish on the scent of the whale trail is a statement in a 1979 Scientific American article by Bernt Würsig, a student of cetacean behavior, suggesting that the dolphins evolved from "land mammals that may have resembled the even-toed ungulates of today such as cattle, pigs and buffaloes." This prospect was just too much for Gish to swallow, and he moved right in:

> It is quite entertaining, starting with cows, pigs, or buffaloes, to attempt to visualize what the intermediates may have looked like. Starting with a cow, one could even imagine one line of descent which prematurely became extinct, due to what might be called "udder failure." (1980, p. iii)

Entertainment did indeed follow, according to an account by Frederick Edwords of a debate between Duane Gish and Kenneth Miller held in Tampa, Florida, in March of 1982 (Edwords, 1983a, pp. 3-4). Gish related Würsig's hypothesis, illustrating it by a cartoon slide showing a smiling dairy cow and subsequent evolutionary stages in which the cow developed whale flukes in place of its hind legs, and front flippers instead of front legs, while retaining the udder. The cartoon series ended with a complete modern whale.

Skipping over some of Gish's recorded commentary on this evolutionary sequence, I repeat the final sentences of his message:

> How did some hairy four-legged mammal get into the water, stick around for eons of time, and just gradually and slowly evolve into a whale which is wonderfully and marvelously designed for life in the water? You see, when it comes right down to a specific case, the whole idea of evolution is an absurdity. (Gish, in Edwords, 1983a, p. 4)

When Dr. Miller's turn came to reply, he produced a humorous slide of the imagined transition from cow to whale, drawn for the occasion by his artist wife. We reproduce it here as Figure 45.1 with the original title "Bossie to Blowhole." Miller then turned to the serious business of examining one of the early fossil whales.

In a 1983 issue of ICR Impact Gish again raised the question of the origin of whales, this time drawing energy from an Associated Press article in the Detroit Free Press, April, 15, 1983, telling of a new and surprising fossil find--a possible missing link between the earliest known whales and land animals early in Eocene time. Gish reacted:

> One should be immediately suspicious of the term "whale" being given to such a creature, whatever it was, since whales are totally incapable of living or breeding on land.
>
> News of this kind, as tentative and unreliable as it might be, is no doubt welcome to evolutionists since there is indeed, as is the case with all other mammalian orders, a huge gap between the order Cetacea (this order includes all creatures known inclusively as "whales"--, whales, dolphins and porpoises) and any supposed ancestral creatures. (1983, p. iii)

Gish then inserts a comforting passage from Edwin H. Colbert's treatise, Evolution of the Vertebrates, evidently

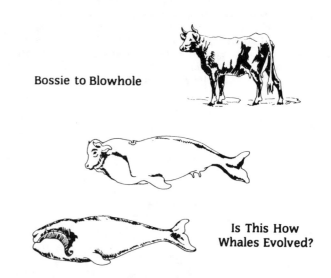

Bossie to Blowhole

Is This How Whales Evolved?

Figure 45.1 This cartoon mimics one thrown on the screen by creationist Dr. Duane Gish in his humorous approach to the problem of ancestry and evolution of the whales. (Drawn by Jody Zanot Miller. From Creation/ Evolution, vol. 3, no. 4, cover illustration. Copyright © 1982 by Jody Zanot Miller. Used by permission.)

to reassure the readers that there could be no such "missing links" or intermediates for the cetaceans. Colbert wrote:

> These mammals must have had an ancient origin, for no intermediate forms are apparent in the fossil record between the whales and the ancestral Cretaceous placentals. Like the bats, the whales (using the term in a general and inclusive sense) appear suddenly in early Tertiary times, fully adapted by profound modifications of the basic mammalian structure for a highly specialized mode of life. Indeed, the whales are even more isolated with relation to other mammals than the bats; they stand quite alone. (1969, p. 336)[1]

Gish repeats this quotation from Colbert in his 1985 revision and update of Evolution: The Fossils Say NO! (1985, p. 80). In the meantime, Colbert's Third Edition, 1980, was carrying a different version of the first sentence of the above quotation. It reads: "These mammals probably arose in early Cenozoic time from primitive carnivore-like ungulates known as mesonychids" (Colbert, 1980, p. 326).[1] In Figure 117, p. 328, of the Third Edition, Colbert shows stages in the possible derivation of the cetacean skull from a terrestrial mesonychid condylarth. Clearly, Colbert is reporting substantial progress in tracing the ancestry of the whale, but we would not be aware of it on simply reading the only quotation Gish gives us.

We will return later to a full report on the finding of the early Eocene fossils remains of a transitional form between land mammals and whales, but first it will be helpful to list those morphological characteristics of modern cetaceans that are unique. We will then look back in time to the earliest of the true cetaceans for signs of some intermediate characteristics suggesting less than complete adaptation to the marine environment.

Cetaceans show some of the same adaptive features as do the extinct marine reptiles, especially the ichthyosaurs. The body became beautifully streamlined with no suggestion of a neck. The skin became bare and smooth; the mammalian hair disappeared. A fishlike tail appeared and provided the forward thrust for swimming, while a dorsal fin appeared to serve as a stabilizer. The hind limbs disappeared, but the front limbs became flippers serving as paddles. The lungs were enlarged and improved for air storage and the nostrils shifted to the top and back of the skull to become the blowhole. In this change, the olfactory system was lost, but the auditory apparatus was changed and improved to receive underwater sound waves. The jaws became greatly enlarged and the brain was pushed far back in the skull.

The modern cetaceans are carnivores; they fall into one of two suborders, depending on the kind of food consumed. One suborder consists of the toothed whales, or Odontoceti, which are primarily fish feeders. The odontocetes include the porpoises and dolphins; these small toothed cetaceans go under the collective name of delphinids. Also included are the killer whale (blackfish), some river dolphins, and the narwhals. The other modern suborder, feeding upon plankton, consists of the whalebone whales, or Mysticiti; their mouths contain transverse plates of fibrous keratin (baleen) that serve to strain the plankton from the seawater.

A third suborder, now extinct, flourished in the Eocene Epoch; they are the Archaeoceti, or archaic whales. It is to the archaeocetes that we must look for information on the ancestry of the cetaceans. The odontocetes and mysticetes did not appear until late Pliocene or earliest Miocene time. The absurdity of the creationists' cartoon showing a modern dairy cow being transformed into a modern whale lies in the false notion that modern, living genera of whales could have evolved recently from a modern ungulate genus. No such relationship was ever in the minds of evolutionists for the history of any living mammalian orders. The laugh in this case is at the expense of the creationists--if they really believe what they are saying.

Let us now look back in time as far as possible to locate the earliest cetaceans. Until the early 1980s, information on whale ancestry was limited to fossil archaeocetes of middle and upper Eocene age. These were small, toothed creatures, comparable in size to the modern porpoises, and already fully adapted to the marine environment. The nostrils had migrated part way back on the top of the skull, but the skull itself and the teeth resembled those of a primitive carnivore (Romer, 1966, pp. 297-99). The skull had a long, low braincase; the forward part of the skull was narrow and pointed with peglike front teeth, as shown in Figure 45.2. The cheek teeth resembled those of the primitive placental mammals.

The body of the primitive cetaceans showed significant differences from that of the modern whales. An example is Basilosaurus (Zeuglodon), an archaic Eocene cetacean, whose skeleton is shown in Figure 45.3. This creature was long and slender--up to 20 m overall length--with a tail section over triple the length of the abdominal section, or ribcase. The hind limbs had been reduced to mere vestiges that no longer protruded from the body; only the forelimbs remained, converted to paddles. In the restoration of the body, shown in Figure 45.4, Basilosaurus is given horizontal tail flukes, but no dorsal fin. These archaic cetaceans became extinct at the close of the Eocene, and there is no alternative but to regard them as ancestral to the odontocetes and mysticites; in every sense of the word they were intermediate between the latter suborders and land-dwelling placental mammals. Creationists simply ignore the archaic cetaceans. Instead,

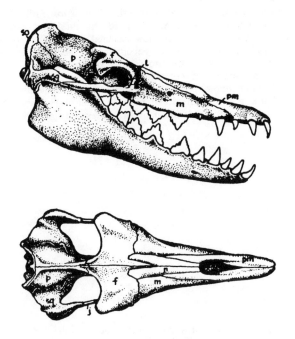

Figure 45.2 Restoration of the skull of an Eocene archaeocete, Prozeuglodon atrox. Length of the skull is about 60 cm. (From Vertebrate Paleontology, 3d ed., by Alfred S. Romer, University of Chicago Press, p. 298, Figure 431. Copyright © 1933, 1945, and 1966 by The University of Chicago. Reprinted by permission.)

---

[1]From Evolution of the Vertebrates, First edition 1955, Second edition 1969, Third edition 1980, by Edwin H. Colbert, John Wiley & Sons, Inc., New York. Copyright © 1955, 1969, 1980 by John Wiley & Sons, Inc. Reprinted by permission.

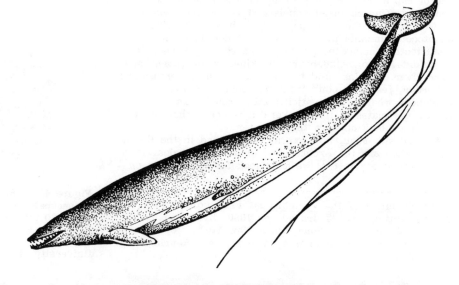

Figure 45.3 Restoration of the skeleton of an archaic Eocene cetacean, <u>Basilosaurus</u> (<u>Zeuglodon</u>). Length about 17 m. (From <u>Vertebrate Paleontology</u>, 3d ed., by Alfred S. Romer, University of Chicago Press, p. 299, Figure 432. Copyright © 1933, 1945, and 1966 by The University of Chicago. Reprinted by permission.)

they try to give the impression that whales came into sudden existence as the familiar sperm whale or blue whale. No such impression can be drawn from Colbert's statement, quoted above, which refers to a gap between all cetaceans, archaic forms included, and placental mammals of the Cretaceous Period.

We return, now, to the 1983 discovery of remains of primitive cetaceans even older than those already described. The locality in which they were discovered is of great geologic interest; it is the "crunch" zone of two colliding continents--peninsular India and southern Asia. As these continental plates drew closer together in the early Cenozoic time, the Tethys Sea between them became progressively narrower; ultimately it disappeared. Strata of early Eocene age deposited in shallow marine water and in coastal deltaic and river plains on the Asiatic side of the seaway became squeezed and uplifted to form a series of low, open folds now located in Pakistan, northwest of the Indus River. The continental beds, representing stream deposits, are bright red, colored by oxides of iron (they are generally known as redbeds) and contain numerous fossils of early mammals, including representatives of bats, rats, primates, creodonts (early carnivores), and ancient ungulates (hoofed herbivorous mammals).

In 1977, vertebrate paleontologist Professor Philip Gingerich organized an expedition to the Pakistan redbed locality to attempt to locate new mammal sites. Continued work in 1978 produced many recognizable mammal fossils, but among them some strange fossils belonging to a carnivorous mammal of some kind (Gingerich, 1983, p. 31). These consisted of the rear part of a cranium, a piece of lower jaw with three pointed premolars, and three unattached sharply pointed molars. These fragments are shown in Figure 45.5 in their correct positions in a restoration of a complete skull. To the skilled eyes of the paleontologists, these fossils were clearly recognizable as belonging to a primitive whale; they named it <u>Pakicetus</u>, after its country of origin. For further interpretation of the fossils, Gingerich's own words are better than any I might attempt in paraphrase:

> Finally, study of the teeth of <u>Pakicetus</u> indicates that this early cetacean fed on fishes as did middle Eocene whales, while study of the cranium indicates that <u>Pakicetus</u> had none of the modifications of the skull and hearing apparatus characteristic of later whales and necessary for a fully aquatic existence. The age, environment of deposition, associated fauna, and functional morphology of <u>Pakicetus</u> indicate not only that it is the oldest and most primitive whale yet discovered, but that it is an important transitional form linking Paleocene carnivorous land mammals and later, more advanced marine whales. Whales apparently made the transition from land to sea in the early Eocene when protocetids like <u>Pakicetus</u> entered the shallow epicontinental seas to feed on abundant planktivorous fishes living there. (1983, p. 144)

The technical account of <u>Pakicetus</u> appears in <u>Science</u>, authored by the four paleontologists who studied the fossil: Philip Gingerich and his student, Neil Wells of the University of Michigan, Donald E. Russell of the French National Museum of Natural History in Paris, and S. M. Ibrahim Shah of the Geological Survey of Pakistan (Gingerich et al., 1983). I find two sentences in this report that point to a possible ancestor of <u>Pakicetus</u>:

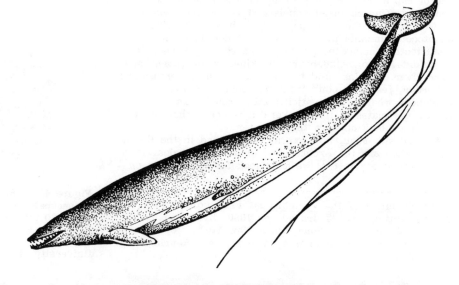

Figure 45.4 An artist's restoration of <u>Basilosaurus</u> (Figure 45.3). (From Edwin H. Colbert, 1980, <u>Evolution of the Vertebrates</u>, 3d ed., John Wiley & Sons, New York, p. 330, Figure 118. Drawn by Lois Darling. Copyright © by Lois M. Darling. Used by permission of the artist.)

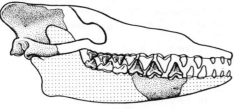

Figure 45.5 Above: An artist's reconstruction of Pakicetus from the early Eocene of Pakistan. Below: Restoration of the skull showing the preserved parts (stipple). (Drawn by Karen Klitz. From P. D. Gingerich, Jour. of Geological Education, vol. 31, p. 140. Copyright ©1983 by the National Association of Geology Teachers. Used by permission.)

Teeth of Pakicetus resemble those of terrestrial mesonychid Condylarthra and are similar to teeth of middle Eocene archeocete Cetacea such as Protocetus and Indocetus. Mesonychids and protocetids are thought to have been carnivorous or piscivorous (fish-eaters), or both. (P. 403)

This sounds formidably technical, but if we can get a handle on the "mesonychid Condylarthra" we may just have found our candidate for an ancestor. Do you suppose it will look like the Borden cow? Search for this candidate takes us to the archaic ungulates. The modern or advanced ungulates are (loosely defined) hoofed mammals that feed on plants. Bossy would be in that group. The primitive ungulates, all extinct, are of other orders (four orders, actually). One order of primitive ungulates is the Order Condylarthra (Romer, 1966, p. 241). Mostly of Paleocene age, they may be the basic stock from which the ungulate orders were derived, but they certainly don't look very much like the ungulates we see around us.

Next, what were the mesonychids like? According to Romer they were a family of primitive placentals showing little in the way of ungulate specialization "and have, in fact, been generally considered in the past to be allied in some fashion to the carnivores" (1966, p. 243).[2] The mesonychids appeared in the Paleocene and continued through the Eocene. An example of an Eocene mesonychid is Mesonyx, whose skeleton is shown in Figure 45.6A. This was a small animal, perhaps 1.5 m long, with wolflike feet having toes like a carnivorous mammal. The skull of Mesonyx was about 30 cm long and rather massive in construction, with mean-looking canine teeth and pointed incisors (Figure 45.6B). No one seeing this animal alive today would mistake it for a dairy cow. Perhaps it would be mistaken for a large wolf or dog. Romer writes: "The habits of these grotesque creatures are difficult to imagine; carrion feeding, mollusk-eating, a diet of some type of tough vegetable matter are among the guesses" (p. 244).[2] If Mesonyx and its Paleocene ancestors were indeed mollusk-eaters, they would have spent some of their time wading in shallow water of rivers and estuaries, heads deeply submerged as they scraped along in the mud and sand. They could also have enjoyed a meal of fish. From that point on, structural adaptations to an aquatic life would readily follow, and the animals would spend more and more time in deeper saline waters but also return at times to dry land where floodplain and deltaic sediments were accumulating. What a far cry from the ridiculous cartoon scenario used by Duane Gish to draw guffaws and cat-calls from a captive audience of fundamentalist creationists! And, best of all, no possibility whatsoever of "udder failure"!

Gish gives a good account of the Pakicetus fossils and Gingerich's appraisal of them but, of course, he is not happy with the paleontologists' conclusions (1983, p. iv). He points out that no fragment of the postcranial skeleton has been found, so we have no idea what the creature's body looked like. He considers the claim that a missing link has been found to be premature. He suggests that Pakicetus will eventually become another of the debunked "missing links," such as Piltdown Man, Nebraska Man, and others. No doubt other persons, including open-minded science students, will share Gish's reluctance to partake of the paleontologists' confidence in interpreting the few fossil fragments thus far available of this possible intermediate genus. But searches will continue for more and better fossils from these same strata, and perhaps from other strata of the same general age elsewhere in the region of the closing Tethys seaway. The book is not closed on this subject.

That cetaceans are living animals makes possible a great deal of additional and independent evidence of their ancestry, through anatomy, embryology, and the newly developed methods of biochemical phylogeny. That the cetaceans are correctly classed as placental mammals seems not to have been questioned by creationists. Their biology textbook, Biology: A Search for Order in Complexity, presents all the mammals in completely conventional terms, cetaceans included (Moore and Slusher, 1970, p. 250). They do not, however, call attention to a remarkable observation that relates the whales to land mammals. In the fetal stage the mysticete whales develop tooth buds, but these are resorbed before they can erupt through the gums (Awbrey and Thwaites, 1981, p. 69). This embryonic development clearly relates the mysticetes to earlier toothed whales and thus to land mammals. The same embryos also have a coat of hair, but this is lost before birth. Here, again is a strong indication of mammalian affinity.

[2]From Vertebrate Paleontology, Third edition, by Alfred S. Romer, University of Chicago Press. Copyright © 1933, 1945, and 1966 by The University of Chicago. Used by permission.

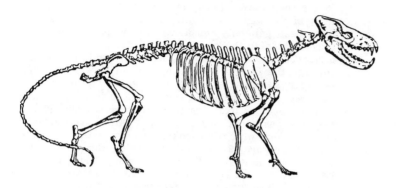

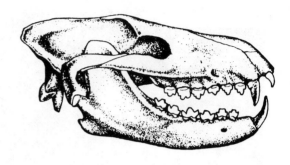

Figure 45.6 Reconstruction of the skeleton and skull of _Mesonyx_, a condylarth of Eocene age. Skeleton about 2 m long; skull about 28 cm long. (From _Vertebrate Paleontology_, 3d ed., by Alfred S. Romer, University of Chicago Press, p. 244, Figure 352; p. 242, Figure 349. Copyright © 1933, 1945, and 1966 by The University of Chicago. Reprinted by permission.)

Vestigial Organs and Their Significance

We turn next to the subject of vestigial organs and their significance. Most of us, if asked to give an example of a vestigial organ, would name the human vermiform appendix or the tonsils. Many years ago, no value was seen in either of these organs, and because they were troublemakers, they were commonly removed in anticipation. They are not, however, acceptable under the title of vestigial organ as the evolutionists define the term. One reason is that the appendix and tonsils occur in nearly all human individuals, whereas a true vestigial organ occurs infrequently or rarely. This requirement was stressed by a Russian zoologist, Alexy Yablokov, whose 1966 book, _Variability of Mammals_, discusses vestigial organs, including those found in whales. Yablokov stresses the point that a true vestigial organ is the expression of a vestigial function, i.e., some useful property of the former structure or organ that is no longer fulfilled. Thus a vestigial structure neither helps nor hinders its possessor in terms of survival; it is usually atrophied--shrunken and simplified in comparison with its earlier useful form.

Evolutionary biologists interpret vestigial structures as strong supporting evidence for evolution. Leading geneticist Theodosius Dobzhansky (now deceased) wrote in 1955: "There is, indeed, no doubt that vestigial rudimentary organs silently proclaim the fact of evolution." Creationists disagree. The nature of this disagreement has been explored by Ernest C. Conrad in an article devoted to the meaning of vestigial structures in cetaceans (1983). Conrad quotes a statement by creationist advocate Robert E. Kofahl from his 1977 book with the intriguing title of _Handy Dandy Evolution Refuter_. Kofahl writes:

> [1] Advancing knowledge and physiology has shown that most of the supposed vestigial organs are useful and even essential. [2] If there are any true vestigial organs, they show loss of structure and design, not the production of something new. [3] But to support the theory of evolution, evidence for the production of new organs is required. (1977, p. 102)

(Note: Numbers added to assist in discussion.)

Point one can be dismissed in the light of the definition of a vestigial organ. The human tonsils may have indeed been found useful, but because everyone has

tonsils, they can scarcely be called vestigial organs. Evolutionists are referring to organs or structures not useful in the present owner.

Point two, as Conrad explains (pp. 9-10), would be quite acceptable if the word "design" were deleted. (One might also simply substitute "function" for "design.") Most certainly, a vestigial structure is not something new, since "new" and "vestigial" are contradictory and mutually exclusive words. "Design" is a creationist concept; it invokes Divine Creation of an organism and its parts for God's purpose, which is God's secret and need not be explained or justified to humans.

Point three is a total misconstruction of the prevailing statement of the theory of descent with modification. Evolution is seen only in retrospect (looking backward in time); its future course cannot be predicted, because it is controlled by chance variation under an extremely complex set of environmental controls. Thus, it would be absurd to interpret a useless and nonfunctioning organ or structure as the precursor to an unknown new organ, not as yet needed. Evolutionary changes make use of existing functional structures, modifying those structures to perform a new function.

So, we now take a look at a set of strange vestigial bones in the modern sperm whale. The pelvic girdle of the sperm whale is a single elongate bone, presently used to support the reproductive organs, and in that sense is not vestigial. In reference to its former function of supporting the bones and muscles of hind limbs, it can be regarded as vestigial. However, every sperm whale has a pelvic bone. The true vestigial bones are rarely present in the sperm whale; they are remnants of leg bones. The occurrence of these relict leg bones has been studied in detail by Yablokov (1966), who describes five kinds of occurrences; these are pictured in Figure 45.7. A single extra bone may be found close to the pelvic bone (B); it is a remnant of the former femur. Additional bones, representing the remnants of other leg bones, ankle bones, and toes, may be present (C through F).

Creationists accept the presence of these unused extra bones, saying they are degenerative structures-- degenerating in accord with the second law of thermodynamics. This particular interpretation is attributed to creation scientist Dr. Gary Parker, biologist on the faculty of the Christian Heritage College (Awbrey and Thwaites, 1981, p. 69). Degenerating from what, we ask? If the whale is a unique, created kind, it must have been deliberately created with a useless or defective organ of some kind. Why would an intelligent and loving God make such an unsatisfactory arrangement in the first place? Obviously, the sperm whales' vestigial limb bones are not to be expected under the creationist hypothesis. They are, however, expected consequences of the evolutionary hypothesis.

On rare occasions, the genes for creating hind limbs, retained by the sperm whale but normally inactive, are reactivated to generate a large part of the lost limb. This event also occurs in horses to produce on rare occasions

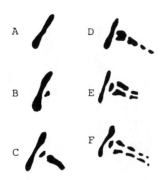

Figure 45.7 Drawings of the pelvic bone and vestigial hind leg bones of the sperm whale. The pelvic bone alone is shown in A. (After Yablokov, 1966, as shown in F. Awbrey and W. Thwaites, 1981, Creation vs. Evolution, Aztec Lecture Notes, San Diego State Univ., p. 69.)

an extra toe, or toes. It occurs in horses when one of the side splints that is a vestigal structure is activated to grow into an extra toe. The phenomenon is called "polydactyly by atavism" (see Gould, 1983, pp. 177-79). Well-documented cases are on record of finding protuberances on the body of a whale that prove to be formed by a femur, tibia, or metatarsal in combination with cartilage forms of other limb bones. Details of several such findings are given by Conrad (1982, pp. 11-13). These remarkable but very rare occurrences strongly support the evolution of the whale from a four-legged terrestrial mammal.

Application of molecular biology to phylogeny of the whales has been discussed by Matthew Landau (1983, pp. 16-17). We have already gone over the principles involved in protein phylogenies, with one example being the amino acid sequences of cytochrome c proteins from 20 different species of organisms (see Figure 36.2). The sixth species listed is the grey whale, for which the amino acid sequence is different in only two positions from that of the pig/bovine/sheep sequence, representing the artiodactyls. The two sequences are exactly alike in 102 of the 104 acids. The phylogenetic relationship is obviously very close. A second comparison is based on the amino sequences of myoglobin (Ayala, 1977, p. 305, Figure 9-21). The sperm whale, dolphin, and porpoise-- grouped together--are separated in the cladogram from the bovine and sheep by only one animal, the seal (Figure 45.8). Again, the phylogenetic relationship is a close one. Thus the data of protein phylogenies are strongly supportive of the hypothesis that the cetaceans and artiodactyls arose from a common Paleocene ancestor.

Horses' Hooves and Tapirs' Toes

It almost seems that the better the fossil record of descent with modification in an animal group, the louder are the creationists' claims that the record is inadequate. Perhaps this perverse relationship of quality of record to level of complaint is only a manifestation of Gish's Law: the more the intermediates, the more the gaps. When it comes to the evolution of the horses (the equids) through the Cenozoic Era, Dr. Gish's complaints gush forth. In his 1980 article "The Origin of the Mammals," in ICR Impact Series, No. 87, the attack is leveled at a familiar museum exhibit and textbook illustration--the stages in evolution of the skull and limb of the horse, from the terrier-sized "dawn horse" of the Eocene Epoch to the extant Dobbin.

Figure 45.9 is a typical exhibit of this kind. The little dawn horse (formerly called Eohippus, but now correctly called Hyracotherium) had three toes in the hind foot, but four in the forefoot. As time passed, a single toe took

over the support of the leg, developing a hoof, while the other toes shrank in size, receding up the limb to become mere splints alongside the metapodial bone called the cannon. Figure 45.10 shows the foot bones in a front view, giving a better idea of the recession and disappearance of the side toes. We also see the great change in dentition as the small molars of Hyracotherium are modified into long molars requiring a lot of vertical jaw space to hold them. Presumably, this dental change reflected the change in diet from soft, green forest foliage to tough, dry grass stems of the prairie. The molars of the full-fledged horse, Equus, are continually ground away, and the teeth move surfaceward during the individual's life to keep the grinding surfaces in a functioning position.

Clearly, the exhibit of horse evolution, as thus presented, gives the impression of a simple linear progression of new forms with time. Earlier I used the analogy of the growth of an unbranching bamboo stalk in the context of presenting the alternative evolutionary model, which is like a shrub or bush, with many forks at all levels. The creationists, as in previous situations, want to be shown transitional forms between each of the four stages shown in the diagram. Gish says:

> To us the family tree of the horse appears to be merely a scenario put together from non-equivalent parts. Nowhere, for example, are there intermediate forms documenting transition from a non-horse ancestor (supposedly a condylarth) with five toes on each foot, to Hyracotherium with four toes on the front foot and three on the rear. Neither are there transitional forms between the four-toed Hyracotherium and the three-toed Myohippus, or between the latter, equipped with browsing teeth, and the three-toed Merychippus, equipped with high-crowned grazing teeth. Finally, the one-toed grazers, such as Equus, appear abruptly with no intermediates showing gradual evolution from the three-toed grazers. (1980, p. iv)

Gish goes on to quote phrases from a modern textbook of evolution (Birdsell, 1975) stressing the rapidity and

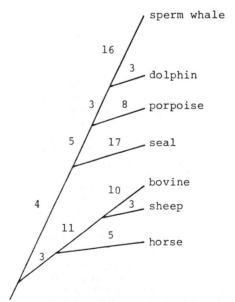

Figure 45.8 A portion of the phylogeny of the globin genes for numerous animals, based on the amino acid sequences in 55 globins. Figures tell the numbers of nucleotide substitutions that have occurred during evolution. (Data of Goodman, 1976, as presented in Dobzhansky et al., 1977, Evolution, W. H. Freeman & Company, p. 304, Figure 9-21.)

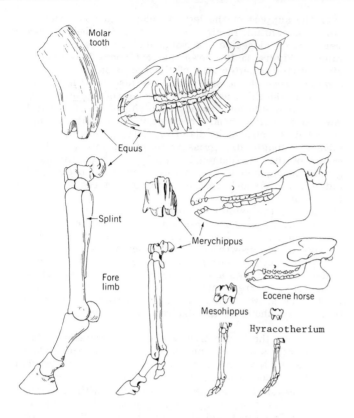

Figure 45.9 Evolution of the horse forelimb, skull, and molar teeth, beginning with Hyracotherium in the Eocene Epoch and ending with the modern horse, Equus. Skull of Equus is shown as cut away to expose the molars. (Forelimbs redrawn from W. D. Matthew, courtesy of the American Museum of Natural History; teeth sketched from photographs prepared by Carl O. Dunbar from collections of the Yale Peabody Museum; skulls redrawn from Edwin H. Colbert, 1980, Evolution of the Vertebrates, 3d ed., John Wiley & Sons, New York, p. 381, Figure 135. Copyright © 1955, 1969, 1980 by John Wiley & Sons, Inc. Reprinted by permission.)

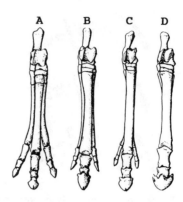

Figure 45.10 Hind feet of the horses (Equidae). A. Hyracotherium (Lower Eocene). B. Miohippus (Oligocene). C. Merychippus (Late Miocene). D. Equus (Recent). (From Vertebrate Paleontology, 3d ed., by Alfred S. Romer, University of Chicago Press, p. 263, Figure 383. Copyright © 1933, 1945, and 1966 by The University of Chicago. Reprinted by permission.)

It should also be noted that in the Rattlesnake Formation of the John Day Country of northeastern Oregon, the three-toed horse Neohipparion is found with the one-toed horse, Pliohippus (S. Nevins, 1974). No transitional forms between the two are found. In other cases "primitive" species of a genus, such as those of Merychippus, are found in geological formations supposedly younger than those containing "advanced" species (J. B. Birdsell, 1975). (1980, p. v)

(See Gish for references.)

Not to be concerned! We need not question Gish's statements because the actual fossil record of horse ancestors--of the Equidae, that is--through Cenozoic time shows a branching evolutionary pattern. Colbert lists the progressive trends that characterized the evolution of horses through the Cenozoic Era, eleven specific points in all, but then he says:

> However, when the horses are considered in their entirety no such picture of uniform evolution emerges. The horses are often cited as an outstanding example of "straight-line evolution" or of "orthogenesis," and it is frequently maintained that these animals evolved with little deviation along a straight path from the little Eocene Hyracotherium or eohippus to the modern horse, Equus. It is true that most of the progressive changes listed above can be followed through time from Hyracotherium to the modern horses, but in middle and late Tertiary times there were various lateral branches of horses that were progressive in some features and conservative in others. When all fossils are taken into account the history of horses in North America is seen to be anything but a simple progression along a single line of development. (1980, pp. 379-83)[1]

Colbert explains that the single-line development actually occurred only at the beginning of their development, from lower Eocene to middle and upper Oligocene (p. 380). This development takes us from Hyracotherium in lower Eocene to Miohippus in upper Oligocene. But by the end of the Oligocene Epoch, this single-line evolution ended: "With the advent of Miocene times there was a branching out of horses on several lines of development, probably as a response to an increase in

abruptness of the changes in the foot structures, in keeping with the current concept of punctuated equilibrium. For evolutionists, this mode of evolution explains why transitional sequences are rarely found in the fossil record. Not only are the changes rapid, but they also occur in small, isolated populations. Gish comments that the continuity demanded by theory cannot be documented from the fossil record. It is not clear whether Gish accepts the reasoning for scarcity of transitional forms, or whether he continues to insist on intermediates, despite the rapidity of changes. But then Gish moves to a different line of attack.

Under the linear, or bamboo-stalk, model of evolution adhered to by the creationists, an "advanced" genus of the evolutionary line cannot be found in strata of the same age as an "archaic" or "primitive" genus. This prohibition follows from the creationists' assumption that as each new genus arises, the one from which it evolved must immediately become extinct. Turning from bamboo stalks to mothers, this is like saying that every human mother must die in childbirth, making it impossible for any mother to be alive at the same time as the child. Such a rule is utter nonsense, of course, whether applied to the chain of life or to an evolutionary succession of genera or species. Nonsense not being denied to humans in the course of an argument, the creationists persist in using it. Gish performs the nonsense maneuver with these statements:

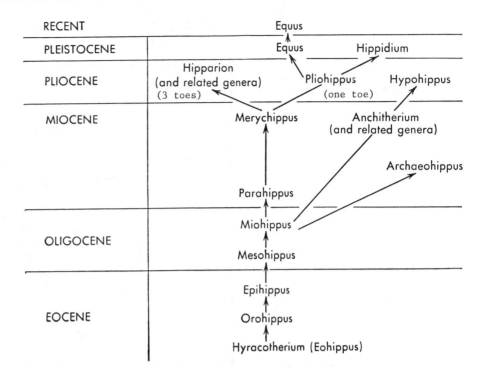

Figure 45.11 Evolution of the horses (Equidae) through the Cenozoic Era. (From Edwin H. Colbert, 1980, <u>Evolution</u> <u>of</u> <u>the</u> <u>Vertebrates</u>, 3d ed., John Wiley & Sons, New York, p. 383. Copyright © 1955, 1969, 1980 by John Wiley & Sons, Inc. Reprinted by permission.)

the variety of environments available to them, and especially because of the spread of early grasses and other flowering ground plants" (1980, pp. 381-82).[1] Colbert goes on to say that <u>Archaeohippus</u>, of Miocene age, remained conservative, showing very little further development of the skull, teeth, and feet beyond the <u>Miohippus</u> stage. He then describes another line of horses, the Miocene and Pliocene anchitheres, that, although eventually reaching the size of the modern horse, "retained the conservative <u>Miohippus</u>-like teeth and functional three-toed feet" (p. 382).[1] Colbert speculates that the anchitheres were forest-living horses, browsing in the deep woodlands (p. 382). Under such conditions they would not have needed to run swiftly over great distances on hard ground, nor would they have needed the huge molars of <u>Equus</u>, used to masticate siliceous grasses. The maintenance of conservative characteristics easily explains why fossils of a three-toed horse would be found in the same strata with those that had progressed to use of a single toe.

Colbert has yet another example of branching lines of horses. At the close of the Miocene, <u>Merychippus</u> gave rise to two groups of horses. One, the hipparions, retained the three-toed feet, while the other, centered on <u>Pliohippus</u>, progressed to become a single-toed equid. So here is another case where three-toed horses could share the same stratigraphic horizon with those having one toe. Figure 45.11 is a table reproduced from Colbert in which you can see the branching pattern of evolution we have been discussing (1980, p. 383).

Gish again performs the nonsense maneuver, but this time in a bizarre setting--among the South American ungulates, which evolved in isolation from the other lands of the earth from early in Cenozoic time to the Pleistocene (1980, pp. iv-v). (See Chapter 32.) From Romer, Gish reproduces an illustration of the hind feet of three of these ungulates, shown here as Figure 45.12 (Romer, 1966, p. 260). All three of the genera belong to the ungulate order of litopterns (Litopterna). From left to right we see <u>Macrauchenia</u>, with three toes; <u>Diadiaphorus</u>, with a single hoofed toe and only the shrunken remnants of two side toes; and <u>Thoatherium</u> in which even the vestiges of the side toes are almost invisible. About this seeming horselike evolutionary pattern, Gish has this to say:

Do they not thus provide another nice, logical evolutionary series? No, not at all, for they do not occur in this sequence at all! <u>Diadiaphorus</u>, the three-toed ungulate with reduced lateral toes, and <u>Thoatherium</u>, the one-toed ungulate, were contemporaries in the Miocene epoch. <u>Macrauchenia</u>, with pes (hind foot) containing three full-sized toes, is not found until the Pliocene epoch, which followed the Miocene according to the geological column. In fact, it is said that the one-toed <u>Thoatherium</u> became extinct in the Miocene before the three-toed <u>Macrauchenia</u> made his (sic) appearance in the Pliocene. (1980, p. v)

(Note: Words in parentheses added by the author.)

Gish senses a kill here. His audience is already convinced he has caught the evolutionists in a checkmate.

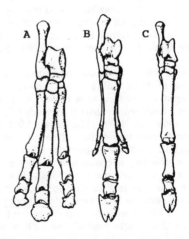

Figure 45.12 The hind-foot bones of litopterns. A. <u>Macrauchenia</u>. B. <u>Diadiaphorus</u>. C. <u>Thoatherium</u>. (From <u>Vertebrate</u> <u>Paleontology</u>, 3d ed., by Alfred S. Romer, University of Chicago Press, p. 260, Figure 378. Copyright © 1933, 1945, and 1966 by The University of Chicago. Reprinted by permission.)

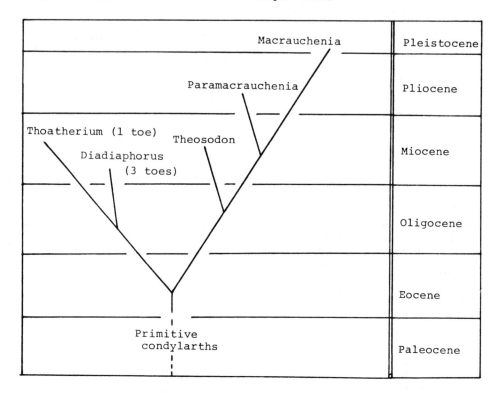

PROTOTHERES
(horse-like)

MACRAUCHENIDS
(tapir-like)

Figure 45.13 Phylogenetic diagram of the litopterns, an order of South American ungulates. (Based on data of A. S. Romer, 1966, and E. H. Colbert, 1980.)

There remains only the coup de grace and perhaps one or two more thrusts besides--just for the fun of it:

> Thus, if evolutionists would permit the fossil evidence and their usual assumptions concerning geological time to be their guide, they should suppose that in South America a one-toed ungulate gave rise to a three-toed ungulate with reduced lateral toes, which then gave rise to an ungulate with three full-sized toes. This is precisely the opposite of the supposed sequence of events that occurred with North American horses. I don't know any evolutionist who suggests such an evolutionary sequence of events, but why not? Perhaps it is because the three-toed to one-toed sequence for North American horses became so popularized in evolutionary circles that no one dare suggest the reverse transition. Of course there is no more real evidence for transitional forms in South America than there is in North America. (1980, p. v)

Why not accept this "reversed" evolutionary sequence? A careful reading of both Colbert and Romer discloses a perfectly good answer; it lies in a closer look at the phylogeny of the litopterns, clearly revealed in the fossil record (Romer, 1966, p. 260). The three hind feet shown in the figure come from two quite different families of litopterns; they had branched apart in Oligocene or earlier time after sharing a common ancestry in the primitive condylarths of Eocene time. Figure 45.13 shows the total picture. At the left is the horselike branch, called the prototheres; to it belong Diadiaphorus with a big center toe and two recessive side toes and Thoatherium with just the single hoofed toe. They were contemporaries in the Miocene Epoch. In Thoatherium, says Romer, "we have an ungulate, about the size of Mesohippus, which seems more horselike than any true horse, for it was single toed with splints more reduced than those of modern equids" (p. 260).[2] But Romer adds a very important observation--that this "pseudohorse" was unprogressive in other respects; for example, it had low-

crowned cheek teeth and the carpus (set of wrist bones) was not well suited to running on one toe. That a three-toed genus and a one-toed genus should be contemporaries is no more unusual, say, than the existence today of two related perissodactyls, the one-toed horse and the tapir (four toes up front, three in the rear). Somewhere back in the Eocene, horses and tapirs had a common ancestor, but the tapir retained much of its primitive character while the horse line changed dramatically. But let's get on with the litopterns.

The second family of litopterns consists of the tapir-like macrauchenids, shown as the righthand branch in Figure 45.13. They had representatives in Miocene time (Theosodon), as well as in Pliocene time (Promacrauchenia). Macrauchenia, living in the Pleistocene, was a very strange animal indeed. The skull, pictured in Figure 45.14, featured a nasal opening very far back on the top of the skull, suggesting that (like the modern hippo) it may have spent most of its time submerged in the fresh water of swamps and rivers. Its proportions were camel-like, as the reconstruction in Figure 45.15 shows, and some paleontologists think it may have had a short proboscis. In any case, the snout was long and narrow. Like the tapirs, it had three functional toes.

The moral to this story of the South American ungulates is the same as for the phylogeny of any group of fossils. Their relative age is established by their stratigraphic position, not by arranging them in some supposed or assumed sequence of bodily features, as the creationists would like to have us do. The principle of stratigrapic succession always has priority, because sediment particles along with all material objects respect the law of gravitation in the earth's surface zone.

Gish devotes four paragraphs to attempting to answer his own question: "Was Hyracotherium (Eohippus) really a horse?" (1980, p. v) His answer, like that of mainstream paleontologists, is "No," so one wonders why there should be an argument. Gish uses quotations from G. A. Kerkut and G. G. Simpson to show that the position of Hyracotherium as ancestor to the next in line of the horse chain is questionable (p. vi). Gish reasons that if

Figure 45.14 Restoration of the skull of the litoptern Macrauchenia, length about 45 cm. Lower figure is view from beneath. (From Vertebrate Paleontology, 3d ed., by Alfred S. Romer, University of Chicago Press, p. 260, Figure 381. Copyright © 1933, 1945, and 1966 by The University of Chicago. Reprinted by permission.)

Figure 45.15 A restoration of the litoptern Macrauchenia. (From Edwin H. Colbert, 1980, Evolution of the Vertebrates, 3d ed., John Wiley & Sons, New York, p. 361. Drawn by Lois Darling. Copyright © by Lois M. Darling. Used by permission of the artist.)

Hyracotherium is not a true horse, then it should not be placed in the bottom position of the horse line of evolution. That much seems reasonable enough, but Gish goes on to draw a completely erroneous conclusion: if Hyracotherium is incorrectly positioned as a horse ancestor, the entire chain of horse evolution above it is rendered invalid.

In case you do not see why the conclusion is untenable, consider the example of a person living today who is able to trace his or her ancestry back four generations, on the basis of entries in the family Bible backed up by a set of birth certificates. Tracing the ancestry back one more generation is difficult and leads to controversy, because some of the individuals involved were immigrants born abroad and the information on their parentage is largely of the hearsay variety, found in recorded recollections. Nevertheless, the excellent record of ancestry back to a particular point in time stands on its own merits.

Chris McGowan, preparing to answer the creationists' arguments about lack of intermediates in the horse series, addresses himself first to Gish's question in these words:

> The first point that has to be made, one that is seldom recognized, is the various fossils in the series are not horses; only the modern horse, Equus caballus, is a real horse. Nobody would make the mistake of calling a zebra a horse, or an ass a horse, even though they are all members if the same genus, Equus. We should therefore not make the mistake of calling any of the fossils in the series horses. When we look at this evolutionary sequence, then, we are not looking at modifications within the "horse kind," as creationists might refer to them, but at major evolutionary changes within a particular group of hoofed mammals. (1984, pp. 142-43)[3]

McGowan traces in great detail the changes from Hyracotherium through Mesohippus to Merychippus. He concludes:

> As far as the evolution of the foot is concerned, the only stage which appears to be missing is the final stage, from an equid like Merychippus, with two small side toes, to the modern horse, which has no side toes at all. However, when we look closely at the anatomy of the modern horse, and when we

consider its embryonic development, we can see that there is no gap at all. (1984, pp. 143-46)[3]

McGowan then details the evolutionary relationship between the splint bones on the posterior surface of the cannon bone and the metapodial bones of Merychippus that support its side toes. He concludes: "This is compelling evidence for descent with modification, and it is very difficult to rationalize the retention of splint bones in the horse other than by evolution" (1984, p. 147).[3]

According to McGowan the evidence from embryonic development goes back to the 1890s, when an anatomist, J. C. Ewart, studied horse embryos ranging in length from 2 cm to 88 cm (p. 147). At an early stage a small bud appears at the lower end of each of the embryonic splint bones. Upon dissection, the buds appear to be embryonic side toes; they are complete with end caps that can be interpreted as embryonic hooves. Actually, in the most advanced stage of the development of these buds, it is possible to distinguish three individual elements, each corresponding to one of the three bones in the side toes of Merychippus. In a later stage, the entire bud disappears, leaving only the splint. This evidence indicates that the modern horse carries within its structural genes the inherited capacity to produce side toes. Although the process of development of the side toes is initiated in the embryo, regulator genes later switch off the structural genes. But, as we have seen in an earlier section dealing with vestigial organs, on rare occasions the regulator genes fail to act and the horse is born with well-developed side toes.

In conclusion, McGowan places the creationists on the defensive by asking: "Why would the Creator give the horse the capability of developing side toes, a capability always exercised during embryonic development but rarely manifest after birth?" (1984, p. 148)[3] The answer lies in theology rather than in science, so we need not pursue it further.

Bats and Rodents--Who Are Their Ancestors?

Creationists are required by their doctrine of created kinds to insist that transitional fossil forms are absent from every phylum, class, and order within all life kingdoms. Ultimately, creation scientists will be required to present evidence of the lack of transitional forms for every one of those taxa that they recognize as "kinds."

Thus far, they have selected only a few kinds as examples of taxa without transitional forms and hence without discernible ancestry. Of those orders of placental mammals the creationists have cited specifically in this connection there remain to be considered in our review the bats, rodents, lagomorphs, and primates. In this section we take up the first three orders named, leaving the primates for two chapters that include the ancestry of the hominoids.

We look first at the bats, as briefly as possible. As mammals of the order Chiroptera, living today, the physiology and habits of bats are well understood. Romer describes these creatures:

> Only in the bats . . . has true flight been developed by mammals. As in the pterosaurs (and in contrast with birds), the wings are formed by webs of skin; but, instead of their being supported by a single elongate finger, nearly the whole hand is involved. The thumb, a clutching organ, is free and clawed; the other four fingers are all utilized in support of the wing membrane; claws are lost on these fingers (except the second in fruit bats); and, as would be expected, the end phalanges are reduced and may be absent. In having the wing expanse broken by the long digits, the bat has evolved a more flexible and less easily damaged wing than that of the pterosaur. In connection with the bat's habit of hanging by the hind legs, those structures, as well as the pelvis, are peculiarly adapted but rather weak. (1966, p. 212)[2]

As Gish has correctly stated, the earliest known bat fossil, _Paleochiropteryx_, is of Eocene age (1978a, pp. 90-93; 1980, p. iii). Colbert states that "they must have experienced an initial stage of very rapid evolution, because the first known bats of Eocene age . . . were highly developed and not greatly different from their modern relatives. There are no known intermediate stages between bats and insectivores" (1980, p.283).[1] Again, Gish has a reason to gloat:

> Well, here he is--the world's oldest known bat. And what is he? One hundred percent bat! The complete absence of any supposed transitional forms between the bat and his alleged ancestor leaves unanswered, on the basis of evolution, such questions as when, from what, and where, and how did bats originate? (1978a, p. 91)

Fortunately, in this case, bats are living creatures, so that full anatomical and biochemical studies can be applied, as they cannot with the pterosaurs. Surely, cladistic analysis of the anatomical characters of the bats can give strong evidence of the bat's position in mammalian phylogeny. I do not find the bat listed among the various molecular phylogenies that have been calculated, but one report on hybridized DNA analysis lists the hedgehog, an insectivore, which lies close to and just below the mouse, but well above the chicken (Ayala, 1977, p. 279). I suspect that it will be only a matter of time before molecular phylogeny will put the bat in place and then we can see how close it is to the insectivores, which are presently thought to be the ancestral group of both bats and rodents (Romer, 1966, pp. 213, 303).

Gish gives two pages to the rodents under the banner heading "Rodents--the most prolific mammal--produce no evidence of evolution" (1978a, pp. 93-94). He explains:

> The order Rodentia should provide evolutionists with another group of animals ideal for evolutionary studies. In number of species and genera, the rodents exceed all other mammalian orders combined. They flourish under almost all conditions. Surely, if

any group of animals could supply transitional forms, this group could. (P. 93)

Gish goes on to recite passages from Romer stating that transitional forms are unknown for rodents in general and for beavers and porcupines in particular (Romer, 1966). He throws in an added sentence to say the same about the lagomorphs. Gish concludes: "Thus we see that the order Rodentia, which should supply an excellent case for evolution, if evolution really did occur, offers powerful evidence against the evolutionary hypthesis" (p. 94).

Colbert also stresses the obscurity of ancestry of these mammal orders, but he also expresses some interesting thoughts on the reasons for the lack of fossil intermediates:

> Our unsatisfactory knowledge of the fossil rodents is to a considerable degree a result of the general lack of interest among paleontologists in these mammals. In the earlier collections of fossil mammals rodents are comparatively rare, since the inconspicuous remains of these small mammals were frequently overlooked by collectors who concentrated on the larger animals. Moreover, large mammals have been given priority during past years in collecting programs for the purposes of impressive museum displays. (1980, p. 313)[1]

Dr. Colbert should know what he is talking about after a long career as Curator of Vertebrate Paleontology in the American Museum of Natural History in New York City. How could rodents, rabbits, insectivores and even early primates compete with dinosaurs and sabre-tooth cats for public attention and benevolent financial support? Colbert notes, however, that paleontologists have been turning in increasing numbers to the study of the rodents and that in years to come we may expect to have much more information about their ancestry.

That Colbert's prediction may be nearing fulfillment is suggested in a _Science News_ article by B. Bower describing some of the newer finds being made in the Wind River Basin of Wyoming by two paleontologists of the Carnegie Museum of Natural History, Leonard Krishtalka and Richard Stucky (Bower, 1984, p. 213). From several quarries in a rich locality in early Eocene strata, a summer's work yielded up fossil bones and eggs of 65 species of mammals, lizards, and frogs. The report says:

> Most of the remains come from small mammals such as primates and rodents. Several of the species are "new to science," says Krishtalka. These include shrew-like animals related to modern hedgehogs, and bat-like animals related to modern bats.

> However they got there, the bones fill in a crucial gap in the fossil record, he explains. They may add significant evidence to the scientific debate about whether evolution occurred gradually or proceeded in rapid bursts followed by long periods of little change.

Study of fossil specimens such as these takes a great deal of time. For every negative pronouncement by the creationists--taking only a few seconds to make and no time whatsoever in research on fossils--a single positive statement by mainstream paleontologists about the ancestry of the rodents, or bats, or what-have-you, may require hundreds or thousands of hours of field and laboratory study. Gish's book title _Evolution? The Fossils Say NO!_ is an absurdity. It is the creationists, not the fossils, who say "NO"! The fossils say "YES"!

Credits

Chapter 46

Mass Extinctions in the Geologic Record

In our earlier review of stratigraphy and the fossil record (Chapter 32) we took note of a number of points in Phanerozoic history at which extinctions of entire taxa at the levels of genus, family, order, and class occurred in a seemingly very short time interval, and typically at the close of an era or a period. The reason for the latter coincidence is simple enough: the eras, periods, and sometimes epochs (as in the Cenozoic) were defined on the basis of extinctions followed quickly by radiations.

The Major Mass Extinctions

Lists of the major extinctions differ from author to author, and we shall not attempt one here. Important mass extinctions of marine life occurred at the close of the Ordovician Period and very late in the Devonian Period (Raup and Sepkoski, 1982). These extinctions are shown in the graph of Figure 46.1 as sharp drops in a curve showing numbers of families of skeletonized marine animals in existence (labeled 1 and 2). These early extinctions occurred, of course, in a time when land animals and plants were either not present or existed in rather limited numbers. Another important marine extinction occurred late in Triassic time (4). A great extinction occurred at the end of Permian time (3), closing the Paleozoic Era, and another at the end of Cretaceous time (5), closing the Mesozoic Era. We will focus on these last two extinctions, both to illustrate what happened and to examine some of the many hypotheses that have been proposed to explain them.

Late Permian extinctions and early Triassic radiations are shown graphically in Figure 46.2 for the higher taxa of selected animal groups, both marine and terrestrial. Each horizontal line represents a single group, usually an order. Several lines end at the Permian/Triassic boundary, showing extinction; several originate at that line, representing radiation of new orders. Note the total extinction of the trilobites (a phylum). Extinctions shown within the bryozoans and brachiopods are for selected groups only. Three orders of stalked echinoderms, including crinoids and blastoids, became extinct. Of the groups shown, only the ceratite cephalopods crossed over the time boundary to persist for any substantial length of time. A recent appraisal of this event by paleontologist J. John Sepkoski, Jr., states that "it stands alone as the single most devastating collapse of the marine ecosystem in the Phanerozoic" (1982, p. 285). He also comments that this extinction was almost certainly not instantaneous but extended over two stages encompassing the final 15 million years of the Permian Period.

Figure 46.3 is a similar diagram for selected orders on either side of the Cretaceous/Tertiary time boundary. Although we have previously used "Cenozoic" in preference to "Tertiary," current literature on this extinction uses the latter name, abbreviating the boundary to "K-T." Particularly striking is the total demise of the ammonites and some other cephalopod orders, and of several reptilian orders including the dinosaurs, marine reptiles, and flying reptiles. Also

striking is the radiation of the several orders of mammals. Notice that the insectivores "jumped the gun," for they were to give rise to all other placental mammalian orders. Not shown in this figure are marine plankton. Several major groups of marine phytoplankton and zooplankton totally disappeared. Most severely affected were plankton of the open ocean surface waters and shallow coastal waters; least affected were those of deep ocean waters (Thierstein, 1982, p. 385). An important casualty (not shown) was the calcareous zooplanktonic group of foraminifera, of which only one species survived from the Cretaceous. It has been estimated that some 75 percent of all Cretaceous species may have become extinct at the K-T boundary (McCartney, 1984, p. 306).

What Caused These Great Extinctions?

Important in considering the cause or causes of great extinctions is the observation that the vascular plants do not show a corresponding history to that of the animals (Newell, 1963, p. 78-79, 81). When the numbers of genera of the three major plant groups--ferns and mosses,

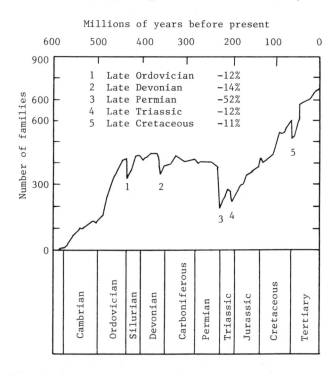

Figure 46.1 Graph of the fluctuation in numbers of families of marine vertebrates and invertebrates through Phanerozoic time. The numbered sharp drops are important extinctions. Percentages show relative drop in numbers of families. (From D. M. Raup and J. J. Sepkoski, Jr., _Science_, vol. 215, p. 1502, Figure 2. Copyright © 1982 by the American Association for the Advancement of Science. Reproduced by permission.)

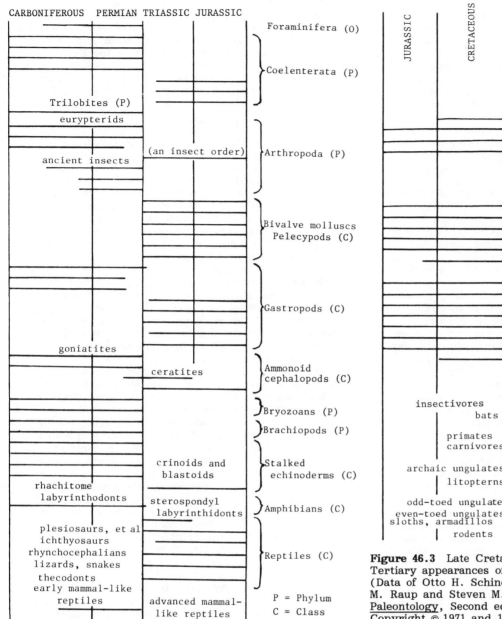

Figure 46.2 Late Permian extinctions and Early Triassic appearances of selected higher taxa of animals. Each horizontal line represents a taxon, usually an order. (Data of Otto H. Schindewolf, 1962. Adapted from David M. Raup and Steven M. Stanley, Principles of Paleontology, Second edition, p. 297, Figure 10-30. Copyright © 1971 and 1978 by W. H. Freeman and Company. Used by permission.)

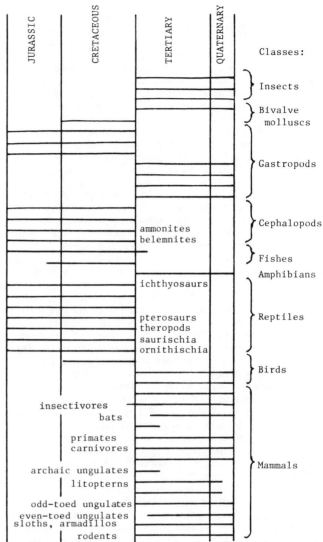

Figure 46.3 Late Cretaceous extinctions and early Tertiary appearances of selected higher taxa of animals. (Data of Otto H. Schindewolf, 1962. Adapted from David M. Raup and Steven M. Stanley, Principles of Paleontology, Second edition, p. 301, Figure 10-32. Copyright © 1971 and 1978 by W. H. Freeman and Company. Used by permission.)

gymnosperms, and angiosperms--are plotted against the time scale from Silurian to the present, no suggestion of an extinction appears at the Permian/Triassic boundary or the K-T boundary. (Refer back to Figure 32.4 showing the evolution of the vascular plants.) One group of gymnosperms, the seed ferns, became extinct at the close of the Jurassic. Note also that the cycads and ginkgoes arose early in the Triassic, immediately after the great Permian/Triassic animal extinction. When we look at the K-T boundary, we find the angiosperms in a condition of rather steady increase in abundance of groups and showing no temporary reversal.

An important animal group that fails to show the effects of either the Permian/Triassic or K-T extinction are the bony fishes (Newell, 1963, p. 85). Why did they

pass unaffected through the same K-T catastrophe that claimed the ammonoids and the marine plankton?

Anyone hardy enough (and perhaps foolish enough, as well) to undertake a complete listing of possible causes of mass extinctions is immediately confronted with a discouraging variety of complications and contradictions. A given single-mechanism explanation may seem to fit well with extinction taking place in a single ecological environment or affecting a single taxonomic group. For example, cooling of the ocean may account nicely for several important extinctions of marine animals and plants (Stanley, 1984), but will that same explanation apply to extinctions of many reptile groups on the lands? Many explanations have been offered for the disappearance of the dinosaurs--for example, that they died off in a great disease epidemic--but the same explanation cannot easily apply to mass extinctions of marine plankton.

How sudden was a great extinction? This is a crucial question because one hypothesis may assume a single catastrophic event in which the extinction is largely accomplished within, say, a few years; another assigns several million years to the event. Is it to be sudden death or slow, lingering death?

Extinction hypotheses that bear the designation of being "catastrophic" can differ in a very important respect. One type begins with a sudden high-energy event, such as the infall of an asteroid or the energy burst of a supernova; this is followed by a succession of consequences, some of which may be delayed or attenuated for a long period of time. The other type begins with some gradual changes, such as slow cooling of the ocean or rifting of continents, the effects of which slowly build and multiply to reach a short-lived climax of catastrophic proportions. You might think of this second program as one of uniformitarianism building up to terminal catastrophism. The general idea is that gradual changes accumulate to a critical point that triggers the extinction event. As an example, suppose that sea level falls globally (perhaps because of a change in rate of spreading apart of lithospheric plates), gradually reducing the width of the continental shelf and causing shallow-water marine animals to be crowded into smaller and smaller coastal strips. Although this worsening condition is tolerated over a long period of time, there comes a critical point at which the situation is intolerable and mass extinction occurs.

The thought of a terrible disaster befalling planet earth from external causes has proved attractive over the years, not only to pseudoscience writers such as Velikovsky, but also to otherwise rather sober-minded scientists. Columbia University geologist James D. Hays in 1971 proposed that extinctions of marine planktonic species have accompanied reversals of polarity of the earth's magnetic field. While studying deep sea cores, Hays found several extinctions of radiolarian species within the record of the past 2.5 m.y., each extinction occurring directly after a magnetic reversal (Hays, 1971). It was not a direct effect of the field reversal, but an indirect one that had been suggested as early as 1963. At the time that the magnetic intensity fell to near zero, the earth's external magnetic field, or magnetosphere, would have virtually disappeared, and with it the shield it provided from high-energy cosmic radiation from outer space. This ionizing radiation would have damaged the genetic materials of exposed life forms, causing a high frequency of mutations, leading to extinction of some species, but increasing the pace of evolution of new species as well. The suggestion quickly met with strong opposition from biologists, who felt they could show that the increased radiation effects would be too small to cause extinctions (Purrett, 1971; Eberhart, 1976). The question was kicked around pro and con for some years afterward, but has rarely been revived in the last decade and may have been put to rest (Plotnick, 1980). Also discussed in the early 1970s was the suggestion that enormous energy fluxes received during reversal times from supernova outbursts might have resulted in drastically lowered surface temperatures that caused mass extinctions (Ruderman, 1974).

Catastrophic Impact Hypotheses

The current cataclysmic "quick fix" hit the scientific community in May, 1979, when geologist Walter Alvarez of the University of California at Berkeley presented to a meeting of the American Geophysical Union a paper reporting a most remarkable discovery. In the thin layer of clay that had previously been zeroed in on by paleontologists as representing the boundary between Cretaceous and Tertiary strata at Gubbio, Italy, there occurs a spectacular jump in the amount of iridium present, as compared with that in strata above and below. Iridium, a noble element, is rare in crustal rocks but is about 10,000 times more abundant in extraterrestrial materials-- meteorites, that is--than in crustal rock. It is also thought to be present in much higher concentrations in the earth's core and could be brought to the surface by

magma rising from very deep in the mantle. Walter Alvarez and his team of research scientists (his father, who is Berkeley nuclear physicist Luis Alvarez; physical chemists Frank Asaro and Helen V. Michel) favored the extraterrestrial origin of the iridium at the K-T boundary. They were uncertain what source to favor but allowed that it could have been a meteoroid or asteroid some 10 km in diameter that impacted the earth, throwing up an enormous cloud of iridium-rich dust. When their paper was published in Science in 1980, it carried a strong positive vote for the asteroid impact hypothesis and additional evidence in the form of similar iridium anomalies found at the K-T boundary in Denmark and New Zealand. In the Denmark sample the iridium level is about 168 times higher than the background level. The iridium and the clay in which it is incorporated were interpreted as the fallout of a great dust cloud that enveloped the earth, blocking out solar radiation for several years, bringing severely cold surface temperatures, and resulting in the direct death of animals, or the cutting off of their food supply because of the cessation of photosynthesis. Marine plankton as well as life on land would have been equally affected. Other writers have suggested that a great heat flash of brief duration immediately followed the impact. It was the heat, they say, that killed the dinosaurs and other animals, rather than severe, prolonged cold (Emiliani, 1980). The scenario is remarkably like that we hear prophesied as the impending nuclear winter.

Introduction of the asteroid impact hypothesis of mass extinction raised a storm of controversy such as has seldom occurred in the earth sciences in our time--even including the heated debate over plate tectonics. I suspect, but dare not say, that the thought of a nuclear physicist and two physical chemists butting in with a powerful mechanism of extinction must have raised a lot of hackles. In my experience I have seen many instances of geophysicists and geochemists turning up novel mechanisms of geologic processes, thereby jolting the complacency of the traditional field geologists. Indeed, without a succession of these jolts we would not now have the theory of plate tectonics as our paradigm.

Objections to the impact hypothesis were numerous and varied. The paleontologists insisted that the extinction of dinosaurs and other groups was gradual, occurring over at least several millions of years. The question is still strongly debated. Why, the paleontologists asked, did the vascular plants show no deleterious effects of the period of darkness and extreme cold? Answer: The plants retained their viability through roots and dormant seeds, quickly reviving when the air cleared and temperatures returned to normal. Why, some asked, have we not found the remains of the great impact crater? Much older impact craters are clearly visible in continental shield rocks. While a few impact craters of about the right age are known, they are judged too small to meet the require- ments. One answer is that the meteorite landed on the deep ocean floor. If it was located on a plate headed toward a subduction boundary, it would have long since been carried down to a location deep in the mantle.

One line of opposition to the meteorite hypothesis was in the form of a suggested alternative source of iridium. Analyses of gases emitted from vents of Hawaiian volcanoes show strikingly large quantities of iridium (see EOS, vol. 65, no. 6, p. 41). Basalt lava of this island chain is thought to originate in a mantle plume that may tap the lower mantle. Could the iridium in the clay layers have originated as concentrations from volcanic dust emitted in brief periods of exceptionally great volcanic activity, perhaps related to periods of stepped-up plate spreading? This alternative was put forward by Charles B. Officer and Charles L. Drake, both of whom were heavily involved in the geophysical research of the early ocean-floor discovery period in which plate tectonics was born. These authors claim that the iridium anomaly was formed over a time interval spanning some 10,000 to 100,000 years, ample time for deep volcanic sources to emit the iridium as volcanic dust (Officer and Drake, 1984). In support of this hypothesis they cite the finding in collected samples of airborne particles erupted by the Hawaiian volcano Kilauea of an enrichment in iridium by a factor of 10,000 to 20,000 over that found in normal basalt rock. They point to the eruption of enormous quantities of basalt at about the time of the K-T boundary (-65 m.y.); these are the Deccan traps of peninsular India. The form of volcanic activity involved here is of the hot-spot type, developed over a deep mantle plume.

We cannot follow this debate further, but you have a sample of the way mainstream science works when a new suggestion stirs the brains of the scientific community it impacts. Nor do we have space to give equal treatment to a current hypothesis of more gradual extinction, which began with climatic change involving a temperature drop capable of leading to basic changes in the processing of atmospheric carbon dioxide and a depletion of calcium carbonate in the ocean (see McCartney, 1984, p. 307).

We also lack space to present the most outrageous catastrophic extinction scenario being debated currently by mainstream scientists--including the astronomers, who have not been able to resist getting into the act. I refer to the suggestion that periodic extinctions can be attributed to occasional periods of bombardment by comets dislodged from the distant Öort region, where comets live peaceably most of the time. The villain is Nemesis, a ghostly companion star to our Sun. Nemesis orbits beyond the cometary region but at intervals sweeps in close to that region, causing some comets to break free and enter orbits that bring them into collision with the inner planets. As I write this, astronomers are busy plotting the necessary orbits and trying to fit the episodes of cometary dislodgement into what may be (or may not be) a regularity of occurrence of extinctions since earliest Phanerozoic time (Simon, 1984b; Weisburd, 1984). Other astronomical hypotheses for periodic extinctions from cometary impacts are also being considered. One version relates the extinction periodicity to the passage of the Solar System through interstellar clouds of gas and dust. Not without reason, this explanantion has been dubbed the "Milky Way" hypothesis (Simon, 1984a).

In closing this brief review of catastrophic hypotheses, I think you will enjoy a philosophical note by vertebrate paleontologist Paul E. Olson, a professor in Columbia University who has participated in the discovery of rich assemblages of reptilian fossils of late Triassic time in strata in Nova Scotia. Here the patterns of extinction appear to fit a catastrophic event comparable to that proposed for the K-T extinction. Olsen writes:

> Scientific theories are not divorced from society's norms and common philosophy. We have lived for 40 years with the concept of nuclear holocaust and the immediate intensity of interest with which the impact theory has been greeted almost certainly has been affected by our fear and fascination with the possibility of our own imminent annihilation. The impact theory fits our rather new view of the ultimate destructive and uncontrollable control by catastrophes unrelated to our day to day activities. The parallel and intertwined development of the impact theory and Nuclear Winter scenarios is a rather striking demonstration of how comfortable the impact theory is with current popular concerns. While I do not want to push the issue too strongly, as Cuvier's catastrophism and Darwin's theory of evolution were during the 19th century, the asteroid impact theory is symptomatic of society's current philosophical milieu. This is quite independent of the correctness of the theory. Hopefully that will be judged by its correspondence to reality as judged by critical tests. (1986, p. 10)

An Embarrassment to the Creationists

Where have the creationists stood in this raging debate? Which side do they root for? The meteorite (or cometary) impact hypothesis is obviously great embarrassment to them because they see themselves as the keepers of catastrophism, lined up against mainstream scientists, whom they regard as the keepers of uniformitarianism. With the annihilation of the clean dichotomy in the opposed viewpoints, what is left for the creationists to oppose?

A search through the larger works of the Institute for Creation Research--Duane Gish's Evolution: The Fossils Say NO! for example, and the earlier Whitcomb/Morris basic work, The Genesis Flood--fails to turn up the term "extinction" in the index of any of them. Their reference to fossil graveyards, a topic we covered earlier, seems to come close to the subject of major extinctions, but not close enough. I did, however, find the literature of the 1979-1982 period on extinction by meteorite impact included in ICR Technical Monograph No. 13 by Stephen A. Austin (1984, pp. 62-70). His volume is titled Catastrophes in Earth History: A Source Book of Geologic Evidence Speculation and Theory. It consists of quotations from mainstream scientists on all sorts of catastrophic events. I gather that the purpose of publishing such a useless collection of quotations is to show that mainstream science relies heavily on catastrophism, even as it purports to be uniformitarian in philosophy. Such has never been the case in modern times, as we have repeatedly stated, but the idea persists among the creationists. The K-T extinction is mentioned in a quotation from K. J. Hsu (1980); the earlier hypothesis of extinction by supernova explosion in a 1977 paper by Clark, McCrea, and Stephenson. (See Austin for exact references.) The following quotation from the Preface reveals what Austin had in mind in assembling the quotations:

> One thing is sure to incite emotional response from geologists, geophysicists and geomorphologists--it is the idea of catastrophes in earth history. Just propose that a regional or global catastrophe left evidence in the geologic record and you will be promptly charged with giving free reign to fantasy. Even worse, your notions may be assigned to a shelf with a whole gamut of suggestions ventured by innumerable "crackpots." Such neglect or censorship has existed in geological science for the last 150 years when "megathinking" was renounced and "microthinking" was standard. However, the situation has changed. Geology contains a rich body of evidence, speculation and theory challenging the notion that the earth evolved to its present configuration simply by the action of gradual processes. This book is about some extraordinary processes that have been discovered, and which challenge our way of thinking about the earth.

There it is again, the parading of that straw person, the creationists' inaccurate version of a modern view of uniformitarianism! Must we be subjected endlessly to this senseless ritual? Also published in 1984 is a book titled Catastrophes and Earth History, The New Uniformitarianism (Berggren and Van Couvering, 1984). It is a collection of essays by mainstream scientists, including several of great distinction in their respective fields. Part One is titled "The concept of catastrophe as a natural agent." We have been over this ground in Chapter 21. Recall my flippant suggestion that the new term for the mainstream view of geology should be either "uniformi-catastrophism" or "cataformitarianism." Those alternatives being absurd, the term "actualism" stands a better chance of wide adoption. Anyhow, the bulk of this rather large collection of essays deals with the K-T extinction event and other extinctions.

The real difference in thought between creationists and mainstream evolutionists lies not in the magnitude of natural catastrophes, but in that the latter group considers them to be natural events. Science deals with natural things. Creationism considers catastrophes to be acts of God; as such, they are supernatural acts and cannot be examined by science. By what stretch of the imagination do Austin and his creation science colleagues think that any rational human will accept the transfer of documented geologic events from the natural realm to the supernatural realm? The creationists require catastrophe in the Flood scenario, for how else could all of stratigraphy and tectonics be encompassed in a one-year period. Natural catastrophe belongs to natural science, Flood catastrophe to the supernatural, and never the two shall meet.

I suggest that it would be prudent for the creationists to learn to live with the great surge of interest shown by the mainstream evolutionists in natural catastrophes. Science sets no limits on the size or scope of natural events it may from time to time evoke to explain what it observes of the real world. On the other hand, the business of creation science is to attempt to demonstrate that extinctions of the geologic record are expected consequences of a set of rigidly constrained and unchangeable postulates that constitute the hypothesis of recent Divine Creation and the ensuing Flood of Noah.

Actually, I did find one substantive creationist article on extinctions; it was written by Kenneth B. Cumming, Chairman of the Biology Department at Christian Heritage College and Research Associate in Bioscience for the ICR (Cumming, 1980). Carefully, the author goes over the subject of extinctions of species occurring in historical times. There follows a review of extinctions of the geologic record. All of the citations and quotations come from works written before the meteorite impact hypothesis was first presented by Alvarez and associates in 1979 and 1980, but the supernova hypothesis is mentioned. Cumming closes with a 1973 statement by paleontologist James Valentine to the effect that mass extinctions are caused by the same natural regulators of diversity that have acted through all geologic time. Cumming considers Valentine's statement to be an expression of uniformitarianism; whereas the creationist view is one of a succession of post-Flood catastrophes to be expected of a decaying system in which entropy (disorder) increases as time passes. In such a regime extinctions are often rapid and devastating, but with very few appearances of new species following the events. Cumming downgrades the documented great radiations that followed mass extinctions in these words: "Subsequent diversification of benthic families, genera, and species since then is probably due to horizontal variation within kinds" (p. iv). As we have pointed out in Chapter 37, this position represents the creationists' shift toward enlarging the definition of kinds to include much higher taxa than they perhaps originally intended. Cumming summarizes the features of two models of extinction, that of the creation model and that of the evolution model, shown here as Table 46.1 (Cumming, p. iv.)

Those of my readers who have developed a full appreciation of the intricacies and inconsistencies of Flood geology will not be content to let Professor Cumming get away with such a vague creationist scenario and fancy

Table 46.1   Comparison of Two Extinction Models

	Creation	Evolution
Initial number of taxa	Many	Few
Number of taxa with time:		
Higher taxa (phylum, class, order)	Decreasing	Increasing
Lower taxa (family, genus, species)	Variable	Variable
Geological events	Catastrophe	Uniformity
Niche replacement with "new" taxa	Limited	Extensive

Data of Kenneth B. Cumming, 1980, Extinction, ICR Impact Series, no. 84, p. iv.

talk about decreasing numbers of higher taxa and niche replacements with new taxa through variations in kinds. Flood geology simply fails to account for the phenomenon of extinction followed by radiation. All organisms that are represented by the fossil record must, according to Flood geology, have been alive on the eve of the Flood, and all must have perished during the single year of the Flood. The entire scenario was that of a single great extinction accomplished in one year. Stratigraphic succession as we observe it requires a sorting agent within the Flood ocean that arranged the fossils in the taxonomic order in which we find them. In Chapter 39, dealing with hydraulic stratigraphy of the Flood, we showed that the observed fossil sequence makes no sense in terms of Flood geology. The same conclusion holds firm, but even more strongly, when it comes to the record of mass extinctions and ensuing radiations. For reasons unknown and perhaps impossible to imagine, the new taxa represented by the Tertiary strata would have had to remain in suspension in the floodwaters while the taxa that failed to pass the K-T boundary would have an opportunity to settle out first. This selective settling must have been perfect, for there are no representatives of the extinct Mesozoic reptilian and other groups above the K-T boundary. If hydraulic stratigraphy was governed by differences in settling velocity, the separation we observe would be totally impossible. Corpses of titanotheres and mastodonts (and of all Tertiary animals) would have had to remain suspended until every last corpse of the Mesozoic reptiles (and of all other Mesozoic organisms, including the Mesozoic marine zooplankton) was laid to rest in bottom sediment. This segregation requirement would have had to be repeated for every one of the major and minor extinctions of the geologic record, and for all occurrences of individual extinctions at various times through the record. Lest the creationists be tempted to attribute the sorting out of taxa to God's personal control, we should remind them that, by their own reckoning, God's role in the Flood was only to start it and stop it; what went on during and after the Flood was controlled by natural processes conforming with laws of nature.

Chapter 47

Living Fossils

A living fossil? What's so remarkable about that? If it's assumed we mean that a fossil has been found that is almost identical with a living species of organism, then most living species are also living fossils. Fossilization can take place in a very short time--perhaps as quickly as a few years or even months. If a footprint is to be classed as a fossil, a human footprint on a freshly congealed lava surface (ouch!) could have been imprinted as recently as within the past few minutes. Chris McGowan states that more than half of the living species of European mammals can be traced back some half-million years without showing any obvious changes (1984, p. 26). Surely fossils exist somewhere for all of them.

Creationists' Views on Living Fossils

So, if a living fossil is simply a living species or genus for which a like fossil can be found, why would creationists Whitcomb and Morris devote four pages to a section headed "Living fossils?" (1961, pp. 176-80) These authors consider living fossils to be cases of supposedly long extinct animals that have suddenly and unexpectedly been found living today (p. 176). A little sorting out of ideas is needed here. Fossils do not become extinct; the taxa they represent can become extinct, but extinction is not a prerequisite of fossilization. How can the event of extinction possibly leave a fossil record? We assume extinction of a taxon has occurred not because we find fossils, but because we do not find fossils. The assumption of extinction always remains in question, because a younger fossil of the taxon may yet be found. That is what seems to have happened a number of times in the history of paleontology; it is a happening to be expected so long as there are more strata to be uncovered, broken apart, and closely examined for indications of past life.

Whitcomb and Morris list several candidates for the dubious distinction of being labeled "living fossil" (1961, pp. 176-80). On their list are the following:

Name	Date of last known fossil appearance
Tuatara (a beakhead reptile	Jurassic
Coelacanth (a cross-opterygian fish)	Cretaceous
Neopilina (a segmented mollusk)	Devonian
Metasequoia (dawn redwood tree)	Miocene

The tuatara lives only in New Zealand. According to the authority quoted by Whitcomb and Morris, "The skeleton of a reptile found in the Jurassic deposits of Europe is so nearly identical with that of the living tuatara that very little change in the bony structure must have taken place during a period of 150,000,000 years" (Charles M. Bogert, 1953, p. 167). The same authority states that no fossil remains of this reptile have been found in rocks younger than early Cretaceous (-140 m.y.).

Whitcomb and Morris find this information difficult to believe:

> The remarkable thing is that a creature which is so apparently out of place in the modern world and which has apparently little selection value in the struggle for existence could have survived the countless vicissitudes of the millions of years that are supposed to have elapsed since all its relatives perished. A few thousands of years of survival under adverse circumstances might be possible, but hardly millions! (1961, p. 176)

The story of the coelacanth is well worth reviewing as an object lesson in scientific discovery: Who knows what will be discovered next? Of the two groups of crossopterygians, or lobe fins, the rhipidistians were the group from which the amphibians arose. The second group, the coelacanths, were far removed from the main line of evolution to the land vertebrates. Coelacanths appeared in the Devonian Period and flourished in the Mesozoic Era. A representative example from the Upper Triassic is Diplurus (Figure 47.1); one from the Cretaceous Period is the genus Macrapoma. Since no post-Cretaceous coelacanth fossils had been found, the group was judged to have become extinct at that time. Then, in 1939 a trawler working off the coast of South Africa dredged up a fish that was quickly recognized as a coelacanth. Paleontologist Edwin H. Colbert describes the find and subsequent events as follows:

> The fish, named Latimeria, is rather large, being about five or six feet in length. In form, it is extraordinarily similar to Macropoma of the Cretaceous period. It has brilliant blue scales, rounded and large, and the lobed paired fins are long and strong. For fourteen years after its discovery, this single specimen was the only known example of Latimeria. Then in December, 1952, a second specimen was caught off the coast of Madagascar, an event that set off a train of excited newspaper and magazine articles. This fish was flown to South Africa in a special airplane, and elaborate plans were made for a detailed study of it. But hardly had the excitement died down when several more specimens of Latimeria were caught, also near Madagascar. Since then, more coelacanths have been taken on frequent occasions in the vicinity of the Comoro Islands, between Madagascar and Africa. A considerable series of specimens of this fascinating fish has been intensively studied in Paris, and large, monographic descriptions of it have been published. Here we have a valuable link

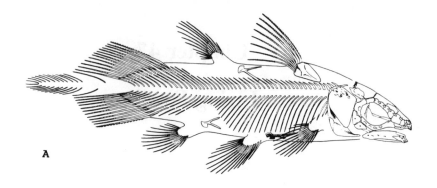

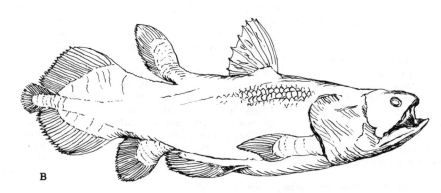

Figure 47.1 A. _Diplurus_, a small freshwater coelacanth from the Upper Triassic of eastern North America. B. A "living fossil," the coelacanth _Latimeria chalumnae_, sketched from photographs. (A. From _Stratigraphy and Life History_, by M. Kay and E. H. Colbert, p. 382, Figure 16-28. Copyright © 1965 by John Wiley & Sons, Inc. Reprinted by permission of John Wiley & Sons.)

with the past that gives us a glimpse of an important group of vertebrates, hitherto known only from fossils. (1980, p. 73)[1]

The segmented mollusk, _Neopilina_, had been thought extinct since Devonian time, but in 1957 several specimens of the same class were dredged up from the Acapulco Trench, where the bottom lies at a depth of about 3,700 m. Whitcomb and Morris quote from an article by biologist Bentley Glass, describing the discovery:

> To zoologists the recently reported discovery by the Galathea Expedition of the extraordinary deep-sea mollusk _Neopilina galatheae_ will seem even more incredible than the famous discovery in recent times of _Latimeria_, the living coelacanth. . . . the new-found mollusk represents a class that existed in the Cambrian to Devonian periods of the Paleozoic, and was supposed to have become extinct about 280 million years ago. (Glass, 1957, p. 158, in Whitcomb and Morris, 1961, p. 1978))

Whitcomb and Morris continue with an account by Harry S. Ladd of another living-fossil discovery, that of a series of primitive crustaceans living in the saltwater that saturates beach sands in New England, and of an even more primitive crustacean living in mud of Long Island Sound (Ladd, 1959, p. 74). The closest known relative of the latter species lived in Middle Devonian time. Whitcomb and Morris comment as follows:

> In view of these and many similar discoveries, one also wonders whether or not many more of the supposedly extinct creatures of geologic history might not also be living in some unexplored region of the globe, especially in the deep oceans. It

_____
[1]From _Evolution of the Vertebrates_, First edition 1955, Second edition 1969, Third edition 1980, by Edwin H. Colbert, John Wiley & Sons, Inc., New York, Copyright © 1955, 1969, 1980 by John Wiley & Sons, Inc. Reprinted by permission.

would not be surprising if even the famous trilobite, perhaps the most important "index fossil" of the earliest period of the Paleozoic, the Cambrian, should turn up one of these days. A creature very similar to it has already been found. (1961, pp. 178-79)

The last sentence is in reference to a 1957 article in _Science Digest_, describing the finding of an invertebrate with certain characters of the trilobites.

Finally, Whitcomb and Morris (p. 179) describe the finding in a remote region of China of stands of the conifer _Metasequoia_, of which fossils are found in Eocene and Miocene strata, but not in younger rocks (Ralph W. Chaney, 1948, p. 490). The creationist authors comment: "Evidently something must have been wrong with the geological record deduced from the Pliocene and Pleistocene strata, which failed to reveal the continued existence of the trees, in spite of their great abundance in the supposedly earlier strata" (Whitcomb and Morris, pp. 179-80).

The Mainstream View of Living Fossils

Many more examples of "living fossils" could be cited, among them the familiar living nautilus and horseshoe crab. An entire volume on the subject of living fossils, published in 1984 and containing 32 case histories by 31 authors, is proof enough that paleontologists are well aware of such occurrences and speculate on their meaning (Eldredge and Stanley, 1984). The usual interpretation is that of extreme stasis in the evolutionary process. I suppose the logic involved is that if stasis can be observed in one particular genus or family as extending over several million years, there is no reason to rule out the possibility that it might have continued in some instances for 100 to 500 million years. If it is assumed that punctuated change is forced in small, isolated populations by strong environmental change, then stasis is demanded in equal strength for large, widespread populations in environments that show very little change. Chris McGowan addresses the problem of stasis from a very broad and basic point of view. He notes: "over a period of time a species undergoes change in response to fluctuating environmental conditions, but these changes are not directional; rather, they fluctuate about the mean. There is, therefore, no permanent change in the species" (1984, pp. 25-26).[2] I infer that the same program of fluctuation about the mean applies also to the environment. As a geologist, I should suppose that almost every important category of environment of sedimentation capable of entombing fossils has been in existence somewhere on the globe continuously since the time that plate tectonics had first produced sizable continents much as they are today. For example, there have always been, since late Precambrian time or earlier, continental shelves and deltas of passive margins, deep trenches, forearc and backarc troughs of subduction boundaries, aggrading floodplains, and fault-block basins within continents, and so forth, in each of which the environment changed little through time. These environments would disappear in one geographical location only to reappear in another. In a specific case, that of the coelacanths, their deep ocean-floor environment could remain essentially unchanged for the entire Phanerozoic Eon, even as ocean basins closed and opened in repetitions of the Wilson cycle. Why then shouldn't the coelacanths correspondingly also show almost no evolutionary change?

McGowan points out that there are many examples of genera that have remained essentially unchanged for millions of years (p. 26). Examples he cites include the brachiopod Lingula, the reptile Sphenodon, the Port Jackson shark Heterodontus, the horseshoe crab Limulus, the crocodile Crocodilus, the Australian lungfish Epiceratodus, the marine turtle Chelonia, and the lizard Lacerta. He continues:

> If we broaden our category to include major groups of organisms, orders, and classes in the classification system, we find that stasis, that is, no change, is the rule rather than the exception. There appear to be no important differences between modern representatives of major groups and their ancient relatives. (1984, pp. 26-27)[2]

Examples given by McGowan are the coelacanth, the jellyfish, the turtles, the sharks, the bony fishes, and the insectivorous mammals. McGowan then comments:

> I can almost see the creationists nodding in agreement and saying that this is exactly what their model predicts, but I hasten to point out these various groups we have mentioned did not appear all at once. (1984, p. 27)[2]

McGowan asserts that stasis might seem to fly in the face of evolution (as the creationists claim), but that it actually makes perfectly good sense. Examples he cites demonstrate the principle that (in my words) might read: "When you have a good thing, stick with it!" Once a good structural adaptation had been made, it continued to be used in that particular group. For the reptiles, the amniotic egg proved highly successful; it continues to be a successful way for reptiles to propagate; why should it have been abandoned if it was successful?

McGowan scores a hit when he compares stasis in the larger taxonomic groups with the progress of technology. For example, he notes that "the present-day internal combustion engine has not undergone significant change since its invention over a century ago, and the same is true of television sets, electric light bulbs and sewing machines." In contrast, a significant change in technology results in the initiation of a new line of machines based on an entirely different principle. In transportation, the invention and development of the jet engine led quickly to a new line of aircraft, but even so, aircraft driven by propellers powered by internal combustion engines continued to fly and to increase in numbers. As I see this

analogy, we might think of the situation as one of stasis in contrast to punctuated evolutionary change. The small airplane remains successful in the environment of low-level, short-distance flying; the big jet plane is well settled into its own environment of high altitudes and long distances. (Should I mention the B-52 bomber?) Each class of flying machine is enjoying its prolonged period of stasis.

Looking back over the creationists' presentation of the subject of living fossils, I find nothing of scientific value or importance in their attempted argument. They report the paleontological details accurately enough, but they fail to make any positive argument to the effect that no genus, family, or suborder could persist for many tens of millions of years with so little change. Their case boils down to a statement such as this: "We just don't believe such persistence is possible, but we can't give any reasons for that belief." In contrast, the mainstream scientists are saying, in effect: "We have a reasonable general hypothesis of evolution--punctuated equilibrium-- that provides for extreme examples of stasis such as the geologic record shows." The evolutionists' hypothesis is subject to modification or even complete replacement, should that be necessitated by new observations. Looking into the future, it seems quite possible that the fossil record may well improve greatly in the matter of the long gaps between the living fossils and the last appearances of the corresponding similar ancestors. Filling of the large blank stretches of geologic time is the only logical direction in which new information can lead under the postulate of descent with modification from a single common ancestor.

Living Dinosaurs--Really!

The "living fossil" story has taken on a new and bizarre twist in the hands of the creationists. As we explained in an earlier section, Flood geology requires that all air-breathing animals represented by the fossil record must have been taken on the Ark and must then have been released at the same time after the floodwaters receded. Thus, there could in theory be a living fossil for every terrestrial animal species of the fossil record. If the living coelacanth could have remained unknown to science until 1938, couldn't there be dinosaurs living today that have escaped detection? The creationists think there may be some dinosaurs living today. In an article in the creationist journal, Ex Nihilo, author Ken Ham is very clear on this matter (1984, pp. 10-11). He says there have been reported sightings of dinosaurs up to the present. He says that published accounts claim that "explorers and natives in Africa have reported sightings of dinosaur-like creatures." He continues: "It certainly would be no embarrassment to a creationist if we discovered a living Tyrannosaurus rex in a jungle. It wouldn't even be surprising if it happened to be plant eater! However, this would be a tremendous embarrassment to an evolutionist."

The Institute for Creation Research keeps close tabs on such reports of sightings of dinosaur-like creatures. A 1985 issue of Acts and Facts carried a report on the Lake Tele expedition to the Congo Basin (vol. 14, no. 1, pp. 3, 5). The expedition, planned for April, 1985, was to attempt to document the existence of a brontosaurus-like animal, called "Mokele Mbembe" by natives, claimed to have been seen by the two expedition leaders in 1982 (p. 5). The news item went on to say:

> This expedition has a very strong implications as far as the theory of evolution is concerned. If this creature is indeed a living dinosaur (another living fossil) and can be documented as such, then the data supporting dinosaur and man tracks in the Paluxy River of Glen Rose, Texas, cannot be taken lightly. It would eliminate 65,000,000 years of

geologic history needed for the theory of evolution and it would also correlate with archaeological and biblical records (Job 40: 15-24).

Sounds fascinating, but I wonder why, if a bona fide living dinosaur is captured and brought back alive or dead to civilization, the find would prove any more embarrassing to mainstream evolutionists or more damaging to their theory of evolution than was the finding of Latimeria in the deep ocean off South Africa? More power to the living fossils! May their numbers increase!

The Case of the Monotremes

Is it conceivable that a living fossil could exist today without any fossil counterpart having been discovered as its early ancestor? This situation is actually recognized in the case of two strange mammals living today in Australia and New Guinea: the platypus or duckbill, Ornitho-rhynchus, and the spiny anteaters, Echidna and Tachyglossus. These are monotremes (order Monotremata), unknown as fossils older than Pleistocene. Their strange appearance is a result of extreme specialization. The platypus uses its ducklike bill for burrowing into the mud of stream beds in search of grubs and worms. It has teeth in early life, but these are shed and replaced in the adults by hard pads. The anteaters have sharp spines covering the body; their toothless jaws are elongated into a snout, well adapted to probing into ant hills.

Aside from these rather superficial features, the monotremes are judged to be very primitive mammals. Colbert describes their primitive features as follows:

> They reproduce by laying eggs, which are hatched in burrows. The young are suckled on milk that is secreted . . . by modified sweat glands that are homologous to the mammae or breasts in the higher mammals. The skeleton and soft anatomy show the persistence of various reptilian characters. For instance, the shoulder girdle is very primitive, with a persistent interclavicle, large coracoids, and no true scapular spine. The cervical ribs are unfused. Various reptilian characters persist in the skull. The rectum and urinogenital system open into a common cloaca as in reptiles, not separately as in mammals. There are no external ears or pinnae as in most other mammals. (1980, p. 256)[1]

Colbert says there is every reason to infer that the monotremes represent a separate line of descent from the mammal-like reptiles that lived in middle Mesozoic time. Clearly the monotremes are not derived from either the marsupials or the placental mammals, both of which groups appear first as fossils in late Cretaceous time. Colbert states: "In many respects the monotremes give us an excellent view in the flesh of mammals intermediate in their stage of evolution between the mammal-like reptiles and the higher mammals" (1980, p. 256).[1] That being the case, I should suppose that their ancestral fossil remains, if any exist, will perhaps some day be discovered in strata as old as early Jurassic or late Triassic. Those fossils would not, however, necessarily show the platypus's ducklike bill or the echidna's long snout, since

these features may have appeared much more recently in the long course of Cenozoic time. Stasis would here be represented by unchanging skeletal features mentioned above--in the skull, shoulder girdle, ribs, and spine--and in retention of the cloaca and the egg-laying habit.

The creationists refer to the monotremes in the context of their argument over the significance of possession or lack of possession of teeth by birds, reptiles, and amphibians. Duane Gish states:

> Some fossil birds had teeth and some did not. That this should be true is not surprising at all since this is true of all other classes of vertebrates--fish, amphibians, reptiles and mammals. Furthermore, following the notion that absence of teeth denotes a more "advanced" state, then the duck-billed platypus and the spiny anteater, mammals that do not have teeth, should be considered more advanced or highly evolved than man, yet in many other ways, as previously mentioned, the duck-billed platypus and spiny anteater could be considered the most primitive of all mammals. Thus, the possession or absence of teeth proves nothing about ultimate ancestry. (1985, p. 114)

Loss of teeth in the monotremes represents a relatively superficial adaptation to specialized feeding habits, as we have noted; it is not a criterion of being more or less "advanced." Thus, Gish's suggestion that the monotremes should be considered more advanced than humans is simply specious and need not be considered seriously.

Gish's point that the lack of pre-Pleistocene fossils rules out the monotremes as descendant from ancestors of the more advanced orders of mammals can also be dismissed (1978a, p. 86). As we have seen, homology based on anatomy and embryology of living animals is a powerful and independent means of establishing valid evolutionary relationships. Interestingly enough, phylogeny of the alpha chain of globin genes locates the echidna just about where it should be if the phylogeny from anatomical homology is valid--between the kangaroo and the chicken (Ayala, 1977c, p. 304, Figure 9-21).

When Gish cites the lack of a monotreme fossil record, he implies that no fossils of ancient monotremes exist, but this implication cannot be sustained. We are permitted no statement other than this: "No fossils of ancient monotremes have yet been found." Let the creationists continue to search for living dinosaurs in the Congo basin and let the mainstream paleontologists continue to search for fossils of monotremes in Mesozoic and younger strata. Let's see who succeeds first!

Credits

1. From Evolution of the Vertebrates, First edition 1955, Second edition 1969, Third edition 1980, by Edwin H. Colbert, John Wiley & Sons, Inc., New York, Copyright © 1955, 1969, 1980 by John Wiley & Sons, Inc. Reprinted by permission.

2. From IN THE BEGINNING © by Chris McGowan. Reprinted by permission of Macmillan of Canada, A Division of Canada Publishing Corporation.

Chapter 48

Out-of-Order Fossils

"Anomalous coexistences within the record" is the name we used for this phenomenon in introducing the creationists' arguments against evolutionary biostratigraphy and phylogeny. To repeat, an anomalous coexistence is the finding or alleged finding in a single stratum or on a single bedding surface of two or more fossil varieties that, in the standard biostratigraphic column, are widely separated in age. As we noted then, creationists think that an out-of-order fossil has the potential of demolishing in one stroke the entire hypothesis of evolutionary descent with modification; they see it as a "sudden death" threat. When one fossil is that of a human and the other of a dinosaur or a trilobite, the stakes are the highest of all. Mainstream science then bets its entire bankroll on one throw of the dice, while the creationists, on the other hand, have everything to win but nothing to lose.

Anomalous Occurrences and Flood Geology

We will be reviewing two kinds of anomalous occurrences. One is the finding of supposed human footprints (or shoeprints) on bedding surfaces that also bear tracks or actual fossil remains of long extinct animals. The other is the finding of human fossil remains enclosed in strata that, by reason of their stratigraphic position and fossil content, are conventionally assigned an age of at least some tens of millions of years.

Humans living in the Cambrian Period? "Preposterous!" you may be tempted to say. The fossil record of the Cambrian, studied intensively for more than a century, reveals only marine invertebrates among the metazoans. If that is your attitude, the creationists will chastise you soundly for exhibiting prejudice and an unscientific attitude. Creationists will point out that a human shoeprint (or sandal print) has been found in Cambrian strata in Utah, and that a trilobite lies inside the outline of the print--proof positive that the human and the trilobite were both alive at the same time.

It all fits nicely into Flood geology in a way we have explained at length in Part Six. All humans except the Noah family perished in the Flood along with all organisms of the entire fossil record. As those hapless beings strove to stay above the rising flood waters, they left many footprints in mud layers accumulating in shallow water. Many other air-breathing land animals were engaged in a similar survival activity. Dinosaurs, too, must have been stepping on trilobites and other invertebrates. The panic scene has been vividly described by creationists John D. Morris (1976, pp. vii-viii) and Ken Ham (1984, pp. 9-10). Keep in mind that Flood geology recognizes no fossil-iferous strata older than about 4,350 years; it further specifies that all fossiliferous strata were deposited in less than one year. This being the case, the possibility exists for the finding of any two known fossil species on the same bedding plane, irrespective of the kingdom, phylum, or lesser taxon to which each belongs. Finding a human footprint on top of a trilobite is actually an expected consequence of Flood geology; and it serves at the same time to falsify the theory of evolution.

A Human Footprint of Cambrian Age?

Where is this human footprint on top of a trilobite? It was found in the Wheeler shale formation of Middle Cambrian age in the Wheeler Amphitheater in the House Range, Millard County, Utah. The finder was William J. Meister. He had split open a joint slab, one of many such fragments that litter the ground at that locality. There on the freshly exposed bedding surface was the sharp outline of the sole of a shoe or perhaps a sandal, such as one sees worn today in tropical countries (Figure 48.1). The sole might have been cut from heavy animal hide. The print of a heel is particularly well defined and (wouldn't you know it?) right there under the heel is the trilobite fossil! In the spring of 1968 Meister and two of his associates brought the specimen to Professor William Lee Stokes of the Department of Geology and Geophysics at the University of Utah in Salt Lake City. Stokes tells of this experience:

> They came to me for a geologic opinion as to the authenticity of their find as a footprint. However, this was not before the story had received international newspaper publicity as a genuine challenge and embarrassment to the geologists. (1986, p. 187)

It seems that the find had been described by newswriter Don Searle in the June 13, 1967, issue of The Deseret News under the title "Puzzling Fossils Unearthed." Professor Stokes explained to his visitors the criteria by which a genuine fossil footprint can be recognized. There will be a rim of material squeezed up around the outline of the foot (or shoe); this is often called "up-push" or "up-squeeze." A single print is not enough, there must be found a trail of prints--right and left--with appropriate stride length. Also, the nature of the process of forming a print is such that the soft bed of sand, clay, or marl receiving the print is typically filled in and covered over with sediment of a different composition and texture, for example, soft mud that settles out from suspension in comparatively calm water. These criteria seem to demand the former existence of a floodplain backswamp, coastal tidal flat, or other shallow depression alternately exposed and flooded, but free from strong water currents. Stokes reports that after he gave his explanation, he pronounced his conclusion:

> My judgment was that since the specimen does not display an elevated rim, and that it seemed impossible to check any additional criteria in the field, that the finders should not publish their find as a proven footprint but should at least leave open the possibility that it might be something else, a natural break or spall, for example. This was not what Meister and his followers wanted to hear; they already had opinions favorable to their theory from a variety of media people, cobblers, ministers and interested neighbors. (p. 188)

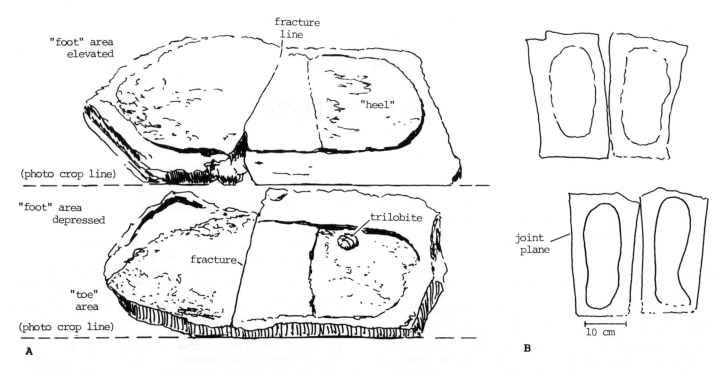

Figure 48.1 A. Drawing from a photograph of the Meister "footprint." Shown are the two rock surfaces as they came apart from a single bedding fracture. Length of impression is 27 cm. B. Sketches of two sets of mirror-image foot-shaped markings revealed by splitting apart of bedding planes of the Wheeler shale. (From photos and drawings by W. L. Stokes. Used by permission.)

What followed is a personal horror story of the kind of misrepresentation and vilification to which a conservative mainstream scientist is often subjected by those who find the scientific opinion unpalatable. It can happen in a completely secular setting as well; for example, when a geologist expresses the considered opinion that the ground-water table is rapidly falling beneath an agricultural valley, threatening agriculture and making unwise the possibility of further urban and industrial development.

Stokes received "the treatment" at the hands of the creationists. One particularly vicious and unwarranted accusation came from creationist Clifford L. Burdick, published in 1980 in Bible-Science Newsletter (vol. 18, no. 2, pp. 3-5). It was a statement to the effect that Stokes was fearful that the Meister "print" would endanger the basic principles of historical geology, causing a decline in his (Stokes's) textbook of historical geology and result in financial losses to both author and publisher (Stokes, 1985, p. 188).

Without going into the great amount of detail needed to analyze the Meister specimen and offer a complete alternative naturalistic explanation, it must suffice here to say that there is such an explanation and Stokes has presented it in detail in his 1986 paper. The phenomenon involved is familiar to all geologists. Joint blocks and slabs of hard shale freshly fallen from a cliff commonly show rectangular faces and sharp edges. Two of the parallel block surfaces are bedding planes; the four surfaces perpendicular to the bedding are joint planes in two parallel sets (Figure 48.2). In some cases only one joint set is well developed, the other faces being rough pressure fractures. Chemical weathering by aqueous solutions from rainwater penetrating the block alters the rock progressively inward along a weathering front that is sharply defined. In many localities this alteration leads to the formation of concentric shells of soft rock that easily fall away. In other cases, the process simply leaves an inner core of unaltered rock that can be exposed when the slab is forcibly split along the bedding. In the

Wheeler shale "print" specimens a thin bedding layer in the core area was detached from one block and adhered to the opposite block, making what superficially resembles the sole of a shoe. In the Meister specimen a trilobite happened to be exposed in the "heel" zone of the depressed area of the block (Figure 48.1A). Somewhat similar shale specimens have been collected in this area and they substantiate the naturalistic explanation of weathering (Figure 48.1B). Nothing in the way of a trail of tracks on a bedding plane in situ has been found. The creationists do not even know which was the "up" direction in the Meister specimen. There is a 50 percent chance that the supposed sole imprint is actually on the underside of a bed, in which case, the human that made it would have had to "walk upside-down on the ceiling," if you can imagine such a physically impossible maneuver!

According to Ernest C. Conrad, opinion within the creationist group concerning the authenticity of the Meister specimen as a human foot or shoe print is divided (1981, p. 31). Creationist Melvin Cook is quoted from a 1970 source as saying: "No intellectually honest individual examining this specimen can reasonably deny its genuine appearance" (Conrad, 1981, p. 31). Less assured are Robert E. Kofahl and Kelly L. Seagraves (1975, p. 54). I gather that enough doubt exists in the minds of the creation science leaders that the case is not now highly touted.

Carboniferous Footprints--Of What?

So let's make our own tracks from the Cambrian to the Carboniferous, where the next set of "human footprints" is to be found. Whitcomb and Morris in The Genesis Flood have a section on "misplaced fossils," beginning with the mysterious Carboniferous figures (1961, pp. 172-76). They appear on exposed rock surfaces in states ranging from Virginia, through Kentucky, Illinois, and Missouri, and to points even farther westward. The figures came to public attention in 1940 through a short article in Scientific American written by Albert G. Ingalls. That is

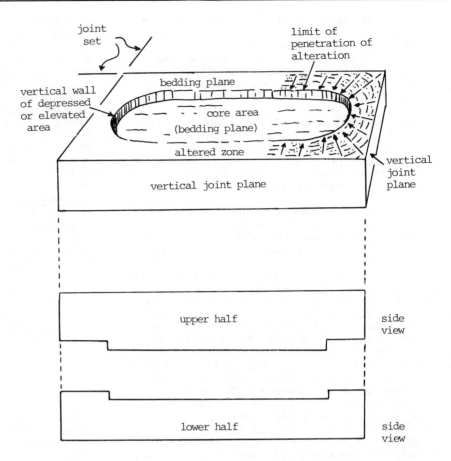

Figure 48.2 Schematic drawing showing how elongate printlike areas can be formed when a joint block is split parallel with the bedding. (A. N. Strahler.)

Whitcomb and Morris's source of information as well as mine at the moment, although Ingalls drops the names of four persons--including an ethnologist, a geologist, and a paleontologist--who have investigated the phenomenon. Ingalls's article leaves much to be desired in the way of solid information but is not lacking in wit and humor. Photographs suggest that shallow depressions in the rock surface are shaped like crudely drawn outlines of toed feet (Figure 48.3). At first glance, they reminded me of petroglyphs scratched by American Indians into the desert varnish of boulders and cliffs throughout the south-western desert region. Whitcomb and Morris select those of Ingalls's statements they want, while omitting those that do not favor the creationists' case. One sentence in the latter group reads: "They look like human footprints and it often has been said, though not by scientists, that they really are human footprints made in the soft mud before it became rock." For comparison, here is the Whitcomb/Morris version: "These prints give every evidence of having been made by human feet, at a time when the rocks were soft mud" (p. 173).

Actually, nothing in Ingalls's article even implies that the features on the rock surface are prints (natural foot impressions) of any sort. We are not told whether they occur on bedding surfaces, on joint surfaces, or on rounded surfaces of weathered outcrops. We are not even told what lithology bears the markings. I suspect that nine out of ten persons seeing the pictures would identify them as drawings or carvings made by Indians. That is one expert interpretation; it was held by David I. Bushnell of the Smithsonian Institution, who suggested that the human foot drawing was a symbol for presence of water nearby, as, for example, in a spring or seep. A second explanation reviewed by Ingalls is that they are indeed foot impressions, but the maker was a Carbon-iferous tetrapod--perhaps a five-toed amphibian. It seems that the animal-track hypothesis was (when the article was written) supported by geologist Professor W. G. Burroughs of Berea College, Kentucky, and by

paleontologist Charles W. Gilmore at the U.S. National Museum. Neither explanation appeals to Whitcomb and Morris, who say:

> Such explanations illustrate the methods by which the uniformitarians can negate even the most plain and powerful evidence in opposition to their philosophy. Nevertheless, it is obvious that it is only the philosophy, and not the objective scientific evidence, what would prevent one from accepting these prints as of true human origin. (1961, p. 173)

Well, Ingalls offers no scientific evidence one way or the other in his article; neither do Whitcomb and Morris. Instead of evidence, Ingalls does have a clever paragraph in his article:

> Nevertheless, asking the scientist for man in the Carboniferous is like asking the historian for Diesel engines in ancient Sumeria. The comparison is no exaggeration but an understatement. If man, or even his ape ancestor, or even that ape ancestor's early mammalian ancestor, existed as far back as in

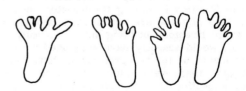

Figure 48.3 Outlines of footlike depressions on surfaces of Carboniferous rocks. (Traced from photos by Albert G. Ingalls, 1940, <u>Scientific American</u>, vol. 162 [January], p. 14.)

the Carboniferous Period in any shape, then the whole science of geology is so completely wrong that all geologists will resign their jobs and take up truck driving. Hence, for the present at least, science rejects the attractive explanation that man made these mysterious prints in the mud of the Carboniferous Period with his feet.

What science does know is that, anyway, unless 2 and 2 are 7, and unless the Sumerians had airplanes and radios and listened to Amos and Andy, these prints were not made by any Carboniferous Period man. What did make them may appear with certainty later. Or it may not.

This must be the perfect nondebate. Neither the evolutionist nor the creationist offers any information of substantive value, let alone the semblance of a scientific argument, but each is steadfastly faithful to its group's respective dogma.

Since this section was written, James Stewart Monroe, chairman of the Geology Department at Central Michigan University, has published a paper titled "Creationism, Human Footprints, and Flood Geology," in which he covers in detail many examples of supposedly human footprints besides those described by Ingalls (Monroe, 1987). Particularly interesting are the Nevada "human footprints," exposed in 1882 during quarrying operations in the State Prison yard at Carson City. These huge tracks, forming trails of alternate right and left tracks, were immediately interpreted as tracks of giant humans. Instead, the tracks are readily interpreted as those of a giant Pleistocene ground sloth (Mylodon or Morotherium), the fossil remains of which were found in essentially the same stratigraphic horizon and described by the noted paleontologist Othniel Charles Marsh in 1883.

The Paluxy Story

For at least four decades, field skirmishes have been fought in the rocky channel of the Paluxy River, near the town of Glen Rose, Texas. Here, the creationist forces dug in their positions on limestone ledges of Lower Cretaceous age. Here, they claimed, are genuine and unmistakable human tracks--"man tracks," they are called--appearing as trails crossing those of gigantic dinosaurs, including many made by a colossal sauropod and others by terrifying beasts of a bipedal carnivorous genus, possibly Acrocanthosaurus. The story of Paluxy has every ingredient an absorbing drama needs--strong personalities in conflict, defamations of character, allegations of cover-up of evidence and perhaps a bit of midnight chisel work, and claims of flood washouts obliterating absolutely convincing evidence that humans cohabited with dinosaurs. During the last decade, forces from both sides converged on the Paluxy riverbed, and the conflict intensified into a full-scale battle.

For perhaps the first time in the creationist-evolutionist conflict, teams of scientific experts from both sides came to a controversial field area to see it for themselves and give their expert appraisals of a varied collection of features indenting the limestone bedding surfaces. Some of these are nondescript cavities, pits, and hollows in limestone bedding surfaces, many of which, taken alone, would scarcely have supported a serious creationist argument for the presence of humans with dinosaurs, but there were other features not easily dismissed. Repeating elongate depressions, fairly uniform in size and occurring more or less evenly spaced in trails that cross the trails of obvious dinosaur tracks, do indeed bear some superficial resemblance to human footprints made in soft, yielding mud. It is on these non-random sets of depressions that we will focus.

The story of the Paluxy had the best of scientific

beginnings, but even so the tinge of chicanery was there from the start. Local talk of "man tracks" in the bed of the Paluxy River was going the rounds at least as early as 1908 among the residents of Glen Rose and its surrounding ranches (Milne and Schafersman, 1983). Authentic dinosaur tracks were found here and described as early as 1917 by E. W. Schuler, a paleontologist, but his work apparently did not attract much attention. Roland T. Bird, a professional collector of dinosaur fossils in the employ of the American Museum of Natural History in New York City, seems not to have been aware of Schuler's earlier find when, in the fall of 1938, he was searching the southwest for new fossil localities. His discovery account appeared in the May, 1939, issue of Natural History, from which I have extracted the following material. By chance, Bird happened to spot in the window of Jack Hill's Indian trading store in Gallup, New Mexico, a startling display--two massive chunks of limestone, each bearing on its surface the deep "imprint" of a "human" foot. Bird wrote:

> I could conceive of no animal that might have made them. It was ridiculous to think they were human footprints. They were too large and bear-like; and yet they weren't like the largest prehistoric bear I could think of, the great Pleistocene cave bear, for the toes were not typical. I felt a keen sense of regret when I told the clerk: "I'm afraid your Jack Hill has found himself a pair of fake footprints." (P. 255)

Conversation with the clerk led to information that dinosaur tracks in the same type of stone and from the same place were on display in a store in Lupten. Wrote Bird:

> The dinosaur footprints were found as represented and like the "mystery tracks," they were fine specimens--too fine. I had every reason to suspect the entire lot had been fashioned by some stone artist, but how they had been so neatly done, how a man could have duplicated the dinosaur tracks at least, without an intimate knowledge of something genuine, there was no means of telling.

The scent led to Glen Rose, where Bird spotted the 20-inch imprint of a three-toed dinosaur on a limestone block embedded in masonry near the courthouse. Inquiry of the local folk soon led Bird to the bed of the Paluxy River, and here, he realized, was excellent material to collect for museum display. But Bird was still mindful of the "man tracks" in Gallup and he soon broached the subject to James Ryals, whose farm bordered the Paluxy channel. To Bird's surprise, Ryals replied, "Oh, you mean the man tracks. Why sure, there used to be a whole trail of them up above the fourth crossing, before the river washed them out." Ryals was persuaded to guide Bird to a point where a single "man track" could be seen. It was submerged in shallow water on the river bed, and Ryals began to clean out the mud that filled it. Reported Bird: "I watched closely as the outline of a foot took form, something about 15 inches long with a curious elongated heel." Bird continues:

> What I saw was discouraging in one sense, enlightening in another. Apparently it had been made by some hitherto unknown dinosaur or reptile. The original mud had been very soft at this point, and the rock had preserved faithfully this element of softness, but the track lacked definition on which to base conclusions. There was only one, and though my eyes itched to see a good one, the overlying ledge covered any possible next print. Ryals said he knew of no others exposed at present. (P. 257)

In 1940, a splendid trail of sauropod tracks was removed under Bird's direction and taken to New York for installation in the American Museum. For several years thereafter, Bird wrote magazine articles about his exploration there, but they contain no mention of seeing other man tracks. (From this point on, we drop the quotation marks from man tracks.) He must have seen many more, because he later gave his own considered opinion as to how such features could have been formed by dinosaurs. According to Laurie Godfrey, his final opinion and explanation were contained in a letter dated 1969 to creationist Mike Turnage (1981, p. 29). Godfrey quotes the following portion of the letter:

> They are definitely, repeat, definitely not human. I am well familiar with all the fossil footprints found in the Glen Rose (Cretaceous) of Central Texas, and have seen those purported to be "human" by farmers lacking any geologic training.
>
> They were made by carnivorous dinosaurs wading through deep mud. When the foot was withdrawn, the sides of the resulting cavity flowed inward leaving an oblong opening only faintly suggestive of the footprint of a man in the eye of the beholder. When one followed such a trail, tracks of the dinosaur were invariably found that showed all the details of a three-toed dinosaur. Anything else "human" exhibited or reported "found" in the area is the product of a very clever prankster with a hammer and chisel.

The first outspoken creationist advocate of the man tracks to arrive from outside the Glen Rose community and exploit the phenomenon seems to have been Clifford L. Burdick, mentioned earlier in connection with the Meister print. According to a footnote by John D. Morris, Burdick had been examining the Paluxy tracks since 1945, when they had first come to the attention of creation scientists in the old Society for the Study of Creation, the Deluge and Related Sciences (Morris, 1976, p. ii). Burdick published a paper on the tracks, titled "When Reptiles Ruled the Earth," in a 1950 issue of Signs of the Times. Henry M. Morris learned of Burdick's work and obtained his photographs of slabs of Glen Rose limestone, one bearing a three-toed dinosaur print, the other two with large "human" prints. The photographs were reproduced in The Genesis Flood, with captions and text strongly supporting the man-track interpretation (Whitcomb and Morris, 1961, pp. 174-75). Whitcomb and Morris raised the antievolution argument most commonly used by creationists, namely, that evolutionists reject the human origin of the prints for the reason that it does not fit their evolutionary time scale. This argument received a thorough workout by John D. Morris (son of Henry M. Morris) in his 1976 ICR Impact Series article "The Paluxy River Tracks." The ICR's major work on the Paluxy tracks has been John Morris's 1980 book, Tracking Those Incredible Dinosaurs: and the People Who Knew Them, in which rough site maps of several track localities are shown, together with photographs of the supposed tracks.

Other creationist groups have shown an interest in the Paluxy man tracks and have sent investigators to the scene. Institutions whose representatives have participated in this investigation include Loma Linda University, Columbia Union College, and Ozark Bible College. The Loma Linda team conducted its field work in 1970 at three major track sites. A noteworthy feature of the Loma Linda report, which appeared in Origins, was that it not only did not support the man-track claims, but even strongly refuted them (Neufeld, 1975).

A significant event, providing a convenient bone of contention, was the making at one site the Paluxy riverbed of a creationist film titled Footprints in Stone by Stanley E. Taylor. Reverend Taylor (since deceased) was

director of the Films for Christ Association. In 1968 Taylor had taken a creationist group to visit the Paluxy locality. At one site, now known as the Taylor Site, they found numerous elongate tracks forming trails and interpreted them as man tracks. The following year and in 1970, Taylor's group returned to the Paluxy riverbed and excavated in the Taylor Site. Here they exposed many of the curious elongate depressions, along with numerous, obvious dinosaur tracks. The Taylor film was released in 1973. The widespread publicity it brought led to a spate of citations in creationist publications, in which the case for authentic man tracks was strongly promoted. The film was reviewed in depth by physical anthropologist Laurie Godfrey, who recounted her experiences in showing the film to a student/faculty audience at the University of Massachusetts in Amherst, where she is a faculty member (1981, pp. 23-25). Godfrey noted: "Even without good resolution, it is possible to tell that the 'man prints' in the film are not genuine human footprints" (p. 26).

Another creationist who has actively investigated and collected in the Paluxy riverbed in recent years is the Reverend Carl E. Baugh. He claims to have found some fifty man prints and, on the basis of these, has named the species that formed them Humanus Bauanthropus. He founded and stocked the Creation Evidence Museum, near Glen Rose; it contained several prints or casts of prints of supposed human origin.

In 1980, a serious field investigation of the Paluxy phenomena from the side of mainstream science was begun by Glen J. Kuban, who holds the B.A. degree in biology from the College of Wooster and is presently employed as a computer programmer. Working largely independently until 1984, Kuban compiled a large volume of field data on the Paluxy tracks, including detailed site maps. This scientific information, together with its interpretation, is contained in his privately published 1986 monograph titled The Texas "Man Track Controversy" (Kuban, 1987).[1]

In the period 1982-1984, a group of four mainstream scientists--self-named the "Raiders of the Lost Tracks"--began investigating the claims made by the creationists for the Paluxy materials. One team member is Laurie Godfrey, mentioned above. Two others are Texas scientists particularly active in opposing censorship of Texas biology textbooks and other intrusions of creationism into Texas schools. Dr. Ronnie J. Hastings, with a Ph.D. in physics, is a public school science instructor residing in Waxahachie, not far from Glen Rose. He has made many visits to the Paluxy track exposures and has familiarized himself with the activities of many local persons on the Paluxy scene. Steven D. Schafersman of the Department of Geology, Rice University, Houston, holds the Ph.D. degree in geology from Rice University. As President of the Texas Council for Science Education and President of the Texas Chapter of The Voice of Reason, he is familiar with creationist activities in the education field. The fourth team member is anthropologist/archaeologist Dr. John R. Cole, formerly of the University of Northern Iowa and presently a staff member of the American Humanist Association and the journal Creation/Evolution; he is affiliated with the State University of New York at Buffalo. To the list of mainstream science activists we can add biologist David H. Milne of Evergreen State College, Olympia, Washington. He coauthored with Schafersman a review of the pros and cons of the Paluxy debate (Milne and Schafersman, 1983).

The "Paluxy Raiders" documented with a video tape record much of what they saw along the Paluxy riverbed. This material was greatly supplemented and finally edited by Hastings into an educational film titled Footprints in the Mind. Copies of the videotape were distributed in 1982 and 1983 for educational viewing, the aim being to present

---

[1]Distributed by the author, P.O. Box 33232, North Royalton, Ohio, 44133.

the evolutionary scientists' appraisal for comparison with the creationists' view, shown in their film, Footprints in Stone. Interest in the Hastings videotape was such that in 1983, a professional version videotape, covering the same material as Footprints in the Mind and featuring the same scientific team, was produced by Cole and made available for classroom use.[2] The culmination of the team effort was publication of a collection of its members' individual articles on the Paluxy materials, appearing in 1985 in the journal Creation/Evolution (Cole and Godfrey, 1985). This collection focused on early man-track claims and Baugh's claims and introduced only briefly a study of the Taylor Site. The concerted educational program of the investigative team represented the desire of mainstream science to state its case as fully and rationally as possible in answer to every aspect of the creationists' claims of man prints in the Paluxy dinosaur track locality.

Many additional interesting details of the investigations by mainstream scientists are given by Ronnie Hastings in an account titled "Tracking Those Incredible Creationists" (1985) and a sequel published a year later (Hastings, 1986a). With this introduction to the history of the Paluxy controversy, we can turn to a critical review of the evidence itself.

Geologic Environment of the Paluxy Tracks

The Paluxy River, a tributary to the Brazos River, has a flow regime to be expected of a rather short stream of steep gradient--torrential bankfull flow in response to heavy rains, but little or no flow in dry summer periods. Throughout the 11-km stretch in which most of the tracks are found, the Paluxy has a winding course entrenched in the Glen Rose Formation of Lower Cretaceous age. The lower member of this formation, which here is exposed in the river bed and lower ledges of its banks, consists of layers of resistant limestone, 15 to 20 cm thick, alternating with soft layers that consist of clay or clay-rich limestone. Rapid erosion of the soft layers in times of river flood undermines the hard layers, causing limestone slabs to collapse and break apart, to be swept downstream. In this way, bedding surfaces of the hard limestone layers are newly exposed to view. It is on these surfaces that the dinosaur tracks are found, and on which the supposed human tracks occur as well.

As deduced from its composition and fossil content, the depositional environment of the Glen Rose Formation probably was that of a tidal flat or shallow lagoon; perhaps it was sequestered from the open ocean by a barrier island or shell reef. The limestone layers accumulated as a lime mud consisting in part of lime shells and shell fragments of a variety of marine invertebrates, including foraminifera and mollusks. Exposed at low tide, the mud flats provided an ideal medium for receiving the dinosaur tracks. Preservation of such tracks also requires special circumstances. The sediment that buries the tracks must be of different texture than the layer beneath it and must be brought to the area without strong current action that would destroy the tracks. In a tidal lagoon, an influx of turbid river water, spreading widely over the tidal basin at high water, would provide a gentle rain of minute particles of clay minerals and finely ground quartz. The clay particles may, on contact with salt water, coagulate (flocculate) into larger particles that sink easily to the bed. Much later, after burial under many such alternating layers of sediment, the lime mud became hardened (lithified) to yield dense limestone layers, while the clay layers remained comparatively soft.

Besides the dinosaur tracks, bedding surfaces exposed

on the Paluxy riverbed bear various kinds of natural, minor irregular relief features. These may be (a) of original depositional origin or (b) produced by erosion processes, either immediately after deposition or during recent exposure to weathering processes under atmospheric exposure and to flow of river water.

Features produced by erosion processes have been interpreted by creationists as man tracks. They can be characterized as elongate shallow depressions of varied and irregular outer limits. The depressed area may contain smaller depressions that appear to be more rounded than elongate and in some cases cuplike. The development of cuplike depressions is a well-known feature of limestone exposed to hydraulic action in river channels. In a paper published in 1929 Philip B. King, a distinguished geologist known for his extensive research on the geology of Texas, described various kinds of grooved and cupped surfaces on limestone exposed in the channel and on benches of Barton Creek, near Austin. Features of this kind develop both from direct abrasion by rock fragments carried by the stream and by dissolution of the calcium carbonate by action of weak acid (carbonic acid) in the water. Laurie Godfrey offers an excellent photograph of this type of depression properly attributed to solution of the limestone (1985, Figure 8). The solution phenomenon is often referred to as karren erosion. This and other forms of erosional depressions are described and explained by Cole, Godfrey, and Schafersman (1985, pp. 38-39). In a related category are burrow casts made in the freshly deposited, soft sediment. (Such features, as well as vertebrate tracks, are of a class called "trace fossils.") A drawing of the casts is shown in Godfrey, 1985, Figure 13. Burrow casts are, however, of minor importance in the man track controversy. We need not spend more words on features so easily explained by solution and hydraulic action or the activities of invertebrates; they are of secondary scientific importance, even though they have been strongly presented by creationists as man tracks. Features of this kind at the State Park Shelf, a facility open to the public, were recognized as natural features by the Loma Linda scientific team (Neufeld, 1975).

This might also be an appropriate point at which to dispense with alleged human tracks that appear to have been carved out or altered in some way by local residents. These are not numerous and are easily recognized as artifacts. The rather gross carvings of "giant" manlike footprints, such as those described by Roland Bird, and appearing in Burdick's photos, reproduced by Whitcomb and Morris, were probably carved out of loose limestone slabs, carried to more convenient locations for tooling. Their authenticity as artifacts, rather than natural features, is no longer seriously doubted. The so-called "Caldwell track," a cast of which was displayed in Baugh's museum and copies of which were distributed as a premium for contributions to the Louisiana Creation Legal Defense Fund, has received wide publicity. (For further descriptions of artificially chiseled "giant" man tracks, see Godfrey, 1985, pp. 19-21 and Figures 3, 4.)

Dinosaur Prints Erroneously Interpreted as Man Prints

We now focus attention on the elongate depressions, arranged in trails, that have been most vigorously offered by the creationists as genuine man tracks. This class of tracks is found in close association on the same bedding surfaces with unquestioned dinosaur tracks. Two aspects of the question are (a) recognition of physical features of the elongate prints that cast serious doubt on their human origin, and (b) recognition of physical features that point strongly to their being a distinctive variety of dinosaur track. The first category of evidence is of a negative form (the proposition is not valid); the second is of a positive form (the proposition is valid). Both are useful

---

[2]John R. Cole and Pia Nicolini, Producers, 1984, The Case of the Texas Footprints. Videotape (VHS or Beta), 27 minutes, distributed by ISHI Films, P.O. Box 2367, Philadelphia, PA 19103.

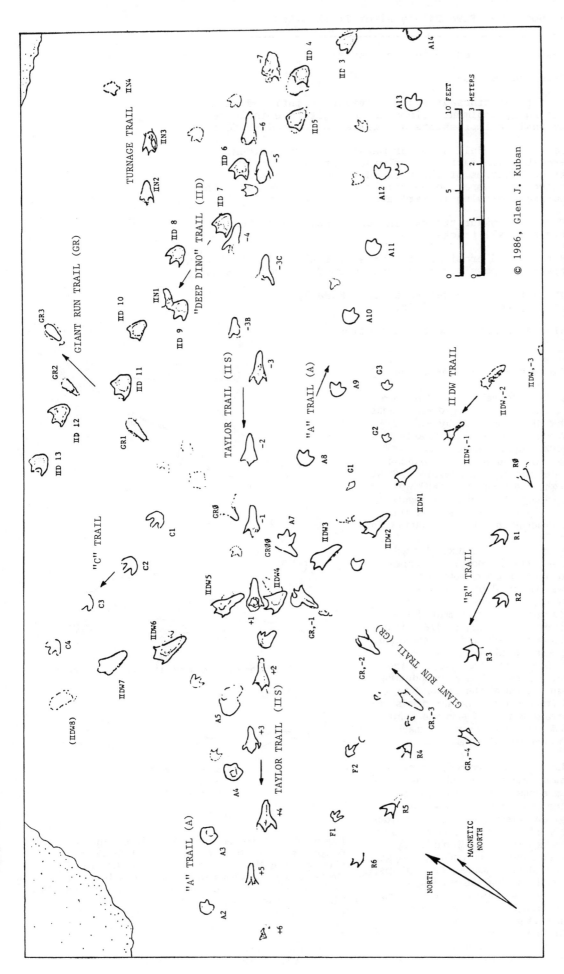

Figure 48.4 Map of main section of the Taylor Site, Paluxy River, Glen Rose, Texas, based on measurements and photographs taken by Glen J. Kuban, Ronnie J. Hastings, and associates in the period from 1980 through 1985. The north bank of the river lies just beyond the upper margin of the map. The Ryals Trail and extension of the IIDW Trail occur just south of the site shown on the map. The track outlines represent the borders of color distinctions and/or relief differences with respect to the surrounding rock surface. (Copyright © 1986 by Glen J. Kuban. Used by permission.)

Table 48.1   Condensed Summary of Five Paluxy River Track Sites

Site:	Features alleged to be man tracks:
State Park Ledge	Erosional marks
Taylor Site	Metatarsal dinosaur tracks, evidently infilled with secondary sediment. Dinosaurian digits are indicated by shallow impressions and dramatic color distinctions.
Baugh Sites[*]	Partial and full dinosaur metatarsal impressions. Probable dinosaur tail or manus marks. Natural irregularities of the rock surface. Evidence of selective tampering on some depressions; at least one outright carving in marl above track surface.
Original McFall Site	Eroded dinosaur tracks, including some elongated (metatarsal) tracks.
Dougherty/ von Däniken Site	Natural and erosional features. One alleged "giant track" is possibly a severely eroded dinosaur track.

[*]The area of the Baugh Sites is referred to as the "McFall Site" in Creation/Evolution, vol. 5, no. 1, pp. 34-35.

forms of argument, but the positive form deserves the greater emphasis.

We must digress at this point to make clear that there are five principal Paluxy sites at which alleged man tracks occur (Kuban, 1987). Table 48.1, prepared by Kuban, summarizes the salient features of each. One, already mentioned, is the State Park Ledge, bearing only erosional features. Another, the Dougherty/von Däniken Site, displays largely natural and erosional features. That leaves three (Taylor, Baugh, and Original McFall) with abundant dinosaur tracks, along with trails of elongate depressions that are offered by the creationists as man tracks. We shall limit our description and interpretation to only one of the three--the Taylor Site.

Rough maps of print sites along the Paluxy channel were prepared by creationist John D. Morris and appear along with photographs in his 1980 book, referred to above. Morris's material was reviewed by Milne and Schafersman (1983). Their paper includes a redrawn, modified, and annotated version of Morris's map of the Taylor Site (Milne and Schafersman, 1983, p. 116, Figure 7). The map shows two dinosaur trails and five chainlike sequences of elongate depressions alleged to be human footprints. Kuban and his field associates remapped the Taylor Site in great detail. Figure 48.4 is a reproduction of Kuban's map of the main portion of the Taylor Site (Kuban, 1987). Tridactyl (three-toed) dinosaur prints are easily recognized. Other trails consist of elongate prints with only weakly developed toe marks, or no toe marks; these are the objects on which attention needs to be focused.

The elongate tracks alleged to be human footprints are quite different from dinosaur tracks we usually see pictured or displayed in museums. Glen Kuban devoted much of his research at the tracked Paluxy sites to measuring, photographing, and casting the different track forms. On the basis of this information he arrived at an explanation of how the tracks originated. Kuban's key to understanding the two general classes of tridactyl tracks lies in his recognition of two distinct modes, or positions, assumed by the dinosaur while making the tracks. This key led Kuban to make new and significant contributions to the science of ichnology (the study of fossil tracks and other markings made by live organisms). By way of background, I refer to Edwin H. Colbert's popular book, Men and Dinosaurs, in which there is a set of drawings by W. A. Parks, showing a Cretaceous bipedal dinosaur

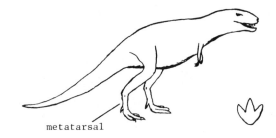

A Digitigrade stance

B Plantigrade resting/feeding position

C Plantigrade walking

Figure 48.5 Postures used by a bipedal dinosaur. The footprints are quite different in digitigrade and plantigrade modes. Sketches (A) and (B) are of a Cretaceous carnivorous dinosaur from Alberta. Plantigrade walking tracks (C) are common in the Paluxy riverbed. The plantigrade resting/feeding type (B) is not found in the Paluxy riverbed, but is known from other track localities. (A. and B. From E. H. Colbert, 1968, Men and Dinosaurs, E. P. Dutton, New York, p. 194, Figure 33; reconstruction attributed to W. A. Parks. C. Drawn by A. N. Strahler and R. J. Hastings.)

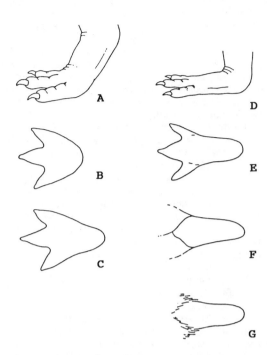

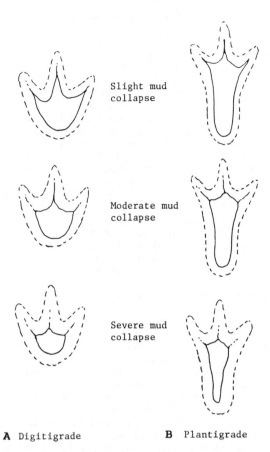

Figure 48.6 Variations of bipedal dinosaur tracks. A. Dinosaur foot in a digitigrade, or "toe-walking" stance. B. Typical digitigrade track. C. Track with a partial metatarsal (sole) impression, misinterpreted by some as a man track overlapping a dinosaur track. D. Dinosaur foot in a plantigrade stance. E. Distinct plantigrade track. F. and G. Plantigrade tracks with indistinct digits, due to mud back-flow (F), erosion (G), or other factors, resulting in shapes that superficially resemble "giant man tracks." (Drawings and legend by Glen Kuban. Copyright © 1986 by Glen J. Kuban. Used by permission.)

Figure 48.7 Tridactyl dinosaur tracks modified by mud collapse following withdrawal of the foot. The plantigrade tracks formed by moderate and severe mud collapse are easily misinterpreted as man tracks, especially when further subdued by erosion or infilling. (Drawn by Glen Kuban. Copyright © 1986 by Glen J. Kuban. Used by permission.)

in four poses: standing, sitting, feeding, and resting (Colbert, 1982, p. 194, Figure 33). These drawings clearly show the feet in two basically different positions. Our Figure 48.5, adapted from those by Parks, will help clarify the positions. In diagram A, the three-toed dinosaur is "walking on its toes" in the <u>digitigrade</u> <u>mode</u>, leaving only the imprint of the toes and the area where they join. Many of the Paluxy trails are of this kind. In diagram B, the animal is crouching down into the <u>plantigrade</u> <u>position</u>, bringing the metatarsal, or "heel" into a horizontal attitude. Apparently, it could also walk in this manner, perhaps to enable it to keep its head down low to search the ground as it moved, as shown in diagram C. Kuban introduced the possibility that the dinosaur could also walk upright with its feet in the plantigrade position. Kuban recognized that the plantigrade walking mode seems best to explain the Taylor trail and others like it. Figure 48.6 shows how Kuban explains the variations in tracks of the Taylor Site according to each of the two walking modes.

In certain track areas, a second factor influenced the appearance of the track: the degree to which soft, wet mud flowed back into the track. Kuban's drawings of the effect of different degrees of mud collapse are shown in Figure 48.7. Notice that collapse occurs more or less uniformly inward around the entire rim of the initial track. As a result, the narrow toe impressions close entirely whereas the wider metatarsal portion becomes a vague, oblong depression. A completely continuous series of track forms is generated by the extent of mud collapse. In the extreme case, the plantigrade track is reduced to a shallow, elongate form that easily inspires a susceptible person to see it as a human footprint. Kuban's painstaking reconstruction from actual prints of the sequence of changes caused by mud collapse is in itself a

convincing line of evidence favoring the dinosaurian origin of all the trails of the Paluxy sites. Keep in mind that other factors of both primary and secondary origin also contribute to the alleged "human" appearance of the elongate dinosaur tracks; these include erosion, imperfect initial impressions, and subsequent infilling with fresh sediment.

This positive form of evidence greatly outweighs the negative line of evidence, based on the total lack in the Paluxy surface depressions of features diagnostic of a human footprint, namely, the familiar heel-and-ball depressions and a set of five toes. Another important consideration is dimensional--track length and stride length. Both dimensions are clearly more typical of dinosaur tracks than of human tracks. That stride length for the elongate (plantigrade) tracks is similar to that of the clearly bipedal (digitigrade) dinosaur tracks is obvious from a close look at the stride lengths of the trails plotted on Kuban's map (Figure 48.4). Measurements of strides and track lengths of elongate tracks made independently by Hastings and Kuban at the Taylor Site fit well among typical dinosaur trail measurements. Laurie Godfrey has treated the anatomical aspects of the comparison between human footprints and those of dinosaurs, as well as the comparison of foot lengths and stride lengths (Godfrey, 1985).

By 1984, Kuban had amassed a large body of convincing evidence that the elongate man tracks are indeed dinosaurian. But there was a final scene yet to be played, in which the creationist claim to the existence of Paluxy man tracks was given the coup de grace.

Shades of Color Close the Story of Paluxy

The players who come on stage to act out this final scene in the Paluxy drama are Kuban and Hastings, who had met along the Paluxy in August 1984. Hastings had been engaged in making casts of tracks under shallow water at the Taylor Site. Then, thanks to a rainless September, the riverbed of the Taylor Site became dry. Kuban, who was nearing the completion of his exhaustive treatise on the Paluxy tracks, came from Ohio to take full advantage of this rare situation to make additional observations. He was joined by Hastings and a group of students. Together, they undertook to clean off the Taylor Site for a better look at the alleged man tracks. Hastings reported: "What we saw was quite a surprise to all of us" (1987a, p. 5).

As a preface to their remarkable discovery, we must go back a month or so to relate a general observation made independently by both Hastings and Kuban. Each had noticed that, as the water level fell and the rock surfaces dried out, there appeared zones of distinctive surface coloration related to the tracks.

Hastings described his discovery of the coloration phenomenon as follows:

> In August of 1984, I had seen a fascinating phenomenon similar to those color patterns on an undocumented dinosaur trail exposed by the drying riverbed nearby. Outlined in the dry limestone was a series of three depressionless tridactyl dinosaur prints. . . . The outline could be seen only because the color of the material inside the track areas was slightly different from that outside the tracks, being reddish-brown and blue-gray in contrast to the surrounding limestone's ivory shade. Kuban (who had independently seen this phenomenon on other trails) and I called this color contrast "discoloration" or, later, "color distinction." (1987a, p. 5)

On closer examination, it was found that the color distinctions also corresponded with at least slight relief differences between the print area and the surrounding rock surface, and with rock texture differences, as well. These findings served to strengthen the evidence from the color phenomena.

At the Taylor Site there is a particularly crucial display of elongate tracks, alleged to be man tracks, but already recognized by Kuban as the plantigrade form of dinosaur tracks. Called the Taylor Trail, it crosses diagonally two obvious dinosaur trails. (Find the Taylor Trail on Kuban's map, Figure 48.4; it runs from right to left across the middle of the map.) What Hastings and Kuban documented at the Taylor Site was that tracks of the Taylor Trail showed color distinctions associated with the shallow depressions. The color area covered the entire track and was seen to extend out from the anterior (front) end of the narrow, elongate "foot" depression to reveal at least one of the three toes of a dinosaur. Some tracks showed two distinct toes splayed to the sides of the anterior end, while several tracks actually showed clearly all three toes of a typical tridactyl print (Figure 48.8). Hastings wrote: "There can be no doubt that the Taylor Trail was made by a bipedal, tridactyl dinosaur which left unusually elongate tracks" (1987a, p. 5).

As the exposure of bedding surfaces to drying continued, previously undocumented dinosaur trails appeared due to the color distinctions. Entire trails were found with very little surface relief but dramatic coloration revealing the full extent of the tracks. Of course, the scientists made a complete photographic record and measurements of the tridactyl prints revealed by the color distinctions. Similar evidences were found on prints of the Turnage Trail, the Giant Run Trail, and the Ryals Trail, all of which had been claimed as man tracks by many creationists.

Hastings and Kuban have obtained evidence from the study of samples of the rock, both within the color-distinct area and across the boundary into the surrounding rock, that infilling by subsequent influx of sediment is the most plausible explanation for the color distinctions. (Infilling should not be confused with mud collapse that, on most Taylor Site tracks, was evidently of less importance in track development than infilling.) The samples conclusively show that the colorations correspond to deeper subsurface features. Texture differences between the color-distinction areas and the surrounding rock are also recognizable by direct observation. (These observations preclude any possibility that the color distinctions were painted on the rock surface by humans.) Impressions made by the dinosaurs' toes were shortly thereafter filled with sediment of a different physical/chemical makeup than the sediment into which the animal had impressed the toeprints. In some cases the color-distinct rock is separated from the surrounding rock by a faint crack, or fissure. Laboratory studies (in progress) of mineral composition and texture of the colored rock from within the print limits should reveal significant differences from that of rock of the surrounding limestone. Recent chemical weathering would explain the reddish iron-oxide compound that now colors the rock surface within the area of some of the tracks.

The color distinctions at the Taylor Site were observed by some interested visitors, among them Kyle Davies, a

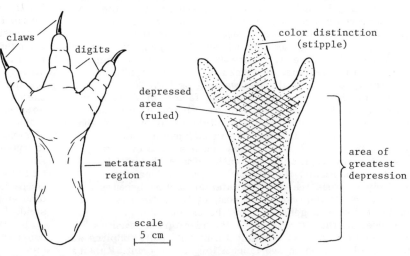

Figure 48.8 Relation of color distinction to elongate, infilled dinosaur tracks of the Taylor Site. (Copyright © 1987 by Glen J. Kuban. Used by permission.)

A Hypothetical reconstruction of underside of foot

B Track with color distinction

paleontologist from the University of Texas at Austin. Creationist Carl Baugh was another visitor. Kuban reports that in 1984 in his presence Baugh remarked, "No one would call these tracks human."

The dry conditions of September 1984 provided a rare opportunity to observe the evidence firsthand, but the opportunity was in imminent danger of being lost as the river rose in October to inundate the rock surfaces and to begin to deposit sediment on the tracks. Accordingly, Kuban telephoned John Morris from the site, urging him to come and inspect the color distinctions and other dinosaurian features of the tracks, or at least send another ICR representative. However, no one came from that organization. Other creationists were contacted as well but did not come. In August 1985, when the colorations had become even more dramatic, Kuban again urged Morris to come, and again Morris declined (Kuban, personal communication).

The importance of the new evidence was so great that, in September 1985, Kuban sent a letter addressed to five members of the ICR staff, reviewing in detail the nature and significance of the color distinctions and other evidence of dinosaurian origin of the tracks. Kuban included photos and slides to illustrate his points. This time a response followed quickly. The ICR delegated John Morris to visit the site and to see the evidence in the company of Kuban. At this time the tracks were covered by a few centimeters of water, but the water was very clear and the color distinctions were still vivid. Morris arrived with four others, one of whom was Paul Taylor, who had helped his father, Stanley Taylor, make the film _Footprints_ in _Stone_. Hastings wrote: "According to Kuban all the visitors were astounded at what they saw" (1986, p. 23). The creationist group examined the tracks and there followed a lengthy discussion at a nearby motel. The visitors attempted to find reasons to question the dinosaurian features, and to hold on to the possibility that some of the tracks might be human, but Kuban patiently reviewed the evidence over and over. Hastings reported: "By November, all the creationists present agreed at least that the Taylor trail was made by a dinosaur" (1986, p. 23) A few weeks later, John Morris and other creationists revisited the Taylor Site. In the company of Kuban, they reexamined the Taylor Site tracks. Later, John Mackay, an Australian creation scientist, came with Taylor to take core samples of the color distinctions.

The Creationists' Response

Late in November 1985, Paul Taylor, John Morris, and other members of the ICR staff met near San Diego to prepare a response. It appeared in the January 1986 issue of ICR _Impact_, in which John Morris wrote:

> In view of these developments, none of the four trails at the Taylor site can today be regarded as unquestionably of human origin. The Taylor Trail appears, obviously, dinosaurian, as do two prints thought to be in the Turnage Trail. The Giant Trail has what appears to be dinosaur prints leading toward it, and some of the Ryals tracks seem to be developing claw features, also. (Morris, J. D., 1986a, p. iii)

The text of Morris's article reads as if creationists of the ICR staff had done most of the original research on the color-distinction phenomena; this is, of course, a gross injustice to Kuban and Hastings. Morris goes on to list six "mysterious" points about the Paluxy tracks that he considers to remain significant. These are posed in the form of questions, implying that the mainstream scientists' interpretation of the evidence is flawed. In reference to the distinctive color that reveals the track outlines, Morris (who calls the color a "reddish stain") states that

such coloration can easily be achieved by applying "readily available chemical agents." The unwarranted suggestion of a fraud having been perpetrated seems clearly evident in Morris's words, but is not explicitly stated. Morris concludes his report with these words:

> Even though it would now be improper for creationists to continue to use the Paluxy data as evidence against evolution, in the light of these questions, there is still much that is not known about the tracks and continued research is in order. We stand committed to truth, and will gladly modify or abandon our previous intepretation of the Paluxy data as the facts dictate. (1986a, p. iv.)

In a letter enclosed with the January 1986 issue of _Impact_, Dr. Henry Morris, President of the ICR, reinforced the ICR conclusions we have quoted above, but he adds: "At the same time, it should be emphasized that this question in no way affects the basic creation/evolution issue. These tracks have always been only illustrative, not definitive, and the over-all scientific case against evolution, which is overwhelmingly strong, is not affected in any way."

Glen Kuban soon published in _Origins Research_ a detailed reply to John Morris's ICR article, in which he takes up each of the six "mysterious points" raised by Morris (Kuban, 1986c). Kuban takes special pains to reiterate evidence that the color distinctions are associated with an underlying rock material and cannot be interpreted as superficial applications of stain as a hoax. Kuban then takes the initiative by showing that Morris's drawing of the typical color distinction is inaccurate.

In the same issue of _Origins Research_, John Morris published a brief statement titled "Follow Up on the Paluxy Mystery" (1986c). He insists that the creationists' man print interpretation was, at the time it appeared, a sound one in view of the data then available. Significantly, the implication that a painting hoax may have been perpetrated would seem to stand retracted by the following statement:

> Many have suggested that someone may have stained the surface surrounding some of the humanlike tracks, to given them a reptilian appearance. However, no evidence of fraud has been found, and some hints of these dinosaur toe stains have now possibly been discerned on photos taken when the prints in question were originally discovered. . . . Evolutionists have long falsely accused creationists of fraudulently carving their tracks (with no evidence), but we must not resort to their tactics by similarly charging them with fraud without clear evidence. (1986c, p. 14)

Also in the same issue of _Origins Research_ there appeared a statement by Films For Christ (FFC), of Mesa, Arizona, explaining that field visits by the FFC staff impressed them with the validity of the color distinctions and their significance; also that they had found good reason to doubt that the colorations were the result of a hoax (Films for Christ Association, 1986, p. 15). Accordingly, they had immediately ceased accepting any new requests for showings of their film _Footprints_ in _Stone_. On the other hand, FFC urges caution in discarding previously gathered information on the tracks earlier regarded as human. They state that "the new evidence gathered certainly requires a re-evaluation and possible reinterpretation of all the data gathered." They add: "We highly recommend that no one represent any of the Paluxy tracks as proven evidence of human existence during the Cretaceous until final, reliable conclusions can be reached regarding new and old data."

Aftershocks from the devastating quake released by Kuban and Hastings continued to be felt in many quarters. The far-reaching temblor evoked an article in Nature, written by Tony Thulborn, president and founder of the Australian Association for the Protection of Evolution (APE) (vol. 320, p.308). John Morris quickly replied with a letter, emphasizing that he had seen no direct evidence of fraud in the color distinctions and expressing his conviction that the interpretation of the elongate tracks as of human origin "was not only valid, but arguably the best" (Nature, vol. 321, p. 722). Responses came in the form of a second wave of letters from the scientists. One that was published in Nature came from fossil-track specialist James O. Farlow of Indiana University-Purdue University at Fort Wayne. Farlow has a wide range of field experience in Texas dinosaur track sites (including Paluxy). What caught my eye in Farlow's letter was this information about color distinctions: "Although unusual, the phenomenon is not 'unprecedented' as Morris states; a splendid bipedal dinosaur trail with footprints delimited by such color distinctions occurs along the Alameda Parkway, near Denver, in Cretaceous Rocks of the Dakota Group; the site is being studied by Martin Lockley of the University of Colorado at Denver. Lockley also has some sauropod footprints, preserved only as colored tracks, occurring in the Morrison Formation (Jurassic) along the Purgatory River in Colorado."

An interesting item of creationist news appeared in the Bible Science Newsletter, July, 1986. It seems that the Bible-Science Association has responded by forming its own "Paluxy Task Force," commissioned with the carrying out of a full-fledged field and laboratory reexamination of the Paluxy tracks and the "recently identified stains." They state: "Our goal is to conduct an independent analysis to determine the exact nature of changes in the evidence at Paluxy."

The Paluxy disclosures enjoyed a remarkably good response from the news media. The New York Times, June 17, 1986, captioned its article by John Noble Wilford "Man Tracks Revealed to Be Dinosaurian." Time Magazine, June 30, 1986, used the headline "Defeat for Strict Creationists."

Have the young-earth creationists really suffered a defeat along the banks of the Paluxy River? If so, do they know it? If they have been defeated, it is to them only a minor and temporary setback. Their army fights on many fronts.

Bones in Stone--The Moab Skeletons

Now, how about those human remains alleged to have been fossilized in strata many tens of millions of years old? Two such cases have recently received the attention of both creationists and mainstream scientists. One is a claim that human remains are enclosed in rocks of Jurassic or Cretaceous age. If this were proved a valid claim, it would invalidate my previous statement that no hominid fossils have been found in Mesozoic strata. The second case is that of a human skeleton claimed to be a fossil enclosed in Miocene strata, giving it an age of some 20 million years.

The case of the alleged Jurassic/Cretaceous human fossils starts back in 1971. That is the point to which I have traced its beginning, thanks to a large file of letters and clippings passed on to me by Dr. Ronnie J. Hastings of Waxahachie (one of the team of Paluxy "Raiders" referred to earlier). The find surfaced as a news report in October(?) 1971 in the Moab, Utah, Times Independent. It is perhaps from this source that creationist Clifford L. Burdick (now familiar to you as a player in the Paluxy drama) learned of the find, investigated it further, and published his own account of it in 1973 in the Creation Science Research Quarterly (Burdick, 1973). In 1975 another account of the fossil find appeared in Desert Magazine, a popular journal widely distributed in the western United States. The author was F. A. Barnes,

Utah Associate Editor of the magazine; his title, "The Case of the Bones in Stone."

According to Barnes, the find was purely accidental (as surely all such amateur discoveries must be) during the course of a field trip of a group of "rockhounds" intent on collecting specimens of the mineral azurite. In Lisbon Valley, near Moab, as they searched over freshly bulldozed land adjacent to an active copper mining operation, one member found what appeared to be a human tooth. Further search turned up several more teeth and even a bone fragment that was obviously from a human skull or jawbone. The ground surface beneath the bulldozed area is described in the article as "semi-rock sand" and "decomposing sandstone." Further search over this material soon disclosed a brown-colored patch, and when this was dug into with a knife blade, the rounded end of a bone was revealed. It had a greenish tinge, apparently staining by a hydrous copper-carbonate compound derived from other copper-bearing minerals in the vicinity.

Why am I relating these minute details accompanying the discovery? The crucial question proves to be whether or not the human fossils--for such they are--were actually found enclosed in solid, undisturbed bedrock, which in this case would be the Morrison Formation of Late Jurassic age, famous throughout the West for its dinosaur fossils, or possibly the overlying Dakota sandstone formation of lowermost Cretaceous. Alternatively, they may have been preserved in weathered mineral matter (called "regolith" by geologists) derived from one of the above-named sandstones, but of comparatively recent age as a surficial deposit overlying or downslope from the undisturbed outcrops of bedrock.

Barnes writes that the leader of the rockhound group, a man named Lin Ottinger, was well aware of the potential value of the human bones as evidence that humans lived in the Age of Dinosaurs, a creationist theme that must have been widely disseminated from the Paluxy controversy. Ottinger ceased further excavation, covered the bone, and the group departed. Ottinger promptly notified Professor W. Lee Stokes at the University of Utah (previously introduced to you in connection with the Cambrian "Meister print"), who in turn referred the case to John P. Marwitt, a graduate student and Ph.D. candidate in anthropology in the same university. About a week later, Marwitt and Ottinger, accompanied by a "natural history television photography team," a reporter, and several other persons arrived on the scene (Barnes, 1975, p. 38). Marwitt began to excavate the buried bone. Barnes describes the matrix as being "neither hard sandstone nor loose sand, but somewhere in between, a kind of semi-rock that forms when loose sand bonds together in the presence of moisture over long periods of time" (p. 38). It was "too hard to penetrate with bare fingers, yet scraped away readily enough with the knife blades and pointed trowels being used for the delicate excavation." Complete excavation yielded two partial human skeletons, each missing the upper part. Barnes states that the anthropologist (Marwitt) summarized the situation as one in which "the bones were obviously human and 'in situ,' that is, in place and not washed or fallen into the stratum where they rested from higher, younger strata" (p. 38). One paragraph by Barnes needs to be given verbatim for careful scrutiny and later comment:

> There was some question as to the exact geological formation in which the bones were found. Mine metallurgist Keith Barrett, of the Big Indian Copper Mine that owned the discovery site, recalled that the rock and sandy soil that had been removed by 'dozer from above the bones had been solid, with no visible caves or crevices. He also remembered that at least 15 feet of material had been removed, including five or six feet of solid rock. This

provided strong, but not conclusive, evidence that the remains are as old as the stratum in which they were found. (P. 38)

Although Barnes's text shows strong bias toward the bedrock enclosure interpretation, he did insert a paragraph explaining that there still remained a possibility that there may have existed some sort of natural or excavated opening into which the humans might have entered or fallen to arrive in a space enclosed on all sides by undisturbed bedrock. The remainder of Barnes's article is given to deploring the alleged failure of the university scientists to have the bones dated by radiometric means. That they did not do so is, according to Barnes, a mystery in itself. Barnes says that Lin Ottinger, growing tired of waiting for more than a year, reclaimed his box of bones. Eventually, the bones were purchased by Carl E. Baugh, for a price (it is said) of $10,000, and they were exhibited in his Creation Evidence Museum near Glen Rose. These are now the type specimens of Baugh's new species, Humanus Bauanthropus.

Going back to Burdick's 1973 paper in CRC Quarterly, I was on the lookout for things Barnes might have picked up for his Desert article, as well as for unwarranted creationist assumptions and conclusions. Burdick says there could be no doubt that the bones were definitely "in place," as there was no sign of disturbance of the surrounding "rock," and he himself had dug "a foot deep at the site" (p. 110). He further states:

It was evident from the location of the find deep within the man-made pit that the bodies were buried at the time of emplacement of the sandstone rock. (P. 110)

The "man-made pit" is, of course, the recent excavation by bulldozer that, Burdick notes earlier, stopped about fifteen feet below the previous surface of the hill. He says that the mine superintendent stated that at least six feet had become hard rock, though still soft enough to bulldoze. Altogether, it seems clear enough that Burdick wished his readers to have no doubt that the skeletons were actually enclosed in strata of Mesozoic age, requiring that the skeletons also be of that age. Referring to Marwitt's alleged disposition of the bones at the university, Burdick says that Marwitt exhibited little interest in them as a museum display. Burdick then states: "Could it be that their association with the Cretaceous rock, presumed to be very old, could be the reason?" (P. 109)

I suspected that there was another side to the Moab story, and that it could best be presented by John Marwitt, who excavated the bones and whose training in anthropology, coupled with field experience in Indian anthropology of the West, put him in the best possible situation to evaluate the geology of the site. Turning to copies of correspondence of Marwitt and others beginning in 1971, I quickly found the other side of the story.

Writing on October 19, 1971, from his new academic post in Ohio, Marwitt replied to a letter of inquiry from a local promotional association in Moab whose director wanted to know more about the content of a local newspaper article reporting the find. Perhaps it was the article in the Moab Times Independent, written by the local news correspondent mentioned by Barnes as being present when Marwitt excavated the bones. Marwitt's letter contains the following paragraph:

In relation to the finds in Lisbon Valley, I tried to impress upon the local reporter that there was no possibility of the finds themselves being 100 million years old. The burials are of completely modern type and were obviously intrusive into the deposits in which they were found, but evidently the reporter in question was looking for a story, not the truth. Since his article was not cleared with any scientfcally responsible person, I must dissociate myself and the University of Utah with any claims made therein.

Marwitt's next correspondence in my file took place in the fall of 1973; it was in reply to an inquiry from Duane E. Jeffery, Assistant Professor of Zoology at Brigham Young University, Provo. It seems that local newspapers had been repeating the Moab story with its creationist slant. Jeffery had also been told of Clifford Burdick's 1973 article in the CRC Quarterly. I suspect that Jeffery was anticipating some questions being directed to him by local creationists and needed to be forearmed. Marwitt's reply, dated November 27, 1973, contains the following paragraphs:

As you probably expected, Burdick's account of the human skeletons found near Moab is not strictly true. The burials had apparently been placed in a rock crevice (a common practice historically and prehistorically) and fell to the bottom. Blasting and bulldozing later removed the overburden and exposed the burials. I took some pains to point out to all concerned, including the "Creation Science Research Society," that in no sense could the human remains be seen as contemporary with the sandstone deposits. The burials were surrounded and covered with loose blowsand; they were not in a rock matrix as implied by Burdick. The bones themselves were not fossilized and there had been no replacement of bone calcium by mineralization. They were soft, friable and partly decayed--in short, of rather recent vintage, probably historic Paiute or Ute. As a matter of fact though, since no associated artifacts were found the burial could possibly be of Euro-American origin.

There is no need to look further for the answer to the creationists' warped version of the Moab skeletons. Marwitt's statements are complete and succinct. There are in my file some letters of more recent dates to and from Marwitt, but nothing in them that needs to be added to the quoted paragraphs. Marwitt was burdened with inquiries, but these seem to have been mostly from academic colleagues and well-wishers. Nevertheless, each time the creationists' version of the Moab skeletons has been aired, it has carried the insinuation that scientists of the University of Utah deliberately suppressed information and failed to carry out the further investigation--such as dating--that the skeletons warranted. Marwitt has fully answered such insinuations. I did, however, learn from Ronnie Hastings of a new development that might now alleviate such pressure: the Moab bones were dated at the University of California, Los Angeles; their age, 200 to 300 years (Creation/ Evolution Newsletter, 1984, vol. 4, no. 6, p. 23).

"Miocene Man" or Modern Woman?

Finally, we come to "Miocene Man," whose discovery story actually takes us back to the early 1800s. Miocene Man has resided in the British Museum since 1812, when "it" was presented to that establishment as the sole remaining skeleton of many like it that were excavated on the Caribbean island of Guadeloupe. According to creationist author Bill Cooper, writing in Ex Nihilo (1983, p. 6), the fossilized bones were taken from a limestone formation near the town of Moule, their removal superintended by one General Ernouf. The rock was extremely hard and the removal task made difficult by the position of the bed below the level of high tide. Enclosed in a block weighing two tons, Miocene Man (or should it really have been "Miocene Woman?") was shipped by the

Royal Navy to the British Museum, where it was on regular display. This exposure ceased when Darwin's theory of descent with modification became popular. Cooper says that he became the first member of the public to see the skeleton since the early 1930s (p. 7). Cooper describes the skeleton as that of a woman (indeed!), about 5 ft 2 in (157 cm) in height. Cooper emphasizes the "evidence of extensive damage caused by tremendous natural forces" (p. 7). The spinal chord had been dislocated and various bones wrenched from their correct positions. At this point, creationist watchers will forecast what Cooper has in mind. We read on:

> The significance of this evidence is the fact that since the bones are still together (even though dislocated and broken), all of these injuries occurred before the body had decayed and before the rock had solidified around the body. The damage is in fact, only consistent with that caused by a fluid mass and/or water exerting a colossal impact similar to that produced by a tidal wave, the sediment then enveloping the body before solidifying into rock. Had the impact been caused by already hardened rock, then the damage to the skeleton would have been of totally different nature. (P. 7)

Well, there you have it: the Flood of Noah "done her in," poor soul! Cooper goes on to say that extremely rapid burial of the corpses is indicated by the very high organic content of the rock in the immediate vicinity of the bones. This organic matter was, he suggests, derived from the flesh of the bodies. If the formation is indeed of Lower Miocene age (-20 m.y.), then (he claims) so are the enclosed skeletons. The remainder of Cooper's article is orthodox creationist argument and includes a quick flashing of the uniformitarianist corpse, as well as a severe admonition as to what happens to humans who refuse to heed the word of God.

There are people who read Ex Nihilo and do not hesitate to take issue with its creationist authors. Two such persons are Michael Howgate and Alan Lewis, the former of the Department of Zoology at University College, London, the latter a geologist. Their report appeared in New Scientist (March 29, 1984, pp. 44-45). It seems that following publication of Cooper's article in Ex Nihilo and subsequently in pamphlet form for a British-based creationist organization, the British Museum was beset with inquiries about Miocene Man, or more correctly, Guadeloupe Woman. Museum representatives thoroughly repudiated accusations of having perpetrated a conspiracy of silence. True, the fossil specimen had been residing on top of a cabinet in the museum basement, but only "after being on public display for many years between 1882 and 1967."

Cooper then gave a public lecture on Guadeloupe Woman. The lecture was under the auspices of the British-based Creation Science Movement (CSM). It was here that our heroes, Howgate and Lewis, made their appearance. They are the sole representatives of Britain's only anticreationist organization, APE (Association for the Protection of Evolution). During the question period following his talk, Cooper was asked by a creationist listener how it was established that the skeleton was of Miocene age. Put on the spot, Cooper was forced to concede that it was his own assessment. (In his paper, Cooper states: "According to conventional geologic dating, the limestone was supposedly laid down some 25 million years ago [Miocene Era].") APE then pressed the attack by suggesting that the skeleton be independently dated by the radiocarbon method, which could give reliable ages at least as far back as the late Pleistocene and thus potentially come up with an age no older than that already established for other fossil humans.

Shortly after this encounter the debate took a new turn when a rival creationist organization, the Biblical Creation Society (BCS), stepped in to censure their colleagues in the CSM. According to the report in New Scientist "experts" from the BCS examined Guadaloupe Woman and repudiated the claim that the fossil is a relic of the Flood. The report suggests that in this strange creationist internal self-censure "the men from APE have been privileged by history to witness the first signs of a Kuhnian paradigm shift in creation 'science'."

In closing this episode, I call your attention to a letter to the editor of New Scientist (April 26, 1984, p. 52). It contributed the following comment on Guadeloupe Woman by the distinguished British geologist (and the last great diluvialist!), Rev. William Buckland, taken from his 1836 volume, Geology and Mineralogy:

> There is no reason to consider these bones to be of high antiquity, as the rock in which they occur is of very recent formation, and so composed of agglutinated fragments of shells and corals which inhabit the adjacent water. (Pp. 104-105)

It does look as if the matter has been closed since 1836, doesn't it? After all, who would have greater authority in such matters than the Last of the Great Diluvialists?

The Guadeloupe Human Skeleton
Plate I in Essay on the Theory of the Earth, Third Edition, 1817, by M. Cuvier, with notes by Professor Jameson. The Guadeloupe fossil is described in detail on pages 253–57. (William Blackwood, London, Publishers. Courtesy of John R. Armstrong.)

FOSSIL
HUMAN SKELETON.
FOUND IN GUADALOUPE

PART VIII

The Rise of Man and Emergence of the Human Mind

Introduction

In this, our eighth part, we continue to focus upon the general problem of origins--the creationistic view in opposition to the naturalistic view. We continue to probe into natural history, focusing on the origin of the human genus, <u>Homo</u>, and carrying the development of that animal to higher levels of consciousness and mental activity that seem to set apart the one extant human species from all other animals, including even the apes that are the human's closest evolutionary relatives.

A search for ancestors of our genus starts far down the tree of mammalian evolution at a point early in Cenozoic time--some 60 million years ago--when the first primates appeared. Following the branching evolutionary tree up through the Cenozoic Era, we look for forks that mark the points of common ancestry with each of the other primate families--prosimians, monkeys, and apes. At each potential fork the creationists form a cordon of resistance to bar our attempts to form a new branch that might eventually lead to humans. Creation scientists will tell us that no such evolutionary tree exists, that we harbor only an illusion. For them, each primate family is seen as a single and separate stalk, planted in the instant of recent creation. The urgency of the creationists' challenge mounts at each evolutionary step, reaching a crescendo at the inferred fork marking the common ancestor of humans and apes.

Somewhere along our climb through the primate tree in the company of the paleontologists, we begin to encounter the anthropologists, who have worked their way down the human branch in search of the common fork with the apes. These two groups work together very nicely, on the whole, but with endless arguments between individuals over the meaning of the often-fragmentary fossil evidence. The creationists are listening closely to what seems like a succession of irreconcilable differences of opinion. It is these diametrically opposed opinions that the creationists use to refute all of the evolutionists' tentative conclusions of common ancestry. They perhaps reason that, as it is with paired elementary particles--one of matter and the other of antimatter--a collision annihilates both. It is the same strategy of creation science we have exposed

repeatedly throughout earlier chapters, beginning with the rise of the invertebrates at the dawn of the Paleozoic Era and progressing through the evolution of all classes and orders of animals to the Present.

Anthropology, as a branch of mainstream science, looks also for naturalistic origins of such human mental phenomena as memory, imagination, and language and their expressions in primitive forms of ritual, religion, and ethics. Archaeology, also a branch of mainstream science, studies the material remains of past cultures, interpreting artifacts to reveal cultural practices.

Creationists already have the answer to questions of the origin of human mental phenomena: they are God's unique endowment to the human kind through special, recent creation of its first two members, Adam and Eve.

In the emerging scientific field of sociobiology disturbing thoughts are being voiced again about the possible subtle but pervasive role of genes in controlling human behavior. As Darwinian natural selection begins to repermeate an area of human history that was recently interpreted almost entirely by theories of cultural dominance, the nature/nurture debate grows louder in the halls of science. The creationists will have none of sociobiology, which they consider as just more evolutionary falsehood.

Religion and ethics fall within those cognitive fields we have recognized as belief fields, using Professor Mario Bunge's epistemological classification, explained in Chapter 1. They are thus apart from science, which is one of the research fields, along with the humanities, mathematics, and applied science. Science is free to examine the origins and content of any of the belief fields. This analytical process views religion and ethics as real phenomena, carefully avoiding value judgments and recommendations as to what particular position is better or worse than another. Are religion and ethics products and by-products of the same biological evolutionary process that produced the human brain itself? In carrying out this investigation we intend no infringement upon the rights of individuals to subscribe to transnatural beliefs of their own choosing.

Chapter 49

Hominoid Evolution

Humans, apes, and monkeys, together with the prosimians (lemurs and tarsiers), are within the Order Primates, or simply primates. This order is the subject of our chapter, but with only a brief mention of the prosimians and monkeys, so that we can concentrate attention on the Superfamily Hominoidea, containing the apes and humans. (The primate taxonomy is shown in Figure 49.6.)

Evolution of the Primates

Figure 49.1 is an evolutionary chart of the order of primates. Notice that the time scale is greatly enlarged for the hominoids. The Order Primates has two suborders: Prosimii (prosimians) and Anthropoidea (anthropoids). "Prosimian" means simply "premonkey." Prosimians are the older of the two suborders and are the stem from which the remaining primates evolved.

According to mainstream paleontology, the prosimians--like all other orders of placental mammals--arose early in the Cenozoic Era from the insectivores, small animals resembling tree shrews of today. Prosimians are represented today by the lemurs, lorises, and tarsiers, which are small tree-dwelling mammals of Africa and Asia. Lemurlike prosimians entered in Paleocene time and were most abundant in the Eocene Epoch, after which they declined to minor status. It is important to take note of this early evolution of the prosimians, because it means that the evolutionary line of humans and the other anthropoids is as old as that of the other mammals, following a parallel course of evolution. However, the important evolutionary radiation of the higher primates was long delayed and occurred only after that of the other mammals had been largely completed.

In Oligocene time there evolved from the prosimians three primate groups: the New World monkeys, the Old World monkeys, and primitive apelike hominoids. This evolution is not yet adequately documented by fossil evidence but seems to have occurred rather rapidly. The hominoids are considered more closely related to the Old World monkeys than to the New World monkeys, because the former are more apelike in development. (Figure 49.1 shows the Old World monkeys to have arisen from primitive hominoids.)

The primates showed a number of important specializations that led to their rapid evolutionary radiation late in the Cenozoic Era. In adapting to forest living, the ability to judge distances accurately at close range led to the development of binocular vision, in which the eyes came forward on the head to lie in the same plane and hence to secure maximum overlap of vision. The senses of smell and hearing were secondary in importance, so that the olfactory equipment underwent some decline. The need to grasp tree limbs firmly led to an evolution of the hands and feet in which a thumb and first toe became opposed to the remaining four digits. Most important of all was the large increase in brain size and the consequent increase in intelligence and in ability to control the limbs. Good binocular vision enabled the hands and feet to be put to use in manipulating food and various objects, ultimately leading to the use of tools by humans.

As you would expect, the creationists do not accept the suggestion that the primates may have evolved from the insectivores. They also question the hypothesis that the prosimians are the ancestors of the anthropoids. Duane T. Gish writes: "Although the primates are supposed to have evolved from an insectivorous ancestor, there are no series of transitional forms connecting primates to insectivores" (1978a, p. 96; 1985, p. 130). He then quotes from a 1969 work by Elwyn Simons, whom he considers one of the world's leading experts in the field of primates: "In spite of recent finds, the time and place of origin of Primates remains shrouded in mystery" (see Gish for reference). From a 1974 work by A. J. Kelso he gleans the comment: "the transition from insectivore to primate is not documented by fossils. The basis of knowledge about the transition is by inference from living forms" (see Gish for reference).

Writing further on this subject, Gish refers to a 1966 report in Science by C. B. G. Campbell to the effect that earlier acceptance by paleontologists of a close relationship between tree shrews (tupaiids) and primates "is unlikely" (Gish, 1980, p. vii).

Continuing on the subject of primate evolution, Gish quotes from Romer's 1966 treatise and Kelso statements that the supposed transition from prosimians to monkeys, both New World monkeys (platyrrhines) and Old World monkeys (within the catarrhines), is not documented by fossil evidence (1978a, pp. 98-99; 1985, p. 130). Gish concludes:

> Already, then, the fossil record has failed twice to produce man's supposed progenitors: antecedents of the entire primate order are lacking and transitional forms between the prosimians, alleged the more "primitive" of the primates, and the catarrhines, or more "advanced" primates, have never been found. (1978a, p. 99; 1985, pp. 133-34)

In 1985 Gish introduced a statement from a 1982 work by R. D. Martin to the effect that "the tree shrew is not on the roster of human ancestors" (Gish, 1985, p. 132). Martin's conclusion was based on the maternal habits and fat content of the milk of living female shrews; Martin found both of these characters different from those of living primates. On the basis of this testimony and that of Kelso, Simons, and others, Gish concludes:

> There is thus no evidence either in the present world or in the world of the past to link primates to any other creatures. Right at the start, then, an evolutionary origin of man is invalidated by empirical evidence. The primates, as a group, stand completely isolated from all other creatures. (1985, p. 132)

In view of Gish's first sentence in the above paragraph, I turned to Edwin H. Colbert for an explanation of the supposed tree-shrew/primate connection. Colbert finds it evident that "the primates arose from their insectivore ancestors at a very early

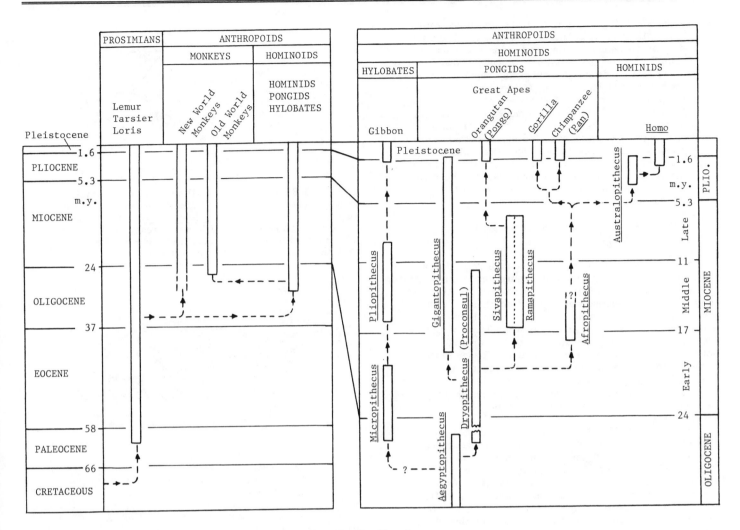

Figure 49.1 Evolutionary chart of the primates. (A. N. Strahler.)

date, perhaps even during the Cretaceous period. Certainly ancestral primates had become defined by the beginning of Cenozoic times, as indicated by their fossil remains in deposits of Paleocene age" (1980, p. 292).[1] The fossils are fragmentary--mostly teeth and jaw fragments. Colbert continues:

> A glimpse of the most ancient primates can be obtained by looking at the modern tree shrews, Tupaia, Ptilocerus, and their relatives, that live in the Orient. These interesting mammals have been the subject of much study, and their relationships have been a matter of debate for many years. They are now commonly regarded as insectivores, but many authorities have believed that they more properly should be included among the primates. Morphologically the tree shrews are so close to the line of demarcation between insectivores and primates that, even if they are included in the former mammalian order, they, nevertheless, give us some insight into the adaptations that were leading from primate eutherian (placental) mammals toward the first primates. (1980, p. 292)[1]

Colbert goes on to explain that the tree shrews, or tupaiids, are partially or dominantly fruit-eating animals.

This suggests the possibility that the earliest primates may have changed over from a carnivorous or insectivorous diet to one composed largely of fruits. Changed behavior patterns required by the diet change would in time, through natural selection, change the morphology and function of the feeding mechanism, and eventually of the whole animal. I was intrigued by Colbert's observation that, as in the primates, the thumb and great toe of the tree shrews are set somewhat apart from the other toes (1980, p. 293).

Colbert regards as particularly significant the fact that the brain of the tree shrew is relatively large while at the same time the olfactory region is small. While this may be simply a case of parallelism, Colbert observes, it is suggestive of the same evolutionary trend to development of a large brain and dominance of vision among the senses that determined the successful direction of primate radiation. These are arboreal adaptations.

As to the fossil record linking the insectivores with primates, remains of primitive forms of the latter are by no means lacking from strata of Paleocene and early Eocene age. Colbert has something to say on this subject, and although rather technical in places, it should be entered here:

> Turning now to the fossil record, we find in the Paleocene deposits of North America and Europe three groups of eutherians (placental mammals) that can be regarded as very primitive primates--namely, the paromomyids, the carpolestids, and the plesiadapids. By some authorities these ancient primates have been considered as constituting a

separate category, possibly of subordinal rank; here, they are classified as among the lemurs, in the broad sense of the word. These fossils, most of which are very fragmentary indeed, show us some very small primates in which the molar teeth were rather generalized, being tribosphenic (shaped to both shear and crush) and low crowned, but in which the incisors were commonly enlarged and procumbent chisels. The postcranial skeletal materials that have been found show that these mammals had claws rather than the nails that are general among the primates. Thus, they appear to have been on a lateral rather than on the direct line of evolution to the later lemurs and tarsiers. (1980, pp. 295-96)[1]

(Note: Words in parentheses added for clarification.)

The groups referred to above can be regarded as "cousins" to the early Cenozoic lemurs, as groups sharing a common ancestor among the insectivores. In this sense, they are true intermediates, even though not in the direct line of descent. Romer also describes the three families and comments that they are forms that "most would agree are definitely on the primate side of the ordinal boundary." In agreement with Colbert, Romer states: "These ancient families are obviously well off the line leading to higher primates" (1966, pp. 217-18).[2] A late Paleocene representative of the plesiadapids is Plesiadapis, for which a skull is available; it is shown here as Figure 49.2.

Recall that in Chapter 41 on fossil evidence of evolution of species we reviewed the research of Philip D. Gingerich, who traced the changes in teeth of the early Eocene primate genus Cantius to show how one species was succeeded by another. These data were presented as a possible example of gradualism. Gingerich (1980), working in cooperation with Kenneth Rose, also studied the evolution of representatives of the carpolestids and plesiadapids in Paleocene strata in Clark's Fork Basin, Wyoming. In the plesiadapids, they traced the evolution from an earlier genus, Pronothodectes, at the base of the stratigraphic column, to an early species of Plesiadapis, then through several other species of that same genus.

In recent years, attention has focused on the early Eocene primate Cantius as a candidate for ancestry of all modern primates. Cantius trigonodus, in particular, has been studied in great detail by Johns Hopkins University paleontologist Robert T. Bakker and his associates (Herbert, 1982). A fossil foot of this primitive primate shows it to have had a strong grasping toe, suited to tree climbing, and eyes facing forward for close-range stereoscopic vision. Bakker considers the acquisition of this tree-climbing facility to be the major evolutionary advance in early primate evolution. He suggests that their new arboreal home gave the Eocene primates a high level of immunity from predators and at the same time strengthened the bonding between parent and adolescent offspring. As the adolescent period lengthened, the brain grew in size and learning capacity increased. An interesting feature of C. trigonodus is that its teeth were very primitive and had not adapted to a diet of tree fruits. In other words, we have here an excellent example of mosaic evolution, making C. trigonodus a true intermediate between ground-dwelling placental mammals and the prosimian primates.

Another intermediate form that fills the gap between the Paleocene primates and the later Lemuroidea (lemurs) is an Eocene lemur, Notharctus, shown in Figure 49.3.

[2]From Vertebrate Paleontology, Third edition, by Alfred S. Romer, University of Chicago Press. Copyright © 1933, 1945, and 1966 by The University of Chicago. Used by permission.

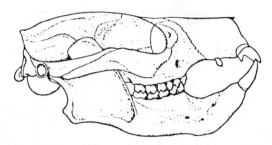

Figure 49.2 Restoration of the skull of Plesiadapis, a late Eocene primate. Note the enlarged incisors. Length about 10 cm. (From Vertebrate Paleontology, 3d ed., by Alfred S. Romer, University of Chicago Press, p. 218, Figure 327. Copyright © 1933, 1945, and 1966 by The University of Chicago. Reprinted by permission.)

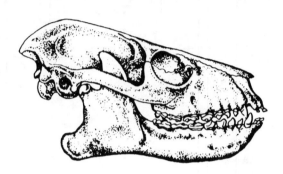

Figure 49.3 Restoration of the skull of Notharctus, an Eocene lemur. Length about 8 cm. (From Vertebrate Paleontology, 3d ed., by Alfred S. Romer, University of Chicago Press, p. 218, Figure 329. Copyright © 1933, 1945, and 1966 by The University of Chicago. Reprinted by permission.)

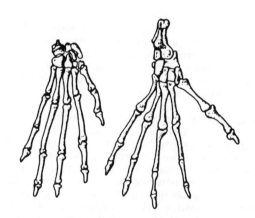

Figure 49.4 Bones of the hand (left) and foot (right) of Notharctus. (From Vertebrate Paleontology, 3d ed., by Alfred S. Romer, University of Chicago Press, p. 195, Figure 304. Copyright © 1933, 1945, and 1966 by The University of Chicago. Reprinted by permission.)

Complete skeletons of this genus have been found in western North America (Colbert, 1980, p. 296). It was a small animal, its skull only about 5 cm in length. The tail was very long; the long, slender legs were capable of a wide range of movements. Notharctus was well adapted to tree climbing; the digits of the hands and feet, shown in Figure 49.4, were long and slender, with thumb and first toe shortened and angled out from the other digits.

Gingerich also made a study of the evolution of Notharctus, which traced the rise of that genus from an older genus, Pelycodus (1976). His raw data consisted of fossils of 255 individuals distributed through a total thickness of some 500 m of early Eocene strata in the Big Horn Basin of Wyoming. His completed study provides a small window of information showing how forms closely related to the insectivores gradually evolved into forms definitely of primate affinity.

C. Loring Brace, Professor of Anthropology at the University of Michigan-Ann Arbor and an authority on human evolution comments on this transition:

> The spectrum of living primates runs from that most modified and aberrant species, Homo sapiens, to prosimian forms that are so little different from non-primate insectivores that scientists have been arguing for a century about their correct classification (Luckett, 1980). The important thing, in reality, is not the "correct" pigeonhole but the fact that they represent a condition intermediate between the two orders and suggest to us the kind of evolutionary change by which primates could have diverged from the generalized mammalian stem. (Brace, 1983, p. 246)[3]

(See Brace for reference to Luckett.)

Perhaps from among the early Eocene fossils, found as recently as 1984 in the Wind River Basin of Wyoming (see Chapter 45, section on bats and rodents), new evidence on the ancestry of the primates will come forth. The news report of these fossil finds makes mention of an important find in the form of "the partial skull of a monkey-like creature, known as shoshonius, that may have been the ancestor of the tarsier, a primate now living in southeast Asia. Five other skulls belong to ancient primate relatives of living monkeys, lemurs, and tarsiers" (Bower, 1984, p. 213). Far from being broken by great information gaps, as the creationists claim, the existing record of evolution of the primates already displays abundant fossil evidence of transitions from insectivores to prosimians.

The close evolutionary connections between prosimians, monkeys, and anthropoids are supported by studies in molecular biology. A phylogeny of the globin genes based on the estimated numbers of nucleotide replacements locates the human adjacent or very close to Old World

monkeys and the slow loris, all much closer to the human than the mouse, dog, and other placental mammals (data of Goodman, 1976, reviewed in Ayala, 1977c, p. 303 and Table 9-21).

Evolution of the Hominoids

The fossil and living hominoids make up the superfamily Hominoidea. Classification of the superfamily is shown in Figure 49.5, its evolution in Figure 49.1. Two families within the Hominoidea are the Pongidae, or apes, and the Hominidae, or humans. The living Pongidae (pongids) include the gibbons, orangutan (Pongo), chimpanzee (Pan), and Gorilla. The Hominidae, or hominids, have only one living genus: Homo (humans).

Evolution of the hominoids began in Oligocene time with the appearance of Aegyptopithecus, literally translated as "Egyptian ape." (The Greek word pithekos means "ape.") Knowledge of this animal is based on several well-preserved jaw fragments and various other bone fragments found since 1977 in the Fayum Depression of Egypt (Science News, 1980, vol. 117, p. 100). Its age is greater than 26 m.y., as dated from an overlying lava flow of that age. As described by Brace, Aegyptopithecus was "an arboreal quadruped about the size of a smallish dog. It had a tail, grasping hands and feet, and a brain that was somewhat enlarged compared with those of the prosimians" (1983, p. 247).[3] An important difference from the Eocene primates we have already described is seen in the dentition of Aegyptopithecus:

> Molar tooth crowns, unlike the earlier Paleocene and Eocene primates, were low and rounded like those of recent fruit-eating primates. All of this is completely monkeylike, but when one looks at the patterns of cusp arrangement on the molar teeth, they are quite different from those of modern monkeys but absolutely indistinguishable from those of modern anthropoid apes--and human beings. (Brace, pp. 246-48)[3]

Gish does not mention the question of the evolutionary transition from Eocene primates to the hominoids (1978a, pp. 96-100). However, the existence of a true intermediate is put strongly on the record by Brace, who states that although the teeth of Aegyptopithecus are those of an ape, its body plan was that of a monkey:

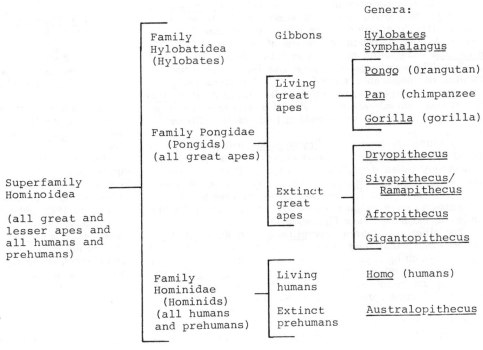

Figure 49.5 Members of the super-family Hominoidea. (Based on data of E. L. Simons, 1977; Roger Lewin, 1982; David Pilbeam, 1984; and others.)

i.e., "it was an orthodox arboreal quadruped" (1983, pp.248-49).[3] And finally, this significant conclusion by Brace: "All told, it provides a splendid representation of the ancestral condition from which modern apes--and humans--descended" (p. 249).[3] Again, we find a case of mosaic evolution that is the mark of a genuine transitional form.

At this point we should take note of the gibbons and their relationship to the other apes (great apes). The gibbons, smallest of the living hominoids, now live in the Malaysian region. Unlike the great apes, the gibbon usually walks erect when on the ground, but is primarily a tree dweller and is famous for performing remarkable acrobatic feats. Because of structural differences the gibbons are sometimes placed in a separate family of hominoids, the Hylobatidae. An earlier member of the family, Pliopithecus, is found in Miocene and Pliocene time; it resembled the living gibbon in cranial structure but had not yet developed the long arms of today's gibbon. A link between Pliopithecus and Aegyptopithecus may exist in a possible intermediate of late Oligocene and early Miocene time called Micropithecus. This gibbon phylogeny is shown in Figure 49.1. Molecular biology provides corroborative evidence of the separateness of the gibbons from the great apes. The early branching of the gibbons from a common ancestor with all other anthropoids is shown clearly in phylogenies based on immunological differences in albumin proteins (Ayala, 1977c, pp 291-92); and on hybridization of DNA (Lewin, 1984b, p. 1181). (We refer later to the DNA evidence. See Figure 49.14.)

The Dryopithecines

Dryopithecus and the closely related African genus Proconsul include early hominoid species generally designated as the common ancestor of all great apes and hominids. They appear in the late Oligocene and extend into the Miocene for a duration of about 20 m.y. The fossil remains, mostly skull and jaw fragments, are widely distributed: Africa, Europe, India, and China. Through early Miocene time Africa was separated from Eurasia by the Tethys Sea, which was connected to the ancestral Mediterranean Sea. As the collision of the two continents became imminent, the Tethys was excluded from what is now the Near East, allowing animal migration between the continents. Thus Proconsul and its descendants were able to spread widely through Eurasia in middle Pliocene time (Pilbeam, 1984, p. 91).

The African representative of the genus, Proconsul africanus, has provided a large sample of the bones of the skeleton in addition to the skull and jawbone (see Figure 49.7A). Excellent specimens of the feet of P. africanus were found in Kenya in 1984 (Bower, 1984, p. 26). The specimens come from five individuals--two adults, an adolescent, a child, and an infant--allowing structural changes during aging to be analyzed. Proconsul was a small animal--about the size of a baboon--walked on four legs, and was a tree-dwelling fruit eater (Pilbeam, 1984, p. 91).

By middle Miocene time, Dryopithecus had given rise to three important genera of great apes; the four genera are collectively called dryopithecines (Brace, 1983, p. 249). One genus, Gigantopithecus, was a huge ape. Only lower jaws have been found, and from these the creature is estimated to have weighed up to 270 kg (600 lbs); its line ended in extinction (see Figure 49.1).

The other two genera, Sivapithecus and Ramapithecus, are so similar that they are often considered as equivalent and representing a single genus. Interpreted as such, we might want to refer to them jointly as Siva/Ramapithecus, but they are commonly referred to simply as "rama-morphs." Ramamorphs were more apelike than Dryo-pithecus (Proconsul), with longer limbs attached to a relatively shorter body. More importantly, the ramamorphs had large cheek teeth covered by thick enamel and set in a robust jaw; this tooth/jaw characteristic was interpreted as a distinctly hominid feature and led to Ramapithecus being first assigned to the hominids.

There are, however, distinct differences between the two ramamorph genera. Sivapithecus (see Figure 49.7B) had a large face and has been described as distinctly apelike, whereas the face of Ramapithecus was small and less apelike, which is another reason why that genus was first considered by some anthropologists as a hominid. Accordingly, until quite recently, Ramapithecus was placed in the family Hominidae (Simons, 1977, p. 31).

Skull and jawbone fossils of Sivapithecus, found in 1981 in Pakistan by anthropologist David Pilbeam of the Harvard Peabody Museum, seemed to him clearly to resemble features of the modern orangutan, and not to resemble as closely the skull of the chimpanzee (Science News, 1982, vol. 121, p. 84). Pilbeam was reported as suggesting that specimens of Ramapithecus are actually females of the genus Sivapithecus (Herbert, 1984, p. 41). Pilbeam and his associates now consider Siva/Ramapithecus as directly ancestral to the orangutan, but not to the chimpanzee and gorilla, and not to any of the hominids (Cartmill, Pilbeam, and Isaac, 1986). He would, therefore, place Siva/Ramapithecus in the family of the apes, or Pongidae, as shown in Figures 49.1 and 49.5). If Siva/Ramapithecus is not ancestral to the hominids or to the chimpanzee and gorilla, those apes and the hominids must be traced all the way back to Dryopithecus, leaving a very long evolutionary gap--on the order of 15 m.y. Commenting on what they call "the fall of Ramapithecus" as a hominid ancestor, paleoanthropology professors Matt Cartmill, David Pilbeam, and Glynn Isaac write: "Most researchers now think that the last common ancestor of humans, chimpanzees, and gorillas was a robust-jawed (and thus far hypothetical) ape with thick dental enamel, which inhabited Africa from 5 to 10 million years ago" (1986, p. 414). They add: "That ancestor is now generally recognized as a tree-dweller with long arms, similar in its habits and general appearance to a modern chimpanzee or orangutan."

New fossil discoveries announced in 1986 by Richard and Maeve Leakey may have provided intermediates to fill the Miocene gap (Bower, 1986, p. 324). The fossil hominoids, found in the Lake Turkana locality of northern Kenya, are in sediments thought to be in the age-range of 16 to 18 m.y. They are of two distinct genera, named Afropithecus and Turkanopithecus. Bower reports as follows on the Leakeys' findings:

> Afropithecus, explains the researchers, displays the characteristics of a variety of hominoids combined in a single, distinctive category. Its palate is shallow, long and narrow and the nasal passage is "remarkably narrow and high." The forehead inclines steeply to a long muzzle. (1986, p. 324)

It looks as if these and similar fossil finds yet to come may successfully fill the void created by discreditation of Ramapithecus as the hominid ancestor of middle Miocene time.

However, the older view of Ramapithecus as a genus quite distinct from a Sivapithecus in several important respects continued to be strongly supported (Herbert, 1984, p. 41). Remains of Sivapithecus were discovered in Africa in 1983 and given an age of 17 m.y.--much older than those from Pakistan. Numerous new fossil specimens collected recently in China seem to show clearly that two genera actually do exist, and that the fossils cannot be interpreted as simply a mixture of female and male remains within one genus (Herbert, 1984, p. 41). The arguments pro and con for both older and newer views are a reassuring sign of mainstream science in a state of excellent health!

The creationists are adamantly opposed to recognizing Ramapithecus as a hominid intermediate, ancestor to the

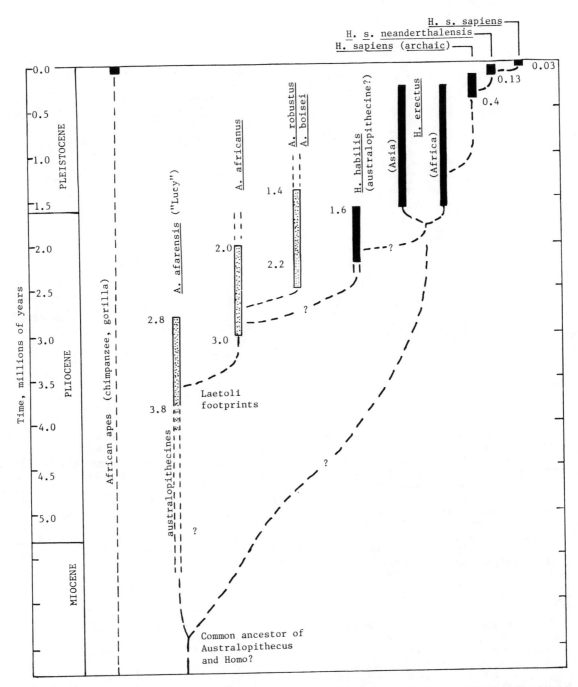

Figure 49.6 Phylogeny of the hominids. (Based on data of B. Rensberger, 1984, <u>Science</u>, vol. 5, no. 2, pp. 29-39; D. Pilbeam, 1984, <u>Scientific American</u>, vol. 250, no. 3, pp. 84-91; and others.)

true and undisputed hominids. Gish gives several pages to the creationist campaign to kick <u>Ramapithecus</u> clear out of the hominid camp (1978, pp. 100-104; 1985, pp.140-43).

Gish is in good company in supporting the nonhominid interpretation of the ramapiths. Mainstream science cultivates debates over alternative interpretations of the same data, but becoming an activist only for one side carries the responsibility of first becoming an authority on the subject under debate.

Australopithecines--The First Hominids

Starting about -4 m.y., which is about in the middle of the Pliocene Epoch, a good record of hominid evolution comes into view--good, that is, in comparison with the gap of some 4 or 5 m.y. (or even 15 m.y. or more) that preceded it. This gap is the last line of defense for the

creationists; it is the moat that surrounds the creationist castle. Nothing means more to them than the belief that God created the human in His own image. Creationists recognize all fossil and living species of the genus <u>Homo</u> as a single created kind, but no earlier hominid can be accepted as ancestral to <u>Homo</u>. The moat must never be crossed by the evolutionists, for if that disaster ever should occur, the castle would be taken and its defenders destroyed.

The first discovery of what mainstream anthropologists recognize as undoubted hominid remains was made in 1924 in limestone quarries at Taungs, near Kimberly, South Africa. The discovery included a nearly complete skull. The fossils were brought to Raymond Dart, a professor of anatomy at the University of Witwatersrand, who named the creature <u>Australopithecus</u> <u>africanus</u>. The name literally translates to "southern ape of Africa." (The prefix "austral" should not be in any way related to

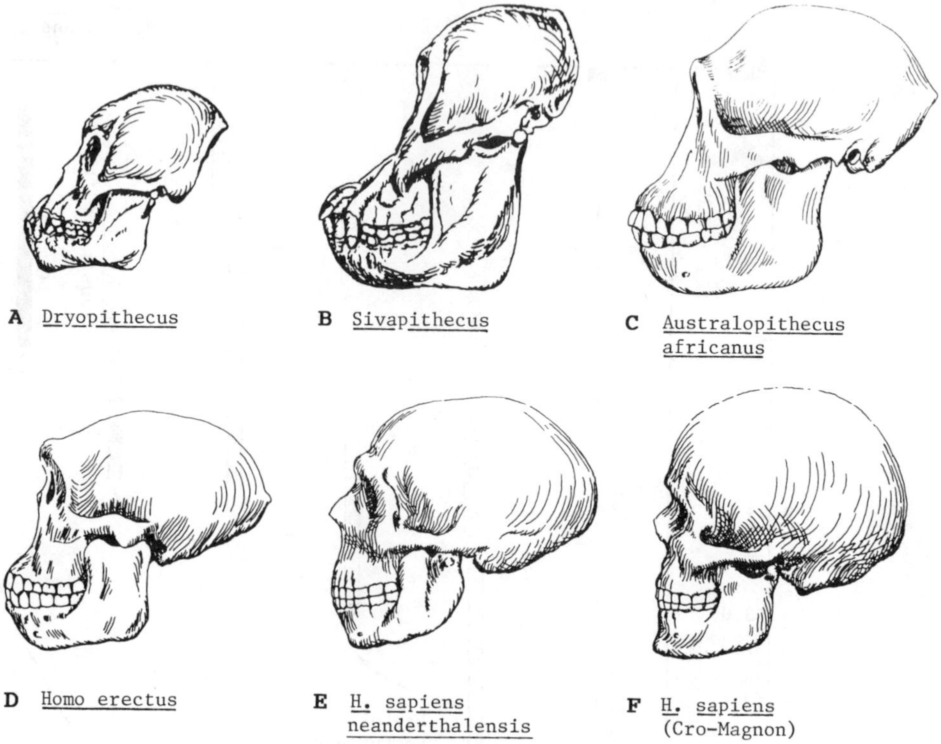

A Dryopithecus   **B** Sivapithecus   **C** Australopithecus africanus

Figure 49.7 Sketches of fossil skulls of six hominoids. (Drawn by A. N. Strahler. A. and B. From The Descent of Hominoids and Hominids, by David Pilbeam, <u>Scientific</u> <u>American</u>, vol. 250, no. 3, p. 85. Copyright © 1984 by Scientific American, Inc. All rights reserved. Used by permission. C-F. Drawn from photographs in the collection of the American Museum of Natural History.)

D Homo erectus   **E** H. sapiens neanderthalensis   **F** H. sapiens (Cro-Magnon)

Australia in this context.) At the time of its discovery, A. africanus was judged to have lived at least one million years ago, but radiometric dating was not available to establish absolute age. As the years passed, many new discoveries of remains of relatives of Australopithicus (conveniently called "australopithecines," or simply "australopiths"), were made at several African sites, ranging from South Africa to Ethiopia. Three distinguished anthropologists who engaged in these discoveries are the late Louis S. B. Leakey, his wife Mary, and, more recently, their son Richard B. Leakey. Radiometric ages were obtained and age ranges established for at least three important species of australopiths (Figure 49.6). Australopithicus africanus has been assigned a range of about -3.0 to -2.0 m.y.

Brain size of A. africanus was about the same as in the living great apes--450 to 550 cubic centimeters. However, in terms of ratio of brain size to body size, measured by the encephalization quotient (EQ), the brain of A. africanus had an EQ of 4.3, much larger than the EQ values of apes, which range between 1.6 and 2.3 (McGowan, 1984, p. 175). The head shape might seem to us more apelike than humanlike, because the crown of the skull is low and the jaws are large and thrust far forward (Figure 49.7C). Molar teeth of the australopiths are exceptionally large as compared with those of humans. A particularly significant feature of the australopithecine jaw is that, when seen in plan view, the teeth form a nicely curved arch, as in humans, whereas in the apes the cheek teeth and canines lie in two straight parallel lines (Figure 49.8). An important feature of the australophith dentition is that the canine teeth did not project above the general level of the other teeth, as they do in the apes and other primates. Whereas the ape jaw shows a gap (called the diastema) between the canine and incisors, no such gap is present in the australopithecine jaw and human jaw (Figure 49.8). McGowan calls attention to a rather remarkable difference between the ape and human: whereas in the apes the canine teeth are the last permanent teeth to appear, in humans the permanent canines appear earlier, along with the incisors and premolars, but the wisdom teeth come along in the late

teens or early twenties (1983, p. 172). When Professor Dart examined the Taungs skull, which was that of a juvenile, he found that the permanent canines were already present, but that the wisdom teeth had not yet erupted. Clearly, then, the Taungs skull was of a hominid, rather than a pongid.

In the australopiths the foramen magnum, which is the hole in the skull through which the spinal chord passes, was located at the bottom of the skull and facing vertically downward, as it is in genus Homo, but not at

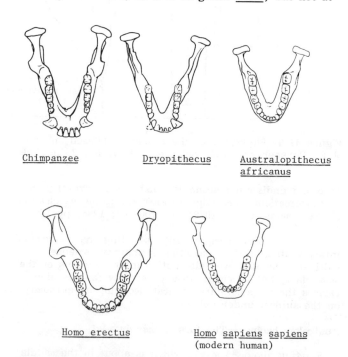

Chimpanzee   Dryopithecus   Australopithecus africanus

Homo erectus   Homo sapiens sapiens (modern human)

Figure 49.8 Drawings of the lower jaw of five hominoids. (From: Bernard Campbell, <u>HUMAN</u> <u>EVOLUTION</u>, Second Edition. Copyright © 1974 by Bernard Campbell. Aldine Publishing Company, New York. Adapted by permission.)

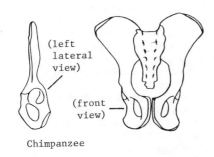

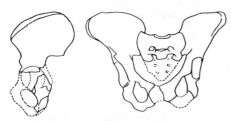

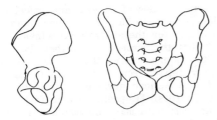

Chimpanzee

Australopithecus africanus

South African Bushman

Figure 49.9 Comparison of the pelvis of three hominoids. (From: Bernard Campbell, HUMAN EVOLUTION, Second Edition. Copyright © 1974 by Bernard Campbell. Aldine Publishing Company, New York. Adapted by permission.)

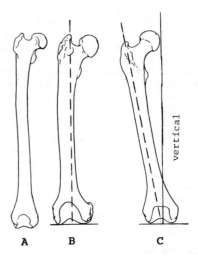

Figure 49.10 Femur of Dryopithecus (A), chimpanzee (B), and modern human (C). Notice that the human tibia is tilted from the vertical with respect to the horizontal line of the two tips of the condyles of the knee joint. (From: Bernard Campbell, HUMAN EVOLUTION, Second Edition. Copyright © 1974 by Bernard Campbell. Aldine Publishing Company, New York. Adapted by permission.)

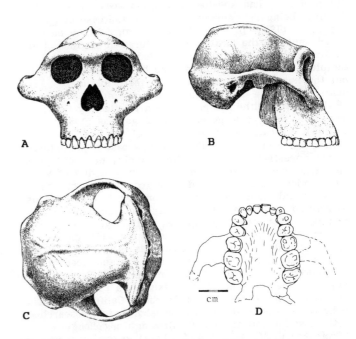

Figure 49.11 Australopithecus boisei (robustus). A-C. Drawings of the reconstructed skull viewed from the front, side, and top. D. Drawing of a photograph of the upper jaw of a specimen of A. boisei from the Olduvai Gorge. Notice the small incisors and canines in contrast with the very large cheek teeth. (A-C. From The Hominids of East Turkana, by A. Walker and R. E. F. Leakey, Scientific American, vol. 239, no. 2, p. 64. Copyright © 1978 by Scientific American, Inc. All rights reserved. Used by permission. D. After a photograph in P. V. Tobias, 1965, Science, vol. 149, p. 26, Figure 3.)

the rear of the skull and angled diagonally downward and backward as it is in the apes and other primates.

Comparison of the australopithecine pelvis with that of modern humans shows them to be "remarkably similar" (Campbell, 1974, p. 158). In particular, the australopithecine ilium is broader than long, whereas in the apes it is narrow and long (Figure 49.9). The lower limb bones together with the pelvis show that the australopiths stood upright and walked on two legs (i.e., they were bipedal). Evidence for this conclusion can be found at the distal end of the femur--it is angled to the shaft, as in modern humans (Figure 49.10). The arms of the australopiths were proportionately a little longer than in modern humans, but their hands were quite similar to those of modern humans.

Another species of australopith, A. robustus, was identified in the late 1930s in southern Africa. It was massively built and had a much wider face than A. africanus. Its huge molars, probably suited to grinding coarse plant material, were operated by exceptionally powerful jaw muscles (Figure 49.11). The skull shows a prominent bony crest, resembling that of the gorilla skull. (A closely related robust species, found in eastern Africa, is A. boisei, placed together with A. robustus in Figure 49.6.) Australopithecus robustus is thought to have evolved from A. africanus, but it became extinct in early Pleistocene time.

Brace summarizes the status of Australopithecus as presently viewed by mainstream anthropologists and paleontologists:

The direct evidence, then, gives us a picture of Australopithecus as a terrestrial biped with an ape-sized brain and possessing ape-sized teeth that, however, occlude in a completely human fashion. The locomotor adaptation is so different from that of a typical pongid--ape--that the creature is placed in the same family, Hominidae, with modern human beings, and referred to informally as a hominid. However, this does not make it a human being. With a brain that is only one-third the capacity of the modern human average and teeth that are double the bulk, the Australopithecines warrant separate generic designation--Australopithecus. The implications of the anatomical, ecological, and archaeological evidence have all been weighed and considered with care (literature citations given here), and despite the differences of opinion among those who have studied the original material, no one doubts the fact that the Australopithecines present that "mosaic of advanced and ancestral characters" that "always" characterizes an evolutionary intermediate (Mayr, 1971, p. 50). Within the spectrum of the Australopithecines, then, we find a picture of an intermediate between the pongid and the hominid condition that is just as convincing as that provided by Archaeopteryx of an intermediate condition between the reptilian and avian condition. (1983, p. 251)[3]

Creationists Demote the Australopiths

Duane Gish devotes several pages to refuting the decision of mainstream anthropologists to place Australopithecus among the hominids (1978a, pp. 104-13; 1985, pp. 144-51). Gish first points out that the australopithecine brain, being small, was the brain of an ape. We have shown that the more meaningful measure of brain size, the EQ, shows that this was not so. Gish says: "Both of them had ape-like skulls and jaws, these features being particularly obvious in the case of A. robustus" (1985, p. 145). Gish is wrong. As we have clearly shown, the location of the foramen magnum of the australopiths is definitely unlike that of the apes and definitely like that of humans. We have already pointed out the humanlike rounded arch of teeth in the australopithecine jaw, unlike that of the apes. Massiveness of the australopithecine cheek teeth is cited by Gish, who says they were as large as those of gorillas (1985, p. 146). We can accept this as one of the mosaic characters of an intermediate, in this case a persistence of the ancestral character. Alternatively, we might explain the large cheek teeth as an adaptation to a diet of very coarse plant materials found near ground level on the savanna and open woodland, in contrast to the more easily masticated leaves and fruits of trees of the denser tropical forests that lay adjacent to the savanna.

Gish places heavy emphasis upon the writings of two distinguished physical anthropologists (1985, pp. 148-51). One is Solly Lord Zuckerman, Professor and Head of the Department of Anatomy of the University of Birmingham in England; the other is Charles Oxnard, a professor of anatomy at the University of Southern California Medical School. Professor Oxnard has collaborated in research with Lord Zuckerman and his Birmingham colleagues (Professor E. H. Ashton and others) in applying methods of biometrics, particularly multivariate analysis in which populations of variates are seen as clusters in three-dimensional and multi-dimensional space. The particular area of the research to which Gish makes reference can be described as "bone-joint-muscle biomechanics" of the primates (Oxnard, 1972, p. viii). These investigations have been primarily concerned with the functioning of the locomotor apparatus of the primates, i.e., how the primates use their limbs for climbing, walking, or running. Of course, when primate fossils are the materials under study, the measurements are limited to solid bones, and in most cases these are individual bones or bone fragments. Corresponding bones of several genera or species are compared in terms of certain measurements of form that may include ratios of dimensions and angles between geometric elements of the bone.

Gish uses one quotation from Lord Zuckerman's 1970 book, Beyond the Ivory Tower, but our purposes will be served by analyzing only the material from Oxnard's works. Professor Oxnard published in 1975 a review

article in the British journal Nature; it bore the rather intriguing title "The place of the australopithecines in human evolution: grounds for doubt." What a marvelous opportunity for the creationists! Another distinguished scientist finds no grounds for evolution! Gish quotes the opening sentence from Oxnard's abstract:

> Although most studies emphasize the similarity of the australopithecines to modern man, and suggest, therefore, that these creatures were bipedal tool-makers at least one form of which (Australopithecus africanus--"Homo habilis", "Homo africanus") was almost directly ancestral to man, a series of multivariate statistical studies of various postcranial fragments suggests other conclusions. (Oxnard, 1975, p. 77, in Gish, 1985, p. 150)

Gish leaves that statement without comment, perhaps in the hope that the words "other conclusions" will mean to the reader that Oxnard finds no evidence of any evolutionary connection between the two groups of hominids. Gish then quotes from a 1974 magazine article by Oxnard:

> Multivariate studies of several anatomical regions, shoulder, pelvis, ankle, foot, elbow, and hand are now available for the australopithecines. These suggest that the common view, that these fossils are similar to modern man or that on those occasions when they depart from a similarity to man they resemble the African great apes, may be incorrect. Most of the fossil fragments are in fact uniquely different from both man and man's nearest living genetic relatives, the chimpanzee and gorilla. . . . To the extent that resemblances exist with living forms, they tend to be with the orangutan. (Oxnard, 1982, p. 242 in Gish, 1985, p. 150)

Get ready, now, for the classic debating maneuver of creation scientists (and of pseudoscientists in general). We can anticipate that Oxnard's words "uniquely different" will be picked up and linked to a fictitious conclusion that was never intended by the author. Here it is:

> Oxnard's conclusions are, then, that Australopithecus is not related to anything living today, man or ape, but was uniquely different. If Oxnard and Lord Zuckerman are correct, certainly Australopithecus was neither ancestral to man nor intermediate between ape and man. (Gish, 1985, p. 151)

Gish's first sentence is unwarranted and clearly a gross perversion of the statements to which he refers. Consider the words "not related to anything living today." No reputable evolutionary scientist would assert that one extinct genus of what is clearly a hominoid is "not related to anything living today." The evolutionary theory of descent with modification requires that every living genus is related to every other genus, living or extinct, of any of the life kingdoms. Oxnard's use of the words "directly ancestral" in the first selection quoted does not rule out an indirect relationship of common ancestry residing in a third party.

In cases such as this, it is usually instructive to go back to the author's paper or book and look for statements that put those carefully selected quotations in their full context. A reading of Oxnard's Nature article gives all the background material we need (1975a, p. 389). First, we go to the abstract, prominently set in bold italic type on the first page; the final sentence reads: "The genus Homo may, in fact, be so ancient as to parallel entirely the genus Australopithecus thus denying the latter a direct place in the human lineage." There you have it loud and clear: parallel but separate evolution in two genera arising from a common ancestor. On page 392 of the Nature article Oxnard summarizes his findings on the relationship of humans to the living primates. He states that the results of his research give "more information about structurally relevant aspects of behaviour than about genetic propinquity." ("Propinquity" means "nearness of blood" or "kinship.") He also says: "This uniqueness for man does not contradict the findings from traditional morphology or from the molecular studies referenced above" (p. 392).

At this point the concept of "uniqueness," as intended by Oxnard, needs some clarification. When the variates (raw data, or individual measurements from the bones) are treated by multivariate analysis, they may appear as clusters or "swarms" in three-dimensional space, each cluster representing the bones of a given genus or species. Visual inspection alone would satisfy the average viewer that each cluster is so clearly removed from the others that they may all be said to be "uniquely different." Statistical tests based on ideal probability distributions of the variates can be applied to evaluate the significance of the observed differences, using a null hypothesis. (We explained this procedure in Chapter 5.) The quality of "uniqueness" referred to by Oxnard simply means that the isolation of the clusters is such that the chance of their appearing as they do by random sampling from a truly homogeneous population of variates is extremely remote. Keeping this in mind, let us read further what Oxnard has to say:

> This uniqueness presumably arises from his (man's) having a totally different behavioural repertoire from any other primates, and therefore unique functions in different anatomical regions. . . . Although all primates can use their forelimbs for manipulation, communication, feeding, and so on, the overall functions of the human forelimb are different from those of other primates because man does not use this member for locomotion. Although almost all primates can walk on two legs, the functions of the human hindlimb are uniquely different because only man does this habitually and also, and perhaps especially, because only man is utterly unable to use the hindlimb in quadripedal locomotion. (P. 392)

And finally, to be sure you have the correct meaning and intent of Oxnard's results, read this summary statement:

> The findings for man provide, therefore, not only additional important information necessary for studying fossils that may be thought to be close to human, but also further confirmation of the functional, more than any other biological, such as genetic, emphasis to the particular morphological results. (P. 392)

In the part of his Nature paper describing the results of multivariate analysis of bone fragments of australo-pithecines in comparison with corresponding bones of other primates, including modern humans, Oxnard finds a uniqueness of the australopithecenes such that he considers that group most unlikely as being a direct part of the human ancestry (p. 394). Further analysis that includes australopithecine fossils seems to reinforce the conclusion that modern humans, living apes, and australopithecines are uniquely different from one another in terms of foot bones (Oxnard, 1973, pp. 159-68). Clearly, Oxnard at no point in his writings abandons the generally accepted basic principles of evolution, nor does he throw out the reconstruction of the mosaic evolution of humans from hominoid ancestors. The thrust of his argument is aimed more to a rejection of direct linear

evolution using the fossils as links in a single chain. In taking this viewpoint, Oxnard is certainly in good company among mainstream anthropologists and paleontologists.

In his critique of the hominid status of Ramapithecus, referred to earlier, Gish finds "even more devastating evidence against the assumption of a hominid status for Ramapithecus" (1978a, p. 103). He returns to this same evidence in his argument against the hominid status of the australopiths (p. 112). The "evidence" comes from the teeth and jaws of a species of baboon living at high altitude in Ethiopia. Read carefully the following statement by Gish:

> This baboon, Theropithecus galada, has incisors and canines which are small relative to those of extant African apes, closely packed and heavily worn cheek teeth, powerful masticatory muscles, and short deep face, and other "man-like" features possessed by Ramapithecus and Australopithecus.[*] Since this animal is nothing but a baboon in every other feature, and living today, it is certain that it has no genetic relationship to man. Yet it has many of the dental and mandibular characteristics used to classify Ramapithecus as a hominid! (P. 103)

([*]References are given to C. J. Jolly, 1970, and D. R. Pilbeam, 1970.)

Gish goes on to assert that because the dental and mandibular characteristics cited above are possessed by a baboon, they cannot then also be diagnostic of hominids. On the surface, this conclusion may seem valid because "diagnostic" implies exclusivity.

Chris McGowan takes up Gish's argument about this species of baboons, otherwise known as "gelada baboons," replying as follows:

> Gish's references to the gelada baboons sounds a little disturbing, because it seemingly destroys the validity of dental characteristics which we have used to distinguish the hominids from the other primates. However, when we look at a gelada baboon we see that only two dental features parallel the hominid condition: the incisors are relatively small compared with the cheek teeth, and they tend to be vertical rather than to point forwards. In all other dental features the gelada baboon compares with non-hominids: the cheek teeth and canines on the left and right sides lie in a straight row, and these rows lie essentially parallel to one another; the distances between the canines and the last molars are approximately the same; the canines are long and project beyond the level of the cheek teeth, especially in males; the canines are pointed; and there is a diastema between the canine and the incisor teeth. The suggestion that the gelada baboon possesses dental fatures that undermine the hominid status for australopithecines is therefore unfounded. (1984, pp. 176-77)[4]

Certainly "close packed and heavily worn cheek teeth" are not exclusively a hominid characteristic. Wear shown on the crowns of teeth is not an inheritable characteristic to begin with and, of course, many grazing herbivores have closely packed cheek teeth that show wear with aging--the modern horse, for example. Many other primates could be said to possess "powerful masticatory muscles." The "short deep face" is certainly not exclusively hominid. A look at the monkeys reveals several with short, deep faces: capuchin, rhesus, colobus, dourocouli, langur, and ouakari. Gish has named in a loose, general way some structural features that hominids may possess, but none that is by itself diagnostic of a hominid to the exclusion of any other primate. McGowan handles the dental and mandibular problem efficiently. Why does Gish overlook cranial capacity, stated as index of cephalization, as a diagnostic feature? What about the basal location and vertical attitude of the foramen magnum--a diagnostic feature of hominids? Surely these two features alone would clearly separate the hominid from the Ethiopean baboon, and also from all other hominoids and anthropoids generally.

Such structural modifications as the high-altitude Ethiopian baboon may actually possess could probably be explained simply enough as an adaption to a specialized environment. In this connection, I came across an interesting statement by Edwin H. Colbert: "Baby baboons have round heads and short noses, as do most of the higher primates, but in the adults the snout becomes very long and dog-like, and the canine teeth are greatly enlarged" (1980, p. 301).[1] The changes in snout and canines are explained by Colbert as an adaptation to ground living. The baby baboons exhibit faces of ancestors formerly adapted to an arboreal life with a diet of fruits, berries, and insects. All that would be needed to modify some high-altitude forest-dwelling baboon to fit Gish's description of the Ethiopian species would be a minor change in the action of genes controlling the maturation process.

Gish's 1985 work, Evolution: The Challenge of the Fossil Record, omits mention of the gelada baboons. Perhaps McGowan's argument has successfully disposed of the matter. I thought the subject worth including here, because large numbers of the 1981 printing of Gish's 1978 predecessor work, NO! must still be circulating among students and teachers.

Lucy--The Oldest Australopith

Since 1973, two African sites--Laetoli in Tanzania and Hadar in Ethiopia--have yielded abundant remains of an older australopith, A. afarensis, in the firmly dated age range of -4 to -3 m.y. The first finds were made in 1973 by members of a joint American-French-Ethiopian expedition under the leadership of Donald C. Johanson in the Awash River valley of Ethiopia, located at the northern end of the African rift valley system. More finds followed a year later at the Hadar locality, not far from the Awash locality. By far the most remarkable find at Hadar, one that brought great public attention to the Afar/Hadar fossils, was that of a female for whom the skeleton was about 40 percent complete. The individual was quickly named Lucy (after the Beatles' song "Lucy in the Sky with Diamonds"), and she soon had a fan club of millions of people eager to learn more about her.

Lucy's remains consist of a few skull parts, a jaw with several teeth, most bones of one arm, some wrist and ankle bones, several vertebrae, and part of the pelvis with the sacrum that connects it to the backbone. An adult, Lucy was a little over one meter tall and may have weighed about 30 kilograms. That Lucy walked on two feet (was bipedal) was inferred from the formation of the pelvis (Figure 49.12).

Australopith fossils of about the same age range from the Laetoli locality in Tanzania, collected and studied by Mary Leakey, proved strikingly similar to those from the Hadar locality. Most of the specimens were, of course, of individuals much larger than Lucy, and it seems that males may have been much larger than females--perhaps as much as twice the size. All were bipedal, thick-boned, and probably strongly muscled.

Brace states that all who have examined the fossil materials from both the Ethiopian and Laetoli sites agree that they are the remains of australopithecines (1983, pp. 252-53). Brace summarizes their evolutionary status as follows:

All also agree that in a series of traits these early Australopithecines are more primitive than those

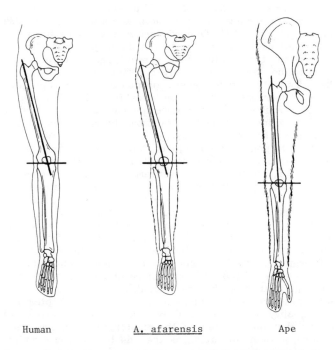

Human   A. afarensis   Ape

Figure 49.12 Diagrams of the legs of three hominoids. The angular relationship of the femur to the lower leg bones (the valgus angle) is critical to bipedal locomotion and is shared by human and australopith. Notice also the similarity of the human and australopith pelvis, both differing markedly from that of the ape. (Drawn by Luba Gudtz. From Roger Lewin, Human Evolution, W. H. Freeman & Company, New York, p. 40. Copyright © 1984 by Blackwell Scientific Publications, Ltd. Used by permission.)

that are dated to approximately 2 million years ago. The legs are relatively short and the arms are relatively long, although the anatomy of the pelvis, femur, tibia, and ankle and foot bones leaves absolutely no doubt that they were erect-walking bipeds. There is nothing to indicate that they could have been knuckle-walkers. Brain size was right in the range of variation of anthropoid apes and possibly, on the average, slightly smaller than in later Australopithecines. The position of entry of the spinal chord into the skull is in-between that characteristic of humans and anthropoid apes. (1983, pp. 252-53)[3]

Brace goes on to say that the jaw of A. afarensis shows the apelike structure, with the sides of the dental arch being relatively straight and parallel (pp. 253-54) (Figure 49.13). There is a noticeable diastema separating the canines from the incisors, and the canine crowns are more robust than those of younger hominids. The unworn canines would project slightly above the line of the rest of the teeth, but they were worn down to that level and flattened, as in the case of the human canines. Brace describes certain other features of the molars that are "typically (but not universally) found in nonhuman primates and quite different from the bicuspid crown forms typical for humans" (p. 254).[3]

Because no shaped stone tools have been found in direct association with the earliest australopiths, it is assumed that they did not have this capability, although they may have used stones and pieces of wood as they found them in the function of tools and weapons. (Crudely shaped broken pebbles have been found in association with australopiths as old as -2.6 m.y. at Hadar, but apparently not with remains of A. afarensis.)

An interesting point is that bipedalism seems to have

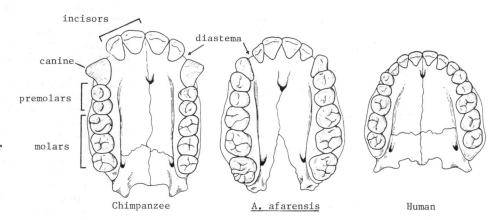

Figure 49.13 Comparison of the upper jaws of three hominoids. (Drawn by Luba Gudtz. From Roger Lewin, Human Evolution, W. H. Freeman & Company, New York, p. 39. Copyright © 1984 by Blackwell Scientific Publications, Ltd. Used by permission.)

preceded the great increase in brain size that was yet to come. A basic factor, often cited as encouraging bipedalism, is that a change of habitat from forest to open grassland (the savanna environment) required living and moving on the ground, rather than in trees, so that walking upright would have been a valuable adapation. It has also been suggested that the australopiths took to walking on two hind limbs because the hands of the forelimbs, no longer needed for climbing trees, had become well adapted to grasping objects, such as stones, sticks, food plants, or game. It would have been most advantageous to be able to move about rapidly while carrying objects in the hands (Campbell, 1974, p. 199).

But not all the experts agree that A. afarensis had forsaken the forest for open ground. Anthropologist Randall L. Susman and anatomist Jack Stern interpret Lucy's bone structure as indicating good adaptation for climbing but that she also had developed efficient locomotion. Gish presents these arguments in considerable detail (1985, pp. 158-62). In rebuttal to Susman and Stern, C. Owen Lovejoy, an anatomist, argues that Lucy's hip is "beautifully adapted for bipedality and poorly adapted for climbing" (Herbert, 1982, p. 116; see Figure 49.12). The argument continued (Lewin, 1983a; 1983b).

An extremely interesting and much-publicized discovery at Laetoli in Tanzania is a collection of hominid footprints in the bedding surfaces of layers of volcanic ash, dated by the potassium-argon method as between -3.5 and -3.8 m.y. (Hay and Leakey, 1982). Parallel tracks of hominid footprints occur beside the prints of a variety of other animals. The hominid prints show that the foot had a rounded heel, an uplifted arch, and a forward-pointing big toe. If the prints are those of A. afarensis, which seems highly likely in view of the age of the strata, their presence would seem to settle the question of whether or not the species walked upright.

Brace offers some thoughts on classification in the closing paragraphs of his section on the australopiths (1983, pp. 254-55). Although he has had an opportunity to handle the A. afarensis fossils from both the Ethiopian and South African sources, and although he agrees with others who find the older species to be more primitive in its traits than A. africanus, he is not convinced that the difference is so great as to warrant their being designated as two distinct species. He goes on:

> Our disagreement is merely a matter of the assignment of names. This is based on judgment of individual scholars and is a trivial matter, but it does point up an issue of fundamental significance. In an evolutionary continuum, change occurs more or less gradually through time. At the early and late ends of such change, everyone agrees that different names are justified, but when one form slowly transforms into another without break, the point where the change of name is to be applied is

a completely arbitrary matter imposed by the namers for their convenience only--it is not something compelled by the data. (1983, pp. 254-55)[3]

Molecular Phylogeny of the Primates

In Chapter 36, on methods of molecular biology used to establish phylogenies based on differences in protein and DNA molecules, we gave three examples of primate phylogenies. One was based on immunological distances, a second on amino acids in carbonic anhydrase, and a third on DNA hybridization. The discussion of DNA hybridization methods included a review of recent research by Charles Sibley and John Ahlquist leading to a phylogenetic tree of the primates with a time scale attached. Their tree is shown here as Figure 49.14. When you compare this tree with that of Sarich and Wilson, based on immunological distances and published in 1967, you will find the two remarkably similar in plan (see Figure 36.1). The important difference with respect to the hominoids is that the albumen immunological tree gives equal rank to human, chimpanzee, and gorilla, whereas the newer DNA tree gives human and chimpanzee equal rank with a common junction and shows the gorilla as having branched off somewhat earlier.

What should perhaps interest us most about the Sibley/Ahlquist tree is the comparatively recent date at which the human is shown to have branched from the chimpanzee: -6 to -8 m.y. Sibley and Ahlquist calibrated

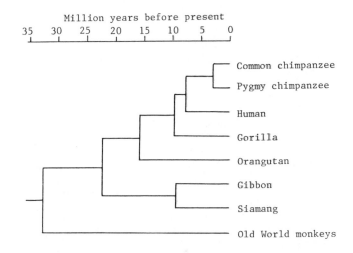

Figure 49.14 The hominoid family tree as calculated from DNA hybridization. Compare with Figure 36.1. (From C. G. Sibley and J. E. Ahlquist, Jour. of Molecular Evolution, vol. 20, p. 12, Figure 6. Copyright © 1984 by Springer-Verlag New York. Used by permission.)

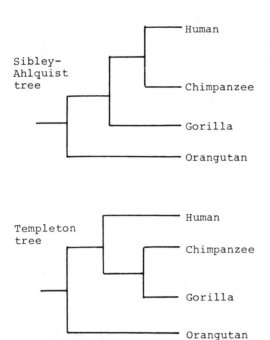

Figure 49.15 Comparison of two phylogenetic trees for the pongids and hominids, based on methods of molecular phylogeny.

their DNA "clock" for the primate phylogeny in part on the basis of fossil evidence of the date of origin of the orangutan, thought to be -13 to -16 m.y. According to science writer Roger Lewin, Vincent Sarich (whose immunological studies in coauthorship with Allan Wilson we presented earlier) favors a somewhat more recent date for the branching of humans from the African apes, specifically, -4.5 m.y. (1984, p. 1181). Both Wilson and Sarich are critical of Sibley and Ahlquist's two major calibration points, i.e., the 80-m.y. age of the separation of ostrich and rhea on opposite sides of the widening

South Atlantic and the 13- to 16-m.y. age of the first orangutan.

To add to the happy confusion on the placement of branches on the anthropoid tree and the age of the human/African ape split, Alan Templeton of Washington University in St. Louis in 1983 published a statistical analysis of data of mitochondrial and globin DNA and other (cleavage map) DNA data. His conclusion is that the chimpanzee and gorilla should be clustered at the ends of two branches from a common fork, with the human on its own separate branch arising earlier from a common fork with the African apes. Templeton's hominoid tree is shown in Figure 49.15 in comparison with that of Sibley and Ahlquist, but without any time scale. Look now at our evolutionary diagram of the primates, Figure 49.1, with special attention focusing on the phylogeny of the great apes and the hominids; it favors Templeton's tree. The early splitting off of the orangutan, the Asiatic ape, is clearly indicated in the molecular phylogenies in marked contrast to the later splitting off of the hominids from the African apes.

Credits

1. From Evolution of the Vertebrates, First edition 1955, Second edition 1969, Third edition 1980, by Edwin H. Colbert, John Wiley & Sons, Inc., New York. Copyright © 1955, 1969, 1980 by John Wiley & Sons, Inc. Reprinted by permission.

2. From Vertebrate Paleontology, Third edition, by Alfred S. Romer, University of Chicago Press. Copyright © 1933, 1945, and 1966 by The University of Chicago. Reprinted by permission.

3. From C. Loring Brace, Humans in Time and Space, pp. 245-82 in Scientists Confront Creationism, Laurie R. Godfrey, ed., W. W. Norton & Company, New York. Copyright © 1983 by Laurie R. Godfrey. Used by permission.

4. From IN THE BEGINNING © 1983 by Chris McGowan. Reprinted by permission of Macmillan of Canada, A Division of Canada Publishing Corporation.

Chapter 50

Human Evolution

The human genus Homo is represented by three species: habilis, erectus, and sapiens (Figure 49.6). Of these, only a subspecies of H. sapiens is alive today. As in the case of the australopiths, the oldest species, H. habilis, was the most recently discovered, but, as we shall find, the scientific community is divided as to whether this species should be in the genus Homo or in the genus Australopithecus. A skull of H. erectus was discovered on the island of Java in 1898 by Eugene Dubois, a Dutch physician, and it was given the name Pithecanthropus erectus, which translates to "ape-man who stood upright." To the public it became known as Java Man. Then, in the 1920s in the deposits of a cave not far from Beijing, China, a Swedish paleontologist, Birger Bohlen, found a single lower molar that could be clearly recognized as belonging to an ancestral human species. It was studied by a professor of anatomy at the Peking Union Medical College, Davidson Black, who gave it the name Sinanthropus pekinensis, meaning "Chinese Man from Peking," or simply Peking Man. A skull of Peking Man was found in 1929. At the time there was no means of dating the deposits that enclosed Java Man and Peking Man, but it is now known that the species to which they belong, H. erectus, goes no farther back than the beginning of the Pleistocene Epoch. So let us get back to a correct chronological order in the phylogeny of the human genus, which may begin with H. habilis.

Homo habilis--The Handy Man

In 1964, Louis Leakey discovered in the Olduvai Gorge in Tanzania the jaw fragments of a hominid dated as -1.75 m.y. He recognized it as more primitive than H. erectus and sharing many features with the the australopiths. Even at the time, there was some doubt as to whether to place it in the genus Australopithecus or in the genus Homo. Leakey gave the creature the name Homo habilis. In 1967, at Kanapoi, Kenya, an elbow joint judged to belong to a hominid of genus Homo was found beneath a volcanic ash layer dated by the potassium/argon method as -2.5 m.y. (Sullivan, 1967). Presumably, it was an older occurrence of H. habilis.

Starting in 1968, Richard Leakey, his wife Dr. Maeve Leakey (an anthropologist), and several other anthropologists began to search for hominid fossils on the eastern shore of Lake Turkana (formerly Lake Rudolph) in northern Kenya. In 1972 at a site called Koobi Fora they found at one point a large number of loose skull fragments that had been exposed by rain-beat erosion. These were reassembled by Maeve Leakey to form a remarkably complete skull (designated ER 1470) with a cranial capacity of about 800 cc--more than that of the gracile australopiths (450 to 550 cc), but less than that of H. erectus (800 to 1,200 cc) (Figure 50.1). Richard Leakey first announced that the age of this fossil was -2.6 m.y. (Sullivan, 1972). That figure has since been revised downward to between -1.6 and -1.8 m.y. Leg bones found at the same site gave convincing evidence that the creature was bipedal. Leakey later used the name

H. habilis for the Koobi Fora skull, the same name his father had used in 1964 for the skull fragments found at Olduvai Gorge. In later years, more remains of the ancient hominid were found in eastern Africa, both in the Lake Turkana area and in the Olduvai Gorge. By 1981 the new species had become reasonably well established in hominid paleontology as Homo habilis; it is entered in Figure 49.6 as having a time range between -2.2 and -1.6 m.y. Found in deposits of the same age are bones of the robust australopith, A. boisei, which lived on later than any of the other australopiths. Clearly, the two hominids were contemporaries in a shared environment.

Homo habilis was not greatly different in outward appearance from A. africanus. Both showed some remarkable similarities with modern humans: the jaws lacked the large projecting canines found in the apes; the pelvis showed adaptation to prolonged walking and

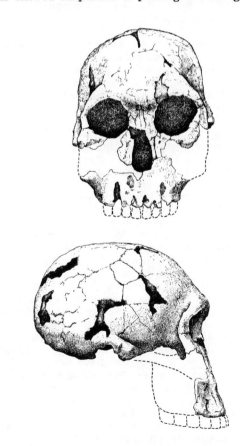

Figure 50.1 Drawings of the skull of Homo habilis, found at Koobi Fora in 1972 and assembled from many small fragments by Maeve Leakey. It is designated as specimen ER 1470. (From The Hominids of East Turkana, by A. Walker and R. E. F. Leakey, Scientific American, vol. 239, no. 2, p. 60. Copyright © 1978 by Scientific American, Inc. All rights reserved. Used by permission.)

running upright. On closer inspection, the ER 1479 skull of H. habilis can be distinguished by its high-vaulted, well-rounded skull, lacking a sagittal crest (Leakey, 1976, p. 177). The hands of H. habilis could easily manipulate objects between the thumb and forefinger. Leakey selected the name "habilis" for its meaning as "adroit" or "skillful" (as in the word "habile"); the appelation "Handy Man" is in popular use for the species. H. habilis had learned to make rudimentary tools by chipping an egg-shaped pebble to form a sharp cutting edge at one end. The species was probably a scavenger as well as a vegetarian, feeding on carcasses or on small animals it was capable of killing. The stone tools would have greatly assisted in dismembering large animals. At one site at Koobi Fora stone tools of this type were "found in association with the partially butchered skeleton of an extinct hippopotamus all silted over in the deposits of an ancient river delta" (Brace, 1983, p. 261; from Isaac, 1976).

Not everyone in the mainstream community of anthropologists and paleontologists was pleased with the Leakeys' designation of the Olduvai and Koobi Fora specimens as members of the genus Homo. Among the dissenters were C. Loring Brace, whose work we have quoted in earlier paragraphs, and Milford H. Wolpoff (Wolpoff and Brace, 1975). Brace states that the jaw and teeth of the Olduvai specimen cannot be distinguished from a typical specimen of A. africanus; he would place the brain size as "either at the small end of H. erectus or the large end of the Australopithecine range of variation" (1983, p. 258).[1] As to Koobi Fora EH 1470, Brace comments that ". . .the 'true Homo' status of the skull was claimed with somewhat more enthusiasm than a careful appraisal of the evidence would support" (p. 260).[1] Brace continues:

> As it happens, the long, somewhat "dish-shaped" face and the enormous molar root sockets look perfectly Australopithecine. The brain case itself looks Australopithecene in form, but, at 750+ cc., it is exactly in-between Australopithecus and Homo. (Reference to Holloway, 1975.)
>
> It is taking nothing away from the significance of ER 1470 to say that it does not support the contentions of its finders. It does fit between two major categories--the creationists would say "kinds"--of hominid, and it provides a gratifying picture of one of the steps by which Australopithecus became transformed into Homo. Because it shows such a mixture of features, it is not surprising that the authorities who have studied it have been unable to agree on a named pigeonhole for its assignment. (Brace, 1983, p. 260)[1]

Brace adds that other specimens (ER 1813 and OH 16) show a reduction in tooth size along with an increase in brain size, something that is to be expected during the course of a transition of one genus into another.

As shown on a conventional type of evolutionary ladder, H. habilis is usually connected to A. africanus, but ancestors of Homo erectus may have branched from the australopiths much earlier. Paleontologist Stephen J. Gould even goes so far as to suggest that the true humans did not arise from the australopiths and that the former may actually be older than any of the latter (1977, p. 60). Gould cites in support of this suggestion the work and conclusions of anthropologist Charles Oxnard (1975b). (This is the same material we covered in Chapter 49 in evaluating Gish's use of Oxnard's statements.) Gould is favorably impressed with Oxnard's data and supports Oxnard's suggestion favoring "the removal of the different members of this relatively small-brained, curiously unique genus Australopithecus into one or more parallel side lines away from a direct link with man" (Oxnard, 1975b).

Gould goes on in his essay to use this interpretation of a prolonged parallel evolution of genera in support of his and Eldredge's model of punctuated evolution, in which a very small, isolated population forms a new species very rapidly, but the old and new species endure as cohabitants for millions of years with little or no visible change. Nothing in what Gould says denies the common ancestry of the coexisting hominid species; the points in time of bifurcations from common ancestral species are simply not yet documented. I think you will enjoy reading Gould's anticipation of the creationists' reaction to his promotion of Oxnard's findings:

> At this point, I confess, I cringe, knowing full well what all the creationists who deluge me with letters must be thinking. "So Gould admits that we can trace no evolutionary ladder among early African hominids; species appear and later disappear, looking no different from their great-grandfathers. Sounds like special creation to me." (Although one might ask why the Lord saw fit to make so many kinds of hominids, and why some of his later productions, H. erectus in particular, look so much more human than the earlier models.) I suggest that the fault is not with evolution itself, but with a false picture of its operation that most of us hold-- namely the ladder; which brings me to the subject of bushes. (Gould, 1977, pp. 61-62)

Well, Gould predicted correctly. As we have seen in the previous chapter, Gish came out with an incorrect use of Oxnard's statements to claim that that Australopithecus is not related to anything living today and is uniquely different from both man and ape (1978a). Gish reiterates this claim in his revised edition of the same work (1985, p. 179). This, as we noted earlier, is tantamount to a flat statement that none of the three have any ancestry whatsoever, let alone a shared ancestry. (Gould's essay was written in 1976 and first published in Natural History; it was thus available to Gish prior to publication of Evolution: the Fossils Say No!")

Gish mentions Homo habilis briefly in connection with Australopithecus (1978a, pp. 110, 112; 1985, p. 150). One reference is in a quotation from Oxnard stating that multivariate statistical studies (reviewed above by Gould) lead to conclusions other than that H. habilis "was almost directly ancestral to man" (1975b). Oxnard's rejection of a "direct" ancestral relation does not exclude existence of indirect ancestral relationship, a point we have made repeatedly before, yet Gish would have us construe Oxnard's words as rejecting any ancestral relationship whatsoever. To see how grossly Gish has misconstrued Oxnard's conclusions arising from multivariate analysis, we need only read the following concluding paragraph from Oxnard's section on the fossil primates:

> We may thus be a considerable way toward discarding ideas such as that in Australopithecus bipedal human walking is established, or that "Homo habilis" is unequivocally a human toolmaker. Rather, we may now be able to search for the actual nature of morphologies and functions relating to a species that is becoming somewhat similar to man, but that is clearly not yet there. Recognition of such possibilities may also allow alternative hypotheses to be explored; for instance, it may yet be shown that Australopithecus, though a close ancestor, may nevertheless have been situated on a side branch. Recent discoveries that push human and prehuman remains back in time may eventually force this particular conclusion upon us. (1973, p. 168)

How, I ask, could there be devised a clearer affirmation of common ancestry of two genera? To suggest parallel evolution of two genera over a long period of time is completely within the framework of the theory of evolution, albeit with a preference for a model that stresses rapid change followed by stasis, i.e., punctuated equilibrium.

A second reference to H. habilis by Gish is in the context of his summary of what he thinks is the correct place of the australopithecines, i.e., that they are really apes (pongids) and not hominids (1987a, pp. 112-13). The reference reads: "It has been argued by Robinson and others that Homo habilis is the same as A. africanus. If this is true, then the above arguments would also apply to this creature." What Gish intends to make clear is simply his belief that H. habilis, too, is an ape. This is a creationists' way of keeping any animal with an apelike countenance on the far side of the moat protecting the specially created "kind" that is the true human. In the folksy vernacular of fundamentalist Protestants I knew in my early years, this reads: "I can tell you for sure, I didn't come from no ape!"

Peking Man and Java Man--Homo erectus

Homo erectus, as we previously noted, was first found in eastern Asia. For Peking Man, correctly titled H. erectus pekinensis to designate it as a subspecies, a continuous record of occupation of the cave at Zhoukou-dian may span as much as 200,000 years (-460,000 to -230,000 yr). The fossils of Peking Man accumulated by Chinese anthropologists from this and other localities over the past 50 years include 6 complete skulls, numerous skull fragments andpieces of mandibles, and more than 150 teeth. The limb fossils are fewer in number, but altogether the collection makes it possible to reconstruct the form and appearance of Peking Man and to document the changes the species underwent. During the possible 200,000 years of cave occupancy, the brain of Peking Man increased in size from about 900 cc to about 1,100 cc, while stone tools in the cave deposits show an evolution from primitive choppers and scrapers to complex arrow and spear points (Rukang and Shenlong, 1983). Charcoal in the lowest layers shows that Peking Man used fire from the earliest time of cave habitation. Age determinations include uranium/thorium ratios, fission-track dating, and paleomagnetism. One major paleomagnetic reversal at -730,000 yr (from Matuyana Epoch to Bruhnes Epoch) is recorded near the base of the cave deposits (Zechun, 1983, p. 298). Pollen analysis gives information on the prevailing climate and the tree species that grew in the area.

In the course of time, remains of H. erectus began to be reported in widely scattered localities in Europe, North Africa, East Africa, and South Africa. Eight subspecies were recognized and these were arranged into five "grades" or levels of evolutionary progress (Howells, 1966). The Java Man skulls, with thick walls and relatively low brain capacity, fall into the lower grades and are judged to be over 700,000 years old.

Following World War II, African members of H. erectus were discovered: at Swartkrans, South Africa, in 1949; at Olduvai Gorge by Louis Leakey in 1961. The former hominid, called Swartkrans Man, is of the first grade, comparable with Java Man. The Olduvai specimen has been dated at -1.25 m.y. Another of the primitive forms, found at the Koobi Fora locality and designated as skull KNMER 3733, is dated at -1.6 m.y. In general, then, the oldest representatives of H. erectus are from Africa, nearly a million years older than those of Europe, China, and Java, although one specimen from Java has been tentatively and with low reliability dated at -1.5 m.y. (Lewin, 1984c, p. 54). A European relative, Heidelberg Man, is of the second grade, while others in Hungary and England (Swanscombe Man) are more advanced and compare in grade with Peking Man. Thus the conclusion has emerged that H. erectus has two separate and parallel branches, one Asian, the other African. The two

branches are shown in Figure 49.6. This hypothesis is supported by the finding in 1963 in the Lantian district of Shensi, China, of a grade-one representative comparable in age with both the Java Man and the Swartkrans Man.

One line of current thought, based on tracing of distinctive types of stone tools, is that H. erectus originated in western Europe, then migrated eastward to China and Java and southward into Africa (Stebbins, 1982, p. 337). An alternative interpretation has the African and Asian branches derived from H. habilis in Africa, migrating from that continent first into eastern Asia, then later into Europe (Andrews, 1984). The oldest representatives are shown in Figure 49.6 as dating back to -1.6 m.y. and to have arisen from H. habilis, or alternatively, if the latter species is an australopithecine, from some much more ancient common ancestor of the two genera.

Homo erectus shows distinct changes in skull configuration from the oldest to the youngest forms, corresponding with an increase in brain capacity. The flattened face and high-domed cranium of the high-grade forms, such as Rhodesian Man and Steinheim Man, give the subspecies a strikingly "modern look," at least to the eye of an amateur. The lower-grade forms are more apelike, with a massive lower jaw, a somewhat flattened cranium, and a prominent brow ridge, as in H. habilis and the australopiths (Figure 49.7D). Whereas these older hominids seem to have changed little if any during their spans, H. erectus shows a definite linear evolution from one subspecies to the next. If this interpretation is correct, it may seem to support gradualism, rather than punctuation but, on the other hand, we may be simply seeing a series of smaller, more frequent punctuations. (More about this later.)

Compared with modern humans, H. erectus was more heavily boned, had more prominent brow ridges, jaws more protruding, and a less strongly developed chin, but may have been no less dexterous than modern humans. There seems to be no question that H. erectus was a capable hunter and killer of large animals, a practice made possible by the crafting of sharp, effective stone blades and points. Cave deposits left by Peking Man contain abundant remains of large mammals, especially deer. The use of fire by Peking Man suggests the possibility that cooking of meat may have been practiced.

The creationists have found H. erectus a bit of a problem. Their alternatives are: (a) throw the creature back to the apes; (b) admit it as a subspecies of the human "kind." Henry Morris opines: "It may well be that Homo erectus was a true man, but somewhat degenerate in size and culture, possibly because of inbreeding, poor diet and a hostile environment" (1974a, p. 174). He thinks there is some question about the humanness of the species because of the small brain size (900-1100 cc.), but he concludes: "However, that is definitely within the range of brain size of modern man, though on the low end of the scale." So, chalk up one vote in favor of admitting H. erectus to the human "kind."

Gish approaches the question with a different strategy, which is to concentrate on Java Man and Peking Man, sketching a rather long, complicated, and cloudy history of the fossils and their interpretation (1978a, pp. 113-16; 1985, pp. 180-87, 190-200). It seems that the Dutch physician, Dubois, who found the skull of Java Man and named it Pithecanthropus erectus, concealed his discovery of two other skulls at Wadjak (the "Wadjak skulls"), presumably because they had large cranial capacities, larger even than the human cranial capacity. Gish considers this concealment an act of dishonesty. I suggest that Gish's strategy here is to imply that Dubois's dishonesty impugns the authenticity of the Java Man skull. Anyhow, Gish claims that at a later time Dubois changed his mind about Java Man being a human and declared instead that it was a large gibbon. (See Gish for references.) Then later, two French scientists, M. Boule and H. M. Valois--the former being Director of the French Institute of Human Paleontology--published their opinion that the Java Man skull was very similar to that of chimpanzees and gibbons. Details of the case are too involved to recite here, but Gish thinks it has been made clear enough that _Pithecanthropus_ was an ape (1978a, p. 116). Chalk up one creationist vote for tossing H. erectus back out across the moat, with strict instructions never to set foot in the human castle again.

Gish offers a minitreatise on Peking Man, the history of which he characterizes as a "tangled web of contradictions" (1978a, pp. 116-34; 1985, pp. 190-200). Episodes in that clouded history correctly include the loss of most of the original Peking Man fossils during the war. Several pages are devoted to the speculations of a Roman Catholic priest, Rev. Patrick O'Connell, contained in his 1969 book titled Science of Today and the Problems of Genesis. O'Connell was in China during the early excavations of the cave at Choukoutien (now Zhoukoudian) and several years subsequently but did not actually examine the site; instead, he based his scenario on local journal accounts. To make a long story short, O'Connell concluded that the human remains found at the cave site were those of Late Paleolithic inhabitants, while the supposed Peking Man skulls were those of large macaques or baboons killed and eaten by the humans. O'Connell's conclusions are welcomed by the creationists because they effectively dispose of H. erectus and leave only a large gap between apes and a created human kind.

Should you read Gish's material on Peking Man or venture so far as to look up O'Connell's book, I would recommend as an antidote that you read a recent article written by two Chinese scientists, Wu Rukang and Lin Shenglong (1983). One of them is on the staff of the Chinese Institute of Vertebrate Paleontology and Paleoanthropology. Also recommended is an article by a professor in the Nanjing Institute of Geography, Liu Zechun (1983). These lucid, well-illustrated accounts give an up-to-date review of the geology of the cave and its deposits and of the fossil materials and their chronology, all based on modern methods of age determination--paleomagnetic as well as radiometric.

Gish inserts several paragraphs on the history of two well-known fraudulent "fossil men" (Nebraska Man and Piltdown Man) to serve as object lessons and cast doubt on the validity of the Peking Man fossils (1978a, pp. 119-21; 1985, pp. 187-90). Boule and Vallois again enter the scene, and again they are reported to have associated the Peking Man fossils with anthropoids that could not have been the ancestors of humans, but instead represented large monkeylike or apelike creatures. And so, Gish casts another vote to kick out one of the H. erectus representatives from the castle of the human "kind."

Do we have a schism here between two creationists from the same institution? Dr. Morris is the spiritual and scientific leader of the Institute for Creation Research; Dr. Gish is coming on strong as the resident scientist judged most versed in modern paleontology. Someone must rule on this issue, which is of grave concern, for at the moment poor Homo erectus, Asiatic style, is floundering around in the creationist moat in grave danger of drowning. This scurvy treatment of a possible human relative is unconscionable and must end!

Finally, Gish turns to Richard Leakey's discovery of remains of both australopiths and H. erectus in Bed II of the Olduvai Gorge strata (1978a, p. 130). Gish asks:

> If Australopithecus and Homo erectus existed contemporaneously, then how could one have been the evolutionary ancestor of the other? And how could either be ancestral to man, when man's artifacts are found at a lower stratigraphic level directly underneath the fossil remains of these creatures? These are very hard questions, indeed. The evolutionary hypothesis for the origin of man

becomes less and less plausible as more and more evidence becomes available. (P. 130)

(Note: Gish, in 1985, twice repeats the above statement in similar wordings, pp. 171, 203.)

The answer to the contemporaneity of the two hominids is one we have given before--more than once--and is quite simply answered by asking whether or not it is possible in a modern human family for grandparents to be alive and well in the same household as their grandchildren. According to ages derived by the potassium/-argon method, A. robustus survived into the early Pleistocene; H. erectus came on the scene at the start of the Pleistocene. There was a span of time when both lived in close proximity. On the other hand, remains of A. robustus go back in time a half-million years before the appearance of H. erectus, so that the former could have been the ancestor of the latter, perhaps with H. habilis as an intermediate. So what is all the fuss about? In regard to the second question, that of a primitive, hutlike habitation being found at lower level, the artifact has already been attributed to the intermediate, H. habilis--a perfectly plausible explanation.

Evolution of the Human Brain

Thus far, we have covered the evolution of the hominid skull, but not its contents--the brain itself. For the fossil hominids, the brain is not preserved, but its surface form is indicated by the inner surface of the skull, and details of that surface are sometimes preserved sufficiently well to enable the major parts of the brain to be outlined. It would be best, however, to start with a comparison between the brain of living humans and those of the great apes.

Side views of the brains of chimpanzee and modern human are shown in Figure 50.2. Apart from size, which is not of concern here, notice first the greater vertical dimension of the human brain and the more convoluted (infolded) cerebral cortex (outer layer). The occipital lobe, which is involved in vision, is proportionately much larger in the chimp than in the human, whereas the other lobes are proportionately larger in the human. The human parietal lobe, involved in sensory integration and association, lies farther back and extends lower down; the frontal lobe, involved in motor behavior and complex aspects of adaptive behavior, and the temporal lobe, involved in memory, are also larger and extend farther toward the back. In the chimp brain a deep infold, the lunar sulcus, sets off the occipital lobe from the rest of the brain, but this feature is less clearly seen in the human brain. Differences such as these are looked for on the inside casts (endocasts) of fossil skulls.

Anthropologist Ralph L. Holloway made a comprehensive comparative study of the endocasts of all available fossil hominid brains (1974). Of special interest to us are his findings on the brain of Australopithecus, which is classified by mainstream paleontology as a hominid, but considered an ape by the creationists. Holloway studied both the gracile form, A. africanus, and the robust form, A. robustus. He also included Homo habilis, which some anthropologists prefer to place with the australopiths. Holloway's conclusion is highly positive:

> Fortunately, no matter what controversy may surround the question of how these early African hominids are related to one another it has very little bearing on the question of their neurological development. The reason is that in each instance where an endocast is available, whether the skull is less than a million years old or more than two million years old, the brain shows the distinctive pattern of hominid neurological organization. (p. 109)

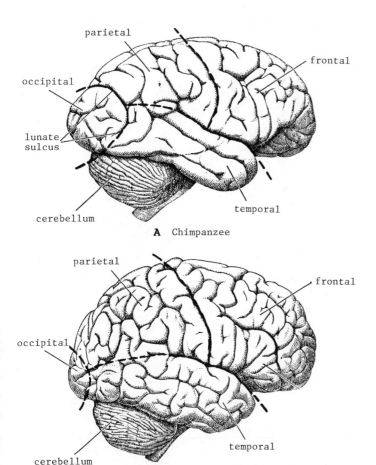

Figure 50.2 Brains of the chimpanzee (A) and human (B), drawn to the same length and showing the approximate outlines of the major lobes. (From The Casts of Hominid Brains, by Ralph L. Holloway, Scientific American, vol. 231, no. 1, p. 109. Copyright © 1974 by Scientific American, Inc. All rights reserved. Used by permission.)

Holloway's description of the differences in the two brains is quite technical in spots, but he notes that the australopithecine frontal lobe is larger and more convoluted than in pongid endocasts. Brain height and form of the temporal lobes show the hominid configurataion. The lunate sulcus, when it can be found, lies in the human position.

I found most interesting Holloway's discussion of the endocranial cast of Richard Leakey's Koobi Fora skull ER 1470, which we have described and assigned to H. habilis. At the time Holloway wrote his article, ER 1470 was a comparatively recent find and its age had been calculated, erroneously, as as -2.6 m.y. This great age was quite a shock to the anthropologists. Upon examining the endocast, Holloway was surprised to find that not only was the brain larger than that of the australopiths, but slightly larger even than in three other specimens of H. habilis. What this meant to Holloway was that Leakey's find of ER 1470 pushed back the history of hominid brain evolution to a time even older than that of the oldest australopiths then dated. Since then, of course, the error in dating ER 1470 has been corrected and an earlier australopith, A. afarensis (Lucy) has been found. Holloway has emphasized that absolute size of these fossil brains is not a criterion of whether or not they are of the hominid type (p. 112). His main point remains firm: the australopiths had brains much like those of modern humans, but unlike those of pongids. Creationists take

note! This is strong corroborative evidence contrary to your assertions that the australopiths were actually apes.

We turn now to brain size and its increase from the australopiths, through the older human species--H. habilis and H. erectus--and finally to H. sapiens. Holloway has devoted a great deal of attention to the rate of increase in brain size after body size has been taken into account (1974, pp. 111-15). First, the relationship between brain size and body size must be established within a given genus or species, taking into account the basic consideration that a larger brain is needed to control a larger body. As individuals go through the growth process, the various structures and organs increase in size to the maximum at full maturity. The study of this changing body geometry falls within the general title of "allometric growth." A particular structure or organ may or may not grow in the same ratio as the whole individual. Equations can be formulated for allometric growth of brain versus body; they have been formulated for various living primate groups--insectivores, prosimians, monkeys, apes, and humans. The equations are alike in basic formulation, but the rates of change of the two variable quantities are not the same. The differences in these change rates can tell us something of value about the evolution of the groups. The equation reads as follows:

$$\text{brain weight} = b \, (\text{body weight})^a,$$

where b is a numerical coefficient and a is an exponent (power) that specifies the rate of increase of brain size with body size.

Endocranial capacity (volume in cubic centimeters) is usually substituted for brain weight, for practical reasons. Figure 50.3 is a graph showing examples from three hominoid groups: living pongids (apes); australopiths, all fossils; humans, both fossil and living (Pilbeam and Gould, 1974). Each group is fitted by the same equation but with different coefficients and exponents. Particularly important here is the exponent, a, which determines the slope (or slant) of the line (regression line) fitted to the plotted points. Each point is a particular individual or (in this case) the average of measurements for individuals of a species. Note that the graph is logarithmically scaled on both axes (log-log plot), making the regression lines appear straight. A slope with value a equal to 1.0 would appear as a regression line inclined 45 degrees from the vertical (or horizontal); it would mean identical rates of change for the two variables, e.g., if one quantity is doubled, so is the other, etc.

Look first at the fitted regression lines for the lower two groups. These have almost the same slope because the exponent a is almost identical for both. For both, the rates of increase in brain size with body size are almost identical. But there is an important difference. The ratio of brain size to body size, expressed by the coefficient b, is not the same. The pongid brain is always smaller relative to its body than for the australopiths. Check this out by comparing the gorilla and the robust australopiths. Brain size is about the same for both (500 cc), but the body weight of the gorilla is over 100 kg while that of the australopiths is a mere 40 to 50 kg. Clearly, the brains of all the australopiths are always much larger in ratio to their body weights than for the apes. This has something to say about the need of the australopiths for proportionately larger brains to carry out their body functioning. It is a difference appropriate to their being placed in different families.

Next, compare the the genus Homo--the true humans-- with the australopiths. First, all of the humans have much larger brains than the australopiths for a given body weight. As you can see, H. habilis, with much the same

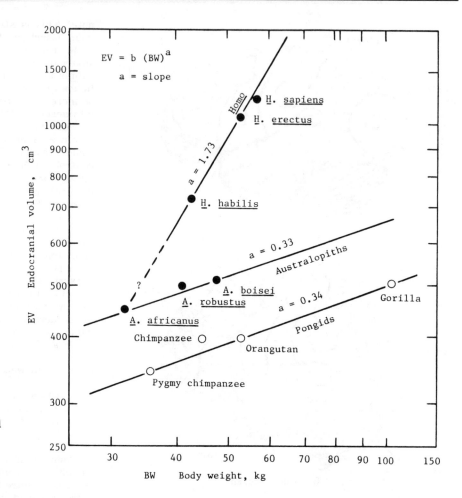

Figure 50.3 Log-log graph of endocranial volume as a function of body weight. (Data from D. Pilbeam and S. J. Gould, 1974, _Science_, vol. 186, pp. 894-95, Tables 1 and 2.)

body weight as the robust australopiths, has a substantially larger brain. The striking feature of the human regression line is that its slope is so much steeper than for the other two. This means that the relative increase (ratio of increase) is much greater for the humans. You might say that their hat sizes were growing out of all proportion to their clothing sizes--they were becoming "swell-headed." (Anthropologists apply the term "cephalization" to the relative growth rate shown by the slope of the regression line.)

The regression analysis described above was carried out by David Pilbeam and Stephen Gould in an attempt to ascertain the phyletic relationship between australopiths and humans, particularly as to a possible linkage between A. africanus and H. habilis, and the correct affiliation of the latter--whether with the australopiths or with the humans (1974). Their conclusion was that H. habilis is indeed right on the human line and of the same genus. It is difficult to argue against that conclusion when looking only at the regression data.

But here comes an interesting point not previously mentioned. In constructing the "human" regression equation, Pilbeam and Gould included A. africanus as one of the points. This was based on an assumption that the australopith is a direct ancestor of H. habilis. On the other hand you can see that if only the three representatives of genus Homo had been used (three points instead of four), the regression line would have been almost identical--passing very close to A. africanus when projected downward. But, as Pilbeam and Gould say:

> In the lineage leading to H. sapiens, brain volume does increase dramatically, but A. africanus was not the ancestor that first showed brain expansion beyond the australopithecine level; that honor must go to H. habilis, one reason for placing this species in Homo. (1974, p. 896)

Pilbeam and Gould carried out a similar study of the size of the cheek teeth of the same three groups used in the brain-size study. Here the results were comparable in revealing the uniqueness of the Homo group, except that the regression line for the human teeth slopes downward toward the right. This means that the human teeth were decreasing in size during the period of increase in body weight. Was their food becoming softer and easier to masticate?

In Chapter 51 we will have more to say about the evolution of the brain in Homo. What evolutionary pressures or influences would have caused such a rapid increase in the size and convolution of the brain?

The Rise of Homo sapiens

Homo sapiens seems to have arisen from H. erectus during a transitional period from -400,000 to -300,000 y. The change may have been gradual, with increasing cranial capacity causing greater bulging of the forehead region and reduction of the brow ridges. This evolving new member of Homo sapiens is designated by the adjective archaic, with the status of a subspecies. These humans knew how to construct dwellings made of branches formed into a domed shape. There then appeared a new subspecies that has long been known as Neanderthal Man, named for a valley (a "thal") in western Germany, the Neanderthal, where a human skull was discovered in 1856 and recognized as an extinct species or race of humans. Originally named Homo neanderthalensis, it is now considered a subspecies of H. sapiens (i.e., H. sapiens neanderthalensis); it spanned the period from -130,000 y. to -35,000 y. Neanderthal Man had a brain actually a bit larger (1,470 cc) than the average for living humans, which ranges from 1,000 to 2,000 cc with an average of 1,330 cc. The Neanderthals were somewhat more heavily boned than living humans, with brow ridges more prominent, and a more elongate skull (see Figure 49.7E). They developed a more advanced method of shaping stone tools, knocking off thin flakes that left razorlike edges; these are known as Mousterian tools. The Neanderthals have left evidence in their burial sites of displays expressing qualities of spirituality and reverence. They may have practiced burial rituals and ceremonies, and perhaps religious rites as well. The question of whether they had a language, as distinct from a system of signalling by sounds and gestures, is not resolved. One authority, Curtis G. Smith, places the start of language among the late Neanderthals, within the last 100,000 y. (Smith, 1985, p. 45, 145).

The creationists welcome Neanderthal Man into the fold of the human "kind." Henry Morris generously overlooks the bad image of the Neanderthals with their stooped posture, brutish character, heavy brow ridges, and crude habits; it was not their fault (1974a, p. 175). After all, says Morris, these features were due to disease, possibly arthritis or rickets. One creationist advocate, Dr. Rush K. Acton, a professor of orthopedic surgery and anatomy at the University of Miami, discusses the paleopathology of the Neanderthals (Acton, 1978). They may have had chronic vitamin D deficiency, which could have produced bone deformities. Paget's disease and congenital syphylis are also considered possibilities; the latter is known to cause the "Olympian brow" such as that of the Neanderthal skull. Acton goes so far as to tell us: "A specialist in venereal diseases in London named D. J. M. Wright examined the collection of Neanderthal bones in the British Museum of Natural History and reported that these bones could be merely modern man affected by congenital syphylis" (p. iv). Gish reinforces the opinions offered by Morris and Acton (1978a, pp. 134-35).

The point is, of course, that creationists do not like to admit to too much evolutionary change--microevolution, they call it--in a created "kind." Free of disease, Neanderthal Man would have been a perfectly good specimen of modern <u>H</u>. <u>sapiens</u>. Dr. Acton speaks for the creationists in his summary view of the Neanderthals:

> Putting aside preconceived notions of evolution or creation, one can clearly see that the evolutionary scientists have provided good evidence to suggest that Neanderthal Man might well represent some of Noah's descendants ravaged by various diseases. How blind man can be to scientific evidence when it conflicts with a compelling need to demonstrate that God does not exist and that the creation did not take place. (p. iv)

Demise of the Neanderthals in Europe and southwest Asia, occurring about -30,000 y., coincided with their replacement by a people essentially similar in all respects to living humans. This newcomer is known as Cro-Magnon Man, named for a locality in France where several skeletons were found. Whether the replacement of Neanderthal Man by Cro-Magnon Man was by conquest and slaughter or by gradual genetic evolution within the Neanderthal population is a question that has been debated. Skeletal remains that have been interpreted as transitional may, instead, be a result of hybridizing.

The Cro-Magnon people had a rich culture, quite different from that of the Neanderthals. The Cro-Magnons were cave dwellers; their cave paintings display advanced artistic capability in producing realistic renditions of animals, as well as in generating abstract symbols. In addition to excellence in the crafting of stone tools, they were skilled at carving bone into ornate art forms.

The creationists capitalize on the anthropologists' admission to the seeming abrupt appearance and demise of the Neanderthals and their replacement by Cro-Magnon Man, apparently from nowhere. Gish writes:

> Neanderthal Man, for example, abruptly appeared in

Europe. Anthropologists have no evidence whatever concerning his origin. He as abruptly disappeared and is replaced by Cro-Magnon Man whose origin is equally mysterious to evolutionists. (1978a, p. 143)

If acting true to form, Gish might have simply concluded that because evolutionists don't know all the answers, their entire theory is wrong. In this case, however, he launches into his own explanation, which is straight evolutionary theory on the lines of punctuated equilibrium. He explains that if small groups of humans became geographically isolated, they would have undergone rapid evolutionary change. Because a high degree of inbreeding would have resulted, genetic traits that were previously suppressed through dilution would have rapidly surfaced. The result would have been the rise of "tribes" or "races." Evidently pleased that creationism can handle the problem of origin of subspecies and races through application of principles of genetics, Gish turns upon the evolutionists, scorning their impotence:

> That evolution theory, in view of known genetic data, produces no satisfactory explanation for the origin of races is evident from the following statement made in 1972 by the famous evolutionist, the late Theodosius Dobzhansky: "It is almost incredible that a century after Darwin, the problem of the origin of racial differences in the human species remains about as baffling as it was in his time." (1978a, p. 145)

> (See Gish for reference.)

At this point, Gish is near the end of his chapter on the origin of the human species; it is time to pull out all the stops:

> In other words, there is no way of correlating the genetic data associated with the various races within an evolutionary framework. It is an amazing thing that evolutionists insist that they can explain how the universe evolved, how life evolved, how fishes, amphibians, reptiles, birds, and mammals evolved, and how apes, monkeys, and men evolved from earlier primates, and yet must admit that they cannot explain the origin of races within the species <u>Homo</u> <u>sapiens</u>! If evolution theory cannot even explain the origin of races in the light of the known scientific evidence, how then can one pretend to use the theory to explain the most profound mysteries of all? Apparently the closer the theory approaches the actual scientific data, the more untenable it becomes. (p. 145)

Microevolution within the Human Species?

Continuing on the subject of the evolution of the races or subspecies of <u>Homo</u> <u>sapiens</u>, the transition in Europe from Cro-Magnon Man to the modern human subspecies, <u>Homo</u> <u>sapiens</u> <u>sapiens</u>, occurring during the final deglaciation following the Wisconsinan glacial stage about -12,000 y. ago, seems to have been as much one of cultural evolution as of genetic evolution. It has been stated by some geneticists that for at least the last 25,000 years the human gene pool has not changed appreciably.

Gish, as we have already seen, shares that opinion, for he has given us an outline of the way in which reshuffling of genes in isolated populations has produced new races. As an example, Gish cites the evolution of the Tasaday people of the interior of Mindanao, an island of the Philippines (1978a, p. 144; 1985, p. 213). He claims that this group became isolated from fellow Filipinos about 500 to 1000 years ago. Relieved of pressures of competition, they reverted or regressed to a more primitive state, which sets them apart as a different race or tribe. Gish may have gotten his information from a 1971 article in National Geographic and its subsequent television airing by NBC, picturing the Tasaday as stone-age people, faithfully preserving the paleolithic culture. That presentation proved a "carefully-orchestrated 20th-century hoax," as recently explained in Geographical Magazine (Leather, 1986, pp. 490-91).

Apparently, Gish is in disagreement with Dr. Morris on the possibility that evolution has taken place in recent centuries. Morris has given us another version:

> If evolution is taking place today, it operates too slowly to be measurable, and, therefore, is outside the realm of empirical science. To transmute one kind of organism into a higher organism would presumably take millions of years, and no team of scientific observers is available to make measurements on any such experiment. (1974a, p. 5)

Granted that the instantaneous rate of evolutionary change in human populations may be difficult or impossible to measure within only a few decades, there is suggestive evidence that increments of evolutionary change--in inheritable traits or features, that is--have taken place in the time span since the introduction of written records, or about the past four millenia. Brace offers two such examples. One is of the early farmers of Jericho inhabiting that city at the time of its destruction. Their faces, jaws, and teeth "were slightly larger and more robust than those of the modern inhabitants of the same area" (1983, p. 271).[1] This conclusion is based on Brace's anthropological studies of the groups being compared. A second example from Brace's own research is that of inhabitants of the London, England, area during the seventeenth century. Brace found that the faces of the farmers of the region surrounding London were more robust than those of average citizens of London. Brace states: "The differences are not dramatic and the magnitude amounts to less than a five percent average change, which is not by itself statistically different, but the fact that the direction of difference is the same in all traits measured is in itself highly important" (p. 271).[1]

The microevolution described by Brace is thought by evolutionary biologists to have led not only to the differentiation of races and subspecies, but also in the past to the rise of new species. This topic is one we have covered in an earlier section on principles of speciation. The problem associated with the rise and demise of such subspecies as the Neanderthals and the Cro-Magnons-- how, when, and why these events occurred--is that the actual genetic changes that produced a distinctly new race or subspecies were usually accompanied by cultural changes that obscure the time and place of the evolutionary change. Migration of a people over long distances and the expansion of a region occupied by a particular race bring that race into a new ecological environment, perhaps very different in its physical and biological aspects from the environment in which the genetic change occurred. Anthropologist Bernard Campbell gives an excellent account of the problems of origin of the several subspecies or geographical races of modern humans (1974, pp. 394-98). He writes:

> The problems introduced by migration and expansion in understanding the relationship between the present pattern of Homo sapiens populations and their environment must involve, in practice, a consideration of their past history and cultural adaptations as well as their anatomical and physiological characters. (P. 395)[2]

Campbell states that anthropologists can account for a certain number of the more obvious inheritable differences among and between the geographical races generally

recognized in modern humans. Examples he gives are "the ratio of area to weight, which is correlated with temperature and relative humidity, or skin and eye color, which is correlated with solar radiation, or nose shape, which is correlated with absolute humidity" (p. 394).[2] He continues:

> These kinds of bodily adaptations are clear enough to make recognizable the sorts of differences that zoologists would classify as subspecific, and, as a result, anthropologists recognize between six and nine different subspecies that have arisen as modern man's adaptations to different ecological situations [reference to Garn, 1961]. Of course, their isolation is incomplete, and gene flow continues between them; otherwise they would be defined as distinct species. Man is a single widely dispersed polytypic species, showing adaptation to a wide range of environments. (1974, pp. 394-95)[2]

The Creationists' View of the Origin of Races and Languages

The creationists' view of the origin of races is, of course, entirely different from what we have just described (Morris, 1974a, pp. 178-85). Humans are a single created kind; they arose from a single male individual of the kind and from him a single female individual was derived. The concept of race could have no meaning at that time. The creationist problem is to have the initial human pair multiply into a large population, and from that initial gene endowment differentiate into races bearing the observed differences. The creationists also accept the basic principles of genetics and heredity used in mainstream biology (see Moore and Slusher, 1970, Chapter 7). Morris has this to say:

> Many varieties of dogs have been developed from one ancestral dog "kind," yet they are still interfertile and capable of reverting back to the ancestral form. Similarly, all different tribal groups among men have developed from the originally created man and woman and are still basically one biological unit. (1974a, p. 180)

Morris then states that the modern evolutionists have essentially the same problem: If all humankind descended from the same ancestors, how did they become differentiated into races, each different in appearance from the other? He continues:

> Creationists also have the same problem of explaining the origin of the different tribal physical characteristics from one common ancestral population. Obviously, segregation into small groups is necessary in either model if distinctive characteristics are to emerge and become stabilized in each group. (P. 181)

Morris then draws upon anthropologist Ralph Linton's 1955 assertion that small, highly inbred groups are ideal sites for the fixation of mutations and consequent speeding up of the evolutionary process. Morris then objects on grounds that mutations are harmful, not helpful. Granted that small, inbreeding populations are favorable to rapid "physiological changes," Morris suggests that an acceptable hypothesis can be found if the term "mutations" in Linton's statement is dropped out and in its place we simply insert "recessive Mendelian characteristics." He suggests that the Creator provided the first humans with a tremendously diverse supply of

---

[2]From: Bernard Campbell, HUMAN EVOLUTION, Second Edition. Copyright © 1974 by Bernard Campbell. (Aldine Publishing Company, New York.) Adapted by permission.

genetic characters, of which the abundant recessive genes would lie dormant until they were activated in small, isolated populations. Morris continues:

> It would be difficult, if not impossible, to prove that a new characteristic in a population was a true mutation rather than a mere recessive characteristic. The difference is that a recessive characteristic is already implicitly present in the structured genetic program for the organism but hitherto hidden. (P. 182)

Morris goes on to say: "A designed genetic structure, however, though previously recessive, might very well have immediate benefits in a given environment" (p. 182). Altogether, Morris's model is quite similar to that popular today in mainstream evolution. (We presented this model in Chapter 35.) The major difference is that the mainstream model requires the continual but slow occurrence of mutations in a large population during periods when it is undergoing little change because the environment is holding constant. The point of difference seems to resolve itself into the source of genetic diversity. Initial creation of that diversity by God is a religious concept that cannot be examined by science. Morris goes on to generate a problem based on his own version of the origin of races:

> Now, the question is, how would it be possible to force the ancestral human population to split up into small, inbreeding groups in order to permit the process of change, whatever it was, to take effect? Since they originated together, it would seem advantageous to the group as a whole to have remained together, or at least in communication and commerce with each other, as such would have discouraged and minimized inbreeding. (1974a, p. 182)

Morris has the perfect solution to his self-made problem: give a different language to each of several subgroups. That would make communication between groups impossible. The cultural barriers thus created would prevent intermarriage. Is Morris so naive as to think difference of language prevents sexual intercourse? Morris is, of course, aware that there is a chicken-and-egg question here. Did reproductive isolation foster the rise of different language, or did the sudden onset of language difference cause the isolation?

Morris does not mention the subject of origin of multiple languages in this part of his book, but we all know what he really has in mind. The story is told in Chapter 11 of Genesis, but you may need to be refreshed as to what happened. After the Flood of Noah, the tribes of Noah's sons populated the earth. In the words of Genesis 11, as translated in The Jerusalem Bible, "Throughout the earth men spoke the same language, with the same vocabulary." These people moved eastward until they found a plain in the land of Shinar (Babylonia). There they decided to build a city, using brick instead of stone, and erecting there a tower with its top reaching heaven. Yahweh (God) came down to see what was going on and didn't like what he saw. Said Yahweh: "Come, let us go down and confuse their language on the spot so that they can no longer understand one another. Yahweh scattered them thence over the whole face of the earth." The tower was, of course, the Tower of Babel.

In the general edition of his textbook, Scientific Creationism, Morris refers to the confusion of tongues at Babel, noting that the Biblical model of earth history does not preclude such events (1974a, p. 215). In his ICR Impact article No. 28, "Language, Creation, and the Inner Man," Morris writes:

> To the creationist, of course, Babel is not only

symbolic but actual. This supernatural confusion of phonologies, with its resultant tribal dispersions throughout the world and its logical genetic consequences in the rapid emergence of distinctive tribal (not "racial"--the Bible knows nothing of the racial categories of evolutionary biology) characteristics, fits all the known facts of philology, ethnology and archaeology beautifully. (1975, p. iv)

Gish seems to recognize the confusion of tongues at Babel, although implicitly, in his creationist version of the cause of skin color differences between races:

> Creationists believe that skin color variations developed as a natural sorting out of preexisting traits during the formation of races as described in the section above. According to this view, blacks tended to migrate into those areas where their black skin offered protection from intense sunlight, while the fair-skinned, blue-eyed Scandinavian race naturally migrated to the far north to escape the more intense ultraviolet light encountered near the equator. (1978a, p. 146; 1985, p. 215)

Presumably, the migrations referred to above followed immediately upon the imposition of different languages at Babel. But how did the segregation into black and white races occur in the first instance? Obviously not by adaptation to the solar radation environment. The black group was already black before it decided to migrate to a high-intensity UV environment; the white group, too, was already white before it trekked toward the polar regions. Do you suppose that Yahweh imposed different skin colors on the various language groups He created at Babel? That seems like a reasonable solution, keeping the creationist explanation of races and languages internally consistent and reducing the problem of causes to its simplest statement.

The creationist version of the origin of races has been reviewed recently by Australian creation scientist John Mackay in the "Science for the Layman" section of Ex Nihilo (Mackay, 1984, pp. 6-12). The discussion follows closely the party line we have seen in the ICR writings of Gish and Morris. There is, however, a somewhat different version of what happened at Babel. Each of the small groups into which the single large breeding group had been split (according to language) contained a wide range in racial traits, including skin color. When a given group moved out into a new environment, selection tended to favor those individuals endowed with traits well adapted to the new environment, while at the same time tending to eliminate those with traits poorly adapted to the environment. Thus by natural selection groups that migrated to a cold, high-latitude climate became entirely light-skinned, as the dark-skinned individuals declined in numbers. In groups that migrated to sunny, hot regions the selection process favored the dark-skinned individuals at the expense of the light-skinned individuals, ultimately yielding an all-black population.

In Chapters 51 and 52 we will speculate about cultural evolution and especially about the origins of morality, ethics, and religion. Do today's humans inherit genes that code for certain traits of social behavior?

Credits

1. From C. Loring Brace, Humans in Time and Space, pp. 245-82 in Scientists Confront Creationism, Laurie R. Godfrey, ed., W. W. Norton & Company, New York. Copyright © 1983 by Laurie R. Godfrey. Used by permission.

2. From: Bernard Campbell, HUMAN EVOLUTION, Second Edition. Copyright © 1974 by Bernard Campbell. (Aldine Publishing Company, New York.) Adapted by permission.

Chapter 51

Evolution of the Human Mind and Culture

To find out what biologists and anthropologists of today can tell us about the evolution of the human mind and culture, including the early origins of ethics and religion, I turned to the writings of Theodosius Dobzhansky, George Gaylord Simpson, and Bernard G. Campbell. In the period 1940-1962, Dobzhansky served as Professor of Biology at Columbia University. His office was only a few doors away from mine, and I saw him occasionally. He was then at the height of his achievements as a geneticist and was particularly known for his experimental studies with fruit flies (Drosophila). In later years, his interests broadened to deal with the evolution of humans and he became involved with theories of human cultural evolution and the future of mankind. He dealt with these subjects in his 1962 work, Mankind Evolving (Dobzhansky, 1962), and later in The Biology of Ultimate Concern (Dobzhansky, 1967). George G. Simpson, paleontologist and zoologist, was long a curator of the American Museum of Natural History and also a professor at Columbia University (1945-1959) and at Harvard University (1959-1970). It was at Columbia, where we were faculty colleagues, that I had an opportunity to meet him. He too, in later years, turned to problems of human evolution, discussed in This View of Life (Simpson, 1964) and Biology and Man (Simpson, 1969). Bernard G. Campbell, formerly Professor of Anthropology at the University of California, Los Angeles, has been introduced in our chapter on hominid evolution; he received his higher education at Cambridge University. I found most informative his major work, Human Evolution: An Introduction to Man's Adaptations (Campbell, 1974; 1985); it has a particularly valuable chapter on culture and society, with a section on the evolution of ethics.

The Human Mental Faculty

Ernst Haeckel, a German biologist and philosopher who earned the reputation of being "Darwin's German bulldog," wrote in his 1874 work, The Evolution of Man, to the effect that there is no fundamental difference between man and the higher animals in their mental faculties. Can such a statement be sustained today?

In chapters 49 and 50 we found that the genus Homo may have entered the scene as far back as 2.2 million years ago, for a short time contemporaneous with Australopithecus africanus, from which Homo habilis is thought to have evolved. From H. habilis, the genus may have evolved into H. erectus, then, about a half-million years ago, into an archaic form of H. sapiens. There followed Neanderthal Man, now considered a subspecies of H. sapiens. Modern H. sapiens, as Cro-Magnon Man, appeared in Europe some 35,000 to 40,000 years ago, rather abruptly replacing Neanderthal Man.

From our standpoint, the most interesting evolutionary development in this sequence is the increase in the size of the brain, in the proportion of the brain that is cerebral cortex, and in the surface area of the brain. From a brain volume of about 600 cm^3 (average) for A. africanus, there was a rapid increase to 950 cm^3

(average) in H. erectus, and brain volume continued increase to an average value of 1330 cm^3 in H. sapiens (range 1000-2000 cm^3) (Campbell, 1974, pp. 272-73). Important in this great increase in brain volume was the disproportionately great increase in volume of the cerebral cortex, and in increased surface area accommodated by infolding of the surface layers of the cerebral hemispheres. Thus, "the surface area of the present human brain is four times as great as that of the gorillas' although the volume has increased by a factor of only 2.5" (Campbell, 1974, p. 273).[1] Particularly meaningful in comparing the human brain with that of other primates and mammals in general is use of the ratio of brain weight to body weight of the individual. One index used to measure this ratio (index of cephalization) gives modern man a three-times higher ratio than that of the three great apes, more than twice that of the monkeys, and fifteen times greater than that of mammals exclusive of the primates.

The above statistics relate to brain size. They might be used to support Haeckel's contention that human mental makeup differs from that of the nearest related mammals only in degree, the difference being quantitative but not qualitative. What, if anything, does increase in brain size have to do with possible qualitative differences? Is there associated with increase in brain size a fundamental new dimension in the mental faculties of H. sapiens, as compared with hominid predecessors and other higher mammals?

To find an answer to such questions, we need to review the consensus held by modern anthropologists about important changes in the mental activities and capabilities of the brain as H. habilis evolved from A. africanus, and subsequently into H. erectus and H. sapiens.

Campbell writes: "In our consideration of the evolution of man and his culture, it is necessary to refer to what has been described as man's unique mental characteristic: conceptual thought" (1975, p. 332).[1] (Footnote on his page reads: "Mental is the adjective of mind, the functional and subjective aspect of the living brain.")

Just what is conceptual thought? To get an answer we need to analyze the path of information entering the mind, and the storage of that information. This is a topic discussed in our investigation of science and the scientific method. Recall from Chapter 1 that a percept, the mental image of the external environment, consists of two types of information (1) input through the senses, and (2) memory of previous experience (Campbell, 1974, p. 332). Primate perception is very special, as compared with that of other organisms, and particularly with that of other higher mammals. Campbell states:

Primates, more than all other mammals (except bats), live in a three-dimensional world; their eyes

---

[1]From: Bernard Campbell, HUMAN EVOLUTION, Second Edition. Copyright © 1974 by Bernard Campbell. (Aldine Publishing Company, New York.) Adapted by permission.

are stereoscopic, and their movements are in all three planes of space. They must have very precise ideas of spatial relationships, for arboreal locomotion involves a knowlege of space far greater than that which may be necessary to ground-living forms.... The integration of spatial data from the senses to form a composite perception of the environment has clearly gone further among primates than among other groups of animals. (P. 333)[1]

Campbell also concludes that the memory component of perception in primates inevitably comes to include, in addition to experiential record of events, some generalizations about spatial relationships. He states: "By manipulation, the higher primate can extract an object from the environment, free it, as it were, from spatial implication, and build up a perception of it as a discrete object, not merely as part of a pattern. In time, the primate will come to perceive the environment not only as a three-dimensional pattern but also as an assemblage of objects" (p. 333).[1]

Campbell speculates on the effects of these special perceptive faculties:

When man began to hunt, his perception evolved accordingly. Using his prime sense, vision, man evolved the ability to identify objects on the move without reference to their relationship to the fixed part of the environment; he saw them as totally separate from their environment. Here was a fundamental improvement in perception and something novel among land animals: a carnivore that hunted by sight.

It is clear that, first, manipulation and, later, hunting came to make man's perceptual world different from that of other primates; indeed, different from that of all other animals. Man's analytic perception, more than any other factor, opened the door to the development of conceptual thought and eventually to his remarkable culture. (P. 334)[1]

Although what Campbell is describing is hypothesis, it shows a rational approach to the linkage between evolution of the unique human perceptive faculty and human culture.

Campbell next turns to concepts. In discussing science and the scientific method, we defined a concept as "an abstraction from the particular to the class" (Campbell, 1975, p. 334).[1] Campbell finds in the evolution of perception the need to imply "some degree of abstraction from experience, from the particular to the class" (p. 334).[1] Although, like perception, such abstraction is not a conscious activity, conceptual thought (which is a conscious activity) may have had its origin in the advanced perceptive process described above. Campbell considers that conceptual thought is an activity limited to man alone. Man alone can have conceptual thoughts about objects that are not concurrently visible to the thinker. In other words, man is uniquely capable of imagination.

Referring to the work of Freedman and Roe (1958), Campbell elucidates further on imagination: "imagination is the consciousness of sets of concepts, which are the classification of experience" (p. 335).[1] Imagination permits humans to escape mentally from the present into the future, in order to plan or devise some act or artifact for future use. And now we come to what Campbell considers another unique mental faculty of humans:

It has been said that what distinguishes man from animals is the length of time through which his consciousness extends. In animals, this dimension is small, stretching a little way into past and future; in man, it grows both qualitatively and quanti-

tatively. The evolution of conceptual thought gives man greater power to live in the past and in the future by abstraction from the past. (P. 335)[1]

Campbell finds in mammals evidence for a classification of experience that can be called "unconscious conceptualization," but in contrast, the conceptual thought engaged in by humans is conscious. He goes on:

Human conceptual thought appears to be characterized particularly by its conscious nature, but it is no doubt the result of a steady process of evolution from less conscious and indeed unconscious concepts in primates. Man's achievement was the fully conscious concept of things he does not possess but needs; the recognition of game, weapons, women, or children as classes brought with it the classification of more and more of man's environment and the possibility of foresight of future needs. Man, leaving behind the narrow limits of present time experienced, entered the broad expanse of past memory and future concepts. (P. 336)[1]

Dobzhansky recognizes the conscious nature described above as self-awareness. For humans, he states:

In point of fact, self-awareness is the most immediate and incontrovertible of all realities. . . . Only by analogy can I infer that other humans have self-awareness--they usually act as I do in situations in which I know that my self-awareness is involved. . . . Human self-awareness obviously differs greatly from any rudiments of mind that may be present in nonhuman animals. The magnitude of the difference makes it a difference in kind, and not one of degree. Owing primarily to this difference, mankind became an extraordinary and unique product of biological evolution. (1977, p. 453)[2]

Dobzhansky adds another special form of human awareness that is unique: death awareness. "There is no indication that individuals of any species other than man know that they will inevitably die."[2] He goes on:

Foreseeing the remote future and planning for future contingencies require capabilities that we know exist only in the human mind. Self-awareness and death-awareness are probably causally related and appeared together in evolution. They appeared because they enhanced the adaptedness of their possessors. The adaptive role of self-awareness is sufficiently obvious, no matter how elusive self-awareness may be. It is an integral part of the complex of adaptations that include the use of symbols, language, and hence acquisition and transmission of culture. (P. 454)[2]

Language, Memory, and Intelligence

Campbell continues with his exposition of the unique qualities of the human mind (1974). In the field of language, he points out that the language of humans is based on the use of symbols. While the symbols must be learned (a cultural phenomenon), the conceptual thought that is required for the formulation of language is a product of evolution. Thus, animals do not communicate with language that uses word symbols; their communication is based on signs (or signals). In animals, "communication is a means by which one organism can bring about change in one or more others; . . . it is the means by which one animal can trigger a response in another. . . . In all animals this means is non-verbal;

only in man is language evolved to supplement non-verbal signals" (p. 344).[1]

Campbell states that all evidence points to the conclusion that the ability to use language is a genetically determined character of modern man. The language itself is part of culture--an artifact--"but this remarkable and uniquely human kind of behavior rests on man's peculiar endowment, which gives him his large, slowly-maturing brain and his ability to verbalize" (p. 357).[1] Summing up the importance and uniqueness of language, Campbell states: "Language has made possible the vast development of man's culture, and brought us the unique human consciousness of ourselves and others" (p. 356).[1]

On the subject of memory, Campbell refers to the work of Penfield and Roberts (1959). They distinguish three levels of memory. Experiential memory is a continuous record of the stream of experience. This form of memory operates at the subconscious level and is found in animals as well as humans. Conceptual memory is "the accumulation of abstract concepts about experience" (Campbell, p. 338).[1] While conceptual memory is found in a simple form in animals, it is highly complex in humans. Humans have the unique ability to experience emotions without reference to directly attached objects. Humans have the ability to store in memory highly abstract concepts based on symbols, as for example in mathematics. A third class is word memory, closely linked with but distinct from conceptual memory. Only humans can possess word memory.

On the subject of intelligence, Campbell states:

> Intelligence has been defined as the relating activity of mind--the ability to realize the connection between discrete objects and events. It clearly involves imagination, which may be considered to be the presentation from the experiential memory of memory traces--knowledge--not obviously directly connected with the immediate experience. It may involve conscious recall of memory, but the significant feature of imaginative thought is that a much broader range of memory traces is brought to the interpretation of day-to-day experience than in the process of simple unconscious learning, for only thus could the imaginative connection between events be realized. (P. 341)[1]

Intelligence cannot exist alone, Campbell states, "it is a capacity that interacts with knowledge (which in man may be considered to be the accumulation of experiential and conceptual memory)" (p. 341).[1] He quotes C. J. Herrick:

> Intelligence integrates knowledge and gives it direction; it is a conscious and "purposively directed mental process with awareness of means and ends." (Herrick, 1956, p. 367)

Campbell's complete picture of the evolution of human mental faculties is carefully reasoned and is based on generally acceptable psychological theory of mental activity. It makes the point that, by the process of biological evolution, the human mind became unique in quality from that of all other primates and all other animals. This hypothesis of mental evolution requires us to refute Haeckel's statement to the effect that human mental activity does not differ qualitatively from that of other animals. The rise of human culture accompanied the evolution of the human brain functions and indeed would not have been possible without the emergence of the unique human ability for conscious conceptualization and the specific phenomenon of imagination extending far into the future, but making use of experiential memory extending far into the past. Ethics as we know it among humans could not have developed prior to the development

of those unique human mental faculties and the ability to communicate by use of language.

Although I felt satisfied with Campbell's and Dobzhansky's analyses of the evolution of the human brain and its unique mental capacities, I decided to seek a third interpretation, which I found in the writings of G. Ledyard Stebbins, an eminent authority on genetics and evolution. He received the Ph.D. degree from Harvard University in 1931 and has fulfilled a long career of teaching and research in the University of California, first at Berkeley and later at Davis. His book, titled Darwin to DNA, Molecules to Humanity, appeared in 1982 and is intended for the general reader (Stebbins, 1982). I found his style extremely lucid, and I strongly recommend the volume to those wishing to fill out a background on modern views of the evolutionary process. For the most part, his views on evolution of human mental functioning parallel those I have just reviewed, but with some variations in terms and some new insights.

How Distinctive Is Humanity?

In answer to the question "How distinctive is humanity?" Stebbins notes the many observed similarities in behavior between apes and humans. Despite seeming likenesses in many respects, Stebbins finds fundamental differences that far overshadow the apparent similarities:

> Given this great degree of similarity, must we conclude that humans are only sophisticated apes endowed with enormous technical prowess but otherwise only quantitatively different fron animals? I think not.
>
> . . . human society must be regarded as containing many novel traits. The family life of humans, including our attitudes toward relatives and friends, could be reasonably well predicted on the basis of a thorough knowledge of chimpanzee and gorilla societies. Nevertheless, these apes have not developed any form of organization that resembles even remotely the human division of labor, disciplined armies, religions, nations, and international organizations. Modern human societies are qualitatively as well as quantitatively different from all existing animal societies. (P. 363)[3]

Stebbins names three distinctively novel human characteristics: artisanship, conscious time-binding, and imaginal thinking. Artisanship is a specific capability not mentioned by Campbell or Dobzhansky. Stebbins considers artisanship to be "the transcendent outcome of evolution in tool-making" (p. 363).[3] He notes that it is present in rudimentary form in chimpanzees, but he adds: "One could not predict the Parthenon, the pyramids, or the Empire State Building from watching pre-Olduvian hominids 'fishing' for termites with sticks" (p. 363).[3]

Conscious time-binding is essentially the capacity for conceptual memory we have already noted. The term was introduced by a Polish philosopher, Korzybski; it refers to the ability of humans to plan far into the future on the basis of experiential memory going far into the past. Stebbins suggests that this ability developed in connection with cooperative hunting and the taking of fire. He calls attention to carved animal bones, interpreted as lunar calendars, as evidence of conscious time-binding.

Imaginal thinking we have covered in some detail under the heading of human imagination and its uniqueness. Stebbins stresses the importance of symbolic language in promoting both conscious time-binding and imaginal thinking.

Stebbins attributes the unique combination of the three human characteristics just cited to biological evolution of the human brain: "From the biological viewpoint, these characteristics are based on nothing but a quantitative increase in the number of cells that form the surface layers of a single organ, the neocortex of the brain. This genetically determined change, interacting with the rudiments of culture that was already present when it began, triggered a direction of cultural evolution that brought novel and transcendent social behavior" (p. 365).[3]

I should also insert a note of clarification here as to Stebbins's use of the word "quantitative." The individual brain cells to which he refers are viewed as essentially similar (qualitatively the same) as those in the brains of the apes and the hominid ancestors of the present human species. This should not be interpreted to mean that the difference between the human brain and the ape brain is therefore merely a quantitative difference. The vast increase in the number of cells in the human neocortex as compared with that of the apes and early man gives to the whole brain as a functioning organ a qualitative uniqueness. Stebbins clarifies this concept in an earlier chapter of his book where he explains how quality differences arise as a result of complex quantitative differences. (See Stebbins, 1982, pp. 147-50.)

Stebbins summarizes the enormous consequences of the evolutionary development of the human brain to its extraordinary level of uniqueness:

> These three qualities are the foundations of the three principal realms of human knowledge. Natural science and engineering are the outgrowth of prescientific artisanship. History, political science, and other branches of the social sciences are fundamentally ways of directing social behavior to avoid disaster and to improve the material state of mankind. The humanities--literature, the arts, and philosophy--are extensions of imaginal thinking. The fact that artisanship, conscious time binding, and imaginal thinking evolved gradually by means of quantitative physical changes that affected a complex of behavioral characteristics does not detract in any way from the novel or transcendent quality of the human way of life. (Pp. 365-66)[3]

What kind of evolutionary pressure induced this remarkable development of the human brain? Science writer Roger Lewin, reporting in Science on a meeting of paleoanthropologists in Cambridge, England, in 1987, reviewed the "group-against-group" hypothesis (Lewin, 1987). It identifies an environmental force judged to have been more effective in human brain evolution than other forces traditionally listed under Darwin's "hostile forces of nature" (Darwin, 1859, Chapter 3).

The group-against-group hypothesis, attributed to Nicholas Humphrey of the University of Cambridge, was expanded upon by Richard Alexander of the University of Michigan. Alexander is quoted as saying: "The only plausible way to account for the striking departure of humans from their predecessors and all other species with respect to mental and socal attributes is to assume that humans uniquely became their own principal hostile forces of nature" (Lewin, 1987, p. 669). Lewin explains:

> The key engine in this evolutionary drive, suggests Alexander, is a positive feedback resulting from the close match between the competitors--human versus human--in the battle for survival. "In social-intellectual-physical competition [members of the same species] are likely to be--as no other competitors or hostile forces can--inevitably no more than a step behind or ahead in any evolving system of strategies and capabilities. Evolutionary races* are thus set in motion that have a severity and centrality as in no other circumstances." The result is a "runaway" evolutionary trajectory that is analogous to the mechanism favored by some biologists for the phenomenon of sexual selection

that, among other things, produced exaggerated features such as the peacock's tail.

(*Here, "race" means a contest, not an ethnic race.)

The central concept expressed by Alexander follows Charles Darwin, whose preview of Chapter 3 reads: "Struggle for life most severe between individuals and varieties of the same species; often severe between species of the same genus" (Darwin, 1859, p. 114). Alexander likes the group-against-group hypothesis because "it can explain any size or complexity of group; it accords with all of recorded human history; it is consistent with the fact that humans alone play competitively group-against-group on a large and complex scale; and it accords with the ecological dominance of the human species" (Lewin, 1987, p. 669). Lewin explains further:

> At the core of this evolutionary explanation is the idea that intense social interaction and manipulation demands unprecedented skills in dealing within one another. Human intelligence and human reflective consciousness are therefore seen as the product of natural selection for dealing with the most challenging things in the human environment: other humans, not, as usually has been assumed, technological exigencies. (P. 669)

Creationists, take note! Darwin's theory of organic evolution by means of natural selection continues to dominate biology and paleontology as the paradigm of choice.

Two Behavioral Controls: Biological and Cultural

Before embarking on an investigation of possible biological origins of ethics and religion, we need to examine the broader topic of the possible interplay of two forms of control over basic human behavior--the biological control through genetic change on the one hand, and the cultural control through information transmission and learning, on the other. Stebbins covers this subject in a remarkably clear fashion (1982). Referring to human evolution of mental faculties, Stebbins states that transcendence of the human animal "was achieved by the development of three distinctive qualities--artisanship, conscious time binding, and imaginal thinking" (p. 369).[3]

The phenomenon on which we are focusing attention is human behavior in its broadest sense--what humans do and how they do it--but with particular emphasis upon human functions in a society of humans. We are concerned with how people act in relation to one another and to the environment and the kinds of artifacts (using the term broadly) they produce. Most particularly, we are interested in how people act when they have available choices of two or more ways in which to act. We want to understand how the choices they make influence the likelihood that they will survive in groups and continue to reproduce themselves.

Changes in human behavioral patterns through time, starting from, say, the earliest species of genus Homo, have, in general, evolved from simpler to more complex and more sophisticated modes. The driving mechanisms of observed changes--biological and cultural--are continuously interacting with a possible built-in feedback mechanism that makes it exceedingly difficult for us to separate their effects. It is to this problem that Stebbins devotes a chapter titled "The genetic and cultural heritage of humanity" (pp. 369-82).

The totality of changes observed in human behavior as time passes can be described as a form of evolution, which simply means a process of change through time. The total picture of change is commonly thought of as consisting of two kinds of evolution--genetic evolution and cultural evolution. These terms can be confusing; they must be carefully defined and their meanings closely scrutinized.

Genetic evolution is responsible for physical changes in the human body and in the capability of the human brain to perform the higher functions we have already examined. Stebbins states that genetic evolution, "established in populations by the action of natural selection modified by chance events, has provided us with a wide range of physical and mental capacities" (pp. 369-70).[3] Cultural evolution makes use of unique mental capabilities in the making of conscious choices and decisions leading to improved ways to act in given situations and in developing new and improved artifacts. Once invented or discovered, these improved behavioral changes are dispersed through the population by imitation and, more effectively, by language. The knowledge itself is passed on to succeeding generations by those same methods of transmission. Nothing in the innovative or inventive process is directly related in any physical sense to human genes. There is, however, an indirect connection to genetic evolution through a feedback process that is acceptable in terms of modern genetics.

The feedback mechanism between culture change and the genetic endowment is thought of as a mechanism of selection--selecting for survival, that is, of those groups whose innovations and inventions improve their chances of survival and reproduction, as compared with groups with lesser inventive capabilities. The feedback circuit is described by Stebbins in these words:

> As humans relied more and more on artisanship for survival, genes that increased the capacity for making better tools, shelters, and clothing and for cooperation through sharing and division of labor spread more and more widely because of their increased adaptive value. At the same time, societies that used their capacities more efficiently could acquire more food and defend themselves better against predators, and thus improved their chances for survival and reproduction. (P. 370)[3]

Let me try to clarify the distinction between the cultural feedback mechanism of genetic change and the basic mechanism of genetic change implied in natural selection as it functions throughout the plant and animal kingdoms. As an example, take the evolutionary change of some part of the human body. At a time when (as the hypothesis goes) early hominids took leave of the dwindling tropical forests of Africa and occupied the expanding savanna grasslands, available foods in the new environment consisted largely of hard seeds or other tough plant tissues. Through mutations, certain individuals received a set of teeth better adapted to masticating these new food varieties, e.g., harder enamel and/or flatter molar crowns. Such individuals would be better adapted to the changed environment, and the probability for survival through adulthood and thus for propagating offspring would be increased. In due time, the genes coding for the improved dental equipment would spread through the population. The important point here, and in all similar cases, is that the individual human had no choice in selecting the improved teeth. No tray of tooth types was set before the individual with instructions to pick the one that seemed most likely to make chewing easier. Evolutionary changes in the human tooth came

about in the same nonconscious framework as the evolutionary changes in the tooth of a beaver or a gopher. Noninvolvement of the thought processes is the essential point. In contrast, cultural feedback involves conscious choice, or at least the possibility of choice. Which of two hunting strategies shall we adopt? Which of two kinds of storage sites for nuts is the safer? Shall we go over to the neighboring tribe and steal some babies for adoption, or is the risk of retaliation too great? Choice in itself could have been a basic factor in inducing genetic improvements in the human brain. The better equipped one individual's brain to conceptualize, to remember, to imagine, and to invent, the better would be the choice made by that individual in a given situation involving survival.

Philosophers point out that ethics always involves choices of action. It is easy to understand how the selection of morals at the family level of interaction could have adaptive significance through the feedback mechanism. The distinction between genetic change without choice and with choice based on the ability to conceptualize is encompassed in the term "coevolutionary circuit," i.e., from genes to culture and back, found in sociobiology (see Edward O. Wilson and Charles Lumsden, 1981).

Stebbins continues to develop the concept of two kinds of selection in evolving humans. He notes that in time they would come to have almost equal importance for human evolution (p. 370). But, in due course, cultural evolution would have become the dominant means of behavioral or social change. As the content of the culture itself became more and more complex, and the store of information required to sustain it became vast, genetic adaptation by increase in mental capacity would have become inadequate to keep pace. This is suggested by the observation that brain size has not increased in modern man as compared with Neanderthal Man. The most important factor, however, would be the increased rate of cultural change in contrast to the slowness of the process of genetic change. There would come a point at which genetic evolution (unguided by deliberate selective breeding) could play no significant role in social change. Cultural evolution would then be required to carry the whole burden of implementing and guiding change, using an effectively unchanging mental capacity. We must wrestle with the problem of extinction by nuclear weaponry using a genetic mental endowment that is no greater than it was in the time of Plato or Aristotle or even inhabitants of an earlier time, when civilizations first arose in Mesopotamia and the Nile Valley.

Credits

1. From: Bernard Campbell, HUMAN EVOLUTION, Second Edition. Copyright © 1974 by Bernard Campbell. (Aldine Publishing Company, New York.) Adapted by permission.

2. From Theodosius Dobzhansky, Evolution of Mankind, Chapter 14, pp. 438-63 in T. Dobzhansky, F. J. Ayala, G. L. Stebbins, and J. W. Valentine, Evolution, W. H. Freeman and Company, San Francisco. Copyright © 1977 by W. H. Freeman and Company. Used by permission.

3. From Darwin to DNA, Molecules to Humanity, by G. Ledyard Stebbins, W. H. Freeman and Company, New York. Copyright © 1982 by W. H. Freeman and Company. Used by permission.

Chapter 52

Biological Origins of Ethics and Religion

As we turn next to consider speculations on the biological origins of ethics and religion, we must keep in focus the uniqueness of the human mind/brain--uniqueness in many functions and capabilities. That uniqueness is a product of evolution within the human genus.

Speculations on the Evolution of Ethics

Writing on the evolution of ethics, Professor Bernard Campbell considers it "one of the most important events in the evolution of man. Ethics arose as a direct result of the appearance of self-awareness in the growing human consciousness" (1975, p. 362).[1] Campbell explains: "Man could see, as a result of his self-consciousness, how many of his activities were directed to satisfy his needs, his basic requirements for life; but he could also see that certain of his actions satisfied only social needs, that they led not to personal satisfaction but to frustration" (p. 362).[1] "Ethics therefore arise when man finds that he has to make conscious choices in a social context" (p. 363).[1] Campbell refers to the speculations of Freedman and Roe that the need for values arose from anxiety, from the conscious frustration of human needs (1958). Campbell states: "Internal conflict and frustration may, however, be among the most important stimuli of cultural progress, of the development of adaptive behavior and technology" (p. 364).[1]

Dobzhansky also fits the rise of ethics into human evolution: "There are two interesting sources of ethics and values--cultural and biological. The ethical standards of every individual are imparted, mainly in childhood and youth, by other members of the society for which the individual is being prepared. Ethics are acquired, not biologically inherited" (1977, p. 455).[2] Yet, although the particular moral values--the "oughts"--are cultural (artifacts), the ability to see ethical issues is a genetically controlled capability. Dobzhansky quotes from C. H. Waddington to the effect that humans are genetically determined "ethicizing beings" and, especially in childhood and youth, "authority acceptors" (1960). Dobzhansky continues: "Every member of a human society must become familiar with the ethical and value systems of his society. Failure to do so makes him a misfit or an outcast; it jeopardizes his and his progeny's success and survival. Therefore natural selection has exerted pressure to insure that every member of the human species comes into the world with a genetic endowment making him an 'ethicizing being'" (p. 455).[2] This, of course, is a purely naturalistic hypothesis--part of the total theory of biologic evolution--and you might be prompted to conclude that ethics does, indeed, have a mechanistic base that can be explained by the scientific method.

That conclusion might, in turn, seem to support the hypothesis that an ethical system is achieved wholly

through natural processes. Dobzhansky addresses himself to that conclusion, for he points out that the ethics of a particular society are not genetically derived: "Anthropologists have ample evidence that different cultures demand different modes of behavior, and that these demands are usually complied with" (p. 456).[2] While some human attitudes and evaluations that seem quite instinctive (inherited) may seem to be based on value judgments, it is more reasonable to suppose that they were developed purely by survival pressures. For example, in the human family, high value is placed on motherhood, while children are cherished and loved. These values directly promote the survival of the family, and they are also displayed to some degree in animals. But, says Dobzhansky:

> This can hardly be said of many other ethics and values that are recognized in most, if not all, human societies as valid. For example, it is wrong to steal, swindle, rob, waylay, or murder other people, especially members of one's own group or society and, by extension, any human being. This is wrong even if so doing is profitable, the misdeed is undetected, and no vengeance or retribution is to be feared. On the contrary, honesty, generosity, and veracity are praiseworthy, especially if they bring hardships to persons who practice them. Human life, that of a stranger no less than that of a relative, is sacred, with the significant exception of war. Life is to be preserved at all costs (including that of incurably ill persons whose existence may be sheer misery). At the summit of ethics, we have the commandments of universal love (including one's enemies), service to others, and resistance to evil. (Pp. 456-57)[2]

It is evident in the above statement Dobzhansky has taken the quantum jump from that which is an empirical concept (survival of the fittest) to a different realm couched in language not amenable to evaluation by empirical science--notice the burst of emotive words, such as "wrong," "misdeed," and "praiseworthy," and even such excessively emotive phrases as "life is sacred" and "universal love."

At this point a digression seems in order to develop a concept of a category of human knowledge that lies outside of those research fields that are areas in which knowledge rests solely on observation of the real world. Recall from Chapter 1 that Mario Bunge's epistemology distinguishes the knowledge of research fields from that of belief fields in which knowledge rests on belief--belief in something that cannot be observed to exist. As we noted in Chapter 6, the dualistic ontological models include two realms: a natural realm and a supernatural realm. The natural realm is the domain of the knowledge of research fields, including empirical science. The supernatural realm falls in the domain of the belief fields of knowledge and, specifically, of the field of religion. We are now in search of an ontological realm that can

---

[1]From: Bernard Campbell, HUMAN EVOLUTION, Second Edition. Copyright © 1975 by Bernard Campbell. (Aldine Publishing Company, New York.) Adapted by permission.

be paired off with the natural realm in a somewhat different dualism from that seen in the science/religion dualism. We need here a realm that includes value assessments made in the field of ethics and morality. We need a place to house the products of human quality judgments, which we often refer to as "values." Examples abound in everyday life, as we repeatedly assign to human acts such adjectives as "good" or "bad" and "right" or "wrong." We are also prone to make quality judgments about products of human imagination in the field of the arts. These are aesthetic values, expressed through such adjectives as "beautiful," "ugly," or "sublime." I propose that all such forms of human value judgments be assigned to the category of transempirical concepts.

I found the term "transempirical" in an essay by Herbert Feigl, Professor of Philosophy at the University of Minnesota. His essay is in a volume titled Moral Problems in Contemporary Society, edited by Paul Kurtz (Kurtz, 1969, pp. 48-64). In discussing his ethical outlook, which he describes as scientific humanism, Feigl introduces the word "transempirical" as a substitute for "supernatural," to be used in describing beliefs that are beyond the scope of scientific analysis. In adopting his term, however, I restrict its usage in the manner indicated earlier in this paragraph, that is, to descriptive value judgments and to aesthetic qualities.

My suggestion is that a transempirical concept exists in only one form: as a product of the human brain. Nevertheless, its naturalistic origin notwithstanding, the basic nature of the transempirical concept is such that its truth-status as a human judgment cannot be confirmed or denied by empirical (scientific) methods. By "cannot be confirmed" or "denied" I mean that if an alternative judgment were to be assigned to the same object of valuation, the alternative would enjoy equal status. Thus, if one individual says "action A is good," equal status applies to another individual's statement that "action A is evil." Another characteristic of transempirical concepts, and perhaps a descriptive criterion as well, is that they arise from emotive mental processes and attitudes, i.e., they are always associated with an individual's feelings about what is desirable to possess or to have accomplished. Recognition of "that which is transempirical" as distinct from "that which is empirical" sets up a dualistic ontology (a dualistic metaphysics) that can operate in total freedom from the supernatural realm.

Returning now to Dobzhansky on ethics, he is well aware of the two realms I have identified. He makes a clear distinction between two kinds of ethics:

> Family ethics are shared by man with the "quasi-ethics" of at least some animals; in animals as well as in men, many family ethics are genetically conditioned dispositions (although in man they may be overcome by an exercise of will). Family ethics can be envisioned as products of natural selection, which established the genetic bases of these ethics in our ancestors as well as in other animal species. Group ethics are products not of biological but of cultural evolution. They confer no advantage and may be disadvantageous to individuals who practice them, although they are indispensable to the maintenance of human societies. Natural selection has not made man inherently evil (as is so readily assumed by believers in original sin or proponents of territorial and other "imperatives"). Whatever proclivities to selfishness and hedonism man may have, he also has a genetically established educability that permits him to counteract these proclivities by means of culturally derived group ethics. Natural selection for educability and plasticity of behavior, rather than for genetically fixed egoism or altruism, has been the dominant directive factor in human evolution. (P. 457)[2]

It seems reasonable to me that group ethics, heavily involved with transempirical concepts, would have developed concurrently with religion in Homo sapiens, and that the moral precepts would have been enforced by religious sanctions. This linkage could have developed as soon as language was capable of communicating abstract ideas, but before the invention of writing. Abstract transempirical value concepts are much easier to hand over to supernatural authority than to justify rationally. Perhaps it was simply a case of following the path of least resistance. Family ethics, being much simpler in its concepts, would have long preceded group ethics, since the former would have been much simpler to comprehend and to communicate.

The third evolutionary biologist whom I named earlier, George Gaylord Simpson, has a lot to say on the subject of "biology and ethics" in a chapter of that title in his volume Biology and Man (Simpson, 1969, pp. 130-48). He describes among biologists "the search for naturalistic ethics, that is, an ethical or moral system rationally related to the nature of things or to the material universe" (p. 130).[3] To my surprise, I found that Simpson feels quite good about the prospects for achieving a naturalistic ethics.

First, however, Simpson refers to what philosophers of ethics call the naturalistic fallacy, an error in logic that takes place when one passes from pure description of nature ("it is" or "it is not") to the conclusion prescribing what should be the case ("it ought to be" or "it ought not to be"). Simpson cites an example of the naturalistic fallacy from biology. The late Alfred Kinsey, a biologist whose later researches turned to human sexual behavior (well known through the "Kinsey Report"), claimed to have discovered that homosexualism was rather more prevalent than previously thought. On the grounds that the phenomenon exists in nature, he recommended that laws prohibiting the practice be modified to reflect the conclusion that it cannot be morally wrong. But Simpson has a point to make:

> It is undoubtedly illogical to conclude that what is therefore ought to be. It is, however, equally illogical to make that the basis for a further conclusion that decision as to what ought to be cannot be based on consideration of what is--in other words, that naturalistic ethics are impossible. (P. 132)[3]

> Let us not be too dismayed if some attempts to set up naturalistic ethics turn out themselves to be fallacious, or if we cannot all agree as to either origin or criteria for naturalistic ethics. It would be enough to go on with if we could conclude that naturalistic ethics are eventually possible. In fact, I believe we can do considerably better than that. (P. 133)[3]

Naturalistic ethics, Simpson notes, is usually considered in the context of biology and evolution. Perhaps he is referring to the writings of Konrad Lorenz of the mid-1960s. As did Dobzhansky, Simpson refers to Waddington's assertion that humans have the genetic predisposition to ethicize and to accept authority. Simpson writes:

> In the evolutionary context, the problem really becomes one of why and how organic evolution produced an animal capable of cultural evolution and of the ethicizing that helps to mediate cultural progress. There is no real doubt, and neither Julian Huxley nor Waddington has doubted, that the capacity or, one can say, the necessity for ethicizing is in fact a biological characteristic of the human species developed by natural selection because it is adaptive for the species. In generally

biological but nontechnical terms, the direction of human adaptation early became one depending on individual flexibility with mainly learned abilities, with alternatives of action, and with consequent responsibility for those actions. (P. 134)[3]

After discussing and rejecting the Freudian theory that involved inheritance of acquired characteristics, Simpson returns to reinforce his view of naturalistic ethics:

It is thus plausible and indeed, I think, practically certain that ethicizing, the capacity and the necessity for some system of ethics, arose in the course of human evolution in a completely natural way. That does not mean that men early and generally recognized the natural origin of the moral sense or that they adopted specific ethical systems that were naturalistic. We all know that they did not. (P. 135)[3]

At this point, Simpson turns to the introduction of systems of revealed or inspired religion that would have taken advantage of the ethicizing and authority-accepting capability in humans. He is, as one might expect, skeptical of the value of religious ethics in terms of promoting survival. He gives two reasons. "In the first place, no given one of the systems of revealed or inspired religion and ethics is necessarily adaptive biologically or specially" (p. 135).[3] Those who promoted and manipulated those ethics may have had ends other than group survival in view. There were "bitter conflicts between rival systems of supposedly revealed ethics," and these "are obviously inadaptive even when either system alone might be sufficiently adaptive for survival" (pp. 135-36).[3] Simpson then points out that ethical systems that might have been adaptive for survival under primitive tribal or pastoral conditions would almost certainly be nonadaptive under today's social and environmental conditions. As a second reason for not placing reliance in traditional religion-based ethical systems, Simpson argues that large segments of society today (presumably atheists and secular humanists) "will not accept fiat as a substitute for reason." Unless based on reason, those systems will not be functionally adaptive today. Simpson concludes: "Therefore, continued human welfare requires ethical systems that are not supernaturalistic but are naturalistic" (p. 137).[3] I presume that what he means is that ethics derived by reason in consideration of present environmental conditions is constructed in such a manner as to promote human survival. As I see it, this does not mean that such ethics will actually pass from the category of acquired culture to that of the genetic endowment by natural selection. Shortness of time and the rapid flux of social change could not permit natural selection to do its work in the Darwinian sense.

Simpson continues with an excellent review of the rise of Social Darwinism and its fallacious assumptions about the meaning of "survival of the fittest." But he soon leaves the biological area of survival ethics and enters the area of value judgments. This is the quantum jump to the transempirical realm I have identified in earlier arguments. He reviews the theory of society as the supraorganism (Gerard; Comte), in which the "good of the state" requires sacrifice of the good of individuals. Nothing in evolutionary biology would, as I see it, point to recognition of social organization on the level of nations. The ethical concept of a nation having intrinsic value is better viewed historically as an artifact of religious origin.

Simpson refers next to "numerous biological phenomena that have been considered general tendencies of evolution and have been involved in proposals of evolutionary ethics" (p. 139).[3] These are technical points and include some of the unique features of living organisms that I

discussed in Chapter 3, for example, the ability of organisms to decrease system entropy, the ability of organic systems to achieve steady states, and the general increase of complexity of organisms through time. Simpson does not think these trends or tendencies are relevant to the question of ethics: "I do not believe there is an 'ought' inherent in any of these processes" (p. 140).[3] In other words, what have often been referred to as biological laws are not in themselves ethical in content.

Taking up next the subject of trend ethics, Simpson makes his final encounter with the quantum step I have referred to earlier--the shift from empirical to transempirical ideas and constructs. Referring to Julian Huxley as an exponent of trend ethics, Simpson states: "An overall evolutionary trend is inferred or postulated. It is then concluded that the continuance of that trend is desirable and that whatever promotes it is ethical" (p. 140).[3] In a quotation from Huxley (not referenced) the word "right" is used repeatedly to describe the supposed goal of further social evolution, which is both to respect human individuality and to encourage its fullest development. Huxley defends his stated evolutionary trend on the grounds that its organic phase would constitute an "improvement" in society. Simpson does not let this assertion get by him, for he states:

"Improvement" is an evaluating word. It means "change for the better," so that when unlimited improvement is taken as the criterion for the ethically good, what is being said is simply that it is good to become better. The argument becomes circular to the point of meaninglessness unless some external, independent criterion of "better" is found and applied. It is surprising and, at first sight, dismaying to find Huxley maintaining that the concepts of progress, of higher and lower, of better and worse, "spring automatically to the mind" and are known "as an immediate and obvious fact." If, after all, we are to take intuitive concepts as ethical principles, then purported derivation of naturalistic ethics from the facts of evolution is either irrelevant or spurious. (P. 140)[3]

Simpson continues with a review of C. H. Waddington's views on evolutionary ethics, which closely resemble those of Julian Huxley and are subject to similar criticisms. Skipping over this rather involved analysis, I turn to Simpson's conclusions. One is "that neither strictly organic nor social evolution necessarily leads to improvement in any sense of the word acceptable for the human situation. In fact either, in their ways that in themselves are blindly amoral, may have results extremely undesirable for mankind" (pp. 145-46). Simpson sees a saving grace in the knowledge that both organic and social evolution are now to some degree, though limited, under human control, and that we can work deliberately for the desirable. Of course, in using the words "desirable" and "undesirable," Simpson stumbles into the morass of the transempirical. Whether "desirable" means "to promote survival of the human race" or anything else, it is a value judgment that is religious in nature, for it refers us to the questions: "Why are we here? What value have we? To whom or what are we valuable other than to ourselves?"

Although Simpson does not come right out and say what I have covered in the foregoing sentence, his concluding statement shows that he is well aware of what is derivable by reason and what is beyond either formal logic or empirical analysis:

Man has risen, not fallen. He can choose to develop his capacities as the highest animal and to try to rise still farther, or he can choose otherwise. The choice is his responsibility, and his alone. . . .

Evolution has no purpose; man must supply this for himself. . . . It is futile to search for an absolute ethical criterion retroactively in what occurred before ethics themselves evolved. (P. 148)[3]

Sociobiology

We made brief mention of sociobiology in Chapter 2 in the context of examples of the ruling hypothesis. The going code word for the controversy we looked at is "nature/nurture," signifying the conflict between the strongly entrenched cultural school of anthropology that claims human behavior is almost entirely acquired, and the sociobiologists, reviving the thesis that behavior is strongly controlled by genes and thus largely a product of biological evolution through natural selection. The writings of George G. Simpson on the origin of ethics, reviewed in this chapter, touch on the message of sociobiology, but his 1969 book, Biology and Man, actually predates the enormous splash that came in the early 1970s with the arrival of a full-blown theory of genetically controlled human behavior that includes moral propensities.

Sociobiology was preceded in the 1960s by the extension of modern ethology (the science of animal behavior) into the area of human behavior and human traits, especially in reference to human aggression. Konrad Lorenz was a leader in modern ethology, and he ventured into what are now called sociobiological areas in his 1966 work, On Aggression. In rereading his Chapter 13 titled "Ecce Homo," I find him saying that our understanding of the bad track record of humanity in managing its affairs can be explained "if one assumes that human behavior, and particularly human social behavior, far from being determined by reason and cultural tradition alone, is still subject to all the laws prevailing in all phylogenetically adapted instinctive behavior" (p. 237). On the subject of aggression, Lorenz says that in 1955 he had written that human aggressive drives "simply derive from the fact that in prehistoric times intra-specific selection bred into man a measure of aggression drive for which in the social order of today he finds no adequate outlet" (p. 243). In the same chapter, however, Lorenz gives full credit to the dominance of culture evolving at a pace that greatly outstripped the snail's pace of biological evolution.

The work of Lorenz and his colleagues seems to have attracted the attention of secular humanists in search of a naturalistic explanation of ethics. For example, in a paper by Delos B. McKown, Professor of Philosophy at Auburn University, Alabama, appearing in The American Rationalist, I found this statement: "It is only natural say modern ethologists and other students of animal behavior for an intelligent, gregarious species like homo sapiens to develop ethics. In short, morality is to be expected and is necessary for the good life, God or no God. Once again the supernatural is denied necessity" (McKown, 1985, p. 85). His first sentence carried a reference to page 235 in Lorenz's 1966 book, cited above. I did not find exactly that statement, but in reference to Immanuel Kant's views on moral law, Lorenz says: "Would he, who did not yet know of the evolution of the world of organisms, be shocked that we consider the moral law within us not as something given, a priori, but as something which has arisen by natural evolution, just like the laws of the heavens?" (1966, p. 235) Thus the ethologists of the mid-1900s seem to have been adopted by the secular humanists as buddies in antitheism, and this affair was to persist through the explosive appearance of modern sociobiology.

The "splash" of the 1970s is credited to two Harvard professors, Edward O. Wilson and Robert L. Trivers. Wilson, a zoologist and entomologist, gained the position of innovator of sociobiology when, in 1975, his book, titled Sociobiology: The New Synthesis, was published by the Harvard University Press. Sometimes in the history of science a particular idea or hypothesis is evolved and discussed for a long time--even decades--before it hits the fan of public attention. What sometimes triggers this splash phenomenon is the invention of a buzzword that catches the public fancy, not only for its intrinsic value in a special field, but also for its wide-ranging implications extending into fields of general human interest. In using "sociobiology" in his book title, and devoting a final chapter to the sociobiology of humans, Wilson brought down a storm of criticism on the "nature" side of the nature/nurture debaters.

Defined as simply "the biology of human behavior," sociobiology was nothing new. Konrad Lorenz and his colleagues Karl von Frisch and Nicholas Tinbergen had seen to that in scientific research that brought them jointly a Nobel prize in medicine and physiology in 1973. Their predecessors in ethological speculation were Herbert Spencer, Charles Darwin, and William James. The early ethologists had already touched off a storm of public indignation through the implications of Social Darwinism, bringing on a severe depression to that line of investigation among anthropologists, as we have already seen.

Most of Wilson's 1975 treatise was devoted to producing a unifying explanation of the systems of social insects. Wrote Jeffrey Saver: "Meticulously extending this approach to the many thousands of social animal species, Wilson in Sociobiology provided an encompassing evolutionary account of the origin and maintenance of the disparate patterns of sociality displayed by birds, primates, insects, and man" (1985, p. 15). There followed in 1978 Wilson's shorter work, On Human Nature; it won him a Pulitzer Prize. He then teamed up with physicist Charles Lumden to produce Genes, Mind, and Culture (1981), in which a serious attempt was made to establish more firmly the genetic base for speculations advanced in On Human Nature. In this work, the authors developed the concept of the coevolutionary circuit, mentioned in Chapter 51.

One critic of Wilson's sociobiology has been John Maynard Smith, a professor of biology in the University of Sussex, England, and himself a leading research worker in the field of animal behavior. I was interested to find that Smith's Scientific American article of 1978, titled "The Evolution of Animal Behavior," contains no mention of Wilson and his publications, but Smith refers to the work of Robert L. Trivers and H. Hare of Harvard University. Smith gives a great deal of attention to the work of W. D. Hamilton of the Imperial College of Science and Technology in London as a pioneer in understanding the evolution of altruistic behavior in animals, contained in papers published in 1964 (Smith, 1978, p. 176). Doesn't it look as if Wilson may have raised the hackles of a number of students of social behavior of animals (ethology) who had gotten along quite well for years without that newfangled word "sociobiology"?

The 1977 Dahlem Workshop on Biology and Morals, organized and managed by molecular biologist Gunther S. Stent of the University of California in Berkeley, had as its purpose to give sociobiology a very close and critical scrutiny. Stent's tone is evident in these words from his introduction to the Dahlem Workshop volume:

As Darwin himself had realized, natural selection operates not only on the physical features of animals, but on their behavior as well. And so it was not far-fetched to envisage that also human behavior, and moral behavior in particular, is the product of selective, evolutionary forces. Thus not only the existence of a moral code, but also of its actual content would be justifiable by discovering what adaptive value, or fitness, it brought to the human species in the evolutionary struggle for survival. The naturalistic view received a further impetus forty years ago upon the rise of ethology, of the study of animal behavior in an evolutionary context poioneered by Konrad Lorenz. And more recently the naturalistic approach has gained wide notoriety under the sponsorship of the ethological specialty that calls itself "sociobiology." (Stent, 1978, p. 14)[4]

John Maynard Smith was a contributor to the Dahlem Workshop. He describes Wilson's first book as "a major, albeit controversial, treatise on social organization in animals" (Smith 1978b, p. 28).[4] Smith places the main source of controversy on Wilson's claim "that an evolutionary approach can illuminate not only animal but also human societies." Smith continues:

> Wilson does not allege that the _differences_ between various human societies are caused by genetic differences between their members. Such differences Wilson accepts as being culturally transmitted, basically because historically documented changes in social organization have occurred too rapidly to be explained by genetic change. What he does claim is that there are universal characteristics of human societies which exist because of universal features of human nature that evolved by natural selection. (P. 28)[4]

The specific areas of morality cited by Wilson and other sociobiologists would seem to fall mostly into what Dobzhansky calls family ethics. For example, incest taboos are repeatedly singled out as part of our genetic makeup. This seems reasonable enough because close inbreeding is recognized as deleterious to a family group. Another alleged genetic trait frequently cited is economic role differentiation between the sexes.

In continuing his critical examination of Wilson's work, Smith is not happy with Wilson's suggestion or inference that the genetic makeup (in the form of alleles) can determine in an individual certain morally relevant traits such as aggressiveness or lack of aggressiveness. Smith observes: "There is no evidence for a gene in man determining how aggressive he will be, or whether he will commit incest, or any other morally relevant trait" (p. 30).[4] On the other hand, Smith accepts a more general statement that "there are genetic differences between people influencing how likely they are to be aggressive in particular circumstances." Unequivocal evidence that such differences actually exist "is hard to come by in man," Smith admits, "but it would be very odd indeed if this assumption were not true."

Smith turns to a second line of criticism of Wilson's general thesis. Wilson treats as innate "human universals" certain genetically determined behavior patterns and implies that these are present in all human beings at all times, regardless of environment, and that "almost all human societies will acquire these characteristics, regardless of initial conditions or conscious intentions" (p. 31).[4] Smith cautions: "I do not know how this implication can be justified, but it certainly cannot be justified by evolutionary theory" (p. 31).[4]

The spate of popular books on the "new" sociobiology that soon appeared included one by R. Dawkins with the title _The Selfish Gene_ (1976). Stent considers this book as "an exemplary case of muddying the waters by lack of terminological hygiene." Stent continues:

> I single out Dawkins' book in this Introduction, not because I want to criticize preemptorily its conclusions, or on the sociobiological approach to morals that it represents, but because its parlance manages to create total confusion with regard to the essential concepts in _both_ the moral _and_ the biological domain. (1978, p. 16)[4]

Stent goes on to explain there is a consensus among the workshop participants "that, whatever else morals may be, or whatever is their biological basis, the concept of morality pertains to an intentional state of an agent" (p. 18). The intentional aspect of morality, Stent holds, gives morality the status of a "non-subject" in biology, i.e., morality is a transempirical concept. Stent says that there are people--behaviorists, for instance--who hold that biology can have nothing to say about intentional, or "mentalistic," states. He continues:

> But such people would not be justified in redefining the concept of morality by simply replacing covert intentionality by its overt consequences as the criterion of the moral component of behavior, just so that biology can be brought to bear on something called "morals" but which is not, in fact, the morals of ordinary discourse. (P. 17)[4]

Dawkins's work got considerable public exposure through his use of expressions such as "altruistic genes" and "selfish genes." So we have good genes and bad genes, respectively. The suggestion here is that intent is not at work, that genes carry out the moral direction. Dawkins considers altruism as behavior that increases the chances of survival of another individual at the expense of the altruistic individual. Selfishness (bad moral value) decreases the survival chances of others. Stent considers Dawkins's notions of altruism and selfishness to "lie outside the domain of moral discourse." "Accordingly," says Stent, "ethologico-evolutionary considerations regarding this trans-ethical concept of 'altruism' can be expected to add very little to the understanding of morality as a biological phenomenon" (p. 17).[4]

Stent scores the sociobiologists for first claiming that evolution has endowed humans with "a selfish, aggressive, and on the whole, rather nasty genome," then, by reason of the awareness of this bad inheritance, they tell us we must struggle to "transcend" that genetic inheritance. "But," adds Stent, "the idea of any organism, including man, transcending its genome is a biological absurdity" (p. 19).[4]

In view of Dawkins's use of such terms and their widespread dissemination in the media, it is little wonder that Wilson's sociobiology has come under attack not only by other biologists, but also by social scientists. Tom Unger, writing on sociobiology in the humanist journal _The American Rationalist_, describes the reaction in these words:

> These revolutionary ideas have been received with something less than open arms by most social science theorists, many of whom are dedicated to the proposition that cultural upbringing, not biological imperatives, almost exclusively shapes human nature. Their opposition to sociobiology has been generally vehement, often openly hostile, and on occasion downright violent. Even personal attacks on Professor Wilson are not uncommon. At a meeting of the American Academy of Arts and Sciences, for example, Wilson's detractors doused him with water. Elsewhere he has been the target of rotten eggs and other projectiles. Already at least three or four anti-sociobiology books have been hastily thrown together. Anthropologist Marshall Sahlins of the University of Chicago, one of sociobiology's most fervent opponents, has summarily dismissed the new science as "genetic capitalism." Jerome Schneewind, a philosopher formerly at Manhattan's Hunter College, has called it "mushy metaphor . . . a souped-up version of Hobbes." Harvard Evolutionary Biologist Richard Lewontin is perhaps the bluntest critic of all.
> . . . . At a conference of the American Anthropological Association in December of 1976, a minority of reactionists went so far as to advocate that

sociobiology books be banned from our state libraries and universities. (Unger, 1984, p. 57)

Unger reflects on the underlying reasons for such outspoken and virulent hostility to the sociobiologists. Keep in mind that Unger is a humanist advocate and that Wilson has become a darling of the secular humanists, elected to the American Academy of Humanism. Unger attributes the opposition largely to anthropocentrism, which has dominated cultural anthropology and is a leading tenet of non-fundamentalist Christians (as well as of the creation scientists) (1984, p. 58). Theistic scientists, those who really consider their religious conviction important along with their science, see the rise of <u>Homo sapiens</u> as the culmination of the Creator's design. Unger comments: "In many ways sociobiology is profoundly at odds with the idea of an anthropocentric teleology. It is even more at odds with the notion of a benevolent design in the universe" (p. 58). Of course, that same statement reads equally well if we simply substitute "Darwinism" for "sociobiology" and set the scene back one century. In that sense, Unger's critique borders on banality, at least for those familiar with the history of Darwinism and the reactions it drew from theistic elements of society. According to Unger, Wilson has fueled the hostility of the religious anthropocentrists by statements such as that dogmatic religion is "the single most dangerous force in the world today" (pp. 58-59). (Reference given to an article in <u>Science Digest</u>, May, 1982, p. 86.) Unger says: "In fact, it is Wilson's hope that sociobiology's ability to present a fully satisfactory, evolutionary account of religion will constitute the ultimate disproof of the traditional assumptions of theology" (p. 59).

Unger's adulation for humanist academician Edward Wilson, whom he refers to as the "father of sociobiology," soars to stratospheric heights in the opinion that Wilson "is now being attacked for many of the same reasons that religionists once attacked Copernicus, Galileo, and Darwin. If history is anything to go by, we may assume with confidence that Wilson will eventually triumph" (p. 59).

I'm all for pursuing the scientific quest for a fully satisfactory naturalistic explanation of religion, making use of both biological and cultural human evolution, but that quest has and will continue to include the efforts of many anthropologists and geneticists who were working on the problem long before Wilson came on the scene and who find much to criticize in sociobiology as presented by Wilson, Dawkins, and others. Let's close this review of sociobiology with some comments from two philosophers who contributed significantly to the Dahlem Workshop.

Philosophy professors T. Nagel of Princeton University and C. Fried of Harvard University were in strong agreement that "ethics is an autonomous subject." Nagel wrote:

> Ethics is a theoretical subject developed through the collective human capacity to submit pre-reflective motivations and behavior patterns to rational criticism and thereby to modify them. The result is an open-ended process of development of social institutions, motives, and methods of moral reasoning. It can be best understood internally, through the methods appropriate to the subject, and its history is part of intellectual and cultural, not of biological history. (1978, p. 221)[4]

Nagel observes: "No one, to my knowledge, has suggested a biological theory of mathematics; yet the biological approach to ethics has aroused a great deal of interest" (p. 222). He goes on:

> Ethics exists on both the behavioral and the theoretical level. Its appearance in some form in every culture and subculture as a pattern of conduct and judgments about conduct is more conspicuous than its treatment by philosophers, political and legal theorists, utopian anarchists and evangelical reformers. Not only is ethical theory and the attempt at ethical discovery less socially conspicuous than common behavioral morality, but the amount of disagreement about ethics at both levels produces doubt that it is a field for relational discovery at all. (P. 222)[4]

Nagel takes the view that the development of ethics must be participated in and accepted by a large number of people and that this requirement makes the pursuit of ethics a "more democratic subject than any science" (p. 224). His rational approach to ethics is close to that expressed by many secular humanists:

> Still, the premise of this view of ethics as a subject for rational development is that motives, like beliefs, can be criticized, justified, and improved-- in other words that there is such a thing as practical reason. This means that we can reason not only, as Hume thought, about the most effective methods of achieving what we want, but also about what we should want, both for ourselves and for others. (P. 224)[4]

In his discussion of methods of ethical inquiry Nagel talks about the principle of moral equality--most modern ethical positions hold that moral claims or rights of all persons are equal (p. 228). This is a conclusion that I would find difficult if not impossible to arrive at by practical reason; it concerns presuppositions about the intrinsic value of humanity that are in the transempirical or supernatural realm.

Nagel's paper closes with a firm stand on sociobiology. He realizes that humans who engage in ethical discussions are "organisms about whom we can learn a great deal from biology." He grants that "their capacity to perform the reflective and critical tasks involved is presumably somehow a function of their organic structure. But it would be as foolish to seek a biological evolutionary explanation of ethics as it would be to seek such an explanation of the development of physics" (p. 229).[4]

Professor Fried emphasizes how close the sociobiologists come to committing the naturalistic fallacy (1978, p. 209). Being of the opinion that ethics is an autonomous subject, he comments: "If it puts me in the camp of those who hold that one cannot derive an 'ought' from an 'is', so that naturalism is indeed a fallacy, then I must comfort myself with the otherwise conflicting company of both Hume and Kant." He goes on:

> The persistence of this view that naturalism is fallacious is in large part due to the fact that the attempts to show that one can derive an "ought" from an "is" rely on some very fishy, fancy, and suspiciously "ought"-like "is"es from which to launch the derivation. (P. 209)[4]

Fried emphasizes that ethics relates to choice, and normative ethics is concerned with guiding choice. About the latter he says: "I take it to be clear that when it is a matter of recommending a choice, no description of past or present states of the world can of itself determine that recommendation" (p. 210). In the choice of goals or ends "no system of statements about the way the world is or has been can of itself compel conclusions in an ethical argument. And of course the propositions of biology, (including sociobiology) do not--in any reputable scientific circles of which I am aware--purport to be anything other than propositions about the way the world is and about the laws which describe the world at some level of abstraction or generality" (p. 210).[4]

A Naturalistic Origin of Religion

If, as evolutionary biologists seem to think is the case, the human capacity to ethicize is a genetic endowment, can it not also be argued that the human capacity to have religious feelings and to generate religious concepts is a genetic endowment? The only essential requirement for producing a religious thought is imagination. We have already seen that imagination is uniquely a mental faculty of humans and that is a product of biological evolution. Imagination enables humans to supply explanations of phenomena, the causes of which are not immediately available.

It is not difficult to reason that imagination is of adaptive value in human evolution, when imagination is used to predict situations that, on the one hand, may be favorable to survival or, on the other hand, unfavorable to survival. To imagine a successful group hunting strategy could permit survival when the lack of that imagination could under similar circumstances result in death by starvation. The ability to imagine a dangerous encounter with a group of animals or of other humans could result in devising an evasive strategy leading to survival.

When we pass from such simple, mundane situations over which humans have some measure of control to imagination about events that seem to have no obvious cause and over which humans have no control, the adaptive role of imagination is difficult to demonstrate and perhaps should be presumed not to exist. For example, an unseen pestilence causing the rapid death of many individuals of both sexes and various ages would have sent the survivors into a state of great emotional distress. The same imagination that served a practical purpose in the circumscribed hunting situation now runs free and uninhibited in seeking an agent on which blame may be placed. Imagined supernatural agents or forces fill the void; they provide an explanation and an object of emotion--fear, anger, or hate, for example. Imagination provides further details of the agent and its relationship to humans. By analogy with human relationships, it can be imagined that the supernatural agent has a plan, purpose, or design in perpetrating the pestilence. It can be imagined that the agent is in the role of an all-powerful ruler or creator or whatever. The imagery thus produced, once accepted by individuals and standardized by a social group, has become a working mythology or a religion and, as such, is a cultural package to which morals are referred for authority and enforcement. In short, religion is an artifact.

It is hard to see in imaginative activity about the supernatural realm any adaptive role that might be expressed in the genetic evolution of humans. Does it relate to survival fitness in any meaningful way? Or is it simply a recreational activity of the brain, indulged in when there were no pressing requirements to produce images for fitness-related situations? I suppose one could argue that a satisfactory supernatural system that accounted for all ulterior questions would bring increased ability of a social group to cope with stressful situations, but on the other hand, one can deduce negative effects. For example, setting out generous helpings of scarce foods for forest spirits to appease their innate hostility, or sacrificing one's son or daughter to appease the wrath of a god could severely diminish the survival capability of the group.

Once a religion has been fed into the culture stream, it can persist indefinitely or can be replaced by other imagination-induced religious systems. This connection between imagination and its elaborate cultural structures is recognized by psychologist N. Bischof of the University of Zurich, a participant in the 1977 Dahlem Workshop:

> Human imagination, capable of dreaming up possible future environments and ego-states, is not bound to a strict down-to-earth level. It can produce mythological ideas that reduce the inherent emotional tensions. Here we find the consoling eschatologies of religions, and also the socialist utopia of a paradise of general solidarity, or the capitalist ideology of the unlimited possibilities of personal development. (1978, p. 66)[4]

Professor Bischof closes his article with a paragraph I find very much to my liking; I pass it on in hopes it will also tickle your fancy:

> Mythological or moral ideas, then, do not evolve because they promote the physical survival of mankind. They are accepted or rejected according to the degree to which they assist in finding and holding on to points of equilibrium in the paradoxical field of human emotion. When they find their niche in this ecology, they then have the chance to endure, the chance that temples and palaces be built for them, virgins consecrated, and human hearts sacrificed in their honor, the chance that they be immortalized in hymns and carved into the face of rocks, the chance that martyrs bear witness to them, and that missionary conquistadores lay the world in ruin so as to make them triumph in millions of subjugated brains. (1978, p. 72)[4]

I found the literature of cultural anthropology so rich and diverse on the subject of the early development of religion as to be almost unassailable. Some general principles of interest are covered by Campbell and deserve review (1975, pp. 359-62). He points out that "in preliterate societies, traditional knowledge (folklore) is preserved in ritual form by, for example, religious institutions. This function of religious and similar institutions is also carried through into literate society" (p. 357).[1] Religion would serve the important function of integration of the social group and promote its cohesion because it would direct feelings toward a symbolic center. Religion would also assist in controlling individual behavior through religious sanctions. Primitive religion would have first been promoted through rituals having deep and abstract meaning. Ritual is seen as a condensed and compact form of social memory. Says Campbell: "Ritual is a special form of language with a very high information content, but, unlike language, the meaning of ritual depends on its social context. A ritual act is a social act--it is society's meditation of traditional knowledge and behavior" (p. 359).[1] Campbell explains further:

> In its remarkable way ritual records knowledge about social origins, about social structure, kinship, and obligations. It records behavior patterns of a fundamental nature, such as hunting, toolmaking, and food preparation. Ritual is, as it were, the DNA of society, the encoded informational basis of culture; it is the memory core of human social achievement. (P. 359)[1]

Myth, the verbal derivative of ritual, followed directly, and later, with the evolution of writing, continued to play a role of promoting social stability. Religion, which in its early stages was predominantly concerned with the supernatural realm, also served the same important social function: to bind society more tightly together. The awesome power (literally) of supernatural forces could be harnessed in support of behavior required for social stability. Many important social activities would then take on a sacred and ritual quality, safely secured in a domain beyond reason. Campbell writes:

> Not only are traditions ritualized, but many important social activities appear to take on a

sacred and ritual quality. An element in the common life of a society may come to have a special significance and in time become an act of religious observance. Especially important in this respect are the rites de passage (birth rites, puberty rites, marriage rites, death rites), which consolidate social roles and social structure, as well as bind members of the society together. But religion does more than that, for it directs social sentiments toward one stable and symbolic center. (P. 360)[1]

The phenomenon of death awareness, referred to earlier, may have led to the primitive religious rituals associated with burial, for which artifacts can be found. The religious concept of continuance of ancestors in a supernatural realm would have emerged and perhaps led to ancestor worship as one of the primitive forms of religion.

Laws and legal codes forming a secular structure would, I suspect, have followed the growth of an entangled web consisting of functional ethics and religion. A written legal code could be derived by one society in power from its own religion to serve the wider function of directing the behavior of subjugate societies with different religions--conquered nations and imported slaves, for example. Campbell observes:

These different means of binding and stabilizing society are clearly factors of overriding importance in the evolution of man. It remains to point out that religious and legal institutions help control human behavior at an individual level for the benefit of the society. What "is done" and "is not done" is a powerful determinant of human behavior; it is dictated by tradition, religion, and law, through each individual's peers and his elders. Religious and legal sanctions control human behavior at its most fundamental level, and the two kinds of sanctions are often so closely allied as to be indistinguishable, as in the Ten Commandments. Among these vital functions of religious and legal institutions falls the control of the way men express their biological needs, by means of sanctions against theft, aggression, adultery, and so on. (Pp. 361-62)[1]

In this chapter I have argued the case that religion is, at its rock-bottom base, a phenomenon of completely naturalistic origin, i.e., it is an artifact. Thus, a naturalistic hypothesis accounts for all belief systems composed of supernatural phenomena. This is not a self-contradictory statement, as it might seem to be at first reading. The relationship between natural and super-natural is fully and precisely accounted for in a monistic ontology--that of mechanistic materialism--which I presented in Chapter 6 on science and religion. Imagination, a physical/chemical functioning of the brain (a neurophysiological activity), produces religious imagery, which is a form of reality amenable to inquiry by empirical science. This view accomplishes an ontological reduction from a dualistic to a monistic system. Where does that put ethics, which seems to require an infusion of either supernatural or transempirical constructs? One component of ethics is directly naturalistic; it is the biological capacity to ethicize. The other is religious, through human imagination. But, when religion is seen as naturalistic in origin, then both ethics and religion are fully tractable to empirical science for investigation as real phenomena. If this general hypothesis is viable, it must accommodate all ethical and religious systems existing in the past, present, and future. The range of imaginative constructs is surely unlimited in the human population. Therefore, we can expect to find limitless variation in both ethical and religious systems the world over through human history.

Religion as the formalization of mental imagery is not my own idea, although I formulated it independently (euphemism for "in ignorance of previously published material saying the same thing"). Besides the reinforcement I discovered in the writings of such scholars as Professors Bischof and Campbell, I came across some relevant passages in an essay by Kenneth E. Boulding with the title "Toward an Evolutionary Theology." Boulding is a professor of economics at the University of Colorado and has published widely read articles on the general environmental mix of population, resources, and technology. (Theology, as I see it, is the structured and dogmatized body of religion comprising a mass of accumulated excrescent artifacts that can be endlessly churned up and discussed in the absence of any great amount of substantive content in the bare religious outline itself.) Boulding finds it necesary to introduce his essay with some statements on the nature and origin of theology and religion. His opening sentence reads thus:

Theology consists of those images in the human mind, and the language describing them, which are relevant to and ultimately derived from the record of human experiences and states of mind which can be classified as "religious." (1984, p. 142)

Boulding goes on to say that it is not easy to draw a line between religious and nonreligious experience. In the middle ground lie the experiences of art and music, human love and sacrifice--not easy to classify. Boulding continues:

Nevertheless, there is a large area of human experience reflected not only in personal testimony, but in a very ancient and complex record, the existence of which can hardly be denied. The traces of flowers on Neanderthal graves, cave paintings, figurines and little statues of household gods, funerary objects, temples, mosques and cathedrals, holy books, a vast descriptive literature of personal religious experience in journals and other writings, the enormous literature of mythology, and even the sober descriptive writings of psychologists like William James, all suggest that there is a religious element of human experience which cannot be denied as a part of the ongoing history of the human race. (P. 142)

If, as Boulding asserts, there is a "religious element of human experience," is that element part of our genetic makeup? If the ability to see ethical issues is a genetically controlled capability, as Waddington proposed, and if, as Dobzhansky added, it follows that "every member of the human species comes into the world with a genetic endowment making him an ethicizing being," then cannot a parallel set of statements be made about the ability of humans to accept as real those religious images (images of the supernatural)? (Dobzhansky, 1977, p. 455) For Waddington's "ethicizing beings" we substitute "religi-cizing beings" with a natural propensity for "religiosity." Both classses fall within the category of authority acceptors.

If fitness for survival was enhanced by accepting authority on ethical matters, could it not also have been enhanced by accepting authority on religious statements? Let us assume a stable primitive society at the level of the tribe consisting of (a) a small group of strong managers with low susceptibility to belief in imagined entities coupled with strong inclination to wield authority, and (b) a large number of subservient individuals with high susceptibility to believe imagined entities coupled with strong inclination to accept authority. This bipolar or dualistic arrangement, seen everywhere today in political, military, industrial, and religious establishments on various scales, is remarkably successful, as compared with egalitarian arrangements generally. Perhaps at a time when culture was extremely primitive and its evolution extremely slow, religiosity became a genetic endowment with alleles coding for a range of intensities from weak to strong. In terms of what modern sociobiology asks us to take seriously, this hypothesis is neither extreme or unreasonable. But, as molecular biologist Gunther Stent says, much of our behavior is controlled by "covert deep structures" inaccessible to direct observation (1975, p. 1053). Moreover there are no records from this formative time in human evolution to give us clues as to how and why selection pressures may have acted to generate the accumulated content of inaccessible genetic information. And thus, says Stent, "We encounter the barrier to an ultimate scientific understanding of man which Descartes had recognized more than three centuries ago" (p. 1057).

A fitting statement with which to end this chapter on the possible biological origins of ethics and religion is also Stent's closing statement:

> From this ultimate insufficiency of the everyday concepts which our brain obliges us to use for science it does not, of course, follow that further study of the mind should cease, no more than it follows from it that one should stop further study of physics. But I think that it is important to give due recognition to this fundamental epistemological limitation to the human sciences, if only as a

safeguard against the psychological prescriptions put forward by those who allege that they have already managed to gain a scientifically validated understanding of man. (1975, p. 1057)

Credits

1. From: Bernard Campbell, HUMAN EVOLUTION, Second Edition. Copyright © 1974 by Bernard Campbell. (Aldine Publishing Company, New York.) Adapted by permission.

2. From Theodosius Dobzhansky, Evolution of Mankind, Chapter 14, pp. 438-63, in T. Dobzhansky, F. J. Ayala, G. L. Stebbins, and J. W. Valentine, Evolution, W. H. Freeman and Company, San Francisco. Copyright © 1977 by W. H. Freeman and Company. Used by permission.

3. From George Gaylord Simpson, 1969, Biology and Man, Chapter 10, Harcourt Brace World, New York. Originally published under the title "Naturalistic Ethics in the Social Sciences" in American Psychologist, vol. 21, pp. 27-36. Copyright 1966 by the American Psychological Association. Reprinted by permission of the publisher.

4. From Morality As a Biological Phenomenon, Gunther S. Stent, ed., Report of the Dahlem Workshop, Berlin, 1977, Abakon Verlagsgesellschaft, Berlin. Copyright © 1978, Dr. Silke Bernhard, Dahlem Konferenzen, Berlin. Used by permission.

PART IX

The Origin of Life on Earth— Naturalistic or Creationistic?

Introduction

The origin of life from nonlife, which we call biopoesis, is a topic we have left to the end in our overview of earth history for two reasons. First, it it involves much speculation and may require us to draw upon nearly everything we know of earth history, biology, and the record of life on earth. Second, it represents that point in earth history at which organic evolution starts and modification with descent begins its unbroken chain. Here, at the start of life, an act of God--a true miracle in the supernatural realm--can be inserted without impugning a completely naturalistic theory of subsequent evolution.

For many deistic scientists, Darwin perhaps included, the initiation of life can be taken as the Creator's last overt and specific act of interference in nature's realm. These persons find satisfaction in having assigned God an act both essential and otherwise impossible for nature to accomplish. As long as God has refrained ever since from influencing the course of evolution, deistic scientists can pursue their research on evolution as true materialists, free of any conflict between science and religion. Because of the time sequencing involved, these deistic scientists cannot be held guilty of Orwellian "double-think," for God's creation and Darwin's natural selection are not in time conflict. In this respect, the unique point in time that life originated shares the same qualities as the unique point in time of origin of the universe. In a cosmic relay race, God and naturalist have run alternate laps as teammates. God ran the first lap, passing the baton to naturalist, who in turn passed it back to God, and thence it was given back to naturalist for the fourth and final lap.

Chapter 53

Philosophical Speculations on Life Origins

The origin of life on earth is a major topic of conflict between creationism and mainstream evolution science. It is a topic about which creation scientists of the Institute for Creation Research have written a great deal, going deeply into modern molecular biology for support of their negative position. Besides the conflicting hypotheses (a) that God created all life along with the entire universe, and (b) that spontaneous mechanistic development of living matter from nonliving matter occurred early in geologic times, what alternative hypotheses have been presented?

Vitalism, Its Ghost Revived

Chapter 4 referred to vitalism as a philosophical doctrine evoking a pervasive and mysterious vital force in living matter. I remarked that vitalism was eliminated in the early 1900s as a serious scientific hypothesis, because it invokes a supernatural (or transnatural) concept and cannot be examined or tested by empirical science. Nevertheless, vitalism seems to have a lot of vitality left and, if it does not have much chance of being revived on a large scale, its ghost does seem to arise once in a while. One scientist who recently revived vitalism is biologist George Wald, whose belief in the guiding role of cosmic consciousness I mentioned in Chapter 1.

Creationists show considerable interest in another biologist who has proposed a form of vitalism. He is Albert Szent-Gyorgyi, Hungarian-born (1893) and recipient of the Nobel prize in 1937 and 1955. In recent years he served as Research Director of the Institute for Muscle Research, located in Massachusetts. Szent-Gyorgyi's vital force goes under the name of "syntropy." In the context of the second law of thermodynamics, syntropy is a universal force that tends to counteract the growth of entropy and disorder. Syntropy tends to cause organisms to progress to higher levels of order and organization. Thus, he says, the second law does not function in living systems. Syntropy was responsible for the origin of life, as explained in this passage from a work by Szent-Gyorgyi:

> Inanimate nature stops at the low level organization of simple molecules. But living systems go on and combine molecules to form macromolecules, macromolecules to form organelles (such as nuclei, mitochondria, chloroplasts, ribosomes, and membranes) and eventually put these all together to form the greatest wonder of creation, a cell, with its astounding inner regulations. Then it goes on putting cells together to form "higher organisms" and increasingly more complex individuals . . . at every step new, more complex individuals . . . at every step, new, more complex and subtle qualities are created, and so in the end we are faced with properties which have no parallel in the inanimate world. (1977, pp. 15-16)

Syntropy, being an ever-present force, continued to drive organic evolution throughout its upward course and

is active today. Syntropy is identified with the teleology that theists consider God's prerogative. Syntropy can be invoked by nontheists and atheists. Consequently, creationists have no use whatsoever for syntropy or any other version of a vitalistic force. Writing in the Institute for Creation Research Impact Series, Jerry Bergman, a psychologist on the faculty of Bowling Green State University in Ohio, rejects syntropy because it is a hypothesis about transnatural things; it sets up a dualism that science can no more entertain than a dualism containing the theistic supernatural realm (1977).

Panspermia--An Attractive Idea

Another version of the origin of life on earth is that it was imported from somewhere else in the universe. This possibility is reviewed in an article in Technology Review, titled "The Cosmic Cradle," written by Robert C. Cowen, Science Editor of the Christian Science Monitor. His opening paragraph reads thus:

> Earth in orbit sweeps up some 16,000 tons of interplanetary matter each year, much of it the remnants of decaying comets. Are new life forms present in this stellar gift? Do viruses evolved in comets or interstellar dust bring novel genes to influence earthly evolution? Did earth's life itself evolve from these cosmic seedlings? (1978, p. 6)

The general hypothesis of an extraterrestrial (ET) origin of earthly life is not new; it goes under the name of panspermia. The idea was put forward by Aristarchus of Samos (c. 310-230 B.C.), best known for having proposed a heliocentric theory of the universe. In modern times, panspermia was proposed by J. B. S. Haldane (1892-1964), a distinguished geneticist known for having applied mathematics to that area of biology. As if one geneticist of renown were not enough to get into the act, no less a celebrity than Francis Crick has suggested a scenario that would do credit to Velikovsky and von Däniken. Crick is codiscoverer with James D. Watson of the double helix of the genetic material DNA, which carries the growth-directive program for all life on earth. Writing in a book titled Life Itself: Its Origin and Nature, Nobel laureate Crick proposed that bacteria carried in an unmanned space rocket reached earth after a journey from another planet within the Milky Way. Those bacteria survived and thrived in the earth's environment as it existed about 3.5 billion years ago. Of course, Crick supposed that the ET bacteria originated in a naturalistic manner from nonliving matter, so the rocket journey merely moves back the problem to a place remote in time and space. One wonders which of the two scenarios required is the less probable: (a) spontaneous origin of life from nonlife on our planet, or (b) a rocket journey of enormous distance leading to a precise landing on a mere speck of an earth at a time when that earth had no other life but was in possession of a physical environment favorable to perpetuating new life. I would opt for the first scenario as being less improbable, particularly since we must multiply the probability of (a) times that of (b). (And whatever became of Occam's razor?)

In our time of resurgence of pseudoscience and its enormous cultist followings and media exposure, eminent scientists have perhaps been unable to resist the urge to get into the limelight. The occasion for an orgy of speculative discourse was set up in 1980 in the form of a colloquium titled "Comets and the Origin of Life," held at the University of Maryland. (See reports in EOS, vol. 61, no. 52, December 1980; Chemical & Engineering News, vol. 58, no. 47, November 1980, p. 36.) The two principal performers were Chandra Wickramasinghe and Sir Fred Hoyle, both of the University of Cardiff, Wales. Both are scientists of impeccable reputation. Hoyle, an astronomer and cosmologist, is widely known for his part in formulating the steady state hypothesis of cosmology. The two scientists proposed that interstellar clouds of dust and gas serve as factories for the production of organic molecules and their combination into living matter as complex as viruses or bacteria. Comets, formed from such clouds that lie within the outer reaches of our Solar System, could have carried the organic molecules and primitive life forms to our planet. This would explain the first life on earth. But to have the speculators stop with only one grandiose idea would be as outrageous as asking them to eat only one potato chip. With many comets operating continuously throughout geologic time, why not suppose that our earth experienced numerous inoculations of such organic materials? Infusions of new genetic materials from time to time could have abruptly and profoundly altered the course of organic evolution. In our own civilization perhaps we have experienced inoculations of new viruses that have caused great disease epidemics throughout history. (I'm reminded of Michael Crichton's Andromeda Strain, 1969.) Not surprisingly, such suggestions have aroused sharp rebuffs from geneticists and epidemiologists.

Panspermia, along with Hoyle's 1983 importation hypothesis in particular, is flatly rejected by the creationists (Oller, 1984). For creationists, where life begins is a question apart from what or who caused it and mechanistic process anywhere in the universe is unacceptable. Evolutionists, on the other hand, will continue to look for a naturalistic origin of life on our planet, where at least the geologic record holding the simplest life forms is available for investigation.

The Creationists' Argument from Design

Creationists "make a market" in the subject of the origin of life (to lift a phrase from the stock brokerage business). It is an easy target because it is on the outer fringe of science and it seems easy enough for them to try to push it entirely out of the science realm. Duane T. Gish of the ICR staff is their point man on biochemistry. He holds the Ph.D. from the University of California at Berkeley and serves as Professor of Biochemistry in the ICR graduate degree program. Writing in ICR Impact Series, Gish claims that scientific hypotheses about the origin of life (biopoesis) do not satisfy the criteria of empirical science (1) because no human observers were present at the origin, (2) it is impossible to reenact the process, and (3) the process, if it did occur, could have left no fossil record (1976a). With the background developed in our Part One on the nature of science, you should be able to make an effective reply to these points.

First, the absence of human observers at about -3.6 b.y. in no way disqualifies a phenomenon from being discussed in a scientific manner. This is a point we have gone into previously. If anyone, including the creationists, took their argument seriously, there would be almost nothing left of the creation/evolution controversy. Second, as emphasized previously, historical natural science cannot reproduce time-bound segments of complex sequential events, but they can be investigated in the light of remaining evidence; hypotheses relating to them can be tested. Third, we cannot assume that the process of origin of life has left no fossil record. Acting negatively on that kind of assumption would stifle all field research in paleontology of the ancient shield rocks. That Dr. Gish cannot take these three points seriously is obvious, because he goes on at great length in his published papers to discuss in detail the scientific hypotheses of biopoesis (Gish, 1976a; 1976b; 1976c). Although creation scientists claim that the subject of origins is beyond the scope of science or of any form of human knowledge, their entire effort is directed to examining origins in what they consider a scientific context.

Fundamentalist creationists as a religious group are obsesssed with the subject of origins. Nothing matters so much to them as the belief that the universe originated as the work of God by special and recent creation as described in the book of Genesis. Recall from Chapter 11 that this postulate is the first tenet of scientific creationism. The second tenet reads: "The phenomenon of biological life did not develop by natural processes from inanimate systems but was specially and supernaturally created by the Creator."

The basic argument persistently put forward by creation scientists against a naturalistic origin of the earliest life in the distant geologic past has a twofold base. First is the argument from design, which they apply to all origins, including the entire universe. Second is the argument from improbability.

The argument from design can be traced back to a set of arguments for the existence of God set forth by Thomas Aquinas (1225-1274) in his most important work, titled Summa Theologica. These arguments have been clearly set forth and analyzed in great detail by philosopher Peter A. Angeles (1980, pp. 17-43). Saint Thomas offered "Five Ways" by means of which he believed he could prove the existence of God. Angeles notes that "all of the Five Ways are elaborations of concepts taken directly from Aristotle and indirectly from Plato" (p. 20). The common theme of these Five Ways is that each describes an object or condition that can be identified in the universe and that each requires a motivating agent. For example (in my words), observed motion requires a 'mover', observed effect requires a 'causer', necessity of existence requires a 'needer', scales of value require an 'evaluator', and, lastly, design requires a designer. It is the last "way" that concerns us here. Angeles quotes St. Thomas's description of the fifth way as follows:

> The fifth way is taken from the governance of the world. We see that things which lack knowledge, such as natural bodies, act for an end, and this is evident from their acting always, or nearly always, in the same way, so as to obtain the best result. Hence it is plain that they achieve their end, not fortuitously, but designedly. Now whatever lacks knowledge cannot move towards an end, unless it be directed by some being endowed with knowledge and intelligence; as the arrow is directed by the archer. Therefore some intelligent being exists by whom all natural things are directed to their end; and this being we call God. (Pp. 19-20)

The argument from design closely follows the argument of the other four ways in requiring an agent (mover, causer, needer, or evaluator); it is this requirement that gets the whole argument into a logical difficulty known as infinite regress. If we require God as the causative or creative agent, then we must ask: "Who caused or created God in the first place?" The answer calls for another godlike agent to create God, and before that yet another such agent, etc., ad infinitum, and so we end up with an infinite regress.

The logic of the infinite regress is unassailable as the arguments are laid out. The only escape is to rule out infinite regress by postulating that there is only one agent in control--i.e., God--and He is eternal. In describing this option, Peter Angeles explains:

> God is Eternal. He is not caused by anything else. There never was a time at which God was not. He always has been. Thus if God is Eternal, the question "What caused God?" is meaningless. (1980, p. 31)

The creationists' dogma is precisely given in the above quoted sentences. The eternal nature of God also rules out any possibility that God is self-caused, in itself a logical impossibility (Angeles, 1980, p. 31). The premise of an eternal God cannot be challenged by science. The message here seems to be that it is futile for nontheists to invoke the infinite regress as an argument against divine causation and design. The creationists have in the eternal God a "waterproof" argument; like a duck's back, its impervious shield sheds all attempts to penetrate it.

The argument from design resurfaced prominently in England in the popular writings of theologian William Paley (1743-1785). It can be found in his popular books arguing for the existence of God. One of these, titled Natural Theology; or, Evidences of the Existence and Attributes of the Deity, was published in 1802 and is often quoted today, perhaps because Paley used what is now called the watchmaker analogy, a favorite of creationist debaters in modern times. Perhaps in no other topic in the creationist lecture circuit are the gullible cultists more easily deceived than by this distorted version of the argument from design and its companion piece, the argument from improbability.

Paley's watchmaker ploy allows the creationist orator to pull from his vest a large gold watch. (No women, I trust, ever took part in this nonsense.) Professor Russell F. Doolittle, Ph.D., Biochemistry, Harvard University, now on the faculty of the University of California, San Diego, calls this oratorical device "the timeworn watch caper" (Doolittle, 1983, p. 86). Professor Doolittle's research lies in areas of protein chemistry, molecular evolution, and the origin of life. He and Dr. Duane Gish of the ICR recorded a debate on "creation/evolution," which was aired on national television in 1982 (see Miller, 1982). Doolittle's 1983 article, titled "Probability and the Origin of Life," is a detailed repudiation of the ICR's arguments for the impossibility of a spontaneous naturalistic origin of life. I recommend it strongly to those with some background knowledge of the biochemistry of the cell (see also Doolittle, 1984).

Paley's orator first asks (rhetorically) if anyone in the audience could tell how the watch originated merely by examining the intricate parts inside it; then quickly follows by asking if anyone would go so far as to assert that the watch formed spontaneously from a collection of inanimate single atoms brought together in one place. This possibility being absurdly remote, the orator confidently concludes: "Every watch must have a watchmaker." This is, of course, the argument from design, backed up by the argument from improbability.

Before going any further, we should be aware that the argument from design, taken by itself, is automatically invalidated on grounds of being a tautology. The word "design" is defined in our dictionaries as a purposeful creation of an intelligent mind, which has to be either a human or a superhuman being. No other meaning is implicit in the definition. So "design" is already linked unambiguously to "designer" by definition--neither could exist in concept without the other. To argue that the discovery of design is proof of the existence of a designer is therefore specious.

If there is to be an effective argument favoring the existence of a designer, it must stem from the nature of the thing identified as the design--i.e., something that is so unique and so complex in its structure or organization as to seem virtually impossible of arising through spontaneous and natural random processes. So we are back to the argument from improbability.

Another good example of the creationist attempt to use the arguments from design and improbability to refute a naturalistic origin of life was that made in 1983 by Dr. Norman L. Geisler, who teaches systematic theology at the Dallas Theological Seminary. He has written many books on theological subjects; he testified for the creationist cause at the Arkansas trial of 1981. Geisler invaded hostile territory to publish an article in Creation/ Evolution with the title "A Scientific Basis for Creation; the Principle of Uniformity" (1983). Perhaps because the stem-wound pocket watch has passed from the scene in favor of the digital microchip timepiece--or to circumvent ennui, or both--Geisler chose for his designed structure the gigantic sculpted faces of four presidents carved in granite high on Mount Rushmore in the Black Hills of South Dakota by Gutzon Borglum. Geisler considers how a viewer might react on first seeing this gross, incongruous scar, while being completely ignorant of its origin. Geisler is certain that the viewer would rightly conclude that the group of sculptured faces requires an intelligent creator and that therefore an intelligent creator existed at some past time. Geisler follows with a list of twelve speculations or musings the viewer might entertain, including one that thrusts his audience into the microchip age. Surely, he argues, we would never infer that there has been an infinite regress of computers, each designed by a previous one. No, he thinks, we would infer that an intelligent designer created all the computers.

Couching the analogy in more modern terms changes nothing in the age-old argument. Geisler tactfully refrains from stomping out the infinite chain of so-called "intelligent computers" by postulating an eternal God as its exterminator.

What is new, if anything, in Geisler's presentation is a brief discourse skirting on the question of the criteria for distinguishing an artifact from a naturally formed thing. Geisler asserts that we know natural forces never produce effects such as the Mount Rushmore faces; therefore, we know such a form is produced by intelligent humans. He quotes from Paley the words: "wherever we see marks of contrivance, we are led for its cause to an intelligent author" (1902). To this, Paley adds a significant sentence: "And this transition of the understanding is found in uniform experience." There you have the key to the distinction being made: uniform experience.

Identification of an artifact (human-made) is a part of the scientific study of archeology. It requires a great deal of accumulated information (experience) based on observation. For one thing, artifacts are always limited to deposits of an age known to yield human remains, i.e., Pleistocene, mostly. Age of the enclosing deposit is established by stratigraphic principles, lithology, and radiometric age determination. The earliest stone tools are often in question by experts because the flaking off of fragments from a stream-rounded pebble by hydraulic impact can produce somewhat similar forms. The Oldowan tools from the Kenya fossil hominid localities, dated at about -2 m.y., are an example (Lewin, 1984, p. 64). In some cases examination of the rock or bone surface under high magnification reveals distinctive abrasion markings of a kind never found on sedimentary particles of pre-Pliocene age. Throughout the heated debates over the age of the earliest humans in North America, there was much argument over the authenticity of alleged stone tools at supposed early-man sites. This is all part of the science of archaeology and has nothing to do with supernatural causes.

Geisler compounds confusion by turning next to

naturally occurring crystals. The remarkable order and symmetry of the crystal lattice and its expression in a beautiful crystal do not lead us to conclude that crystals are the work of a creator, because we can observe crystal growth as a natural physical process.

In Chapter 13 we showed that the creationists' attempts to fit the orderliness of crystal lattices into the necessity for ever-increasing disorder (increasing entropy) led them into serious difficulties. Here, orderliness of crystalline structure is accepted as naturalistic because, as Paley stated, the understanding of it is found in uniform experience.

Geisler uses the subject of naturalistic origin of crystals as a base from which to launch his major thrust, involving the argument from improbability. He refers to the genetic code held by strands of DNA, saying that the information in the DNA of even a one-celled organism is comparable to the information contained in an encyclopedia. He notes that the codons in DNA resemble the combinations of letters devised by humans to convey information through written language. These considerations lead Geisler to conclude that an intelligent agent was required to produce every living organism. Geisler goes on to add that when we discover that the human brain contains more genetic information than the world's largest libraries, we would be inclined to reject the idea that the brain could have emerged without intelligent intervention.

Frederick Edwords has answered Dr. Geisler's arguments in the same issue of Creation/Evolution (a journal of which Edwords is the editor) (1983b). Edwords emphasizes a significant point, namely, that Geisler uses the principle of uniformitarianism as the base of his argument. Projecting back today's experience with intelligent design (Mount Rushmore) into the distant past is accepted by Geisler as a sound procedure. Now, as we have seen several times in earlier chapters, creationists do not hesitate to espouse uniformitarianism ad hoc (when they need it for a special case), momentarily throwing to the winds their espoused doctrine of catastrophism. This maneuver is clear throughout Geisler's argument. Edwords exposes the difficulty, noting that Geisler uses the principle of uniformitarianism to demonstrate a supernatural designer. "This," Edwords says, "is contradictory: arguing from naturalism to prove supernaturalism. In logic, a conclusion is never allowed to refute its premise, yet this is precisely what Geisler seeks to accomplish" (p. 7). Edwords points out that Geisler's entire argument is nothing more than the assertion that "the design evident in artifacts is also present in life forms." The point is that knowing how a particular artifact was made throws no light whatsoever on the origin of a natural object of great geologic age. Geisler might just as well have claimed that knowing how a microcomputer is designed and made by an intelligent human mind will throw light on how the sun originated. There remains, however, the argument from improbability, and that is the topic to which we turn to next. What about Geisler's claim that DNA shows the "marks of contrivance" by a superior intelligence, a supernatural designer?

The Argument from Improbability

The creationists' attack on a naturalistic origin of life features the argument from improbability. To understand this argument we can use a familiar illustration--a calculation of the probability of a person being dealt a perfect bridge hand of thirteen spades. The odds can be calculated as one in 635,013,559,600 (Doolittle, 1983,

p. 94). In contrast to a deck of 52 playing cards, a living cell consists of millions of atoms. In the writings of mainstream scientists there are many probability statements that can be lifted out of context to show the extremely minute probability that the atoms of a single cell will spontaneously come together to form a typical living cell. One of these can be found in a book titled Energy Flow in Biology, by biologist H. J. Morowitz (1968, p. 7); there we read that the probability is less than one in 10 raised to the tenth power raised to the eleventh power. That would expand to one followed by 100 billion zeros. The calculation is based on the assumption that perfect randomness exists in the derivation of a population of combinations and that the individual variates (atoms, in this case) are entirely independent of each other. Those conditions apply to molecules of a monatomic gas in the closed-container experiment used to illustrate the second law of thermodynamics (see Chapter 13). The argument from improbability is simply that odds of that order of magnitude effectively rule out random chance and can lead us to only one conclusion: the cell must reflect the work of a supernatural creator.

Applied to the origin of a cell of living matter from nonliving matter, however, an entirely different set of conditions must be stipulated. First, the basic components of living matter are atoms of various elements and groupings of elements into ions and simple molecules. These units of matter do not necessarily behave independently of one another when in close proximity. Because of positive and negative charges, chemically similar or unlike units may attract one another. They may have a strong tendency to join together by chemical bonds that are difficult to break. Take the example of the snowflake, or any crystalline solid forming from a solution or melt as energy is lost from the system. Crystals of marvelous orderliness arise from a prior disordered state --witness the perfect diamond or sapphire. Even creationists will agree that mineral crystallization is a natural process in conformity with laws of physics and chemistry, carried out without the aid of a supernatural agent. The calculated probability that the carbon atoms of a perfect 10-karat diamond could have taken their places by pure chance, in the absence of interatomic forces in exact order in the crystal lattice from an initial amorphous state as free carbon atoms, is so small that we can only conclude (using the creationists' argument) that it was impossible and never happened.

Second, the creationists' argument based on extremely small probabilities of chance of assembly of a cell or other unit of living matter fails completely to include the possibility that the end result would come about by a succession of small steps, each one a relatively simple increase in complexity of form or function, and each new step being capable of maintaining itself with the available supplies of matter and energy.

Third, the creationists fail to take into account the enormous spans of time available for the steps to be accomplished. Creationists, you must remember, recognize no span of time longer than 10,000 years. They believe that God created not only all matter and energy, but also time itself. Time did not exist before the creation. This is the absolute truth, revealed in the Bible, and for the creationists it rules out the possibility that there could have been, say, 200,000,000 years available for nonliving molecules to advance by small steps to the level of complexity of a simple cell capable of replicating itself.

In the next chapter, we review the history of modern scientific thought as to possible ways in which life could have arisen in a naturalistic manner from nonliving matter.

Chapter 54

Biochemical Speculations on Life's Beginning

Before launching into our biochemical speculations about a naturalistic beginning of life on earth, perhaps the first thing we can do is to establish the requirements for a very simple form of life that can serve as the precursor of the simplest life forms in existence today. The minimum system must consist of organic molecules capable of synthesizing themselves and other organic molecules that make up the mass and structure of the organism and of carrying out endless replications of that same organism. The organism must be an open system of both energy and matter, with the apparatus needed to process and store energy and to expend the energy in constructing and degrading molecules as required to sustain the organism. It will be of no value to imagine an organism that uses different molecules and different chemical processes from those found in life today, for in that case we would simply postpone the explanation of what actually is.

A remarkable feature of life on earth today is that all cells of all living things are alike in using the DNA molecule as the carrier of genetic information and supplier of specific information on the assembly of chains of amino acids that constitute the various proteins, including the enzymes. All living cells use RNA to read the DNA code, transcribe it, and use it in building the protein molecules. All living cells produce adenosine triphosphate (ATP), a molecule capable of supplying the energy for cell functions. We should, therefore, go to the simplest of the one-celled organisms for the list of essential molecules whose origin must be traced by scientific hypothesis (even when it is largely a speculative statement) in steps from nonliving to living matter.

According to a leading modern classification of earthly life-forms developed by Robert H. Whittaker, there are five kingdoms: Monera, Protoctista, Fungi, Animalia (animals), and Plantae (plants) (Margulis and Schwartz, 1982). Of the five kingdoms of earthly life, two consist of one-celled individuals (Monera and Protoctista). Of these, the Monera, commonly called "bacteria," have the simpler form of cell, referred to in Chapter 34 as the prokaryotic cell. Thus, the members of the Monera are prokaryotes, whereas all members of the other four kingdoms are eukaryotes, having the eukaryotic type of cell. The prokaryotic cell is comparatively small and has a thin outer wall and membrane. The genetic material of the cell consists of a central mass of complex molecules (the nucleoid) including DNA, but it lacks the enclosing membrane that surrounds the nucleus, found in all prokaryotic cells. The prokaryotic cell reproduces by simple division (fission); it lacks the more complex apparatus and division process (mitosis) of cells with a true nucleus. Most prokaryote cells have a slender "tail," called a flagellum, capable of giving them mobility in a fluid medium.

According to the Whittaker five-kingdom system of phylogeny and the subdivision of phyla adopted by Margulis and Schwartz, the Monera consist of 16 phyla, with a total of more than 5000 species (1982). In looking over the descriptions of these phyla and the many genera they include, one cannot help but be impressed not only by the great diversity of form they show, but also especially by the varied internal chemistry systems by which they process nutrients and energy. That the phyla of the bacteria are very different among themselves must be highly significant from the standpoint of their earliest origins, for they could obviously adapt to widely different environmental conditions and nutrient sources. Some bacteria are anaerobic, living in an environment devoid of oxygen; some are aerobic, requiring oxygen. Autotrophic bacteria (autotrophs) synthesize organic molecules in one of two ways. One is by the use of chemical energy; these bacteria are chemautotrophs and they perform chemosynthesis. The other is by photosynthesis, using the energy of sunlight. Chemosynthesis is limited to one group of bacteria and is not carried out by members of the other kingdoms. (Photosynthesis is, of course, carried out by the green plants, which are also autotrophs for that reason.) One kind of bacteria can oxidize sulfides of nitrogen and manganese; another can deposit iron sulfides. While several of the phyla are autotrophs, others are heterotrophs; they must derive their sustenance from organic matter already produced by other organisms. One phylum of bacteria can take atmospheric nitrogen and "fix" it into nitrogen compounds useful to other organisms. Thus, the bacteria as a group both take in and give out all of the gases of the atmosphere, including nitrogen, oxygen, carbon dioxide and monoxide, hydrogen, methane, ammonia, nitrous oxide, and several gases that contain sulfur.

The Monera include one phylum in which we shall be particularly interested: the Cyanobacteria, or blue-green bacteria. The name "blue-green algae" has been widely used for this phylum, which was formerly placed in the plant kingdom. Correctly used, the term "algae" refers to certain phyla within the Protoctista. It is important to keep this usage in mind when reading about the evolutionary succession of the one-celled organisms in early geologic time. The cyanobacteria are photo-synthesizers, a function also found among the true algae and the plants.

The great diversity in chemical functioning within the Monera seems to suggest successful "experimentation" of the earliest bacteria in occupying a wide range of environments. What we are most interested in here is the apparatus they all share in common (and with all earthly life), which is the DNA/RNA combination capable of directing the production of all kinds of other molecules and of passing on that information, exact and fully intact, to following generations of individuals.

As with all life forms on earth, the bacteria carry out various chemical activities within the cell. These activities go under the general name of metabolism; it consists of synthesizing organic compounds from simpler, nonorganic molecules, and of breaking down organic molecules previously synthesized. Metabolism stores chemical energy within the cell when energy surpluses are available and releases that energy when energy sources are reduced or cut off. Metabolism also includes the production of molecules that make up the physical structure of the cell, causing the cell to grow in volume until it must subdivide into two separate cells. Biochemical activity requires the presence of catalysts, molecules that affect the rate at which the chemical reactions occur but do not themselves participate in the reaction. In organisms, the catalysts are enzymes. (See Chapter 34.)

Besides explaining how complex protein molecules arose, any hypothesis of origin of life must explain the coming into existence of the DNA and RNA molecules that hold the code for the production of protein chains. The extremely large DNA molecules take the form of paired chains of nucleotides coiled into a double helix, as explained in Chapter 34. The RNA molecules copy information from the DNA molecules and, in turn, serve as templates to program the synthesis of the protein molecules that are to serve as enzymes. The problem of a naturalistic origin of life is therefore reduced to explaining how proteins, RNA, and DNA came into existence. Only after this problem is solved, in the sense that a reasonable hypothesis is set up, can we proceed to the next step, that of the origin of the simplest cell.

Now that geneticists have tied in the older knowledge of genetics ("microscope knowledge") with modern molecular biology, we are better prepared to attack some issues in the debate between creation scientists and mainstream evolution scientists. The creationists have no scientific hypothesis of their own to debate, so it is up to the proponents of naturalistic evolution to send up their hypotheses like reconnaissance helicopters, to be shot at by the creationists. Creationists can call the shots as they please; they play the role of guerrillas, lying in wait concealed and opening fire when a kill looks easy.

The Oldest Fossils

Intensive searches for fossils of great age have been made by paleontologists in the oldest known rocks of the continental crust. Precambrian time is all of geologic time prior to the beginning of the Cambrian Period, which began the Phanerozoic Eon approximately 600 million years ago. We need to keep in mind that Precambrian time from the start of Archean time, for which a rock record exists, endured at least six times longer than the elapsed time from the start of the Cambrian Period to the present. The oldest crustal rocks have a whole-rock age of about 3.8 billion years (-3.8 b.y.). An older age (-4.2 b.y.) of individual minerals grains (zircons) within rocks of nearly that age was reported in 1983 from a locality in western Australia.

Microscopic structures that can be interpreted as showing actual living cells are found in rock dated about -3.5 b.y. These may have been primitive forms of the cyanobacteria, which are prokaryotes of the moneran kingdom. This interpretation is suggested by their association with layered, moundlike structures that resemble similar structures being produced today by cyanobacteria in shallow marine waters. The structures are called stromatolites, and they can be clearly identified going back in geologic time well into the Precambrian.

Fossil or fossil-like structures in rocks dating to -3.5 b.y. have been reported in two localities. One locality is in the Barberton area of Swaziland in southern Africa. Here, in 1975, Harvard paleontologist Elso S. Barghoorn collected fossils from the Archean Fig Tree formation, a siliceous rock (chert) dated at -3.5 b.y. The fossils are spherical bodies resembling modern prokaryotic cells; some of the bodies appear in contact in pairs, as if completing fission (Knoll and Barghoorn, 1977). Another find is from North Pole in Western Australia, not far from Port Hedland (Geotimes, vol. 25, no. 9, 1980). Microscopic chainlike filaments found there in association with stromatolites are taken to represent linked bacterial cells (Groves et al., 1981). Since then, more specimens have been found in the same region. A New Scientist news article (July 31, 1986) stated: "There is circumstantial

evidence to suggest that the microorganisms produced oxygen by photosynthesis. Morphologically, they are indistinguishable from modern oxygen-producing cyanobacteria." Claims that prokaryotes were in existence earlier than -3.5 b.y. are based on the finding of carbon deposits in sedimentary rocks of older ages. The presence of carbon suggests that life was present but is not direct evidence.

Younger, but less controversial microfossils of what appear to be bacterial cells have been found at a number of localities with rock ages of -3.2 to -2.7 b.y. (Schopf, 1973). It is generally agreed that the prokaryotes, as ancestral forms of the cyanobacteria, were the only life forms for a period of some 2.0 to 2.5 b.y. The first appearance of eukaryotic cells was about -1.0 b.y., or perhaps as early as -1.5 b.y. (Cloud, 1976; 1983).

We can expect a great deal of new evidence to appear as to the nature of the oldest life forms. Hypotheses of the origin of that life can be formulated on the tentative assumption that it was occurring in the time-frame of -3.6 to -3.5 b.y. The next question to raise is: What environmental conditions prevailed at that time? Was the earth's surface cold or hot? Did the earth have an atmosphere and ocean, and if so, what were their chemical composition?

Keeping in mind that answers offered now to these questions are highly speculative, but that they are constrained by some firm evidence from chemistry of crustal rocks and meteorites, a working statement is possible and has the form of a tentative consensus for much that has been written about the planetary environment in which life originated.

Early History of the Planetary Environment

Modern explanations of the origin of the planets of our Solar System favor the condensation hypothesis, described in Chapter 16. Recall that the planets grew within a contracting cloud of dust and gases that formed the solar nebula, from which the sun itself was being formed. Condensation of gases into liquid and solid particles occurred along with the clotting together of those particles. By numerous collisions the particles accumulated into larger objects, and these in turn joined together to form the planets, their satellites, and numerous smaller objects (asteroids, comets) orbiting the sun. There are many variations of the condensation hypothesis, and the details are strongly debated as to their merits and weaknesses.

By about -4.7 b.y. accretion of earth and the other planets was largely complete, although for hundreds of millions of years there was to continue the infall of solid objects of many sizes (as we infer from the surfaces of the Moon and Mars). It is thought that the accretion occurred very rapidly, nearly the entire accumulation being accomplished in only about 200 m.y. This seems like a very long time but is short in comparison with the four billion years of Precambrian time that was to follow. During the accretion process, the growing earth as a whole remained in the solid state but with local melting of parts of the outer rock layer from the intense heat of impact of large masses. If the earth had a primeval atmosphere when accretion was complete, it was rapidly eroded away and dispersed into outer space by the force of the solar wind of energetic particles that began to emanate from the sun as that star went radioactive and its surface became intensely hot. If that were the case, how did the earth obtain its present atmosphere and oceans?

Origin of the Atmosphere and Oceans

The earth's atmosphere and ocean consist largely of chemical substances of a general class called volatiles. These substances can remain in the gaseous or liquid state at temperatures prevailing widely over the earth's

Table 54.1   Volatiles of the Earth's Atmosphere and Hydrosphere Compared with Volcanic Gases.

	Volatiles of atmosphere and hydrosphere	Gases from basaltic lava of Hawaiian volcanoes
Water (H_2O)	92.8	57.8
Carbon (C), as CO_2	5.1	23.5
Chlorine (Cl_2)	1.7	0.1
Nitrogen (N_2)	0.24	5.7
Sulfur (S_2)	0.13	12.6
Hydrogen (H_2)	0.07	0.04
Argon, A	trace	0.3

Figures in table represent percentages by weight.

Data of W. W. Rubey, 1953, Bull., Geol. Soc .Amer., vol. 62, p. 1137.

surface. A list of the principal volatiles of the present atmosphere and oceans (including all other free water on the land surfaces and held in a shallow rock zone) is given in Table 54.1. The first thing to note about this list is that the elements it contains include the four most abundant nutrients of life on earth today: hydrogen, carbon, oxygen, and nitrogen. (Other essential nutrient elements are abundant in minerals found in rocks near the earth surface; they are not volatiles. See Table 34.1.)

There is general agreement among geologists and geochemists today that the volatiles of the atmosphere and oceans came from the earth's crust, largely through eruption of volcanoes, in a process called outgassing. Comparison of the volatiles listed above with those in samples of gas emanating from molten volcanic rock (magma) shows a certain similarity in relative amounts of each substance.

Outgassing may have been very rapid in the first billion years following earth accretion. It is reasonable to suppose that by about -3.9 b.y., extensive deep oceans existed and atmospheric storms were producing abundant precipitation, which fell upon barren land surfaces of exposed rock or decomposed rock particles (regolith).

It is thought that the primitive atmosphere was very rich in carbon dioxide (CO_2)--perhaps from 500 to 100 times greater in concentration than in today's atmosphere (Monastersky, 1986). Carbon dioxde gas readily dissolves in water to produce carbonic acid. Acidity of rainwater in the primitive atmosphere may therefore have been much higher than it is today under natural conditions. Acid in rainwater was reacting with minerals to release many kinds of ions, for example, ions of magnesium (Mg), iron (Fe), calcium (Ca), sodium (Na), and phosphorus (P). As these ions entered the oceans, they combined with ions of chlorine and sulfur, forming the "salts" of seawater.

Thus, the list of available nutrients of modern life was largely completed. But the chemical environment of the oceans was perhaps lacking in the one vital nutrient of life as we know it today--free oxygen, represented by the oxygen molecule, O_2. Not that oxygen was not present on the planet for, as an element, oxygen comprises about 30 percent of the entire earth, nearly all of it locked up in rock-forming mineral compounds. One hypothesis, heavily

supported by many scientists who speculate on the origin of life, is that the earth's atmosphere and ocean of -3.7 b.y. were almost totally deficient in free oxygen, and that the lack was to continue until about -2.0 b.y.

The Primitive Atmosphere

If free molecular oxygen (O_2) had been lacking in the atmosphere at -3.9 b.y., there could have been no ozone layer in the upper atmosphere. Ozone is a molecule consisting of three atoms of oxygen (O_3). The high-altitude ozone layer now absorbs much of the ultraviolet radiation of the sun's spectrum, allowing very little to reach the earth's surface. Ultraviolet radiation is severely damaging to living cells and would have been lethal to the earliest cells if they had been exposed to it. The consequences of the zero-oxygen postulate are that the earliest life did not make use of oxygen in its metabolism, and that it must have been restricted to environments protected from ultraviolet radiation, i.e., deep below the ocean surface or inside masses of mineral matter. The lack of free oxygen in the atmosphere of -3.9 b.y. has been challenged by mainstream scientists, and the issue is open to debate (Clemmey and Badham, 1982). If free oxygen had been present in substantial quantities, it would have tended to destroy by oxidation those unprotected primitive organic molecules. (See Chapter 32, "Spread of Life to the Lands," for a discussion of early atmospheric oxygen levels.)

The common, or typical, description of the atmosphere at about -3.9 b.y. reads that it was composed principally of water vapor (H_2O), carbon dioxide (CO_2), nitrogen (N_2), carbon monoxide (CO), hydrogen sulfide (H_2S), and hydrogen (H_2). Nitrogen can combine with hydrogen to form ammonia (NH_3), which is essential to the formation of many organic compounds. When carbon (C) is also included in forming molecules with nitrogen and hydrogen, there result compounds known as amines and amides. The group NH_2 (or NH) is an essential part of the structure of the amino acids that make up proteins, and of nucleotides of DNA and RNA. The intense ultraviolet radiation of the time could have provided the energy needed to form ammonia, which would thus have been available for synthesis into organic molecules. Methane (CH_4) is another compound often included in the list of early atmospheric components.

The Oparin-Haldane Theory of Life

In the 1920s, two scientists working independently came up with the first hypothesis specifically proposing the formation of organic molecules in the primitive atmosphere described above, through the application of natural sources of energy. One was a distinguished Soviet biochemist, academician Aleksandr I. Oparin. In 1924 he published a book titled The Origin of Life, outlining his version of the way in which naturally produced organic molecules led to the formation of proteinlike compounds (Oparin, 1968, p. 34). In 1929, J. B. S. Haldane, a British geneticist, published a short essay titled "The Origin of Life" (reprinted in Bernal, 1976, pp. 242-49). He wrote: "when ultra-violet light acts on a mixture of water, carbon dioxide, and ammonia, a vast variety of organic substances are made, including sugars and apparently some of the materials from which proteins are built up." He noted that this process had already been demonstrated in the laboratory of a British colleague. Haldane continued as follows:

> In this present world, such substances, if left about, decay--that is to say, they are destroyed by micro-organisms. But before the origin of life they must have accumulated till the primitive oceans reached the consistency of hot dilute soup. Today an organism must trust to luck, skill, or strength

to obtain its food. The first precursors of life found food available in considerable quantities, and had no competitors in the struggle for existence. As the primitive atmosphere contained little or no oxygen, they must have obtained the energy which they needed for growth by some other process than oxidation--in fact, by fermentation. For, as Pasteur put it, fermentation is life without oxygen. If this was so, we should expect that high organisms like ourselves would start life as anaerobic beings, just as we start as single cells. This is the case.

The first living or half-living things were probably large molecules synthesized under the influence of the Sun's radiation, and only capable of reproduction in the particularly favourable medium in which they originated. Each presumably required a variety of highly specialized molecules before it could reproduce itself, and it depended on chance for a supply of them. (From Bernal, 1968, pp. 246-47)

Haldane emphasized that his conclusions were speculative, and that they will remain so until living creatures have been synthesized in the laboratory. He knew the goal was far distant in time, which for him ran out a decade after the Crick-Watson discovery of the double helix had opened up a molecular view of life that not only greatly complicated the problem of origin of life, but also made possible the conception of detailed and specific hypotheses of the necessary steps in that scenario.

Similarities in the independent hypotheses of Oparin and Haldane led to their being joined under the name of the Oparin-Haldane theory of origin of life or, less formally, the "hot-soup theory."

In 1953, in order to test experimentally the hot-soup theory, Stanley L. Miller, then a graduate student working under the direction of Nobel chemist Harold C. Urey at the University of Chicago, set up an apparatus in which a mixture of methane, hydrogen, and ammonia was combined with water vapor produced by boiling water in a flask. The gaseous mixture was passed through a glass-tube circuit feeding into and out of a large flask fitted with two tungsten electrodes. As the gaseous mixture moved through the large flask, spark discharges were passed between the exposed electrodes. The water vapor then condensed and was caught in a trap at the base of the system. Miller allowed the experiment to run continuously for one week, after which the gases were pumped out and the liquid residue extracted for chemical analysis. Twenty new organic compounds were identified; they included several amino acids. Miller was surprised to find that "the major products were not themselves a random selection of organic compounds but included a surprising number of substances that occur in living organisms" (Miller and Orgel, 1974, p. 84).

Similar experiments, repeated many times and with many variations, have consistently yielded positive results, often with large yields of organic molecules. Of the 20 amino acids, 18 have been synthesized in this manner, along with some chains of the amino acids resembling protein molecules but, as yet, the nucleotides of DNA and RNA have not been obtained.

Energy for Life

In applying the results of these experiments to possible natural conditions in early earth history, sources of energy need to be considered. Assuming the atmospheric composition to have been as described earlier, by far the most abundant and reliable energy source would be the ultraviolet rays of the sun, free to reach the earth's surface in the absence of an ozone layer. Another possibility is that of geothermal energy, which is

heat at the earth's surface at points where recent volcanic activity has occurred, or that lie over deep masses of hot magma that has not reached the surface. Commonly, geothermal localities have hot springs, where boiling water (consisting almost entirely of cycled, heated rainwater) feeds surface pools. These must have been numerous throughout early earth history, but they occur as small, isolated areas. Another high-energy source that has been considered is from lightning discharges; they would also be highly localized.

The laboratory experiments have used closed systems, so far as the movement of matter is concerned, and this gives them little resemblance to natural geologic systems, which are open in terms of both energy and matter. Nevertheless, they tend to substantiate the general hypothesis that a large variety of small organic molecules can arise spontaneously from a small number of simple ingredients, given a suitable energy source.

An interesting new turn in thinking about biopoesis has followed on the discovery of unusual organisms on the deep ocean floor at points where jets of hot, mineral-ladened water issue from the floors of mid-oceanic rifts, where the crust is actively spreading apart. (This setting and its hydrothermal activity were described in our earlier section on ore deposits and the Flood, Chapter 25.) This environment is completely lacking in solar energy, but there is ample energy in the heat of the emerging sulfur-rich water at temperatures over 300 C. In the water jets is found a primitive type of methane-generating bacteria, the Archaebacteria, thought to have evolved early in Precambrian time. It seems probable that lithospheric plates were in motion over the earth before the end of the first billion years following earth accretion. Seafloor spreading would have been essential to the formation of deep, flat-floored ocean basins, while plate subduction would have been required to create the high-standing continents. If so, it is probable that the sites of hot water springs (hydrothermal activity) were abundantly available as early as -3.9 b.y. It has even been suggested that new life has been forming more or less continually, or at various intervals, at such localities through all geologic time following the development of the first deep ocean basins.

Creationists' Arguments against the Hot-Soup Hypothesis

Creation scientists point out that the primitive earth atmosphere postulated under the Oparin-Haldane scenario and generally favored by mainstream scientists was devised to satisfy the need of the hot-soup environment for synthesis of organic compounds. The postulate of no free oxygen is contrived, they say, through the need to protect the earliest organic molecules from being destroyed by oxidation. Duane Gish refers to dissenting opinions by mainstream scientists with respect to a no-oxygen atmosphere (1976a, p. ii). He cites a reference to an opinion by R. T. Brinkman that the action of ultraviolet solar radiation on molecules of water vapor would have been able to produce "a significant quantity of oxygen very early in the earth's history." Another reference is to a conviction expressed by C. F. Davidson to the effect there is no evidence for assuming that the primitive atmosphere was any different from that of today. (References are given in Gish's paper.) To this I add my own earlier reference to a recent paper (Clemmey and Badham, 1982) making the same protest and arguing for the presence of oxygen.

The ICR scientists were quick to pick up Clemmey and Badham's paper, which appeared in the March 1982 issue of Geology. Rushing into print with an issue of the Impact Series, dated July 1982, Steven A. Austin, Ph.D., ICR's graduate staff member in charge of their geology program, reviewed Clemmey and Badham's geological arguments suggesting the presence of oxygen in the early

Precambrian time. But the pro-oxygen argument, even if accepted as describing a more likely state of the primeval atmosphere, is by no means fatal to a scenario of naturalistic biopoesis. Even if substantial amounts of oxygen had been present, there could have existed oxygen-free (anoxic) environments in which organic molecules could have been preserved. One such environment would be in masses of water-saturated clays deposited in the tidal zone; another would be in muds of the deep, hydrothermal environments of seafloor spreading rifts. Today, with an atmospheric level of molecular oxygen at 21 percent, anoxic environments exist, inhabited by anaerobic bacteria.

Another argument put forward by Gish is that the assumption of the presence of methane and ammonia in the primitive atmosphere is untenable in the light of known facts (1982a, pp. ii-iii). Gish cites statements by Philip H. Abelson, a mainstream scientist who has worked on the problem of biopoesis, to the effect that there is no evidence that methane was present, and that the dissociation (breakdown) of ammonia by ultraviolet radiation would have prevented ammonia accumulation. While Abelson's point may be valid, it can be negative only if methane and ammonia were necessary atmospheric ingredients for the formation of organic compounds. Stanley Miller reported that, when running subsequent laboratory expriments with his sparking flask, "Amino acids are also produced if the original CH_4-NH_3-N_2 mixture is replaced by other reducing mixtures--for example, CO-N_2-H_2" (Miller and Orgel, 1974, p. 86). This latter mixture agrees perfectly with the primitive atmospheric composition postulated by Abelson.

Gish reviews Stanley Miller's 1953 experiment and later experiments of the same type (1976a, pp. iii-iv). About these findings, Gish states:

> The first thing that must be emphasized about these results is that while the production of these compounds is a vital necessity in any origin of life scheme, success at this stage is many orders of magnitude easier to achieve than success at the next stage, which would include arranging these subunits in the precise order required for biologically active proteins, DNA and RNA. Furthermore, bringing these large biologically active molecules together into a coordinated functional system required for a living cell is again many orders of magnitude more difficult and less likely. In other words, even if these results are accepted uncritically, they are trivial in view of the immensity of the overall problem. (P. iv)

The above paragraph might have come from a mainstream scientist showing a proper appreciation of the enormity of the problem of naturalistic biopoesis. I find fault only with the final sentence, which describes the experimental work as "trivial." The experimental results are by no means trivial. They are first steps, and they can be pursued to develop hypotheses about more advanced levels of molecular organization.

Gish continues with an argument fully respected by mainstream scientists, namely, that the Miller-type experimental apparatus is an artifact embodying technical features that do not resemble natural open systems. For example, the apparatus traps out the organic products, preventing them from being continuously destroyed by recycling through the spark chamber. Gish goes on to develop the argument that high environmental temperatures tend to destroy organic molecules, and that the rate of destruction would far exceed the rates of production even in the ocean. The argument in general may very well be sound, but it cannot be an excuse for rejecting the possibility of naturalistic biopoesis, for it categorically rules out all possibility that there might exist natural environments in which organic molecules,

once formed, could become sequestered and could accumulate. An example might be the marine tidal environment of a tidal flat, where the salt water of each flood tide mixes with an influx of colloidal organic and mineral particles brought in by freshwater streams. As flocculation takes place, the aggregates sink to the bottom, where they accumulate and are soon removed from exposure to the solar rays at low tide by burial. Here, organic molecules bound to clay crystal lattices might find refuge from the destructive environment in which they were created. This is only one suggestion to show that one must not accept a general negative assumption as grounds to dismiss the entire hypothesis. When the creationists dismiss naturalistic biopoesis on sweeping generalities about improbability of organic synthesis occurring in combination with high probability of destruction of those substances in a hostile environment, they not only exhibit a nonscientific attitude, but they are also coming close to stating their final clinching argument that, because naturalistic biopoesis was impossible, Divine Creation of life must, therefore, be true.

Prebiotic Structures

In the second of his series of three _Impact_ articles, Gish takes up a much-discussed hypothesis of origin of pre-life (prebiotic) structures that may have functioned somewhat in the manner of primitive cells (1976b). The hypothesis has been put forward by Sidney W. Fox, Director and Professor of the University of Miami's Institute for Molecular and Cellular Evolution in Coral Gables, Florida. He is known for his research into the determination of the amino acid sequences in proteins, and for his laboratory experiments in synthesizing amino acid chains (see Fox and Dose, 1977).

Fox's laboratory experiments are unusual in investigating the effects of heating dry amino acids at temperatures well above the boiling point of water, mixing the heated acids with hot water, and allowing the solution to cool. Upon cooling, there emerge in the solution microscopic spherical objects of almost perfectly circular outline, but of varying diameters; these are called _microspheres_ and they contain proteinlike materials (proteinoids). The microspheres (also called _coacervates_) have a superficial resemblance to bacterial cells, but, as geneticist G. Ledyard Stebbins points out, the majority of spheres show no chemical activity and only a few can carry out even weakly the function of enzymes (1982, p. 176). Most important, says Stebbins, they cannot replicate ordered sequences of amino acids, as RNA is capable of doing in the cell of a living organism. Stebbins concludes: "Each proteinlike sphere is a separate unit, doomed to exist for a short time and perish without leaving a trace" (p. 176).[1] Gish attacks Fox's suggestion that microspheres similar to those produced in the laboratory might have been the precursors of true cells (1976b). I gather that one thrust of Fox's argument seems to be that the material within the microsphere would be physically isolated from the environment, and this isolation might have allowed the contents to escape destruction. Otherwise, as Stebbins makes clear, such nonliving microspheres are simply a dead end.

Gish concentrates on the observation that Fox's synthetic products are random sequences of amino acids and could have no value as enzymes. Fox has already fared badly at the hands of the mainstream scientists, who are never hesitant to exert their prerogative as peers to criticize other researchers in the same field. This, of course, is the way science functions, and this is the exposure process that gives science its strength. For Gish to claim that Fox has a weak case for explaining biopoesis is merely to restate what the scientific community has already said, but it gives an opportunity to the creation scientists to imply that naturalistic biopoesis as a whole is an unsupported and untenable concept, because one particular hypothesis, among several important working hypotheses currently being promoted by their authors, proves vulnerable or inadequate.

In his third article on the origin of life, Gish selects another hypothesis based on laboratory production of a structure that might be regarded as a prebiotic cell (1976c). He is referring to the work of A. I. Oparin, whom we mentioned earlier as originating the hot-soup synthesis of organic molecules. Oparin and his colleagues were able to obtain microspheres (called coacervate droplets) consisting of macromolecules, including DNA and RNA. Those macromolecules were supplied as ingredients; they were not synthesized during the experiment. Oparin claimed that the droplets are open systems in equilibrium with the surrounding medium, and that they are capable of growth by taking in small organic molecules and joining them together in chains (polymerization) (1968, Chapter 5). However, nothing was observed that can be suggestive of a link between coacervate droplets and more advanced prebiotic and biotic structures. Like the Fox microspheres, the coacervate droplets are perhaps best viewed as just another dead end.

In dismissing Oparin's suggestion that coacervate droplets might have been the prebiotic stage that led to living matter, Gish is only repeating the conclusion of the mainstream scientists, as put very clearly by Stebbins (1982, p. 176). Gish can only return to exploitation of the low probabilities of obtaining complex macromolecules by pure chance through fully random sequential arrangements of a given number of independent variates. He flatly asserts that no mechanism has been proposed for ordered sequences of molecular units in enzymes, DNA, and RNA, nor could any such mechanism exist under natural conditions (1976c, p. ii). Contrary to that assertion, an ordering mechanism has been suggested, and to see what it might be, we return to the problem of the macromolecules.

Problem of the Macromolecules

The problem of origin of the giant molecules (macro-molecules) found in many proteins and in DNA and RNA (the nucleic acids) still lies largely in the realm of carefully considered speculation. Debate has occurred on the question of which came first: the DNA and RNA templates coding for the amino acid sequences of the protein molecules, or (b) the protein chains themselves. Or did they all evolve concurrently in gradual steps?

Suppose you were an anthropologist studying works of art of the Navajo Indians in the 1890s, and that you visited the rug weavers in their traditional surroundings. You carefully documented all details of the process of preparation and dyeing of the wool yarn, the construction and use of the looms, and the design of the patterns. You wrote all of this down and published it as book of instructions: "How to make Navajo rugs." Obviously you could not have written that book in total ignorance of the product. The art of rug weaving has accumulated and changed over centuries by a process of cultural evolution. True, the instructions were always there--passed down from generation to generation by word of mouth and visual observation--but each change in instructions was preceded by an innovation based on creative imagination and perfected by experimentation. So you might conclude that product evolution comes first; instructions for replication lag a bit behind. Perhaps a more useful conclusion is that the process of rug weaving observed by the anthropologist is extremely advanced in technology and incorporates a high level of information, none of which necessarily reveals anything about the origins of the process. Graham Cairns-Smith makes the point that even the most primitive, one-celled forms of life we observe today as nucleotide/protein enzyme systems are highly advanced products of evolution (1985, pp. 90-91).

He calls them "high-tech" forms. What we should be looking for, he says, are "low-tech" systems of great simplicity with which the first evolutionary steps were carried out, leading to the basic framework for life to emerge.

Geneticist G. Ledyard Stebbins states that life could not evolve until the appearance of protein chains capable of replication (1982, pp. 176-77). He then proceeds to tackle the question of which came first, the proteins or the nucleic acids, or whether they both developed in close relationship with each other. He states:

> Several geneticists have suggested that nucleic acids with informational content appeared first and that proteins were built around them. This is unlikely, however, because the possibility is low that a nucleic acid consisting of randomly arranged nucleotide units would code for a protein chain having a definite function. Moreover, the relationship between triplets of nucleotides and single amino acids that makes the present genetic code work is highly complex and is almost certainly the end result of a long course of evolution. More direct relationships between nucleotides and single amino acids that are sufficiently precise for information to be transferred have been suggested, but the structural relationships between these two kinds of molecules are such that a direct functional relationship is highly improbable. Finally, DNA molecules in modern living systems cannot replicate themselves without the aid of a battery of enzymes, and simpler methods of replication are difficult to imagine. Few biochemists believe that the original macromolecule was DNA or RNA. (P. 178)[1]

Stebbins finds more attractive the hypothesis that functional proteins appeared first. He gives three reasons. First, experiments show that amino acids can be polymerized (joined up into chains) more easily than nucleotides. Second, in the primordial "soup" amino acids of proteins were probably much more abundant than nucleotides (the building blocks of DNA and RNA). Third, there must have existed among the millions of short protein chains present some kinds that could serve as enzymes.

Professor Russell F. Doolittle, a biochemist whose special interests are in protein chemistry, molecular evolution, and the origin of life, observes that "the most puzzling problem remains the invention of coding whereby the information in a polynucleotide sequence (RNA) is transferred into a corresponding polypeptide (protein molecule)" (Doolittle, 1983, p. 90).[2] Despite Stebbins's preference for starting with the protein molecules, Doolittle would suggest we begin with the DNA or RNA molecules. He writes: "The first critical step is the formation of a few short strings of nucleotides, perhaps under the catalytic influence of certain metal ions" (p. 90).[2] Such nucleotide strings, or strands, he notes, "have an intrinsic ability to associate with complimentary strands." Once a few such strands "had formed in in some warm little primeval pool, self-assembly properties would come into play." Doolittle goes on to show that the recombination of two short strands that fit only along a short stretch of the length of each can lead to longer strands. This process he calls "bootstrapping." But Doolittle admonishes us: "It wasn't until protein manufacture was coupled to the information in poly-nucleotide sequences that the roots were struck for genuine living systems" (p. 93).[2] That coupling, he admits, "is certainly the most challenging problem in understanding the origin of life." The foregoing suggestions by Stebbins and Doolittle--one opting for starting with the proteins and the other for starting with polynucleotides--are really nothing more than statements of the problem itself.

Does this sound too much like a creationist complaining about the impossibility of naturalistic biopoesis? Duane Gish, in his article referred to earlier, develops the theme of difficulties of explaining the naturalistic development of macromolecules (1976a, pp. v-vi). He says that this is a problem dwarfing all earlier problems (p. vi). I'm sure that any mainstream scientist will agree that the problem of the macromolecules is orders of magnitude more difficult to solve than the synthesis of amino acids and small protein chains. Having given up hope of seeing any scientific progress in coping with the macromolecules, Gish closes his paper with a return to the argument from the second law of thermodynamics. He suggests that there is a "thermodynamic barrier" to polymerization (1976a, p. vii). Destruction and decay of the macromolecules is an inevitable response to the universal trend to increasing entropy and increasing disorder. A macromolecule, says Gish, is just like an automobile. It can run uphill as long as it has gas, but when the gas is gone, it runs back down to the bottom of the hill (p. vii). How, I ask, can a fundamentalist creationist liken God's marvelous creation of the DNA macromolecule to a Detroit lemon with a faulty gas gauge?

The replication of DNA and RNA requires a large supply of the four nucleotides in the surrounding medium, and the polymerization (chaining) of these free-floating nucleotides requires the assistance of catalysts. For nearly two decades after the discovery of DNA and RNA, all these catalysts were thought to be protein enzymes. The extraordinary accuracy of replication of DNA was also thought to require the participation of a group of replicating enzymes. Even the synthesis of the nucleotide itself from the units of which it is composed (phosphate unit, ribose, and amine) would require the intervention of enzymes. This essential role of enzymes suggests that highly complex, large protein molecules must have evolved at an earlier time. But protein molecules have no means of directly replicating themselves, and the probabilities of their forming spontaneously in vast numbers by chance encounters of amino acids in just the right order seems almost infinitely small.

The synthesis of proteins by ribosomal transfer of RNA also requires the intervention of a catalyst. If these catalysts are protein enzymes, they must have evolved earlier by chance as random sequences of amino acids that just happened to be correctly equipped to form specific catalytic functions on specific organic molecules--in this case amino acid polymerization--all in anticipation of an RNA that had not yet appeared. The problem took on a new twist with the discovery that the enzyme ribonuclease P, involved in the splicing of ribosomal RNA, consists largely of RNA. In other words, RNA has a self-splicing capability. In a report by Roger Lewin on this important development, published in Science, the closing paragraph reads as follows:

> For those interested in the origin of life, the existence of RNA catalysts offers an intriguing glimpse of a former, more primitive age when the full range of metabolic and genetic machinery had yet to evolve. If, as now seems certain, RNA molecules can perform a range of catalytic functions in addition to being carriers of information, the old origins conundrum of "protein before DNA or DNA before protein?" is mercifully eschewed. (1984d, p. 267)

The case for RNA evolving before both DNA and protein has strengthened substantially. Confirmation of the capacity of RNA to replicate itself in the absence of protein enzymes seems to have put RNA securely in the "driver's seat," so to speak. Roger Lewin reported: "For those who like to speculate about the beginnings of life, these new results strongly support the hypothesis that RNA, not DNA, could have been the start of it all" (1986,

p. 545). For those familiar with the three forms of RNA and their functions, Lewin's explanation may be meaningful: "The new data all come from the realm of RNA splicing, that is the removal of intervening sequences, or introns, from the precursors of various RNA molecules, including transfer RNA, messenger RNA, and ribosomal RNA" (1986, p. 545). Harvard biologist Walter Gilbert, writing in Nature, speculates as follows on the origin of the macromolecules:

> One can contemplate an RNA world, containing only RNA molecules that serve to catalyze the synthesis of themselves. . . . The first stage of evolution proceeds, then, by RNA molecules performing the catalytic activities necessary to assemble themselves from a nucleotide soup. The RNA molecules evolve in self-replicating patterns, using recombination and mutation to explore new functions and to adapt to new niches.
>
> At the next stage, RNA molecules began to synthesize proteins, first by developing RNA adapter molecules that can bind activated amino acids and then by arranging them according to an RNA template using other RNA molecules such as the RNA core of the ribosome. This process would make the first proteins, which would simply be better enzymes than their RNA counterparts. (Gilbert, 1986, p. 618)

Gilbert concludes his scenario by adding that DNA finally appeared on the scene, obtaining its genetic code by reverse transcription from the genetic DNA molecules. Since then, double-stranded DNA has provided a stable store of information capable of self-correction of errors, and also capable of mutation and recombination. RNA then became relegated to an intermediate role, which we explained in Chapter 34.

In looking back over this discussion of the origin of macromolecules, I sense that there is an enormous gap in both time and evolutionary change between the two scenarios we have entertained, namely, the early hot-soup period in which amino acids and small protein molecules could have arisen spontaneously, and the later "high-tech" period when a system of interacting all-organic macromolecules had evolved. With what speculations can we fill that void?

Crystals of Clay as Our Ancestors?

Chemist Graham Cairns-Smith of the University of Glasgow, mentioned earlier, suggests that we should not be looking to the "high-tech" macromolecular apparatus of the modern cell for explanations of the origin of life. He writes:

> Of course, it might still be that the first organisms were made from molecules similar to those in organisms today, but this should be seen as an assumption with no special warrant.
>
> There are indeed good reasons for doubt. They arise from that high-tech interdependent complexity of central biochemistry. The first organisms could not have been like this. They must have been "low tech" machines of the kind that are fairly easily put together and for which there are simple versions of work, more or less (spears, not machine guns). There is a difference in approach here that might lead you to suspect the first organisms would have been made differently from today's, and with different materials. (1985, p. 91)

Cairns-Smith gives a good example of what he means. If you were developing a hypothesis for the origin of the modern electronic pocket calculator, you would not search the debris of ancient civilizations for batteries, transistors, or microchips; you would be finding there, instead, the wooden beads of an abacus. Moreover, you would never find an abacus bead in the mechanism of your pocket calculator. By the same token, Cairns-Smith does not think amino acids should be our starting point: "We should doubt whether amino acids or any other of the now critical biochemicals would have been at all useful at the start."

To develop this concept further, Cairns-Smith considers the example of an ancient arch of stones, freestanding on a flat plain. How could it have come into existence? Not likely by a fortuitous occurrence such as stones being dropped from above and just happening to form an arch. More likely, the stones were laid down on a mound of other rocks (or of earth) and that mound subsequently removed from beneath the arch. In other words, the mound served as a scaffolding. My own analogy would be that of the "lost-wax" process used in castings of beautifully sculptured small objects of metal. The prototype is first carved in a mass of hard wax, which is then surrounded by a ceramic clay shell. Once the shell has hardened, it is heated, melting the wax, which pours out from a hole in the shell. We now have a perfect mold that can receive molten gold or silver, but to retrieve the casting we must break up and discard the mold. This analogy has some pertinent points about it. First the object originated in materials of lowly and common status--clay and wax--but these have disappeared without a trace, leaving only a rare and valuable object. Of course, I was careful to mention clay in this analogy, because clay is the lowly and common inorganic material to which Cairns-Smith turns in seeking the origin of life. Were crystals of clay our ancestors? Were they the first "genes"?

This new idea appeals to me as a geologist with some knowledge of common clays governing soil fertility. Chemically highly active because of the negative electrical charges their surfaces bear, colloidal particles of crystalline clay store and exchange the base cations that plants need as macronutrients. The crystals we are talking about are not pretty and they occur in unenticing situations, such as dark agricultural soils and in odorous tidal mudflats. They belong to a group called the clay minerals. A nice clean example of a clay mineral is pure kaolin, a white substance used in making the finest ceramics. Perhaps more important from the standpoint of possible connection to the earliest macromolecules are those clay minerals found in dark soils and muds. They are derived from common silicate minerals that make up igneous rocks, such as granite or basalt. During chemical weathering minerals like mica, for example, take up water into their crystal structure, which comes in the form of innumerable thin parallel plates, easily pried apart with a sharp knife. (The structure is known as perfect cleavage in one plane.) The weathering process yields individual thin plates or scales, each of which is a set of layers of atoms in strictly prescribed arrangements called lattice layers. The scales are of colloidal or smaller size and bear negative surface electrical charges capable of attracting and holding molecules of water and such positively charged ions as iron, magnesium, calcium, aluminum, and hydrogen.

Figure 54.1 has diagrams of the lattice-layer structures of mica and illite, the latter a common clay mineral (named for the University of Illinois, where its structure was first analyzed). These are ball-and-stick models in which individual atoms or atom groups are represented by balls; the sticks represent bonds. An edge-on view of illite shows sheets of bonded units. There are two sheets of silicon-oxygen tetrahedrons, the upper one being inverted (upside down) with respect to the lower one. Between the tetrahedral sheets is a sheet of octahedrons consisting of aluminum atoms bonded to either oxygen atoms or

hydroxyl (OH) groups. The oxygens also serve as points of the adjacent silicon-oxygen tetrahedrons. Thus, the complete layer consists of three sheets. Between layers is an interlayer space occupied by potassium ions of positive charge. These have weak bonds with the adjacent layers, resulting in easy penetration by aqueous solutions and easy separation of the layers.

The next point to consider is how a lattice-layer structure such as that of illite might carry information in the form of a code. In this case, the usual central silicon atom of each silicon-oxygen tetrahedron can be substituted for by an aluminum atom. Presence of the aluminum atom in this location gives the surface of the layer a negative charge. We have here the makings of a binary code system. Along a given line of tetrahedrons, aluminums can alternate with silicons in random sequences. Like dots and dashes in the Morse code, or "0" and "1" in the binary code used in computers, the sequence can hold information. The nice thing about the clay lattice structure (or any crystal lattice) is that the distances between atoms are exactly fixed. Barring defects in the lattice, the spaces between silicon/aluminum positions run in uniform strings of hundreds or thousands of linked tetrahedrons. This feature makes a wonderful template for holding and transferring information.

Many forms of crystal lattices are available in nature, and they are prone to bearing many kinds of structural defects, of which the random atom substitutions are only one type. Our next step is to consider that as new lattice layers are created from atoms and ions in solution, the new layer may pick up the same information sequence as occurs in the old layer. This form of copying has actually been observed in the case of formation of new lattice layers of the clay mineral smectite (Cairns-Smith, 1985, pp. 97-98). The copying is attributed to the control by charge density differences due to the alternations of silicons and aluminums in the tetrahedrons.

I think by now you know what I am leading up to. Is it possible that each lattice layer can act as a sort of inorganic DNA molecule, carying a genetic code and capable of exact replication? Are there such things as "inorganic genes"? Cairns-Smith leaps forward at this point in these words:

> Clearly there are further observations and experimental clarifications to be made of the big question: Do mineral crystal genes exist? At this point I can only answer "Quite possibly" and go on to the next question: Could mineral crystal genes evolve? The answer to this, it seems to me, is "Yes, they could hardly help it." (1985, p. 98)

And how might mineral crystal genes "evolve"? In the case of the illite lattice layers, we might suppose that a new layer would, for some reason that is accidental, depart from the information sequence of the older layer next to which it forms. Now a "mutation" has occurred in the clay "gene" and the new sequence is replicated, replacing the previous one. As if this is not enough of a brain-busting idea, a bigger one is yet to follow logically upon it: the mutant "gene" might possess better survival value than the previous "gene" and a whole new population of lattice layers would form. Here we have evolution by natural selection! In what respect might "survival" be enhanced by such a mutation? Perhaps the mutant "gene" is less subject to being dissolved by acids or better able to resist abrasion. Cairns-Smith elaborates on this topic and it makes fascinating reading (1985, p. 99). We must, however, get on with the story.

The next question is: How can clay "genes" give rise to genes made up of organic macromolecules such as the nucleic acids? The answer may lie in the ability of the clay "genes" to attract organic molecules, such as nucleotides and amino acids. There might conceivably be specific attractions for specific sites. Thus, the sequence

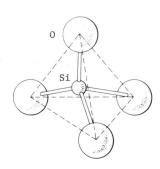

Expanded silicon–oxygen tetrahedron

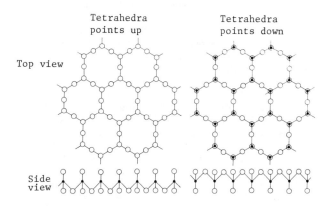

Sheets of silicon–oxygen tetrahedra in mica

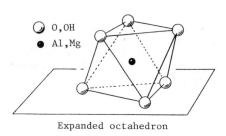

Expanded octahedron

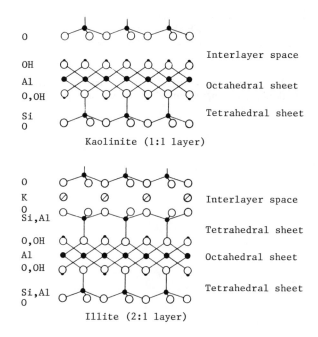

Figure 54.1 Lattice-layer structure of mica and illite. (Copyright © 1977, 1982 by Arthur N. Strahler.)

of nucleotides might always be the same for a given portion of the clay code. Binding of nucleotides into chains would be a requirement, requiring some sorts of catalysts, which might be metal ions adsorbed by the clay surface.

Also necessary at some stage would be the ability of the polymerized nucleotide to detach itself to become a free unit. There would be no problem of the availability of large numbers of identical clay "genes" (better called templates). From here on, the problems of evolution are not so difficult. The nucleic acid strand released from the clay surface could attract another strand, joining together as a DNA molecule. Bootstrapping could then allow the molecule to grow in length and complexity.

Essentially the same process could be visualized for the construction of protein chains. The clay template could serve as the factory for proteins, but this function would cease when it was taken over by DNA and RNA. Now the entire system has "gone biological" and life has taken off as a new and independent self-organizing and self-perpetuating system.

All of the above may sound great as speculation, but are there any observational or experimental data that suggest Cairns-Smith may be on the right track, at least? Indeed there are! (See Stebbins, 1982, pp. 181-82; Cairns-Smith, 1985, p. 100.) Scientists of NASA's Ames Research Center were able to achieve the binding of amino acids to clay particle surfaces. They found that when a

mixture of various kinds of amino acids included types not found in organisms, the clay crystals selected only those kinds used in the construction of proteins. (Amino acids come in two styles, expressed by the term "handedness." Only those that have the counterclockwise form of handedness are found in nature.) It was also found that metal cations--zinc and copper--exerted selective catalytic effects on the amino acids. Thus, the amino acids could be polymerized without the need of organic catalysts.

Building of amino acid chains and nucleic acids requires an input of energy. Apparently, clays are uniquely capable of storing and releasing energy. For example, it has been found that kaolinite is capable of absorbing energy of natural background radioactivity, storing this energy, and releasing it during such physical changes as wetting and drying (Cairnes-Smith, 1985, p. 100). To summarize, Stebbins writes:

> Thus experiments apparently show that prebiotic evolution was possible, that at its very beginning it already depended on interactions between evolving systems and their environment, and that various kinds of selection played a dominant role in making possible the sequence of events that led to life. We can expect that further investigations will reveal more about the nature of replication of prebiotic systems. (1982, p. 182)[2]

Mainstream science never closes the book on any problem of origins, no matter how bleak the outlook for finding evidence that can be brought to bear upon it. In the case of biopoesis, the closed doors have begun to open, revealing many surprises and new avenues of search. We close with this assessment by Professsor Doolittle:

> Many interesting problems concerning the origin of life remain to be explored, including the general area of bioenergetics and the encapsulation of systems by membranes. In both these areas we have reasonable scenarios that can serve as starting points for study. With this essay I have tried to show that the arguments that are raised about the improbability of the origin of life, particularly those concerning functional proteins, are often naive and misdirected. The forerunners of today's proteins were not formed from a random collection of amino acids any more than cells were the result of a simple aggregation of atoms. Life on earth developed in stages, each of which was built upon the stabilizing, catalytic, or replicative power of the stage before it.
>
> If I have made my point, the next time you hear creationists railing about the "impossibility" of making a particular protein, whether hemoglobin or ribonuclease or cytochrome c, you can smile wryly and know that they are nowhere near a consideration of the real issues. Comfort yourself also with the fact that a mere thirty years ago (before the Watson and Crick era) no one had the slightest inkling how proteins were genetically coded. Given the rapid rate of progress in our understanding of molecular biology, I have no doubt that satisfactory explanations of the problems posed here will soon be forthcoming. (1983, p. 96)[2]

Credits

1. From Darwin to DNA, Molecules to Humanity, by G. Ledyard Stebbins, W. H. Freeman and Company, New York. Copyright © 1982 by W. H. Freeman and Company. Used by permission.

2. From Russell F. Doolittle, 1983, Probability and the Origin of Life, pp. 85-97 in Scientists Confront Creationism, Laurie R. Godfrey, ed., W. W. Norton & Company, New York. Copyright © 1983 by Laurie R. Godfrey. Used by permission.

Summation and Verdict—
Creation Science Assessed

Now that we have examined every facet of creation science, from its origins and tenets through its arguments in favor of special recent creation and the cataclysmic events of the one-year worldwide Flood, it is appropriate to reflect on this question: Does creation science qualify for inclusion with mainstream science as one and the same field of human knowledge, or is it, on the other hand, pseudoscience? Inextricably woven into this question is another: Is creation science an expression of religion dealing with the supernatural realm rather than with the natural realm?

Early chapters in this book laid the foundation for assessing these questions. First, we examined the body of science, its methods and goals, and the nature of scientific hypotheses and theories. Second, we arrived at a clear distinction between science and religion. Third, we looked closely at pseudoscience, using examples that did not include creation science, but which enabled us to recognize specific features of pseudoscience. Chapter 8, to which we now refer for guidance, set down the characteristics of the scientific community, its practitioners, and its self-imposed guidelines. Most important was the establishment of the essential components of a cognitive field and the comparative referencing of both science and pseudoscience to those components, following the guidance of Professor Mario Bunge.

Recall, first, Bunge's contention that, whereas there exist numerous important and respected cognitive fields that are nonscientific--the belief fields, he calls them--the crucial point is that "any cognitive field that, though nonscientific, is advertised as scientific will be said to be pseudoscientific" (Bunge, 1984, p. 39).[1] To repeat a few lines from Chapter 8, a claim put forward by its adherents that a belief field is science is a fraudulent claim; it is a misrepresentation of the real nature of that belief field. Our charge, then, is that the fundamentalist creationist view of the universe, based on the literal interpretation of the book of Genesis and clearly a belief field, when presented by its adherents as if it were, instead, a research field under the name of "creation science," constitutes pseudoscience.

As explained in Chapter 8, Bunge describes and defines science in terms of twelve conditions, all of which must be satisfied. He further stipulates that if any cognitive field fails to satisfy all of these conditions, it should be judged as nonscientific. Bunge follows with a second list; it gives those conditions that will identify a cognitive field as pseudoscience. (At this point, you may wish to refer to the two lists of conditions given in Chapter 8.) It is to the second list that I refer here by number. Whereas in Chapter 8 each point was directed at the two historical scenarios of earth history used as examples (Velikovsky's Worlds in Collision and von Däniken's Chariots of the Gods), we now direct each point at creation science.

Point 1. [Little change occurs in the components in the course of time.] Keeping in mind that the single unique component of creation science is its set of tenets (listed in Chapter 11), that component has not changed in the slightest through the modern period of an organized body of creation scientists (Chapter 10). Rooted in a single book of the Old Testament and strictly limited to one specific English version of that religious document, the basic hypothesis of creation science--recent and rapid creation followed by a catastrophic Flood--cannot be modified in the light of any external observation of the real world. No change in the tenets of creation science can be induced from external (empirical) sources, because to make any change would be, in effect, an admission that the tenets were previously faulty, i.e., in error. If changed, a revised tenet of creation science would, in turn, become vulnerable to further change; hence, no tenet currently held could be accepted as eternally true. To avoid this logical self-destruct process, no change can be tolerated, no matter what empirical science comes up with in the way of new information.

Contrast this inflexible position of creation science with that of mainstream science, which has a set of tentative theories and hypotheses, all of them vulnerable to the impact of new information based on observation. Not only is forced change in those tentative models to be expected, it is to be diligently sought after. This single, glaring contrast between mainstream science and creation science is by itself sufficient reason to declare the latter to be pseudoscience.

Point 2. [There is no research community as such. Instead, the cognitive community consists of believers, who, although calling themselves scientists, conduct no scientific research.] Without question, creation scientists form a community of believers. First and foremost in their outlook is the inerrancy of the Scriptures. To become enrolled in its membership, all inductees of that community take a pledge of belief in the religious tenets of creation science. All this is freely stated in the literature disseminated by the Institute for Creation Research and allied organizations. One has only to read the ICR's series Impact and Acts and Facts or the Australian journal Ex Nihilo to find the repeated statements of religious belief that underpin the only acceptable view of origin of the universe and everything in it.

Turning to the matter of "research community" and "scientific research," members of the ICR stoutly maintain that they are, indeed, a scientific research community and that they do, in fact, carry on scientific research and publish the results of such research. Mainstream scientists will counter with the charge that the actual quantity of scientific research produced by all of the creation scientists is minuscule, that it fails to meet standards of the mainstream scientific community through peer review and open journal publication. The charge has merit. Serious research articles supporting the hypothesis of recent and sudden creation are almost totally lacking in the journals and treatises of mainstream science. Gentry's published papers on pleochroic halos may be an exception (Chapter 17).

In this connection, consider the following statement that appeared in the September, 1985, issue of Science 85:

> A three-year data-base search of 4,000 scientific publications--focusing on the names of people associated with the Institution for Creation Research and on phrases and keywords such as "creationism"--didn't turn up a single paper. A follow-up study of 68 journals found that only 18 of 135,000 total manuscript submissions concerned scientific creationism, and all 18 were rejected. Reasons cited included "flawed arguments," "ramblings," and a "high-school theme quality." (Vol. 6, no. 7, p. 11)

I frequently visit the library of the University of California in Santa Barbara. Perhaps it is typical of a major university library, and its science collection representative of the status of mainstream science, with emphasis on modern science. One has only to walk past aisle after aisle filled with science publications, numbering in the tens of thousands, to gain some measure of respect for the enormousness of accumulated scientific information open to peer review, dissent, and debate, to revision, and to the prospect of summary discard. Where, pray tell, is the library's collection of creation science publications? Certainly not in with the mainstream science collection. Instead, I discovered it filed with "religion" in a distant region of the same floor as science. I estimate this collection to occupy scarcely two shelves, each about one meter in length. Here, I found bound copies of the Quarterly Journal of the Creation Research Society and a small collection of miscellaneous creationist books. The core area of creation science items seems to merge at either end with purely religious materials. Does this enormous disparity in quantity of publications carry a message? I think it does. Although I have tried to acquire just about every recent publication on creation science I see advertised, my collection occupies only one shelf of a total of 60-odd shelves in my personal library, which is limited to science, highly selective, and continually weeded of obsolete or redundant materials.

The published literature of creation science, almost without exception, exists apart from the literature of mainstream science. Publication of articles and technical monographs on creation science appear almost exclusively in publications owned and controlled by creationist organizations sponsored by fundamentalist religious groups. Here the publications are sequestered from peer review by mainstream science. The articles often contain within the body of their texts references to religious dogma and religious goals. A single issue of the Australian journal Ex Nihilo, for example, is a mix of articles on religion and science, spanning a range in level from lessons for children to so-called "technical" papers. This mixture of overtly religious matter with scientific topics is a hallmark of the creation science educational publications offered to the general public.

As to the quality of "scientific" research conducted by creation scientists, it differs strikingly from that seen in mainstream science. First, it is directed entirely at the two functions: (a) attacking the so-called "evolutionary" view of the cosmos, and (b) supporting recency and rapidity of universal creation and the Flood catastrophe. The first function is entirely negative, or destructive, one in which the "research" activity is limited almost entirely to a search of the published research of mainstream science. Gleaned from this activity is a long list of alleged inconsistencies in the published or oral

statements and of disagreements among the mainstream scientists. As we have seen repeatedly, these supposed flaws of science are typically documented by out-of-context quotations--often using citations of statements long since discredited and superseded by new findings. Propositions and principles alleged to be held by mainstream scientists are shown to be untenable, using the fallacious straw-person argument, as in the oft-repeated allegation that geologists continue to adhere to an outmoded and discredited concept of uniformitarianism. Article after article in the creation science literature consists of this abuse of second-hand material in an attempt to discredit mainstream science. To call the production of such a literature "scientific research" is a travesty on the meaning of that term.

Creation scientists will claim that they have a positive program of field research looking for new evidence to support recent and rapid formation of geologic materials, such as petroleum and coal. In comparison with field research programs in mainstream geology and geophysics, these creationist projects are trivial in scope and almost devoid of substantive results. Another form of creationist field activity consists of observing the effects of catastrophic processes, such as volcanic eruptions, using such observations as arguments against their own version of the doctrine of uniformitarianism. Actually, the mainstream geologists are always interested in catastrophic processes, as witness the enormous scientific effort mounted on the occasion of the recent eruption of Mount St. Helens and the thousands of pages of scientific reports that have resulted. Creation scientists simply do not participate in mainstream science research, whether it relates to controversial questions of origin or not, nor do they produce independently a comparable body of scientific results. It seems fair to say that in terms of results, there is no research community as such in the cognitive field of creation science. Thus, on the second point alone, creation science is clearly recognized as pseudoscience.

Point 3. [The host society supports the cognitive community for practical reasons (because the cognitive field is good business) or tolerates it while relegating it beyond the border of its official culture.] The pertinence of this condition depends on what is meant by "host society." If the host society is the entire body of conservative, right-wing Christian fundamentalists (some 40 million strong), then creation science is indeed "good business" in aiding the broad frontal attack on nontheistic, naturalistic evolutionary science with its alleged evil effects on society. If the host society is the entire American public, tolerance in the sense of widespread indifference seems to prevail. What is our "official culture"? If, at the moment, it is the Reagan administration's view of religion, ethics, and morality, creation science is not only within the borders of that official culture, but also a respected member in the innermost circle of that culture. Perhaps we should just move on to another point.

Points 4 and 5. [The domain teems with unreal or at least not certifiably real entities; the general outlook includes either an ontology countenancing immaterial entities or processes or an epistemology making room for arguments from authority.] Both the domain of creation science and its general outlook deal in unreal or nonmaterial entities or processes, i.e., with the supernatural, which is the Judeo-Christian God as creator. While there is nothing for science to criticize about a religious belief in God's existence and creative role, the wedding of that religious belief with naturalistic science to produce an incongruous dualism, proffered as science, can only be regarded as pseudoscience.

Point 6. [The formal background is usually modest. Logic is not always respected.] As noted in Chapter 8, the formal background of logic and mathematics plays little part in delving into origins, which are in the realm of historical science. On the other hand, logic is certainly not respected by the creation scientists in their arguments against mainstream evolutionary science. Time and again we have seen the non-sequitur chain used in the creationist argument that reads: "Science is in disarray; therefore science is in error; therefore recent creation is the truth." Another of the egregious errors of logic is the claim that manifestation of design proves the existence of a designer. Judged solely on grounds of the abuse of logic in presenting its case against mainstream science, creation science qualifies admirably as pseudoscience.

Point 7. [The specific background is small or nil; a pseudoscience learns little or nothing from other cognitive fields. Likewise, it contributes little or nothing to the development of other cognitive fields.] Bunge's reference here is to borrowed, specific scientific laws, principles, and methods. Because the Creator had his own unique and mysterious ways of creating time, matter, and energy, the specific background of recent creation in terms of mainstream science is simply nil. By the same token, science cannot contribute to an understanding of the ways in which God created everything; that knowledge is forbidden to humans. Chalk up another clear indicator that creation science is pseudoscience.

Point 8. (Does not apply meaningfully to our analysis.)

Point 9. [The fund of knowledge is practically stagnant and contains numerous untestable or even false hypotheses in conflict with well-confirmed scientific hypotheses; it contains no universal and well-confirmed hypotheses.] The matter of stagnation has been fully covered under Point 1. Recent special creation and the catastrophic Flood are untestable with reference to the asserted causation, which is supernatural. The part that is testable is a simple statement that the universe with all its present configurations and behavioral laws came into existence suddenly (within six days) at a point in time 4000 years ago (more or less) and was followed some 1600 years later by a one-year worldwide marine inundation that saw the deposition of all fossiliferous rocks of the earth. I suppose we could regard that statement as a universal hypothesis, since it accounts for everything that needs to be accounted for. A scientific hypothesis does not need to explain the phenomenon it describes. The causation, or hypothesis that God did it all, cannot be tested, falsified, or verified. On the other hand, the postulated event itself can be probed by use of a set of deductions, serving as tests. Several of our chapters contain the results of such scientific testing of the assertion of recent and rapid origin of various terrestrial phenomena (for example, the limits to rates of accumulation of rhythmites, evaporites, and coral reefs). The diagnosis of creation science as pseudoscience rests here on the untestability (total inaccessibility) of the hypothesis of causation by a supernatural entity.

Point 10. (Does not apply to historical science.)

Point 11. (Not closely applicable to our discussion.)

Point 12. [No other field of knowledge, except possibly another pseudoscience, overlaps the stated cognitive field. This means that the field is isolated and free from control of other cognitive fields.] Certainly mainstream science does not overlap the religious doctrine of special recent creation by a supernatural being. In that respect the creationist tenet is completely isolated from all research fields.

Any one of the points we have reviewed can by itself serve to identify creation science as pseudoscience. Points 1 and 2, singly or together, provide positive identification of creation science as pseudoscience. That conclusion, then, is our clear verdict.

Pseudosciences, according to Professor Bunge, share with religions and political ideologies membership in the set of belief fields (1984, p. 38). Special recent creation finds a comfortable and secure place among the religions. On this matter Judge William Overton may have spoken for the majority of Americans in the text of his 1982 decision striking down the Arkansas Balanced Treatment Act (Act 590 of 1981). In Judge Overton's own words we find some fitting lines with which to close this book. He is here expanding upon his major conclusion to the effect that "the evidence is overwhelming that both the purpose and effect of Act 590 is the advancement of religion in the public schools" (Overton, 1982, p. 937). Overton repeats the definitions of "creation-science" and "evolution science" found in Section 4 of the act (p. 938). Let us first give the full text of that definition of "creation-science," bearing in mind that it was composed by the creationists. We do not need to concern ourselves with their definition of "evolution science," as it would not be acceptable to mainstream evolutionary scientists (nor was it acceptable to Judge Overton) (p. 938). The text of section 4a reads as follows:

> 4(a) "Creation-science" means the scientific evidences for creation and inferences from those scientific evidences. Creation-science includes the scientific evidences and related inferences that indicate: (1) Sudden creation of the universe, energy, and life from nothing; The insufficiency of mutation and natural selection in bringing about development of all living kinds from a single organism; (3) Changes only within fixed limits of originally created kinds of plants and animals; (4) Separate ancestry for man and apes; (5) Explanation of the earth's geology by catastrophism, including the occurrence of a worldwide flood; and (6) A relatively recent inception of the earth and living kinds.

Of the above definition of "creation-science," Overton says: "Section 4a is unquestionably a statement of religion, with the exception of 4(a)2, which is a negative thrust aimed at what creationists understand to be the theory of evolution." Overton continues:

> Both the concepts and wording of Section 4(a) convey an inescapable religiosity. Section 4(a) describes "sudden creation of the universe, energy, and life from nothing." Every theologian who testified, including the defense witnesses, expressed the opinion that the statement referred to a supernatural creation which was performed by God.

> The argument that creation from nothing in 4(a) does not involve a supernatural deity has no evidentiary or rational support. To the contrary, "creation out of nothing" is a concept unique to Western religions. In traditional Western religious thought, the conception of a creator of the world is a conception of God. Indeed, creation of the world "out of nothing" is the ultimate religious statement because God is the only actor.

> The facts that creation science is inspired by the Book of Genesis and that Section 4(a) is consistent with a literal interpretation of Genesis leave no doubt that a major effect of the Act is the advancement of particular religious beliefs. (P. 938)

At the very end of his text Overton states:

> The Court closes this opinion with a thought expressed eloquently by the great Justice Frankfurter: We renew our conviction that "we have staked the very existence of our country on the faith that complete separation between the state and religion is best for the state and best for religion. (Everson v. Board of Education, 330 U.S. at 59) (P. 942)

It seems, then, that the welfare of mainstream science depends on the strict enforcement of two great freedoms: Freedom OF Religion and Freedom FROM Religion. If either freedom were to be abridged, the other would also fall, and with their collapse the community of scientists would lose its freedom of choice--the choice of what to examine--and its freedom to express divergent opinions about the phenomena it observes.

Credit

1. From Mario Bunge, What is Pseudoscience? Skeptical Inquirer, vol. 9, no. 1, pp. 36-46. Copyright © 1984 by the Committee for the Scientific Investigation of Claims of the Paranormal. Used by permission of the author and publisher.

ADDENDUM

On June 19, 1987, the United States Supreme Court, by a 7-to-2 vote, held unconstitutional the 1981 Louisiana law, known as the Act for Balanced Treatment for Creation-Science and Evolution-Science in Public School Instruction (Creationism Act). Sponsored by state senator Bill Keith, the bill was easily passed by the legislature, and the law was upheld by the Louisiana Supreme Court. It was then remanded back to the U.S. District Court where, in 1985, Federal Judge Adrian Duplantier ruled the law unconstitutional on First Amendment grounds. Later in 1985, the U.S. Fifth Circuit Court of Appeals upheld Duplantier's ruling, an action that led the Louisiana attorney general to take the case to the U.S. Supreme Court. There, the law was defended by Wendell R. Bird, serving as special counsel for the state of Louisiana, and assailed by Jay Topkis of the American Civil Liberties Union, representing a consortium of Louisiana educators, parents and religious groups.

The Court's opinion, written by Justice William J. Brennan, Jr., noted that requiring schools to teach creation science with evolution does not advance academic freedom. He added: "Furthermore, the goal of basic 'fairness' is hardly furthered by the Act's discriminatory preference for the teaching of creation science against the teaching of evolution." Justice Brennan's text included the following paragraphs:

> The preeminent purpose of the Louisiana Legislature was clearly to advance the religious viewpoint that a supernatural being created humankind. The term "creation science" was defined as embracing this particular religious doctrine by those responsible for the passage of the Creationism Act.

> The legislative history documents that the Act's primary purpose was to change the science curriculum of public schools in order to provide persuasive advantage to a particular religious doctrine that rejects the factual basis of evolution in its entirety. The sponsor of the Creationism Act, Senator Keith, explained during the legislative hearings that his disdain for the theory of evolution resulted from the support that evolution supplied to views contrary to his own religious beliefs.

> In this case, the purpose of the Creationism Act was to restructure the science curriculum to conform with a particular religious viewpoint. Out of many possible science subjects taught in the public schools, the legislature chose to affect the teaching of the one scientific theory that historically has been opposed by certain religious sects.

> But because the primary purpose of the Creationism Act is to endorse a particular religious doctrine, the Act furthers religion in violation of the Establishment Clause.

> The Louisiana Creationism Act advances a religious doctrine by requiring either the banishment of the theory of evolution from public school classrooms or the presentation of a religious viewpoint that rejects evolution in its entirely. The Act violates the Establishment Clause of the First Amendment because it seeks to employ the symbolic and financial support of government to achieve a religious purpose. The judgment of the Court of Appeals therefore is affirmed.

The dissenting opinion was written by Justice Antonin Scalia, joined by Chief Justice William H. Rehnquist. The following review of the dissenting opinion is excerpted from a New York Times article (June 20, 1987) by Stuart Taylor, Jr.

> Justice Scalia, in his dissent, said the limited evidence before the Court meant that "we can only guess" at the meaning of the law but that it appeared not to "require the presentation of religious doctrine." Rather, he said, it was intended to foster the "academic freedom" of schoolchildren, by legislators who sincerely believed they were being "indoctrinated" in the false belief that "evolution is proven fact" and "that science has proven their religious beliefs false."

> These legislators believed, Justice Scalia said, that "the evidence for evolution is far less compelling than we have been led to believe," that "teachers have been brainwashed by an entrenched scientific establishment," and that "creation science" consists not of religious precepts but of "scientific data supporting the theory that life abruptly appeared on earth." While noting that he did not necessarily agree with these views, Justice Scalia said they were supported by affidavits of scientific and other experts with "impressive" academic credentials, which he reviewed in detail.

> The Court had no basis in the record of the case, Justice Scalia said, for assuming that there is no evidence to support the idea of creationism as science, that the scientific evidence for evolution is beyond challenge or that the state's scheme "will amount to no more than a presentation of the Book of Genesis." Justice Scalia said the fact that religious feeling helped propel the law through the legislature did not make it unconstitutional. "Political activism by the religiously motivated is part of our heritage," he said.

> More broadly, he attacked the Court's "embarrassing Establishment Clause jurisprudence" as a whole as inconsistent, unprincipled, and "fundamentally unsound."

Data sources: The New York Times, June 20, 1987; Colin Norman, 1987, Science, vol. 236, p. 1620.

References Cited

Acton, Rush K., 1978, Bone disease simulating ancient age in "pre-human" fossils. ICR Impact Series, no. 59, pp. i-iv.

Ager, Derek V., 1981, The Nature of the Stratigraphic Record. 2d ed., Macmillan Press, London, 122 pp.

Akridge, Russell, 1980, The sun is shrinking. ICR Impact Series, no. 82, pp. i-iv.

Albritton, Claude C., Jr., 1963, The Fabric of Geology. Addison-Wesley Publ. Co., Reading, Mass., 372 pp.

Alfven, Hannes, and Gustaf Arrhenius, 1976, Evolution of the Solar System. NASA SP-345, Nat. Space and Aeronautics Administration, Washington, D.C., 599 pp.

Allen, Benjamin Franklin, 1942, The geologic age of the Mississippi River. Bull., Deluge Soc. and Related Sciences, vol. 2, no. 2, pp. 37-62.

------ 1972, The geologic age of the Mississippi River. Creation Research Soc. Quarterly, vol. 9, no. 2, pp. 96-114.

Alvarez, Luis W., Walter Alvarez, Frank Asaro, and Helen V. Michel, 1980, Extraterrestrial cause for the Cretaceous-Tertiary extinction. Science, vol. 208, pp. 1095-1108. See also Letters and authors' reply, Science, vol. 211, pp. 648-56.

Anderson, R. Y., W. E. Dean, D. W. Kirkland, and H. I. Snyder, 1972, Permian Castile varved evaporite sequence, West Texas and New Mexico. Bull., Geological Soc. of America, vol. 83, pp. 59-86.

Amato, Ivan, 1987, Tics in the tocs of molecular clocks. Science News, vol. 131, pp. 74-75.

Andrews, Peter, 1984, The descent of man. New Scientist, vol.102, no.1408, pp. 24-25.

Angeles, Peter A., 1980, The Problem of God: A Short Introduction. Prometheus Books, Buffalo, N.Y., 156 pp.

Asimov, Isaac, 1977, Foreword. Pp. 7-15 in Goldsmith, 1977.

Atkins, Kenneth R., 1972, Physics--Once Over Lightly. John Wiley & Sons, Inc., New York, 370 pp.

Austin, Steven A., 1980, Origin of limestone caverns. ICR Impact Series, no. 79, pp. i-iv.

------ 1982, Did the early earth have a reducing atmosphere? ICR Impact Series, no. 109, pp. i-iv.

------ 1983, Did landscapes evolve? ICR Impact Series, no. 118, pp. i-iv.

------ 1984, Catastrophes in Earth History: A Source Book of Geologic Evidence, Speculation and Theory. ICR Technical Monograph 13. Institute for Creation Research, El Cajon, Calif., 318 pp.

Awbrey, Frank T., 1983, Space dust, the moon's surface, and the age of the cosmos. Creation/Evolution, vol. 4, no. 3, pp. 21-29.

Awbrey, Frank, and William M. Thwaites, 1981, Evolution Vs. Creation. Aztec Lecture Notes, San Diego State University (San Diego, Calif., 92182), 77 pp.

------ eds., 1984, Evolutionists Confront Creationists. Proceedings of the 63rd Annual Meeting of the Pacific Division, American Assoc. for the Advancement of Science, vol. 1, Part 3, San Francisco, 213 pp.

Ayala, Francisco J., 1977a, The genetic structure of populations. Pp. 20-56 in Dobzhansky et al., 1977.

------ 1977b, The origin of hereditary variation. Pp. 57-94 in Dobzhansky et al., 1977.

------ 1977c, Phylogenies and macromolecules. Pp. 262-313 in Dobzhansky et al., 1977.

------ 1977d, Philosophical issues. Pp. 474-516 in Dobzhansky et al., 1977.

Ayrton, Stephen, 1980, High fluid pressure, isothermal surfaces, and the initiation of nappe movement. Geology, vol. 8, pp. 172-74. See also reply on p. 406.

Bainbridge, William Sims, and Rodney Stark, 1981, Superstitions: old and new. Pp. 46-59 in Frazier, 1981.

Bambach, Richard K., 1983, Responses to creationism. Science, vol. 220, pp. 851-53.

Bambach, Richard K., C. R. Scotese, and A. M. Ziegler, 1980, Before Pangaea: The geographies of the Paleozoic world. American Scientist, vol. 68, pp. 26-38.

Barnes, F. A., 1975, The case of the bones in stone. Desert Magazine, vol. 38, no. 2, pp. 36-39.

Barnes, Thomas G., 1973, Origin and destiny of the earth's magnetic field. ICR Technical Monograph, no. 4, pp. 1-64.

------ 1981, Depletion of the earth's magnetic field. ICR Impact Series, no. 100, pp. i-iv.

------ 1982, Young age for the moon and earth. ICR Impact Series, no. 110, pp. i-iv.

------ 1983, The earth's magnetic age: the Achilles Heel of evolution. ICR Impact Series, no. 122, pp. i-iv.

Bauer, Henry H., 1981, Passions and purposes: A perspective. Pp. 409-12 in Frazier, 1981.

Bayly, Brian, 1968, Introduction to Petrology. Prentice-Hall, Englewood Cliffs, N.J., 317 pp.

Beaty, Chester B., 1978, The causes of glaciation. American Scientist, vol. 66, pp. 452-59.

Berggren, W. A., and John A. Van Couvering, eds., 1984, Catastrophes and Earth History, The New Uniformitarianism. Princeton Univ. Press, Princeton, N.J., 465 pp.

Bergman, Jerry, 1977, Albert Szent-Georgyi's theory of syntropy and creationism. ICR Impact Series, no. 54, pp. i-iv.

Berkner, L. V., and L. C. Marshall, 1964, Pp. 102-26 in P. J. Brancazio and A. G. W. Cameron, eds., 1964, The Origin and Evolution of the Atmosphere and Oceans. John Wiley & Sons, New York.

------ 1965, History of major atmospheric components. Pp. 1215-26 in Symposium on the Evolution of the Earth's Atmosphere, Proc., Nat. Acad. Sciences, vol. 53, no.6.

Bernal, J. D., 1967, The Origin of Life. Weidenfeld and Nicolson, London, 345 pp.

Biederman, Edwin W., Jr., Crude-oil composition and migration. Pp. 212-20 in Fairbridge and Bourgeois, 1978.

Bird, Roland T., 1939, Thunder in his footsteps. Natural History, vol. 43, pp. 255-61, 302.

Birdsell, J. B., 1975, Human Evolution. Rand McNally College Publ. Co., Chicago, 546 pp.

Bischof, N., 1978, On the phylogeny of human morality. Pp. 53-55 in Stent, 1978.

Blank, Richard G., and Stanley V. Margolis, 1975, Pliocene climatic and glacial history of Antarctica as revealed by southeast Indian Ocean deep-sea cores. Bull., Geological Soc. of America, vol. 86, pp. 1058-66.

Blitz, Leo, 1982, Giant molecular-cloud complexes in the galaxy. Scientific American, vol. 246, no. 4, pp. 84-94.

Blitz, Leo, Michel Fich, and Shrinivas Kulkarni, 1983, The new Milky Way. Science, vol. 220, pp. 1233-40.

Boardman, William, Robert F. Koontz, and Henry Morris, 1973, Science and Creation. Creation-Science Research Center, San Diego, Calif.

Bogert, Charles M., 1953, The tuatara: Why is it a lone survivor? Scientific Monthly, vol. 76, pp. 163-70.

Boulding, Kenneth E., 1984, Toward an evolutionary theology. Pp. 142-48 in Montagu, 1984.

Bower, Bruce, 1984, Fossils may clarify mammal evolution. Science News, vol. 126, p. 213.

------ 1985, A "mosaic ape" takes shape. Science News, vol. 127, pp. 26-27.

------ 1986, Fossil finds diversify ancient apes. Science News, vol. 130, p. 324.

Brace, C. Loring, 1983, Humans in time and space. Pp. 245-82 in Godfrey, 1983.

Bradley, J. P., D. E. Brownlee, and P. Fraundorf, 1984, Discovery of nuclear tracks in interplanetary dust. Science, vol. 226, pp. 1432-34.

Bradley, Wilmot H., 1929, The varves and climate of the Green River Epoch. U.S Geological Survey Professional Paper, no. 158-E, pp. 87-110.

------ 1948, Limnology and the Eocene lakes of the Rocky Mountain region. Bull., Geological. Soc. of America, vol. 59, pp. 635-48.

Braitsch, Otto, and David J. Kinsman, 1978, Marine evaporites--diagenesis and metamorphism. Pp. 464-68 in Fairbridge and Bourgeois, 1978.

Brandt, John C., ed., 1981, Comets. Readings from Scientific American, W. H. Freeman and Co., San Francisco, 92 pp.

Bridgman, Percy W., 1936, The Nature of Physical Theory. Princeton Univ. Press, Princeton, N.J., 138 pp.

Briggs, Louis I., 1978, Evaporite facies. Pp. 300-303 in Fairbridge and Bourgeois, 1978.

Britten, Roy J., 1986, Rates of DNA sequence evolution differ between taxonomic groups. Science, vol. 231, pp. 1393-98.

Broad, William J., 1981, Creationists limit scope of evolution case. Science, vol. 211, pp. 1331-32.

Broecker, Wallace S., 1966, Absolute dating and the astronomical theory of glaciation. Science, vol. 151, pp. 299-304.

Broecker, W. S., D. L. Thurber, J. Goddard, T. Ku, R. K. Matthews, and K. J. Mesolella, 1968, Milankovich hypothesis supported by precise dating of coral reefs and deep-sea sediments. Science, vol. 159, pp. 297-300.

Broecker, W. S., T. Takahashi, H. J. Simpson, and T. H. Peng, 1979, Fate of fossil fuel carbon dioxide and the global carbon budget. Science, vol. 206, pp. 409-18.

Brooks, C. E. P., 1949, Climate Through the Ages. McGraw-Hill Book Co., New York, 395 pp.

Brooks, W. K. 1984, The origin of the oldest fossils and the discovery of the bottom of the ocean. Jour. of Geology, vol. 2, pp. 455-79.

Brush, Stephen G., 1982, Finding the age of the earth by physics or by faith? Jour. of Geological Education, vol. 30, pp. 34-58.

------ 1983, Ghosts from the nineteenth century: creationist arguments for a young earth. Pp. 49-84 in Godfrey, 1983.

Bucha, V., 1971, pp. 57-117 in H. N. Michael and E. K. Ralph, 1971, Dating Techniques for the Archaeologist, MIT Press, Cambridge, Mass.

Bucher, Walter H., 1941, The nature of geological inquiry and the training required for it. Amer. Inst. of Mining and Metallurgical Engineers, Tech. Publ., no. 1377, 6 pp.

Bunge, Mario, 1983a, Exploring the world. D. Reidel, Dordrecht and Boston, 404 pp.

------ 1983b, Understanding the World. D. Reidel, Dordrecht and Boston, 296 pp.

------ 1984, What is pseudoscience? The Skeptical Inquirer, vol. 9, no. 1, pp. 36-46.

Burdick, Clifford L., 1973, Discovery of human skeletons in Cretaceous formation. Creation Research Society Quarterly, vol. 10, pp. 109-10.

Byers, Charles W., 1982, Stratigraphy--the fall of continuity. Jour. of Geological Education, vol. 30, pp. 215-21.

Cairns-Smith, A. Graham, 1971, The Life Puzzle: On Crystals and Organisms and on the Possibility of a Crystal As an Ancestor, Univ. of Toronto Press, Toronto, 165 pp.

------ 1985, The first organisms. Scientific American, vol. 252, no.6, pp. 90-100.

Campbell, Bernard G., 1974, Human Evolution: An Introduction to Man's Adaptations. Second Edition, Aldine Publishing Co., Chicago, 469 pp. (Third edition, 1985.)

Campbell, Marius R., 1914, The Glacier National Park. U.S. Geological Survey Bulletin, no. 600, pp. 1-54.

Carlston, Charles W., 1965, The relation of free meander geometry to stream discharge and its geomorphic implications. Amer. Jour. of Science, vol. 263, pp. 864-85.

Carroll, Robert C., 1981, Plesiosaur ancestors from the Upper Permian of Madagascar. Philosophical Transactions of the Royal Society of London, vol. 293, pp. 315-83.

Carrozi, Albert V., 1984, Glaciology and the Ice Age. Jour. of Geological Education, vol. 32, pp. 158-70.

Carter, William E., and Douglas S. Robertson, 1986, Studying the earth by very-long-baseline interferometer. Scientific American, vol. 255, no. 5, pp. 46-54.

Cartmill, Matt, David Pilbeam, and Glynn Isaac, 1986, One hundred years of paleoanthropology. American Scientist, vol. 74, pp. 410-20.

Caster, Kenneth E., 1984, Ediacaran fossils. Science, vol. 223, pp. 1129-30.

Chamberlin, Thomas C., 1904, The methods of the earth-sciences. Popular Science Monthly, vol. 66, pp. 66-75.

Chappell, John, 1974, Geology of coral terraces, Huon Peninsula, New Guinea; a study of Quaternary tectonic movements and sea-level changes. Bull., Geological Soc. of America, vol. 85, pp. 553-70.

Chappell, John, and H. A. Polach, 1976, Holocene sea-level change and coral-reef growth at Huon Peninsula, Papua, New Guinea. Bull., Geological Soc. of America, vol. 87, pp. 235-40.

Chappell, J., and H. H. Veeh, 1978, Late Quaternary tectonic movements and sea-level changes at Timor and Atauro Island. Bull., Geological Soc. of America, vol. 89, pp. 356-68.

Charig, A. J., F. Greenaway, A. C. Milner, C. A. Walker, and P. J. Whybrow, 1986, Archaeopteryx is not a forgery. Science, vol. 232, pp. 622-26.

Chave, K. E., S. V. Smith, and K. J. Roy, 1972, Calcium carbonate production by coral reefs. Marine Geology, vol. 12, no. 2, pp. 123-40.

Chen, A., 1983, Signs of first intergalactic cloud spotted. Science News, vol. 123, p. 148.

Cheney, Ralph W., 1948, Metasequoia discovery. American Scientist, vol. 36, pp. 490-94.

Christodoulidis, D. C., D. E. Smith, R. Kelenkiewicz, S. M. Klosko, and P. J. Dunn, 1985, Observing plate motions and deformations from satellite laser ranging, Jour. of Geophysical Research, vol. 90, pp. 9249-63.

Clemmey, Harry, and Nick Badham, 1982, Oxygen in the Precambrian atmosphere: an evaluation of the geologic evidence. Geology, vol. 10, pp. 141-46.

Cloud, Preston E., 1973, Possible stratotype sequences for the basal Paleozoic in North America. Amer. Jour. of Science, vol. 273, pp. 193-206.

------ 1976, Beginnings of biospheric evolution and their biogeochemical consequences. Paleobiology, vol. 2, pp. 351-87.

------ 1977, "Scientific creationism"--A new Inquisition brewing? The Humanist, vol. 37, no. 1, pp. 6-16.

------ 1983, The biosphere. Scientific American, vol. 249, no. 3, pp. 176-89.

Cloud, Preston, and Martin F. Glaessner, 1982, The Ediacarian Period and System: Metazoa inherit the earth. Science, vol. 218, pp. 783-92.

Cloud, Preston E., Jr., and C. A. Nelson, 1966, Phanerozoic-Cryptozoic and related transitions: New evidence. Science, vol. 154, pp. 766-70.

Colbert, Edwin H., 1949, Evolutionary growth rates in the dinosaurs. Scientific Monthly, vol. 69, p. 71.

------ 1965, The Age of Reptiles. W. W. Norton and Co., New York, 228 pp.

------ 1968, Men and Dinosaurs: The Search in Field and Laboratory. E. P. Dutton and Co., New York, 283 pp.

------ 1969, Evolution of the Vertebrates, 2d ed. John Wiley & Sons, New York, 535 pp.

------ 1980, Evolution of the Vertebrates, 3d ed. John Wiley & Sons, New York, 510 pp.

Cole, John R., 1981, Misquoted scientists respond. Creation/Evolution, vol. 2, no. 4, pp. 34-44.

Cole, John R. and Laurie R. Godfrey, eds., 1985, The Paluxy River footprint mystery--solved. Creation/Evolution, vol. 5, no. 1, pp. 1-56.

Cole, J. R., L. R. Godfrey, and S. D. Schafersman, 1985, Mantracks? The fossils say No!. Creation/Evolution, vol. 5, no. 1, pp. 37-45.

Connor, Carol Waite, 1985, Sixty-five volcanic events recorded in a single coal bed. Bull., American Assoc. of Petroleum Geologists, vol. 69, p. 246.

Conrad, Ernest C., 1981, Tripping over a trilobite: A study of the Meister tracks. Creation/Evolution, vol. 2, no. 4, pp. 30-33.

------ 1983, True vestigial structures in whales and dolphins. Creation/Evolution, vol. 3, no. 4, pp. 8-13.

Cook, Melvin A., 1968, Do radiological clocks need repair? Creation Research Soc. Quarterly, vol. 5, p. 70.

------ 1970, William J. Meister discovery of human footprint with trilobites in a Cambrian formation of western Utah. Pp. 185-86 in Walter E. Lammerts, ed., 1970, Why Not Creationism? Presbyterian and Reformed Publishing Co., Philadelphia.

Cooper, Bill, 1983, Human fossils from Noah's Flood. Ex Nihilo, vol. 5, no. 3 (International), pp. 6-9.

Cowen, Robert C., 1978, The cosmic cradle. Technology Review, vol. 80, no. 5, pp. 6-7, 19.

Cracraft, Joel, 1983, Systematics, comparative biology, and creationism. Pp. 163-91 in Godfrey, 1983.

------ 1984, The significance of the data of systematics and paleontology for the evolution-creationism controversy. Pp. 189-205 in Awbrey and Thwaites, 1984.

Crick, Francis, 1982, Life Itself: Its Origin and Nature. W. W. Norton, New York, 192 pp.

Crittenden, Max D., Jr., 1963, Effective viscosity of the earth derived from isostatic loading of Pleistocene Lake Bonneville. Jour. of Geophysical Research, vol. 68, pp. 5517-30.

Croll, James, 1875, Climate and Time in Their Geologic Relationships. A Theory of Secular Changes of the Earth's Climate. Daldy, Isbister, and Co., London, 577 pp.

Crompton, A. W., and Pamela Parker, 1978, Evolution of the mammalian masticatory apparatus. American Scientist, vol. 66, pp. 192-201

Cronin, Thomas M., 1985, Speciation and stasis in marine ostracoda: climatic modulation of evolution. Science, vol. 227, pp. 60-63.

Cumming, Kenneth B., 1980, Extinction. ICR Impact Series, no. 84, pp. i-iv.

Dalrymple, G. Brent, 1983, Can the earth be dated from decay of its magnetic field? Jour. of Geological Education, vol. 31, pp. 124-33.

------ 1984, How old is the earth? A reply to "scientific" creationism. Pp. 66-131 in Awbrey and Thwaites, 1984.

Daly, Reginald, 1972, The cause of the ice age. Creation Research Soc. Quarterly, vol. 9, no. 4, pp. 210-17.

Damon, Paul E., 1965, Correlation and chronology of ore deposits and volcanic rocks. Annual Progress Report, C00-689-50, Contract AT(11-1)-689. Research Division, U.S. Atomic Energy Commission: Arizona State University. Geochronology Labs, 157 pp.

Dansgaard, W., et al., 1982, A new Greenland deep ice core. Science, vol. 218, pp. 1273-77.

Darwin, Charles, 1859, The Origin of Species by Means of Natural Selection. John Murray, London. (Published by Penguin Books, 1968, reprinted 1986, with an introduction and bibliography by J. W. Burrow, 477 pp.)

Davidheiser, Bolton, 1979, Evolution and Christian Faith. Baker Book House, Grand Rapids, Mich.

Davies, J. T., 1973, The Scientific Approach. Academic Press, London and New York, 185 pp.

Davis, William Morris, 1899, The geographical cycle. Pp. 249-78 in W. M. Davis, 1909.

------ 1902, River terraces in New England. Pp. 514-86 in W. M. Davis, 1909.

------ 1909, Geographical Essays. Ginn and Co., Boston, 777 pp.

Dawkins, Richard, 1976, The Selfish Gene. Oxford Univ. Press, New York, 224 pp.

Deevey, Edward S., Jr., 1960, Population. Scientific American, vol. 203, no. 5, pp. 194-204.

Defant, Albert, 1961, Physical Oceanography, vol. 1. Pergamon Press, Oxford, 729 pp.

Degens, Egon T., and David A. Ross, 1976, Strata-bound metalliferous deposits found in or near active rifts. Pp. 165-202 in K. H. Wolf, ed., 1976, Handbook of Strata-Bound Ore Deposits, vol. 4, Elsevier, Amsterdam, 429 pp.

DeYoung, Donald B., 1979, The moon: A faithful witness in the sky. ICR Impact Series, no. 68, pp. i-iv.

Dietz, Robert S., 1983, Gish's Law. Geotimes, vol. 28, no. 8, pp. 11-12.

Dillon, William P., and Robert N. Oldale, 1978, Late Quaternary sea-level curve: reinterpretation based on glaciotectonic influence. Geology, vol. 6, pp. 56-60.

Dillow, J. C., 1981, The Waters Above. Moody Press, 479 pp.

------ 1983, The vertical temperature structure of the pre-flood vapor canopy. Creation Research Soc. Quarterly, vol. 20, pp. 7-14.

Dobzhansky, Theodosius, 1955, Evolution, Genetics, and Man. John Wiley & Sons, New York, 398 pp.

------ 1962, Mankind Evolving. Yale Univ. Press, New Haven, 381 pp.

------ 1967, The Biology of Ultimate Concern. New American Library, New York, 152 pp.

Dobzhansky, T., F. J. Ayala, G. L. Stebbins, and J. W. Valentine, 1977, Evolution. W. H. Freeman & Co., San Francisco, 572 pp.

Dohnanyi, J. S., 1972, Interplanetary objects in review: Statistics of their masses and dynanics. Icarus, vol. 17, pp. 1-48.

Donn, William L., and David M. Shaw, 1977, Model of climate evolution based on continental drift and polar wandering. Bull., Geological Soc. of America, vol. 88, pp. 390-96.

Doolittle, Russell F., 1983, Probability and the origin of life. Pp. 85-97 in Godfrey, 1983.

------ 1984, Some rebutting comments to creationist views on the origin of life. Pp. 153-63 in Awbrey and Thwaites, 1984.

Dorn, Ronald I., 1983, Cation-ratio dating: a new rock varnish age-determination technique. Quaternary Research, vol. 20, pp. 49-73.

Dorn, R. I., and T. M. Oberlander, 1981, Microbial origin of desert varnish. Science, vol. 213, pp. 1245-47.

------ 1982, Rock varnish. Progress in Physical Geography, vol. 6, no. 3, pp. 317-66.

Dorn, Ronald I., and David S. Whitley, 1984, Chronometric and relative age determination of petroglyphs in the western United States. Annals, Assoc. of American Geographers, vol. 74, pp. 308-22.

Dorn, Ronald I., et al., 1986, Cation-ratio and accelerator radiocarbon dating of rock varnish on Mojave artifacts and landforms. Science, vol. 231, pp. 830-33.

Dudley, H. D., 1976, The Morality of Nuclear Planning? Kronos Press, Glassboro, N.J.

Dunbar, Carl O., and John Rodgers, 1957, Principles of Stratigraphy. John Wiley & Sons, New York, 356 pp.

Dunbar, Carl O., and Karl M. Waage, 1969, Historical Geology, 3d ed. John Wiley & Sons, New York, 556 pp.

Dunham, D. W., S. Sofia, A. D. Fiala, D. Herald, and P. M. Muller, 1980, Observations of a probable change in the solar radius between 1715 and 1979. Science, vol. 210, pp. 1243-44.

Dury, George H., 1958, Tests of a general theory of misfit streams. Inst. of British Geographers, Trans. and Papers, Publication No. 25, pp. 105-18.

------ 1960, Misfit streams: problems in interpretation, discharge, and distribution. The Geographical Review, vol. 50, pp. 221-42.

------ 1965, Theoretical implications of underfit streams. U.S. Geological Survey Professional Paper, no. 542-C.

------ 1980, Neocatastrophism? A further look. Progress in Physical Geography, vol. 4, no. 3, pp. 391-413.

Dutch, Steven I., 1982a, Notes on the nature of fringe science. Jour. of Geological Education, vol. 30, pp. 6-13.

------ 1982b, A critique of creationist cosmology. Jour. of Geological Education, vol. 30, pp. 27-33.

Dutton, Clarence E., 1882, Tertiary History of the Grand Canyon District. U.S. Geological Survey Monographs, vol. 2, Government Printing Office, Washington, D.C., 264 pp.

Eberhart, Jonathan, 1976, Of life and death and magnetism. Science News, vol. 109, p. 204.

Ebert, James D., Chairman, Committee on Science and Creationism, 1984, Science and Creationism: A View from the National Academy of Sciences. National Academy Press, Washington, D.C., 28 pp.

Eddy, John A., and Aram A. Boornazian, 1979, Secular decrease in solar diameter, 1863-1953 (Abstract). Bull. American Astronomical Soc., vol. 11, p. 437.

Edmond, John M., and Karen Von Damm, 1983, Hot springs on the ocean floor. Scientific American, vol. 248, no. 4, pp. 78-93.

Edwards, Frederick, 1982, The dilemma of the horned dinosaurs. Creation/Evolution, vol. 3, no. 3, pp. 1-11.

------ 1983, An answer to Dr. Geisler--from the perspective of philosophy. Creation/Evolution, vol. 4, no. 3, pp. 6-12.

------ 1983a, Those amazing animals: the whales and dolphins. Creation/Evolution, vol. 3, no. 4, pp. 1-7.

------ 1983b, Searching for Noah's Ark. The Humanist, vol. 43, no. 6, p. 35.

Ehrlich, Paul, and L. C. Birch, 1967, Evolutionary history and population biology. Nature, vol. 214, pp. 349-52.

Eicher, Don L., 1976, Geologic Time, 2d ed. Prentice-Hall, Englewood Cliffs, N.J., 150 pp.

Eldredge, Niles, and Stephen J. Gould, 1972, Punctuated equilibria: an alternative to phyletic gradualism. Pp. 82-115 in T. J. M. Schopf, ed., 1972, Models in Paleobiology. Freeman, Cooper and Co., San Francisco, 250 pp.

Eldredge, Niles, and Steven M. Stanley, eds., 1984, Living Fossils. Springer-Verlag. New York, 291 pp.

Emiliani, Cesare, 1980, Death and renovation at the end of the Mesozoic. EOS, vol. 61, no. 1, pp. 505-6.

Ericson, David B., and Geosta Wollin, 1968, Pleistocene climates and chronology in deep-sea sediments. Science, vol. 162, pp. 1227-34.

Fackerell, Edward, 1984, The age of the astronomical universe. Ex Nihilo Technical Jour., vol. 1, pp. 87-94.

Fair, Charles, 1974, The New Nonsense: The End of Rational Consensus. Simon and Schuster, New York, 287 pp.

Fairbridge, R. W., and J. Bourgeois, eds., 1978, The Encyclopedia of Sedimentology. Dowden, Hutchinson, and Ross, Stroudsburg, Pa., 901 pp.

Farrand, William R., 1961, Frozen mammoths and modern geology. Science, vol. 133, pp. 729-35.

------ 1962, Postglacial rebound in North America. American Jour. of Science, vol. 260, pp. 181-98.

Fassett, James E., and J. S. Hinds, 1971, Geology and fuel resources of the Fruitland Formation and Kirtland Shale of the San Juan Basin, New Mexico and Colorado. U.S. Geological Survey Professional Paper, no. 676, Washington, D.C., 76 pp.

Faul, Henry, and Carol Faul, 1983, It Began with a Stone: A History of Geology from The Stone Age to the Age of Plate Tectonics. John Wiley & Sons, New York, 270 pp.

Feigl, Herbert, 1953, The scientific outlook: naturalism and humanism. Pp. 8-18 in Feigl and Brodbeck, 1953. (First published in American Quarterly, vol. 1, 1949.)

------ 1969, Ethics, religion, and scientific humanism. Pp. 48-64 in Kurtz, 1969.

Feigl, H., and M. Brodbeck, 1953, Readings in the Philosophy of Science. Appleton-Century-Crofts, New York, 811 pp.

Fenn, John B., 1982, Engines, Energy, and Entropy. W. H. Freeman and Co., New York, 293 pp.

Fenneman, Nevin M., 1931, Physiography of the Western United States. McGraw-Hill Book Co., New York, 534 pp.

Ferm, Vergilius, 1936, First Adventures in Philosophy. Charles Scribner's Sons, New York, 548 pp.

Fezer, Karl D., 1984, (Editor's comments.) Creation/Evolution Newsletter, vol. 4, no. 6, pp. 4-5.

Films for Christ Association, 1986, Footprints in stone: The current situation. Origins Research, vol. 9, no. 1, p. 15.

Finlow-Bates, T., 1979, Cyclicity in the lead-zinc-silver-bearing sediments at Mount Isa mine, Queensland, Australia, and rates of sulfide accumulation. Economic Geology, vol. 74, pp. 1408-19.

Fisher, D., 1969, Dating the spreading sea floor. New Scientist, vol. 44, pp. 185-87.

Fisher, Lloyd W., 1934, Growth of stalactites. American Mineralogist, vol. 19, pp. 429-31.

Fisk, Harold N., 1944, Geological Investigation of the Alluvial Valley of the Lower Mississippi River. Mississippi River Commission, Vicksburg, Miss., 78 pp.

Fisk, Harold N., and E. McFarlan, Jr., 1955, Late Quaternary deltaic deposits of the Mississippi River. Geological Soc. of America Special Paper, no. 62, pp. 279-302.

Fitch, W. M., and E. Margoliash, 1967, Construction of phylogenetic trees. Science, vol. 155, pp. 279-84.

Flint, Richard F., 1957, Glacial and Pleistocene Geology. John Wiley & Sons, New York, 553 pp.

------ 1971, Glacial and Quaternary Geology. John Wiley & Sons, New York, 892 pp.

Fox, Sidney W., and Klaus Dose, 1977, Molecular Evolution and the Origin of Life. Marcel Dekker, Inc., New York and Basel, 370 pp.

Frank, Philipp, 1946, Foundations of Physics. International Encyclopedia of Unified Science, vol. 1, no. 7, The University of Chicago Press, 78 pp.

Freedman, L. Z., and A. Roe, 1958, Evolution and human behavior. Pp. 455-79 in A. Roe and G. G. Simpson, Behavior in Evolution. Yale Univ. Press, New Haven, 557 pp.

Freske, Stanley, 1980, Evidence supporting a great age for the universe. Creation/Evolution, vol. 1, no. 2, pp. 34-39.

------ 1981, Creationist misunderstanding, misrepresentation, and misuse of the Second Law of Thermodynamics. Creation/Evolution, vol. 2, no. 2, pp. 8-16.

Fried, C., 1978, Biology and ethics: normative implications. Pp. 209-20 in Stent, 1978.

Friedman, Gerald M., and Sanders, John E., 1978, Principles of Sedimentation. John Wiley & Sons, New York, 972 pp.

Fritz, William J., 1980a, Reinterpretation of the depositional environment of the Yellowstone "fossil forests." Geology, vol. 8, pp. 309-13.

------ 1980b, Stumps transported and deposited upright by Mount St. Helens mud flows. Geology, vol. 8, pp. 586-88.

Funkhouser, J. G., and J. J. Naughton, 1968, Radiogenic helium and argon in ultramafic inclusions in Hawaii. Jour. of Geophysical Research, vol. 73, pp. 4601-7.

Gans, Carl, ed., 1969, Biology of the Reptilia, vol. 1, Morphology A. Academic Press, London and New York, 373 pp.

Gardner, Martin, 1957, Fads and Fallacies in the Name of Science. Dover Publications, New York, 363 pp.

Gardner, T. W., 1975, The history of part of the Colorado River and its rivers: an experimental study. Four Corners Geological Soc. Guidebook, 9th Field Conference, Canyonlands, pp. 87-95.

Garner, H. F., 1974, The Origin of Landscapes: Synthesis of Geomorphology. Oxford Univ. Press, New York, 734 pp.

Gascoyne, M., G. J. Benjamin, and H. P. Schwartz, 1979, Sea-level lowering during the Illinoian glaciation: evidence from a Bahama "blue hole." Science, vol. 205, pp. 806-8.

Geisler, Norman L., 1983, A scientific basis for creation; the principle of uniformity. Creation/Evolution, vol. 4, no. 3, pp. 1-6.

Gentry, Robert V., 1984, Radioactive halos in a radiochronological and cosmological perspective. Pp. 38-65 in Awbrey and Thwaites, 1984.

Gilbert, Grove K., 1890, Lake Bonneville. U.S. Geological Survey, Monograph 1, Government Printing Office, Washington, D.C., 438 pp.

Gilbert, Walter, 1986, The RNA world. Nature, vol. 319, February 20, p. 619.

Gingerich, Philip D., 1976, Paleontology and phylogeny: patterns of evolution of the species level in early Tertiary mammals. American Jour. of Science, vol. 276, no. 1, pp. 1-28.

------ 1980, Evolutionary patterns in early Cenozoic mammals. Annual Review of Earth and Planetary Sciences, vol. 8, pp. 407-24.

------ 1983, Evidence for evolution from the vertebrate fossil record. Jour. of Geological Education, vol. 31, pp. 140-44.

Gingerich, P. D., N. A. Wells, D. E. Russell, and S. M. Ibrahim Shah, 1983, Origin of whales in epicontinental remnant seas: new evidence from the early Eocene of Pakistan. Science, vol. 220, pp. 403-6.

Gish, Duane, 1974, The solar system--new discoveries produce new mysteries. ICR Impact Series, no. 15, pp. i-iv.

------ 1976a, Origin of life: critique of early stage chemical evolution theories. ICR Impact Series, no. 31, pp. i-iv.

------ 1976b, Origin of life; The Fox thermal model of the origin of life. ICR Impact Series, no. 33, pp. i-iv.

------ 1976c, The origin of life: Theories on the origin of biological order. ICR Impact Series, no. 37, pp. i-iv.

------ 1977, Dinosaurs, Those Terrible Lizards. Creation-Life Publishers, San Diego, Calif., 62 pp.

------ 1978a, Evolution: The Fossils Say No! Public School Edition, Creation-Life Publishers, San Diego, Calif., 189 pp.

------ 1978b, Thermodynamics and the origin of life (Part II). ICR Impact Series, no. 58, pp. i-iv.

------ 1980, The origin of mammals. ICR Impact Series, no. 87, pp. i-viii.

------ 1981, The mammal-like reptiles. ICR Impact Series, no. 102, pp. i-viii.

------ 1982, Letters to editor. Science 82, vol. 3, no. 1, p. 16.

------ 1983, Creating a missing link; a tale about a whale. ICR Impact Series, no. 123, pp. i-iv.

------ 1985, Evolution: The Challenge of the Fossil Record. Creation-Life Publishers, Master Books Division, El Cajon, Calif., 278 pp.

Glass, Bentley, 1957, New missing link discovered. Science, vol. 126, pp. 158-59.

Glass, Billy P., 1982, Introduction to Planetary Geology. Cambridge Univ. Press, Cambridge, 469 pp.

Godfrey, Laurie R., 1981, An analysis of the creationist film, Footprints in Stone. Creation/Evolution, vol. 2, no. 4, pp. 23-30.

------ ed., 1983, Scientists Confront Creationism. W. W. Norton and Co., New York, 324 pp.

------ 1983a, Creationism and gaps in the fossil record. Pp. 193-218 in Godfrey, 1983.

------ 1985, Foot notes of an anatomist. Creation/Evolution, vol. 5, no. 1, pp. 16-36.

Goldberg, Edward D., 1961, Chemistry in the oceans. Pp. 583-97 in Mary Sears, ed., Oceanography, Publ. No. 67, American Assoc. for the Advancement of Science, Washington, D.C., 654 pp.

Goldreich, Peter, 1972, Tides and the Earth-Moon system. Scientific American, vol. 226, no. 4, pp. 43-52.

Goldsmith, D., ed., 1977, Scientists Confront Velikovsky. Cornell Univ. Press, Ithaca, N.Y., 183 pp.

Goldstein, S. J., Jr., J .D. Trasco, and T. J. Ogburn III, 1973, On the velocity of light three centuries ago. Astronomical Jour., vol. 78, no. 1, pp. 122-25.

Gottlieb, L. D., 1973, Genetic differentiation, sympatric speciation, and the origin of a diploid species of Stephanomeria. American Jour. of Botany, vol. 60, pp. 545-53.

Gould, Stephen J., 1975, Catastrophes and steady-state earth. Natural History, vol. 84, no. 2, pp. 14-18.

------ 1977, Ever Since Darwin: Reflections in Natural History. W. W. Norton & Co., New York, 285 pp.

------ 1980, The Panda's Thumb: More Reflections in Natural History. W. W. Norton & Co., New York, 343 pp.

------ 1982, Darwinism and the expansion of evolutionary theory. Science, vol. 216, pp. 380-87.

------ 1983, Hens' Teeth and Horses' Toes. W. W. Norton and Co., New York, 413 pp.

------ 1987, Darwinism defined: the difference between fact and theory. Discover, vol. 8, no. 1, pp. 64-70.

Gretener, P. E., 1977, On the character of thrust faults with particular reference to the basal tongues. Bull., Canadian Petroleum Geology, vol. 25, pp. 110-22.

------ 1980, More on pore pressure and overthrusts. In Proceedings, Conference on Thrust and Nappe Tectonics, Blackwell Scientific Publishers, London and Oxford,   pp.

Gribbin, John, and Omar Sattaur, 1984, The school-children's eclipse. Science 84, vol. 5, no. 4, pp. 51-56.

Grootes, P. M., 1978, Carbon-14 time scale extended: comparison of chronologies. Science, vol. 200, pp. 11-21.

Groves, David I., John S. R. Dunlop, and Roger Buick, 1981, An early habitat of life. Scientific American, vol. 245, no. 4, pp. 64-73.

Guisti, Ennio V., 1978, Hydrogeology of the karst of Puerto Rico. U.S. Geological Survey Professional Paper, no. 1012, 68 pp.

Gutenberg, Beno, 1941, Changes in sea level, post-glacial uplift, and mobility of the earth's interior. Bull, Geological Soc. of America, vol, 52, pp. 721-72.

Guth, Alan H., and Paul J. Steinhardt, 1984, The inflationary universe. Scientific American, vol. 250, no. 5, pp. 116-28.

Guth, Peter L., L. V. Hodges, and J. H. Willemin, 1982, Limitations on the role of pore pressure in gravity sliding. Bull. Geological Soc. of America, vol. 93, p. 611.

Habgood, John, 1982, Evolution and the doctrine of creation. Insight, Wycliffe College, Toronto, Canada, no. 13, pp. 1-9.

Haldane, J. B. S., 1929, The origin of life. Pp. 242-49 in Bernal, 1967. (Originally published in The Rationalist Annual, 1929.)

Hall, J. M., and P. T. Robinson, 1979, Deep crustal drilling in the North Atlantic Ocean. Science, vol. 204, pp. 573-86.

Ham, Ken, 1984, What happened to the dinosaurs? Ex Nihilo, vol. 7, no. 2, pp. 6-11.

Hardin, Garrett, 1984, "Scientific creationism"--marketing deception as truth. Pp. 159-66 in Montagu, 1984.

Harrison, E.R., 1980, The paradox of the dark night sky. Mercury, vol. 9, no. 4, pp. 83-89.

Harwit, Martin, 1973, Astrophysical Concepts. John Wiley & Sons, New York, 561 pp.

Hastings, Ronnie J., 1985, Tracking those incredible creationists. Creation/Evolution, vol. 5, no. 1, pp. 5-15.

------ 1986, Tracking those incredible creationists--the trail continues. Creation/Evolution, vol. 6, no. 1, pp. 19-27.

------ 1987, "Creation physics" and the speed of light. (Unpublished manuscript.)

------ 1987a, New observations on Paluxy tracks confirm their dinosaurian origin. Jour. of Geological Education, vol. 35, pp. 4-15.

Hay, Richard L., and Mary Leakey, 1982, The fossil footprints of Laetolil. Scientific American, vol. 246, no. 4, pp. 50-57.

Hays, James D., 1971, Faunal extinctions and reversals of the earth's magnetic field. Bull., Geological Soc. of America, vol. 82, pp. 2433-47.

Hays, J. D., John Imbrie, and N. J. Shackleton, 1976, Variations in earth's orbit: pacemaker of the ice ages. Science, vol. 194, pp. 1121-32.

Heezen, Bruce C., and D. J. Fornari, (no year), Geologic map of the Pacific Ocean. World Geological Atlas, UNESCO, Paris, France.

Heiskanen, W. A., and F. A. Vening Meinesz, 1959, The Earth and Its Gravity Field. McGraw-Hill Book Co., New York, 470 pp.

Helwig, James, and Gerald A. Hall, 1974, Steady state trenches? Geology, vol. 2, pp. 309-16.

Hempel, C. G., and P. Oppenheim, 1953, The logic of explanation. Pp. 319-52 in Feigl and Brodbeck, 1953.

Henderson, G. H. and F. W. Sparks, 1939, A quantitative study of halos, 4, New types of halos. Proceedings of the Royal Society, A, pp. 173, 393-403.

Hendry, Allan, 1979, The UFO Handbook; A Guide to Investigating, Evaluating and Reporting UFO Sightings. Doubleday & Co., Garden City, N.Y., 297 pp.

Herbert, Wray, 1982a, Fossil raises question about earliest primates. Science News, vol. 121, p. 372.

----- 1982b, Was Lucy a climber? Dissenting views on ancient bones. Science News, vol. 122, p. 116.

------, 1984, The living link? Science News, vol. 125, p. 41.

Herman, Yvonne, and David M. Hopkins, 1980, Arctic Ocean climate in late Cenozoic time. Science, vol. 209, pp. 557-62.

Herrick, C. J., 1956, The Evolution of Human Nature. Harper & Bros., New York, 506 pp.

Hildemann, W. H., 1982, "Creative evolution." Letters to the Editor, Science, vol. 215, p. 1182.

Hilgard, E. W., 1869-1870, Report on the geologic age of the Mississippi River delta. Report of the U.S. Army Engineers, 1869-1870.

Hitch, Charles J., 1982, Dendrochronology and serendipity. American Scientist, vol. 70, pp. 300-305.

Hoblitt, R. P., D. R. Crandell, and D. R. Mullineaux, 1980, Mount St. Helens eruptive behavior during the past 1,500 yr. Geology, vol. 8, pp. 555-59.

Hoffstetter, Robert, and Jean-Pierre Gasc, 1969, Vertebrae and ribs of modern reptiles. Pp. 201-310 in Gans, 1969.

Holloway, Ralph L., 1974, The casts of fossil hominid brains. Scientific American, vol. 231, no. 1, pp. 106-15.

Holmes, Arthur, 1978, Principles of Geology, 3d ed., revised by Doris L. Holmes. John Wiley & Sons, New York, 730 pp.

Holmes, William H., 1878 (1883), Report on the Geology of the Yellowstone National Park. U.S. Geological and Geographical Survey of the Territories, Annual Report 12, Part 2, pp. 1-57.

Hooke, Robert, 1665, Micrographia. Martyn & Allestry, London. (Reprinted, 1961, by Dover Books, New York.)

Horie, Shoji, 1978, Lacustrine sedimentation. Pp. 421-27 in Fairbridge and Bourgeois, 1978.

Hospers, John, 1980, Law. Pp. 104-11 in Klemke, Hollinger, and Kline, 1980.

Howgate, Michael E., 1985, Archaeopteryx counterview. British Jour. of Photography, vol. 132, p. 348.

Hoyle, Fred, 1983, The Intelligent Universe. Michael Joseph, London, 256 pp.

Hoyle, F., Wichramasinghe, and R. S. Watkins, 1985, Archaeopteryx. British Jour. of Photography, vol. 132, pp. 693-694, 703.

Hsu, Kenneth J., 1972, When the Mediterranean dried up. Scientific American, vol. 227, no. 6, pp. 27-36.

------ 1978, When the Black Sea was drained. Scientific American, vol. 238, no. 5, pp. 53-63.

Hubbert, M. King, and W. W. Rubey, 1959, Role of fluid pressure in mechanics of overthrust faulting. Bull., Geological Soc. of America, vol. 70, pp. 115-66.

Hulse, Joseph H., and David Spurgeon, 1974, Triticale. Scientific American, vol. 231, no. 2, pp. 72-80.

Humphreys, A. A., 1869-1870, U.S. Engineers Report, 1869-1870.

Humphreys, A. A., and H. L. Abbott, 1876, Report on the physics and hydraulics of the Mississippi River. U.S. Army Corps of Engineers, Professional Paper no. 13, pp. 92-95.

Hunt, J. M., 1979, Petroleum Geochemistry and Geology. W. H. Freeman and Co., San Francisco, 617 pp.

Ingalls, Albert G., 1940, The Carboniferous mystery. Scientific American, vol. 162, January, p. 14.

Institute for Creation Research, 1981-1882, Graduate School Catalog. Institute for Creation Research, El Cajon, Calif., 48 pp.

Jackson, M. P. A., and S. J. Seni, 1983, Geometry and evolution of salt structures in a marginal rift basin of the Gulf of Mexico, east Texas. Geology, vol. 11, pp. 131-35.

Jacobs, J. A., 1963, The Earth's Core and Geomagnetism. Pergamon Press, the Macmillan Co., New York, 137 pp.

------ 1975, The Earth's Core. Academic Press, New York, London, 253 pp.

------ 1983, Reversals of the Earth's Magnetic Field. Adam Hilger, Ltd., Bristol, 230 pp.

Jastrow, Robert, 1980, Have astronomers found God? Reader's Digest, vol. 117 (699), pp. 49-53.

Jeuneman, Frederick B., 1972, Will the real monster please stand up? Industrial Research, September, p. 15.

Judson, Sheldon, 1968, Erosion of the land, or What's happening to our continents? American Scientist, vol. 56, pp. 356-74.

Jennings, J. N., 1971, Karst. The MIT Press, Cambridge, Mass., 252 pp.

Jukes, Thomas H., 1983, Molecular evidence for evolution. Pp. 117-38 in Godfrey, 1983.

Jukes, Thomas H., and W. Richard Holmquist, 1972, Evolutionary clock: nonconstancy of rate in different species. Science, vol. 177, pp. 530-32.

Kahn, Peter G. H., and Stephen M. Pompea, 1978, Nautiloid growth and dynamical evolution of the Earth-Moon system. Nature, vol. 275, no. 5681, pp. 606-11.

Kay, Marshall, and Edwin H. Colbert, 1965, Stratigraphy and Life History. John Wiley & Sons, New York, 736 pp.

Keith, M. L., 1971, Ocean-floor convergence: A contrary view of global tectonics. Jour. of Geology, vol. 80, pp. 249-76.

Kemp, T. S., 1982, Mammal-like Reptiles and the Origin of Mammals. Academic Press, London, 363 pp.

Kennett, James P., and Robert C. Thunnell, 1975, Global increase in Quaternary explosive volcanism. Science, vol. 187, pp. 497-503.

Kermack, K. A., Frances Mussett, and H. W. Rigney, 1973, The lower jaw of Morganucodon. Zoological Jour. of the Linnaean Soc., vol. 53, pp. 87-175.

Kerr, Richard A., 1978, Climate control: How large a role for orbital variations? Science, vol. 201, pp. 144-46.

------ 1978, Seismic reflection profiling: A new look at the deep crust. Science, vol. 199, pp. 672-74.

------ 1982a, New evidence fuels Antarctic ice debate. Science, vol. 216, pp. 973-74.

------ 1982b, Planetary rings explained and unexplained. Science, vol. 218, pp. 141-44.

------ 1982c, Where was the moon eons ago? Science, vol. 221, p. 1166.

------ 1984, Making the Moon from a big splash. Science, vol. 226, pp. 1060-61.

------ 1986, Plate tectonics is the key to the distant past. Science, vol. 234, pp. 670-72.

------ 1987, Milankovitch climate cycles through the ages. Science, vol. 235, pp. 973-94.

Kieth, M. S., and G. M. Anderson, 1963, Radiocarbon dating: fictitious results with mollusk shells. Science, vol. 141, p. 634.

King, Philip B., 1929, Corrosion and corrasion on Barton Creek, Texas. Jour. of Geology, vol. 35, pp. 631-38.

Klemke, E. D., Robert Hollinger, and A. David Kline, eds., 1980, Introductory Readings in the Philosophy of Science, Prometheus Books, Buffalo, N.Y., 373 pp.

Kitcher, Philip, 1982, Abusing Science: The Case Against Creationism. The MIT Press, Cambridge, Mass., 213 pp.

Klass, Philip J., 1968, UFOs--Identified. Random House, New York, 290 pp.

Klotz, J. W., 1970, Genes, Genesis, and Evolution. Concordia Publishing Co., St. Louis, Mo.

Knaub, Clete, and Gary Parker, 1982, Molecular evolution? ICR Impact Series, no. 114, pp. i-iv.

Knoll, Andrew H., and Elso S. Barghoorn, 1977, Archaean microfossils showing cell division from the Swaziland System of South Africa. Science, vol. 198, pp. 396-98.

Kofahl, Robert E., 1977, Handy Dandy Evolution Refuter. Beta Books, San Diego, Calif.

------ 1981, Letters to editor. Science, vol. 212, p. 873.

Kofahl, Robert E., and Kelly L. Seagraves, 1975, The Creation Explanation. Harold Shaw Publishers, Wheaton, Ill.

Kolb, Charles R., and Jack R. Van Lopik, 1966, Depositional environments of the Mississippi River deltaic plain--southeastern Louisiana. Pp. 18-61 in M. L. Shirley and J. A. Ragsdale, eds., 1966, Deltas in Their Geologic Framework, Houston Geological Society, 251 pp.

Kröner, Alfred, 1985, Evolution of the Archaean continental crust. Annual Reviews of Earth and Planetary Sciences, vol. 13, pp. 49-74.

Kuban, Glen J., 1986a, A summary of the Taylor Site evidence. Creation/Evolution, vol. 6, no. 1, pp. 10-18.

------ 1986b, The Taylor Site "man tracks." Origins Research, vol. 9, no. 1, pp. 1, 7-9.

------ 1986c, Review of ICR Impact article 151. Origins Research, vol. 9, no. 1, pp. 10-15.

------1987, The Texas "Mantrack" Controversy. (Privately published monograph. P.O. Box 33232, North Royalton, Oh., 44113.)

Kuhn, Thomas S., 1962, The Structure of Scientific Revolutions. The University of Chicago Press, 172 pp.

Kulp, J. Laurence, 1950, Flood geology. Jour., Amer. Scientific Affiliation, vol. 2, pp. 1-15.

Kurtz, Paul, 1969, Moral Problems in Contemporary Society; Essays in Humanistic Ethics. Prometheus Books, Buffalo, N.Y., 301 pp.

LaBonte, Barry J., and Robert Howard, 1981, Measurement of solar radius changes. Science, vol. 214, pp. 907-9.

Ladd, Harry S., 1959, Ecology, paleontology and stratigraphy. Science, vol. 129, pp. 69-78.

Ladd, Harry S., and M. Grant Gross, 1967, Drilling on Midway Atoll, Hawaii. Science, vol. 156, pp. 1088-94.

Laferriere, A. P., D. E. Hattin, and A. W. Archer, 1987, Effects of climate, tectonics, and sea-level changes on rhythmic bedding patterns in the Niobrara Formation (Upper Cretaceous), U.S. Western Interior. Geology, vol. 15, pp. 233-36.

LaHaye, Tim F., and John D. Morris, 1976, The Ark on Ararat. Thomas S. Nelson, Nashville, and Creation-Life Publishers, San Diego (joint publishers), 275 pp.

Lammerts, Walter, 1961, Personal communication to Dr. Henry M. Morris. Pp. 189-91 in Whitcomb and Morris, 1961.

Landau, Matthew, 1983, Whales: Can evolution account for them? Creation/Evolution, vol. 3, no. 4, pp. 14-19.

Landis, E. R., and Paul Averitt, 1978, Coal. Pp. 165-67 in Fairbridge and Bourgeois, 1978.

Langseth, Marcus, 1977, The seafloor and the earth's heat engine. Lamont-Doherty Geological Observatory Yearbook, vol. 4, pp. 41-44.

Langston, Wann, Jr., 1974, Nonmammalian Comanchian tetrapods. Geoscience and Man, vol. 8, pp. 39-55.

------ 1981, Pterosaurs. Scientific American, vol. 244, no. 2, pp. 122-36.

Laplace, Pierre Simon, Marquis de, 1812, Introduction to the Analytical Theory of Probability. (English tr., 1886), Paris, p. vi.

Leakey, Richard E., 1976, Hominids in Africa. American Scientist, vol. 64, pp. 174-78.

Leather, Derek, 1986, The tale of the Tasaday. Geographical Magazine, vol. 58, no. 10, pp. 490-91.

Leopold, Luna B., and M. Gordon Wolman, 1960, River meanders. Bull., Geological Soc. of America, vol. 71, pp. 769-94.

Le Pichon, Xavier, 1968, Sea-floor spreading and continental drift. Jour. of Geophysical Research, vol. 73, pp. 3661-97.

Leslie, J. G., 1984, Mutation and design in the genome. Ex Nihilo, vol. 6, no. 4, pp. 38-45.

Levorsen, A. I., 1967, Geology of Petroleum, 2d ed., W. H. Freeman, San Francisco, 724 pp.

Lewin, Roger, 1980, Evolutionary theory under fire. Science, vol. 210, pp. 883-87.

------ 1981, No gap here in the fossil record. Science, vol. 214, pp. 645-46.

------ 1982a, Biology is not postage stamp collecting. Science, vol. 216, pp. 718-20.

------ 1982b, Molecules come to Darwin's aid. Science, vol. 216, pp. 1091-92.

------ 1983a, Fossil Lucy grows younger again. Science, vol. 219, pp. 43-44.

------ 1983b, How did vertebrates take to the air? Science, vol. 221, pp. 38-39.

------ 1983c, Do ape-size legs mean ape-like gait? Science, vol. 221, pp. 537-38.

------ 1983d, Extinctions and the history of life. Science, vol. 221, pp. 935-37.

------ 1984a, Alien beings here on earth. Science, vol. 223, p. 39.

------ 1984b, DNA reveals surprises in human family tree. Science, vol. 226, pp. 1179-82.

------ 1984c, Human Evolution: An Illustrated Introduction. W. H. Freeman and Co., New York, 104 pp.

------ 1984d, First true RNA catalyst found. Science, vol. 223, pp. 266-67.

------ 1985, How does half a bird fly? Science, vol. 230, pp. 530-31.

------ 1985a, Molecular clocks scrutinized. Science, vol. 228, p. 571.

------ 1986, RNA catalysis gives fresh perspective on the origin of life. Science, vol. 231, pp. 545-46.

------ 1987, The origin of the modern human mind. Science, vol. 236, pp. 668-70.

Lightman, Alan P., 1983, Weighing the odds. Science 83, vol. 4, no. 10, pp. 21-22.

Lingenfelter, R. E., 1963, Production of carbon 14 by cosmic ray neutrons. Reviews of Geophysics, vol. 1, no. 1, pp. 35-55.

Lively, R. S., 1983, Late Quaternary U-series speleothem growth record from southeastern Minnesota, Geology, vol. 11, pp. 259-62.

Lorenz, Konrad, 1966, On Aggression. Harcourt Brace & World, New York, 306 pp.

Lotka, A. J., 1922, Contribution to the energetics of evolution. Proc., Nat. Acad. Sci., vol. 8, pp. 147-55.

Mac Aoidh, Caoimhin, 1983, The new fossil record. Geotimes, vol. 28, no. 3, p. 12.

Macdonald, Gordon A., and Agatin T. Abbott, 1970, Volcanoes in the Sea: The Geology of Hawaii. Univ. of Hawaii Press, Honolulu, 441 p.

Mackay, John, 1983, Mt. St. Helens: Key to rapid coal formation? Ex Nihilo, vol. 2, no. 1 (International), pp. 6-8.

------ 1984, The origin of races. Ex Nihilo, vol. 6, no. 4, pp. 6-12.

Mahard, Richard H., 1942, The origin and significance of intrenched meanders. Jour. of Geomorphology, vol. 5, pp. 32-44.

Malmgren, Björn A., W. A. Berggren, and G. P. Lohmann, 1984, Species formation through punctuated gradualism in planktonic foraminifera. Science, pp. 317-19.

Margulis, Lynn, 1984, Early Life. Jones and Bartlett Publishers, Boston, 160 pp.

Margulis, Lynn, and Karlene V. Schwartz, 1982, Five Kingdoms: An Illustrated Guide to the Phyla of Life on Earth. W. H. Freeman and Co., San Francisco, 338 pp.

Marsh, Frank L., 1947, Evolution, Creation, and Science. Review and Herald Publishing Assoc., Washington, D.C.

Marshall, Eliot, 1983, A controversy on Samoa comes of age. Science, vol. 219, pp. 1042-45.

Matthews, Ralph W., 1982, Radiometric dating and the age of the earth. Ex Nihilo, vol. 5, no. 1, pp. 41-44.

Mayr, Ernst, 1970, Population, Species, and Evolution. Belknap Press of Harvard Univ. Press, Cambridge, Mass.

------ 1971, Evolution vs. special creation. The American Biology Teacher, vol. 33, no. 1, pp. 49-50.

------ 1982, The Growth of Biological Thought. Belknap Press of Harvard Univ. Press, Cambridge, Mass., 974 pp.

McAlester, A. Lee, 1968, The History of Life. Prentice-Hall, Englewood Cliffs, N.J., 151 pp.

McCartney, Kevin, 1984, The Cretaceous/Tertiary extinction. Jour. of Geological Education, vol. 32, pp. 306-09.

McCrea, W. H., 1968, Cosmology after half a century. Science, vol. 160, pp. 1295-99.

McDonald, K. L., and R. H. Gunst, 1967, An analysis of the earth's magnetic field from 1835 to 1965. ESSA Tech. Report IER 46-IES 1, U.S. Government Printing Office, Washington, D.C., 87 pp.

McGowan, Chris, 1984, In the Beginning: A Scientist Shows Why the Creationists Are Wrong. Prometheus Books, Buffalo, N.Y., 208 pp.

McKee, Edwin D., 1945, Cambrian History of the Grand Canyon Region. Carnegie Institution of Washington, Publ. 563, Part I., 232 pp.

McKee, Edwin D., and Edwin H. McKee, 1972, Pliocene uplift of the Grand Canyon region--time of drainage adjustment. Bull., Geological Soc. of America, vol. 83, pp. 1923-32.

McKee, Edwin H., R. F. Wilson, W. J. Breed, and C. S. Breed, eds., 1967, Evolution of the Colorado River in Arizona. Bull., Museum of Northern Arizona, no. 44, pp. 1-67.

McKown, Delos B., 1985, The real culprit behind religious conflicts in public education. The American Rationalist, vol. 29, no. 6, pp. 84-86.

Medawar, Peter, 1967, Mathematical Challenges to the Neo-Darwinism Interpretation of Evolution. Wistar Institute Press, Philadelphia.

Meier, M. F., P. J. Carpenter, and R. J. Janda, 1981, Hydrologic effects of Mount St. Helens' 1980 eruptions. EOS, vol. 62, no. 33, pp. 625-26.

Menard, W. H., 1984, Evolution of ridges by asymmetrical spreading. Geology, vol. 12, pp. 177-80.

Merrill, Ronald T., and Michael W. McElhinny, 1983, The Earth's Magnetic Field. Academic Press, London, New York, 410 pp.

Merton, Robert K., 1973, The Sociology of Science; Theoretical and Empirical Investigations. The University of Chicago Press, 605 pp.

Meyerhoff, A. A., 1970, Continental drift, I., II. Jour. of Geology, vol. 78, pp. 1-51, 406-44.

Meyerhoff, A. A., and Curt Teichert, 1971, Continental drift, III. Jour. of Geology, vol. 79, pp. 285-321.

Meyerhoff, A. A., and H. A. Meyerhoff, 1972a, Continental drift, IV. Jour. of Geology, vol. 80, pp. 34-60.

------, and H. A. Meyerhoff, 1972b, "The new global tectonics": Age of linear magnetic anomalies of ocean basins. Bull., American Assoc. of Petroleum Geologists, vol. 56, pp. 337-59.

Meyerhoff, A. A., H. A. Meyerhoff, and R. S. Briggs, 1972, Continental drift, V. Jour. of Geology, vol. 80, pp. 663-92.

Miller, Dan C., 1980, Potential hazards from future eruptions of Mount Shasta volcano, northern California. U.S. Geological Survey Bulletin, no. 1503, 43 pp.

Miller, Kenneth, 1982, Answers to the standard creationist arguments. Creation/Evolution, vol. 3, no. 1, pp. 1-13.

Miller, Stanley L., and Leslie E. Orgel, 1974, The Origins of Life on Earth. Prentice-Hall, Englewood Cliffs, N.J., 229 pp.

Milne, David H., 1981, How to debate with creationists--and "Win." The American Biology Teacher, vol. 43, no. 5, pp. 235-45.

------ 1984, Creationists, population growth, bunnies, and the Great Pyramid. Creation/Evolution, vol. 4, no. 4, pp. 1-5.

Milne, David H., and Steven D. Schafersman, 1983, Dinosaur tracks, erosion marks and midnight chisel work (but no human footprints) in the Cretaceous limestone of the Paluxy River bed, Texas. Jour. of Geological Education, vol. 31, pp. 111-23.

Mintz, Leigh W., 1977, Historical Geology: The Science of a Dynamic Earth. 2d ed., Charles E. Merrill Publ. Co., Columbus, Ohio, 588 pp.

Mixter, Russell L., 1950, Creation and Evolution. American Scientific Affiliation, Monograph 2.

Monastersky, R., 1986, Reining in a runaway theory. Science News, vol. 130, p. 374.

Monroe, James Stewart, 1987, Creationism, human footprints, and flood geology. Jour. of Geological Education, vol. 35, pp. 93-103.

Montagu, Ashley, ed., 1984, Science and Creationism, Oxford Univ. Press, New York, 415 pp.

Moon, Parry, and Domina E. Spencer, 1953, Binary stars and the velocity of light. Jour. of the Optical Society of America, vol. 43, pp. 635-41.

Moore, E. S., 1940, Coal; Its Properties, Analysis, Classification, Extraction, Uses and Distribution, 2d ed., John Wiley & Sons, New York, 473 pp.

Moore, George W., 1968, Speleothems. Pp. 1040-41 in R. W. Fairbridge, ed., 1968, Encyclopedia of Geomorphology. Reinhold Book Corp., New York, 1295 pp.

Moore, J. Casey, et al., 1982, Geology and tectonic evolution of a juvenile accretionary terrane along a truncated convergent margin. Bull., Geological Soc. of America, vol. 93, pp. 847-61.

Moore, John N., and Harold S. Slusher, eds., 1970, Biology, A Search for Order in Complexity. Zondervan Publishing House, Grand Rapids, Mich., 595 pp.

Moore, Robert A., 1983, The impossible voyage of Noah's Ark. Creation/Evolution, vol. 4, no. 1, pp. 1-43.

Morner, Nils-Axel, 1978, Varves and varved clays. Pp. 841-43 in Fairbridge and Bourgeois, 1978.

Morowitz, Harold J., 1968, Energy Flow in Biology. Academic Press, New York, 179 pp.

------ 1982, Navels of Eden. Science 82, vol. 3, no.2, pp. 20-22.

Morris, Henry M., 1946, 1978, That You Might Believe. (Revised ed., 1978) Creation-Life Publishers, San Diego.

------ 1968, Science versus scientism. Pp. 9-13 in Morris, H. M., et al., A Symposium on Creation. Baker Book House, Grand Rapids, Mich.

------ 1973?, Evolution, thermodynamics, and entropy. ICR Impact Series, no. 3 (pages not numbered).

------ ed., 1974a, Scientific Creationism. Creation-Life Publishers, San Diego. (Public School Edition, 217 pp.; General Edition, 277 pp.) Second edition, revised, 1985.

------ 1974b The young earth. ICR Impact Series, no. 17, pp. i-iv.

------ 1974c, The Troubled Waters of Evolution. Creation-Life Publishers, San Diego.

------ 1975b, Language, creation, and the inner man. ICR Impact Series, no. 28, pp. i-iv.

------ 1976a, Up with catastrophism! ICR Impact Series, no. 38, pp. i-iv.

------ 1976b, Entropy and open systems. ICR Impact Series, no. 40, pp. i-iv.

------ 1977, Circular reasoning in evolutionary geology. ICR Impact Series, no. 48, pp. i-iv.

------ 1978a, The day-age theory revisited. ICR Impact Series, no. 50, pp. i-iv.

------ 1978b, Thermodynamics and the origin of life. ICR Impact Series, no. 57, pp. i-iv.

------ 1983, Those remarkable floating rock formations. ICR Impact Series, no. 119, pp. i-iv.

------ 1984a, Evolution ex nihilo. ICR Impact Series, no. 135, pp. i-iv.

------ 1984b, A History of Modern Creationism. Master Book Publishers, San Diego, Calif., 382 pp.

Morris, John D., 1976, The Paluxy River tracks. ICR Impact Series, no. 35, pp. i-viii.

------ 1980, Tracking Those Incredible Dinosaurs and the People Who Knew Them. Creation-Life Publishers, San Diego, Calif., 239 pp.

------ 1986a, The Paluxy River mystery. ICR Impact Series, no. 151, pp. i-iv.

------ 1986b, The Paluxy River mystery. Letter in Nature, vol. 321 (June), p. 722.

------ 1986c, Follow up on the Paluxy mystery. Origins Research, vol. 9, no. 1, p. 14.

Morris, Simon Conway, and H.B. Whittington, 1979, The animals of the Burgess Shale. Scientific American, vol. 241, no. 1, pp. 122-33.

Morrison, David, 1977, Planetary astronomy and Velikovsky's catastrophism. Pp. 145-76 in Goldsmith, 1977.

Morrison, Roger B., 1975, Predecessors of Great Salt Lake. Geological Society of America, Abstracts with Programs, vol. 7, no. 6, p. 1206.

Morrison, Roger B., and John C. Frye, 1965, Correlation of the middle and late Quaternary successions of the Lake Lahontan Lake Bonneville, Rocky Mountain (Wasatch Range), southern Great Plains, and eastern midwest areas. Nevada Bur. Mines, Report 9, University of Nevada, 45 pp.

Nagel, Ernest, 1961, The Structure of Science: Problems in the Logic of Scientific Explanation. Harcourt, Brace & World, New York, 618 pp.

Nagel, T., 1978, Ethics as an autonomous theoretical subject. Pp. 221-32 in Stent, 1978.

Nelkin, Dorothy, 1982, The Creation Controversy; Science or Scripture in the Schools. W. W. Norton & Co., New York, 242 pp.

Neufeld, Berney, 1975, Dinosaur tracks and giant men. Origins, vol. 2, no. 2, pp. 64-67.

Nevins, Stuart E., 1976, Continental drift, plate tectonics, and the Bible. ICR Impact Series, no. 32, pp. i-iv.

------ 1976, The origin of coal. ICR Impact Series, no. 41, pp. i-iv.

Newell, Norman D., 1959, Adequacy of the fossil record. Jour. of Paleontology, vol. 33, pp. 488-99.

------ 1963, Crises in the history of life. Scientific American, vol. 208, no. 2, pp. 77-92.

------ 1982, Creation and Evolution: Myth or Reality? Columbia Univ. Press, New York, 199 pp.

------ 1984, Why Scientists Believe in Evolution. American Geological Institute, Washington, D.C., 13 pp.

Newton, Robert, 1969, Secular accelerations of the earth and moon. Science, vol. 166, pp. 825-31.

Ninkovich, Dragoslav, and William L. Donn, 1976, Explosive Cenozoic volcanism and climatic implications. Science, vol. 194, pp. 899-906

Noble, C. S., and J. J. Naughton, 1968, Deep-ocean basalts: Inert gas contents and uncertainties in age dating. Science, vol. 162, pp. 265-67.

Numbers, Ronald L., 1982, Creationism in 20th-Century America. Science, vol. 218, pp. 538-44.

Oberg, James, 1981, Ideas in collision. Pp. 401-8 in Frazier, 1981.

Odum, H. T., and R. C. Pinkerton, 1955, Time's speed regulator: the optimum efficiency for maximum power output in physical and biological systems. Amer. Scientist, vol. 43, no. 2, pp. 331-43.

Officer, Charles B., and Charles L. Drake, 1984, Terminal Cretaceous events. Science, vol. 227, pp. 1161-67.

Oller, John W., Jr., 1984, Not according to Hoyle. ICR Impact Series, no. 138, pp. i-iv.

Olsen, Paul E., 1986a, Impact theory: Is the past the key to the future? Lamont-Doherty Geological Observatory Yearbook 1984-86 (Columbia University), pp. 5-10.

------ 1986b, A 40-million-year lake record of early Mesozoic orbital climatic forcing. Science, vol. 234, pp. 842-48.

Omohundro, John T., 1981, Von Däniken's chariots: a primer in the art of cooked science. Pp. 307-31 in Frazier, 1981.

Oparin, Aleksandr I., 1968, Genesis and Evolutionary Development of Life. Tr. by Eleanor Maass, Academic Press, New York, 203 pp.

O'Rourke, J. E., 1976, Pragmatism versus materialism in stratigraphy. American Jour. of Science, vol. 276, pp. 47-55.

Ostrom, H. John, 1970, Archaeopteryx: notice of a "new" specimen. Science, vol. 170, pp. 537-38.

------ 1979, Bird flight: How did it begin? American Scientist, vol. 67, pp. 46-56.

Overton, William R., 1982, Creationism in the schools; The decision in McLean versus the Arkansas Board of Education. Science, vol. 215, pp. 934-43.

Oxnard, Charles E., 1973, Form and Pattern in Human Evolution. Univ. of Chicago Press, Chicago and London, 218 pp.

------ 1975a, The place of the australopithecines in human evolution: grounds for doubt? Nature, vol. 258, pp. 389-95.

----- 1975b, Uniqueness and Diversity in Human Evolution; Morphometric Studies of Australopithecines. Univ. of Chicago Press, Chicago, 133 pp.

Pannella, G., C. MacClintock, and M. N. Thompson, 1968, Paleontological evidence of variations in length of synodic month since late Cambrian. Science, vol. 162, pp. 792-96

Patterson, Colin, 1984, Letter in reply to Steven Binkley, June 17, 1982. Creation/Evolution Newsletter, vol. 4, no. 6, pp. 4-5.

Patterson, John W., 1983, Thermodynamics and evolution. Pp. 99-116 in Godfrey, 1983.

------ 1984, Thermodynamics and probability. Pp. 132-52 in Awbrey and Thwaites, 1984.

Pecker, J. C., A. P. Roberts, and J. P. Vigier, 1972, Non-velocity redshifts and photon-photon interactions. Nature, vol. 237, pp. 227-29.

Penfield, W., and L. Roberts, 1959, Speech and Brain Mechanisms. Princeton Univ. Press, Princeton.

Pennington, Wayne D., 1983, Role of shallow phase changes in the subduction of oceanic crust. Science, vol. 220, pp. 1045-47.

Pensée Editors, 1976, Velikovsky Reconsidered. Doubleday and Co., Garden City, N.Y., 260 pp.

Pettijohn, F. J., 1957, Sedimentary Rocks, 2d ed., John Wiley & Sons, New York, 718 pp.

Petto, Andrew J., 1982, The turtle: Evolutionary dilemma or creationist shell game? Creation/Evolution, vol. 3, no. 4, pp. 20-29.

Pierce, William G., 1957, Heart Mountain and South Fork detachment thrusts of Wyoming. Bull., American Assoc. of Petroleum Geologists, vol. 41, pp. 591-626.

------ 1963, Tectonic denudation as exemplified by the Heart Mountain fault, Wyoming. Transactions, XXIII International Geological Congress, vol. 3, pp. 191-97.

------ 1979, Clastic dikes of Heart Mountain fault breccia, northwestern Wyoming, and their significance. U.S. Geological Survey Professional Paper, no. 1133, 25 pp.

Pilbeam, David, 1984, The descent of hominoids and hominids, Scientific American, vol. 250, no. 3 pp. 84-96.

Pilbeam, David, and Stephen J. Gould, 1974, Size and scaling in human evolution. Science, vol. 186, pp. 892-901.

Plotnick, Roy E., 1980, Relationship between biological extinctions and geomagnetic reversals. Geology, vol. 8, pp. 578-81.

Pollack, James B., and Jeffry N. Cuzzi, 1981, Rings in the solar system. Scientific American, vol. 245, no. 5, pp. 105-29.

Popper, Karl R., 1959, The Logic of Scientific Discovery. Hutchinson, London, and Basic Books, New York, 480 pp. (Translation of Logik der Forschung, 1934.)

------ 1976, Unended Quest: An Intellectual Autobiography. Open Court Publishing Co., La Salle, Ill., 255 pp.

------ 1978, Natural selection and the emergence of mind. Dialectica, vol. 32, no. 3-4, pp. 339-55.

------ 1980, Science: conjectures and refutations. Pp. 19-34 in Klemke, Hillinger, and Kline, 1980.

Porter, S. C., M. Stuiver, and I. C. Yang, l977, Chronology of Hawaiian glaciers. Science, vol. 195, pp. 61-63.

Potter, Russell M., and George R. Rossman, 1977, Desert varnish: the importance of clay minerals. Science, vol. 196, pp. 1446-48.

Price, George McCready, 1923, The New Geology. Pacific Press, Mountain View, Calif., 706 pp.

Purrett, Louise, 1971, Magnetic reversals and biological extinctions. Science News, vol. 100, p. 300.

Ralph, Elizabeth K., and Henry N. Michael, 1974, Twenty-five years of radiocarbon dating. American Scientist, vol. 62, pp. 553-60.

Ramberg, Hans, 1963, Experimental study of gravity tectonics by means of centrifugal models. Bull., Geological Institute, University of Uppsala, vol. 62, pp. 1-97.

Ramm, Bernard, 1954, The Christian View of Science and Scripture. Wm. B. Erdmans Publ. Co., Grand Rapids, Mich., 255 pp.

Rampino, Michael R., Stephen Self, and Rhodes W. Fairbridge, 1979, Can rapid climatic change cause volcanic eruptions? Science, vol. 206, pp. 826-30.

Ransom, C. J., 1976, The Age of Velikovsky. Kronos Press, Glassboro, N.J., 274 pp.

Raup, David M., 1983, The geological and paleontological arguments of creationism. Pp. 147-62 in Godfrey, 1983.

Raup, David M., and Steven M. Stanley, 1971, Principles of Paleontology. W. H. Freeman and Co., San Francisco, 388 pp.

Raup, David M., and J. John Sepkoski, Jr., 1982, Mass extinctions in the marine fossil record. Science, vol. 215, pp. 1501-2.

Rensberger, Boyce, 1983, Margaret Mead on becoming human: The nature-nurture debate. Science 83, vol. 4, no. 3, pp. 28-46.

Retallack, George; William J. Fritz, 1981, Comment and reply on "Reinterpretation of the depositional environment of the Yellowstone fossil forests." Geology, vol. 9, pp. 52-54.

Richards, J. R., 1975, Lead isotope data on three north Australian galena localities. Mineralium Deposita, vol. 10, pp. 287-301.

Romer, Alfred S., 1966, Vertebrate Paleontology, 3d ed. Univ. of Chicago Press, 468 pp.

Rona, Peter A., and D. F. Gray, 1980, Structural behavior of fracture zones symmetric and asymmetric about a spreading axis. Bull., Geological Soc. of America, vol. 91, pp. 485-94.

Root-Bernstein, Robert S., 1981, Letters to editor. Science, vol. 212, p. 1446.

------ 1984, Ignorance versus knowledge in the evolutionist-creationist controversy. Pp. 8-24 in Awbrey and Thwaites, 1984.

Ross, C. P., and Richard Rezak, 1959, Rocks and Fossils of Glacier National Park, U.S. Geological Survey Professional Paper, no. 294-K, pp. 401-39.

Ross, David A., 1972, Red Sea hot brine area: revisited. Science, vol. 175, pp. 1455-57.

Ruddiman, William F., 1984, Ice-age thermal response and climatic role of the surface Atlantic Ocean, 40 degrees N to 63 degrees N. Bull., Geological Soc. of America, vol. 95, pp. 381-96.

Ruddiman, William F. and Andrew McIntyre, 1981, Oceanic mechanisms for amplification of the 23,000-year ice-volume cycle. Science, vol. 212, pp. 617-27.

Ruddiman, W. F., A. McIntyre, and J. Hays, 1979, Causes and mechanisms of climate change. Lamont-Doherty Yearbook, vol. 6, pp. 27-30.

Ruderman, M. A., 1974, Possible consequences of nearby supernova explosions for atmospheric ozone and terrestrial life. Science, vol. 184, pp. 1079-81.

Rukang, Wu, and Lin Shenlong, 1983, Peking Man. Scientific American, vol. 248, pp. 86-94.

Runcorn, S. K., 1966, Corals as paleontological clocks. Scientific American, vol. 215, no. 4, pp. 26-33.

Russell, Richard J., 1957, The instability of sea level. American Scientist, vol. 45, pp. 414-30.

Rybka, Theodore W., 1982, Consequences of time dependent nuclear decay indices on half lives. ICR Impact Series, no. 106, pp. i-iv.

Sagan, Carl, 1977, An Analysis of Worlds in Collision. Pp. 41-104 in Goldsmith, 1977.

Saller, Arthur, 1984, Petrologic and geochemical constraints on the origin of subsurface dolomite, Enewetak Atoll. Geology, vol. 12, pp. 217-20.

Saunders, David R., and R. Roger Harkins, 1968, UFOs? Yes!: Where the Condon Committee Went Wrong. World Publishing Co., New York, 256 pp.

Saver, Jeffrey, 1985, An interview with E. O. Wilson on sociobiology and religion. Free Inquiry, vol. 5, no. 2, pp. 15-22.

Schadewald, Robert J., 1982, Six "Flood" arguments creationists can't answer. Creation/Evolution, vol. 3, no. 3, pp. 12-17.

------ 1983, The evolution of Bible-science. Pp. 283-99 in Godfrey, 1983.

Schafersman, Steven D., 1983, Fossils, stratigraphy, and evolution: consideration of a creationist argument. Pp. 219-44 in Godfrey, 1983.

Schmidt, Victor A., 1982, Magnetostratigraphy of sediments in Mammoth Cave, Kentucky. Science, vol. 217, pp. 827-29.

Scholl, D. W., et al., 1970, Peru-Chile Trench sediments and sea-floor spreading. Bull., Geological Soc. of America, vol. 81, p. 1339-60.

Schopf, J. William, 1973, The evolution of the earliest cells. Scientific American, vol. 239, no. 3, pp. 111-38.

Schramm, David N., 1974, The age of the elements. Scientific American, vol. 230, no. 1, pp. 69-77.

Schuchert, Charles, and Carl O. Dunbar, 1933, A Textbook of Geology; Part II Historical Geology. John Wiley & Sons, New York. 551 pp.

Schumm, Stanley A., 1963, The disparity between present rates of denudation and orogeny. U.S. Geological Survey Professional Paper, no. 454-H, 13 pp.

------ 1967, Meander wavelength of alluvial rivers. Science, vol. 157, pp. 1549-50.

------ 1968, River adjustment to altered hydrologic regimen, Murrumbidgee River and paleochannels, Australia. U.S. Geological Survey Professional Paper, no. 598.

------ 1977, The Fluvial System. John Wiley & Sons, New York, 338 pp.

Sclater, John G., and Robert L. Fisher, 1974, Evolution of the East Indian Ocean. Bull., Geological Soc. of America, vol. 85, pp. 683-702.

Scotese, C. R., 1984, Paleozoic paleomagnetism and the assembly of Pangaea. Pp. 1-10 in Van der Voo, R., C. R. Scotese, and N. Bonhommet, eds., 1984, Plate Reconstruction from Paleozoic Paleomagnetism. Geodynamics Series, vol. 12, American Geophysical Union, Washington, D.C., 136 pp.

Scriven, M., 1959, Explanation and prediction in evolutionary theory. Science, vol. 130, pp. 477-82.

Sepkoski, J. John, Jr., 1982. Mass extinctions in the Phanerozoic oceans. Pp. 283-89 in Silver and Schultz, 1982.

Setterfield, Barry, 1981, The velocity of light and the age of the universe. Ex Nihilo, vol. 4, no. 1, pp. 38-48.

------ 1982, The velocity of light and the age of the universe. Ex Nihilo, vol. 1, no. 1 (International Edition), pp. 53-93.

------ 1983, The velocity of light and the age of the universe. Ex Nihilo, vol. 1, no. 3 (International Edition), pp. 41-46.

------ 1984a, C decay and the red-shift. Ex Nihilo Technical Jour., vol. 1, pp. 71-86.

------ 1984b, The age of the astronomical universe--a reply. Ex Nihilo Technical Jour., vol. 1, pp. 95-104.

Seyfert Carl K., and Leslie A. Sirkin, 1979, Earth History and Plate Tectonics, 2d ed., Harper & Row, Publishers, New York, 600 pp.

Shackleton, N. J., and N. D. Opdyke, 1973, Oxygen isotope and paleomagnetic stratigraphy of equatorial Pacific core V28-238. Quaternary Research, vol. 3, pp. 39-55.

Shapiro, Irwin I., 1980, Is the sun shrinking? Science, vol. 208, pp. 51-53.

Sharp, Robert P., 1940, Ep-Archean and Ep-Algonkian erosion surfaces, Grand Canyon, Arizona. Bull., Geological Soc. of America, vol. 51, pp. 1235-70.

Shea, James H., 1982, Twelve fallacies of uniformitarianism. Geology, vol. 10, pp. 455-60.

------ 1983, Creationism, uniformitarianism, geology and science. Jour. of Geological Education, vol. 31, pp. 105-10.

Shepard, Francis P., 1963, Submarine Geology, 2d ed. Harper & Row, New York, 557 pp.

------ 1973, Submarine Geology, 3d ed. Harper & Row, New York, 517 pp.

Shepherd, R. G., 1972, Incised river meanders: evolution in simulated bedrock. Science, vol. 178, pp. 409-11.

Shibaoka, M., J. D. Saxby, and G. H. Taylor, 1978, Hydrocarbon generation in Gippsland Basin, Australia-- Comparison with Cooper Basin, Australia. Bull., American Assoc. of Petroleum Geologists, vol. 62, no. 7, pp. 1151-58.

Shu, Frank H., 1973, Spiral structure, dust clouds, and star formation. American Scientist, vol. 61, pp. 524-36.

Shuler, E. W., 1917, Dinosaur tracks in the Glen Rose limestone near Glen Rose, Texas. American Jour. of Science, vol. 44, no. 262, pp. 294-98.

Sibley, Charles, and Jon Ahlquist, 1984, The phylogeny of the hominoid primates as indicated by DNA-DNA hybridization. Jour. of Molecular Evolution, vol. 20, pp. 2-15.

Silk, Joseph, 1980, The Big Bang: The Creation and Evolution of the Universe. W. H. Freeman and Co., San Francisco, 394 pp.

Silver, Leon T., and Peter H. Schultz, eds., 1982, Geological implications of impacts of large asteroids and comets on the earth. Geological Society of America Special Paper, no. 190, Boulder, Colorado, 528 pp.

Simon, Cheryl, 1984a, Mass extinctions and sister stars. Science News, vol. 125, p. 116.

------ 1984b, Death star. Science News, vol. 125, pp. 250-52.

Simons, Elwyn L., 1977, Ramapithecus. Scientific American, vol. 236, no. 5, pp. 28-35.

Simpson, George Gaylord, 1944, Tempo and Mode in Evolution. Columbia Univ. Press, New York, 237 pp.

------ 1963, Historical science. Pp. 24-48 in Albritton, 1963.

------ 1964, This View of Life. Harcourt, Brace & World, New York, 308 pp.

------ 1969, Biology and Man. Harcourt, Brace & World, New York, 175 pp.

Skehan, James W., 1983, Theological basis for a Judeo-Christian position on creationism. Jour. of Geological Education, vol. 31, pp. 307-14.

Slusher, Harold S., 1976, Some recent developments having to do with time. Pp. 278-85 in H. M. Morris and D. T. Gish, eds., The Battle for Creation. Creation-Life Publishers, San Diego, 321 pp.

------ 1978, The origin of the universe: an examination of the big-bang and steady-state cosmogenies. ICR Technical Monograph, no. 8, Institute for Creation Research, San Diego, 50 pp.

------ 1980, Age of the cosmos. ICR Technical Monograph, no. 9, Institute for Creation Research, San Diego, 76 pp.

------ 1981, Critique of radiometric dating. 2d ed., ICR Technical Monograph, no. 2, Institute for Creation Research, San Diego, 58 pp.

Smart, J. S., 1979, Determinism and randomness in fluvial geomorphology. EOS, vol. 60, no. 36, pp. 651-55.

Smith, Curtis G., 1985, Ancestral Voices: Language and the Evolution of Consciousness. Prentice-Hall, Englewood Cliffs, N. J., 178 pp.

Smith, Guy M., 1985, Source of marine magnetic anomalies: some results from DSDP Leg 83. Geology, vol. 13, pp. 162-65.

Smith, John Maynard, 1978a, The evolution of human behavior. Scientific American, vol. 239, no. 3, pp. 176-91.

------ 1978b, The concepts of sociobiology. Pp. 23-34 in Stent, 1978.

Snelling, Andrew, 1982, The recent origin of Bass Strait oil and gas. Ex Nihilo, vol. 5, no. 2 (International Edition, vol. 1, no. 2), pp. 43-46.

------ 1983a, What about continental drift? Have the continents moved apart? Ex Nihilo (International Edition), vol. 2, no. 1, pp. 14-16.

------ 1983b, Creationist geology: the Precambrian. Ex Nihilo, vol. 6, no. 2 (International Edition, vol. 2, no. 2) pp. 42-46.

------ 1984, The recent, rapid formation of the Mount Isa orebodies during Noah's Flood. Ex Nihilo, vol. 6, no. 3 (International Edition, vol. 2, no. 3), pp. 40-46.

Snelling, Andrew, and John Mackay, 1984, Coal volcanism and Noah's Flood. Ex Nihilo Technical Journal, vol. 1, pp. 11-29.

Snyder, Walter S., 1978, Manganese deposited by submarine hot springs in chert-greenstone complexes, western United States. Geology, vol. 6, pp. 741-44.

Sofia, S., J. O'Keefe, J. R. Lesh, and A. S. Endal, 1979, Solar constant: constraints on possible variations derived from solar diameter measurements. Science, vol. 204, pp. 1306-8.

Soroka, Leonard G., and Charles L. Nelson, 1983, Physical constraints on the Noachian Deluge. Jour. Geological Education, vol. 31, pp. 135-39.

Sozansky, V. I., 1973, Origin of salt deposits in deep-water basins of Atlantic Ocean. Bull., American Assoc. of Petroleum Geologists, vol. 57, pp. 589-90.

Spera, Frank, 1980, Thermal evolution of plutons: a parameterized approach. Science, vol. 207, pp. 299-301.

Stanley, Steven M., 1973, An ecological theory for the sudden origin of multicellular life in the Late Precambrian. Proc. of the National Academy of Sciences, vol. 70, pp. 1486-89.

------ 1979, Macroevolution: Pattern and Process. W. H. Freeman and Co., San Francisco.

------ 1984, Mass extinctions in the ocean. Scientific American, vol. 250, no. 6, pp. 64-72.

Stebbins, G. Ledyard, 1977, Patterns of speciation. Pp. 195-232 in Dobzhansky, 1977.

------ 1982, Darwin to DNA, Molecules to Humanity. W. H. Freeman and Co., San Francisco, 491 pp.

Stent, Gunther S., 1975, Limits to scientific understanding of man. Science, vol. 187, pp. 1052-57.

------ ed., 1978, Morality As a Biological Phenomenon. Report of the Dahlem Workshop, Berlin, 1977, Abakon Verlagsgesellschaft, Berlin, 323 pp.

Stephenson, F. Richard, 1982, Historical eclipses. Scientific American, vol. 274, no. 4, pp. 170-83.

Stokes, William Lee, 1986, Alleged human footprint from Middle Cambrian strata, Millard County, Utah. Jour. of Geological Education, vol. 34, pp. 187-90.

Storer, Norman W., 1977, The sociological context of the Velikovsky controversy. Pp. 29-39 in Goldsmith, 1977.

Story, Ronald, 1976, The Space Gods Revealed. Harper & Row, New York, 139 pp.

------ 1980, Guardians of the Universe? St. Martin's Press, New York, 207 pp.

Strahler, Arthur N., 1971, The Earth Sciences, 2d ed. Harper & Row, New York, 824 pp.

------ 1980, Systems theory in physical geography. Physical Geography, vol. 1, no. 1, pp. 1-27.

------ 1981, Physical Geology. Harper & Row, New York, 612 pp.

------ 1983, Toward a broader perspective in the evolutionist-creationist debate. Jour. of Geological Education, vol. 31, pp. 87-94.

Strahler, A. N., and A. H. Strahler, 1974, Introduction to Environmental Science. Hamilton Publishing Company, Santa Barbara, Calif. (John Wiley & Sons, New York), 633 pp.

Strom, Stephen E., and Karen M. Strom, 1979, The evolution of disk galaxies. Scientific American, vol. 240, no. 4, pp. 72-82.

Stuiver, Minze, 1976, First Miami conference on isotope climatology and paleoclimatology. EOS, vol. 57, no. 1, pp. 830-36.

Stuiver, Minze, and Paul D. Quay, 1980, Changes in atmospheric carbon-14 attributed to a variable sun. Science, vol. 207, pp. 11-19.

Suess, H. E., 1982, Personal communication cited as source of Figure 1, p. 14, in Ellen M. Druffel, 1982, Banded corals: changes in oceanic carbon-14 during the Little Ice Age. Science, vol. 218, pp. 13-19.

Sullivan, Walter, 1967, Bone found in Kenya indicates man is 2.5 million years old. The New York Times, January 14, 1967.

------ 1972, Skull pushes back man's origin. The New York Times, November 10, 1972.

Sunderland, Luther D., and Gary E. Parker, 1982, Evolution? Prominent scientist reconsiders. ICR Impact Series, no. 108, pp. i-iv.

Sykes, Lynn R., 1967, Mechanism of earthquakes and nature of faulting on the mid-oceanic ridges, Jour. of Geophysical Research, vol. 72, pp. 2131-53.

Szent-Gyorgyi, Albert, 1977, Drive in living matter to perfect itself. Synthesis 1, vol. 1, no. 1, pp. 14-26.

Tappan, Helen, and A. R. Loeblich, Jr., 1970, Geobiologic implications of fossil phytoplankton evolution and time-space distribution. Geological Soc. of America Special Paper, no.127, pp. 247-340.

Tanner, W.F., 1973, Deep-sea trenches and the compression assumption. Bull., American Assoc.of Petroleum Geologists, vol. 57, pp. 2195-2206.

Taylor, G. Jeffrey, 1985, Lunar origin meeting favors impact theory. Geotimes, vol. 30, no. 4, pp. 16-17.

Teilhard de Chardin, P., 1966, Man's Place in Nature. (Tr. by Rene Hague.) Harper & Row, New York, 124 pp.

Templeton, Alan R., 1983, Phylogenetic inference from restriction endonuclease cleavage site maps with particular reference to the evoution of humans and the apes. Evolution, vol. 37, pp. 221-44.

Terzaghi, 1950, Mechanism of landslides. Pp. 83-123 in Sidney Paige, Chairman, 1950, Application of Geology to Engineering Practice, Berkey Volume, Geological Society of America, New York, 327 pp.

Thierstein, Hans R., 1982, Terminal Cretaceous plankton extinctions: a critical assessment. Pp. 385-399 in Silver and Schultz, 1982.

Thompson, L. G., Wayne L. Hamilton, and Colin Bull, 1975, Climatological implications of microparticle concentrations in the ice core from "Byrd" Station, western Antarctica. Jour. of Glaciology, vol. 14, pp. 433-44.

Thomsen Dietrick E., 1983a, The new inflationary nothing cosmology. Science News, vol. 123, pp. 108-9.

------ 1983b, A knowing universe seeking to be known. Science News, vol. 123, p. 124.

Thrailkill, J. V., 1972, Carbonate chemistry of aquifer and stream water in Kentucky. Jour. of Hydrology, vol. 16, pp. 93-104.

Thurston, Diana R., 1978, Chert and flint. Pp. 119-24 in Fairbridge and Bourgeois, 1978.

Thwaites, William M., 1983, An answer to Dr. Geisler--from the perspective of biology. Creation/Evolution, vol. 4, no. 3, pp. 13-20.

Thwaites, William M. and Frank T. Awbrey, 1982, As the world turns: Can creationists keep time? Creation/Evolution, vol. 3, no. 3, pp. 18-22.

Trefil, J. S., 1978, A consumer's guide to pseudo-science. Saturday Review, April 29, pp. 16-21.

Trop, Moshe, 1983, Is the Archaeopteryx a fake? Creation Research Soc Quarterly, vol. 20, pp. 120-21.

Trowbridge, Arthur C., 1930, Building the Mississippi delta. Bull., American Assoc. of Petroleum Geologists, vol. 38. pp. 167-92.

Tryon, Edward, 1984, What made the world? New Scientist, vol. 101, March 8, p. 14.

Unger, Tom, 1984, The sociobiology debate: What is it really all about? The American Rationalist, vol. 29, no. 4, pp. 57-59.

Uyeda, Seiya, 1978, The New View of the Earth. W. H. Freeman & Co., San Francisco, 217 pp.

Van den Bergh, Sidney, 1981, Size and age of the universe. Science, vol. 213, pp. 825-30.

Van Huene, R.E., 1972, Structure of the continental margin and tectonism at the eastern Aleutian Trench. Bull. ,Geological Soc. of America, vol. 83, pp. 3613-26.

Vardiman, Larry, 1984, The sky has fallen. ICR Impact Series, no. 128, pp. i-iv.

Vawter, Lisa, and Wesley M. Brown, 1986, Nuclear and mitochondrial comparisons reveal extreme rate variation in the molecular clock. Science, vol. 234, pp. 194-6.

Veatch, A. C., and P. A. Smith, 1939, Atlantic submarine valleys of the United States and the Congo submarine valley. Geological Soc. of America Special Paper, no. 7, 101 pp.

Veeh, H. H., and John Chappell, 1970, Astronomical theory of climatic change: support from New Guinea. Science, vol. 167, pp. 862-65.

Velikovsky, Immanuel, 1950, Worlds in Collision. Doubleday and Company, Garden City, N.Y., 401 pp.

Vines, Gail, 1985, Strange case of Archaeopteryx "fraud." New Scientist, March 14, p. 3.

von Bertalanffy, Ludwig, 1950, The theory of open systems in physics and biology. Science, vol. 111, pp. 23-29.

von Däniken, Erich, 1969, Chariots of the Gods? Unsolved Mysteries of the Past (Tr. by Michael Heron). G. P. Putnam's Sons, New York, 188 pp.

------ 1971, Gods from Outer Space. G. P. Putnam's Sons, New York, 190 pp.

------ 1973, The Gold of the Gods. G. P. Putnam's Sons, New York, 210 pp.

Waddington, Conrad. H., 1960, The Ethical Animal. Allen and Unwin, London, 231 pp.

Wahr, John, 1985, The earth's rotation rate. American Scientist, vol. 73, pp. 41-46.

Waldrop, M. Mitchell, 1984, Before the beginning. Science 84, vol. 5, no. 1, pp. 45-51.

Wartofsky, Marx W., 1968, Conceptual Foundations of Scientific Thought; An Introduction to the Philosophy of Science. Macmillan Company, New York, 560 pp.

Washburn, Sherwood L., 1978, The evolution of man. Scientific American, vol. 239, no. 3, pp. 194-208.

Watkins, R. S., F. Hoyle, N. C. Wickramsinghe, J. Watkins, R. Rabilizirof, and L. M. Spetner, 1985, Archaeopteryx--a photographic study. British Journal of Photography, vol. 132, pp. 264-66, 358-59, 367, 469-70.

Weber, Christopher G., 1980a, The fatal flaws of flood geology. Creation/Evolution, vol. 1, no. 1, pp. 24-37.

------ 1980b, Common creationist attacks on geology. Creation/Evolution, vol. 1, no. 2, pp. 10-25.

------ 1982, Answers to creationist attacks on carbon-14 dating. Creation/Evolution, vol. 3, no. 2, pp. 23-29.

Weisburd, S., 1984, Sister star scenario: sound or shot? Science News, vol. 126, p. 279.

------ 1986, Oldest bird and longest dinosaur. Science News, vol. 130, p. 103.

Weller, J. Marvin, 1960, Stratigraphic Principles and Practice. Harper & Row, New York, 725 pp.

Wesson, Paul S., 1972, Objections to continental drift and plate tectonics. Jour. of Geology, vol. 80, pp. 185-97.

West, Susan, 1978, Moon history in a sea shell. Science News, vol. 114, pp. 426-28.

Westphal, Frank, 1976, The dermal armour of some Triassic placodont reptiles. Pp. 31-41 in Bellairs, A. d'A., and C. Barry Cox, eds., 1976, Morphology and Biology of Reptiles, Linnean Society Symposium Series, No. 3, Academic Press, London and New York, 290 pp.

Whitcomb, John C., and Henry M. Morris, 1961, The Genesis Flood. Presbyterian and Reformed Publishing Co., Philadelphia, 518 pp.

White, Andrew D., 1896, A History of the Warfare of Science with Theology, vol. I. Reprinted in 1978 by Peter Smith, Gloucester, Mass., 415 pp.

Whitelaw, Robert L., 1968, Radiocarbon confirms biblical creation. Creation Research Soc. Quarterly, vol. 5, p. 80.

Willemin, J. H., P. L. Guth, and K. V. Hodges, 1980, Comment and reply on "High fluid pressure, isothermal surfaces, and the initiation of nappe movement." Geology, vol. 8, pp. 405-6.

Willis, Bailey, 1902, Stratigraphy and structure, Lewis and Livingstone Ranges, Montana. Bull., Geological Soc. of America, vol. 13, pp. 305-52.

Wilson, Clifford, 1972, Crash Go the Chariots. Lancer Books, New York, 126 pp.

Wilson, Edward O., 1975, Sociobiology: The New Synthesis. Harvard Univ. Press, Cambridge, Mass., 697 pp.

------ 1978, On Human Nature. Harvard Univ. Press, Cambridge, Mass., 260 pp.

Wilson, Edward O., and Charles Lumden, 1981, Genes, Mind, and Culture; The Evolutionary Process. Harvard Univ. Press, Cambridge, Mass., 248 pp.

Wolpoff, Milford H., and C. Loring Brace, 1975, Allometry and early hominids. Science, vol. 189, pp. 61-63.

Woodmorappe, J., 1979, Radiometric geochronology reappraised. Creation Research Soc. Quarterly, vol. 16, pp. 102-29, 47.

Woodruff, F., S. M. Savin, and R. G. Douglas, 1981, Miocene stable isotope record: a detailed deep Pacific Ocean study and its paleoclimatic implications. Science, vol. 212, pp. 665-68.

Wszolek, P. C., and A. L. Burlingame, 1978, Petroleum--origin and evolution. Pp. 565-74 in Fairbridge and Bourgeois, 1978.

Wursig, Bernd, 1979, Dolphins. Scientific American, vol. 240, no. 3, pp. 136-48.

Wysong, Randy L., 1976, The Creation-Evolution Controversy. Inquiry Press, Midland, Mich., 455 pp.

Yablokov, Alexy, 1966, Variability of Mammals. Nauka Publishers, Moscow.

York, Derek, 1979, Pleochroic halos and geochronology. EOS, vol. 60, no. 33, pp. 617-18.

Young, Davis A., 1977, Creation and the Flood. Baker Book House, 217 pp.

------ 1982, Christianity and the Age of the Earth. Zondervan Publishing House, Grand Rapids, Mich.

Yuretich, Richard F., 1984, Yellowstone fossil forests: new evidence for burial in place. Geology, vol. 12, pp. 159-62.

Zangerl, Rainer, 1969, The turtle shell. Pp. 311-39 in Gans, 1969.

Zechun, Liu, 1983, Peking Man's cave yields new finds. The Geographical Magazine, vol. 55, no. 6, pp. 297-300.

Zeilik, Michael, 1982, Astronomy; The Evolving Universe, 2d ed. Harper and Row, New York, 623 pp.

Zeisel, Hans, 1981, Letters to editor. Science, vol. 212, p. 873.

Zenger, Donald H., 1986, Lyell and episodicity. Jour. of Geological Education, vol. 34, pp. 10-13.

Ziman, John, 1980, What is science? Pp. 35-54 in Klemke, Hollinger, and Kline, 1980.

Name Index*

---

*Letter "q" indicates a quotation.

Subject Index*